Formulas from Geometry

Formulas for Area (A), Perimeter (P), Circumference (C), and Volume (V):

Square

$A = s^2$

$P = 4s$

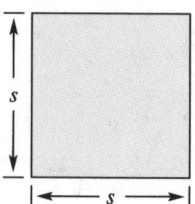

Rectangle

$A = lw$

$P = 2l + 2w$

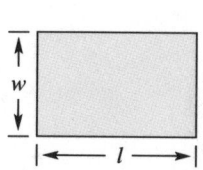

Circle

$A = \pi r^2$

$C = 2\pi r$

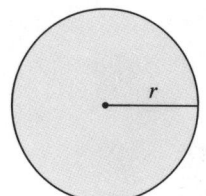

Triangle

$A = \frac{1}{2}bh$

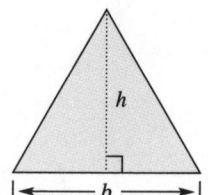

Trapezoid

$A = \frac{1}{2}h(b_1 + b_2)$

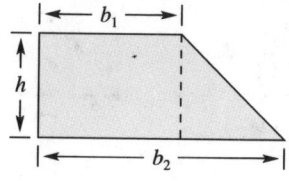

Parallelogram

$A = bh$

$P = 2a + 2b$

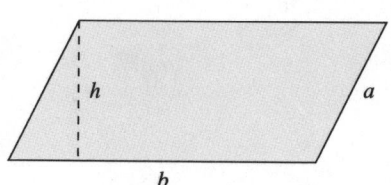

Pythagorean Theorem

$a^2 + b^2 = c^2$

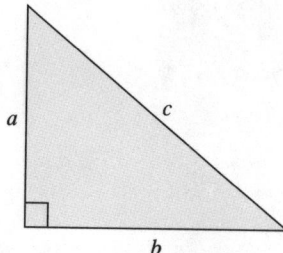

Cube

$V = s^3$

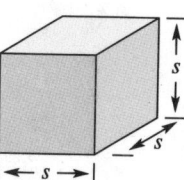

Rectangular Solid

$V = lwh$

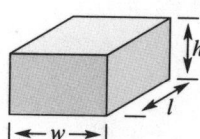

Circular Cylinder

$V = \pi r^2 h$

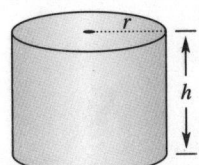

Sphere

$V = \frac{4}{3}\pi r^3$

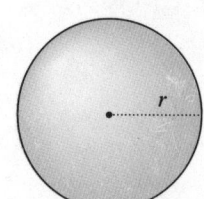

Instructor's Annotated Edition
Intermediate Algebra
Graphs and Functions

Second Edition

Roland E. Larson
The Pennsylvania State University
The Behrend College

Robert P. Hostetler
The Pennsylvania State University
The Behrend College

Carolyn F. Neptune
Johnson County Community College

with the assistance of

David E. Heyd
The Pennsylvania State University
The Behrend College

HOUGHTON MIFFLIN COMPANY   Boston   New York

Editor in Chief—Mathematics: Charles Hartford
Managing Editor: Catherine B. Cantin
Senior Associate Editor: Maureen Brooks
Associate Editor: Michael Richards
Assistant Editor: Carolyn Johnson
Supervising Editor: Karen Carter
Art Supervisor: Gary Crespo
Marketing Manager: Sara Whittern
Associate Marketing Manager: Ros Kane
Marketing Assistant: Carolyn Lipscomb
Cover design: Henry Rachlin
Composition and Art: Meridian Creative Group

Photo credits: 1, Dale E. Boyer/Photo Researchers, Inc.; 36, Edward L. Miller/Stock Boston; 103, Kerbs/Monkmeyer Press; 144, J. Carrini/The Image Works; 183, Merlin D. Tuttle/Photo Researchers, Inc.; 211, Corbis/Bettmann; 251, David J. Sams/Stock Boston; 265, J. Pickerell/The Image Works; 347, NASA; 363, Corbis/Bettmann; 407, Bob Daemmrich/Stock Boston; 477, Bob Daemmrich/Stock Boston; 555, NASA; 581, Gamma-Liaison; 643, R.T. Loesch/Stock Boston; 692, Nita Winters/The Image Works; 711, Amy C. Etra/PhotoEdit.

Trademark Acknowledgments: TI is a registered trademark of Texas Instruments, Inc. Casio is a registered trademark of Casio, Inc. Sharp is a registered trademark of Sharp Electronics Corp. Hewlett-Packard is a registered trademark.

Printed in the U.S.A.

Library of Congress Catalog Card Number: 97-72512

ISBN: 0-395-87773-3

123456789—DC—01 00 99 98 97

Contents

Preface

A word from the authors ...

Welcome to *Intermediate Algebra: Graphs and Functions,* Second Edition! In the revision of this early graphing and functions text, we focused on fine-tuning our student-oriented approach and incorporating the best aspects of reform in a meaningful yet easy-to-use manner. We hope you will be as excited about the Second Edition as we are after you take a look at it.

A student-oriented approach ...

Some students must take intermediate algebra more than once because they do not know how to study mathematics. The Second Edition helps students break out of this cycle by outlining a straightforward program of study with continual reinforcement and progressive confidence building.

This practical approach begins with Strategies for Success, a new feature found at the beginning of each chapter. In addition to outlining the key skills to be covered in the chapter, this checklist provides page references for the various study tools in the chapter. The Chapter Summary at the end of each chapter reinforces the Strategies for Success with a comprehensive list of the skills covered in the chapter, section references, and a correlation to the Review Exercises for guided practice.

Throughout each chapter there are many opportunities for students to assess their progress: at the end of each section (Warm-Ups and section exercises); in the middle of each chapter (Mid-Chapter Quiz); and at the end of each chapter (Review Exercises, Chapter Test, and Cumulative Test). The test items and text exercises are carefully crafted and graded in difficulty to give students a higher comfort level with algebraic skill building and problem solving.

These study tools reinforce the message that mathematics is a continuing story that requires constant synthesis and review. Along the way, students are guided by Study Tips that address special cases, expand on concepts, and help them avoid common errors.

Foremost for us, this text was written to be read and understood by students—not to be merely a source of homework assignments. We paid careful attention to the presentation—using precise mathematical language, innovative full-color design for emphasis and clarity, and a level of exposition that appeals to students—to create an effective teaching and learning tool.

... that incorporates the best aspects of reform

We wholeheartedly embrace many of the features of the mathematics reform movement. Our First Edition led the way in developing many innovative learning techniques and our Second Edition is maintaining the pace.

In the Second Edition, we have increased the coverage of technology and integrated it throughout the text at point of use. Students are encouraged to use a graphing utility as a tool for exploration, discovery, and problem solving. Our reviewers were pleased to notice that technology is always introduced in support of the concepts, rather than being the central focus of the text.

We introduce graphing and functions early and integrate them throughout, always stressing visualization. We have increased the emphasis on real-life applications, problem solving, conceptual exercises (for example, Think About It and Section Projects), and motivational features (for example, Explorations and Group Activities). In addition, we added three new sections on modeling data: Section 3.2, Modeling Data with Linear Functions; Section 6.6, Modeling Data with Quadratic Functions; and Section 9.6, Modeling Data (with exponential and logarithmic functions). This new material gives students the opportunity to focus on generating, exploring, and analyzing data.

We hope you will enjoy the Second Edition. It is a readable text that incorporates the best aspects of reform. The straightforward approach and effective study tools should appeal to your students.

Roland E. Larson

Robert P. Hostetler

Carolyn F. Neptune

Introduction to Graphs and Functions

2

Strategies for Success

SKILLS *When you have completed this chapter, make sure you are able to:*
○ Create graphs that represent real-life data
○ Sketch graphs of equations and find the x- and y-intercepts
○ Sketch graphs of lines using slopes and y-intercepts
○ Evaluate functions and find their domains
○ Identify and sketch transformations of graphs of functions

TOOLS *Use these study tools to master the skills above:*
○ Mid-Chapter Quiz (page 145)
○ Chapter Summary (page 176)
○ Review Exercises (page 177)
○ Chapter Test (page 181)
○ Cumulative Test (page 182)

In Exercise 79 of Section 2.1, you will graphically represent the number of male and female sports participants in the United States.

Section Topics

Each section begins with a list of important topics that are covered in that section. These topics are also the subsection titles and can be used for easy reference and review by students.

Definitions, Rules, and Formulas

All of the important definitions, rules, and formulas are highlighted for emphasis. Each is also titled for easy reference.

Features of the Text

Chapter Openers NEW

Each chapter opens with *Strategies for Success.* This new checklist outlines the key skills to be covered in the chapter and gives a list of the study tools in the chapter—with page references that will help students master the key skills. Each chapter opener also contains a list of the section topics, as well as a photo referring students to an interesting exercise in the chapter.

SECTION 5.2 Rational Exponents and Radicals   **357**

5.2 Rational Exponents and Radicals

Roots and Radicals • Rational Exponents •
Radicals and Calculators • Radical Functions

Roots and Radicals

In Section 1.1, you reviewed the use of radical notation to represent nth roots of real numbers. Recall from that section that b is called an nth root of a if $a = b^n$. Also recall that the principal nth root of a real number is defined as follows.

NOTE "Having the same sign" means that the principal nth root of a is positive if a is positive and negative if a is negative. For example, $\sqrt{4} = 2$ and $\sqrt[3]{-8} = -2$.

Principal nth Root of a Number

Let a be a real number that has at least one (real number) nth root. The **principal nth root of a** is the nth root that has the same sign as a, and it is denoted by the **radical symbol**

$$\sqrt[n]{a}. \qquad \text{Principal } n\text{th root}$$

The positive integer n is the **index** of the radical, and the number a is the **radicand**. If $n = 2$, omit the index and write $\sqrt{a}$ rather than $\sqrt[2]{a}$.

Therefore, $\sqrt{49} = 7$, $\sqrt[3]{1000} = 10$, and $\sqrt[5]{-32} = -2$.

You need to be aware of the following properties of nth roots. (Remember that for nth roots, n is an integer that is greater than or equal to 2.)

Properties of nth Roots

1. If a is a positive real number and n is *even*, then a has exactly two (real) nth roots, which are denoted by $\sqrt[n]{a}$ and $-\sqrt[n]{a}$.

2. If a is any real number and n is *odd*, then a has only one (real) nth root, which is denoted by $\sqrt[n]{a}$.

3. If a is a negative real number and n is *even*, then a has no (real) nth root.

Integers such as 1, 4, 9, 16, 49, and 81 are called **perfect squares** because they have integer square roots. Similarly, integers such as 1, 8, 27, 64, and 125 are called **perfect cubes** because they have integer cube roots.

Explorations NEW

Before students are exposed to selected topics, Explorations invite them to discover concepts and patterns on their own, often taking advantage of the power of technology. This active participation by students strengthens their intuition and their critical-thinking skills and makes it more likely that they will remember the results. These new, optional boxed features can be omitted if the instructor desires with no loss of continuity in the coverage of material.

---

112   CHAPTER 2 Introduction to Graphs and Functions

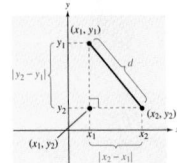

FIGURE 2.13 Distance between two points

To develop a general formula for the distance between two points, let (x_1, y_1) and (x_2, y_2) represent two points in the plane (that do not lie on the same horizontal or vertical line). With these two points, a right triangle can be formed, as shown in Figure 2.13. Note that the third vertex of the triangle is (x_2, y_1). Because (x_1, y_1) and (x_2, y_1) lie on the same vertical line, the length of the vertical side of the triangle is $|y_2 - y_1|$. Similarly, the length of the horizontal side is $|x_2 - x_1|$. By the Pythagorean Theorem, the square of the distance between (x_1, y_1) and (x_2, y_2) is

$$d^2 = |x_2 - x_1|^2 + |y_2 - y_1|^2.$$

Because the distance d must be positive, you can choose the positive square root and write

$$d = \sqrt{|x_2 - x_1|^2 + |y_2 - y_1|^2}.$$

Finally, replacing $|x_2 - x_1|^2$ and $|y_2 - y_1|^2$ by the equivalent expressions $(x_2 - x_1)^2$ and $(y_2 - y_1)^2$ gives you the **Distance Formula.**

Exploration

Plot the points $A(-1, -3)$ and $B(5, 2)$ and sketch the line segment from A to B. How could you verify that point $C(2, -0.5)$ is the midpoint of the segment? Why is it not sufficient to show that the distances from A to C and from C to B are equal? Find a formula for the coordinates of the midpoint of the line segment connecting (x_1, y_1) and (x_2, y_2).

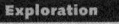

The Distance Formula

The distance d between two points (x_1, y_1) and (x_2, y_2) is

$$d = \sqrt{(x_2 - x_1)^2 + (y_2 - y_1)^2}.$$

Note that for the special case in which the two points lie on the same vertical or horizontal line, the Distance Formula still works. For instance, applying the Distance Formula to the points $(2, -2)$ and $(2, 4)$ produces

$$d = \sqrt{(2 - 2)^2 + [4 - (-2)]^2} = \sqrt{6^2} = 6,$$

which is the same result obtained in Example 6.

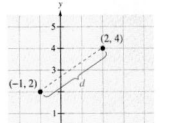

EXAMPLE 7   Finding the Distance Between Two Points

Find the distance between the points $(-1, 2)$ and $(2, 4)$, as shown in Figure 2.14.

Solution

Let $(x_1, y_1) = (-1, 2)$ and $(x_2, y_2) = (2, 4)$, and apply the Distance Formula.

$d = \sqrt{[2 - (-1)]^2 + (4 - 2)^2}$	Substitute coordinates of points.
$= \sqrt{3^2 + 2^2}$	Simplify.
$= \sqrt{13}$	Simplify.
≈ 3.61	Use a calculator.

---

324   CHAPTER 4 Systems of Linear Equations and Inequalities

EXAMPLE 4   Sketching the Graph of a Linear Inequality

Use the slope-intercept form of a linear equation as an aid in sketching the graph of the inequality

$$2x - 3y \le 15.$$ Original linear inequality

Solution

To begin, rewrite the inequality in slope-intercept form.

$2x - 3y \le 15$	Original inequality
$-3y \le -2x + 15$	Subtract 2x from both sides.
$y \ge \dfrac{2}{3}x - 5$	Divide both sides by −3 and reverse the inequality symbol.

From this form, you can conclude that the solution is the half-plane lying *on* or *above* the line

$$y = \frac{2}{3}x - 5.$$

The graph is shown in Figure 4.19.

FIGURE 4.19

Technology

A graphing utility can be used to graph an inequality. The actual keystrokes used depend on the graphing utility, but here is an example of how to graph

$$3x + 2y < 4$$

on a *TI-83*.

1. Solve the inequality for y to obtain $y < -\frac{3}{2}x + 2$.
2. Press [Y=] and enter $-(3/2)X + 2$ for Y₁.
3. Move the cursor to the left of Y₁.
4. Press [ENTER] until the ◣ icon appears.
5. Press [GRAPH].

The graph is shown at the left. Try using a graphing utility to graph the following inequalities.

(a) $2x + 3y \ge 3$ (b) $x - 2y \le 2$

Graphics

Visualization is a critical problem-solving skill. Graphing is introduced in Chapter 2, and from that point on, students are encouraged to use graphs to reinforce algebraic or numeric solutions, to interpret data, and to explore concepts. The numerous figures in this text—all computer generated for accuracy—help students develop these skills.

Examples

Each of the nearly 900 examples was carefully chosen to illustrate a particular concept or problem-solving technique and to enhance students' understanding. Students are taught a five-step strategy in the spirit of the AMATYC and NCTM standards, which starts with constructing a verbal model and ends with checking the answer. The examples are titled for easy reference, and comments adjacent to the solutions offer additional explanations.

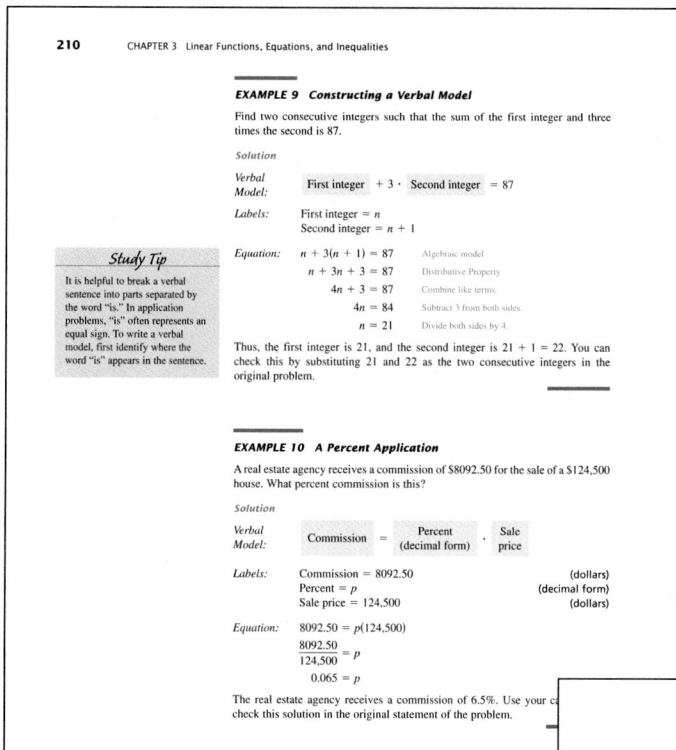

210 CHAPTER 3 Linear Functions, Equations, and Inequalities

EXAMPLE 9 Constructing a Verbal Model

Find two consecutive integers such that the sum of the first integer and three times the second is 87.

Solution

Verbal Model: First integer + 3 · Second integer = 87

Labels: First integer = n
Second integer = $n + 1$

Equation:
$n + 3(n + 1) = 87$ Algebraic model
$n + 3n + 3 = 87$ Distributive Property
$4n + 3 = 87$ Combine like terms.
$4n = 84$ Subtract 3 from both sides.
$n = 21$ Divide both sides by 4.

Thus, the first integer is 21, and the second integer is $21 + 1 = 22$. You can check this by substituting 21 and 22 as the two consecutive integers in the original problem.

Study Tip

It is helpful to break a verbal sentence into parts separated by the word "is." In application problems, "is" often represents an equal sign. To write a verbal model, first identify where the word "is" appears in the sentence.

EXAMPLE 10 A Percent Application

A real estate agency receives a commission of $8092.50 for the sale of a $124,500 house. What percent commission is this?

Solution

Verbal Model: Commission = Percent (decimal form) · Sale price

Labels: Commission = 8092.50   (dollars)
Percent = p (decimal form)
Sale price = 124,500   (dollars)

Equation:
$8092.50 = p(124,500)$
$\dfrac{8092.50}{124,500} = p$
$0.065 = p$

The real estate agency receives a commission of 6.5%. Use your ca[...] check this solution in the original statement of the problem.

Applications

A rich and varied selection of real-world applications are integrated throughout the text in examples and in exercises. These applications offer students a constant review of problem-solving skills and emphasize the relevance of the mathematics. Many of the applications use current, real data, and are titled for easy reference.

Group Activities   NEW

Group Activities appear at the end of each section. They encourage students to think, talk, and write about mathematics in a peer-assisted learning environment.

SECTION 6.6 Modeling Data with Quadratic Functions **459**

Application

EXAMPLE 4 Finding a Quadratic Model

The total amounts A (in millions of tons) of solid waste materials recycled in the United States in selected years from 1980 through 1993 are shown below. Decide whether a linear model or a quadratic model better fits the data. Then use the model to predict the amount that will be recycled in the year 2000. In the list of data points (t, A), t represents the year, with $t = 0$ corresponding to 1980. (Source: Franklin Associates, Ltd.)

(0, 14.5), (5, 16.4), (6, 18.3), (7, 20.1), (8, 23.5), (9, 29.9), (10, 32.9), (11, 37.3), (12, 41.5), (13, 45.0)

Solution

Begin by entering the data into a calculator or computer that has least squares regression programs. Then run the regression programs for linear and quadratic models.

Linear: $y = ax + b$ $a = 2.624$ $b = 6.684$
Quadratic: $y = ax^2 + bx + c$ $a = 0.243$ $b = -0.655$ $c = 14.037$

From the graphs in Figure 6.27, you can see that the quadratic model fits better.

(The correlation coefficient is $r = 0.994$, which implies that the model is a good fit to the data.) From this model, you can predict the amount that will be recycled in the year 2000 to be

$A = 0.243(20)^2 - 0.655(20) + 14.037 = 98.137$ million tons,

which is more than two times the amount recycled in 1993.

Linear

Quadratic
FIGURE 6.27

Group Activities **Problem Solving**

Modeling Data The amounts y (in gallons) of bottled water consumed in the United States in the years 1980 through 1993 are listed below. The data is given as ordered pairs of the form (t, y), where t is the year, with $t = 0$ representing 1980. Create a scatter plot of the data. With others in your group, decide which type of model best fits the data. Then find the model.

(0, 2.4), (1, 2.7), (2, 3.0), (3, 3.4), (4, 4.0), (5, 4.5), (6, 5.0), (7, 5.7), (8, 6.5), (9, 7.4), (10, 8.0), (11, 8.0), (12, 8.2), (13, 9.2)

Technology   NEW

Students are encouraged to use a graphing utility as a tool for exploration, discovery, and problem solving. Many opportunities to visualize concepts, to discover alternative approaches, to execute computations or programs, and to verify the results of other solution methods using technology are integrated throughout the text at point of use. However, students are not required to have access to a graphing utility to use this text effectively. In addition to describing the benefits of using technology, the text also pays special attention to its possible misuse or misinterpretation.

---

denominators and is called the **least common denominator** (or **LCD**) of the original rational expressions. Once the rational expressions have been written with like denominators, you can simply add or subtract the rational expressions using the rule given at the beginning of this section.

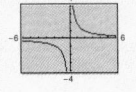

Technology

You can use a graphing utility to check your results when adding or subtracting rational expressions. In Example 5, for instance, try graphing the equations

$$y_1 = \frac{7}{6x} + \frac{5}{8x}$$

and

$$y_2 = \frac{43}{24x}$$

in the same viewing rectangle. If the two graphs coincide, as shown below, you can conclude that the solution checks.

EXAMPLE 5   Adding with Unlike Denominators

Add the rational expressions: $\dfrac{7}{6x} + \dfrac{5}{8x}$.

Solution

By factoring the denominators, $6x = 2 \cdot 3 \cdot x$ and $8x = 2^3 \cdot x$, you can conclude that the least common denominator is $2^3 \cdot 3 \cdot x = 24x$.

$$\frac{7}{6x} + \frac{5}{8x} = \frac{7(4)}{6x(4)} + \frac{5(3)}{8x(3)} \qquad \text{Rewrite fractions using least common denominator.}$$
$$= \frac{28}{24x} + \frac{15}{24x} \qquad \text{Like denominators}$$
$$= \frac{28 + 15}{24x} \qquad \text{Add fractions.}$$
$$= \frac{43}{24x} \qquad \text{Simplified form}$$

EXAMPLE 6   Subtracting with Unlike Denominators

Subtract the rational expressions: $\dfrac{3}{x-3} - \dfrac{5}{x+2}$.

Solution

The only factors of the denominators are $(x-3)$ and $(x+2)$. Therefore, the least common denominator is $(x-3)(x+2)$.

$$\frac{3}{x-3} - \frac{5}{x+2} = \frac{3(x+2)}{(x-3)(x+2)} - \frac{5(x-3)}{(x-3)(x+2)}$$
$$= \frac{3x+6}{(x-3)(x+2)} - \frac{5x-15}{(x-3)(x+2)}$$
$$= \frac{(3x+6) - (5x-15)}{(x-3)(x+2)}$$
$$= \frac{3x+6-5x+15}{(x-3)(x+2)}$$
$$= \frac{-2x+21}{(x-3)(x+2)}$$

NOTE In Example 1, the solutions are rational numbers, which means that the equation could have been solved by factoring. Try solving the equation by factoring.

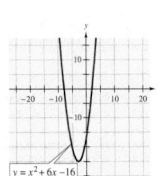

FIGURE 6.6

Solving Equations by the Quadratic Formula

When using the Quadratic Formula, remember that *before* the formula can be applied, you must first write the quadratic equation in standard form.

EXAMPLE 1   The Quadratic Formula: Two Distinct Solutions

$$x^2 + 6x = 16 \qquad \text{Original equation}$$
$$x^2 + 6x - 16 = 0 \qquad \text{Write in standard form.}$$
$$x = \frac{-b \pm \sqrt{b^2 - 4ac}}{2a} \qquad \text{Quadratic Formula}$$
$$x = \frac{-6 \pm \sqrt{6^2 - 4(1)(-16)}}{2(1)} \qquad \text{Substitute: } a = 1, b = 6, c = -16.$$
$$x = \frac{-6 \pm \sqrt{100}}{2} \qquad \text{Simplify.}$$
$$x = \frac{-6 \pm 10}{2} \qquad \text{Simplify.}$$
$$x = 2 \text{ or } x = -8 \qquad \text{Solutions}$$

The solutions are 2 and −8. Check these in the original equation. Or, try using a graphic check, as shown in Figure 6.6.

Study Tip

If the leading coefficient of a quadratic equation is negative, we suggest that you begin by multiplying both sides of the equation by −1, as shown in Example 2. This will produce a positive leading coefficient, which is less cumbersome to work with.

EXAMPLE 2   The Quadratic Formula: Two Distinct Solutions

$$-x^2 - 4x + 8 = 0 \qquad \text{Leading coefficient is negative.}$$
$$x^2 + 4x - 8 = 0 \qquad \text{Multiply both sides by } -1.$$
$$x = \frac{-b \pm \sqrt{b^2 - 4ac}}{2a} \qquad \text{Quadratic Formula}$$
$$x = \frac{-4 \pm \sqrt{4^2 - 4(1)(-8)}}{2(1)} \qquad \text{Substitute: } a = 1, b = 4, c = -8.$$
$$x = \frac{-4 \pm \sqrt{48}}{2} \qquad \text{Simplify.}$$
$$x = \frac{-4 \pm 4\sqrt{3}}{2} \qquad \text{Simplify.}$$
$$x = \frac{2(-2 \pm 2\sqrt{3})}{2} \qquad \text{Divide out common factor.}$$
$$x = -2 \pm 2\sqrt{3} \qquad \text{Solutions}$$

Some students may be tempted to reduce $(-4 \pm 4\sqrt{3})/2$ to $-2 \pm 4\sqrt{3}$. Ask them what is wrong with this reasoning.

The solutions are $-2 + 2\sqrt{3}$ and $-2 - 2\sqrt{3}$. Check these in the original equation.

Notes

Many instructional notes accompany definitions, rules, and examples to give additional insight or describe generalizations.

Study Tips   NEW

Study Tips help students avoid common errors, address special cases, and expand upon concepts. They appear in the margin at point of use.

Exercises

The nearly 7000 section exercises contain numerous computational and applied problems dealing with a wide range of topics. The exercise sets are designed to build competence, skill, and understanding; each exercise set is graded in difficulty to allow students to gain confidence as they progress. Each pair of consecutive problems is similar, with the answers to the odd-numbered problems given at the end of the text. Detailed solutions to all odd-numbered exercises are given in the *Student Study and Solutions Guide*.

Warm-Ups

For each text section (except for Section 1.1), there is a corresponding set of ten Warm-Up exercises in the Appendix, as indicated by an icon. The Warm-Ups enable students to review and practice the previously learned skills necessary to master the new skills presented in the section. Answers to Warm-Ups appear in the Appendix as well.

SECTION 6.4 Applications of Quadratic Equations   **441**

6.4 Exercises

See Warm-Up Exercises, p. A45

1. *Unit Analysis* Describe the units of the product.
$$\frac{9 \text{ dollars}}{\text{hour}} \cdot (20 \text{ hours})$$

2. *Unit Analysis* Describe the units of the product.
$$\frac{20 \text{ feet}}{\text{minute}} \cdot \frac{1 \text{ minute}}{60 \text{ seconds}} \cdot (45 \text{ seconds})$$

Number Problems In Exercises 3–6, find two positive integers that satisfy the requirement.

3. The product of two consecutive integers is 8 less than 10 times the smaller integer.

4. The product of two consecutive integers is 80 more than 15 times the larger integer.

5. The product of two consecutive even integers is 50 more than 3 times the larger integer.

6. The product of two consecutive odd integers is 22 less than 15 times the smaller integer.

Dimensions of a Rectangle In Exercises 7–16, complete the table of widths, lengths, perimeters, and areas of rectangles.

	Width	Length	Perimeter	Area
7.	$0.75l$	l	42 in.	
8.	w	$1.5w$	40 m	
9.	w	$2.5w$		250 ft²
10.	w	$1.5w$		216 cm²
11.	$\frac{1}{3}l$	l		192 in.²
12.	$\frac{3}{4}l$	l		2700 in.²
13.	w	$w+3$	54 km	
14.	$l-6$	l	108 ft	
15.	$l-20$	l		12,000 m²
16.	w	$w+5$		500 ft²

17. *Lumber Storage Area* A retail lumberyard plans to store lumber in a rectangular region adjoining the sales office (see figure). The region will be fenced on three sides and the fourth side will be bounded by the wall of the office building. Find the dimensions of the region if 350 feet of fencing is available and the area of the region is to be 12,500 square feet.

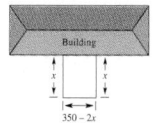

Building

$350 - 2x$

18. *Fencing the Yard* You have 100 feet of fencing. Do you have enough to enclose a rectangular region whose area is 630 square feet? Is there enough to enclose a circular region of area 630 square feet? Explain.

19. *Fencing the Yard* A family has built a fence around three sides of their property. In total, they used 550 feet of fencing. By their calculations, the lot is one acre (43,560 square feet). Is this correct? Explain your reasoning.

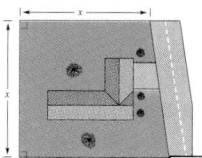

---

202   CHAPTER 3 Linear Functions, Equations, and Inequalities

3.2 Exercises

See Warm-Up Exercises, p. A41

1. *Falling Object* In an experiment, students measured the speed s (in meters per second) of a falling object t seconds after it was released. The results are given in the table.

t	0	1	2	3	4
s	0	11.0	19.4	29.2	39.4

A model for the data is $s = 9.7t + 0.4$.

(a) Plot the data and graph the model on the same set of coordinate axes.

(b) Create a table showing the given data and the approximations given by the model.

(c) Use the model to predict the speed of the object after falling 5 seconds.

(d) Interpret the slope in the context of the problem.

2. *Cable TV* The average monthly basic rate R (in dollars) for cable TV for the years 1989 through 1994 in the United States is given in the table. *(Source: Paul Kagan Associates, Inc.)*

Year	1989	1990	1991
R	15.21	16.78	18.10

Year	1992	1993	1994
R	19.08	19.39	21.62

A model for the data is $R = 1.17t + 16.61$, where t is the time in years, with $t = 0$ corresponding to 1990.

(a) Plot the data and graph the model on the same set of coordinate axes.

(b) Create a table showing the given data and the approximations given by the model.

(c) Use the model to predict the average monthly basic rate for cable TV for the year 2000.

(d) Interpret the slope in the context of the problem.

3. *Property Tax* The property tax in a township is directly proportional to the assessed value of the property. The tax on property with an assessed value of $17,072 is $1067.

(a) Find a mathematical model that gives the tax T in terms of the assessed value v.

(b) Use the model to find the tax on property with an assessed value of $11,500.

(c) Determine the tax rate.

4. *Revenue* The total revenue R is directly proportional to the number of units sold x. When 25 units are sold, the revenue is $6225.

(a) Find a mathematical model that gives the revenue R in terms of the number of units sold x.

(b) Use the model to find the revenue when 32 units are sold.

(c) Determine the price per unit.

5. *The English and Metric Systems* The label on a roll of tape gives the amount of tape in inches and centimeters. These amounts are 500 and 1270.

(a) Use the information on the label to find a mathematical model that relates inches to centimeters.

(b) Use part (a) to convert 15 inches to centimeters.

(c) Use part (a) to convert 650 centimeters to inches.

(d) Use a graphing utility to graph the model in part (a). Use the graph to confirm the results in parts (b) and (c).

6. *The English and Metric Systems* The label on a bottle of soft drink gives the amount in liters and fluid ounces. These amounts are 2 and 67.63.

(a) Use the information on the label to find a mathematical model that relates liters to fluid ounces.

(b) Use part (a) to convert 27 liters to fluid ounces.

(c) Use part (a) to convert 32 fluid ounces to liters.

(d) Use a graphing utility to graph the model in part (a). Use the graph to confirm the results in parts (b) and (c).

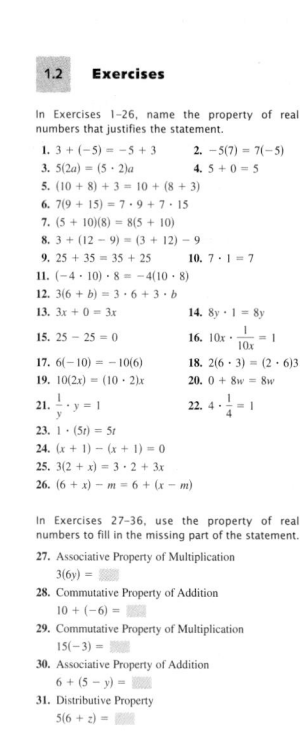

SECTION 1.2   Properties of Real Numbers   **25**

1.2   Exercises

See Warm-Up Exercises, p. A39

In Exercises 1–26, name the property of real numbers that justifies the statement.

1. $3 + (-5) = -5 + 3$ **2.** $-5(7) = 7(-5)$
3. $5(2a) = (5 \cdot 2)a$ **4.** $5 + 0 = 5$
5. $(10 + 8) + 3 = 10 + (8 + 3)$
6. $7(9 + 15) = 7 \cdot 9 + 7 \cdot 15$
7. $(5 + 10)(8) = 8(5 + 10)$
8. $3 + (12 - 9) = (3 + 12) - 9$
9. $25 + 35 = 35 + 25$ **10.** $7 \cdot 1 = 7$
11. $(-4 \cdot 10) \cdot 8 = -4(10 \cdot 8)$
12. $3(6 + b) = 3 \cdot 6 + 3 \cdot b$
13. $3x + 0 = 3x$ **14.** $8y \cdot 1 = 8y$
15. $25 - 25 = 0$ **16.** $10x \cdot \frac{1}{10x} = 1$
17. $6(-10) = -10(6)$ **18.** $2(6 \cdot 3) = (2 \cdot 6)3$
19. $10(2x) = (10 \cdot 2)x$ **20.** $0 + 8w = 8w$
21. $\frac{1}{y} \cdot y = 1$ **22.** $4 \cdot \frac{1}{4} = 1$
23. $1 \cdot (5t) = 5t$
24. $(x + 1) - (x + 1) = 0$
25. $3(2 + x) = 3 \cdot 2 + 3x$
26. $(6 + x) - m = 6 + (x - m)$

In Exercises 27–36, use the property of real numbers to fill in the missing part of the statement.

27. Associative Property of Multiplication
$3(6y) = \blacksquare$
28. Commutative Property of Addition
$10 + (-6) = \blacksquare$
29. Commutative Property of Multiplication
$15(-3) = \blacksquare$
30. Associative Property of Addition
$6 + (5 - y) = \blacksquare$
31. Distributive Property
$5(6 + z) = \blacksquare$

32. Distributive Property
$-3(4 + x) = \blacksquare$
33. Commutative Property of Addition
$25 + (-x) = \blacksquare$
34. Additive Inverse
$13x - 13x = \blacksquare$
35. Multiplicative Identity
$(x + 8) \cdot 1 = \blacksquare$
36. Additive Identity
$(8x) + 0 = \blacksquare$

True or False? In Exercises 37–40, decide whether the statement is true or false. Explain.

37. $-6x + 6x = 0$ **38.** $-9 + 5 = -5 + 9$
39. $6(7 + 2) = 6(7) + 2$
40. $-4(8 + 1) = -4(8) - 4(1)$

41. *Think About It* Does every real number have a multiplicative inverse? Explain.
42. *Think About It* What is the additive inverse of a real number? Give an example of the Additive Inverse Property?

In Exercises 43–50, give (a) the additive inverse and (b) the multiplicative inverse of the quantity.

43. 10 **44.** 18
45. -16 **46.** -52
47. $6z,\ z \neq 0$ **48.** $2y,\ y \neq 0$
49. $x + 1,\ x \neq -1$ **50.** $y - 4,\ y \neq 4$

In Exercises 51–58, rewrite the expression using the Associative Property of Addition or the Associative Property of Multiplication.

51. $(x + 5) - 3$ **52.** $(z$
53. $32 + (-4 + y)$ **54.** $15 +$
55. $3(4 \cdot 5)$ **56.** $(10$
57. $6(2y)$ **58.** $8(3x$

Graphing Utilities

Many exercises in the text can be solved using technology; however, the symbol 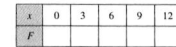 identifies all exercises in which students are specifically instructed to use a graphing utility. Students are encouraged to use scientific and graphing calculators to discover patterns, to experiment, to calculate, and to create graphic models.

True or False   NEW

To help students understand the logical structure of algebra, a set of True or False questions is included toward the end of selected exercise sets. These questions help students focus on concepts, common errors, and the correct statements of definitions and rules.

Think About It   NEW

These exercises are thought-provoking, conceptual problems that help students grasp underlying theories.

---

130   CHAPTER 2   Introduction to Graphs and Functions

In Exercises 63–66, explain how the *x*-intercepts of the graph correspond to the solutions of the polynomial equation when $y = 0$.

63. $y = x^2 - 9$ **64.** $y = x^2 - 4x + 4$

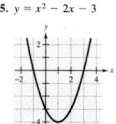   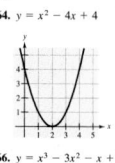

65. $y = x^2 - 2x - 3$ **66.** $y = x^3 - 3x^2 - x + 3$

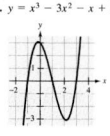

In Exercises 67–70, use a graphing utility to graph the equation and find any *x*-intercepts of the graph. Verify algebraically that any *x*-intercepts are solutions of the polynomial equation when $y = 0$.

67. $y = \frac{1}{2}x - 2$
68. $y = -3x + 6$
69. $y = x^2 - 6x$
70. $y = x^2 - 11x + 28$

In Exercises 71–80, use a graphing utility to solve the equation graphically.

71. $7 - 2(x - 1) = 0$ **72.** $2x - 1 = 3(x + 1)$
73. $4 - x^2 = 0$ **74.** $x^2 + 2x = 0$
75. $x^2 - 2x + 1 = 0$ **76.** $1 - (x - 2)^2 = 0$
77. $2x^2 + 5x - 12 = 0$ **78.** $(x - 2)^2 - 9 = 0$
79. $x^3 - 4x = 0$ **80.** $2 + x - 2x^2 - x^3 = 0$

81. *Hooke's Law* The force F (in pounds) to stretch a spring x inches from its natural length is given by
$$F = \tfrac{4}{3}x, \quad 0 \leq x \leq 12.$$

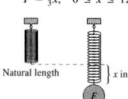

Natural length $\}$ x in.

(a) Use the model to complete the table.

x	0	3	6	9	12
F					

(b) Sketch a graph of the model.
(c) Use the graph in part (b) to determine how the length of the spring changes each time the force is doubled.

82. *Dairy Farms* The number of farms in the United States with milk cows has been decreasing. The numbers of farms N (in thousands) for the years 1988 through 1994 are given in the table.

Year	1988	1989	1990	1991	1992	1993	1994
N	216	203	193	181	171	159	150

A model for this data is
$$N = -11.0t + 192.9,$$
where t is the time in years, with $t = 0$ corresponding to 1990. (*Source: U.S. Department of Agriculture*)

(a) Use a graphing utility to plot the data and graph the model.
(b) How well does the model represent the data? Explain your reasoning.
(c) Use the model to predict the number of farms with milk cows in 1997.
(d) Explain why the model may not be accurate in the future.

Section Projects   NEW

Section Projects appear at the end of every exercise set. These extended applications are often multi-part exercises that make use of real data to develop critical-thinking and problem-solving skills. Section Projects are designed for individual or group assignments.

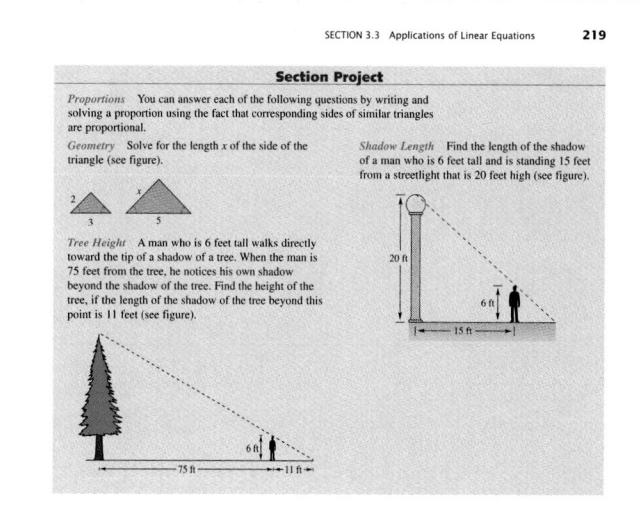

SECTION 3.3 Applications of Linear Equations   **219**

Section Project

Proportions You can answer each of the following questions by writing and solving a proportion using the fact that corresponding sides of similar triangles are proportional.

Geometry Solve for the length x of the side of the triangle (see figure).

Tree Height A man who is 6 feet tall walks directly toward the tip of a shadow of a tree. When the man is 75 feet from the tree, he notices his own shadow beyond the shadow of the tree. Find the height of the tree, if the length of the shadow of the tree beyond this point is 11 feet (see figure).

Shadow Length Find the length of the shadow of a man who is 6 feet tall and is standing 15 feet from a streetlight that is 20 feet high (see figure).

98   CHAPTER 1   Concepts of Elementary Algebra

Chapter Summary

After studying this chapter, you should have acquired the following skills. These skills are keyed to the Review Exercises that begin on page 99. Answers to odd-numbered Review Exercises are given in the back of the book.

• Plot real numbers on a real number line and compare them by using inequality symbols. *(Section 1.1)*	**Review Exercises 1–4**
• Evaluate expressions containing operations with real numbers. *(Section 1.1)*	**Review Exercises 5–24, 47**
• Identify the rule of algebra that is illustrated by an equation. *(Sections 1.2, 1.3)*	**Review Exercises 25–30**
• Expand expressions using the Distributive Property. *(Sections 1.2, 1.3)*	**Review Exercises 31–34**
• Simplify expressions by removing symbols of grouping. *(Section 1.3)*	**Review Exercises 35–38**
• Simplify expressions by applying the properties of exponents. *(Section 1.3)*	**Review Exercises 39–44**
• Solve problems involving geometry. *(Sections 1.3, 1.4)*	**Review Exercises 45, 46, 129, 130**
• Use expressions or equations to solve real-life problems. *(Sections 1.1, 1.8)*	**Review Exercises 48, 127, 128**
• Interpret graphs representing real-life data. *(Sections 1.1–1.4, 1.7)*	**Review Exercises 49, 50**
• Simplify expressions by performing arithmetic operations. *(Section 1.4)*	**Review Exercises 51–66**
• Multiply polynomials using the special product formulas. *(Section 1.4)*	**Review Exercises 67–74**
• Factor expressions completely. *(Sections 1.5, 1.6)*	**Review Exercises 75–96**
• Solve linear equations. *(Section 1.7)*	**Review Exercises 97–104, 115, 116, 120, 121, 123, 124**
• Solve literal equations. *(Section 1.7)*	**Review Exercises 105, 106**
• Solve polynomial equations. *(Section 1.8)*	**Review Exercises 107–114, 117–119, 122, 125, 126**
• Use a calculator to evaluate expressions containing operations with real numbers. *(Section 1.1)*	**Review Exercise 131**

Chapter Summary   NEW

The Chapter Summary reviews the skills covered in the chapter. Section references make this an effective study tool, and correlation to the Review Exercises offers guided practice.

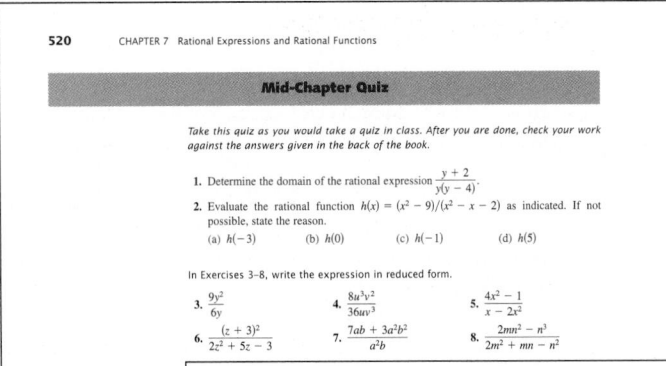

520   CHAPTER 7   Rational Expressions and Rational Functions

Mid-Chapter Quiz

Take this quiz as you would take a quiz in class. After you are done, check your work against the answers given in the back of the book.

1. Determine the domain of the rational expression $\dfrac{y+2}{y(y-4)}$.

2. Evaluate the rational function $h(x) = (x^2 - 9)/(x^2 - x - 2)$ as indicated. If not possible, state the reason.
   (a) $h(-3)$ (b) $h(0)$ (c) $h(-1)$ (d) $h(5)$

In Exercises 3–8, write the expression in reduced form.

3. $\dfrac{9y^2}{6y}$ 4. $\dfrac{8u^3v^2}{36uv^3}$ 5. $\dfrac{4x^2-1}{x-2x^2}$

6. $\dfrac{(z+3)^2}{2z^2+5z-3}$ 7. $\dfrac{7ab+3a^2b^2}{a^2b}$ 8. $\dfrac{2mn^2-n^3}{2m^2+mn-n^2}$

---

550   CHAPTER 7   Rational Expressions and Rational Functions

Review Exercises

In Exercises 1–4, find the domain of the rational expression.

1. $\dfrac{3y}{y-8}$ 2. $\dfrac{t+4}{t+12}$

3. $\dfrac{u}{u^2-7u+6}$ 4. $\dfrac{x-12}{x(x^2-16)}$

In Exercises 5–12, simplify the rational expression.

5. $\dfrac{6x^4y^2}{15xy^2}$ 6. $\dfrac{2(y^3z)^2}{28(yz^2)^2}$

7. $\dfrac{5b-15}{30b-120}$ 8. $\dfrac{4a}{10a^2+26a}$

9. $\dfrac{9x-9y}{y-x}$ 10. $\dfrac{x+3}{x^2-x-12}$

11. $\dfrac{x^2-5x}{2x^2-50}$ 12

23. $\dfrac{x^2-7x}{x+1} \div \dfrac{x^2-14x+49}{x^2-1}$

24. $\left(\dfrac{6x}{y^2}\right)^2 \div \left(\dfrac{3x}{y}\right)^3$

25. $\dfrac{4}{9} - \dfrac{11}{9}$ 26. $\dfrac{2(3y+4)}{2y+1} + \dfrac{3-y}{2y+1}$

27. $\dfrac{15}{16} - \dfrac{5}{24} - 1$ 28. $-\dfrac{3}{8} + \dfrac{7}{6} - \dfrac{1}{12}$

29. $\dfrac{1}{x+5} + \dfrac{3}{x-12}$ 30. $\dfrac{2}{x-10} + \dfrac{3}{4-x}$

31. $5x + \dfrac{2}{x-3} - \dfrac{3}{x+2}$

32. $4 - \dfrac{4x}{x+6} + \dfrac{7}{x-5}$

CHAPTER TEST   **553**

Chapter Test

Take this test as you would take a test in class. After you are done, check your work against the answers given in the back of the book.

1. Find the domain of the rational expression $\dfrac{3y}{y^2-25}$.

2. Simplify the rational expression $\dfrac{2-x}{3x-6}$.

In Exercises 3–11, perform the operation(s) and simplify.

3. $\dfrac{4z^3}{5} \cdot \dfrac{25}{12z^2}$ 4. $\dfrac{y^2+8y+16}{2(y-2)} \cdot \dfrac{8y-16}{(y+4)^3}$

5. $(4x^2-9) \cdot \dfrac{2x+3}{2x^2-x-3}$ 6. $\dfrac{(2xy^2)^3}{15} \div \dfrac{12x^3}{21}$

7. $\dfrac{\left(\dfrac{3x}{x+2}\right)}{}$ 8. $\dfrac{\left(9x-\dfrac{1}{x}\right)}{}$ 9. $2x + \dfrac{1-4x^2}{}$

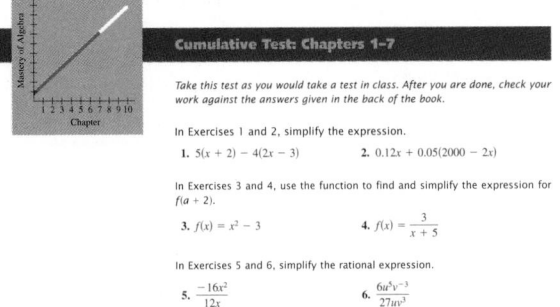

Cumulative Test: Chapters 1–7

Take this test as you would take a test in class. After you are done, check your work against the answers given in the back of the book.

In Exercises 1 and 2, simplify the expression.

1. $5(x+2) - 4(2x-3)$ 2. $0.12x + 0.05(2000 - 2x)$

In Exercises 3 and 4, use the function to find and simplify the expression for $f(a+2)$.

3. $f(x) = x^2 - 3$ 4. $f(x) = \dfrac{3}{x+5}$

In Exercises 5 and 6, simplify the rational expression.

5. $\dfrac{-16x^2}{12x}$ 6. $\dfrac{6u^4v^{-3}}{27uv^3}$

In Exercises 7–9, perform the operation and simplify. (Assume that all variables

---

Mid-Chapter Quiz   NEW

Each chapter contains a Mid-Chapter Quiz. This feature allows students to perform a self-assessment midway through the chapter. Answers to Mid-Chapter Quizzes appear at the end of the text.

Review Exercises

The Review Exercises at the end of each chapter offer students an opportunity for additional practice. Answers to all the odd-numbered Review Exercises appear at the end of the text.

Chapter Test

Each chapter contains an end-of-chapter test for students to assess their progress. Answers appear at the end of the text.

Cumulative Test

In this edition, Cumulative Tests have been placed at the end of each chapter (except Chapter 1). These tests reinforce the message that is presented throughout the text—that mathematics is a continuing story and requires constant synthesis and review. Answers appear at the end of the text.

Acknowledgments

We would like to thank the many people who have helped us prepare the Second Edition of this text. Their encouragement, criticisms, and suggestions have been invaluable to us.

Second Edition Reviewers: Mary K. Alter, University of Maryland, College Park; Mary Jean Brod, University of Montana; Martin Brown, Jefferson Community College; John W. Burns, Mt. San Antonio College; Bradd Clark, University of Southwestern Louisiana; D. J. Clark, Portland Community College; Linda Crabtree, Metropolitan Community College; Patricia Dalton, Montgomery College; Paul A. Dirks, Miami-Dade Community College; Ingrid Holzner, University of Wisconsin—Milwaukee; Donald Patrick Kinney, University of Minnesota; Barbara C. Kistler, Lehigh Carbon Community College; Antonio M. Lopez, Jr., Loyola University; David Lunsford, Grossmont College; Giles Wilson Maloof, Boise State University; Eric Preibisius, Cuyamaca College; William Radulovich, Florida Community College; Jack Rotman, Lansing Community College; Minnie W. Shuler, Gulf Coast Community College; Judith D. Smalling, St. Petersburg Junior College; James Tarvin, Grossmont College; Gwen Terwilliger, University of Toledo; Frances Ventola, Brookdale Community College; Jo Fitzsimmons Warner, Eastern Michigan University.

First Edition Reviewers: Lionel Geller, Dawson College; William Grimes, Central Missouri State University; Rosalyn T. Jones, Albany State College; Debra A. Landre, San Joaquin Delta College; Myrna F. Manly, El Camino Community College; James I. McCullough, Arapahoe Community College; Katherine McLain, Cosumnes River College; Karen S. Norwood, North Carolina State University; Nora I. Schukei, University of South Carolina at Beaufort; Kay Stroope, Phillips County Community College.

A special thanks to all the people at Houghton Mifflin Company who worked with us on the Second Edition, especially Chris Hoag, Cathy Cantin, Maureen Brooks, Michael Richards, Carolyn Johnson, Karen Carter, Rachel Wimberly, Carrie Lipscomb, Gary Crespo, Henry Rachlin, Sara Whittern, and Ros Kane.

We would also like to thank the staff at Larson Texts, Inc., who assisted with proofreading the manuscript; preparing and proofreading the art package; and checking and typesetting the supplements.

A special note of thanks goes to all the students who have used the first edition of the text.

On a personal level, we are grateful to our spouses, Deanna Gilbert Larson, Eloise Hostetler, and Harold Neptune, for their love, patience, and support. Also, a special thanks goes to R. Scott O'Neil.

If you have suggestions for improving this text, please feel free to write to us. Over the past two decades we have received many useful comments from both instructors and students, and we value these very much.

Roland E. Larson
Robert P. Hostetler
Carolyn F. Neptune

Supplements

Intermediate Algebra: Graphs and Functions, Second Edition by Larson, Hostetler, and Neptune is accompanied by a comprehensive supplements package. Most items are keyed directly to the text.

Printed Resources for the Instructor

Instructor's Annotated Edition by Larson, Hostetler, and Neptune
- Includes the entire student edition of the text
- Answers to all exercises and tests
- Teaching tips at point of use
- Additional examples and exercises with answers at point of use

Instructor's Guide by Carolyn F. Neptune, Johnson County Community College
- Detailed solutions to all even-numbered Section Exercises
- Transparency Masters

Test Item File by David C. Falvo, The Pennsylvania State University, The Behrend College
- Over 4,000 test items keyed to the text by section and organized by objective
- Six Chapter Tests per chapter
- Questions given in both multiple-choice and fill-in formats
- Answers to all test items and to chapter tests
- Also available as a computerized test bank

Printed Resources for the Student

Student Study and Solutions Guide by Carolyn F. Neptune, Johnson County Community College
- Step-by-step solutions to all odd-numbered Section Exercises, all Review Exercises, and all Mid-Chapter Quiz, Chapter Test, and Cumulative Test problems

Graphing Technology Guide: Algebra by Benjamin N. Levy and Laurel Technical Services
- Keystroke instructions for a wide variety of Texas Instruments, Casio, Sharp, and Hewlett-Packard graphing calculators, including the most current models
- Examples with step-by-step solutions
- Extensive graphics screen output
- Technology tips

Media Resources for the Instructor

Computerized Testing
- Test-generating software for Windows and Macintosh
- Over 4,000 test items
- Also available as a printed test bank

Media Resources for the Student

Videotape Series by Dana Mosely
- Comprehensive section-by-section coverage
- Detailed explanation of important concepts
- Numerous examples and applications, often illustrated by computer-generated graphics

Tutorial Software
- Interactive tutorial software with comprehensive section-by-section coverage
- Diagnostic feedback
- Additional examples
- Chapter self-tests
- Glossary

How to Study Algebra

Studying Mathematics Studying mathematics is a linear process: The material you learn each day builds upon material you learned previously. There are no shortcuts—you must keep up with the coursework every day.

Making a Plan Make your own course plan right now! A good rule of thumb is to study two to four hours for every hour in class. After your first major test, you will know if your efforts were sufficient. If you did not make the grade you wanted, then you should increase your study time, improve your study efficiency, or both.

Preparing for Class Before class, review your notes from the previous class. Then, read the portion of the text that is to be covered, paying special attention to the definitions and rules that are highlighted. This takes self-discipline, but it pays off because you will benefit much more from your instructor's presentation.

Attending Class Attend every class. Arrive on time with your text, a pen or pencil and paper for notes, and your calculator. If you must miss a class, get the notes from another student, go to your tutor for help, or view the appropriate mathematics videotape. You *must* learn the material that was covered in the missed class before attending the next class.

Participating in Class As you read the text before class, write down any questions you may have about the material. Ask your instructor these questions during class. This way, you will understand the material better, and you will be prepared to do your homework.

Taking Notes During class, take notes on definitions, examples, concepts, and rules. Focus on the instructor's cues to identify important material. Then, as soon after class as possible, review your notes and add any explanations that are necessary to make your notes understandable *to you.*

Doing the Homework Learning algebra is like learning to play the piano or basketball. You cannot develop skills just by watching someone do it; you must do it yourself. The best time to do your homework is right after class, when the concepts are still fresh in your mind. This increases your chances of retaining the information in long-term memory.

Finding a Study Partner When you get stuck on a problem, it may help to work with a partner. Even if you feel you are giving more help than you are getting, you will find that teaching others is an excellent way to learn.

Building a Math Library Build a library of books that can help you with your math courses. Consider using the *Student Study and Solutions Guide* for this text. As you will probably take other math courses after this one, we suggest that you keep the text. It will be a valuable reference book. Adding computer software and math videotapes is another way to build your math library.

Assessing Your Progress   Refer to the *Strategies for Success* checklist at the beginning of each chapter, and use the *Tools and Skills* lists to track your progress through the material. Halfway through the chapter, take the Mid-Chapter Quiz as if you were in class, then check your answers in the back of the text.

Keeping Up with the Work   Don't let yourself fall behind in the course. If you are having trouble, seek help immediately—from your instructor, a math tutor, your study partner, or additional study aids such as videotapes and software tutorials. Remember: If you have trouble with one section of your algebra text, there's a good chance that you will have trouble with later sections unless you take steps to improve your understanding.

Keeping Your Skills Sharp   In the Appendix we have included short sets of *Warm-Up Exercises* that correspond to each section in the text (except Section 1.1). These exercises will help you review skills that you learned in previous exercises and retain them in long-term memory. These sets are designed to take only a few minutes to solve. We suggest working the entire set before you start each new exercise set. (All of the *Warm-Up Exercises* are answered in the Appendix as well.)

Getting Stuck   Every math student has had this experience: You work a problem and cannot solve it, or the answer you get does not agree with the one given in the text. When this happens, you might ask for help, take a break to clear your thoughts, sleep on it, rework the problem, or reread the section in the text. Avoid getting frustrated or spending too much time on a single problem.

Checking Your Work   In algebra, you can check your work by testing your solution. One way is to plug the answer into the equation, then solve it to see if the numbers on each side of the equation are equal. Another way is to rework the problem on a separate sheet of paper and compare your answers. If they don't match, compare each step of the solutions to find the error. If you work on your "checking skills" as well as your "solving skills," your test scores should improve.

Preparing for Tests   Cramming for an algebra test seldom works. If you have kept up with the work and followed the suggestions given here, you should be almost ready for the test. To prepare for the chapter test, go over the Mid-Chapter Quiz, work the Review Exercises, and review the Skills list at the beginning of the chapter and the Summary at the end. Set aside an hour to take the sample Chapter Test and another for the sample Cumulative Test. Analyze the results of your Chapter and Cumulative Tests to locate and correct test-taking errors.

Taking a Test   Most instructors do not recommend studying right up to the minute the test begins. This practice tends to make people anxious. The best cure for test-taking anxiety is to prepare well in advance. Once the test has begun, read the directions carefully and work at a reasonable pace. (You might want to read the entire test first, then work the problems in the order in which you feel most comfortable.) People who hurry tend to make careless errors. If you finish early, take a few moments to clear your thoughts and then go over your work.

Learning from Mistakes   When your test is returned to you, go over any errors you might have made. This will help you avoid repeating some systematic or conceptual errors. Don't dismiss an error as just a "dumb mistake." Take advantage of any mistakes by hunting for ways to improve your test-taking skills.

What Is Algebra?

To some, algebra is manipulating symbols or performing mathematical operations with letters instead of numbers. To others, it is factoring, solving equations, or solving word problems. And to still others, algebra is a mathematical language that can be used to model real-world problems. In fact, algebra is all of these!

As you study this text, it is helpful to view algebra from the "big picture"—to see how the various rules, operations, and strategies fit together.

The rules of arithmetic are generalized through the use of symbols and letters to form the basic rules of algebra. The algebra rules are used to *rewrite* algebraic expressions and equations in new, more useful forms. The ability to write algebraic expressions and equations is needed in all three major components of algebra—*simplifying* expressions, *solving* equations, and *graphing* functions. The following chart shows how this text fits into the "big picture" of algebra.

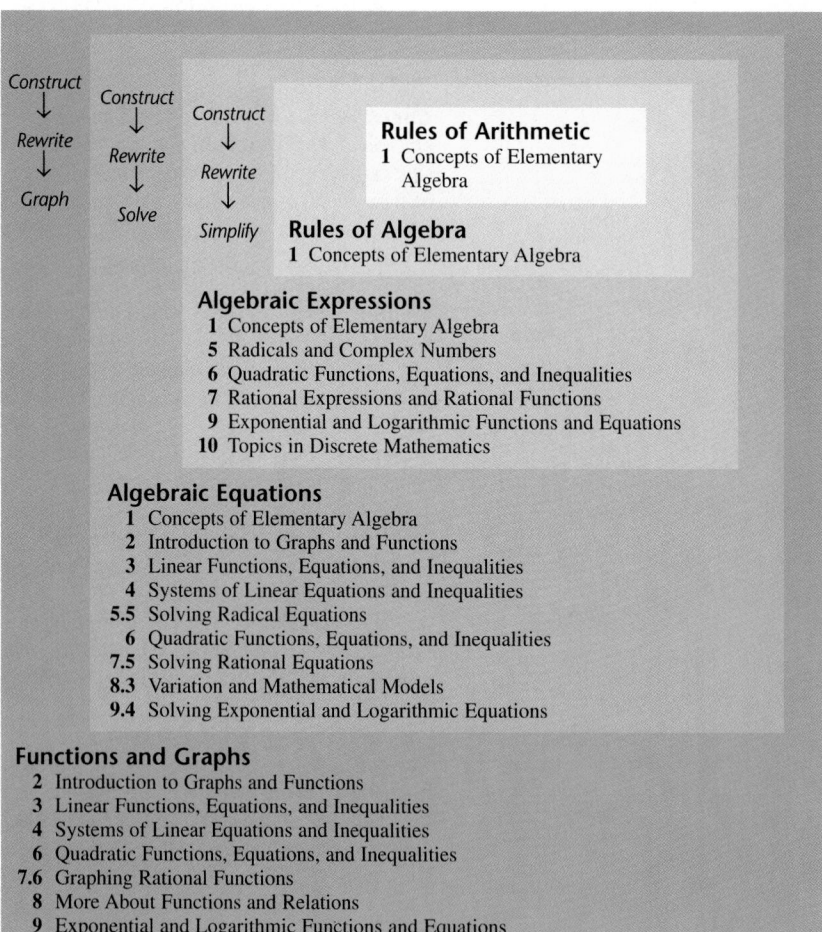

Construct → *Rewrite* → *Graph*

Construct → *Rewrite* → *Solve*

Construct → *Rewrite* → *Simplify*

Rules of Arithmetic
1 Concepts of Elementary Algebra

Rules of Algebra
1 Concepts of Elementary Algebra

Algebraic Expressions
1 Concepts of Elementary Algebra
5 Radicals and Complex Numbers
6 Quadratic Functions, Equations, and Inequalities
7 Rational Expressions and Rational Functions
9 Exponential and Logarithmic Functions and Equations
10 Topics in Discrete Mathematics

Algebraic Equations
1 Concepts of Elementary Algebra
2 Introduction to Graphs and Functions
3 Linear Functions, Equations, and Inequalities
4 Systems of Linear Equations and Inequalities
5.5 Solving Radical Equations
6 Quadratic Functions, Equations, and Inequalities
7.5 Solving Rational Equations
8.3 Variation and Mathematical Models
9.4 Solving Exponential and Logarithmic Equations

Functions and Graphs
2 Introduction to Graphs and Functions
3 Linear Functions, Equations, and Inequalities
4 Systems of Linear Equations and Inequalities
6 Quadratic Functions, Equations, and Inequalities
7.6 Graphing Rational Functions
8 More About Functions and Relations
9 Exponential and Logarithmic Functions and Equations

Concepts of Elementary Algebra

Strategies for Success

SKILLS *When you have completed this chapter, make sure you are able to:*

○ Simplify expressions by removing symbols of grouping and combining like terms

○ Add, subtract, and multiply polynomials

○ Solve linear and literal equations

○ Factor expressions completely and solve polynomial equations by factoring

○ Use expressions or equations to solve real-life problems

TOOLS *Use these study tools to master the skills above:*

○ Mid-Chapter Quiz   (page 54)

○ Chapter Summary   (page 98)

○ Review Exercises   (page 99)

○ Chapter Test   (page 102)

In Exercise 88 in Section 1.7, you will solve a linear equation to find how long it takes for a drop of water to travel from the base of a water fountain to the maximum spraying height.

| **1.1** | **Operations with Real Numbers** |

Real Numbers ▪ Order and Distance ▪ Absolute Value ▪ Operations with Real Numbers ▪ Positive Integer Exponents ▪ Roots ▪ Order of Operations ▪ Calculators

Real Numbers

A **set** is a collection of objects. For instance, the set $\{1, 2, 3\}$ contains the three numbers 1, 2, and 3. In this text, a *pair* of braces $\{\quad\}$ always indicates that we are describing the members of a set. Parentheses $(\quad)$ and brackets $[\quad]$ are used to represent other ideas.

The set of numbers that is used in arithmetic is the set of **real numbers.** The term *real* distinguishes real numbers from *imaginary* numbers—a type of number that you will study later in this text. One of the most commonly used **subsets** of real numbers is the set of **natural numbers** or **positive integers**

$$\{1, 2, 3, 4, \ldots\}. \qquad \text{Set of natural numbers}$$

The three dots indicate that the pattern continues (the set also contains the numbers 5, 6, 7, and so on).

Positive integers can be used to describe many quantities that you encounter in everyday life—you might be taking four classes this term, or you might be paying $420 per month for rent. But even in everyday life, positive integers cannot describe some concepts accurately. For instance, you could have a zero balance in your checking account, or the temperature could be $-10°$ (10 degrees below zero). To describe such quantities, you need to expand the set of positive integers to include **zero** and the **negative integers.** The positive integers and zero make up the set of **whole numbers.** The expanded set containing the whole numbers and the negative integers is called the set of **integers,** which is written as follows.

Zero
$$\{\ldots, -3, -2, -1, 0, 1, 2, 3, \ldots\}$$
Negative integers   Positive integers

The set of integers is a subset of the set of real numbers, which is another way of saying that every integer is a real number. Even when the entire set of integers is used, there are still many quantities in everyday life that cannot be described accurately. The costs of many items are not in whole-dollar amounts, but in parts of dollars, such as $1.19 or $39.98. You might work $8\frac{1}{2}$ hours, or you might miss the first *half* of a movie. To describe quantities, you can expand the set of integers to include **fractions.** The expanded set is called the set of **rational numbers.** Formally, a real number is called **rational** if it can be written as the ratio p/q of two integers, where $q \neq 0$ (the symbol $\neq$ means **does not equal**).

Historical Note: The Egyptians had no symbol for zero. The earliest evidence we have of zero in Babylonia is from 150 A.D. The Mayans in Mexico were the first to use zero as both a number and a placeholder.

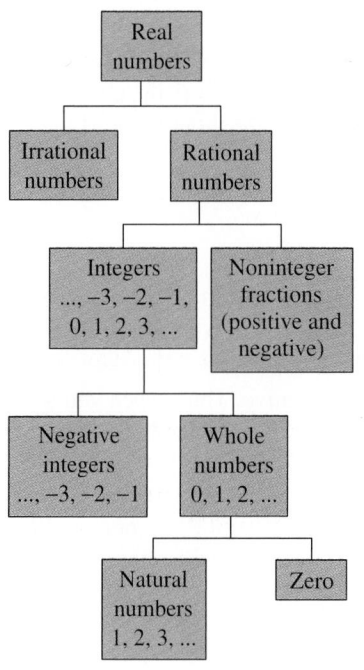

FIGURE 1.1   Subsets of real numbers

Here are some examples of rational numbers.

$$2 = \frac{2}{1}, \quad \frac{1}{3} = 0.333 \ldots, \quad \frac{1}{8} = 0.125, \quad \text{and} \quad \frac{125}{111} = 1.126126 \ldots$$

Real numbers that cannot be written as ratios of two integers are called **irrational.** For instance, the numbers

$$\sqrt{2} = 1.4142135 \ldots \quad \text{and} \quad \pi = 3.1415926 \ldots$$

are irrational. The decimal representation of a rational number is either **terminating** or **repeating.** For instance, the decimal representation of $\frac{1}{4} = 0.25$ is terminating, and the decimal representation of $\frac{4}{11} = 0.363636 \cdots = 0.\overline{36}$ is repeating. (The line over 36 indicates which digits repeat.)

The decimal representation of an irrational number neither terminates nor repeats. When performing operations with such numbers, you usually use a decimal approximation that has been **rounded.*** For instance, rounded to four decimal places, the decimal approximations of $\frac{2}{3}$ and π are

$$\frac{2}{3} \approx 0.6667 \quad \text{and} \quad \pi \approx 3.1416.$$

The symbol $\approx$ means **equals approximately.**

Figure 1.1 shows several commonly used subsets of real numbers and their relationships to each other.

EXAMPLE 1   Identifying Real Numbers

Which of the numbers in the following set are (a) natural numbers, (b) integers, (c) rational numbers, and (d) irrational numbers?

$$\left\{ -7, \, -\sqrt{3}, \, -1, \, -\frac{1}{5}, \, 0, \, \frac{3}{4}, \, \sqrt{2}, \, \pi, \, 5 \right\}$$

Solution

(a) Natural numbers: $\{5\}$

(b) Integers: $\{-7, \, -1, \, 0, \, 5\}$

(c) Rational numbers: $\left\{ -7, \, -1, \, -\frac{1}{5}, \, 0, \, \frac{3}{4}, \, 5 \right\}$

(d) Irrational numbers: $\left\{ -\sqrt{3}, \, \sqrt{2}, \, \pi \right\}$

Remind students that every integer can be written as *both* a fraction and a repeating decimal. For example, $18 = \frac{18}{1} = 18.000 \ldots$

---

*The rounding rule we use in this text is to round *up* if the succeeding digit is 5 or more and round *down* if the succeeding digit is 4 or less. For example, if you wanted to round 7.35 to one decimal place, you would round up to 7.4. Similarly, if you wanted to round 2.364 to two decimal places, you would round down to 2.36.

Order and Distance

A **real number line** can be used to picture the real numbers. It consists of a horizontal line with a point (the **origin**) labeled as 0. Numbers to the left of 0 are **negative** and numbers to the right of 0 are **positive,** as shown in Figure 1.2.

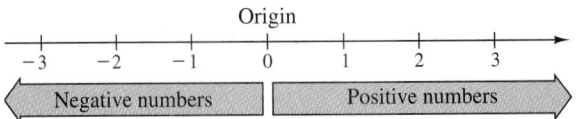

FIGURE 1.2

The real number zero is neither positive nor negative. Thus, when you want to talk about real numbers that might be positive or zero, you should use the term **nonnegative real number.**

Each point on the real number line corresponds to exactly one real number, and each real number corresponds to exactly one point on the real number line, as shown in Figure 1.3. When you draw the point (on the real number line) that corresponds to a real number, you are **plotting** the real number.

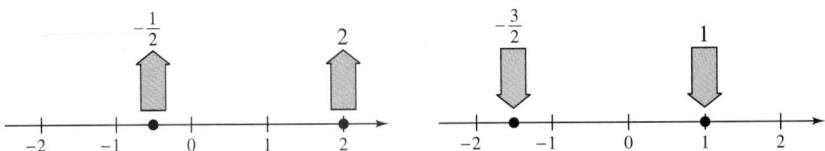

FIGURE 1.3

The real number line provides you with a way of comparing any two real numbers. For instance, if you choose any two (different) numbers on the real number line, one of the numbers must be to the left of the other number. The number to the left is **less than** the number to the right. Similarly, the number to the right is **greater than** the number to the left, as shown in Figure 1.4.

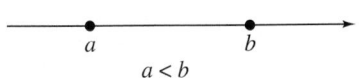

FIGURE 1.4 a is to the left of b.

Definition of Order on the Real Number Line

If the real number a lies to the left of the real number b on the real number line, then we say that a is **less than** b and write

   $a < b.$

This relationship can also be described by saying that b is **greater than** a and by writing $b > a$. The symbol $a \leq b$ means that a is **less than or equal to** b, and the symbol $b \geq a$ means that b is **greater than or equal to** a. The symbols $<$, $>$, $\leq$, and $\geq$ are called **inequality symbols.**

When you are asked to **order** two numbers, you are simply being asked to say which of the two numbers is greater.

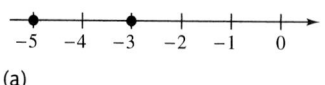

(a)

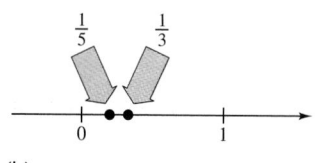

(b)

FIGURE 1.5

EXAMPLE 2 Ordering Real Numbers

Determine the correct inequality symbol ($<$ or $>$) between each pair.

(a) -3 ▢ -5 (b) $\frac{1}{5}$ ▢ $\frac{1}{3}$

Solution

(a) Because -3 lies to the right of -5, it follows that -3 is *greater than* -5.

$$-3 > -5 \qquad \text{See Figure 1.5(a).}$$

(b) Because $\frac{1}{5}$ lies to the left of $\frac{1}{3}$, it follows that $\frac{1}{5}$ is *less than* $\frac{1}{3}$.

$$\frac{1}{5} < \frac{1}{3} \qquad \text{See Figure 1.5(b).}$$

Once you know how to represent real numbers as points on the real number line, it is natural to talk about the **distance between two real numbers.** Specifically, if a and b are two real numbers such that $a \le b$, the distance between a and b is defined as $b - a$.

NOTE Note from this definition that if $a = b$, the distance between a and b is zero. If $a < b$, the distance between a and b is positive.

Definition of Distance Between Two Real Numbers

If a and b are two real numbers such that $a \le b$, then the **distance between a and b** is

(Distance between a and b) $= b - a$.

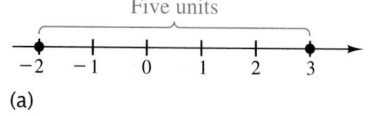

(a)

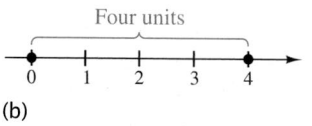

(b)

FIGURE 1.6

EXAMPLE 3 Finding the Distance Between Two Real Numbers

Find the distance between the real numbers in each pair.

(a) -2 and 3   (b) 0 and 4

Solution

(a) The distance between -2 and 3 is

$$3 - (-2) = 3 + 2 = 5. \qquad \text{See Figure 1.6(a).}$$

(b) The distance between 0 and 4 is

$$4 - 0 = 4. \qquad \text{See Figure 1.6(b).}$$

Absolute Value

The distance between a real number a and 0 (the origin) is called the **absolute value** of a. A pair of vertical bars $|\ \ |$ are used to denote absolute value. For example, $|-8|$ denotes the distance between -8 and 0. Thus, $|-8| = 8$.

NOTE   Be sure you see from this definition that the absolute value of a real number is never negative. For instance, if $a = -3$, then

$$|a| = |-3| = -(-3) = 3 = -a.$$

Definition of Absolute Value of a Real Number

The **absolute value** of a real number a is defined as the distance between a and 0 on the real number line.

Rule	Example
1. If $a \geq 0$, then $|a| = a$.	$|3| = 3$
2. If $a < 0$, then $|a| = -a$.	$|-2| = -(-2) = 2$

The absolute value of any number is either positive or zero. Moreover, zero is the only number whose absolute value is zero. That is, $|0| = 0$.

Two real numbers are **opposites** of each other if they lie the same distance from, but on opposite sides of, zero. For example, -2 is the opposite of 2. Because opposite numbers lie the same distance from zero on the real number line, they have the same absolute value. Thus, $|5| = 5$ and $|-5| = 5$.

For any two real numbers a and b, exactly one of the following must be true: $a < b$, $a = b$, or $a > b$. This property is called the **Law of Trichotomy.**

EXAMPLE 4   *Comparing Real Numbers*

Determine the correct symbol ($<$, $>$, or $=$) between each pair.

(a) $|-9|$ ▦ $|9|$ (b) $|-2|$ ▦ 1 (c) -4 ▦ $-|-4|$

Solution

(a) $|-9| = |9|$, because both are equal to 9.

(b) $|-2| > 1$, because $|-2| = 2$ and 2 is greater than 1.

(c) $-4 = -|-4|$, because both numbers are equal to -4.

NOTE   Here is an example of using absolute value to find the distance between two real numbers. The distance between -2 and 1 is

$$|-2 - 1| = |-3| = 3.$$

When we defined the distance between two real numbers a and b as $b - a$, the definition included the restriction $a \leq b$. Using absolute value, you can generalize this definition as follows. If a and b are *any* two real numbers, the distance between a and b is

(Distance between a and b) $= |b - a| = |a - b|$.

Operations with Real Numbers

There are four basic operations of arithmetic: addition, subtraction, multiplication, and division.

The result of **adding** two real numbers is called the **sum** of the two numbers, and the two real numbers are called **terms** of the sum. **Subtraction** of one real number from another can be described as *adding the opposite* of the second number to the first number. For instance,

$$7 - 5 = 7 + (-5) = 2$$

and

$$10 - (-13) = 10 + 13 = 23.$$

The result of subtracting one real number from another is called the **difference** of the two numbers.

EXAMPLE 5   Adding and Subtracting Real Numbers

(a) $-84 + 14 = -70$

(b) $6 + (-13) + 10 = 3$

(c) $-13.8 - 7.02 = -13.8 + (-7.02) = -20.82$

(d) $\dfrac{1}{5} - \dfrac{2}{5} = \dfrac{1-2}{5} = -\dfrac{1}{5}$

(e) To add two fractions with *unlike denominators*, you must first rewrite one (or both) of the fractions so that they have the same denominator.

$$\frac{3}{8} + \frac{5}{12} = \frac{3(3)}{8(3)} + \frac{5(2)}{12(2)} \qquad \text{Common denominator is 24.}$$

$$= \frac{9}{24} + \frac{10}{24} \qquad \text{Fractions have like denominators.}$$

$$= \frac{9+10}{24} \qquad \text{Add numerators.}$$

$$= \frac{19}{24}$$

The result of **multiplying** two real numbers is called their **product,** and each of the two numbers is called a **factor** of the product. The product of zero and any other number is zero. For instance, if you multiply 0 and 4, you obtain $(0)(4) = 0$.

Multiplication is denoted in a variety of ways. For instance,

$$7 \times 3, \quad 7 \cdot 3, \quad 7(3), \quad (7)3, \quad \text{and} \quad (7)(3)$$

all denote the product of "7 times 3," which you know is 21.

EXAMPLE 6   *Multiplying Real Numbers*

(a) $(-5)(-7) = 35$

(b) $(-1.2)(0.4) = -0.48$

(c) To find the product of more than two numbers, find the product of their absolute values. If there is an *even* number of negative factors, the product is positive. If there is an *odd* number of negative factors, the product is negative. For instance, in the following product there are two negative factors, so the product must be positive, and you can write

$$5(-3)(-4)(7) = 420.$$

(d) To multiply two fractions, multiply their numerators and their denominators. For instance, the product of $\frac{2}{3}$ and $\frac{4}{5}$ is

$$\left(\frac{2}{3}\right)\left(\frac{4}{5}\right) = \frac{(2)(4)}{(3)(5)}$$

$$= \frac{8}{15}.$$

The **reciprocal** of a nonzero real number a is defined as the number by which a must be multiplied to obtain 1. For instance, the reciprocal of 3 is $\frac{1}{3}$ because

$$3\left(\frac{1}{3}\right) = 1.$$

Similarly, the reciprocal of $-\frac{4}{5}$ is $-\frac{5}{4}$ because

$$-\frac{4}{5}\left(-\frac{5}{4}\right) = 1.$$

In general, the reciprocal of a/b is b/a. Note that the reciprocal of a positive number is positive, and the reciprocal of a negative number is negative. Also, be sure you see that zero does not have a reciprocal because there is no number that can be multiplied by zero to obtain 1.

To divide one real number by a second (nonzero) real number, multiply the first number by the reciprocal of the second number. The result of dividing two real numbers is called the **quotient** of the two numbers. The quotient of a and b can be written as

$$a \div b, \quad a/b, \quad \frac{a}{b}, \quad \text{or} \quad b\overline{)a}.$$

The number a is called the **numerator** (or **dividend**), and the number b is called the **denominator** (or **divisor**).

EXAMPLE 7   *Dividing Real Numbers*

NOTE   Division by zero is *not* defined. For instance, the expression $\frac{3}{0}$ is undefined.

(a) $-30 \div 5 = -30\left(\frac{1}{5}\right) = -\frac{30}{5} = -6$

(b) $-\frac{9}{14} \div -\frac{1}{3} = -\frac{9}{14}\left(-\frac{3}{1}\right) = \frac{27}{14}$

(c) $\frac{5}{16} \div 2\frac{3}{4} = \frac{5}{16} \div \frac{11}{4} = \frac{5}{16}\left(\frac{4}{11}\right) = \frac{5(4)}{4(4)(11)} = \frac{5}{44}$

Remind students that $-a/b$ can be written as either $(-a)/b$ or $a/(-b)$.

(d) $\dfrac{-\frac{2}{3}}{\frac{3}{5}} = -\frac{2}{3} \div \frac{3}{5} = -\frac{2}{3}\left(\frac{5}{3}\right) = -\frac{10}{9}$

The answers to parts (b) and (d) are written as *improper fractions*. They could also have been written as *mixed numbers*. For instance,

$$\frac{27}{14} = 1\frac{13}{14} \qquad \text{and} \qquad -\frac{10}{9} = -1\frac{1}{9}.$$

Positive Integer Exponents

Let n be a positive integer and let a be a real number. Then the product of n factors of a is given by

$$a^n = \underbrace{a \cdot a \cdot a \cdots a}_{n \text{ factors}}.$$

In the **exponential form** a^n, a is called the **base** and n is the **exponent.** When you write the exponential form a^n, you are **raising a to the nth power.**

When a number is raised to the *first* power, you usually do not write the exponent 1. For instance, you would usually write 5 rather than 5^1.

EXAMPLE 8   *Evaluating Exponential Expressions*

(a) $(-3)^4 = (-3)(-3)(-3)(-3) = 81$

(b) $-3^4 = -(3)(3)(3)(3) = -81$

(c) $-(-3)^4 = -(-3)(-3)(-3)(-3) = -81$

(d) $\left(\frac{2}{5}\right)^3 = \left(\frac{2}{5}\right)\left(\frac{2}{5}\right)\left(\frac{2}{5}\right) = \frac{8}{125}$

(e) $(-2)^5 = (-2)(-2)(-2)(-2)(-2) = -32$

(f) $-2^5 = -(2)(2)(2)(2)(2) = -32$

Roots

When you multiply a number by itself, you are **squaring** the number. For instance, $5^2 = 5 \cdot 5 = 25$. To undo this "squaring operation," you can take the **square root** of 25. Similarly, you can undo a cubing operation by taking a **cube root.**

Number	Equal Factors	Root
$25 = 5^2$	$5 \cdot 5$	5 (square root)
$25 = (-5)^2$	$(-5)(-5)$	-5 (square root)
$9 = 3^2$	$3 \cdot 3$	3 (square root)
$-27 = (-3)^3$	$(-3)(-3)(-3)$	-3 (cube root)
$16 = 2^4$	$2 \cdot 2 \cdot 2 \cdot 2$	2 (fourth root)

In general, the **nth root** of a number is defined as follows.

Definition of nth Root of a Number

Let a and b be real numbers and let n be a positive integer such that $n \geq 2$. If

$$a = b^n,$$

then b is an **nth root of a.** If $n = 2$, the root is a **square root,** and if $n = 3$, the root is a **cube root.**

Historical Note: In the 16th century, as many as four different symbols were used to indicate a square root. Other symbols were used for cube roots.

By applying this definition, you can see that some numbers have more than one nth root. For example, both 5 and -5 are square roots of 25 because $25 = 5^2$ and $25 = (-5)^2$. Similarly, both 2 and -2 are fourth roots of 16 because $16 = 2^4$ and $16 = (-2)^4$. To avoid ambiguity, if a is a real number, the **principal nth root of a** is defined as the nth root that has the same sign as a, and it is denoted by the **radical symbol**

$$\sqrt[n]{a}. \qquad \text{Principal } n\text{th root}$$

NOTE The number -4 has no real square root because there is no real number that can be squared to produce -4. Thus, $\sqrt{-4}$ is not a real number. This same comment can be made about any negative number. That is, if a is negative, $\sqrt{a}$ is not a real number.

The positive integer n is called the **index** of the radical, and the number a is called the **radicand.** If $n = 2$, we omit the index and write $\sqrt{a}$ rather than $\sqrt[2]{a}$. "Having the same sign" means that the principal nth root of a is positive if a is positive and negative if a is negative. For example, $\sqrt{4} = 2$ and $\sqrt[3]{-8} = -2$.

EXAMPLE 9 Finding the Principal nth Root of a Number

(a) $\sqrt{36} = 6$ (b) $\sqrt{0} = 0$ (c) $\sqrt[3]{8} = 2$

(d) $\sqrt[3]{-27} = -3$ (e) $\sqrt[4]{256} = 4$

Order of Operations

An established **order of operations** helps avoid confusion when one is evaluating numerical expressions.

Order of Operations

1. Perform operations that occur within grouping symbols such as parentheses or brackets.

2. Evaluate powers and roots.

3. Perform multiplications and divisions from left to right.

4. Perform additions and subtractions from left to right.

EXAMPLE 10   Evaluating Expressions Using Order of Operations

This is a good time to demonstrate informally the use of the calculator. Have students do each of the following by hand and then check their answers on the calculator.

$3 + 8 \div 2 + 1 = 8$

$5 - 2^2 \cdot 3 + 7 = 0$

(a) $-4 + 2(-2 + 5)^2 = -4 + 2(3)^2$ Add within parentheses.

$\qquad\qquad\qquad = -4 + 2(9)$ Evaluate the power.

$\qquad\qquad\qquad = -4 + 18$ Multiply.

$\qquad\qquad\qquad = 14$ Add.

(b) $(-4 + 2)(-2 + 5)^2 = (-2)(3)^2$ Add within parentheses.

$\qquad\qquad\qquad = (-2)(9)$ Evaluate the power.

$\qquad\qquad\qquad = -18$ Multiply.

(c) $5 - 12 \div 3 - 7 = 5 - 4 - 7$ Divide first.

$\qquad\qquad\qquad = -6$ Subtract from left to right.

(d) $(-3)(-2)^2 = (-3)(4)$ Evaluate the power.

$\qquad\qquad\qquad = -12$ Multiply.

(e) $(-3)(-2^2) = (-3)(-4)$ Evaluate the power.

$\qquad\qquad\qquad = 12$ Multiply.

The symbols for absolute value, radicals, and fraction bars are given the same order of operation as a grouping symbol. Here are two examples.

$$5 - |3 - 4| = 5 - |-1| = 5 - 1 = 4$$

$$2 - \frac{9 - 3}{2} = 2 - \frac{6}{2} = 2 - 3 = -1$$

Calculators

This book includes many examples and exercises that are best done with a calculator. Most of these can be done with a scientific calculator *or* a graphing calculator. (Some require a graphing calculator.)

EXAMPLE 11 *Evaluating Expressions on a Calculator*

Expression	Keystrokes	Display	
(a) $-4 - 5$	4 [+/−] [−] 5 [=]	-9	Scientific
$-4 - 5$	[(−)] 4 [−] 5 [ENTER]	-9	Graphing
(b) $-3^2 + 4$	3 [x^2] [+/−] [+] 4 [=]	-5	Scientific
$-3^2 + 4$	[(−)] 3 [x^2] [+] 4 [ENTER]	-5	Graphing
(c) $(-3)^2 + 4$	[(] 3 [+/−] [)] [x^2] [+] 4 [=]	13	Scientific
$(-3)^2 + 4$	[(] [(−)] 3 [)] [x^2] [+] 4 [ENTER]	13	Graphing
(d) $7 - \sqrt{5}$	7 [−] 5 [√] [=]	4.764	Scientific
$7 - \sqrt{5}$	7 [−] [√] 5 [ENTER]	4.764	Graphing

NOTE In Example 11(d), the display on most calculators will show more digits. For instance, your calculator might display 4.763932023. Rounded to three decimal places, the result is 4.764.

On a scientific calculator, notice the difference between the change sign key [+/−] and the subtraction key [−]. On a graphing calculator, the negation key [(−)] and the subtraction key [−] may *not* perform the same operations.

Incorrectly negating a number on the calculator is one of the most common mistakes students make. Be sure to stress proper use of calculators.

EXAMPLE 12 *Evaluating Expressions on a Calculator*

Expression	Keystrokes	Display	
(a) $24 \div 2^3$	24 [÷] 2 [y^x] 3 [=]	3	Scientific
$24 \div 2^3$	24 [÷] 2 [^] 3 [ENTER]	3	Graphing
(b) $(24 \div 2)^3$	[(] 24 [÷] 2 [)] [y^x] 3 [=]	1728	Scientific
$(24 \div 2)^3$	[(] 24 [÷] 2 [)] [^] 3 [ENTER]	1728	Graphing
(c) $\dfrac{5}{4 + 3 \cdot 2}$	5 [÷] [(] 4 [+] 3 [×] 2 [)] [=]	0.5	Scientific
$\dfrac{5}{4 + 3 \cdot 2}$	5 [÷] [(] 4 [+] 3 [×] 2 [)] [ENTER]	0.5	Graphing

----

Historical Note: Technologically, we have made significant progress in calculator development. Blaise Pascal (1623–1662), the French mathematician, invented the first mechanical calculator in 1642 at the young age of 18. He wanted to help his father, an accountant of sorts, to compute more easily. The Pascaline calculator is currently on display at the Museum of Technology in Paris.

When keystrokes are given in this text, remember that the keystroke sequences listed may not agree precisely with the steps required by *your* calculator. So be sure you are familiar with the use of the keys on your own calculator.

Many calculators follow the same order of operations that we use in algebra. Some calculators, however, use a different order of operations—so be sure that you understand the order of operations used by your own calculator. If you are unsure of the order of operations used by your calculator, you should insert enough parentheses to make sure the calculator performs the operations in the order you intend.

EXAMPLE 13   *Using Parentheses with a Calculator*

Suppose you live in a state that charges 6% sales tax. You are buying two items that cost $24.95 and $36.95. The sales clerk uses a calculator to compute your sales tax as follows.

24.95 ⊞ 36.95 ⊠ .06 [ENTER]

From the calculator display, the sales tax appears to be $27.17. Both you and the sales clerk realize that this is too much. What went wrong?

Solution

The order of operations used by the calculator was

$$24.95 + (36.95)(0.06) \approx 24.95 + 2.22 = 27.17. \qquad \text{Calculator operations}$$

The correct sales tax of $3.71 can be obtained by using the following calculator steps.

⦅ 24.95 ⊞ 36.95 ⦆ ⊠ .06 [ENTER]

Group Activities   **Calculator Operations**

Example 13 describes a real-life situation in which a calculator could be used incorrectly because its user did not understand the calculator's order of operations. With others in your group, describe a different real-life example in which a calculator could give an incorrect result. Use your own calculator to demonstrate both the incorrect result and the correct result.

1.1 Exercises

In Exercises 1 and 2, determine which of the real numbers are (a) natural numbers, (b) integers, (c) rational numbers, and (d) irrational numbers.

1. $\left\{-10, -\sqrt{5}, -\frac{2}{3}, -\frac{1}{4}, 0, \frac{5}{8}, 1, \sqrt{3}, 4, 2\pi, 6\right\}$
2. $\left\{-\frac{7}{2}, -\sqrt{6}, -\frac{\pi}{2}, -\frac{3}{8}, 0, \sqrt{15}, \frac{10}{3}, 8, 245\right\}$

In Exercises 3–6, list the numbers satisfying the specified requirements.

3. The integers between -5.8 and 3.2
4. The even integers between -2.1 and 10.5
5. The odd integers between 0 and 3π
6. All prime numbers between 0 and 25

In Exercises 7 and 8, locate the real numbers on the real number line.

7. (a) 3 (b) $\frac{5}{2}$ (c) $-\frac{7}{2}$ (d) -5.2
8. (a) 8 (b) $\frac{4}{3}$ (c) -6.75 (d) $-\frac{9}{2}$

In Exercises 9 and 10, approximate the two numbers and order them.

9.

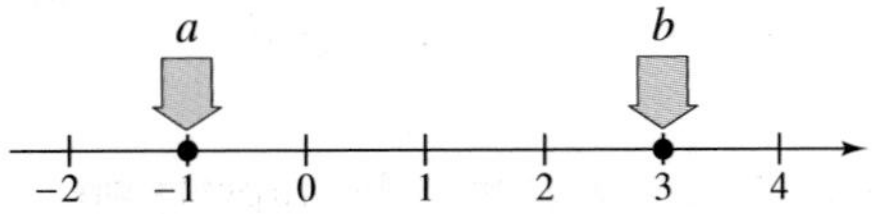

10.

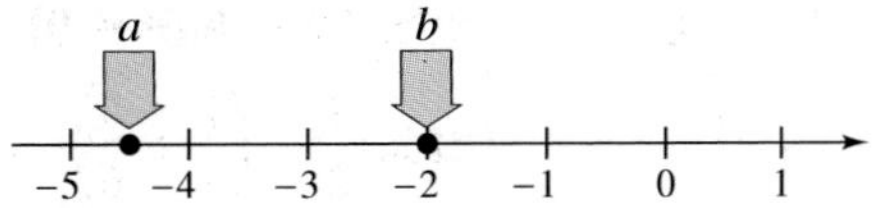

In Exercises 11–16, plot the two real numbers on the real number line and place the correct inequality symbol (< or >) between the two numbers.

11. 2 ___ 5
12. 8 ___ 3
13. -7 ___ -2
14. -2 ___ -5
15. $-\frac{2}{3}$ ___ $-\frac{10}{3}$
16. $\frac{11}{4}$ ___ π

17. *Think About It* Is there a difference between saying that a real number is positive and saying that a real number is nonnegative? Explain your answer.
18. Does -8 or 6 lie farther from the real number -4? Explain your answer.

In Exercises 19–24, write the statement using inequality notation.

19. x is negative.
20. y is more than 25.
21. u is at least 16.
22. z is greater than 2 and no more than 10.
23. The price p of a coat will be less than \$225 during the sale.
24. The tire pressure p must be at least 30 pounds per square inch and no more than 35 pounds per square inch.

In Exercises 25–30, find the distance between the two real numbers.

25. 4 and 10
26. 75 and 20
27. 18 and -32
28. -54 and 32
29. -35 and 0
30. 0 and 35

In Exercises 31–36, evaluate the quantity.

31. $|10|$
32. $|62|$
33. $-|3.5|$
34. $|-6|$
35. $-|-25|$
36. $-\left|\frac{3}{4}\right|$

In Exercises 37–40, place the correct symbol (<, >, or =) between the two real numbers.

37. $|-6|$ ___ $|2|$
38. $|150|$ ___ $|-310|$
39. $\left|-\frac{3}{4}\right|$ ___ $-\left|\frac{4}{5}\right|$
40. $-|-16.8|$ ___ $-|16.8|$

In Exercises 41–44, find the opposite and the absolute value of the number.

41. 14

42. -22.5

43. $-\frac{5}{4}$

44. π

In Exercises 45–48, plot the number and its opposite on the real number line.

45. -3

46. 3.5

47. $\frac{5}{3}$

48. $-\frac{3}{4}$

In Exercises 49–66, perform the indicated operation(s).

49. $13 + 32$

50. $16 + 84$

51. $-13 + 32$

52. $16 + (-84)$

53. $-7 - 15$

54. $-5 + (-52)$

55. $5.8 - 6.2 + 1.1$

56. $46.08 - 35.1 - 16.25$

57. $\frac{3}{4} - \frac{1}{4}$

58. $\frac{5}{6} + \frac{7}{6}$

59. $\frac{5}{8} + \frac{1}{4} - \frac{5}{6}$

60. $\frac{3}{11} + \frac{-5}{2}$

61. $5\frac{3}{4} + 7\frac{3}{8}$

62. $8\frac{1}{2} - 24\frac{2}{3}$

63. $-(-11.325) + |34.625|$

64. $|-16.25| - 54.78$

65. $-|-15.667| - 12.333$

66. $-\left|-15\frac{2}{3}\right| - 12\frac{1}{3}$

In Exercises 67–72, find the product without using a calculator.

67. $5(-6)$

68. $-7(3)$

69. $6.3(5.1)$

70. $(-4.4)(-3.2)$

71. $\left(-\frac{5}{8}\right)\left(-\frac{4}{5}\right)$

72. $\frac{2}{3}\left(-\frac{18}{5}\right)\left(-\frac{5}{6}\right)$

In Exercises 73–78, find the quotient.

73. $\frac{-18}{-3}$

74. $-\frac{30}{-15}$

75. $-\frac{4}{5} \div \frac{8}{25}$

76. $\frac{-11/12}{5/24}$

77. $5\frac{3}{4} \div 2\frac{1}{8}$

78. $-3\frac{5}{6} \div -2\frac{2}{3}$

In Exercises 79 and 80, find the quotient and round the result to two decimal places. (A calculator may be useful.)

79. $\dfrac{25.5}{6.325}$

80. $\dfrac{265.45}{25.6}$

In Exercises 81 and 82, write the quantity as a repeated multiplication problem.

81. $(-3)^4$

82. $\left(\frac{2}{3}\right)^3$

In Exercises 83 and 84, write the repeated multiplication problem using exponent notation.

83. $(-5)(-5)(-5)(-5)$

84. $-(5 \times 5 \times 5 \times 5 \times 5 \times 5)$

In Exercises 85–90, evaluate the exponential expression.

85. $(-4)^3$

86. $(-3)^4$

87. -5^2

88. -3^5

89. $\left(-\frac{7}{8}\right)^2$

90. $\left(\frac{2}{3}\right)^4$

In Exercises 91–98, find the root.

91. $\sqrt{81}$

92. $\sqrt{100}$

93. $\sqrt[3]{125}$

94. $\sqrt[3]{64}$

95. $\sqrt[3]{-8}$

96. $-\sqrt[4]{1}$

97. $-\sqrt[3]{-125}$

98. $\sqrt[5]{32}$

In Exercises 99–104, evaluate the expression.

99. $16 - 5(6 - 10)$

100. $72 - 8(6^2 \div 9)$

101. $\dfrac{3^2 - 5}{12} - 3\frac{1}{6}$

102. $\dfrac{3}{2} - \left(\dfrac{1}{3} \div \dfrac{5}{6}\right)$

103. $0.2(6 - 10)^3 + 85$

104. $\dfrac{5^3 - 50}{-15} + 27$

In Exercises 105–112, evaluate the expression using a calculator. Round the result to two decimal places.

105. $5.6[13 - 2.5(-6.3)]$

106. $35(1032 - 4650)$

107. $5^6 - 3(400)$

108. $5(100 - 3.6^4) \div 4.1$

109. $\dfrac{500}{(1.055)^{20}}$

110. $\dfrac{265.45}{25.6}$

111. $\sqrt{9^2 + 7.5^2}$

112. $\sqrt{9.4^2 - 4(2)(6)}$

113. *Think About It*   Is it true that $3 \cdot 4^2 = 12^2$? Explain.

114. *Company Profits*   The annual profit for a company (in millions of dollars) is shown in the bar graph. Complete the table showing the increase or decrease in profit from the preceding year.

Year	Yearly gain or loss
1993	
1994	
1995	
1996	
1997	

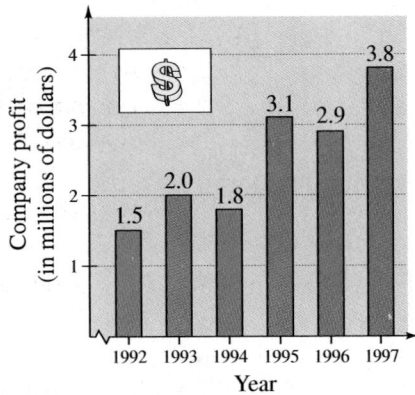

115. *Balance in an Account*   During one month, you made the following transactions in your checking account.

Initial Balance:	$2618.68
Deposit:	$1236.45
Withdrawal:	$25.62
Withdrawal:	$455.00
Withdrawal:	$125.00
Withdrawal:	$715.95

Find the balance at the end of the month. (Disregard any interest that may have been earned.)

116. *Savings Plan*   Suppose you decide to save $50 per month for 18 years. How much money will you set aside during the 18 years?

In Exercises 117 and 118, determine the unknown fractional part of the circle graph.

117.

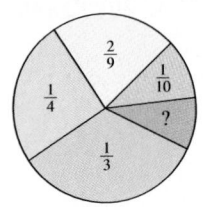

118.

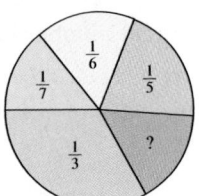

Area   In Exercises 119–122, find the area of the figure. (The area of a rectangle is $A = lw$, and the area of a triangle is $A = \frac{1}{2}bh$.)

119.

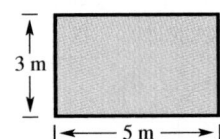

120.

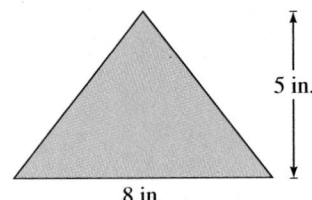

121.

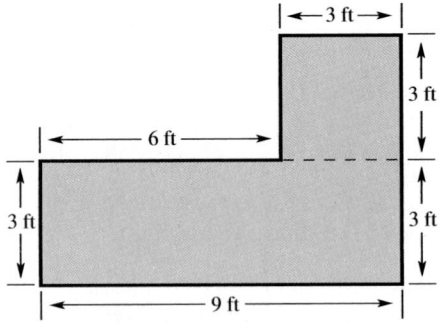

122.

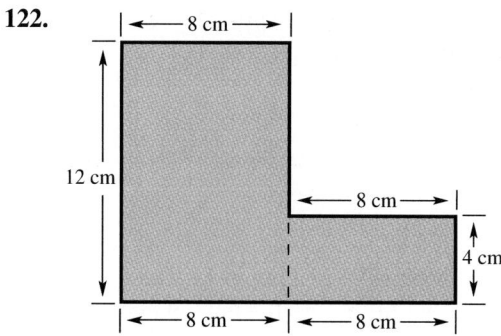

Volume In Exercises 123 and 124, use the following information to answer the questions. A bale of hay is a rectangular solid having the dimensions shown in the figure and weighs approximately 50 pounds. (The volume of a rectangular solid is $V = lwh$.)

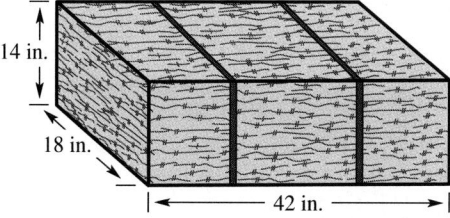

123. Find the volume of a bale of hay in cubic feet if the volume of a rectangular solid is the product of its length, width, and height. (Use the fact that $1728 \text{ in.}^3 = 1 \text{ ft}^3$.)

124. Approximate the number of bales in a ton of hay. Then approximate the volume of a stack of baled hay that weighs 12 tons.

True or False? In Exercises 125–130, determine whether the statement is true or false. Explain your reasoning.

125. Every integer is a rational number.

126. Every rational number is an integer.

127. The reciprocal of every nonzero integer is an integer.

128. The reciprocal of every nonzero rational number is a rational number.

129. $\frac{2}{3} + \frac{3}{2} = \frac{2+3}{3+2} = 1$

130. If a negative real number is raised to the 11th power, the result will be positive.

In Exercises 131 and 132, a calculator was used incorrectly to evaluate the expression. The keystrokes and display are shown. Describe the error and correct it.

131. *Expression:* $523 - (145 - 136)$

Keystrokes: 523 [−] 145 [−] 136 [ENTER]

Display: 242

132. *Expression:* $\frac{126 + 37}{4}$

Keystrokes: 126 [+] 37 [÷] 4 [ENTER]

Display: 135.25

Section Project

UPC Codes Packaged products sold in the United States have a Universal Product Code (UPC) or bar code as shown below. The following algorithm is used on the first 11 digits and the result should equal the 12th digit, called the check digit.

1. Add the numbers in the odd-numbered positions together. Multiply by 3.

2. Add the numbers in the even-numbered positions together.

3. Add the results of Steps 1 and 2.

4. Subtract the result of Step 3 from the next highest multiple of 10.

0 25215 04658 2

First digit   Check digit

Using the algorithm on this UPC produces $(0 + 5 + 1 + 0 + 6 + 8)(3) + (2 + 2 + 5 + 4 + 5) = 78$. The next highest multiple of 10 is 80, so $80 - 78 = 2$, which is the check digit.

(a) Does a UPC of 0 76737 20012 9 check? Explain.

(b) Does a UPC of 0 41800 48700 3 check? Explain.

1.2   Properties of Real Numbers

Mathematical Systems ▪ Basic Properties of Real Numbers ▪
Additional Properties of Real Numbers

Mathematical Systems

In this section, we will review the properties of real numbers. These properties make up the third component of what is called a **mathematical system.** These three components are a set of numbers, operations with the set of numbers, and properties of the numbers (and operations).

Figure 1.7 is a diagram that represents different mathematical systems. Note that the set of numbers for the system can vary. The set can consist of whole numbers, integers, rational numbers, or real numbers. (Later, you will see that the set can also consist of algebraic expressions.)

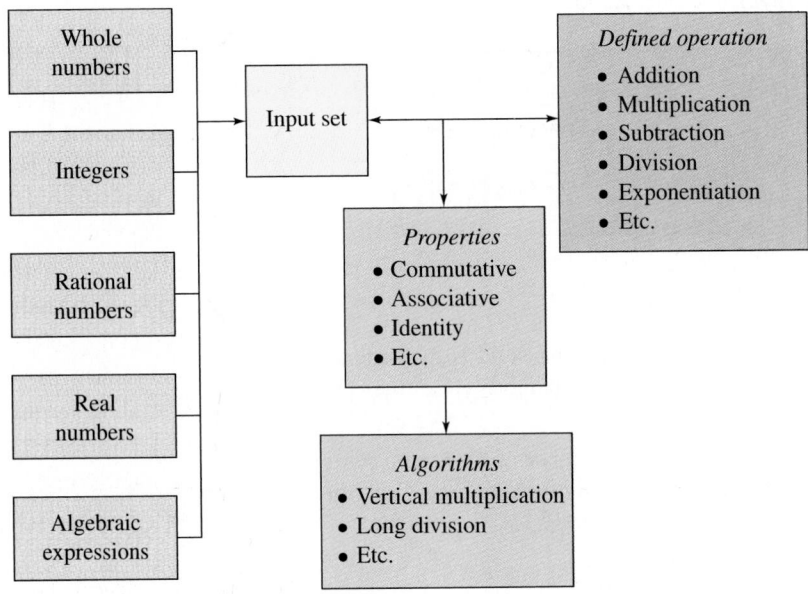

FIGURE 1.7

Basic Properties of Real Numbers

For the mathematical system that consists of the set of real numbers together with the operations of addition, subtraction, multiplication, and division, the resulting properties are called the **properties of real numbers.** In the following list, we give a verbal description of each property, as well as one or two examples.

Properties of Real Numbers

Let a, b, and c be real numbers.

Property	*Verbal Description*
Closure Property of Addition	The sum of two real numbers is a real number.
$a + b$ is a real number	Example: $1 + 5 = 6$, and 6 is a real number
Closure Property of Multiplication	The product of two real numbers is a real number.
ab is a real number	Example: $7 \cdot 3 = 21$, and 21 is a real number
Commutative Property of Addition	Two real numbers can be added in either order.
$a + b = b + a$	Example: $2 + 6 = 6 + 2$
Commutative Property of Multiplication	Two real numbers can be multiplied in either order.
$a \cdot b = b \cdot a$	Example: $3 \cdot (-5) = -5 \cdot 3$
Associative Property of Addition	When three real numbers are added, it makes no difference which two are added first.
$(a + b) + c = a + (b + c)$	Example: $(1 + 7) + 4 = 1 + (7 + 4)$
Associative Property of Multiplication	When three real numbers are multiplied, it makes no difference which two are multiplied first.
$(ab)c = a(bc)$	Example: $(4 \cdot 3) \cdot 9 = 4 \cdot (3 \cdot 9)$
Distributive Property	Multiplication distributes over addition.
$a(b + c) = ab + ac$ $(b + c)a = ba + ca$	Examples: $2(3 + 4) = 2 \cdot 3 + 2 \cdot 4$ $(3 + 4)2 = 3 \cdot 2 + 4 \cdot 2$
Additive Identity Property	The sum of zero and a real number equals the number itself.
$a + 0 = 0 + a = a$	Example: $4 + 0 = 0 + 4 = 4$
Multiplicative Identity Property	The product of 1 and a real number equals the number itself.
$a \cdot 1 = 1 \cdot a = a$	Example: $5 \cdot 1 = 1 \cdot 5 = 5$
Additive Inverse Property	The sum of a real number and its opposite is 0.
$a + (-a) = 0$	Example: $5 + (-5) = 0$
Multiplicative Inverse Property	The product of a nonzero real number and its reciprocal is 1.
$a \cdot \dfrac{1}{a} = 1, \quad a \neq 0$	Example: $7 \cdot \dfrac{1}{7} = 1$

NOTE Why are the operations of subtraction and division not listed in the preceding collection? It is because they fail to possess many of the properties described in the list. For instance, subtraction and division are not commutative. To see this, consider $4 - 3 \neq 3 - 4$ and $15 \div 5 \neq 5 \div 15$. Similarly, the examples $8 - (6 - 2) \neq (8 - 6) - 2$ and $20 \div (4 \div 2) \neq (20 \div 4) \div 2$ illustrate the fact that subtraction and division are not associative.

EXAMPLE 1 Identifying Properties of Real Numbers

Name the property of real numbers that justifies each statement.

(a) $4(a + 3) = 4 \cdot a + 4 \cdot 3$

(b) $6 \cdot \frac{1}{6} = 1$

(c) $-3 + (2 + b) = (-3 + 2) + b$

(d) $(b + 8) + 0 = b + 8$

Solution

(a) This statement is justified by the Distributive Property.

(b) This statement is justified by the Multiplicative Inverse Property.

(c) This statement is justified by the Associative Property of Addition.

(d) This statement is justified by the Additive Identity Property.

EXAMPLE 2 Identifying Properties of Real Numbers

The area of the rectangle in Figure 1.8 can be represented in two ways: as the area of a single rectangle, or as the sum of the areas of the two rectangles. Find this area in both ways. What property of real numbers does this demonstrate?

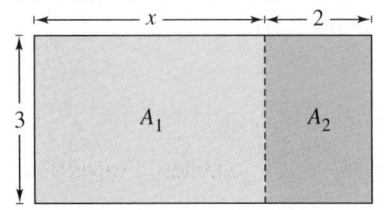

FIGURE 1.8

Solution

The area of the single rectangle with width = 3 and length = $x + 2$ is

$$A = 3(x + 2).$$

The areas of the two rectangles are $A_1 = 3(x)$ and $A_2 = 3(2)$. Because the area of the single rectangle and the sum of the areas of the two rectangles are equal, you can say that

$$A = A_1 + A_2$$
$$3(x + 2) = 3(x) + 3(2)$$
$$= 3x + 6.$$

This is an example of the Distributive Property.

EXAMPLE 3   *Using the Properties of Real Numbers*

Complete each statement using the specified property of real numbers.

(a) Multiplicative Identity Property

$(4a)1 = $ ▨

(b) Associative Property of Addition

$(b + 8) + 3 = $ ▨

(c) Additive Inverse Property

$0 = 5c + $ ▨

(d) Distributive Property

$4 \cdot b + 4 \cdot 5 = $ ▨

Solution

Review the difference between the Commutative Properties and the Associative Properties.

(a) By the Multiplicative Identity Property, you can write

$(4a)1 = 4a.$

(b) By the Associative Property of Addition, you can write

$(b + 8) + 3 = b + (8 + 3).$

(c) By the Additive Inverse Property, you can write

$0 = 5c + (-5c).$

(d) By the Distributive Property, you can write

$4 \cdot b + 4 \cdot 5 = 4(b + 5).$

Additional Properties of Real Numbers

Once you have determined the basic properties of a mathematical system (called the **axioms** of the system), you can go on to develop other properties of the system. These additional properties are often called **theorems,** and the formal arguments that justify the theorems are called **proofs.** The list on page 22 summarizes several additional properties of real numbers.

EXAMPLE 4   *Proof of a Property of Equality*

Prove that if $a + c = b + c$, then $a = b$. (Use the Addition Property of Equality.)

Solution

$a + c = b + c$	Given equation
$(a + c) + (-c) = (b + c) + (-c)$	Addition Property of Equality
$a + [c + (-c)] = b + [c + (-c)]$	Associative Property of Addition
$a + 0 = b + 0$	Additive Inverse Property
$a = b$	Additive Identity Property

Additional Properties of Real Numbers

Let a, b, and c be real numbers.

Properties of Equality

Addition Property of Equality

   If $a = b$, then $a + c = b + c$.

Multiplication Property of Equality

   If $a = b$, then $ac = bc, c \neq 0$.

Cancellation Property of Addition

   If $a + c = b + c$, then $a = b$.

Cancellation Property of Multiplication

   If $ac = bc$ and $c \neq 0$, then $a = b$.

Verbal Description

Adding a real number to both sides of a true equation produces another true equation.

Multiplying both sides of a true equation by a nonzero real number produces another true equation.

Subtracting a real number from both sides of a true equation produces another true equation.

Dividing both sides of a true equation by a nonzero real number produces another true equation.

Properties of Zero

Multiplication Property of Zero

   $0 \cdot a = 0$

Division Property of Zero

   $\dfrac{0}{a} = 0, \ a \neq 0$

Division by Zero Is Undefined

   $\dfrac{a}{0}$ is undefined.

Verbal Description

The product of zero and any real number is zero.

If zero is divided by any *nonzero* real number, the result is zero.

We do not define division by zero.

Properties of Negation

Multiplication by -1

   $(-1)a = -a$
   $(-1)(-a) = a$

Placement of Minus Signs

   $-(ab) = (-a)(b) = (a)(-b)$

Product of Two Opposites

   $(-a)(-b) = ab$

Verbal Description

The opposite of a real number a can be obtained by multiplying the real number by -1.

The opposite of the product of two numbers is equal to the product of one of the numbers and the opposite of the other.

The product of the opposites of two real numbers is equal to the product of the two real numbers.

EXAMPLE 5 *Proof of a Property of Negation*

Prove that

$$(-1)a = -a.$$

(You may use any of the properties of equality and properties of zero.)

Solution

At first glance, it is a little difficult to see what you are being asked to prove. However, a good way to start is to consider carefully the definitions of each of the three numbers in the equation.

$$a = \text{given real number}$$
$$-1 = \text{the additive inverse of } 1$$
$$-a = \text{the additive inverse of } a$$

By showing that $(-1)a$ has the same properties as the additive inverse of a, you will be showing that $(-1)a$ must be the additive inverse of a.

$$
\begin{aligned}
(-1)a + a &= (-1)a + (1)(a) && \text{Multiplicative Identity Property} \\
&= (-1 + 1)a && \text{Distributive Property} \\
&= (0)a && \text{Additive Inverse Property} \\
&= 0 && \text{Multiplication Property of Zero}
\end{aligned}
$$

Because you have shown that $(-1)a + a = 0$, you can now use the fact that $-a + a = 0$ to conclude that $(-1)a + a = -a + a$. From this, you can complete the proof as follows.

$$
\begin{aligned}
(-1)a + a &= -a + a && \text{Shown in first part of proof} \\
(-1)a &= -a && \text{Cancellation Property of Addition}
\end{aligned}
$$

There are many other properties of real numbers that we could have included in the list. Many of these are properties that you use—even if you don't happen to know their formal names. Here are a couple.

If $a = b$, then $b = a$.   Symmetric Property of Equality

If $a = b$ and $b = c$, then $a = c$.   Transitive Property of Equality

The list of additional properties of real numbers forms a very important part of algebra. Knowing the names of the properties is not especially important, but knowing how to use each property is extremely important. The next two examples show how several of the properties are used to solve common problems in algebra.

EXAMPLE 6   Applying the Properties of Real Numbers

In the list of equations below, each equation can be justified on the basis of the previous equation using one of the properties of real numbers. The property of real numbers that justifies each step is listed.

$b + 2 = 6$	Given equation
$(b + 2) + (-2) = 6 + (-2)$	Addition Property of Equality
$b + [2 + (-2)] = 4$	Associative Property of Addition
$b + 0 = 4$	Additive Inverse Property
$b = 4$	Additive Identity Property

EXAMPLE 7   Applying the Properties of Real Numbers

In the list of equations below, each equation can be justified on the basis of the previous equation using one of the properties of real numbers. The property of real numbers that justifies each step is listed.

$3a = 9$	Given equation
$\left(\dfrac{1}{3}\right)(3a) = \left(\dfrac{1}{3}\right)(9)$	Multiplication Property of Equality
$\left(\dfrac{1}{3} \cdot 3\right)(a) = 3$	Associative Property of Multiplication
$(1)(a) = 3$	Multiplicative Inverse Property
$a = 3$	Multiplicative Identity Property

Group Activities   You Be the Instructor

One of the most common errors in algebra is illustrated in the following *incorrect* use of the Distributive Property.

$$-3(4 + x) = -12 + 3x$$

With others in your group, make a list of other common errors in algebra. Suppose you were teaching an algebra class. What would you say to your class to help them avoid the errors in your list?

1.2 Exercises

See Warm-Up Exercises, p. A39

In Exercises 1–26, name the property of real numbers that justifies the statement.

1. $3 + (-5) = -5 + 3$

2. $-5(7) = 7(-5)$

3. $5(2a) = (5 \cdot 2)a$

4. $5 + 0 = 5$

5. $(10 + 8) + 3 = 10 + (8 + 3)$

6. $7(9 + 15) = 7 \cdot 9 + 7 \cdot 15$

7. $(5 + 10)(8) = 8(5 + 10)$

8. $3 + (12 - 9) = (3 + 12) - 9$

9. $25 + 35 = 35 + 25$

10. $7 \cdot 1 = 7$

11. $(-4 \cdot 10) \cdot 8 = -4(10 \cdot 8)$

12. $3(6 + b) = 3 \cdot 6 + 3 \cdot b$

13. $3x + 0 = 3x$

14. $8y \cdot 1 = 8y$

15. $25 - 25 = 0$

16. $10x \cdot \dfrac{1}{10x} = 1$

17. $6(-10) = -10(6)$

18. $2(6 \cdot 3) = (2 \cdot 6)3$

19. $10(2x) = (10 \cdot 2)x$

20. $0 + 8w = 8w$

21. $\dfrac{1}{y} \cdot y = 1$

22. $4 \cdot \dfrac{1}{4} = 1$

23. $1 \cdot (5t) = 5t$

24. $(x + 1) - (x + 1) = 0$

25. $3(2 + x) = 3 \cdot 2 + 3x$

26. $(6 + x) - m = 6 + (x - m)$

In Exercises 27–36, use the property of real numbers to fill in the missing part of the statement.

27. Associative Property of Multiplication

$3(6y) = $

28. Commutative Property of Addition

$10 + (-6) = $

29. Commutative Property of Multiplication

$15(-3) = $

30. Associative Property of Addition

$6 + (5 - y) = $

31. Distributive Property

$5(6 + z) = $

32. Distributive Property

$-3(4 + x) = $

33. Commutative Property of Addition

$25 + (-x) = $

34. Additive Inverse

$13x - 13x = $

35. Multiplicative Identity

$(x + 8) \cdot 1 = $

36. Additive Identity

$(8x) + 0 = $

True or False? In Exercises 37–40, decide whether the statement is true or false. Explain.

37. $-6x + 6x = 0$

38. $-9 + 5 = -5 + 9$

39. $6(7 + 2) = 6(7) + 2$

40. $-4(8 + 1) = -4(8) - 4(1)$

41. *Think About It* Does every real number have a multiplicative inverse? Explain.

42. *Think About It* What is the additive inverse of a real number? Give an example of the Additive Inverse Property?

In Exercises 43–50, give (a) the additive inverse and (b) the multiplicative inverse of the quantity.

43. 10

44. 18

45. -16

46. -52

47. $6z, z \neq 0$

48. $2y, y \neq 0$

49. $x + 1, x \neq -1$

50. $y - 4, y \neq 4$

In Exercises 51–58, rewrite the expression using the Associative Property of Addition or the Associative Property of Multiplication.

51. $(x + 5) - 3$

52. $(z - 6) + 10$

53. $32 + (-4 + y)$

54. $15 + (3 + x)$

55. $3(4 \cdot 5)$

56. $(10 \cdot 8) \cdot 5$

57. $6(2y)$

58. $8(3x)$

In Exercises 59–62, identify the property of real numbers that justifies each step.

59.
$$x + 5 = 3 \qquad \text{Given}$$
$$(x + 5) + (-5) = 3 + (-5)$$
$$x + [5 + (-5)] = -2$$
$$x + 0 = -2$$
$$x = -2$$

60.
$$x - 8 = 20 \qquad \text{Given}$$
$$(x - 8) + 8 = 20 + 8$$
$$x + (-8 + 8) = 28$$
$$x + 0 = 28$$
$$x = 28$$

61.
$$2x - 5 = 6 \qquad \text{Given}$$
$$(2x - 5) + 5 = 6 + 5$$
$$2x + (-5 + 5) = 11$$
$$2x + 0 = 11$$
$$2x = 11$$
$$\tfrac{1}{2}(2x) = \tfrac{1}{2}(11)$$
$$\left(\tfrac{1}{2} \cdot 2\right)x = \tfrac{11}{2}$$
$$1 \cdot x = \tfrac{11}{2}$$
$$x = \tfrac{11}{2}$$

62.
$$3x + 4 = 10 \qquad \text{Given}$$
$$(3x + 4) + (-4) = 10 + (-4)$$
$$3x + [4 + (-4)] = 6$$
$$3x + 0 = 6$$
$$3x = 6$$
$$\tfrac{1}{3}(3x) = \tfrac{1}{3}(6)$$
$$\left(\tfrac{1}{3} \cdot 3\right)x = 2$$
$$1 \cdot x = 2$$
$$x = 2$$

In Exercises 63–68, rewrite the expression using the Distributive Property.

63. $20(a + 5)$

64. $-3(y + 8)$

65. $5(3x + 4)$

66. $6(2x + 5)$

67. $(x + 6)(-2)$

68. $(z + 10)(12)$

In Exercises 69–74, the right side of the equation is *not* equal to the left side. Change the right side so that it *does* equal the left side.

69. $3(x + 5) \neq 3x + 5$

70. $4(x + 2) \neq 4x + 2$

71. $-2(x + 8) \neq -2x + 16$

72. $-9(x + 4) \neq -9x + 36$

73. $3\left(\tfrac{0}{3}\right) \neq 1$

74. $6\left(\tfrac{1}{6}\right) \neq 0$

In Exercises 75–80, use the Distributive Property to perform the arithmetic mentally. For example, suppose you work in an industry where the wage is $14 per hour with "time and a half" for overtime. Thus, your hourly wage for overtime is

$$14(1.5) = 14\left(1 + \tfrac{1}{2}\right)$$
$$= 14 + 7$$
$$= \$21.$$

75. $16(1.75)$

76. $15\left(1\tfrac{2}{3}\right)$

77. $7(62) = 7(60 + 2)$

78. $5(49) = 5(50 - 1)$

79. $9(6.98) = 9(7 - 0.02)$

80. $12(19.95) = 12(20 - 0.05)$

81. Prove that if $ac = bc$ and $c \neq 0$, then $a = b$.

82. Prove that $(-1)(-a) = a$.

83. *Investigation* Suppose you define a new mathematical operation using the symbol $\odot$. This operation is defined as $a \odot b = 2 \cdot a + b$.

(a) Is this operation commutative? Explain.

(b) Is this operation associative? Explain.

Dividends In Exercises 84–87, the dividend paid per share of common stock by the H.J. Heinz Company for the years 1990 through 1996 is approximated by the model

Dividend per share $= 0.08t + 0.54$.

In this model, the dividend per share is measured in dollars and t represents the year, with $t = 0$ corresponding to 1990 (see figure). *(Source: H.J. Heinz Company Annual Report 1996)*

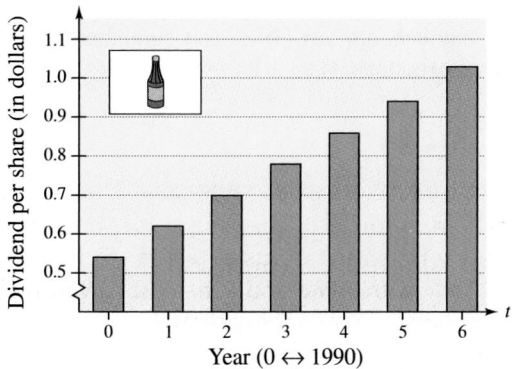

84. Use the graph to approximate the dividend paid in 1994.

85. Use the model to approximate the annual increase in the dividend paid per share.

86. Use the model to forecast the dividend per share in 1998.

87. In 1994, the actual dividend paid per share of common stock was $0.86. Compare this with the approximation given by the model.

Section Project

Area of a Rectangle (a) The figure shows two adjoining rectangles. Demonstrate the Distributive Property by filling in the blanks to express the total area of the two rectangles in two ways.

☐ (☐ + ☐) = ☐ + ☐

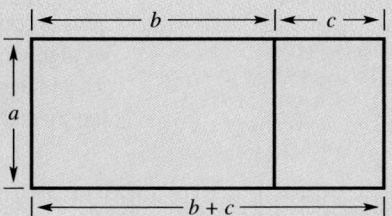

(b) The figure shows two adjoining rectangles. Demonstrate the "subtraction version" of the Distributive Property by filling in the blanks to express the area of the left rectangle in two ways.

☐ (☐ − ☐) = ☐ − ☐

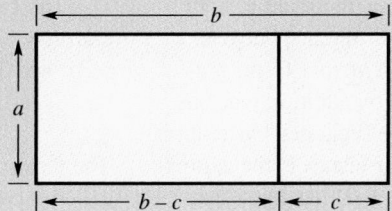

1.3	**Algebraic Expressions**

Algebraic Expressions ▪ Basic Rules of Algebra ▪
Properties of Exponents ▪ Simplifying Algebraic Expressions ▪
Evaluating Algebraic Expressions

Algebraic Expressions

Historical Note: The French mathematician François Viéte (1540–1603) was the first to use letters to represent numbers. He used vowels to represent unknown quantities and consonants to represent known quantities.

This section marks our transition from arithmetic to algebra. A basic characteristic of algebra is the use of letters (or combinations of letters) to represent numbers. The letters used to represent the numbers are **variables,** and combinations of letters and numbers are **algebraic expressions.** Here are a few examples.

$$3x, \quad x + 2, \quad \frac{x}{x^2 + 1}, \quad 2x - 3y$$

Definition of an Algebraic Expression

A collection of letters (called **variables**) and real numbers (called **constants**) combined using the operations of addition, subtraction, multiplication, and division is called an **algebraic expression.**

NOTE It is important to understand the difference between a *term* and a *factor*. Terms are separated by addition whereas factors are separated by multiplication. For instance, the expression $4x(x + 2)$ has three factors: 4, x, and $(x + 2)$.

The **terms** of an algebraic expression are those parts that are separated by *addition*. For example, the algebraic expression $x^2 - 3x + 6$ has three terms: x^2, $-3x$, and 6. Note that $-3x$ is a term, rather than $3x$, because

$$x^2 - 3x + 6 = x^2 + (-3x) + 6. \qquad \text{Think of subtraction as a form of addition.}$$

The terms x^2 and $-3x$ are called the **variable terms** of the expression, and 6 is called the **constant term** of the expression. The numerical factor of a variable term is called the **coefficient** of the variable term. For instance, the coefficient of the variable term $-3x$ is -3, and the coefficient of the variable term x^2 is 1. (The constant term of an expression is also considered to be a coefficient.)

EXAMPLE 1   *Identifying Terms and Coefficients*

Algebraic Expression	*Terms*	*Coefficients*
(a) $5x - \frac{1}{3}$	$5x, -\frac{1}{3}$	$5, -\frac{1}{3}$
(b) $4y + 6x - 9$	$4y, 6x, -9$	$4, 6, -9$

Basic Rules of Algebra

The properties of real numbers that were discussed in Section 1.2 can be used to rewrite algebraic expressions. The following list is similar to those given in Section 1.2, except that the examples involve algebraic expressions. In other words, the properties are true for variables and algebraic expressions as well as for real numbers.

Basic Rules of Algebra

Let a, b, and c be real numbers, variables, or algebraic expressions.

Property	*Example*
Commutative Property of Addition	
$a + b = b + a$	$5x + x^2 = x^2 + 5x$
Commutative Property of Multiplication	
$ab = ba$	$(3 + x)x^3 = x^3(3 + x)$
Associative Property of Addition	
$(a + b) + c = a + (b + c)$	$(-x + 6) + 3x^2 = -x + (6 + 3x^2)$
Associative Property of Multiplication	
$(ab)c = a(bc)$	$(5x \cdot 4y)(6) = (5x)(4y \cdot 6)$
Distributive Property	
$a(b + c) = ab + ac$	$2x(4 + 3x) = 2x \cdot 4 + 2x \cdot 3x$
$(a + b)c = ac + bc$	$(y + 6)y = y \cdot y + 6 \cdot y$
Additive Identity Property	
$a + 0 = 0 + a = a$	$4y^2 + 0 = 0 + 4y^2 = 4y^2$
Multiplicative Identity Property	
$a \cdot 1 = 1 \cdot a = a$	$(-5x^3)(1) = (1)(-5x^3) = -5x^3$
Additive Inverse Property	
$a + (-a) = 0$	$4x^2 + (-4x^2) = 0$
Multiplicative Inverse Property	
$a \cdot \dfrac{1}{a} = 1, a \neq 0$	$(x^2 + 1)\left(\dfrac{1}{x^2 + 1}\right) = 1$

NOTE   Because subtraction is defined as "adding the opposite," the Distributive Property is also true for subtraction. For instance, the "subtraction form" of $a(b + c) = ab + ac$ is

$$a(b - c) = a[b + (-c)]$$
$$= ab + a(-c)$$
$$= ab - ac.$$

In addition to these basic rules, the properties of equality, zero, and negation given in the previous section are also valid for algebraic expressions. The next example illustrates the use of a variety of the basic rules and properties.

EXAMPLE 2   Identifying the Basic Rules of Algebra

Identify the rule of algebra used in each equation.

(a) $(5x^2)3 = 3(5x^2)$

(b) $(3x^2 + x) - (3x^2 + x) = 0$

(c) $(y - 6)3 + (y - 6)y = (y - 6)(3 + y)$

(d) $(5 + x^2) + 4x^2 = 5 + (x^2 + 4x^2)$

(e) $5x \cdot \dfrac{1}{5x} = 1, x \neq 0$

Solution

(a) This equation illustrates the Commutative Property of Multiplication. In other words, you obtain the same result whether you multiply $5x^2$ by 3, or 3 by $5x^2$.

(b) This equation illustrates the Additive Inverse Property. In terms of subtraction, this property simply states that when any expression is subtracted from itself the result is zero.

The concept in Example 2(c) can sometimes be difficult for students.

(c) This equation illustrates the Distributive Property in reverse order.

$$ab + ac = a(b + c) \qquad \text{Distributive Property}$$

$$(y - 6)3 + (y - 6)y = (y - 6)(3 + y)$$

Note in this case that $a = y - 6$, $b = 3$, and $c = y$.

(d) This equation illustrates the Associative Property of Addition. In other words, to form the sum $5 + x^2 + 4x^2$, it doesn't matter whether 5 and x^2 are added first or x^2 and $4x^2$ are added first.

(e) This equation illustrates the Multiplicative Inverse Property. Note that it is important that x be a nonzero number. If x were zero, the reciprocal of x would be undefined.

Exploration

Discovering Properties of Exponents   Try writing each of the following as a single power of 2. Explain how you obtained your answer. Then generalize your procedure by completing the statement "*When you multiply exponential expressions that have the same base, you. . . .*"

a. $2^2 \cdot 2^3$ b. $2^4 \cdot 2^1$ c. $2^5 \cdot 2^2$ d. $2^3 \cdot 2^4$ e. $2^1 \cdot 2^5$

Properties of Exponents

Historical Note: Originally, Arabian mathematicians used their words for colors to represent quantities (*cosa, censa, cubo*). These words were eventually abbreviated to *co, ce, cu*. René Descartes (1596–1650) simplified this even further by introducing the symbols *x, x²,* and *x³*.

When multiplying two exponential expressions that have the *same base,* you add exponents. To see why this is true, consider the product $a^3 \cdot a^2$. Because the first expression represents $a \cdot a \cdot a$ and the second represents $a \cdot a$, the product of the two expressions represents $a \cdot a \cdot a \cdot a \cdot a$, as follows.

$$a^3 \cdot a^2 = \underbrace{(a \cdot a \cdot a)}_{\substack{\text{Three} \\ \text{Factors}}} \cdot \underbrace{(a \cdot a)}_{\substack{\text{Two} \\ \text{Factors}}} = \underbrace{(a \cdot a \cdot a \cdot a \cdot a)}_{\substack{\text{Five} \\ \text{Factors}}} = a^{3+2} = a^5$$

Properties of Exponents

Let m and n be positive integers, and let a and b represent real numbers, variables, or algebraic expressions.

Property	*Example*
1. $a^m \cdot a^n = a^{m+n}$	$x^5(x^4) = x^{5+4} = x^9$
2. $(ab)^m = a^m \cdot b^m$	$(2x)^3 = 2^3(x^3) = 8x^3$
3. $(a^m)^n = a^{mn}$	$(x^2)^3 = x^{2 \cdot 3} = x^6$
4. $\dfrac{a^m}{a^n} = a^{m-n},\ m > n,\ a \neq 0$	$\dfrac{x^6}{x^2} = x^{6-2} = x^4,\ x \neq 0$
5. $\left(\dfrac{a}{b}\right)^m = \dfrac{a^m}{b^m},\ b \neq 0$	$\left(\dfrac{x}{2}\right)^3 = \dfrac{x^3}{2^3} = \dfrac{x^3}{8}$

NOTE The first and second properties can be extended to products involving three or more factors. For example,

$$a^m \cdot a^n \cdot a^k = a^{m+n+k}$$

and

$$(abc)^m = a^m b^m c^m.$$

EXAMPLE 3   *Illustrating the Properties of Exponents*

(a) To multiply exponential expressions that have the *same base,* add exponents.

$$x^2 \cdot x^4 = \underbrace{x \cdot x}_{\text{2 factors}} \cdot \underbrace{x \cdot x \cdot x \cdot x}_{\text{4 factors}} = \underbrace{x \cdot x \cdot x \cdot x \cdot x \cdot x}_{\text{6 factors}} = x^{2+4} = x^6$$

(b) To raise the product of two factors to the *same power,* raise each factor to the power and multiply the results.

$$(3x)^3 = \underbrace{3x \cdot 3x \cdot 3x}_{\text{3 factors}} = \underbrace{3 \cdot 3 \cdot 3}_{\text{3 factors}} \cdot \underbrace{x \cdot x \cdot x}_{\text{3 factors}} = 3^3 \cdot x^3 = 27x^3$$

(c) To raise an exponential expression to a power, multiply exponents.

$$(x^3)^2 = \underbrace{(x \cdot x \cdot x)}_{\text{3 factors}} \cdot \underbrace{(x \cdot x \cdot x)}_{\text{3 factors}} = \underbrace{(x \cdot x \cdot x \cdot x \cdot x \cdot x)}_{\text{6 factors}} = x^{3 \cdot 2} = x^6$$

EXAMPLE 4   Illustrating the Properties of Exponents

(a) To divide exponential expressions that have the *same base,* subtract exponents.

$$\frac{x^4}{x^2} = \frac{\overbrace{x \cdot x \cdot x \cdot x}^{4 \text{ factors}}}{\underbrace{x \cdot x}_{2 \text{ factors}}} = x^{4-2} = x^2$$

(b) To raise the quotient of two expressions to the *same power,* raise each expression to the power and divide the results.

$$\left(\frac{x}{3}\right)^3 = \frac{x}{3} \cdot \frac{x}{3} \cdot \frac{x}{3} = \frac{\overbrace{x \cdot x \cdot x}^{3 \text{ factors}}}{\underbrace{3 \cdot 3 \cdot 3}_{3 \text{ factors}}} = \frac{x^3}{3^3} = \frac{x^3}{27}$$

EXAMPLE 5   Applying Properties of Exponents

Use the properties of exponents to simplify each expression.

It may be helpful to have students clearly identify the base of an exponential expression, which will help them distinguish between expressions such as -2^4 and $(-2)^4$.

(a) $(x^2y^4)(3x)$ (b) $-2(y^2)^3$ (c) $(-2y^2)^3$

Solution

(a) $(x^2y^4)(3x) = 3(x^2 \cdot x)(y^4) = 3x^3y^4$

(b) $-2(y^2)^3 = (-2)(y^{2 \cdot 3}) = -2y^6$

(c) $(-2y^2)^3 = (-2)^3(y^2)^3 = -8y^6$

EXAMPLE 6   Applying Properties of Exponents

Use the properties of exponents to simplify each expression.

(a) $\dfrac{14a^5b^3}{7a^2b^2}$ (b) $\left(\dfrac{x^2}{2y}\right)^3$ (c) $\dfrac{x^{n+2}y^{3n}}{x^2y^n}$

Solution

(a) $\dfrac{14a^5b^3}{7a^2b^2} = 2(a^{5-2})(b^{3-2}) = 2a^3b$

(b) $\left(\dfrac{x^2}{2y}\right)^3 = \dfrac{(x^2)^3}{(2y)^3} = \dfrac{x^6}{2^3y^3} = \dfrac{x^6}{8y^3}$

(c) $\dfrac{x^{n+2}y^{3n}}{x^2y^n} = x^{(n+2)-2}y^{3n-n} = x^ny^{2n}$

Simplifying Algebraic Expressions

One common use of the basic rules of algebra is to rewrite an algebraic expression in a simpler form. To **simplify** an algebraic expression generally means to *remove symbols of grouping* such as parentheses or brackets and *combine like terms*.

Two or more terms of an algebraic expression can be combined only if they are *like terms*. In an algebraic expression, two terms are said to be **like terms** if they are both constant terms or if they have the same variable factor(s). For example, the terms

$$4x \quad \text{and} \quad -2x \qquad\qquad \text{Like terms}$$

are like terms because they have the same variable factor. Similarly,

$$2x^2y, \quad -x^2y, \quad \text{and} \quad \tfrac{1}{2}(x^2y) \qquad\qquad \text{Like terms}$$

are like terms. Note that $4x^2y$ and $-x^2y^2$ are not like terms because their variable factors are different.

To combine like terms in an algebraic expression, simply add their respective coefficients and attach the common variable factor. This is actually an application of the Distributive Property, as shown in Example 7.

EXAMPLE 7   *Combining Like Terms*

Simplify each expression by combining like terms.

(a) $2x + 3x - 4$ (b) $-3 + 5 + 2y - 7y$ (c) $5x + 3y - 4x$

Solution

(a)
$$\begin{aligned}
2x + 3x - 4 &= (2 + 3)x - 4 &&\text{Distributive Property}\\
&= 5x - 4 &&\text{Simplest form}
\end{aligned}$$

(b)
$$\begin{aligned}
-3 + 5 + 2y - 7y &= (-3 + 5) + (2 - 7)y &&\text{Distributive Property}\\
&= 2 - 5y &&\text{Simplest form}
\end{aligned}$$

(c)
$$\begin{aligned}
5x + 3y - 4x &= 5x - 4x + 3y &&\text{Commutative Property}\\
&= (5x - 4x) + 3y &&\text{Associative Property}\\
&= (5 - 4)x + 3y &&\text{Distributive Property}\\
&= x + 3y &&\text{Simplest form}
\end{aligned}$$

As you gain experience with the rules of algebra, you may want to combine some of the steps in your work. For instance, you might feel comfortable listing only the following steps to solve Example 7(c).

$$\begin{aligned}
5x + 3y - 4x &= (5x - 4x) + 3y &&\text{Group like terms.}\\
&= x + 3y &&\text{Combine like terms.}
\end{aligned}$$

EXAMPLE 8   Combining Like Terms

(a) $7x + 7y - 4x - y = (7x - 4x) + (7y - y)$ Group like terms.

$\qquad\qquad\qquad\quad = 3x + 6y$ Combine like terms.

(b) $2x^2 + 3x - 5x^2 - x = (2x^2 - 5x^2) + (3x - x)$ Group like terms.

$\qquad\qquad\qquad\qquad = -3x^2 + 2x$ Combine like terms.

(c) $3xy^2 - 4x^2y^2 + 2xy^2 + (xy)^2 = (3xy^2 + 2xy^2) + (-4x^2y^2 + x^2y^2)$

$\qquad\qquad\qquad\qquad\qquad\qquad = 5xy^2 - 3x^2y^2$

A set of parentheses preceded by a *minus* sign can be removed by changing the sign of each term inside the parentheses. For instance,

$$8x - (5x - 4) = 8x - 5x + 4 = 3x + 4.$$

A set of parentheses preceded by a *plus* sign can be removed without changing the signs of the terms inside the parentheses. For instance,

$$8x + (5x - 4) = 8x + 5x - 4 = 13x - 4.$$

EXAMPLE 9   Removing Symbols of Grouping

$3(x - 5) - (2x - 7) = 3x - 15 - 2x + 7$ Distributive Property

$\qquad\qquad\qquad\quad = (3x - 2x) + (-15 + 7)$ Group like terms.

$\qquad\qquad\qquad\quad = x - 8$ Combine like terms.

When removing symbols of grouping, combine like terms within the innermost symbols of grouping first, as shown in the next example.

EXAMPLE 10   Removing Symbols of Grouping

(a) $5x - 2x[3 + 2(x - 7)] = 5x - 2x[3 + 2x - 14]$ Remove parentheses.

$\qquad\qquad\qquad\qquad = 5x - 2x[2x - 11]$ Combine like terms in brackets.

$\qquad\qquad\qquad\qquad = 5x - 4x^2 + 22x$ Remove brackets.

$\qquad\qquad\qquad\qquad = -4x^2 + 27x$ Combine like terms.

(b) $-3x(5x^4) + (2x)^5 = -15x^5 + (2^5)(x^5)$

$\qquad\qquad\qquad\qquad = -15x^5 + 32x^5$

$\qquad\qquad\qquad\qquad = 17x^5$

Evaluating Algebraic Expressions

To **evaluate** an algebraic expression, substitute numerical values for each of the variables in the expression. Here are some examples.

A common error is to confuse the operation of subtraction with the substitution of a negative number. Point out the use of parentheses when substituting a numerical value for a variable in an expression.

Expression	*Value of Variable*	*Substitute*	*Value of Expression*
$3x + 2$	$x = 2$	$3(2) + 2$	$6 + 2 = 8$
$4x^2 + 2x - 1$	$x = -1$	$4(-1)^2 + 2(-1) - 1$	$4 - 2 - 1 = 1$
$2x(x + 4)$	$x = -2$	$2(-2)(-2 + 4)$	$2(-2)(2) = -8$

EXAMPLE 11   Evaluating Algebraic Expressions

Evaluate each algebraic expression when $x = -2$ and $y = 5$.

(a) $2y - 3x$ (b) $5 + x^2$ (c) $5 - x^2$

Solution

(a) When $x = -2$ and $y = 5$, the expression $2y - 3x$ has a value of
$$2(5) - 3(-2) = 10 + 6 = 16.$$

(b) When $x = -2$, the expression $5 + x^2$ has a value of
$$5 + (-2)^2 = 5 + 4 = 9.$$

(c) When $x = -2$, the expression $5 - x^2$ has a value of
$$5 - (-2)^2 = 5 - 4 = 1.$$

EXAMPLE 12   Evaluating Algebraic Expressions

Evaluate each algebraic expression when $x = 2$ and $y = -1$.

(a) $x^2 - 2xy + y^2$ (b) $|y - x|$ (c) $\dfrac{2xy}{5x + y}$

Additional example: Evaluate $-x^4$ when $x = -1$.
Solution: $-(-1)^4 = -1$
(Note that $-(-1)^4 \neq +1^4$.)

Solution

(a) When $x = 2$ and $y = -1$, the expression $x^2 - 2xy + y^2$ has a value of
$$2^2 - 2(2)(-1) + (-1)^2 = 4 + 4 + 1 = 9.$$

(b) When $x = 2$ and $y = -1$, the expression $|y - x|$ has a value of
$$|-1 - 2| = |-3| = 3.$$

(c) When $x = 2$ and $y = -1$, the expression $2xy/(5x + y)$ has a value of
$$\frac{2(2)(-1)}{5(2) + (-1)} = \frac{-4}{10 - 1} = -\frac{4}{9}.$$

According to the National Education Association, more than half of the adults in the United States read at least one book for enjoyment each year. Reading books to children increases the likelihood that the children will read for enjoyment as adults.

EXAMPLE 13   *Creating a Real-Life Model*

For the years 1985 through 1992, the number B (in millions) of children's books that were sold in the United States can be modeled by $B = 20.8t + 84.9$, where $t = 5$ represents 1985. During the same years, the average price p of a children's book can be modeled by $p = 0.32t + 2.54$. These two models can be used to write a model for the total sales (in millions of dollars) of children's books for 1985 through 1992. Use the model to find the total sales in 1988.   *(Source: Book Industry Trends)*

Solution

Let S represent the yearly sales in millions of dollars.

$$\begin{array}{ccc} \boxed{\begin{array}{c}\text{Total yearly}\\\text{sales}\end{array}} & = & \boxed{\begin{array}{c}\text{Number sold}\\\text{each year}\end{array}} \cdot \boxed{\begin{array}{c}\text{Average price per}\\\text{book each year}\end{array}} \end{array}$$

$$S = (20.8t + 84.9)(0.32t + 2.54)$$

To find the total sales in 1988, substitute 8 for t in this model.

$S = (20.8t + 84.9)(0.32t + 2.54)$	Total sales model
$= (20.8 \cdot 8 + 84.9)(0.32 \cdot 8 + 2.54)$	Substitute 8 for t.
$= (251.3)(5.1)$	Perform operations within parentheses.
$= 1281.63$	Multiply.

The total sales of children's books in 1988 was about \$1282 million (or about \$1.3 billion). The bar graph presented in Figure 1.9 shows the total sales of children's books from 1985 through 1992.

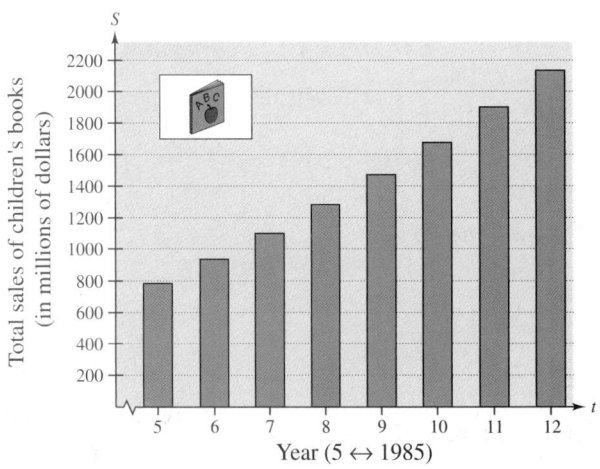

Year (5 ↔ 1985)

FIGURE 1.9

Group Activities   **Unit Analysis**
_____

In Example 13, the *unit* for the model for the yearly sales of children's books is *millions of dollars per year.* With others in your group, explain how the following "unit analysis" diagram can be used to determine this unit.

$$\frac{\text{Millions of books}}{\text{Year}} \cdot \frac{\text{dollars}}{\text{book}} = \frac{\text{millions of dollars}}{\text{year}}$$

Then give other examples in which you can use similar unit analyses.

1.3   Exercises

See Warm-Up Exercises, p. A39

1. *Think About It* Explain the difference between terms and factors in an algebraic expression.
2. *Think About It* Explain the difference between constants and variables in an algebraic expression.

In Exercises 3–10, identify the terms of the algebraic expression.

3. $10x + 5$

4. $-16t^2 + 48$

5. $-3y^2 + 2y - 8$

6. $25z^3 - 4.8z^2$

7. $4x^2 - 3y^2 - 5x + 2y$

8. $14u^2 + 25uv - 3v^2$

9. $x^2 - 2.5x - \dfrac{1}{x}$

10. $\dfrac{3}{t^2} - \dfrac{4}{t} + 6$

In Exercises 11–14, determine the coefficient of the term.

11. $5y^3$

12. $4x^6$

13. $-\frac{3}{4}t^2$

14. $-8.4x$

In Exercises 15–24, identify the rule of algebra that is illustrated by the equation.

15. $4 - 3x = -3x + 4$

16. $(10 + x) - y = 10 + (x - y)$

17. $-5(2x) = (-5 \cdot 2)x$

18. $(x - 2)(3) = 3(x - 2)$

19. $(x + 5) \cdot \dfrac{1}{x + 5} = 1, \quad x \neq -5$

20. $(x^2 + 1) - (x^2 + 1) = 0$

21. $5(y^3 + 3) = 5y^3 + 5 \cdot 3$

22. $10x^3y + 0 = 10x^3y$

23. $(16t^4) \cdot 1 = 16t^4$

24. $-32(u^2 - 3u) = -32u^2 + 96u$

In Exercises 25–30, use the property to rewrite the expression.

25. (a) Distributive Property

$5(x + 6) = $ ▢

(b) Commutative Property of Multiplication

$5(x + 6) = $ ▢

26. (a) Distributive Property

$6x + 6 = $ ▢

(b) Commutative Property of Addition

$6x + 6 = $ ▢

27. (a) Commutative Property of Multiplication

$6(xy) = $ ▨

(b) Associative Property of Multiplication

$6(xy) = $ ▨

28. (a) Additive Identity Property

$3ab + 0 = $ ▨

(b) Commutative Property of Addition

$3ab + 0 = $ ▨

29. (a) Additive Inverse Property

$4t^2 + (-4t^2) = $ ▨

(b) Commutative Property of Addition

$4t^2 + (-4t^2) = $ ▨

30. (a) Associative Property of Addition

$(3 + 6) + (-9) = $ ▨

(b) Additive Inverse Property

$9 + (-9) = $ ▨

True or False? In Exercises 31–34, decide whether the statement is true or false. Explain.

31. $(3a)b = 3(ab)$

32. $7x + (6 - y) = (7x + 6) - y$

33. $5(y^3 + 2) = 5y^3 + 2$

34. $2x(5 - y) = 2x(5) + 2x(-y)$

In Exercises 35–40, use the definition of exponents to write the expression as a repeated multiplication.

35. $x^3 \cdot x^4$

36. $z^2 \cdot z^5$

37. $(-2x)^3$

38. $(2y)^3$

39. $\left(\dfrac{y}{5}\right)^4$

40. $\left(\dfrac{3}{t}\right)^5$

In Exercises 41–44, write the expression using exponential notation.

41. $(5x)(5x)(5x)(5x)$

42. $(y \cdot y \cdot y)(y \cdot y \cdot y \cdot y)$

43. $(x \cdot x \cdot x)(y \cdot y \cdot y)$

44. $(-9t)(-9t)(-9t)(-9t)(-9t)(-9t)$

In Exercises 45–76, use the properties of exponents to simplify the expression.

45. $3^3y^4 \cdot y^2$

46. $6^2x^3 \cdot x^5$

47. $(-4x)^2$

48. $(-4x)^3$

49. $(-5z^2)^3$

50. $(-5z^3)^2$

51. $(2xy)(3x^2y^3)$

52. $(-5a^2b^3)(2ab^4)$

53. $\dfrac{3^7x^5}{3^3x^3}$

54. $\dfrac{2^4y^5}{2^2y^3}$

55. $\dfrac{(2xy)^5}{6(xy)^3}$

56. $\dfrac{4^3(ab)^6}{4(ab)^2}$

57. $(5y)^2(-y^4)$

58. $(3y)^3(2y^2)$

59. $-5z^4(-5z)^4$

60. $(-6n)(-3n^2)$

61. $(-2a)^2(-2a)^2$

62. $(-2a^2)(-8a)$

63. $\dfrac{(2x)^4y^2}{2x^3y}$

64. $(-z^2)(z^2)$

65. $\dfrac{6(a^3b)^3}{(3ab)^2}$

66. $\dfrac{(-3c^5d^3)^2}{(-2cd)^3}$

67. $-\left(\dfrac{2x^4}{5y}\right)^2$

68. $-\left(\dfrac{3a^3}{2b^5}\right)^3$

69. $\dfrac{x^{n+1}}{x^n}$

70. $\dfrac{a^{m+3}}{a^3}$

71. $(x^n)^4$

72. $(a^3)^k$

73. $x^{n+1} \cdot x^3$

74. $y^{m-2} \cdot y^2$

75. $\dfrac{r^{n+2}s^{m+4}}{r^{n+1}s}$

76. $\left(\dfrac{x^{2n}y^{m+4}}{x^ny^3}\right)^2$

77. *Think About It* Write, from memory, the rules of exponents.

78. *Think About It* Explain the difference between $(2x)^3$ and $2x^3$.

In Exercises 79–90, simplify the expression by combining like terms.

79. $3x + 4x$

80. $-2x^2 + 4x^2$

81. $9y - 5y + 4y$

82. $8y + 7y - y$

$$\dot{2}x(xy + x^2) + \frac{6x^4}{3x} = 2x^2y + 4x^3$$

83. $3x - 2y + 5x + 20y$

84. $-2a + \frac{1}{3}b - 7a - b$

85. $8z^2 + \frac{3}{2}z - \frac{5}{2}z^2 + 10$

86. $-5y^3 + 3y - 6y^2 + 8y^3 + y - 4$

87. $2uv + 5u^2v^2 - uv - (uv)^2$

88. $3m^2n^2 - 4mn - n(5m) + 2(mn)^2$

89. $5(ab)^2 + 2ab - 4ab$

90. $3xy - xy + 8$

oz^5

$a^{12}b^4$

In Exercises 91–104, simplify the algebraic expression.

91. $10(x - 3) + 2x - 5$

92. $3(x^2 + 1) + x^2 - 6$

93. $-3(y^2 + 3y - 1) + 2(y - 5)$

94. $5(a + 6) - 4(a^2 - 2a - 1)$

95. $4[5 - 3(x^2 + 10)]$

96. $2x(5x^2) - (x^3 + 5)$

97. $2[3(b - 5) - (b^2 + b + 3)]$

98. $x(x^2 + 3) - 3(x + 4)$

99. $y^2(y + 1) + y(y^2 + 1)$

100. $2ab(b^2 - 3) - ab(b^2 + 2)$

101. $x(xy^2 + y) - 2xy(xy + 1)$

102. $z^2(z^4 - z^2) + 4z^3(z + 1)$

103. $-2a(3a^2)^3 + \frac{9a^8}{3a}$ ✓

104. $5y^3 + \frac{4y^5}{2y^2} - (7y)y^2$

Geometry In Exercises 105 and 106, write an expression for the area of the region. Then simplify the expression.

105.

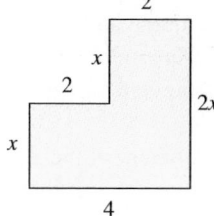

106.

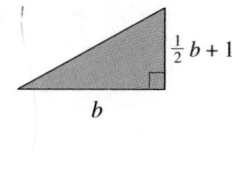

In Exercises 107–118, evaluate the algebraic expression for the specified values of the variable(s). If not possible, state the reason.

Expression	Value	
107. $5 - 3x$	(a) $x = \frac{2}{3}$	(b) $x = 5$
108. $\frac{3}{2}x - 2$	(a) $x = 6$	(b) $x = -3$
109. $10 - |x|$	(a) $x = 3$	(b) $x = -3$
110. $2x^2 + 5x - 3$	(a) $x = \frac{1}{2}$	(b) $x = -3$
111. $\dfrac{x}{x^2 + 1}$	(a) $x = 0$	(b) $x = 3$
112. $5 - \dfrac{3}{x}$	(a) $x = 0$	(b) $x = -6$
113. $3x + 2y$	(a) $x = 1, \; y = 5$	
	(b) $x = -6, \; y = -9$	
114. $x^2 - xy + y^2$	(a) $x = 2, \; y = -1$	
	(b) $x = -3, \; y = -2$	
115. $\dfrac{x}{x - y}$	(a) $x = 0, \; y = 10$	
	(b) $x = 4, \; y = 4$	
116. $|y - x|$	(a) $x = 2, \; y = 5$	
	(b) $x = -2, \; y = -2$	
117. rt	(a) $r = 40, \; t = 5\frac{1}{4}$	
	(b) $r = 35, \; t = 4$	
118. Prt	(a) $P = \$5000, \; r = 0.085,$	
	$t = 10$	
	(b) $P = \$750, \; r = 0.07,$	
	$t = 3$	

Geometry In Exercises 119 and 120, find an expression for the area of the figure. Then evaluate the expression for the given value of the variable.

119. $b = 15$

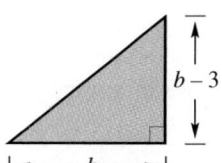

120. $h = 12$

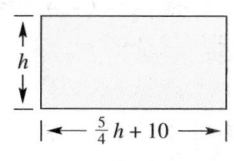

Using a Model   In Exercises 121 and 122, use the model, which approximates the annual sales (in millions of dollars) of exercise equipment in the United States from 1987 through 1994 (see figure).

Sales $= 201.90t - 174.25$

In this formula, $t = 7$ represents 1987. *(Source: National Sporting Goods Association)*

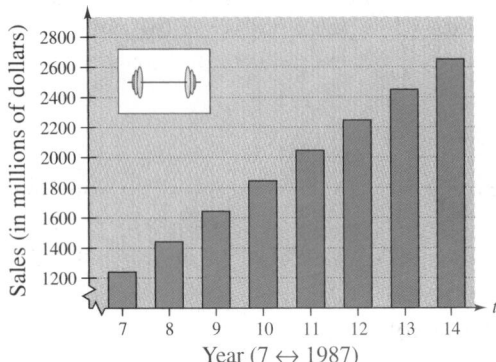

121. Graphically approximate the sales of exercise equipment in 1988. Then use the model to confirm your estimate algebraically.

122. Graphically approximate the sales of exercise equipment in 1993. Then use the model to confirm your estimate algebraically.

123. *Geometry*   The area of the trapezoid with parallel bases of lengths b_1 and b_2, and height h, in the figure is $\frac{1}{2}(b_1 + b_2)h$. Use the Distributive Property to show that the area can also be expressed as $b_1h + \frac{1}{2}(b_2 - b_1)h$.

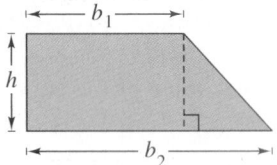

124. *Area of a Trapezoid*   Use both formulas given in Exercise 123 to find the area of a trapezoid with $b_1 = 7$, $b_2 = 12$, and $h = 3$.

125. *Amount of Roofing Material*   The roof shown in the figure is made up of two trapezoids and two triangles. Find the total area of the roof.

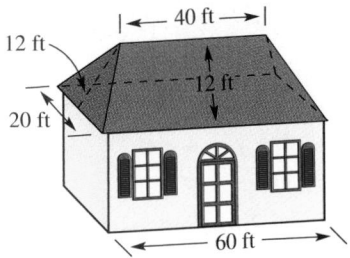

Figure for 125

Section Project

(a) Complete the table by evaluating $2x - 5$.

x	-1	0	1	2	3	4
$2x - 5$						

(b) From the table in part (a), determine the increase in the value of the expression for each one-unit increase in x.

(c) Complete the table by using the table feature of a graphing utility to evaluate $3x + 2$.

x	-1	0	1	2	3	4
$3x + 2$						

(d) From the table in part (c), determine the increase in the value of the expression for each one-unit increase in x.

(e) Use the results in parts (a) through (d) to make a conjecture about the increase in the algebraic expression $7x + 4$ for each one-unit increase in x. Use the table feature of a graphing utility to confirm your result.

(f) Use the results in parts (a) through (d) to make a conjecture about the increase in the algebraic expression $-3x + 1$ for each one-unit increase in x. Use the table feature of a graphing utility to confirm your result.

(g) In general, what does the coefficient of the x term represent in the expressions above?

1.4 Operations with Polynomials

Basic Definitions ▪ Adding and Subtracting Polynomials ▪
Multiplying Polynomials ▪ Special Products ▪ Applications

Basic Definitions

An algebraic expression containing only terms of the form ax^k, where a is any real number and k is a nonnegative integer, is called a **polynomial in one variable** or simply a **polynomial.** Here are some examples of polynomials in one variable.

$$3x - 8, \quad x^4 + 3x^3 - x^2 - 8x + 1, \quad x^3 + 5, \quad \text{and} \quad 9x^5$$

In the term ax^k, a is called the **coefficient,** and k the **degree,** of the term. Note that the degree of the term ax is 1, and the degree of a constant term is zero. Because a polynomial is an algebraic *sum,* the coefficients take on the signs between the terms. For instance,

$$x^3 - 4x^2 + 3 = (1)x^3 + (-4)x^2 + (0)x + 3$$

has coefficients 1, -4, 0, and 3. Polynomials are usually written in order of descending powers of the variable. This is referred to as **standard form.** For example, the standard form of $3x^2 - 5 - x^3 + 2x$ is

$$-x^3 + 3x^2 + 2x - 5. \qquad \text{Standard form}$$

The **degree of a polynomial** is defined as the degree of the term with the highest power, and the coefficient of this term is called the **leading coefficient** of the polynomial. For instance, the polynomial $-3x^4 + 4x^2 + x + 7$ is of fourth degree and its leading coefficient is -3.

Definition of a Polynomial in x

Let $a_n, \ldots, a_2, a_1, a_0$ be real numbers and let n be a *nonnegative integer.* A **polynomial in x** is an expression of the form

$$a_n x^n + a_{n-1}x^{n-1} + \cdots + a_2 x^2 + a_1 x + a_0,$$

where $a_n \neq 0$. The polynomial is of **degree n,** and the number a_n is called the **leading coefficient.** The number a_0 is called the **constant term.**

NOTE The following are *not* polynomials, for the reasons stated.

$2x^{-1} + 5$ Exponent in $2x^{-1}$ is *not* nonnegative.

$x^3 + 3x^{1/2}$ Exponent in $3x^{1/2}$ is *not* an integer.

EXAMPLE 1   *Identifying Leading Coefficients and Degrees*

Polynomial	Standard Form	Degree	Leading Coefficient
(a) $5x^2 - 2x^7 + 4 - 2x$	$-2x^7 + 5x^2 - 2x + 4$	7	-2
(b) $16 - 8x^3$	$-8x^3 + 16$	3	-8
(c) 10	10	0	10
(d) $5 + x^4 - 6x^3$	$x^4 - 6x^3 + 5$	4	1

NOTE   The prefix *mono* means one, the prefix *bi* means two, and the prefix *tri* means three.

A polynomial with only one term is a **monomial.** Polynomials with two *unlike* terms are **binomials,** and those with three *unlike* terms are **trinomials.** Here are some examples.

Monomial: $5x^3$ *Binomial:* $-4x + 3$ *Trinomial:* $2x^2 + 3x - 7$

EXAMPLE 2   *Evaluating a Polynomial*

Find the value of $x^3 - 5x^2 + 6x - 3$ when $x = 4$.

Solution

When $x = 4$, you have the following.

$$
\begin{aligned}
\text{Value} &= x^3 - 5x^2 + 6x - 3 && \text{Original polynomial}\\
&= 4^3 - 5(4^2) + 6(4) - 3 && \text{Replace } x \text{ by 4.}\\
&= 64 - 80 + 24 - 3 && \text{Evaluate terms.}\\
&= 5 && \text{Simplify.}
\end{aligned}
$$

Adding and Subtracting Polynomials

To add two polynomials, simply combine like terms. This can be done in either a horizontal or a vertical format, as shown in Examples 3 and 4.

EXAMPLE 3   *Adding Polynomials Horizontally*

$$
\begin{aligned}
&(2x^3 + x^2 - 5) + (x^2 + x + 6) && \text{Original polynomials}\\
&= (2x^3) + (x^2 + x^2) + (x) + (-5 + 6) && \text{Group like terms.}\\
&= 2x^3 + 2x^2 + x + 1 && \text{Combine like terms.}
\end{aligned}
$$

EXAMPLE 4 *Using a Vertical Format to Add Polynomials*

To use a vertical format to add polynomials, align the terms of the polynomials by their degrees, as follows.

$$
\begin{array}{r}
5x^3 + 2x^2 - x + 7 \\
3x^2 - 4x + 7 \\
-x^3 + 4x^2 \quad\ \ - 8 \\
\hline
4x^3 + 9x^2 - 5x + 6
\end{array}
$$

To subtract one polynomial from another, change the sign of each of the terms of the polynomial that is being subtracted and then add the resulting like terms.

EXAMPLE 5 *Subtracting Polynomials Horizontally*

Use a horizontal format to subtract $x^3 + 2x^2 - x - 4$ from $3x^3 - 5x^2 + 3$.

Solution

$$
\begin{aligned}
(3x^3 - 5x^2 + 3) &- (x^3 + 2x^2 - x - 4) && \text{Original polynomials} \\
&= 3x^3 - 5x^2 + 3 - x^3 - 2x^2 + x + 4 && \text{Change signs and add.} \\
&= (3x^3 - x^3) + (-5x^2 - 2x^2) + (x) + (3 + 4) && \text{Group like terms.} \\
&= 2x^3 - 7x^2 + x + 7 && \text{Combine like terms.}
\end{aligned}
$$

Be especially careful to get the correct signs when you are subtracting one polynomial from another. One of the most common mistakes in algebra is to forget to change signs correctly when subtracting one expression from another.

EXAMPLE 6 *Using a Vertical Format to Subtract Polynomials*

Use a vertical format to perform the operations.

$$(4x^4 - 2x^3 + 5x^2 - x + 8) - (3x^4 - 2x^3 + 3x - 4)$$

Solution

$$
\begin{array}{ll}
\begin{array}{r}
(4x^4 - 2x^3 + 5x^2 - x + 8) \\
-(3x^4 - 2x^3 \quad\quad + 3x - 4) \\
\hline
\end{array}
\Longrightarrow
\begin{array}{r}
4x^4 - 2x^3 + 5x^2 - \ x + \ 8 \\
-3x^4 + 2x^3 \quad\quad\ - 3x + \ 4 \\
\hline
x^4 \quad\quad + 5x^2 - 4x + 12
\end{array}
\end{array}
$$

Multiplying Polynomials

The simplest type of polynomial multiplication involves a monomial multiplier. The product is obtained by direct application of the Distributive Property.

EXAMPLE 7 *Finding Products with Monomial Multipliers*

You might illustrate the use of the Commutative Property of Multiplication.

$$(2x - 7)(-3x) = (-3x)(2x - 7)$$
$$= -3x(2x) - 3x(-7)$$
$$= -6x^2 + 21x$$

(a) $(2x - 7)(-3x) = 2x(-3x) - 7(-3x)$ Distributive Property

$\qquad\qquad\qquad\; = -6x^2 + 21x$ Properties of exponents

(b) $4x^2(3x - 2x^3 + 1) = 4x^2(3x) - 4x^2(2x^3) + 4x^2(1)$ Distributive Property

$\qquad\qquad\qquad\quad = 12x^3 - 8x^5 + 4x^2$ Properties of exponents

$\qquad\qquad\qquad\quad = -8x^5 + 12x^3 + 4x^2$ Standard form

---

To multiply two binomials, you can use both (left and right) forms of the Distributive Property. For example, if you treat the binomial $(2x + 7)$ as a single quantity, you can multiply $(3x - 2)$ by $(2x + 7)$ as follows.

$$(3x - 2)(2x + 7) = 3x(2x + 7) - (2)(2x + 7)$$
$$= (3x)(2x) + (3x)(7) - (2)(2x) - (2)(7)$$
$$= 6x^2 + 21x - 4x - 14$$

Product of First terms	Product of Outer terms	Product of Inner terms	Product of Last terms

$$= 6x^2 + 17x - 14$$

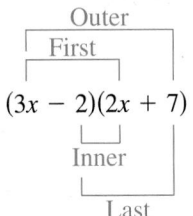

Outer

First

$(3x - 2)(2x + 7)$

Inner

Last

FOIL Diagram

With practice, you should be able to multiply two binomials without writing out all of the steps above. In fact, the four products in the boxes above suggest that you can put the product of two binomials in the FOIL form in just one step. This is called the **FOIL Method.** Note that the words *first, outer, inner,* and *last* refer to the positions of the terms in the original product.

EXAMPLE 8 *Multiplying Binomials (Distributive Property)*

Use the Distributive Property to multiply $x + 2$ by $x - 3$.

Solution

$$(x + 2)(x - 3) = x(x - 3) + 2(x - 3)$$ Distributive Property
$$= x^2 - 3x + 2x - 6$$ Distributive Property
$$= x^2 - x - 6$$ Combine like terms.

EXAMPLE 9   *Multiplying Binomials (FOIL Method)*

Use the FOIL method to multiply $3x + 4$ by $2x + 1$.

Solution

$$
\begin{array}{c}
\quad\quad\quad\quad\quad\quad \text{F}\quad\ \ \text{O}\quad\ \text{I}\quad\ \ \text{L} \\
(3x + 4)(2x + 1) = 6x^2 + 3x\ + 8x\ + 4 \\
 = 6x^2 + 11x + 4 \qquad \text{Combine like terms.}
\end{array}
$$

To multiply two polynomials that have three or more terms, you can use the same basic principle that you use when multiplying monomials and binomials. That is, *each term of one polynomial must be multiplied by each term of the other polynomial.* This can be done using either a horizontal or a vertical format.

EXAMPLE 10   *Multiplying Polynomials (Horizontal Format)*

$$
\begin{aligned}
&(4x^2 - 3x - 1)(2x - 5) \\
&= 4x^2(2x - 5) - 3x(2x - 5) - 1(2x - 5) \qquad \text{Distributive Property}\\
&= 8x^3 - 20x^2 - 6x^2 + 15x - 2x + 5 \qquad \text{Distributive Property}\\
&= 8x^3 - 26x^2 + 13x + 5 \qquad \text{Combine like terms.}
\end{aligned}
$$

When multiplying two polynomials, it is best to write each in standard form before using either the horizontal or vertical format. This is illustrated in the next example.

EXAMPLE 11   *Multiplying Polynomials (Vertical Format)*

With a vertical format, line up like terms in the same vertical columns, much as you align digits in whole-number multiplication.

$$
\begin{array}{r}
4x^2 + \ x\ -\ 2 \qquad \text{Standard form}\\
\times \quad\quad\quad -x^2 + 3x + \ 5 \qquad \text{Standard form}\\
\hline
20x^2 + 5x - 10 \qquad\ \ 5(4x^2 + x - 2)\\
12x^3 + \ 3x^2 - 6x \qquad\ \ 3x(4x^2 + x - 2)\\
-4x^4 - \ \ x^3 + \ 2x^2 \qquad\ \ -x^2(4x^2 + x - 2)\\
\hline
-4x^4 + 11x^3 + 25x^2 - \ x - 10
\end{array}
$$

EXAMPLE 12 Multiplying Polynomials

$$
\begin{aligned}
(x - 4)^3 &= (x - 4)(x - 4)(x - 4) \\
&= [(x - 4)(x - 4)](x - 4) && \text{Associative Property} \\
&= (x^2 - 4x - 4x + 16)(x - 4) && \text{Find } (x - 4)^2. \\
&= (x^2 - 8x + 16)(x - 4) && \text{Combine like terms.} \\
&= x^2(x - 4) - 8x(x - 4) + 16(x - 4) && \text{Distributive Property} \\
&= x^3 - 4x^2 - 8x^2 + 32x + 16x - 64 && \text{Distributive Property} \\
&= x^3 - 12x^2 + 48x - 64 && \text{Combine like terms.}
\end{aligned}
$$

EXAMPLE 13 An Area Model for Multiplying Polynomials

Use an area model to show that $(x + 2)(2x + 1) = 2x^2 + 5x + 2$.

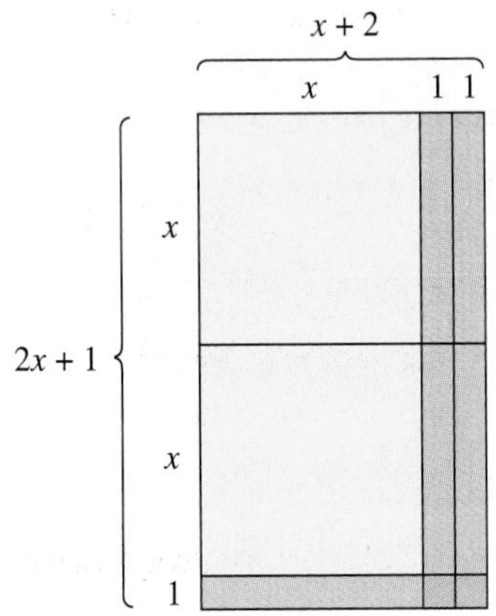

Solution

Think of a rectangle whose sides are $x + 2$ and $2x + 1$. The area of this rectangle is

$$(x + 2)(2x + 1). \qquad \text{Area} = \text{(width)(length)}$$

Another way to find the area is to add the areas of the rectangular parts, as shown in Figure 1.10. There are two squares whose sides are x, five rectangles whose sides are x and 1, and two squares whose sides are 1. The total area of these nine rectangles is

$$2x^2 + 5x + 2. \qquad \text{Area} = \text{sum of rectangular areas}$$

Because each method must produce the same area, you can conclude that

$$(x + 2)(2x + 1) = 2x^2 + 5x + 2.$$

FIGURE 1.10

Special Products

Some binomial products have special forms that occur frequently in algebra. For instance, the product $(x + 3)(x - 3)$ is called the **product of the sum and difference of two terms.** With such products, the two middle terms cancel, as follows.

$$
\begin{aligned}
(x + 3)(x - 3) &= x^2 - 3x + 3x - 9 && \text{Sum and difference of two terms} \\
&= x^2 - 9 && \text{Product has no middle term.}
\end{aligned}
$$

Another common type of product is the **square of a binomial.** With this type of product, the middle term is always twice the product of the terms in the binomial.

$$(2x + 5)^2 = (2x + 5)(2x + 5) \qquad \text{Square of a binomial}$$
$$= 4x^2 + 10x + 10x + 25 \qquad \text{Outer and inner terms are equal.}$$
$$= 4x^2 + 20x + 25 \qquad \text{Middle term is twice the product of the terms in the binomial.}$$

Special Products

Let u and v be real numbers, variables, or algebraic expressions. Then the following formulas are true.

Sum and Difference of Two Terms *Example*

$(u + v)(u - v) = u^2 - v^2$ $(3x - 4)(3x + 4) = (3x)^2 - 4^2$
$= 9x^2 - 16$

Square of a Binomial *Example*

$(u + v)^2 = u^2 + 2uv + v^2$ $(4x + 9) = (4x)^2 + 2(4x)(9) + 9^2$
$= 16x^2 + 72x + 81$

$(u - v)^2 = u^2 - 2uv + v^2$ $(x - 6)^2 = x^2 - 2(x)(6) + 6^2$
$= x^2 - 12x + 36$

NOTE When a binomial is squared, the resulting middle term is always *twice* the product of the two terms.

$$(x + y)^2 = x^2 + 2(xy) + y^2$$

Terms   Twice the product
of the terms

EXAMPLE 14 *Finding Special Products*

(a) $(3x - 2)(3x + 2) = (3x)^2 - 2^2$ Special Product
$= 9x^2 - 4$ Simplify.

(b) $(2x - 7)^2 = (2x)^2 - 2(2x)(7) + 7^2$ Special Product
$= 4x^2 - 28x + 49$ Simplify.

(c) $[(a - 2) + b]^2 = (a - 2)^2 + 2(a - 2)b + b^2$ Special Product
$= a^2 - 4a + 4 + 2ab - 4b + b^2$ Simplify.

Applications

There are many applications that require the evaluation of polynomials. One commonly used second-degree polynomial is called a **position polynomial.** This polynomial has the form

$$-16t^2 + v_0 t + s_0, \qquad \text{Position polynomial}$$

where t is the time, measured in seconds. The value of this polynomial gives the height (in feet) of a free-falling object above the ground, assuming no air resistance. The coefficient of t, v_0, is called the **initial velocity** of the object, and the constant term, s_0, is called the **initial height** of the object. If the initial velocity is positive, the object was projected upward (at $t = 0$), and if the initial velocity is negative, the object was projected downward.

EXAMPLE 15   *Finding the Height of a Free-Falling Object*

An object is thrown downward from the top of a 200-foot building. The initial velocity is -10 feet per second. Use the position polynomial

$$-16t^2 - 10t + 200$$

to find the height of the object when $t = 1$, $t = 2$, and $t = 3$ (see Figure 1.11).

Solution

When $t = 1$, the height of the object is

$$\begin{aligned} \text{Height} &= -16(1^2) - 10(1) + 200 \\ &= -16 - 10 + 200 \\ &= 174 \text{ feet.} \end{aligned}$$

When $t = 2$, the height of the object is

$$\begin{aligned} \text{Height} &= -16(2^2) - 10(2) + 200 \\ &= -64 - 20 + 200 \\ &= 116 \text{ feet.} \end{aligned}$$

When $t = 3$, the height of the object is

$$\begin{aligned} \text{Height} &= -16(3^2) - 10(3) + 200 \\ &= -144 - 30 + 200 \\ &= 26 \text{ feet.} \end{aligned}$$

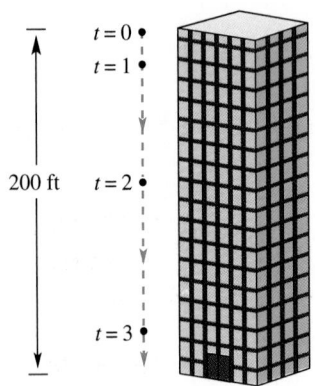

FIGURE 1.11

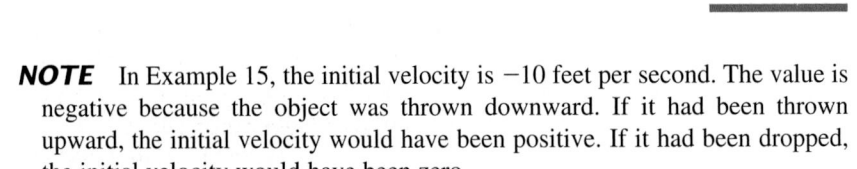

NOTE   In Example 15, the initial velocity is -10 feet per second. The value is negative because the object was thrown downward. If it had been thrown upward, the initial velocity would have been positive. If it had been dropped, the initial velocity would have been zero.

EXAMPLE 16 *Using Polynomial Models*

The numbers of pounds of poultry P and of beef B consumed per person in the United States from 1985 to 1993 (see Figure 1.12) can be modeled by

$$P = 1.4t + 28.8 \qquad \text{Poultry (pounds per person)}$$
$$B = 0.16t^2 - 4.62t + 94.69, \qquad \text{Beef (pounds per person)}$$

where $t = 5$ represents 1985. Find a model that represents the total amount T of poultry *and* beef consumed from 1985 to 1993. Estimate the total amount T consumed in 1991. *(Source: U.S. Department of Agriculture)*

Solution

The sum of the two polynomial models would be

$$(1.4t + 28.8) + (0.16t^2 - 4.62t + 94.69) = 0.16t^2 - 3.22t + 123.49.$$

The model for the total consumption of poultry and beef is

$$T = 0.16t^2 - 3.22t + 123.49. \qquad \text{Total (pounds per person)}$$

Using this model, and substituting $t = 11$, you can estimate the 1991 consumption to be

$$T = 0.16(11^2) - 3.22(11) + 123.49 = 107.43 \text{ pounds per person.}$$

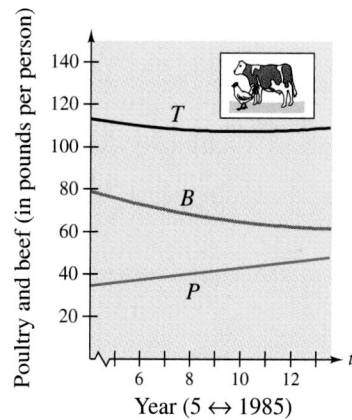

Poultry and beef (in pounds per person)

Year ($5 \leftrightarrow 1985$)

FIGURE 1.12

Group Activities

The Position Polynomial

Consider the following polynomials in the variable t to represent the height above ground of an object moving vertically up or down through the air.

$$-16t^2 + 80t + 30, \quad -16t^2 - 80t + 30, \quad -16t^2 + 30, \quad -16t^2 + 30t$$

Discuss each question in your group. Then summarize each answer with a short paragraph, giving reasons for your conclusions.

a. Which polynomial represents the height of an object thrown *downward?*

b. Which polynomial represents the height of an object that is *dropped* from a height of 30 feet?

c. Which polynomial represents the height of an object thrown *upward* from ground level?

1.4   **Exercises**

See Warm-Up Exercises, p. A39

1. *Think About It*   Explain why $x^2 - 3\sqrt{x}$ is not a polynomial.

2. *Think About It*   Explain the difference between the degree of a term in a polynomial and the degree of a polynomial.

3. *Think About It*   What operation separates the terms of a polynomial? What operation separates the factors of a term?

4. *Think About It*   Is every trinomial a second-degree polynomial? Explain.

In Exercises 5–12, write the polynomial in standard form, and find its degree and leading coefficient.

5. $10x - 4$

6. $3x^2 + 8$

7. $5 - 3y^4$

8. $-3x^3 - 2x^2 - 3$

9. $6t + 4t^5 - t^2 + 3$

10. $8z - 16z^2$

11. -4

12. $v_0 t - 16t^2$ (v_0 is a constant.)

In Exercises 13–16, determine whether the polynomial is a monomial, binomial, or trinomial.

13. $12 - 5y^2$

14. t^3

15. $x^3 + 2x^2 - 4$

16. $2u^7 - 9u^3$

In Exercises 17–20, give an example of a polynomial in one variable satisfying the conditions. (*Note:* There is more than one correct answer.)

17. A monomial of degree 3

18. A trinomial of degree 4 and leading coefficient -2

19. A binomial of degree 2 and leading coefficient 8

20. A monomial of degree 0

In Exercises 21–24, find the value of the polynomial at the given values of the variable.

21. $x^3 - 12x$

 (a) $x = -2$

 (b) $x = 0$

 (c) $x = 2$

 (d) $x = 4$

22. $\frac{1}{4}x^4 - 2x^2$

 (a) $x = -2$

 (b) $x = 0$

 (c) $x = 2$

 (d) $x = 3$

23. $x^4 - 4x^3 + 16x - 16$

 (a) $x = -1$

 (b) $x = 0$

 (c) $x = 2$

 (d) $x = \frac{5}{2}$

24. $3t^4 + 4t^3$

 (a) $t = -1$

 (b) $t = -\frac{2}{3}$

 (c) $t = 0$

 (d) $t = 1$

In Exercises 25–30, perform the addition using a horizontal format.

25. $(2x^2 - 3) + (5x^2 + 6)$

26. $(3x^3 - 2x + 8) + (3x - 5)$

27. $(x^2 - 3x + 8) + (2x^2 - 4x) + 3x^2$

28. $(5y + 6) + (4y^2 - 6y - 3)$

29. $(3x^2 + 8) + (7 - 5x^2)$

30. $(20s - 12s^2 - 32) + (15s^2 + 6s)$

In Exercises 31–34, perform the addition using a vertical format.

31. $(5x^2 - 3x + 4) + (-3x^2 - 4)$

32. $(4x^3 - 2x^2 + 8x) + (4x^2 + x - 6)$

33. $(2b - 3) + (b^2 - 2b) + (7 - b^2)$

34. $(v^2 + v - 3) + (4v + 1) + (2v^2 - 3v)$

35. *Think About It* Can two third-degree polynomials be added to produce a second-degree polynomial? If so, give an example.

36. *Think About It* Describe the method for subtracting polynomials.

In Exercises 37–42, perform the subtraction using a horizontal format.

37. $(3x^2 - 2x + 1) - (2x^2 + x - 1)$

38. $(5y^4 - 2) - (3y^4 + 2)$

39. $(8x^3 - 4x^2 + 3x) - [(x^3 - 4x^2 + 5) + (x - 5)]$

40. $(5y^2 - 2y) - [(y^2 + y) - (3y^2 - 6y + 2)]$

41. $(10x^3 + 15) - (6x^3 - x + 11)$

42. $(y^2 + 3y^4) - (y^4 - y^2)$

In Exercises 43–46, perform the subtraction using a vertical format.

43. $(x^2 - x + 3) - (x - 2)$

44. $(0.2t^4 - 5t^2) - (-t^4 + 0.3t^2 - 1.4)$

45. $(-2x^3 - 15x + 25) - (2x^3 - 13x + 12)$

46. $(z^2 - 5) - (z^2 - 5)$

In Exercises 47–54, perform the operations.

47. $(4x^2 + 5x - 6) - (2x^2 - 4x + 5)$

48. $(13x^3 - 9x^2 + 4x - 5) - (5x^3 + 7x + 3)$

49. $(10x^2 - 11) - (-7x^3 - 12x^2 - 15)$

50. $(15y^4 - 18y - 18) - (-11y^4 - 8y - 8)$

51. $5s - [6s - (30s + 8)]$

52. $3x^2 - 2[3x + (9 - x^2)]$

53. $2(t^2 + 12) - 5(t^2 + 5) + 6(t^2 + 5)$

54. $-10(v + 2) + 8(v - 1) - 3(v - 9)$

In Exercises 55–72, perform the multiplication and simplify.

55. $(-2a^2)(-8a)$

56. $(-6n)(3n^2)$

57. $2y(5 - y)$

58. $5z(2z - 7)$

59. $4x^3(2x^2 - 3x + 5)$

60. $3y^2(-3y^2 + 7y - 3)$

61. $-2x^2(5 + 3x^2 - 7x^3)$

62. $-3a^2(11a - 3)$

63. $(x + 7)(x - 4)$

64. $(y - 2)(y + 3)$

65. $(2x + y)(3x + 2y)$

66. $(2x - y)(3x - 2y)$

67. $\left(4y - \frac{1}{3}\right)(12y + 9)$

68. $\left(5t - \frac{3}{4}\right)(2t - 16)$

69. $-3x(-5x)(5x + 2)$

70. $4t(-3t)(t^2 - 1)$

71. $5a(a + 2) - 3a(2a - 3)$

72. $(2t - 1)(t + 1) + 3(2t - 5)$

In Exercises 73–80, perform the multiplication using a horizontal format.

73. $(x^3 - 3x + 2)(x - 2)$

74. $(t + 3)(t^2 - 5t + 1)$

75. $(u + 5)(2u^2 + 3u - 4)$

76. $(x^2 + 4)(x^2 - 2x - 4)$

77. $(2x^2 - 3)(2x^2 - 2x + 3)$

78. $(x - 1)(x^2 - 4x + 6)$

79. $(a + 5)^3$

80. $(y - 2)^3$

In Exercises 81–84, perform the multiplication using a vertical format.

81.
$$7x^2 - 14x + 9$$
$$\times \qquad\quad x + 3$$

82.
$$4x^4 - 6x^2 + 9$$
$$\times \qquad 2x^2 + 3$$

83.
$$-x^2 + 2x - 1$$
$$\times \qquad 2x + 1$$

84.
$$2s^2 - 5s + 6$$
$$\times \qquad 3s - 4$$

In Exercises 85–100, multiply.

85. $(a - 6c)(a + 6c)$ **86.** $(8n - m)(8n + m)$

87. $\left(2x - \frac{1}{4}\right)\left(2x + \frac{1}{4}\right)$ **88.** $\left(\frac{2}{3}x + 7\right)\left(\frac{2}{3}x - 7\right)$

89. $(0.2t + 0.5)(0.2t - 0.5)$

90. $(4a - 0.1b)(4a + 0.1b)$

91. $(x + 5)^2$ **92.** $(x + 9)^2$

93. $(5x - 2)^2$ **94.** $(3x - 8)^2$

95. $(2a + 3b)^2$ **96.** $(4x - 5y)^2$

97. $[(x + 2) - y]^2$ **98.** $[(x - 4) + y]^2$

99. $[u - (v - 3)]^2$ **100.** $[2z + (y + 1)]^2$

In Exercises 101–106, perform the operations and simplify.

101. $(x + 3)(x - 3) - (x^2 + 8x - 2)$

102. $(k - 8)(k + 8) - (k^2 - k + 3)$

103. $5y(y - 4) + (y - 6)^2$

104. $(b + 1)^2 - 4b(b + 5)$

105. $(t + 3)^2 - (t - 3)^2$

106. $(a + 6)^2 + (a - 6)^2$

107. *Think About It*   What is the degree of the product of two polynomials of degrees m and n?

108. *True or False?*   Decide whether the statement is true or false. Explain.

   (a) The product of two monomials is a monomial.

   (b) The product of two binomials is a binomial.

Geometry   In Exercises 109 and 110, find the area of the shaded portion of the figure.

109.

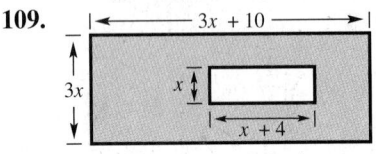

110.

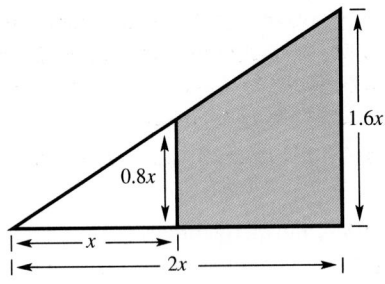

Geometrical Modeling   In Exercises 111 and 112, use the area model to write two different expressions for the total area. Then equate the two expressions and name the algebraic property that is illustrated.

111.

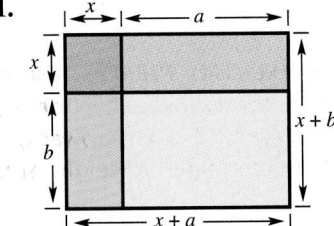

112.

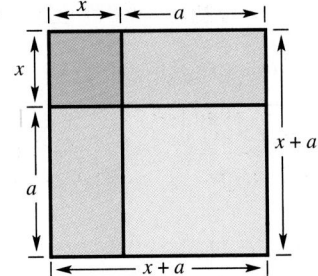

113. *Dimensions of a Rectangle*   The length of a rectangle is one and one half times its width w. Find (a) the perimeter and (b) the area of the rectangle.

114. *Area of a Triangle*   The base of a triangle is $3x$ and its height is $x + 5$. Find the area A of the triangle.

115. *Compounded Interest*   After 2 years, an investment of \$1000 compounded annually at an interest rate of r will yield an amount $1000(1 + r)^2$. Find this product.

116. *Compounded Interest*   After 2 years, an investment of $1000 compounded annually at an interest rate of 9.5% will yield an amount $1000(1 + 0.095)^2$. Find this product.

117. *Find a Pattern*   Perform the following multiplications.

(a) $(x - 1)(x + 1)$

(b) $(x - 1)(x^2 + x + 1)$

(c) $(x - 1)(x^3 + x^2 + x + 1)$

From the pattern formed by these products, can you predict the result of $(x - 1)(x^4 + x^3 + x^2 + x + 1)$?

118. *Evaluating Expressions*   Verify that $(x + y)^2$ is not equal to $x^2 + y^2$ by evaluating both expressions when $x = 3$ and $y = 4$.

Free-Falling Object   In Exercises 119–122, use the position polynomial to determine whether the free-falling object was dropped, thrown upward, or thrown downward. Also determine the height of the object at time $t = 0$.

119. $-16t^2 + 100$ **120.** $-16t^2 + 50t$

121. $-16t^2 - 24t + 50$ **122.** $-16t^2 + 32t + 300$

123. *Free-Falling Object*   An object is thrown upward from the top of a 200-foot building (see figure). The initial velocity is 40 feet per second. Use the position polynomial

$$\text{Height} = -16t^2 + 40t + 200$$

to find the height of the object when $t = 1$, $t = 2$, and $t = 3$.

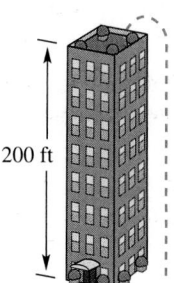

200 ft

In Exercises 124–127, use a calculator to perform the operations.

124. $8.04x^2 - 9.37x^2 + 5.62x^2$

125. $-11.98y^3 + 4.63y^3 - 6.79y^3$

126. $(4.098a^2 + 6.349a) - (11.246a^2 - 9.342a)$

127. $(27.433k^2 - 19.018k) + (-14.61k^2 + 3.814k)$

Section Project

Consumption of Milk   The per capita consumption (average consumption per person) of whole milk and lowfat milk in the United States between 1980 and 1993 can be approximated by these two polynomial models.

$$y = 17.01 - 0.60t \qquad \text{Whole milk}$$

$$y = 8.92 + 0.47t - 0.01t^2 \qquad \text{Lowfat milk}$$

In these models, y represents the average consumption per person in gallons and t represents the year, with $t = 0$ corresponding to 1980. *(Source: U.S. Department of Agriculture)*

(a) Find a polynomial model that represents the per capita consumption of milk (of both types) during this time period. Use this model to find the per capita consumption of milk in 1985 and in 1990.

(b) During the given period, the per capita consumption of whole milk was decreasing and the per capita consumption of lowfat milk was increasing (see figure). Was the combined per capita consumption of milk increasing or decreasing?

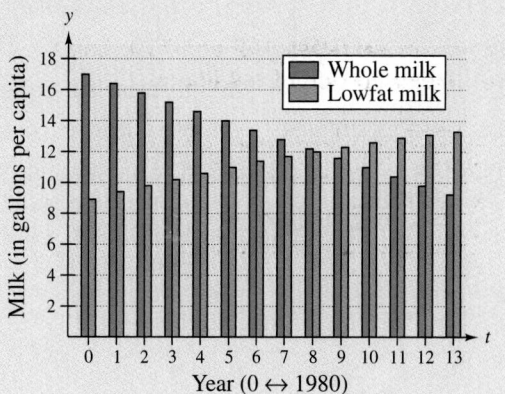

Mid-Chapter Quiz

Take this quiz as you would take a quiz in class. After you are done, check your work against the answers given in the back of the book.

In Exercises 1–6, evaluate the expression. Write fractions in reduced form.

1. $16 + (-84)$

2. $\frac{5}{8} + \frac{1}{4} - \frac{5}{6}$

3. $|-16.25| + 54.78$

4. $\frac{3}{4} \div \frac{15}{8}$

5. $-\left(-\frac{3}{2}\right)^3$

6. $\frac{2^4 - 6}{5} - \sqrt{16}$

In Exercises 7–9, name the property of real numbers demonstrated.

7. $5(x + 3) = 5 \cdot x + 5 \cdot 3$

8. $6y \cdot 1 = 6y$

9. $(z + 3) - (z + 3) = 0$

10. Write the expression $(5x)(5x)(5x)(5x)$ using exponential notation.

In Exercises 11–22, perform the operations and/or simplify the expression.

11. $4x^2 \cdot x^3$

12. $(-2x)^4$

13. $\left(\frac{y^2}{3}\right)^3$

14. $\frac{18x^2y^3}{12xy}$

15. $4x^2 - 3xy + 5xy - 5x^2$

16. $3[x - 2x(x^2 + 1)]$

17. $3(x - 5) + 4x$

18. $(v - 3) - 3(2v + 5)$

19. $(6r + 5s)(6r - 5s)$

20. $(x + 1)(x^2 - x + 1)$

21. $2z(z + 5) - 7(z + 5)$

22. $(v - 3)^2 - (v + 3)^2$

2x + 5

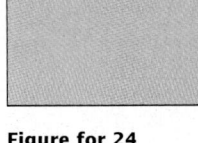

x + 3

Figure for 24

23. Evaluate the expression $10x^2 - |5x|$ when (a) $x = 5$ and (b) $x = -1$.

24. Find an expression for the perimeter of the rectangle shown at the left.

25. The midyear financial statement of a company showed a profit of $1,415,322.62. At the close of the year, the financial statement showed a profit for the year of $916,489.26. Find the profit or loss of the company for the second 6 months of the year.

| 1.5 | **Factoring Polynomials** |

Factoring Polynomials with Common Factors ▪ Factoring by Grouping ▪ Factoring the Difference of Two Squares ▪ Factoring the Sum or Difference of Two Cubes ▪ Factoring Completely

Factoring Polynomials with Common Factors

Now we will switch from the process of multiplying polynomials to the *reverse* process—**factoring polynomials.** This section and the next section deal only with polynomials that have integer coefficients. Remember that in Section 1.4 you used the Distributive Property to *multiply* and *remove* parentheses, as follows.

Multiply ▷

$$3x(4 - 5x) = 12x - 15x^2 \qquad \text{Distributive Property}$$

In this and the next section, you will use the Distributive Property in the reverse direction to *factor* and *create* parentheses.

Factor ▷

$$12x - 15x^2 = 3x(4 - 5x) \qquad \text{Distributive Property}$$

Factoring an expression (by the Distributive Property) changes a *sum of terms* into a *product of factors.* Later you will see that this is an important strategy for solving equations and for simplifying algebraic expressions.

To be efficient in factoring, you need to understand the concept of the **greatest common factor.** Recall from arithmetic that every integer can be factored into a product of prime numbers. The **greatest common factor** of two or more integers is the greatest integer that divides evenly into each integer. For example, the greatest common factor of 18 and 42 is 6, as shown at the left.

NOTE   To find the greatest common factor of two integers or two expressions, begin by writing each as a product of prime factors. The greatest common factor is the product of the *common* prime factors. For instance, from the factorizations

$$18 = 2 \cdot 3 \cdot 3$$
$$42 = 2 \cdot 3 \cdot 7$$

it follows that the greatest common factor is $2 \cdot 3$ or 6.

EXAMPLE 1   *Finding the Greatest Common Factor*

Find the greatest common factor of $6x^5$, $30x^4$, and $12x^3$.

Solution

From the factorizations

$$6x^5 = 2 \cdot 3 \cdot x \cdot x \cdot x \cdot x \cdot x = (6x^3)(x^2)$$
$$30x^4 = 2 \cdot 3 \cdot 5 \cdot x \cdot x \cdot x \cdot x = (6x^3)(5x)$$
$$12x^3 = 2 \cdot 2 \cdot 3 \cdot x \cdot x \cdot x = (6x^3)(2),$$

you can conclude that the greatest common factor is $6x^3$.

Consider the three terms given in Example 1 as terms of the polynomial

$$6x^5 + 30x^4 + 12x^3.$$

NOTE If a polynomial in x (with integer coefficients) has a greatest common monomial factor of the form ax^n, the following statements must be true.

1. The coefficient a of the greatest common monomial factor must be the greatest integer that *divides* each of the coefficients in the polynomial.

2. The variable factor x^n of the greatest common monomial factor has the *least* power of x of all terms of the polynomial.

The greatest common factor, $6x^3$, of these terms is called the **greatest common monomial factor** of the polynomial. When you use the Distributive Property to remove this factor from each term of the polynomial, you are **factoring out** the greatest common monomial factor.

$$6x^5 + 30x^4 + 12x^3 = 6x^3(x^2) + 6x^3(5x) + 6x^3(2) \quad \text{Factor each term.}$$
$$= 6x^3(x^2 + 5x + 2) \quad \text{Factor out common monomial factor.}$$

EXAMPLE 2 Factoring Out a Greatest Common Monomial Factor

Factor out the greatest common monomial factor from $24x^3 - 32x^2$.

Solution

The greatest common factor of $24x^3$ and $32x^2$ is $8x^2$. Thus, you can factor the given polynomial as follows.

$$24x^3 - 32x^2 = (8x^2)(3x) - (8x^2)(4) \quad \text{Factor each term.}$$
$$= 8x^2(3x - 4) \quad \text{Factor out common monomial factor.}$$

We usually consider the greatest common monomial factor of a polynomial to have a positive coefficient. However, sometimes it is convenient to factor a negative number out of a polynomial, as shown in the next example.

EXAMPLE 3 A Negative Common Monomial Factor

Factor the polynomial $-3x^2 + 12x - 18$ in two ways.

(a) Factor out a 3.
(b) Factor out a -3.

Solution

(a) To factor out the common monomial factor of 3, write the following.
$$-3x^2 + 12x - 18 = 3(-x^2) + 3(4x) + 3(-6) \quad \text{Factor out each term.}$$
$$= 3(-x^2 + 4x - 6) \quad \text{Factored form}$$

(b) To factor out the common monomial factor of -3, write the following.
$$-3x^2 + 12x - 18 = -3(x^2) + (-3)(-4x) + (-3)(6) \quad \text{Factor out each term.}$$
$$= -3(x^2 - 4x + 6) \quad \text{Factored form}$$

Factoring by Grouping

There are occasions when a common factor of a polynomial is not simply a monomial. For instance, the polynomial $x^2(2x - 3) + 4(2x - 3)$ has the common *binomial* factor $(2x - 3)$. Factoring out this common factor produces

$$x^2(2x - 3) + 4(2x - 3) = (2x - 3)(x^2 + 4).$$

This type of factoring is part of a more general procedure called **factoring by grouping.**

Point out that when factoring a polynomial with four or more terms, students first should try factoring by grouping.

Note that the common factor must be exactly the same. For example, $x(x - 3) - 5(x + 3)$ cannot be factored.

EXAMPLE 4   A Common Binomial Factor

Factor the polynomial $5x^2(6x - 5) - 2(6x - 5)$.

Solution

Each of the terms of this polynomial has a binomial factor of $(6x - 5)$. Factoring this binomial out of each term produces the following.

$$5x^2(6x - 5) - 2(6x - 5) = (6x - 5)(5x^2 - 2)$$

In Example 4, the given polynomial was already grouped, and so it was easy to determine the common binomial factor. In practice, you will have to do the grouping as well as the factoring. To see how this works, consider the expression $x^3 - 3x^2 - 5x + 15$ and try to *factor* it. Note first that there is no common monomial factor to take out of all four terms. But suppose you *group* the first two terms together and the last two terms together. Then you have the following.

$$
\begin{aligned}
x^3 - 3x^2 - 5x + 15 &= (x^3 - 3x^2) - (5x - 15) && \text{Group terms.}\\
&= x^2(x - 3) - 5(x - 3) && \text{Factor out common}\\
& && \text{factor in each group.}\\
&= (x - 3)(x^2 - 5) && \text{Factor out common}\\
& && \text{binomial factor.}
\end{aligned}
$$

NOTE   In Example 5, the polynomial is factored by grouping the first and second terms and the third and fourth terms. You could just as easily have grouped the first and third terms and the second and fourth terms, as follows.

$$
\begin{aligned}
&x^3 - 5x^2 + x - 5\\
&= (x^3 + x) - (5x^2 + 5)\\
&= x(x^2 + 1) - 5(x^2 + 1)\\
&= (x^2 + 1)(x - 5)
\end{aligned}
$$

EXAMPLE 5   Factoring by Grouping

Factor the polynomial $x^3 - 5x^2 + x - 5$.

Solution

$$
\begin{aligned}
x^3 - 5x^2 + x - 5 &= (x^3 - 5x^2) + (x - 5) && \text{Group terms.}\\
&= x^2(x - 5) + 1(x - 5) && \text{Factor out common}\\
& && \text{factor in each group.}\\
&= (x - 5)(x^2 + 1) && \text{Factored form}
\end{aligned}
$$

You can check this by multiplying out the factors and comparing the result with the original expression.

Factoring the Difference of Two Squares

Some polynomials have special forms that you should learn to recognize so that they can be factored easily. Here are some examples of forms that you should be able to recognize by the time you have completed this section.

$$x^2 - 9 = (x + 3)(x - 3)$$ Difference of two squares

$$x^3 + 8 = (x + 2)(x^2 - 2x + 4)$$ Sum of two cubes

$$x^3 - 1 = (x - 1)(x^2 + x + 1)$$ Difference of two cubes

One of the easiest special polynomial forms to recognize and to factor is the form $u^2 - v^2$, called a **difference of two squares.**

Difference of Two Squares

Let u and v be real numbers, variables, or algebraic expressions. Then the expression $u^2 - v^2$ can be factored using the following pattern.

$$u^2 - v^2 = (u + v)(u - v)$$

Difference   Opposite signs

To recognize perfect squares, look for coefficients that are squares of integers and for variables raised to *even* powers. Here are some examples.

Consider reinforcing the pattern using parentheses.

$$81y^4 - 49 = (\quad)^2 - (\quad)^2$$
$$= (9y^2)^2 - (7)^2$$
$$= (9y^2 + 7)(9y^2 - 7)$$

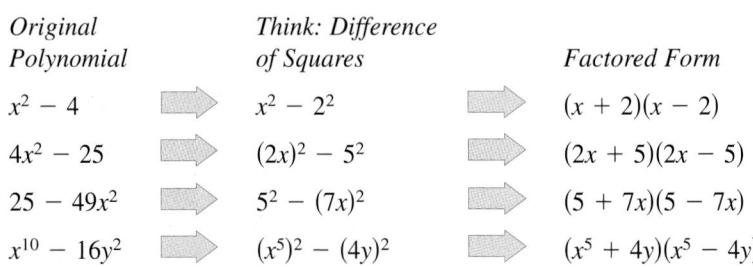

Original Polynomial	*Think: Difference of Squares*	*Factored Form*
$x^2 - 4$	$x^2 - 2^2$	$(x + 2)(x - 2)$
$4x^2 - 25$	$(2x)^2 - 5^2$	$(2x + 5)(2x - 5)$
$25 - 49x^2$	$5^2 - (7x)^2$	$(5 + 7x)(5 - 7x)$
$x^{10} - 16y^2$	$(x^5)^2 - (4y)^2$	$(x^5 + 4y)(x^5 - 4y)$

EXAMPLE 6   *Factoring the Difference of Two Squares*

Factor the polynomial $x^2 - 64$.

Solution

Because x^2 and 64 are both perfect squares, you can recognize this polynomial as the difference of two squares. Therefore, the polynomial factors as follows.

$$x^2 - 64 = x^2 - 8^2$$ Write as difference of two squares.

$$= (x + 8)(x - 8)$$ Factored form

EXAMPLE 7   *Factoring the Difference of Two Squares*

Factor the polynomial $49x^2 - 81y^2$.

Solution

Because $49x^2$ and $81y^2$ are both perfect squares, you can recognize this polynomial as the difference of two squares. Therefore, the polynomial factors as follows.

$$49x^2 - 81y^2 = (7x)^2 - (9y)^2 \qquad \text{Write as difference of two squares.}$$
$$= (7x + 9y)(7x - 9y) \qquad \text{Factored form}$$

Remember that the rule $u^2 - v^2 = (u + v)(u - v)$ applies to polynomials or expressions in which u and v are themselves expressions. The next example illustrates this possibility.

EXAMPLE 8   *Factor the Difference of Two Squares*

Factor the polynomial $(x + 2)^2 - 9$.

Extension example:
$(3x - 1)^2 - (x + 6)^2$
 $= [(3x - 1) + (x + 6)][(3x - 1) - (x + 6)]$
 $= [3x - 1 + x + 6][3x - 1 - x - 6]$
 $= (4x + 5)(2x - 7)$

Solution

$$(x + 2)^2 - 9 = (x + 2)^2 - 3^2 \qquad \text{Write as difference of two squares.}$$
$$= [(x + 2) + 3][(x + 2) - 3] \qquad \text{Factored form}$$
$$= (x + 5)(x - 1) \qquad \text{Simplify.}$$

Factoring the Sum or Difference of Two Cubes

The last type of special factoring discussed in this section is the sum or difference of two cubes. The patterns for these two special forms are summarized as follows. In these patterns, pay particular attention to the signs of the terms.

Sum or Difference of Two Cubes

Let u and v be real numbers, variables, or algebraic expressions. Then the expressions $u^3 + v^3$ and $u^3 - v^3$ can be factored as follows.

Like signs   Like signs

$$u^3 + v^3 = (u + v)(u^2 - uv + v^2) \qquad u^3 - v^3 = (u - v)(u^2 + uv + v^2)$$

Unlike signs   Unlike signs

NOTE   Remember that you can check a factoring result by multiplying. For instance, you can check the results in Example 9 as follows.

(a)
$$
\begin{array}{r}
x^2 + 5x + 25 \\
x - 5 \\
\hline
-5x^2 - 25x - 125 \\
x^3 + 5x^2 + 25x \\
\hline
x^3 \qquad\qquad - 125
\end{array}
$$

(b)
$$
\begin{array}{r}
4y^2 - 2y + 1 \\
2y + 1 \\
\hline
4y^2 - 2y + 1 \\
8y^3 - 4y^2 + 2y \\
\hline
8y^3 \qquad\qquad + 1
\end{array}
$$

EXAMPLE 9   *Factoring Sums and Differences of Cubes*

Factor each polynomial.

(a) $x^3 - 125$ (b) $8y^3 + 1$

Solution

(a) This polynomial is the difference of two cubes, because x^3 is the cube of x and 125 is the cube of 5. Therefore, you can factor the polynomial as follows.

$$
\begin{aligned}
x^3 - 125 &= x^3 - 5^3 && \text{Write as difference of two cubes.}\\
&= (x - 5)(x^2 + 5x + 5^2) && \text{Factored form}\\
&= (x - 5)(x^2 + 5x + 25) && \text{Simplify.}
\end{aligned}
$$

(b) This polynomial is the sum of two cubes, because $8y^3$ is the cube of $2y$ and 1 is its own cube. Therefore, you can factor the polynomial as follows.

$$
\begin{aligned}
8y^3 + 1 &= (2y)^3 + 1^3 && \text{Write as sum of two cubes.}\\
&= (2y + 1)[(2y)^2 - (2y)(1) + 1^2] && \text{Factored form}\\
&= (2y + 1)(4y^2 - 2y + 1) && \text{Simplify.}
\end{aligned}
$$

Factoring Completely

Exploration

Find a formula for completely factoring $u^6 - v^6$ using the formulas from this section. Use your formula to factor completely $x^6 - 1$ and $x^6 - 64$.

Sometimes the difference of two squares can be hidden by the presence of a common monomial factor. Remember that with *all* factoring techniques, you should first remove any common monomial factors.

EXAMPLE 10   *Removing a Common Monomial Factor First*

Factor the polynomial $125x^2 - 80$ completely.

Solution

The polynomial $125x^2 - 80$ has a common monomial factor of 5. After removing this factor, the remaining polynomial is the difference of two squares.

$$
\begin{aligned}
125x^2 - 80 &= 5(25x^2 - 16) && \text{Remove common monomial factor.}\\
&= 5[(5x)^2 - 4^2] && \text{Write as difference of two squares.}\\
&= 5(5x + 4)(5x - 4) && \text{Factored form}
\end{aligned}
$$

Remind students always to check first for a common monomial factor.

The polynomial in Example 10 is said to be **completely factored** because none of its factors can be factored further using integer coefficients.

Which of the following is factored completely and correctly?

(a) $8x^3 - 64 = 8(x^3 - 8)$
$\qquad = 8(x - 2)(x^2 + 2x + 4)$

(b) $8x^3 - 64 = (2x - 4)(4x^2 + 8x + 16)$

Discuss the results.

EXAMPLE 11 *Removing a Common Monomial Factor First*

Factor the polynomial $3x^3 + 81$ completely.

Solution

$$3x^3 + 81 = 3(x^3 + 27) \qquad \text{Remove common monomial factor.}$$
$$= 3(x^3 + 3^3) \qquad \text{Write as sum of two cubes.}$$
$$= 3(x + 3)(x^2 - 3x + 9) \qquad \text{Factored form}$$

To factor a polynomial completely, always check to see whether the factors obtained might themselves be factorable. That is, can any of the factors be factored? For instance, after factoring the polynomial $x^4 - 16$ once as the difference of two squares,

$$x^4 - 16 = (x^2)^2 - 4^2 = (x^2 + 4)(x^2 - 4),$$

you can see that the second factor is itself the difference of two squares. Thus, to factor the polynomial *completely,* you must continue factoring, as follows.

$$x^4 - 16 = (x^2 + 4)(x^2 - 4) = (x^2 + 4)(x + 2)(x - 2)$$

Other instances of "repeated factoring" are given in the next example.

NOTE Note in Example 12 that the *sum of two squares* does not factor further. A second-degree polynomial that is the sum of two squares, such as $x^2 + y^2$ or $9m^2 + 1$, is *not factorable,* which means that it is not factorable *using integer coefficients.*

EXAMPLE 12 *Factoring Completely*

Factor (a) $x^4 - y^4$ and (b) $81m^4 - 1$ completely.

Solution

(a) Recognizing $x^4 - y^4$ as a difference of two squares, you can write
$$x^4 - y^4 = (x^2)^2 - (y^2)^2 = (x^2 + y^2)(x^2 - y^2).$$

Note that the second factor $(x^2 - y^2)$ is itself a difference of two squares and you therefore obtain
$$x^4 - y^4 = (x^2 + y^2)(x^2 - y^2) = (x^2 + y^2)(x + y)(x - y).$$

(b) Recognizing $81m^4 - 1$ as a difference of two squares, you can write
$$81m^4 - 1 = (9m^2)^2 - (1)^2 = (9m^2 + 1)(9m^2 - 1).$$

Note that the second factor $(9m^2 - 1)$ is itself a difference of two squares and you therefore obtain
$$81m^4 - 1 = (9m^2 + 1)(9m^2 - 1) = (9m^2 + 1)(3m + 1)(3m - 1).$$

Group Activities **Factoring Higher-Degree Polynomials**

Have each person in your group create a cubic polynomial by multiplying three polynomials of the form

$$(x - a)(x + a)(x + b).$$

Then write the result in standard form on a piece of paper and give it to another person in the group to factor. When each person has factored his or her polynomial, compare the results with the original factors used to create the polynomials.

1.5 Exercises

See Warm-Up Exercises, p. A39

1. *Think About It* Explain what is meant by saying that a polynomial is in factored form.

2. *Think About It* Explain how the word *factor* can be used as a noun and as a verb.

In Exercises 3–14, find the greatest common factor of the expressions.

3. 48, 90

4. 36, 150, 100

5. $3x^2, 12x$

6. $27x^4, 18x^3$

7. $30z^2, 12z^3$

8. $45y, 150y^3$

9. $28b^2, 14b^3, 42b^5$

10. $16x^2y, 84xy^2, 36x^2y^2$

11. $42(x + 8)^2, 63(x + 8)^3$

12. $66(3 - y), 44(3 - y)^2$

13. $4x(1 - z)^2, x^2(1 - z)^3$

14. $2(x + 5), 8(x + 5)$

In Exercises 15–30, factor out the greatest common monomial factor. (Some of the polynomials have no common monomial factor other than 1 or −1.)

15. $8z - 8$

16. $5x + 5$

17. $24x^2 - 18$

18. $14z^3 + 21$

19. $2x^2 + x$

20. $-a^3 - 4a$

21. $21u^2 - 14u$

22. $36y^4 + 24y^2$

23. $11u^2 + 9$

24. $16x^2 - 3y^3$

25. $3x^2y^2 - 15y$

26. $4uv + 6u^2v^2$

27. $28x^2 + 16x - 8$

28. $9 - 27y - 15y^2$

29. $14x^4 + 21x^3 + 9x^2$

30. $17x^5y^3 - xy^2 + 34y^2$

In Exercises 31–38, factor a negative real number from the polynomial and then write the polynomial factor in standard form.

31. $10 - 5x$

32. $32 - 4x^4$

33. $7 - 14x$

34. $15 - 5x$

35. $8 + 4x - 2x^2$

36. $12x - 6x^2 - 18$

37. $2t - 15 - 4t^2$

38. $16 + 32s^2 - 5s^4$

In Exercises 39–42, factor the expression.

39. $\frac{3}{2}x + \frac{5}{4} = \frac{1}{4}(\quad)$

40. $\frac{1}{3}x - \frac{5}{6} = \frac{1}{6}(\quad)$

41. $\frac{5}{8}x + \frac{5}{16}y = \frac{5}{16}(\quad)$

42. $\frac{7}{12}u - \frac{21}{8}v = \frac{7}{24}(\quad)$

In Exercises 43–48, factor the polynomial by factoring out the common binomial factor.

43. $2y(y - 3) + 5(y - 3)$

44. $7t(s + 9) - 6(s + 9)$

45. $5t(t^2 + 1) - 4(t^2 + 1)$

46. $3a(a^2 - 3) + 10(a^2 - 3)$

47. $a(a + 6) - a^2(a + 6)$

48. $(5x + y)(x - y) - 5x(x - y)$

In Exercises 49–60, factor the expression by grouping.

49. $y^2 - 6y + 2y - 12$

50. $y^2 + 3y + 4y + 12$

51. $x^2 + 25x + x + 25$

52. $x^2 - 7x + x - 7$

53. $x^3 + 2x^2 + x + 2$

54. $t^3 - 11t^2 + t - 11$

55. $a^3 - 4a^2 + 2a - 8$

56. $3s^3 + 6s^2 + 5s + 10$

57. $z^4 + 3z^3 - 2z - 6$

58. $4u^4 - 2u^3 - 6u + 3$

59. $cd + 3c - 3d - 9$

60. $u^2 + uv - 4u - 4v$

In Exercises 61–76, factor the difference of two squares.

61. $x^2 - 64$

62. $y^2 - 144$

63. $16y^2 - 9z^2$

64. $9z^2 - 25w^2$

65. $x^2 - 4y^2$

66. $81a^2 - b^6$

67. $a^8 - 36$

68. $y^{10} - 64$

69. $100 - 9y^2$

70. $625 - 49x^2$

71. $a^2b^2 - 16$

72. $u^2v^2 - 25$

73. $(a + 4)^2 - 49$

74. $(x - 3)^2 - 4$

75. $81 - (z + 5)^2$

76. $100 - (y - 3)^2$

In Exercises 77–82, factor the sum or difference of two cubes.

77. $x^3 - 8$

78. $t^3 - 27$

79. $y^3 + 64z^3$

80. $z^3 + 125w^3$

81. $8t^3 - 27$

82. $27s^3 + 64$

In Exercises 83–88, factor the algebraic expression completely.

83. $8 - 50x^2$

84. $a^3 - 16a$

85. $y^4 - 81x^4$

86. $u^4 - 256v^4$

87. $2x^3 - 54$

88. $5y^3 - 625$

Mental Math In Exercises 89 and 90, evaluate the quantity mentally using the sample as a model.

$$48 \cdot 52 = (50 - 2)(50 + 2) = 50^2 - 2^2 = 2496$$

89. $79 \cdot 81$

90. $18 \cdot 22$

In Exercises 91 and 92, factor the expression. (Assume $n > 0$.)

91. $4x^{2n} - 25$

92. $81 - 16y^{4n}$

Think About It In Exercises 93 and 94, show all the different groupings that can be used to factor the polynomial completely. Carry out the various factorizations to show that they yield the same result.

93. $3x^3 + 4x^2 - 3x - 4$

94. $6x^3 - 8x^2 + 9x - 12$

95. *Simple Interest* The total amount of money from a principal of P invested at $r\%$ simple interest for t years is given by $P + Prt$. Factor this expression.

96. *Revenue and Price* The revenue for selling x units of a product at a price of p dollars per unit is given by xp. For a particular commodity, the revenue is

$$R = 800x - 0.25x^2.$$

Factor the expression for the revenue and determine an expression for the price in terms of x.

97. *Geometry*   The area of a rectangle of length l is given by $45l - l^2$. Factor this expression to determine the width of the rectangle. Draw a diagram to illustrate your answer.

98. *Geometry*   The area of a rectangle of width w is given by $32w - w^2$. Factor this expression to determine the length of the rectangle. Draw a diagram to illustrate your answer.

99. *Geometry*   The surface area of a right circular cylinder is $S = 2\pi r^2 + 2\pi rh$ (see figure). Factor the expression for the surface area.

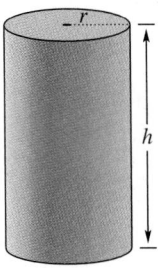 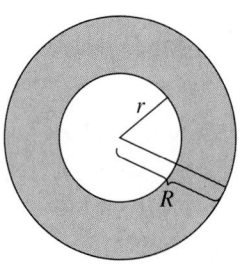

Figure for 99	Figure for 100

100. *Product Design*   A washer on the drive train of a car has an inside radius of r centimeters and an outside radius of R centimeters (see figure). Find the area of one of the flat surfaces of the washer and express the area in factored form.

101. *Chemical Reaction*   The rate of change of some chemical reaction is

$$kQx - kx^2,$$

where Q is the amount of the original substance, x is the amount of substance formed, and k is a constant of proportionality. Factor the expression for this rate of change.

Section Project

The cube shown in the figure is formed by four solids: I, II, III, IV.

(a) Explain how you could determine the following expressions for volume.

	Volume
Entire cube	a^3
Solid I	$(a - b)a^2$
Solid II	$(a - b)ab$
Solid III	$(a - b)b^2$
Solid IV	b^3

(b) Add the volumes of solids I, II, and III. Factor the result to show that the total volume can be expressed as $(a - b)(a^2 + ab + b^2)$.

(c) Explain why the total volume of solids I, II, and III can also be expressed as $a^3 - b^3$. Then explain how the figure can be used as a geometric model for the *difference of two cubes* factoring pattern.

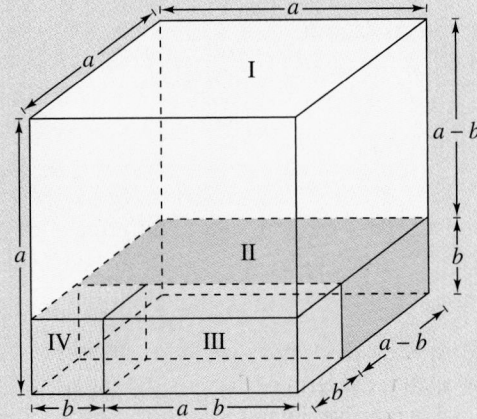

1.6	**Factoring Trinomials**

Factoring Trinomials of the Form $x^2 + bx + c$ ▪ Factoring Trinomials of the Form $ax^2 + bx + c$ ▪ Factoring Trinomials by Grouping (Optional) ▪ Factoring Perfect Square Trinomials ▪ Summary of Factoring

Factoring Trinomials of the Form $x^2 + bx + c$

Try covering the factored forms in the left-hand column below. Can you determine the factored forms from the trinomial forms?

Factored Form   **F**   **O**   **I**   **L**   *Trinomial Form*

$$(x - 1)(x + 4) = \; x^2 + 4x - \; x - 4 = \; x^2 + 3x - 4$$
$$(x - 3)(x - 2) = \; x^2 - 2x - 3x + 6 = \; x^2 - 5x + 6$$
$$(3x + 5)(x + 1) = 3x^2 + 3x + 5x + 5 = 3x^2 + 8x + 5$$

Your goal here is to factor trinomials of the form $x^2 + bx + c$. To begin, consider the following factorization.

$$x^2 + bx + c = (x + m)(x + n)$$

By multiplying the right-hand side, you obtain the following.

$$(x + m)(x + n) = x^2 + nx + mx + mn$$
$$= x^2 + \underbrace{(m + n)}_{\substack{\text{Sum of} \\ \text{terms} \\ \downarrow}}x + \underbrace{mn}_{\substack{\text{Product of} \\ \text{terms} \\ \downarrow}}$$
$$= x^2 + \boxed{b}\,x + \boxed{c}$$

Thus, to *factor* a trinomial $x^2 + bx + c$ into a product of two binomials, you must find factors of c with a sum of b.

NOTE In Example 1, note that when the constant term of the trinomial is positive, its factors must have *like* signs; otherwise, its factors have *unlike* signs.

EXAMPLE 1 *Factoring Trinomials*

Factor the trinomials (a) $x^2 - 2x - 8$ and (b) $x^2 - 5x + 6$.

Solution

(a) $x^2 - 2x - 8 = (x + \boxed{})(x + \boxed{})$ Think: You need factors of -8 with a sum of -2.
$$= (x + 2)(x - 4)$$ $(2)(-4) = -8, 2 - 4 = -2$

(b) $x^2 - 5x + 6 = (x + \boxed{})(x + \boxed{})$ Think: You need factors of 6 with a sum of -5.
$$= (x - 2)(x - 3)$$ $(-2)(-3) = 6, -2 - 3 = -5$

When factoring a trinomial of the form $x^2 + bx + c$, if you have trouble finding two factors of c with a sum of b, it may be helpful to list all of the distinct pairs of factors and then choose the appropriate pair from the list. For instance, consider the trinomial

$$x^2 - 5x - 24.$$

Factors of -24	Sum of Factors
$(1)(-24)$	$1 - 24 = -23$
$(-1)(24)$	$-1 + 24 = 23$
$(2)(-12)$	$2 - 12 = -10$
$(-2)(12)$	$-2 + 12 = 10$
$(3)(-8)$	$3 - 8 = -5$
$(-3)(8)$	$-3 + 8 = 5$
$(4)(-6)$	$4 - 6 = -2$
$(-4)(6)$	$-4 + 6 = 2$

For this trinomial, you have $c = -24$ and $b = -5$. Thus, you need to find two factors of -24 with a sum of -5, as shown at the left. With experience, you will be able to narrow this list down *mentally* to only two or three possibilities whose sums can then be tested to determine the correct factorization, which is $x^2 - 5x - 24 = (x + 3)(x - 8)$.

EXAMPLE 2   Factoring a Trinomial

Factor the trinomial $x^2 - 7x - 18$.

Solution

To factor this trinomial, you need to find two factors of -18 with a sum of -7.

$$x^2 - 7x - 18 = (x + \quad)(x + \quad)$$
$$= (x + 2)(x - 9)$$

Think: You need factors of -18 with a sum of -7.
$(2)(-9) = -18, 2 - 9 = -7$

Applications of algebra sometimes involve trinomials that have a common monomial factor. To factor such trinomials completely, first factor out the common monomial factor. Then try to factor the resulting trinomial by the methods given in this section. For instance, the trinomial

$$5x^2 - 5x - 30 = 5(x^2 - x - 6) = 5(x - 3)(x + 2)$$

is factored completely.

EXAMPLE 3   Factoring Completely

Factor the trinomial $4x^3 - 8x^2 - 60x$ completely.

Solution

Remind students to check whether a common factor can be removed before trying other methods of factoring.

This trinomial has a common monomial factor of $4x$. Thus, you should start the factoring process by factoring $4x$ out of each term.

$$4x^3 - 8x^2 - 60x = 4x(x^2 - 2x - 15)$$
$$= 4x(x + \quad)(x + \quad)$$
$$= 4x(x + 3)(x - 5)$$

Factor out common factor.
Think: You need factors of -15 with a sum of -2.
$(3)(-5) = -15, 3 - 5 = -2$

Factoring Trinomials of the Form $ax^2 + bx + c$

To factor a trinomial whose leading coefficient is not 1, use the following pattern.

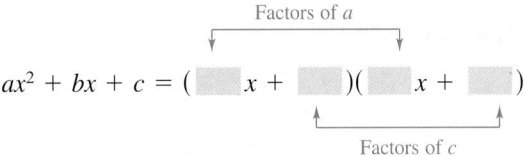

Factors of a

$$ax^2 + bx + c = (x +)(x +)$$

Factors of c

The goal is to find a combination of factors of a and c such that the outer and inner products add up to the middle term bx. For instance, in the trinomial $6x^2 + 17x + 5$, you have $a = 6$, $c = 5$, and $b = 17$. After some experimentation, you can determine that the factorization is

$$6x^2 + 17x + 5 = (2x + 5)(3x + 1).$$

EXAMPLE 4   *Factoring a Trinomial of the Form $ax^2 + bx + c$*

Factor the trinomial $6x^2 + 5x - 4$.

Solution

For this trinomial, you have $ax^2 + bx + c = 6x^2 + 5x - 4$, which implies that $a = 6$, $c = -4$, and $b = 5$. The possible factors of 6 are $(1)(6)$ and $(2)(3)$, and the possible factors of -4 are $(-1)(4)$, $(1)(-4)$, and $(2)(-2)$. By trying the *many* different combinations of these factors, you obtain the following list.

$$(x + 1)(6x - 4) = 6x^2 + 2x - 4$$
$$(x - 1)(6x + 4) = 6x^2 - 2x - 4$$
$$(x + 4)(6x - 1) = 6x^2 + 23x - 4$$
$$(x - 4)(6x + 1) = 6x^2 - 23x - 4$$
$$(x + 2)(6x - 2) = 6x^2 + 10x - 4$$
$$(x - 2)(6x + 2) = 6x^2 - 10x - 4$$
$$(2x + 1)(3x - 4) = 6x^2 - 5x - 4$$
$$(2x - 1)(3x + 4) = 6x^2 + 5x - 4 \quad \Longleftarrow \quad \text{Correct factorization}$$
$$(2x + 4)(3x - 1) = 6x^2 + 10x - 4$$
$$(2x - 4)(3x + 1) = 6x^2 - 10x - 4$$
$$(2x + 2)(3x - 2) = 6x^2 + 2x - 4$$
$$(2x - 2)(3x + 2) = 6x^2 - 2x - 4$$

Thus, you can conclude that the correct factorization is

$$6x^2 + 5x - 4 = (2x - 1)(3x + 4).$$

To help shorten the list of *possible* factorizations of a trinomial of the form $ax^2 + bx + c$, use the following guidelines.

Guidelines for Limiting Possible Trinomial Factorizations

1. If the trinomial has a *common monomial factor,* you should remove the monomial factor before trying to find binomial factors. For instance, the trinomial $12x^2 + 10x - 8$ has a common factor of 2. By removing this common factor, you obtain $12x^2 + 10x - 8 = 2(6x^2 + 5x - 4)$.

2. Do not switch the signs of the factors of c unless the middle term is correct except in sign. In Example 4, after determining that $(x + 4)(6x - 1)$ is not the correct factorization, for instance, it is unnecessary to test $(x - 4)(6x + 1)$.

3. Do not use binomial factors that have a common monomial factor. Such a factor cannot be correct, because the trinomial has no common monomial factor. (Any common monomial factor was already removed in Step 1.) For instance, in Example 4, it is unnecessary to test $(x + 1)(6x - 4) = 6x^2 + 2x - 4$ because the factor $(6x - 4)$ has a common factor of 2.

With these suggestions, you could shorten the list given in Example 4 to the following three possible factorizations.

$$(x + 4)(6x - 1) = 6x^2 + 23x - 4$$
$$(2x + 1)(3x - 4) = 6x^2 - 5x - 4$$
$$(2x - 1)(3x + 4) = 6x^2 + 5x - 4 \quad \longleftarrow \quad \text{Correct factorization}$$

Do you see why you can cut the list from 12 possible factorizations to only three?

EXAMPLE 5   *Factoring a Trinomial of the Form* $ax^2 + bx + c$

Factor the trinomial $2x^2 - x - 21$.

Solution

For this trinomial, you have $a = 2$, which factors as $(1)(2)$, and $c = -21$, which factors as $(1)(-21)$, $(-1)(21)$, $(3)(-7)$, or $(-7)(3)$.

$$(2x + 1)(x - 21) = 2x^2 - 41x - 21$$
$$(2x + 21)(x - 1) = 2x^2 + 19x - 21$$
$$(2x + 3)(x - 7) = 2x^2 - 11x - 21$$
$$(2x + 7)(x - 3) = 2x^2 + x - 21 \qquad \text{Middle term has incorrect sign.}$$
$$(2x - 7)(x + 3) = 2x^2 - x - 21 \quad \longleftarrow \quad \text{Correct factorization}$$

NOTE   Remember that if the middle term is correct except in sign, you need only change the signs of the factors of c, as in Example 5.

EXAMPLE 6 *Factoring a Trinomial of the Form ax² + bx + c*

Factor the trinomial

$$3x^2 + 11x + 10.$$

Solution

For this trinomial, you have $a = 3$, which factors as $(1)(3)$, and $c = 10$, which factors as $(1)(10)$ or $(2)(5)$. You can test the possible factors as follows.

$$(x + 10)(3x + 1) = 3x^2 + 31x + 10$$
$$(x + 1)(3x + 10) = 3x^2 + 13x + 10$$
$$(x + 5)(3x + 2) = 3x^2 + 17x + 10$$
$$(x + 2)(3x + 5) = 3x^2 + 11x + 10 \quad \Longleftarrow \quad \text{Correct factorization}$$

Therefore, the correct factorization is

$$3x^2 + 11x + 10 = (x + 2)(3x + 5).$$

Remember that if a trinomial has a common monomial factor, the common monomial factor should be removed first. This is illustrated in the next example.

EXAMPLE 7 *Factoring Completely*

Factor the following trinomial completely.

$$8x^2y - 60xy + 28y$$

Solution

Begin by factoring out the common monomial factor $4y$.

$$8x^2y - 60xy + 28y = 4y(2x^2 - 15x + 7)$$

Now, for the new trinomial $2x^2 - 15x + 7$, you have $a = 2$ and $c = 7$. The possible factorizations of this trinomial are as follows.

$$(2x - 7)(x - 1) = 2x^2 - 9x + 7$$
$$(2x - 1)(x - 7) = 2x^2 - 15x + 7 \quad \Longleftarrow \quad \text{Correct factorization}$$

Therefore, the complete factorization of the original trinomial is

$$8x^2y - 60xy + 28y = 4y(2x^2 - 15x + 7)$$
$$= 4y(2x - 1)(x - 7).$$

When factoring a trinomial with a negative leading coefficient, first factor -1 out of the trinomial, as demonstrated in Example 8.

EXAMPLE 8   A Trinomial with a Negative Leading Coefficient

Progress check: Factor the following completely.

(a) $x^2 + 8x + 12$ (b) $6y^2 - 13y + 6$

(c) $2x^3 - 18x$

Answers: (a) $(x + 2)(x + 6)$

(b) $(2y - 3)(3y - 2)$

(c) $2x(x + 3)(x - 3)$

Factor the trinomial

$$-3x^2 + 16x + 35.$$

Solution

This trinomial has a negative leading coefficient, so you should begin by factoring (-1) out of the trinomial.

$$-3x^2 + 16x + 35 = (-1)(3x^2 - 16x - 35)$$

Now, for the new trinomial $3x^2 - 16x - 35$, you have $a = 3$ and $c = -35$. The possible factorizations of this trinomial are as follows.

$$(3x - 1)(x + 35) = 3x^2 + 104x - 35$$
$$(3x - 35)(x + 1) = 3x^2 - 32x - 35$$
$$(3x - 7)(x + 5) = 3x^2 + 8x - 35$$
$$(3x - 5)(x + 7) = 3x^2 + 16x - 35 \qquad \text{Middle term has incorrect sign.}$$
$$(3x + 5)(x - 7) = 3x^2 - 16x - 35 \quad \Longleftarrow \quad \text{Correct factorization}$$

Thus, the correct factorization is

$$-3x^2 + 16x + 35 = (-1)(3x + 5)(x - 7).$$

Alternative forms of this factorization include

$$(3x + 5)(-x + 7)$$

and

$$(-3x - 5)(x - 7).$$

Not all trinomials are factorable using only integers. For instance, to factor

$$x^2 + 3x + 5$$

you need factors of 5 that add up to 3. This is not possible, because the only integer factors of 5 are 1 and 5, and their sum is not 3. Such a trinomial is *not factorable.* Try factoring

$$2x^2 - 3x + 2$$

to see that it is also not factorable. Watch for other trinomials that are not factorable in the exercises for this section.

Factoring Trinomials by Grouping (Optional)

In this section, you have seen that factoring a trinomial can involve quite a bit of trial and error. An alternative technique that some people like is to use *factoring by grouping*. For instance, suppose you rewrite the trinomial $2x^2 + x - 15$ as

$$2x^2 + x - 15 = 2x^2 + 6x - 5x - 15.$$

Then, by grouping the first and second terms and the third and fourth terms, you can factor the polynomial as follows.

$$
\begin{aligned}
2x^2 + x - 15 &= 2x^2 + 6x - 5x - 15 && \text{Rewrite middle term.} \\
&= (2x^2 + 6x) - (5x + 15) && \text{Group terms.} \\
&= 2x(x + 3) - 5(x + 3) && \text{Factor groups.} \\
&= (x + 3)(2x - 5) && \text{Distributive Property}
\end{aligned}
$$

The key to this method of factoring is knowing how to rewrite the middle term. In general, *to factor a trinomial $ax^2 + bx + c$ by grouping, choose factors of the product ac that add up to b and use these factors to rewrite the middle term.* This technique is illustrated in Example 9.

EXAMPLE 9 *Factoring a Trinomial by Grouping*

Use factoring by grouping to factor the trinomial

$$2x^2 + 5x - 3.$$

Solution

In the trinomial $2x^2 + 5x - 3$, you have $a = 2$ and $c = -3$, which implies that the product ac is -6. Now, because -6 factors as $(6)(-1)$, and $6 - 1 = 5 = b$, you can rewrite the middle term as $5x = 6x - x$. This produces the following.

$$
\begin{aligned}
2x^2 + 5x - 3 &= 2x^2 + 6x - x - 3 && \text{Rewrite middle term.} \\
&= (2x^2 + 6x) - (x + 3) && \text{Group terms.} \\
&= 2x(x + 3) - (x + 3) && \text{Factor groups.} \\
&= (x + 3)(2x - 1) && \text{Distributive Property}
\end{aligned}
$$

Therefore, the trinomial factors as

$$2x^2 + 5x - 3 = (x + 3)(2x - 1).$$

What do you think of this optional technique? Some people think that it is more efficient than the trial-and-error process, especially when the coefficients a and c have many factors.

Factoring Perfect Square Trinomials

A **perfect square trinomial** is the square of a binomial. For instance,

$$x^2 + 6x + 9 = (x + 3)^2$$

is the square of the binomial $(x + 3)$, and

$$4x^2 - 20x + 25 = (2x - 5)^2$$

is the square of the binomial $(2x - 5)$. Perfect square trinomials come in two forms: one in which the middle term is positive, and the other in which the middle term is negative.

Perfect Square Trinomials

Let u and v represent real numbers, variables, or algebraic expressions. Then the following perfect square trinomials can be factored as indicated.

$$u^2 + 2uv + v^2 = (u + v)^2 \qquad\qquad u^2 - 2uv + v^2 = (u - v)^2$$

Same sign   Same sign

To recognize a perfect square trinomial, remember that the first and last terms must be perfect squares and positive, and that the middle term must be twice the product of u and v. (The middle term can be positive or negative.)

EXAMPLE 10   *Factoring Perfect Square Trinomials*

(a) $x^2 - 4x + 4 = x^2 - 2(x)(2) + 2^2 = (x - 2)^2$

(b) $16y^2 + 24y + 9 = (4y)^2 + 2(4y)(3) + 3^2 = (4y + 3)^2$

(c) $9x^2 - 30xy + 25y^2 = (3x)^2 - 2(3x)(5y) + (5y)^2$
$$= (3x - 5y)^2$$

EXAMPLE 11   *Removing a Common Monomial Factor First*

(a) $3x^2 - 30x + 75 = 3(x^2 - 10x + 25)$ Remove common monomial factor.

$\qquad\qquad\qquad\ = 3(x - 5)^2$ Factor as perfect square trinomial.

(b) $16y^3 + 80y^2 + 100y = 4y(4y^2 + 20y + 25)$ Remove common monomial factor.

$\qquad\qquad\qquad\qquad\ = 4y(2y + 5)^2$ Factor as perfect square trinomial.

Summary of Factoring

Although the basic factoring techniques have been discussed one at a time, from this point on you must decide which technique to apply to any given problem situation. The following guidelines should assist you in this selection process.

Ask students to write a paragraph about what it means to "factor completely."

Guidelines for Factoring Polynomials

1. Factor out any common factors.

2. Factor according to one of the special polynomial forms: difference of squares, sum or difference of cubes, or perfect square trinomials.

3. Factor trinomials, $ax^2 + bx + c$, using the methods for $a = 1$ or $a \neq 1$.

4. Factor by grouping—for polynomials with four terms.

5. Check to see if the factors themselves can be further factored.

6. Check the results by multiplying the factors.

EXAMPLE 12   *Factoring Polynomials*

(a) $3x^2 - 108 = 3(x^2 - 36)$ Factor out common factor.

$\qquad\qquad\quad = 3(x + 6)(x - 6)$ Difference of two squares

(b) $4x^3 - 32x^2 + 64x = 4x(x^2 - 8x + 16)$ Factor out common factor.

$\qquad\qquad\qquad\quad = 4x(x - 4)^2$ Perfect square trinomial

(c) $6x^3 + 27x^2 - 15x = 3x(2x^2 + 9x - 5)$ Factor out common factor.

$\qquad\qquad\qquad\quad = 3x(2x - 1)(x + 5)$ Factor trinomial.

(d) $x^3 - 3x^2 - 4x + 12 = (x^3 - 3x^2) - (4x - 12)$ Group terms.

$\qquad\qquad\qquad\quad = x^2(x - 3) - 4(x - 3)$ Factor out common factors.

$\qquad\qquad\qquad\quad = (x - 3)(x^2 - 4)$ Distributive Property

$\qquad\qquad\qquad\quad = (x - 3)(x + 2)(x - 2)$ Difference of two squares

(e) $x^2 + 6x + 9 - y^2 = (x^2 + 6x + 9) - y^2$ Group terms.

$\qquad\qquad\qquad\quad = (x + 3)^2 - y^2$ Perfect square trinomial

$\qquad\qquad\qquad\quad = [(x + 3) + y][(x + 3) - y]$ Difference of two squares

$\qquad\qquad\qquad\quad = (x + 3 + y)(x + 3 - y)$ Simplify.

Group Activities   **Extending the Concept**

With your group, determine how you could factor the following expressions.

$$(a + b)^2 - 7(a + b) + 12$$
$$(m - n)^2 - 4(m - n) - 32$$
$$3(w + k)^2 + 19(w + k) - 14$$

Then factor the following expressions.

$$x^2 + 2xy + y^2 + 6x + 6y + 8$$
$$c^2 - 2cd + d^2 + 8c - 8d - 20$$
$$2r^2 + 4rt + 2t^2 - 15r - 15t + 18$$

1.6   **Exercises**

See Warm-Up
Exercises, p. A39

1. *Error Analysis*   Identify the error.

$$9x^2 - 9x - 54 = (3x + 6)(3x - 9)$$
$$= 3(x + 2)(x - 3)$$

2. *Think About It*   Is $x(x + 2) - 2(x + 2)$ completely factored? Explain.

In Exercises 3–8, fill in the missing factor.

3. $x^2 + 5x + 4 = (x + 4)(\quad)$
4. $a^2 + 2a - 8 = (a + 4)(\quad)$
5. $y^2 - y - 20 = (y + 4)(\quad)$
6. $y^2 + 6y + 8 = (y + 4)(\quad)$
7. $z^2 - 6z + 8 = (z - 4)(\quad)$
8. $z^2 + 2z - 24 = (z - 4)(\quad)$

In Exercises 9–18, factor the algebraic expression.

9. $x^2 + 4x + 3$
10. $x^2 - 10x + 24$
11. $y^2 + 7y - 30$
12. $m^2 - 3m - 10$
13. $t^2 - 4t - 21$
14. $x^2 + 4x - 12$

15. $x^2 - 20x + 96$
16. $u^2 + 5uv + 6v^2$
17. $x^2 - 2xy - 35y^2$
18. $a^2 - 21ab + 110b^2$

In Exercises 19–24, fill in the missing factor.

19. $5x^2 + 18x + 9 = (x + 3)(\quad)$
20. $5x^2 + 19x + 12 = (x + 3)(\quad)$
21. $5a^2 + 12a - 9 = (a + 3)(\quad)$
22. $5c^2 + 11c - 12 = (c + 3)(\quad)$
23. $2y^2 - 3y - 27 = (y + 3)(\quad)$
24. $3y^2 - y - 30 = (y + 3)(\quad)$

In Exercises 25–34, factor the algebraic expression, if possible. Some of the expressions are not factorable using integer coefficients.

25. $3x^2 + 4x + 1$
26. $5x^2 + 7x + 2$
27. $8t^2 - 6t - 5$
28. $6b^2 + 19b - 7$
29. $2a^2 - 13a + 20$
30. $24x^2 - 14xy - 3y^2$
31. $20x^2 + x - 12$
32. $10x^2 + 9xy - 9y^2$
33. $2u^2 + 9uv - 35v^2$
34. $r^2 - 9rs - 9s^2$

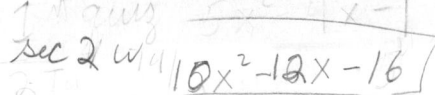

In Exercises 35–38, factor the trinomial.

35. $-2x^2 - x + 6$

36. $-6x^2 + 5x + 6$

37. $1 - 11x - 60x^2$

38. $2 + 5x - 12x^2$

In Exercises 39–44, factor the trinomial by grouping (see Example 9).

39. $3x^2 + 10x + 8$

40. $2x^2 + 9x + 9$

41. $6x^2 + x - 2$

42. $6x^2 - x - 15$

43. $15x^2 - 11x + 2$

44. $12x^2 - 28x + 15$

In Exercises 45–56, factor the perfect square trinomial.

45. $x^2 + 4x + 4$

46. $z^2 + 6z + 9$

47. $a^2 - 12a + 36$

48. $y^2 - 14y + 49$

49. $25y^2 - 10y + 1$

50. $4z^2 + 28z + 49$

51. $9b^2 + 12b + 4$

52. $4x^2 - 4x + 1$

53. $4x^2 - 4xy + y^2$

54. $m^2 + 6mn + 9n^2$

55. $u^2 + 8uv + 16v^2$

56. $4y^2 + 20yz + 25z^2$

In Exercises 57–80, factor the algebraic expression completely.

57. $3x^5 - 12x^3$

58. $20y^2 - 45$

59. $10t^3 + 2t^2 - 36t$

60. $16z^2 - 56z + 49$

61. $4x(3x - 2) + (3x - 2)^2$

62. $6 - x - 6x^2 + x^3$

63. $36 - (z + 3)^2$

64. $3t^3 - 24$

65. $54x^3 - 2$

66. $v^3 + 3v^2 + 5v$

67. $27a^3 - 3ab^2$

68. $8m^3n + 20m^2n^2 - 48mn^3$

69. $x^3 + 2x^2 - 16x - 32$

70. $x^3 - 7x^2 - 4x + 28$

71. $x^3 - 6x^2 - 9x + 54$

72. $x^3 + 10x^2 - 16x - 160$

73. $x^2 - 10x + 25 - y^2$

74. $9y^2 + 12y + 4 - z^2$

75. $a^2 - 2ab + b^2 - 16$

76. $x^2 + 14xy + 49y^2 - 4a^2$

77. $x^8 - 1$

78. $x^4 - 16y^4$

79. $b^4 - 216b$

80. $3y^3 - 192$

In Exercises 81–86, find two real numbers b such that the algebraic expression is a perfect square trinomial.

81. $x^2 + bx + 81$

82. $x^2 + bx + 16$

83. $9y^2 + by + 1$

84. $36z^2 + bz + 1$

85. $4x^2 + bx + 9$

86. $16x^2 + bxy + 25y^2$

In Exercises 87–92, find a real number c such that the algebraic expression is a perfect square trinomial.

87. $x^2 + 8x + c$

88. $x^2 + 12x + c$

89. $y^2 - 6y + c$

90. $z^2 - 20z + c$

91. $16a^2 + 40a + c$

92. $9t^2 - 12t + c$

In Exercises 93 and 94, fill in the missing number.

93. $x^2 + 12x + 50 = (x + 6)^2 + $ ▨

94. $x^2 + 10x + 22 = (x + 5)^2 + $ ▨

In Exercises 95–100, find all integers b such that the trinomial can be factored.

95. $x^2 + bx + 18$

96. $x^2 + bx + 14$

97. $x^2 + bx - 21$

98. $x^2 + bx - 7$

99. $5x^2 + bx + 8$

100. $3x^2 + bx - 10$

In Exercises 101–106, find two integers c such that the trinomial can be factored. (There are many correct answers.)

101. $x^2 + 6x + c$

102. $x^2 + 9x + c$

103. $x^2 - 3x + c$

104. $x^2 - 12x + c$

105. $t^2 - 4t + c$

106. $s^2 + s + c$

In Exercises 107 and 108, evaluate the quantity mentally using the sample as a model.

$$29^2 = (30 - 1)^2 = 30^2 - 2(30)(1) + 1^2$$
$$= 900 - 60 + 1$$
$$= 841$$

107. 52^2

108. 39^2

In Exercises 109 and 110, write, in factored form, an expression for the shaded portion of the figure.

109.

110.

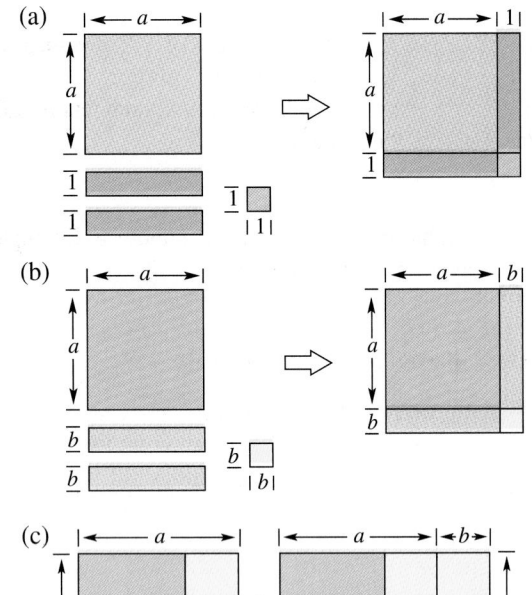

(d)

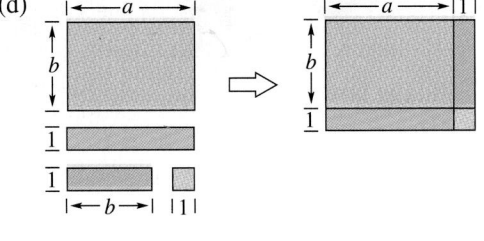

In Exercises 111–114, match the geometric factoring model with the correct factoring formula. [The models are labeled (a), (b), (c), and (d).]

111. $a^2 - b^2 = (a + b)(a - b)$

112. $a^2 + 2ab + b^2 = (a + b)^2$

113. $a^2 + 2a + 1 = (a + 1)^2$

114. $ab + a + b + 1 = (a + 1)(b + 1)$

(a)

(b)

(c)

Section Project

Factor the trinomial and represent the result with a geometric factoring model. The sample shows the factoring of $x^2 + 3x + 2 = (x + 1)(x + 2)$.

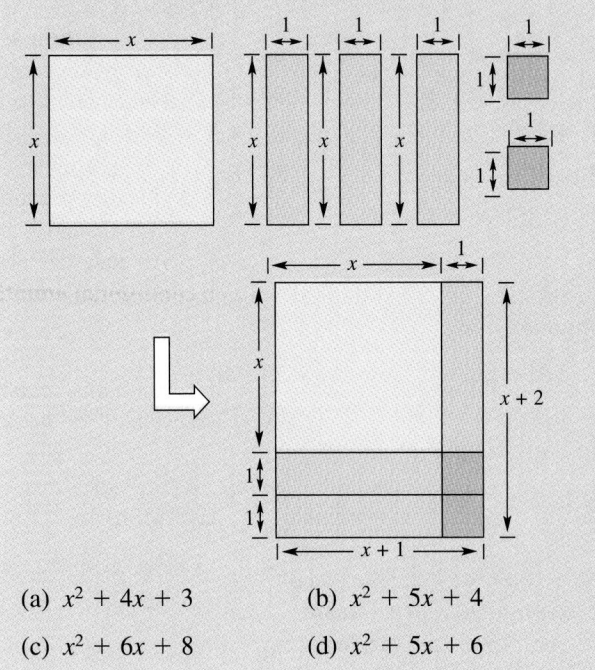

(a) $x^2 + 4x + 3$ (b) $x^2 + 5x + 4$

(c) $x^2 + 6x + 8$ (d) $x^2 + 5x + 6$

1.7	**Solving Linear Equations**

Introduction ▪ Solving Linear Equations in Standard Form ▪
Solving Linear Equations in Nonstandard Form ▪ Literal Equations

Introduction

An **equation** is a statement that equates two mathematical expressions. Some examples are

$$x = 4, \quad 4x + 5 = 17, \quad 2x - 8 = 2(x - 4), \quad \text{and} \quad x^2 - 16 = 0.$$

Remind students to *simplify* expressions and *solve* equations.

Solving an equation involving x means finding all values of x for which the equation is true. Such values are **solutions** and are said to **satisfy** the equation. For instance, 3 is a solution of $4x + 5 = 17$ because $4(3) + 5 = 17$ is a true statement.

The **solution set** of an equation is the set of all solutions of the equation. Sometimes, an equation will have the set of all real numbers as its solution set. Such an equation is an **identity.** For instance, the equation

$$2x - 8 = 2(x - 4) \qquad \text{Identity}$$

is an identity because the equation is true for all real values of x. Try values such as 0, 1, -2, and 5 in this equation to see that each one is a solution.

An equation whose solution set is not the entire set of real numbers is called a **conditional equation.** For instance, the equation

$$x^2 - 16 = 0 \qquad \text{Conditional equation}$$

is a conditional equation because it has only two solutions, 4 and -4. Example 1 shows how to **check** whether a given value is a solution.

NOTE When checking a solution, we suggest you write a question mark over the equal sign to indicate that you are uncertain whether the "equation" is true.

EXAMPLE 1 *Checking a Solution of an Equation*

Decide whether -3 is a solution of $-3x - 5 = 4x + 16$.

Solution

$-3x - 5 = 4x + 16$	Original equation
$-3(-3) - 5 \overset{?}{=} 4(-3) + 16$	Substitute -3 for x.
$9 - 5 \overset{?}{=} -12 + 16$	Simplify.
$4 = 4$	Solution checks. ✔

Because both sides turn out to be the same number, you can conclude that -3 is a solution of the original equation. Try checking to see whether -2 is a solution. You should be able to decide that it is not.

It is helpful to think of an equation as having two sides that are "in balance." Consequently, when you try to solve an equation, you must be careful to maintain that balance by performing the same operation(s) on both sides.

Two equations that have the same set of solutions are **equivalent.** For instance, the equations $x = 3$ and $x - 3 = 0$ are equivalent because both have only one solution—the number 3. When any one of the four techniques in the following list is applied to an equation, the resulting equation is *equivalent* to the original equation.

Forming Equivalent Equations: Properties of Equality

	Original	*Equivalent*
1. *Simplify Each Side:* Remove symbols of grouping, combine like terms, or reduce fractions on one or both sides of the equation.	$3x - x = 8$	$2x = 8$
2. *Apply the Addition Property of Equality:* Add (or subtract) the same quantity to (from) *both* sides of the equation.	$x - 3 = 5$	$x = 8$
3. *Apply the Multiplication Property of Equality:* Multiply (or divide) *both* sides of the equation by the same nonzero quantity.	$3x = 9$	$x = 3$
4. *Interchange Sides:* Interchange the two sides of the equation.	$7 = x$	$x = 7$

Point out that students should not multiply or divide both sides of an equation by a variable, as the value of that variable could be zero.

NOTE   The goal when you *solve an equation* is to get the *variable* on one side of the equal sign and everything else on the other side.

When solving an equation, you can use any of the four techniques for forming equivalent equations to eliminate terms or factors in the equation. For example, to solve the equation $x + 4 = 2$, you need to get rid of the term $+4$ on the left side. This is accomplished by subtracting 4 from both sides.

$x + 4 = 2$	Original equation
$x + 4 - 4 = 2 - 4$	Subtract 4 from both sides.
$x + 0 = -2$	Combine like terms.
$x = -2$	Simplify.

Although this solution involved subtracting 4 from both sides, you could just as easily have added -4 to both sides. Both techniques are legitimate—which one you decide to use is a matter of personal preference.

Solving Linear Equations in Standard Form

The most common type of equation in one variable is a **linear equation.**

Definition of Linear Equation

A **linear equation** in one variable x is an equation that can be written in the standard form

$$ax + b = 0,$$

where a and b are real numbers with $a \neq 0$.

A linear equation in one variable is also called a **first-degree equation** because its variable has an implied exponent of 1. Some examples of linear equations in the standard form $ax + b = 0$ are $3x + 2 = 0$ and $5x - 4 = 0$.

Remember that to *solve* an equation in x means to find the values of x that satisfy the equation. For a linear equation in the standard form $ax + b = 0$, the goal is to **isolate** x by rewriting the standard equation in the form

$$x = \text{(a number)}.$$

Beginning with the original equation, you write a sequence of equivalent equations, each having the same solution as the original equation.

EXAMPLE 2 *Solving a Linear Equation in Standard Form*

$4x - 12 = 0$	Original equation
$4x - 12 + 12 = 0 + 12$	Add 12 to both sides.
$4x = 12$	Combine like terms.
$\dfrac{4x}{4} = \dfrac{12}{4}$	Divide both sides by 4.
$x = 3$	Simplify.

It appears that the solution is 3. You can check this as follows.

Check

$4x - 12 = 0$	Original equation
$4(3) - 12 \overset{?}{=} 0$	Substitute 3 for x.
$12 - 12 \overset{?}{=} 0$	Simplify.
$0 = 0$	Solution checks. ✓

Study Tip

Be sure you see that solving an equation such as the one in Example 2 has two basic steps. The first step is to *find* the solution(s). The second step is to *check* that each solution you find actually satisfies the original equation. You can improve your accuracy in algebra by developing the habit of checking each solution.

NOTE Equations in one variable that are not linear may have two or more solutions. For instance, the nonlinear equation $x^2 = 4$ has two solutions: 2 and -2.

You know that 3 is a solution of the equation in Example 2, but at this point you might be asking, "How can I be sure that the equation does not have other solutions?" The answer is that a linear equation in one variable always has *exactly one* solution. You can prove this with the following steps.

$$ax + b = 0 \qquad \text{Original equation, with } a \neq 0$$
$$ax + b - b = 0 - b \qquad \text{Subtract } b \text{ from both sides.}$$
$$ax = -b \qquad \text{Combine like terms.}$$
$$\frac{ax}{a} = \frac{-b}{a} \qquad \text{Divide both sides by } a.$$
$$x = \frac{-b}{a} \qquad \text{Simplify.}$$

It is clear that the last equation has only one solution, $x = -b/a$. Moreover, because the last equation is equivalent to the given equation, you can conclude that every linear equation in one variable has exactly one solution.

Solving Linear Equations in Nonstandard Form

Linear equations often occur in nonstandard forms that contain symbols of grouping or like terms that are not combined. Here are some examples.

$$2x - 3 = 5, \quad x + 2 = 2x - 6, \quad 6(y - 1) + 4y = 3(7y + 1)$$

Remember that the goal is still to isolate the variable.

EXAMPLE 3   Solving a Linear Equation in Nonstandard Form

Solve the equation $2x - 3 = -5$.

Solution

$$2x - 3 = -5 \qquad \text{Original equation}$$
$$2x - 3 + 3 = -5 + 3 \qquad \text{Add 3 to both sides.}$$
$$2x = -2 \qquad \text{Combine like terms.}$$
$$\frac{2x}{2} = \frac{-2}{2} \qquad \text{Divide both sides by 2.}$$
$$x = -1 \qquad \text{Simplify.}$$

The solution is -1. Check this in the original equation.

As you gain experience in solving linear equations, you will probably find that you can perform some of the solution steps in your head. For instance, you might solve the equation given in Example 3 by performing two of the steps mentally, and writing only the following three steps.

$2x - 3 = -5$	Original equation
$2x = -2$	Add 3 to both sides.
$x = -1$	Divide both sides by 2.

Remember, however, that you should not skip the final step—checking your solution. You may find your calculator useful for checking solutions.

NOTE In Example 4, we solved the equation by isolating the variable on the left side. You could just as easily have isolated the variable on the right side, as follows.

$$x + 2 = 2x - 6$$
$$-x + x + 2 = -x + 2x - 6$$
$$2 = x - 6$$
$$2 + 6 = x - 6 + 6$$
$$8 = x$$

EXAMPLE 4 Solving a Linear Equation in Nonstandard Form

$x + 2 = 2x - 6$	Original equation
$-2x + x + 2 = -2x + 2x - 6$	Add $-2x$ to both sides.
$-x + 2 = -6$	Combine like terms.
$-x + 2 - 2 = -6 - 2$	Subtract 2 from both sides.
$-x = -8$	Combine like terms.
$(-1)(-x) = (-1)(-8)$	Multiply both sides by -1.
$x = 8$	Simplify.

The solution is 8. Check this in the original equation.

Sometimes students add or subtract terms from both sides of an equation before simplifying the original sides. Stress the usefulness of simplifying each side completely first to make the equation easier to manipulate.

EXAMPLE 5 Solving a Linear Equation in Nonstandard Form

$6(y - 1) + 4y = 3(7y + 1)$	Original equation
$6y - 6 + 4y = 21y + 3$	Distributive Property
$10y - 6 = 21y + 3$	Combine like terms.
$10y - 21y - 6 = 21y - 21y + 3$	Subtract $21y$ from both sides.
$-11y - 6 = 3$	Combine like terms.
$-11y - 6 + 6 = 3 + 6$	Add 6 to both sides.
$-11y = 9$	Combine like terms.
$\dfrac{-11y}{-11} = \dfrac{9}{-11}$	Divide both sides by -11.
$y = -\dfrac{9}{11}$	Simplify.

NOTE In most cases, it helps to remove symbols of grouping as a first step to solving an equation. This is illustrated in Example 5.

The solution is $-\frac{9}{11}$. Check this in the original equation.

If a linear equation contains fractions, we suggest that you first *clear the equation of fractions* by multiplying both sides of the equation by the least common denominator (LCD) of the fractions.

Discuss the idea of trying alternative strategies for completing a mathematical task. Note the text suggestion here. In addition, another approach to Example 6 that students may suggest is to begin by finding a common denominator and then combining fractions. Point out how this process is generally less efficient than the text method.

EXAMPLE 6 Solving a Linear Equation that Contains Fractions

Solve the equation $\dfrac{x}{18} + \dfrac{3x}{4} = 2$.

Solution

$$\frac{x}{18} + \frac{3x}{4} = 2 \qquad\qquad \text{Original equation}$$

$$36\left(\frac{x}{18} + \frac{3x}{4}\right) = 36(2) \qquad\qquad \text{Multiply both sides by LCD of 36.}$$

$$36 \cdot \frac{x}{18} + 36 \cdot \frac{3x}{4} = 36(2) \qquad\qquad \text{Distributive Property.}$$

$$2x + 27x = 72 \qquad\qquad \text{Simplify.}$$

$$29x = 72 \qquad\qquad \text{Combine like terms.}$$

$$\frac{29x}{29} = \frac{72}{29} \qquad\qquad \text{Divide both sides by 29.}$$

$$x = \frac{72}{29} \qquad\qquad \text{Simplify.}$$

The solution is $\frac{72}{29}$. Check this in the original equation.

The next example shows how to solve a linear equation involving decimals. The procedure is basically the same, but the arithmetic can be messier.

NOTE A different approach to Example 7 would be to begin by multiplying both sides of the equation by 100. This would clear the equation of decimals to produce

$12x + 9(5000 - x) = 51,300.$

Try solving this equation to see that you obtain the same solution.

EXAMPLE 7 Solving a Linear Equation Involving Decimals

$$0.12x + 0.09(5000 - x) = 513 \qquad \text{Original equation}$$

$$0.12x + 450 - 0.09x = 513 \qquad \text{Distributive Property}$$

$$(0.12x - 0.09x) + 450 = 513 \qquad \text{Group like terms.}$$

$$0.03x + 450 - 450 = 513 - 450 \qquad \text{Subtract 450 from both sides.}$$

$$0.03x = 63 \qquad \text{Combine like terms.}$$

$$\frac{0.03x}{0.03} = \frac{63}{0.03} \qquad \text{Divide both sides by 0.03.}$$

$$x = 2100 \qquad \text{Simplify.}$$

The solution is 2100. Check this in the original equation.

Challenge problem: If $3x - 7 = 17$, what is $2x + 1$?

Answer: 17

When solving equations that are in nonstandard form, you should be aware that it is possible for the equation to have *no solution*. This is demonstrated in Example 8.

EXAMPLE 8   An Equation with No Solution

Solve the equation $2x - 4 = 2(x - 3)$.

Solution

$2x - 4 = 2(x - 3)$	Original equation
$2x - 4 = 2x - 6$	Distributive Property
$-4 = -6$	Subtract $2x$ from both sides.

NOTE   An equation that has no solution is said to have an **empty solution set,** which is denoted by { } or Ø.

Because the last equation has no solution, you can conclude that the original equation also has no solution.

Literal Equations

A **literal equation** is an equation that has more than one variable. For instance, $5x + 2y = 7$ is a literal equation because it has two variables, x and y. The word *literal* comes from the Latin word for "letter." To **solve a literal equation** for one of its variables means to write an equivalent equation in which the "solved variable" is isolated on one side of the equation.

NOTE   There is no general agreement as to the "best" or "simplest" way to write solutions of literal equations. In Example 9(b), we rewrote the solution

$$t = \frac{8 - 5s}{-2} \text{ as } t = \frac{5s - 8}{2}$$

because the latter form seems to be simpler. This, however, is partly a matter of personal preference. For instance, you might prefer to keep the solution in the first form *or* you might prefer to write the solution as $t = \frac{5}{2}s - 4$. The actual form that you use is not particularly important. It is important, however, to realize that many forms can be used *and* that you are able to convert from one form to another.

EXAMPLE 9   Solving a Literal Equation

Solve the equation $5s - 2t = 8$ (a) for s and (b) for t.

Solution

(a) You can solve the equation for s as follows.

$5s - 2t = 8$	Original equation
$5s = 8 + 2t$	Add $2t$ to both sides.
$s = \dfrac{8 + 2t}{5}$	Divide both sides by 5.

(b) You can solve the equation for t as follows.

$5s - 2t = 8$	Original equation
$-2t = 8 - 5s$	Subtract $5s$ from both sides.
$t = \dfrac{8 - 5s}{-2}$	Divide both sides by -2.
$t = \dfrac{5s - 8}{2}$	Simplify.

EXAMPLE 10   Solving a Literal Equation

Solve the literal equation $3y + 2x - 4 = 5y - 3x + 2$ for y.

Solution

$$3y + 2x - 4 = 5y - 3x + 2 \qquad \text{Original equation}$$

$$-2y + 2x - 4 = -3x + 2 \qquad \text{Subtract } 5y \text{ from both sides.}$$

$$-2y - 4 = -5x + 2 \qquad \text{Subtract } 2x \text{ from both sides.}$$

$$-2y = -5x + 6 \qquad \text{Add 4 to both sides.}$$

$$y = \frac{-5x + 6}{-2} \qquad \text{Divide both sides by } -2.$$

$$y = \frac{5x - 6}{2} \qquad \text{Simplify.}$$

NOTE   In Example 11, notice that it is not necessary to isolate the variable on the left side of the equation. Sometimes it is more convenient to isolate the variable on the right side.

Point out that this answer is *not* the same as $A/1 + A/rt$.

EXAMPLE 11   Solving a Literal Equation

Solve the literal equation $A = P + Prt$ for P.

Solution

Notice how the Distributive Property is used to factor P out of the terms on the right side of the equation.

$$A = P + Prt \qquad \text{Original equation}$$

$$A = P(1 + rt) \qquad \text{Distributive Property}$$

$$\frac{A}{1 + rt} = P \qquad \text{Divide both sides by } 1 + rt.$$

Group Activities   **Extending the Concept**

Analyzing and Interpreting Equations   Classify each of the following equations as an identity, a conditional equation, or an equation with no solution. Compare your conclusions with those of the rest of your group, and discuss the reasons for each conclusion.

a. $2x - 3 = -4 + 2x$

b. $x + 0.05x = 37.75$

c. $5(2 + x) + 3 = 5x + 13$

Discuss possible realistic situations in which the equations you classified as an identity and a conditional equation might apply.

sec 4 quiz

$$3 - (4 + x) = \frac{x}{2}$$

1.7 Exercises

See Warm-Up
Exercises, p. A40

In Exercises 1–6, determine whether the given values of the variable are solutions of the equation.

Equation	Values	
1. $3x - 7 = 2$	(a) $x = 0$	(b) $x = 3$
2. $5x + 9 = 4$	(a) $x = -1$	(b) $x = 2$
3. $3 - 2x = 21$	(a) $x = -3$	(b) $x = -9$
4. $10x - 3 = 7x$	(a) $x = 0$	(b) $x = -1$
5. $3x + 3 = 2(x - 4)$	(a) $x = -11$	(b) $x = 5$
6. $7x - 1 = 5(x - 5)$	(a) $x = 2$	(b) $x = -2$

7. *Think About It* Explain the difference between (a) an expression and an equation and (b) a conditional equation and an identity.

8. *Think About It* What is meant by equivalent equations? Give an example of two equivalent equations.

In Exercises 9–12, identify the equation as a conditional equation, an identity, or an equation with no solution.

9. $3(x - 1) = 3x$ **10.** $2x + 8 = 6x$

11. $5(x + 3) = 2x + 3(x + 5)$

12. $\frac{2}{3}x + 4 = \frac{1}{3}x + 12$

In Exercises 13–16, determine whether the equation is linear. If not, state why.

13. $3x + 4 = 10$ **14.** $x^2 + 3 = 8$

15. $\frac{4}{x} - 3 = 5x$ **16.** $3(x - 2) = 4x$

In Exercises 17–20, justify each step of the solution.

17.
$$3x + 15 = 0$$
$$3x + 15 - 15 = 0 - 15$$
$$3x = -15$$
$$\frac{3x}{3} = \frac{-15}{3}$$
$$x = -5$$

18.
$$7x - 21 = 0$$
$$7x - 21 + 21 = 0 + 21$$
$$7x = 21$$
$$\frac{7x}{7} = \frac{21}{7}$$
$$x = 3$$

19.
$$-2x + 5 = 12$$
$$-2x + 5 - 5 = 12 - 5$$
$$-2x = 7$$
$$\frac{-2x}{-2} = \frac{7}{-2}$$
$$x = -\frac{7}{2}$$

20.
$$25 - 3x = 10$$
$$25 - 3x - 25 = 10 - 25$$
$$-3x = -15$$
$$\frac{-3x}{-3} = \frac{-15}{-3}$$
$$x = 5$$

In Exercises 21–56, solve the equation and check the result. (If not possible, state the reason.)

21. $3x = 12$ **22.** $8z - 10 = 0$

23. $23x - 4 = 42$ **24.** $15x - 18 = 27$

25. $7 - 8x = 13x$ **26.** $2s - 16 = 34s$

27. $15t = 0$ **28.** $6a + 2 = 6a$

29. $-8t + 7 = -8t$ **30.** $4x = -12x$

31. $4x - 7 = x + 11$ **32.** $5x + 3 = 2x - 21$

33. $2 - 3x = 10 + x$ **34.** $4 + 2x = 7 - 4x$

35. $8(x - 8) = 24$ **36.** $6(x + 2) = 30$

37. $5 - (2y - 4) = 15$ **38.** $26 - (3x - 10) = 6$

39. $8x - 3(x - 2) = 12$ **40.** $2(x + 3) = 7(x + 3)$

41. $3(x + 5) - 2x = 3 - (4x - 2)$

42. $4x + 5(2x - 4) = 8 - 10(x + 1)$

Sec 2 quiz

$$4x + 10x - 20 = 8 - 10x - 10$$
$$14x - 20 = -2 - 10x$$
$$24x = 18 \qquad x = \frac{18}{24} = \frac{3}{4}$$

43. $-25(x - 100) = 16(x - 100)$

44. $12 = 6(y + 1) - 8(1 - y)$

45. $\dfrac{u}{5} = 10$ **46.** $-\dfrac{z}{2} = 7$

47. $t - \frac{2}{5} = \frac{3}{2}$ **48.** $z + \frac{1}{15} = -\frac{3}{10}$

49. $\dfrac{t + 4}{14} = \dfrac{2}{7}$ **50.** $\dfrac{11x}{6} + \dfrac{1}{3} = 0$

51. $\dfrac{t}{5} - \dfrac{t}{2} = 1$ **52.** $\dfrac{t}{6} + \dfrac{t}{8} = 1$

53. $\dfrac{4u}{3} = \dfrac{5u}{4} + 6$ **54.** $\dfrac{-3x}{4} - 4 = \dfrac{x}{6}$

55. $0.3x + 1.5 = 8.4$ **56.** $16.3 - 0.2x = 7.1$

In Exercises 57–60, solve the equation and round the result to two decimal places. (A calculator may be helpful.)

57. $1.234x + 3 = 7.805$ **58.** $325x - 4125 = 612$

59. $\dfrac{x}{10.625} = 2.850$ **60.** $2x + \dfrac{4.7}{4} = \dfrac{3}{2}$

In Exercises 61–68, solve the equation (a) for x and (b) for y.

61. $2x - 3y = 6$ **62.** $-5x + 3y = 15$

63. $7x + 4 = 10y - 7$

64. $12(x - 2) + 7(y + 1) = 25$

65. $4[2x - 3(x + 2y)] = 0$

66. $-3[2(x + 4y) - (x - y)] = 1$

67. $\dfrac{x}{2} + \dfrac{y}{5} = 1$ **68.** $\dfrac{3x}{4} + \dfrac{y}{6} = 5$

In Exercises 69–74, verify that the solution can be written in the rewritten form.

	Solution	Rewritten Form
69.	$\dfrac{y - 5}{-1}$	$5 - y$
70.	$u + 3uv$	$u(1 + 3v)$
71.	$\dfrac{h + 4\pi}{2}$	$\dfrac{h}{2} + 2\pi$
72.	$\dfrac{2t - 5}{-2}$	$\dfrac{5}{2} - t$

	Solution	Rewritten Form
73.	$\dfrac{3x - 7}{6}$	$\dfrac{1}{2}x - \dfrac{7}{6}$
74.	$\dfrac{-7(3 - x)}{2}$	$\dfrac{7}{2}(x - 3)$

In Exercises 75–86, solve for the indicated variable.

75. Solve for R

Ohm's Law: $E = IR$

76. Solve for C

Markup: $S = C + rC$

77. Solve for L

Discount: $S = L - rL$

78. Solve for h

Area of a Triangle: $A = \frac{1}{2}bh$

79. Solve for b

Area of a Trapezoid: $A = \frac{1}{2}(a + b)h$

80. Solve for h

Volume of a Right Circular Cylinder: $V = \pi r^2 h$

81. Solve for r

Investment at Simple Interest: $A = P + Prt$

82. Solve for P

Investment at Compound Interest: $A = P\left(1 + \dfrac{r}{n}\right)^{nt}$

83. Solve for n

Arithmetic Progression: $S = \dfrac{n}{2}(a_1 + a_n)$

84. Solve for n

Arithmetic Progression: $L = a + (n - 1)d$

85. Solve for r

Geometric Progression: $S = \dfrac{a_1}{1 - r}$

86. Solve for m_2

Newton's Law of Universal Gravitation: $F = \dfrac{km_1m_2}{r^2}$

87. *Investigation*   The length of a rectangle is t times its width (see figure). Thus, the perimeter P is given by $P = 2w + 2(tw)$, where w is the width of the rectangle. The perimeter of the rectangle is 1200 meters.

(a) Complete the table of widths, lengths, and areas of the rectangle by solving the given equation.

t	1	1.5	2	3	4	5
Width						
Length						
Area						

(b) Use the table to write a short paragraph describing the relationship among the width, length, and area of a rectangle that has a *fixed* perimeter.

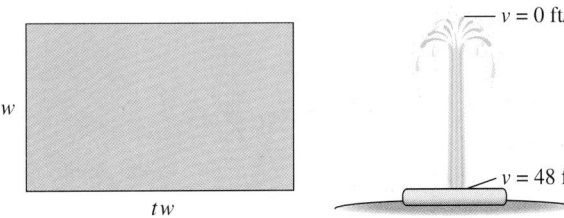

Figure for 87

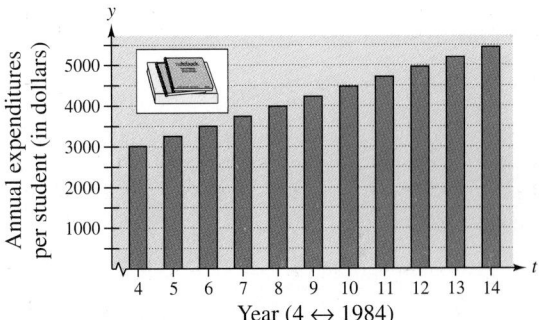

Figure for 88

88. *Maximum Height of a Fountain* Consider the water fountain shown in the figure. The initial velocity of the stream of water is 48 feet per second. The velocity v of the water at any time t (in seconds) is then given by $v = 48 - 32t$. Find the time for a drop of water to travel from the base to the maximum height of the fountain. (*Hint:* The maximum height is reached when $v = 0$.)

89. *Maximum Height of an Object* The velocity v of an object projected vertically upward with an initial velocity of 64 feet per second is given by $v = 64 - 32t$, where t is the time in seconds. When does the object reach its maximum height? Explain.

90. *Work Rate Problem* Two people can complete a task in t hours, where t must satisfy the equation

$$\frac{t}{10} + \frac{t}{15} = 1.$$

Find the required time t.

91. *Using a Model* The average annual expenditures y (in dollars) per student for U.S. public schools from 1984 to 1994 can be modeled by

$$y = 2039.83 + 242.83t,$$

where t represents the year, with $t = 4$ corresponding to 1984 (see figure). Determine algebraically and graphically during which year the expenditures reached \$4000. (*Source: National Education Association*)

Section Project

Using a Two-Part Model Use the following *two-part* model, which approximates the number of cable television subscribers y (in thousands) in the United States between 1970 and 1993.

$$y = 4193 + 1099.4t, \quad 0 \le t \le 9$$
$$y = -15{,}731 + 3210t, \quad 10 \le t \le 23$$

In this model, t represents the year, with $t = 0$ corresponding to 1970. (*Source: Corporation for Public Broadcasting*)

(a) Make a table that shows the number of subscribers from 1970 through 1993.

(b) Draw a bar graph of the data in the table.

(c) Which year had 10,789.4 (thousand) subscribers?

(d) Which year had 42,049 (thousand) subscribers?

(e) In parts (c) and (d), did you use the model, the table, or the bar graph? Could you have used any of the three? Explain.

1.8	**Solving Equations by Factoring**

Quadratic Equations and the Zero-Factor Property ▪
Solving Quadratic Equations by Factoring ▪
Solving Higher-Degree Equations by Factoring

Quadratic Equations and the Zero-Factor Property

In Section 1.7, you studied techniques for solving first-degree polynomial equations (linear equations). In this section, you will study a technique for solving **second-degree** polynomial equations (*quadratic* equations) and *higher-degree* polynomial equations.

Definition of a Quadratic Equation

A **quadratic equation in x in standard form** is an equation that can be written in the form

$$ax^2 + bx + c = 0, \qquad \text{Quadratic equation}$$

where a, b, and c are real numbers with $a \neq 0$.

In Section 1.6, you reviewed techniques for factoring trinomials of the form

$$ax^2 + bx + c. \qquad \text{Trinomial}$$

These skills can be combined with the **Zero-Factor Property** to *solve* quadratic equations.

Zero-Factor Property

Let u and v represent real numbers, variables, or algebraic expressions. If u and v are factors such that

$$uv = 0,$$

then $u = 0$ or $v = 0$. This property applies to three or more factors as well.

This is a good time to justify the time students spend on learning how to factor.

NOTE The Zero-Factor Property is just another way of saying that the only way the product of two (or more) real numbers can be zero is if one (or more) of the real numbers is zero.

Solving Quadratic Equations by Factoring

The Zero-Factor Property is the primary property that is used to solve equations in algebra. For instance, to solve the equation

$$(x - 2)(x + 3) = 0,$$

you can use the Zero-Factor Property to conclude that either $(x - 2)$ or $(x + 3)$ must be zero. Setting the first factor equal to zero implies that $x = 2$ is a solution. That is,

$$x - 2 = 0 \quad \Longrightarrow \quad x = 2.$$

Similarly, setting the second factor equal to zero implies that $x = -3$ is a solution. That is,

$$x + 3 = 0 \quad \Longrightarrow \quad x = -3.$$

Thus, the equation $(x - 2)(x + 3) = 0$ has exactly two solutions: 2 and -3.

In each of the examples that follow, note how factoring skills are combined with the Zero-Factor Property to solve equations.

NOTE Factoring and the Zero-Factor Property allow you to solve a quadratic equation by converting it into two *linear* equations (which you already know how to solve). This is a common strategy of algebra—to break down a given problem into simpler parts, each solvable by previously learned methods.

EXAMPLE 1 Using Factoring to Solve an Equation

Solve the quadratic equation $x^2 - x - 12 = 0$.

Solution

Begin by checking to see that the right side of the equation is zero. Next, factor the left side of the equation. Finally, apply the Zero-Factor Property to find the solutions.

$x^2 - x - 12 = 0$	Original equation in standard form
$(x + 3)(x - 4) = 0$	Factor left side of equation.
$x + 3 = 0 \quad \Longrightarrow \quad x = -3$	Set 1st factor equal to 0.
$x - 4 = 0 \quad \Longrightarrow \quad x = 4$	Set 2nd factor equal to 0.

Thus, the given equation has two solutions: -3 and 4.

Check First Solution

$x^2 - x - 12 = 0$	Original equation
$(-3)^2 - (-3) - 12 \stackrel{?}{=} 0$	Replace x by -3.
$9 + 3 - 12 \stackrel{?}{=} 0$	
$0 = 0$	Solution checks. ✓

Use the same technique to check the second solution.

Encourage students to get into the habit of checking their answers. It will help them to remember to do so later when checking is critical, and it is a useful study skill when doing exercises or taking tests.

To use the Zero-Factor Property, a quadratic equation *must* be written in **standard form.** That is, the quadratic must be on the left side of the equation and zero must be the only term on the right side of the equation. For instance, to write the equation $x^2 - 3x = 18$ in standard form, you must subtract 18 from both sides of the equation, as follows.

$x^2 - 3x = 18$	Original equation (nonstandard form)
$x^2 - 3x - 18 = 18 - 18$	Subtract 18 from both sides.
$x^2 - 3x - 18 = 0$	Standard form

To solve this equation, factor the left side as $(x + 3)(x - 6)$ and then form the linear equations $x + 3 = 0$ and $x - 6 = 0$. The solutions of these two linear equations are -3 and 6. The general strategy for solving a quadratic equation by factoring is summarized in the following diagram.

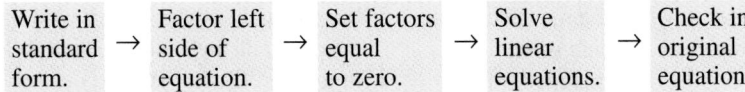

Write in standard form. $\rightarrow$ Factor left side of equation. $\rightarrow$ Set factors equal to zero. $\rightarrow$ Solve linear equations. $\rightarrow$ Check in original equation.

EXAMPLE 2   *Solving a Quadratic Equation by Factoring*

Solve the quadratic equation

$$3x^2 + 5x = 12.$$

Solution

$3x^2 + 5x = 12$	Original equation
$3x^2 + 5x - 12 = 0$	Write in standard form.
$(3x - 4)(x + 3) = 0$	Factor left side of equation.
$3x - 4 = 0 \quad\Longrightarrow\quad x = \frac{4}{3}$	Set 1st factor equal to 0.
$x + 3 = 0 \quad\Longrightarrow\quad x = -3$	Set 2nd factor equal to 0.

Therefore, the solutions are $\frac{4}{3}$ and -3. The first solution is checked as follows.

Check First Solution

$3x^2 + 5x = 12$	Original equation
$3\left(\frac{4}{3}\right)^2 + 5\left(\frac{4}{3}\right) \overset{?}{=} 12$	Replace x by $\frac{4}{3}$.
$\frac{16}{3} + \frac{20}{3} \overset{?}{=} 12$	Simplify.
$\frac{36}{3} = 12$	Solution checks. ✔

The second solution can be checked in a similar way.

As students gain experience with solving equations, encourage them to perform some steps manually. For instance, students can be shown how to solve $3x - 4 = 0$ or $x + 3 = 0$ mentally.

   After converting a quadratic equation to standard form, you should check to see whether the left side of the equation has a common *numerical* factor. If it does, you can divide both sides of the equation by this factor without "losing" any of the solutions. For instance, each of the equations

$$5x^2 + 30x + 40 = 0 \qquad \text{Original equation}$$
$$5(x^2 + 6x + 8) = 0 \qquad \text{Factor out common monomial.}$$
$$x^2 + 6x + 8 = 0 \qquad \text{Divide both sides by 5.}$$

has the same solutions. Be sure, however, that you *do not divide both sides of an equation by a variable factor.* This will usually result in "losing" one or more of the solutions.

EXAMPLE 3   *Equations with Common Monomial Factors*

Solve each equation.

(a) $2x^2 - 2x - 24 = 0$

(b) $2x^2 + 10x = 0$

Solution

(a) $2x^2 - 2x - 24 = 0$ Original equation

      $2(x^2 - x - 12) = 0$ Factor out common monomial.

      $2(x - 4)(x + 3) = 0$ Factor trinomial.

                  $2 \neq 0$ Set 1st factor equal to 0.

              $x - 4 = 0 \implies x = 4$ Set 2nd factor equal to 0.

              $x + 3 = 0 \implies x = -3$ Set 3rd factor equal to 0.

The solutions are 4 and -3. Check these in the original equation. Notice that you do not obtain a solution by setting a *constant* factor equal to zero.

(b) $2x^2 + 10x = 0$ Original equation

      $(2x)(x + 5) = 0$ Factor out common monomial.

              $2x = 0 \implies x = 0$ Set 1st factor equal to 0.

           $x + 5 = 0 \implies x = -5$ Set 2nd factor equal to 0.

The solutions are 0 and -5. Check these in the original equation. Notice that you obtain a solution of zero by setting a factor of the form kx equal to zero.

NOTE   In Example 3(a), notice that you could divide both sides of the equation by the *constant* factor 2 without losing a solution. In Example 3(b), however, dividing both sides of the equation by the *variable* factor $2x$ would result in losing the solution $x = 0$.

In Examples 1, 2, and 3, the given equations each involved a second-degree polynomial and each had *two different* solutions. Sometimes you will encounter a second-degree polynomial equation that has only one (repeated) solution. This occurs when the left side of the equation is a perfect square trinomial, as shown in Example 4.

EXAMPLE 4   *A Quadratic Equation with a Repeated Solution*

Solve the quadratic equation

$$x^2 - 6x + 11 = 2.$$

Solution

$x^2 - 6x + 11 = 2$	Original equation
$x^2 - 6x + 9 = 0$	Write in standard form.
$(x - 3)(x - 3) = 0$	Factor.
$x - 3 = 0$ or $x - 3 = 0$	Set factors equal to 0.
$x = 3$	Repeated solution is 3.

Note that even though the left side of this equation has two factors, the two factors are the same. Thus, you can conclude that the only solution of the equation is 3. (This solution is called a **repeated solution.**) Check this solution in the original equation.

Be sure you see that the Zero-Factor Property can be applied only to a product that is equal to *zero*. For instance, the following solution is incorrect.

$$x^2 - x = 6$$
$$x(x - 1) = 6$$
$$x = 6 \quad\Rightarrow\quad x = 6$$
$$x - 1 = 6 \quad\Rightarrow\quad x = 7$$

Instead, you must first write the equation in standard form and then factor the left side, as follows.

$$x^2 - x = 6$$
$$x^2 - x - 6 = 0$$
$$(x - 3)(x + 2) = 0$$
$$x - 3 = 0 \quad\Rightarrow\quad x = 3$$
$$x + 2 = 0 \quad\Rightarrow\quad x = -2$$

Try checking that 6 and 7 are *not* solutions of $x^2 - x = 6$, and that 3 and -2 *are* solutions.

EXAMPLE 5   *Solving a Quadratic Equation in Nonstandard Form*

Solve the quadratic equation

$$(x + 2)(x + 4) = 3.$$

Solution

$(x + 2)(x + 4) = 3$	Original equation
$x^2 + 6x + 8 = 3$	Multiply factors.
$x^2 + 6x + 5 = 0$	Write in standard form.
$(x + 1)(x + 5) = 0$	Factor.
$x + 1 = 0$ ⟹ $x = -1$	Set 1st factor equal to 0.
$x + 5 = 0$ ⟹ $x = -5$	Set 2nd factor equal to 0.

Therefore, the equation has two solutions: -1 and -5. Check these solutions in the original equation.

Some quadratic equations do not have solutions that are real numbers. For instance, there is no real number that satisfies the equation $x^2 = -4$. The reason that this equation has no real solution is that there is no real number that can be multiplied by itself to produce -4. This type of equation will be discussed further in Chapter 5.

You might remember from an earlier course in algebra that a polynomial equation can have *at most* as many solutions as its degree. For instance, a second-degree equation can have zero, one, or two real solutions, but it cannot have three or more solutions.

You might again discuss Example 4, a quadratic equation with degree 2. We can expect at most two real solutions. Point out that we have two solutions, but they just happen to be the same so we don't need to repeat them.

EXAMPLE 6   *Solving Linear and Quadratic Equations*

Describe the technique you would use to solve each equation.

(a) $7x + 1 = 2x - 3$

(b) $7x + 1 = 2x^2 - 3$

Solution

(a) This is a linear equation, so you would solve it by isolating x. After doing this, you would find that the solution is $-\frac{4}{5}$.

(b) This is a quadratic equation. To solve it, you would write the equation in standard form, factor the quadratic, and apply the Zero-Factor Property to conclude that the solutions are 4 and $-\frac{1}{2}$.

Solving Higher-Degree Equations by Factoring

The Zero-Factor Property can be used to solve polynomial equations of degree 3 or higher. To do this, use the same strategy you used with quadratic equations. That is, write the equation in standard form with the polynomial on the left and zero on the right. Then factor the left side of the equation. Finally, set each factor equal to zero to obtain the solutions.

EXAMPLE 7   Solving a Polynomial Equation with Three Factors

Solve $3x^3 = 15x^2 + 18x$.

Solution

$$3x^3 = 15x^2 + 18x$$ Original equation

$$3x^3 - 15x^2 - 18x = 0$$ Write in standard form.

$$3x(x^2 - 5x - 6) = 0$$ Factor out common monomial.

$$3x(x - 6)(x + 1) = 0$$ Factor.

$$3x = 0 \implies x = 0$$ Set 1st factor equal to 0.

$$x - 6 = 0 \implies x = 6$$ Set 2nd factor equal to 0.

$$x + 1 = 0 \implies x = -1$$ Set 3rd factor equal to 0.

There are three solutions: 0, 6, and -1. Check these in the original equation.

EXAMPLE 8   Solving a Polynomial Equation with Four Factors

Solve $x^4 + x^3 - 4x^2 - 4x = 0$.

Solution

$$x^4 + x^3 - 4x^2 - 4x = 0$$ Original equation

$$x(x^3 + x^2 - 4x - 4) = 0$$ Factor out common monomial.

$$x[x^2(x + 1) - 4(x + 1)] = 0$$ Factor by grouping.

$$x[(x + 1)(x^2 - 4)] = 0$$ Distributive Property

$$x(x + 1)(x + 2)(x - 2) = 0$$ Factor.

$$x = 0 \implies x = 0$$ Set 1st factor equal to 0.

$$x + 1 = 0 \implies x = -1$$ Set 2nd factor equal to 0.

$$x + 2 = 0 \implies x = -2$$ Set 3rd factor equal to 0.

$$x - 2 = 0 \implies x = 2$$ Set 4th factor equal to 0.

There are four solutions: 0, -1, -2, and 2. Check these in the original equation.

[handwritten at top:] quiz $2x^2 + 72 = 24x$
$x^2 - 12x + 36 = 0$
$(x - 6)^2 = 0$ $x = 6$

Group Activities

Technology

Solving Equations with a Graphing Calculator   Some graphing calculators, such as the *TI-82* or *TI-83*, have a built-in "equation solver." For instance, to solve for one solution of the equation $x^2 - 3x + 2 = 0$ on either of these calculators, you can enter

solve($X^2 - 3x + 2$,X,Guess),

where "Guess" is a number that you have guessed to be the solution. In your group, use the *solve* feature of a graphing calculator to solve each of the following equations.

a. $x^2 + 4x + 3 = 0$ b. $3x^2 - x - 2 = 0$ c. $2x^2 + 11x - 6 = 0$

In your group, discuss this solution technique. Is it more efficient than solving the equations "by hand"? Explain your reasoning.

1.8   Exercises

[handwritten:] quiz Sec 8   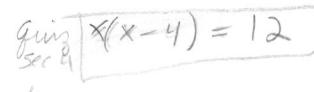 $x(x - 4) = 12$

See Warm-Up
Exercises, p. A40

In Exercises 1–10, use the Zero-Factor Property to solve the equation.

1. $2x(x - 8) = 0$
2. $z(z + 6) = 0$
3. $(y - 3)(y + 10) = 0$
4. $(s - 16)(s + 15) = 0$
5. $25(a + 4)(a - 2) = 0$
6. $17(t - 3)(t + 8) = 0$
7. $(x - 3)(2x + 1)(x + 4) = 0$
8. $\frac{1}{5}x(x - 2)(3x + 4) = 0$
9. $4x(2x - 3)(2x + 25) = 0$
10. $(y - 39)(2y + 7)(y + 12) = 0$

In Exercises 11–48, solve the polynomial equation by factoring.

11. $x^2 - 3x - 10 = 0$
12. $x^2 + x = 12$
13. $y^2 + 20 = 9y$
14. $y^2 + 12y + 27 = 0$
15. $3x^2 + 9x = 0$
16. $5y - y^2 = 0$

17. $x^2 - 25 = 0$
18. $x^2 - 81 = 0$
19. $3x^2 - 300 = 0$
20. $4b^3 - 36b = 0$
21. $m^2 - 8m = -16$
22. $a^2 + 4a + 4 = 0$
23. $4z^2 + 9 = 12z$
24. $2x^2 + 24x + 72 = 0$
25. $7 + 13x - 2x^2 = 0$
26. $11 + 32y - 3y^2 = 0$
27. $x(x - 3) = 10$
28. $s(s + 4) = 96$ *[handwritten:]* 6·16 3·32 12·8
29. $y(y + 6) = 72$
30. $x(x - 2) = 15$
31. $x(x + 2) - 10(x + 2) = 0$
32. $x(x - 15) + 3(x - 15) = 0$
33. $(t - 2)^2 - 16 = 0$
34. $(x + 4)^2 - 49 = 0$
35. $6t^3 - t^2 - t = 0$
36. $3u^3 - 5u^2 - 2u = 0$
37. $x^3 - 19x^2 + 84x = 0$
38. $x^3 + 18x^2 + 45x = 0$
39. $x^2(x - 25) - 16(x - 25) = 0$
40. $y^2(y + 250) - (y + 250) = 0$

41. $z^2(z + 2) - 4(z + 2) = 0$

42. $16(3 - u) - u^2(3 - u) = 0$

43. $c^3 - 3c^2 - 9c + 27 = 0$

44. $v^3 + 4v^2 - 4v - 16 = 0$

45. $a^3 + 2a^2 - 9a - 18 = 0$

46. $x^3 - 2x^2 - 4x + 8 = 0$

47. $x^4 - 5x^3 - 9x^2 + 45x = 0$

48. $2x^4 + 6x^3 - 50x^2 - 150x = 0$

49. *True or False?* If

$$(2x - 5)(x + 4) = 1,$$

then $2x - 5 = 1$ or $x + 4 = 1$. Explain.

50. *Think About It* Is it possible for a quadratic equation to have only one solution? Explain.

51. *Think About It* What is the maximum number of solutions of an nth-degree polynomial equation? Give an example of a third-degree equation that has only one real number solution.

52. *Investigation* Solve the equation

$$3(x + 4)^2 + (x + 4) - 2 = 0$$

in the following two ways.

(a) Let $u = x + 4$, and solve the resulting equation for u. Then find the corresponding values of x that are solutions of the original equation.

(b) Expand and collect like terms in the original equation, and solve the resulting equation for x.

(c) Which method is easier? Explain.

Investigation In Exercises 53 and 54, solve the equation using either method (or both methods) described in Exercise 52.

53. $3(x + 6)^2 - 10(x + 6) - 8 = 0$

54. $8(x + 2)^2 - 18(x + 2) + 9 = 0$

55. Let a and b be real numbers such that $a \neq 0$. Find the solutions of $ax^2 + bx = 0$.

56. Let a be a nonzero real number. Find the solutions of $ax^2 - ax = 0$.

Think About It In Exercises 57 and 58, find a quadratic equation with the given solutions.

57. $x = -3, x = 5$ 58. $x = 1, x = 6$

59. *Number Problem* The sum of a positive number and its square is 240. Find the number.

60. *Number Problem* The sum of a positive number and its square is 72. Find the number.

61. *Free-Falling Object* An object is dropped from a weather balloon 6400 feet above the ground (see figure). Find the time t for the object to reach the ground by solving the equation $-16t^2 + 6400 = 0$.

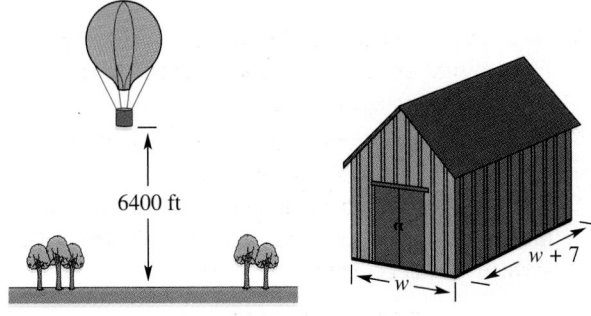

6400 ft

Figure for 61   **Figure for 63**

62. *Free-Falling Object* An object is thrown upward from a height of 64 feet with an initial velocity of 48 feet per second. Find the time t for the object to reach the ground by solving the equation $-16t^2 + 48t + 64 = 0$.

63. *Geometry* The rectangular floor of a storage shed has an area of 330 square feet. The length of the floor is 7 feet more than its width (see figure). Find the dimensions of the floor.

64. *Geometry* The outside dimensions of a picture frame are 28 centimeters and 23 centimeters (see figure). The area of the exposed part of the picture is 414 square centimeters. Find the width w of the frame.

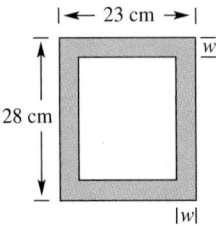

$|\leftarrow$ 23 cm $\rightarrow|$

28 cm

$|w|$

65. *Geometry* The triangular cross section of a machined part must have an area of 48 square inches (see figure). Find the base and height of the triangle if the height is $1\frac{1}{2}$ times the base.

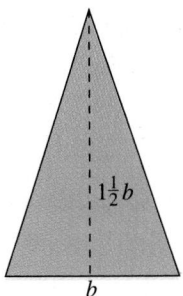

$1\frac{1}{2}b$

b

66. *Geometry* The height of a triangle is 4 inches less than its base. Find the base and height of the triangle if its area is 70 square inches.

67. *Geometry* An open box with a square base is to be constructed from 880 square inches of material (see figure). What should the dimensions of the base be if the height of the box is to be 6 inches? (*Hint:* The surface area is given by $S = x^2 + 4xh$.)

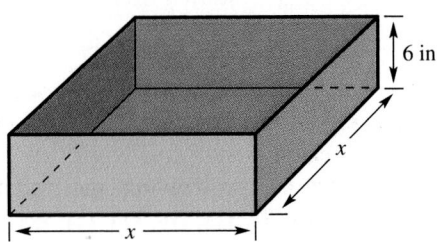

6 in.

x

x

68. *Break-Even Analysis* The revenue R from the sale of x units of a product is given by $R = 90x - x^2$. The cost of producing x units of the product is $C = 200 + 60x$. How many units of the product must be produced and sold in order to break even?

69. *Numerical Reasoning* Consider the product $P = (x + 5)(x - 4)$.

(a) Complete the table.

x	3	4	5	6	7	8
P						

(b) Use the table to determine how P changes for each one-unit increase in x.

(c) Use your answer to part (b) to solve the equation $(x + 5)(x - 4) = 70$.

Section Project

Geometry An open box is to be made from a piece of material that is 5 meters long and 4 meters wide. The box is made by cutting squares of dimension x from the corners and turning up the sides (see figure). The volume V of a rectangular solid is the product of its length, width, and height.

(a) Show that the volume is given by
$$V = (5 - 2x)(4 - 2x)x.$$

(b) Determine the values of x for which $V = 0$.

(c) Use the table feature of a graphing utility to complete the table.

x	0.25	0.50	0.75	1.00	1.25	1.50	1.75
V							

(d) Use the table to determine x if $V = 3$. Verify the result analytically.

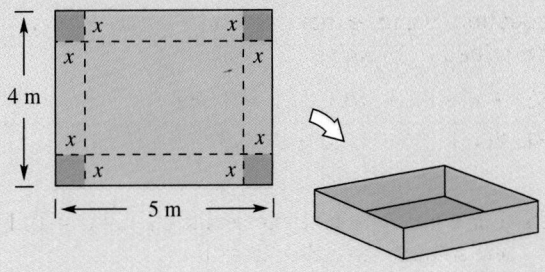

x

x

x

x

x

x

x

x

4 m

5 m

Chapter Summary

After studying this chapter, you should have acquired the following skills. These skills are keyed to the Review Exercises that begin on page 99. Answers to odd-numbered Review Exercises are given in the back of the book.

- Plot real numbers on a real number line and compare them by using inequality symbols. *(Section 1.1)*

  Review Exercises 1–4

- Evaluate expressions containing operations with real numbers. *(Section 1.1)*

  Review Exercises 5–24, 47

- Identify the rule of algebra that is illustrated by an equation. *(Sections 1.2, 1.3)*

  Review Exercises 25–30

- Expand expressions using the Distributive Property. *(Sections 1.2, 1.3)*

  Review Exercises 31–34

- Simplify expressions by removing symbols of grouping. *(Section 1.3)*

  Review Exercises 35–38

- Simplify expressions by applying the properties of exponents. *(Section 1.3)*

  Review Exercises 39–44

- Solve problems involving geometry. *(Sections 1.3, 1.4)*

  Review Exercises 45, 46, 129, 130

- Use expressions or equations to solve real-life problems. *(Sections 1.1, 1.8)*

  Review Exercises 48, 127, 128

- Interpret graphs representing real-life data. *(Sections 1.1–1.4, 1.7)*

  Review Exercises 49, 50

- Simplify expressions by performing arithmetic operations. *(Section 1.4)*

  Review Exercises 51–66

- Multiply polynomials using the special product formulas. *(Section 1.4)*

  Review Exercises 67–74

- Factor expressions completely. *(Sections 1.5, 1.6)*

  Review Exercises 75–96

- Solve linear equations. *(Section 1.7)*

  Review Exercises 97–104, 115, 116, 120, 121, 123, 124

- Solve literal equations. *(Section 1.7)*

  Review Exercises 105, 106

- Solve polynomial equations. *(Section 1.8)*

  Review Exercises 107–114, 117–119, 122, 125, 126

- Use a calculator to evaluate expressions containing operations with real numbers. *(Section 1.1)*

  Review Exercise 131

Review Exercises

In Exercises 1–4, plot each real number as a point on the real number line and place the correct inequality symbol (< or >) between the real numbers.

1. $-\frac{1}{8}$ ▯ 3

2. -2 ▯ -8

3. $-\frac{8}{5}$ ▯ $-\frac{2}{5}$

4. 8.4 ▯ $-\pi$

In Exercises 5–24, evaluate the expression. If it is not possible, state the reason. Write all fractions in reduced form.

5. $340 - 115 + 5$

6. $|-96| - |134|$

7. $120(-5)(7)$

8. $(-16)(-15)(-4)$

9. $\frac{-56}{-4}$

10. $\frac{85}{0}$

11. $\frac{4}{21} + \frac{7}{21}$

12. $\frac{21}{16} - \frac{13}{16}$

13. $-\frac{5}{6} + 1$

14. $\frac{21}{32} + \frac{11}{24}$

15. $8\frac{3}{4} - 6\frac{5}{8}$

16. $-2\frac{9}{10} + 5\frac{3}{20}$

17. $\frac{3}{8} \cdot \frac{-2}{15}$

18. $\frac{5}{21} \cdot \frac{21}{5}$

19. $-\frac{7}{15} \div -\frac{7}{30}$

20. $-\frac{2}{3} \div \frac{4}{15}$

21. $(-6)^3$

22. $-(-3)^4$

23. $120 - (5^2 \cdot 4)$

24. $45 - 45 \div 3^2$

In Exercises 25–30, identify the rule of algebra that is illustrated by the equation.

25. $2x - 2x = 0$

26. $5 + (4 - y) = (5 + 4) - y$

27. $(u - v)(2) = 2(u - v)$

28. $(x + y) + 0 = x + y$

29. $ab \cdot \dfrac{1}{ab} = 1$

30. $x(yz) = (xy)z$

This is a good place for students to review "How to Study Algebra" on pages xx–xxi.

Urge students to use the Mid-Chapter Quiz, Review Exercises, and Chapter Test to review and practice the skills discussed in this chapter and to assess their mastery of these skills.

In Exercises 31–34, expand the expression by using the Distributive Property.

31. $y(3y - 10)$

32. $x(3x + 4y)$

33. $-(-u + 3v)$

34. $(5 - 3j)(-4)$

In Exercises 35–38, simplify the algebraic expression.

35. $3x - (y - 2x)$

36. $30x - (10x + 80)$

37. $3[b + 5(b - a)]$

38. $-2t[8 - (6 - t)] + 5t$

In Exercises 39–48, simplify the expression by using the rules of exponents.

39. $x^2 \cdot x^3 \cdot x$

40. $y^3(-2y^2)$

41. $(xy)(-3x^2y^3)$

42. $3uv(-2uv^2)^2$

43. $(-2a^2)^3(8a)$

44. $2(a - b)^4(a - b)^2$

45. $-(u^2v)^2(-4u^3v)$

46. $(12x^2y)(3x^2y^4)$

47. $\dfrac{120u^5v^3}{15u^3v}$

48. $-\dfrac{(-2x^2y^3)^2}{-3xy^2}$

In Exercises 49 and 50, write expressions for the perimeter and area of the region, and then simplify.

49.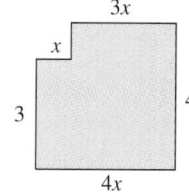

50.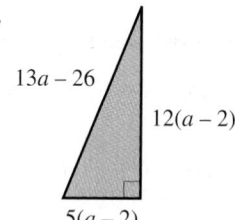

51. Perform the indicated operations and simplify.

$$6 \cdot 10^3 + 9 \cdot 10^2 + 1 \cdot 10^1$$

52. *Total Charge* You purchase a product and make a down payment of $387 plus 12 monthly payments of $68.00. What is the total amount you pay for the product?

Graphical Interpretation In Exercises 53 and 54, use the figure, which shows the average weekly overtime for production workers in August for the years 1982, 1986, 1990, and 1994. *(Source: U.S. Department of Labor)*

Weekly Overtime for Production Workers

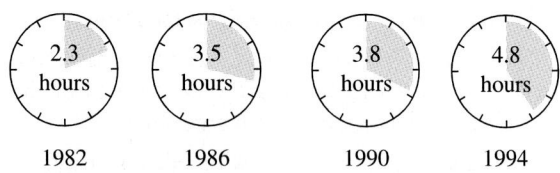

| 2.3 hours | 3.5 hours | 3.8 hours | 4.8 hours |
| 1982 | 1986 | 1990 | 1994 |

53. Determine the increase in average weekly overtime during the 4 years from 1990 to 1994.

54. During which 4-year period did the average weekly overtime increase the least?

In Exercises 55–68, perform the operations and simplify.

55. $(5x + 3x^2) + (x - 4x^2)$

56. $\left(\frac{1}{2}x + \frac{2}{3}\right) + \left(4x + \frac{1}{3}\right)$

57. $(-x^3 - 3x) - 4(2x^3 - 3x + 1)$

58. $(7z^2 + 6z) - 3(5z^2 + 2z)$

59. $3y^2 - [2y - 3(y^2 + 5)]$

60. $(16a^3 + 5a) - 5[a + (2a^3 - 1)]$

61. $(-2x)^3(x + 4)$

62. $3y(-4y)(y - 2)$

63. $(2z + 3)(3z - 5)$

64. $(6t + 1)(t - 11)$

65. $(5x + 3)(3x - 4)$

66. $(3y^2 + 2)(4y^2 - 5)$

67. $(2x^2 - 3x + 2)(2x + 3)$

68. $(5s^2 + 4s - 3)(4s - 5)$

In Exercises 69–74, find the product by using the special product formulas.

69. $(4x - 7)^2$

70. $(8 - 3x)^2$

71. $(5u - 8)(5u + 8)$

72. $(7a + 4)(7a - 4)$

73. $[(u - 3) + v][(u - 3) - v]$

74. $[(m - 5) + n]^2$

In Exercises 75–96, factor the expression completely.

75. $6x^2 + 15x^3$

76. $8y - 12y^4$

77. $u^2 - 10uv + 25v^2$

78. $u^3 - 1$

79. $9a^2 - 100$

80. $x^2 - 11x + 24$

81. $a^2 + 7a - 18$

82. $3x^2 + 23x - 8$

83. $27x^3 + 64$

84. $y^3 + 4y^2 - y - 4$

85. $4t^3 + 7t^2 - 2t$

86. $x^2 - 40x + 400$

87. $v^3 - 2v^2 - 4v + 8$

88. $(y - 3)^2 - 16$

89. $4u^2 - 28u + 49$

90. $28(x + 5) - 70(x + 5)^2$

91. $6h^3 - 23h^2 - 13h$

92. $35y^3 - 10y^2 - 25y$

93. $x^4 + 7x^3 - 9x^2 - 63x$

94. $(u - 9v)(u - v) + v(u - 9v)$

95. $x^2 + 18x + 81 - 4y^2$

96. $250a^3 - 2b^3$

In Exercises 97–104, solve the linear equation.

97. $4y - 6(y - 5) = 2$ **98.** $7x + 2(7 - x) = 8$

99. $1.4t + 2.1 = 0.9t - 2$

100. $8(x - 2) = 3(x - 2)$

101. $\frac{4}{5}x - \frac{1}{10} = \frac{3}{2}$ **102.** $\frac{1}{4}s + \frac{3}{8} = \frac{5}{2}$

103. $\dfrac{v - 20}{-8} = 2v$ **104.** $x + \dfrac{2x}{5} = 1$

In Exercises 105 and 106, solve the equation for the specified variable.

105. $V = \pi r^2 h$, solve for h.

106. $S = 2\pi r^2 + 2\pi rh$, solve for h.

In Exercises 107–114, solve the equation by factoring.

107. $10x(x - 3) = 0$ **108.** $3x(4x + 7) = 0$

109. $v^2 - 100 = 0$ **110.** $(x + 3)^2 - 25 = 0$

111. $3s^3 - 2s - 8 = 0$

112. $2y^3 + 2y^2 - 24y = 0$

113. $z(5 - z) + 36 = 0$

114. $b^3 - 6b^2 - b + 6 = 0$

In Exercises 115–126, solve the equation.

115. $8 - 5t = 20 + t$ **116.** $3y + 14 = y + 20$

117. $3y^2 - 48 = 0$ **118.** $x^2 - 121 = 0$

119. $7 + 13x - 2x^2 = 0$

120. $6[x - (5x - 7)] = 4 - 5x$

121. $2(x + 7) - 9 = 5(x - 4)$

122. $11 + 32y - 3y^2 = 0$

123. $\frac{1}{3}x + 1 = \frac{1}{12}x - 4$ **124.** $1.2(x - 3) = 10.8$

125. $(x - 4)(x + 5) = 10$ **126.** $(u - 8)(u + 10) = 40$

127. *Free-Falling Object*   An object is thrown upward from a height of 48 feet with an initial velocity of 32 feet per second. Find the time t for the object to reach the ground by solving the equation

$$-16t^2 + 32t + 48 = 0.$$

128. *Calculator Experiment*   Use a calculator to calculate 14^4 in two ways.

Scientific	*Graphing*
(a) 14 ⎡y^x⎤ 4 ⎡=⎤	(a) 14 ⎡∧⎤ 4 ⎡ENTER⎤
(b) 14 ⎡x^2⎤ ⎡x^2⎤	(b) 14 ⎡x^2⎤ ⎡x^2⎤ ⎡ENTER⎤

Why do these methods give the same result?

129. *Geometry*   The width of a rectangle is three-fourths of its length. Find the dimensions of the rectangle if its area is 432 square inches.

130. The figure shows a square with sides of length x, within which is a smaller square with sides of length y.

(a) Remove the smaller square from the larger square. What is the area of the remaining figure?

(b) After removing the smaller square, slide and rotate the remaining top rectangle so that it fits against the right side of the figure. What are the dimensions of the resulting rectangle and what special product formula have you demonstrated geometrically?

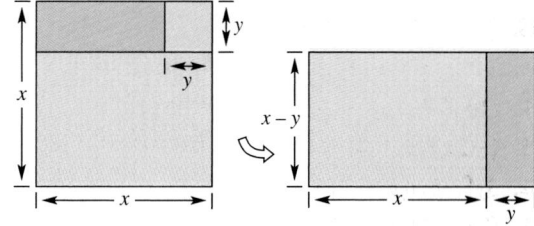

The symbol ▦ indicates an exercise in which you are instructed to use a calculator or graphing utility. The solutions of other exercises may also be facilitated by the use of appropriate technology.

Chapter Test

Take this test as you would take a test in class. After you are done, check your work against the answers given in the back of the book.

In Exercises 1–4, evaluate the expression.

1. $\frac{2}{3} + \left(-\frac{7}{6}\right)$

2. $\frac{5}{18} \div \frac{15}{8}$

3. $\left(-\frac{3}{5}\right)^3$

4. $\sqrt{25} + 3(36 \div 18)$

5. Name the property of real numbers demonstrated by $(-3 \cdot 5) \cdot 6 = -3(5 \cdot 6)$.

6. Give the additive inverse of the quantity $5x$.

In Exercises 7–12, simplify the expression.

7. $(3x^2y)(-xy)^2$

8. $3x^2 - 2x - 5x^2 + 7x - 1$

9. $(16 - y^2) - (16 + 2y + y^2)$

10. $-2(2x^4 - 5) + 4x(x^3 + 2x - 1)$

11. $4t - [3t - (10t + 7)]$

12. $2y\left(\frac{y}{4}\right)^2$

In Exercises 13–16, multiply the polynomials and simplify.

13. $(2x - 3y)(x + 5y)$

14. $(2s - 3)(3s^2 - 4s + 7)$

15. $(4x - 3)^2$

16. $[4 - (a + b)][4 + (a + b)]$

In Exercises 17–20, factor completely.

17. $18y^2 - 12y$

18. $5x^3 - 10x^2 - 6x + 12$

19. $9u^2 - 6u + 1$

20. $6x^2 - 26x - 20$

In Exercises 21–26, solve the equation.

21. $6x - 5 = 19$

22. $15 - 7(1 - x) = 3(x + 8)$

23. $\frac{2x}{3} = \frac{x}{2} + 4$

24. $3y^2 - 5y = 12$

25. $(y + 2)^2 - 9 = 0$

26. $2x^3 + 10x^2 + 8x = 0$

Figure for 27

27. Find the area of the shaded region shown at the left.

28. Solve the literal equation $5a + 2b - 10 = 8a + 7$ for b.

Introduction to Graphs and Functions

Strategies for Success

SKILLS *When you have completed this chapter, make sure you are able to:*

○ Create graphs that represent real-life data
○ Sketch graphs of equations and find the *x*- and *y*-intercepts
○ Sketch graphs of lines using slopes and *y*-intercepts
○ Evaluate functions and find their domains
○ Identify and sketch transformations of graphs of functions

TOOLS *Use these study tools to master the skills above:*

○ Mid-Chapter Quiz   (page 145)
○ Chapter Summary   (page 176)
○ Review Exercises   (page 177)
○ Chapter Test   (page 181)
○ Cumulative Test   (page 182)

In Exercise 79 of Section 2.1, you will graphically represent the number of male and female sports participants in the United States.

2.1	**Describing Data Graphically**
	The Rectangular Coordinate System ▪ Comparing Different Types of Graphs ▪ Ordered Pairs as Solutions ▪ The Distance Formula

The Rectangular Coordinate System

Historical Note: Descartes introduced his analytical geometry in 1637. Most of the terminology we use comes from the Cartesian-coordinate system (linear, quadratic, etc.). Prior to this time there was not a distinct tie between algebra and geometry.

Just as you can represent real numbers by points on the real number line, you can represent **ordered pairs** of real numbers by points in a plane. This plane is called a **rectangular coordinate system** or the **Cartesian plane,** after the French mathematician René Descartes.

A rectangular coordinate system is formed by two real lines intersecting at right angles, as shown in Figure 2.1. The horizontal number line is usually called the **x-axis,** and the vertical number line is usually called the **y-axis.** (The plural of axis is *axes*.) The point of intersection of the two axes is called the **origin,** and the axes separate the plane into four regions called **quadrants.**

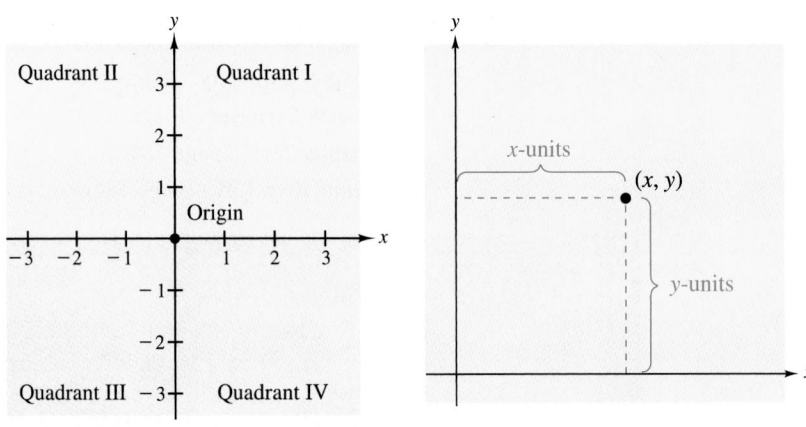

FIGURE 2.1   **FIGURE 2.2**

NOTE   The signs of the coordinates tell you which quadrant the point lies in. For instance, if x and y are positive, the point (x, y) lies in Quadrant I.

Each point in the plane corresponds to an *ordered pair* (x, y) of real numbers x and y, called the **coordinates** of the point. The first number (or **x-coordinate**) tells how far to the left or right the point is from the vertical axis, and the second number (or **y-coordinate**) tells how far up or down the point is from the horizontal axis, as shown in Figure 2.2.

Locating a given point in a plane is called **plotting** the point. Example 1 shows how this is done.

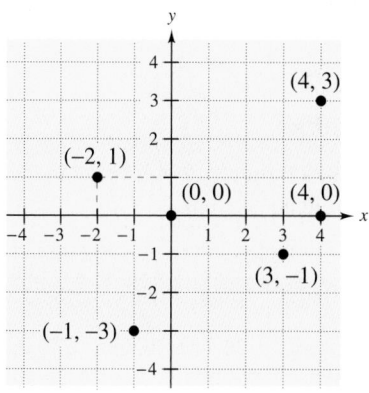

FIGURE 2.3

Point out the sign of *x* and *y* in each of the four quadrants. Discuss questions similar to: If $(-a, b)$ is in quadrant IV, in which quadrant would we find $(a, -b)$?

Students tend to think that $(-a, b)$ must be in quadrant II. This discussion will help them realize that "$-a$" does not necessarily represent a negative value.

EXAMPLE 1   *Plotting Points on a Rectangular Coordinate System*

Plot the points $(-2, 1)$, $(4, 0)$, $(3, -1)$, $(4, 3)$, $(0, 0)$, and $(-1, -3)$ on a rectangular coordinate system.

Solution

The point $(-2, 1)$ is two units to the *left* of the vertical axis and one unit *above* the horizontal axis.

Two units to the left of   One unit above the
the vertical axis   horizontal axis

$$(-2, 1)$$

Similarly, the point $(4, 0)$ is four units to the *right* of the vertical axis and *on* the horizontal axis. (It is on the horizontal axis because its *y*-coordinate is 0.) The other four points can be plotted in a similar way, as shown in Figure 2.3.

In Example 1, you were given the coordinates of several points and asked to plot the points on a rectangular coordinate system. Example 2 looks at the reverse problem. That is, you are given points on a rectangular coordinate system and are asked to determine their coordinates.

EXAMPLE 2   *Finding Coordinates of Points*

Determine the coordinates of each of the points shown in Figure 2.4.

Solution

Point *A* lies two units to the *right* of the vertical axis and one unit *below* the horizontal axis. Therefore, point *A* must be given by the ordered pair $(2, -1)$. The coordinates of the other four points can be determined in a similar way; the results are summarized as follows.

Point	Position	Coordinates
A	2 units *right*, 1 unit *down*	$(2, -1)$
B	1 unit *left*, 5 units *up*	$(-1, 5)$
C	0 units *right*, 2 units *up*	$(0, 2)$
D	3 units *left*, 2 units *down*	$(-3, -2)$
E	2 units *right*, 4 units *up*	$(2, 4)$

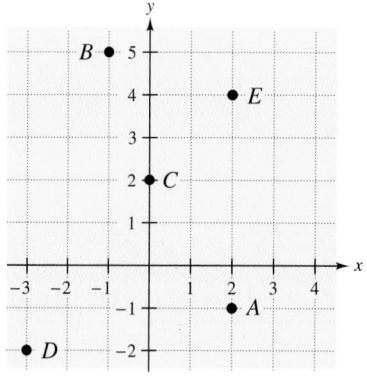

FIGURE 2.4

Notice that because point *C* lies on the *y*-axis, it has an *x*-coordinate of 0.

The primary value of a rectangular coordinate system is that it allows you to visualize relationships between two variables. Today, Descartes's ideas are commonly used in virtually every scientific and business-related field.

EXAMPLE 3   *Representing Data Graphically*

The population (in millions) of California from 1979 through 1994 is listed in the table. Plot these points on a rectangular coordinate system. *(Source: U.S. Bureau of Census)*

Year	1979	1980	1981	1982	1983	1984	1985	1986
Population	23.3	23.7	24.3	24.8	25.3	26.4	27.0	27.0

Year	1987	1988	1989	1990	1991	1992	1993	1994
Population	27.7	28.3	28.6	29.8	30.4	30.9	31.2	31.4

Solution

Begin by choosing which variable will be plotted on the horizontal axis and which will be plotted on the vertical axis. For this data, it seems natural to plot the years on the horizontal axis (which means that the population must be plotted on the vertical axis). Next, use the data in the table to form ordered pairs. For instance, the first three ordered pairs are (1979, 23.3), (1980, 23.7), and (1981, 24.3). All 16 points are shown in Figure 2.5. Note that the break in the x-axis indicates that the numbers between 0 and 1979 have been omitted. The break in the y-axis indicates that the numbers between 0 and 21 have been omitted.

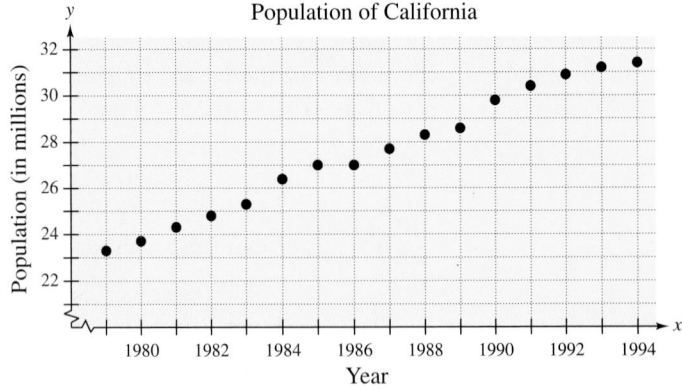

FIGURE 2.5

Comparing Different Types of Graphs

In Example 3, the data for the population of California was represented graphically by points on a rectangular coordinate system. This type of graph is called a **scatter plot.** This is only one of the many ways to represent data graphically. The same data could be represented by a **bar graph** or by a **line graph,** as shown in Figures 2.6 and 2.7.

NOTE When data is presented graphically without a table, the exact values are not evident. However, you can use the graph to estimate the values.

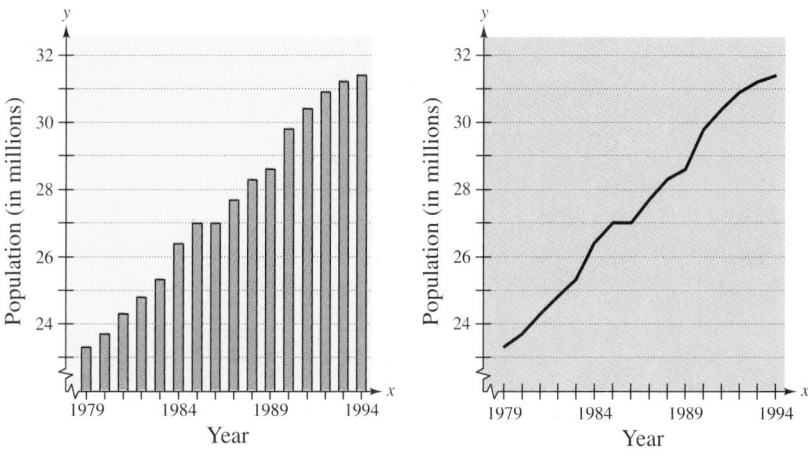

FIGURE 2.6 Bar graph   **FIGURE 2.7** Line graph

In addition to bar and line graphs, you can also use **line plots, circle graphs,** and several other types of graphical representations. Examples of a line plot and a circle graph are shown in Figures 2.8 and 2.9.

Study Tip

Line plots are used to organize data. For instance, the line plot in Figure 2.8 represents the digits in the telephone numbers of ten students in a class.

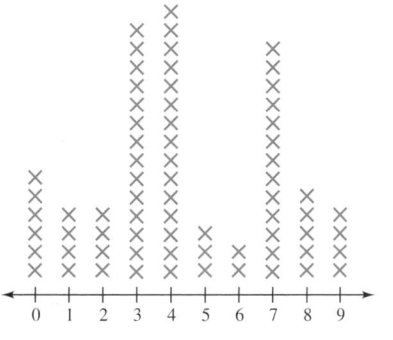

FIGURE 2.8 Line plot

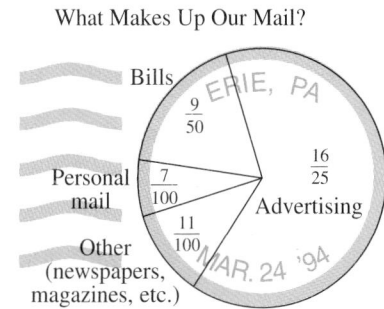

FIGURE 2.9 Circle graph

Ordered Pairs as Solutions

In Example 3, the relationship between the year and the population was given by a **table of values.** In mathematics, the relationship between the variables x and y is often given by an equation. From the equation, you can construct your own table of values.

EXAMPLE 4   Constructing a Table of Values

Construct a table of values for $y = 3x + 2$. Then plot the solution points on a rectangular coordinate system. Choose x-values of -3, -2, -1, 0, 1, 2, and 3.

Solution

For each x-value, you must calculate the corresponding y-value. For example, if you choose $x = 1$, the y-value is

$$y = 3(1) + 2 = 5.$$

The ordered pair $(x, y) = (1, 5)$ is a **solution point** (or **solution**) of the equation.

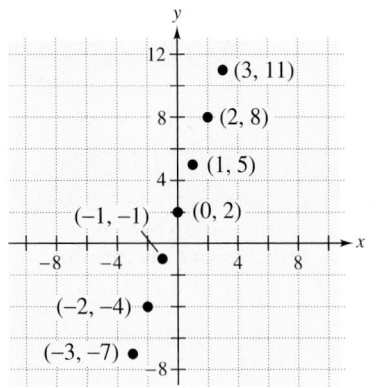

Choose x	Calculate y from $y = 3x + 2$	Solution point
$x = -3$	$y = 3(-3) + 2 = -7$	$(-3, -7)$
$x = -2$	$y = 3(-2) + 2 = -4$	$(-2, -4)$
$x = -1$	$y = 3(-1) + 2 = -1$	$(-1, -1)$
$x = 0$	$y = 3(0) + 2 = 2$	$(0, 2)$
$x = 1$	$y = 3(1) + 2 = 5$	$(1, 5)$
$x = 2$	$y = 3(2) + 2 = 8$	$(2, 8)$
$x = 3$	$y = 3(3) + 2 = 11$	$(3, 11)$

FIGURE 2.10

Once you have constructed a table of values, you can get a visual idea of the relationship between the variables x and y by plotting the solution points on a rectangular coordinate system, as shown in Figure 2.10.

When making a table of values for an equation, it is helpful first to solve the equation for y. Here is an example.

$$5x + 3y = 4 \qquad \text{Original equation}$$
$$3y = -5x + 4 \qquad \text{Subtract } 5x \text{ from both sides.}$$
$$y = -\frac{5}{3}x + \frac{4}{3} \qquad \text{Divide both sides by 3.}$$

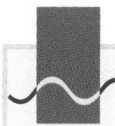

~ Technology

Creating a Table with a Graphing Utility

Some graphing utilities, such as the *TI–82* and *TI–83*, have built-in programs that can create tables. Suppose, for instance, that you want to create a table of values for the equation $2x^2 - 3y = 5$. To begin, solve the equation for y.

$$2x^2 - 3y = 5 \qquad \text{Original equation}$$
$$-3y = -2x^2 + 5 \qquad \text{Subtract } 2x^2 \text{ from both sides.}$$
$$y = \frac{2}{3}x^2 - \frac{5}{3} \qquad \text{Divide both sides by } -3.$$

Now, using the equation $y = \frac{2}{3}x^2 - \frac{5}{3}$, you can use the following steps.

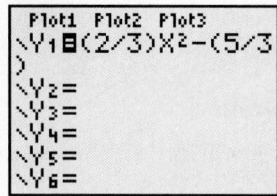

1. Y=

1. Press the $\boxed{Y=}$ key and enter the equation as Y1.

2. Press $\boxed{\text{2nd}}$ $\boxed{\text{TBLSET}}$ and enter the values of TblMin (or TblStart) and ΔTbl. The value of TblMin is the beginning x-value you want displayed in the table, and the value of ΔTbl is the increment for the x-values in the table. For instance, using TblMin=–3 and ΔTbl=1, the table has x-values of $-3, -2, -1$, and so on.

2. TblSet

3. Press $\boxed{\text{2nd}}$ $\boxed{\text{TABLE}}$ to obtain the table of values. You can use the cursor keys to view x- and y-values that are not shown on the default screen.

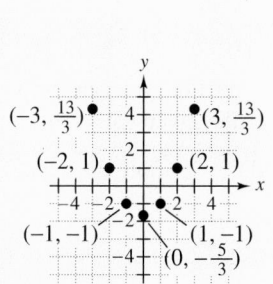

3. Table

Sample Screens from a *TI-83*.

You can confirm the values shown on the graphing utility's screen by substituting the appropriate x-values, as shown in the table below. (Note that the graphing utility lists decimal approximations of fractional values. For instance, $-\frac{5}{3}$ is listed as -1.667.)

Choose x	Calculate y from $y = \frac{2}{3}x^2 - \frac{5}{3}$	Solution point
$x = -3$	$y = \frac{2}{3}(-3)^2 - \frac{5}{3} = \frac{13}{3}$	$\left(-3, \frac{13}{3}\right)$
$x = -2$	$y = \frac{2}{3}(-2)^2 - \frac{5}{3} = 1$	$(-2, 1)$
$x = -1$	$y = \frac{2}{3}(-1)^2 - \frac{5}{3} = -1$	$(-1, -1)$
$x = 0$	$y = \frac{2}{3}(0)^2 - \frac{5}{3} = -\frac{5}{3}$	$\left(0, -\frac{5}{3}\right)$
$x = 1$	$y = \frac{2}{3}(1)^2 - \frac{5}{3} = -1$	$(1, -1)$
$x = 2$	$y = \frac{2}{3}(2)^2 - \frac{5}{3} = 1$	$(2, 1)$
$x = 3$	$y = \frac{2}{3}(3)^2 - \frac{5}{3} = \frac{13}{3}$	$\left(3, \frac{13}{3}\right)$

After creating the table, you can plot the points, as shown at the left.

In the next example, you are given several ordered pairs and are asked to determine whether they are solutions of the given equation. To do this, you need to substitute the values of x and y into the equation. If the substitution produces a true equation, the ordered pair (x, y) is a solution.

EXAMPLE 5 Verifying Solutions of an Equation

NOTE When a substitution of (x, y) produces a true equation, the ordered pair (x, y) is said to **satisfy** the equation.

Which of the ordered pairs are solutions of $x^2 - 2y = 6$?

(a) $(2, 1)$ (b) $(0, -3)$ (c) $(-2, -5)$ (d) $\left(1, -\frac{5}{2}\right)$

Solution

(a) For the ordered pair $(2, 1)$, substitute $x = 2$ and $y = 1$ into the equation.

$$x^2 - 2y = 6 \qquad \text{Original equation}$$
$$(2)^2 - 2(1) \overset{?}{=} 6 \qquad \text{Substitute 2 for } x \text{ and 1 for } y.$$
$$2 \neq 6 \qquad \text{Is not a solution} \ ✗$$

Because the substitution does not satisfy the given equation, you can conclude that the ordered pair $(2, 1)$ *is not* a solution of the given equation.

(b) For the ordered pair $(0, -3)$, substitute $x = 0$ and $y = -3$ into the equation.

$$x^2 - 2y = 6 \qquad \text{Original equation}$$
$$(0)^2 - 2(-3) \overset{?}{=} 6 \qquad \text{Substitute 0 for } x \text{ and } -3 \text{ for } y.$$
$$6 = 6 \qquad \text{Is a solution} \ ✓$$

Because the substitution satisfies the given equation, you can conclude that the ordered pair $(0, -3)$ *is* a solution of the given equation.

(c) For the ordered pair $(-2, -5)$, substitute $x = -2$ and $y = -5$ into the equation.

$$x^2 - 2y = 6 \qquad \text{Original equation}$$
$$(-2)^2 - 2(-5) \overset{?}{=} 6 \qquad \text{Substitute } -2 \text{ for } x \text{ and } -5 \text{ for } y.$$
$$14 \neq 6 \qquad \text{Is not a solution} \ ✗$$

Because the substitution does not satisfy the given equation, you can conclude that the ordered pair $(-2, -5)$ *is not* a solution of the given equation.

(d) For the ordered pair $\left(1, -\frac{5}{2}\right)$, substitute $x = 1$ and $y = -\frac{5}{2}$ into the equation.

$$x^2 - 2y = 6 \qquad \text{Original equation}$$
$$(1)^2 - 2\left(-\frac{5}{2}\right) \overset{?}{=} 6 \qquad \text{Substitute 1 for } x \text{ and } -\frac{5}{2} \text{ for } y.$$
$$6 = 6 \qquad \text{Is a solution} \ ✓$$

Because the substitution satisfies the given equation, you can conclude that the ordered pair $\left(1, -\frac{5}{2}\right)$ *is* a solution of the given equation.

The Distance Formula

You know from Section 1.1 that the distance d between two points a and b on the real number line is simply $d = |b - a|$. The same "absolute value rule" is used to find the distance between two points that lie on the same *vertical or horizontal line*, as shown in Example 6.

EXAMPLE 6   *Finding Horizontal and Vertical Distances*

(a) Find the distance between the points $(2, -2)$ and $(2, 4)$.

(b) Find the distance between the points $(-3, -2)$ and $(2, -2)$.

Solution

(a) Because the x-coordinates are equal, you can visualize a vertical line through the points $(2, -2)$ and $(2, 4)$, as shown in Figure 2.11. The distance between these two points is the absolute value of the difference of their y-coordinates.

$$\text{Vertical distance} = |4 - (-2)| \qquad \text{Subtract } y\text{-coordinates.}$$
$$= 6 \qquad \text{Evaluate absolute value.}$$

(b) Because the y-coordinates are equal, you can visualize a horizontal line through the points $(-3, -2)$ and $(2, -2)$, as shown in Figure 2.11. The distance between these two points is the absolute value of the difference of their x-coordinates.

$$\text{Horizontal distance} = |2 - (-3)| \qquad \text{Subtract } x\text{-coordinates.}$$
$$= 5 \qquad \text{Evaluate absolute value.}$$

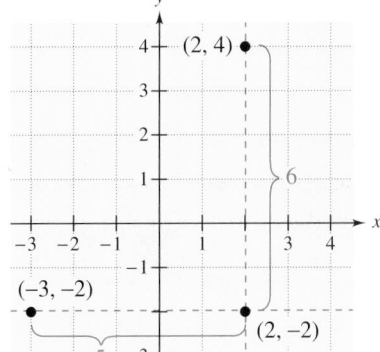

FIGURE 2.11

NOTE   In Figure 2.11, note that the horizontal distance between the points $(-3, -2)$ and $(2, -2)$ is the absolute value of the difference of the x-coordinates, and the vertical distance between the points $(2, -2)$ and $(2, 4)$ is the absolute value of the difference of the y-coordinates.

The technique applied in Example 6 can be used to develop a general formula for finding the distance between two points in the plane. This general formula will work for any two points, even if they do not lie on the same vertical or horizontal line. To develop the formula, you use the **Pythagorean Theorem,** which states that for a right triangle, the hypotenuse c and sides a and b are related by the formula

$$a^2 + b^2 = c^2, \qquad \text{Pythagorean Theorem}$$

as shown in Figure 2.12. (The converse is also true. That is, if $a^2 + b^2 = c^2$, the triangle is a right triangle.)

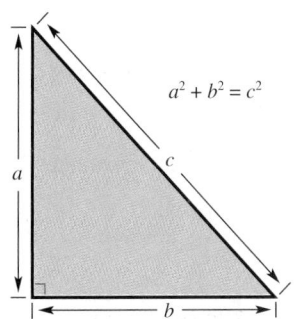

FIGURE 2.12   Pythagorean Theorem

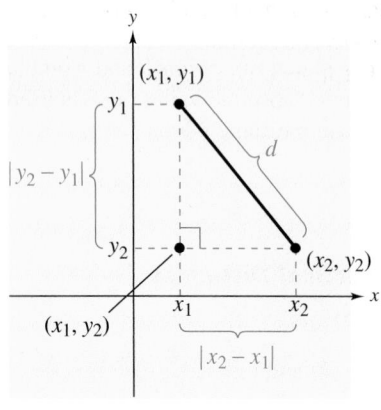

FIGURE 2.13 Distance between two points

To develop a general formula for the distance between two points, let (x_1, y_1) and (x_2, y_2) represent two points in the plane (that do not lie on the same horizontal or vertical line). With these two points, a right triangle can be formed, as shown in Figure 2.13. Note that the third vertex of the triangle is (x_1, y_2). Because (x_1, y_1) and (x_1, y_2) lie on the same vertical line, the length of the vertical side of the triangle is $|y_2 - y_1|$. Similarly, the length of the horizontal side is $|x_2 - x_1|$. By the Pythagorean Theorem, the square of the distance between (x_1, y_1) and (x_2, y_2) is

$$d^2 = |x_2 - x_1|^2 + |y_2 - y_1|^2.$$

Because the distance d must be positive, you can choose the positive square root and write

$$d = \sqrt{|x_2 - x_1|^2 + |y_2 - y_1|^2}.$$

Finally, replacing $|x_2 - x_1|^2$ and $|y_2 - y_1|^2$ by the equivalent expressions $(x_2 - x_1)^2$ and $(y_2 - y_1)^2$ gives you the **Distance Formula.**

Exploration

Plot the points $A(-1, -3)$ and $B(5, 2)$ and sketch the line segment from A to B. How could you verify that point $C(2, -0.5)$ is the midpoint of the segment? Why is it not sufficient to show that the distances from A to C and from C to B are equal? Find a formula for the coordinates of the midpoint of the line segment connecting (x_1, y_1) and (x_2, y_2).

The Distance Formula

The distance d between two points (x_1, y_1) and (x_2, y_2) is

$$d = \sqrt{(x_2 - x_1)^2 + (y_2 - y_1)^2}.$$

Note that for the special case in which the two points lie on the same vertical or horizontal line, the Distance Formula still works. For instance, applying the Distance Formula to the points $(2, -2)$ and $(2, 4)$ produces

$$d = \sqrt{(2 - 2)^2 + [4 - (-2)]^2} = \sqrt{6^2} = 6,$$

which is the same result obtained in Example 6.

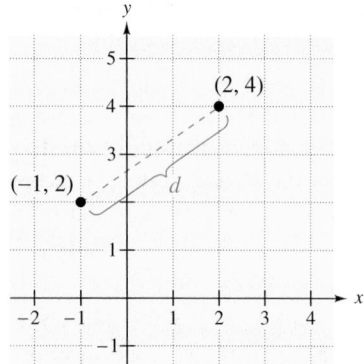

FIGURE 2.14

EXAMPLE 7 *Finding the Distance Between Two Points*

Find the distance between the points $(-1, 2)$ and $(2, 4)$, as shown in Figure 2.14.

Solution

Let $(x_1, y_1) = (-1, 2)$ and $(x_2, y_2) = (2, 4)$, and apply the Distance Formula.

$$d = \sqrt{[2 - (-1)]^2 + (4 - 2)^2} \qquad \text{Substitute coordinates of points.}$$
$$= \sqrt{3^2 + 2^2} \qquad \text{Simplify.}$$
$$= \sqrt{13} \qquad \text{Simplify.}$$
$$\approx 3.61 \qquad \text{Use a calculator.}$$

Encourage students to follow the development of the Distance Formula and to understand *why* the formula works. This understanding is critical to many areas in algebra and geometry.

The Distance Formula has many applications in mathematics. For instance, the next example shows how you can use the Distance Formula and the converse of the Pythagorean Theorem to decide whether three points form the vertices of a right triangle.

EXAMPLE 8   *An Application of the Distance Formula*

Show that the points $(1, 2)$, $(3, 1)$, and $(4, 3)$ are vertices of a right triangle.

Solution

The three points are plotted in Figure 2.15. Using the Distance Formula, you can find the lengths of the three sides of the triangle.

$$d_1 = \sqrt{(3 - 1)^2 + (1 - 2)^2} = \sqrt{4 + 1} = \sqrt{5}$$
$$d_2 = \sqrt{(4 - 3)^2 + (3 - 1)^2} = \sqrt{1 + 4} = \sqrt{5}$$
$$d_3 = \sqrt{(4 - 1)^2 + (3 - 2)^2} = \sqrt{9 + 1} = \sqrt{10}$$

Because

$$d_1^2 + d_2^2 = 5 + 5$$
$$= 10$$
$$= d_3^2,$$

you can conclude from the converse of the Pythagorean Theorem that the triangle is a right triangle.

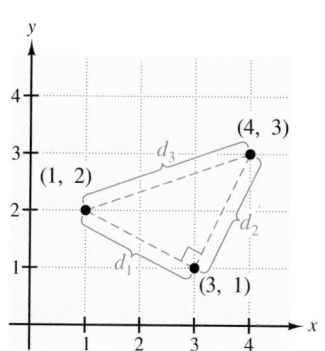

FIGURE 2.15

Group Activities

Extending the Concept

Determining Collinearity   Three or more points are **collinear** if they all lie on the same line. Use the following steps to determine if the set of points $\{A(3, 1), B(5, 4), C(9, 10)\}$ and the set of points $\{A(2, 2), B(4, 3), C(5, 4)\}$ are collinear.

a. For each set of points, use the Distance Formula to find the distances from A to B, from B to C, and from A to C. What relationship exists among these distances for each set of points?

b. Plot each set of points on a rectangular coordinate system. Do all the points of either set appear to lie on the same line?

c. Compare your conclusions from part (a) with the conclusions you made from the graphs in part (b). Make a general statement about how to use the Distance Formula to determine collinearity.

2.1 Exercises

See Warm-Up Exercises, p. A40

1. *Think About It* Discuss the significance of the word *order* when referring to an ordered pair (x, y).

2. *Think About It* What is the x-coordinate of any point on the y-axis? What is the y-coordinate of any point on the x-axis?

In Exercises 3–6, plot the points on a rectangular coordinate system.

3. $(4, 3), (-5, 3), (3, -5)$

4. $(-2, 5), (-2, -5), (3, 5)$

5. $\left(\frac{5}{2}, -2\right), \left(-2, \frac{1}{4}\right), \left(\frac{3}{2}, -\frac{7}{2}\right)$

6. $\left(-\frac{2}{3}, 3\right), \left(\frac{1}{4}, -\frac{5}{4}\right), \left(-5, -\frac{7}{4}\right)$

In Exercises 7 and 8, approximate the coordinates of the points.

7.
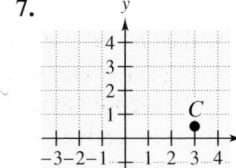

8.
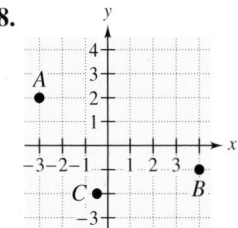

In Exercises 9–12, plot the points and connect them with line segments to form the figure. (*Note:* A *rhombus* is a parallelogram whose sides are all the same length.)

9. *Triangle:* $(-1, 2), (2, 0), (3, 5)$

10. *Rectangle:* $(7, 0), (9, 1), (4, 6), (6, 7)$

11. *Parallelogram:* $(4, 0), (6, -2), (0, -4), (-2, -2)$

12. *Rhombus:* $(-3, -3), (-2, -1), (-1, -2), (0, 0)$

In Exercises 13–18, determine the quadrant(s) in which the point is located. Explain.

13. $(-3, -5)$

14. $(4, -2)$

15. $(x, 4)$

16. $(-10, y)$

17. $(x, y), \quad xy < 0$

18. $(x, y), \quad x > 0, y > 0$

In Exercises 19–22, find the coordinates of the point.

19. The point is located five units to the left of the y-axis and two units above the x-axis.

20. The point is located ten units to the right of the y-axis and four units below the x-axis.

21. The point is on the positive x-axis ten units from the origin.

22. The point is on the negative y-axis five units from the origin.

In Exercises 23–26, plot the points whose coordinates are given in the table.

23. *Exam Score* The table gives the times x in hours invested in studying for five different algebra exams and the resulting exam scores y.

x	5	2	3	6.5	4
y	81	71	88	92	86

24. *Price of Stock* The year-end market prices y per share of common stock in PepsiCo, Inc. for the years 1987 through 1993 are given in the table. The years are given by x. (*Source: PepsiCo, Inc. 1993 Annual Report*)

x	1987	1988	1989	1990	1991	1992	1993
y	$\$11\frac{1}{4}$	$\$13\frac{1}{8}$	$\$21\frac{3}{8}$	$\$25\frac{3}{4}$	$\$33\frac{3}{4}$	$\$42\frac{1}{4}$	$\$41\frac{7}{8}$

25. *Fuel Efficiency* The table gives various speeds x of a car in miles per hour and the corresponding approximate fuel efficiencies y in miles per gallon.

x	50	55	60	65	70
y	28	26.4	24.8	23.4	22

26. *Average Temperature* The table gives the average temperature y (degrees Fahrenheit) for Duluth, Minnesota, for each month of a year, with $x = 1$ representing January. *(Source: NOAA)*

x	1	2	3	4	5	6
y	7.0	12.3	24.4	38.6	50.8	59.8

x	7	8	9	10	11	12
y	66.1	63.7	54.2	43.7	28.4	12.8

In Exercises 27–32, determine whether the ordered pairs are solutions of the equation.

27. $y = 3x + 8$
  (a) $(3, 17)$ (b) $(-1, 10)$
  (c) $(0, 0)$ (d) $(-2, 2)$

28. $5x - 2y + 50 = 0$
  (a) $(-10, 0)$ (b) $(-5, 5)$
  (c) $(0, 25)$ (d) $(20, -2)$

29. $y = \frac{7}{8}x$
  (a) $\left(\frac{8}{7}, 1\right)$ (b) $\left(4, \frac{7}{2}\right)$
  (c) $(0, 0)$ (d) $(-16, 14)$

30. $y = \frac{5}{8}x - 2$
  (a) $(0, 0)$ (b) $(2, 2)$
  (c) $(-4, -7)$ (d) $(32, 49)$

31. $4y - 2x + 1 = 0$
  (a) $(0, 0)$ (b) $\left(\frac{1}{2}, 0\right)$
  (c) $\left(-3, -\frac{7}{4}\right)$ (d) $\left(1, -\frac{3}{4}\right)$

32. $y = 10x - 7$
  (a) $(2, 10)$ (b) $(-2, -27)$
  (c) $(5, 43)$ (d) $(1, 5)$

In Exercises 33–36, complete the table of values. Then plot the solution points on a rectangular coordinate system.

33.

x	-2	0	2	4	6
$y = 5x - 1$					

34.

x	-2	0	2	4	6
$y = \frac{3}{4}x + 2$					

35.

x	-4	$\frac{2}{5}$	4	8	12
$y = -\frac{5}{2}x + 4$					

36.

x	-6	-3	0	$\frac{3}{4}$	10
$y = \frac{4}{3}x - \frac{1}{3}$					

In Exercises 37 and 38, use the table feature of a graphing utility to complete the table of values.

37.

x	-2	0	2	4	6
$y = 4x^2 + x - 2$					

38.

x	-2	0	2	4	6
$y = \frac{4}{3}x - \frac{1}{3}$					

39. *Making a Conjecture* Plot the points $(2, 1)$, $(-3, 5)$, and $(7, -3)$ on a rectangular coordinate system. Then change the sign of the x-coordinate of each point and plot the three new points on the same rectangular coordinate system. What conjecture can you make about the location of a point when the sign of its x-coordinate is changed?

40. *Making a Conjecture* Plot the points $(2, 1)$, $(-3, 5)$, and $(7, -3)$ on a rectangular coordinate system. Then change the sign of the y-coordinate of each point and plot the three new points on the same rectangular coordinate system. What conjecture can you make about the location of a point when the sign of its y-coordinate is changed?

41. *Numerical Interpretation* The cost C of producing x units is given by $C = 28x + 3000$. Use a table to help write a description of the relationship between x and C.

42. *Numerical Interpretation* When an employee produces x units per hour, the hourly wage y is given by $y = 0.75x + 8$. Use a table to help write a description of the relationship between x and y.

The symbol ⊞ indicates an exercise in which you are instructed to use a calculator or graphing utility. The solutions of other exercises may also be facilitated by the use of appropriate technology.

Shifting a Graph   In Exercises 43 and 44, the figure is shifted to a new location in the plane. Find the coordinates of the vertices of the figure in its new location.

43.

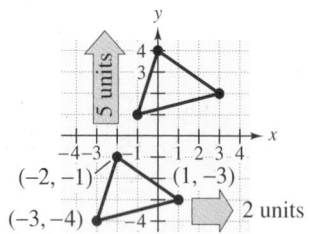

44.

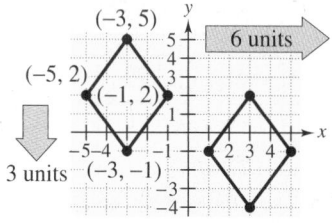

In Exercises 45–52, plot the points and find the distance between them. State whether the points lie on a horizontal or a vertical line.

45. $(3, -2), (3, 5)$

46. $(-2, 8), (-2, 1)$

47. $\left(\frac{1}{2}, \frac{7}{8}\right), \left(\frac{11}{2}, \frac{7}{8}\right)$

48. $\left(\frac{3}{4}, 1\right), \left(\frac{3}{4}, -10\right)$

49. $(3, 2), (10, 2)$

50. $(-120, -2), (130, -2)$

51. $\left(-3, \frac{3}{2}\right), \left(-3, \frac{9}{4}\right)$

52. $\left(\frac{1}{3}, -4\right), \left(\frac{1}{3}, \frac{5}{2}\right)$

In Exercises 53–60, find the distance between the points.

53. $(1, 3), (5, 6)$

54. $(3, 10), (15, 5)$

55. $(0, 0), (12, -9)$

56. $(-5, 0), (3, 15)$

57. $(-2, -3), (4, 2)$

58. $(-5, 4), (10, -3)$

59. $(1, 3), (3, -2)$

60. $\left(\frac{1}{2}, 1\right), \left(\frac{3}{2}, 2\right)$

In Exercises 61–64, determine whether the points are vertices of a right triangle.

61. $(2, 3), (2, 6), (6, 3)$

62. $(2, 4), (1, 1), (7, -1)$

63. $(8, 3), (5, 2), (1, 9)$

64. $(2, 4), (-1, 6), (-3, 1)$

Geometry   In Exercises 65 and 66, find the perimeter of the triangle with the given vertices.

65. $(-2, 0), (0, 5), (1, 0)$

66. $(-5, -2), (-1, 4), (3, -1)$

Midpoint   In Exercises 67–70, plot the points and the midpoint of the line segment joining the points. The coordinates of the midpoint of the line segment joining the points (x_1, y_1) and (x_2, y_2) are

$$\text{Midpoint} = \left(\frac{x_1 + x_2}{2}, \frac{y_1 + y_2}{2}\right).$$

67. $(-2, 0), (4, 8)$

68. $(-3, -2), (7, 2)$

69. $(1, 6), (6, 3)$

70. $(2, 7), (9, -1)$

Quiz Scores   In Exercises 71 and 72, use the following scores from a math class with 30 students. The scores are for two 25-point quizzes.

Quiz #1   20, 15, 14, 20, 16, 19, 10, 21, 24, 15, 15, 14, 15, 21, 19, 15, 20, 18, 18, 22, 18, 16, 18, 19, 21, 19, 16, 20, 14, 12

Quiz #2   22, 22, 23, 22, 21, 24, 22, 19, 21, 23, 23, 25, 24, 22, 22, 23, 23, 23, 23, 22, 24, 23, 22, 24, 21, 24, 16, 21, 16, 14

71. Construct a line plot for quiz #1. Which score occurred with the greatest frequency?

72. Construct a line plot for quiz #2. Which score occurred with the greatest frequency?

Exam Scores In Exercises 73 and 74, use the following scores from a math class with 30 students. The scores are for two 100-point exams.

Exam #1 77, 100, 77, 70, 83, 89, 87, 85, 81, 84, 81, 78, 89, 78, 88, 85, 90, 92, 75, 81, 85, 100, 98, 81, 78, 75, 85, 89, 82, 75

Exam #2 76, 78, 73, 59, 70, 81, 71, 66, 66, 73, 68, 67, 63, 67, 77, 84, 87, 71, 78, 78, 90, 80, 77, 70, 80, 64, 74, 68, 68, 68

73. Construct a line plot for exam #1. Which score occurred with the greatest frequency?

74. Construct a line plot for exam #2. Which score occurred with the greatest frequency?

75. *Recreational Vehicles* The lists give the numbers (in thousands) of recreational vehicles sold and their retail values (in millions of dollars) in the United States for the year 1992. Organize the data in each list graphically. *(Source: Recreation Vehicle Industry Association)*

Total number

Motorized homes	226.3
Travel trailers	102.5
Folding camping trailers	43.3
Truck campers	10.6

Retail Value

Motorized homes	6963
Travel trailers	1523
Folding camping trailers	189
Truck campers	99

76. *Travel to the United States* The places of origin and the numbers of travelers (in millions) to the United States in 1993 are as follows: Canada, 17.3; Mexico, 9.8; Europe, 8.6; Latin America, 3.7; Other, 6.3. Construct a horizontal bar graph for the data. *(Source: U.S. Travel and Tourism Administration)*

77. *Snowfall* The list gives the seasonal snowfall (in inches) at Erie, Pennsylvania, starting with the 1960/61 winter and ending with the 1994/95 winter. The amounts are listed in order by year. Organize the data graphically. *(Source: National Oceanic and Atmospheric Association)*

69.6, 42.5, 75.9, 115.9, 92.9, 84.8, 68.6, 107.9, 79.7, 85.6, 120.0, 92.3, 53.7, 68.6, 66.7, 66.0, 111.5, 142.8, 76.5, 55.2, 89.4, 71.3, 41.2, 110.0, 106.3, 124.9, 68.2, 103.5, 76.5, 114.9, 59.6, 104.8, 108.5, 131.3, 53.7

78. *Fruit Crops* The list gives the cash receipts (in millions of dollars) from fruit crops for farmers in 1993. Construct a bar graph for the data. *(Source: U.S. Department of Agriculture)*

Apples	1364	Oranges	1337
Cherries	221	Peaches	398
Grapefruit	256	Pears	247
Grapes	2000	Plums and Prunes	264
Lemons	178	Strawberries	747

79. *Sports Participants* The list gives the numbers of males and females (in millions) over the age of seven who participated in popular sports activities in the year 1993 in the United States. Construct a double bar graph for the data. *(Source: National Sporting Goods Association)*

Activity	Male	Female
Exercise walking	21.1	43.4
Swimming	27.7	33.6
Bicycling	24.6	23.4
Camping	23.2	19.5
Bowling	20.7	20.6
Basketball	21.3	8.3
Running	11.4	8.9
Aerobic exercising	3.5	21.4

80. *College Attendance* The table shows the enrollments in a liberal arts college for the years 1988 through 1995. Construct a line graph for the data.

Year	1988	1989	1990	1991
Enrollment	1675	1704	1710	1768

Year	1992	1993	1994	1995
Enrollment	1833	1918	1967	1972

81. *Oil Imports* The table shows the amounts of crude oil imported into the United States (in millions of barrels) for the years 1985 through 1994. Construct a line graph for the data and state what information the graph reveals. *(Source: U.S. Energy Information Administration)*

Year	1985	1986	1987	1988	1989
Oil imports	1168	1525	1706	1864	2133

Year	1990	1991	1992	1993	1994
Oil imports	2151	2110	2226	2477	2565

82. *Federal Income* The list gives the receipts (in billions of dollars) for the federal government of the United States for the years 1972 through 1994. Construct a line graph for the data. What can you conclude? *(Source: U.S. Office of Management and Budget)*

1972	207.3	1983	600.6
1973	230.8	1984	666.5
1974	263.2	1985	734.1
1975	279.1	1986	769.1
1976	298.1	1987	854.1
1977	355.6	1988	909.0
1978	399.6	1989	990.7
1979	463.3	1990	1031.3
1980	517.1	1991	1054.3
1981	599.3	1992	1090.5
1982	617.8	1993	1153.5
		1994	1257.7

83. *Housing Construction* A house is 30 feet wide and the ridge of the roof is 7 feet above the tops of the walls (see figure). Find the length of the rafters if they overhang the edges of the walls by 2 feet.

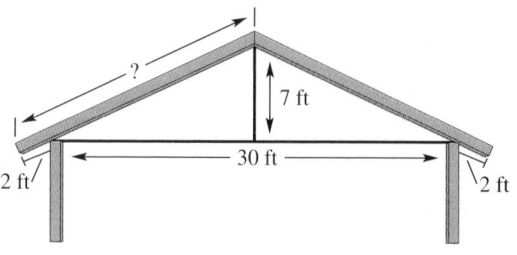

Section Project

Find the coordinates of (x, y), the lengths of the vertical and horizontal sides of the right triangle, and the length of the hypotenuse.

(a)

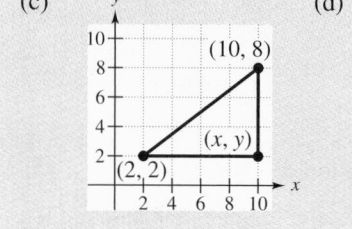

(b)

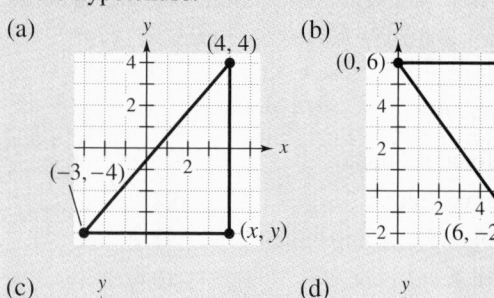

(c)

(d)

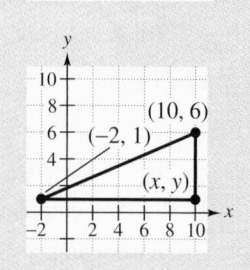

2.2 | **Graphs of Equations**

The Graph of an Equation • Intercepts: An Aid to Sketching Graphs • The Connection Between Solutions and *x*-Intercepts • Real-Life Application of Graphs

The Graph of an Equation

You might begin this section by noting that when we graph *equations involving two variables, we are looking for the set of all solutions—as well as for trends and a visual representation of the relationship.*

In Section 2.1, you saw that the solutions of an equation in x and y can be represented by points on a rectangular coordinate system. The set of *all* solution points of an equation is called its **graph.** In this section, you will study a basic technique for sketching the graph of an equation—the **point-plotting method.**

EXAMPLE 1   *Sketching the Graph of an Equation*

Sketch the graph of $3x - y = 2$.

Solution

*NOTE   The equation in Example 1 is an example of a **linear equation** in two variables—it is of first degree in both variables and its graph is a straight line. You will study this type of equation in Section 2.3.*

To begin, solve the equation for y to obtain $y = 3x - 2$. Next, create a table of values. The choice of x-values to use in the table is somewhat arbitrary. However, the more x-values you choose, the easier it will be to recognize a pattern.

x	-2	-1	0	1	2	3
$y = 3x - 2$	-8	-5	-2	1	4	7
Solution point	$(-2, -8)$	$(-1, -5)$	$(0, -2)$	$(1, 1)$	$(2, 4)$	$(3, 7)$

Now, plot the points, as shown in Figure 2.16(a). It appears that all six points lie on a line, so complete the sketch by drawing a line through the points, as shown in Figure 2.16(b).

As mentioned in Example 1, the choice of x-values is arbitrary. Emphasize to students that if they chose different values for x, they would still generate the same line. Have them re-graph y = 3x − 2 by putting x = 4, 5, 6, 7, 8 in their tables. After they have finished, ask them to compare their graphs to the graph in Figure 2.16.

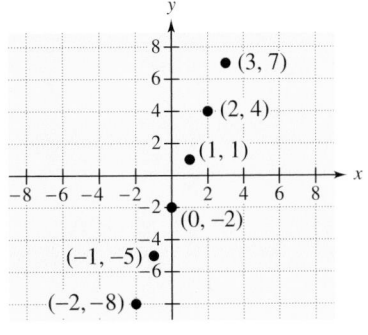

(a)

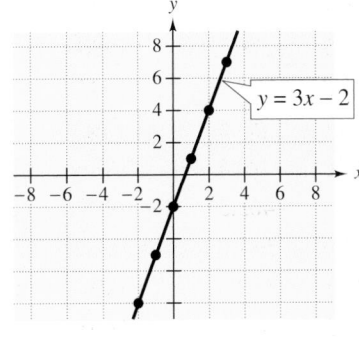

(b)

FIGURE 2.16

NOTE By drawing a line (curve) through the plotted points, we are implying that every point on this line (curve) is a solution point of the given equation and, conversely, that every solution point is on the line (curve).

The Point-Plotting Method of Sketching a Graph

1. If possible, rewrite the equation by isolating one of the variables.

2. Make a table of values showing several solution points.

3. Plot these points on a rectangular coordinate system.

4. Connect the points with a smooth curve or line.

EXAMPLE 2   Sketching the Graph of a Nonlinear Equation

Sketch the graph of $-x^2 + 2x + y = 0$.

Solution

Begin by solving the equation for y to obtain $y = x^2 - 2x$. Next, create a table of values.

x	-2	-1	0	1	2	3	4
$y = x^2 - 2x$	8	3	0	-1	0	3	8
Solution point	$(-2, 8)$	$(-1, 3)$	$(0, 0)$	$(1, -1)$	$(2, 0)$	$(3, 3)$	$(4, 8)$

Now plot the seven solution points, as shown in Figure 2.17(a). Finally, connect the points with a smooth curve, as shown in Figure 2.17(b).

Study Tip

It is possible to have an equation with only one variable—for example, $y = 3$. When you make a table of values for the graph of $y = 3$, you will notice that y is 3 regardless of the value of x. This produces a horizontal line, as shown below.

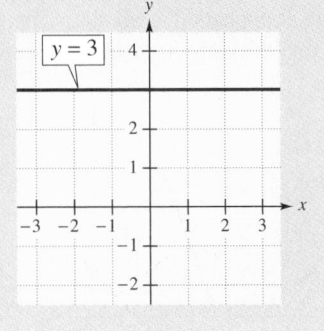

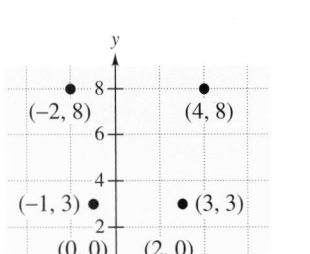

(a)

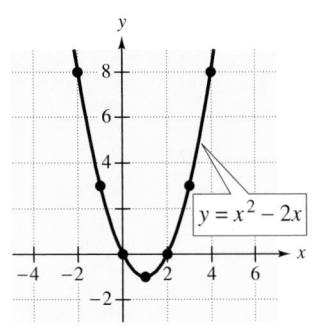

(b)

FIGURE 2.17

NOTE The graph of the equation given in Example 2 is called a **parabola.** You will study this type of graph in detail in Section 6.5.

Example 3 looks at the graph of an equation that involves an absolute value. Remember that to find the absolute value of a number you disregard the sign of the number. For instance,

$$|-5| = 5, \quad |2| = 2, \quad \text{and} \quad |0| = 0.$$

EXAMPLE 3   *The Graph of an Absolute Value Equation*

Sketch the graph of

$$y = |x - 2|.$$

Solution

This equation is already written in a form with y isolated on the left. So begin by creating a table of values. Be sure that you understand how the absolute value is evaluated. For instance, when $x = -2$, the value of y is

$$y = |-2 - 2| = |-4| = 4,$$

and when $x = 3$, the value of y is

$$y = |3 - 2| = |1| = 1.$$

x	-2	-1	0	1	2	3	4	5
$y = |x - 2|$	4	3	2	1	0	1	2	3
Solution point	$(-2, 4)$	$(-1, 3)$	$(0, 2)$	$(1, 1)$	$(2, 0)$	$(3, 1)$	$(4, 2)$	$(5, 3)$

Next, plot the points, as shown in Figure 2.18(a). It appears that the points lie in a "V-shaped" pattern, with the point $(2, 0)$ lying at the bottom of the "V." Following this pattern, connect the points to form the graph shown in Figure 2.18(b).

Remind students that a graph is a visual representation of a relationship.

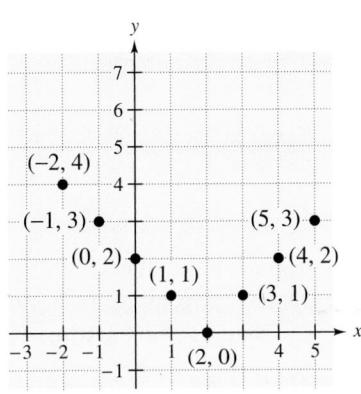

(a)

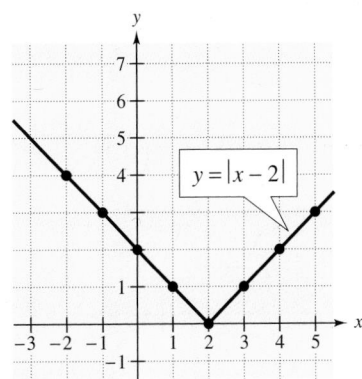

(b)

FIGURE 2.18

Technology

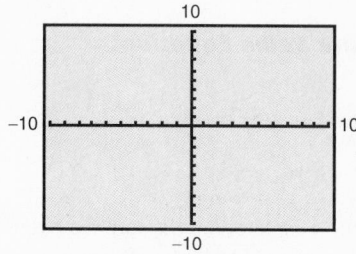

Standard viewing rectangle

A **viewing rectangle** for a graph is a rectangular portion of the coordinate plane. A viewing rectangle is determined by six values: the minimum x-value, the maximum x-value, the x-scale, the minimum y-value, the maximum y-value, and the y-scale. When you enter these six values into a graphing utility, you are setting the **range** or **window.** Some graphing utilities have a standard viewing rectangle, as shown at the left.

By choosing different viewing rectangles for a graph, it is possible to obtain very different impressions of the graph's shape. For instance, below are four different viewing rectangles for the graph of

$$y = -x^2 + 12x - 6.$$

Of these, the view in part (d) is most complete.

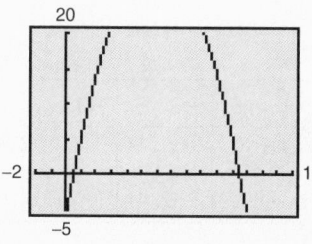

(a)

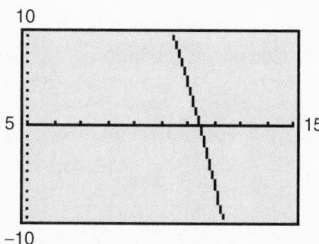

(b)

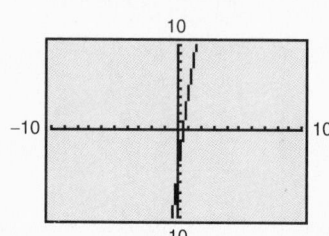

(c)

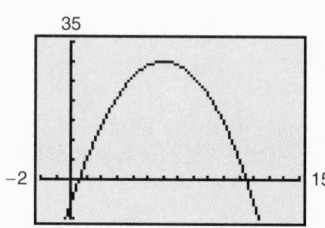

(d)

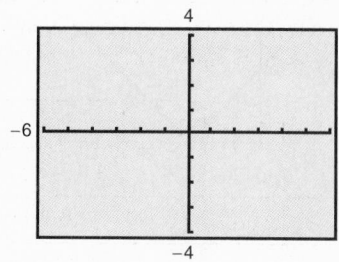

Square setting

On most graphing utilities, the display screen is two-thirds as high as it is wide. On such screens, you can obtain a graph with a true geometric perspective by using a **square setting**—one in which

$$\frac{Y_{max} - Y_{min}}{X_{max} - X_{min}} = \frac{2}{3}.$$

One such setting is shown at the left. Notice that the x and y tick marks are equally spaced on a square setting, but not on a standard one.

Intercepts: An Aid to Sketching Graphs

Two types of solution points that are especially useful are those having zero as the x-coordinate and those having zero as the y-coordinate. Such points are called **intercepts** because they are the points at which the graph intersects, respectively, the y- and x-axes.

Definitions of Intercepts

The point $(a, 0)$ is called an **x-intercept** of the graph of an equation if it is a solution point of the equation. To find the x-intercepts, let $y = 0$ and solve the equation for x.

The point $(0, b)$ is called a **y-intercept** of the graph of an equation if it is a solution point of the equation. To find the y-intercepts, let $x = 0$ and solve the equation for y.

EXAMPLE 4   *Finding the Intercepts of a Graph*

Find the intercepts and sketch the graph of

$$y = 2x - 3.$$

Solution

Find the x-intercept by letting $y = 0$ and solving for x.

$$y = 2x - 3$$
$$0 = 2x - 3$$
$$\tfrac{3}{2} = x$$

Find the y-intercept by letting $x = 0$ and solving for y.

$$y = 2x - 3$$
$$y = 2(0) - 3$$
$$y = -3$$

Therefore, the graph has one x-intercept, the point $\left(\tfrac{3}{2}, 0\right)$, and one y-intercept, the point $(0, -3)$. To sketch the graph of the equation, create a table of values. (Include the intercepts in the table.) Finally, using the solution points given in the table, sketch the graph of the equation, as shown in Figure 2.19.

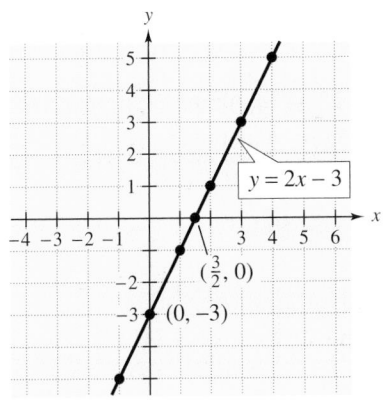

FIGURE 2.19

x	-1	0	1	$\frac{3}{2}$	2	3	4
$y = 2x - 3$	-5	-3	-1	0	1	3	5
Solution point	$(-1, -5)$	$(0, -3)$	$(1, -1)$	$\left(\frac{3}{2}, 0\right)$	$(2, 1)$	$(3, 3)$	$(4, 5)$

It is possible for a graph to have no intercepts or several intercepts. For instance, consider the three graphs in Figure 2.20.

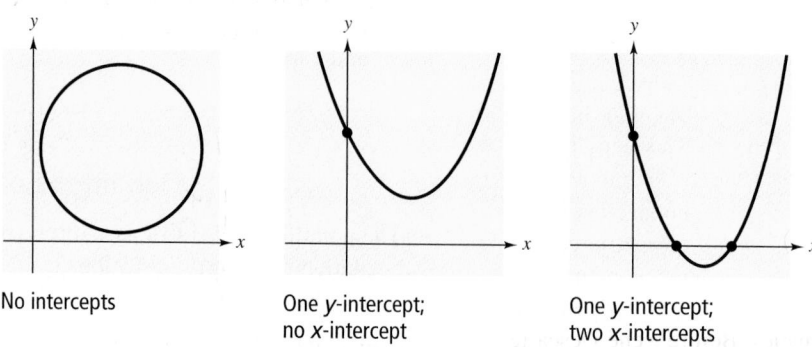

No intercepts

One *y*-intercept;
no *x*-intercept

One *y*-intercept;
two *x*-intercepts

FIGURE 2.20

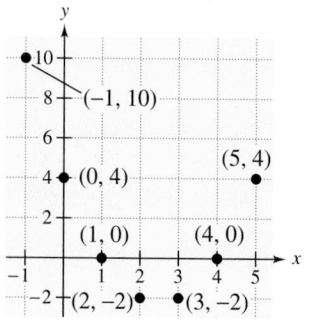

(a)

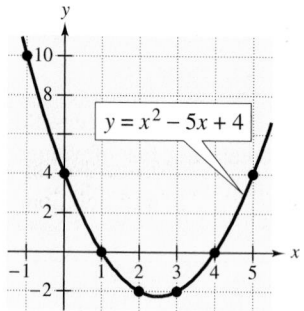

(b)

FIGURE 2.21

EXAMPLE 5   A Graph that Has Two x-Intercepts

Find the intercepts and sketch the graph of $y = x^2 - 5x + 4$.

Solution

Find the x-intercepts by letting	*Find the y-intercept by letting*
y = 0 and solving for x.	*x = 0 and solving for y.*

$$y = x^2 - 5x + 4 \qquad\qquad y = x^2 - 5x + 4$$
$$0 = x^2 - 5x + 4 \qquad\qquad y = 0^2 - 5(0) + 4$$
$$0 = (x - 4)(x - 1) \qquad\qquad y = 4$$
$$x - 4 = 0 \Longrightarrow x = 4$$
$$x - 1 = 0 \Longrightarrow x = 1$$

Therefore, the graph has two *x*-intercepts, the points $(4, 0)$ and $(1, 0)$, and one *y*-intercept, the point $(0, 4)$. To sketch the graph of the equation, create a table of values. (Include the intercepts in the table.)

x	-1	0	1	2	3	4	5
$y = x^2 - 5x + 4$	10	4	0	-2	-2	0	4
Solution point	$(-1, 10)$	$(0, 4)$	$(1, 0)$	$(2, -2)$	$(3, -2)$	$(4, 0)$	$(5, 4)$

Next, plot the points given in the table, as shown in Figure 2.21(a). Then connect them with a smooth curve, as shown in Figure 2.21(b).

The Connection Between Solutions and x-Intercepts

In Chapter 1, we emphasized the importance of checking the solutions of equations algebraically. *You should continue to do this.* However, there is also a way to check a solution *graphically*.

NOTE This connection between algebra and geometry represents one of the most wonderful discoveries ever made in mathematics. Before René Descartes introduced the coordinate plane in 1637, mathematicians had no easy way of "seeing" a solution of an algebraic equation.

Using a Graphic Check of a Solution

The solution of an equation involving one variable x can be checked graphically with the following steps.

1. Write the equation so that all nonzero terms are on one side and zero is on the other side.

2. Sketch the graph of $y =$ (nonzero terms).

3. The solution of the one-variable equation is the x-intercept of the graph of the two-variable equation.

EXAMPLE 6   Using a Graphic Check of a Solution

Solve the equation $3x + 1 = -8$. Check your solution graphically and algebraically.

Solution

$$3x + 1 = -8 \qquad \text{Original equation}$$
$$3x = -9 \qquad \text{Subtract 1 from both sides.}$$
$$x = -3 \qquad \text{Divide both sides by 3.}$$

Graphic Check

Rewrite the equation so that all nonzero terms are on the left side.

$$3x + 1 = -8 \qquad \text{Original equation}$$
$$3x + 9 = 0 \qquad \text{Add 8 to both sides.}$$

Now, sketch the graph of $y = 3x + 9$, as shown in Figure 2.22. Notice that the x-intercept is -3 (where $3x + 9 = 0$), which checks with the algebraic solution.

Algebraic Check

Substitute -3 for x in the original equation.

$$3(-3) + 1 \overset{?}{=} -8 \qquad \text{Substitute } -3 \text{ for } x.$$
$$-9 + 1 \overset{?}{=} -8 \qquad \text{Simplify.}$$
$$-8 = -8 \qquad \text{Solution checks. } \checkmark$$

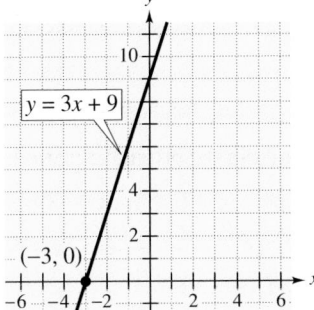

$y = 3x + 9$

$(-3, 0)$

FIGURE 2.22

EXAMPLE 7   *Using a Graphic Check of a Solution*

Solve the equation $x^2 = 4$. Check your solution graphically and algebraically.

Solution

$x^2 = 4$	Original equation
$x^2 - 4 = 0$	Subtract 4 from both sides.
$(x + 2)(x - 2) = 0$	Factor as difference of two squares.
$x + 2 = 0$ ⟹ $x = -2$	Set 1st factor equal to 0.
$x - 2 = 0$ ⟹ $x = 2$	Set 2nd factor equal to 0.

Graphic Check

Rewrite the equation so that all nonzero terms are on the left side.

$x^2 = 4$	Original equation
$x^2 - 4 = 0$	Subtract 4 from both sides.

Now sketch the graph of $y = x^2 - 4$, as shown in Figure 2.23. Notice that the x-intercepts are -2 and 2 (where $x^2 - 4 = 0$), which checks with the algebraic solution.

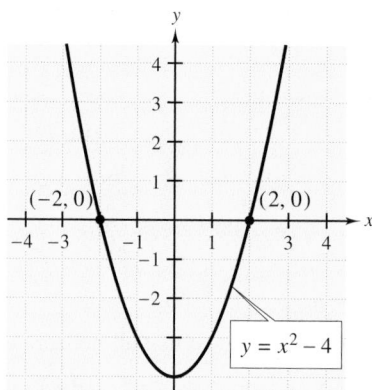

$y = x^2 - 4$

FIGURE 2.23

Algebraic Check

Check the first solution, $x = -2$, by substituting -2 for x in the original equation.

$(-2)^2 \overset{?}{=} 4$	Substitute -2 for x.
$4 = 4$	Solution checks. ✓

Check the second solution, $x = 2$, in a similar manner.

$(2)^2 \overset{?}{=} 4$	Substitute 2 for x.
$4 = 4$	Solution checks. ✓

Exploration

Using a Graphing Utility
Use a graphing utility to perform a graphic check for the solution found in Example 7. After you have found a viewing rectangle that shows both x-intercepts, try using the zoom feature to get a better look at each intercept. From the graph alone, could you have found the solutions to the original equation? Explain your reasoning.

Real-Life Application of Graphs

Have students discuss graphs they come in contact with in everyday life and how graphs complement and enhance reports and presentations.

Newspapers and news magazines frequently use graphs to show real-life relationships between variables. Example 8 shows how such a graph can help you visualize the concept of **straight-line depreciation.**

EXAMPLE 8   *Straight-Line Depreciation*

Your small business buys a new printing press for $65,000. For income tax purposes, you decide to depreciate the printing press over a 10-year period. At the end of the 10 years, the salvage value of the printing press is expected to be $5000. Find an equation that relates the depreciated value of the printing press to the number of years. Then sketch the graph of the equation.

Solution

The total depreciation over the 10-year period is $65,000 - 5000 = \$60,000$. Because the same amount is depreciated each year, it follows that the annual depreciation is $60,000/10 = \$6000$. Thus, after 1 year, the value of the printing press is

Value after 1 year $= 65,000 - (1)6000 = \$59,000$.

By similar reasoning, you can see that the value after 2 years is

Value after 2 years $= 65,000 - (2)6000 = \$53,000$.

Let y represent the value of the printing press after t years and follow the pattern determined for the first 2 years to obtain

$$y = 65,000 - 6000t.$$

A sketch of the graph of this equation is shown in Figure 2.24.

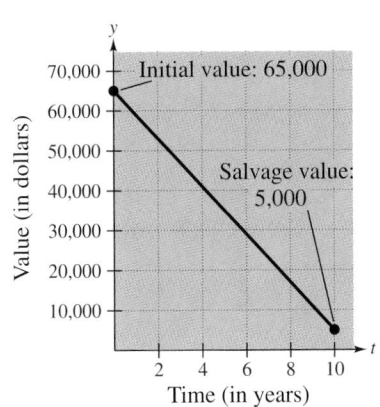

FIGURE 2.24 Straight-line depreciation

Group Activities   ### Extending the Concept

Straight-Line Depreciation   In Example 8, suppose that you depreciated the printing press over 8 years instead of 10 years. Write an equation that represents the depreciated value of the printing press during the 8-year period. Then graph both depreciation models on the same rectangular coordinate system and compare the results. What are the advantages and disadvantages of each model?

2.2   **Exercises**

See Warm-Up Exercises, p. A40

In Exercises 1–6, match the equation with its graph. [The graphs are labeled (a), (b), (c), (d), (e), and (f).]

(a)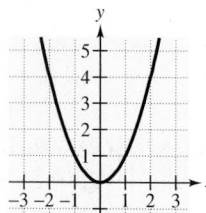

(b)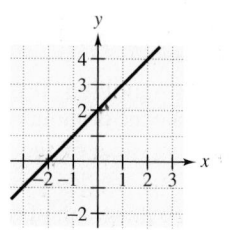

(c)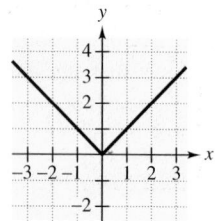

(d)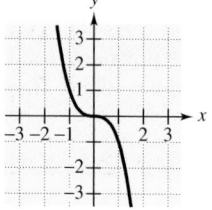

(e)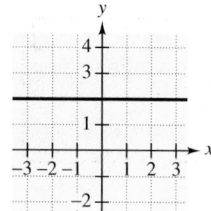

(f)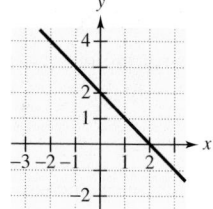

1. $y = 2$

2. $y = 2 + x$

3. $y = 2 - x$

4. $y = x^2$

5. $y = -x^3$

6. $y = |x|$

In Exercises 7–10, complete the table and use the results to sketch the graph of the equation.

7. $2x + y = 3$

x	-4			2	4
y		7	3		

8. $2x - 3y = 6$

x	-3	0		3	
y			$-\frac{2}{3}$		5

9. $y = 4 - x^2$

x		-1		2	
y	0		4		-5

10. $y = \frac{1}{2}x^3 - 4$

x	-1		1		3
y		-4		0	

In Exercises 11–18, sketch the graph of the equation.

11. $y = 3x$

12. $y = \frac{1}{3}x$

13. $y = 2x - 3$

14. $y = -x + 2$

15. $y = x^2 - 1$

16. $y = -x^2$

17. $y = |x| - 1$

18. $y = |x - 1|$

In Exercises 19–22, graphically estimate the intercepts of the graph. Then check your results algebraically.

19. $y = x^2 + 3$

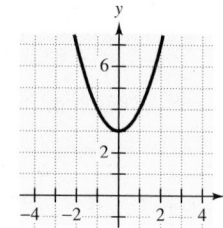

20. $x = 3$

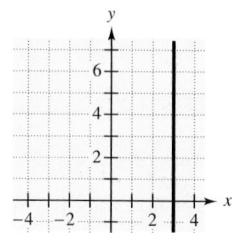

21. $y = |x - 2|$

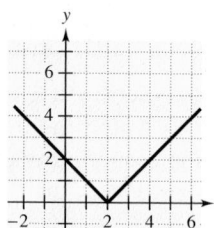

22. $y = (x - 3)(x - 4)$

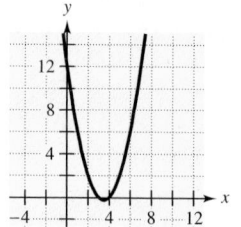

In Exercises 23–30, find the *x*- and *y*-intercepts (if any) of the graph of the equation.

23. $x + 2y = 10$

24. $3x - 2y + 12 = 0$

25. $y = (x + 1)^2$

26. $y = (x + 5)(x - 5)$

27. $y = \frac{3}{4}x + 15$

28. $y = 12 - \frac{2}{5}x$

29. $y = |x|$

30. $y = x^3$

In Exercises 31–40, sketch the graph of the equation and show the coordinates of three solution points (including intercepts).

31. $y = 3 - x$

32. $y = x - 3$

33. $y = 4$

34. $x = -6$

35. $4x + y = 3$

36. $y - 2x = -4$

37. $y = x^2 - 4$

38. $y = 1 - x^2$

39. $y = |x + 2|$

40. $y = |x| + 2$

In Exercises 41–50, use a graphing utility to graph the equation. Use the standard setting.

41. $y = 2x - 6$

42. $y = x - 2$

43. $y = x^2 - 3$

44. $y = 6 - x^2$

45. $y = 1 - x^3$

46. $y = x^4 - 4$

47. $y = \sqrt{x + 4}$

48. $y = 2\sqrt{2x + 3}$

49. $y = |x| - 6$

50. $y = |x - 6|$

In Exercises 51–54, use a graphing utility to graph the equation. Begin by using the standard setting. Then graph the equation a second time using the specified setting. Which setting is better? Explain.

51. $y = \frac{1}{2}x - 10$

52. $y = -2x + 25$

Xmin = -1
Xmax = 20
Xscl = 1
Ymin = -15
Ymax = 5
Yscl = 1

Xmin = -1
Xmax = 20
Xscl = 1
Ymin = -2
Ymax = 28
Yscl = 1

53. $y = 2x^3 - 5x^2$

54. $y = 6\sqrt{x} + 4$

Xmin = -5
Xmax = 5
Xscl = 1
Ymin = -5
Ymax = 5
Yscl = 1

Xmin = -5
Xmax = 5
Xscl = 1
Ymin = -1
Ymax = 20
Yscl = 2

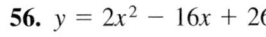

In Exercises 55–58, find a setting on a graphing utility such that the graph of the equation agrees with the given graph.

55. $y = -3x + 15$

56. $y = 2x^2 - 16x + 26$

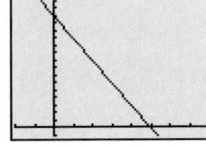

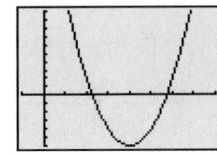

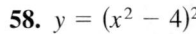

57. $y = x^3 - 3x + 2$

58. $y = (x^2 - 4)^2$

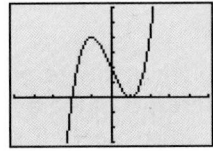

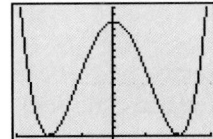

In Exercises 59–62, explain how to use a graph to verify that $y_1 = y_2$. Then identify the rule of algebra that is illustrated.

59. $y_1 = \frac{1}{2}(x - 4)$

$y_2 = \frac{1}{2}x - 2$

60. $y_1 = 3x - x^2$

$y_2 = -x^2 + 3x$

61. $y_1 = x + (2x - 1)$

$y_2 = (x + 2x) - 1$

62. $y_1 = 2x + 0$

$y_2 = 2x$

In Exercises 63–66, explain how the x-intercepts of the graph correspond to the solutions of the polynomial equation when $y = 0$.

63. $y = x^2 - 9$

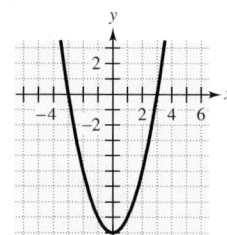

64. $y = x^2 - 4x + 4$

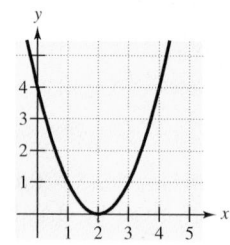

65. $y = x^2 - 2x - 3$

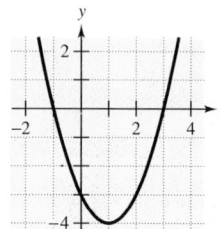

66. $y = x^3 - 3x^2 - x + 3$

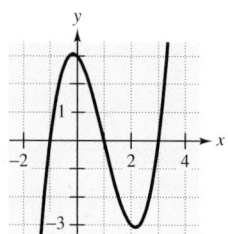

In Exercises 67–70, use a graphing utility to graph the equation and find any x-intercepts of the graph. Verify algebraically that any x-intercepts are solutions of the polynomial equation when $y = 0$.

67. $y = \frac{1}{2}x - 2$

68. $y = -3x + 6$

69. $y = x^2 - 6x$

70. $y = x^2 - 11x + 28$

In Exercises 71–80, use a graphing utility to solve the equation graphically.

71. $7 - 2(x - 1) = 0$

72. $2x - 1 = 3(x + 1)$

73. $4 - x^2 = 0$

74. $x^2 + 2x = 0$

75. $x^2 - 2x + 1 = 0$

76. $1 - (x - 2)^2 = 0$

77. $2x^2 + 5x - 12 = 0$

78. $(x - 2)^2 - 9 = 0$

79. $x^3 - 4x = 0$

80. $2 + x - 2x^2 - x^3 = 0$

81. *Hooke's Law*   The force F (in pounds) to stretch a spring x inches from its natural length is given by

$$F = \tfrac{4}{3}x, \quad 0 \le x \le 12.$$

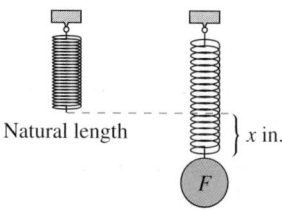

Natural length $\Big\} x$ in.

(a) Use the model to complete the table.

x	0	3	6	9	12
F					

(b) Sketch a graph of the model.

(c) Use the graph in part (b) to determine how the length of the spring changes each time the force is doubled.

82. *Dairy Farms*   The number of farms in the United States with milk cows has been decreasing. The numbers of farms N (in thousands) for the years 1988 through 1994 are given in the table.

Year	1988	1989	1990	1991	1992	1993	1994
N	216	203	193	181	171	159	150

A model for this data is

$$N = -11.0t + 192.9,$$

where t is the time in years, with $t = 0$ corresponding to 1990. (*Source: U.S. Department of Agriculture*)

(a) Use a graphing utility to plot the data and graph the model.

(b) How well does the model represent the data? Explain your reasoning.

(c) Use the model to predict the number of farms with milk cows in 1997.

(d) Explain why the model may not be accurate in the future.

83. *Straight-Line Depreciation*   A manufacturing plant purchases a new molding machine for \$225,000. The depreciated value y after t years is given by

$$y = 225{,}000 - 20{,}000t, \quad 0 \le t \le 8.$$

Sketch the graph of this model.

84. *Maximum Area*   A rectangle of length l and width w has a perimeter of 12 meters.

  (a) Show that the width of the rectangle is $w = 6 - l$ and its area is $A = l(6 - l)$.

  (b) Sketch the graph of the equation for the area.

  (c) Use the graph in part (b) to estimate the dimensions of the rectangle that yield a maximum area.

85. *Investigation*   Graph the equations $y = x^2 + 1$ and $y = -(x^2 + 1)$ on the same set of coordinate axes. Explain how the graph of an equation changes if the expression for y is multiplied by -1. Justify your answer by giving additional examples.

Investigation   In Exercises 86 and 87, graph the equations on the same set of coordinate axes. What conclusions can you make by comparing the graphs?

86. (a) $y = x^2$

  (b) $y = (x - 2)^2$

  (c) $y = (x + 2)^2$

  (d) $y = (x + 4)^2$

87. (a) $y = x^2$

  (b) $y = x^2 - 2$

  (c) $y = x^2 - 4$

  (d) $y = x^2 + 4$

Section Project

Misleading Graphs   Graphs can help us visualize relationships between two variables, but they can also be misused to imply results that are not correct. In each pair of graphs below, both graphs represent the *same* data.

(a) Which graph is misleading and why?

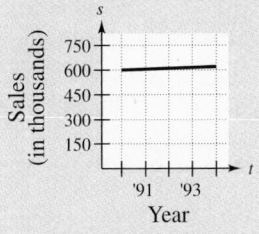
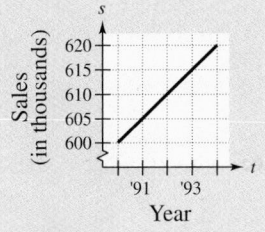

(b) Which graph is misleading and why?

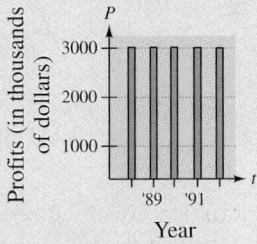
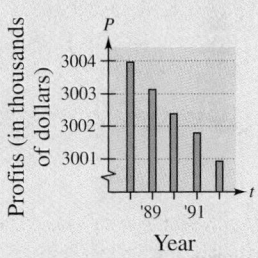

(c) Look for an example of a misleading graph in a newspaper or magazine. Explain how the graph could represent the data more accurately.

<table>
<tr><td>**2.3**</td><td>## Slope: An Aid to Graphing Lines</td></tr>
</table>

The Slope of a Line ▪ Slope as a Graphing Aid ▪
Slope as a Rate of Change ▪ Parallel and Perpendicular Lines

The Slope of a Line

Historical Note: The French verb meaning to mount, to climb, or to rise is *monter*. Because Descartes was largely responsible for the development of analytical geometry, his use of *m*—short for *monter*—to indicate the slope became the accepted term among European mathematicians.

The **slope** of a nonvertical line is the number of units the line rises or falls vertically for each unit of horizontal change from left to right. For example, the line in Figure 2.25 rises two units for each unit of horizontal change from left to right, and we say that this line has a slope of $m = 2$.

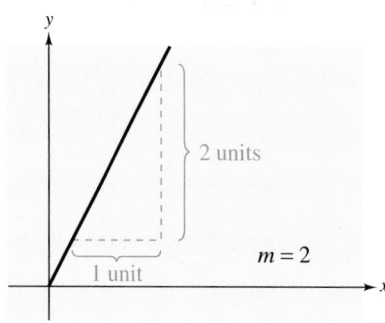

FIGURE 2.25

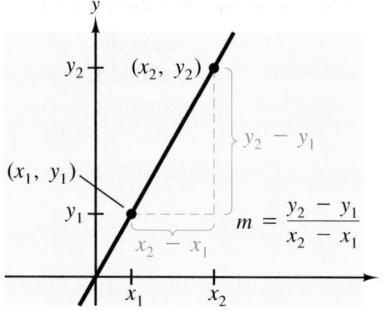

FIGURE 2.26

Definition of the Slope of a Line

The **slope** m of the nonvertical line passing through the points (x_1, y_1) and (x_2, y_2) is

$$m = \frac{y_2 - y_1}{x_2 - x_1} = \frac{\text{change in } y}{\text{change in } x},$$

where $x_1 \neq x_2$ (see Figure 2.26).

It may be helpful to show an example of each of the correct forms. The slope of the line through $(-1, 3)$ and $(4, -2)$ is

$$m = \frac{-2 - 3}{4 - (-1)} = \frac{-5}{5} = -1 \quad \text{or}$$

$$m = \frac{3 - (-2)}{-1 - 4} = \frac{5}{-5} = -1.$$

When the formula for slope is used, the *order of subtraction* is important. Given two points on a line, you are free to label either of them (x_1, y_1) and the other (x_2, y_2). However, once this is done, you must form the numerator and denominator using the same order of subtraction.

$$m = \frac{y_2 - y_1}{x_2 - x_1} \qquad m = \frac{y_1 - y_2}{x_1 - x_2} \qquad m = \frac{y_2 - y_1}{x_1 - x_2}$$

Correct   Correct   Incorrect

EXAMPLE 1   *Finding the Slope of a Line Through Two Points*

Find the slope of the line passing through each pair of points.

(a) $(3, 4)$ and $(1, -2)$

(b) $(-2, 4)$ and $(3, 4)$

(c) $(2, -3)$ and $(0, 1)$

Solution

NOTE   You can use slope to determine if three points are collinear. Consider any three points A, B, and C. If the slope of the line through points A and B is the same as the slope of the line through points B and C, the three points are collinear.

(a) Let $(x_1, y_1) = (3, 4)$ and $(x_2, y_2) = (1, -2)$.

$$m = \frac{y_2 - y_1}{x_2 - x_1}$$ ⟸ Difference in y-values

    ⟸ Difference in x-values

$$m = \frac{-2 - 4}{1 - 3} = \frac{-6}{-2} = 3$$

(b) The slope of the line through $(-2, 4)$ and $(3, 4)$ is

$$m = \frac{4 - 4}{3 - (-2)} = \frac{0}{5} = 0.$$

(c) The slope of the line through $(2, -3)$ and $(0, 1)$ is

$$m = \frac{1 - (-3)}{0 - 2} = \frac{4}{-2} = -2.$$

The graphs of the three lines are shown in Figure 2.27.

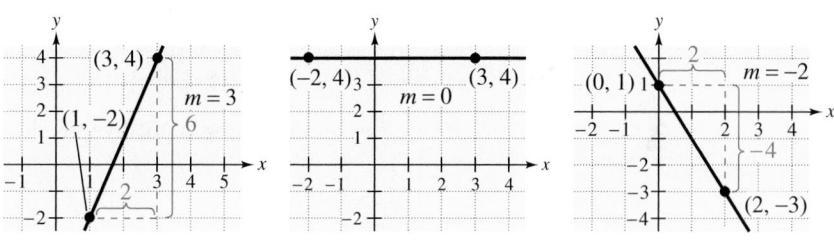

(a) Positive slope   (b) Zero slope   (c) Negative slope

FIGURE 2.27

The definition of slope does not apply to vertical lines. For instance, consider the points $(3, 1)$ and $(3, 3)$ on the vertical line shown in Figure 2.28. Applying the formula for slope, you have

$$\frac{3 - 1}{3 - 3} = \frac{2}{0}.$$ Undefined

Because division by zero is not defined, the slope of a vertical line is not defined.

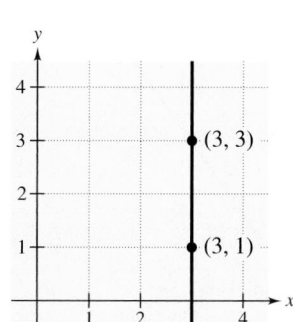

FIGURE 2.28   Slope is undefined.

Problem: Given the points $(2w, 3)$ and $(w, -2)$, what is the value of w if the slope of the line connecting these two points is equal to 2?

Solution: $\dfrac{-2 - 3}{w - 2w} = \dfrac{-5}{-w} = \dfrac{5}{w} = 2$,

so $w = \dfrac{5}{2}$.

From the slopes of the lines shown in Figures 2.27 and 2.28, you can make the following generalizations about the slope of a line.

1. A line with positive slope $(m > 0)$ *rises* from left to right.
2. A line with negative slope $(m < 0)$ *falls* from left to right.
3. A line with zero slope $(m = 0)$ is *horizontal*.
4. A line with undefined slope is *vertical*.

EXAMPLE 2   *Using Slope to Describe Lines*

Describe the line through each pair of points.

(a) $(2, -1), (2, 3)$ (b) $(-2, 4), (3, 1)$ (c) $(1, 3), (4, 3)$ (d) $(-1, 1), (2, 5)$

Solution

(a) $m = \dfrac{3 - (-1)}{2 - 2} = \dfrac{4}{0}$ Undefined slope [Figure 2.29(a)]

Because the slope is undefined, the line is vertical.

(b) $m = \dfrac{1 - 4}{3 - (-2)} = -\dfrac{3}{5} < 0$ Negative slope [Figure 2.29(b)]

Because the slope is negative, the line falls from left to right.

(c) $m = \dfrac{3 - 3}{4 - 1} = \dfrac{0}{3} = 0$ Zero slope [Figure 2.29(c)]

Because the slope is zero, the line is horizontal.

(d) $m = \dfrac{5 - 1}{2 - (-1)} = \dfrac{4}{3} > 0$ Positive slope [Figure 2.29(d)]

Because the slope is positive, the line rises from left to right.

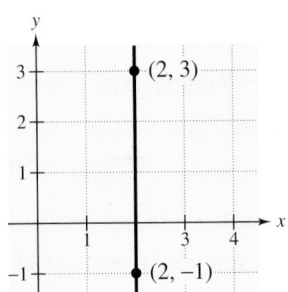

(a) Vertical line: undefined slope

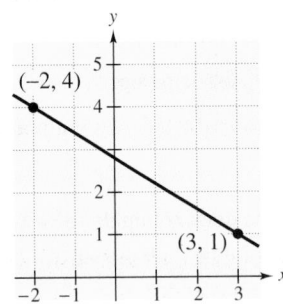

(b) Line falls: negative slope

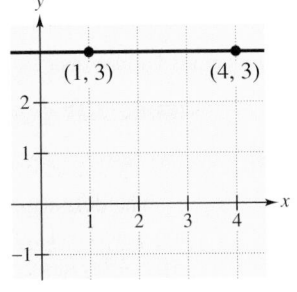

(c) Horizontal line: zero slope

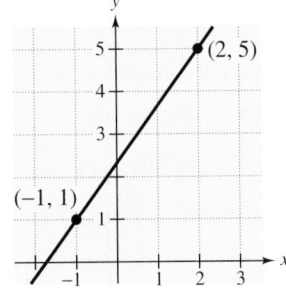

(d) Line rises: positive slope

FIGURE 2.29

Any two points on a nonvertical line can be used to calculate its slope. This is demonstrated in the next example.

EXAMPLE 3   *Finding the Slope of a Line*

Sketch the graph of the line given by $2x + 3y = 6$. Then find the slope of the line. (Choose two different pairs of points on the line and show that the same slope is obtained from either pair.)

Solution

Begin by solving the given equation for y.

$$2x + 3y = 6 \qquad \text{Original equation}$$
$$3y = -2x + 6 \qquad \text{Subtract } 2x \text{ from both sides.}$$
$$\frac{3y}{3} = \frac{-2x + 6}{3} \qquad \text{Divide both sides by 3.}$$
$$y = -\frac{2}{3}x + 2 \qquad \text{Simplify.}$$

Then construct a table of values, as shown below.

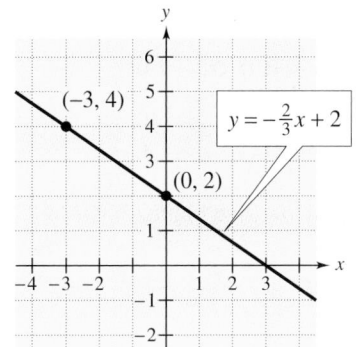

x	-3	0	3	6
$y = -\frac{2}{3}x + 2$	4	2	0	-2
Solution point	$(-3, 4)$	$(0, 2)$	$(3, 0)$	$(6, -2)$

From the solution points shown in the table, sketch the graph of the line, as shown in Figure 2.30. To calculate the slope of the line using two different sets of points, first use the points $(-3, 4)$ and $(0, 2)$, and obtain a slope of

$$m = \frac{2 - 4}{0 - (-3)} = -\frac{2}{3}.$$

Next, use the points $(3, 0)$ and $(6, -2)$ to obtain a slope of

$$m = \frac{-2 - 0}{6 - 3} = -\frac{2}{3}.$$

Try some other pairs of points on the line to see that you obtain a slope of $m = -\frac{2}{3}$ regardless of which two points you use.

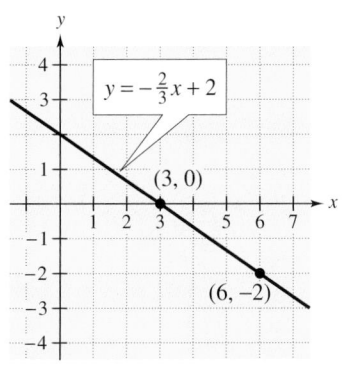

FIGURE 2.30

a. Use a graphing utility to graph the following equations in the same viewing rectangle. How are the equations similar? How are the graphs similar? What does this imply?

$$y = 2x + 4$$
$$y = x + 4$$
$$y = -2x + 4$$
$$y = -x + 4$$

b. Use a graphing utility to graph the following equations in the same viewing rectangle. How are the equations similar? How are the graphs similar? What does this imply?

$$y = 2x + 4$$
$$y = 2x + 2$$
$$y = 2x - 1$$
$$y = 2x - 3$$

Technology

Setting the viewing rectangle on a graphing utility affects the appearance of its slope. When you are using a graphing utility, remember that you cannot judge whether a slope is steep or shallow *unless* you use a square setting.

Slope as a Graphing Aid

You have seen that, before creating a table of values for an equation, you should first solve the equation for y. When you do this for a linear equation, you obtain some very useful information. Consider the results of Example 3. When the equation $2x + 3y = 6$ is solved for y, you see that $y = -\frac{2}{3}x + 2$.

Observe that the coefficient of x is the slope of the graph of this equation (see Example 3). Moreover, the constant term, 2, gives the y-intercept of the graph.

$$y = -\frac{2}{3}x + \boxed{2}$$

Slope y-intercept $(0, 2)$

This form is called the **slope-intercept** form of the equation of the line.

Slope-Intercept Form of the Equation of a Line

The graph of the equation

$$y = mx + b$$

is a line with a slope of m and a y-intercept of $(0, b)$. (See Figure 2.31.)

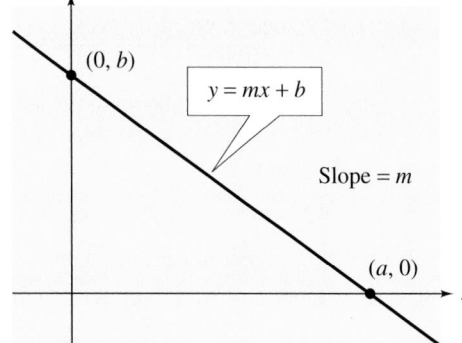

FIGURE 2.31

NOTE   When you substitute 0 for x in the slope-intercept form of the equation of a line, $y = mx + b$, you obtain $y = m(0) + b$ or $y = b$. Thus, $(0, b)$ is the point where the line crosses the y-axis.

So far, you have been plotting several points to sketch the equation of a line. However, now that you can recognize equations of lines, you don't have to plot as many points—two points are enough. (You might remember from geometry that *two points are all that are necessary to determine a line*.)

EXAMPLE 4   *Using the Slope and y-Intercept to Sketch a Line*

Use the slope and y-intercept to sketch the graph of each equation.

(a) $y = \dfrac{5}{2}x - 3$

(b) $y = -3x + 5$

Solution

(a) The equation is already in slope-intercept form.

$$y = mx + b$$

$$y = \frac{5}{2}x - 3 \qquad \text{Slope-intercept form}$$

Thus, the slope of the line is $m = \frac{5}{2}$ and the y-intercept is $(0, b) = (0, -3)$. Now you can sketch the graph of the line as follows. First, plot the y-intercept. Then, using a slope of $\frac{5}{2}$

$$m = \frac{5}{2} = \frac{\text{change in } y}{\text{change in } x},$$

locate a second point on the line by moving two units to the right and five units up (or five units up and two units to the right). Finally, obtain the graph by drawing a line through the two points, as shown in Figure 2.32.

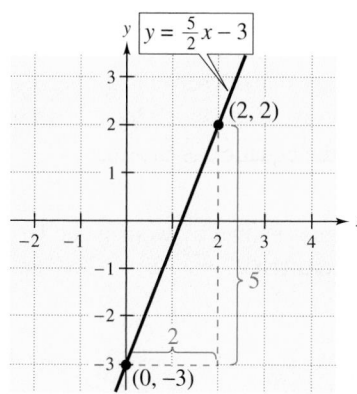

FIGURE 2.32

(b) The equation is already in slope-intercept form.

$$y = mx + b$$

$$y = -3x + 5 \qquad \text{Slope-intercept form}$$

Thus, the slope of the line is $m = -3$ and the y-intercept is $(0, b) = (0, 5)$. Now you can sketch the graph of the line as follows. First, plot the y-intercept. Then, using a slope of -3

$$m = \frac{-3}{1} = \frac{\text{change in } y}{\text{change in } x},$$

locate a second point on the line by moving one unit to the right and three units down (or three units down and one unit to the right). Finally, obtain the graph by drawing a line through the two points, as shown in Figure 2.33.

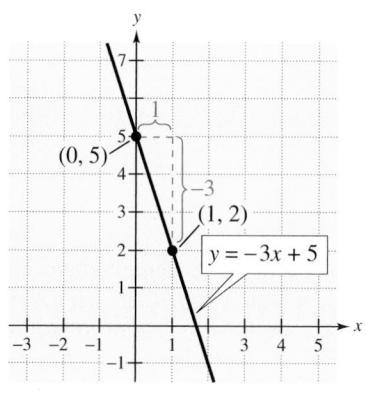

FIGURE 2.33

Slope as a Rate of Change

In real-life problems, slope is often used to describe a **constant rate of change** or an **average rate of change.** In such cases, units of measure are assigned, such as miles per hour or dollars per year.

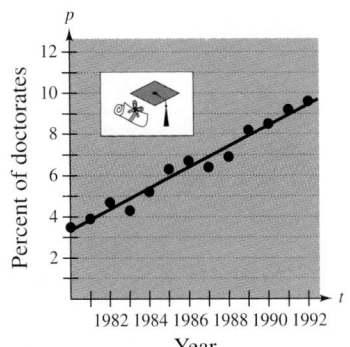

FIGURE 2.34

EXAMPLE 5   *Slope as a Rate of Change*

In 1980, 3.5% of the doctorates in engineering were awarded to women. By 1992, 9.6% were being earned by women. Find the average rate of change in the percent of engineering doctorates earned by women from 1980 to 1992.   *(Source: National Research Council)*

Solution

Let p represent the percent of engineering doctorates earned by women and let t represent the year. The two given data points are represented by (t_1, p_1) and (t_2, p_2).

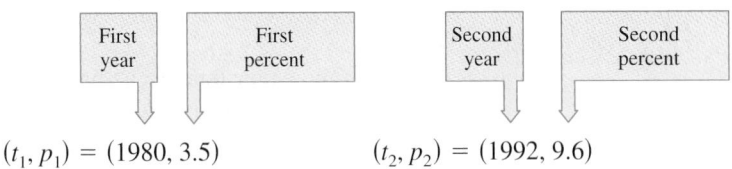

$$(t_1, p_1) = (1980, 3.5) \qquad (t_2, p_2) = (1992, 9.6)$$

Now, use the formula for slope to find the average rate of change.

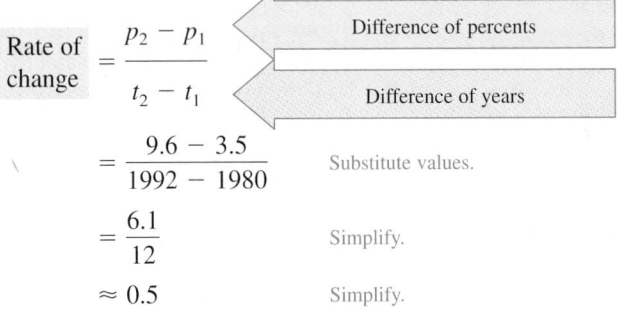

$$\text{Rate of change} = \frac{p_2 - p_1}{t_2 - t_1}$$

$$= \frac{9.6 - 3.5}{1992 - 1980} \qquad \text{Substitute values.}$$

$$= \frac{6.1}{12} \qquad \text{Simplify.}$$

$$\approx 0.5 \qquad \text{Simplify.}$$

From 1980 through 1992, the *average rate of change* in the percent of engineering doctorates awarded to women was about $\frac{1}{2}\%$ per year. (The exact changes in percent varied from one year to the next, as shown in the scatter plot in Figure 2.34.)

Discuss the pros and cons of predicting what percentage of engineering doctorates will be awarded to women in the years 2000, 2010, and 2040.

NOTE   Make sure you understand that the answer in Example 5 is $\frac{1}{2}\%$ or 0.005, not 50% or 0.5, because you were finding the *average rate of change in the percent.*

Parallel and Perpendicular Lines

You know from geometry that two lines in a plane are *parallel* if they do not intersect. What this means in terms of their slopes is suggested in Example 6.

EXAMPLE 6   Lines that Have the Same Slope

On the same set of coordinate axes, sketch the lines given by $y = 2x$ and $y = 2x - 3$.

Solution

For the line given by

$$y = 2x,$$

the slope is $m = 2$ and the y-intercept is $(0, 0)$. For the line given by

$$y = 2x - 3,$$

the slope is also $m = 2$ and the y-intercept is $(0, -3)$. The graphs of these two lines are shown in Figure 2.35.

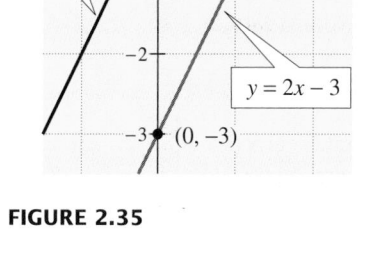

FIGURE 2.35

In Example 6, notice that the two lines have the same slope *and* appear to be parallel. The following rule states that this is always the case.

NOTE The phase "if and only if" in this rule is used in mathematics as a way to write two statements in one. The first statement says that *if two distinct nonvertical lines have the same slope, they must be parallel.* The second statement says that *if two distinct nonvertical lines are parallel, they must have the same slope.*

Parallel Lines

Two distinct nonvertical lines are parallel if and only if they have the same slope.

Another rule from geometry is that two lines in a plane are *perpendicular* if they intersect at right angles. In terms of slope, this means that two nonvertical lines are perpendicular if their slopes are negative reciprocals of each other.

Perpendicular Lines

Consider two nonvertical lines whose slopes are m_1 and m_2. The two lines are perpendicular if and only if their slopes are *negative reciprocals* of each other. That is,

$$m_1 = -\frac{1}{m_2}, \qquad \text{or equivalently,} \qquad m_1 \cdot m_2 = -1.$$

EXAMPLE 7 *Parallel or Perpendicular?*

Is each pair of lines parallel, perpendicular, or neither?

(a) $y = -2x + 4, y = \frac{1}{2}x + 1$ (b) $y = \frac{1}{3}x + 2, y = \frac{1}{3}x - 3$

Solution

(a) The first line has a slope of $m_1 = -2$, and the second line has a slope of $m_2 = \frac{1}{2}$. Because these slopes are negative reciprocals of each other, the two lines must be perpendicular, as shown in Figure 2.36.

(b) Each of these two lines has a slope of $m = \frac{1}{3}$. Therefore, the two lines must be parallel, as shown in Figure 2.37.

 Technology

Use a graphing utility to confirm the results in Example 7. To do so on a *TI-82* or *TI-83,* use the following steps.

1. Choose an appropriate viewing rectangle. (For lines that appear perpendicular, use a square setting.)

2. Enter the equations into Y_1 and Y_2.

3. Press the GRAPH key.

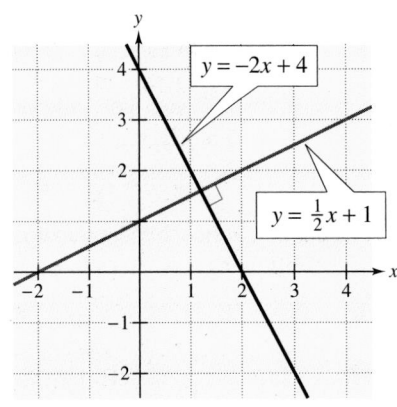

FIGURE 2.36

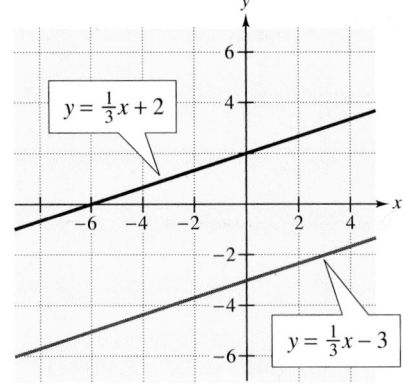

FIGURE 2.37

Group Activities   **Extending the Concept**

Interpreting Slope   Write a linear equation for the given verbal model. Identify the slope and then interpret the slope in the real-life setting.

1. $\dfrac{\text{Total pay}}{\text{per hour}} = \dfrac{\text{Piecework}}{\text{rate}} \cdot \dfrac{\text{Number}}{\text{of pieces}} + \dfrac{\text{Fixed}}{\text{hourly rate}}$

2. $\dfrac{\text{Total}}{\text{cost}} = \dfrac{\text{List}}{\text{price}} + \dfrac{\text{Tax}}{\text{rate}} \cdot \dfrac{\text{List}}{\text{price}}$

2.3 **Exercises**

See Warm-Up
Exercises, p. A40

In Exercises 1–6, estimate the slope of the line from its graph.

1.
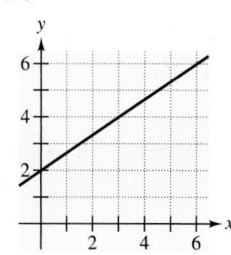

2.
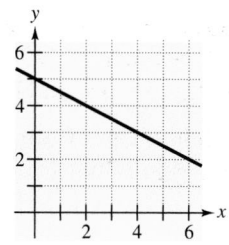

3.
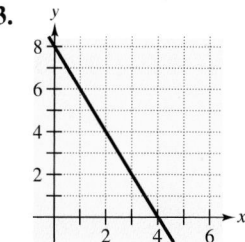

4.
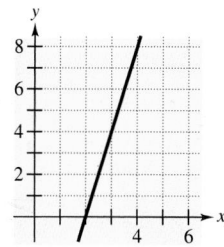

5.
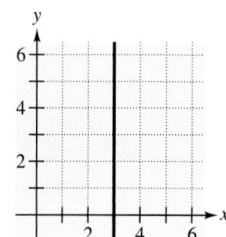

6.
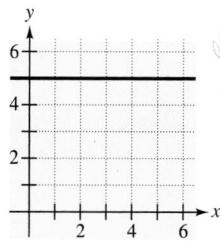

In Exercises 7 and 8, identify the line in the figure that has the specified slope m.

7. (a) $m = \frac{3}{4}$ (b) $m = 0$ (c) $m = -3$

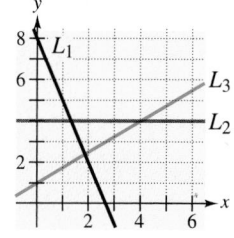

8. (a) $m = -\frac{5}{2}$ (b) m is undefined.   (c) $m = 2$

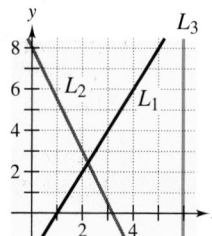

In Exercises 9–26, plot the points and find the slope (if possible) of the line passing through them. State whether the line is rising, falling, horizontal, or vertical.

9. $(0, 0), (7, 5)$ **10.** $(0, 0), (-3, -4)$

11. $(-2, -3), (6, 1)$ **12.** $(0, -4), (6, 0)$

13. $(0, 12), (8, 0)$ **14.** $(3, 6), (5, -2)$

15. $(-5, -3), (-5, 4)$ **16.** $(0, -8), (-5, 0)$

17. $(2, -5), (7, -5)$ **18.** $(-2, 1), (-4, -3)$

19. $\left(\frac{3}{4}, 2\right), \left(5, -\frac{5}{2}\right)$ **20.** $\left(-\frac{3}{2}, -\frac{1}{2}\right), \left(\frac{5}{8}, \frac{1}{2}\right)$

21. $(4.2, -1), (-4.2, 6)$ **22.** $(3.4, 0), (3.4, 1)$

23. $(0, 4.5), (3, 4.5)$ **24.** $(2.5, -2), (4.75, 5.25)$

25. $(-3, -5), (3, 5)$ **26.** $(-5, 0), (0, 6)$

27. *Think About It*   In your own words, give interpretations of a negative slope, a zero slope, and a positive slope.

28. *Think About It*   Can any pair of points on a line be used to calculate the slope of the line? Explain.

In Exercises 29 and 30, solve for x so that the line through the points has the given slope.

29. $(4, 5), (x, 7); m = -\frac{2}{3}$

30. $(x, -2), (5, 0); m = \frac{3}{4}$

In Exercises 31 and 32, solve for y so that the line through the points has the given slope.

31. $(-3, y), (9, 3); m = \frac{3}{2}$

32. $(-3, 20), (2, y); m = -6$

In Exercises 33–42, a point on a line and the slope of the line are given. Find two additional points on the line. (There are many correct answers.)

33. $(5, 2)$
$m = 0$

34. $(-4, 3)$
m is undefined.

35. $(3, -4)$
$m = 3$

36. $(-1, -5)$
$m = 2$

37. $(0, 3)$
$m = -1$

38. $(-2, 6)$
$m = -3$

39. $(-5, 0)$
$m = \frac{4}{3}$

40. $(-1, 1)$
$m = -\frac{3}{4}$

41. $(4, 2)$
m is undefined.

42. $(-2, -2)$
$m = 0$

In Exercises 43–46, sketch the graph of a line through the point $(3, 2)$ having the given slope.

43. $m = 3$

44. $m = \frac{3}{2}$

45. $m = -\frac{1}{3}$

46. $m = 0$

In Exercises 47–52, sketch the graph of a line through the point $(0, 1)$ having the given slope.

47. $m = 2$

48. $m = 0$

49. m is undefined.

50. $m = -1$

51. $m = -\frac{4}{3}$

52. $m = \frac{2}{3}$

Think About It   In Exercises 53 and 54, determine which line is "steeper."

53. $m_1 = 3, m_2 = -4$

54. $m_1 = 4, m_2 = 5$

In Exercises 55–58, plot the x- and y-intercepts and sketch the graph of the line.

55. $2x - y + 4 = 0$

56. $3x + 5y + 15 = 0$

57. $-5x + 2y - 20 = 0$

58. $3x - 5y - 15 = 0$

In Exercises 59–70, write the equation of the line in slope-intercept form and sketch the line. Use a graphing utility to confirm your sketch.

59. $3x - y - 2 = 0$

60. $x - y - 5 = 0$

61. $x + y = 0$

62. $x - y = 0$

63. $3x + 2y - 2 = 0$

64. $x - 2y - 2 = 0$

65. $x - 4y + 2 = 0$

66. $8x + 6y - 3 = 0$

67. $y - 2 = 0$

68. $y + 4 = 0$

69. $x - 0.2y - 1 = 0$

70. $0.5x + 0.6y - 3 = 0$

In Exercises 71–74, determine whether the lines L_1 and L_2 passing through the given pairs of points are parallel, perpendicular, or neither.

71. $L_1: (1, 3), (2, 1)$
$L_2: (0, 0), (4, 2)$

72. $L_1: (-3, -3), (1, 7)$
$L_2: (0, 4), (5, -2)$

73. $L_1: (-2, 0), (4, 4)$
$L_2: (1, -2), (4, 0)$

74. $L_1: (-5, 3), (3, 0)$
$L_2: (1, 2), \left(3, \frac{22}{3}\right)$

In Exercises 75–78, use a graphing utility to graph the pair of equations in the same viewing rectangle. Are the lines parallel, perpendicular, or neither? (Use the *square* setting so the slopes of the lines appear visually correct.)

75. $y = \frac{1}{2}x - 2$
$y = \frac{1}{2}x + 3$

76. $y = 3x - 2$
$y = 3x + 1$

77. $y = \frac{3}{4}x - 3$
$y = -\frac{4}{3}x + 1$

78. $y = -\frac{2}{3}x - 5$
$y = \frac{3}{2}x + 1$

In Exercises 79–82, use a graphing utility to graph the three equations in the same viewing rectangle. Describe the relationships among the graphs. (Use the *square* setting so the slopes of the lines appear visually correct.)

79. $y_1 = 3x$

$y_2 = -3x$

$y_3 = \frac{1}{3}x$

80. $y_1 = \frac{3}{4}x$

$y_2 = -\frac{4}{3}x$

$y_3 = \frac{4}{3}$

81. $y_1 = \frac{1}{4}x$

$y_2 = \frac{1}{4}x - 2$

$y_3 = \frac{1}{4}x + 3$

82. $y_1 = 2x$

$y_2 = 2x - 5$

$y_3 = 2x + \frac{3}{2}$

83. *Think About It* Is it possible for two lines with positive slopes to be perpendicular to each other? Explain.

84. *Geometry* The length and width of a rectangular flower garden are 40 feet and 30 feet (see figure). A walkway of width x surrounds the garden.

(a) Write the outside perimeter P of the walkway in terms of x.

(b) Use a graphing utility to graph the model for the perimeter in part (a).

(c) Determine the slope of the graph in part (b). For each additional 1-foot increase in the width of the walkway, determine the increase in its outside perimeter.

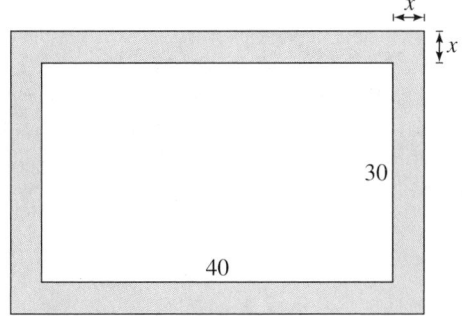

85. *Road Grade* When driving down a mountain road, you notice warning signs indicating that it is a "12% grade." This means that the slope of the road is $-\frac{12}{100}$. Over a stretch of road, your elevation drops by 2000 feet. What is the horizontal change in your position?

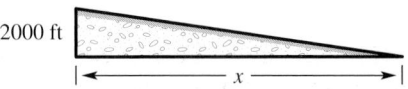

86. *Simple Interest* An inheritance of $8000 is invested in two different mutual funds. One fund pays 6% simple interest and the other pays $7\frac{1}{2}$% simple interest.

(a) If x dollars is invested in the fund paying 6%, how much is invested in the fund paying $7\frac{1}{2}$%?

(b) Use the result in part (a) to write the annual interest I in terms of x.

(c) Use a graphing utility to graph the equation in part (b) over an appropriate interval for x.

(d) Explain why the slope of the line in part (b) is negative.

87. *Income per Share* The graph gives the income per share of common stock for American Home Products Corporation for the years 1986 through 1995. Use the slope of each segment to determine the years when income increased most rapidly. *(Source: American Home Products Corporation 1995 Annual Report)*

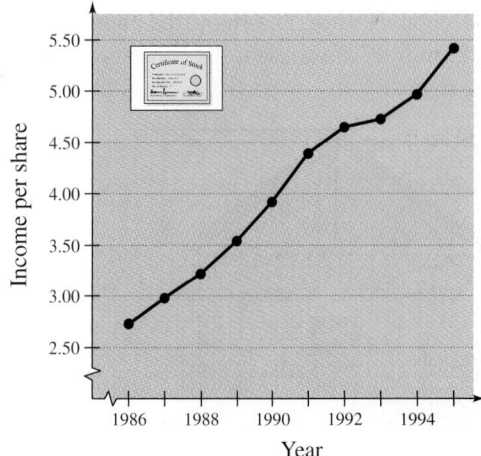

88. *Leaning Tower of Pisa*   When it was built, the Leaning Tower of Pisa in Italy was 180 feet tall. Since then, one side of the base has sunk 1 foot into the ground, causing the top of the tower to lean 16 feet off center (see figure). Approximate the slope of the side of the tower.

The tower leans because it was built on a layer of unstable soil—a mixture of clay, sand, and water.

89. *Height of an Attic*   The slope, or pitch, of a roof is such that it rises (or falls) 3 feet for every 4 feet of horizontal distance. Determine the maximum height in the attic of the house if the house is 30 feet wide (see figure).

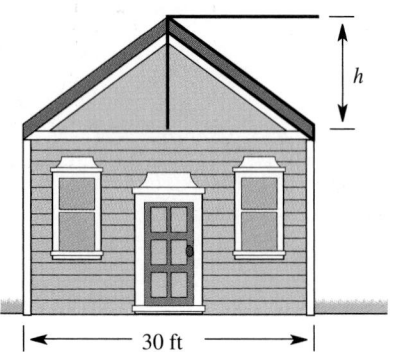

30 ft

Section Project

Spending Your Paycheck   Consider the ordered pair (x, y), where y represents your pocket money and x is the time in days, with $x = 0$ corresponding to payday. On payday, you have $200. Eight days later, you have $46.

(a) Plot the two points modeling the given information and sketch a line segment connecting the points.

(b) Determine the slope of the line segment in part (a).

(c) At what average rate did you spend your paycheck over the 8-day period?

(d) On the next payday, you again have $200 of pocket money. Three days later, you have $95. Plot these two points and sketch a line segment connecting the points.

(e) Determine the slope of the line segment in part (d).

(f) At what average rate did you spend your paycheck over this 3-day period?

(g) Explain how slope is related to an average rate of change in this example.

Mid-Chapter Quiz

Take this quiz as you would take a quiz in class. After you are done, check your work against the answers given in the back of the book.

In Exercises 1 and 2, (a) plot the points on a rectangular coordinate system, (b) find the distance between the points, and (c) find the slope of the line through the points.

1. $(-1, 5), (3, 2)$

2. $(-3, -2), (2, 10)$

3. Find the coordinates of the point that lies ten units to the right of the y-axis and three units below the x-axis.

4. Determine whether the ordered pairs are solution points of the equation $4x - 3y = 10$.

(a) $(2, 1)$ (b) $(1, -2)$ (c) $(2.5, 0)$ (d) $\left(2, -\frac{2}{3}\right)$

5. Find the x- and y-intercepts of the graph of the equation $6x - 8y + 48 = 0$. Plot the intercepts and sketch the graph of the line.

6. The earnings per share for General Electric Company for the years 1991, 1992, 1993, 1994, and 1995 were $2.27, $2.41, $2.45, $3.46, and $3.90. Construct a line graph for the data. *(Source: General Electric Company)*

7. The data at the right gives software applications and sales (in millions of dollars) for North America for 1994. Construct a horizontal bar graph for the data. *(Source: Software Publisher's Association)*

Software	Sales
Entertainment	716
Home Education	522
Finance	411
Word Processors	1031
Spreadsheets	830
Databases	351

Data for 7

In Exercises 8–11, sketch the graph of the equation and show the coordinates of three solution points (including intercepts).

8. $y = 2x - 3$

9. $y = 5$

10. $y = 6x - x^2$

11. $y = |x| - 3$

In Exercises 12 and 13, write the equation in slope-intercept form. Then sketch the graph of the line.

12. $3x + y - 6 = 0$

13. $8x - 6y = 30$

14. Find a setting on a graphing utility such that the graph of the equation $y = x^3 - 5x + 5$ is approximately the same as the graph shown at the right.

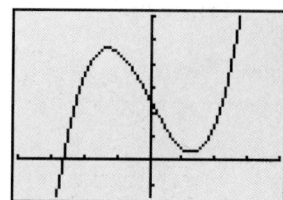

Figure for 14

15. Sketch the line passing through $(0, -6)$ with a slope of $m = \frac{4}{3}$.

16. Write the equation $5x + 3y - 9 = 0$ in slope-intercept form. Find the slope of a line that is perpendicular to this line.

| **2.4** | **Relations, Functions, and Function Notation** |

Relations ▪ Functions ▪ Function Notation ▪
Finding the Domain and Range of a Function ▪ Application

Relations

Many everyday occurrences involve two quantities that are paired or matched with each other by some rule of correspondence. The mathematical term for such a correspondence is a **relation.**

Definition of a Relation

A **relation** is any set of ordered pairs. The set of first components in the ordered pairs is the **domain of the relation,** and the set of second components is the **range of the relation.**

EXAMPLE 1   Analyzing a Relation

Find the domain and range of $\{(0, 1), (1, 3), (2, 5), (3, 5), (0, 3)\}$.

Solution

The domain is the set of all first components of the relation, and the range is the set of all second components.

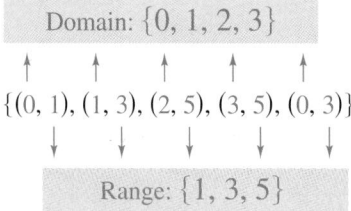

A graphical representation is given in Figure 2.38.

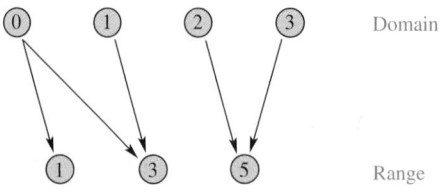

FIGURE 2.38

Functions

Historical Note: Although the concept of function dates back to the Babylonians as early as 2000 B.C., the concept was fine-tuned in the mid-1600s. Descartes is often credited as the first to use the term *function*.

In modeling real-life situations, you will work with a special type of relation called a function. A **function** is a relation in which no two ordered pairs have the same first component and different second components. For instance, (2, 3) and (2, 4) could not be ordered pairs of a function.

Definition of a Function

A **function** f from a set A to a set B is a rule of correspondence that assigns to each element x in set A exactly one element y in set B.

Set A is called the **domain** (or set of inputs) **of the function** f, and set B contains the **range** (or set of outputs) **of the function.**

The rule of correspondence for a function establishes a set of "input-output" ordered pairs of the form (x, y), where x is an input and y is the corresponding output. In some cases, the rule may generate only a finite set of ordered pairs, whereas in other cases the rule may generate an infinite set of ordered pairs.

EXAMPLE 2 *Input-Output Ordered Pairs for Functions*

Ask students to identify functions from real-life examples. Use relationships such as family members, age, social security numbers, and telephone numbers to illustrate functions and nonfunctional relations.

(a) For the function that pairs the year from 1991 through 1994 with the winner of the Super Bowl, each ordered pair is of the form (year, winner).

{(1991, Giants), (1992, Redskins), (1993, Cowboys), (1994, Cowboys)}

(b) For the function given by $y = x - 2$, each ordered pair is of the form (x, y).

{All points on the graph of $y = x - 2$}

(c) For the function that pairs the positive integers that are less than 7 with their squares, each ordered pair is of the form (n, n^2).

{(1, 1), (2, 4), (3, 9), (4, 16), (5, 25), (6, 36)}

(d) For the function that pairs each real number with its square, each ordered pair is of the form (x, x^2).

{All points (x, x^2), where x is a real number}

NOTE In Example 2, the sets in parts (a) and (c) have finite numbers of ordered pairs, whereas the sets in parts (b) and (d) have infinite numbers of ordered pairs.

Characteristics of a Function

1. Each element in the domain A must be matched with an element in the range, which is contained in the set B.

2. Some elements in the set B may not be matched with any element in the domain A.

3. Two or more elements of the domain may be matched with the same element in the range.

4. No element of the domain is matched with two different elements in the range.

EXAMPLE 3   Testing for Functions Represented by Ordered Pairs

Let $A = \{a, b, c\}$ and let $B = \{1, 2, 3, 4, 5\}$. Which represents a function from A to B?

Remind students that not all elements in set B must be used and that those elements from set B that are matched to elements from the domain, set A, form the set called the range. Have students identify the range in (a) and (c) in Example 3.

(The answers are (a) {2, 3, 4} and (c) {1}.)

(a) $\{(a, 2), (b, 3), (c, 4)\}$

(b) $\{(a, 4), (b, 5)\}$

(c)

(d)

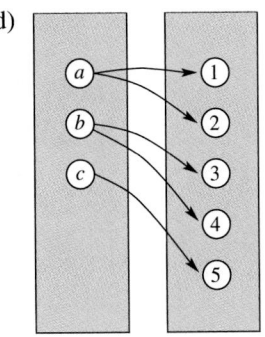

Solution

(a) This set of ordered pairs *does* represent a function from A to B. Each element of A is matched with exactly one element of B.

(b) This set of ordered pairs *does not* represent a function from A to B. Not all elements of A are matched with an element of B.

(c) This diagram *does* represent a function from A to B. It does not matter that each element of A is matched with the same element in B.

(d) This diagram *does not* represent a function from A to B. The element a in A is matched with *two* elements, 1 and 2, in B. This is also true of element b.

Representing functions by sets of ordered pairs is a common practice in the study of *discrete mathematics,* which deals mainly with finite sets of data or with finite subsets of the set of real numbers. In algebra, however, it is more common to represent functions by equations or formulas involving two variables. For instance, the equation

$$y = x^2 \qquad \text{Squaring function}$$

represents the variable y as a function of the variable x. The variable x is the **independent variable** and the variable y is the **dependent variable.** In this context, the domain of the function is the set of all *allowable* real values for the independent variable x, and the range of the function is the *resulting* set of all values taken on by the dependent variable y.

EXAMPLE 4   Testing for Functions Represented by Equations

Which of the equations represents y as a function of x?

(a) $y = x^2 + 1$ (b) $x - y^2 = 2$ (c) $-2x + 3y = 4$

Solution

(a) From the equation

$$y = x^2 + 1$$

you can see that for each value of x there corresponds just one value of y. For instance, when $x = 1$, the value of y is $1^2 + 1 = 2$. Therefore, y *is* a function of x.

(b) By writing the equation $x - y^2 = 2$ in the form

$$y^2 = x - 2$$

you can see that some values of x correspond to *two* values of y. For instance, when $x = 3$, $y^2 = 3 - 2 = 1$ and y can be 1 or -1. Hence, the solution points $(3, 1)$ and $(3, -1)$ show that y *is not* a function of x.

(c) By writing the equation $-2x + 3y = 4$ in the form

$$y = \frac{2}{3}x + \frac{4}{3}$$

you can see that for each value of x there corresponds just one value of y. For instance, when $x = 2$, the value of y is $\frac{4}{3} + \frac{4}{3} = \frac{8}{3}$. Therefore, y is a function of x.

NOTE   An equation that defines y as a function of x may or may not also define x as a function of y. For instance, the equation in part (a) does not define x as a function of y, but the equations in parts (b) and (c) do.

Function Notation

NOTE   Often y and $f(x)$ are used to represent the same quantity. That is, each represents the value of a function for a given value of x.

When an equation is used to represent a function, it is convenient to name the function so that it can be easily referenced. For example, the function $y = x^2 + 1$ in Example 4(a) can be given the name "f" and written in **function notation** as

$$f(x) = x^2 + 1.$$

Function Notation

In the notation $f(x)$:

   f is the **name** of the function,

   x is the **domain** (or input) value, and

   $f(x)$ is a **range** (or output) value y for a given x.

The symbol $f(x)$ is read as *the value of f at x* or simply *f of x.*

Emphasize that $f(x)$ does *not* mean "f times x." We are simply giving the function the name f and indicating in which variable the function is written.

The process of finding the value of $f(x)$ for a given value of x is called **evaluating a function.** This is accomplished by substituting a given x-value (input) into the equation to obtain the value of $f(x)$ (output). Here is an example.

Function	*x-Value*	*Function Value*
$f(x) = 3 - 4x$	$x = -1$	$f(-1) = 3 - 4(-1) = 3 + 4 = 7$

Although f is often used as a convenient function name and x as the independent variable, you can use other letters. For instance, the equations

$$f(x) = 2x^2 + 5, \quad f(t) = 2t^2 + 5, \quad \text{and} \quad g(s) = 2s^2 + 5$$

all define the same function. In fact, the letters used are simply "placeholders" and this same function is well described by the form

$$f(\;\;\;\;) = 2(\;\;\;\;)^2 + 5$$

where the parentheses are used in place of a letter. To evaluate $f(-2)$, simply place -2 in each set of parentheses, as follows.

$$f(-2) = 2(-2)^2 + 5$$
$$= 8 + 5$$
$$= 13$$

When evaluating a function, you are not restricted to substituting only numerical values into the parentheses. For instance, the value of $f(3x)$ is

$$f(3x) = 2(3x)^2 + 5$$
$$= 18x^2 + 5.$$

EXAMPLE 5   Evaluating a Function

Let $g(x) = 3x - x^2$ and find each of the following.

(a) $g(1)$

(b) $g(x + 1)$

(c) $g(x) + g(1)$

Solution

(a) Replacing x with 1 produces $g(1) = 3(1) - (1)^2 = 3 - 1 = 2$.

(b) Replacing x with $x + 1$ produces

$$
\begin{aligned}
g(x + 1) &= 3(x + 1) - (x + 1)^2 \\
&= 3x + 3 - (x^2 + 2x + 1) \\
&= 3x + 3 - x^2 - 2x - 1 \\
&= -x^2 + x + 2.
\end{aligned}
$$

Mention the importance of this concept in calculus.

(c) Using the result of part (a) for $g(1)$, you have

$$g(x) + g(1) = (3x - x^2) + 2 = -x^2 + 3x + 2.$$

Note that $g(x + 1) \neq g(x) + g(1)$. In general, $g(a + b)$ is not equal to $g(a) + g(b)$.

Sometimes a function is defined by more than one equation. An illustration of this is given in Example 6.

EXAMPLE 6   A Function Defined by Two Equations

NOTE   Functions such as the one in Example 6 that are defined by two or more equations are called **piecewise functions.**

Evaluate the function given by

$$
f(t) = \begin{cases} t^2 + 1, & t < 0 \\ t - 2, & t \geq 0 \end{cases}
$$

at (a) $t = -1$, (b) $t = 0$, and (c) $t = 1$.

Solution

Point out that the absolute value function can be piecewise defined as follows.

$$
f(x) = \begin{cases} x, & \text{if } x \geq 0 \\ -x, & \text{if } x < 0 \end{cases}
$$

(a) Because $t = -1 < 0$, you use $f(t) = t^2 + 1$ to obtain

$$f(-1) = (-1)^2 + 1 = 2.$$

(b) Because $t = 0 \geq 0$, you use $f(t) = t - 2$ to obtain

$$f(0) = 0 - 2 = -2.$$

(c) Because $t = 1 \geq 0$, you use $f(t) = t - 2$ to obtain

$$f(1) = 1 - 2 = -1.$$

Finding the Domain and Range of a Function

The domain of a function may be explicitly described along with the function, or it may be *implied* by the expression used to define the function. The **implied domain** is the set of all real numbers (inputs) that yield real number values for the function. For instance, the function given by

$$f(x) = \frac{1}{x^2 - 9} \qquad \text{Domain: all } x \neq \pm 3$$

has an implied domain that consists of all real values of x other than $x = \pm 3$. These two values are excluded from the domain because division by zero is undefined. Another common type of implied domain is that used to avoid even roots of negative numbers. For instance, the function given by

$$f(x) = \sqrt{x} \qquad \text{Domain: all } x \geq 0$$

is defined only for $x \geq 0$. Therefore, its implied domain is the set of all real numbers $x \geq 0$. More will be said about the domains of square root functions in Chapter 5.

EXAMPLE 7   *Finding the Domain and Range of a Function*

Find the domain and range of each function.

(a) f: $\{(-3, 0), (-1, 2), (0, 4), (2, 4), (4, -1)\}$

(b) Area of a circle: $A = \pi r^2$

Solution

(a) The domain of f consists of all first coordinates in the set of ordered pairs. The range consists of all second coordinates in the set of ordered pairs. Thus, the domain is

$$\text{Domain} = \{-3, -1, 0, 2, 4\}$$

and the range is

$$\text{Range} = \{0, 2, 4, -1\}.$$

(b) For the area of a circle, you must choose positive values for the radius r. Thus, the domain is the set of all real numbers r such that $r > 0$. The range is therefore the set of all real numbers A such that $A > 0$.

Note in Example 7(b) that the domain of a function can be implied by a physical context. For instance, from the equation $A = \pi r^2$, you would have no reason to restrict r to positive values. However, because you know this function represents the area of a circle, you can conclude that the radius must be positive.

Application

EXAMPLE 8 Finding an Equation to Represent a Function

Is the area of a square a *function* of the length of one of its sides? If so, find an equation that represents this function.

Solution

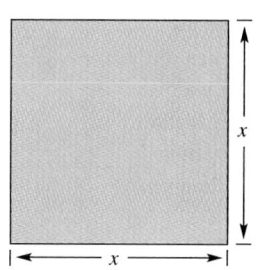

FIGURE 2.39

Figure 2.39 shows a typical square. For this square, let the variable A represent the area of the square, and let the variable x represent the length of any one of its sides. (Note that by definition all sides of the square have the same length.) Because the area of a square is completely determined from the length of its side, you can see that A *is* a function of x. The equation that represents this function is

$$A = x^2.$$

Group Activities   **Extending the Concept**

Determining Relationships that Are Functions Compile a list of statements describing relationships in everyday life. For each statement, identify the dependent and independent variables and discuss whether the statement *is* a function or *is not* a function and why. Here are two examples.

a. In the statement, "The number of ceramic tiles required to floor a kitchen is a function of the floor's area," the dependent variable is the required number of ceramic tiles and the independent variable is the area of the floor. This statement *is* a mathematically correct use of the word "function" because for each possible floor area there corresponds exactly one number of tiles needed to do the job.

b. In the statement, "Interest rates are a function of economic conditions," the dependent variable is interest rates and the independent variable is economic conditions. This statement *is not* a mathematically correct use of the word "function" because "economic conditions" is ambiguous; it is difficult to tell if one set of economic conditions would always result in the same interest rates.

2.4   **Exercises**

See Warm-Up
Exercises, p. A40

In Exercises 1–4, give the domain and range of the relation. Then draw a graphic representation.

1. $\{(-2, 0), (0, 1), (1, 4), (0, -1)\}$

2. $\{(3, 10), (4, 5), (6, -2), (8, 3)\}$

3. $\{(0, 0), (4, -3), (2, 8), (5, 5), (6, 5)\}$

4. $\{(-3, 6), (-3, 2), (-3, 5)\}$

In Exercises 5–8, write a set of ordered pairs that represents the rule of correspondence.

5. In a given week, a salesperson travels a distance d in t hours at an average speed of 50 miles per hour. The travel times for each day are 3 hours, 2 hours, 8 hours, 6 hours, and $\frac{1}{2}$ hour.

6. The cubes of all positive integers less than 8

7. The winners of the World Series from 1992 through 1996

8. The men inaugurated president of the United States in 1969, 1973, 1977, 1981, 1989, 1993, and 1997

In Exercises 9–16, determine if the relation is a function.

9. Domain   Range

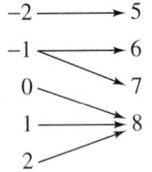

10. Domain   Range

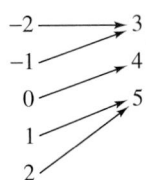

11.

Domain	Range
CBS	60 Minutes
	Dr. Quinn
	Dan Rather
ABC	Family Matters
	Step by Step
	Peter Jennings

12.

Domain	Range
(Year)	(Single women in the labor force, in millions)
1987	13.5
1988	13.8
1989	14.0
1990	14.1
1991	
1992	

(Source: U.S. Bureau of Labor Statistics)

13.

Input value	0	1	2	3	4
Output value	0	1	4	9	16

14.

Input value	0	1	2	1	0
Output value	1	8	12	15	20

15.

Input value	4	7	9	7	4
Output value	2	4	6	8	10

16.

Input value	0	2	4	6	8
Output value	5	5	5	5	5

In Exercises 17 and 18, determine which sets of ordered pairs represent functions from *A* to *B*.

17. $A = \{0, 1, 2, 3\}$ and $B = \{-2, -1, 0, 1, 2\}$

(a) $\{(0, 1), (1, -2), (2, 0), (3, 2)\}$

(b) $\{(0, -1), (2, 2), (1, -2), (3, 0), (1, 1)\}$

(c) $\{(0, 0), (1, 0), (2, 0), (3, 0)\}$

(d) $\{(0, 2), (3, 0), (1, 1)\}$

18. $A = \{1, 2, 3\}$ and $B = \{9, 10, 11, 12\}$

   (a) $\{(1, 10), (3, 11), (3, 12), (2, 12)\}$

   (b) $\{(1, 10), (2, 11), (3, 12)\}$

   (c) $\{(1, 10), (1, 9), (3, 11), (2, 12)\}$

   (d) $\{(3, 9), (2, 9), (1, 12)\}$

19. *Think About It* Explain the difference between a relation and a function.

20. *Think About It* Is it true that every relation is a function? Is it true that every function is a relation? Explain.

In Exercises 21–24, show that both ordered pairs are solutions of the equation and explain why this implies that *y* is not a function of *x*.

21. $x^2 + y^2 = 25$, $(0, 5)$, $(0, -5)$

22. $x^2 + 4y^2 = 16$, $(0, 2)$, $(0, -2)$

23. $|y| = x + 2$, $(1, 3)$, $(1, -3)$

24. $|y - 2| = x$, $(2, 4)$, $(2, 0)$

Interpreting a Graph In Exercises 25 and 26, use the graph, which shows the numbers of high school and college students in the United States. *(Source: U.S. National Center for Education Statistics)*

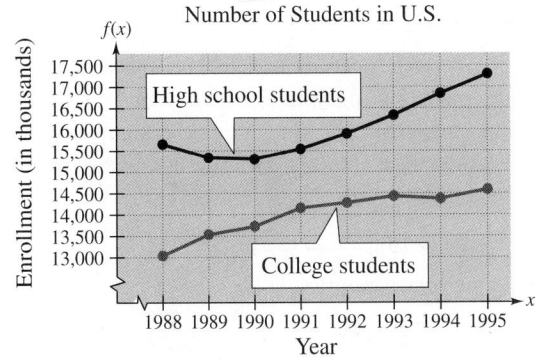

Number of Students in U.S.

25. Is the high school enrollment a function of the year? Is the college enrollment a function of the year? Explain.

26. Let $f(x)$ represent the number of high school students in year x. Find $f(1993)$.

In Exercises 27–30, fill in the blanks using the function and the specified value of the independent variable.

27. $f(x) = 3x + 5$

   (a) $f(2) = 3(\ \ ) + 5$

   (b) $f(-2) = 3(\ \ ) + 5$

   (c) $f(k) = 3(\ \ ) + 5$

   (d) $f(k + 1) = 3(\ \ ) + 5$

28. $f(x) = 3 - x^2$

   (a) $f(0) = 3 - (\ \ )^2$

   (b) $f(-3) = 3 - (\ \ )^2$

   (c) $f(m) = 3 - (\ \ )^2$

   (d) $f(t + 2) = 3 - (\ \ )^2$

29. $f(x) = \dfrac{x}{x + 2}$

   (a) $f(3) = \dfrac{(\ \ )}{(\ \ ) + 2}$

   (b) $f(-4) = \dfrac{(\ \ )}{(\ \ ) + 2}$

   (c) $f(s) = \dfrac{(\ \ )}{(\ \ ) + 2}$

   (d) $f(s - 2) = \dfrac{(\ \ )}{(\ \ ) + 2}$

30. $f(x) = \sqrt{x + 8}$

   (a) $f(1) = \sqrt{(\ \ ) + 8}$

   (b) $f(-4) = \sqrt{(\ \ ) + 8}$

   (c) $f(h) = \sqrt{(\ \ ) + 8}$

   (d) $f(h - 8) = \sqrt{(\ \ ) + 8}$

In Exercises 31–42, evaluate the function as indicated, and simplify.

31. $f(x) = 12x - 7$

   (a) $f(3)$ (b) $f\left(\frac{3}{2}\right)$

   (c) $f(a) + f(1)$ (d) $f(a + 1)$

32. $f(x) = 3 - 7x$

   (a) $f(-1)$ (b) $f\left(\frac{1}{2}\right)$

   (c) $f(t) + f(-2)$ (d) $f(w)$

33. $f(x) = \sqrt{x + 5}$

   (a) $f(-1)$ (b) $f(4)$

   (c) $f\left(\frac{16}{3}\right)$ (d) $f(5z)$

34. $g(x) = 8 - |x - 4|$

   (a) $g(0)$ (b) $g(8)$

   (c) $g(16) - g(-1)$ (d) $g(x - 2)$

35. $f(x) = \dfrac{3x}{x - 5}$

   (a) $f(0)$ (b) $f\left(\frac{5}{3}\right)$

   (c) $f(2) - f(-1)$ (d) $f(x + 4)$

36. $g(x) = \dfrac{|x + 1|}{x + 1}$

   (a) $g(2)$ (b) $g\left(-\frac{1}{3}\right)$

   (c) $g(-4)$ (d) $g(3) + g(-5)$

37. $f(x) = \begin{cases} x + 8, & x < 0 \\ 10 - 2x, & x \geq 0 \end{cases}$

   (a) $f(4)$ (b) $f(-10)$

   (c) $f(0)$ (d) $f(6) - f(-2)$

38. $f(x) = \begin{cases} -x, & x \leq 0 \\ x^2 - 3x, & x > 0 \end{cases}$

   (a) $f(0)$ (b) $f\left(-\frac{3}{2}\right)$

   (c) $f(4)$ (d) $f(-2) + f(25)$

39. $f(x) = 2x + 5$

   (a) $\dfrac{f(x + 2) - f(2)}{x}$ (b) $\dfrac{f(x - 3) - f(3)}{x}$

40. $f(x) = x^2 + 4$

   (a) $f(x + 1) - f(1)$ (b) $\dfrac{f(x - 5) - f(x)}{5}$

41. $g(x) = 1 - x^2$

   (a) $g(2.2)$ (b) $\dfrac{g(2.2) - g(2)}{0.2}$

42. $f(x) = x^3$

   (a) $f(x + 2)$ (b) $\dfrac{f(x + 2) - f(x)}{2}$

In Exercises 43–46, find the domain and range of the function.

43. $f: \{(0, 0), (2, 1), (4, 8), (6, 27)\}$

44. $f: \left\{\left(-3, -\frac{17}{2}\right), \left(-1, -\frac{5}{2}\right), (4, 2), (10, 11)\right\}$

45. Circumference of a circle: $C = 2\pi r$

46. Area of a square of side s: $A = s^2$

In Exercises 47–62, find the domain of the function.

47. $f(x) = \dfrac{2x}{x - 3}$ **48.** $g(x) = \dfrac{x + 5}{x + 4}$

49. $g(x) = \dfrac{5x}{x^2 - 3x + 2}$ **50.** $h(x) = 4x - 3$

51. $f(t) = \dfrac{t + 3}{t(t + 2)}$ **52.** $g(s) = \dfrac{s - 2}{(s - 6)(s - 10)}$

53. $h(x) = \dfrac{9}{x^2 + 1}$ **54.** $f(x) = \dfrac{4}{x^2 - 4}$

55. $f(x) = \sqrt{x - 2}$ **56.** $f(x) = \sqrt{x - 5}$

57. $H(x) = \sqrt{x}$ **58.** $G(x) = \sqrt[3]{x}$

59. $f(x) = \sqrt[3]{x + 1}$ **60.** $h(x) = \sqrt[4]{x}$

61. $f(t) = t^2 + 4t - 1$ **62.** $f(x) = |x + 3|$

63. *Geometry* Express the perimeter P of a square as a function of the length x of a side.

64. *Geometry* Express the surface area S of a cube as a function of the length x of one of its edges.

65. *Geometry* An open box is to be made from a square piece of material 24 inches on a side by cutting equal squares from the corners and turning up the sides (see figure). Write the volume V of the box as a function of x.

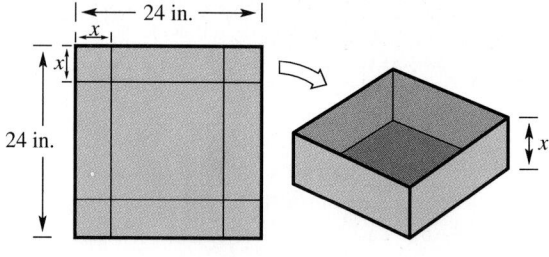

66. *Geometry* Strips of width x are cut from the four sides of a square that is 32 inches on a side (see figure). Write the area A of the remaining square as a function of x.

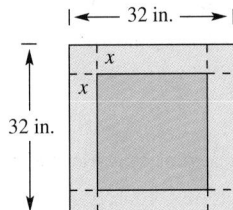

|←— 32 in. —→|

x

x

32 in.

67. *Geometry* Strips of width x are cut from two adjacent sides of a square that is 32 inches on a side (see figure). Write the area A of the remaining square as a function of x.

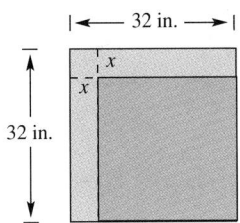

|←— 32 in. —→|

x

x

32 in.

68. *Cost* The inventor of a new game believes that the variable cost of producing the game is $1.95 per unit and the fixed costs are $8000. Write the total cost C as a function of x, the number of games produced.

69. *Profit* The marketing department of a business has determined that the profit for selling x units of a product is approximated by the model

$$P(x) = 50\sqrt{x} - 0.5x - 500.$$

Find (a) $P(1600)$ and (b) $P(2500)$.

70. *Distance* A plane is flying at a speed of 230 miles per hour. Express the distance d traveled by the plane as a function of the time t in hours.

71. *Safe Load* A solid rectangular beam has a height of 6 inches and a width of 4 inches. The safe load S of the beam with the load at the center is a function of its length L and is approximated by the model

$$S(L) = \frac{128{,}160}{L},$$

where S is measured in pounds and L is measured in feet. Find (a) $S(12)$ and (b) $S(16)$.

Think About It In Exercises 72 and 73, determine whether the statements use the word *function* in ways that are *mathematically* correct.

72. (a) The sales tax on a purchased item is a function of the selling price.

(b) Your score on the next algebra exam is a function of the number of hours you study the night before the exam.

73. (a) The amount in your savings account is a function of your salary.

(b) The speed at which a free-falling baseball strikes the ground is a function of the height from which it was dropped.

Section Project

Wages A wage earner is paid $12.00 per hour for regular time and time-and-a-half for overtime. The weekly wage function is

$$W(h) = \begin{cases} 12h, & 0 < h \le 40 \\ 18(h - 40) + 480, & h > 40 \end{cases}$$

where h represents the number of hours worked in a week. Evaluate the weekly wage function for the following values of h.

(a) $h = 30$ (b) $h = 40$ (c) $h = 45$ (d) $h = 50$

Could you use values of h for which $h < 0$ in this model? Why or why not? If the company increased the regular work week to 45 hours, what would the new weekly wage function be?

<table>
<tr><td>**2.5**</td><td>**Graphs of Functions**</td></tr>
</table>

The Graph of a Function ▪ Graphs of Linear Functions ▪
The Vertical Line Test ▪ Finding Domains and Ranges Graphically ▪
Graphs of Piecewise Functions

The Graph of a Function

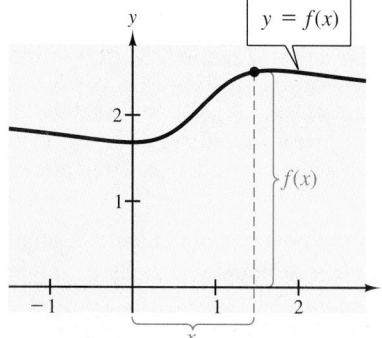

Consider a function f whose domain and range are the set of real numbers. The
graph of f is the set of ordered pairs $(x, f(x))$, where x is in the domain of f.

$$x = x\text{-coordinate of the ordered pair}$$
$$f(x) = y\text{-coordinate of the ordered pair}$$

Figure 2.40 shows a typical graph of such a function.

FIGURE 2.40

EXAMPLE 1 Sketching the Graph of a Function

Sketch the graph of $f(x) = x^2 - 4x + 3$.

Solution

One way to sketch the graph is to begin by making a table of values.

x	-1	0	1	2	3	4	5
$f(x)$	8	3	0	-1	0	3	8

Next, plot the seven points given by the table.

$$(-1, 8), (0, 3), (1, 0), (2, -1), (3, 0), (4, 3), (5, 8)$$

Finally, connect the points with a smooth curve, as shown in Figure 2.41.

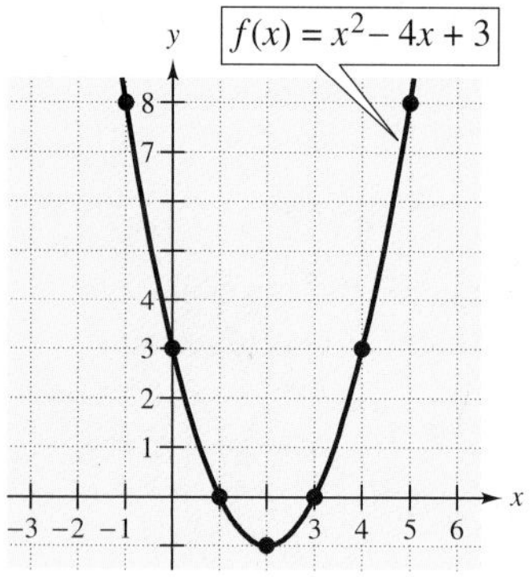

FIGURE 2.41

Exploration

Sketching Graphs of Functions   A graphing utility can be a big help in sketching
the graph of a function. The difficulty, however, is to find a viewing rectangle that
provides a good view of the overall characteristics of the function. With others in your
group, use a graphing utility to find a good viewing rectangle for each of the
following functions. How are the three graphs similar?

a. $f(x) = -2x^2 + 4x + 12$ b. $f(x) = x^2 - 5x$ c. $f(x) = x^2 - 24$

Stress the interchangeability of $f(x)$
and y. Point out that the ordered
pair (x, y) can also be written as
$(x, f(x))$.

Graphs of Linear Functions

There are three basic strategies for sketching the graph of a function.

1. Use the point-plotting method.
2. Use a graphing utility.
3. Use an algebraic approach.

The first two strategies are addressed on the facing page: the point-plotting method is illustrated in Example 1, and the use of a graphing utility is discussed in the Exploration feature. In the third strategy—*an algebraic approach*—you write the function in a form that allows you to recognize its graph. You might recognize the graph to be a line, a parabola, a V-shaped graph, or some other shape. For instance, in Section 2.3 you learned that the graph of an equation of the form $y = mx + b$ is a line with a slope of m and a y-intercept of $(0, b)$. This type of equation is called a **linear function.**

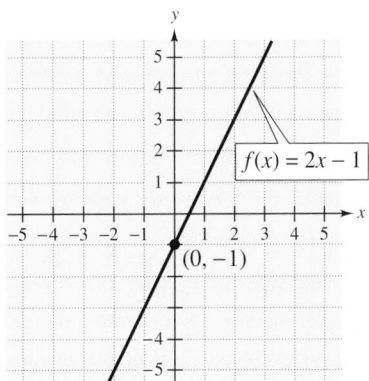

(a)

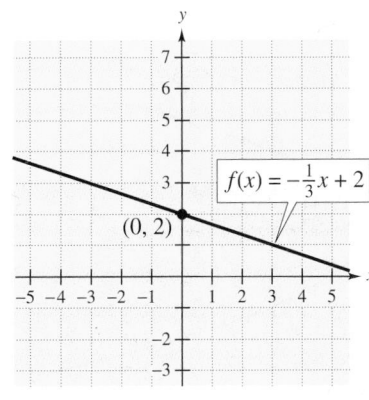

(b)

FIGURE 2.42

Definition of a Linear Function

A **linear function** of x is one that can be written in the form

$$f(x) = mx + b,$$

where $m \neq 0$. The graph of a linear function is a line with a slope of m and a y-intercept of $(0, b)$.

EXAMPLE 2   *Graphing Linear Functions*

Sketch the graphs of (a) $f(x) = 2x - 1$ and (b) $f(x) = -\frac{1}{3}x + 2$.

Solution

(a) This is a linear function.

Slope is 2. y-intercept is $(0, -1)$.

$$f(x) = 2x - 1$$

Its graph is a line with a slope of 2 and a y-intercept of $(0, -1)$, as shown in Figure 2.42 (a).

(b) This is a linear function.

Slope is $-\frac{1}{3}$. y-intercept is $(0, 2)$.

$$f(x) = -\frac{1}{3}x + 2$$

Its graph is a line with a slope of $-\frac{1}{3}$ and a y-intercept of $(0, 2)$, as shown in Figure 2.42(b).

The Vertical Line Test

By the definition of a function, at most one y-value corresponds to a given x-value. This implies that any vertical line can intersect the graph of a function at most once.

Study Tip

The Vertical Line Test provides you with an easy way to determine whether an equation represents y as a function of x. If the graph of an equation has the property that no vertical line intersects the graph at two (or more) points, then the equation represents y as a function of x. On the other hand, if you can find a vertical line that intersects the graph at two (or more) points, then the equation does not represent y as a function of x.

Vertical Line Test for Functions

A set of points on a rectangular coordinate system is the graph of y as a function of x if and only if no vertical line intersects the graph at more than one point.

EXAMPLE 3   Using the Vertical Line Test

Decide whether each equation represents y as a function of x.

(a) $y = |x| - 2$ (b) $x = y^2 - 1$

Solution

(a) From the graph of the equation in Figure 2.43(a), you can see that every vertical line intersects the graph at most once. Therefore, by the Vertical Line Test, the equation *does* represent y as a function of x.

(b) From the graph of the equation in Figure 2.43(b), you can see that a vertical line intersects the graph twice. Therefore, by the Vertical Line Test, the equation *does not* represent y as a function of x.

Ask students to consider whether any lines would fail the vertical line test.

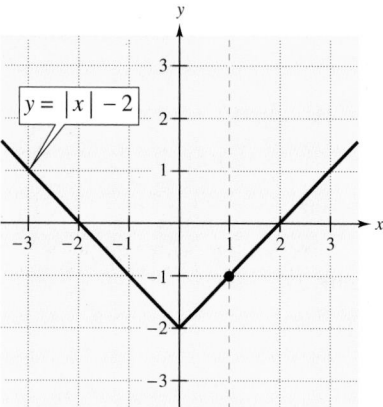

(a) Graph of a function of x; vertical line intersects once.

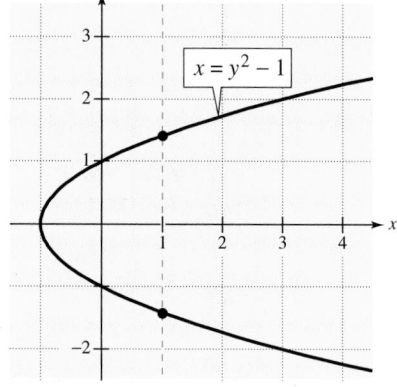

(b) Not a graph of a function of x; vertical line intersects twice.

FIGURE 2.43

EXAMPLE 4 *Testing for y as a Function of x*

Which of the following represent y as a function of x? Explain your reasoning.

(a) $x^2 = y - 5$ (b) $|y| = x - 5$ (c) $y = 2$

Solution

(a) Begin by rewriting the equation in the form

$$y = x^2 + 5.$$

Because you are able to isolate y, it appears that the equation does represent y as a function of x. You can confirm this conclusion by the Vertical Line Test. Notice in Figure 2.44(a) that no vertical line intersects the graph more than once. Therefore, this equation *does* represent y as a function of x.

(b) The graph of this equation is shown in Figure 2.44(b). Note that it is possible for a vertical line to intersect the graph at more than one point. Therefore, this equation *does not* represent y as a function of x. Remember that if an equation represents y as a function of x, then no two different solution points of the equation can have the same x-value. For this equation, you can see that the points $(6, 1)$ and $(6, -1)$ are both solution points of the equation. Because these two points have the same x-value, it follows that the equation $|y| = x - 5$ does not represent y as a function of x.

Ask students to determine the domain and range of the function $y = 2$.

(The answers are domain, all real numbers, and range {2}.)

(c) At first glance, you might be tempted to say that this equation does not represent y as a function of x, because x does not appear in the equation. However, in Figure 2.44(c), you can see that the graph passes the Vertical Line Test. Hence, the equation $y = 2$ *does* represent y as a function of x. This type of function is called a **constant function.**

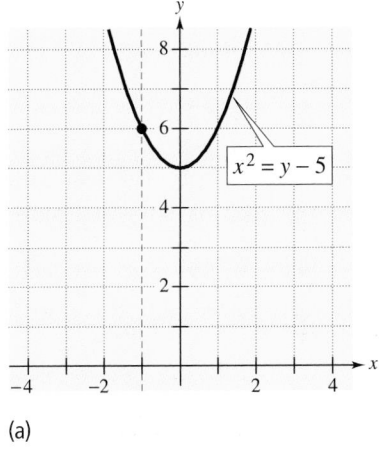

(a)

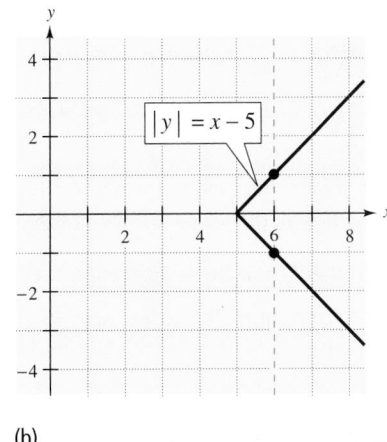

(b)

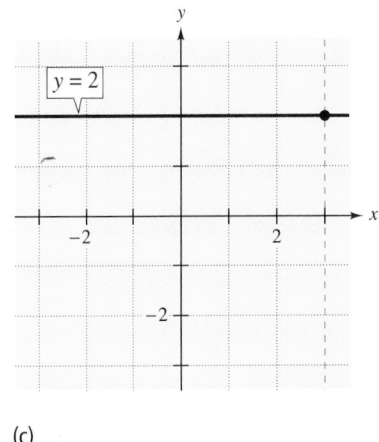

(c)

FIGURE 2.44

Finding Domains and Ranges Graphically

Recall from Section 2.4 that the *domain* of a function of x is the set of all x-values for which the function is defined. Unless specified differently, the domain of a linear function is the set of all real numbers.

If you want to restrict the domain of a function, you can write a restriction to the right of the equation. For instance, the domain of the function

$$f(x) = 4x + 5, \qquad x \geq 0$$

is the set of all nonnegative real numbers. As x varies over the domain of a function, the values of $f(x)$ form the *range* of the function. The graph of a function can help you describe its domain and its range.

EXAMPLE 5 *Finding a Function's Domain and Range*

Describe the domain and range of $f(x) = |x - 3|$.

Solution

The graph of this function is V-shaped, as shown in Figure 2.45. The domain of the function is the set of all real numbers. In other words, you can substitute *any* real number for x and obtain a corresponding y-value.

$$\text{Domain} = \{x: x \text{ is a real number.}\}$$

From the graph of the function, it appears that the y-values are never negative. This is confirmed by the fact that the absolute value of a number can never be negative. The range of this function is the set of all nonnegative real numbers.

$$\text{Range} = \{y: y \geq 0\}$$

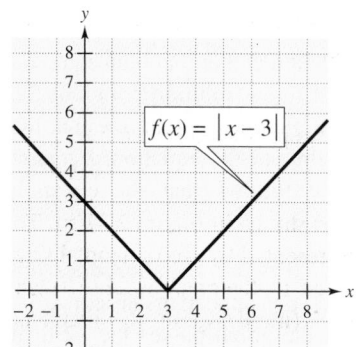

FIGURE 2.45

EXAMPLE 6 *Finding a Function's Domain and Range*

Describe the domain and range of $f(x) = \sqrt{4 - x^2}$.

Solution

The graph of this function is a half circle, as shown in Figure 2.46. The domain of the function is the set of all real numbers between -2 and 2, including -2 and 2.

$$\text{Domain} = \{x: -2 \leq x \leq 2\}$$

From the graph of the function, you can see that the y-values vary from 0 to 2. This implies that the range can be written as follows.

$$\text{Range} = \{y: 0 \leq y \leq 2\}$$

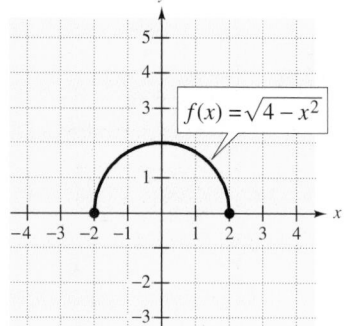

FIGURE 2.46

Graphs of Piecewise Functions

In Example 6 in Section 2.4, you saw that a *piecewise* function is defined by two or more equations. To sketch the graph of a piecewise function, you must consider each of the equations used to define the function.

EXAMPLE 7 *Sketching Piecewise Functions*

Sketch the graph of each function.

(a) $f(x) = \begin{cases} 2x + 1, & x \le 0 \\ x - 3, & x > 0 \end{cases}$

(b) $f(x) = \begin{cases} 2, & x \le -2 \\ -x, & -2 < x < 1 \\ -1, & x \ge 1 \end{cases}$

Solution

(a) The graph is made up of parts of two lines, as shown in Figure 2.47(a). To the left of 0 (including 0), the graph is represented by the line

$$y = 2x + 1. \qquad \text{Left portion of graph}$$

To the right of 0 (not including 0), the graph is represented by the line

$$y = x - 3. \qquad \text{Right portion of graph}$$

(b) The graph is made up of parts of three lines, as shown in Figure 2.47(b). To the left of -2 (including -2), the graph is represented by the horizontal line $y = 2$. Between -2 and 1 (not including -2 and 1), the graph is represented by the line $y = -x$. Finally, to the right of 1 (including 1), the graph is represented by the horizontal line $y = -1$.

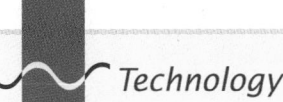 *Technology*

Most graphing utilities can graph a piecewise function. For instance, to graph the function in Example 7(a) on a *TI-82* or *TI-83*, you can enter the following.

$Y_1 = 2x + 1(x \le 0) +$
$\qquad\qquad x - 3(x > 0)$

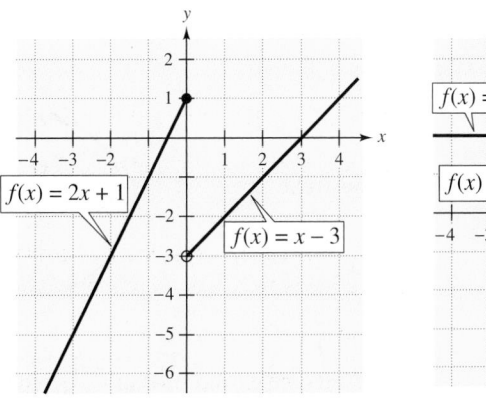

(a)

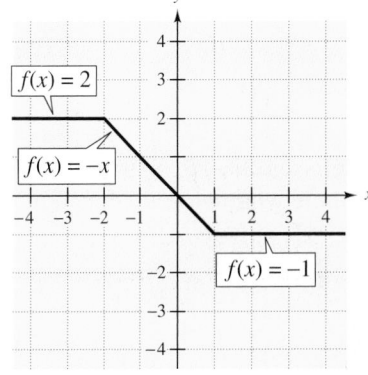

(b)

FIGURE 2.47

Group Activities **Exploring With Technology**

Recognizing Linear Functions One person in your group should enter a linear function with *integer coefficients* into a graphing utility. The person should then graph the function using a viewing rectangle that shows all intercepts of the function. From looking at *only* the graph of the function, the other members of the group should try to determine the equation of the linear function.

Take turns graphing functions and asking the other group members to find the equation that was graphed.

2.5 Exercises

See Warm-Up Exercises, p. A41

In Exercises 1–6, use function notation to write y as a function of x.

1. $3x - y + 10 = 0$ **2.** $x + 2y - 8 = 0$

3. $0.2x + 0.8y - 4.5 = 0$ **4.** $\frac{2}{3}x - \frac{3}{4}y = 0$

5. $y - 4 = 0$ **6.** $y - 3 = \frac{2}{3}(x + 2)$

In Exercises 7–10, find the intercepts of the linear function and use the intercepts to sketch the graph of the function.

7. $f(x) = 2x - 6$ **8.** $g(x) = -3x + 6$

9. $g(x) = -\frac{3}{4}x + 1$ **10.** $f(x) = 0.8x + 2$

In Exercises 11–16, use the Vertical Line Test to determine whether y is a function of x.

11. $y = \frac{1}{3}x^3$ **12.** $y = x^2 - 2x$

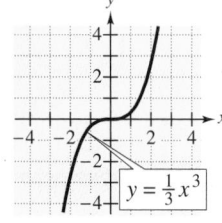

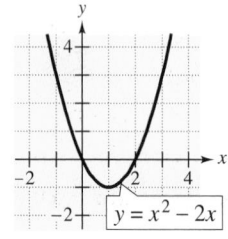

13. $y^2 = x$ **14.** $y = |x|$

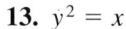

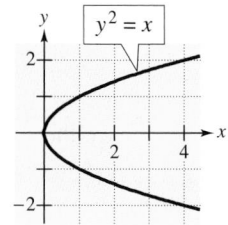

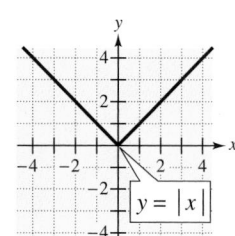

15. $x^2 + y^2 = 16$ **16.** $x - 2y^2 = 0$

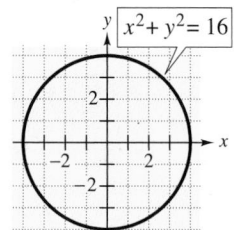

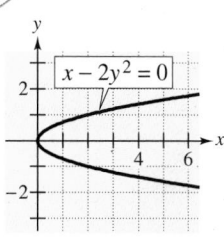

17. *Think About It* Does the graph in Exercise 11 represent x as a function of y? Explain.

18. *Think About It* Does the graph in Exercise 12 represent x as a function of y? Explain.

In Exercises 19–22, sketch the graph of the equation. Does the graph represent *y* as a function of *x*?

19. $3x - 5y = 15$ **20.** $y = x^2 + 2$

21. $y^2 = x + 1$ **22.** $x = y^4$

In Exercises 23–26, find the domain and range of the function.

23. $f(x) = \sqrt{x^2 - 4}$ **24.** $g(x) = \frac{3}{2}|x - 2|$

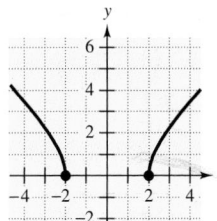 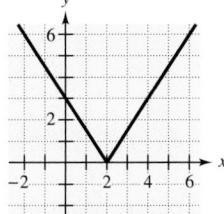

25. $g(x) = \sqrt{9 - x^2}$ **26.** $h(x) = \dfrac{3|x - 3|}{x - 3}$

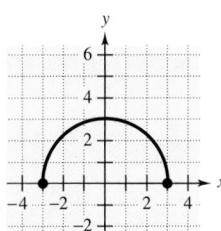 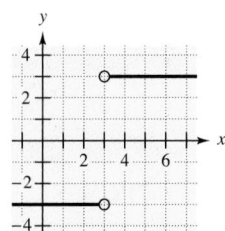

In Exercises 27–30, use a graphing utility to graph the function and find its domain and range.

27. $g(x) = 1 - x^2$ **28.** $f(x) = |x + 1|$

29. $f(x) = \sqrt{x - 2}$ **30.** $h(t) = \sqrt{4 - t^2}$

In Exercises 31–36, match the function with its graph. [The graphs are labeled (a), (b), (c), (d), (e), and (f).]

(a)

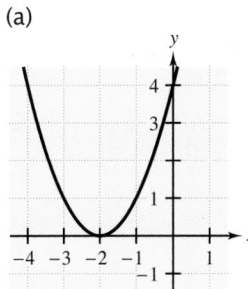

(b)

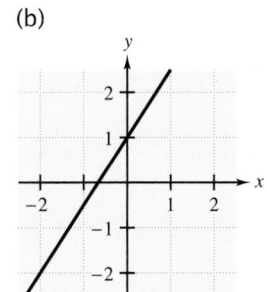

(c)

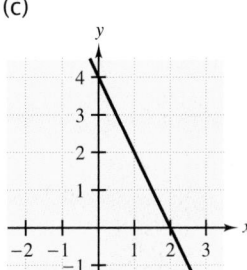

(d)

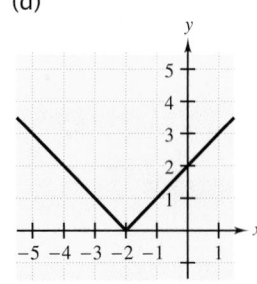

(e)

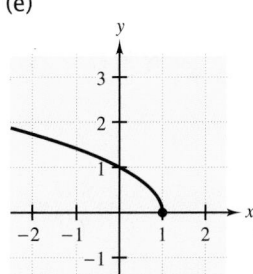

(f)

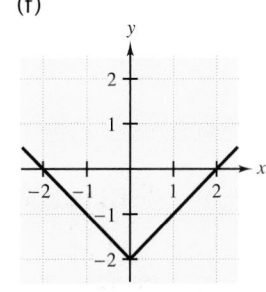

31. $f(x) = 4 - 2x$ **32.** $f(x) = \frac{3}{2}x + 1$

33. $g(x) = \sqrt{1 - x}$ **34.** $g(x) = (x + 2)^2$

35. $h(x) = |x| - 2$ **36.** $h(x) = |x + 2|$

In Exercises 37–58, sketch the graph of the function and use the graph to determine its domain and range.

37. $g(x) = 4$ **38.** $f(x) = -2$

39. $f(x) = 2x - 7$ **40.** $f(x) = 3 - 2x$

41. $g(x) = \frac{1}{2}x^2$ **42.** $h(x) = \frac{1}{4}x^2 - 1$

43. $f(x) = -(x - 1)^2$ **44.** $g(x) = (x + 2)^2 + 3$

45. $K(s) = |s - 4| + 1$ **46.** $Q(t) = 1 - |t + 1|$

47. $f(t) = \sqrt{t - 2}$

48. $h(x) = \sqrt{4 - x}$

49. $g(s) = \frac{1}{2}s^3$

50. $f(x) = x^3 - 4$

51. $f(x) = 6 - 3x,\qquad 0 \le x \le 2$

52. $f(x) = \frac{1}{3}x - 2,\qquad 6 \le x \le 12$

53. $h(x) = x^3,\qquad -2 \le x \le 2$

54. $h(x) = x(6 - x),\qquad 0 \le x \le 6$

55. $f(t) = \begin{cases} \sqrt{4 + t}, & t < 0 \\ \sqrt{4 - t}, & t \geq 0 \end{cases}$

56. $f(x) = \begin{cases} -x & x < 0 \\ x^2 - 4 & x \geq 0 \end{cases}$

57. $h(x) = \begin{cases} 4 - x^2 & x \leq 2 \\ x - 2, & x > 2 \end{cases}$

58. $f(x) = \begin{cases} x^2, & x < 1 \\ x^2 - 3x + 2, & x \geq 1 \end{cases}$

In Exercises 59 and 60, select the viewing rectangle that shows the most complete graph of the function.

59. $f(x) = -(x^2 - 20x + 50)$

Xmin = 0	Xmin = 0	Xmin = 15
Xmax = 10	Xmax = 20	Xmax = 30
Xscl = 1	Xscl = 2	Xscl = 2
Ymin = 0	Ymin = -10	Ymin = -10
Ymax = 30	Ymax = 60	Ymax = 60
Yscl = 2	Yscl = 6	Yscl = 5

60. $f(x) = x^4 - 10x^3$

Xmin = -10	Xmin = -5	Xmin = -3
Xmax = 10	Xmax = 5	Xmax = 12
Xscl = 1	Xscl = 1	Xscl = 1
Ymin = -10	Ymin = -10	Ymin = -1200
Ymax = 10	Ymax = 20	Ymax = 400
Yscl = 1	Yscl = 2	Yscl = 100

61. *Geometry* The perimeter of a rectangle is 200 meters.

(a) Show that the area of the rectangle is given by $A = x(100 - x)$, where x is its length.

(b) Use a graphing utility to graph the area function in part (a).

(c) Approximate the value of x that yields the largest value of A. Interpret the results.

62. *Profit* The profit P when x units of a product are sold is given by $P(x) = 0.47x - 100$ for x in the interval $0 \leq x \leq 1000$.

(a) Use a graphing utility to graph the profit function over the specified domain.

(b) Approximately how many units must be sold for the company to break even ($P = 0$)?

(c) Approximately how many units must be sold for the company to make a profit of $300?

Section Project

Aircraft Orders   The table gives the number N of new civil jet aircraft orders for U.S. aircraft manufacturers for the years 1989 through 1994.   *(Source: Aerospace Industries Association of America)*

Year	1989	1990	1991	1992	1993	1994
N	1015	670	280	231	31	79

A model for this data is given by

$$N = 48.7t^2 - 1309.1t + 8861.5,$$

where t is the time in years, with $t = 9$ corresponding to 1989.

(a) What is the domain of the function?

(b) Use a graphing utility to make a scatter plot of the data and graph the model in the same viewing rectangle.

(c) For which year does the model most accurately estimate the actual data? For which year is it least accurate?

2.6 | **Transformations of Functions**

Graphs of Basic Functions ■
Transformations of Graphs of Functions

Graphs of Basic Functions

NOTE Be sure you can recognize each of the graphs below. If you have a graphing utility, try using it to confirm the shape of each graph.

To become good at sketching the graphs of functions, it helps to be familiar with the graphs of some basic functions. The functions shown in Figure 2.48, and variations of them, occur frequently in applications.

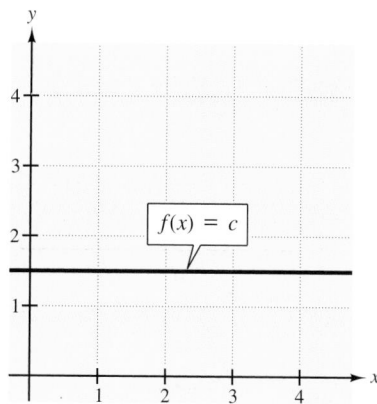

(a) Constant function

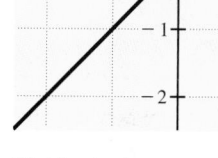

(b) Identity function

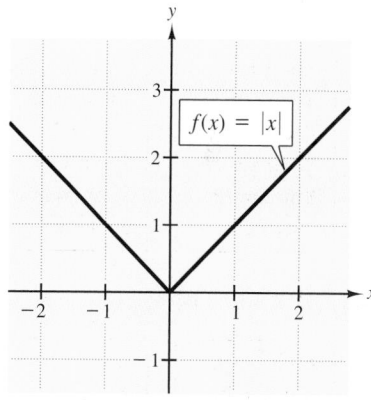

(c) Absolute value function

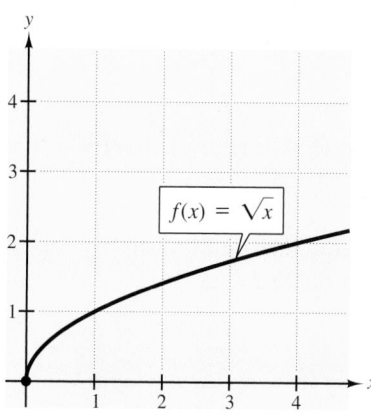

(d) Square root function

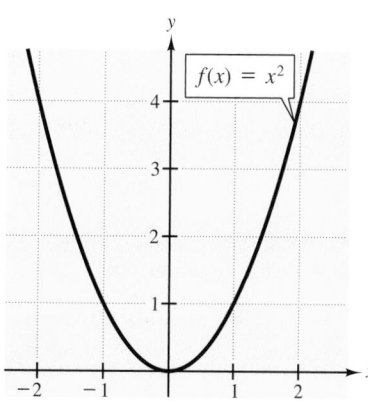

(e) Squaring function

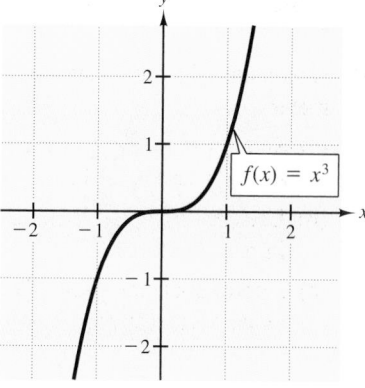

(f) Cubing function

FIGURE 2.48

Transformations of Graphs of Functions

Many functions have graphs that are simple transformations of the basic graphs shown in Figure 2.48. For example, you can obtain the graph of $h(x) = x^2 + 1$ by shifting the graph of $f(x) = x^2$ *upward* one unit, as shown in Figure 2.49(a). In function notation, h and f are related as follows.

$$h(x) = x^2 + 1 = f(x) + 1 \qquad \text{Upward shift of 1}$$

Similarly, you can obtain the graph of $g(x) = (x - 1)^2$ by shifting the graph of $f(x) = x^2$ to the *right* one unit, as shown in Figure 2.49(b). In this case, the functions g and f have the following relationship.

$$g(x) = (x - 1)^2 = f(x - 1) \qquad \text{Right shift of 1}$$

Have students do additional graphs $u(x) = x^2 - 1$ and $v(x) = (x + 1)^2$ and compare them to the graphs in Figure 2.49. Have students distinguish between $u(x) = x^2 - 1$ and $g(x) = (x - 1)^2$ by comparing the graphs of each one.

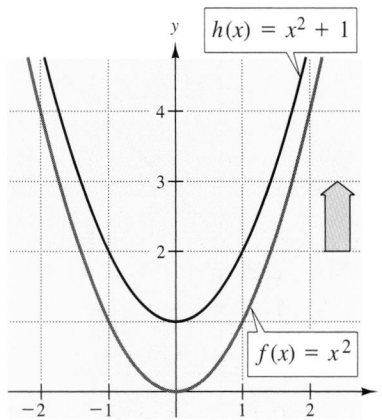

(a) Vertical shift: one unit up

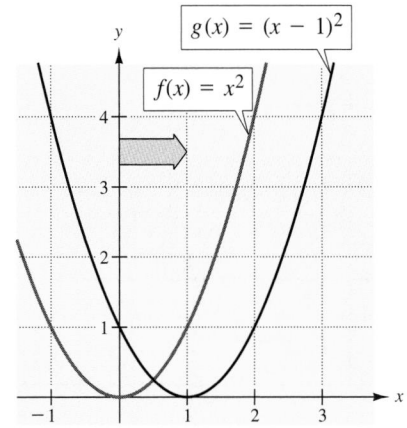

(b) Horizontal shift: one unit right

FIGURE 2.49

The various types of **horizontal and vertical shifts** of the graphs of functions are summarized as follows.

Vertical and Horizontal Shifts

Let c be a positive real number. **Vertical and horizontal shifts** of the graph of the function $y = f(x)$ are represented as follows.

Let students discover these relationships by graphing several functions on the same plane.

1. Vertical shift c units **upward:** $h(x) = f(x) + c$

2. Vertical shift c units **downward:** $h(x) = f(x) - c$

3. Horizontal shift c units to the **right:** $h(x) = f(x - c)$

4. Horizontal shift c units to the **left:** $h(x) = f(x + c)$

EXAMPLE 1   *Shifts of the Graphs of Functions*

Use the graph of $f(x) = x^2$ to sketch the graph of each function.

(a) $g(x) = x^2 - 2$

(b) $h(x) = (x + 3)^2$

Solution

(a) Relative to the graph of $f(x) = x^2$, the graph of $g(x) = x^2 - 2$ represents a *downward shift* of two units, as shown in Figure 2.50(a).

(b) Relative to the graph of $f(x) = x^2$, the graph of $h(x) = (x + 3)^2$ represents a *left shift* of three units, as shown in Figure 2.50(b).

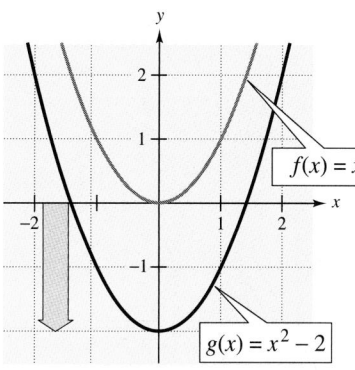 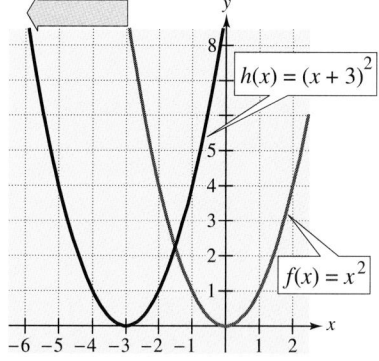

(a) Vertical shift: two units down   (b) Horizontal shift: three units left

FIGURE 2.50

Exploration

Combining Vertical and Horizontal Shifts   The graph of each of the following can be formed by shifting the graph of one of the six basic functions *twice*—once vertically and once horizontally. Describe the graph of each function and use your description to sketch the graph. If you have a graphing utility, use it to confirm your result.

a. $f(x) = (x - 3)^2 - 1$ b. $f(x) = |x + 1| - 2$

c. $f(x) = (x - 1)^3 + 2$ d. $f(x) = \sqrt{x + 4} + 1$

e. $f(x) = (x + 2)^2 + 2$ f. $f(x) = |x - 1| - 4$

Some graphs can be obtained from a *combination* of vertical and horizontal shifts, as shown in part (b) of the next example.

EXAMPLE 2   Shifts of the Graphs of Functions

Use the graph of $f(x) = x^3$ to sketch the graph of each function.

(a) $g(x) = x^3 + 2$ (b) $h(x) = (x - 1)^3 + 2$

Solution

(a) Relative to the graph of $f(x) = x^3$, the graph of

$$g(x) = x^3 + 2$$

represents an *upward shift* of two units, as shown in Figure 2.51(a).

(b) Relative to the graph of $f(x) = x^3$, the graph of

$$h(x) = (x - 1)^3 + 2$$

represents a *right shift* of one unit, followed by an *upward shift* of two units, as shown in Figure 2.51(b).

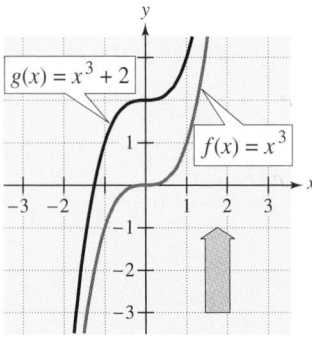

(a) Vertical shift: two units up

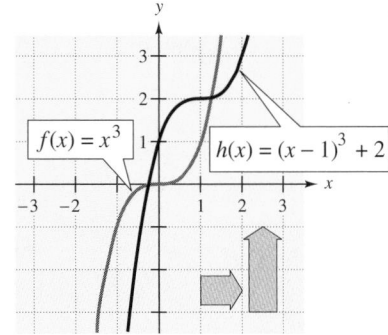

(b) Horizontal shift: one unit right
Vertical shift: two units up

FIGURE 2.51

NOTE   When you are combining horizontal and vertical shifts, it doesn't matter which you do first. For instance, in the graph shown in Figure 2.51(b), you would obtain the same result with either of the following: (a) First shift the graph of $f(x) = x^3$ to the right one unit, then shift this graph up two units; (b) first shift the graph of $f(x) = x^3$ up two units, then shift this graph to the right one unit.

The second basic type of transformation is called a **reflection.** For instance, if you imagine that the x-axis represents a mirror, the graph of $h(x) = -x^2$ is the mirror image (or reflection) of the graph of $f(x) = x^2$, as shown in Figure 2.52.

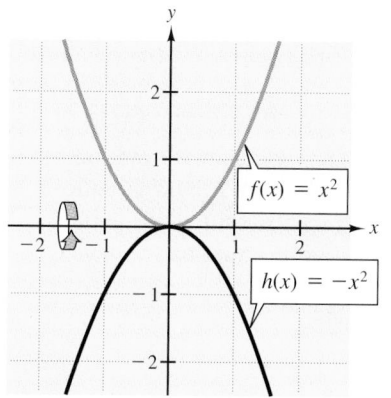

FIGURE 2.52 Reflection

Reflections in the Coordinate Axes

Reflections, in the coordinate axes, of the graph of $y = f(x)$ are represented as follows.

1. **Reflection in the x-axis:** $h(x) = -f(x)$

2. **Reflection in the y-axis:** $h(x) = f(-x)$

EXAMPLE 3 Reflections of the Graphs of Functions

Use the graph of $f(x) = \sqrt{x}$ to sketch the graph of each function.

(a) $g(x) = -\sqrt{x}$ (b) $h(x) = \sqrt{-x}$

Solution

(a) Relative to the graph of $f(x) = \sqrt{x}$, the graph of $g(x) = -\sqrt{x}$ represents a *reflection in the x-axis,* as shown in Figure 2.53(a).

(b) Relative to the graph of $f(x) = \sqrt{x}$, the graph of $h(x) = \sqrt{-x}$ represents a *reflection in the y-axis,* as shown in Figure 2.53(b).

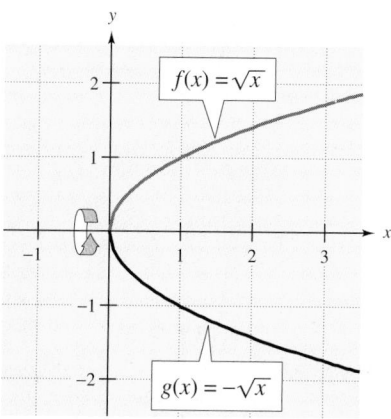

(a) Reflection in x-axis

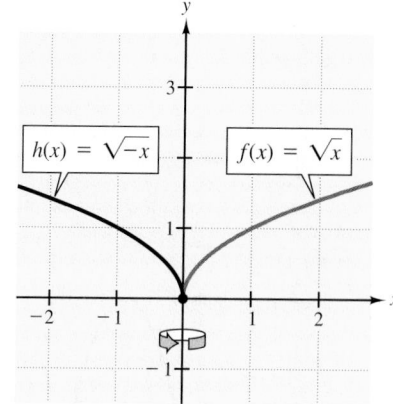

(b) Reflection in y-axis

FIGURE 2.53

Group Activities | **Graphical Reasoning**

Graphs of nine functions are shown below. Each is a transformation of one of the six basic graphs shown on page 167. With others in your group, write an equation for each function. Explain your reasoning. If you have a graphing utility, use it to confirm your equation.

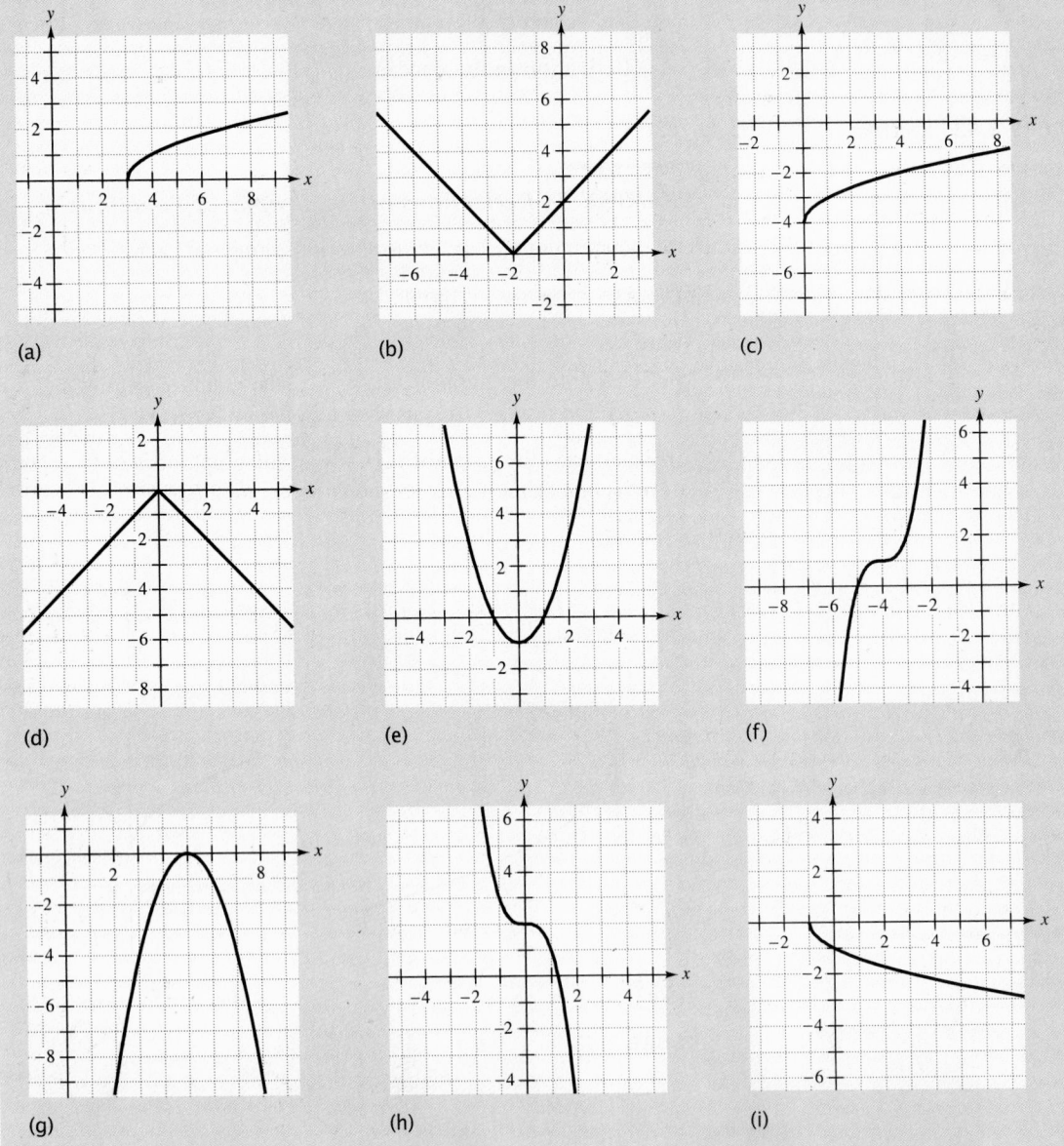

(a)

(b)

(c)

(d)

(e)

(f)

(g)

(h)

(i)

2.6   **Exercises**

See Warm-Up Exercises, p. A41

In Exercises 1–12, sketch the graphs of the three functions *by hand* on the same rectangular coordinate system. Verify your result with a graphing utility.

1. $f(x) = x$
$g(x) = x - 3$
$h(x) = 3x$

2. $f(x) = \frac{1}{2}x$
$g(x) = \frac{1}{2}x + 1$
$h(x) = \frac{1}{2}(x - 1)$

3. $f(x) = x^2$
$g(x) = x^2 + 3$
$h(x) = (x - 3)^2$

4. $f(x) = x^2$
$g(x) = x^2 - 2$
$h(x) = (x - 2)^2 + 1$

5. $f(x) = -x^2$
$g(x) = -x^2 + 2$
$h(x) = -(x - 1)^2$

6. $f(x) = (x + 1)^2$
$g(x) = (x + 1)^2 + 2$
$h(x) = -(x + 1)^2 + 4$

7. $f(x) = x^2$
$g(x) = \left(\frac{1}{2}x\right)^2$
$h(x) = (2x)^2$

8. $f(x) = x^2$
$g(x) = \left(\frac{1}{4}x\right)^2 + 3$
$h(x) = -\left(\frac{1}{4}x\right)^2$

9. $f(x) = |x|$
$g(x) = |x| - 2$
$h(x) = |x - 2|$

10. $f(x) = |x|$
$g(x) = \frac{1}{2}|x|$
$h(x) = 1 - |x|$

11. $f(x) = \sqrt{x}$
$g(x) = \sqrt{x} + 2$
$h(x) = \sqrt{x - 2}$

12. $f(x) = \sqrt{x}$
$g(x) = \sqrt{x + 3}$
$h(x) = \sqrt{x - 2} + 1$

In Exercises 13–26, identify the basic function, and any transformation shown in the graph. Write the equation for the graphed function.

13.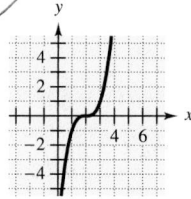

14.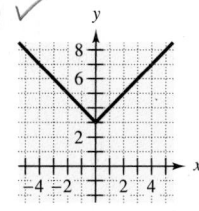

15.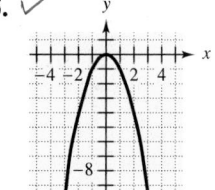

16.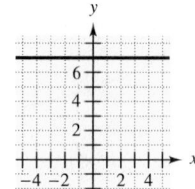

17.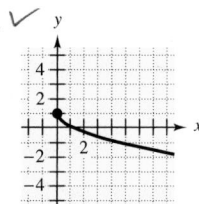

18.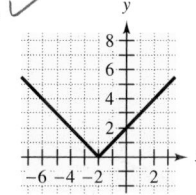

19.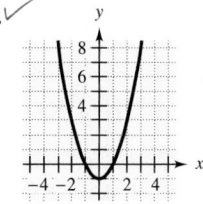

20.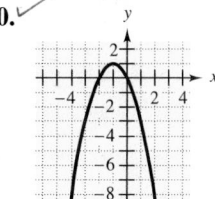

21.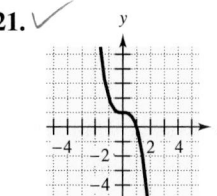

22.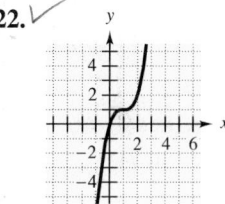

23.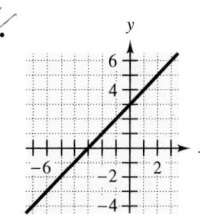

24.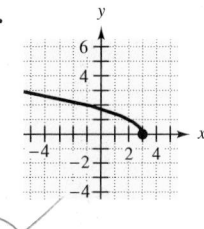

25.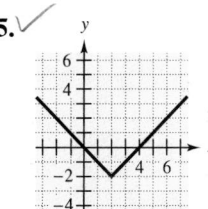

26.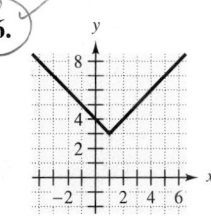

In Exercises 27–32, use the graph of $f(x) = \sqrt{x}$ to write a function that represents the graph.

27.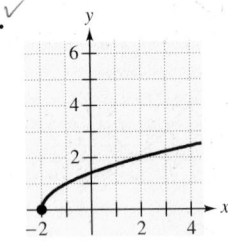

28.

29.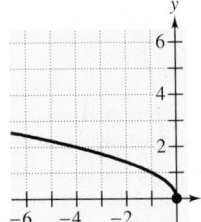

30.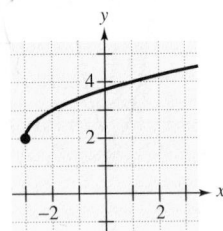

31.

32.

In Exercises 33–40, identify the transformation of the graph of $f(x) = x^3$ and sketch the graph of h.

33. $h(x) = x^3 + 3$
34. $h(x) = x^3 - 5$
35. $h(x) = (x - 3)^3$
36. $h(x) = (x + 2)^3$
37. $h(x) = (-x)^3$
38. $h(x) = -x^3$
39. $h(x) = 2 - (x - 1)^3$
40. $h(x) = (x + 2)^3 - 3$

In Exercises 41–46, identify the transformation of the graph of $f(x) = |x|$ and use a graphing utility to graph h.

41. $h(x) = |x - 5|$
42. $h(x) = |x + 3|$
43. $h(x) = |x| - 5$
44. $h(x) = |-x|$
45. $h(x) = -|x|$
46. $h(x) = 5 - |x|$

In Exercises 47–50, use a graphing utility to graph the two functions in the same viewing rectangle. Describe the graph of g relative to the graph of f.

47. $f(x) = x^3 - 3x^2$
$g(x) = f(x + 2)$

48. $f(x) = x^3 - 3x^2 + 2$
$g(x) = f(x - 1)$

49. $f(x) = x^3 - 3x^2$
$g(x) = f(-x)$

50. $f(x) = x^3 - 3x^2 + 2$
$g(x) = -f(x)$

In Exercises 51 and 52, use the graph of $f(x) = x^3 - 3x^2$ to write an equation for the function g shown in the graph.

51.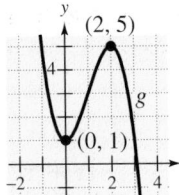

52.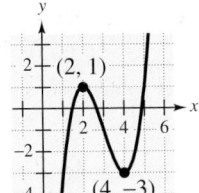

53. Use the graph of f to sketch the graphs.
(a) $y = f(x) + 2$
(b) $y = -f(x)$
(c) $y = f(x - 2)$
(d) $y = f(x + 3)$
(e) $y = f(x) - 1$
(f) $y = f(-x)$

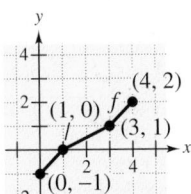

Figure for 53

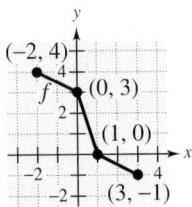

Figure for 54

54. Use the graph of f to sketch the graphs.
(a) $y = f(x) - 1$
(b) $y = f(x + 1)$
(c) $y = f(x - 1)$
(d) $y = -f(x - 2)$
(e) $y = f(-x)$
(f) $y = f(x) + 2$

55. *Investigation*   Use a graphing utility to graph each function. Describe any similarities and differences you observe among the graphs.

(a) $y = x$ (b) $y = x^2$

(c) $y = x^3$ (d) $y = x^4$

(e) $y = x^5$ (f) $y = x^6$

56. *Conjecture*   Use the results in Exercise 55 to make a conjecture about the graphs of $y = x^7$ and $y = x^8$. Use a graphing utility to verify your conjecture.

57. Use the results in Exercise 55 to sketch the graph of $y = (x - 3)^5$ by hand. Use a graphing utility to verify your graph.

58. Use the results in Exercise 55 to sketch the graph of $y = (x + 1)^6$ by hand. Use a graphing utility to verify your graph.

Conjecture   In Exercises 59–62, use the results in Exercises 55–58 to make a conjecture about the shape of the graph of the function. Use a graphing utility to verify your conjecture.

59. $f(x) = x^2(x - 6)^2$ **60.** $f(x) = x^3(x - 6)^2$

61. $f(x) = x^2(x - 6)^3$ **62.** $f(x) = x^3(x - 6)^3$

63. *Civilian Population of the United States*   For 1950 through 1996, the civilian population P (in thousands) of the United States can be modeled by

$$P_1(t) = -5.55t^2 + 2680t + 153{,}121,$$

where $t = 0$ represents 1950. *(Source: U.S. Bureau of the Census)*

(a) Use a graphing utility to graph the function over the appropriate domain.

(b) In the transformation of the population function $P_2(t) = -5.55(t + 20)^2 + 2680(t + 20) + 153{,}121,$ $t = 0$ corresponds to what calendar year? Explain.

(c) Use a graphing utility to graph P_2 over the appropriate domain.

Section Project

Graphical Reasoning   An electronically controlled thermostat in a home is programmed to lower the temperature automatically during the night. The temperature in the house T, in degrees Fahrenheit, is given in terms of t, the time in hours on a 24-hour clock.

(a) Explain why T is a function of t.

(b) Find $T(4)$ and $T(15)$.

(c) Suppose the thermostat were reprogrammed to produce a temperature H where $H(t) = T(t - 1)$. Explain how this would change the temperature in the house.

(d) Suppose the thermostat were reprogrammed to produce a temperature H where $H(t) = T(t) - 1$. Explain how this would change the temperature in the house.

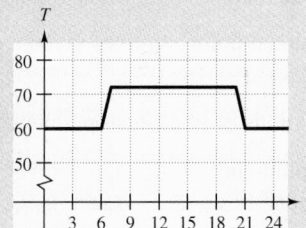

Chapter Summary

After studying this chapter, you should have acquired the following skills. These skills are keyed to the Review Exercises that begin on page 177. Answers to odd-numbered Review Exercises are given in the back of the book.

- Create graphs that represent real-life data. *(Section 2.1)*   **Review Exercises 1, 2**
- Plot points that form lines or polygons. *(Section 2.1)*   **Review Exercises 3–6**
- Determine the quadrants in which points are located. *(Section 2.1)*   **Review Exercises 7–10**
- Determine whether ordered pairs are solutions of equations. *(Section 2.1)*   **Review Exercises 11, 12**
- Match equations with their graphs. *(Section 2.2)*   **Review Exercises 13–16**
- Sketch graphs of equations and label the x- and y-intercepts. *(Section 2.2)*   **Review Exercises 17–22**
- Find the slopes of lines through pairs of points. *(Section 2.3)*   **Review Exercises 23–28**
- Find missing values so that sets of points are collinear. *(Sections 2.1, 2.3)*   **Review Exercises 29, 30**
- Find additional points on lines given a point on each line and the slope. *(Section 2.3)*   **Review Exercises 31–36**
- Write equations of lines in slope-intercept form and sketch the lines. *(Section 2.3)*   **Review Exercises 37–40**
- Use a graphing utility to graph pairs of lines to determine whether they are parallel, perpendicular, or neither. *(Section 2.3)*   **Review Exercises 41–44**
- Decide whether relations are functions. *(Section 2.4)*   **Review Exercises 45–48**
- Estimate the intercepts of graphs, check them algebraically, and determine whether the graphs represent y as a function of x. *(Sections 2.2, 2.4, 2.5)*   **Review Exercises 49–52**
- Graph equations and use the Vertical Line Test to determine whether they are functions. *(Section 2.5)*   **Review Exercises 53–58**
- Evaluate functions and simplify. *(Section 2.4)*   **Review Exercises 59–66**
- Find the domains of functions. *(Section 2.4)*   **Review Exercises 67–70**
- Select viewing rectangles on a graphing utility that show the most complete graphs of functions. *(Section 2.5)*   **Review Exercises 71, 72**
- Sketch graphs of functions and check results using a graphing utility. *(Section 2.5)*   **Review Exercises 73–84**
- Identify and sketch transformations of graphs. *(Section 2.6)*   **Review Exercises 85–88**
- Solve real-life problems modeled by functions. *(Sections 2.4, 2.5)*   **Review Exercises 89–92**

Review Exercises

1. *Weather*   The normal daily maximum and minimum temperatures for each month for the city of Chicago are given in the table. Make a double line graph for the data. During what time of the year is there the greatest difference between the daily maximum and minimum temperatures?   *(Source: NOAA)*

Month	Jan	Feb	Mar	Apr	May	Jun
Max	29.0	33.5	45.8	58.6	70.1	79.6
Min	12.9	17.2	28.5	38.6	47.7	57.5

Month	Jul	Aug	Sep	Oct	Nov	Dec
Max	83.7	81.8	74.8	63.3	48.4	34.0
Min	62.6	61.6	53.9	42.2	31.6	19.1

2. *CompuServe Revenues*   The revenues (in millions of dollars) for CompuServe for the years 1990 through 1994 are given in the table. Create a bar graph for the data.   *(Source: H & R Block 1995 Annual Report)*

Year	1990	1991	1992	1993	1994
Revenue	206.7	251.6	280.9	315.4	429.9

In Exercises 3 and 4, plot the points on a rectangular coordinate system. Are the points collinear or do they form a triangle?

3. $(0, -3), \left(\frac{5}{2}, 5\right), (-2, -4)$

4. $\left(1, -\frac{3}{2}\right), \left(-2, 2\frac{3}{4}\right), (5, 10)$

In Exercises 5 and 6, plot the points and connect them with line segments to form the figure.

5. *Right triangle:* $(1, 1), (12, 9), (4, 20)$

6. *Parallelogram:* $(0, 0), (7, 1), (8, 4), (1, 3)$

In Exercises 7–10, determine the quadrant(s) in which the point is located.

7. $(2, -6)$ **8.** $(-4.8, -2)$

9. $(4, y)$ **10.** $(x, y), \ xy > 0$

In Exercises 11 and 12, determine whether the ordered pairs are solution points of the given equation.

11. $y = 4 - \frac{1}{2}x$

  (a) $(4, 2)$ (b) $(-1, 5)$

  (c) $(-4, 0)$ (d) $(8, 0)$

12. $3x - 2y + 18 = 0$

  (a) $(3, 10)$ (b) $(0, 9)$

  (c) $(-4, 3)$ (d) $(-8, 0)$

In Exercises 13–16, match the equation with its graph. [The graphs are labeled (a), (b), (c), and (d).]

(a)
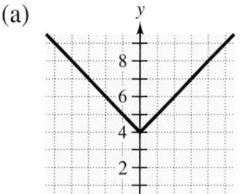

(b)
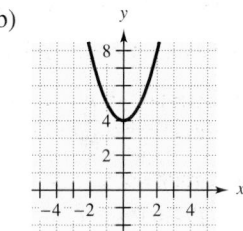

(c)
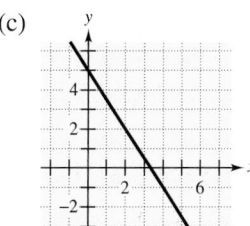

(d)
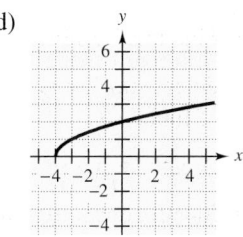

13. $y = 5 - \frac{3}{2}x$ **14.** $y = x^2 + 4$

15. $y = |x| + 4$ **16.** $y = \sqrt{x + 4}$

In Exercises 17–22, sketch the graph of the equation, and label the x- and y-intercepts.

17. $y = 6 - \frac{1}{3}x$ **18.** $y = \frac{3}{4}x - 2$

19. $3y - 2x - 3 = 0$ **20.** $3x + 4y + 12 = 0$

21. $x = |y - 3|$ **22.** $y = 1 - x^3$

In Exercises 23–28, find the slope of the line through the points.

23. $(-1, 1), (6, 3)$ **24.** $(-2, 5), (3, -8)$

25. $(-1, 3), (4, 3)$ **26.** $(7, 2), (7, 8)$

27. $(0, 6), (8, 0)$ **28.** $(0, 0), \left(\frac{7}{2}, 6\right)$

In Exercises 29 and 30, find t so that the three points are collinear.

29. $(-3, -3), (0, t), (1, 3)$

30. $(2, 1), (1, t), (8, 3)$

In Exercises 31–36, a point on a line and the slope of the line are given. Find two additional points on the line. (There are many correct answers.)

31. $(2, -4)$ **32.** $\left(-4, \frac{1}{2}\right)$

   $m = -3$ $m = 2$

33. $(3, 1)$ **34.** $\left(-3, -\frac{3}{2}\right)$

   $m = \frac{5}{4}$ $m = -\frac{1}{3}$

35. $(3, 7)$ **36.** $(7, -2)$

   m is undefined. $m = 0$

In Exercises 37–40, write the equation of the line in slope-intercept form and sketch the line.

37. $5x - 2y - 4 = 0$

38. $x - 3y - 6 = 0$

39. $x + 2y - 2 = 0$

40. $y - 6 = 0$

In Exercises 41–44, use a graphing utility to graph the equations in the same viewing rectangle. Are the lines parallel, perpendicular, or neither?

41. $y = \frac{3}{2}x + 1$ **42.** $y = 2x - 5$

   $y = \frac{2}{3}x - 1$ $y = 2x + 3$

43. $y = \frac{3}{2}x - 2$ **44.** $y = -0.3x - 2$

   $y = -\frac{2}{3}x + 1$ $y = 0.3x + 1$

In Exercises 45–48, determine if the relation is a function.

45. Domain   Range

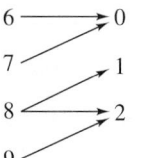

46. Domain   Range

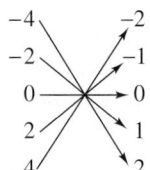

47. Domain   Range

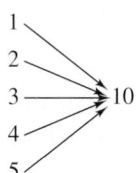

48. Domain   Range

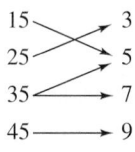

In Exercises 49–52, graphically estimate the inter-cepts. Then check your estimate algebraically. Does the graph represent *y* as a function of *x*?

49. $9y^2 = 4x^3$

50. $y = 4x^3 - x^4$

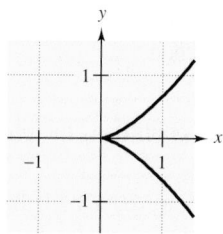

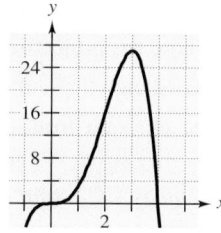

51. $y = x^2(x - 3)$

52. $x^3 + y^3 - 6xy = 0$

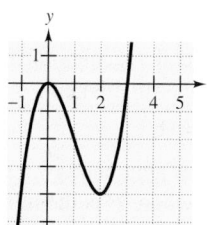

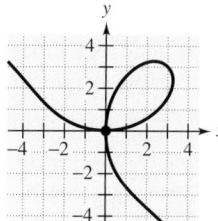

In Exercises 53–58, sketch the graph of the equation and use the Vertical Line Test to determine whether *y* is a function of *x*.

53. $y = 6 - \frac{2}{3}x$

54. $y = \frac{5}{2}x - 4$

55. $2y - 4x + 3 = 0$

56. $3x + 2y + 12 = 0$

57. $x - y^4 = 0$

58. $|y| = 2 - x$

In Exercises 59–66, evaluate the function at the specified values of the independent variable, and simplify.

59. $f(x) = 4 - \frac{5}{2}x$

(a) $f(-10)$ (b) $f(\frac{2}{5})$

(c) $f(t) + f(-4)$ (d) $f(x + h)$

60. $h(x) = x(x - 8)$

(a) $h(8)$ (b) $h(10)$

(c) $h(-3)$ (d) $h(t + 4)$

61. $f(t) = \sqrt{5 - t}$

(a) $f(-4)$ (b) $f(5)$

(c) $f(3)$ (d) $f(5z)$

62. $g(x) = \dfrac{|x + 4|}{4}$

(a) $g(0)$ (b) $g(-8)$

(c) $g(2) - g(-5)$ (d) $g(x - 2)$

63. $f(x) = \begin{cases} -3x, & x \le 0 \\ 1 - x^2, & x > 0 \end{cases}$

(a) $f(2)$ (b) $f\left(-\frac{2}{3}\right)$

(c) $f(1)$ (d) $f(4) - f(3)$

64. $h(x) = \begin{cases} x^3, & x \le 1 \\ (x - 1)^2 + 1, & x > 1 \end{cases}$

(a) $h(2)$ (b) $h\left(-\frac{1}{2}\right)$

(c) $h(0)$ (d) $h(4) - h(3)$

65. $f(x) = 3 - 2x$

(a) $\dfrac{f(x + 2) - f(2)}{x}$ (b) $\dfrac{f(x - 3) - f(3)}{x}$

66. $f(x) = 7x + 10$

(a) $\dfrac{f(x + 1) - f(1)}{x}$ (b) $\dfrac{f(x - 5) - f(5)}{x}$

In Exercises 67–70, find the domain of the function.

67. $h(x) = 4x^2 - 7$

68. $g(s) = \dfrac{s + 1}{(s - 1)(s + 5)}$

69. $f(x) = \sqrt{x - 2}$

70. $f(x) = |x - 6| + 10$

In Exercises 71 and 72, select the viewing rectangle on a graphing utility that shows the most complete graph of the function.

71. $f(x) = x^4 - 2x^3$

Xmin = 0	Xmin = -5	Xmin = -3
Xmax = 20	Xmax = 5	Xmax = 3
Xscl = 1	Xscl = 1	Xscl = 1
Ymin = -10	Ymin = -100	Ymin = -3
Ymax = 10	Ymax = 100	Ymax = 5
Yscl = 1	Yscl = 10	Yscl = 1

72. $f(x) = 5x\sqrt{16 - x^2}$

Xmin = -6	Xmin = -20	Xmin = -10
Xmax = 6	Xmax = 20	Xmax = 10
Xscl = 1	Xscl = 4	Xscl = 2
Ymin = -50	Ymin = -600	Ymin = -20
Ymax = 50	Ymax = 600	Ymax = 20
Yscl = 10	Yscl = 100	Yscl = 4

In Exercises 73–84, sketch the graph of the function. Use a graphing utility to confirm your result.

73. $g(x) = \frac{1}{8}x^2$

74. $y = 4 - (x - 3)^2$

75. $y = (x - 2)^2$

76. $h(x) = 9 - (x - 2)^2$

77. $y = \frac{1}{2}x(2 - x)$

78. $f(t) = \sqrt{\frac{t}{2}}$

79. $y = 8 - 2|x|$

80. $f(x) = |x + 1| - 2$

81. $g(x) = \frac{1}{4}x^3$, $-2 \le x \le 2$

82. $h(x) = x(4 - x)$, $0 \le x \le 4$

83. $f(x) = \begin{cases} 2 - (x - 1)^2, & x < 1 \\ 2 + (x - 1)^2, & x \ge 1 \end{cases}$

84. $f(x) = \begin{cases} 2x, & x \le 0 \\ x^2 + 1, & x > 0 \end{cases}$

In Exercises 85–88, identify the transformation of the graph of $f(x) = x^4$ and sketch a graph of h.

85. $h(x) = -x^4$

86. $h(x) = x^4 + 2$

87. $h(x) = (x - 1)^4$

88. $h(x) = 1 - x^4$

89. *Path of a Projectile*   The height y (in feet) of a projectile is given by

$$y = -\frac{1}{16}x^2 + 5x$$

where x is the horizontal distance (in feet) from where the projectile was launched.

(a) Sketch the path of the projectile.

(b) How far from the launch point does the projectile strike the ground?

90. *Velocity of a Ball*   The velocity of a ball thrown upward from ground level is given by $v = -32t + 80$, where t is the time in seconds and v is the velocity in feet per second.

(a) Find the velocity when $t = 2$.

(b) Find the time when the ball reaches its maximum height. (*Hint:* Find the time when $v = 0$.)

(c) Find the velocity when $t = 3$.

91. *Power Generation*   The power generated by a wind turbine is given by $P = kw^3$, where P is the number of kilowatts produced at a wind speed of w miles per hour and k is the constant of proportionality.

(a) Find k if $P = 1000$ when $w = 20$.

(b) Find the output for a wind speed of 25 miles per hour.

92. *Wire Length*   A wire 100 inches long is to be cut into four pieces to form a rectangle whose shortest side has a length of x. Express the area A of the rectangle as a function of x. Use a graphing utility to graph the function.

Chapter Test

Take this test as you would take a test in class. After you are done, check your work against the answers given in the back of the book.

1. The numbers (in millions) of votes cast for the Democratic candidates for president in 1980, 1984, 1988, and 1992 were 35.5, 37.6, 41.8, and 44.9. Create a bar graph for the data.

2. Determine the quadrant in which the point (x, y) lies if $x > 0$ and $y < 0$.

3. Find the distance between the points $(0, 9)$ and $(3, 1)$.

4. Find the x- and y-intercepts of the graph of the equation $y = -3(x + 1)$.

5. Sketch the graph of the equation $y = \frac{1}{2}x - 4$.

6. Find the slope (if possible) of the line passing through each pair of points.
   (a) $(-4, 7), (2, 3)$
   (b) $(3, -2), (3, 6)$

7. Plot the x- and y-intercepts, and sketch the graph, of the line given by the equation $2x + 5y - 10 = 0$.

8. Write the equation $2x - 4y = 12$ in slope-intercept form and sketch the graph of the line. Find the slope of the line perpendicular to the given line.

9. Use the Vertical Line Test to determine whether the equation $y^2(4 - x) = x^3$, shown at the right, represents y as a function of x.

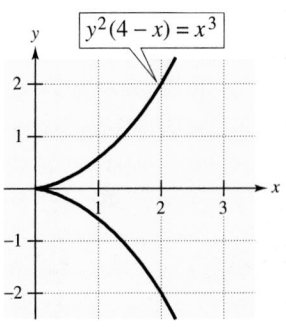

Figure for 9

10. Evaluate (if possible) the function $g(x) = x/(x - 3)$ for the indicated values of the independent variable.
    (a) $g(2)$
    (b) $g\left(\frac{7}{2}\right)$
    (c) $g(3)$
    (d) $g(x + 2)$

11. Let $f(x) = \begin{cases} 3x - 1, & x < 5 \\ x^2 + 4, & x \geq 5 \end{cases}$

    and find the following.
    (a) $f(10)$
    (b) $f(-8)$
    (c) $f(5)$
    (d) $f(0)$

12. Find the domain of each function.
    (a) $h(t) = \sqrt{t + 9}$
    (b) $f(x) = \dfrac{x + 1}{x - 4}$

13. Sketch the graph of the function $f(x) = x^2 + 3$.

14. Use the graph of $f(x) = |x|$ to write an equation for each graph. Use a graphing utility to verify your result.

(a)
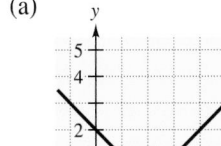

(b)
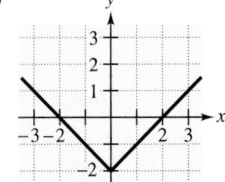

(c)
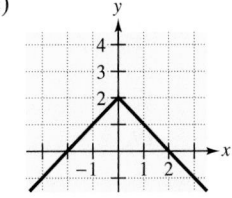

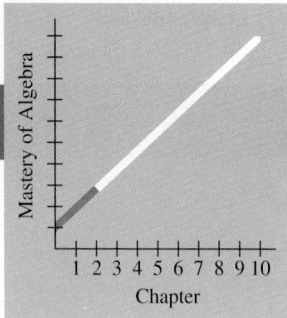

Mastery of Algebra

Chapter

Take this test as you would take a test in class. After you are done, check your work against the answers given in the back of the book.

1. Evaluate $-\frac{8}{45} \div \frac{12}{25}$.

2. Evaluate the expression $\dfrac{a^2 - 2ab}{a + b}$ for $a = -4$ and $b = 7$.

In Exercises 3 and 4, perform the operations and simplify.

3. (a) $(2a^2b)^3(-ab^2)^2$ (b) $3x(x^2 - 2) - x(x^2 + 5)$

4. (a) $t(3t - 1) - 2t(t + 4)$ (b) $[2 + (x - y)]^2$

In Exercises 5 and 6, solve the equation.

5. (a) $12 - 5(3 - x) = x + 3$ (b) $1 - \dfrac{x + 2}{4} = \dfrac{7}{8}$

6. (a) $y^2 - 64 = 0$ (b) $2t^2 - 5t - 3 = 0$

7. Solve for y in the equation $2x - 3y + 9 = 0$.

8. Given the line $y = -4x + 7$, find (a) the slope of the line, (b) the slope of a line parallel to the given line, and (c) the slope of a line perpendicular to the given line.

9. Name the property of real numbers that justifies the statement.

(a) $8x \cdot \dfrac{1}{8x} = 1$ (b) $5 + (-3 + x) = (5 - 3) + x$

10. Find and simplify the expression for the perimeter of the figure shown at the left.

11. Factor $y^3 - 3y^2 - 9y + 27$ by grouping.

12. Factor $3x^2 - 8x - 35$.

13. Determine whether the equation $x - y^3 = 0$ represents y as a function of x.

14. Find the domain of the function $f(x) = \sqrt{x - 2}$.

15. Given $f(x) = x^2 - 3x$, find (a) $f(4)$ and (b) $f(c + 3)$.

16. Find the slope of the line passing through $(-4, 0)$ and $(4, 6)$.

In Exercises 17–20, sketch a graph of the equation. Use a graphing utility to check the result.

17. $4x + 3y - 12 = 0$ **18.** $y = -(x - 2)^2$

19. $y = |x + 3| + 2$ **20.** $y = x^3 - 1$

x

x

$x + 5$

$2x + 1$

$x + 1$

$2x + 5$

Figure for 10

182

Linear Functions, Equations, and Inequalities

Strategies for Success

SKILLS *When you have completed this chapter, make sure you are able to:*

○ Write the equation of a line given a description of its graph
○ Find the best-fitting lines through sets of points
○ Write linear equations to model real-life situations and solve
○ Solve linear inequalities and sketch the graphs of their solution sets
○ Solve absolute value equations and inequalities

TOOLS *Use these study tools to master the skills above:*

In the Section Project of Section 3.5, you will solve inequalities to determine if a bat can hear a human's voice and if a human can hear a bat's voice.

3.1	**Writing Equations of Lines**

Point-Slope Equation of a Line ▪ Horizontal and Vertical Lines ▪
Summary of Equations of Lines ▪ Application

Point-Slope Equation of a Line

There are two basic types of problems in coordinate geometry.

1. Given an algebraic equation, sketch its graph.

2. Given a description of a graph, write its equation.

The first type of problem can be thought of as moving from algebra to geometry, whereas the second type can be thought of as moving the other way—from geometry to algebra. So far in the text, you have been working primarily with the first type of problem. In this section, you will look at the second type.

In this section, you will learn that if you know the slope of a line *and* you also know the coordinates of one point on the line, you can find an equation for the line. Before giving a general formula for doing this, let's look at an example.

Discuss the number of lines having a slope of $\frac{4}{3}$. Then discuss the number of lines that pass through $(-2, 1)$. Point out that only one line satisfies both conditions.

EXAMPLE 1   *Writing an Equation of a Line*

A line has a slope of $\frac{4}{3}$ and passes through the point $(-2, 1)$. Find an equation of this line.

Solution

Begin by sketching the line, as shown in Figure 3.1. You know that the slope of a line is the same through any two points on the line. Thus, to find an equation of the line, let (x, y) represent *any* point on the line. Using the representative point (x, y) and the given point $(-2, 1)$, it follows that the slope of the line is

$$m = \frac{y - 1}{x - (-2)}.$$

 Difference in *y*-values
Difference in *x*-values

Because the slope of the line is $m = \frac{4}{3}$, this equation can be rewritten as follows.

$$\frac{4}{3} = \frac{y - 1}{x + 2}$$ Slope formula

$$4(x + 2) = 3(y - 1)$$ Cross multiply.

$$4x + 8 = 3y - 3$$ Distributive Property

$$4x - 3y = -11$$ Subtract 8 and 3*y* from both sides.

An equation of the line is $4x - 3y = -11$.

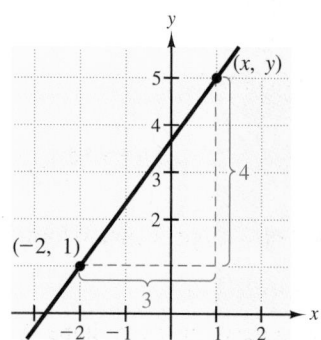

FIGURE 3.1

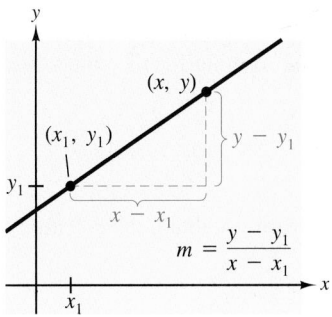

FIGURE 3.2

The procedure in Example 1 can be used to derive a *formula* for the equation of a line, given its slope and a point on the line. In Figure 3.2, let (x_1, y_1) be a given point on the line whose slope is m. If (x, y) is any *other* point on the line, it follows that

$$\frac{y - y_1}{x - x_1} = m.$$

This equation in variables x and y can be rewritten in the form

$$y - y_1 = m(x - x_1),$$

which is called the **point-slope form** of the equation of a line.

Point-Slope Form of the Equation of a Line

The **point-slope form** of the equation of the line that passes through the point (x_1, y_1) and has a slope of m is

$$y - y_1 = m(x - x_1).$$

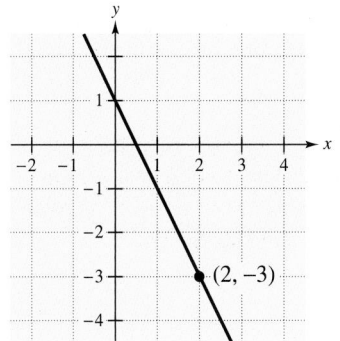

FIGURE 3.3

EXAMPLE 2 The Point-Slope Form of the Equation of a Line

Find an equation of the line that passes through the point $(2, -3)$ and has a slope of -2.

Solution

Use the point-slope form with $(x_1, y_1) = (2, -3)$ and $m = -2$.

$$y - y_1 = m(x - x_1) \qquad \text{Point-slope form}$$
$$y - (-3) = -2(x - 2) \qquad \text{Substitute } y_1 = -3, x_1 = 2, \text{ and } m = -2.$$
$$y + 3 = -2x + 4 \qquad \text{Simplify.}$$
$$y = -2x + 1 \qquad \text{Subtract 3 from both sides.}$$

The graph of this line is shown in Figure 3.3.

In Example 2, notice that the final equation is written in slope-intercept form

$$y = mx + b. \qquad \text{Slope-intercept form}$$

You can use this form to check your work. First, observe that the slope is $m = -2$. Then substitute the coordinates of the given point $(2, -3)$ to see that the equation is satisfied.

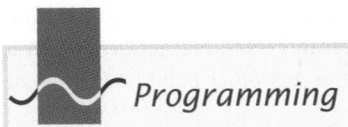

Programming

The *TI-82/TI-83* program listed below uses the two-point form to find an equation of a line. (Programs for other calculator models are given in the appendix.) After the coordinates of two points are entered, the program outputs the slope and *y*-intercept of the line that passes through the points.

```
PROGRAM: TWOPTFM
:Disp "ENTER X1, Y1"
:Input X
:Input Y
:Disp "ENTER X2, Y2"
:Input C
:Input D
:(D-Y)/(C-X)→M
:M*(-X) + Y→B
:Disp "SLOPE="
:Disp M
:Disp "Y-INT="
:Disp B
```

The point-slope form can be used to find the equation of a line passing through two points (x_1, y_1) and (x_2, y_2). First, use the formula for the slope of a line passing through two points.

$$m = \frac{y_2 - y_1}{x_2 - x_1}$$

Then, once you know the slope, use the point-slope form to obtain the equation

$$y - y_1 = \frac{y_2 - y_1}{x_2 - x_1}(x - x_1). \qquad \text{Two-point form}$$

This is sometimes called the **two-point form** of the equation of a line.

EXAMPLE 3   An Equation of a Line Passing Through Two Points

Find an equation of the line that passes through the points $(4, 2)$ and $(-2, 3)$.

Solution

Let $(x_1, y_1) = (4, 2)$ and $(x_2, y_2) = (-2, 3)$. Then apply the formula for the slope of a line passing through two points as follows.

$$m = \frac{y_2 - y_1}{x_2 - x_1} = \frac{3 - 2}{-2 - 4} = \frac{1}{-6} = -\frac{1}{6}$$

Now, using the point-slope form, you can find the equation of the line.

$$y - y_1 = m(x - x_1) \qquad \text{Point-slope form}$$

$$y - 2 = -\frac{1}{6}(x - 4) \qquad \text{Substitute } y_1 = 2, x_1 = 4, \text{ and } m = -\tfrac{1}{6}.$$

$$y - 2 = -\frac{1}{6}x + \frac{2}{3} \qquad \text{Simplify.}$$

$$y = -\frac{1}{6}x + \frac{8}{3} \qquad \text{Add 2 to both sides.}$$

The graph of this line is shown in Figure 3.4.

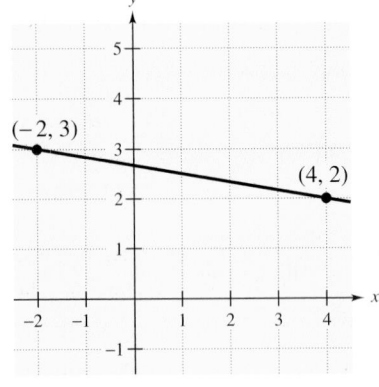

FIGURE 3.4

In Example 3, it does not matter which of the two points is labeled (x_1, y_1) and which is labeled (x_2, y_2). Try switching these labels to

$$(x_1, y_1) = (-2, 3) \qquad \text{and} \qquad (x_2, y_2) = (4, 2)$$

and reworking the problem to see that you obtain the same equation.

Horizontal and Vertical Lines

From the slope-intercept form of the equation of a line, you can see that a horizontal line ($m = 0$) has an equation of the form

$$y = (0)x + b \qquad \text{or} \qquad y = b. \qquad \text{Horizontal line}$$

This is consistent with the fact that each point on a horizontal line through $(0, b)$ has a y-coordinate of b, as shown in Figure 3.5. Similarly, each point on a vertical line through $(a, 0)$ has an x-coordinate of a, as shown in Figure 3.6. Hence, a vertical line has an equation of the form

$$x = a. \qquad \text{Vertical line}$$

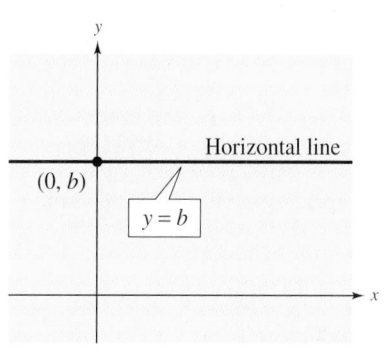

FIGURE 3.5

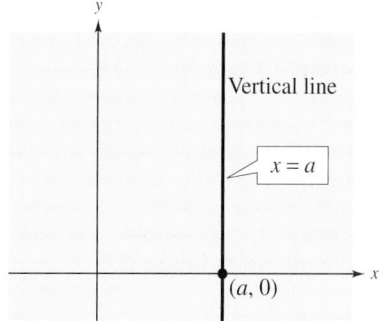

FIGURE 3.6

EXAMPLE 4   *Writing Equations of Horizontal and Vertical Lines*

Write an equation for each line.

(a) Vertical line through $(-2, 4)$

(b) Line passing through $(-2, 3)$ and $(3, 3)$

(c) Line passing through $(-1, 2)$ and $(-1, 3)$

Solution

(a) Because the line is vertical and passes through the point $(-2, 4)$, you know that every point on the line has an x-coordinate of -2. Thus, the equation is
$$x = -2.$$

(b) The line through $(-2, 3)$ and $(3, 3)$ is horizontal, and every point on the line has a y-coordinate of 3. Thus, its equation is
$$y = 3.$$

(c) The line through $(-1, 2)$ and $(-1, 3)$ is vertical, and every point on the line has an x-coordinate of -1. Thus, its equation is
$$x = -1.$$

Summary of Equations of Lines

The equation of a vertical line cannot be written in slope-intercept form because the slope of a vertical line is undefined. However, *every* line has an equation that can be written in the **general form**

$$ax + by + c = 0, \qquad \text{General form}$$

where a and b are not *both* zero.

Summary of Equations of Lines

Writing Assignment: Have students describe what happens to the graph of $y = mx + b$ when the value of the coefficient on x is increased.

1. Slope of line through (x_1, y_1) and (x_2, y_2): $\quad m = \dfrac{y_2 - y_1}{x_2 - x_1}$

2. General form of equation of line: $\quad ax + by + c = 0$

3. Equation of vertical line: $\quad x = a$

4. Equation of horizontal line: $\quad y = b$

5. Slope-intercept form of equation of line: $\quad y = mx + b$

6. Point-slope form of equation of line: $\quad y - y_1 = m(x - x_1)$

7. Parallel lines (equal slopes): $\quad m_1 = m_2$

8. Perpendicular lines (negative reciprocal slopes): $\quad m_2 = -\dfrac{1}{m_1}$

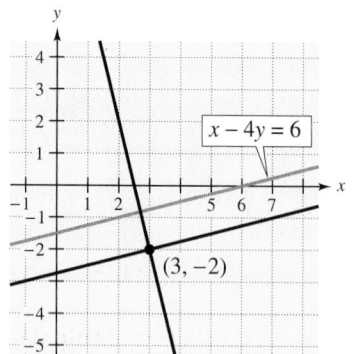

FIGURE 3.7

An application: To avoid midair collisions, flight controllers instruct pilots to take off and land in directions parallel to other planes.

EXAMPLE 5   Parallel and Perpendicular Lines

Find equations of the lines that pass through the point $(3, -2)$ and are (a) parallel and (b) perpendicular to the line $x - 4y = 6$, as shown in Figure 3.7.

Solution

Write the equation in slope-intercept form to determine the slope of the line.

$$x - 4y = 6 \qquad \text{Original equation}$$

$$-4y = -x + 6 \qquad \text{Subtract } x \text{ from both sides.}$$

$$y = \tfrac{1}{4}x - \tfrac{3}{2} \qquad \text{Slope-intercept form}$$

Therefore, the line has a slope of $\tfrac{1}{4}$.

(a) Any line parallel to the given line must also have a slope of $\tfrac{1}{4}$. The line through $(3, -2)$ with slope $m = \tfrac{1}{4}$ has the following equation.

$$y - y_1 = m(x - x_1) \qquad \text{Point-slope form}$$

$$y - (-2) = \tfrac{1}{4}(x - 3) \qquad \text{Substitute } y_1 = -2, x_1 = 3, \text{ and } m = \tfrac{1}{4}.$$

$$y + 2 = \tfrac{1}{4}x - \tfrac{3}{4} \qquad \text{Distributive Property}$$

$$y = \tfrac{1}{4}x - \tfrac{11}{4} \qquad \text{Equation of parallel line}$$

(b) Any line perpendicular to the given line must have a slope of -4. The line through $(3, -2)$ with slope $m = -4$ has the following equation.

$$y - y_1 = m(x - x_1) \qquad \text{Point-slope form}$$

$$y - (-2) = -4(x - 3) \qquad \text{Substitute } y_1 = -2, x_1 = 3, \text{ and } m = -4.$$

$$y + 2 = -4x + 12 \qquad \text{Distributive Property}$$

$$y = -4x + 10 \qquad \text{Equation of perpendicular line}$$

Application

Linear equations are used frequently as mathematical models in business. The next example gives you one idea of how useful such a model can be.

EXAMPLE 6   An Application: Total Sales

The total sales of a new computer software company were $500,000 for the second year and $1,000,000 for the fourth year. Using only this information, what would you estimate the total sales to be during the fifth year?

Solution

To solve this problem, use a *linear model*, with y representing the total sales (in thousands of dollars) and t representing the year. That is, in Figure 3.8, let $(2, 500)$ and $(4, 1000)$ be two points on the line representing the total sales for the company. The slope of the line passing through these points is

$$m = \frac{1000 - 500}{4 - 2} = \frac{500}{2} = 250.$$

Now, using the point-slope form, the equation of the line is

$$y - y_1 = m(t - t_1) \qquad \text{Point-slope form}$$
$$y - 500 = 250(t - 2) \qquad \text{Substitute } y_1 = 500,\ t_1 = 2, \text{ and } m = 250.$$
$$y - 500 = 250t - 500 \qquad \text{Distributive Property}$$
$$y = 250t. \qquad \text{Linear model for sales}$$

Finally, estimate the total sales during the fifth year $(t = 5)$ to be

$$y = 250(5) = \$1250 \text{ thousand} = \$1,250,000.$$

The estimation method illustrated in Example 6 is called **linear extrapolation.** Note in Figure 3.9 that for linear extrapolation, the estimated point lies to the right of the given points. When the estimated point lies *between* two given points, the procedure is called **linear interpolation.**

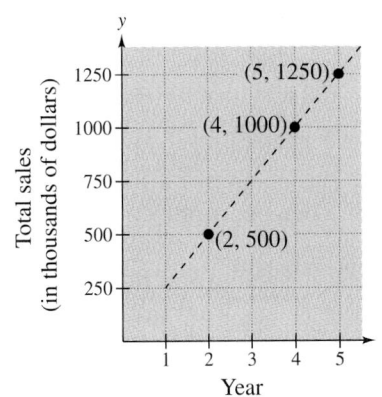

FIGURE 3.8

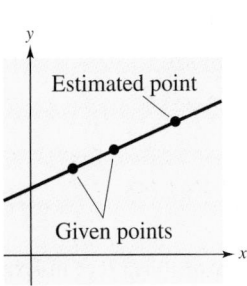

Linear extrapolation

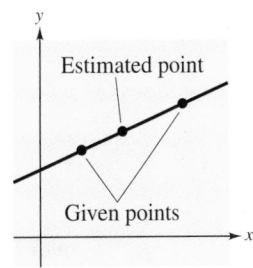

Linear interpolation

FIGURE 3.9

EXAMPLE 7   Slope of a Sewer Line

A city ordinance requires a minimum drop of 0.6% in a home sewer line running to the main sewer line (along the street). This means that for each foot from the house to the street, the sewer line must drop 0.006 foot, as shown in Figure 3.10.

(a) Write an equation that expresses the minimum vertical drop y in terms of the horizontal length x of the pipe.

(b) How much drop is needed for a house that is 200 feet from the main line?

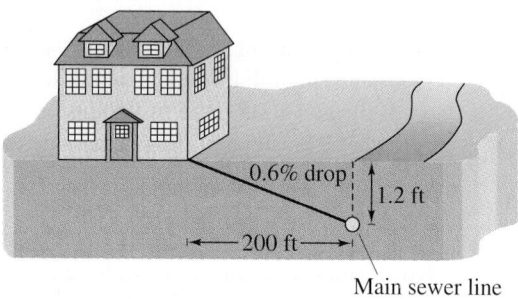

0.6% drop
1.2 ft
200 ft
Main sewer line

FIGURE 3.10

Solution

(a) In this case the slope is

$$m = -0.006,$$

because the line *drops* 0.006 foot vertically for each foot of horizontal length. Because a pipe of zero length would have a zero drop, it follows that $(x_1, y_1) = (0, 0)$ is a point that satisfies the equation relating x and y. Therefore, using the point-slope form, you can write the following equation.

$$y - y_1 = m(x - x_1)$$
$$y - 0 = -0.006(x - 0)$$
$$y = -0.006x$$

(b) The drop in a line for a house that is 200 feet from the street would have to be at least

$$y = -0.006(200)$$
$$= -1.2 \text{ feet.}$$

NOTE   The answer of -1.2 feet in Example 7(b) indicates a *drop* in the line, whereas a positive answer would have indicated a *rise*.

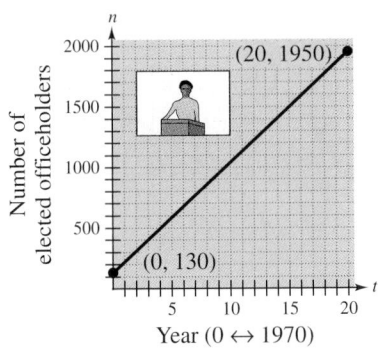

EXAMPLE 8   *Writing a Linear Model*

In 1970, 130 African-American women held elected offices in the United States. By 1990, the number had increased to 1950. Write a linear model for the number n of African-American women who held elected offices between 1970 and 1990. (Let $t = 0$ represent 1970.)   *(Source: Joint Center for Political and Economic Studies)*

Solution

One way to create such a model is to interpret the given information as two points on a line. Using $n = 130$ in 1970 ($t = 0$), you can determine that one point on the line is $(0, 130)$. Using $n = 1950$ in 1990 ($t = 20$), you can determine that a second point on the line is $(20, 1950)$. (See Figure 3.11.) Thus, the slope of the line is

$$m = \frac{1950 - 130}{20 - 0} = \frac{1820}{20} = 91.$$

Because the n-intercept of the line is $(0, 130)$, you can conclude that an equation of the line is

$n = mt + b$ Slope-intercept form—use t and n.

$n = 91t + 130.$ Substitute $m = 91$ and $b = 130$.

FIGURE 3.11

In 1973, Barbara Jordan became the first African-American woman from a southern state to serve in the United States Congress. From 1970 to 1990, the number of African-American women holding elective office increased at a rate that was approximately linear.

Group Activities   **Problem Solving**

Mathematical Modeling   Your manager asks you to make sense of the following set of data, in which y represents the percent of households with personal computer owners and x represents the year from 1983 through 1994, with $x = 3$ corresponding to 1983. You think that a mathematical model will help you understand the trend in the data and may be useful in predicting what could happen in the future. Plot the data. Do you think that a linear model would represent the data well? If so, find the equation of the best-fitting line. Interpret the meaning of the slope in the context of the data. Use your model to predict the percent of households with personal computer owners in 1999 (assuming that the trend continues).   *(Source: Electronic Industries Association Consumer Electronic U.S. Sales, 1982–1994)*

x	3	4	5	6	7	8	9	10	11	12	13	14
y	7	13	15	16	20	21	22	27	29	34	35	37

3.1   **Exercises**

See Warm-Up
Exercises, p. A41

In Exercises 1–6, determine the slope and the *y*-intercept of the line.

1. $y = \frac{2}{3}x - 2$ **2.** $y = -5x + 12$

3. $3x - 2y = 0$ **4.** $y = -2(6x - 1)$

5. $5x - 2y + 24 = 0$ **6.** $3x + 4y - 16 = 0$

In Exercises 7–22, write an equation of the line that passes through the given point and has the specified slope. When possible, write the equation in slope-intercept form. Sketch the line.

7. $(0, 0)$ **8.** $(0, 0)$
$\quad m = -\frac{1}{2}$ $\quad m = -2$

9. $(0, -4)$ **10.** $(0, 9)$
$\quad m = 3$ $\quad m = -\frac{1}{3}$

11. $(5, 6)$ **12.** $(3, 7)$
$\quad m = 2$ $\quad m = 4$

13. $(-8, 1)$ **14.** $(6, -9)$
$\quad m = \frac{3}{4}$ $\quad m = \frac{5}{2}$

15. $(5, -3)$ **16.** $(2, -6)$
$\quad m = \frac{2}{3}$ $\quad m$ is undefined.

17. $(-8, 5)$ **18.** $(-3, -7)$
$\quad m = 0$ $\quad m = \frac{5}{4}$

19. $(2, -1)$ **20.** $(4, -1)$
$\quad m$ is undefined. $\quad m = 0$

21. $\left(\frac{3}{4}, \frac{5}{2}\right)$ **22.** $\left(-\frac{3}{2}, \frac{1}{2}\right)$
$\quad m = \frac{4}{3}$ $\quad m = -3$

In Exercises 23–26, match the equation with its graph. [The graphs are labeled (a), (b), (c), and (d).]

23. $y = \frac{2}{3}x + 2$ **24.** $y = \frac{2}{3}x - 2$

25. $y = -\frac{3}{2}x + 2$ **26.** $y = -3x + 2$

(a)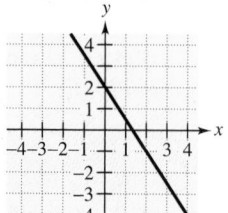

(b)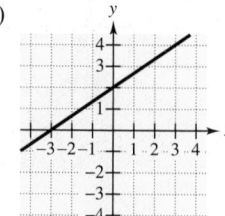

(c)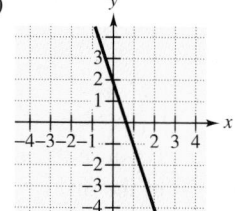

(d)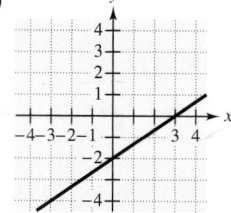

In Exercises 27–42, write an equation of the line that passes through the two points. When possible, write the equation in slope-intercept form.

27. $(0, 0), (2, 3)$ **28.** $(0, 0), (3, -5)$

29. $(7, 3), (5, -5)$ **30.** $(-9, 1), (-6, 10)$

31. $(-5, 2), (5, -2)$ **32.** $(4, -3), (-1, -3)$

33. $(-2, 12), (6, 12)$ **34.** $(-3, -2), (6, -8)$

35. $(-2, 3), (5, 0)$ **36.** $(5, 4), (-3, 5)$

37. $(1, -2), (1, 8)$ **38.** $\left(4, \frac{7}{2}\right), \left(-1, \frac{2}{3}\right)$

39. $(-5, 0.6), (3, -3.4)$ **40.** $(7.5, 2), (7.5, 9)$

41. $\left(\frac{3}{2}, 3\right), \left(\frac{9}{2}, -4\right)$ **42.** $\left(2, -8\right), \left(6, \frac{8}{3}\right)$

In Exercises 43–46, write an equation of the line passing through the two points. Use function notation to write *y* as a function of *x*. Use a graphing utility to graph the linear function.

43. $(-2, 2), (4, 5)$ **44.** $(0, 10), (5, 0)$

45. $(-2, 3), (4, 3)$ **46.** $(-6, -3), (4, 3)$

The symbol ⊞ indicates an exercise in which you are instructed to use a calculator or graphing utility. The solutions of other exercises may also be facilitated by the use of appropriate technology.

47. *Think About It* Can any pair of points on a line be used to determine an equation of the line? Explain.

48. *Think About It* Write, from memory, the point-slope form, the slope-intercept form, and the general form of an equation of a line.

In Exercises 49–56, write equations of the lines through the point (a) parallel to the given line and (b) perpendicular to the given line.

49. $(2, 1)$

$6x - 2y = 10$

50. $(-8, 3)$

$4x - y = 8$

[handwritten: $a \ y = 4x + b$ $3 = -32 + b$ $b = 35$]

51. $(-6, 2)$

$3x + 2y = 4$

52. $(-3, 4)$

$x + 6y = 12$

53. $(1, -7)$

$4x - 3y = 9$

54. $(-5, 4)$

$5x + 4y = 24$

55. $(-1, 2)$

$y = -5$

56. $(3, -4)$

$x = 10$

In Exercises 57–64, use a graphing utility on a square setting to graph both lines in the same viewing rectangle. Determine whether the lines are parallel, perpendicular, or neither. Verify your answer algebraically.

57. $y = -0.6x + 1$

$y = \frac{5}{3}x - 2$

58. $y = \frac{1}{2}(4x - 3)$

$y = \frac{1}{4}(4x + 3)$

59. $y = 0.6x + 1$

$y = \frac{1}{5}(3x + 22)$

60. $y = \frac{3}{2}x - 4$

$y = -\frac{2}{3}x + 1$

61. $y = 3(x - 2)$

$y = 3x + 2$

62. $y = \frac{3}{2}(2x - 3)$

$y = -\frac{1}{3}(x + 3)$

63. $x + 2y - 3 = 0$

$-2x - 4y + 1 = 0$

64. $3x - 4y - 1 = 0$

$4x + 3y + 2 = 0$

65. *Cost* The cost (in dollars) of producing x units of a certain product is given by $C = 20x + 5000$. Use this equation to complete the table.

x	0		100		1000
C		6000		15,000	

66. *Temperature Conversion* The relationship between the Fahrenheit and Celsius temperature scales is given by $C = \frac{5}{9}F - \frac{160}{9}$. Use this equation to complete the table.

F	$-20°$		$20°$		$100°$
C		$-17.8°$		$0°$	$100°$

67. *Sales Commission* The salary for a sales representative is $2500 per month plus a 3% commission of total monthly sales. Write a linear function giving the salary S in terms of the monthly sales M.

68. *Reimbursed Expenses* A sales representative is reimbursed $175 per day for lodging and meals plus $0.32 per mile driven. Write a linear function giving the daily cost C to the company in terms of x, the number of miles driven.

69. *Discount Price* A store is offering a 30% discount on all items in its inventory.

(a) Write a linear function giving the sale price S for an item in terms of its list price L.

(b) Use the function in part (a) to find the sale price of an item that has a list price of $135.

70. *Straight-Line Depreciation* A small business purchases a photocopier for $7400. After 4 years, its depreciated value will be $1500.

(a) Assuming straight-line depreciation, write a linear function giving the value V of the copier in terms of the time t.

(b) Use the function in part (a) to find the value of the copier after 2 years.

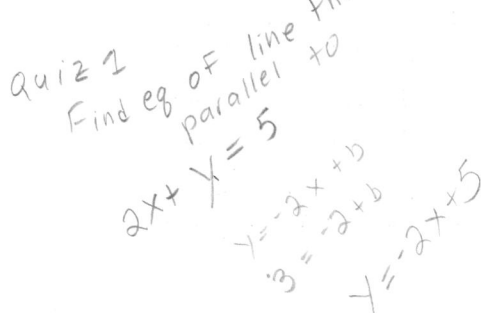

[handwritten: quiz 1
Find eq of line thru (1,3)
parallel to
$2x + y = 5$
$y = -2x + b$
$3 = -2 + b$
$3 =$
$y = -2x + 5$]

71. *Rental Occupancy* A real estate office handles an apartment complex with 50 units. When the rent is $450 per month, all 50 units are occupied. However, when the rent is $525 per month, the average number of occupied units drops to 45. Assume that the relationship between the monthly rent p and the demand x is linear.

(a) Write a linear function giving the demand x in terms of the rent p.

(b) Use a graphing utility to graph the function in part (a).

(c) *Linear Extrapolation* Use the function in part (a) to predict the number of units occupied if the rent is raised to $570.

(d) *Linear Interpolation* Use the function in part (a) to estimate the number of units occupied if the rent is $480.

72. *Soft-Drink Sales* When soft drinks sold for $0.80 per can at football games, approximately 6000 cans were sold. When the price was raised to $1.00 per can, the demand dropped to 4000. Assume that the relationship between the price p and the demand x is linear.

(a) Write a linear function giving the demand x in terms of the price p.

(b) Use a graphing utility to graph the function in part (a).

(c) *Linear Extrapolation* Use the function in part (a) to predict the number of cans of soft drink sold if the price is raised to $1.10.

(d) *Linear Interpolation* Use the function in part (a) to estimate the number of cans of soft drink sold if the price is $0.90.

73. *College Enrollment* A small college had an enrollment of 1500 students in 1980. During the next 10 years, the enrollment increased by approximately 60 students per year.

(a) Write a linear function giving the enrollment N in terms of the year t. (Let $t = 0$ represent 1980.)

(b) Use a graphing utility to graph the function in part (a).

(c) *Linear Extrapolation* Use the function in part (a) to predict the enrollment in the year 2000.

(d) *Linear Interpolation* Use the function in part (a) to estimate the enrollment in 1985.

Section Project

The intercepts of a line are $(a, 0)$ and $(0, b)$, $a \neq 0$, $b \neq 0$.

(a) Find the slope of the line.

(b) Write the equation of the line in slope-intercept form.

(c) Using the result in part (b), clear the equation of fractions and show that the equation can be written in the form $bx + ay = ab$.

(d) Use the result in part (c) to show that the equation of the line can be written in the form $\dfrac{x}{a} + \dfrac{y}{b} = 1$.

Use the result in part (d) to find an equation of the line with the indicated intercepts.

(i) x-intercept: $(3, 0)$
    y-intercept: $(0, 2)$

(ii) x-intercept: $(-6, 0)$
     y-intercept: $(0, 2)$

(iii) x-intercept: $\left(-\frac{5}{6}, 0\right)$
      y-intercept: $\left(0, -\frac{7}{3}\right)$

(iv) x-intercept: $\left(-\frac{8}{3}, 0\right)$
     y-intercept: $(0, -4)$

3.2   Modeling Data with Linear Functions

Introduction ▪ Direct Variation ▪ Rates of Change ▪ Scatter Plots

Introduction

The primary objective of applied mathematics is to find equations or **mathematical models** that describe real-world phenomena. In developing a mathematical model to represent actual data, you should strive for two (often conflicting) goals—accuracy and simplicity. That is, you want the model to be simple enough to be workable, yet accurate enough to produce meaningful results.

EXAMPLE 1   A Mathematical Model

The total annual amounts of advertising expenses y (in billions of dollars) in the United States from 1984 through 1995 are given in the table.   *(Source: McCann Erickson)*

Year	1984	1985	1986	1987	1988	1989	1990	1991	1992	1993	1994	1995
y	88.1	94.8	102.1	109.7	118.1	123.9	128.6	126.4	131.3	138.1	150.0	161.5

A linear model that approximates this data is

$$y = 5.92t + 66.45, \qquad 4 \le t \le 15,$$

where y represents the advertising expenses (in billions of dollars) and t represents the year, with $t = 4$ corresponding to 1984. Plot the actual data *and* the model on the same graph. How closely does the model represent the data?

Solution

The actual data is plotted in Figure 3.12, along with the graph of the linear model. From the figure, it appears that the model is a "good fit" for the actual data. You can see how well the model fits by comparing the actual values of y with the values of y given by the model (these are labeled y^* in the table below).

t	4	5	6	7	8	9	10	11	12	13	14	15
y	88.1	94.8	102.1	109.7	118.1	123.9	128.6	126.4	131.3	138.1	150.0	161.5
y^*	90.1	96.1	102.0	107.9	113.8	119.7	125.7	131.6	137.5	143.4	149.3	155.3

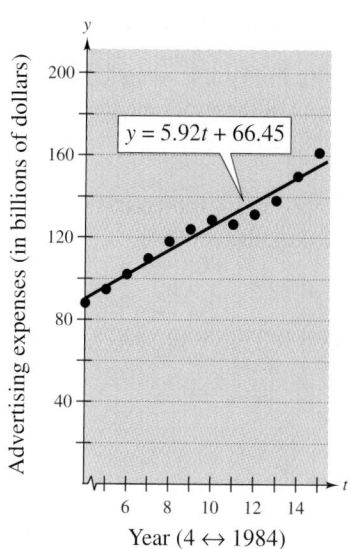

$y = 5.92t + 66.45$

Year (4 ↔ 1984)

FIGURE 3.12

Direct Variation

Direct variation is compared with other types of variation in Section 8.3.

There are two basic types of linear models. The more general model has a y-intercept that is nonzero: $y = mx + b, b \neq 0$. The simpler one, $y = mx$, has a y-intercept that is zero. In the simpler model, y is said to **vary directly** as x, or to be **proportional** to x.

Direct Variation

The following statements are equivalent.

1. y **varies directly** as x.

2. y is **directly proportional** to x.

3. $y = mx$ for some nonzero constant m.

m is the **constant of variation** or the **constant of proportionality**.

EXAMPLE 2   State Income Tax

In Pennsylvania, the state income tax is directly proportional to *gross income*. Suppose you were working in Pennsylvania and your state income tax deduction was $31.50 for a gross monthly income of $1500.00. Find a mathematical model that gives the Pennsylvania state income tax in terms of the gross income.

Solution

If you let y represent the state income tax in dollars and x represent the gross income in dollars, you know that y and x are related by the equation

$$y = mx.$$

You are given $y = 31.50$ when $x = 1500$. By substituting these values into the equation $y = mx$, you can find the value of m.

$y = mx$	Direct variation model
$31.50 = m(1500)$	Substitute $y = 31.50$ and $x = 1500$.
$0.021 = m$	Divide both sides by 1500.

Thus, the equation (or model) for state income tax in Pennsylvania is $y = 0.021x$. In other words, Pennsylvania has a state income tax rate of 2.1% of the gross income. The graph of this equation is shown in Figure 3.13.

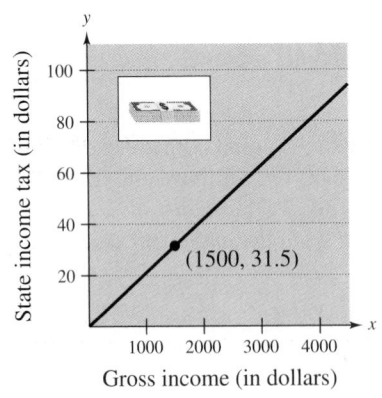

State income tax (in dollars) / Gross income (in dollars) — (1500, 31.5)

FIGURE 3.13

Most measurements in the English system and the metric system are directly proportional. The next example shows how to use a direct proportion to convert between miles per hour and kilometers per hour.

EXAMPLE 3   The English and Metric Systems

You are traveling at a rate of 64 miles per hour. You switch your speedometer reading to metric units and notice that the speed is 103 kilometers per hour. Use this information to find a mathematical model that relates miles per hour to kilometers per hour.

Solution

If you let y represent the speed in miles per hour and x represent the speed in kilometers per hour, you know that y and x are related by the equation

$$y = mx.$$

You are given $y = 64$ when $x = 103$. By substituting these values into the equation $y = mx$, you can find the value of m.

$$y = mx \qquad \text{Direct variation model}$$
$$64 = m(103) \qquad \text{Substitute } y = 64 \text{ and } x = 103.$$
$$\frac{64}{103} = m \qquad \text{Divide both sides by 103.}$$
$$0.62136 \approx m \qquad \text{Use a calculator.}$$

Thus, the conversion factor from kilometers per hour to miles per hour is approximately 0.62136, and the model is

$$y = 0.62136x.$$

The graph of this equation is shown in Figure 3.14.

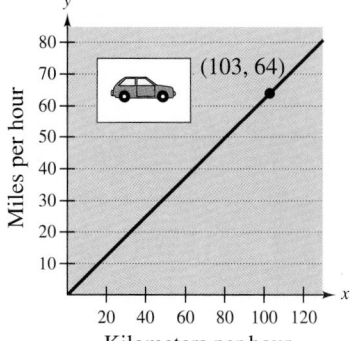

FIGURE 3.14

Once you have found a model that converts speeds from kilometers per hour to miles per hour, you can use the model to convert other speeds from the metric system to the English system, as shown in the table.

Kilometers per hour	20.0	40.0	60.0	80.0	100.0	120.0
Miles per hour	12.4	24.9	37.3	49.7	62.1	74.6

NOTE   The conversion equation $y = 0.62136x$ can be approximated by the simpler equation $y = \frac{5}{8}x$. For instance, to convert 40 kilometers per hour, divide by 8 and multiply by 5 to obtain 25 miles per hour.

Rates of Change

A second common type of linear model is one that involves a known rate of change. In the linear equation

$$y = mx + b,$$

It is important for students to grasp the concept of slope as a rate of change.

you know that m represents the slope of the line. In real-life problems, the slope can often be interpreted as the **rate of change** of y with respect to x. Rates of change should always be listed in appropriate units of measure.

EXAMPLE 4   *Height of a Mountain Climber*

A mountain climber is climbing up a 500-foot cliff. By 1 P.M., the mountain climber has climbed 115 feet up the cliff. By 4 P.M., the climber has reached a height of 280 feet, as shown in Figure 3.15. Find the average rate of change of the climber and use this rate of change to find the equation that relates the height of the climber to the time. Use the model to estimate the time when the climber will reach the top of the cliff.

Solution

Let y represent the height of the climber and let t represent the time. Then the two points that represent the climber's two positions are

$$(t_1, y_1) = (1, 115) \qquad \text{and} \qquad (t_2, y_2) = (4, 280).$$

Thus, the average rate of change of the climber is

$$\text{Average rate of change} = \frac{y_2 - y_1}{t_2 - t_1}$$

$$= \frac{280 - 115}{4 - 1}$$

$$= 55 \text{ feet per hour.}$$

Thus, an equation that relates the height of the climber to the time is

$$\begin{aligned} y - y_1 &= m(t - t_1) && \text{Point-slope form} \\ y - 115 &= 55(t - 1) && \text{Substitute } y_1 = 115, t_1 = 1, \text{ and } m = 55. \\ y &= 55t + 60. && \text{Linear model} \end{aligned}$$

To find the time when the climber reaches the top of the cliff, let $y = 500$ and solve for t to obtain

$$\begin{aligned} 500 &= 55t + 60 \\ t &= 8. \end{aligned}$$

Thus, continuing at the same rate, the climber will reach the top of the cliff at 8 P.M.

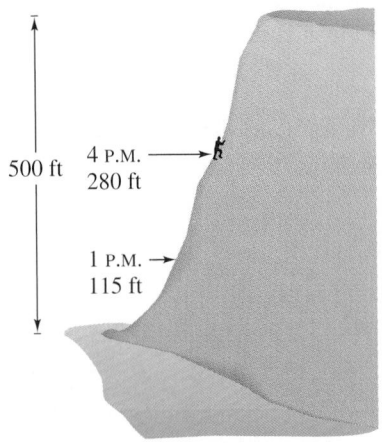

500 ft

4 P.M.
280 ft

1 P.M.
115 ft

FIGURE 3.15

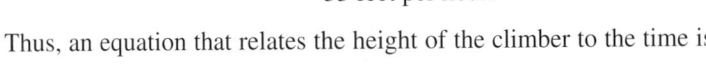

EXAMPLE 5   *Population of Hampton, Virginia*

Between 1980 and 1990, the population of the city of Hampton, Virginia, increased at an average rate of approximately 1100 people per year. In 1980, the population was 123,000. Find a mathematical model that gives the population of Hampton in terms of the year, and use the model to estimate the population in 1992.   *(Source: U.S. Bureau of the Census)*

Solution

Let y represent the population of Hampton, and let t represent the calendar year, with $t = 0$ corresponding to 1980. Letting $t = 0$ correspond to 1980 is convenient because you were given the population in 1980. Now, using the rate of change of 1100 people per year, you have

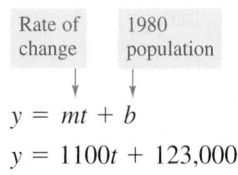

$$y = mt + b$$
$$y = 1100t + 123,000.$$

Using this model, you can estimate the 1992 population to be

$$1992 \text{ population} = 1100(12) + 123,000 = 136,200.$$

The graph is shown in Figure 3.16. (In this particular example, the linear model is quite good—the actual population of Hampton, Virginia, in 1992 was 137,000.)

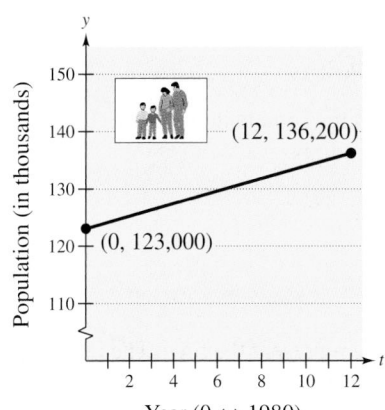

FIGURE 3.16

In Example 5, note that in the linear model the population changed by the same *amount* each year [see Figure 3.17(a)]. If the population had changed by the same *percent* each year, the model would have been exponential, not linear [see Figure 3.17(b)]. (You will study exponential models in Chapter 9.)

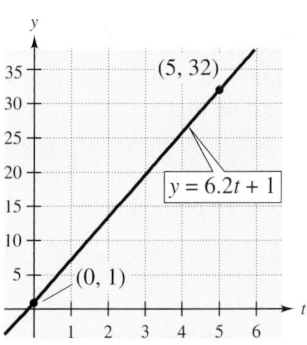

(a) Linear model changes by same amount each year.

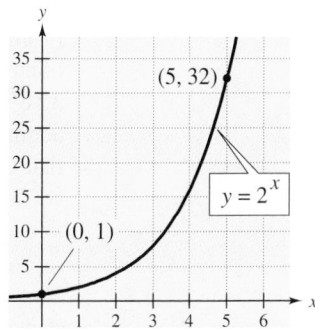

(b) Exponential model changes by same percent each year.

FIGURE 3.17

Technology

Most graphing utilities have built-in programs that will calculate the equation of the best-fitting line for a collection of points. For instance, to find the best-fitting line on a *TI-82* or *TI-83*, enter the *x*- and *y*-values with the STAT (Edit) menu so that the *x*-values are in L1 and the *y*-values are in L2. Then, from the STAT (Calc) menu, choose "LinReg(ax + b)." The graph below shows the best-fitting line for the data in Example 6.

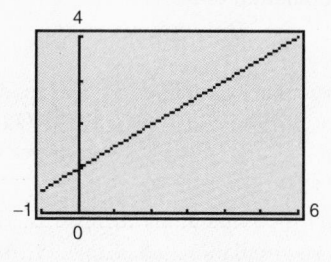

Scatter Plots

Another type of linear modeling is a graphical approach that is commonly used in statistics. To find a mathematical model that approximates a set of actual data points, plot the points on a rectangular coordinate system. This collection of points is called a **scatter plot.** Once the points have been plotted, try to find the line that most closely represents the plotted points. (In this section, we will rely on a visual technique for fitting a line to a set of points. If you take a course in statistics, you will encounter *regression analysis* formulas that can fit a line to a set of points.)

EXAMPLE 6   Fitting a Line to a Set of Points

The scatter plot in Figure 3.18(a) shows 35 different points in the plane. Find the equation of a line that approximately fits these points.

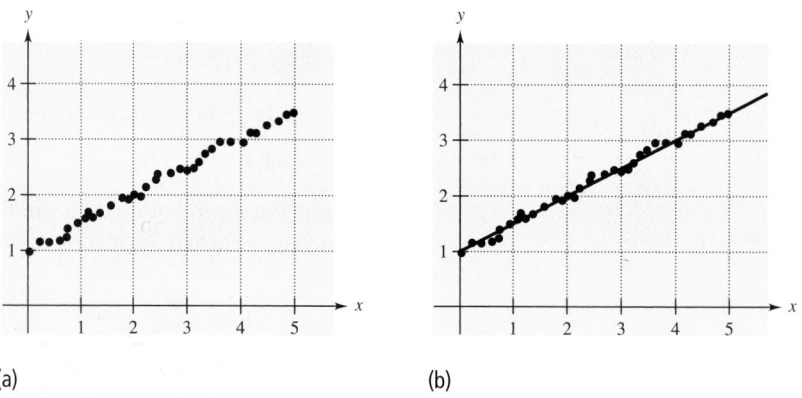

(a)   (b)

FIGURE 3.18

Solution

From Figure 3.18(a), you can see that there is no line that *exactly* fits the given points. The points, however, do appear to resemble a linear pattern. Figure 3.18(b) shows a line that appears to best describe the given points. (Notice that about as many points lie above the line as below it.) From this figure, you can see that the best-fitting line has a *y*-intercept at about (0, 1) and has a slope of about $\frac{1}{2}$. Thus, the equation of the line is

$$y = \frac{1}{2}x + 1.$$

If you had been given the coordinates of the 35 points, you could have checked the accuracy of this model by constructing a table that compared the actual *y*-values with the *y*-values given by the model.

EXAMPLE 7 *Prize Money at the Indianapolis 500*

The total prize money p (in millions of dollars) awarded at the Indianapolis 500 race from 1985 through 1993 is given in the table. Construct a scatter plot that represents the data and find a linear model that approximates the data. *(Source: Indianapolis Motor Speedway Hall of Fame)*

Year	1985	1986	1987	1988	1989	1990	1991	1992	1993
p	$3.27	$4.00	$4.49	$5.03	$5.72	$6.33	$7.01	$7.53	$7.68

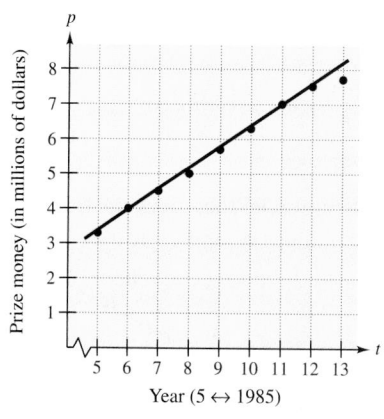

FIGURE 3.19

Solution

Let $t = 5$ represent 1985. The scatter plot for the points is shown in Figure 3.19. From the scatter plot, draw a line that approximates the data. Then, to find the equation of the line, approximate two points *on the line*: $(6, 4)$ and $(11, 7)$. The slope of this line is

$$m \approx \frac{p_2 - p_1}{t_2 - t_1} = \frac{7 - 4}{11 - 6} = \frac{3}{5} = 0.6.$$

Using the point-slope form, you can determine that the equation of the line is

$$p - 4 = 0.6(t - 6) \qquad \text{Point-slope form}$$
$$p = 0.6t + 0.4. \qquad \text{Linear model}$$

To check this model, compare the actual p-values with the p-values given by the model (these are labeled $p*$ below).

t	5	6	7	8	9	10	11	12	13
p	$3.27	$4.00	$4.49	$5.03	$5.72	$6.33	$7.01	$7.53	$7.68
$p*$	$3.4	$4.0	$4.6	$5.2	$5.8	$6.4	$7.0	$7.6	$8.2

Group Activities   **Data Research, Modeling, and Interpretation**

An Experiment   As a preparation for this activity, go to the library, or some other reference source, and find some data that you think describes a linear relationship. Plot the points and find the equation of a line that approximately represents the points. Then present your data and conclusions to your group.

3.2 Exercises

See Warm-Up Exercises, p. A41

1. *Falling Object*   In an experiment, students measured the speed s (in meters per second) of a falling object t seconds after it was released. The results are given in the table.

t	0	1	2	3	4
s	0	11.0	19.4	29.2	39.4

A model for the data is $s = 9.7t + 0.4$.

(a) Plot the data and graph the model on the same set of coordinate axes.

(b) Create a table showing the given data and the approximations given by the model.

(c) Use the model to predict the speed of the object after falling 5 seconds.

(d) Interpret the slope in the context of the problem.

2. *Cable TV*   The average monthly basic rate R (in dollars) for cable TV for the years 1989 through 1994 in the United States is given in the table.   *(Source: Paul Kagan Associates, Inc.)*

Year	1989	1990	1991
R	15.21	16.78	18.10

Year	1992	1993	1994
R	19.08	19.39	21.62

A model for the data is $R = 1.17t + 16.61$, where t is the time in years, with $t = 0$ corresponding to 1990.

(a) Plot the data and graph the model on the same set of coordinate axes.

(b) Create a table showing the given data and the approximations given by the model.

(c) Use the model to predict the average monthly basic rate for cable TV for the year 2000.

(d) Interpret the slope in the context of the problem.

3. *Property Tax*   The property tax in a township is directly proportional to the assessed value of the property. The tax on property with an assessed value of $17,072 is $1067.

(a) Find a mathematical model that gives the tax T in terms of the assessed value v.

(b) Use the model to find the tax on property with an assessed value of $11,500.

(c) Determine the tax rate.

4. *Revenue*   The total revenue R is directly proportional to the number of units sold x. When 25 units are sold, the revenue is $6225.

(a) Find a mathematical model that gives the revenue R in terms of the number of units sold x.

(b) Use the model to find the revenue when 32 units are sold.

(c) Determine the price per unit.

5. *The English and Metric Systems*   The label on a roll of tape gives the amount of tape in inches and centimeters. These amounts are 500 and 1270.

(a) Use the information on the label to find a mathematical model that relates inches to centimeters.

(b) Use part (a) to convert 15 inches to centimeters.

(c) Use part (a) to convert 650 centimeters to inches.

(d) Use a graphing utility to graph the model in part (a). Use the graph to confirm the results in parts (b) and (c).

6. *The English and Metric Systems*   The label on a bottle of soft drink gives the amount in liters and fluid ounces. These amounts are 2 and 67.63.

(a) Use the information on the label to find a mathematical model that relates liters to fluid ounces.

(b) Use part (a) to convert 27 liters to fluid ounces.

(c) Use part (a) to convert 32 fluid ounces to liters.

(d) Use a graphing utility to graph the model in part (a). Use the graph to confirm the results in parts (b) and (c).

Civilian Labor Force   In Exercises 7 and 8, use the graph, which shows the total civilian labor force N (in millions) in the United States from 1983 through 1994. *(Source: U.S. Bureau of Labor Statistics)*

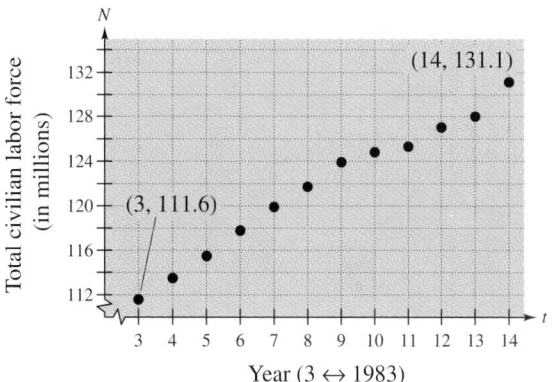

Year (3 ↔ 1983)

7. Using the data for 1983 through 1994, write a linear model for the total civilian labor force, letting $t = 3$ represent 1983. Use the model to predict N in 1998. Use a graphing utility to graph the model and confirm the result.

8. In 1990, 6.9 million of the labor force was unemployed. Approximate the percent of the labor force that was unemployed in 1990.

In Exercises 9–12, a scatter plot is given. Determine whether the data appears linear. If so, determine the sign of the slope of a best-fitting line.

9.
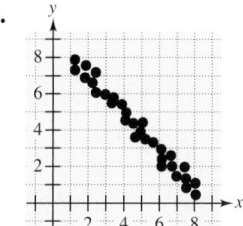

10.
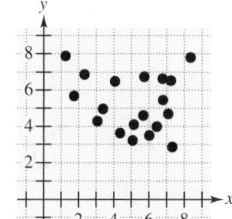

11.
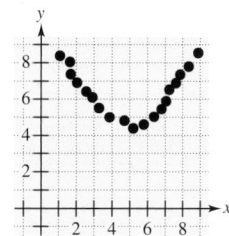

12.
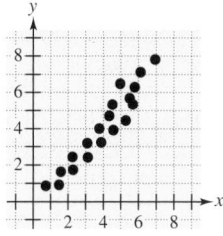

Rate of Change   In Exercises 13–16, you are given the dollar value of a product in 1998 *and* the rate at which the value of the product is expected to change during the next 5 years. Use this information to write a linear equation that gives the dollar value V of the product in terms of the year t. Use a graphing utility to graph the function. (Let $t = 8$ represent 1998.)

	1998 Value	*Rate*
13.	$2540	$125 increase per year
14.	$156	$4.50 increase per year
15.	$20,400	$2000 increase per year
16.	$245,000	$5600 increase per year

Think About It   In Exercises 17–20, match the description with its graph. Determine the slope and interpret its meaning in the context of the problem. [The graphs are labeled (a), (b), (c), and (d).]

(a)
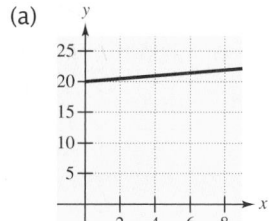

(b)
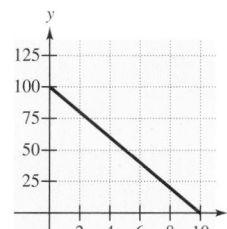

(c)
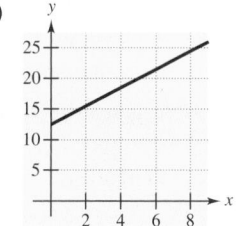

(d)
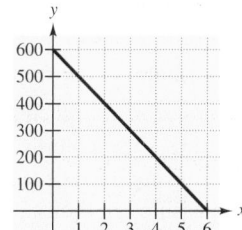

17. A person is paying $10 per week to a friend to repay a $100 loan.

18. An employee is paid $12.50 per hour plus $1.50 for each unit produced per hour.

19. A sales representative receives $20 per day for food plus $0.25 for each mile traveled.

20. A word processor that was purchased for $600 depreciates $100 per year.

21. *Investigation*   An instructor gives 20-point quizzes and 100-point tests in a mathematics course. The average quiz and test scores for six students given as ordered pairs (x, y), where x is the average quiz score and y is the average test score, are $(18, 87)$, $(10, 55)$, $(19, 96)$, $(16, 79)$, $(13, 76)$, and $(15, 82)$.

(a) Plot the points.

(b) Use a ruler to sketch the best-fitting line through the points.

(c) Find an equation for the line sketched in part (b).

(d) Use part (c) to estimate the average test score for a person with an average quiz score of 17.

(e) Describe the changes in parts (a) through (d) that would result if the instructor added 4 points to each average test score.

22. *Holders of Mortgage Debts*   The table gives the amount of mortgage debt (in billions of dollars) held by savings institutions x and commercial banks y for the years 1989 through 1993 in the United States. *(Source: The Federal Reserve Bulletin)*

Year	1989	1990	1991	1992	1993
x	910	802	705	628	598
y	767	845	876	895	940

(a) Plot the points.

(b) Use a ruler to sketch the best-fitting line through the points.

(c) Find an equation for the line sketched in part (b).

(d) Interpret the slope in the context of the problem.

23. *Advertising and Sales*   The table gives the advertising expenditures x and sales volume y for a company for six randomly selected months. Both are measured in thousands of dollars.

Month	1	2	3	4	5	6
x	2.4	1.6	2.0	2.6	1.4	1.6
y	202	184	220	240	180	164

(a) Plot the points.

(b) Use a ruler to sketch the best-fitting line through the points.

(c) Find an equation for the line sketched in part (b).

(d) Interpret the slope in the context of the problem.

In Exercises 24–27, use a ruler to sketch the best-fitting line through the set of points, and find an equation of the line.

24.

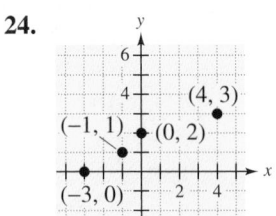

25.

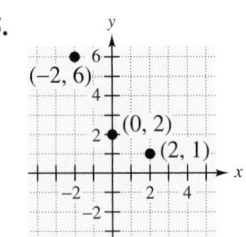

26.

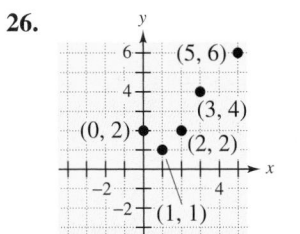

27.

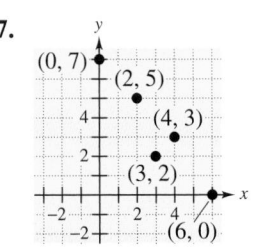

Section Project

Finding a Pattern   Complete the table. The entries in the third row are the differences between consecutive entries in the second row. Describe the third row's pattern.

(a)

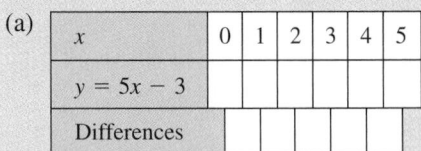

x	0	1	2	3	4	5
$y = 5x - 3$						
Differences					

(b)

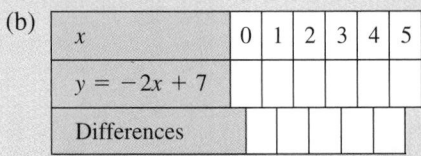

x	0	1	2	3	4	5
$y = -2x + 7$						
Differences					

Find m and b such that the equation $y = mx + b$ yields the table. What does m represent? What does b represent?

(c)

x	0	1	2	3	4	5
$y = mx + b$	3	7	11	15	19	23

(d)

x	0	1	2	3	4	5
$y = mx + b$	-1	-7	-13	-19	-25	-31

3.3	**Applications of Linear Equations**

Translating Phrases ▪ Hidden Operations ▪ Mathematical Modeling ▪
Rates, Ratios, and Proportions

Translating Phrases

In this section, you will study ways to *construct* algebraic expressions. When you translate a verbal sentence or phrase into an algebraic expression, watch for key words and phrases that indicate the four different operations of arithmetic.

Translating Key Words and Phrases

Key Words and Phrases	*Verbal Description*	*Algebraic Expression*
Addition:		
Sum, plus, greater than, increased by, more than, exceeds, total of	The sum of 5 and x	$5 + x$
	Seven more than y	$y + 7$
Subtraction:		
Difference, minus, less than, decreased by, subtracted from, reduced by, the remainder	b is subtracted from 4.	$4 - b$
	Three less than z	$z - 3$
Multiplication:		
Product, multiplied by, twice, times, percent of	Two times x	$2x$
Division:		
Quotient, divided by, ratio, per	The ratio of x and 8	$\dfrac{x}{8}$

EXAMPLE 1   Translating Verbal Phrases

(a) *Verbal Description:* Seven more than three times x

   Algebraic Expression: $3x + 7$

(b) *Verbal Description:* Four less than the product of 6 and n

   Algebraic Expression: $6n - 4$

EXAMPLE 2 Translating Verbal Phrases

(a) *Verbal Description:* Eight added to the product of 2 and n
 Algebraic Expression: $2n + 8$

(b) *Verbal Description:* Four times the sum of y and 9
 Algebraic Expression: $4(y + 9)$

(c) *Verbal Description:* Three times the ratio of x and 7
 Algebraic Expression: $3\left(\dfrac{x}{7}\right)$

In Examples 1 and 2, the verbal description specified the name of the variable. In most real-life situations, however, the variables are not specified and it is your task to assign variables to the *appropriate* quantities.

EXAMPLE 3 Translating Verbal Phrases

This approach makes translation easier for students when they are solving application problems.

(a) *Verbal Description:* The sum of 7 and a number
 Label: The number $= x$
 Algebraic Expression: $7 + x$

(b) *Verbal Description:* Four decreased by the product of 2 and a number
 Label: The number $= x$
 Algebraic Expression: $4 - 2x$

A good way to learn algebra is to do it both forward and backward. For instance, the next example translates algebraic expressions into verbal form. Keep in mind that other key words can be used to describe the operations in each expression.

EXAMPLE 4 Translating Expressions into Verbal Phrases

Without using a variable, write a verbal description for each expression.

(a) $5x - 10$ (b) $\dfrac{3 + x}{4}$

Solution

(a) 10 less than the product of 5 and a number

(b) The sum of 3 and some number, all divided by 4

Hidden Operations

When verbal phrases are translated into algebraic expressions, products are often overlooked. Watch for hidden products in the next three examples.

EXAMPLE 5 *Discovering Hidden Products*

A cash register contains x quarters. Write an expression for this amount of money in dollars.

Solution

| *Verbal Model:* | Value of coin | $\cdot$ | Number of coins |

Labels: Value of coin = 0.25   (dollars per quarter)
              Number of coins = x (quarters)

Expression: $0.25x$

EXAMPLE 6 *Discovering Hidden Products*

A cash register contains n nickels and d dimes. Write an expression for this amount of money in cents.

Solution

| *Verbal Model:* | Value of nickel | $\cdot$ | Number of nickels | $+$ | Value of dime | $\cdot$ | Number of dimes |

Labels: Value of nickel = 5   (cents per nickel)
              Number of nickels = n (nickels)
              Value of dime = 10   (cents per dime)
              Number of dimes = d (dimes)

The concept of multiplying value by quantity is an important one as it can be applied later to a variety of applications.

Expression: $5n + 10d$

In Example 6, the final expression $5n + 10d$ is measured in cents. This makes "sense" in the following way.

$$\frac{5 \text{ cents}}{\text{nickel}} \cdot n \text{ nickels} + \frac{10 \text{ cents}}{\text{dime}} \cdot d \text{ dimes}$$

Note that the nickels and dimes "cancel," leaving cents as the unit of measure for each term. This technique is called *unit analysis,* and it can be very helpful in determining the final unit of measure.

NOTE   Using unit analysis, you can see that the expression in Example 7 has *miles* as its unit of measure.

$$12\,\frac{\text{miles}}{\text{hour}} \cdot t\,\text{hours}$$

Encourage students to choose arbitrary values for a specific case before generalizing with variables. This is a valuable problem-solving strategy, which will ensure that their approach is feasible.

EXAMPLE 7   Discovering Hidden Products

Write an expression showing how far a person can ride a bicycle in t hours if the person travels at a constant rate of 12 miles per hour.

Solution

For this problem, use the formula (distance) = (rate)(time).

Verbal Model: | Rate | · | Time |

Labels: Rate = 12   (miles per hour)
             Time = t (hours)

Expression: $12t$

In mathematics it is useful to know how to represent certain types of integers algebraically.

Two integers are called **consecutive integers** if they differ by 1. Hence, for any integer n, the next two larger consecutive integers are $n + 1$ and $(n + 1) + 1$, or $n + 2$. Thus, you can denote three consecutive integers by n, $n + 1$, and $n + 2$.

Labels for Integers

Let n represent an integer.

1. $\{n, n + 1, n + 2, \ldots\}$ denotes a set of *consecutive* integers.

2. If n is an even integer, then $\{n, n + 2, n + 4, \ldots\}$ denotes a set of *consecutive even integers.*

3. If n is an odd integer, then $\{n, n + 2, n + 4, \ldots\}$ denotes a set of *consecutive odd integers.*

Mathematical Modeling

So far in this section, you have translated *verbal phrases* into *algebraic expressions.* When algebra is used to solve real-life problems, you usually have to carry the process one step further—by translating *verbal sentences* into *algebraic equations.* As discussed in Section 3.2, the process of translating phrases or sentences into algebraic expressions or equations is called **mathematical modeling.**

A good approach is to use a *verbal model* by using the given verbal description of the problem. Then, after assigning labels to the unknown quantities in the verbal model, you can form a *mathematical model* or *algebraic equation.*

Verbal models help students organize and picture relationships, which can then be translated into equations.

Verbal description ⟹ Verbal model ⟹ Assign labels ⟹ Algebraic equation

EXAMPLE 8   *Mathematical Modeling*

Write an algebraic equation that represents the following problem. Then solve the equation and answer the question.

> *You have accepted a job at an annual salary of $27,630. This salary includes a year-end bonus of $750. If you are paid twice a month, what will your gross pay be for each paycheck?*

Solution

Because there are 12 months in a year and you will be paid twice a month, it follows that you will receive 24 paychecks during the year. Construct an algebraic equation for this problem as follows. Begin with a verbal model, then assign labels, and finally form an algebraic equation.

Verbal Model:	$\boxed{\text{Income for year}} = \boxed{\dfrac{24}{\text{paychecks}}} + \boxed{\text{Bonus}}$	
Labels:	Income for year = 27,630	(dollars)
	Amount of each paycheck = x	(dollars)
	Bonus = 750	(dollars)

Equation:

$$27{,}630 = 24x + 750$$
$$27{,}630 - 750 = 24x + 750 - 750$$
$$26{,}880 = 24x$$
$$\frac{26{,}880}{24} = \frac{24x}{24}$$
$$1120 = x$$

Each paycheck will be $1120. Check this in the original statement of the problem, as follows.

$$\boxed{\text{Income for year}} = 24(1120) + 750 = 26{,}880 + 750 = \$27{,}630$$

EXAMPLE 9 *Constructing a Verbal Model*

Find two consecutive integers such that the sum of the first integer and three times the second is 87.

Solution

Verbal Model:

$$\boxed{\text{First integer}} + 3 \cdot \boxed{\text{Second integer}} = 87$$

Labels:

First integer $= n$
Second integer $= n + 1$

Equation:

$n + 3(n + 1) = 87$	Algebraic model
$n + 3n + 3 = 87$	Distributive Property
$4n + 3 = 87$	Combine like terms.
$4n = 84$	Subtract 3 from both sides.
$n = 21$	Divide both sides by 4.

Thus, the first integer is 21, and the second integer is $21 + 1 = 22$. You can check this by substituting 21 and 22 as the two consecutive integers in the original problem.

EXAMPLE 10 *A Percent Application*

A real estate agency receives a commission of $8092.50 for the sale of a $124,500 house. What percent commission is this?

Solution

Verbal Model:

$$\boxed{\text{Commission}} = \boxed{\begin{array}{c}\text{Percent}\\\text{(decimal form)}\end{array}} \cdot \boxed{\begin{array}{c}\text{Sale}\\\text{price}\end{array}}$$

Labels:

Commission $= 8092.50$ (dollars)
Percent $= p$ (decimal form)
Sale price $= 124,500$ (dollars)

Equation:

$$8092.50 = p(124,500)$$
$$\frac{8092.50}{124,500} = p$$
$$0.065 = p$$

The real estate agency receives a commission of 6.5%. Use your calculator to check this solution in the original statement of the problem.

EXAMPLE 11   *The First Transcontinental Railroad*

The Central Pacific Company averaged 8.75 miles of track per month. The Union Pacific Company averaged 20 miles of track per month. The drawing below shows the two companies meeting in Promontory, Utah, as the track was completed. Use the additional information in the caption to answer the following. When was the track completed? How many miles of track did each company build?

In 1862, the United States Congress granted construction rights to two railroad companies to build a railroad connecting Omaha, Nebraska, with Sacramento, California. The entire railroad was about 1590 miles long. The Central Pacific Company began building eastward from Sacramento in 1863. Twenty-four months later, the Union Pacific Company began building westward from Omaha.

Solution

Verbal Model:

$$\underbrace{\boxed{\text{Total miles of track}}}_{} = \underbrace{\boxed{\text{Miles per month}} \cdot \boxed{\text{Number of months}}}_{\text{Central Pacific}} + \underbrace{\boxed{\text{Miles per month}} \cdot \boxed{\text{Number of months}}}_{\text{Union Pacific}}$$

Labels:

Total length of track $= 1590$	(miles)
Central Pacific rate $= 8.75$	(miles per month)
Central Pacific time $= t$	(months)
Union Pacific rate $= 20$	(miles per month)
Union Pacific time $= t - 24$	(months)

Equation:

$1590 = 8.75t + 20(t - 24)$	Linear model
$1590 = 8.75t + 20t - 480$	Distributive Property
$2070 = 28.75t$	Add 480 to both sides.
$72 = t$	Divide both sides by 28.75.

The construction took 72 months (6 years) from the time the Central Pacific Company began. Thus, the track was completed in 1869. The numbers of miles of track built by the two companies were as follows.

Central Pacific: (8.75 miles per month)(72 months) = 630 miles

Union Pacific: (20 miles per month)(48 months) = 960 miles

Rates, Ratios, and Proportions

In real-life applications, the quotient a/b is called a **rate** if a and b have different units, and is called a **ratio** if a and b have the same unit. Note the *order* implied by a ratio. The ratio of a to b means a/b, whereas the ratio of b and a means b/a.

EXAMPLE 12   Expressing Rates and Ratios

Point out the differences between rate and ratio in this problem.

(a) You have driven 110 miles in 2 hours. Your average rate for the trip can be expressed as

$$\text{Rate} = \frac{110 \text{ miles}}{2 \text{ hours}} = 55 \text{ miles per hour.}$$

(b) You have driven 110 miles in 2 hours and your friend has driven 100 miles in the same length of time. The ratio of the distance you have traveled to the distance your friend has traveled is

$$\text{Ratio} = \frac{110 \text{ miles}}{100 \text{ miles}} = \frac{11}{10}.$$

When comparing two *measurements* by means of a ratio, be sure to use the same unit of measurement in both the numerator and the denominator.

EXAMPLE 13   Comparing Measurements

Find ratios for comparing the relative sizes of each of the following. Use the same unit of measurement in the numerator and the denominator.

(a) 200 cents to 5 dollars   (b) 28 months to $1\frac{1}{2}$ years

Solution

(a) Because 200 cents is the same as 2 dollars, the ratio is

$$\frac{200 \text{ cents}}{5 \text{ dollars}} = \frac{2 \text{ dollars}}{5 \text{ dollars}} = \frac{2}{5}.$$

Note that if you had converted 5 dollars to 500 cents, you would have obtained the same ratio, because

$$\frac{200 \text{ cents}}{5 \text{ dollars}} = \frac{200 \text{ cents}}{500 \text{ cents}} = \frac{200}{500} = \frac{2}{5}.$$

(b) Because $1\frac{1}{2}$ years $= 18$ months, the ratio is

$$\frac{28 \text{ months}}{1\frac{1}{2} \text{ years}} = \frac{28 \text{ months}}{18 \text{ months}} = \frac{28}{18} = \frac{14}{9}.$$

Another comparison: Find a ratio to compare the relative size of 3 pounds to 10 ounces.

$\frac{3 \text{ lb}}{10 \text{ oz}} = \frac{48 \text{ oz}}{10 \text{ oz}} = \frac{24}{5}$

A **proportion** is a statement that equates two ratios. For example, if the ratio of a to b is the same as the ratio of c to d, you can write the proportion as $a/b = c/d$. In typical problems, you know three of the values and need to find the fourth. The quantities a and d are called the **extremes** of the proportion, and the quantities b and c are called the **means** of the proportion. In a proportion, the product of the extremes is equal to the product of the means. Rewriting a proportion in the form $ad = bc$ is called **cross multiplying.**

EXAMPLE 14   Solving a Proportion

The ratio of 8 to x is the same as the ratio of 5 to 2. What is x?

Solution

$$\frac{8}{x} = \frac{5}{2} \qquad \text{Set up proportion.}$$

$$16 = 5x \qquad \text{Cross multiply.}$$

$$\frac{16}{5} = x \qquad \text{Divide both sides by 5.}$$

Thus, $x = \frac{16}{5}$. Check this in the original statement of the problem.

EXAMPLE 15   An Application of Proportion

You are driving from Arizona to New York, a trip of 2750 miles. You begin the trip with a full tank of gas and after traveling 424 miles, you refill the tank for $22.00. How much should you plan to spend on gasoline for the entire trip?

Solution

Verbal Model:
$$\frac{\text{Dollars for trip}}{\text{Dollars for tank}} = \frac{\text{Miles for trip}}{\text{Miles for tank}}$$

Labels:
    Cost of gas for entire trip $= x$ (dollars)
    Cost of gas for tank $= 22$ (dollars)
    Miles for entire trip $= 2750$ (miles)
    Miles for tank $= 424$ (miles)

Proportion:
$$\frac{x}{22} = \frac{2750}{424}$$

$$x = 22\left(\frac{2750}{424}\right)$$

$$x \approx 142.69$$

You should plan to spend approximately $142.69 for gasoline on the trip. Check this in the original statement of the problem.

Study Tip

You can write a proportion in several ways. Just be sure to put like quantities in similar positions on each side of the proportion.

The following list summarizes a strategy for modeling and solving real-life problems.

Strategy for Solving Word Problems

1. Ask yourself what you need to know to solve the problem. Then *write a verbal model* that will give you what you need to know.

2. *Assign labels* to each part of the verbal model—numbers to the known quantities and letters (or expressions) to the variable quantities.

3. Use the labels to *write an algebraic model* based on the verbal model.

4. *Solve* the resulting algebraic equation.

5. *Answer* the original question and *check* that your answer satisfies the original problem as stated.

In previous mathematics courses, you studied several other problem-solving strategies, such as *drawing a diagram, making a table, looking for a pattern,* and *solving a simpler problem.* Each of these strategies can also help you to solve problems in algebra.

Group Activities

Extending the Concept

Checking the Sensibility of an Answer When you solve problems related to real-life situations, you should always ask yourself whether your answers make sense. In your group, discuss why the following answers are suspicious and decide what more reasonable answers might be.

(a) A problem asks you to find the life expectancy of an American male born in 1970, and you use a formula to obtain a preliminary answer of 124 years.

(b) A problem asks you to find the Celsius temperature equivalent to 80° Fahrenheit, and you obtain a preliminary answer of −24°C.

(c) A problem asks you to find the diameter of a household electrical extension cord, and you obtain a preliminary answer of 5.0 inches.

(d) A problem asks you to find the floor area of a gymnasium, and you obtain a preliminary answer of 300 square feet.

3.3 Exercises

*See Warm-Up
Exercises, p. A42*

In Exercises 1–14, translate the statement into an algebraic expression.

1. The sum of 8 and a number n

2. Six less than a number n

3. Fifteen decreased by three times a number n

4. Six less than four times a number n

5. One-third of a number n

6. Seven-fifths of a number n

7. Thirty percent of the list price L

8. Forty percent of the cost C

9. The quotient of a number x and 6

10. The ratio of y to 3

11. The sum of 3 and four times a number x, all divided by 8

12. The product of a number y and 10, decreased by 35

13. The absolute value of the difference between a number n and 5

14. The absolute value of the quotient of y and 4

In Exercises 15–22, write a verbal description of the algebraic expression without using the variable.

15. $3x + 2$

16. $4x - 5$

17. $8(x - 5)$

18. $-3(x + 2)$

19. $\dfrac{y}{8}$

20. $\dfrac{4x}{5}$

21. $\dfrac{x + 10}{3}$

22. $25 + \dfrac{x}{6}$

In Exercises 23–36, write an algebraic expression that represents the specified quantity in the verbal statement and simplify if possible.

23. The amount of money (in dollars) represented by n quarters

24. The amount of money (in cents) represented by m dimes and n quarters

25. The distance traveled in t hours at an average speed of 55 miles per hour

26. The distance traveled in 5 hours at an average speed of r miles per hour

27. The time required to travel 100 miles at an average speed of r miles per hour

28. The average rate of speed for a journey of 360 miles in t hours

29. The amount of antifreeze in a cooling system containing y gallons of coolant that is 45% antifreeze

30. The amount of water in q quarts of food product that is 65% water

31. The amount of wage tax due for a taxable income of I dollars that is taxed at a rate of 1.25%

32. The amount of sales tax on a purchase valued at L dollars if the tax rate is 6%

33. The sale price of a coat that has a list price of L dollars if it is a "20% off" sale

34. The total cost for a family to stay one night at a campground if the charge is $18 for the parents plus $3 for each of the n children

35. The total hourly wage for an employee when the base pay is $8.25 per hour and an additional $0.60 is paid for each of q units produced per hour

36. The total hourly wage for an employee when the base pay is $11.65 per hour and an additional $0.80 is paid for each of q units produced per hour

In Exercises 37–42, solve the proportion.

37. $\dfrac{x}{6} = \dfrac{2}{3}$

38. $\dfrac{y}{36} = \dfrac{6}{7}$

39. $\dfrac{5}{4} = \dfrac{t}{6}$

40. $\dfrac{7}{8} = \dfrac{x}{2}$

41. $\dfrac{y + 5}{6} = \dfrac{y - 2}{4}$

42. $\dfrac{z - 3}{3} = \dfrac{z + 8}{12}$

In Exercises 43–46, express as a ratio. Use the same unit in both the numerator and denominator, and write your answer in reduced form.

43. 36 inches to 48 inches

44. 5 pounds to 24 ounces

45. 40 milliliters to 1 liter

46. 125 centimeters to 2 meters

Geometry   In Exercises 47 and 48, write expressions for the perimeter and area of the region. Then simplify the expressions.

47.
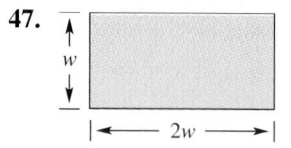

48.
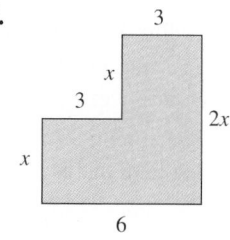

Geometry   In Exercises 49 and 50, write an expression for the area of the region.

49.
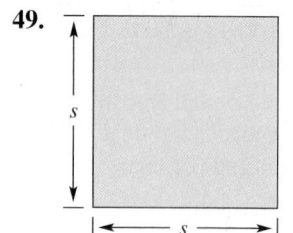

50.
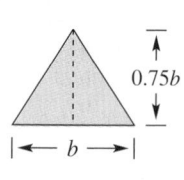

51. *Advertising Space*   An advertising banner has a width of w and a length of $6w$, where w is measured in meters (see figure). Find an algebraic expression that represents the area of the banner. What is the unit of measure for the area?

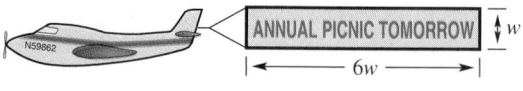

52. *Billiard Table*   The top of a billiard table has a length of l and a width of $l - 6$, where l is measured in feet (see figure). Find an algebraic expression that represents the area of the top of the billiard table. What is the unit of measure for the area?

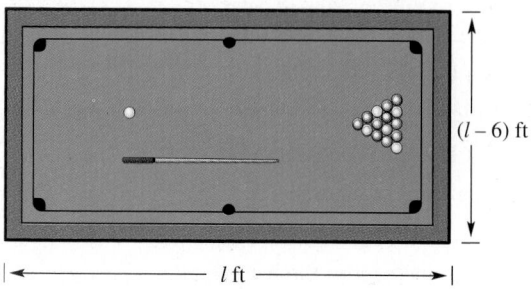

Number Problems   In Exercises 53–58, solve the number problem.

53. The sum of three consecutive even integers is 138. Find the integers.

54. Eight times the sum of a number and 6 is 128. What is the number?

55. When the sum of a number and 18 is divided by 5, the quotient is 12. Find the number.

56. Find a number such that the sum of that number and 30 is 82.

57. Find a number such that six times the difference of the number and 12 is 300.

58. The difference of five times an odd integer and three times the next consecutive odd integer is 24. Find the integers.

59. *Geometry*   The "Slow Moving Vehicle" sign has the shape of a hexagon surrounding an equilateral triangle. The triangle has a perimeter of 129 centimeters. Find the length of each side of the triangle.

60. *Geometry* The length of a rectangle is three times its width. The perimeter of the rectangle is 64 inches. Find the dimensions of the rectangle.

61. *Repair Time* The bill for the repair of an automobile is $380. Included in this bill is a charge of $275 for parts. If the remainder of the bill is for labor at a rate of $35 per hour, how many hours were spent in repairing the car? $275 + 35x = 380$ $35x = 105$ $x = 3$

62. *Units Produced* You have a job on an assembly line for which you are paid $10 per hour plus $0.75 per unit produced. Find the number of units produced in an 8-hour day if your earnings are $146.

63. *Monthly Rent* If you spend 15% of your monthly income of $2800 for rent, what is your monthly rent payment? $.15(2800) = 420$

64. *Pension Fund* Your employer withholds $6\frac{1}{2}\%$ of your gross income for your retirement. Determine the amount withheld each month if the gross monthly income is $3800.

65. *Company Layoff* Because of slumping sales, a small company laid off 25 of its 160 employees. What percent of the work force was laid off? $25 = \frac{x}{100} \cdot 160$ $x = \frac{25}{1.6}$ 15.625% $15\%\%$

66. *Real Estate Commission* A real estate agency receives a commission of $9100 for the sale of a $130,000 house. What percent commission is this?

67. *Defective Parts* A quality control engineer reported that 1.5% of a sample of parts were defective. Find the size of the sample if the engineer detected three defective parts. $.015x = 3$ $x = \frac{3}{.015} = 200$

68. *Price Inflation* The price of a new van is approximately 115% of what it was 3 years ago. What was the approximate price 3 years ago if the current price is $25,750?

69. *Monthly Expenses* The expenses for a small company for January are shown in the circle graph. What percent of the total monthly expenses is each budget item?

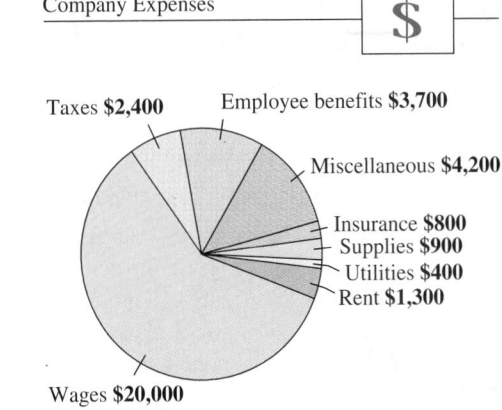

Company Expenses

Taxes **$2,400**
Employee benefits **$3,700**
Miscellaneous **$4,200**
Insurance **$800**
Supplies **$900**
Utilities **$400**
Rent **$1,300**
Wages **$20,000**

70. *Population* The populations of six counties are shown in the circle graph. What percent of the total population is each county's population?

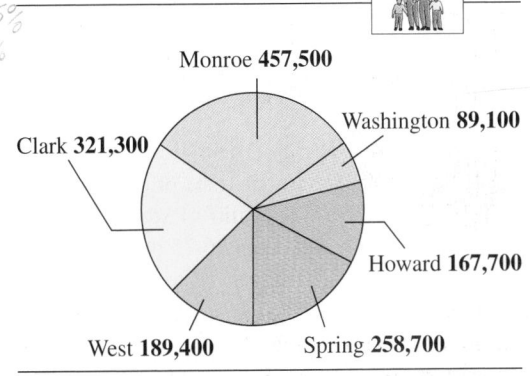

County Populations

Monroe **457,500**
Washington **89,100**
Clark **321,300**
Howard **167,700**
West **189,400**
Spring **258,700**

71. *Think About It*   In the 1970 census, the population of a city was 60,000. The population grew 4% during the decade of the seventies and grew 6% during the decade of the eighties.

(a) Use the given information to approximate the populations of the city at the time of the 1980 census and at the time of the 1990 census.

(b) Use the results in part (a) to approximate the percent increase in the population between the 1970 and the 1990 census. Explain why it isn't 10%.

72. *Think About It*   Is it true that $\frac{1}{2}\% = 50\%$? Explain.

73. *Compression Ratio*   The compression ratio of a cylinder is the ratio of its expanded volume to its compressed volume (see figure). The expanded volume of one cylinder of a small diesel engine is 425 cubic centimeters, and its compressed volume is 20 cubic centimeters. Find the compression ratio of this cylinder.

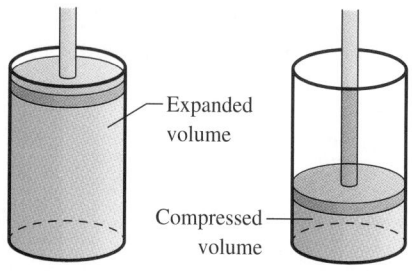

Expanded volume

Compressed volume

74. *Price-Earnings Ratio*   The ratio of the price of a stock to its earnings is called the price-earnings ratio. Find the price-earnings ratio of stock that sells for $56.25 per share and earns $6.25 per share.

75. *State Income Tax*   You have $12.50 of state tax withheld from your paycheck per week when your gross pay is $625. Find the ratio of tax to gross pay. (Write this as a percent.)

76. *Gear Ratio*   The gear ratio of two gears is the number of teeth in one gear to the number of teeth in the other gear (see figure). If two gears in a gearbox have 60 teeth and 40 teeth, find the gear ratio.

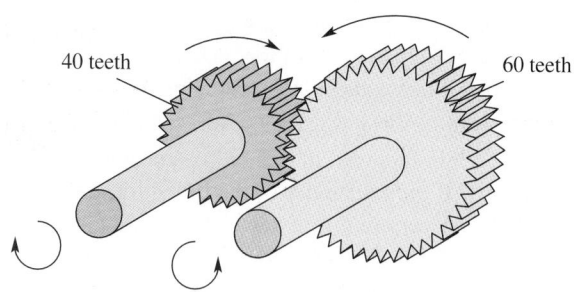

40 teeth   60 teeth

Figure for 76

77. *Property Tax*   The taxes on property with an assessed value of $75,000 are $1125. Find the taxes on property with an assessed value of $120,000.

78. *Public Opinion Poll*   In a public opinion poll, 870 people from a sample of 1500 indicated that they would vote for a specific candidate. Assuming this poll to be a correct indicator of the electorate, how many votes can the candidate expect to receive from a total of 80,000 votes cast?

79. *Fuel Mixture*   The gasoline-to-oil ratio of a two-cycle engine is 40 to 1. Determine the amount of gasoline required to produce a mixture that contains $\frac{1}{2}$ pint of oil. $\frac{40}{1} = \frac{x}{\frac{1}{2}}$ $x = 20\text{ pts}$

80. *Recipe Proportions*   Three cups of flour are required to make one batch of cookies. How many cups are required to make $3\frac{1}{2}$ batches?

81. *Map Scale*   Use the map scale in the figure to approximate the straight-line distance from Los Angeles to San Francisco. ≈ 330 30

0   60
miles

San Francisco

CALIFORNIA

Los Angeles

82. *Spring Length*   A force of 32 pounds stretches a spring 6 inches. Determine the number of pounds of force required to stretch it 9 inches.

Section Project

Proportions   You can answer each of the following questions by writing and solving a proportion using the fact that corresponding sides of similar triangles are proportional.

Geometry   Solve for the length x of the side of the triangle (see figure).

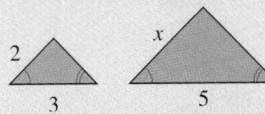

Tree Height   A man who is 6 feet tall walks directly toward the tip of a shadow of a tree. When the man is 75 feet from the tree, he notices his own shadow beyond the shadow of the tree. Find the height of the tree, if the length of the shadow of the tree beyond this point is 11 feet (see figure).

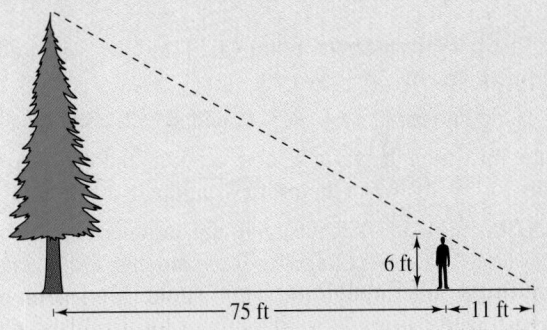

Shadow Length   Find the length of the shadow of a man who is 6 feet tall and is standing 15 feet from a streetlight that is 20 feet high (see figure).

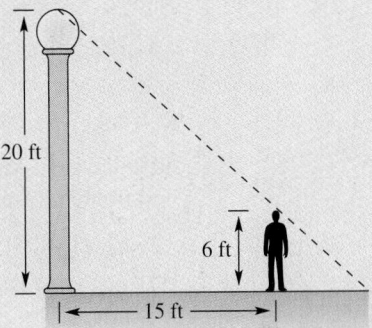

Mid-Chapter Quiz

Take this quiz as you would take a quiz in class. After you are done, check your work against the answers given in the back of the book.

In Exercises 1–4, write an equation of the line passing through the point with the given slope.

1. $\left(0, -\frac{3}{2}\right), m = 2$

2. $(4, 7), m = \frac{1}{2}$

3. $\left(\frac{5}{2}, 6\right), m = -\frac{3}{4}$

4. $(-3.5, -1.8), m = 3$

In Exercises 5–8, write an equation of the line passing through the points. Use a graphing utility to graph the equation.

5. $(2, 1), (4, 5)$

6. $(0, 0.8), (3, -2.3)$

7. $(3, -1), (10, -1)$

8. $\left(4, \frac{5}{3}\right), (4, 8)$

9. Write equations of the lines that pass through the point $(3, 5)$ and are (a) parallel to and (b) perpendicular to the line given by $2x - 3y = 1$.

10. Find the x- and y-intercepts of the line $3x - 2y - 9 = 0$, and sketch its graph.

11. A company produces a product for which the variable cost is $5.60 and the fixed costs are $24,000. The product sells for $9.20. Write the profit as a linear function of x.

12. The points $(75, 2.3)$, $(82, 3.2)$, $(90, 3.6)$, and $(65, 2.3)$ give the entrance exam scores x and the grade-point averages y after 1 year of college for four students. Plot the points. Use a ruler to sketch the best-fitting line through the points. Find an equation of the line and use the equation to estimate the grade-point average for a student with an exam score of 85.

13. Translate the following statement into an algebraic expression.

"The product of a number n and 5 is decreased by 8."

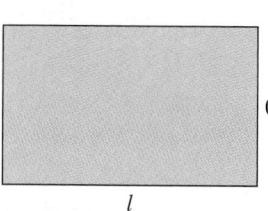

$0.6l$

l

Figure for 14

14. Write algebraic expressions for the perimeter and area of the rectangle shown at the left. Simplify the expressions.

15. Write an algebraic expression for the sum of three consecutive even integers, the first of which is n.

16. A quality control engineer for a manufacturer finds one defective unit in a sample of 300. At this rate, what is the expected number of defective units in a shipment of 600,000?

3.4	**Business and Scientific Problems**

Business Problems · Mixture Problems · Rate Problems · Formulas

Business Problems

Many business problems can be represented by mathematical models involving the sum of a fixed term and a variable term. The variable term is often a *hidden product* in which one of the factors is a percent or some other type of rate. Watch for these occurrences in the discussion and examples that follow.

The **markup** on a consumer item is the difference between the **cost** a retailer pays for an item and the **price** at which the retailer sells the item. A verbal model for this relationship is as follows.

Selling price	=	Cost	+	Markup

Markup is a hidden product.

The markup is the product of the **markup rate** and the cost.

Markup	=	Markup rate	·	Cost

EXAMPLE 1 *Finding the Markup Rate*

A clothing store sells a pair of jeans for $42. If the cost of the jeans is $16.80, what is the markup rate?

Solution

Verbal Model:

Selling price	=	Cost	+	Markup

Labels:

Selling price = 42	(dollars)
Cost = 16.80	(dollars)
Markup rate = p	(percent in decimal form)
Markup = $p(16.80)$	(dollars)

Point out the alternative of first multiplying both sides of the equation by 10 to eliminate the use of decimals.

Equation:

$$42 = 16.8 + p(16.8)$$
$$42 - 16.8 = p(16.8)$$
$$25.2 = p(16.8)$$
$$\frac{25.2}{16.8} = p$$
$$1.5 = p$$

Because $p = 1.5$, it follows that the markup rate is 150%. Check this in the original statement of the problem.

The model for a **discount** is similar to that for a markup.

EXAMPLE 2   Finding the Discount and the Discount Rate

A compact disc player is marked down from its list price of $820 to a sale price of $574. What is the discount rate?

Solution

Verbal Model:

| Discount | = | Discount rate | · | List price |

Labels: Discount = 820 − 574 = 246   (dollars)
            List price = 820   (dollars)
            Discount rate = p (percent in decimal form)

Equation: $246 = p(820)$

$$\frac{246}{820} = p$$

$$0.30 = p$$

The discount rate is 30%. Check this in the original statement of the problem.

EXAMPLE 3   Finding the Hours of Labor

An auto repair bill of $338 lists $170 for parts and the rest for labor. If the labor charge is $28 per hour, how many hours did it take to repair the auto?

Solution

Verbal Model:

| Total bill | = | Price of parts | + | Price of labor |

Labels: Total bill = 338   (dollars)
            Price of parts = 170   (dollars)
            Hours of labor = x (hours)
            Hourly rate for labor = 28   (dollars per hour)
            Price of labor = 28x (dollars)

Equation: $338 = 170 + 28x$

$$168 = 28x$$

$$\frac{168}{28} = x$$

$$6 = x$$

It took 6 hours to repair the auto. Check this in the original statement of the problem.

Mixture Problems

Many real-life problems involve combinations of two or more quantities that make up new or different quantities. Such problems are called **mixture problems.** They are usually composed of the sum of two or more "hidden products" that fit the following verbal model.

$$\boxed{\text{First rate}} \cdot \boxed{\text{Amount}} + \boxed{\text{Second rate}} \cdot \boxed{\text{Amount}} = \boxed{\text{Final rate}} \cdot \boxed{\text{Final amount}}$$

EXAMPLE 4   A Mixture Problem

Study Tip

When you set up a verbal model, be sure to check that you are working with the *same type of units* in each part of the model. In Example 4, for instance, note that each of the three parts of the verbal model measures cost. (If two parts measured cost and the other part measured pounds, you would know that the model was incorrect.)

A nursery wants to mix two types of lawn seed; one type sells for $10 per pound and the other type sells for $15 per pound. To obtain 20 pounds of a mixture at $12 per pound, how many pounds of each type of seed are needed?

Solution

Verbal Model: $\boxed{\text{Total cost of \$10 seed}} + \boxed{\text{Total cost of \$15 seed}} = \boxed{\text{Total cost of \$12 seed}}$

Labels:

Cost of $10 seed $= 10$	(dollars per pound)
Pounds of $10 seed $= x$	(pounds)
Cost of $15 seed $= 15$	(dollars per pound)
Pounds of $15 seed $= 20 - x$	(pounds)
Cost of $12 seed $= 12$	(dollars per pound)
Pounds of $12 seed $= 20$	(pounds)

Equation:

$$10x + 15(20 - x) = 12(20)$$
$$10x + 300 - 15x = 240$$
$$-5x = -60$$
$$x = 12$$

The mixture should contain 12 pounds of the $10 seed and $20 - 12 = 8$ pounds of the $15 seed.

Note that each of these total costs is determined by multiplying the value by the quantity as in Section 3.3.

Remember that when you have found a solution, you should always go back to the original statement of the problem and check to see that the solution makes sense—both algebraically and from a common sense point of view. For instance, you can check the result of Example 4 as follows.

$$\overbrace{\left(\frac{\$10 \text{ per}}{\text{pound}}\right)\left(\frac{12}{\text{pounds}}\right)}^{\$10 \text{ seed}} + \overbrace{\left(\frac{\$15 \text{ per}}{\text{pound}}\right)\left(\frac{8}{\text{pounds}}\right)}^{\$15 \text{ seed}} = \overbrace{\left(\frac{\$12 \text{ per}}{\text{pound}}\right)\left(\frac{20}{\text{pounds}}\right)}^{\$12 \text{ seed}}$$

$$\$120 + \$120 = \$240$$

EXAMPLE 5   A Solution Mixture Problem

A pharmacist needs to strengthen a 20% alcohol solution so that it contains 36% alcohol. How much pure alcohol should be added to 240 milliliters of the 20% solution (see Figure 3.20)?

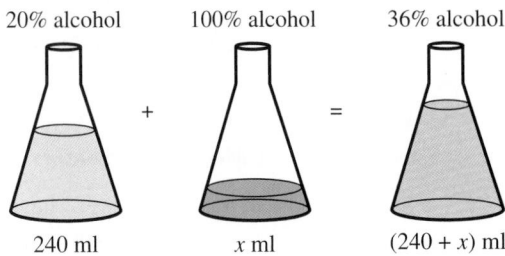

20% alcohol   100% alcohol   36% alcohol

240 ml x ml $(240 + x)$ ml

FIGURE 3.20

Solution

NOTE   In Example 5, the original solution was strengthened by adding pure alcohol. To dilute the solution, you could add a solution containing 0% alcohol. How much solution containing 0% alcohol would you need to add to the original solution to dilute it to a 15% solution?

Verbal Model:

Amount of alcohol in original solution	+	Amount of alcohol in pure solution	=	Amount of alcohol in final solution

Labels:

Original solution: Percent alcohol $= 0.20$ (decimal form)
            Amount of alcohol $= 240$ (milliliters)

Pure alcohol: Percent alcohol $= 1.00$ (decimal form)
            Amount of alcohol $= x$ (milliliters)

Final solution: Percent alcohol $= 0.36$ (decimal form)
            Amount of alcohol $= 240 + x$ (milliliters)

Equation:

$$0.20(240) + 1.00(x) = 0.36(240 + x)$$
$$48 + x = 86.4 + 0.36x$$
$$0.64x = 38.4$$
$$x = \frac{38.4}{0.64}$$
$$= 60 \text{ milliliters}$$

Thus, the pharmacist should add 60 milliliters of pure alcohol to the original solution. Check this solution in the original problem.

Rate Problems

Time-dependent problems such as those involving distance can be classified as **rate problems.** They fit the verbal model

Distance $=$ Rate $\cdot$ Time .

For instance, if you travel at a constant (or average) rate of 55 miles per hour for 45 minutes, the total distance you travel is given by

$$\left(55 \frac{\text{miles}}{\text{hour}}\right)\left(\frac{45}{60} \text{ hour}\right) = 41.25 \text{ miles.}$$

As with all problems involving applications, be sure to check that the units in the verbal model make sense. For instance, in this problem the rate is given in *miles per hour*. Therefore, in order for the solution to be given in *miles*, you must convert the time (from minutes) to *hours*. In the model, you can think of the two "hours" as canceling, as follows.

$$\left(55 \frac{\text{miles}}{\text{hour}}\right)\left(\frac{45}{60} \text{ hour}\right) = 41.25 \text{ miles}$$

EXAMPLE 6   Distance-Rate Problems

Students are traveling in two cars to a football game 150 miles away. The first car leaves on time and travels at an average speed of 48 miles per hour. The second car starts $\frac{1}{2}$ hour later and travels at an average speed of 58 miles per hour. At these speeds, how long will it take the second car to catch up to the first car?

Solution

Verbal Model:	Distance $=$ Rate $\cdot$ Time	

Caution students *not* to express a half hour as 30 *minutes* when using rates expressed in miles per *hour*.

Labels: $\quad d_1 =$ Distance of first car $= 48t$ (miles)
$\quad d_2 =$ Distance of second car $= 58\left(t - \frac{1}{2}\right)$ (miles)

Equation: $\quad d_1 = d_2$

$$48t = 58\left(t - \frac{1}{2}\right)$$
$$48t = 58t - 29$$
$$29 = 10t$$
$$2.9 = t$$

After the first car travels for 2.9 hours, the second car catches up to it. Thus, it takes the second car $2.9 - 0.5 = 2.4$ hours to catch up to the first car.

In work problems, the **rate of work** is the *reciprocal* of the time needed to do the entire job. For instance, if it takes 5 hours to complete a job, the per-hour work rate is

$$\frac{1}{5} \text{ job per hour.}$$

EXAMPLE 7   *Work-Rate Problem*

Consider two machines in a paper manufacturing plant. Machine 1 can complete one job (2000 pounds of paper) in 5 hours. Machine 2 is newer and can complete one job in 2 hours. How long will it take the two machines working together to complete one job?

Solution

| *Verbal Model:* | $\boxed{\dfrac{\text{Work}}{\text{done}}}$ $=$ $\boxed{\begin{array}{c}\text{Portion done}\\\text{by machine 1}\end{array}}$ $+$ $\boxed{\begin{array}{c}\text{Portion done}\\\text{by machine 2}\end{array}}$ |

Labels: Work done by both machines $= 1$ (job)
Time for each machine $= t$ (hours)
Rate for machine 1 $= \frac{1}{5}$ (job per hour)
Rate for machine 2 $= \frac{1}{2}$ (job per hour)

The equation can also be written as $1/t = 1/5 + 1/2$. In this case, $1/t =$ the portion of the work done together in one hour; $1/5 =$ the portion of the work done by Machine 1 in one hour; and $1/2 =$ the portion of the work done by Machine 2 in one hour.

Equation:

$$1 = \frac{1}{5}t + \frac{1}{2}t$$

$$10(1) = 10\left(\frac{1}{5}t + \frac{1}{2}t\right) \qquad \text{Multiply both sides by LCD of 10.}$$

$$10 = 2t + 5t$$

$$10 = 7t$$

$$\frac{10}{7} = t$$

It will take $\frac{10}{7}$ hours (or about 1.43 hours) for both machines to complete the job. Check this solution in the original statement of the problem.

Note in Example 7 that the "2000 pounds" of paper was unnecessary information. We simply represented the 2000 pounds as "one complete job." This type of unnecessary information in an applied problem is sometimes called a *red herring*. The 150 miles given in Example 6 was also a red herring.

Formulas

Many common types of geometric, scientific, and investment problems use ready-made equations, called **formulas.** Knowing formulas such as those in the following lists will help you translate and solve a wide variety of real-life problems involving perimeter, area, volume, temperature, interest, and distance.

Programming

You can use a programmable calculator to solve simple interest problems using the program below. The program may be entered into a *TI-82* or *TI-83*. Programs for other calculator models may be found in the appendix.

```
PROGRAM:SIMPINT
:Fix 2
:Disp "PRINCIPAL"
:Input P
:Disp "INTEREST RATE"
:Disp "IN DECIMAL FORM"
:Input R
:Disp "NUMBER OF YEARS"
:Input T
:PRT→I
:Disp "THE INTEREST IS"
:Disp I
:Float
```

Use the program and the Guess, Check, and Revise method to find P when I = $3330, r = 6%, and t = 3 years.

Common Formulas for Area, Perimeter, and Volume

Square
$A = s^2$
$P = 4s$

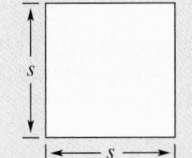

Rectangle
$A = lw$
$P = 2l + 2w$

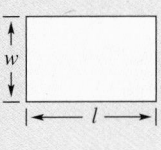

Circle
$A = \pi r^2$
$C = 2\pi r$

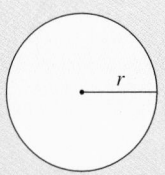

Triangle
$A = \frac{1}{2}bh$

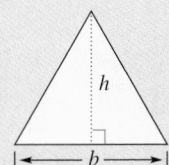

Cube
$V = s^3$

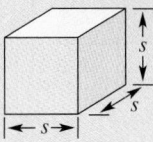

Rectangular Solid
$V = lwh$

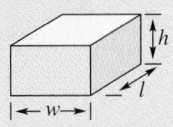

Circular Cylinder
$V = \pi r^2 h$

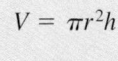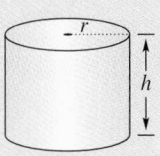

Sphere
$V = \frac{4}{3}\pi r^3$

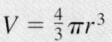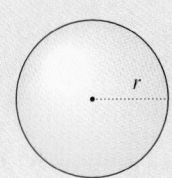

Miscellaneous Common Formulas

Temperature: F = degrees Fahrenheit, C = degrees Celsius

$$F = \frac{9}{5}C + 32$$

Simple Interest: I = interest, P = principal, r = interest rate, t = time

$$I = Prt$$

Distance: d = distance traveled, r = rate, t = time

$$d = rt$$

When working with applied problems, you often need to rewrite one of the common formulas. For instance, the formula $P = 2l + 2w$ (the perimeter of a rectangle) can be rewritten or solved for w in the following manner.

$$P = 2l + 2w \qquad \text{Original formula}$$

$$P - 2l = 2w \qquad \text{Subtract } 2l \text{ from both sides.}$$

$$\frac{P - 2l}{2} = w \qquad \text{Divide both sides by 2.}$$

EXAMPLE 8   Using a Geometric Formula

A rectangular plot has an area of 120,000 square feet. The plot is 300 feet wide. How long is it?

Solution

In a problem such as this, it is helpful to begin by drawing a diagram, as shown in Figure 3.21. In this diagram, label the width of the rectangle as $w = 300$ feet, and label the unknown length as l.

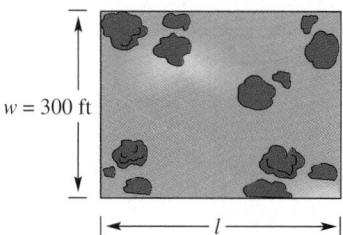

FIGURE 3.21

Now, to solve for the unknown length, use the following steps.

Verbal Model:	Area	$=$	Length	$\cdot$	Width

Labels:	Area $= 120{,}000$	(square feet)
	Length $= l$	(feet)
	Width $= 300$	(feet)

Equation:

$$120{,}000 = l(300)$$

$$\frac{120{,}000}{300} = l$$

$$400 = l$$

The length of the rectangular plot is 400 feet. You can check this by multiplying 300 feet by 400 feet to obtain the area of 120,000 square feet.

EXAMPLE 9   *Simple Interest*

A deposit of $8000 earned $300 in interest in 6 months. What was the annual interest rate for this account?

Solution

Verbal Model:

$$\boxed{\text{Interest}} \; = \; \boxed{\text{Principal}} \; \cdot \; \boxed{\text{Rate}} \; \cdot \; \boxed{\text{Time}}$$

Labels:

Interest = 300	(dollars)
Principal = 8000	(dollars)
Time = $\frac{1}{2}$	(year)
Annual interest rate = r	(percent in decimal form)

Equation:

$$300 = 8000(r)\left(\frac{1}{2}\right)$$

$$\frac{2(300)}{8000} = r$$

$$0.075 = r$$

The annual interest rate is $r = 0.075$ (or 7.5%). Check this in the original statement of the problem.

Remind students that time must be expressed in *years*. Six months should be interpreted as $\frac{1}{2}$ year (not 6).

Group Activities

Communicating Mathematically

Translating a Formula   Use the information provided in the following statement to write a mathematical formula for the 10-second pulse count.

"The target heart rate is the heartbeat rate a person should have during aerobic exercise to get the full benefit of the exercise for cardiovascular conditioning. . . . Using the American College of Sports Medicine method to calculate one's target heart rate, an individual should subtract his or her age from 220, then multiply by the desired intensity level (as a percent—sedentary persons may want to use 60% and highly fit individuals may want to use 85% to 95%) of the workout. Then divide the answer by 6 for a 10-second pulse count. (The 10-second pulse count is useful for checking whether the target heart rate is being achieved during the workout. One can easily check one's pulse—at the wrist or side of the neck—by counting the number of beats in 10 seconds.)" *(Source: Aerobic Fitness Association of America)*

Use the formula you have written to find your own 10-second pulse count.

3.4   **Exercises**

See Warm-Up
Exercises, p. A42

In Exercises 1–8, find the missing quantities. (Assume the markup rate is a percent based on the cost.)

	Cost	Selling Price	Markup	Markup Rate
1.	$45.97	$64.33		
2.	$84.20	$113.67		
3.		$250.80	$98.80	
4.		$603.72	$184.47	
5.		$26,922.50	$4672.50	
6.		$16,440.50	$3890.50	
7.	$225.00			85.2%
8.	$732.00			$33\frac{1}{3}$%

In Exercises 9–16, find the missing quantities. (Assume the discount rate is a percent based on the list price.)

	List Price	Sale Price	Discount	Discount Rate
9.	$49.95	$25.74		
10.	$119.00	$79.73		
11.	$300.00		$189.00	
12.	$345.00		$134.55	
13.	$95.00			65%
14.		$15.92		20%
15.		$893.10	$251.90	
16.		$257.32	$202.18	

17. *Energy Use*   The circle graph shows the sources of the approximately 83.96 quadrillion British thermal units of energy consumed in the United States in 1993. How many quadrillion Btu were obtained from coal? *(Source: Energy Information Administration)*

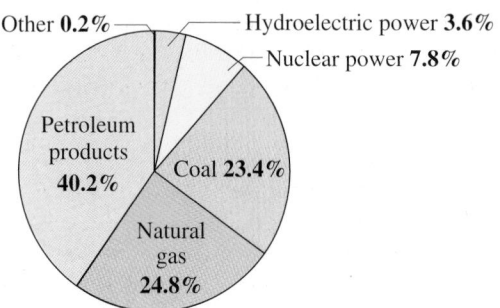

18. *Generation of Electricity*   Use the circle graph to approximate the percent of electricity generated in the United States by hydroelectric plants in 1993. *(Source: Energy Information Administration)*

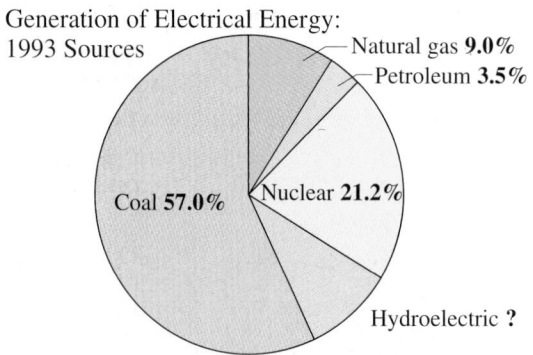

19. *Comparison Shopping*   A department store is offering a discount of 20% on a sewing machine with a list price of $279.95. A mail-order catalog has the same machine for $228.95 plus $4.32 for shipping. Which is the better buy?

d.s. .8(279.98) = 223.96

m.o. 228.95 + 4.32 = 233.27

20. *Comparison Shopping* A hardware store is offering a discount of 15% on a 4000-watt generator with a list price of $699.99. A mail-order company has the same generator for $549.95 plus $14.32 for shipping. Which is the better buy?

21. *Labor Charges* An auto repair bill of $216.37 lists $136.37 for parts and the rest for labor. If the labor rate is $32 per hour, how many hours did it take to repair the auto?

$216.37 = 136.37 + 32x$
$32x = 80$ $x = \frac{80}{32} = 2\frac{1}{2}$

22. *Labor Charges* An appliance repair store charges $35 for the first half hour of a service call. For each additional half hour of labor there is a charge of $18. Find the length of a service call for which the charge is $89.

23. *Long Distance Rates* The weekday rate for a telephone call is $0.75 for the first minute plus $0.55 for each additional minute. Determine the length of a call that costs $5.15. What would have been the cost of the call if it had been made during the weekend when there is a 60% discount?

a) $.75 + .55x = 5.15$ 9 min
$.55(x-1) = 4.4$
$x - 1 = 8$

b) $4(5.15)$
$= 2.06$

24. *Overtime Hours* Last week you earned $740. If you are paid $14.50 per hour for the first 40 hours and $20 for each hour over 40, how many hours of overtime did you work?

25. *Tip Rate* A customer left a total of $10 for a meal that costs $8.45. Determine the tip rate.

$10 - 8.45 = 1.55$
$1.55 = \frac{x}{100} \cdot 8.45$
$x = \frac{155}{8.45}$
$= 18.34\%$

26. *Tip Rate* A customer left a total of $40 for a meal that costs $34.73. Determine the tip rate.

27. *Commission Rate* Determine the commission rate for an employee who earned $450 in commissions for sales of $5000.

$450 = \frac{x}{100} \cdot 5000$ $x = \frac{450}{50} = 9\%$

28. *Insurance Premiums* The annual insurance premium for a policyholder is $862. Find the annual premium if the policyholder must pay a 20% surcharge because of an accident.

29. *Amount Financed* A customer bought a lawn tractor that cost $4450 plus 6% sales tax. Find the amount of the sales tax and the total bill. Find the amount financed if a downpayment of $1000 was made.

$(4450).06 = 267$
total: $4450 + 267 = 4717$
financed 3717

30. *Weekly Pay* The weekly salary of an employee is $250 plus a 6% commission on the employee's total sales. Find the weekly pay for a week in which the sales are $5500.

31. *Number of Stamps* You have a set of 70 stamps with a total value of $20.96. If the set includes 20¢ stamps and 32¢ stamps, find the number of each type.

32. *Coin Problem* A person has 50 coins in dimes and quarters with a combined value of $10.25. Determine the number of coins of each type.

33. *Ticket Sales* Ticket sales for a play total $2200. There are three times as many adult tickets sold as child tickets, and the prices of the tickets for adults and children are $6 and $4. Find the number of child tickets sold.

34. *Nut Mixture* A grocer mixes two kinds of nuts that cost $3.88 per pound and $4.88 per pound, to make 100 pounds of a mixture that costs $4.13 per pound. How many pounds of each kind of nut are in the mixture?

35. *Simple Interest* An inheritance of $40,000 is divided into two investments earning 8% and 10% simple interest. (There is more risk in the 10% fund.) Your objective is to obtain a total annual interest income of $3500 from the investments.

(a) Use the program on page 227 to find the interest accumulated after 3 years on $40,000 for each interest rate.

(b) What is the smallest amount you can invest at 10% in order to meet your objective?

36. *Simple Interest* Four thousand dollars is divided into two investments earning $6\frac{1}{2}\%$ and 9% simple interest. (There is more risk in the 9% fund.) Your objective is to obtain a total annual interest income of $300 from the investments.

(a) Use the program on page 227 to find the interest accumulated after 3 years on $4000 for each interest rate.

(b) What is the smallest amount you can invest at 9% in order to meet your objective?

Mixture Problem In Exercises 37–40, find the number of units of solutions 1 and 2 needed to obtain the desired amount and concentration of the final solution.

	Solution 1	Solution 2	Final Solution	Amount of Solution
37.	20%	60%	40%	100 gal
38.	50%	75%	60%	10 L
39.	15%	60%	45%	24 qt
40.	60%	80%	75%	55 gal

41. *Antifreeze Coolant* The cooling system on a truck contains 5 gallons of coolant that is 40% antifreeze. How much must be withdrawn and replaced with 100% antifreeze to bring the coolant in the system to 50% antifreeze?

42. *Opinion Poll* Fourteen hundred people were surveyed in an opinion poll. Political candidates A and B received approximately the same preference, but candidate C was preferred by twice the number of people as either A or B. Determine the number in the sample that preferred candidate C.

Distance In Exercises 43–48, determine the unknown distance, rate, or time.

	Distance, $d = rt$	Rate, $r = \dfrac{d}{t}$	Time, $t = \dfrac{d}{r}$
43.	2.275 mi	650 mi/hr	$3\frac{1}{2}$ hr
44.		45 ft/sec	10 sec
45.	1000 km	110 km/hr	9.09 hrs
46.	250 ft	32 ft/sec	
47.	1000 ft		$\frac{3}{2}$ sec
48.	385 mi		7 hr

49. *Flying Distance* Two planes leave an airport at approximately the same time and fly in opposite directions. How far apart are the planes after $1\frac{1}{3}$ hours if their speeds are 480 miles per hour and 600 miles per hour?

$480 \cdot \frac{4}{3} + 600 \cdot \frac{4}{3} = 640 + 800$
$= 1440 \text{ mi}$

50. *Average Speed* An Olympic runner completes a 5000-meter race in 13 minutes and 20 seconds. What is the average speed of the runner?

51. *Travel Time* Determine the time for the space shuttle to travel a distance of 5000 miles in orbit when its average speed is 17,000 miles per hour (see figure).

5000 miles

$\frac{5000}{17000} = \frac{5}{17} \text{ hr.}$
$\approx .2941 \text{ hr} \cdot 60 \frac{m}{hr}$
$\approx 17.65 \text{ min}$

52. *Travel Time* On the first part of a 317-mile trip, a sales representative averaged 58 miles per hour. The sales representative averaged only 52 miles per hour on the last part of the trip because of an increased volume of traffic (see figure). Find the amount of driving time at each speed if the total time was 5 hours and 45 minutes.

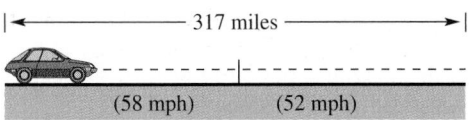

|←——————— 317 miles ———————→|

(58 mph)   (52 mph)

53. *Travel Time* On the first part of a 280-mile trip, a sales representative averaged 63 miles per hour. The sales representative averaged only 54 miles per hour on the last part of the trip because of an increased volume of traffic.

(a) Express the total time for the trip as a function of the distance x traveled at an average speed of 63 miles per hour.

(b) Use a graphing utility to graph the time function. What is the domain of the function?

(c) Approximate the number of miles traveled at 63 miles per hour if the total time was 4 hours and 45 minutes.

54. *Speed of Light* Determine the time for light to travel from the sun to the earth. The distance between the sun and the earth is 93,000,000 miles and the speed of light is 186,282.369 miles per second (see figure).

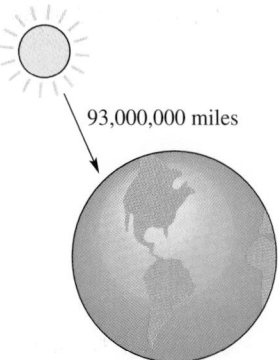

93,000,000 miles

55. *Work Rate* Determine the work rate for each task.

(a) A printer can print eight pages per minute.

(b) A machine shop can produce 30 units in 8 hours.

56. *Work-Rate Problem* You can complete a typing project in 5 hours, and a friend estimates that it would take him 8 hours. What fractional part of the project can be completed by each typist in 1 hour? If you both work on the project, in how many hours can it be completed?

57. *Work-Rate Problem* It takes 30 minutes for a pump to remove an amount of water from a basement. A larger pump can remove the same amount of water in half the time. If both pumps were operating, how long would it take to remove the water?

58. *Work-Rate Problem* It takes 90 minutes to remove an amount of water from a basement when two pumps of different sizes are operating. It takes 2 hours to remove the same amount of water when only the larger pump is running. How long would it take to remove the same amount if only the smaller pump were running?

59. *Height of a Picture Frame* A rectangular picture frame has a perimeter of 3 feet. The width of the frame is 0.62 times its height. Find the height of the frame.

60. *Area of a Square* The figure shows three squares. The perimeter of square I is 12 inches and the area of square II is 36 square inches. Find the area of square III.

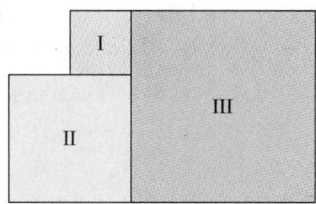

61. *Average Wage* The average hourly wage for bus drivers at public schools in the United States between 1980 and 1994 can be approximated by the linear model

$$y = 5.41 + 0.370t,$$

where y represents the hourly wage (in dollars) and t represents the year, with $t = 0$ corresponding to 1980 (see figure). (*Source: Educational Research Service*)

(a) During which year was the average hourly wage $8.78?

(b) What was the average annual hourly raise for bus drivers during this 15-year period? Explain how you determined your answer.

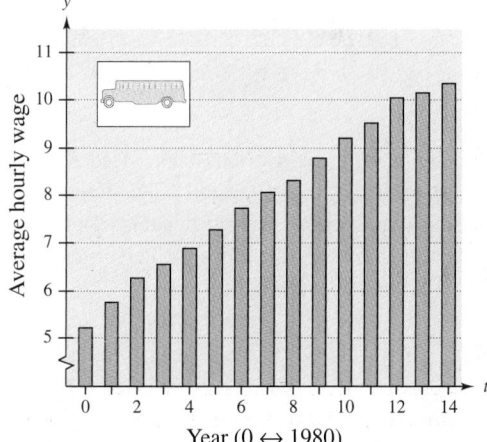

Year (0 ↔ 1980)

62. *Average Wage* The average hourly wage for cafeteria workers at public schools in the United States between 1980 and 1994 can be approximated by the linear model

$$y = 3.96 + 0.282t,$$

where y represents the hourly wage (in dollars) and t represents the year, with $t = 0$ corresponding to 1980 (see figure). *(Source: Educational Research Service)*

(a) During which year was the average hourly wage $6.23?

(b) What was the average annual hourly raise for cafeteria workers during this 15-year period? Explain how you determined your answer.

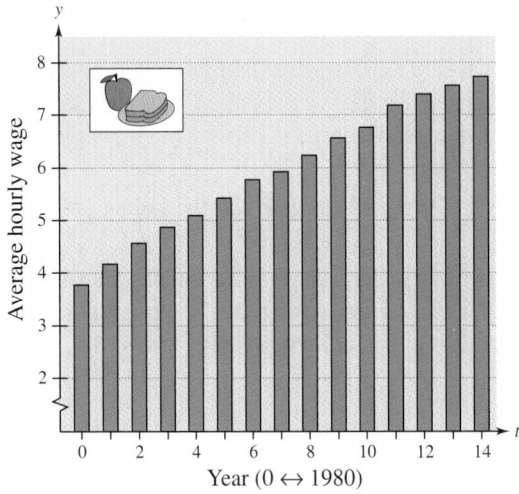

Year (0 ↔ 1980)

63. *Comparing Wage Increases* Use the information given in Exercises 61 and 62 to determine which of the two groups' average wages was increasing at a greater annual rate during the 15-year period from 1980 to 1994. Explain.

Geometry In Exercises 64 and 65, (a) write a function for the area of the region, (b) use a graphing utility to graph the function, and (c) approximate the value of x if the area of the region is 200 square units.

64. **65.**

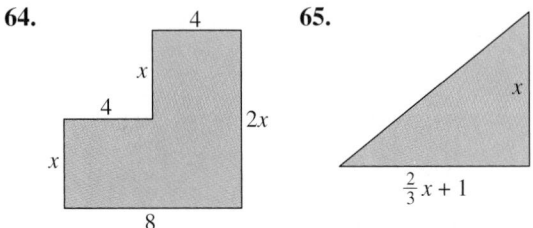

Section Project

Feed Mixture A rancher must purchase 500 bushels of a feed mixture for cattle and is considering oats and corn, which cost $2.70 and $4.00 per bushel, respectively. Complete the table, where x is the number of bushels of oats in the mixture.

Oats x	Corn $500 - x$	Price/bushel of the mixture
0		
100		
200		
300		
400		
500		

(a) How does the increase in the number of bushels of oats affect the number of bushels of corn in the mixture?

(b) How does the increase in the number of bushels of oats affect the price per bushel of the mixture?

(c) If there were an equal number of bushels of oats and corn in the mixture, how would the price of the mixture be related to the price of each component?

3.5	**Linear Inequalities in One Variable**

Intervals on the Real Line ▪ Properties of Inequalities ▪
Solving a Linear Inequality ▪ Application

Intervals on the Real Line

In this section you will study **algebraic inequalities,** which are inequalities that contain one or more variable terms. Some examples are

$$x \leq 4, \qquad x \geq -3, \qquad x + 2 < 7, \qquad \text{and} \qquad 4x - 6 < 3x + 8.$$

As with an equation, you **solve** an inequality in the variable x by finding all values of x for which the inequality is true. Such values are called **solutions** and are said to **satisfy** the inequality. The set of all solutions of an inequality is the **solution set** of the inequality. The **graph** of an inequality is obtained by plotting its solution set on the real number line. Often, these graphs are intervals—either bounded or unbounded.

Historical Note: The first recorded use of the inequality symbols < and > is in Artis Analyticae Praxis, published in 1631 and written by the English mathematician Thomas Harriot (1560–1621).

Bounded Intervals on the Real Number Line

Let a and b be real numbers such that $a < b$. The following intervals on the real number line are called **bounded intervals.** The numbers a and b are the **endpoints** of each interval.

Notation	Interval Type	Inequality	Graph
$[a, b]$	Closed	$a \leq x \leq b$	
(a, b)	Open	$a < x < b$	
$[a, b)$	—	$a \leq x < b$	
$(a, b]$	—	$a < x \leq b$	

NOTE In the list at the right, note that a closed interval contains both of its endpoints and an open interval does not contain either of its endpoints.

Assure students that mastering the use of interval notation will help them in the future.

The **length** of the interval $[a, b]$ is the distance between its endpoints: $b - a$. The lengths of $[a, b]$, (a, b), $(a, b]$, and $[a, b)$ are the same. The reason that these four types of intervals are called bounded is that each has a finite length. An interval that *does not* have a finite length is **unbounded** (or **infinite**).

Historical Note: John Wallis (1616–1703), an English mathematician, was the first to use ∞ as a symbol for infinity. He was a man of many talents: he was the king's chaplain; he devised a system for teaching deaf-mutes; and he was an expert in cryptology.

Unbounded Intervals on the Real Number Line

Let a and b be real numbers. The following intervals on the real number line are called **unbounded intervals.**

Notation	Interval Type	Inequality	Graph
$[a, \infty)$	—	$x \geq a$	
(a, ∞)	Open	$x > a$	
$(-\infty, b]$	—	$x \leq b$	
$(-\infty, b)$	Open	$x < b$	
$(-\infty, \infty)$	Entire real line		

NOTE   The symbols ∞ (**positive infinity**) and $-\infty$ (**negative infinity**) do not represent real numbers. They are simply convenient symbols used to describe the unboundedness of an interval such as $(1, \infty)$.

Study Tip

When using interval notation or graphing inequalities, parentheses are used to denote the inequality symbols $<$ and $>$, and brackets are used to denote the inequality symbols $\leq$ and $\geq$.

EXAMPLE 1   Graphs of Inequalities

(a) The graph of $-3 < x \leq 1$ is a bounded interval.

(b) The graph of $0 < x < 2$ is a bounded interval.

(c) The graph of $-3 < x$ is an unbounded interval.

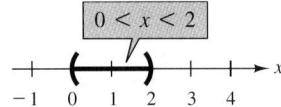

(d) The graph of $x \leq 2$ is an unbounded interval.

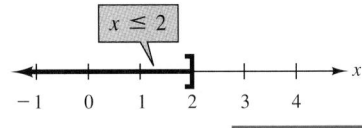

NOTE   In Example 1(c), the inequality $-3 < x$ can also be written as

$x > -3.$

In other words, saying "-3 is less than x" is the same thing as saying "x is greater than -3."

Properties of Inequalities

Solving a linear inequality is much like solving a linear equation. To isolate the variable, you make use of **properties of inequalities.** These properties are similar to the properties of equality, but there are two important exceptions. When both sides of an inequality are multiplied or divided by a negative number, the direction of the inequality symbol must be reversed. Here is an example.

$$-2 < 5 \qquad \text{Original inequality}$$

$$(-3)(-2) > (-3)(5) \qquad \text{Multiply both sides by } -3 \text{ and reverse the inequality symbol.}$$

$$6 > -15 \qquad \text{Simplify.}$$

Two inequalities that have the same solution set are **equivalent.** The following list describes operations that can be used to create equivalent inequalities.

Make sure students understand that if the original statement is an inequality (or an equation), they may perform only operations that produce an equivalent inequality (or equation).

NOTE   These properties remain true if the symbols $<$ and $>$ are replaced by $\leq$ and $\geq$. Moreover, a, b, and c can represent real numbers, variables, or expressions. Note that you cannot multiply or divide both sides of an inequality by zero.

Properties of Inequalities

1. *Addition and Subtraction Properties*

   Adding the same quantity to, or subtracting the same quantity from, both sides of an inequality produces an equivalent inequality.

   If $a < b$, then $a + c < b + c$.
   If $a < b$, then $a - c < b - c$.

2. *Multiplication and Division Properties: Positive Quantities*

   Multiplying or dividing both sides of an inequality by a *positive* quantity produces an equivalent inequality.

   If $a < b$ and c is positive, then $ac < bc$.
   If $a < b$ and c is positive, then $\dfrac{a}{c} < \dfrac{b}{c}$.

3. *Multiplication and Division Properties: Negative Quantities*

   Multiplying or dividing both sides of an inequality by a *negative* quantity produces an equivalent inequality in which the inequality symbol is reversed.

   If $a < b$ and c is negative, then $ac > bc$.
   If $a < b$ and c is negative, then $\dfrac{a}{c} > \dfrac{b}{c}$.

4. *Transitive Property*

   Consider three quantities of which the first is less than the second, and the second is less than the third. It follows that the first quantity must be less than the third quantity.

   If $a < b$ and $b < c$, then $a < c$.

Solving a Linear Inequality

An inequality in one variable is **linear** if it can be written in one of the following forms.

$$ax + b \leq 0, \qquad ax + b < 0, \qquad ax + b \geq 0, \qquad ax + b > 0$$

As you study the following examples, pay special attention to the steps in which the inequality symbol is reversed. Remember that when you multiply or divide an inequality by a negative number, you must reverse the inequality symbol.

EXAMPLE 2   Solving a Linear Inequality

$x + 6 < 9$	Original inequality
$x + 6 - 6 < 9 - 6$	Subtract 6 from both sides.
$x < 3$	Simplify.

The solution set consists of all real numbers that are less than 3. The interval notation for the solution set is $(-\infty, 3)$. The graph is shown in Figure 3.22.

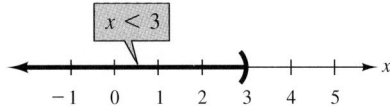

FIGURE 3.22

EXAMPLE 3   Solving a Linear Inequality

$8 - 3x \leq 20$	Original inequality
$8 - 8 - 3x \leq 20 - 8$	Subtract 8 from both sides.
$-3x \leq 12$	Simplify.
$\dfrac{-3x}{-3} \geq \dfrac{12}{-3}$	Divide both sides by -3 and reverse the inequality symbol.
$x \geq -4$	Simplify.

Writing assignment: Why can't you multiply (or divide) both sides of an equation or an inequality by zero?

The solution set consists of all real numbers that are greater than or equal to -4. The interval notation for the solution set is $[-4, \infty)$. The graph is shown in Figure 3.23.

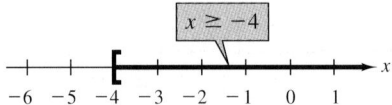

FIGURE 3.23

 *Technology*

Most graphing utilities can sketch the graph of a linear inequality. For instance, the following steps show how to sketch the graph of $5x + 2 < 9x - 4$ on a *TI-82* or *TI-83*.

[Y=] 5 [X,T,θ] [+] 2 [2nd] [TEST] 5 9
[X,T,θ] [-] 4 [ZOOM] 6

The graph produced by these steps is shown below. Notice that the graph occurs as an interval *above* the *x*-axis.

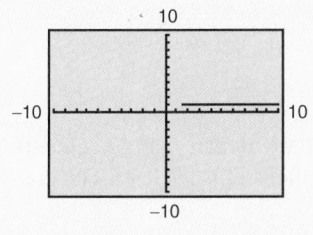

EXAMPLE 4   Solving a Linear Inequality

$$5x + 2 < 9x - 4 \qquad \text{Original inequality}$$
$$5x + 2 - 9x < 9x - 9x - 4 \qquad \text{Subtract } 9x \text{ from both sides.}$$
$$-4x + 2 < -4 \qquad \text{Combine like terms.}$$
$$-4x + 2 - 2 < -4 - 2 \qquad \text{Subtract 2 from both sides.}$$
$$-4x < -6 \qquad \text{Simplify.}$$
$$\frac{-4x}{-4} > \frac{-6}{-4} \qquad \text{Divide both sides by } -4 \text{ and reverse the inequality symbol.}$$
$$x > \frac{3}{2} \qquad \text{Simplify.}$$

The solution set consists of all real numbers that are greater than $\frac{3}{2}$. The interval notation for the solution set is $\left(\frac{3}{2}, \infty\right)$. The graph is shown in Figure 3.24.

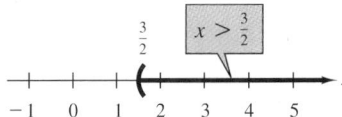

FIGURE 3.24

EXAMPLE 5   Solving a Linear Inequality

$$\frac{2x}{3} + 12 < \frac{x}{6} + 18 \qquad \text{Original inequality}$$
$$6 \cdot \left(\frac{2x}{3} + 12\right) < 6 \cdot \left(\frac{x}{6} + 18\right) \qquad \text{Multiply both sides by LCD of 6.}$$
$$4x + 72 < x + 108 \qquad \text{Distributive Property}$$
$$4x - x < 108 - 72 \qquad \text{Subtract } x \text{ and 72 from both sides.}$$
$$3x < 36 \qquad \text{Combine like terms.}$$
$$x < 12 \qquad \text{Divide both sides by 3.}$$

NOTE   When one or more of the terms in a linear inequality involve constant denominators, it helps to multiply both sides of the inequality by the least common denominator. This clears the inequality of fractions, as shown in Example 5.

The solution set consists of all real numbers that are less than 12. The interval notation for the solution set is $(-\infty, 12)$. The graph is shown in Figure 3.25.

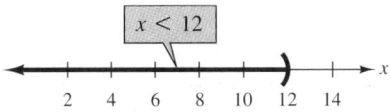

FIGURE 3.25

Sometimes it is convenient to write two inequalities as a **double inequality.** For instance, you can write the two inequalities $-4 \le 5x - 2$ and $5x - 2 < 7$ more simply as

$$-4 \le 5x - 2 < 7.$$

This form allows you to solve the two given inequalities together. Try solving this double inequality. You should find that the solution is $-\frac{2}{5} \le x < \frac{9}{5}$.

EXAMPLE 6   *Solving a Double Inequality*

Write the inequalities $-7 \le 5x - 2$ and $5x - 2 < 8$ as a double inequality. Then solve the double inequality to find the set of all real numbers that satisfy *both* inequalities.

NOTE   When performing an operation on a double inequality, you must apply the operation to *all* three parts of the inequality, as shown in Example 6.

Solution

$-7 \le 5x - 2 < 8$	Original inequality
$-7 + 2 \le 5x - 2 + 2 < 8 + 2$	Add 2 to all three parts.
$-5 \le 5x < 10$	Simplify.
$\dfrac{-5}{5} \le \dfrac{5x}{5} < \dfrac{10}{5}$	Divide all three parts by 5.
$-1 \le x < 2$	Simplify.

The solution set consists of all real numbers that are greater than or equal to -1 and less than 2. The interval notation for the solution set is $[-1, 2)$. The graph is shown in Figure 3.26.

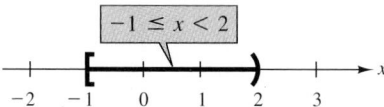

FIGURE 3.26

Show students how they can obtain this solution by graphing $-1 \le x$ and $x < 2$. The intersection of these two inequalities is $-1 \le x < 2$. See Figure 3.26.

The double inequality in Example 6 could have been solved in two parts as follows.

$-7 \le 5x - 2$	and	$5x - 2 < 8$
$-5 \le 5x$		$5x < 10$
$-1 \le x$		$x < 2$

The solution set consists of all real numbers that satisfy *both* inequalities. In other words, the solution set is the set of all values of x for which $-1 \le x < 2$.

Application

Linear inequalities in real-life problems arise from statements that involve phrases such as "at least," "no more than," "minimum value," and so on.

EXAMPLE 7   Translating Verbal Statements

Verbal Statement	*Inequality*	
(a) x is at most 3.	$x \leq 3$	"at most" means "less than or equal to."
(b) x is no more than 3.	$x \leq 3$	
(c) x is at least 3.	$x \geq 3$	"at least" means "greater than or equal to."
(d) x is more than 3.	$x > 3$	
(e) x is less than 3.	$x < 3$	
(f) x is a minimum of 3.	$x \geq 3$	

To solve real-life problems involving inequalities, you can use the same "verbal-model approach" you use with equations.

EXAMPLE 8   Finding the Maximum Width of a Package

An overnight delivery service will not accept any package whose combined length and girth (perimeter of a cross section) exceeds 132 inches. Suppose that you are sending a rectangular package that has square cross sections. If the length of the package is 68 inches, what is the maximum width of the sides of its square cross sections?

Solution

Begin by making a sketch. In Figure 3.27, notice that the length of the package is 68 inches, and each side is x inches wide.

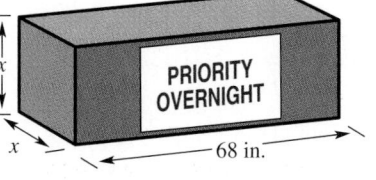

FIGURE 3.27

Verbal Model:

$$\boxed{\text{Length}} \; + \; \boxed{\text{Girth}} \; \leq \; \boxed{132 \text{ inches}}$$

Labels:
Width of a side $= x$ (inches)
Length $= 68$ (inches)
Girth $= 4x$ (inches)

Inequality:
$$68 + 4x \leq 132$$
$$4x \leq 64$$
$$x \leq 16$$

The width of each side of the package must be less than or equal to 16 inches.

EXAMPLE 9   *Comparing Costs*

A subcompact car can be rented from Company A for $240 per week with no extra charge for mileage. A similar car can be rented from Company B for $100 per week, plus 25 cents for each mile driven. How many miles must you drive in a week so that the rental fee for Company B is more than that for Company A?

Solution

Verbal Model:

$$\boxed{\text{Weekly cost for Company B}} > \boxed{\text{Weekly cost for Company A}}$$

Labels: Number of miles driven in one week $= m$ (miles)
Weekly cost for Company A $= 240$ (dollars)
Weekly cost for Company B $= 100 + 0.25m$ (dollars)

Inequality: $100 + 0.25m > 240$

$$0.25m > 140$$

$$m > 560$$

The car from Company B is more expensive if you drive more than 560 miles in a week. A table helps confirm this conclusion.

Miles driven	520	530	540	550	560	570
Company A	$240.00	$240.00	$240.00	$240.00	$240.00	$240.00
Company B	$230.00	$232.50	$235.00	$237.50	$240.00	$242.50

Group Activities   Communicating Mathematically

Problem Posing   Suppose your group owns a small business and must choose between two carriers for long distance telephone service. Create realistic data for cost of the first minute of a call and cost per additional minute for each carrier, and decide what question(s) would be most helpful to ask when making such a choice. Solve the problem your group has created. Write a short memo to your company's business manager outlining the situation, explaining your mathematical solution, and summarizing your recommendations.

3.5 **Exercises**

See Warm-Up
Exercises, p. A42

1. *Think About It* Describe the differences between properties of equality and properties of inequality.

2. *Think About It* Give an example of "reversing an inequality symbol."

In Exercises 3–6, determine whether the given values of x satisfy the inequality.

Inequality   *Values*

3. $7x - 10 > 0$ (a) $x = 3$ (b) $x = -2$
                            (c) $x = \frac{5}{2}$ (d) $x = \frac{1}{2}$

4. $3x + 2 < \dfrac{7x}{5}$ (a) $x = 0$ (b) $x = 4$
                            (c) $x = -4$ (d) $x = -1$

5. $0 < \dfrac{x + 5}{6} < 2$ (a) $x = 10$ (b) $x = 4$
                            (c) $x = 0$ (d) $x = -6$

6. $-2 < \dfrac{3 - x}{2} \le 2$ (a) $x = 0$ (b) $x = 3$
                            (c) $x = 9$ (d) $x = -12$

In Exercises 7–10, match the inequality with its graph. [The graphs are labeled (a), (b), (c), and (d).]

(a)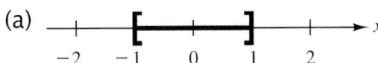

(b)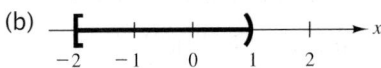

(c)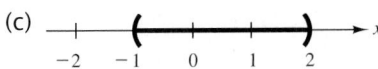

(d)

7. $-1 < x \le 2$ 8. $-1 \le x \le 1$
9. $-1 < x < 2$ 10. $-2 \le x < 1$

In Exercises 11–16, sketch the graph of the inequality. Write the interval notation for the inequality.

11. $-5 < x \le 3$ 12. $4 > x \ge 1$
13. $\frac{3}{2} \ge x > 0$ 14. $-7 < x \le -3$
15. $-\frac{15}{4} < x < -\frac{5}{2}$ 16. $-\pi > x > -5$

In Exercises 17–22, use a graphing utility to graph the inequality. Write the interval notation for the inequality.

17. $x \le -2$ 18. $x > 6$
19. $-2 < x \le 4$ 20. $-10 < x < -5$
21. $9 \ge x \ge 3$ 22. $x \le -1$ or $x > 1$

In Exercises 23–52, solve the inequality and sketch the solution on the real number line.

23. $x + 7 \le 9$ 24. $z - 4 > 0$
25. $4x < 22$ 26. $2x > 5$
27. $-9x \ge 36$ 28. $-6x \le 24$
29. $-\frac{3}{4}x < -6$ 30. $-\frac{1}{5}x > -2$
31. $2x - 5 > 9$ 32. $3x + 4 \le 22$
33. $5 - x \le -2$ 34. $1 - y \ge -5$
35. $5 - 3x < 7$ 36. $12 - 5x > 5$
37. $3x - 11 > -x + 7$
38. $21x - 11 \le 6x + 19$
39. $16 < 4(y + 2) - 5(2 - y)$
40. $4[z - 2(z + 1)] < 2 - 7z$
41. $-3(y + 10) \ge 4(y + 10)$
42. $2(4 - z) \ge 8(1 + z)$
43. $\dfrac{x}{6} - \dfrac{x}{4} \le 1$
44. $\dfrac{x + 3}{6} + \dfrac{x}{8} \ge 1$
45. $0 < 2x - 5 < 9$
46. $-4 \le 2 - 3(x + 2) < 11$

47. $-3 < \dfrac{2x-3}{2} < 3$

48. $0 \le \dfrac{x-5}{2} < 4$

49. $1 > \dfrac{x-4}{-3} > -2$

50. $-\dfrac{2}{3} < \dfrac{x-4}{-6} \le \dfrac{1}{3}$

51. $\dfrac{2}{5} < x + 1 < \dfrac{4}{5}$

52. $-1 < -\dfrac{x}{6} < 1$

In Exercises 53–58, rewrite the statement using inequality notation.

53. x is nonnegative.

54. y is more than -2.

55. z is at least 2.

56. m is at least 4.

57. n is no more than 16.

58. x is at least 450 but no more than 500.

In Exercises 59–62, write a verbal description of the inequality.

59. $x \ge \frac{5}{2}$

60. $t < 4$

61. $0 < z \le \pi$

62. $-4 \le t \le 4$

63. *Think About It* If $t < 8$, then $-t$ must be in what interval?

64. *Think About It* If $-3 < x \le 10$, then $-x$ must be in what interval?

65. Four times a number n must be at least 12 and no more than 30. What interval contains this number?

66. Five times a number n must be at least 15 and no more than 45. What interval contains this number?

67. *Travel Budget* A student group has $4500 budgeted for a field trip. The cost of transportation for the trip is $1900. To stay within the budget, all other costs C must be no more than what amount?

68. *Monthly Budget* You have budgeted $1800 per month for your total expenses. The cost of rent per month is $600 and the cost of food is $350. To stay within your budget, all other costs C must be no more than what amount?

69. *Comparing Temperatures* The average temperature in Miami is greater than the average temperature in Washington D.C., and the average temperature in Washington D.C. is greater than the average temperature in New York. How does the average temperature in Miami compare with the average temperature in New York?

70. *Comparing Elevations* The elevation (above sea level) of San Francisco is less than the elevation of Dallas, and the elevation of Dallas is less than the elevation of Denver. How does the elevation of San Francisco compare with the elevation of Denver?

71. *Operating Costs* A utility company has a fleet of vans. The annual operating cost per van is

$$C = 0.28m + 2900,$$

where m is the number of miles traveled by a van in a year. What number of miles will yield an annual operating cost that is less than $10,000?

72. *Operating Costs* A fuel company has a fleet of trucks. The annual operating cost per truck is

$$C = 0.58m + 7800,$$

where m is the number of miles traveled by a truck in a year. What number of miles will yield an annual operating cost that is less than $25,000?

Profit In Exercises 73 and 74, the revenue R for selling x units and the cost C of producing x units of a product are given. In order to obtain a profit, the revenue must be greater than the cost. For what values of x will this product produce a profit? Use a graphing utility to graph the inequality to confirm the result.

73. $R = 89.95x$
$C = 61x + 875$

74. $R = 105.45x$
$C = 78x + 25,850$

75. *Long Distance Charges* The cost of a long distance telephone call is $0.96 for the first minute and $0.75 for each additional minute. If the total cost of the call cannot exceed $5, find the interval of time that is available for the call. Use a graphing utility to graph the inequality to confirm the result.

76. *Long Distance Charges*   The cost of a long distance telephone call is $1.45 for the first minute and $0.95 for each additional minute. If the total cost of the call cannot exceed $15, find the interval of time that is available for the call. Use a graphing utility to graph the inequality to confirm the result.

77. *Distance*   If you live 5 miles from school and your friend lives 3 miles from you, the distance d that your friend lives from school is in what interval?

78. *Distance*   If you live 10 miles from work and work is 5 miles from school, the distance d that you live from school is in what interval?

79. *Hourly Wage*   You must select one of two plans for payment when working for a company. The first plan pays a straight $12.50 per hour. The second plan pays $8.00 per hour plus $0.75 per unit produced per hour. Write an inequality yielding the number of units that must be produced per hour so that the second plan gives the greater hourly wage. Solve the inequality.

80. *Monthly Wage*   You must select one of two plans for payment when working for a company. The first plan pays a straight $3000 per month. The second plan pays $1000 per month plus a commission of 4% of your gross sales. Write an inequality yielding the gross sales per month so that the second plan gives the greater monthly wage. Solve the inequality.

Air Pollutant Emissions   In Exercises 81 and 82, use the following equation, which models the amount of air pollutant emissions of carbon monoxide in the continental United States from 1987 to 1993 (see figure).

$$y = 5.84 - 0.31t, \quad -3 \le t \le 3$$

In this model, y represents the amount of pollutant in parts per million and t represents the year, with $t = 0$ corresponding to 1990. *(Source: U.S. Environmental Protection Agency)*

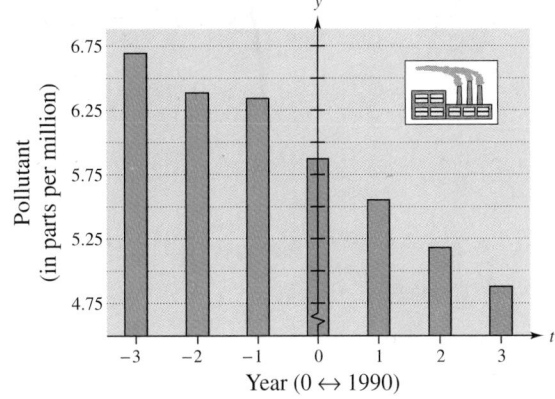

Year (0 ↔ 1990)

81. During which years between 1987 and 1993 was the air pollutant emission of carbon monoxide greater than 6 parts per million?

82. During which years between 1987 and 1993 was the air pollutant emission of carbon monoxide less than 5 parts per million?

Section Project

Bats   The range of a human's voice frequency h (in cycles per second) is about $85 \le h \le 1100$. The range of a human's hearing frequency H is about $20 \le H \le 20,000$.

(a) The relationship between a human's voice frequency h and a bat's voice frequency b is

$$h = 85 + \frac{203}{22,000}(b - 10,000).$$

Find the range of a bat's voice frequency.

(b) The relationship between a human's hearing frequency H and a bat's hearing frequency B is

$$H = 20 + \frac{999}{5950}(B - 1000).$$

Find the range of a bat's hearing frequency.

(c) Use your answer in part (a) to determine whether a human could hear a bat's voice.

(d) Use your answer in part (b) to determine whether a bat could hear a human's voice.

3.6 Absolute Value Equations and Inequalities

Solving Equations Involving Absolute Value ▪ Solving Inequalities Involving Absolute Value ▪ Connections with Graphs of Absolute Value Functions

Solving Equations Involving Absolute Value

Consider the **absolute value equation**

$$|x| = 3.$$

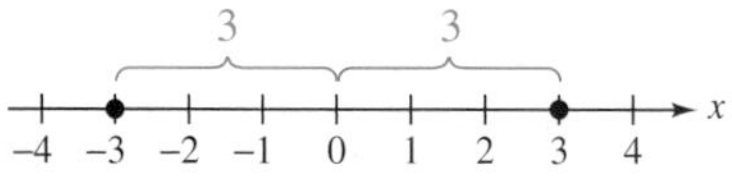

FIGURE 3.28

The only solutions of this equation are -3 and 3, because these are the only two real numbers whose distance from zero is 3. (See Figure 3.28.) In other words, the absolute value equation $|x| = 3$ has exactly two solutions: $x = -3$ and $x = 3$.

Sometimes you can explain the concept of solving equations of the form $|x| = a$ with a variation of the question, "Of what number can you take the absolute value to get a?"

Solving an Absolute Value Equation

Let x be a variable or an algebraic expression and let a be a real number such that $a \geq 0$. The solutions of the equation $|x| = a$ are given by $x = -a$ and $x = a$. That is,

$$|x| = a \quad \Rightarrow \quad x = -a \quad \text{or} \quad x = a.$$

EXAMPLE 1 Solving Absolute Value Equations

Solve each absolute value equation.

(a) $|x| = 10$ (b) $|x| = 0$ (c) $|y| = -1$

Solution

(a) This equation is equivalent to the two linear equations

$$x = -10 \quad \text{or} \quad x = 10. \qquad \text{Equivalent linear equations}$$

Thus, the absolute value equation has two solutions: -10 and 10.

(b) This equation is equivalent to the two linear equations

$$x = -0 \quad \text{or} \quad x = 0. \qquad \text{Equivalent linear equations}$$

Because both equations are equivalent, you can conclude that the absolute value equation has only one solution: 0.

(c) This absolute value equation has *no solution* because it is not possible for the absolute value of a real number to be negative.

Study Tip

The strategy for solving absolute value equations is to *rewrite* the equation in *equivalent forms* that can be solved by previously learned methods. This is a common strategy in mathematics. That is, when you encounter a new type of problem, you try to rewrite the problem so that it can be solved by techniques you already know.

EXAMPLE 2   Solving an Absolute Value Equation

Solve $|3x + 4| = 10$.

Solution

$\|3x + 4\| = 10$	Original equation
$3x + 4 = -10$ or $3x + 4 = 10$	Equivalent equations
$3x + 4 - 4 = -10 - 4$ $3x + 4 - 4 = 10 - 4$	Subtract 4 from both sides.
$3x = -14$ $3x = 6$	Simplify.
$x = -\dfrac{14}{3}$ $x = 2$	Divide both sides by 3.

The solutions are $-\frac{14}{3}$ and 2. Check these in the original equation.

Exploration

Solve $|5x - 6| = -2$ using the technique shown in Example 2. Substitute the resulting solution(s) into the original equation. Do the results check? Why or why not?

NOTE   When you are solving absolute value equations, remember that it is possible that they have no solution. For instance, the equation

$$|3x + 4| = -10$$

has no solution because the absolute value of a real number cannot be negative. Do not make the mistake of trying to solve such an equation by writing the "equivalent" linear equations as $3x + 4 = -10$ and $3x + 4 = 10$. These equations have solutions, but they are both extraneous.

The equation in the next example is not given in the **standard form**

$$|ax + b| = c, \qquad c \geq 0.$$

Notice that the first step in solving such an equation is to write it in standard form.

EXAMPLE 3   An Absolute Value Equation in Nonstandard Form

Solve $|2x - 1| + 3 = 8$.

Solution

$\|2x - 1\| + 3 = 8$	Original equation
$\|2x - 1\| = 5$	Standard form
$2x - 1 = -5$ or $2x - 1 = 5$	Equivalent equations
$2x = -4$ $2x = 6$	Add 1 to both sides.
$x = -2$ $x = 3$	Divide both sides by 2.

The solutions are -2 and 3. Check these in the original equation.

If two algebraic expressions are equal in absolute value, they must either be *equal* to each other or be the *opposites* of each other. Thus, you can solve equations of the form

$$|ax + b| = |cx + d|$$

by forming the two linear equations

Expressions opposite

Expressions equal

$$ax + b = -(cx + d) \quad \text{and} \quad ax + b = cx + d.$$

EXAMPLE 4   Solving an Equation Involving Absolute Values

Solve $|3x - 4| = |7x - 16|$.

Solution

$$|3x - 4| = |7x - 16| \qquad \text{Original equation}$$

$3x - 4 = -(7x - 16)$ or $3x - 4 = 7x - 16$ Equivalent equations

$\quad 3x - 4 = -7x + 16 \qquad\qquad 3x = 7x - 12$

$\qquad\quad 10x = 20 \qquad\qquad\qquad -4x = -12$

$\qquad\qquad x = 2 \qquad\qquad\qquad\quad x = 3$ Solutions

The solutions are 2 and 3. Check these in the original equation.

EXAMPLE 5   Solving an Equation Involving Absolute Values

NOTE   When solving equations of the form
$$|ax + b| = |cx + d|,$$
it is possible that one of the resulting equations will not have a solution. Note this occurrence in Example 5.

Solve $|x + 5| = |x + 11|$.

Solution

By equating the expression $(x + 5)$ to the opposite of $(x + 11)$, you obtain

$$x + 5 = -(x + 11)$$
$$x + 5 = -x - 11$$
$$x = -x - 16$$
$$2x = -16$$
$$x = -8.$$

However, by setting the two expressions equal to each other, you obtain

$$x + 5 = x + 11$$
$$x = x + 6$$
$$0 = 6,$$

which makes no sense. Therefore, the original equation has only one solution: -8. Check this solution in the original equation.

Solving Inequalities Involving Absolute Value

To see how to solve inequalities involving absolute value, consider the following comparisons.

$$|x| = 2 \qquad\qquad |x| < 2 \qquad\qquad |x| > 2$$
$$x = -2 \text{ or } x = 2 \qquad -2 < x < 2 \qquad x < -2 \text{ or } x > 2$$

These comparisons suggest the following rule for solving inequalities involving absolute value.

Solving an Absolute Value Inequality

Let x be a variable or an algebraic expression and let a be a real number such that $a > 0$.

1. The solutions of $|x| < a$ are all values of x that lie between $-a$ and a. That is,

$$|x| < a \quad \text{if and only if} \quad -a < x < a.$$

2. The solutions of $|x| > a$ are all values of x that are *less than* $-a$ or *greater than* a. That is,

$$|x| > a \quad \text{if and only if} \quad x < -a \text{ or } x > a.$$

These rules are also valid if $<$ is replaced by $\leq$ and $>$ is replaced by $\geq$.

EXAMPLE 6   *Solving an Absolute Value Inequality*

Solve $|x - 5| < 2$.

Solution

$|x - 5| < 2$	Original inequality
$-2 < x - 5 < 2$	Equivalent double inequality
$-2 + 5 < x - 5 + 5 < 2 + 5$	Add 5 to all three parts.
$3 < x < 7$	Simplify.

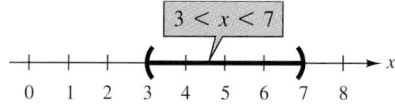

FIGURE 3.29

The solution set consists of all real numbers that are greater than 3 and less than 7. The interval notation for this solution set is $(3, 7)$. The graph is shown in Figure 3.29.

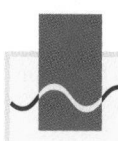

Technology

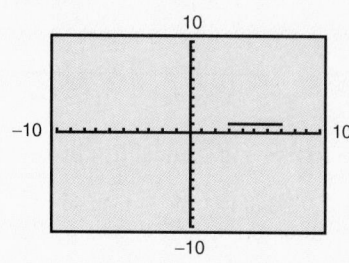

Most graphing utilities can sketch the graph of an absolute value inequality. For instance, the following steps show how to sketch the graph of

$$|x - 5| < 2$$

on a *TI-82* or *TI-83*.

TI-82: $\boxed{\text{Y=}}$ $\boxed{\text{2nd}}$ $\boxed{\text{ABS}}$ $\boxed{(}$ $\boxed{\text{X,T,}\Theta}$ $\boxed{-}$ 5 $\boxed{)}$ $\boxed{\text{2nd}}$ $\boxed{\text{TEST}}$ 5 2 $\boxed{\text{ZOOM}}$ 6

TI-83: $\boxed{\text{Y=}}$ $\boxed{\text{MATH}}$ $\boxed{\blacktriangleright}$ $\boxed{\text{ENTER}}$ $\boxed{\text{X,T,}\Theta,n}$ $\boxed{-}$ 5 $\boxed{)}$ $\boxed{\text{2nd}}$ $\boxed{\text{TEST}}$ 5 2 $\boxed{\text{ZOOM}}$ 6

The graph produced by these steps is shown at the left. Notice that the graph occurs as an interval *above* the x-axis. Compare this result with Example 6.

EXAMPLE 7   Solving an Absolute Value Inequality

Solve $|3x - 4| \ge 5$.

Solution

$|3x - 4| \ge 5$		Original inequality
$3x - 4 \le -5$ or $\quad 3x - 4 \ge 5$		Equivalent inequalities
$3x - 4 + 4 \le -5 + 4 \quad 3x - 4 + 4 \ge 5 + 4$		Add 4 to both sides.
$3x \le -1 \qquad\qquad 3x \ge 9$		Simplify.
$\dfrac{3x}{3} \le \dfrac{-1}{3} \qquad\qquad \dfrac{3x}{3} \ge \dfrac{9}{3}$		Divide both sides by 3.
$x \le -\dfrac{1}{3} \qquad\qquad x \ge 3$		Simplify.

The solution set consists of all real numbers that are less than or equal to $-\frac{1}{3}$ or greater than or equal to 3. The interval notation for this solution set is $\left(-\infty, -\frac{1}{3}\right] \cup [3, \infty)$. The symbol $\cup$ is called a **union** symbol, and it is used to denote the combining of two sets. The graph is shown in Figure 3.30.

$x \le -\frac{1}{3}$ $x \ge 3$

FIGURE 3.30

EXAMPLE 8 *Solving an Absolute Value Inequality*

$$\left|2 - \frac{x}{3}\right| \leq 0.01 \qquad \text{Original inequality}$$

$$-0.01 \leq 2 - \frac{x}{3} \leq 0.01 \qquad \text{Equivalent double inequality}$$

$$-2.01 \leq -\frac{x}{3} \leq -1.99 \qquad \text{Subtract 2 from all three parts.}$$

$$6.03 \geq x \geq 5.97 \qquad \text{Multiply all three parts by } -3 \text{ and reverse both inequality symbols.}$$

$$5.97 \leq x \leq 6.03 \qquad \text{Solution set in standard form}$$

The solution set consists of all real numbers that are greater than or equal to 5.97 and less than or equal to 6.03. The interval notation for this solution set is [5.97, 6.03]. The graph is shown in Figure 3.31.

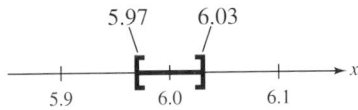

FIGURE 3.31

The pit organs of a rattlesnake can detect temperature changes as small as 0.005°F. By moving its head back and forth, a rattlesnake can detect warm prey, even in the dark.

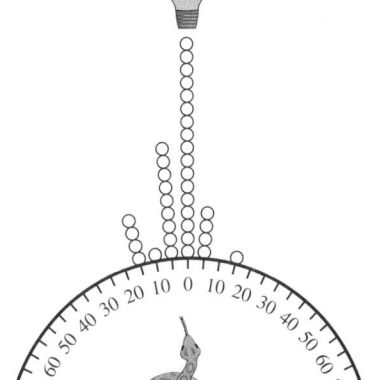

FIGURE 3.32

EXAMPLE 9 *Creating a Model*

To test the accuracy of a rattlesnake's "pit-organ sensory system," a biologist blindfolded a rattlesnake and presented the snake with a warm "target." Of 36 strikes, the snake was on target 17 times. In fact, the snake was within 5 degrees of the target for 30 of the strikes. Let A represent the number of degrees by which the snake is off target. Then $A = 0$ represents a strike that is aimed directly at the target. Positive values of A represent strikes to the right of the target and negative values of A represent strikes to the left of the target. Use the diagram shown in Figure 3.32 to write an absolute value inequality that describes the interval in which the 36 strikes occurred.

Solution

From the diagram, you can see that the snake was never off by more than 15 degrees in either direction. As a compound inequality, this can be represented by

$$-15 \leq A \leq 15.$$

As an absolute value inequality, the interval in which the strikes occurred can be represented by

$$|A| \leq 15.$$

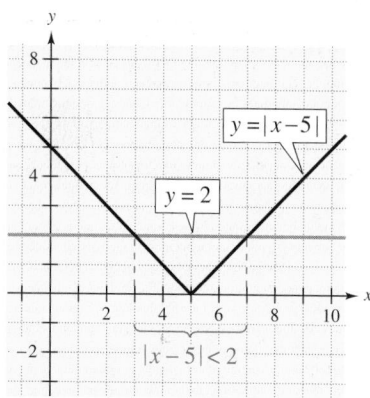

FIGURE 3.33

NOTE Try using Figure 3.33 to solve the inequality $|x - 5| > 2$ graphically. From the graph, can you see why the solution of this inequality involves *two* separate linear inequalities?

Connections with Graphs of Absolute Value Functions

Each of the absolute value equations and inequalities in this section has contained only *one* variable. This section concludes with a look at a technique that uses the graphs of equations with *two* variables to perform a graphic check of solutions. The basic procedure is illustrated in Example 10.

EXAMPLE 10 Performing a Graphic Check

Solve the inequality $|x - 5| < 2$. Then show how the graph of $y = |x - 5|$ can be used to check your solution graphically.

Solution

The algebraic solution of this inequality is shown in Example 6. From that example, you know that the solution is $3 < x < 7$. To check this solution graphically, sketch the graphs of

$$y = |x - 5| \quad \text{and} \quad y = 2$$

on the same set of coordinate axes, as shown in Figure 3.33. In the figure, notice that for all values of x between 3 and 7, the graph of $y = |x - 5|$ lies *below* the line given by $y = 2$. Thus, you can conclude that $|x - 5| < 2$ for all x such that $3 < x < 7$.

Note that each of the previous examples involving absolute value inequalities could be solved analytically and graphically.

Group Activities **You Be the Instructor**

Error Analysis Suppose you are teaching a class in algebra and one of your students hands in the following solution. What is wrong with this solution? What could you say to help your students avoid this type of error?

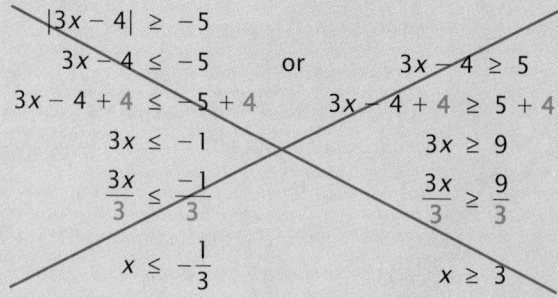

QUIZ $|3x-2| = |x+1|$

SECTION 3.6 Absolute Value Equations and Inequalities   **253**

3.6   **Exercises**

See Warm-Up
Exercises, p. A42

In Exercises 1–4, determine whether the value is a solution of the equation.

Equation	*Value*
1. $|4x + 5| = 10$	$x = -3$
2. $|2x - 16| = 10$	$x = 3$
3. $|6 - 2w| = 2$	$w = 4$
4. $|\frac{1}{2}t + 4| = 8$	$t = 6$

In Exercises 5–8, transform the absolute value equation into two linear equations.

5. $|x - 10| = 17$ **6.** $|7 - 2t| = 5$

7. $|4x + 1| = \frac{1}{2}$ **8.** $|22k + 6| = 9$

In Exercises 9–34, solve the equation. (Some equations have no solution.)

9. $|t| = 45$ **10.** $|s| = 16$

11. $|h| = 0$ **12.** $|x| = -82$

13. $|x - 16| = 5$ **14.** $|z - 100| = 100$

15. $|2s + 3| = 25$ **16.** $|7a + 6| = 8$

17. $|32 - 3y| = 16$ **18.** $|3 - 5x| = 13$

19. $|3x + 4| = -16$ **20.** $|20 - 5t| = 50$

21. $|5x - 3| + 8 = 22$ **22.** $4|x + 5| = 9$

23. $|4 - 3x| = 0$ **24.** $|3x - 2| = -5$

25. $|\frac{2}{3}x + 4| = 9$ **26.** $|\frac{3}{2} - \frac{4}{5}x| = 1$

27. $|0.32x - 2| = 4$

28. $|3.2 - 1.054x| = 2$

29. $|x + 8| = |2x + 1|$

30. $|10 - 3x| = |x + 7|$

31. $|45 - 4x| = |32 - 3x|$

32. $|5x + 4| = |3x + 25|$

33. $|x + 2| = |3x - 1|$

34. $|x - 2| = |2x - 15|$

Think About It   In Exercises 35 and 36, write a single equation that is equivalent to the two equations.

35. $2x + 3 = 5, \ 2x + 3 = -5$

36. $4x - 6 = 7, \ 4x - 6 = -7$

In Exercises 37–44, determine whether the *x*-values are solutions of the inequality.

Inequality	*Values*	
37. $|x| < 3$	(a) $x = 2$	(b) $x = -4$
	(c) $x = 4$	(d) $x = -1$
38. $|x| \leq 5$	(a) $x = -7$	(b) $x = -4$
	(c) $x = 4$	(d) $x = 9$
39. $|x| \geq 3$	(a) $x = 2$	(b) $x = -4$
	(c) $x = 4$	(d) $x = -1$
40. $|x| > 5$	(a) $x = -7$	(b) $x = -4$
	(c) $x = 4$	(d) $x = 9$
41. $|x - 7| < 3$	(a) $x = 9$	(b) $x = -4$
	(c) $x = 11$	(d) $x = 6$
42. $|x - 3| \leq 5$	(a) $x = 16$	(b) $x = 3$
	(c) $x = -2$	(d) $x = -3$
43. $|x - 7| \geq 3$	(a) $x = 9$	(b) $x = -4$
	(c) $x = 11$	(d) $x = 6$
44. $|x - 3| > 5$	(a) $x = 16$	(b) $x = 3$
	(c) $x = -2$	(d) $x = -3$

In Exercises 45–48, transform the absolute value inequality into a double inequality or two separate inequalities.

45. $|y + 5| < 3$

46. $|6x + 7| \leq 5$

47. $|7 - 2h| \geq 9$

48. $|8 - x| > 25$

In Exercises 49–52, sketch a graph that shows the real numbers that satisfy the statement.

49. All real numbers greater than -2 *and* less than 5

50. All real numbers greater than or equal to 3 *and* less than 10

51. All real numbers less than or equal to 4 *or* greater than 7

52. All real numbers less than -6 *or* greater than or equal to 6

In Exercises 53–68, solve the inequality and sketch the solution on a number line.

53. $|y| < 4$

54. $|x| < 6$

55. $|y - 2| \leq 4$

56. $|x - 3| \leq 6$

57. $|y + 2| < 4$

58. $|x + 3| < 6$

59. $|y| \geq 4$

60. $|x| \geq 6$

61. $|y - 2| > 4$

62. $|x - 3| > 6$

63. $|y + 2| \geq 4$

64. $|x + 3| \geq 6$

65. $|2x| < 14$

66. $|4z| \leq 9$

67. $\left|\dfrac{y}{3}\right| \leq 3$

68. $\left|\dfrac{t}{2}\right| < 4$

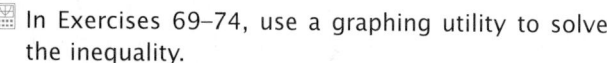 In Exercises 69–74, use a graphing utility to solve the inequality.

69. $|3x + 2| < 4$

70. $|2x - 1| \leq 3$

71. $|x - 5| + 3 \leq 5$

72. $|a + 1| - 4 < 0$

73. $|2x + 3| > 9$

74. $|7r - 3| > 11$

In Exercises 75–84, solve the inequality. (If not possible, state the reason.)

75. $|0.2x - 3| < 4$

76. $|1.5t - 8| \leq 16$

77. $\dfrac{|x + 2|}{10} \leq 8$

78. $\dfrac{|y - 16|}{4} < 30$

79. $|6t + 15| \geq 30$

80. $|3t + 1| > 5$

81. $\dfrac{|s - 3|}{5} > 4$

82. $\dfrac{|a + 6|}{2} \geq 16$

83. $\left|\dfrac{z}{10} - 3\right| > 8$

84. $\left|\dfrac{x}{8} + 1\right| < 0$

In Exercises 85–88, match the inequality with its graph. [The graphs are labeled (a), (b), (c), and (d).]

(a)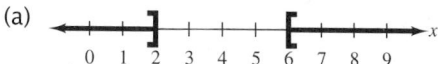

(b)

(c)

(d)

85. $|x - 4| \leq 4$

86. $|x - 4| < 1$

87. $\frac{1}{2}|x - 4| > 4$

88. $|2(x - 4)| \geq 4$

In Exercises 89–94, write an absolute value inequality that represents the interval.

89.

90.

91.

92.

93.

94.

95. *Temperature* The operating temperature of an electronic device must satisfy the inequality

$$|t - 72| < 10,$$

where t is given in degrees Fahrenheit. Sketch the graph of the solution of the inequality.

96. *Time Study* A time study was conducted to determine the length of time required to perform a particular task in a manufacturing process. The times required by approximately two-thirds of the workers in the study satisfied the inequality

$$\left|\frac{t - 15.6}{1.9}\right| < 1,$$

where t is the time in minutes. Sketch the graph of the solution of the inequality.

97. *Body Temperature* Physicians consider an adult's body temperature to be normal if it is between $97.6°$F and $99.6°$F. Write an absolute value inequality that describes this normal temperature range.

98. *Accuracy of Measurements* In woodshop class, you must cut several pieces of wood to within $\frac{3}{16}$ inch of the teacher's specifications. Let $(s - x)$ represent the difference between the specification s and the measured length x of a cut piece.

(a) Write an absolute value inequality that describes the values of x that are within the specifications.

(b) The length of one piece of wood is specified to be $s = 5\frac{1}{8}$ inches. Describe the acceptable lengths for this piece.

99. *Think About It* When you buy a 16-ounce bag of chips, you probably don't expect to get *precisely* 16 ounces. Suppose that the actual weight w (in ounces) of a "16-ounce" bag of chips is given by $|w - 16| \leq \frac{1}{2}$. If you buy four 16-ounce bags, what is the greatest amount you can expect to get? What is the least? Explain.

Section Project

The graph of the inequality

$$|x - 3| < 2$$

can be described as *all real numbers that are within 2 units of 3*. Give a similar description of $|x - 4| < 1$.

The graph of the inequality

$$|y - 1| > 3$$

can be described as *all real numbers that are greater than 3 units from 1*. Give a similar description of $|y + 2| > 4$.

Write an absolute value inequality that represents each verbal statement. Then solve the inequality and sketch the graph of the solution.

(a) The set of all real numbers x whose distance from 0 is less than 3

(b) The set of all real numbers x whose distance from 0 is more than 2

(c) The set of all real numbers x whose distance from 5 is more than 6

(d) The set of all real numbers x whose distance from 16 is less than 5

Chapter Summary

After studying this chapter, you should have acquired the following skills. These skills are keyed to the Review Exercises that begin on page 257. Answers to odd-numbered Review Exercises are given in the back of the book.

- Write the equation of a line given its slope and a point on the line.   *(Section 3.1)*   **Review Exercises 1–8**
- Write the equation of a line passing through two points.   *(Section 3.1)*   **Review Exercises 9–14**
- Write the equations of lines through a given point that are parallel and perpendicular to a given line.   *(Section 3.1)*   **Review Exercises 15–18**
- Graph pairs of lines using a graphing utility to determine if the lines are parallel, perpendicular, or neither.   *(Section 3.1)*   **Review Exercises 19–22**
- Solve real-life problems modeled by linear equations.   *(Sections 3.1, 3.2)*   **Review Exercises 24–26**
- Find linear equations or inequalities to solve real-life problems.   *(Sections 3.1–3.5)*   **Review Exercises 23, 27, 28, 33, 34, 67–90**
- Determine if real-life data represented by scatter plots appears linear.   *(Section 3.2)*   **Review Exercises 29, 30**
- Find the best-fitting lines through sets of points.   *(Section 3.2)*   **Review Exercises 31, 32**
- Translate verbal phrases into algebraic expressions.   *(Section 3.3)*   **Review Exercises 35–40, 45–50**
- Translate algebraic expressions into verbal phrases.   *(Section 3.3)*   **Review Exercises 41–44**
- Compare measurements using ratios.   *(Section 3.3)*   **Review Exercises 51–54**
- Solve proportions.   *(Section 3.3)*   **Review Exercises 55–58**
- Translate real-life situations into proportions and solve.   *(Section 3.3)*   **Review Exercises 59–62, 65, 66**
- Solve problems involving geometry.   *(Section 3.3)*   **Review Exercises 63, 64, 91, 92**
- Solve absolute value equations.   *(Section 3.6)*   **Review Exercises 93–96**
- Solve inequalities and sketch the solutions on the real number line.   *(Sections 3.5, 3.6)*   **Review Exercises 97–110**
- Translate verbal phrases into inequalities.   *(Sections 3.5, 3.6)*   **Review Exercises 111–114**
- Write absolute value inequalities that represent intervals.   *(Section 3.6)*   **Review Exercises 115, 116**

Review Exercises

In Exercises 1–8, write an equation of the line passing through the point with the specified slope.

1. $(1, -4)$
$m = 2$

2. $(-5, -5)$
$m = 3$

3. $(-1, 4)$
$m = -4$

4. $(5, -2)$
$m = -2$

5. $\left(\frac{5}{2}, 4\right)$
$m = -\frac{2}{3}$

6. $\left(-2, -\frac{4}{3}\right)$
$m = \frac{3}{2}$

7. $(7, 8)$
m is undefined.

8. $(-6, 5)$
$m = 0$

In Exercises 9–14, write an equation of the line passing through the two points.

9. $(-6, 0), \quad (0, -3)$

10. $(-2, -3), \quad (4, 6)$

11. $(0, 10), \quad (6, 10)$

12. $(-10, 2), \quad (4, -7)$

13. $\left(\frac{4}{3}, \frac{1}{6}\right), \quad \left(4, \frac{7}{6}\right)$

14. $\left(\frac{5}{2}, 0\right), \quad \left(\frac{5}{2}, 5\right)$

In Exercises 15–18, find equations of the lines passing through the point that are (a) parallel and (b) perpendicular to the given line.

15. $(-1, 5)$
$2x + 4y = 1$

16. $\left(\frac{3}{5}, -\frac{4}{5}\right)$
$3x + y = 2$

17. $\left(\frac{3}{8}, 3\right)$
$4x - 3y = 12$

18. $(12, 1)$
$5x = 3$

In Exercises 19–22, use a graphing utility on a square setting to graph the lines. Decide whether the lines are parallel, perpendicular, or neither.

19. $y = 2x - 3$
$y = \frac{1}{2}(4 - x)$

20. $y = \frac{2}{3}x + 3$
$y = \frac{2}{9}(3x - 4)$

21. $2x - 3y - 5 = 0$
$x + 2y - 6 = 0$

22. $4x + 3y - 6 = 0$
$3x - 4y - 8 = 0$

23. *Cost and Profit* A company produces a product for which the variable cost is $8.55 per unit and the fixed costs are $25,000. The product is sold for $12.60, and the company can sell all it produces.

(a) Write the cost C as a linear function of x, the number of units produced.

(b) Write the profit P as a linear function of x, the number of units sold.

24. *Velocity of a Ball* The velocity of a ball thrown upward from ground level is given by $v = -32t + 80$, where t is the time in seconds and v is the velocity in feet per second.

(a) Find the velocity when $t = 2$.

(b) Find the time when the ball reaches its maximum height. (*Hint:* Find the time when $v = 0$.)

(c) Find the velocity when $t = 3$.

25. *Sales* A sales manager of a company wants to know if there is a relationship between sales and the years of experience for sales personnel who have been with the company 4 or fewer years. The table gives the years of experience x for eight of the company's sales personnel and monthly sales y in thousands of dollars.

x	1.5	1.0	0.3	3.0
y	46.7	32.9	19.2	48.4

x	4.0	0.5	2.5	1.8
y	51.2	28.5	53.4	35.5

A model for this data is $y = 8.22x + 24.47$.

(a) Plot the data and graph the model on the same set of coordinate axes.

(b) Create a table showing the given data and the approximations given by the model.

(c) Use the model to predict the sales for a person who has been with the company for 3.5 years.

(d) Interpret the slope in the context of the problem.

26. *Stress Test*   A machine part was tested by bending it x centimeters ten times per minute until it failed (y = time until failure, in hours). The results are recorded in the table.

x	3	6	9	12	15
y	61	56	53	55	48

x	18	21	24	27	30
y	35	36	33	28	23

A model for this data is $y = -1.43x + 66.44$.

(a) Plot the data and graph the model on the same set of coordinate axes.

(b) Create a table showing the given data and the approximations given by the model.

(c) Use the model to predict the time until failure for $x = 20$.

(d) Interpret the slope in the context of the problem.

27. *Hooke's Law*   Hooke's Law states that the force F required to compress or stretch a spring (within its elastic limits) is proportional to the distance d that the spring is compressed or stretched from its original length. A force of 60 kilograms stretches a spring 4.0 centimeters.

(a) Find a mathematical model that gives F in terms of d.

(b) Use the model in part (a) to find the elongation of the spring if a force of 80 kilograms is applied.

(c) Use the model in part (a) to find the force on the spring if the force stretches the spring 3.2 centimeters.

28. *The English and Metric Systems*   The label on a soft drink can gives the amount of soft drink in ounces and grams. These amounts are 12 and 340.19.

(a) Use the information on the label to find a mathematical model that relates ounces to grams.

(b) Use part (a) to convert 32 ounces to grams.

(c) Use part (a) to convert 500 grams to ounces.

29. *Carcinogens*   The following ordered pairs give the exposure index x of a carcinogenic substance and the cancer mortality rate y per 100,000 population. The higher the index level is, the higher the level of contamination.

(3.50, 150.1), (3.58, 133.1), (4.42, 132.9),

(2.26, 116.7), (2.63, 140.7), (4.85, 165.5),

(12.65, 210.7), (7.42, 181.0), (9.35, 218.4)

(a) Create a scatter plot for the data.

(b) Does the relationship between x and y appear to be approximately linear? Explain.

30. *Quiz Scores*   The following ordered pairs give the scores of two consecutive 15-point quizzes for a class of 18 students.

(7, 13), (9, 7), (14, 4), (15, 15), (10, 15), (9, 7),

(14, 11), (14, 15), (8, 10), (9, 10), (15, 9),

(10, 11), (11, 14), (7, 14), (11, 10), (14, 11),

(10, 15), (9, 6)

(a) Create a scatter plot for the data.

(b) Does the relationship between consecutive quiz scores appear to be approximately linear? If not, give some possible explanations.

In Exercises 31 and 32, use a ruler to sketch the best-fitting line through the set of points and find an equation of the line.

31.   **32.**

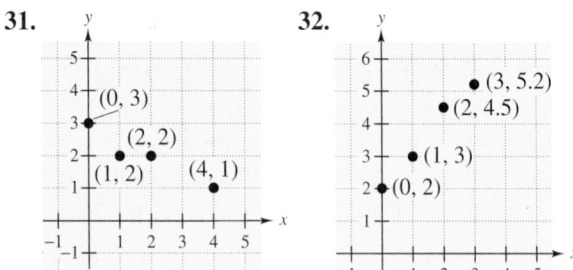

33. *Prices of Homes* The table gives the median sales prices y (in thousands of dollars) of existing one-family homes for the years 1991 through 1994. Let x represent the time in years, with $x = 1$ corresponding to 1991. *(Source: National Association of Realtors)*

Year	1991	1992	1993	1994
x	1	2	3	4
y	100.3	103.7	106.8	109.8

(a) Plot the points.

(b) Use a ruler to sketch the best-fitting line through the points.

(c) Find an equation for the line sketched in part (b).

(d) Interpret the slope in the context of the problem.

(e) Use part (c) to predict the median sales price of an existing one-family home in 2000.

34. *Agriculture* The table gives the yield per acre y of wheat (in bushels) when x pounds of fertilizer (in hundreds) is applied per acre.

x	1.0	1.5	2.0	2.5
y	32	41	48	53

(a) Plot the points.

(b) Use a ruler to sketch the best-fitting line through the points.

(c) Find an equation for the line sketched in part (b).

(d) Interpret the slope in the context of the problem.

(e) Use the model in part (c) to predict the yield when $x = 1.2$.

In Exercises 35–40, translate the phrase into a mathematical expression. (Let n represent the arbitrary real number.)

35. Two hundred decreased by three times a number

36. One hundred increased by the product of 15 and a number

37. The sum of the square of a number and 49

38. The absolute value of the sum of a number and 10

39. The absolute value of the quotient of a number and 5

40. The sum of a number and the square of the number

In Exercises 41–44, write a verbal description of the algebraic expression without using the variable.

41. $2y + 7$

42. $5u - 3$

43. $\dfrac{x - 5}{4}$

44. $-3(a - 10)$

In Exercises 45–50, write an algebraic expression that represents the quantity given by the verbal statement.

45. The amount of income tax on a taxable income of I dollars when the tax rate is 18%

46. The distance traveled in 8 hours at an average speed of r miles per hour

47. The area of a rectangle whose length is l inches and whose width is five units less than the length

48. The sum of three consecutive odd integers, the first of which is n

49. The cost of 30 acres of land if the price per acre is p dollars

50. The time it takes to copy z pages if the copy rate is eight pages per minute

In Exercises 51–54, express as a ratio. (Use the same unit for the numerator and denominator.)

51. 16 feet to 4 yards

52. 3 quarts to 5 pints

53. 45 seconds to 5 minutes

54. 3 meters to 150 centimeters

In Exercises 55–58, solve the proportion.

55. $\dfrac{7}{8} = \dfrac{y}{4}$

56. $\dfrac{x}{16} = \dfrac{5}{12}$

57. $\dfrac{b}{15} = \dfrac{5}{6}$

58. $\dfrac{x + 1}{3} = \dfrac{x - 1}{2}$

59. *Property Tax*   The tax on a property with an assessed value of $80,000 is $1350. Find the tax on a property with an assessed value of $110,000.

60. *Recipe Enlargement*   One and a half cups of milk is required to make one batch of pudding. How much is required to make $2\frac{1}{2}$ batches?

61. *Map Scale*   One-third inch represents 50 miles on a map. Approximate the distance between two cities that are $3\frac{1}{4}$ inches apart on the map.

62. *Fuel Mixture*   The gasoline-to-oil ratio is 50 to 1 for a lawn mower engine. Determine the amount of gasoline required if $\frac{1}{2}$ pint of oil is used in producing the mixture.

Similar Triangles   In Exercises 63 and 64, solve for the length *x* of the side of a triangle by using the fact that corresponding sides of similar triangles are proportional. (*Note:* Assume that the triangles are similar.)

63.

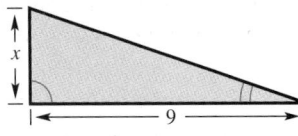

64.

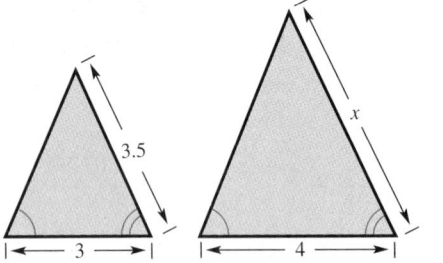

65. *Height of a Building*   You want to determine the height of a building. To do this, you measure the building's shadow and find that it is 20 feet long. You are 6 feet tall and your shadow is $1\frac{1}{2}$ feet long. How tall is the building?

66. *Height of a Flagpole*   You want to determine the height of a flagpole. To do this, you measure the flagpole's shadow and find that it is 30 feet long. You also measure a 5-foot lamppost's shadow and find that it is 3 feet long (see figure). How tall is the flagpole?

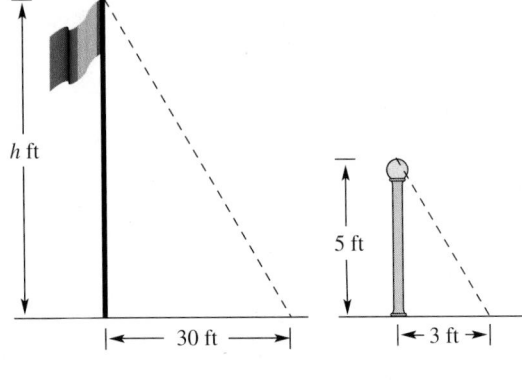

(Not to scale)

67. *Census Costs*   Use the information in the figure to determine the percent increase in the per capita cost of census taking from the 1980 census to the 1990 census. Using a 1990 population of 250 million, determine the total cost of the 1990 census. (*Source: U.S. Bureau of Census*)

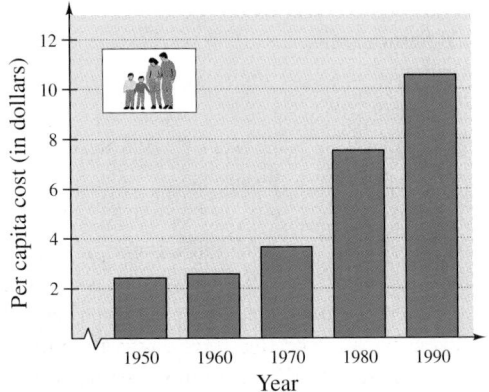

68. *Revenue Increase*   The revenues for a corporation in millions of dollars for the years 1995 and 1996 were $4521.4 and $4679.0. Determine the percent increase in revenue from 1995 to 1996.

69. *Instant Replays* Complete the table on instant replays in the National Football League. *(Source: National Football League)*

Year	Reviewed	Reversed	Percent reversed
1986	374	38	
1987	490	47	
1988	537	53	
1989	492	65	

70. *Price Increase* The manufacturer's suggested retail price for a certain truck model is $25,750. Estimate the price of a comparably equipped truck for the next model year if it is projected that truck prices will increase by $5\frac{1}{2}\%$.

71. *Retail Price* A camera that costs a retailer $259.95 is marked up by 35%. Find the price to the consumer.

72. *Markup Rate* A calculator selling for $175.00 costs the retailer $95.00. Find the markup rate.

73. *Sale Price* The list price of a coat is $259. Find the sale price of the coat if it is reduced by 25%.

74. *Comparison Shopping* A mail-order catalog has an attaché case with a list price of $99.97 plus $4.50 for shipping and handling. A local department store offers the same attaché case for $125.95. The department store has a special 20% off sale. Which is the better buy?

75. *Sales Goal* The weekly salary of an employee is $150 plus a 6% commission on total sales. The employee needs a minimum salary of $650 per week. How much must be sold to produce this salary?

76. *Opinion Poll* Twelve hundred people were surveyed in an opinion poll. Political candidates A and B received the same preference, but candidate C was preferred by one and one third times the number of people as either A or B. Determine the number in the sample that preferred candidate C.

77. *Mixture Problem* Determine the number of liters of a 30% saline solution and the number of liters of a 60% saline solution that are required to make 10 liters of a 50% saline solution.

78. *Mixture Problem* Determine the number of gallons of a 25% alcohol solution and the number of gallons of a 50% alcohol solution that are required to make 8 gallons of a 40% alcohol solution.

79. *Travel Time* Determine the time for a bus to travel 330 miles if its average speed is 52 miles per hour.

80. *Average Speed* An Olympic cross-country skier completed the 15-kilometer event in 41 minutes and 20 seconds. What was the average speed of the skier?

81. *Distance* Determine the distance an Air Force jet can travel in $2\frac{1}{3}$ hours if its average speed is 1200 miles per hour.

82. *Average Speed* For 2 hours of a 400-mile trip, your average speed was 40 miles per hour. Determine the average speed that must be maintained for the remainder of the trip if you want the average speed for the entire trip to be 50 miles per hour.

83. *Work-Rate Problem* Find the time for two people working together to complete half a task if it takes them 8 hours and 10 hours to complete the entire task working individually.

84. *Work-Rate Problem* Find the time for two people working together to complete a task if it takes them 4.5 hours and 6 hours working individually.

85. *Simple Interest* Find the total simple interest you will earn on a $1000 corporate bond that matures in 4 years and has an 8.5% interest rate.

86. *Simple Interest* Find the annual simple interest rate on a certificate of deposit that pays $37.50 per year in interest on a principal of $500.

87. *Simple Interest* Find the principal required to have an annual interest income of $20,000 if the annual simple interest rate on the principal is 9.5%.

88. *Simple Interest* A corporation borrows 3.25 million dollars for 2 years to modernize one of its manufacturing facilities. If it pays an annual simple interest rate of 12%, what will be the total principal and interest that must be repaid?

89. *Simple Interest*   An inheritance of $50,000 is divided between two investments earning 8.5% and 10% simple interest. (The 10% investment has greater risk.) Determine the minimum amount that can be invested at 10% if you want an annual interest income of $4700 from the investment.

90. *Simple Interest*   You invest $1000 in a certificate of deposit that has an annual simple interest rate of 7%. After 6 months the interest is computed and added to the principal. During the second 6 months the interest is computed using the original investment plus the interest earned during the first 6 months. What is the total interest earned during the first year of the investment?

91. *Dimensions of a Rectangle*   The area of the rectangle is 48 square inches. Find the dimensions of the rectangle.

6 in.

x

92. *Dimensions of a Rectangle*   The perimeter of the rectangle is 110 feet. Find x.

$5x$

$4x + 10$

In Exercises 93–96, solve the equation.

93. $|x - 2| - 2 = 4$ **94.** $|2x + 3| = 7$

95. $|3x - 4| = |x + 2|$ **96.** $|5x + 6| = |2x - 1|$

In Exercises 97–110, solve the inequality and sketch the solution on the real number line.

97. $5x + 3 > 18$ **98.** $-11x \geq 44$

99. $\frac{1}{3} - \frac{1}{2}y < 12$ **100.** $3(2 - y) \geq 2(1 + y)$

101. $-4 < \dfrac{x}{5} \leq 4$ **102.** $-13 \leq 3 - 4x < 13$

103. $5 > \dfrac{x + 1}{-3} > 0$ **104.** $12 \geq \dfrac{x - 3}{2} > 1$

105. $|2x - 7| < 15$ **106.** $|5x - 1| < 9$

107. $|x - 4| > 3$ **108.** $|t + 3| > 2$

109. $|b + 2| - 6 \geq 1$ **110.** $\left|\dfrac{t}{3}\right| < 1$

In Exercises 111–114, write an inequality for the given statement.

111. z is no more than 10.

112. x is nonnegative.

113. y is at least 7 but less than 14.

114. The volume V is less than 27 cubic feet.

In Exercises 115 and 116, write an absolute value inequality that represents the interval.

115.

116.

Chapter Test

Take this test as you would take a test in class. After you are done, check your work against the answers given in the back of the book.

x	y
1	2
2	3
3	5
4	6

Data for 4

1. Write an equation of the line passing through $(25, -15)$ and $(75, 10)$.

2. Write an equation of the line passing through $(10, 2)$ and perpendicular to the line $5x + 3y - 9 = 0$.

3. After 4 years, a $26,000 car has a depreciated value of $10,000. Write a linear function giving the value V of the car in terms of t, the number of years. Use the function to find the time t when the value of the car is $16,000.

4. Sketch a scatter plot of the data shown at the right. Then find the equation of the line that you think best fits the data.

5. One afternoon a meteorologist observed that the temperature rose at an average rate of 3 degrees per hour between noon and 5 P.M. The temperature at noon was 72° F. Write a mathematical model that gives the afternoon temperature T in terms of the hour t, with $t = 0$ corresponding to noon.

6. Write an expression for the perimeter of the figure shown at lower right.

7. When the product of a number n and 5 is decreased by 8, the result is 27. Find the number.

8. The sum of two consecutive even integers is 54. Find the integers.

9. The tax on a property with an assessed value of $90,000 is $1200. Estimate the tax on a property with an assessed value of $110,000.

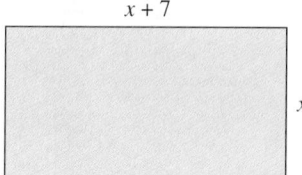

Figure for 6

10. The bill (including parts and labor) for the repair of a home appliance was $165. The cost for parts was $85. How many hours were spent repairing the appliance if the cost of labor was $16 per half hour?

11. Two solutions (10% concentration and 40% concentration) are mixed to create 100 liters of a 30% solution. Determine the numbers of liters of the 10% solution and the 40% solution that are required.

12. Two cars start at a given time and travel in the same direction at average speeds of 40 miles per hour and 55 miles per hour. How much time will elapse before the two cars are 10 miles apart?

13. Find the principal required to earn $300 in simple interest in 2 years if the annual interest rate is 7.5%.

14. Use the inequality notation to denote the phrase, "t is at least 8."

15. Solve $|x + 6| - 3 = 8$.   16. Solve $|4x - 1| = |2x + 7|$.

In Exercises 17–20, solve the inequality and sketch the solution on the real number line.

17. $1 + 2x > 7 - x$

18. $0 \le \dfrac{1 - x}{4} < 2$

19. $|x - 3| \le 2$

20. $|x + 4| > 1$

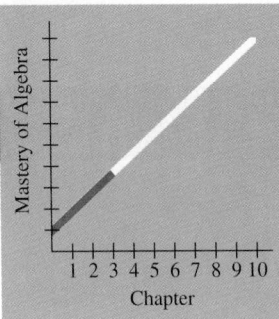

Mastery of Algebra

1 2 3 4 5 6 7 8 9 10
Chapter

Take this test as you would take a test in class. After you are done, check your work against the answers given in the back of the book.

In Exercises 1–4, evaluate the expression.

1. $5(57 - 33)$

2. $-\frac{4}{15} \times \frac{15}{16}$

3. $(12 - 15)^3$

4. $\left(\frac{5}{8}\right)^2$

In Exercises 5 and 6, simplify the expression.

5. $2(x + 5) - 3 - (2x - 3)$

6. $4 - 2[3 + 4(x + 1)]$

In Exercises 7 and 8, factor the expression.

7. $x^2 - 3x + 2$

8. $4x^2 - 28x + 49$

9. Solve by factoring: $3x^2 + 9x - 12 = 0$.

10. Evaluate the function $f(x) = 3 - 2x$.

(a) $f(5)$

(b) $f(x + 3) - f(3)$

11. Determine the slope of a line perpendicular to the line $3x + 7y = 20$.

12. Write an equation of the line passing through the two points $(-2, 7)$ and $(5, 5)$.

13. The inventor of a new game believes that the variable cost for producing the game is $5.75 per unit and the fixed costs are $12,000. Let x represent the number of games produced. Express the total cost C as a function of x.

14. The annual auto insurance premium for a policyholder is $1225. Because of a driving violation, the premium is increased by 15%. What is the new premium?

15. The triangles shown at the left are similar. Solve for x by using the fact that corresponding sides of similar triangles are proportional.

16. A combined total of $24,000 is invested in two bonds that pay 7.5% and 9% simple interest. The total annual interest is $1935. How much is invested in each bond?

In Exercises 17 and 18, solve the inequality and sketch the solution.

17. $7 - 3x > 4 - x$

18. $|x - 2| \geq 3$

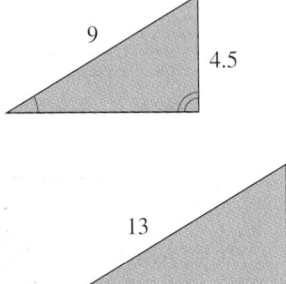

9

4.5

13

x

Figure for 15

Systems of Linear Equations and Inequalities

Strategies for Success

SKILLS *When you have completed this chapter, make sure you are able to:*

○ Solve systems of linear equations by substitution, by elimination, by Gaussian elimination, by using matrices, and graphically

○ Evaluate the determinant of a matrix

○ Sketch graphs of solutions of linear inequalities and systems of linear inequalities

○ Model real-life situations using systems of linear equations and inequalities and solve

TOOLS *Use these study tools to master the skills above:*

In the Section Project of Section 4.1, you will see how a system of linear equations can be used to model the slanting bases of the two walls that form the Vietnam Veterans Memorial.

265

4.1	**Systems of Linear Equations in Two Variables**

Systems of Equations ▪ Solving Systems of Equations by Graphing ▪ Solving Systems of Equations by Substitution ▪ The Method of Elimination ▪ Application

Systems of Equations

Up to this point in the text, most problems have involved a single equation with one or two variables. Many problems in business and science involve a **system of equations** that consists of two or more equations, each involving two or more variables. A **solution** of a system of equations in two variables x and y is an ordered pair (x, y) of real numbers that satisfies *each* equation in the system. When you find the set of all solutions of the system of equations, you are **solving the system of equations.**

EXAMPLE 1 Checking Solutions of a System of Equations

Ask students how they intuitively view a common solution for two equations. Stress that a solution to a system of equations must satisfy each equation in the system.

Check whether (a) the ordered pair $(3, 3)$ and (b) the ordered pair $(4, 2)$ are solutions of the following system of equations.

$$\begin{cases} x + y = 6 & \text{Equation 1} \\ 2x - 5y = -2 & \text{Equation 2} \end{cases}$$

Solution

(a) To determine whether the ordered pair $(3, 3)$ is a solution of the given system of equations, you must substitute $x = 3$ and $y = 3$ into *each* of the given equations. Substituting into Equation 1 produces

$$3 + 3 = 6. \qquad \text{Replace } x \text{ by 3 and } y \text{ by 3.}$$

Similarly, substituting into Equation 2 produces

$$2(3) - 5(3) = -9 \neq -2. \qquad \text{Replace } x \text{ by 3 and } y \text{ by 3.}$$

Because the ordered pair $(3, 3)$ fails to check in *both* equations, you can conclude that it *is not* a solution of the given system of equations.

(b) By substituting the coordinates of the ordered pair $(4, 2)$ into the two given equations, you can determine that this pair is a solution of Equation 1,

$$4 + 2 = 6, \qquad \text{Replace } x \text{ by 4 and } y \text{ by 2.}$$

and is also a solution of Equation 2,

$$2(4) - 5(2) = -2. \qquad \text{Replace } x \text{ by 4 and } y \text{ by 2.}$$

Therefore, $(4, 2)$ *is* a solution of the given system of equations.

Solving Systems of Equations by Graphing

You can gain insight about the location and number of solutions of a system of equations by sketching the graph of each equation in the same coordinate plane. The solutions of the system correspond to the **points of intersection** of the graphs.

EXAMPLE 2   *The Graphical Method of Solving a System*

Use the graphical method to solve the following system of equations.

$$\begin{cases} 2x + 3y = 7 & \text{Equation 1} \\ 2x - 5y = -1 & \text{Equation 2} \end{cases}$$

Solution

Because both equations in the given system are linear, you know that they have graphs that are straight lines. To sketch these lines, write each equation in slope-intercept form, as follows.

$$y = -\frac{2}{3}x + \frac{7}{3} \qquad \text{Equation 1}$$

$$y = \frac{2}{5}x + \frac{1}{5} \qquad \text{Equation 2}$$

The lines corresponding to these two equations are shown in Figure 4.1. From this figure, it appears that the two lines intersect in a single point, and that the coordinates of the point are approximately (2, 1). A check will show that this point is a solution of each of the original equations.

FIGURE 4.1

NOTE   In Example 2, a graphical approach found the *exact* solution, because it had integer coordinates. Even if the coordinates are messy, however, the graphical approach can still produce a good approximation of the solution.

Exploration

Exploring Solutions of Linear Systems   In your group, find each of the following systems of equations and graph them using a graphing utility. One person in the group should try to find and graph a system that has exactly one solution. Another should try to find and graph a system that has no solution. Another should try to find and graph a system that has more than one solution. Discuss your results within your group. Form group conclusions about the number of solutions that a system of linear equations in two variables can have.

A system of linear equations can have exactly one solution, infinitely many solutions, or no solution. To see why this is true, consider the following graphical interpretations of a system of two linear equations in two variables.

Graphical Interpretation of Solutions

For a system of two linear equations in two variables, the number of solutions is given by one of the following.

Number of Solutions	*Graphical Interpretation*
1. Exactly one solution	The two lines intersect at one point.
2. Infinitely many solutions	The two lines coincide (are identical).
3. No solution	The two lines are parallel.

These three possibilities are shown in Figure 4.2.

Use the system $3x - 2y = -4$, $9x + 4y = 13$ to point out the possible inaccuracy or need for approximation when using the graphical method.

Solution: $\left(\frac{1}{3}, \frac{5}{2}\right)$

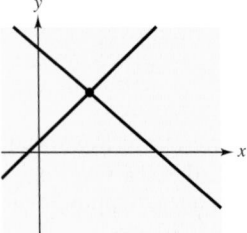

Consistent

Two lines that intersect: single point of intersection

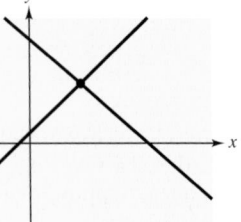

Consistent

Two lines that coincide: infinitely many points of intersection

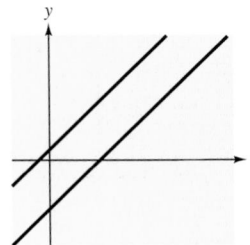

Inconsistent

Two parallel lines: no point of intersection

FIGURE 4.2

A system of linear equations is called **consistent** if it has at least one solution, and it is called **inconsistent** if it has no solution.

 *Technology*

Use a graphing utility to graph each of the following systems. From the graph, how many solutions does each system have?

a. $\begin{cases} x - 4y = 6 \\ \frac{1}{2}x - 2y = 3 \end{cases}$ b. $\begin{cases} x + y = 4 \\ x - y = 4 \end{cases}$ c. $\begin{cases} -3x + y = 2 \\ -3x + y = -5 \end{cases}$

Solving Systems of Equations by Substitution

One way to solve a system of two equations in two variables algebraically is to convert the system to *one* equation in *one* variable by an appropriate substitution. This procedure is illustrated in Example 3.

EXAMPLE 3   *The Method of Substitution: One-Solution Case*

Solve the following system of equations.

$$\begin{cases} -x + y = 3 & \text{Equation 1} \\ 3x + y = -1 & \text{Equation 2} \end{cases}$$

Solution

Begin by solving for y in Equation 1.

$$y = x + 3 \qquad \text{Solve for } y \text{ in Equation 1.}$$

Next, substitute this expression for y into Equation 2.

$$3x + y = -1 \qquad \text{Equation 2}$$
$$3x + (x + 3) = -1 \qquad \text{Replace } y \text{ by } x + 3.$$
$$4x + 3 = -1 \qquad \text{Combine like terms.}$$
$$4x = -4$$
$$x = -1 \qquad \text{Solve for } x.$$

At this point, you know that the x-coordinate of the solution is -1. To find the y-coordinate, back-substitute the x-value into the revised Equation 1.

$$y = x + 3 \qquad \text{Revised Equation 1}$$
$$y = -1 + 3 \qquad \text{Replace } x \text{ by } -1.$$
$$y = 2 \qquad \text{Solve for } y.$$

Thus, the solution is $(-1, 2)$. Check to see that it satisfies both original equations. You can also check the solution graphically, as shown in Figure 4.3.

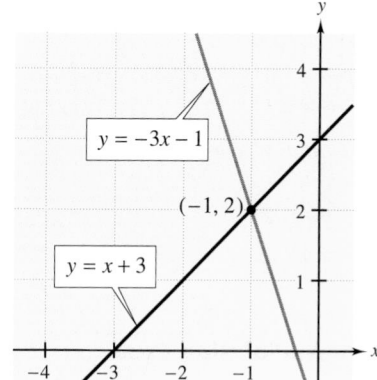

$y = -3x - 1$

$(-1, 2)$

$y = x + 3$

FIGURE 4.3

NOTE   The term **back-substitute** implies that you work backwards. After finding a value for one of the variables, substitute that value back into one of the equations in the original (or revised) system to find the value of the other variable.

When using the method of substitution, it does not matter which variable you choose to solve for first. Whether you solve for y first or x first, you will obtain the same solution. When making your choice, you should choose the variable that is easier to work with. For instance, in the system

$$\begin{cases} 3x - 2y = 1 & \text{Equation 1} \\ x + 4y = 3 & \text{Equation 2} \end{cases}$$

it is easier to solve for x first (in Equation 2).

The steps for using the method of substitution to solve a system of two equations involving two variables are summarized as follows.

The Method of Substitution

To solve a system of two equations in two variables, use the following steps.

1. Solve one of the equations for one variable in terms of the other variable.

2. Substitute the expression found in Step 1 into the other equation to obtain an equation in one variable.

3. Solve the equation obtained in Step 2.

4. Back-substitute the solution from Step 3 into the expression obtained in Step 1 to find the value of the other variable.

5. Check the solution to ensure that it satisfies *both* of the original equations.

EXAMPLE 4   *The Method of Substitution: No-Solution Case*

Solve the following system of equations.

$$\begin{cases} 2x - 2y = 0 & \text{Equation 1} \\ x - y = 1 & \text{Equation 2} \end{cases}$$

Solution

Begin by solving for y in Equation 2.

$$y = x - 1 \qquad \text{Solve for } y \text{ in Equation 2.}$$

Next, substitute this expression for y into Equation 1.

$$2x - 2y = 0 \qquad \text{Equation 1}$$
$$2x - 2(x - 1) = 0 \qquad \text{Replace } y \text{ by } x - 1.$$
$$2x - 2x + 2 = 0$$
$$2 = 0 \qquad \text{False statement}$$

Because the substitution process produced a false statement ($2 = 0$), you can conclude that the original system of equations has no solution. This is confirmed graphically in Figure 4.4, which shows that the two lines are parallel.

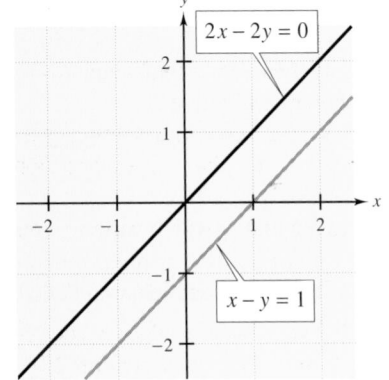

FIGURE 4.4

The Method of Elimination

A third method of solving a system of linear equations is called the **method of elimination.** The key step in the method of elimination is to obtain, for one of the variables, coefficients that differ only in sign, so that by *adding* the two equations this variable will be eliminated. This system contains such coefficients for x.

$$
\begin{array}{ll}
2x + 7y = 16 & \text{Equation 1} \\
\underline{-2x - 3y = -8} & \text{Equation 2} \\
 4y = 8 & \text{Add equations.}
\end{array}
$$

Caution students to avoid the common error of forgetting to find the value for *both* variables.

Note that by adding the two equations, you eliminate the variable x and obtain a single equation in y. Solving this equation for y produces $y = 2$, which can be back-substituted into one of the original equations to solve for x. Thus, the ordered pair $(1, 2)$ is a solution of the system. The method of elimination is summarized as follows.

The Method of Elimination

To use the **method of elimination** to solve a system of two linear equations in x and y, use the following steps.

1. Write each equation in the form $Ax + By = C$.

2. Obtain coefficients for x (or y) that differ only in sign by multiplying all terms of one or both equations by suitable constants.

3. Add the equations to eliminate one variable and solve the resulting equation.

4. Back-substitute the value obtained in Step 3 into either of the original equations and solve for the other variable.

5. Check your solution in *both* of the original equations.

To obtain coefficients (for one of the variables) that differ only in sign, you often need to multiply one or both of the equations by a suitable constant. For instance, in the system

$$
\begin{cases}
2x + 3y = 10 & \text{Equation 1} \\
4x - y = 5 & \text{Equation 2}
\end{cases}
$$

you can multiply both sides of Equation 1 by -2 (which will produce x-coefficients that differ only in sign) or you can multiply both sides of Equation 2 by 3 (which will produce y-coefficients that differ only in sign). This is further demonstrated in the next example.

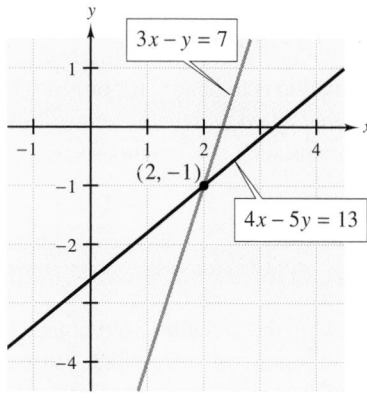

FIGURE 4.5

Technology

Use a graphing utility to graph the system of equations in Example 5(b). Use the zoom and trace features to estimate the solution to the system. Then go through Example 5(b) to see how close your graphical estimation was to the algebraic solution.

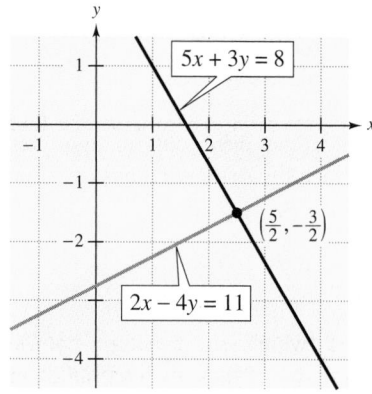

FIGURE 4.6

EXAMPLE 5   The Method of Elimination

Solve each linear system.

(a) $\begin{cases} 4x - 5y = 13 \\ 3x - y = 7 \end{cases}$

(b) $\begin{cases} 5x + 3y = 8 \\ 2x - 4y = 11 \end{cases}$

Solution

(a) For this system, you can obtain coefficients of y that differ only in sign by multiplying Equation 2 by -5.

$$\begin{cases} 4x - 5y = 13 \\ 3x - y = 7 \end{cases} \implies \begin{array}{ll} 4x - 5y = 13 & \text{Equation 1} \\ \underline{-15x + 5y = -35} & \text{Multiply Equation 2 by } -5. \\ -11x = -22 & \text{Add equations.} \end{array}$$

Thus, you can see that $x = 2$. By back-substituting this value of x into Equation 2, you can solve for y, as follows.

$$\begin{array}{ll} 3x - y = 7 & \text{Equation 2} \\ 3(2) - y = 7 & \text{Replace } x \text{ by 2.} \\ -y = 1 & \\ y = -1 & \text{Solve for } y. \end{array}$$

Therefore, the solution is $(2, -1)$. Check to see that this solution satisfies both original equations. The solution is verified graphically in Figure 4.5.

(b) You can obtain coefficients of y that differ only in sign by multiplying Equation 1 by 4 and Equation 2 by 3, in order to eliminate the y-term in both equations.

$$\begin{cases} 5x + 3y = 8 \\ 2x - 4y = 11 \end{cases} \implies \begin{array}{ll} 20x + 12y = 32 & \text{Multiply Equation 1 by 4.} \\ \underline{6x - 12y = 33} & \text{Multiply Equation 2 by 3.} \\ 26x = 65 & \text{Add equations.} \end{array}$$

From this equation, you can determine that $x = \frac{5}{2}$. By back-substituting this value of x into Equation 2, you can solve for y, as follows.

$$\begin{array}{ll} 2x - 4y = 11 & \text{Equation 2} \\ 2\left(\frac{5}{2}\right) - 4y = 11 & \text{Replace } x \text{ by } \frac{5}{2}. \\ -4y = 6 & \\ y = -\frac{3}{2} & \text{Solve for } y. \end{array}$$

Therefore, the solution is $\left(\frac{5}{2}, -\frac{3}{2}\right)$. Check this solution in both of the original equations. The solution is verified graphically in Figure 4.6.

EXAMPLE 6   *The Method of Elimination: No-Solution Case*

Solve the following system of linear equations.

$$\begin{cases} 3x + 9y = 8 & \text{Equation 1} \\ 2x + 6y = 7 & \text{Equation 2} \end{cases}$$

Solution

To obtain coefficients of x that differ only in sign, multiply Equation 1 by 2 and Equation 2 by -3.

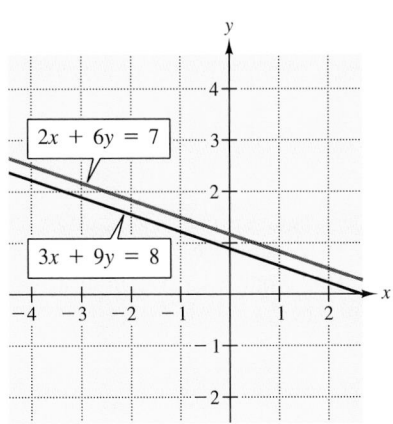

$2x + 6y = 7$

$3x + 9y = 8$

FIGURE 4.7

$$\begin{cases} 3x + 9y = 8 \\ 2x + 6y = 7 \end{cases} \quad \begin{array}{rr} 6x + 18y = & 16 & \text{Multiply Equation 1 by 2.} \\ -6x - 18y = & -21 & \text{Multiply Equation 2 by } -3. \\ \hline 0 = & -5 & \text{False statement} \end{array}$$

Because there are no values of x and y for which $0 = -5$, you can conclude that the system is inconsistent and has no solution. The lines corresponding to the two equations of this system are shown in Figure 4.7. Note that the two lines are parallel, and therefore have no point of intersection.

Example 7 shows how the method of elimination works with a system that has infinitely many solutions.

EXAMPLE 7   *The Method of Elimination: Many-Solutions Case*

Solve the following system of linear equations.

$$\begin{cases} -2x + 6y = 3 & \text{Equation 1} \\ -12y = -6 - 4x & \text{Equation 2} \end{cases}$$

Solution

To begin, write Equation 2 in standard form. Then, to obtain coefficients of x that differ only in sign, multiply Equation 1 by 2.

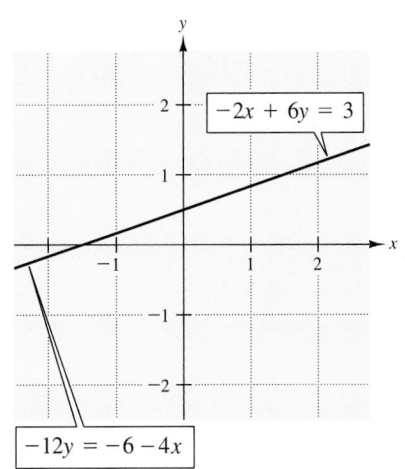

$-2x + 6y = 3$

$-12y = -6 - 4x$

FIGURE 4.8

$$\begin{cases} -2x + 6y = 3 \\ 4x - 12y = -6 \end{cases} \quad \begin{array}{rr} -4x + 12y = & 6 & \text{Multiply Equation 1 by 2.} \\ 4x - 12y = & -6 & \text{Equation 2} \\ \hline 0 = & 0 & \text{Add equations.} \end{array}$$

Because the two equations turned out to be equivalent, you can conclude that the system has infinitely many solutions. The solution set consists of all points (x, y) lying on the line $-2x + 6y = 3$, as shown in Figure 4.8.

The next example shows how the method of elimination works with a system of linear equations having decimal coefficients.

EXAMPLE 8   Solving a Linear System Having Decimal Coefficients

Solve the following system of linear equations.

$$\begin{cases} 0.02x - 0.05y = -0.38 & \text{Equation 1} \\ 0.03x + 0.04y = 1.04 & \text{Equation 2} \end{cases}$$

Solution

Point out that when multiplying an equation by a non-zero constant, all terms on each side of the equation are multiplied by this constant. The resulting equation is equivalent to the original, and therefore the solution set of the system is not affected.

Because the coefficients in this system have two decimal places, it is convenient to begin by multiplying each equation by 100. (This produces a system in which the coefficients are all integers.)

$$\begin{cases} 2x - 5y = -38 & \text{Revised Equation 1} \\ 3x + 4y = 104 & \text{Revised Equation 2} \end{cases}$$

Now, to obtain coefficients of x that differ only in sign, multiply Equation 1 by 3 and Equation 2 by -2.

$$\begin{cases} 2x - 5y = -38 \\ 3x + 4y = 104 \end{cases} \implies$$

$$\begin{aligned} 6x - 15y &= -114 & \text{Multiply Equation 1 by 3.} \\ -6x - 8y &= -208 & \text{Multiply Equation 2 by } -2. \\ \hline -23y &= -322 & \text{Add equations.} \end{aligned}$$

Thus, the value of y is

$$y = \frac{-322}{-23} = 14.$$

Back-substituting this value into revised Equation 2 produces the following.

$$\begin{aligned} 3x + 4y &= 104 & \text{Revised Equation 2} \\ 3x + 4(14) &= 104 & \text{Replace } y \text{ by } 14. \\ 3x &= 48 \\ x &= 16 & \text{Solve for } x. \end{aligned}$$

Technology

Use a graphing utility to graph the system of linear equations in Example 8. Use the zoom and trace features to approximate the solution.

Therefore, the solution is $(16, 14)$. You can check this solution in both of the original equations in the system, as follows.

Check

$$0.02(16) - 0.05(14) = 0.32 - 0.70 = -0.38 \;\checkmark$$
$$0.03(16) + 0.04(14) = 0.48 + 0.56 = 1.04 \;\checkmark$$

Application

EXAMPLE 9   A Mixture Problem

A company with two stores buys six large delivery vans and five small delivery vans. The first store receives four of the large vans and two of the small vans for a total cost of $160,000. The second store receives two of the large vans and three of the small vans for a total cost of $128,000. What is the cost of each type of van?

Solution

The two unknowns in this problem are the costs of the two types of vans.

*Verbal
Model:* $4\left(\begin{array}{c}\text{Cost of}\\\text{large van}\end{array}\right) + 2\left(\begin{array}{c}\text{Cost of}\\\text{small van}\end{array}\right) = \$160{,}000$

$2\left(\begin{array}{c}\text{Cost of}\\\text{large van}\end{array}\right) + 3\left(\begin{array}{c}\text{Cost of}\\\text{small van}\end{array}\right) = \$128{,}000$

Labels: $x = $ Cost of large van   (dollars)
$y = $ Cost of small van   (dollars)

*System of
Equations:* $\begin{cases} 4x + 2y = 160{,}000 & \text{Equation 1}\\ 2x + 3y = 128{,}000 & \text{Equation 2} \end{cases}$

To solve this system of linear equations, you can use the method of elimination. To obtain coefficients of x that differ only in sign, multiply Equation 2 by -2.

$\begin{cases} 4x + 2y = 160{,}000 \\ 2x + 3y = 128{,}000 \end{cases} \quad\Longrightarrow\quad \begin{array}{r} 4x + 2y = 160{,}000 \\ -4x - 6y = -256{,}000 \\ \hline -4y = -96{,}000 \end{array}$

Thus, the cost of each small van is $y = \$24{,}000$. By back-substituting this value into Equation 1, you can find the cost of each large van, as follows.

$$4x + 2y = 160{,}000 \qquad \text{Equation 1}$$
$$4x + 2(24{,}000) = 160{,}000 \qquad \text{Replace } y \text{ by } 24{,}000.$$
$$4x = 112{,}000$$
$$x = 28{,}000 \qquad \text{Solve for } x.$$

Thus, the cost of each large van is $x = \$28{,}000$, and the cost of each small van is $y = \$24{,}000$. Check this solution in the original problem.

Study Tip

When solving application problems, make sure your answers make sense. For instance, a negative result for the x- or y-value in Example 9 would not make sense.

EXAMPLE 10   An Application

A total of $12,000 was invested in two funds paying 6% and 8% simple interest. If the interest for 1 year is $880, how much of the $12,000 was invested in each fund?

Solution

Verbal Model:

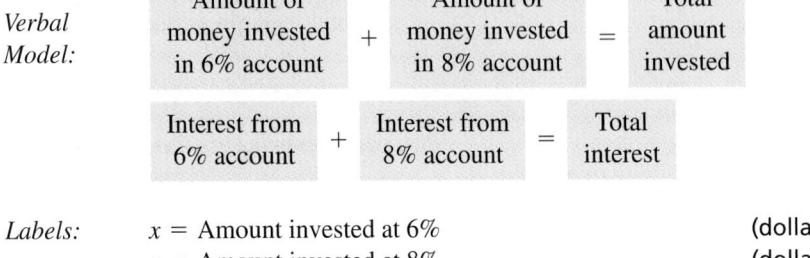

Labels: $x =$ Amount invested at 6%   (dollars)
            $y =$ Amount invested at 8%   (dollars)

System of Equations: $\begin{cases} x + y = 12{,}000 & \text{Equation 1} \\ 0.06x + 0.08y = 880 & \text{Equation 2} \end{cases}$

To solve this system, solve Equation 1 for y to obtain $y = 12{,}000 - x$. Next, substitute this expression into Equation 2 to obtain

$$0.06x + 0.08(12{,}000 - x) = 880 \qquad \text{Replace } y \text{ by } 12{,}000 - x.$$
$$0.06x + 960 - 0.08x = 880$$
$$-0.02x = -80$$
$$x = 4000. \qquad \text{Solve for } x.$$

Back-substituting this value for x into revised Equation 1 produces

$$y = 12{,}000 - 4000 = 8000.$$

Therefore, you can conclude that $4000 was invested in the 6% fund and $8000 was invested in the 8% fund.

Additional problem: For what value of k does the following system have no solution?

$2x - 3y = 4$

$4x - ky = 1$

Answer: $k = 6$

The total cost C of producing x units of a product usually has two components—the initial cost and the cost per unit. When enough units have been sold so that the total revenue R equals the total cost, the sales have reached the **break-even point.** You can find this break-even point by setting C equal to R and solving for x. In other words, the break-even point corresponds to the point of intersection of the cost and revenue graphs.

EXAMPLE 11   *An Application: Break-Even Analysis*

A small business invests $14,000 in equipment to produce a product. Each unit of the product costs $0.80 to produce and is sold for $1.50. How many items must be sold before the business breaks even?

Solution

The total cost C of producing x units is

$$\overset{\text{Cost per unit}}{\overbrace{}} \quad \overset{\text{Initial cost}}{\overbrace{}}$$
$$C = 0.80x + 14{,}000$$

and the revenue R obtained by selling x units is

$$\overset{\text{Price per unit}}{\overbrace{}}$$
$$R = 1.50x.$$

Because the break-even point occurs when $R = C$, you have

$$1.50x = 0.80x + 14{,}000 \qquad R = C$$
$$0.7x = 14{,}000$$
$$x = 20{,}000. \qquad \text{Solve for } x.$$

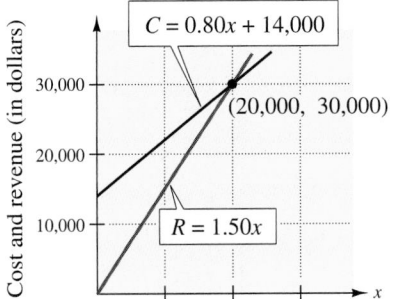

FIGURE 4.9

Therefore, the business must sell 20,000 items before it breaks even. Note in Figure 4.9 that sales less than the break-even point correspond to a loss for the small business, whereas sales greater than the break-even point correspond to a profit.

Group Activities

Writing Systems of Equations

Suppose you are tutoring another student and want to create several systems of equations that the student can use for practice. To begin, you want to find some equations that have relatively simple solutions. For instance, ask each person in your group to try to write a system of equations that has $(3, -2)$ as a solution.

Now, compare your systems. Discuss how you obtained your systems. Did everyone in the group write the same system? How many systems of equations have $(3, -2)$ as a solution? Explain the reasoning behind your answer.

4.1 Exercises

See Warm-Up Exercises, p. A42

1. *Think About It* If you graph a system of equations and the graphs do not intersect, what can you conclude about the solution of the system?

2. *Think About It* You graph a system of equations on a graphing utility and see only one line. Describe how the system could have (a) a unique solution, (b) infinitely many solutions, or (c) no solution.

In Exercises 3–6, determine whether each ordered pair is a solution of the system of linear equations.

3. $\begin{cases} x + 2y = 9 \\ -2x + 3y = 10 \end{cases}$ (a) $(1, 4)$
   (b) $(3, -1)$

4. $\begin{cases} 5x - 4y = 34 \\ x - 2y = 8 \end{cases}$ (a) $(0, 3)$
   (b) $(6, -1)$

5. $\begin{cases} -2x + 7y = 46 \\ y = -3x \end{cases}$ (a) $(-3, 2)$
   (b) $(-2, 6)$

6. $\begin{cases} -5x - 2y = 23 \\ 4y = -x - 19 \end{cases}$ (a) $(-3, -4)$
   (b) $(3, 7)$

In Exercises 7–12, use the graphs of the equations to determine whether the system has any solutions. Find any solutions that exist.

7. $\begin{cases} x + y = 4 \\ x + y = -1 \end{cases}$

8. $\begin{cases} -x + y = 5 \\ x + 2y = 4 \end{cases}$

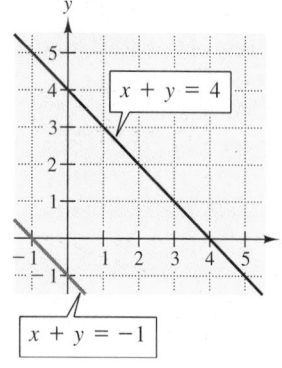

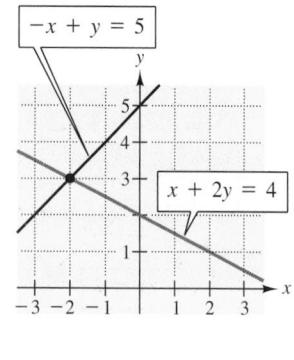

9. $\begin{cases} 5x - 3y = 4 \\ 2x + 3y = 3 \end{cases}$

10. $\begin{cases} 2x - y = 4 \\ -4x + 2y = -12 \end{cases}$

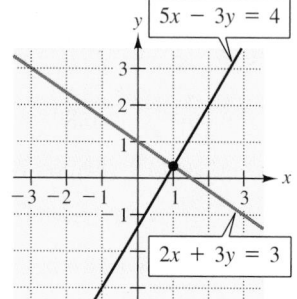

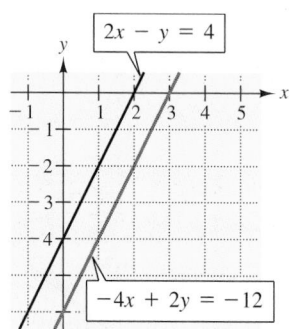

11. $\begin{cases} -x + 2y = 5 \\ 2x - 4y = -10 \end{cases}$

12. $\begin{cases} x = 5 \\ y = -3 \end{cases}$

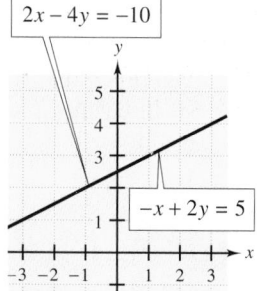

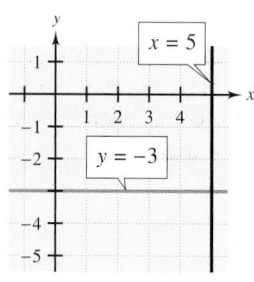

In Exercises 13–18, sketch the graphs of the equations and approximate any solutions of the system of linear equations.

13. $\begin{cases} -2x + y = 1 \\ x - 3y = 2 \end{cases}$

14. $\begin{cases} 5x - 6y = -30 \\ 5x + 4y = 20 \end{cases}$

15. $\begin{cases} 2x - 5y = 20 \\ 4x - 5y = 40 \end{cases}$

16. $\begin{cases} 5x + 3y = 24 \\ x - 2y = 10 \end{cases}$

17. $\begin{cases} x - y = -3 \\ 2x - y = 6 \end{cases}$

18. $\begin{cases} x = 4 \\ y = 3 \end{cases}$

The symbol ▦ indicates an exercise in which you are instructed to use a calculator or graphing utility. The solutions of other exercises may also be facilitated by the use of appropriate technology.

⊞ In Exercises 19–24, use a graphing utility to graph the equations and approximate any solutions of the system of linear equations.

19. $\begin{cases} -5x + 3y = 15 \\ x + y = 1 \end{cases}$

20. $\begin{cases} 9x + 4y = 8 \\ 7x - 4y = 8 \end{cases}$

21. $\begin{cases} 5x + 4y = 35 \\ -x + 3y = 12 \end{cases}$

22. $\begin{cases} 5x - 4y = 0 \\ -3x + 8y = 14 \end{cases}$

23. $\begin{cases} 4x - y = 3 \\ 6x + 2y = 1 \end{cases}$

24. $\begin{cases} x - 6y = 2 \\ 2x + 3y = 9 \end{cases}$

In Exercises 25–28, decide whether the system is consistent or inconsistent.

25. $\begin{cases} 4x - 5y = 3 \\ -8x + 10y = -6 \end{cases}$

26. $\begin{cases} 4x - 5y = 3 \\ -8x + 10y = 14 \end{cases}$

27. $\begin{cases} -2x + 5y = 3 \\ 5x + 2y = 8 \end{cases}$

28. $\begin{cases} x + 10y = 12 \\ -2x + 5y = 2 \end{cases}$

⊞ In Exercises 29–32, use a graphing utility to graph the equations in the system. Use the graphs to determine whether the system is consistent or inconsistent. If the system is consistent, determine the number of solutions.

29. $\begin{cases} \frac{1}{3}x - \frac{1}{2}y = 1 \\ -2x + 3y = 6 \end{cases}$

30. $\begin{cases} x + y = 5 \\ x - y = 5 \end{cases}$

31. $\begin{cases} -2x + 3y = 6 \\ x - y = -1 \end{cases}$

32. $\begin{cases} 2x - 4y = 9 \\ x - 2y = 4.5 \end{cases}$

⊞ In Exercises 33–36, use a graphing utility to graph the equations in the system. The graphs appear parallel. Yet, from the slope-intercept forms of the lines, you find that the slopes are not equal and the graphs intersect. Find the point of intersection of the two lines.

33. $\begin{cases} x - 100y = -200 \\ 3x - 275y = 198 \end{cases}$

34. $\begin{cases} 35x - 33y = 0 \\ 12x - 11y = 92 \end{cases}$

35. $\begin{cases} 3x - 25y = 50 \\ 9x - 100y = 50 \end{cases}$

36. $\begin{cases} x + 40y = 80 \\ 2x + 150y = 195 \end{cases}$

In Exercises 37–42, solve the system by the method of substitution.

37. $\begin{cases} x - 2y = 0 \\ 3x + 2y = 8 \end{cases}$

38. $\begin{cases} x - y = 0 \\ 5x - 2y = 6 \end{cases}$

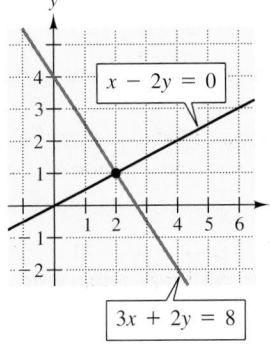

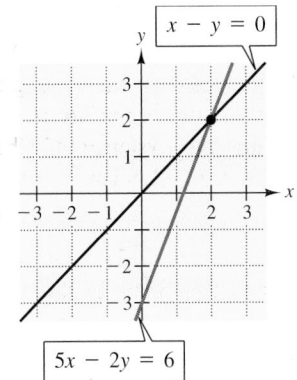

39. $\begin{cases} x = 4 \\ x - 2y = -2 \end{cases}$

40. $\begin{cases} y = 2 \\ x - 6y = -6 \end{cases}$

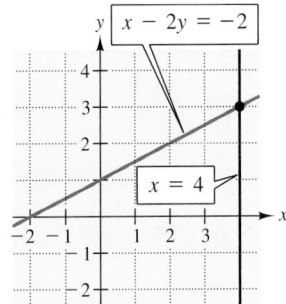

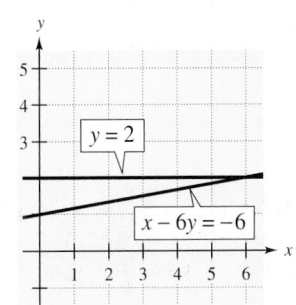

41. $\begin{cases} 7x + 8y = 24 \\ x - 8y = 8 \end{cases}$

42. $\begin{cases} x - 3y = -2 \\ 5x + 3y = 17 \end{cases}$

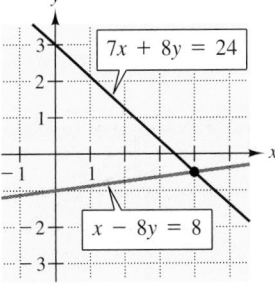

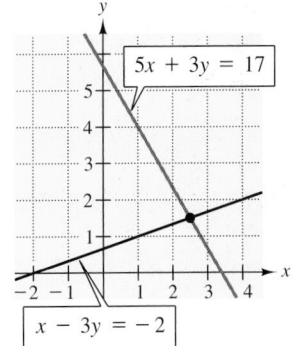

In Exercises 43–52, solve the system by the method of substitution.

43. $\begin{cases} x + y = 3 \\ 2x - y = 0 \end{cases}$

44. $\begin{cases} -x + y = 5 \\ x - 4y = 0 \end{cases}$

45. $\begin{cases} x + y = 2 \\ x - 4y = 12 \end{cases}$

46. $\begin{cases} x - 2y = -1 \\ x - 5y = 2 \end{cases}$

47. $\begin{cases} x + 6y = 19 \\ x - 7y = -7 \end{cases}$

48. $\begin{cases} x - 5y = -6 \\ 4x - 3y = 10 \end{cases}$

49. $\begin{cases} 2x + 5y = 29 \\ 5x + 2y = 13 \end{cases}$

50. $\begin{cases} -13x + 16y = 10 \\ 5x + 16y = -26 \end{cases}$

51. $\begin{cases} 4x - 14y = -15 \\ 18x - 12y = 9 \end{cases}$

52. $\begin{cases} 5x - 24y = -12 \\ 17x - 24y = 36 \end{cases}$

In Exercises 53–60, solve the system by the method of elimination. Identify and label each line with its equation and label the point of intersection.

53. $\begin{cases} -x + 2y = 1 \\ x - y = 2 \end{cases}$

54. $\begin{cases} 3x + y = 3 \\ 2x - y = 7 \end{cases}$

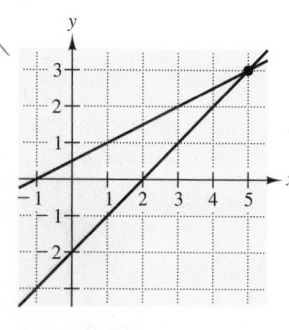

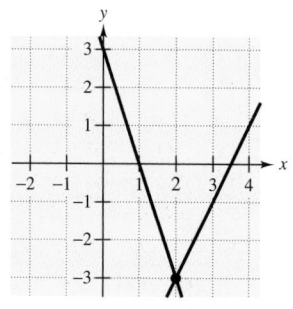

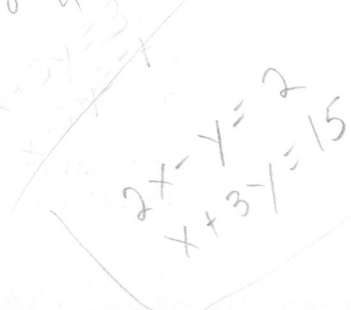

55. $\begin{cases} x + y = 0 \\ 3x - 2y = 10 \end{cases}$

56. $\begin{cases} -x + 2y = 2 \\ 3x + y = 15 \end{cases}$

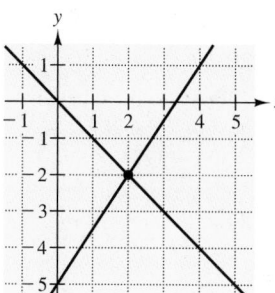

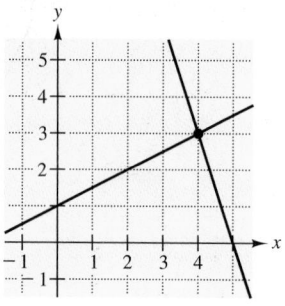

57. $\begin{cases} x - y = 1 \\ -3x + 3y = 8 \end{cases}$

58. $\begin{cases} 3x + 4y = 2 \\ 0.6x + 0.8y = 1.6 \end{cases}$

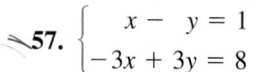

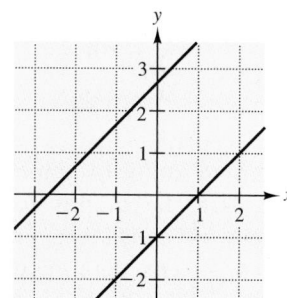

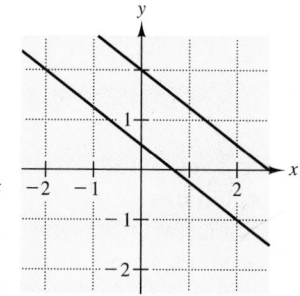

59. $\begin{cases} x - 3y = 5 \\ -2x + 6y = -10 \end{cases}$

60. $\begin{cases} x - 4y = 5 \\ 5x + 4y = 7 \end{cases}$

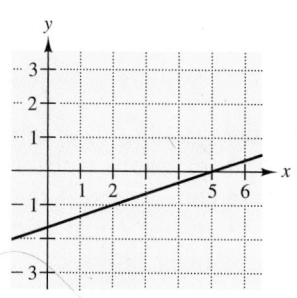

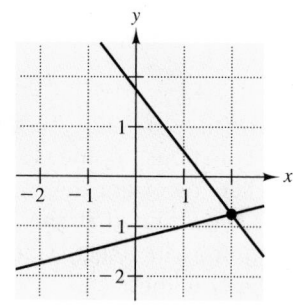

In Exercises 61–80, solve the system by the method of elimination.

61. $\begin{cases} 3x - 2y = 5 \\ x + 2y = 7 \end{cases}$

62. $\begin{cases} -x + 2y = 9 \\ x + 3y = 16 \end{cases}$

63. $\begin{cases} 4x + y = -3 \\ -4x + 3y = 23 \end{cases}$

64. $\begin{cases} -3x + 5y = -23 \\ 2x - 5y = 22 \end{cases}$

65. $\begin{cases} -x - 3y = 2 \\ 3x - 7y = 4 \end{cases}$

66. $\begin{cases} 7r - s = -25 \\ 2r + 5s = 14 \end{cases}$

67. $\begin{cases} 2u + 3v = 8 \\ 3u + 4v = 13 \end{cases}$

68. $\begin{cases} 4x - 3y = 25 \\ -3x + 8y = 10 \end{cases}$

69. $\begin{cases} 12x - 5y = 2 \\ -24x + 10y = 6 \end{cases}$

70. $\begin{cases} -2x + 3y = 9 \\ 6x - 9y = -27 \end{cases}$

71. $\begin{cases} \frac{2}{3}r - s = 0 \\ 10r + 4s = 19 \end{cases}$

72. $\begin{cases} x - y = -\frac{1}{2} \\ 4x - 48y = -35 \end{cases}$

73. $\begin{cases} 0.7u - v = -0.4 \\ 0.3u - 0.8v = 0.2 \end{cases}$

74. $\begin{cases} 0.15x - 0.35y = -0.5 \\ -0.12x + 0.25y = 0.1 \end{cases}$

75. $\begin{cases} 5x + 7y = 25 \\ x + 1.4y = 5 \end{cases}$

76. $\begin{cases} 12b - 13m = 2 \\ -6b + 6.5m = -2 \end{cases}$

77. $\begin{cases} 2x = 25 \\ 4x - 10y = 0.52 \end{cases}$

78. $\begin{cases} 6x - 6y = 25 \\ 3y = 11 \end{cases}$

79. $\begin{cases} \frac{3}{2}x - y = 4 \\ -x + \frac{2}{3}y = -1 \end{cases}$

80. $\begin{cases} 12x - 3y = 6 \\ 4x - y = 2 \end{cases}$

In Exercises 81–90, solve the system by any convenient method.

81. $\begin{cases} \frac{3}{2}x + 2y = 12 \\ \frac{1}{4}x + y = 4 \end{cases}$

82. $\begin{cases} 4x + y = -2 \\ -6x + y = 18 \end{cases}$

83. $\begin{cases} y = 5x - 3 \\ y = -2x + 11 \end{cases}$

84. $\begin{cases} 3x + 2y = 5 \\ y = 2x + 13 \end{cases}$

85. $\begin{cases} 2x - y = 20 \\ -x + y = -5 \end{cases}$

86. $\begin{cases} 3x - 2y = -20 \\ 5x + 6y = 32 \end{cases}$

87. $\begin{cases} 3y = 2x + 21 \\ x = 50 - 4y \end{cases}$

88. $\begin{cases} x + 2y = 4 \\ \frac{1}{2}x + \frac{1}{3}y = 1 \end{cases}$

89. $\begin{cases} 2x + 3y = 0 \\ 3x + 5y = -1000 \end{cases}$

90. $\begin{cases} 0.4u + 1v = 800 \\ 0.7u + 2v = 1850 \end{cases}$

In Exercises 91 and 92, determine the value of k such that the system of linear equations is inconsistent.

91. $\begin{cases} 5x - 10y = 40 \\ -2x + ky = 30 \end{cases}$

92. $\begin{cases} 12x - 18y = 5 \\ -18x + ky = 10 \end{cases}$

93. *True or False?* It is possible for a consistent system of linear equations to have exactly two solutions. Explain.

94. *Think About It* How can you recognize that a system of linear equations has no solution? Give an example.

In Exercises 95 and 96, find a system of linear equations that has the given solution. (There are many correct answers.)

95. $\left(3, -\frac{3}{2}\right)$

96. $(-8, 12)$

Dimensions of a Rectangle In Exercises 97 and 98, find the dimensions of the rectangle meeting the specified conditions.

Perimeter	Condition
97. 220 meters	Length is 120% of the width.
98. 280 feet	Width is 75% of the length.

99. *Gasoline Mixture* Twelve gallons of regular unleaded gasoline plus 8 gallons of premium unleaded gasoline cost $23.08. The price of premium unleaded is 11 cents more per gallon than the price of regular unleaded. Find the price per gallon for each grade of gasoline.

100. *Ticket Sales* Eight hundred tickets were sold for a theater production and the receipts for the performance were $8600. The tickets for adults and students sold for $12.50 and $7.50, respectively. How many of each kind of ticket were sold?

101. *Hay Mixture* How many tons of hay at $125 per ton must be purchased with hay at $75 per ton to have 100 tons of hay with a value of $90 per ton?

102. *Alcohol Mixture* How many liters of a 40% alcohol solution must be mixed with a 65% solution to obtain 20 liters of a 50% solution?

$3y = 100 - 8y + 21$
$11y = 121$ $y = 11$ $x = 50 - 44 = 6$

103. *Simple Interest* A combined total of $20,000 is invested in two bonds that pay 8% and 9.5% simple interest. The annual interest is $1675. How much is invested in each bond?

104. *Break-Even Analysis* A small business invests $8000 in equipment to produce a product. Each unit of the product costs $1.20 to produce and is sold for $2.00. How many items must be sold before the business breaks even?

105. *Rope Length* You must cut a rope that is 160 inches long into two pieces so that one piece is four times as long as the other. Find the length of each piece.

106. *Driving Distance* Two people share the driving for a trip of 300 miles. One person drives three times as far as the other. Find the distance that each person drives.

107. *Air Speed* An airplane flying into a headwind travels the 3000-mile flying distance between two cities in 6 hours and 15 minutes. On the return flight, the distance is traveled in 5 hours. Find the speed of the plane in still air and the speed of the wind, assuming that both remain constant throughout the round trip.

108. *Best-Fitting Line* The slope and y-intercept of the line $y = mx + b$ that best fits the three non-collinear points $(0, 0)$, $(1, 1)$ and $(2, 3)$ are given by the solution of the following system of linear equations.

$$\begin{cases} 3b + 3m = 4 \\ 3b + 5m = 7 \end{cases}$$

(a) Solve the system and find the equation of the best-fitting line.

(b) Plot the three points and sketch the graph of the best-fitting line.

109. *U.S. Aircraft Industry* The total employment (in thousands) in aircraft industries in the United States from 1992 through 1994 is given in the table. *(Source: U.S. Bureau of Labor Statistics)*

Year	1992	1993	1994
Employment	758	666	588

(a) Plot the data given in the table, where $x = 0$ corresponds to 1990.

(b) The line $y = mx + b$ that best fits the data is given by the solution of the following system.

$$\begin{cases} 3b + 9m = 2012 \\ 9b + 29m = 5866 \end{cases}$$

Solve the system and find the equation of this line. Sketch the graph of the line on the same set of coordinate axes used in part (a).

(c) Explain the meaning of the slope of the line in the context of the problem.

Section Project

Vietnam Veterans Memorial "The Wall" in Washington, D.C., designed by Maya Ling Lin when she was a student at Yale University, has two vertical, triangular sections of black granite with a common side (see figure). The top of each section is level with the ground. The bottoms of the two sections can be modeled by the equations $y = \frac{2}{25}x - 10$ and $y = -\frac{5}{61}x - 10$ when the x-axis is superimposed on the top of the wall. Each unit in the coordinate system represents 1 foot. How deep is the memorial at the point where the two sections meet? How long is each section?

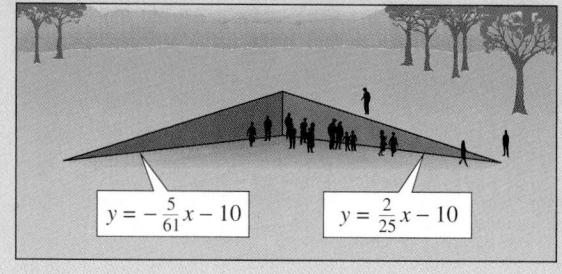

$y = -\frac{5}{61}x - 10$ $y = \frac{2}{25}x - 10$

4.2	**Systems of Linear Equations in Three Variables**

Systems of Linear Equations in Three Variables · Row-Echelon Form · The Method of Gaussian Elimination · Application

Systems of Linear Equations in Three Variables

Each of the systems of linear equations discussed in Section 4.1 had only two variables. In real-life applications, many systems have more than two variables. Here are two examples.

$$\begin{cases} 2x + 5y + z = 0 \\ 3x + 2y - 4z = 0 \\ y + z = 0 \end{cases} \qquad \begin{cases} 3x - 2y + z = 9 \\ x + y - 2z = -8 \\ -x - y + 3z = 12 \end{cases}$$

A **solution** of such a system is an **ordered triple,** (x, y, z), of real numbers that satisfies each equation in the system.

EXAMPLE 1   Checking Solutions

Which of the ordered triples is a solution of the first linear system shown above?

(a) $(4, -2, 2)$ (b) $(-2, -1, 1)$ (c) $(0, 0, 0)$

Solution

(a) To check the ordered triple $(4, -2, 2)$, substitute 4 for x, -2 for y, and 2 for z in each of the original equations. Because $(4, -2, 2)$ is a solution of *each* equation, it *is* a solution of the system.

$$2(4) + 5(-2) + (2) = 8 - 10 + 2 = 0 \quad \checkmark$$
$$3(4) + 2(-2) - 4(2) = 12 - 4 - 8 = 0 \quad \checkmark$$
$$(-2) + (2) = -2 + 2 = 0 \quad \checkmark$$

(b) To check the ordered triple $(-2, -1, 1)$, substitute -2 for x, -1 for y, and 1 for z in each of the original equations. Because $(-2, -1, 1)$ is not a solution of *each* equation, it *is not* a solution of the system.

$$2(-2) + 5(-1) + (1) = -4 - 5 + 1 = -8 \quad \times$$
$$3(-2) + 2(-1) - 4(1) = -6 - 2 - 4 = -12 \quad \times$$
$$(-1) + (1) = -1 + 1 = 0 \quad \checkmark$$

(c) To check the ordered triple $(0, 0, 0)$, substitute 0 for x, y, and z in each of the original equations. When you do this, each equation is satisfied, so $(0, 0, 0)$ *is* a solution of the system.

One way to solve a system of linear equations in three variables is to add multiples of the equations to each other to eliminate one of the variables. Your goal is to reduce the system to one that has only two equations and two variables. This procedure is demonstrated in Example 2.

EXAMPLE 2   Solving a System of Three Linear Equations

Solve the following linear system.

$$\begin{cases} 3x - 2y + z = 9 \\ x + y - 2z = -8 \\ -x - 2y + 3z = 13 \end{cases}$$

Equation 1
Equation 2
Equation 3

Solution

There are many ways to solve this system. One way is to try to eliminate the variable x. Multiply Equation 2 by -3 and add the result to Equation 1.

$$\begin{array}{ll} 3x - 2y + z = 9 & \text{Equation 1} \\ \underline{-3x - 3y + 6z = 24} & -3 \text{ times Equation 2} \\ -5y + 7z = 33 & \text{New Equation 1} \end{array}$$

Then add Equation 2 and Equation 3.

$$\begin{array}{ll} x + y - 2z = -8 & \text{Equation 2} \\ \underline{-x - 2y + 3z = 13} & \text{Equation 3} \\ -y + z = 5 & \text{New Equation 2} \end{array}$$

You now have a new system of linear equations that has only two equations and two variables. You can solve this system using the techniques in Section 4.1.

$$\begin{cases} -5y + 7z = 33 \\ -y + z = 5 \end{cases} \implies \begin{array}{ll} -5y + 7z = 33 & \text{New Equation 1} \\ \underline{5y - 5z = -25} & \text{Multiply new Equation 2 by } -5. \\ 2z = 8 & \text{Add equations.} \end{array}$$

Thus, you can see that $z = 4$. By back-substituting this value of z into new Equation 1, you can solve for y.

$$-5y + 7(4) = 33 \implies y = -1$$

Now, knowing that $y = -1$ and $z = 4$, you can back-substitute these values into any of the original three equations to solve for x.

$$3x - 2(-1) + (4) = 9 \implies x = 1$$

Thus, $(1, -1, 4)$ is the solution of the system. Check this in each equation in the original system.

Row-Echelon Form

Consider the following systems of linear equations.

$$\begin{cases} x - 2y + 2z = 9 \\ -x + 3y = -4 \\ 2x - 5y + z = 10 \end{cases} \qquad \begin{cases} x - 2y + 2z = 9 \\ y + 2z = 5 \\ z = 3 \end{cases}$$

Which one of these two systems do you think is easier to solve? After comparing the two systems, it should be clear that it is easier to solve the system on the right. The system on the right is said to be in **row-echelon form,** which means that it has a "stair-step" pattern with leading coefficients of 1.

Back-substitution can be used to solve a system in row-echelon form, as shown in Example 3.

EXAMPLE 3   Solving a System in Row-Echelon Form

Solve the following system of linear equations.

$$\begin{cases} x - 2y + 2z = 9 & \text{Equation 1} \\ y + 2z = 5 & \text{Equation 2} \\ z = 3 & \text{Equation 3} \end{cases}$$

Solution

From Equation 3, you know the value of z. To solve for y, substitute $z = 3$ into Equation 2 to obtain

$$\begin{aligned} y + 2(3) &= 5 && \text{Substitute 3 for } z. \\ y &= -1. && \text{Solve for } y. \end{aligned}$$

Finally, substitute $y = -1$ and $z = 3$ into Equation 1 to obtain

$$\begin{aligned} x - 2(-1) + 2(3) &= 9 && \text{Substitute } -1 \text{ for } y \text{ and 3 for } z. \\ x &= 1. && \text{Solve for } x. \end{aligned}$$

The solution is $x = 1$, $y = -1$, and $z = 3$, which can also be written as the ordered triple $(1, -1, 3)$. You can check this in the original system of equations, as follows.

Check

$$\begin{aligned} (1) - 2(-1) + 2(3) &= 9 \quad \checkmark \\ (-1) + 2(3) &= 5 \quad \checkmark \\ (3) &= 3 \quad \checkmark \end{aligned}$$

The Method of Gaussian Elimination

Two systems of equations are **equivalent** if they have the same solution set. To solve a system that is not in row-echelon form, you must first convert it to an *equivalent* system that is in row-echelon form. To see how this is done, let's take another look at the method of elimination, as applied to a system of two linear equations.

EXAMPLE 4   The Method of Elimination

Solve the following system of linear equations.

$$\begin{cases} 3x - 2y = -1 & \text{Equation 1} \\ x - y = 0 & \text{Equation 2} \end{cases}$$

The initial goal is a coefficient of 1 for x in the first equation. Point out that the easiest way to achieve the goal is to switch the first and second equations.

Solution

$$\begin{cases} x - y = 0 \\ 3x - 2y = -1 \end{cases}$$
You can interchange two equations in the system.

$$\begin{aligned} -3x + 3y &= 0 \\ 3x - 2y &= -1 \\ \hline y &= -1 \end{aligned}$$
Multiply the first equation by -3.

You can add the multiple of the first equation to the second equation to obtain a new second equation.

$$\begin{cases} x - y = 0 \\ y = -1 \end{cases}$$
New system in row-echelon form

Now, using back-substitution, you can determine that the solution is $y = -1$ and $x = -1$, which can be written as the ordered pair $(-1, -1)$. Check this in the original system of equations.

Operations that Produce Equivalent Systems

Each of the following **row operations** on a system of linear equations produces an *equivalent* system of linear equations.

1. Interchange two equations.

2. Multiply one of the equations by a nonzero constant.

3. Add a multiple of one of the equations to another equation to replace the latter equation.

As shown in Example 4, rewriting a system of linear equations in row-echelon form usually involves a *chain* of equivalent systems, each of which is obtained by using one of the three basic row operations. This process is called **Gaussian elimination,** after the German mathematician Carl Friedrich Gauss (1777–1855). This method of Gaussian elimination easily adapts to computer use for solving systems of linear equations with dozens of variables.

EXAMPLE 5   Using Gaussian Elimination to Solve a System

Solve the following system of linear equations.

$$\begin{cases} x - 2y + 2z = 9 & \text{Equation 1} \\ -x + 3y = -4 & \text{Equation 2} \\ 2x - 5y + z = 10 & \text{Equation 3} \end{cases}$$

Solution

> Stress the importance of using a systematic procedure, especially for more complicated problems.

There are many ways to begin, but we suggest working from the upper left corner, saving the x in the upper left position and eliminating the other x's from the first column.

$$\begin{cases} x - 2y + 2z = 9 \\ y + 2z = 5 \\ 2x - 5y + z = 10 \end{cases}$$

Adding the first equation to the second equation produces a new second equation.

$$\begin{cases} x - 2y + 2z = 9 \\ y + 2z = 5 \\ -y - 3z = -8 \end{cases}$$

Adding -2 times the first equation to the third equation produces a new third equation.

Now that all but the first x have been eliminated from the first column, go to work on the second column. (You need to eliminate y from the third equation.)

$$\begin{cases} x - 2y + 2z = 9 \\ y + 2z = 5 \\ -z = -3 \end{cases}$$

Adding the second equation to the third equation produces a new third equation.

Finally, you need a coefficient of 1 for z in the third equation.

$$\begin{cases} x - 2y + 2z = 9 \\ y + 2z = 5 \\ z = 3 \end{cases}$$

Multiplying the third equation by -1 produces a new third equation.

This is the same system that was solved in Example 3, and, as in that example, you can conclude that the solution is $x = 1$, $y = -1$, and $z = 3$. The solution can be written as the ordered triple $(1, -1, 3)$. Check this solution in all three equations of the original system.

EXAMPLE 6 *Using Gaussian Elimination to Solve a System*

Solve the following system of linear equations.

$$\begin{cases} 2x + 4y + z = 6 & \text{Equation 1} \\ 3x + y + 3z = 17 & \text{Equation 2} \\ -5x - 8y + z = -10 & \text{Equation 3} \end{cases}$$

Solution

Additional problem:

$5x - 2y + 8z = -4$

$x + 3y - 6z = 9$

$2x + y + 4z = 6$

Answer: $x = 0$, $y = 4$, and $z = \frac{1}{2}$

$$\begin{cases} x + 2y + \frac{1}{2}z = 3 \\ 3x + y + 3z = 17 \\ -5x - 8y + z = -10 \end{cases}$$

Multiplying the first equation by $\frac{1}{2}$ produces a new first equation.

$$\begin{cases} x + 2y + \frac{1}{2}z = 3 \\ -5y + \frac{3}{2}z = 8 \\ -5x - 8y + z = -10 \end{cases}$$

Adding -3 times the first equation to the second equation produces a new second equation.

$$\begin{cases} x + 2y + \frac{1}{2}z = 3 \\ -5y + \frac{3}{2}z = 8 \\ 2y + \frac{7}{2}z = 5 \end{cases}$$

Adding 5 times the first equation to the third equation produces a new third equation.

$$\begin{cases} x + 2y + \frac{1}{2}z = 3 \\ y - \frac{3}{10}z = -\frac{8}{5} \\ 2y + \frac{7}{2}z = 5 \end{cases}$$

Multiplying the second equation by $-\frac{1}{5}$ produces a new second equation.

$$\begin{cases} x + 2y + \frac{1}{2}z = 3 \\ y - \frac{3}{10}z = -\frac{8}{5} \\ \frac{41}{10}z = \frac{41}{5} \end{cases}$$

Adding -2 times the second equation to the third equation produces a new third equation.

$$\begin{cases} x + 2y + \frac{1}{2}z = 3 \\ y - \frac{3}{10}z = -\frac{8}{5} \\ z = 2 \end{cases}$$

Multiplying the third equation by $\frac{10}{41}$ produces a new third equation.

Now that the system of equations is in row-echelon form, you can see that $z = 2$. By back-substituting into the second equation, you can determine the value of y.

$$y - \frac{3}{10}(2) = -\frac{8}{5} \implies y = -1$$

By back-substituting $y = -1$ and $z = 2$ into the first equation, you can solve for x.

$$x + 2(-1) + \frac{1}{2}(2) = 3 \implies x = 4$$

Therefore, the solution is $x = 4$, $y = -1$, and $z = 2$, which can be written as the ordered triple $(4, -1, 2)$. Check this in the original system of equations.

The next example involves an inconsistent system—one that has no solution. The key to recognizing an inconsistent system is that, at some stage in the elimination process, you obtain an absurdity such as $0 = -2$.

EXAMPLE 7   An Inconsistent System

Solve the following system of linear equations.

$$\begin{cases} x - 3y + z = 1 & \text{Equation 1} \\ 2x - y - 2z = 2 & \text{Equation 2} \\ x + 2y - 3z = -1 & \text{Equation 3} \end{cases}$$

Solution

$$\begin{cases} x - 3y + z = 1 \\ 5y - 4z = 0 \\ x + 2y - 3z = -1 \end{cases}$$

Adding -2 times the first equation to the second equation produces a new second equation.

$$\begin{cases} x - 3y + z = 1 \\ 5y - 4z = 0 \\ 5y - 4z = -2 \end{cases}$$

Adding -1 times the first equation to the third equation produces a new third equation.

$$\begin{cases} x - 3y + z = 1 \\ 5y - 4z = 0 \\ 0 = -2 \end{cases}$$

Adding -1 times the second equation to the third equation produces a new third equation.

Because the third "equation" is a false statement, you can conclude that this system is inconsistent and therefore has no solution. Moreover, because this system is equivalent to the original system, you can conclude that the original system also has no solution.

As with a system of linear equations in two variables, the solution(s) of a system of linear equations in more than two variables must fall into one of three categories.

The Number of Solutions of a Linear System

For a system of linear equations, exactly one of the following is true.

1. There is exactly one solution.

2. There are infinitely many solutions.

3. There is no solution.

EXAMPLE 8   A System with Infinitely Many Solutions

Solve the following system of linear equations.

$$\begin{cases} x + y - 3z = -1 & \text{Equation 1} \\ y - z = 0 & \text{Equation 2} \\ -x + 2y = 1 & \text{Equation 3} \end{cases}$$

Solution

Begin by rewriting the system in row-echelon form.

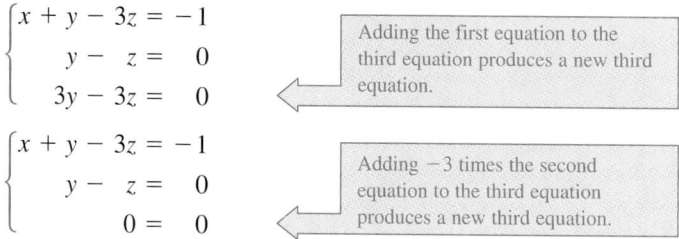

$$\begin{cases} x + y - 3z = -1 \\ y - z = 0 \\ 3y - 3z = 0 \end{cases}$$

Adding the first equation to the third equation produces a new third equation.

$$\begin{cases} x + y - 3z = -1 \\ y - z = 0 \\ 0 = 0 \end{cases}$$

Adding -3 times the second equation to the third equation produces a new third equation.

This means that Equation 3 depends on Equations 1 and 2 in the sense that it gives us no additional information about the variables. Thus, the original system is equivalent to the system

$$\begin{cases} x + y - 3z = -1 \\ y - z = 0. \end{cases}$$

In this last equation, solve for y in terms of z to obtain $y = z$. Then, back-substituting for y in the previous equation produces $x = 2z - 1$. Finally, letting $z = a$, the solutions to the given system are all of the form

$$x = 2a - 1, \qquad y = a, \qquad \text{and} \qquad z = a$$

where a is a real number. Thus, every ordered triple of the form

$$(2a - 1, a, a), \qquad a \text{ is a real number}$$

is a solution of the system.

In Example 8, there are other ways to write the same infinite set of solutions. For instance, the solutions could have been written as

$$\left(b, \tfrac{1}{2}(b + 1), \tfrac{1}{2}(b + 1)\right), \qquad b \text{ is a real number.}$$

Try convincing yourself of this by substituting $a = 0$, $a = 1$, $a = 2$, and $a = 3$ into the solution listed in Example 8. Then substitute $b = -1$, $b = 1$, $b = 3$, and $b = 5$ into the solution listed above. In both cases, you should obtain the same ordered triples. Thus, when comparing descriptions of an infinite solution set, keep in mind that there is more than one way to describe the set.

Application

EXAMPLE 9   An Application: Moving Object

The height at time t of an object that is moving in a (vertical) line with constant acceleration a is given by the **position equation**

$$s = \tfrac{1}{2}at^2 + v_0t + s_0.$$

You might ask students to account for the fact that the object reached a height of 164 feet <u>twice</u>.

The height s is measured in feet, the acceleration a is measured in feet per second squared, the time t is measured in seconds, v_0 is the initial velocity (at time $t = 0$), and s_0 is the initial height. Find the values of a, v_0, and s_0 if $s = 164$ feet at 1 second, $s = 180$ feet at 2 seconds, and $s = 164$ feet at 3 seconds.

Solution

By substituting the three values of t and s into the position equation, you obtain three linear equations in a, v_0, and s_0.

When $t = 1$, $s = 164$: $\tfrac{1}{2}a(1)^2 + v_0(1) + s_0 = 164$ Equation 1

When $t = 2$, $s = 180$: $\tfrac{1}{2}a(2)^2 + v_0(2) + s_0 = 180$ Equation 2

When $t = 3$, $s = 164$: $\tfrac{1}{2}a(3)^2 + v_0(3) + s_0 = 164$ Equation 3

By multiplying Equation 1 and Equation 3 by 2, this system can be rewritten as

$$\begin{cases} a + 2v_0 + 2s_0 = 328 \\ 2a + 2v_0 + s_0 = 180 \\ 9a + 6v_0 + 2s_0 = 328 \end{cases}$$

and you can apply elimination to obtain

$$\begin{cases} a + 2v_0 + 2s_0 = 328 \\ -2v_0 - 3s_0 = -476 \\ 2s_0 = 232. \end{cases}$$

Explain to students how the need to understand the concepts of velocity and acceleration led to new developments in calculus by Newton and Leibniz.

From the third equation $s_0 = 116$, so back-substituting into the second equation yields

$$-2v_0 - 3(116) = -476 \qquad \text{Replace } s_0 \text{ by 116 in second equation.}$$
$$-2v_0 = -128$$
$$v_0 = 64. \qquad \text{Solve for } v_0.$$

Finally, back-substituting $s_0 = 116$ and $v_0 = 64$ into the first equation yields

$$a + 2(64) + 2(116) = 328 \qquad \text{Replace } s_0 \text{ by 116 and } v_0 \text{ by 64 in first equation.}$$
$$a = -32. \qquad \text{Solve for } a.$$

Thus, the position equation for this object is $s = -16t^2 + 64t + 116$.

EXAMPLE 10 Data Analysis: Curve-Fitting

Find a quadratic equation $y = ax^2 + bx + c$ whose graph passes through the points $(-1, 3)$, $(1, 1)$, and $(2, 6)$.

Solution

Because the graph of $y = ax^2 + bx + c$ passes through the points $(-1, 3)$, $(1, 1)$, and $(2, 6)$, you can write the following.

When $x = -1, y = 3$: $\quad a(-1)^2 + b(-1) + c = 3$

When $x = 1, y = 1$: $\quad a(1)^2 + b(1) + c = 1$

When $x = 2, y = 6$: $\quad a(2)^2 + b(2) + c = 6$

This produces the following system of linear equations.

$$\begin{cases} a - b + c = 3 & \text{Equation 1} \\ a + b + c = 1 & \text{Equation 2} \\ 4a + 2b + c = 6 & \text{Equation 3} \end{cases}$$

The solution of this system is $a = 2$, $b = -1$, and $c = 0$ (try solving this on your own). Thus, the quadratic equation is $y = 2x^2 - x$. This graph, as shown in Figure 4.10, is called a **parabola.**

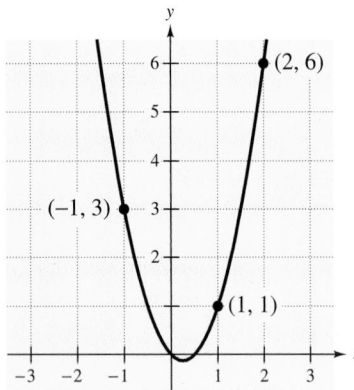

FIGURE 4.10

Group Activities

Problem Solving

Fitting a Quadratic Model The data in the table represents the U.S. government's annual net receipts y (in billions of dollars) from individual income taxes for the year x from 1990 through 1992, where $x = 0$ corresponds to 1990. *(Source: U.S. Department of the Treasury)*

x	0	1	2
y	467	468	476

Use a system of three linear equations to find a quadratic model that fits the data. According to your model, what were the annual net receipts from individual income taxes in 1993? The actual annual net receipts were $510 billion. How does the value you obtained compare? Suppose you had been involved in planning the 1993 federal budget and had used this model to estimate how much federal income could be expected from 1993 individual income taxes. When you review the actual 1993 tax receipts and see that the model wasn't completely accurate, how do you evaluate the model's predictive performance? Are you satisfied with it? Why or why not?

4.2   Exercises

See Warm-Up
Exercises, p. A43

In Exercises 1 and 2, determine whether each ordered triple is a solution of the system of linear equations.

1. $\begin{cases} x + 3y + 2z = 1 \\ 5x - y + 3z = 16 \\ -3x + 7y + z = -14 \end{cases}$

(a) $(0, 3, -2)$ (b) $(12, 5, -13)$
(c) $(1, -2, 3)$ (d) $(-2, 5, -3)$

2. $\begin{cases} 3x - y + 4z = -10 \\ -x + y + 2z = 6 \\ 2x - y + z = -8 \end{cases}$

(a) $(-2, 4, 0)$ (b) $(0, -3, 10)$
(c) $(1, -1, 5)$ (d) $(7, 19, -3)$

In Exercises 3–6, use back-substitution to solve the system of linear equations.

3. $\begin{cases} x - 2y + 4z = 4 \\ 3y - z = 2 \\ z = -5 \end{cases}$

4. $\begin{cases} 5x + 4y - z = 0 \\ 10y - 3z = 11 \\ z = 3 \end{cases}$

5. $\begin{cases} x - 2y + 4z = 4 \\ y = 3 \\ y + z = 2 \end{cases}$

6. $\begin{cases} x = 10 \\ 3x + 2y = 2 \\ x + y + 2z = 0 \end{cases}$

7. *Think About It*   Describe a linear system of equations that is in row-echelon form.

8. Are the following two systems of linear equations equivalent? Give reasons for your answer.

$\begin{cases} x + 3y - z = 6 \\ 2x - y + 2z = 1 \\ 3x + 2y - z = 2 \end{cases}$ $\begin{cases} x + 3y - z = 6 \\ -7y + 4z = 1 \\ -7y - 4z = -16 \end{cases}$

In Exercises 9 and 10, perform the row operation and write the equivalent system of linear equations.

9. Add Equation 1 to Equation 2.

$\begin{cases} x - 2y + 3z = 5 & \text{Equation 1} \\ -x + 3y - 5z = 4 & \text{Equation 2} \\ 2x - 3z = 0 & \text{Equation 3} \end{cases}$

What did this operation accomplish?

10. Add -2 times Equation 1 to Equation 3.

$\begin{cases} x - 2y + 3z = 5 & \text{Equation 1} \\ -x + 3y - 5z = 4 & \text{Equation 2} \\ 2x - 3z = 0 & \text{Equation 3} \end{cases}$

What did this operation accomplish?

In Exercises 11–36, use one or more of the methods illustrated in this section to solve the system of linear equations.

11. $\begin{cases} x + z = 4 \\ y = 2 \\ 4x + z = 7 \end{cases}$

12. $\begin{cases} x = 3 \\ -x + 3y = 3 \\ y + 2z = 4 \end{cases}$

13. $\begin{cases} x + y + z = 6 \\ 2x - y + z = 3 \\ 3x - z = 0 \end{cases}$

14. $\begin{cases} x + y + z = 2 \\ -x + 3y + 2z = 8 \\ 4x + y = 4 \end{cases}$

15. $\begin{cases} x + y + z = -3 \\ 4x + y - 3z = 11 \\ 2x - 3y + 2z = 9 \end{cases}$

16. $\begin{cases} x - y + 2z = -4 \\ 3x + y - 4z = -6 \\ 2x + 3y - 4z = 4 \end{cases}$

17. $\begin{cases} x + 2y + 6z = 5 \\ -x + y - 2z = 3 \\ x - 4y - 2z = 1 \end{cases}$

18. $\begin{cases} x + 6y + 2z = 9 \\ 3x - 2y + 3z = -1 \\ 5x - 5y + 2z = 7 \end{cases}$

19. $\begin{cases} 2x + 2z = 2 \\ 5x + 3y = 4 \\ 3y - 4z = 4 \end{cases}$

20. $\begin{cases} 6y + 4z = -12 \\ 3x + 3y = 9 \\ 2x - 3z = 10 \end{cases}$

21. $\begin{cases} x + y + 8z = 3 \\ 2x + y + 11z = 4 \\ x + 3z = 0 \end{cases}$ **22.** $\begin{cases} 2x + y + 3z = 1 \\ 2x + 6y + 8z = 3 \\ 6x + 8y + 18z = 5 \end{cases}$

23. $\begin{cases} 2x - 4y + z = 0 \\ 3x + 2z = -1 \\ -6x + 3y + 2z = -10 \end{cases}$

24. $\begin{cases} 3x - y - 2z = 5 \\ 2x + y + 3z = 6 \\ 6x - y - 4z = 9 \end{cases}$

25. $\begin{cases} y + z = 5 \\ 2x + 4z = 4 \\ 2x - 3y = -14 \end{cases}$

26. $\begin{cases} 5x + 2y = -8 \\ z = 5 \\ 3x - y + z = 9 \end{cases}$

27. $\begin{cases} 2x + 6y - 4z = 8 \\ 3x + 10y - 7z = 12 \\ -2x - 6y + 5z = -3 \end{cases}$

28. $\begin{cases} x + 2y - 2z = 4 \\ 2x + 5y - 7z = 5 \\ 3x + 7y - 9z = 10 \end{cases}$

29. $\begin{cases} x - 2y - z = 3 \\ 2x + y - 3z = 1 \\ x + 8y - 3z = -7 \end{cases}$

30. $\begin{cases} 2x + y - z = 4 \\ y + 3z = 2 \\ 3x + 2y = 4 \end{cases}$

31. $\begin{cases} 3x + y + z = 2 \\ 4x + 2z = 1 \\ 5x - y + 3z = 0 \end{cases}$

32. $\begin{cases} 2x + 3z = 4 \\ 5x + y + z = 2 \\ 11x + 3y - 3z = 0 \end{cases}$

33. $\begin{cases} 0.2x + 1.3y + 0.6z = 0.1 \\ 0.1x + 0.3z = 0.7 \\ 2x + 10y + 8z = 8 \end{cases}$

34. $\begin{cases} 0.3x - 0.1y + 0.2z = 0.35 \\ 2x + y - 2z = -1 \\ 2x + 4y + 3z = 10.5 \end{cases}$

35. $\begin{cases} x + 4y - 2z = 2 \\ -3x + y + z = -2 \\ 5x + 7y - 5z = 6 \end{cases}$

36. $\begin{cases} 2x + z = 1 \\ 5y - 3z = 2 \\ 6x + 20y - 9z = 11 \end{cases}$

Think About It In Exercises 37 and 38, find a system of linear equations in three variables with integer coefficients that has the given point as a solution. (*Note:* There are many correct answers.)

37. $(4, -3, 2)$ **38.** $(5, 7, -10)$

Curve-Fitting In Exercises 39–44, find the equation of the parabola $y = ax^2 + bx + c$ that passes through the points.

39. $(0, -4), (1, 1), (2, 10)$ **40.** $(0, 5), (1, 6), (2, 5)$

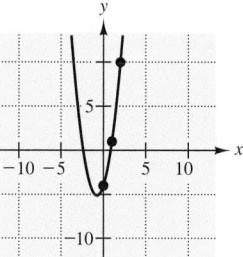 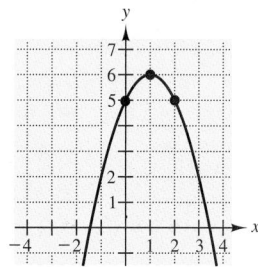

41. $(1, 0), (2, -1), (3, 0)$ **42.** $(1, 2), (2, 1), (3, -4)$

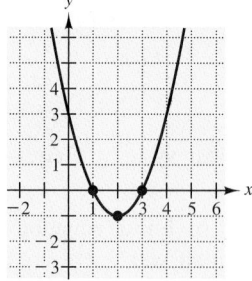 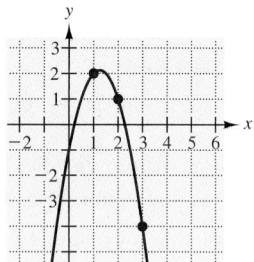

43. $(-1, -3), (1, 1), (2, 0)$
44. $(-1, -1), (1, 1), (2, -4)$

Curve-Fitting In Exercises 45–50, find the equation of the circle $x^2 + y^2 + Dx + Ey + F = 0$ that passes through the points.

45. $(0, 0), (2, -2), (4, 0)$

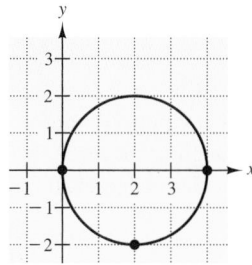

46. $(0, 0), (0, 6), (-3, 3)$

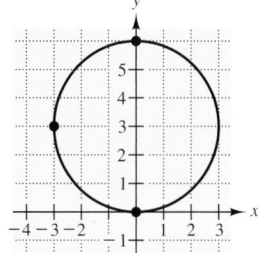

47. $(3, -1), (-2, 4), (6, 8)$

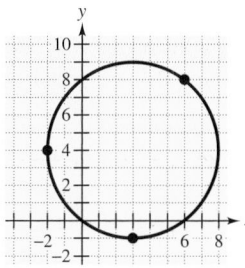

48. $(0, 0), (0, 2), (3, 0)$

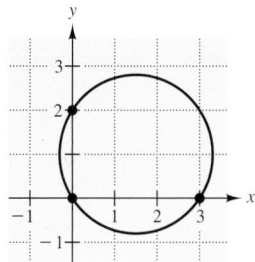

49. $(-3, 5), (4, 6), (5, 5)$

50. $(5, 13), (17, 5), (10, 12)$

Vertical Motion In Exercises 51–54, find the position equation $s = \frac{1}{2}at^2 + v_0 t + s_0$ for an object that has the indicated heights at the specified times.

51. $s = 128$ feet at $t = 1$ second
   $s = 80$ feet at $t = 2$ seconds
   $s = 0$ feet at $t = 3$ seconds

52. $s = 48$ feet at $t = 1$ second
   $s = 64$ feet at $t = 2$ seconds
   $s = 48$ feet at $t = 3$ seconds

53. $s = 32$ feet at $t = 1$ second
   $s = 32$ feet at $t = 2$ seconds
   $s = 0$ feet at $t = 3$ seconds

54. $s = 10$ feet at $t = 0$ second
   $s = 54$ feet at $t = 1$ second
   $s = 46$ feet at $t = 3$ seconds

55. *Crop Spraying* A mixture of 12 gallons of chemical A, 16 gallons of chemical B, and 26 gallons of chemical C is required to kill a certain destructive crop insect. Commercial spray X contains 1, 2, and 2 parts of these chemicals. Spray Y contains only chemical C. Spray Z contains only chemicals A and B in equal amounts. How much of each type of commercial spray is needed to get the desired mixture?

56. *Investigation* The total numbers of sides and diagonals of regular polygons with three, four, and five sides are three, six, and ten, as shown in the figure. Find the function $y = ax^2 + bx + c$ that fits the data. Then check to see if it gives the correct answer for a polygon with six sides.

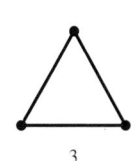 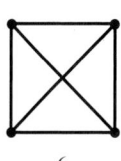 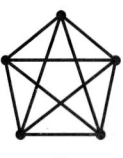 

   3   6   10   15

57. *School Orchestra* The table shows the percents of each section of the North High School orchestra that were chosen to participate in the city orchestra, the county orchestra, and the state orchestra. Thirty members of the city orchestra, 17 members of the county orchestra, and 10 members of the state orchestra are from North High. How many members are in each section of North High's orchestra?

Orchestra	String	Wind	Percussion
City orchestra	40%	30%	50%
County orchestra	20%	25%	25%
State orchestra	10%	15%	25%

Section Project

Chemistry A chemist needs 10 liters of a 25% acid solution. It is to be mixed from three solutions whose concentrations are 10%, 20%, and 50%. How many liters of each solution will satisfy the following?

(a) Use 2 liters of the 50% solution.

(b) Use as little as possible of the 50% solution.

(c) Use as much as possible of the 50% solution.

4.3 **Matrices and Linear Systems**

Matrices ▪ Elementary Row Operations ▪
Using Matrices to Solve a System of Linear Equations

Matrices

In this section, you will study a streamlined technique for solving systems of linear equations. This technique involves the use of a rectangular array of real numbers called a **matrix.** (The plural of matrix is *matrices.*) Here is an example of a matrix.

$$\begin{bmatrix} 3 & -2 & 4 & 1 \\ 0 & 1 & -1 & 2 \\ 2 & 0 & -3 & 0 \end{bmatrix}$$

Students may initially confuse rows with columns. Ask them to think of the (vertical) columns in front of the Parthenon.

This matrix has three rows and four columns, which means that its **order** is 3×4, which is read as "3 by 4." Each number in the matrix is an **entry** of the matrix.

EXAMPLE 1   Examples of Matrices

The following matrices have the indicated orders.

(a) Order: 2×3

$$\begin{bmatrix} 1 & -2 & 4 \\ 0 & 1 & -2 \end{bmatrix}$$

(b) Order: 2×2

$$\begin{bmatrix} 0 & 0 \\ 0 & 0 \end{bmatrix}$$

(c) Order: 3×2

$$\begin{bmatrix} 1 & -3 \\ -2 & 0 \\ 4 & -2 \end{bmatrix}$$

A matrix with the same number of rows as columns is called a **square matrix.** For instance, the 2×2 matrix in part (b) is square.

A matrix derived from a system of linear equations (each written in standard form) is the **augmented matrix** of the system. Moreover, the matrix that is derived from the coefficients of the system (but that does not include the constant terms) is the **coefficient matrix** of the system. Here is an example.

System	Coefficient Matrix	Augmented Matrix

$$\begin{cases} x - 4y + 3z = 5 \\ -x + 3y - z = -3 \\ 2x - 4z = 6 \end{cases} \qquad \begin{bmatrix} 1 & -4 & 3 \\ -1 & 3 & -1 \\ 2 & 0 & -4 \end{bmatrix} \qquad \begin{bmatrix} 1 & -4 & 3 & \vdots & 5 \\ -1 & 3 & -1 & \vdots & -3 \\ 2 & 0 & -4 & \vdots & 6 \end{bmatrix}$$

Note the use of 0 for the missing y-variable in the third equation, and also note the fourth column of constant terms in the augmented matrix.

When forming either the coefficient matrix or the augmented matrix of a system, you should begin by vertically aligning the variables in the equations.

System	*Align Variables*	*Form Augmented Matrix*

$$\begin{cases} x + 3y = 9 \\ -y + 4z = -2 \\ x - 5z = 0 \end{cases} \qquad \begin{aligned} x + 3y &= 9 \\ -y + 4z &= -2 \\ x - 5z &= 0 \end{aligned} \qquad \left[\begin{array}{ccc:c} 1 & 3 & 0 & 9 \\ 0 & -1 & 4 & -2 \\ 1 & 0 & -5 & 0 \end{array}\right]$$

EXAMPLE 2   *Forming Coefficient and Augmented Matrices*

Form the coefficient matrix and the augmented matrix for each system of linear equations.

(a) $\begin{cases} -x + 5y = 2 \\ 7x - 2y = -6 \end{cases}$ (b) $\begin{cases} 3x + 2y - z = 1 \\ x + 2z = -3 \\ -2x - y = 4 \end{cases}$ (c) $\begin{cases} x = 3y - 1 \\ 2y - 5 = 9x \end{cases}$

Solution

System	*Coefficient Matrix*	*Augmented Matrix*

(a) $\begin{cases} -x + 5y = 2 \\ 7x - 2y = -6 \end{cases}$ $\begin{bmatrix} -1 & 5 \\ 7 & -2 \end{bmatrix}$ $\left[\begin{array}{cc:c} -1 & 5 & 2 \\ 7 & -2 & -6 \end{array}\right]$

(b) $\begin{cases} 3x + 2y - z = 1 \\ x + 2z = -3 \\ -2x - y = 4 \end{cases}$ $\begin{bmatrix} 3 & 2 & -1 \\ 1 & 0 & 2 \\ -2 & -1 & 0 \end{bmatrix}$ $\left[\begin{array}{ccc:c} 3 & 2 & -1 & 1 \\ 1 & 0 & 2 & -3 \\ -2 & -1 & 0 & 4 \end{array}\right]$

(c) $\begin{cases} x - 3y = -1 \\ -9x + 2y = 5 \end{cases}$ $\begin{bmatrix} 1 & -3 \\ -9 & 2 \end{bmatrix}$ $\left[\begin{array}{cc:c} 1 & -3 & -1 \\ -9 & 2 & 5 \end{array}\right]$

Be sure students understand that all equations must be written in standard form before they are translated to the augmented matrix.

EXAMPLE 3   *Forming Linear Systems from Their Matrices*

Write the system of linear equations that is represented by each matrix.

(a) $\left[\begin{array}{cc:c} 3 & -5 & 4 \\ -1 & 2 & 0 \end{array}\right]$ (b) $\left[\begin{array}{cc:c} 1 & 3 & 2 \\ 0 & 1 & -3 \end{array}\right]$ (c) $\left[\begin{array}{ccc:c} 2 & 0 & -8 & 1 \\ -1 & 1 & 1 & 2 \\ 5 & -1 & 7 & 3 \end{array}\right]$

Solution

(a) $\begin{cases} 3x - 5y = 4 \\ -x + 2y = 0 \end{cases}$ (b) $\begin{cases} x + 3y = 2 \\ y = -3 \end{cases}$ (c) $\begin{cases} 2x - 8z = 1 \\ -x + y + z = 2 \\ 5x - y + 7z = 3 \end{cases}$

Elementary Row Operations

In Section 4.2, you studied three operations that can be used on a system of linear equations to produce an equivalent system: (1) interchange two equations, (2) multiply an equation by a nonzero constant, and (3) add a multiple of an equation to another equation. In matrix terminology, these three operations correspond to **elementary row operations.**

Elementary Row Operations

Any of the following **elementary row operations** performed on an augmented matrix will produce a matrix that is row-equivalent to the original matrix. Two matrices are **row-equivalent** if one can be obtained from the other by a sequence of elementary row operations.

1. Interchange two rows.

2. Multiply a row by a nonzero constant.

3. Add a multiple of a row to another row.

EXAMPLE 4   Elementary Row Operations

(a) Interchange the first and second rows.

Original Matrix $\qquad\qquad\qquad\qquad$ *New Row-Equivalent Matrix*

$$
\begin{bmatrix} 0 & 1 & 3 & 4 \\ -1 & 2 & 0 & 3 \\ 2 & -3 & 4 & 1 \end{bmatrix}
\qquad
\begin{matrix} R_2 \\ R_1 \end{matrix}
\begin{bmatrix} -1 & 2 & 0 & 3 \\ 0 & 1 & 3 & 4 \\ 2 & -3 & 4 & 1 \end{bmatrix}
$$

(b) Multiply the first row by $\frac{1}{2}$.

Original Matrix $\qquad\qquad\qquad\qquad$ *New Row-Equivalent Matrix*

$$
\begin{bmatrix} 2 & -4 & 6 & -2 \\ 1 & 3 & -3 & 0 \\ 5 & -2 & 1 & 2 \end{bmatrix}
\qquad
\tfrac{1}{2}R_1 \rightarrow
\begin{bmatrix} 1 & -2 & 3 & -1 \\ 1 & 3 & -3 & 0 \\ 5 & -2 & 1 & 2 \end{bmatrix}
$$

(c) Add -2 times the first row to the third row.

Original Matrix $\qquad\qquad\qquad\qquad$ *New Row-Equivalent Matrix*

$$
\begin{bmatrix} 1 & 2 & -4 & 3 \\ 0 & 3 & -2 & -1 \\ 2 & 1 & 5 & -2 \end{bmatrix}
\qquad
-2R_1 + R_3 \rightarrow
\begin{bmatrix} 1 & 2 & -4 & 3 \\ 0 & 3 & -2 & -1 \\ 0 & -3 & 13 & -8 \end{bmatrix}
$$

NOTE In Section 4.2, Gaussian elimination was used with back-substitution to solve a system of linear equations. Example 5 demonstrates the matrix version of Gaussian elimination. The two methods are essentially the same. The basic difference is that with matrices you do not need to keep writing the variables.

EXAMPLE 5 *Solving a System of Linear Equations*

Linear System

$$\begin{cases} x - 2y + 2z = 9 \\ -x + 3y \quad\quad = -4 \\ 2x - 5y + z = 10 \end{cases}$$

Associated Augmented Matrix

$$\begin{bmatrix} 1 & -2 & 2 & \vdots & 9 \\ -1 & 3 & 0 & \vdots & -4 \\ 2 & -5 & 1 & \vdots & 10 \end{bmatrix}$$

Add the first equation to the second equation.

$$\begin{cases} x - 2y + 2z = 9 \\ y + 2z = 5 \\ 2x - 5y + z = 10 \end{cases}$$

Add the first row to the second row $(R_1 + R_2)$.

$$R_1 + R_2 \rightarrow \begin{bmatrix} 1 & -2 & 2 & \vdots & 9 \\ 0 & 1 & 2 & \vdots & 5 \\ 2 & -5 & 1 & \vdots & 10 \end{bmatrix}$$

Add -2 times the first equation to the third equation.

$$\begin{cases} x - 2y + 2z = 9 \\ y + 2z = 5 \\ -y - 3z = -8 \end{cases}$$

Add -2 times the first row to the third row $(-2R_1 + R_3)$.

$$-2R_1 + R_3 \rightarrow \begin{bmatrix} 1 & -2 & 2 & \vdots & 9 \\ 0 & 1 & 2 & \vdots & 5 \\ 0 & -1 & -3 & \vdots & -8 \end{bmatrix}$$

Add the second equation to the third equation.

$$\begin{cases} x - 2y + 2z = 9 \\ y + 2z = 5 \\ -z = -3 \end{cases}$$

Add the second row to the third row $(R_2 + R_3)$.

$$R_2 + R_3 \rightarrow \begin{bmatrix} 1 & -2 & 2 & \vdots & 9 \\ 0 & 1 & 2 & \vdots & 5 \\ 0 & 0 & -1 & \vdots & -3 \end{bmatrix}$$

Multiply the third equation by -1.

$$\begin{cases} x - 2y + 2z = 9 \\ y + 2z = 5 \\ z = 3 \end{cases}$$

Multiply the third row by -1.

$$-R_3 \rightarrow \begin{bmatrix} 1 & -2 & 2 & \vdots & 9 \\ 0 & 1 & 2 & \vdots & 5 \\ 0 & 0 & 1 & \vdots & 3 \end{bmatrix}$$

Now use back-substitution to find that the solution is $x = 1$, $y = -1$, and $z = 3$.

NOTE The last matrix in Example 5 is in **row-echelon form**. The term *echelon* refers to the stair-step pattern formed by the nonzero elements of the matrix.

Definition of Row-Echelon Form of a Matrix

A matrix in **row-echelon form** has the following properties.

1. All rows consisting entirely of zeros occur at the bottom of the matrix.

2. For each row that does not consist entirely of zeros, the first nonzero entry is 1 (called a **leading 1**).

3. For two successive (nonzero) rows, the leading 1 in the higher row is farther to the left than the leading 1 in the lower row.

Using Matrices to Solve a System of Linear Equations

Gaussian Elimination with Back-Substitution

To use matrices and Gaussian elimination to solve a system of linear equations, use the following steps.

1. Write the augmented matrix of the system of linear equations.

2. Use elementary row operations to rewrite the augmented matrix in row-echelon form.

3. Write the system of linear equations corresponding to the matrix in row-echelon form, and use back-substitution to find the solution.

When you perform Gaussian elimination with back-substitution, we suggest that you operate from *left to right by columns,* using elementary row operations to obtain zeros in all entries directly below the leading 1's.

EXAMPLE 6   *Gaussian Elimination with Back-Substitution*

Solve the following system of linear equations.

$$\begin{cases} 2x - 3y = -2 \\ x + 2y = 13 \end{cases}$$

Solution

$$\begin{bmatrix} 2 & -3 & \vdots & -2 \\ 1 & 2 & \vdots & 13 \end{bmatrix}$$ Augmented matrix for system of linear equations

$$\begin{matrix} R_2 \\ R_1 \end{matrix} \begin{bmatrix} 1 & 2 & \vdots & 13 \\ 2 & -3 & \vdots & -2 \end{bmatrix}$$ First column has leading 1 in upper left corner.

$$-2R_1 + R_2 \rightarrow \begin{bmatrix} 1 & 2 & \vdots & 13 \\ 0 & -7 & \vdots & -28 \end{bmatrix}$$ First column has a zero under its leading 1.

$$-\tfrac{1}{7}R_2 \rightarrow \begin{bmatrix} 1 & 2 & \vdots & 13 \\ 0 & 1 & \vdots & 4 \end{bmatrix}$$ Second column has leading 1 in second row.

The system of linear equations that corresponds to the (row-echelon) matrix is

$$\begin{cases} x + 2y = 13 \\ y = 4. \end{cases}$$

Using back-substitution, you can find that the solution of the system is $x = 5$ and $y = 4$. Check this solution in the original system of linear equations.

EXAMPLE 7   *Gaussian Elimination with Back-Substitution*

Solve the following system of linear equations.

$$\begin{cases} 3x + 3y = 9 \\ 2x - 3z = 10 \\ 6y + 4z = -12 \end{cases}$$

Solution

$$\begin{bmatrix} 3 & 3 & 0 & \vdots & 9 \\ 2 & 0 & -3 & \vdots & 10 \\ 0 & 6 & 4 & \vdots & -12 \end{bmatrix}$$ Augmented matrix for system of linear equations

$\frac{1}{3}R_1 \rightarrow \begin{bmatrix} 1 & 1 & 0 & \vdots & 3 \\ 2 & 0 & -3 & \vdots & 10 \\ 0 & 6 & 4 & \vdots & -12 \end{bmatrix}$ First column has leading 1 in upper left corner.

$-2R_1 + R_2 \rightarrow \begin{bmatrix} 1 & 1 & 0 & \vdots & 3 \\ 0 & -2 & -3 & \vdots & 4 \\ 0 & 6 & 4 & \vdots & -12 \end{bmatrix}$ First column has zeros under its leading 1.

$-\frac{1}{2}R_2 \rightarrow \begin{bmatrix} 1 & 1 & 0 & \vdots & 3 \\ 0 & 1 & \frac{3}{2} & \vdots & -2 \\ 0 & 6 & 4 & \vdots & -12 \end{bmatrix}$ Second column has leading 1 in second row.

$-6R_2 + R_3 \rightarrow \begin{bmatrix} 1 & 1 & 0 & \vdots & 3 \\ 0 & 1 & \frac{3}{2} & \vdots & -2 \\ 0 & 0 & -5 & \vdots & 0 \end{bmatrix}$ Second column has a zero under its leading 1.

$-\frac{1}{5}R_3 \rightarrow \begin{bmatrix} 1 & 1 & 0 & \vdots & 3 \\ 0 & 1 & \frac{3}{2} & \vdots & -2 \\ 0 & 0 & 1 & \vdots & 0 \end{bmatrix}$ Third column has leading 1 in third row.

The system of linear equations that corresponds to the (row-echelon) matrix is

$$\begin{cases} x + y \phantom{+ \frac{3}{2}z} = 3 \\ y + \frac{3}{2}z = -2 \\ z = 0. \end{cases}$$

Using back-substitution, you can find that the solution is

$$x = 5, \quad y = -2, \quad \text{and} \quad z = 0.$$

Check this in the original system, as follows.

$$\begin{array}{rcl} 3(5) + 3(-2) & = & 9 \quad \checkmark \\ 2(5) - 3(0) & = & 10 \quad \checkmark \\ 6(-2) + 4(0) & = & -12 \quad \checkmark \end{array}$$

EXAMPLE 8 A System with No Solution

Solve the following system of linear equations.

$$\begin{cases} 6x - 10y = -4 \\ 9x - 15y = 5 \end{cases}$$

Solution

$$\begin{bmatrix} 6 & -10 & \vdots & -4 \\ 9 & -15 & \vdots & 5 \end{bmatrix}$$ Augmented matrix for system of linear equations

$\frac{1}{6}R_1 \rightarrow$ $$\begin{bmatrix} 1 & -\frac{5}{3} & \vdots & -\frac{2}{3} \\ 9 & -15 & \vdots & 5 \end{bmatrix}$$ First column has leading 1 in upper left corner.

$-9R_1 + R_2 \rightarrow$ $$\begin{bmatrix} 1 & -\frac{5}{3} & \vdots & -\frac{2}{3} \\ 0 & 0 & \vdots & 11 \end{bmatrix}$$ First column has a zero under its leading 1.

The "equation" that corresponds to the second row of the matrix is $0 = 11$. Because this is a false statement, the system of equations has no solution.

EXAMPLE 9 A System with Infinitely Many Solutions

Solve the following system of linear equations.

$$\begin{cases} 12x - 6y = -3 \\ -8x + 4y = 2 \end{cases}$$

Solution

$$\begin{bmatrix} 12 & -6 & \vdots & -3 \\ -8 & 4 & \vdots & 2 \end{bmatrix}$$ Augmented matrix for system of linear equations.

$\frac{1}{12}R_1 \rightarrow$ $$\begin{bmatrix} 1 & -\frac{1}{2} & \vdots & -\frac{1}{4} \\ -8 & 4 & \vdots & 2 \end{bmatrix}$$ First column has leading 1 in upper left corner.

$8R_1 + R_2 \rightarrow$ $$\begin{bmatrix} 1 & -\frac{1}{2} & \vdots & -\frac{1}{4} \\ 0 & 0 & \vdots & 0 \end{bmatrix}$$ First column has a zero under its leading 1.

Because the second row of the matrix is all zeros, you can conclude that the system of equations has an infinite number of solutions, represented by all points (x, y) on the line $x - \frac{1}{2}y = -\frac{1}{4}$. Because this line can be written as

$$x = -\frac{1}{4} + \frac{1}{2}y,$$

you can write the solution set as

$$\left(-\frac{1}{4} + \frac{1}{2}a, a\right), \quad \text{where } a \text{ is any real number.}$$

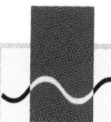

Technology

Graphing Utilities and Matrices

Most graphing utilities are capable of working with matrices. For example, to write the matrix shown at the left in row-echelon form using a *TI-82* or *TI-83*, use the following steps.

$$\begin{bmatrix} 1 & -2 & 2 & \vdots & 9 \\ -1 & 3 & 0 & \vdots & -4 \\ 2 & -5 & 1 & \vdots & 10 \end{bmatrix}$$

1. Use MATRX (EDIT) to enter the matrix.

2. To write the matrix in row-echelon form, use the elementary row operations that appear in the MATRX (MATH) menu, as follows.

row+([A],1,2) ENTER	Add row 1 to row 2.
STO▶ [A] ENTER	Store the result as matrix A.
*row+(−2,[A],1,3) ENTER	Multiply row 1 by −2 and add to row 3.
STO▶ [A] ENTER	Store the result as matrix A.
row+([A],2,3) ENTER	Add row 2 to row 3.
STO▶ [A] ENTER	Store the result as matrix A.
*row(−1,[A],3) ENTER	Multiply row 3 by −1.
STO▶[A] ENTER	Store the result as matrix A.

3. The result that is displayed on the screen should be as follows.

$$\begin{bmatrix} [\; 1 & -2 & 2 & 9 \;] \\ [\; 0 & 1 & 2 & 5 \;] \\ [\; 0 & 0 & 1 & 3 \;] \end{bmatrix}$$

At this point, you can use back-substitution to solve the corresponding system of equations.

Group Activities

Exploring with Technology

Analyzing Solutions to Systems of Equations Use a graphing utility to graph each system of equations given in Example 6, Example 8, and Example 9. Verify the solution given in each example and explain how you may reach the same conclusion by using the graph. Summarize how you may conclude that a system has a unique solution, no solution, or infinitely many solutions when you use Gaussian elimination.

4.3 Exercises

See Warm-Up Exercises, p. A43

In Exercises 1–4, determine the order of the matrix.

1. $\begin{bmatrix} 3 & -2 \\ -4 & 0 \\ 2 & -7 \end{bmatrix}$ **2.** $\begin{bmatrix} 4 & 0 & -5 \\ -1 & 8 & 9 \\ 0 & -3 & 4 \end{bmatrix}$

3. $\begin{bmatrix} 5 & -8 & 32 \\ 7 & 15 & 28 \end{bmatrix}$ **4.** $\begin{bmatrix} -2 & 5 \\ 0 & -1 \end{bmatrix}$

In Exercises 5–8, form the augmented matrix for the system of linear equations.

5. $\begin{cases} 4x - 5y = -2 \\ -x + 8y = 10 \end{cases}$ **6.** $\begin{cases} 8x + 3y = 25 \\ 3x - 9y = 12 \end{cases}$

7. $\begin{cases} x + 10y - 3z = 2 \\ 5x - 3y + 4z = 0 \\ 2x + 4y = 6 \end{cases}$ **8.** $\begin{cases} 9x - 3y + z = 13 \\ 12x - 8z = 5 \end{cases}$

In Exercises 9–12, write the system of linear equations represented by the augmented matrix. (Use variables *x*, *y*, *z*, and *w*.)

9. $\begin{bmatrix} 4 & 3 & \vdots & 8 \\ 1 & -2 & \vdots & 3 \end{bmatrix}$

10. $\begin{bmatrix} 9 & -4 & \vdots & 0 \\ 6 & 1 & \vdots & -4 \end{bmatrix}$

11. $\begin{bmatrix} 1 & 0 & 2 & \vdots & -10 \\ 0 & 3 & -1 & \vdots & 5 \\ 4 & 2 & 0 & \vdots & 3 \end{bmatrix}$

12. $\begin{bmatrix} 5 & 8 & 2 & 0 & \vdots & -1 \\ -2 & 15 & 5 & 1 & \vdots & 9 \\ 1 & 6 & -7 & 0 & \vdots & -3 \end{bmatrix}$

In Exercises 13–16, fill in the blank(s) by using elementary row operations to form a row-equivalent matrix.

13. $\begin{bmatrix} 1 & 4 & 3 \\ 2 & 10 & 5 \end{bmatrix}$ **14.** $\begin{bmatrix} 3 & 6 & 8 \\ 4 & -3 & 6 \end{bmatrix}$

$\begin{bmatrix} 1 & 4 & 3 \\ 0 & \blacksquare & -1 \end{bmatrix}$ $\begin{bmatrix} 3 & 6 & 8 \\ 1 & -9 & \blacksquare \end{bmatrix}$

15. $\begin{bmatrix} 1 & 1 & 4 & -1 \\ 3 & 8 & 10 & 3 \\ -2 & 1 & 12 & 6 \end{bmatrix}$

$\begin{bmatrix} 1 & 1 & 4 & -1 \\ 0 & 5 & \blacksquare & \blacksquare \\ 0 & 3 & \blacksquare & \blacksquare \end{bmatrix}$

$\begin{bmatrix} 1 & 1 & 4 & -1 \\ 0 & 1 & & \\ 0 & 3 & 20 & 4 \end{bmatrix}$

16. $\begin{bmatrix} 2 & 4 & 8 & 3 \\ 1 & -1 & -3 & 2 \\ 2 & 6 & 4 & 9 \end{bmatrix}$

$\begin{bmatrix} 1 & \blacksquare & \blacksquare & \blacksquare \\ 1 & -1 & -3 & 2 \\ 2 & 6 & 4 & 9 \end{bmatrix}$

$\begin{bmatrix} 1 & 2 & 4 & \frac{3}{2} \\ 0 & \blacksquare & -7 & \frac{1}{2} \\ 0 & 2 & \blacksquare & \blacksquare \end{bmatrix}$

In Exercises 17–22, convert the matrix to row-echelon form. (*Note:* There is more than one correct answer.)

17. $\begin{bmatrix} 1 & 2 & 3 \\ 2 & -1 & -4 \end{bmatrix}$ **18.** $\begin{bmatrix} 1 & 3 & 6 \\ -4 & -9 & 3 \end{bmatrix}$

19. $\begin{bmatrix} 4 & 6 & 1 \\ -2 & 2 & 5 \end{bmatrix}$

20. $\begin{bmatrix} 3 & 2 & 6 \\ 2 & 3 & -3 \end{bmatrix}$

21. $\begin{bmatrix} 1 & 1 & 0 & 5 \\ -2 & -1 & 2 & -10 \\ 3 & 6 & 7 & 14 \end{bmatrix}$

22. $\begin{bmatrix} 1 & 2 & -1 & 3 \\ 3 & 7 & -5 & 14 \\ -2 & -1 & -3 & 8 \end{bmatrix}$

In Exercises 23–26, use the matrix capabilities of a graphing utility to write the matrix in row-echelon form. (*Note:* There is more than one correct answer.)

23. $\begin{bmatrix} 1 & -1 & -1 & 1 \\ 4 & -4 & 1 & 8 \\ -6 & 8 & 18 & 0 \end{bmatrix}$

24. $\begin{bmatrix} 1 & -3 & 0 & -7 \\ -3 & 10 & 1 & 23 \\ 4 & -10 & 2 & -24 \end{bmatrix}$

25. $\begin{bmatrix} 1 & 1 & -1 & 3 \\ 2 & 1 & 2 & 5 \\ 3 & 2 & 1 & 8 \end{bmatrix}$ **26.** $\begin{bmatrix} 1 & -3 & -2 & -8 \\ 1 & 3 & -2 & 17 \\ 1 & 2 & -2 & -5 \end{bmatrix}$

In Exercises 27–30, write the system of linear equations represented by the augmented matrix. Then use back-substitution to find the solution. (Use variables x, y, and z.)

27. $\begin{bmatrix} 1 & -2 & \vdots & 4 \\ 0 & 1 & \vdots & -3 \end{bmatrix}$

28. $\begin{bmatrix} 1 & 5 & \vdots & 0 \\ 0 & 1 & \vdots & -1 \end{bmatrix}$

29. $\begin{bmatrix} 1 & -1 & 2 & \vdots & 4 \\ 0 & 1 & -1 & \vdots & 2 \\ 0 & 0 & 1 & \vdots & -2 \end{bmatrix}$

30. $\begin{bmatrix} 1 & 2 & -2 & \vdots & -1 \\ 0 & 1 & 1 & \vdots & 9 \\ 0 & 0 & 1 & \vdots & -3 \end{bmatrix}$

In Exercises 31–44, use matrices to solve the system of linear equations.

31. $\begin{cases} x + 2y = 7 \\ 3x + y = 8 \end{cases}$ **32.** $\begin{cases} 2x + 6y = 16 \\ 2x + 3y = 7 \end{cases}$

33. $\begin{cases} 6x - 4y = 2 \\ 5x + 2y = 7 \end{cases}$ **34.** $\begin{cases} 2x - y = -0.1 \\ 3x + 2y = 1.6 \end{cases}$

35. $\begin{cases} -x + 2y = 1.5 \\ 2x - 4y = 3 \end{cases}$ **36.** $\begin{cases} x - 3y = 5 \\ -2x + 6y = -10 \end{cases}$

37. $\begin{cases} x - 2y - z = 6 \\ y + 4z = 5 \\ 4x + 2y + 3z = 8 \end{cases}$ **38.** $\begin{cases} x - 3z = -2 \\ 3x + y - 2z = 5 \\ 2x + 2y + z = 4 \end{cases}$

39. $\begin{cases} x + y - 5z = 3 \\ x - 2z = 1 \\ 2x - y - z = 0 \end{cases}$ **40.** $\begin{cases} 2y + z = 3 \\ -4y - 2z = 0 \\ x + y + z = 2 \end{cases}$

41. $\begin{cases} 2x + 4y = 10 \\ 2x + 2y + 3z = 3 \\ -3x + y + 2z = -3 \end{cases}$

42. $\begin{cases} 2x - y + 3z = 24 \\ 2y - z = 14 \\ 7x - 5y = 6 \end{cases}$

43. $\begin{cases} x - 3y + 2z = 8 \\ 2y - z = -4 \\ x + z = 3 \end{cases}$

44. $\begin{cases} 2x + 3z = 3 \\ 4x - 3y + 7z = 5 \\ 8x - 9y + 15z = 9 \end{cases}$

In Exercises 45–50, use the matrix capabilities of a graphing utility to solve the system of equations.

45. $\begin{cases} -2x - 2y - 15z = 0 \\ x + 2y + 2z = 18 \\ 3x + 3y + 22z = 2 \end{cases}$

46. $\begin{cases} 2x + 4y + 5z = 5 \\ x + 3y + 3z = 2 \\ 2x + 4y + 4z = 2 \end{cases}$

47. $\begin{cases} 2x + 4z = 1 \\ x + y + 3z = 0 \\ x + 3y + 5z = 0 \end{cases}$

48. $\begin{cases} 3x + y - 2z = 2 \\ 6x + 2y - 4z = 1 \\ -3x - y + 2z = 1 \end{cases}$

49. $\begin{cases} x + 3y = 2 \\ 2x + 6y = 4 \\ 2x + 5y + 4z = 3 \end{cases}$

50. $\begin{cases} 2x + 2y + z = 8 \\ 2x + 3y + z = 7 \\ 6x + 8y + 3z = 22 \end{cases}$

51. *Think About It*   Describe the row-echelon form of an augmented matrix that corresponds to a system of linear equations that is inconsistent.

52. *Think About It*   Describe the row-echelon form of an augmented matrix that corresponds to a system of linear equations that has an infinite number of solutions.

53. *Simple Interest*   A corporation borrowed $1,500,000 to expand its product line. Some of the money was borrowed at 8%, some at 9%, and the remainder at 12%. The annual interest payment to the lenders was $133,000. If the amount borrowed at 8% was 4 times the amount borrowed at 12%, how much was borrowed at each rate?

54. *Investments*   An inheritance of $16,000 was divided among three investments yielding a total of $990 in simple interest per year. The interest rates for the three investments were 5%, 6%, and 7%. Find the amount placed in each investment if the 5% and 6% investments were $3000 and $2000 less than the 7% investment.

Curve-Fitting   In Exercises 55 and 56, find the equation of the parabola $y = ax^2 + bx + c$ that passes through the points.

55.

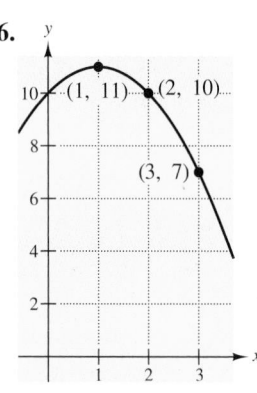

56.

57. *Think About It*   What is meant when it is said that two augmented matrices are row-equivalent?

58. *Think About It*   Give an example of a matrix in row-echelon form.

59. *Mathematical Modeling*   A videotape of the path of a ball thrown by a baseball player was analyzed on a television set with a grid covering the screen (see figure). The tape was paused three times and the coordinates of the ball were measured each time. The coordinates were approximately $(0, 6)$, $(25, 18.5)$, and $(50, 26)$. (The x-coordinate is the horizontal distance in feet from the player and the y-coordinate is the height in feet of the ball above the ground.)

(a) Find the equation $y = ax^2 + bx + c$ that passes through the three points.

(b) Use a graphing utility to graph the model in part (a). Use the graph to approximate the maximum height of the ball and the point at which it struck the ground.

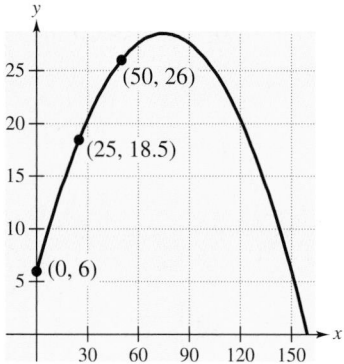

60. *Newsprint Production*   The table gives the number y, in millions of short tons, of newsprint produced in the years 1990 through 1992 in the United States. (*Source: American Paper Institute*)

Year	1990	1991	1992
y	6.6	6.8	7.1

(a) Find the equation $y = at^2 + bt + c$ that passes through the three points, with $t = 0$ corresponding to 1990.

(b) Use a graphing utility to graph part (a).

(c) Use the model in part (a) to predict newsprint production in 1999 if the trend continues.

61. *Private Savings* The bar graph gives the gross private savings y (in billions of dollars) in the years 1992 through 1994 in the United States. *(Source: U.S. Bureau of Economic Analysis)*

(a) Estimate the three points from the bar graph. Find the equation $y = at^2 + bt + c$ that passes through the three points, with $t = 2$ corresponding to 1992.

(b) Use a graphing utility to graph the model in part (a).

(c) Use the model in part (a) to predict gross private savings in the year 2000 if the trend continues.

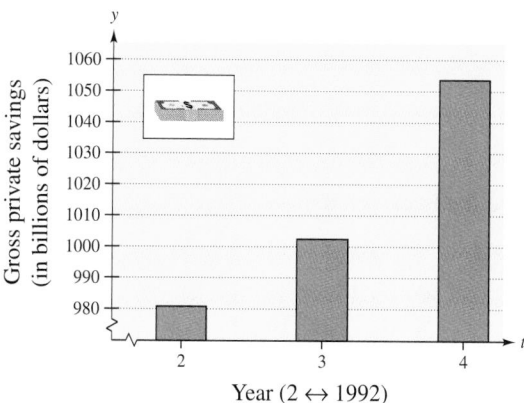

Year (2 ↔ 1992)

62. *Rewriting a Fraction* The fraction

$$\frac{2x^2 - 9x}{(x - 2)^3}$$

can be written as a sum of three fractions as follows.

$$\frac{2x^2 - 9x}{(x - 2)^3} = \frac{A}{x - 2} + \frac{B}{(x - 2)^2} + \frac{C}{(x - 2)^3}$$

The numbers $A, B,$ and C are the solutions of the system

$$\begin{cases} 4A - 2B + C = 0 \\ -4A + B = -9 \\ A = 2. \end{cases}$$

Write the expression as the sum of three fractions.

63. *Rewriting a Fraction* The fraction

$$\frac{1}{(x^3 - x)}$$

can be written as a sum of three fractions as follows.

$$\frac{1}{x^3 - x} = \frac{A}{x} + \frac{B}{x + 1} + \frac{C}{x - 1}$$

The numbers $A, B,$ and C are the solutions of the system

$$\begin{cases} A + B + C = 0 \\ -B + C = 0 \\ -A = 1. \end{cases}$$

Write the expression as the sum of three fractions.

Section Project

Investment Portfolio Consider an investment portfolio totaling $500,000 that is to be allocated among the following types of investments: certificates of deposit, municipal bonds, blue-chip stocks, and growth or speculative stocks. How much should be allocated to each type of investment? If there is more than one solution, determine several possible ways to allocate the investments.

(a) The certificates of deposit pay 10% annually, and the municipal bonds pay 8% annually. Over a 5-year period, the investor expects the blue-chip stocks to return 12% annually, and expects the growth stocks to return 13% annually. The investor wants a combined annual return of 10% and also wants to have only one-fourth of the portfolio invested in stocks.

(b) The certificates of deposit pay 9% annually, and the municipal bonds pay 5% annually. Over a 5-year period, the investor expects the blue-chip stocks to return 12% annually, and expects the growth stocks to return 14% annually. The investor wants a combined annual return of 10% and also wants to have only one-fourth of the portfolio invested in stocks.

Mid-Chapter Quiz

Take this quiz as you would take a quiz in class. After you are done, check your work against the answers given in the back of the book.

In Exercises 1 and 2, solve the system of linear equations graphically.

1. $\begin{cases} x - 2y = 0 \\ 2x + y = 5 \end{cases}$

2. $\begin{cases} y = \frac{1}{3}(1 - 2x) \\ y = \frac{1}{3}(5x - 13) \end{cases}$

In Exercises 3 and 4, solve the system by the method of substitution. Use a graphing utility to check your solution.

3. $\begin{cases} 5x - y = 32 \\ 6x - 9y = 18 \end{cases}$

4. $\begin{cases} 0.2x + 0.7y = 8 \\ -x + 2y = 15 \end{cases}$

In Exercises 5–7, use elimination or Gaussian elimination to solve the linear system.

5. $\begin{cases} x + 10y = 18 \\ 5x + 2y = 42 \end{cases}$

6. $\begin{cases} 3x + 11y = 38 \\ 7x - 5y = -34 \end{cases}$

7. $\begin{cases} x + 4z = 17 \\ -3x + 2y - z = -20 \\ x - 5y + 3z = 19 \end{cases}$

8. Use the matrix capabilities of a graphing utility to solve the system of linear equations.

$\begin{cases} x - 3y + z = -3 \\ 3x + 2y - 5z = 18 \\ y + z = -1 \end{cases}$

In Exercises 9–13, set up a system of linear equations that models the problem. (It is not necessary to solve the system.)

9. The linear system of equations has the unique solution $(10, -12)$. (There are many correct answers.)

10. The linear system of equations has the unique solution $(2, -3, 1)$. (There are many correct answers.)

11. Two people share the driving on a 300-mile trip. One person drives three times as far as the other. Find the distance each person drives.

12. Twenty gallons of a 30% brine solution are obtained by mixing a 20% solution with a 50% solution. Let x represent the number of gallons of the 20% solution and let y represent the number of gallons of the 50% solution. How many gallons of each solution are required?

13. Find the equation $y = ax^2 + bx + c$ that passes through the points $(1, 2)$, $(-1, -4)$, and $(2, 8)$.

4.4	**Determinants and Linear Systems**

The Determinant of a Matrix ▪ Cramer's Rule ▪
Applications of Determinants

The Determinant of a Matrix

Historical Note: An ancient Chinese method of solving systems of equations involved the use of bamboo rods to represent coefficients. In 1683, Seki Kowa (1642–1708), the Japanese mathematician, advanced this method by rearranging the rods in a way that resembles our use of determinant notation today.

Associated with each square matrix is a real number called its **determinant.** The use of determinants arose from special number patterns that occur during the solution of systems of linear equations. For instance, the system

$$\begin{cases} a_1 x + b_1 y = c_1 \\ a_2 x + b_2 y = c_2 \end{cases}$$

has a solution given by

$$x = \frac{c_1 b_2 - c_2 b_1}{a_1 b_2 - a_2 b_1} \qquad \text{and} \qquad y = \frac{a_1 c_2 - a_2 c_1}{a_1 b_2 - a_2 b_1}$$

provided that $a_1 b_2 - a_2 b_1 \neq 0$. Note that the denominator of each fraction is the same. We call this denominator the **determinant** of the coefficient matrix of the system.

$$\textit{Coefficient Matrix} \qquad \textit{Determinant}$$

$$A = \begin{bmatrix} a_1 & b_1 \\ a_2 & b_2 \end{bmatrix} \qquad \det(A) = a_1 b_2 - a_2 b_1$$

The determinant of the matrix A can also be denoted by vertical bars on both sides of the matrix, as indicated in the following definition.

Definition of the Determinant of a 2 × 2 Matrix

$$\det(A) = \begin{vmatrix} a_1 & b_1 \\ a_2 & b_2 \end{vmatrix} = a_1 b_2 - a_2 b_1$$

Historical Note: In a letter written in 1693 to Guillaume de L'Hôpital, Gottfried Wilhelm von Leibniz (1646–1716) gave a written notation for determinants in what is considered to be the first formal use of them.

A convenient method for remembering the formula for the determinant of a 2 × 2 matrix is shown in the following diagram.

$$\det(A) = \begin{vmatrix} a_1 & b_1 \\ a_2 & b_2 \end{vmatrix} = a_1 b_2 - a_2 b_1$$

Note that the determinant is given by the difference of the products of the two diagonals of the matrix.

EXAMPLE 1   The Determinant of a 2 × 2 Matrix

Find the determinant of each matrix.

(a) $A = \begin{bmatrix} 2 & -3 \\ 1 & 4 \end{bmatrix}$ (b) $B = \begin{bmatrix} -1 & 2 \\ 2 & -4 \end{bmatrix}$ (c) $C = \begin{bmatrix} 1 & 3 \\ 2 & 5 \end{bmatrix}$

Solution

NOTE   Notice in Example 1 that the determinant of a matrix can be positive, zero, or negative.

(a) $\det(A) = \begin{vmatrix} 2 & -3 \\ 1 & 4 \end{vmatrix} = 2(4) - 1(-3) = 8 + 3 = 11$

(b) $\det(B) = \begin{vmatrix} -1 & 2 \\ 2 & -4 \end{vmatrix} = (-1)(-4) - 2(2) = 4 - 4 = 0$

(c) $\det(C) = \begin{vmatrix} 1 & 3 \\ 2 & 5 \end{vmatrix} = 1(5) - 2(3) = 5 - 6 = -1$

Technology

A graphing utility can be used to evaluate the determinant of a square matrix. To find the determinant of

$\begin{bmatrix} 2 & -3 \\ 1 & 4 \end{bmatrix}$

first use the matrix feature to enter the matrix, and then evaluate the determinant, as follows.

det [A] ENTER

Check the result against Example 1(a). Evaluate the determinant of the 3 × 3 matrix that follows Example 1 using a graphing utility. Then evaluate the determinant by expanding by minors to check the result.

One way to evaluate the determinant of a 3 × 3 matrix, called **expanding by minors,** allows you to write the determinant of a 3 × 3 matrix in terms of three 2 × 2 determinants. The **minor** of an entry in a 3 × 3 matrix is the determinant of the 2 × 2 matrix that remains after deletion of the row and column in which the entry occurs. Here are two examples.

Determinant	*Entry*	*Minor of Entry*	*Value of Minor*
$\begin{vmatrix} 1 & -1 & 3 \\ 0 & 2 & 5 \\ -2 & 4 & -7 \end{vmatrix}$	1	$\begin{vmatrix} 2 & 5 \\ 4 & -7 \end{vmatrix}$	$2(-7) - 4(5) = -34$
$\begin{vmatrix} 1 & -1 & 3 \\ 0 & 2 & 5 \\ -2 & 4 & -7 \end{vmatrix}$	-1	$\begin{vmatrix} 0 & 5 \\ -2 & -7 \end{vmatrix}$	$0(-7) - (-2)(5) = 10$

Expanding by Minors

$$\det(A) = \begin{vmatrix} a_1 & b_1 & c_1 \\ a_2 & b_2 & c_2 \\ a_3 & b_3 & c_3 \end{vmatrix}$$

$$= a_1(\text{minor of } a_1) - b_1(\text{minor of } b_1) + c_1(\text{minor of } c_1)$$

$$= a_1 \begin{vmatrix} b_2 & c_2 \\ b_3 & c_3 \end{vmatrix} - b_1 \begin{vmatrix} a_2 & c_2 \\ a_3 & c_3 \end{vmatrix} + c_1 \begin{vmatrix} a_2 & b_2 \\ a_3 & b_3 \end{vmatrix}$$

This pattern is called **expanding by minors** along the first row. A similar pattern can be used to expand by minors along any row or column.

$$\begin{bmatrix} + & - & + \\ - & + & - \\ + & - & + \end{bmatrix}$$

FIGURE 4.11   Sign pattern for 3×3 matrix

The *signs* of the terms used in expanding by minors follow the alternating pattern shown in Figure 4.11. For instance, the signs used to expand by minors along the second row are $-, +, -$, as follows.

$$\det(A) = \begin{vmatrix} a_1 & b_1 & c_1 \\ a_2 & b_2 & c_2 \\ a_3 & b_3 & c_3 \end{vmatrix}$$

$$= -a_2(\text{minor of } a_2) + b_2(\text{minor of } b_2) - c_2(\text{minor of } c_2)$$

EXAMPLE 2   Finding the Determinant of a 3 × 3 Matrix

Find the determinant of $A = \begin{bmatrix} -1 & 1 & 2 \\ 0 & 2 & 3 \\ 3 & 4 & 2 \end{bmatrix}$.

Solution

By expanding by minors along the *first column*, you obtain the following.

$$\det(A) = \begin{vmatrix} -1 & 1 & 2 \\ 0 & 2 & 3 \\ 3 & 4 & 2 \end{vmatrix}$$

$$= (-1)\begin{vmatrix} 2 & 3 \\ 4 & 2 \end{vmatrix} - (0)\begin{vmatrix} 1 & 2 \\ 4 & 2 \end{vmatrix} + (3)\begin{vmatrix} 1 & 2 \\ 2 & 3 \end{vmatrix}$$

$$= (-1)(4 - 12) - (0)(2 - 8) + (3)(3 - 4)$$

$$= 8 - 0 - 3$$

$$= 5$$

Study Tip

Note in the expansion in Example 2 that a zero entry will always yield a zero term when expanding by minors. Thus, when you are evaluating the determinant of a matrix, you should choose to expand along the row or column that has the most zero entries. This idea is reinforced in Example 3.

EXAMPLE 3   Finding the Determinant of a 3 × 3 Matrix

Find the determinant of $A = \begin{bmatrix} 1 & 2 & 1 \\ 3 & 0 & 2 \\ 4 & 0 & -1 \end{bmatrix}$.

Solution

By expanding by minors along the *second column*, you obtain the following.

$$\det(A) = \begin{vmatrix} 1 & 2 & 1 \\ 3 & 0 & 2 \\ 4 & 0 & -1 \end{vmatrix}$$

$$= -(2)\begin{vmatrix} 3 & 2 \\ 4 & -1 \end{vmatrix} + (0)\begin{vmatrix} 1 & 1 \\ 4 & -1 \end{vmatrix} - (0)\begin{vmatrix} 1 & 1 \\ 3 & 2 \end{vmatrix}$$

$$= -(2)(-3 - 8) + 0 - 0$$

$$= 22$$

Explain how the sign pattern in Figure 4.11 is used in Examples 2 and 3.

Cramer's Rule

So far in this chapter, you have studied three methods for solving a system of linear equations: substitution, elimination (with equations), and elimination (with matrices). We now look at one more method, called **Cramer's Rule,** which is named after Gabriel Cramer (1704–1752). This rule uses determinants to write the solution of a system of linear equations.

Cramer's Rule

NOTE Cramer's Rule is not as general as the elimination method because Cramer's Rule requires that the coefficient matrix of the system be square *and* that the system have exactly one solution.

1. For the system of linear equations

$$\begin{cases} a_1 x + b_1 y = c_1 \\ a_2 x + b_2 y = c_2 \end{cases}$$

the solution is given by

$$x = \frac{D_x}{D} = \frac{\begin{vmatrix} c_1 & b_1 \\ c_2 & b_2 \end{vmatrix}}{\begin{vmatrix} a_1 & b_1 \\ a_2 & b_2 \end{vmatrix}}, \qquad y = \frac{D_y}{D} = \frac{\begin{vmatrix} a_1 & c_1 \\ a_2 & c_2 \end{vmatrix}}{\begin{vmatrix} a_1 & b_1 \\ a_2 & b_2 \end{vmatrix}}$$

provided that $D \neq 0$.

2. For the system of linear equations

$$\begin{cases} a_1 x + b_1 y + c_1 z = d_1 \\ a_2 x + b_2 y + c_2 z = d_2 \\ a_3 x + b_3 y + c_3 z = d_3 \end{cases}$$

Historical Note: Gabriel Cramer invented determinants independently and in 1750 published his method of using them to solve linear systems of equations. He did not, however, use the notation that we use today.

the solution is given by

$$x = \frac{D_x}{D} = \frac{\begin{vmatrix} d_1 & b_1 & c_1 \\ d_2 & b_2 & c_2 \\ d_3 & b_3 & c_3 \end{vmatrix}}{\begin{vmatrix} a_1 & b_1 & c_1 \\ a_2 & b_2 & c_2 \\ a_3 & b_3 & c_3 \end{vmatrix}}, \qquad y = \frac{D_y}{D} = \frac{\begin{vmatrix} a_1 & d_1 & c_1 \\ a_2 & d_2 & c_2 \\ a_3 & d_3 & c_3 \end{vmatrix}}{\begin{vmatrix} a_1 & b_1 & c_1 \\ a_2 & b_2 & c_2 \\ a_3 & b_3 & c_3 \end{vmatrix}},$$

$$z = \frac{D_z}{D} = \frac{\begin{vmatrix} a_1 & b_1 & d_1 \\ a_2 & b_2 & d_2 \\ a_3 & b_3 & d_3 \end{vmatrix}}{\begin{vmatrix} a_1 & b_1 & c_1 \\ a_2 & b_2 & c_2 \\ a_3 & b_3 & c_3 \end{vmatrix}}$$

provided that $D \neq 0$.

EXAMPLE 4   *Using Cramer's Rule for a 2 × 2 System*

Writing assignment: Have students describe the different methods of solving linear systems of equations. Ask them to comment on the advantages and disadvantages of each method.

Use Cramer's Rule to solve the following system of linear equations.

$$\begin{cases} 4x - 2y = 10 \\ 3x - 5y = 11 \end{cases}$$

Solution

Begin by finding the determinant of the coefficient matrix.

$$D = \begin{vmatrix} 4 & -2 \\ 3 & -5 \end{vmatrix} = -20 - (-6) = -14$$

$$x = \frac{D_x}{D} = \frac{\begin{vmatrix} 10 & -2 \\ 11 & -5 \end{vmatrix}}{-14} = \frac{(-50) - (-22)}{-14} = \frac{-28}{-14} = 2$$

$$y = \frac{D_y}{D} = \frac{\begin{vmatrix} 4 & 10 \\ 3 & 11 \end{vmatrix}}{-14} = \frac{44 - 30}{-14} = \frac{14}{-14} = -1$$

The solution is $(2, -1)$. Check this in the original system of equations.

EXAMPLE 5   *Using Cramer's Rule for a 3 × 3 System*

NOTE   When using Cramer's Rule, remember that the method *does not* apply if the determinant of the coefficient matrix is zero. For instance, the system

$$\begin{cases} 6x - 10y = -4 \\ 9x - 15y = 5 \end{cases}$$

has no solution (see Example 8 in Section 4.3), and the determinant of the coefficient matrix of this system is zero.

Use Cramer's Rule to solve the following system of linear equations.

$$\begin{cases} -x + 2y - 3z = 1 \\ 2x + \quad\ \ z = 0 \\ 3x - 4y + 4z = 2 \end{cases}$$

Solution

The determinant of the coefficient matrix is $D = 10$.

$$x = \frac{D_x}{D} = \frac{\begin{vmatrix} 1 & 2 & -3 \\ 0 & 0 & 1 \\ 2 & -4 & 4 \end{vmatrix}}{10} = \frac{8}{10} = \frac{4}{5}$$

$$y = \frac{D_y}{D} = \frac{\begin{vmatrix} -1 & 1 & -3 \\ 2 & 0 & 1 \\ 3 & 2 & 4 \end{vmatrix}}{10} = \frac{-15}{10} = -\frac{3}{2}$$

$$z = \frac{D_z}{D} = \frac{\begin{vmatrix} -1 & 2 & 1 \\ 2 & 0 & 0 \\ 3 & -4 & 2 \end{vmatrix}}{10} = \frac{-16}{10} = -\frac{8}{5}$$

The solution is $\left(\frac{4}{5}, -\frac{3}{2}, -\frac{8}{5}\right)$. Check this in the original system of equations.

Applications of Determinants

In addition to Cramer's Rule, determinants have many other practical applications. For instance, you can use a determinant to find the area of a triangle whose vertices are given by three points on a rectangular coordinate system.

Area of a Triangle

The area of a triangle with vertices (x_1, y_1), (x_2, y_2), and (x_3, y_3) is

$$\text{Area} = \pm \frac{1}{2} \begin{vmatrix} x_1 & y_1 & 1 \\ x_2 & y_2 & 1 \\ x_3 & y_3 & 1 \end{vmatrix}$$

where the symbol $(\pm)$ indicates that the appropriate sign should be chosen to yield a positive area.

EXAMPLE 6 Finding the Area of a Triangle

Find the area of the triangle whose vertices are $(2, 0)$, $(1, 3)$, and $(3, 2)$, as shown in Figure 4.12.

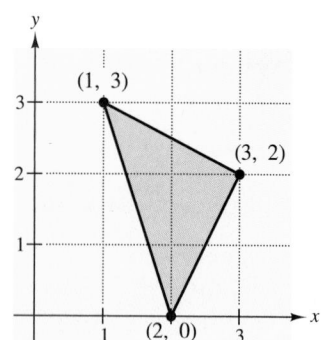

FIGURE 4.12

Solution

Choose $(x_1, y_1) = (2, 0)$, $(x_2, y_2) = (1, 3)$, and $(x_3, y_3) = (3, 2)$. To find the area of the triangle, evaluate the determinant

$$\begin{vmatrix} x_1 & y_1 & 1 \\ x_2 & y_2 & 1 \\ x_3 & y_3 & 1 \end{vmatrix} = \begin{vmatrix} 2 & 0 & 1 \\ 1 & 3 & 1 \\ 3 & 2 & 1 \end{vmatrix}$$

$$= 2 \begin{vmatrix} 3 & 1 \\ 2 & 1 \end{vmatrix} - 0 \begin{vmatrix} 1 & 1 \\ 3 & 1 \end{vmatrix} + 1 \begin{vmatrix} 1 & 3 \\ 3 & 2 \end{vmatrix}$$

$$= 2(1) - 0 + 1(-7)$$

$$= -5.$$

Using this value, you can conclude that the area of the triangle is

$$\text{Area} = -\frac{1}{2} \begin{vmatrix} 2 & 0 & 1 \\ 1 & 3 & 1 \\ 3 & 2 & 1 \end{vmatrix} = -\frac{1}{2}(-5) = \frac{5}{2}.$$

NOTE   To see the benefit of the "determinant formula," try finding the area of the triangle in Example 6 using the standard formula: $\text{Area} = \frac{1}{2}(\text{base})(\text{height})$.

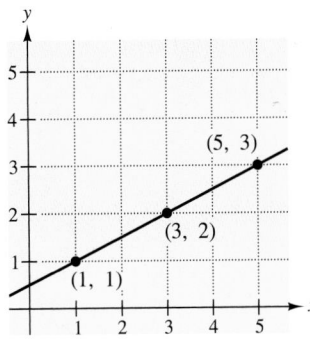

FIGURE 4.13

Suppose the three points in Example 6 had been on the same line. What would have happened had we applied the area formula to three such points? The answer is that the determinant would have been zero. Consider for instance, the three collinear points $(1, 1)$, $(3, 2)$, and $(5, 3)$, as shown in Figure 4.13. The area of the "triangle" that has these three points as vertices is

$$\frac{1}{2}\begin{vmatrix} 1 & 1 & 1 \\ 3 & 2 & 1 \\ 5 & 3 & 1 \end{vmatrix} = \frac{1}{2}\left(1\begin{vmatrix} 2 & 1 \\ 3 & 1 \end{vmatrix} - 1\begin{vmatrix} 3 & 1 \\ 5 & 1 \end{vmatrix} + 1\begin{vmatrix} 3 & 2 \\ 5 & 3 \end{vmatrix} \right)$$

$$= \frac{1}{2}[-1 - (-2) + (-1)] = 0.$$

This result is generalized as follows.

Test for Collinear Points

Three points (x_1, y_1), (x_2, y_2), and (x_3, y_3) are collinear (lie on the same line) if and only if

$$\begin{vmatrix} x_1 & y_1 & 1 \\ x_2 & y_2 & 1 \\ x_3 & y_3 & 1 \end{vmatrix} = 0.$$

EXAMPLE 7 *Testing for Collinear Points*

Determine whether the points $(-2, -2)$, $(1, 1)$, and $(7, 5)$ lie on the same line. (See Figure 4.14.)

Solution

Letting $(x_1, y_1) = (-2, -2)$, $(x_2, y_2) = (1, 1)$, and $(x_3, y_3) = (7, 5)$, you have

$$\begin{vmatrix} x_1 & y_1 & 1 \\ x_2 & y_2 & 1 \\ x_3 & y_3 & 1 \end{vmatrix} = \begin{vmatrix} -2 & -2 & 1 \\ 1 & 1 & 1 \\ 7 & 5 & 1 \end{vmatrix}$$

$$= -2\begin{vmatrix} 1 & 1 \\ 5 & 1 \end{vmatrix} - (-2)\begin{vmatrix} 1 & 1 \\ 7 & 1 \end{vmatrix} + 1\begin{vmatrix} 1 & 1 \\ 7 & 5 \end{vmatrix}$$

$$= -2(-4) - (-2)(-6) + 1(-2) = -6.$$

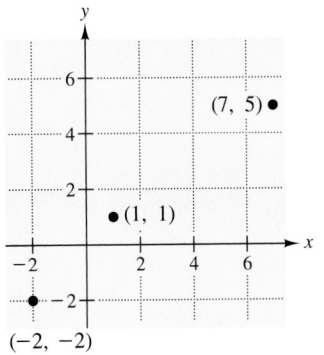

FIGURE 4.14

Because the value of this determinant *is not* zero, you can conclude that the three points *do not* lie on the same line.

A test for collinear points can be adapted to finding the equation of the line passing through two points, as shown on the next page.

Two-Point Form of the Equation of a Line

An equation of the line passing through the distinct points (x_1, y_1) and (x_2, y_2) is given by

$$\begin{vmatrix} x & y & 1 \\ x_1 & y_1 & 1 \\ x_2 & y_2 & 1 \end{vmatrix} = 0.$$

EXAMPLE 8   Finding an Equation of a Line

Find an equation of the line passing through $(-2, 1)$ and $(3, -2)$.

Solution

$$\begin{vmatrix} x & y & 1 \\ -2 & 1 & 1 \\ 3 & -2 & 1 \end{vmatrix} = 0$$

$$x\begin{vmatrix} 1 & 1 \\ -2 & 1 \end{vmatrix} - y\begin{vmatrix} -2 & 1 \\ 3 & 1 \end{vmatrix} + 1\begin{vmatrix} -2 & 1 \\ 3 & -2 \end{vmatrix} = 0$$

$$3x + 5y + 1 = 0$$

Therefore, an equation of the line is $3x + 5y + 1 = 0$.

Group Activities   Extending the Concept

Determinant of a 3 × 3 Matrix   There is an alternative method for evaluating the determinant of a 3×3 matrix A. (This method works *only* for 3×3 matrices.) To apply this method, copy the first and second columns of A to form fourth and fifth columns. The determinant of A is then obtained by adding the products of three diagonals and subtracting the products of three diagonals.

$$A = \begin{bmatrix} 0 & 2 & 1 \\ 3 & -1 & 2 \\ 4 & -4 & 1 \end{bmatrix}$$

$$|A| = 0 + 16 - 12 - (-4) - 0 - 6 = 2$$

Subtract these products.

Add these products.

Try using this technique to find the determinants of the matrices in Examples 2 and 3. Do you think this method is easier than expanding by minors?

See Warm-Up
Exercises, p. A43

4.4   **Exercises**

In Exercises 1–12, find the determinant of the matrix.

1. $\begin{bmatrix} 2 & 1 \\ 3 & 4 \end{bmatrix}$

2. $\begin{bmatrix} -3 & 1 \\ 5 & 2 \end{bmatrix}$

3. $\begin{bmatrix} 5 & 2 \\ -6 & 3 \end{bmatrix}$

4. $\begin{bmatrix} 2 & -2 \\ 4 & 3 \end{bmatrix}$

5. $\begin{bmatrix} 5 & -4 \\ -10 & 8 \end{bmatrix}$

6. $\begin{bmatrix} 4 & -3 \\ 0 & 0 \end{bmatrix}$

7. $\begin{bmatrix} 2 & 6 \\ 0 & 3 \end{bmatrix}$

8. $\begin{bmatrix} -2 & 3 \\ 6 & -9 \end{bmatrix}$

9. $\begin{bmatrix} -7 & 6 \\ \frac{1}{2} & 3 \end{bmatrix}$

10. $\begin{bmatrix} \frac{2}{3} & \frac{5}{6} \\ 14 & -2 \end{bmatrix}$

11. $\begin{bmatrix} 0.3 & 0.5 \\ 0.5 & 0.3 \end{bmatrix}$

12. $\begin{bmatrix} -1.2 & 4.5 \\ 0.4 & -0.9 \end{bmatrix}$

In Exercises 13–16, evaluate the determinant of the matrix six different ways by expanding by minors along each row and column.

13. $\begin{bmatrix} 2 & 3 & -1 \\ 6 & 0 & 0 \\ 4 & 1 & 1 \end{bmatrix}$

14. $\begin{bmatrix} 10 & 2 & -4 \\ 8 & 0 & -2 \\ 4 & 0 & 2 \end{bmatrix}$

15. $\begin{bmatrix} 1 & 1 & 2 \\ 3 & 1 & 0 \\ -2 & 0 & 3 \end{bmatrix}$

16. $\begin{bmatrix} 2 & 1 & 3 \\ 1 & 4 & 4 \\ 1 & 0 & 2 \end{bmatrix}$

In Exercises 17–30, evaluate the determinant of the matrix. Expand by minors along the row or column that appears to make the computation easiest.

17. $\begin{bmatrix} 2 & 4 & 6 \\ 0 & 3 & 1 \\ 0 & 0 & -5 \end{bmatrix}$

18. $\begin{bmatrix} 2 & 3 & 1 \\ 0 & 5 & -2 \\ 0 & 0 & -2 \end{bmatrix}$

19. $\begin{bmatrix} -2 & 2 & 3 \\ 1 & -1 & 0 \\ 0 & 1 & 4 \end{bmatrix}$

20. $\begin{bmatrix} 3 & 2 & 2 \\ 2 & 2 & 2 \\ -4 & 4 & 3 \end{bmatrix}$

21. $\begin{bmatrix} 1 & 4 & -2 \\ 3 & 6 & -6 \\ -2 & 1 & 4 \end{bmatrix}$

22. $\begin{bmatrix} 2 & -1 & 0 \\ 4 & 2 & 1 \\ 4 & 2 & 1 \end{bmatrix}$

23. $\begin{bmatrix} -3 & 2 & 1 \\ 4 & 5 & 6 \\ 2 & -3 & 1 \end{bmatrix}$

24. $\begin{bmatrix} -3 & 4 & 2 \\ 6 & 3 & 1 \\ 4 & -7 & -8 \end{bmatrix}$

25. $\begin{bmatrix} 1 & 4 & -2 \\ 3 & 2 & 0 \\ -1 & 4 & 3 \end{bmatrix}$

26. $\begin{bmatrix} 6 & 8 & -7 \\ 0 & 0 & 0 \\ 4 & -6 & 22 \end{bmatrix}$

27. $\begin{bmatrix} 0.1 & 0.2 & 0.3 \\ -0.3 & 0.2 & 0.2 \\ 5 & 4 & 4 \end{bmatrix}$

28. $\begin{bmatrix} -0.4 & 0.4 & 0.3 \\ 0.2 & 0.2 & 0.2 \\ 0.3 & 0.2 & 0.2 \end{bmatrix}$

29. $\begin{bmatrix} x & y & 1 \\ 3 & 1 & 1 \\ -2 & 0 & 1 \end{bmatrix}$

30. $\begin{bmatrix} x & y & 1 \\ -2 & -2 & 1 \\ 1 & 5 & 1 \end{bmatrix}$

In Exercises 31–42, use a graphing utility to evaluate the determinant of the matrix.

31. $\begin{bmatrix} 5 & -3 & 2 \\ 7 & 5 & -7 \\ 0 & 6 & -1 \end{bmatrix}$

32. $\begin{bmatrix} -\frac{1}{2} & -1 & 6 \\ 8 & -\frac{1}{4} & -4 \\ 1 & 2 & 1 \end{bmatrix}$

33. $\begin{bmatrix} 35 & 15 & 70 \\ -8 & 20 & 3 \\ -5 & 6 & 20 \end{bmatrix}$

34. $\begin{bmatrix} 3 & -1 & 2 \\ 1 & -1 & 2 \\ -2 & 3 & 10 \end{bmatrix}$

35. $\begin{bmatrix} 0.4 & 0.3 & 0.3 \\ -0.2 & 0.6 & 0.6 \\ 3 & 1 & 1 \end{bmatrix}$

36. $\begin{bmatrix} \frac{1}{2} & \frac{3}{2} & \frac{1}{2} \\ 4 & 8 & 10 \\ -2 & -6 & 12 \end{bmatrix}$

37. $\begin{bmatrix} 2 & -3 & 3 \\ \frac{3}{4} & 1 & -\frac{1}{4} \\ 12 & 3 & -\frac{1}{2} \end{bmatrix}$

38. $\begin{bmatrix} 1 & 3 & \frac{2}{3} \\ -\frac{3}{2} & \frac{1}{2} & 5 \\ 5 & 2 & \frac{4}{5} \end{bmatrix}$

39. $\begin{bmatrix} \frac{3}{2} & -\frac{3}{4} & 1 \\ 10 & 8 & 7 \\ 12 & -4 & 12 \end{bmatrix}$ **40.** $\begin{bmatrix} 0.3 & -0.2 & 0.5 \\ 0.6 & 0.4 & -0.3 \\ 1.2 & 0 & 0.7 \end{bmatrix}$

41. $\begin{bmatrix} 0.2 & 0.8 & -0.3 \\ 0.1 & 0.8 & 0.6 \\ -10 & -5 & 1 \end{bmatrix}$

42. $\begin{bmatrix} 250 & -125 & 60 \\ -125 & 200 & -50 \\ 60 & -50 & 150 \end{bmatrix}$

In Exercises 43–58, use Cramer's Rule to solve the system of linear equations. (If not possible, state the reason.)

43. $\begin{cases} x + 2y = 5 \\ -x + y = 1 \end{cases}$ **44.** $\begin{cases} 2x - y = -10 \\ 3x + 2y = -1 \end{cases}$

45. $\begin{cases} 3x + 4y = -2 \\ 5x + 3y = 4 \end{cases}$ **46.** $\begin{cases} 18x + 12y = 13 \\ 30x + 24y = 23 \end{cases}$

47. $\begin{cases} 20x + 8y = 11 \\ 12x - 24y = 21 \end{cases}$ **48.** $\begin{cases} 13x - 6y = 17 \\ 26x - 12y = 8 \end{cases}$

49. $\begin{cases} -0.4x + 0.8y = 1.6 \\ 2x - 4y = 5 \end{cases}$

50. $\begin{cases} -0.4x + 0.8y = 1.6 \\ 0.2x + 0.3y = 2.2 \end{cases}$

51. $\begin{cases} 3u + 6v = 5 \\ 6u + 14v = 11 \end{cases}$ **52.** $\begin{cases} 3x_1 + 2x_2 = 1 \\ 2x_1 + 10x_2 = 6 \end{cases}$

53. $\begin{cases} 4x - y + z = -5 \\ 2x + 2y + 3z = 10 \\ 5x - 2y + 6z = 1 \end{cases}$

54. $\begin{cases} 4x - 2y + 3z = -2 \\ 2x + 2y + 5z = 16 \\ 8x - 5y - 2z = 4 \end{cases}$

55. $\begin{cases} 3x + 4y + 4z = 11 \\ 4x - 4y + 6z = 11 \\ 6x - 6y = 3 \end{cases}$

56. $\begin{cases} 14x_1 - 21x_2 - 7x_3 = 10 \\ -4x_1 + 2x_2 - 2x_3 = 4 \\ 56x_1 - 21x_2 + 7x_3 = 5 \end{cases}$

57. $\begin{cases} 3a + 3b + 4c = 1 \\ 3a + 5b + 9c = 2 \\ 5a + 9b + 17c = 4 \end{cases}$

58. $\begin{cases} 2x + 3y + 5z = 4 \\ 3x + 5y + 9z = 7 \\ 5x + 9y + 17z = 13 \end{cases}$

In Exercises 59–68, solve the system of linear equations using a graphing utility and Cramer's Rule.

59. $\begin{cases} -3x + 10y = 22 \\ 9x - 3y = 0 \end{cases}$ **60.** $\begin{cases} 3x + 7y = 3 \\ 7x + 25y = 11 \end{cases}$

61. $\begin{cases} 4x - y = -2 \\ -2x + y = 3 \end{cases}$ **62.** $\begin{cases} 4x + 8y = 6 \\ 8x + 26y = 19 \end{cases}$

63. $\begin{cases} x + y - z = 2 \\ 6x + 4y + 3z = 4 \\ 3x + 6z = -3 \end{cases}$

64. $\begin{cases} 3x + 2y - 5z = -10 \\ 6x - z = 8 \\ -y + 3z = -2 \end{cases}$

65. $\begin{cases} 3x + y + z = 6 \\ x - 4y + 2z = -1 \\ x - 3y + z = 0 \end{cases}$

66. $\begin{cases} x - y + 2z = 6 \\ -2x + 3y - z = -7 \\ 3x + 2y + 2z = 5 \end{cases}$

67. $\begin{cases} 3x - 2y + 3z = 8 \\ x + 3y + 6z = -3 \\ x + 2y + 9z = -5 \end{cases}$

68. $\begin{cases} 6x + 4y - 8z = -22 \\ -2x + 2y + 3z = 13 \\ -2x + 2y - z = 5 \end{cases}$

69. *Think About It* Is it possible to find the determinant of a 2×3 matrix? Explain your answer.

70. *Think About It* What conditions must be met in order to use Cramer's Rule to solve a system of linear equations?

Area of a Triangle   In Exercises 71–74, use a determinant to find the area of the triangle with the given vertices.

71. $(0, 3), (4, 0), (8, 5)$

72. $(-4, 2), (1, 5), (4, -4)$

73. $(-2, 1), (3, -1), (1, 6)$

74. $\left(0, \frac{1}{2}\right), \left(\frac{5}{2}, 0\right), (4, 3)$

Area of a Region   In Exercises 75 and 76, find the area of the shaded region of the figure.

75.

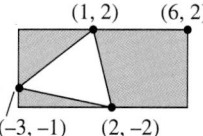

(1, 2)   (6, 2)
(−3, −1)   (2, −2)

76.

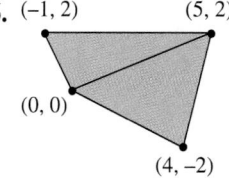

(−1, 2)   (5, 2)
(0, 0)
(4, −2)

Collinear Points   In Exercises 77–80, determine whether the points are collinear.

77. $(-1, 11), (0, 8), (2, 2)$

78. $(-1, -1), (1, 9), (2, 13)$

79. $\left(-2, \frac{1}{3}\right), (2, 1), \left(3, \frac{1}{5}\right)$

80. $\left(0, \frac{1}{2}\right), \left(1, \frac{7}{6}\right), \left(9, \frac{13}{2}\right)$

Equation of a Line   In Exercises 81–84, use a determinant to find the equation of the line through the points.

81. $(0, 0), (5, 3)$

82. $(-4, 3), (2, 1)$

83. $(10, 7), (-2, -7)$

84. $\left(-\frac{1}{2}, 3\right), \left(\frac{5}{2}, 1\right)$

 Curve-Fitting   In Exercises 85–88, use Cramer's Rule to find the equation $y = ax^2 + bx + c$ that passes through the points. Use a graphing utility to plot the points and graph the model.

85. $(0, 1), (1, -3), (-2, 21)$

86. $(2, 3), \left(-1, \frac{9}{2}\right), (-2, 9)$

87. $(1, -1), (-1, -5), \left(\frac{1}{2}, \frac{1}{4}\right)$

88. $(-2, 6), (1, 9), (3, 1)$

89. *U.S. Exports and Imports*   The table gives the merchandise exports y_1 and the merchandise imports y_2 (in billions of dollars) for the years 1992 through 1994 in the United States.   *(Source: U.S. Bureau of Census)*

Year	1992	1993	1994
y_1	448.2	465.1	512.7
y_2	532.7	580.7	663.8

(a) Find the model of the form $y_1 = a_1 t^2 + b_1 t + c_1$ for exports. Let $t = 0$ represent 1990.

(b) Find the model of the form $y_2 = a_2 t^2 + b_2 t + c_2$ for imports. Let $t = 0$ represent 1990.

(c) Use a graphing utility to graph the models in parts (a) and (b).

(d) Find a model for the merchandise trade balance (merchandise exports – merchandise imports).

90. *Electrical Networks*   When Kirchhoff's Laws are applied to the electrical network in the figure, the currents I_1, I_2, and I_3 are the solution of the system

$$
\begin{aligned}
I_1 - I_2 + I_3 &= 0 \\
3I_1 + 2I_2 &= 7 \\
2I_2 + 4I_3 &= 8.
\end{aligned}
$$

Find the currents.

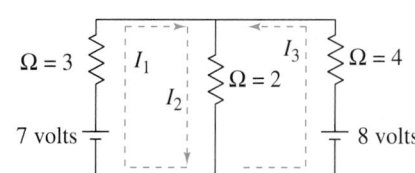

91. (a) Use Cramer's Rule to solve the system of linear equations.

$$\begin{cases} kx + (1 - k)y = 1 \\ (1 - k)x + \quad\quad ky = 3 \end{cases}$$

   (b) For what value(s) of k will the system be inconsistent?

In Exercises 92 and 93, solve the equation.

92. $\begin{vmatrix} 5 - x & 4 \\ 1 & 2 - x \end{vmatrix} = 0$

93. $\begin{vmatrix} 4 - x & -2 \\ 1 & 1 - x \end{vmatrix} = 0$

Section Project

Area of a Region   A large region of forest has been infested with gypsy moths. The region is roughly triangular, as shown in the figure. (Dimensions are in miles.)

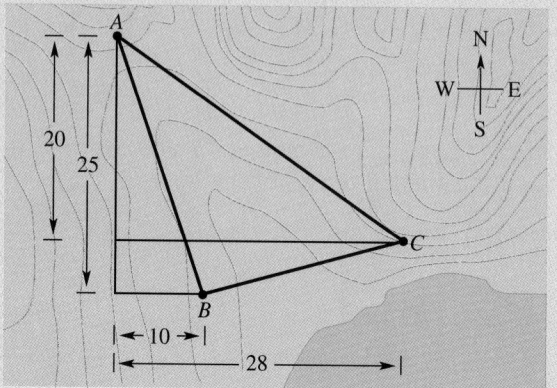

(a) Choose an appropriate point on the figure to use as the origin of a rectangular coordinate system, and determine the coordinates of each vertex of the triangular region. Use a determinant to approximate the number of square miles in this region.

(b) Choose another point on the figure to use as the origin and determine the new coordinates of the vertices. Use a determinant to approximate the number of square miles in the region using the new coordinates.

(c) Do you think the choice of the origin affects the approximation of the area of the figure? Why or why not?

<table>
<tr><td>**4.5**</td><td>**Graphs of Linear Inequalities in Two Variables**</td></tr>
</table>

Linear Inequalities in Two Variables ▪ The Graph of a Linear Inequality in Two Variables ▪ Application

Linear Inequalities in Two Variables

A **linear inequality** in variables x and y is an inequality that can be written in one of the following forms (where a and b are not both zero).

$$ax + by < c, \qquad ax + by > c, \qquad ax + by \le c, \qquad ax + by \ge c$$

Here are some examples.

$$x - y > -3, \qquad 4x - 3y \le 7, \qquad x < 2, \qquad y \ge -4$$

An ordered pair (x_1, y_1) is a **solution** of a linear inequality in x and y if the inequality is true when x_1 and y_1 are substituted for x and y.

EXAMPLE 1 *Verifying Solutions of Linear Inequalities*

Determine whether each point is a solution of $2x - 3y \ge -2$.

(a) $(0, 0)$ (b) $(2, 2)$ (c) $(0, 1)$

Solution

(a) To determine whether the point $(0, 0)$ is a solution of the inequality, substitute the coordinates of the point into the inequality, as follows.

$$2x - 3y \ge -2 \qquad \text{Original inequality}$$
$$2(0) - 3(0) \overset{?}{\ge} -2 \qquad \text{Replace } x \text{ by 0 and } y \text{ by 0.}$$
$$0 \ge -2 \qquad \text{Inequality is satisfied.}$$

Because the inequality is satisfied, the point $(0, 0)$ *is* a solution.

(b) By substituting the coordinates of the point $(2, 2)$ into the inequality, you obtain

$$2(2) - 3(2) = -2.$$

Because -2 is greater than or equal to -2, you can conclude that the point $(2, 2)$ *is* a solution of the inequality.

(c) By substituting the coordinates of the point $(0, 1)$ into the inequality, you obtain

$$2(0) - 3(1) = -3.$$

Because -3 is less than -2, the point $(0, 1)$ *is not* a solution of the inequality.

Exploration

Sketch the graph of $4x - 3y = 12$. Choose several points to the left and right of the line and evaluate $4x - 3y$ at each point. Which points satisfy the inequality $4x - 3y < 12$? Which points satisfy the inequality $4x - 3y > 12$? What can you conclude about the graph of the solution of a linear inequality?

The Graph of a Linear Inequality in Two Variables

The **graph** of an inequality is the collection of all solution points of the inequality. To sketch the graph of a linear inequality such as

$$4x - 3y < 12 \qquad \text{Original linear inequality}$$

begin by sketching the graph of the *corresponding linear equation*

$$4x - 3y = 12. \qquad \text{Write corresponding linear equation.}$$

This graph is made with a *dashed* line for the inequalities $<$ and $>$ and with a *solid* line for the inequalities $\leq$ and $\geq$. The graph of the equation (corresponding to a given linear inequality) separates the plane into two regions, called **half-planes.** In each half-plane, one of the following *must* be true.

1. All points in the half-plane are solutions of the inequality.

2. No point in the half-plane is a solution of the inequality.

Thus, you can determine whether the points in an entire half-plane satisfy the inequality by simply testing *one* point in the region. This graphing procedure is summarized as follows.

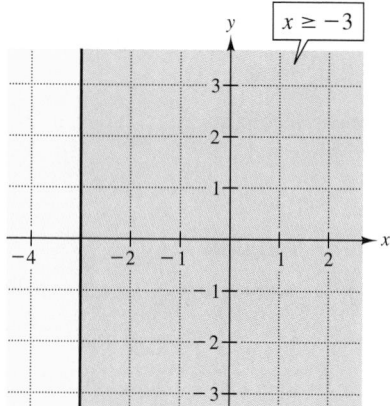

FIGURE 4.15

Sketching the Graph of a Linear Inequality in Two Variables

1. Replace the inequality sign by an equal sign, and sketch the graph of the resulting equation. (Use a dashed line for $<$ or $>$ and a solid line for $\leq$ or $\geq$.)

2. Test one point in each of the half-planes formed by the graph in Step 1. If the point satisfies the inequality, shade the entire half-plane to denote that every point in the region satisfies the inequality.

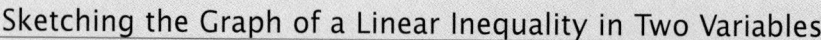

EXAMPLE 2 Sketching the Graph of a Linear Inequality

Sketch the graph of each linear inequality.

(a) $x \geq -3$ (b) $y < 4$

Solution

(a) The graph of the corresponding equation $x = -3$ is a vertical line. The points that satisfy the inequality $x \geq -3$ are those lying to the right of (or on) this line, as shown in Figure 4.15.

(b) The graph of the corresponding equation $y = 4$ is a horizontal line. The points that satisfy the inequality $y < 4$ are those lying below this line, as shown in Figure 4.16.

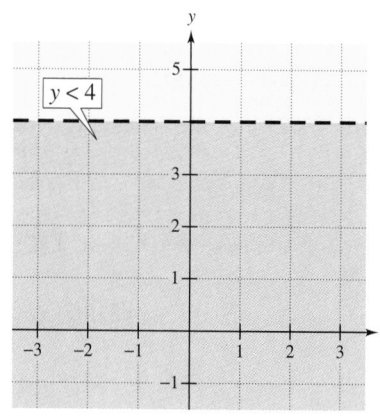

FIGURE 4.16

EXAMPLE 3   *Sketching the Graph of a Linear Inequality*

Sketch the graph of the linear inequality

$$x + y > 3. \qquad \text{Original linear inequality}$$

Solution

The graph of the corresponding equation

$$x + y = 3 \qquad \text{Write corresponding linear equation.}$$

is a line, as shown in Figure 4.17. Because the origin $(0, 0)$ *does not* satisfy the inequality, the graph consists of the half-plane lying *above* the line. (Try checking a point above the line. Regardless of which point you choose, you will see that it satisfies the inequality.)

Point out that initially trying the point $(0, 0)$ is usually the easiest, but should be done only if the line does not pass through the origin.

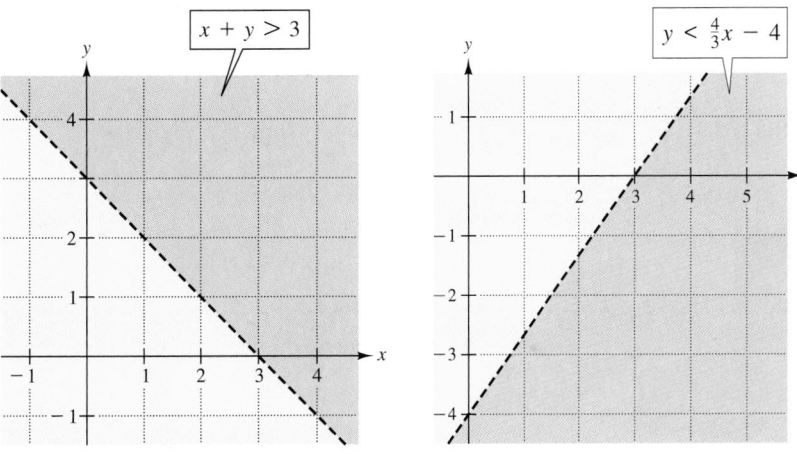

FIGURE 4.17   **FIGURE 4.18**

For a linear inequality in two variables, you can sometimes simplify the graphing procedure by writing the inequality in *slope-intercept* form. For instance, by writing $x + y > 3$ in the form $y > -x + 3$, you can see that the solution points lie *above* the line $y = -x + 3$, as shown in Figure 4.17. Similarly, by writing the inequality

$$4x - 3y > 12 \qquad \text{Original linear inequality}$$

in the form

$$y < \frac{4}{3}x - 4 \qquad \text{Rewrite in slope-intercept form.}$$

you can see that the solutions lie *below* the line $y = \frac{4}{3}x - 4$. See Figure 4.18.

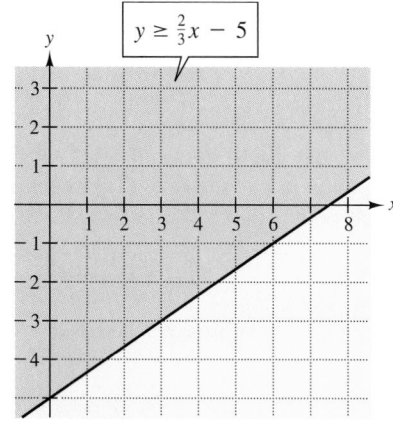

FIGURE 4.19

EXAMPLE 4 Sketching the Graph of a Linear Inequality

Use the slope-intercept form of a linear equation as an aid in sketching the graph of the inequality

$$2x - 3y \leq 15. \qquad \text{Original linear inequality}$$

Solution

To begin, rewrite the inequality in slope-intercept form.

$$2x - 3y \leq 15 \qquad \text{Original inequality}$$

$$-3y \leq -2x + 15 \qquad \text{Subtract } 2x \text{ from both sides.}$$

$$y \geq \frac{2}{3}x - 5 \qquad \text{Divide both sides by } -3 \text{ and reverse the inequality symbol.}$$

From this form, you can conclude that the solution is the half-plane lying *on* or *above* the line

$$y = \frac{2}{3}x - 5.$$

The graph is shown in Figure 4.19.

Technology

A graphing utility can be used to graph an inequality. The actual keystrokes used depend on the graphing utility, but here is an example of how to graph

$$3x + 2y < 4$$

on a *TI-83*.

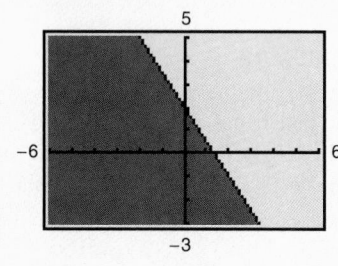

1. Solve the inequality for y to obtain $y < -\frac{3}{2}x + 2$.

2. Press $\boxed{Y=}$ and enter $-(3/2)X + 2$ for Y_1.

3. Move the cursor to the left of Y_1.

4. Press $\boxed{\text{ENTER}}$ until the ◣ icon appears.

5. Press $\boxed{\text{GRAPH}}$.

The graph is shown at the left. Try using a graphing utility to graph the following inequalities.

(a) $2x + 3y \geq 3$ (b) $x - 2y \leq 2$

Application

EXAMPLE 5   *Finding Ocean Treasure*

You are on a treasure-diving ship that is hunting for gold and silver coins. Objects collected by the divers are placed in a wire basket. One of the divers signals you to reel in the basket. It feels as if it contains no more than 50 pounds of material. If gold coins weigh about $\frac{1}{2}$ ounce each and silver coins weigh about $\frac{1}{4}$ ounce each, what are the different amounts of coins that you could be reeling in?

Solution

At 16 ounces per pound, the basket contains 800 ounces.

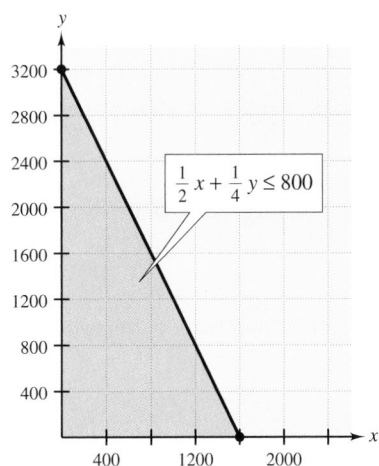

Verbal Model:	Weight per gold coin	$\cdot$	Number of gold coins	$+$	Weight per silver coin	$\cdot$	Number of silver coins	$\leq$	Weight in basket

Labels:	Weight per gold coin $= \frac{1}{2}$	(ounce per coin)
	Number of gold coins $= x$	(coins)
	Weight per silver coin $= \frac{1}{4}$	(ounce per coin)
	Number of silver coins $= y$	(coins)
	Weight in basket $= 800$	(ounces)

Inequality: $\frac{1}{2}x + \frac{1}{4}y \leq 800$ Linear model

One solution is $(1600, 0)$: all gold coins. Another solution is $(0, 3200)$: all silver coins. Unfortunately, there are many other solutions including $(0, 0)$: 50 pounds of something, but no gold or silver coins. All possible solutions are shown in Figure 4.20.

FIGURE 4.20

Group Activities

Discovering Through Technology

Write a linear inequality in x and y. Then use a graphing utility to sketch its graph. Show the graph to the other people in your group. Ask them to write the inequality, using only the graph. Continue this activity until each person in the group has had a chance to graph at least one linear inequality. Finally, as a group, summarize the strategy you used to write an inequality from its graph.

4.5 **Exercises**

See Warm-Up Exercises, p. A43

In Exercises 1–6, match the linear inequality with its graph. [The graphs are labeled (a), (b), (c), (d), (e), and (f).]

(a)
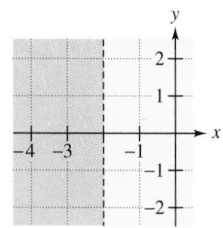

(b)
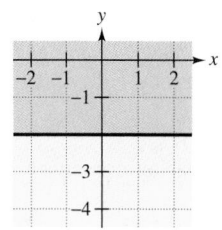

(c)
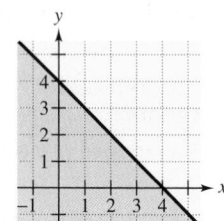

(d)
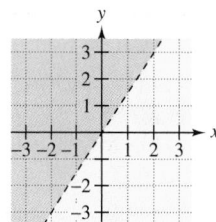

(e)
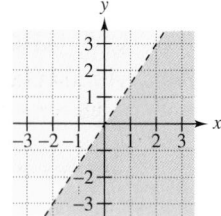

(f)
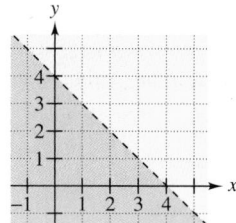

1. $y \geq -2$

2. $x < -2$

3. $3x - 2y < 0$

4. $3x - 2y > 0$

5. $x + y < 4$

6. $x + y \leq 4$

In Exercises 7–14, determine whether the points are solutions of the inequality.

Inequality	*Points*	
7. $x - 2y < 4$	(a) $(0, 0)$	(b) $(2, -1)$
	(c) $(3, 4)$	(d) $(5, 1)$
8. $x + y < 3$	(a) $(0, 6)$	(b) $(4, 0)$
	(c) $(0, -2)$	(d) $(1, 1)$

Inequality	*Points*	
9. $3x + y \geq 10$	(a) $(1, 3)$	(b) $(-3, 1)$
	(c) $(3, 1)$	(d) $(2, 15)$
10. $-3x + 5y \geq 6$	(a) $(2, 8)$	(b) $(-10, -3)$
	(c) $(0, 0)$	(d) $(3, 3)$
11. $y > 0.2x - 1$	(a) $(0, 2)$	(b) $(6, 0)$
	(c) $(4, -1)$	(d) $(-2, 7)$
12. $y < -3.5x + 7$	(a) $(1, 5)$	(b) $(5, -1)$
	(c) $(-1, 4)$	(d) $\left(0, \frac{4}{3}\right)$
13. $y \leq 3 - |x|$	(a) $(-1, 4)$	(b) $(2, -2)$
	(c) $(6, 0)$	(d) $(5, -2)$
14. $y \geq |x - 3|$	(a) $(0, 0)$	(b) $(1, 2)$
	(c) $(4, 10)$	(d) $(5, -1)$

In Exercises 15–32, sketch the graph of the solution of the linear inequality.

15. $x \geq 2$

16. $x < -3$

17. $y < 5$

18. $y > 2$

19. $y > \frac{1}{2}x$

20. $y \leq 2x$

21. $y \leq x + 1$

22. $y > 4 - x$

23. $y - 1 > -\frac{1}{2}(x - 2)$

24. $y - 2 < -\frac{2}{3}(x - 1)$

25. $\frac{x}{3} + \frac{y}{4} \leq 1$

26. $\frac{x}{2} + \frac{y}{6} \geq 1$

27. $x - 2y \geq 6$

28. $3x + 5y \leq 15$

29. $3x - 2y \geq 4$

30. $x + 3y \leq 5$

31. $0.2x + 0.3y < 2$

32. $x - 0.75y > 6$

In Exercises 33–40, use a graphing utility to graph the solution of the linear inequality.

33. $y \geq \frac{3}{4}x - 1$

34. $y \leq 9 - 1.5x$

35. $y \leq -\frac{2}{3}x + 6$

36. $y \geq \frac{1}{4}x + 3$

37. $x - 2y - 4 \geq 0$

38. $2x + 4y - 3 \leq 0$

39. $2x + 3y - 12 \leq 0$

40. $x - 3y + 9 \geq 0$

In Exercises 41–46, write an inequality for the shaded region shown in the figure.

41.
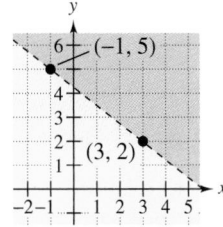

42.
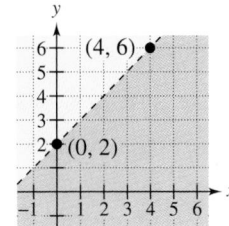

43.
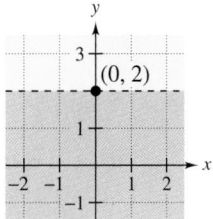

44.
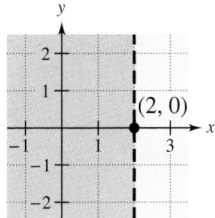

45.
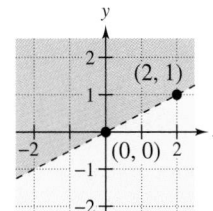

46.
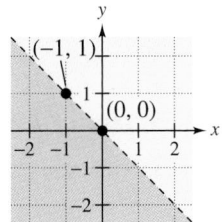

47. *Think About It*   Explain the meaning of the term *half-plane*. Give an example of an inequality whose graph is a half-plane.

48. *Think About It*   How does the solution set of $x - y > 1$ differ from the solution set of $x - y \geq 1$?

49. *Think About It*   After graphing the boundary, explain how you determine which half-plane is the solution set of a linear inequality.

50. *Think About It*   Explain the difference between graphing the solution of the inequality $x \leq 3$ on the real number line and on the rectangular coordinate system.

51. *Geometry*   The perimeter of a rectangle of length x and width y cannot exceed 500 feet. Write a linear inequality for this constraint. Use a graphing utility to graph the inequality.

52. *Storage Space*   A warehouse for storing chairs and tables has 1000 square feet of floor space. Each chair requires 10 square feet and each table requires 15 square feet. Write a linear inequality for this space constraint if x is the number of chairs and y is the number of tables stored. Sketch the graph of the inequality.

53. *Weekly Pay*   You have two part-time jobs. One is at a grocery store, which pays $9 per hour, and the other is mowing lawns, which pays $6 per hour. Between the two jobs you want to earn at least $150 per week. Write a linear inequality that shows the different numbers of hours you can work at each job, and sketch the graph of the inequality. From the graph, find several ordered pairs with positive integer coordinates that are solutions of the inequality.

54. *Getting a Workout*   The maximum heart rate r (in beats per minute) of a person in normal health is related to the person's age A (in years). The relationship between r and A is given by the inequality $r \leq 220 - A$.

(a) Sketch the graph of the inequality with A measured along the horizontal axis and r measured along the vertical axis.

(b) Physiologists recommend that during a workout a person strive to increase his or her heart rate to 75% of the maximum rate for the person's age. Sketch the graph of $r = 0.75(220 - A)$ on the same set of coordinate axes used in part (a).

Section Project

Pizza and Soda Pop   You and some friends go out for pizza. Together you have $26. You want to order two large pizzas with cheese at $8 each. Each additional topping costs $0.40, and each small soft drink costs $0.80. Write a linear inequality that represents the number of toppings x and drinks y that your group can afford. Sketch the graph of the inequality. What are the coordinates for an order of six soft drinks and two large pizzas with cheese, each with three additional toppings? Is this a solution of the inequality? (Assume there is no sales tax.)

4.6   Systems of Inequalities and Linear Programming

Systems of Linear Inequalities in Two Variables ▪
Linear Programming ▪ Application

Systems of Linear Inequalities in Two Variables

Many practical problems in business, science, and engineering involve **systems of linear inequalities.** This type of system arises in problems that have *constraint* statements that contain phrases such as "more than," "less than," "at least," "no more than," "a minimum of," and "a maximum of." A **solution** of a system of linear inequalities in x and y is a point (x, y) that satisfies each inequality in the system. For instance, the point $(2, 4)$ is a solution of the following system because $x = 2$ and $y = 4$ satisfy each of the inequalities in the system.

$$\begin{cases} x + y \le 10 \\ 3x - y \le 2 \end{cases}$$

To sketch the graph of a system of inequalities in two variables, first sketch (on the same coordinate system) the graph of each individual inequality. The **solution set** is the region that is *common* to every graph in the system.

EXAMPLE 1   *Graphing a System of Linear Inequalities*

Sketch the graph of the following system of linear inequalities.

$$\begin{cases} 2x - y \le 5 \\ x + 2y \ge 2 \end{cases}$$

Solution

Begin by sketching the graph of each inequality. The graph of $2x - y \le 5$ consists of all points on and above the line $y = 2x - 5$. The graph of $x + 2y \ge 2$ consists of all points on and above the line $y = -\frac{1}{2}x + 1$. The graph of the *system* of linear inequalities consists of the tan wedge-shaped region that is common to the two half-planes (representing the individual inequalities), as shown in Figure 4.21.

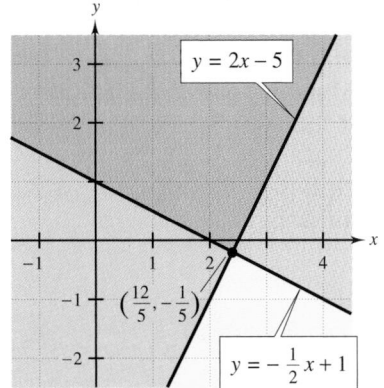

FIGURE 4.21

In Figure 4.21, note that the two borderlines of the region

$$y = 2x - 5 \quad \text{and} \quad y = -\frac{1}{2}x + 1$$

intersect at the point $\left(\frac{12}{5}, -\frac{1}{5}\right)$. Such a point is called a **vertex** of the region. The region shown in the figure has only one vertex. Some regions, however, have several vertices. When you are sketching the graph of a system of linear inequalities, it is helpful to find and label any vertices of the region.

Graphing a System of Linear Inequalities

1. Sketch the line that corresponds to each inequality. (Use dashed lines for inequalities with $<$ or $>$ and solid lines for inequalities with $\le$ or $\ge$.)

2. Lightly shade the half-plane that is the graph of each linear inequality. (Colored pencils may help you distinguish the different half-planes.)

3. The graph of the system is the intersection of the half-planes. (If you have used colored pencils, it's the region that is shaded with *every* color.)

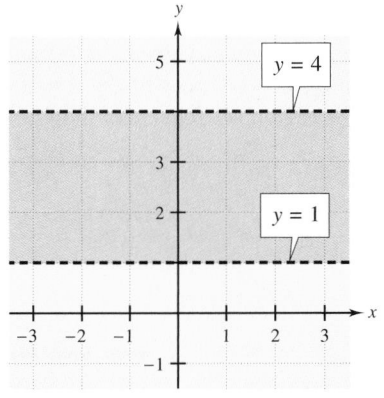

FIGURE 4.22

EXAMPLE 2   *Graphing a System of Linear Inequalities*

Sketch the graph of the following system of linear inequalities.

$$\begin{cases} y < 4 \\ y > 1 \end{cases}$$

Solution

The graph of the first inequality is the half-plane *below* the horizontal line

$$y = 4. \qquad \text{Upper boundary}$$

The graph of the second inequality is the half-plane *above* the horizontal line

$$y = 1. \qquad \text{Lower boundary}$$

The graph of the system is the horizontal band that lies *between* the two horizontal lines (where $y < 4$ *and* $y > 1$), as shown in Figure 4.22.

~~*Technology*

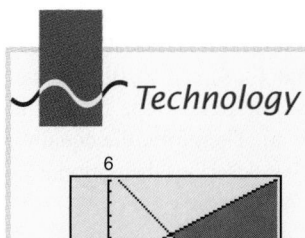

A graphing utility can be used to graph a system of linear inequalities. To graph

$$\begin{cases} 4y < 2x - 6 \\ x + y \ge 7 \end{cases}$$

on a *TI-83*, use steps similar to those shown on page 324. The graph is shown at the left. The region that is shaded by both inequalities is the solution of the system. Try using a graphing utility to graph

$$\begin{cases} 3x + y < 1 \\ -2x - 2y < 8. \end{cases}$$

EXAMPLE 3 Graphing a System of Linear Inequalities

Sketch the graph of the following system of linear inequalities, and label the vertices.

$$\begin{cases} x + y \leq 5 \\ 3x + 2y \leq 12 \\ x \qquad \geq 0 \\ \qquad y \geq 0 \end{cases}$$

Solution

Begin by sketching the half-planes represented by the four linear inequalities. The graph of $x + y \leq 5$ is the half-plane lying on and below the line $y = -x + 5$, the graph of $3x + 2y \leq 12$ is the half-plane lying on and below the line $y = -\frac{3}{2}x + 6$, the graph of $x \geq 0$ is the half-plane lying on and to the right of the y-axis, and the graph of $y \geq 0$ is the half-plane lying on and above the x-axis. As shown in Figure 4.23, the region that is common to all four of these half-planes is a four-sided polygon. The vertices of the region are found as follows.

Vertex A: (0, 5)	Vertex B: (2, 3)	Vertex C: (4, 0)	Vertex D: (0, 0)
Solution of the system	Solution of the system	Solution of the system	Solution of the system
$\begin{cases} x + y = 5 \\ x \quad = 0 \end{cases}$	$\begin{cases} x + y = 5 \\ 3x + 2y = 12 \end{cases}$	$\begin{cases} 3x + 2y = 12 \\ y = 0 \end{cases}$	$\begin{cases} x = 0 \\ y = 0 \end{cases}$

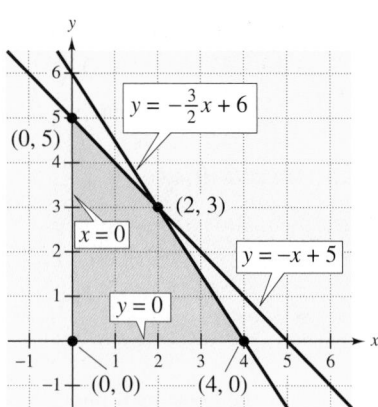

FIGURE 4.23

EXAMPLE 4 Finding the Boundaries of a Region

Find a system of inequalities that defines the region shown in Figure 4.24.

Solution

Three of the boundaries of the region are horizontal or vertical—they are easy to find. To find the diagonal boundary line, use the techniques from Section 3.1 to find the equation of the line passing through the points (4, 4) and (6, 0). You can use the slope formula to find $m = -2$, and then use the point-slope form with point (6, 0) and $m = -2$ to obtain $y - 0 = -2(x - 6)$. Therefore, the equation is $y = -2x + 12$. The system of linear inequalities that describes the region is as follows.

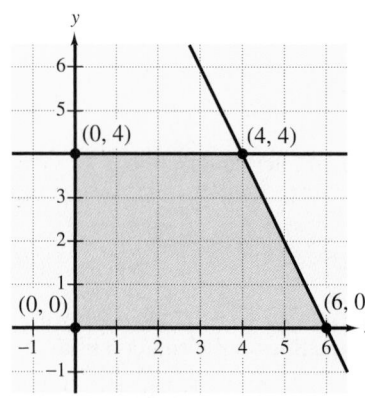

FIGURE 4.24

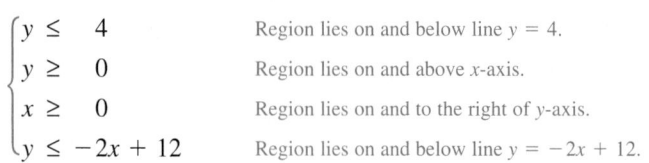

$y \leq 4$	Region lies on and below line $y = 4$.
$y \geq 0$	Region lies on and above x-axis.
$x \geq 0$	Region lies on and to the right of y-axis.
$y \leq -2x + 12$	Region lies on and below line $y = -2x + 12$.

Linear Programming

Systems of linear inequalities are used extensively in business and economics to solve **optimization problems.** The word "optimize" means to find the greatest or least. Many optimization problems can be solved using **linear programming.** A two-variable linear programming problem consists of the following.

1. An **objective function** that expresses the quantity to be maximized (or minimized).

2. A system of **constraint linear inequalities** whose solution set represents the set of **feasible solutions.**

The solution of a linear programming problem is found by determining which point in the set of feasible solutions yields the optimal value of the objective function. For example, consider a linear programming problem in which you are asked to maximize the value of

$$C = ax + by \qquad \text{Objective function}$$

subject to a set of constraints that determine the region indicated in Figure 4.25. It can be shown that if there is an optimal solution, it must occur at one of the vertices of the region. In other words, *you can find the maximum value by testing C at each of the vertices,* as illustrated in Example 5.

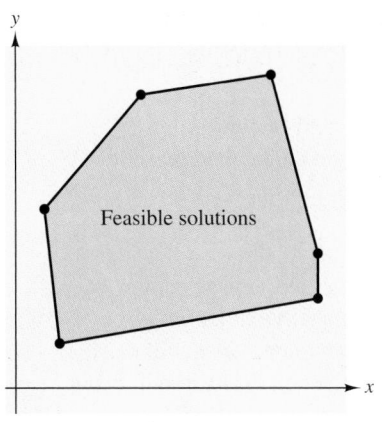

Feasible solutions

FIGURE 4.25

EXAMPLE 5   *Solving a Linear Programming Problem*

Find the minimum value and maximum value of

$$C = 4x + 5y \qquad \text{Objective function}$$

subject to the following constraints.

$$\begin{cases} x & \geq 0 \\ & y \geq 0 \\ x + y \leq 6 \end{cases} \qquad \text{Constraints}$$

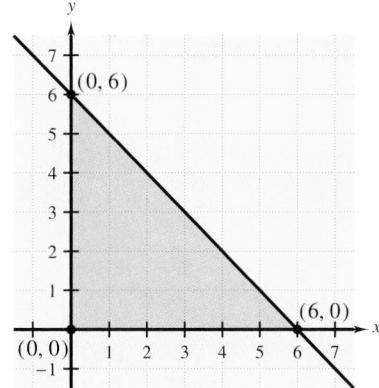

FIGURE 4.26

NOTE   In Example 5, try evaluating C at other *feasible points* in the graph of the constraints. No matter which point you choose, the value of C will be greater than or equal to 0 and less than or equal to 30.

Solution

The graph of the constraint inequalities is shown in Figure 4.26. The three vertices are $(0, 0)$, $(6, 0)$, and $(0, 6)$. To find the minimum and maximum values of C, evaluate $C = 4x + 5y$ at each of the three vertices.

At $(0, 0)$: $C = 4(0) + 5(0) = 0$ Minimum value of C

At $(6, 0)$: $C = 4(6) + 5(0) = 24$

At $(0, 6)$: $C = 4(0) + 5(6) = 30$ Maximum value of C

The minimum value of C is 0. It occurs when $x = 0$ and $y = 0$. The maximum value of C is 30. It occurs when $x = 0$ and $y = 6$.

Guidelines for Solving a Linear Programming Problem

To solve a linear programming problem involving two variables, use the following steps.

1. Sketch the region corresponding to the system of constraints. (The points inside or on the boundary of the region are called feasible solutions.)

2. Find the vertices of the region.

3. Test the objective function at each of the vertices and select the values of the variables that optimize the objective function. For a bounded region, both a minimum and a maximum value will exist. (For an unbounded region, *if* an optimal solution exists, it will occur at a vertex.)

These guidelines will work whether the objective function is to be maximized *or* minimized. For instance, in Example 5 the same test was used to find the maximum value of C and the minimum value of C.

EXAMPLE 6   *Solving a Linear Programming Problem*

Find the minimum value and maximum value of

$$C = 3x + 4y \qquad \text{Objective function}$$

subject to the following constraints.

$$\begin{cases} x \geq 2 \\ x \leq 5 \\ y \geq 1 \\ y \leq 6 \end{cases} \qquad \text{Constraints}$$

Solution

The graph of the constraint inequalities is shown in Figure 4.27. The four vertices are $(2, 1)$, $(2, 6)$, $(5, 1)$, and $(5, 6)$. To find the minimum and maximum values of C, evaluate $C = 3x + 4y$ at each of the four vertices.

At $(2, 1)$: $C = 3(2) + 4(1) = 10$ Minimum value of C
At $(2, 6)$: $C = 3(2) + 4(6) = 30$
At $(5, 1)$: $C = 3(5) + 4(1) = 19$
At $(5, 6)$: $C = 3(5) + 4(6) = 39$ Maximum value of C

The minimum value of C is 10. It occurs when $x = 2$ and $y = 1$. The maximum value of C is 39. It occurs when $x = 5$ and $y = 6$.

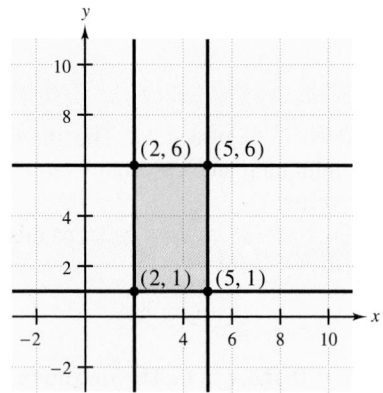

FIGURE 4.27

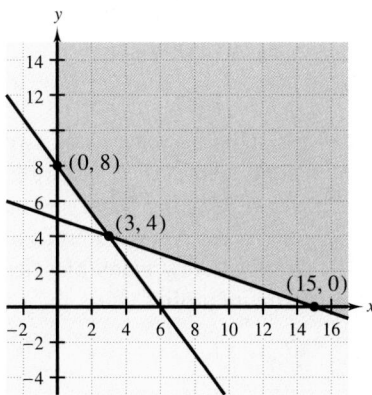

FIGURE 4.28

EXAMPLE 7   *Solving a Linear Programming Problem*

Find the minimum value and maximum value of

$$C = 6x + 7y \qquad \text{Objective function}$$

subject to the following constraints.

$$\begin{cases} x & \geq \ 0 \\ & y \geq \ 0 \\ 4x + 3y \geq 24 \\ x + 3y \geq 15 \end{cases} \qquad \text{Constraints}$$

Solution

The graph of the constraint inequalities is shown in Figure 4.28. The three vertices are $(0, 8)$, $(3, 4)$, and $(15, 0)$. To find the minimum and maximum values of C, evaluate $C = 6x + 7y$ at each of the three vertices.

At $(0, 8)$: $C = 6(0) \ + 7(8) = 56$

At $(3, 4)$: $C = 6(3) \ + 7(4) = 46$ Minimum value of C

At $(15, 0)$: $C = 6(15) + 7(0) = 90$

The minimum value of C is 46. It occurs when $x = 3$ and $y = 4$. There is no maximum value (the graph of the constraints is unbounded).

EXAMPLE 8   *Solving a Linear Programming Problem*

Find the maximum value of

$$C = 3x + 2y \qquad \text{Objective function}$$

subject to the following constraints.

$$\begin{cases} x + 2y \leq 4 \\ x - \ y \leq 1 \\ x \quad\ \geq 0 \\ \quad\ \ y \geq 0 \end{cases} \qquad \text{Constraints}$$

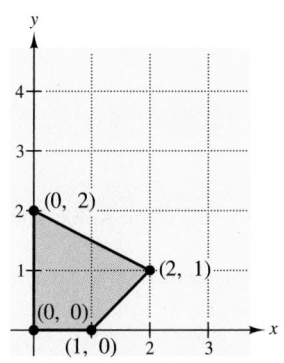

FIGURE 4.29

Solution

The constraints form the region shown in Figure 4.29. At the four vertices of this region, the objective function has the following values.

At $(0, 0)$: $C = 3(0) + 2(0) = 0$

At $(1, 0)$: $C = 3(1) + 2(0) = 3$

At $(2, 1)$: $C = 3(2) + 2(1) = 8$ Maximum value of C

At $(0, 2)$: $C = 3(0) + 2(2) = 4$

Thus, the maximum value of C is 8, and this value occurs when $x = 2$ and $y = 1$.

Application

EXAMPLE 9   Finding the Maximum Profit

You own a bicycle manufacturing plant and can assemble bicycles using two processes. The hours of unskilled labor, machine time, and skilled labor *per bicycle* are given below. You can use up to 4200 hours of unskilled labor and up to 2400 hours each of machine time and skilled labor. Process A earns a profit of $45 per bike, and process B earns a profit of $50 per bike. How many bicycles should you assemble by each process to obtain a maximum profit?

	Unskilled labor	Machine time	Skilled labor
Hours for process A	3	1	2
Hours for process B	3	2	1

Solution

Let a and b represent the numbers of bicycles assembled by the two processes. Because you want a maximum profit P, the objective function is

$$P = 45a + 50b.$$

Profit: $45 per bike for process A
$50 per bike for process B

The constraints are as follows.

$$\begin{cases} 3a + 3b \le 4200 \\ a + 2b \le 2400 \\ 2a + b \le 2400 \\ a \qquad \ge \quad 0 \\ \qquad b \ge \quad 0 \end{cases}$$

Unskilled labor: Up to 4200 hours
Machine time: Up to 2400 hours
Skilled labor: Up to 2400 hours
Cannot produce a negative amount
Cannot produce a negative amount

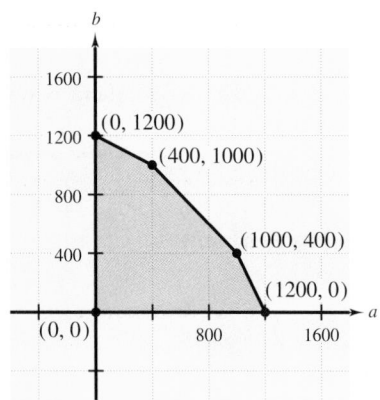

FIGURE 4.30

The region that represents the feasible solutions is shown in Figure 4.30. The profits at the vertices of the region are as follows.

At $(0, 1200)$: $P = \$60,000$

At $(400, 1000)$: $P = \$68,000$ Maximum profit

At $(1000, 400)$: $P = \$65,000$

At $(1200, 0)$: $P = \$54,000$

At $(0, 0)$: $P = \quad \$0$

The maximum profit is obtained by making 400 bicycles by process A and 1000 bicycles by process B.

Group Activities

Writing a Linear Programming Problem

In your group, perform the following steps.

- Draw a coordinate plane on a piece of paper.
- Draw a line segment from (0, 0) to (4, 0).
- Draw a line segment from (4, 0) to (3, 2).
- Draw a line segment from (3, 2) to (0, 3).
- Draw a line segment from (0, 3) to (0, 0).
- Shade the quadrilateral formed by the four line segments.

Now, with others in your group, write a linear programming problem whose constraints determine the region you shaded. Then find an objective function that has a maximum at the vertex (3, 2).

4.6 Exercises

See Warm-Up Exercises, p. A43

In Exercises 1–6, match the system of linear inequalities with its graph. [The graphs are labeled (a), (b), (c), (d), (e), and (f).]

(a)

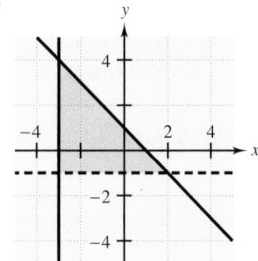

(b)

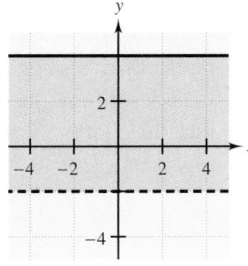

(c)

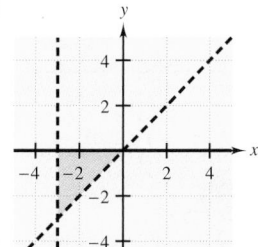

(d)

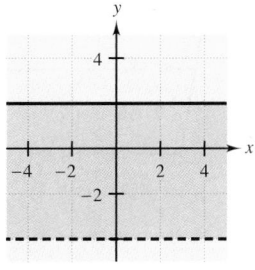

(e)

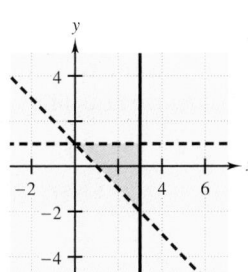

(f)

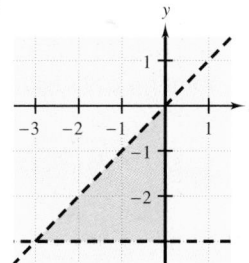

1. $\begin{cases} y > x \\ x > -3 \\ y \le 0 \end{cases}$

2. $\begin{cases} y \le 4 \\ y > -2 \end{cases}$

3. $\begin{cases} y < x \\ y > -3 \\ x \le 0 \end{cases}$

4. $\begin{cases} x \le 3 \\ y < 1 \\ y > -x + 1 \end{cases}$

5. $\begin{cases} y > -1 \\ x \ge -3 \\ y \le -x + 1 \end{cases}$

6. $\begin{cases} y > -4 \\ y \le 2 \end{cases}$

In Exercises 7–20, sketch a graph of the solution of the system of linear inequalities.

7. $\begin{cases} x < 3 \\ x > -2 \end{cases}$

8. $\begin{cases} y > -1 \\ y \le 2 \end{cases}$

9. $\begin{cases} y > -5 \\ x \le 2 \\ y \le x + 2 \end{cases}$

10. $\begin{cases} y \ge -1 \\ x \le 2 \\ y \le x + 2 \end{cases}$

11. $\begin{cases} x + y \le 1 \\ -x + y \le 1 \\ y \ge 0 \end{cases}$

12. $\begin{cases} 3x + 2y < 6 \\ x \ge 0 \\ y \ge 0 \end{cases}$

13. $\begin{cases} x + y \le 5 \\ x \ge 2 \\ y \ge 0 \end{cases}$

14. $\begin{cases} 2x + y \ge 2 \\ x \le 2 \\ y \le 1 \end{cases}$

15. $\begin{cases} -3x + 2y < 6 \\ x - 4y > -2 \\ 2x + y < 3 \end{cases}$

16. $\begin{cases} x - 7y > -36 \\ 5x + 2y > 5 \\ 6x + 5y > 6 \end{cases}$

17. $\begin{cases} 2x + y < 2 \\ 6x + 3y > 2 \end{cases}$

18. $\begin{cases} x - 2y < -6 \\ 5x - 3y > -9 \end{cases}$

19. $\begin{cases} x \ge 1 \\ x - 2y \le 3 \\ 3x + 2y \ge 9 \\ x + y \le 6 \end{cases}$

20. $\begin{cases} x + y \le 4 \\ x + y \ge -1 \\ x - y \ge -2 \\ x - y \le 2 \end{cases}$

In Exercises 21–28, use a graphing utility to graph the solution of the system of linear inequalities.

21. $\begin{cases} 2x - 3y \le 6 \\ y \le 4 \end{cases}$

22. $\begin{cases} 6x + 3y \ge 12 \\ y \le 4 \end{cases}$

23. $\begin{cases} 2x - 2y \le 5 \\ y \le 6 \end{cases}$

24. $\begin{cases} y \ge -2 \\ y \le 8 \end{cases}$

25. $\begin{cases} 2x + y \le 2 \\ y \ge -4 \end{cases}$

26. $\begin{cases} x - 2y \ge -6 \\ y \le 6 \end{cases}$

27. $\begin{cases} 2x + 3y \ge 12 \\ y \ge 2 \end{cases}$

28. $\begin{cases} x - y \ge -4 \\ y \le 1 \end{cases}$

In Exercises 29–32, write a system of linear inequalities that describes the shaded region.

29.

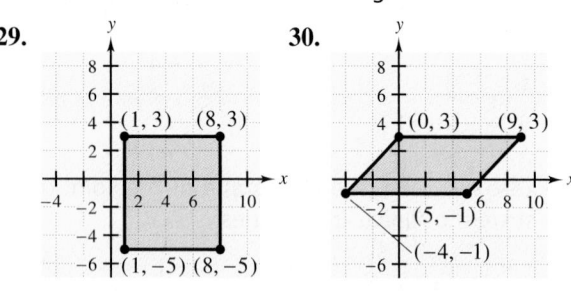

30.

31.

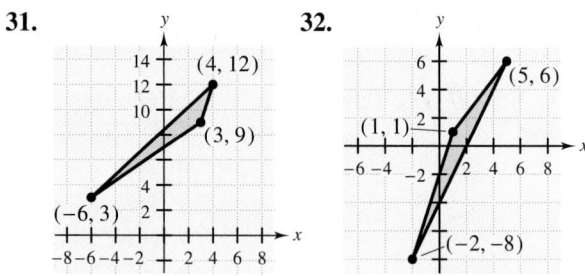

32.

33. *Swimming Safety*   The figure shows a cross section of a roped-off swimming area at a beach. Write a system of linear inequalities describing the cross section. (Each unit in the coordinate system represents 1 foot.)

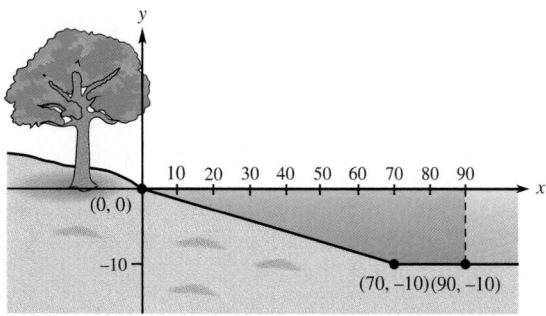

34. *A View of the Chorus*   The figure shows the chorus platform on a stage. Write a system of linear inequalities describing the part of the audience that can see the full chorus. (Each unit in the coordinate system represents 1 meter.)

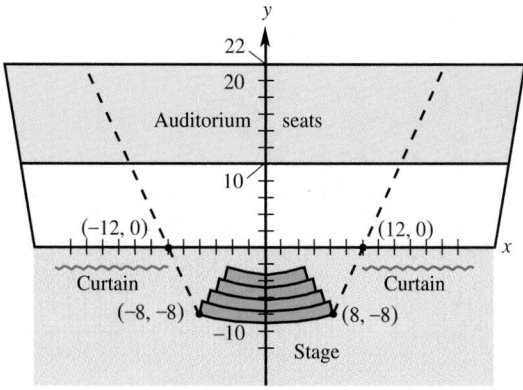

35. *Investment*   A person plans to invest up to $20,000 in two different interest-bearing accounts. Each account is to contain at least $5000. Moreover, one account should have at least twice the amount in the other account. Write a system of linear inequalities describing the various amounts that can be deposited into each account, and sketch the graph of the system.

36. *Roasting a Turkey*   The time t (in minutes) that it takes to roast a turkey weighing p pounds at 350°F is given by the following inequalities.

> For a turkey up to 6 pounds: $t \geq 20p$
>
> For a turkey over 6 pounds: $t \geq 15p + 30$

Sketch the graphs of these inequalities. What are the coordinates for a 12-pound turkey that has been roasting for 3 hours and 40 minutes? Is this turkey fully cooked?

37. *Concert Ticket Sales*   Two types of tickets are to be sold for a concert. One type costs $15 per ticket and the other type costs $25 per ticket. The promoter of the concert must sell at least 15,000 tickets, including at least 8000 of the $15 tickets and at least 4000 of the $25 tickets. Moreover, the gross receipts must total at least $275,000 in order for the concert to be held. Write a system of linear inequalities describing the different numbers of tickets that can be sold. Use a graphing utility to graph the system.

38. *Diet Supplement*   A dietician is asked to design a special diet supplement using two different foods. Each ounce of food X contains 20 units of calcium, 15 units of iron, and 10 units of vitamin B. Each ounce of food Y contains 10 units of calcium, 10 units of iron, and 20 units of vitamin B. The minimum daily requirements in the diet are 280 units of calcium, 160 units of iron, and 180 units of vitamin B. Write a system of linear inequalities describing the different amounts of food X and food Y that can be used in the diet. Use a graphing utility to graph the system.

In Exercises 39–50, find the minimum and maximum values of the objective function subject to the constraints. (For each exercise, the graph of the region determined by the constraints is provided.)

39. Objective function: $C = 4x + 5y$

$$\text{Constraints:} \begin{cases} x & \geq 0 \\ & y \geq 0 \\ x + y \leq 6 \end{cases}$$

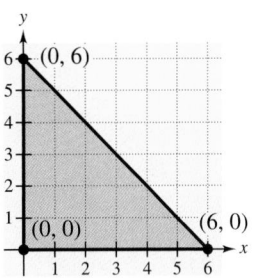
Figure for 39

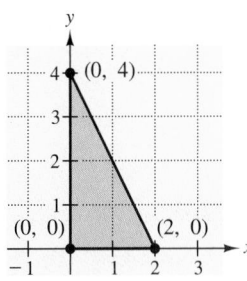
Figure for 40

40. Objective function: $C = 2x + 8y$

$$\text{Constraints:} \begin{cases} x & \geq 0 \\ & y \geq 0 \\ 2x + y \leq 4 \end{cases}$$

41. Objective function: $C = 10x + 6y$

Constraints: (See Exercise 39.)

42. Objective function: $C = 7x + 3y$

Constraints: (See Exercise 40.)

43. Objective function: $C = 3x + 2y$

$$\text{Constraints:} \begin{cases} x & \geq 0 \\ & y \geq 0 \\ x + 3y \leq 15 \\ 4x + y \leq 16 \end{cases}$$

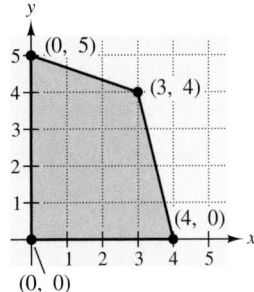

44. Objective function: $C = 4x + 3y$

Constraints: $\begin{cases} x & \geq & 0 \\ 2x + 3y & \geq & 6 \\ 3x - 2y & \leq & 9 \\ x + 5y & \leq & 20 \end{cases}$

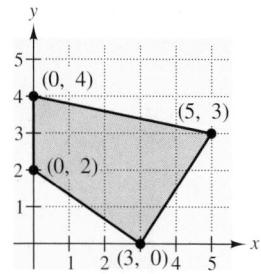

45. Objective function: $C = 5x + 0.5y$

Constraints: (See Exercise 43.)

46. Objective function: $C = x + 6y$

Constraints: (See Exercise 44.)

47. Objective function: $C = 10x + 7y$

Constraints: $\begin{cases} 0 \leq x \leq 60 \\ 0 \leq y \leq 45 \\ 5x + 6y \leq 420 \end{cases}$

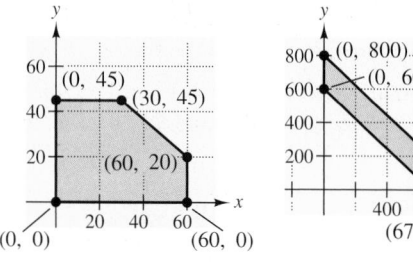

Figure for 47   **Figure for 48**

48. Objective function: $C = 50x + 35y$

Constraints: $\begin{cases} x & \geq & 0 \\ & y \geq & 0 \\ 8x + 9y & \leq & 7200 \\ 8x + 9y & \geq & 5400 \end{cases}$

49. Objective function: $C = 25x + 35y$

Constraints: (See Exercise 47.)

50. Objective function: $C = 16x + 20y$

Constraints: (See Exercise 48.)

In Exercises 51–58, sketch the region determined by the constraints. Then find the minimum and maximum values of the objective function subject to the constraints.

51. Objective function: $C = -2x + y$

Constraints: $\begin{cases} x \geq -5 \\ x \leq 4 \\ y \geq -1 \\ y \leq 3 \end{cases}$

52. Objective function: $C = 5x + 4y$

Constraints: $\begin{cases} x \leq 2 \\ x \geq -4 \\ y \geq 1 \\ y \leq 6 \end{cases}$

53. Objective function: $C = x + 4y$

Constraints: $\begin{cases} x & \geq & 0 \\ x & \leq & 12 \\ & y \leq & 10 \\ 2x + 3y & \geq & 24 \end{cases}$

54. Objective function: $C = 6x + 2y$

Constraints: $\begin{cases} x & \geq & 0 \\ x & \leq & 5 \\ & y \geq & 0 \\ 4x - y & \geq & 1 \end{cases}$

55. Objective function: $C = 6x + 10y$

Constraints: $\begin{cases} x & \geq & 0 \\ & y \geq & 0 \\ 2x + 5y & \leq & 10 \end{cases}$

56. Objective function: $C = 7x + 8y$

Constraints: $\begin{cases} x & \geq & 0 \\ & y \geq & 0 \\ x + \frac{1}{2}y & \leq & 4 \end{cases}$

57. Objective function: $C = 9x + 4y$

Constraints: (See Exercise 55.)

58. Objective function: $C = 7x + 2y$

Constraints: (See Exercise 56.)

In Exercises 59–62, use a graphing utility to graph the region determined by the constraints. Then find the minimum and maximum values of the objective function subject to the constraints.

59. Objective function: $C = 4x + 5y$

Constraints: $\begin{cases} x && \geq 0 \\ & y & \geq 0 \\ 4x + 3y & \geq 27 \\ x + y & \geq 8 \\ 3x + 5y & \geq 30 \end{cases}$

60. Objective function: $C = 4x + 5y$

Constraints: $\begin{cases} x && \geq 0 \\ & y & \geq 0 \\ 2x + 2y & \leq 10 \\ x + 2y & \leq 6 \end{cases}$

61. Objective function: $C = 2x + 7y$

Constraints: (See Exercise 59.)

62. Objective function: $C = 2x - y$

Constraints: (See Exercise 60.)

63. *Maximum Profit* A fruit grower has 150 acres of land available to raise two crops, A and B. It takes 1 day to trim an acre of crop A and 2 days to trim an acre of crop B, and there are 240 days per year available for trimming. It takes 0.3 day to pick an acre of crop A and 0.1 day to pick an acre of crop B, and there are 30 days per year available for picking. Find the number of acres of each fruit that should be planted to maximize profit, assuming that the profit is $140 per acre for crop A and $235 per acre for crop B.

64. *Investments* An investor has up to $450,000 to invest in two types of investments. Type A investments pay 6% annually and type B investments pay 10% annually. To have a well-balanced portfolio, the investor imposes the following conditions. At least one-half of the total portfolio is to be allocated to type A investments and at least one-fourth of the total portfolio to type B investments. How much should be allocated to each type of investment to obtain a maximum return?

65. *Minimum Cost* A farming cooperative mixes two brands of cattle feed. Brand X costs $25 per bag and contains two units of nutrient A, two units of nutrient B, and two units of nutrient C. Brand Y costs $20 per bag and contains one unit of nutrient A, nine units of nutrient B, and three units of nutrient C. The minimum requirements of nutrients A, B, and C are 12 units, 36 units, and 24 units, respectively. Find the number of bags of each brand that should be mixed to produce the required mixture having a minimum cost.

Section Project

Maximum Profit A manufacturer produces two models of bicycles. The amounts of time (in hours) required for assembling, painting, and packaging each model are as follows.

	Assembly	Painting	Packaging
Model A	2	4	1
Model B	2.5	1	0.75

The total amounts of time available for assembly, painting, and packaging are 4000 hours, 4800 hours, and 1500 hours, respectively. The profits per unit for the two models are $45 (model A) and $50 (model B). How many of each model should be produced to obtain a maximum profit?

Chapter Summary

After studying this chapter, you should have acquired the following skills. These skills are keyed to the Review Exercises that begin on page 341. Answers to odd-numbered Review Exercises are given in the back of the book.

- Solve systems of linear equations graphically. *(Section 4.1)*

  Review Exercises 1–10

- Solve systems of linear equations by the method of substitution. *(Section 4.1)*

  Review Exercises 11–16

- Solve systems of linear equations by the method of elimination or Gaussian elimination. *(Sections 4.1, 4.2)*

  Review Exercises 17–22

- Solve systems of linear equations using matrices. *(Section 4.3)*

  Review Exercises 23–26

- Solve systems of linear equations using the matrix capabilities of a graphing utility. *(Section 4.3)*

  Review Exercises 27–30

- Evaluate the determinant of a 2×2 matrix. *(Section 4.4)*

  Review Exercises 31, 32

- Evaluate the determinant of a 3×3 matrix by expanding by minors. *(Section 4.4)*

  Review Exercises 33, 34

- Evaluate the determinant of a 3×3 matrix using any appropriate method. *(Section 4.4)*

  Review Exercises 35, 36

- Evaluate the determinant of a 3×3 matrix using the matrix capabilities of a graphing utility. *(Section 4.4)*

  Review Exercises 37, 38

- Solve systems of linear equations using Cramer's Rule. *(Section 4.4)*

  Review Exercises 39–44

- Use determinants to find areas of triangles and equations of lines. *(Section 4.4)*

  Review Exercises 45–52

- Create systems of linear equations having given solutions. *(Section 4.1)*

  Review Exercises 53, 54

- Model real-life situations with linear equations and inequalities, and solve. *(Sections 4.1–4.3, 4.5, 4.6)*

  Review Exercises 55–62, 64, 79, 80, 85, 86

- Find equations of parabolas and circles passing through points. *(Sections 4.2–4.4)*

  Review Exercises 63, 65, 66

- Sketch graphs of solutions of linear inequalities. *(Section 4.5)*

  Review Exercises 67–72

- Sketch graphs of solutions of systems of linear inequalities. *(Section 4.6)*

  Review Exercises 73–76

- Create systems of linear inequalities to describe regions. *(Section 4.6)*

  Review Exercises 77, 78

- Find the maximum and minimum values of objective functions subject to constraints. *(Section 4.6)*

  Review Exercises 81–84

Review Exercises

In Exercises 1–6, sketch the graphs of the equations and approximate any solutions of the systems of linear equations.

1. $\begin{cases} x + y = 2 \\ x - y = 0 \end{cases}$

2. $\begin{cases} 2x = 3(y - 1) \\ y = x \end{cases}$

3. $\begin{cases} x - y = 3 \\ -x + y = 1 \end{cases}$

4. $\begin{cases} x + y = -1 \\ 3x + 2y = 0 \end{cases}$

5. $\begin{cases} 2x - y = 0 \\ -x + y = 4 \end{cases}$

6. $\begin{cases} x = y + 3 \\ x = y + 1 \end{cases}$

In Exercises 7–10, use a graphing utility to graph the equations and approximate any solutions of the system of linear equations.

7. $\begin{cases} 5x - 3y = 3 \\ 2x + 2y = 14 \end{cases}$

8. $\begin{cases} 8x + 5y = 1 \\ 3x - 4y = 18 \end{cases}$

9. $\begin{cases} x - 3y = -1 \\ -3x + 2y = -4 \end{cases}$

10. $\begin{cases} 2x + y = 4 \\ -x - y = -1 \end{cases}$

In Exercises 11–16, solve the system of linear equations by the method of substitution.

11. $\begin{cases} 2x + 3y = 1 \\ x + 4y = -2 \end{cases}$

12. $\begin{cases} 3x - 7y = 10 \\ -2x + y = -14 \end{cases}$

13. $\begin{cases} -5x + 2y = 4 \\ 10x - 4y = 7 \end{cases}$

14. $\begin{cases} 5x + 2y = 3 \\ 2x + 3y = 10 \end{cases}$

15. $\begin{cases} 3x - 7y = 5 \\ 5x - 9y = -5 \end{cases}$

16. $\begin{cases} 24x - 4y = 20 \\ 6x - y = 5 \end{cases}$

In Exercises 17–22, solve the system of equations by the method of elimination or Gaussian elimination.

17. $\begin{cases} x + y = 0 \\ 2x + y = 0 \end{cases}$

18. $\begin{cases} 4x + y = 1 \\ x - y = 4 \end{cases}$

19. $\begin{cases} 2x - y = 2 \\ 6x + 8y = 39 \end{cases}$

20. $\begin{cases} 0.2x + 0.3y = 0.14 \\ 0.4x + 0.5y = 0.20 \end{cases}$

21. $\begin{cases} -x + y + 2z = 1 \\ 2x + 3y + z = -2 \\ 5x + 4y + 2z = 4 \end{cases}$

22. $\begin{cases} 2x + 3y + z = 10 \\ 2x - 3y - 3z = 22 \\ 4x - 2y + 3z = -2 \end{cases}$

In Exercises 23–26, use matrices to solve the system of linear equations.

23. $\begin{cases} 5x + 4y = 2 \\ -x + y = -22 \end{cases}$

24. $\begin{cases} 2x - 5y = 2 \\ 3x - 7y = 1 \end{cases}$

25. $\begin{cases} x + 2y + 6z = 4 \\ -3x + 2y - z = -4 \\ 4x + 2z = 16 \end{cases}$

26. $\begin{cases} 2x_1 + 3x_2 + 3x_3 = 3 \\ 6x_1 + 6x_2 + 12x_3 = 13 \\ 12x_1 + 9x_2 - x_3 = 2 \end{cases}$

In Exercises 27–30, use the matrix capabilities of a graphing utility to solve the system of equations.

27. $\begin{cases} 0.2x - 0.1y = 0.07 \\ 0.4x - 0.5y = -0.01 \end{cases}$

28. $\begin{cases} 2x + y = 0.3 \\ 3x - y = -1.3 \end{cases}$

29. $\begin{cases} 5x - 3y + 2z = 2 \\ 2x + 2y - 3z = 3 \\ x - 7y + 8z = -4 \end{cases}$

30. $\begin{cases} 3x + 2y + 5z = 4 \\ 4x - 3y - 4z = 1 \\ -8x + 2y + 3z = 0 \end{cases}$

In Exercises 31 and 32, find the determinant of the matrix.

31. $\begin{bmatrix} 7 & 10 \\ 10 & 15 \end{bmatrix}$

32. $\begin{bmatrix} -3.4 & 1.2 \\ -5 & 2.5 \end{bmatrix}$

In Exercises 33 and 34, evaluate the determinant of the matrix six different ways by expanding by minors along each row and column.

33. $\begin{bmatrix} 8 & 6 & 3 \\ 6 & 3 & 0 \\ 3 & 0 & 2 \end{bmatrix}$ **34.** $\begin{bmatrix} 7 & -1 & 10 \\ -3 & 0 & -2 \\ 12 & 1 & 1 \end{bmatrix}$

In Exercises 35 and 36, find the determinant of the matrix using any appropriate method.

35. $\begin{bmatrix} 8 & 3 & 2 \\ 1 & -2 & 4 \\ 6 & 0 & 5 \end{bmatrix}$ **36.** $\begin{bmatrix} 4 & 0 & 10 \\ 0 & 10 & 0 \\ 10 & 0 & 34 \end{bmatrix}$

In Exercises 37 and 38, use the matrix capabilities of a graphing utility to evaluate the determinant.

37. $\begin{bmatrix} 2 & -5 & 0 \\ 4 & 7 & 0 \\ -7 & 25 & 3 \end{bmatrix}$ **38.** $\begin{bmatrix} 8 & 7 & 6 \\ -4 & 0 & 0 \\ 5 & 1 & 4 \end{bmatrix}$

In Exercises 39–44, solve the system of linear equations using Cramer's Rule. (If not possible, state the reason.)

39. $\begin{cases} 7x + 12y = 63 \\ 2x + 3y = 15 \end{cases}$ **40.** $\begin{cases} 12x + 42y = -17 \\ 30x - 18y = 19 \end{cases}$

41. $\begin{cases} 3x - 2y = 16 \\ 12x - 8y = -5 \end{cases}$ **42.** $\begin{cases} 4x + 24y = 20 \\ -3x + 12y = -5 \end{cases}$

43. $\begin{cases} -x + y + 2z = 1 \\ 2x + 3y + z = -2 \\ 5x + 4y + 2z = 4 \end{cases}$

44. $\begin{cases} 2x_1 + x_2 + 2x_3 = 4 \\ 2x_1 + 2x_2 = 5 \\ 2x_1 - x_2 + 6x_3 = 2 \end{cases}$

Area of a Triangle In Exercises 45–48, use a determinant to find the area of the triangle with the given vertices.

45. $(1, 0), (5, 0), (5, 8)$ **46.** $(-4, 0), (4, 0), (0, 6)$
47. $(1, 2), (4, -5), (3, 2)$ **48.** $\left(\frac{3}{2}, 1\right), \left(4, -\frac{1}{2}\right), (4, 2)$

Equation of a Line In Exercises 49–52, use a determinant to find the equation of the line through the points.

49. $(-4, 0), (4, 4)$
50. $(2, 5), (6, -1)$
51. $\left(-\frac{5}{2}, 3\right), \left(\frac{7}{2}, 1\right)$
52. $(-0.8, 0.2), (0.7, 3.2)$

In Exercises 53 and 54, create a system of equations having the given solution. (*Note:* There are many correct answers.)

53. $\left(\frac{2}{3}, -4\right)$
54. $(-10, 12)$

55. *Break-Even Analysis* A small business invests $25,000 in equipment to produce a product. Each unit of the product costs $3.75 to produce and is sold for $5.25. How many items must be sold before the business breaks even?

56. *Dimensions of a Rectangle* The perimeter of a rectangle is 480 meters, and its length is 150% of its width. Find the dimensions of the rectangle.

57. *Acid Mixture* One hundred gallons of a 60% acid solution is obtained by mixing a 75% solution with a 50% solution. How many gallons of each solution must be used to obtain the desired mixture?

58. *Rope Length* Suppose you must cut a rope that is 128 inches long into two pieces such that one piece is three times longer than the other. Find the length of each piece.

59. *Cassette Tape Sales* You are the manager of a music store and are going over receipts for the previous week's sales. Six hundred and fifty cassette tapes of two different types were sold. One type of cassette sold for $9.95, and the other sold for $14.95. The total cassette receipts were $7717.50. The cash register that was supposed to record the number of each type of cassette sold malfunctioned. Can you recover the information? If so, how many of each type of cassette were sold?

60. Flying Speeds Two planes leave Pittsburgh and Philadelphia at the same time, each flying to the other city. Because of the wind, one plane flies 25 miles per hour faster than the other. Find the ground-speed of each plane if the cities are 275 miles apart and if the planes pass one another after 40 minutes.

61. Investments An inheritance of $20,000 is divided among three investments yielding $1780 in interest per year. The interest rates for the three investments are 7%, 9%, and 11%. Find the amount placed in each investment if the second and third are $3000 and $1000 less than the first, respectively.

62. Number Problem The sum of three positive numbers is 68. The second number is four greater than the first, and the third is twice the first. Find the three numbers.

63. Curve-Fitting Find the equation

$$y = ax^2 + bx + c$$

that passes through the points $(0, -6)$, $(1, -3)$, and $(2, 4)$.

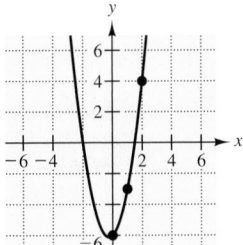

64. Mathematical Modeling A child throws a softball over a garage. The locations of the eaves and the peak of the roof are given by $(0, 10)$, $(15, 15)$, and $(30, 10)$.

(a) Find the equation $y = ax^2 + bx + c$ for the path of the ball if the ball follows a path 1 foot over the eaves and the peak of the roof.

(b) Use a graphing utility to graph the path of the ball in part (a).

(c) From the graph, estimate how far from the edge of the garage the child is standing if the ball is at a height of 5 feet when it leaves his hand.

Curve-Fitting In Exercises 65 and 66, find an equation of the circle $x^2 + y^2 + Dx + Ey + F = 0$ that passes through the points.

65. $(2, 2), (5, -1),$ **66.** $(4, 2), (1, 3), (-2, -6)$
$(-1, -1)$

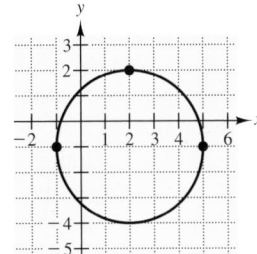

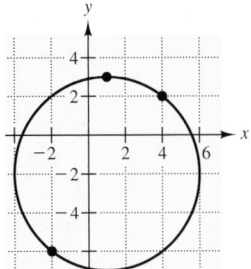

In Exercises 67–72, sketch the graph of the solution of the linear inequality in two variables. Use a graphing utility to confirm your result.

67. $x - 2 \geq 0$ **68.** $y + 3 < 0$

69. $2x + y < 1$ **70.** $3x - 4y > 2$

71. $x \leq 4y - 2$ **72.** $(y - 3) \geq \frac{2}{3}(x - 5)$

In Exercises 73–76, sketch a graph of the solution of the system of linear inequalities.

73. $\begin{cases} x + 2y \leq 160 \\ 3x + y \leq 180 \\ x \quad\quad \geq 0 \\ \quad\quad y \geq 0 \end{cases}$ **74.** $\begin{cases} 2x + 3y \leq 24 \\ 2x + y \leq 16 \\ x \quad\quad \geq 0 \\ \quad\quad y \geq 0 \end{cases}$

75. $\begin{cases} 3x + 2y \geq 24 \\ x + 2y \geq 12 \\ 2 \leq x \leq 15 \\ y \leq 15 \end{cases}$ **76.** $\begin{cases} 2x + y \geq 16 \\ x + 3y \geq 18 \\ 0 \leq x \leq 25 \\ 0 \leq y \leq 25 \end{cases}$

In Exercises 77 and 78, derive a system of linear inequalities to describe the region.

77. Parallelogram with vertices at $(1, 5)$, $(3, 1)$, $(6, 10)$, and $(8, 6)$

78. Triangle with vertices at $(1, 2)$, $(6, 7)$, and $(8, 1)$

In Exercises 79 and 80, determine a system of linear inequalities that models the description, and sketch a graph of the solution of the system.

79. *Fruit Distribution*   A Pennsylvania fruit grower has up to 1500 bushels of apples that are to be divided between markets in Harrisburg and Philadelphia. These two markets need at least 400 bushels and 600 bushels, respectively.

80. *Inventory Costs*   A warehouse operator has up to 24,000 square feet of floor space in which to store two products. Each unit of product I requires 20 square feet of floor space and costs $12 per day to store. Each unit of product II requires 30 square feet of floor space and costs $8 per day to store. The total storage cost per day cannot exceed $12,400.

81. Maximize the objective function:

$$C = 3x + 4y.$$

Constraints: $\begin{cases} x & \geq 0 \\ y \geq 0 \\ 2x + 5y \leq 50 \\ 4x + y \leq 28 \end{cases}$

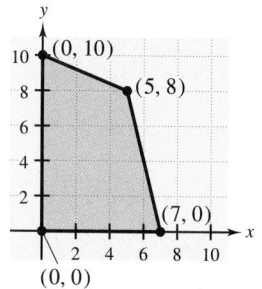

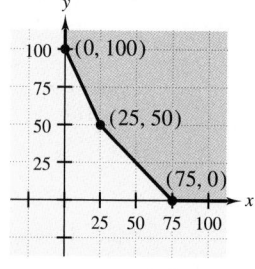

Figure for 81   **Figure for 82**

82. Minimize the objective function:

$$C = 10x + 7y.$$

Constraints: $\begin{cases} x & \geq 0 \\ y \geq 0 \\ 2x + y \geq 100 \\ x + y \geq 75 \end{cases}$

In Exercises 83 and 84, use a graphing utility to graph the region determined by the constraints. Then find the minimum or maximum value of the objective function subject to the constraints.

83. Minimize the objective function $C = 1.75x + 2.25y.$

Constraints: $\begin{cases} x & \geq 0 \\ y \geq 0 \\ 2x + y \geq 25 \\ 3x + 2y \geq 45 \end{cases}$

84. Maximize the objective function $C = 50x + 70y.$

Constraints: $\begin{cases} x & \geq 0 \\ y \geq 0 \\ x + 2y \leq 1500 \\ 5x + 2y \leq 3500 \end{cases}$

85. *Maximum Profit*   A manufacturer produces products A and B, yielding profits of $18 and $24 per unit, respectively. Each product must go through two processes, for which the required times per unit are shown in the table.

Process	Hours for product A	Hours for product B	Hours available per day
I	4	2	24
II	1	2	9

Find the daily production level for each unit to maximize profit.

86. *Minimum Cost*   A pet supply company mixes two brands of dry dog food. Brand X costs $15 per bag and contains eight units of nutrient A, one unit of nutrient B, and two units of nutrient C. Brand Y costs $30 per bag and contains two units of nutrient A, one unit of nutrient B, and seven units of nutrient C. Each bag of dog food must contain at least 16 units, 5 units, and 20 units of nutrients A, B, and C, respectively. Find the numbers of bags of brands X and Y that should be mixed to produce a mixture meeting the minimum nutritional requirements and having a minimum cost.

Chapter Test

Take this test as you would take a test in class. After you are done, check your work against the answers given in the back of the book.

In Exercises 1–6, use the indicated method to solve the system.

1. Substitution:

$$\begin{cases} 5x - y = 6 \\ 4x - 3y = -4 \end{cases}$$

2. Graphing Utility:

$$\begin{cases} x - 2y = -3 \\ 2x + 3y = 22 \end{cases}$$

3. Elimination:

$$\begin{cases} x + 2y - 4z = 0 \\ 3x + y - 2z = 5 \\ 3x - y + 2z = 7 \end{cases}$$

4. Matrices:

$$\begin{cases} x - 3y + z = -3 \\ 3x + 2y - 5z = 18 \\ y + z = -1 \end{cases}$$

5. Cramer's Rule:

$$\begin{cases} 2x - 7y = 7 \\ 3x + 7y = 13 \end{cases}$$

6. Any Method:

$$\begin{cases} 4x + y + 2z = -4 \\ 3y + z = 8 \\ -3x + y - 3z = 5 \end{cases}$$

7. Evaluate the determinant of the matrix. $\begin{bmatrix} 3 & -2 & 0 \\ -1 & 5 & 3 \\ 2 & 7 & 1 \end{bmatrix}$

8. Find the equation $y = ax^2 + bx + c$ that passes through the points $(0, 4)$, $(1, 3)$, and $(2, 6)$.

9. Use a determinant to find the area of the triangle with vertices $(0, 0)$, $(5, 4)$, and $(6, 0)$.

10. Sketch the graph of the inequality $x + 2y \leq 4$.

In Exercises 11 and 12, use a graphing utility to graph the solution of the system of linear inequalities.

11. $\begin{cases} 3x - y < 4 \\ x > 0 \\ y > 0 \end{cases}$

12. $\begin{cases} x + y < 6 \\ 2x + 3y > 9 \\ x \geq 0 \\ y \geq 0 \end{cases}$

13. Find the minimum and maximum values of the objective function subject to the indicated constraints (see figure).

Objective function: $C = 5x + 11y$ Constraints: $\begin{cases} x \geq 0 \\ y \geq 0 \\ x + 3y \leq 12 \\ 3x + 2y \leq 15 \end{cases}$

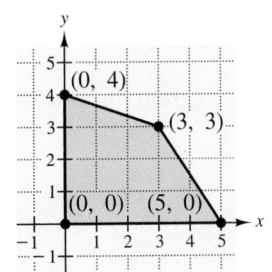

Figure for 13

14. Two people share the driving for a 200-mile trip. One person drives four times as far as the other. Write a system of linear equations that models this problem. Find the distance each person drives.

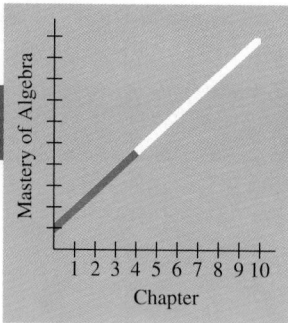

Mastery of Algebra

1 2 3 4 5 6 7 8 9 10
Chapter

Take this test as you would take a test in class. After you are done, check your work against the answers given in the back of the book.

In Exercises 1 and 2, simplify the expression by using the properties of exponents.

1. $(-3a^5)^2 \cdot (-6a^8)$

2. $\dfrac{(2x^3y)^4}{6x^2y^3}$

In Exercises 3 and 4, factor the expression completely.

3. $x^3 - 3x^2 - x + 3$

4. $y^3 - 64$

In Exercises 5–7, solve the equation.

5. $x + \dfrac{x}{2} = 4$

6. $8(x - 1) + 14 = 3(x + 7)$

7. $x^2 + x - 42 = 0$

8. Use a graphing utility to graph each function and identify the transformation of the graph of $f(x) = x^5$.

(a) $g(x) = x^5 - 2$ (b) $g(x) = (x - 2)^5$ (c) $g(x) = -x^5$

9. The length and width of the rectangle shown in the figure are x centimeters and y centimeters. The perimeter of the rectangle is 500 centimeters.

(a) Write y in terms of x.

(b) Write the area A of the rectangle as a function of x.

10. Write an equation of the line that passes through the point $(7, -2)$ and is parallel to the line $5x - y = 8$.

11. Solve the inequality $-16 < 6x + 2 \le 5$ and sketch the graph of the solution set on the real number line.

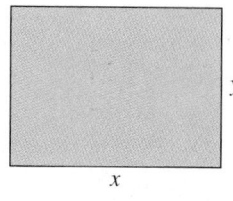

y

x

Figure for 9

In Exercises 12 and 13, use a graphing utility to graph the equations in the same viewing rectangle, and use the graph to approximate any points of intersection. Find the solution of the system algebraically.

12. $\begin{cases} x + 2y = 8 \\ 2x - 4y = -4 \end{cases}$

13. $\begin{cases} x + y = 2 \\ 0.2x + y = 6 \end{cases}$

14. Solve the system: $\begin{cases} 2x + y - 2z = 1 \\ x - z = 1 \\ 3x + 3y + z = 12 \end{cases}$

346

Radicals and Complex Numbers

5

Strategies for Success

SKILLS *When you have completed this chapter, make sure you are able to:*

○ Simplify and evaluate expressions involving integer and rational exponents

○ Simplify and evaluate expressions involving radicals

○ Add, subtract, multiply, and divide radical expressions

○ Solve radical equations

○ Add, subtract, multiply, and divide complex numbers

TOOLS *Use these study tools to master the skills above:*

In the Section Project of Section 5.1, you will use an equation with rational exponents, Kepler's Third Law, to determine the relationship of the period of each planet in our solar system and the planet's mean distance from the sun.

347

5.1 | **Integer Exponents and Scientific Notation**

Integer Exponents ▪ Scientific Notation

Integer Exponents

So far in the text, all exponents have been positive integers. In this section, the definition of an exponent is extended to include zero and negative integers. If a is a real number such that $a \neq 0$, then a^0 is defined as 1. Moreover, if m is an integer, then a^{-m} is defined as the reciprocal of a^m.

Definitions of Zero Exponents and Negative Exponents

Let a be a real number such that $a \neq 0$, and let m be an integer.

1. $a^0 = 1$ 2. $a^{-m} = \dfrac{1}{a^m}$

These definitions are consistent with the properties of exponents given in Section 1.3. For instance, consider the following.

$$x^0 \cdot x^m = x^{0+m} = x^m = 1 \cdot x^m$$

(x^0 is the same as 1.)

EXAMPLE 1 *Zero Exponents and Negative Exponents*

Rewrite each expression without using zero exponents or negative exponents.

(a) 3^0 (b) 0^0 (c) 2^{-1} (d) 3^{-2} (e) $\left(\frac{1}{2}\right)^{-2}$

Solution

(a) $3^0 = 1$ Definition of zero exponents

(b) 0^0 is undefined. Zero cannot have a zero exponent.

(c) $2^{-1} = \dfrac{1}{2^1} = \dfrac{1}{2}$ Definition of negative exponents

(d) $3^{-2} = \dfrac{1}{3^2} = \dfrac{1}{9}$ Definition of negative exponents

(e) $\left(\dfrac{1}{2}\right)^{-2} = \dfrac{1}{\left(\frac{1}{2}\right)^2} = \dfrac{1}{\left(\frac{1}{4}\right)} = 4$ Definition of negative exponents

The following properties of exponents are valid for all integer exponents, including integer exponents that are zero or negative. (The first five properties were listed in Section 1.3.)

Properties of Exponents

Let m and n be integers, and let a and b represent real numbers, variables, or algebraic expressions.

Property	Example
1. $a^m \cdot a^n = a^{m+n}$	$x^4 \cdot x^3 = x^{4+3} = x^7$
2. $(ab)^m = a^m \cdot b^m$	$(3x)^2 = 3^2 \cdot x^2 = 9x^2$
3. $(a^m)^n = a^{mn}$	$(x^3)^3 = x^{3 \cdot 3} = x^9$
4. $\dfrac{a^m}{a^n} = a^{m-n}, \quad a \neq 0$	$\dfrac{x^3}{x} = x^{3-1} = x^2, \quad x \neq 0$
5. $\left(\dfrac{a}{b}\right)^m = \dfrac{a^m}{b^m}, \quad b \neq 0$	$\left(\dfrac{x}{3}\right)^2 = \dfrac{x^2}{3^2} = \dfrac{x^2}{9}$
6. $\left(\dfrac{a}{b}\right)^{-m} = \left(\dfrac{b}{a}\right)^m, \quad \begin{matrix} a \neq 0 \\ b \neq 0 \end{matrix}$	$\left(\dfrac{x}{3}\right)^{-2} = \left(\dfrac{3}{x}\right)^2 = \dfrac{3^2}{x^2} = \dfrac{9}{x^2}, \quad x \neq 0$
7. $a^{-m} = \dfrac{1}{a^m}, \quad a \neq 0$	$x^{-2} = \dfrac{1}{x^2}, \quad x \neq 0$
8. $a^0 = 1, \quad a \neq 0$	$(x^2 + 1)^0 = 1$

EXAMPLE 2 Using Properties of Exponents

(a) $2x^{-1} = 2(x^{-1}) = 2\left(\dfrac{1}{x^1}\right) = \dfrac{2}{x}$

(b) $(2x)^{-1} = \dfrac{1}{(2x)^1} = \dfrac{1}{2x}$

(c) $\dfrac{3}{x^{-2}} = \dfrac{3}{\left(\dfrac{1}{x^2}\right)} = 3\left(\dfrac{x^2}{1}\right) = 3x^2$

(d) $\dfrac{1}{(3x)^{-2}} = \dfrac{1}{\left[\dfrac{1}{(3x)^2}\right]} = \dfrac{1}{\left(\dfrac{1}{3^2 x^2}\right)} = \dfrac{1}{\left(\dfrac{1}{9x^2}\right)} = (1)\left(\dfrac{9x^2}{1}\right) = 9x^2$

EXAMPLE 3 Using Properties of Exponents

Rewrite each expression using only positive exponents. (For each expression, assume that $x \neq 0$ and $y \neq 0$.)

(a) $(-5x^{-3})^2$ (b) $-\left(\dfrac{7x}{y^2}\right)^{-2}$

Solution

(a) $(-5x^{-3})^2 = (-5)^2(x^{-3})^2$ Property 2

$\qquad\qquad\quad = 25x^{-6}$ Property 3

$\qquad\qquad\quad = \dfrac{25}{x^6}$ Property 7

(b) $-\left(\dfrac{7x}{y^2}\right)^{-2} = -\left(\dfrac{y^2}{7x}\right)^2$ Property 6

$\qquad\qquad\quad = -\dfrac{(y^2)^2}{(7x)^2}$ Property 5

$\qquad\qquad\quad = -\dfrac{y^4}{49x^2}$ Property 3

EXAMPLE 4 Using Properties of Exponents

Rewrite each expression using only positive exponents. (For each expression, assume that $x \neq 0$ and $y \neq 0$.)

(a) $\left(\dfrac{8x^{-1}y^4}{4x^3y^2}\right)^{-3}$ (b) $\dfrac{3xy^0}{x^2(5y)^0}$

Solution

(a) $\left(\dfrac{8x^{-1}y^4}{4x^3y^2}\right)^{-3} = \left(\dfrac{2y^2}{x^4}\right)^{-3}$ Properties 1, 4, and 7

$\qquad\qquad\qquad = \left(\dfrac{x^4}{2y^2}\right)^3$ Property 6

$\qquad\qquad\qquad = \dfrac{(x^4)^3}{(2y^2)^3}$ Property 5

$\qquad\qquad\qquad = \dfrac{x^{12}}{8y^6}$ Property 3

(b) $\dfrac{3xy^0}{x^2(5y)^0} = \dfrac{3x(1)}{x^2(1)} = \dfrac{3}{x}$ Properties 1, 7, and 8

NOTE Here is another way to simplify the expression in Example 4(a).

$$\left(\frac{8x^{-1}y^4}{4x^3y^2}\right)^{-3} = \frac{8^{-3}x^3y^{-12}}{4^{-3}x^{-9}y^{-6}}$$

$$= \frac{4^3x^3x^9y^6}{8^3y^{12}}$$

$$= \frac{64x^{12}y^6}{512y^{12}}$$

$$= \frac{x^{12}}{8y^6}$$

EXAMPLE 5   Using Properties of Exponents

Simplify each expression using only positive exponents. (For each expression, assume that $x \neq 0$ and $y \neq 0$.)

(a) $(5x^{-2}y^{-5})^{-2}(3xy^{-4})$ (b) $\dfrac{4x^{-5}y^3}{(7x^{-2}y)^{-1}}$

Solution

(a) $(5x^{-2}y^{-5})^{-2}(3xy^{-4}) = (5^{-2}x^4y^{10})(3xy^{-4}) = \dfrac{x^4y^{10}(3x)}{5^2y^4} = \dfrac{3x^5y^6}{25}$

(b) $\dfrac{4x^{-5}y^3}{(7x^{-2}y)^{-1}} = \dfrac{4x^{-5}y^3}{7^{-1}x^2y^{-1}} = \dfrac{4y^3 \cdot 7y}{x^2x^5} = \dfrac{28y^4}{x^7}$

Scientific Notation

Exponents provide an efficient way of writing and computing with very large (or very small) numbers. For instance, a drop of water contains more than 33 billion billion molecules—that is, 33 followed by 18 zeros.

> 33,000,000,000,000,000,000

It is convenient to write such numbers in **scientific notation.** This notation has the form $c \times 10^n$, where $1 \leq c < 10$ and n is an integer. Thus, the number of molecules in a drop of water can be written in scientific notation as

> $3.3 \times 10,000,000,000,000,000,000 = 3.3 \times 10^{19}$.

The *positive* exponent 19 indicates that the number being written in scientific notation is *large* (10 or more) and that the decimal point has been moved 19 places. A *negative* exponent in scientific notation indicates that the number is *small* (less than 1).

Have students write 2600, 0.00026, and 2.6 in scientific notation.

Answers: 2.6×10^3, 2.6×10^{-4}, and 2.6×10^0

EXAMPLE 6   Writing Scientific Notation

Write each real number in scientific notation.

(a) 0.0000684   (b) 937,200,000

Solution

(a) $0.0000684 = 6.84 \times 10^{-5}$ (b) $937,200,000.0 = 9.372 \times 10^8$

Five places   Eight places

EXAMPLE 7 *Writing Decimal Notation*

Convert each number from scientific notation to decimal notation.

(a) 2.486×10^2 (b) 1.81×10^{-6}

Solution

(a) $2.486 \times 10^2 = 248.6$ (b) $1.81 \times 10^{-6} = 0.00000181$

Two places   Six places

EXAMPLE 8 *Using Scientific Notation with a Calculator*

Discuss how to operate a calculator using scientific notation. Remind students that the number on the right of the display is the exponent of base 10.

Use a scientific or graphing calculator to find $65{,}000 \times 3{,}400{,}000{,}000$.

Solution

6.5 [EXP] 4 [×] 3.4 [EXP] 9 [=]   Scientific

6.5 [EE] 4 [×] 3.4 [EE] 9 [ENTER]   Graphing

The calculator display should read [2.21 14], which implies that

$$(6.5 \times 10^4)(3.4 \times 10^9) = 2.21 \times 10^{14} = 221{,}000{,}000{,}000{,}000.$$

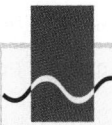 *Technology*

Using Scientific Notation

Most scientific and graphing calculators automatically switch to scientific notation when they are showing large (or small) numbers that exceed the display range. Try multiplying $86{,}500{,}000 \times 6000$. If your calculator follows standard conventions, its display should be

[5.19 11] or [5.19 E 11].

This means that $c = 5.19$ and the exponent of 10 is $n = 11$, which implies that the number is 5.19×10^{11}.

To *enter* numbers in scientific notation, your calculator should have an exponential entry key labeled [EE] or [EXP]. If you were to perform the preceding multiplication using scientific notation, you could begin by writing

$$86{,}500{,}000 \times 6000 = (8.65 \times 10^7)(6.0 \times 10^3)$$

and then entering the following.

8.65 [EXP] 7 [×] 6 [EXP] 3 [=]   Scientific

8.65 [EE] 7 [×] 6 [EE] 3 [ENTER]   Graphing

In each case, you get 5.19×10^{11}.

EXAMPLE 9   *Using Scientific Notation*

Evaluate $\dfrac{(2{,}400{,}000{,}000)(0.00000345)}{(0.00007)(3800)}$.

Solution

Begin by rewriting each number in scientific notation and simplifying.

$$\frac{(2{,}400{,}000{,}000)(0.00000345)}{(0.00007)(3800)} = \frac{(2.4 \times 10^9)(3.45 \times 10^{-6})}{(7.0 \times 10^{-5})(3.8 \times 10^3)}$$

$$= \frac{(2.4)(3.45)(10^3)}{(7)(3.8)(10^{-2})}$$

$$= \frac{(8.28)(10^5)}{26.6}$$

$$\approx 0.3112782(10^5)$$

$$= 31{,}127.82$$

EXAMPLE 10   *Average Amount Spent on Golf Equipment*

The U.S. population in 1994 was 260 million. Use the information shown in Figure 5.1 to find the average amount that an American spent on golf equipment in 1994.   *(Source: National Sporting Goods Association)*

Solution

From the bar graph shown in Figure 5.1, you know that the total amount spent on golf equipment in 1994 was about $1.3 billion. To find the average amount spent on golf equipment in 1994, divide the total amount spent by the number of people.

$$\text{Average amount per person} = \frac{\text{Total amount}}{\text{Number of people}}$$

$$= \frac{1.3 \text{ billion}}{260 \text{ million}}$$

$$= \frac{1.3 \times 10^9}{2.6 \times 10^8}$$

$$= \frac{1.3}{2.6} \times 10^1$$

$$= 5$$

Thus, the average amount spent per person was $5.00.

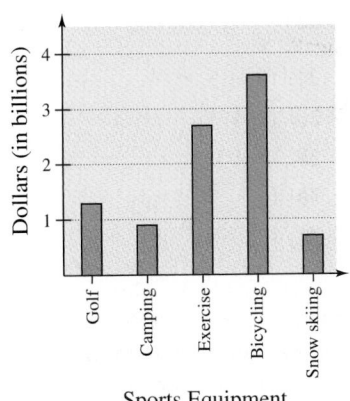

FIGURE 5.1

Group Activities   **Communicating Mathematically**

Developing a Mathematical Method   Discuss why scientific notation is used and give examples of its usefulness. Develop an easy-to-use saying, description, or method for converting numbers to and from scientific notation. Each member of your group should demonstrate the method to the rest of the group, using one of the following as an example.

a. 0.0000042   b. 293,600,000,000

c. 3.1×10^{-6} d. 5.12×10^{11}

Discuss how to tell which of two numbers written in scientific notation is larger. Describe a comparison method and test it by deciding which is larger: 7×10^{51} or 8×10^{50}.

5.1   **Exercises**

See Warm-Up Exercises, p. A44

1. *Think About It*   Discuss any differences between the expressions $(-2x)^4$ and $-2x^4$.

2. *True or False?*   The number 32.5×10^5 is written in scientific notation. Explain.

In Exercises 3–32, evaluate the expression.

3. 5^{-2}

4. 2^{-4}

5. -10^{-3}

6. -20^{-2}

7. $(-3)^{-5}$

8. 25^0

9. $\dfrac{1}{4^{-3}}$

10. $\dfrac{1}{-8^{-2}}$

11. $\dfrac{1}{(-2)^{-5}}$

12. $-\dfrac{1}{6^2}$

13. $\left(\frac{2}{3}\right)^{-1}$

14. $\left(\frac{4}{5}\right)^{-3}$

15. $\left(\frac{3}{16}\right)^0$

16. $\left(-\frac{5}{8}\right)^{-2}$

17. $27 \cdot 3^{-3}$

18. $4^2 \cdot 4^{-3}$

19. $\dfrac{3^4}{3^{-2}}$

20. $\dfrac{5^{-1}}{5^2}$

21. $\dfrac{10^3}{10^{-2}}$

22. $\dfrac{10^{-5}}{10^{-6}}$

23. $(4^2 \cdot 4^{-1})^{-2}$

24. $(5^3 \cdot 5^{-4})^{-3}$

25. $(2^{-3})^2$

26. $(-4^{-1})^{-2}$

27. $2^{-3} + 2^{-4}$

28. $4 - 3^{-2}$

29. $\left(\frac{3}{4} + \frac{5}{8}\right)^{-2}$

30. $\left(\frac{1}{2} - \frac{2}{3}\right)^{-1}$

31. $(5^0 - 4^{-2})^{-1}$

32. $(32 + 4^{-3})^0$

In Exercises 33–68, rewrite the expression using only positive exponents, and simplify. (Assume that any variables in the expression are nonzero.)

33. $y^4 \cdot y^{-2}$

34. $x^{-2} \cdot x^{-5}$

35. $z^5 \cdot z^{-3}$

36. $t^{-1} \cdot t^{-6}$

37. $\dfrac{1}{x^{-6}}$

38. $\dfrac{x^{-3}}{y^{-1}}$

39. $\dfrac{a^{-6}}{a^{-7}}$

40. $\dfrac{6u^{-2}}{15u^{-1}}$

41. $\dfrac{(4t)^0}{t^{-2}}$

42. $\dfrac{(5u)^4}{(5u)^4}$

43. $(2x^2)^{-2}$

44. $(4a^{-2}b^3)^{-3}$ quit

45. $(-3x^{-3}y^2)(4x^2y^{-5})$

46. $(5s^5t^{-5})\left(\dfrac{3s^{-2}}{50t^{-1}}\right)$

47. $(3x^2y^{-2})^{-2}$

48. $(-4y^{-3}z)^{-3}$

49. $\dfrac{6^2x^3y^{-3}}{12x^{-2}y}$

50. $\dfrac{2^{-4}y^{-1}z^{-3}}{4^{-2}yz^{-3}}$

51. $\left(\dfrac{3u^2v^{-1}}{3^3u^{-1}v^3}\right)^{-2}$

52. $\left(\dfrac{5^2x^3y^{-3}}{125xy}\right)\left(\dfrac{5x}{3y}\right)^{-1}$

53. $[(2x^{-3}y^{-2})^2]^{-2}$

54. $\left[\left(\dfrac{2x^2}{4y}\right)^{-3}\right]^2$

55. $(4m)^3\left(\dfrac{4}{3m}\right)^{-2}$

56. $\left(\dfrac{3z^2}{x}\right)^{-2}$

57. $(5x^2y^4)^3(xy^{-5})^{-3}$

58. $(s^4t)^{-2}(s^4t)^2$

59. $\left(\dfrac{x}{10}\right)^{-1}$

60. $\left(\dfrac{4}{z}\right)^{-2}$

61. $\left(\dfrac{a^{-2}}{b^{-2}}\right)\left(\dfrac{b}{a}\right)^3$

62. $\left(\dfrac{a^{-3}}{b^{-3}}\right)\left(\dfrac{b}{a}\right)^3$

63. $[(x^{-4}y^{-6})^{-1}]^2$

64. $(ab)^{-2}(a^2b^2)^{-1}$

65. $\dfrac{(2a^{-2}b^4)^3}{(10a^3b)^2}$

66. $\dfrac{(5x^2y^{-5})^{-1}}{2x^{-5}y^4}$

67. $(2x^3y^{-1})^{-3}(4xy^{-6})$

68. $x^5(3x^0y^4)(7y)^0$

In Exercises 69 and 70, find an expression for the area of the figure.

69.

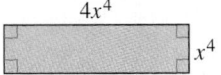

70.

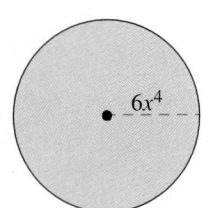

In Exercises 71–80, write the number in scientific notation.

71. 3,600,000

72. 98,100,000

73. 0.00381

74. 0.0007384

75. *Land Area of Earth:* 57,500,000 square miles

76. *Ocean Area of Earth:* 139,400,000 square miles

77. *Light Year:* 9,461,000,000,000,000 meters

78. *Thickness of Soap Bubble:* 0.0000001 meter

79. *Relative Density of Hydrogen:* 0.0000899

80. *One Micron (Millionth of Meter):* 0.00003937 inch

In Exercises 81–90, write the number in decimal notation.

81. 6×10^7

82. 5.05×10^{12}

83. 1.359×10^{-7}

84. 8.6×10^{-9}

85. *1996 Wal-Mart Sales:* $\$9.36 \times 10^{10}$

86. *Number of Air Sacs in Lungs:* 3.5×10^8

87. *Interior Temperature of Sun:* 1.3×10^7 degrees Celsius

88. *Width of Air Molecule:* 9.0×10^{-9} meter

89. *Charge of Electron:* 4.8×10^{-10} electrostatic units

90. *Width of Human Hair:* 9.0×10^{-4} meter

In Exercises 91–100, evaluate without a calculator. Write the result in scientific notation.

91. $(2 \times 10^9)(3.4 \times 10^{-4})$

92. $(5 \times 10^4)^3$

93. $\dfrac{3.6 \times 10^9}{9 \times 10^5}$

94. $\dfrac{2.5 \times 10^{-3}}{5 \times 10^2}$

95. $(4,500,000)(2,000,000,000)$

96. $\dfrac{64,000,000}{0.00004}$

97. $(6.5 \times 10^6)(2 \times 10^4)$

98. $(5 \times 10^8)^3$

99. $\dfrac{3.6 \times 10^{12}}{6 \times 10^5}$

100. $\dfrac{72,000,000,000}{0.00012}$

In Exercises 101–104, evaluate with a calculator. Write the answer in scientific notation, $c \times 10^n$, with c rounded to two decimal places.

101. $\dfrac{1.357 \times 10^{12}}{(4.2 \times 10^2)(6.87 \times 10^{-3})}$

102. $(8.67 \times 10^4)^7$

103. $\dfrac{(0.0000565)(2,850,000,000,000)}{0.00465}$

104. $\dfrac{(5,000,000)^3(0.000037)^2}{(0.005)^4}$

105. *Light Year* One light year (the distance light can travel in 1 year) is approximately 9.46×10^{15} meters. Approximate the time for light to travel from the sun to the earth if that distance is approximately 1.49×10^{11} meters.

106. *Distance to a Star* Determine the distance (in meters) to the star Alpha Andromeda if it is 90 light years from the earth (see figure). See Exercise 105 for the definition of a light year.

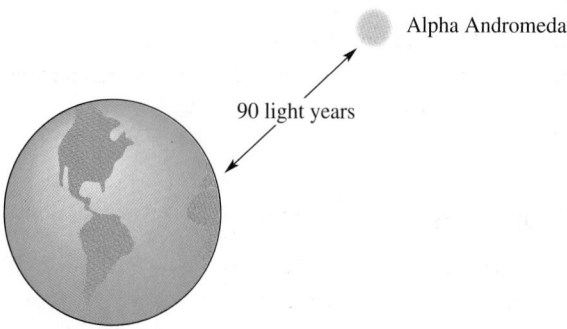

Alpha Andromeda

90 light years

107. *Masses of Earth and Sun* The masses of the earth and the sun are approximately 5.975×10^{24} kg and 1.99×10^{30} kg, respectively. The mass of the sun is approximately how many times that of the earth?

108. *Federal Debt* In September 1996, the estimated population of the United States was 266 million people, and the estimated federal debt was 5240 billion dollars. Use these two numbers to determine the amount each person would have to pay to eliminate the debt. *(Source: U.S. Bureau of Census and U.S. Office of Management and Budget)*

Section Project

Kepler's Third Law In 1619, Johannes Kepler, a German astronomer, discovered that the period T (in years) of each planet in our solar system is related to the planet's mean distance R (in astronomical units) from the sun by the equation

$$\frac{T^2}{R^3} = k.$$

Test Kepler's equation for the nine planets in our solar system, using the table. What value do you get for k for each planet? Are the values of k all approximately the same? (Astronomical units relate a planet's period and mean distance to Earth's period and mean distance.)

Planet	T	R
Mercury	0.241	0.387
Venus	0.615	0.723
Earth	1.000	1.000
Mars	1.881	1.523
Jupiter	11.861	5.203
Saturn	29.457	9.541
Uranus	84.008	19.190
Neptune	164.784	30.086
Pluto	248.350	39.508

| 5.2 | **Rational Exponents and Radicals** |

Roots and Radicals • Rational Exponents •
Radicals and Calculators • Radical Functions

Roots and Radicals

In Section 1.1, you reviewed the use of radical notation to represent nth roots of real numbers. Recall from that section that b is called an nth root of a if $a = b^n$. Also recall that the principal nth root of a real number is defined as follows.

NOTE "Having the same sign" means that the principal nth root of a is positive if a is positive and negative if a is negative. For example, $\sqrt{4} = 2$ and $\sqrt[3]{-8} = -2$.

Principal nth Root of a Number

Let a be a real number that has at least one (real number) nth root. The **principal nth root of a** is the nth root that has the same sign as a, and it is denoted by the **radical symbol**

$$\sqrt[n]{a}. \qquad \text{Principal } n\text{th root}$$

The positive integer n is the **index** of the radical, and the number a is the **radicand.** If $n = 2$, omit the index and write $\sqrt{a}$ rather than $\sqrt[2]{a}$.

Therefore, $\sqrt{49} = 7$, $\sqrt[3]{1000} = 10$, and $\sqrt[5]{-32} = -2$.

You need to be aware of the following properties of nth roots. (Remember that for nth roots, n is an integer that is greater than or equal to 2.)

Properties of nth Roots

1. If a is a positive real number and n is *even*, then a has exactly two (real) nth roots, which are denoted by $\sqrt[n]{a}$ and $-\sqrt[n]{a}$.

2. If a is any real number and n is *odd*, then a has only one (real) nth root, which is denoted by $\sqrt[n]{a}$.

3. If a is a negative real number and n is *even*, then a has no (real) nth root.

Integers such as 1, 4, 9, 16, 49, and 81 are called **perfect squares** because they have integer square roots. Similarly, integers such as 1, 8, 27, 64, and 125 are called **perfect cubes** because they have integer cube roots.

EXAMPLE 1   *Evaluating Radical Expressions*

Evaluate each radical expression.

(a) $\sqrt{144}$ (b) $-\sqrt{81}$ (c) $-\sqrt[3]{-64}$ (d) $\sqrt[4]{1}$

Solution

(a) $\sqrt{144} = 12$ (b) $-\sqrt{81} = -9$

(c) $-\sqrt[3]{-64} = -(-4) = 4$ (d) $\sqrt[4]{1} = 1$

EXAMPLE 2   *Classifying Perfect nth Powers*

State whether each number is a perfect square, a perfect cube, both, or neither.

(a) 81   (b) 64   (c) 32

Solution

(a) 81 is a perfect square because $9^2 = 81$. It is not a perfect cube.

(b) 64 is a perfect square because $8^2 = 64$, and it is also a perfect cube because $4^3 = 64$.

(c) 32 is not a perfect square or a perfect cube. (It is a perfect 5th because $2^5 = 32$.)

Raising a number to the *n*th power and taking the principal *n*th root of a number can be thought of as *inverse* operations. Here are two examples.

$$\left(\sqrt{4}\right)^2 = 2^2 = 4 \quad \text{and} \quad \sqrt{2^2} = \sqrt{4} = 2$$

$$\left(\sqrt[3]{27}\right)^3 = 3^3 = 27 \quad \text{and} \quad \sqrt[3]{3^3} = \sqrt[3]{27} = 3$$

Inverse Properties of *n*th Powers and *n*th Roots

Let a be a real number, and let n be an integer such that $n \geq 2$.

1. If a has a principal *n*th root, then
$$\left(\sqrt[n]{a}\right)^n = a.$$

2. If n is *odd*, then
$$\sqrt[n]{a^n} = a.$$
   If n is *even*, then
$$\sqrt[n]{a^n} = |a|.$$

EXAMPLE 3   *Evaluating Radical Expressions*

Evaluate each radical expression.

(a) $\sqrt[3]{5^3}$ (b) $\sqrt[3]{(-2)^3}$ (c) $\left(\sqrt{7}\right)^2$ (d) $\sqrt{(-3)^2}$ (e) $\sqrt{-3^2}$

Solution

(a) Because the index of the radical is odd, you can write
$$\sqrt[3]{5^3} = 5.$$

(b) Because the index of the radical is odd, you can write
$$\sqrt[3]{(-2)^3} = -2.$$

(c) Using the inverse property of powers and roots, you can write
$$\left(\sqrt{7}\right)^2 = 7.$$

(d) Because the index of the radical is even, you must include absolute value signs, and write
$$\sqrt{(-3)^2} = |-3| = 3.$$

(e) Because $\sqrt{-3^2} = \sqrt{-9}$ is an even root of a negative number, its value is not a real number.

Rational Exponents

Definition of Rational Exponents

Let a be a real number, and let n be an integer such that $n \geq 2$. If the principal nth root of a exists, we define $a^{1/n}$ to be
$$a^{1/n} = \sqrt[n]{a}.$$
If m is a positive integer that has no common factor with n, then
$$a^{m/n} = (a^{1/n})^m = \left(\sqrt[n]{a}\right)^m \quad \text{and} \quad a^{m/n} = (a^m)^{1/n} = \sqrt[n]{a^m}.$$

NOTE   The numerator of a rational exponent denotes the *power* to which the base is raised, and the denominator denotes the *root* to be taken.

Ask students to simplify $27^{4/3}$ using both methods and to decide which is easier. Students should be familiar with both evaluation techniques.

It does not matter in which order the two operations are performed, provided the nth root exists. Here is an example.

$$8^{2/3} = \left(\sqrt[3]{8}\right)^2 = 2^2 = 4 \qquad \text{Cube root, then second power}$$
$$8^{2/3} = \sqrt[3]{8^2} = \sqrt[3]{64} = 4 \qquad \text{Second power, then cube root}$$

The properties of exponents listed in Section 5.1 also apply to rational exponents (provided the roots indicated by the denominators exist). We relist the properties here, with different examples.

Properties of Exponents

Let r and s be rational numbers, and let a and b be real numbers, variables, or algebraic expressions.

Property	*Example*
1. $a^r \cdot a^s = a^{r+s}$	$4^{1/2} \cdot 4^{1/3} = 4^{5/6}$
2. $(ab)^r = a^r \cdot b^r$	$(2x)^{1/2} = 2^{1/2} \cdot x^{1/2}$
3. $(a^r)^s = a^{rs}$	$(x^3)^{1/3} = x$
4. $\dfrac{a^r}{a^s} = a^{r-s}, \quad a \neq 0$	$\dfrac{x^2}{x^{1/2}} = x^{2-(1/2)} = x^{3/2}, \quad x \neq 0$
5. $\left(\dfrac{a}{b}\right)^r = \dfrac{a^r}{b^r}, \quad b \neq 0$	$\left(\dfrac{x}{3}\right)^{1/3} = \dfrac{x^{1/3}}{3^{1/3}}$
6. $\left(\dfrac{a}{b}\right)^{-r} = \left(\dfrac{b}{a}\right)^r, \quad \begin{matrix} a \neq 0 \\ b \neq 0 \end{matrix}$	$\left(\dfrac{x}{25}\right)^{-1/2} = \left(\dfrac{25}{x}\right)^{1/2} = \dfrac{5}{x^{1/2}}, \quad x \neq 0$
7. $a^{-r} = \dfrac{1}{a^r}, \quad a \neq 0$	$9^{-1/2} = \dfrac{1}{9^{1/2}} = \dfrac{1}{3}$
8. $a^0 = 1, \quad a \neq 0$	$(10x^3)^0 = 1$

EXAMPLE 4   Evaluating Expressions with Rational Exponents

(a) $8^{4/3} = \left(\sqrt[3]{8}\right)^4 = 2^4 = 16$

(b) $(4^2)^{3/2} = \left(\sqrt{4^2}\right)^3 = 4^3 = 64$

(c) $25^{-3/2} = \dfrac{1}{25^{3/2}} = \dfrac{1}{\left(\sqrt{25}\right)^3} = \dfrac{1}{5^3} = \dfrac{1}{125}$

(d) $\left(\dfrac{64}{125}\right)^{2/3} = \left(\sqrt[3]{\dfrac{64}{125}}\right)^2 = \left(\dfrac{\sqrt[3]{64}}{\sqrt[3]{125}}\right)^2 = \left(\dfrac{4}{5}\right)^2 = \dfrac{16}{25}$

NOTE In parts (e) and (f) of Example 4, be sure that you see the distinction between the expressions $-9^{1/2}$ and $(-9)^{1/2}$.

(e) $-9^{1/2} = -\sqrt{9} = -3$

(f) $(-9)^{1/2} = \sqrt{-9}$ is not a real number.

EXAMPLE 5   *Using Properties of Exponents*

Rewrite each expression using rational exponents.

(a) $x\sqrt[4]{x^3}$

(b) $\dfrac{\sqrt[3]{x^2}}{\sqrt{x^3}}$

Solution

(a) $x\sqrt[4]{x^3} = x(x^{3/4}) = x^{1+(3/4)} = x^{7/4}$

(b) $\dfrac{\sqrt[3]{x^2}}{\sqrt{x^3}} = \dfrac{x^{2/3}}{x^{3/2}} = x^{(2/3)-(3/2)} = x^{-5/6} = \dfrac{1}{x^{5/6}}$

EXAMPLE 6   *Using Properties of Exponents*

Use the properties of exponents to simplify each expression.

Stress to students that the basic rules of exponents apply even if the exponents are negative or fractions. Give a comparison example such as:

$x^2 \cdot x^4 = x^6$

$x^{-2} \cdot x^4 = x^2$

$x^{1/2} \cdot x^{1/4} = x^{3/4}$

(a) $x^{3/5} \cdot x^{2/3}$ (b) $(x^{3/5})^{2/3}$

(c) $(6x^{3/5}y^{-3/2})^2,$ $y \neq 0$

Solution

(a) $x^{3/5} \cdot x^{2/3} = x^{(3/5)+(2/3)} = x^{(9+10)/15} = x^{19/15}$

(b) $(x^{3/5})^{2/3} = x^{(3/5)(2/3)} = x^{2/5}$

(c) $(6x^{3/5}y^{-3/2})^2 = 6^2 x^{(3/5)2} y^{(-3/2)2} = 36x^{6/5}y^{-3} = \dfrac{36x^{6/5}}{y^3}$

EXAMPLE 7   *Using Properties of Exponents*

Use the properties of exponents to simplify each expression.

(a) $\sqrt{\sqrt[3]{x}}$

(b) $\dfrac{(2x-1)^{4/3}}{\sqrt[3]{2x-1}}$

Solution

(a) $\sqrt{\sqrt[3]{x}} = \sqrt{x^{1/3}} = (x^{1/3})^{1/2} = x^{1/6}$

(b) $\dfrac{(2x-1)^{4/3}}{\sqrt[3]{2x-1}} = \dfrac{(2x-1)^{4/3}}{(2x-1)^{1/3}} = (2x-1)^{(4/3)-(1/3)} = 2x-1$

Radicals and Calculators

There are two methods of evaluating radicals on most calculators.

1. Use a radical key such as $\boxed{\sqrt{}}$ or $\boxed{\sqrt[3]{}}$.

2. Convert to exponential form and then use your calculator's exponential key $\boxed{y^x}$ or $\boxed{\wedge}$.

Technology

Most graphing utilities have an nth root key such as $\boxed{\sqrt[x]{}}$. If your graphing utility has such a key, use it to confirm the result in Example 8(b).

Alternatively, in Example 8(b), you might show that $\sqrt[5]{25} = 25^{1/5} = 25^{0.2} \approx 1.904$.

EXAMPLE 8   *Evaluating Roots with a Calculator*

Evaluate each of the following. Round the result to three decimal places.

(a) $\sqrt{5}$ (b) $\sqrt[5]{25}$

(c) $\sqrt[3]{-4}$ (d) $(1.4)^{-2/5}$

Solution

(a) $5\,\boxed{\sqrt{}}$ Scientific

      $\boxed{\sqrt{}}\,5\,\boxed{\text{ENTER}}$ Graphing

The display is 2.236068. Rounded to three decimal places, you get $\sqrt{5} \approx 2.236$.

(b) First rewrite the expression as $\sqrt[5]{25} = 25^{1/5}$. Then use one of the following keystroke sequences.

         $25\,\boxed{y^x}\,\boxed{(}\,1\,\boxed{\div}\,5\,\boxed{)}\,\boxed{=}$ Scientific

         $25\,\boxed{\wedge}\,\boxed{(}\,1\,\boxed{\div}\,5\,\boxed{)}\,\boxed{\text{ENTER}}$ Graphing

The display is 1.9036539. Rounded to three decimal places, you get $\sqrt[5]{25} \approx 1.904$.

(c) If your calculator does not have a cube root key, use the fact that

$$\sqrt[3]{-4} = \sqrt[3]{(-1)(4)} = \sqrt[3]{-1}\,\sqrt[3]{4} = -\sqrt[3]{4}$$

and attach the negative sign of the radicand as the last keystroke.

         $4\,\boxed{y^x}\,\boxed{(}\,1\,\boxed{\div}\,3\,\boxed{)}\,\boxed{=}\,\boxed{+/-}$ Scientific

         $\boxed{\sqrt[3]{}}\,\boxed{(-)}\,4\,\boxed{\text{ENTER}}$ Graphing

The display is -1.5874011. Rounded to three decimal places, you get $\sqrt[3]{-4} \approx -1.587$.

(d) $1.4\,\boxed{y^x}\,\boxed{(}\,2\,\boxed{+/-}\,\boxed{\div}\,5\,\boxed{)}\,\boxed{=}$ Scientific

    $1.4\,\boxed{\wedge}\,\boxed{(}\,\boxed{(-)}\,2\,\boxed{\div}\,5\,\boxed{)}\,\boxed{\text{ENTER}}$ Graphing

The display is 0.8740752. Rounded to three decimal places, you get $(1.4)^{-2/5} \approx 0.874$.

The *Olympias* is a reconstruction of a trireme (a Greek galley ship). The ship's triple set of oars was operated by volunteers.

EXAMPLE 9 *Finding the Speed of a Ship*

The speed *s* (in knots) of the *Olympias* was found to be related to the power *P* (in kilowatts) generated by the rowers according to the model

$$s = \sqrt[3]{\frac{100P}{3}}.$$

(a) Make a table of values that relates the power and speed. Then use the table to sketch a graph of the model.

(b) The volunteer crew was able to generate maximum power of 10.5 kilowatts. What was the ship's greatest speed? *(Source: Scientific American)*

Solution

(a) Use a calculator to make a table of values, as shown below. From the table, you can sketch the graph of the equation, as shown in Figure 5.2.

P	0	2	4	6	8	10	12	14
s	0	4.1	5.1	5.8	6.4	6.9	7.4	7.8

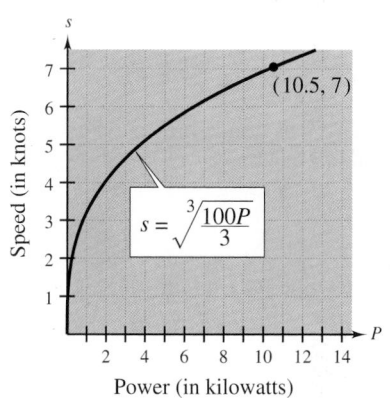

FIGURE 5.2

(b) To find the greatest speed, use the model and substitute 10.5 for *P*.

$$s = \sqrt[3]{\frac{100P}{3}} \qquad \text{Cube root model}$$

$$= \sqrt[3]{\frac{100(10.5)}{3}} \qquad \text{Substitute 10.5 for } P.$$

$$\approx 7 \qquad \text{Use a calculator.}$$

Thus, the greatest speed attained by the *Olympias* was about 7 knots (about 8 miles per hour).

Radical Functions

The **domain** of the radical $\sqrt[n]{x}$ is the set of all real numbers such that x has a principal nth root.

Domain of a Radical

Let n be an integer that is greater than or equal to 2.

1. If n is odd, the domain of $\sqrt[n]{x}$ is the set of all real numbers.

2. If n is even, the domain of $\sqrt[n]{x}$ is the set of all nonnegative real numbers.

EXAMPLE 10   Finding the Domain of a Radical Function

Describe the domain of each function.

(a) $f(x) = \sqrt{x}$ (b) $f(x) = \sqrt[3]{x}$

(c) $f(x) = \sqrt{x^2}$ (d) $f(x) = \sqrt{x^3}$

Solution

Have students evaluate $(-4)^{3/2}$ on their calculators. Ask students why their displays give an E or Error message.

(a) The domain of
$$f(x) = \sqrt{x}$$
is the set of all nonnegative real numbers. For instance, 2 is in the domain, but -2 is not because $\sqrt{-2}$ is not a real number.

(b) The domain of
$$f(x) = \sqrt[3]{x}$$
is the set of all real numbers because for any real number x, the expression $\sqrt[3]{x}$ is a real number.

(c) The domain of
$$f(x) = \sqrt{x^2}$$
is the set of all real numbers because for any real number x, the expression x^2 is a nonnegative real number.

(d) The domain of
$$f(x) = \sqrt{x^3}$$
is the set of all nonnegative real numbers. For instance, 1 is in the domain, but -1 is not because $\sqrt{(-1)^3} = \sqrt{-1}$ is not a real number.

 *Technology*

A graphing utility can be used to investigate the graphs of functions of the form

$$y = x^n, \qquad x \geq 0, \qquad n > 0.$$

After some experimentation, you can see that graphs of functions of this form fall into three categories.

1. If $n = 1$, the graph of $y = x^n$ is a straight line.

2. If $n > 1$, the graph of $y = x^n$ curves upward. (This is called *concave up*.)

3. If $n < 1$, the graph of $y = x^n$ curves downward. (This is called *concave down*.)

Several examples are shown in the calculator screens below. Try confirming this observation by sketching graphs of other functions of the form $y = x^n$.

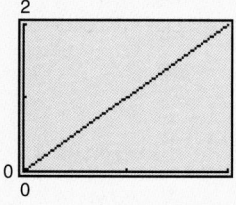  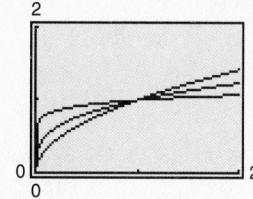

$y = x^n, \quad n = 1$ $y = x^n, \quad n > 1$ $y = x^n, \quad n < 1$

Group Activities

Exploring with Technology

Describing Domains and Ranges   In your group, discuss the domain and range of each of the following functions. Use a graphing utility to verify your conclusions.

a. $y = x^{3/2}$ b. $y = x^2$ c. $y = x^{1/3}$

d. $y = (\sqrt{x})^2$ e. $y = x^{-4/5}$ f. $y = x^5$

5.2   **Exercises**

*See Warm-Up
Exercises, p. A44*

In Exercises 1–6, complete the statement.

1. Because $7^2 = 49$, ▨ is a square root of 49.
2. Because $24.5^2 = 600.25$, ▨ is a square root of 600.25.
3. Because $4.2^3 = 74.088$, ▨ is a cube root of 74.088.
4. Because $6^4 = 1296$, ▨ is a fourth root of 1296.
5. Because $45^2 = 2025$, 45 is called the ▨ of 2025.
6. Because $12^3 = 1728$, 12 is called the ▨ of 1728.

7. *Think About It*   In your own words, define an *n*th root of a number.
8. *Think About It*   Is it true that $\sqrt{2} = 1.414$? Explain.

In Exercises 9–34, evaluate without a calculator. (If not possible, state the reason.)

9. $\sqrt{64}$
10. $-\sqrt{100}$
11. $\sqrt{-100}$
12. $\sqrt{144}$
13. $\sqrt{81}$
14. $\sqrt{-64}$
15. $-\sqrt{\frac{4}{9}}$
16. $\sqrt{\frac{9}{16}}$
17. $\sqrt{0.09}$
18. $\sqrt{0.36}$
19. $\sqrt{0.16}$
20. $-\sqrt{0.0009}$
21. $\sqrt{49 - 4(2)(-15)}$
22. $\sqrt{\frac{75}{3}}$
23. $\sqrt[3]{125}$
24. $\sqrt[3]{-8}$
25. $\sqrt[3]{1000}$
26. $\sqrt[3]{64}$
27. $\sqrt[3]{-\frac{1}{64}}$
28. $-\sqrt[3]{0.008}$
29. $\sqrt[4]{81}$
30. $\sqrt[5]{32}$
31. $-\sqrt[4]{-625}$
32. $-\sqrt[4]{\frac{1}{625}}$
33. $\sqrt[5]{-0.00243}$
34. $\sqrt[6]{64}$

In Exercises 35–38, determine whether the square root is a rational or irrational number.

35. $\sqrt{6}$
36. $\sqrt{\frac{9}{16}}$
37. $\sqrt{900}$
38. $\sqrt{72}$

In Exercises 39–44, fill in the missing description.

Radical Form	Rational Exponent Form
39. $\sqrt{16} = 4$	▨
40. $\sqrt[4]{81} = 3$	▨
41. $\sqrt[3]{27^2} = 9$	▨
42. ▨	$125^{1/3} = 5$
43. ▨	$256^{3/4} = 64$
44. ▨	$27^{2/3} = 9$

In Exercises 45–56, evaluate without a calculator.

45. $25^{1/2}$
46. $49^{1/2}$
47. $-36^{1/2}$
48. $-121^{1/2}$
49. $16^{3/4}$
50. $-\left(\frac{1}{125}\right)^{2/3}$
51. $32^{-2/5}$
52. $81^{-3/4}$
53. $\left(\frac{8}{27}\right)^{2/3}$
54. $\left(\frac{256}{625}\right)^{1/4}$
55. $\left(\frac{121}{9}\right)^{-1/2}$
56. $\left(\frac{27}{1000}\right)^{-4/3}$

▦ In Exercises 57–66, use a calculator to approximate the quantity accurate to four decimal places. (If not possible, state the reason.)

57. $\sqrt{73}$
58. $\sqrt{-532}$
59. $\frac{8 - \sqrt{35}}{2}$
60. $\frac{-5 + \sqrt{3215}}{10}$
61. $1698^{-3/4}$
62. $962^{2/3}$
63. $\sqrt[4]{342}$
64. $\sqrt[3]{159}$
65. $\sqrt[3]{545^2}$
66. $\sqrt[5]{-35^3}$

The symbol ▦ indicates an exercise in which you are instructed to use a calculator or graphing utility. The solutions of other exercises may also be facilitated by the use of appropriate technology.

In Exercises 67–94, simplify the expression.

67. $\sqrt{t^2}$

68. $\sqrt[3]{z^3}$

69. $\sqrt[3]{y^9}$

70. $\sqrt[4]{a^8}$

71. $\sqrt[3]{t^6}$

72. $\sqrt[4]{z^4}$

73. $\sqrt{x^8}$

74. $\sqrt[5]{y^{15}}$

75. $(3^{1/4} \cdot 3^{3/4})$

76. $(2^{1/2})^{2/3}$

77. $\dfrac{2^{1/5}}{2^{6/5}}$

78. $\dfrac{5^{-3/4}}{5}$

79. $\left(\frac{2}{3}\right)^{5/3} \cdot \left(\frac{2}{3}\right)^{1/3}$

80. $(4^{1/3})^{9/4}$

81. $x^{2/3} \cdot x^{7/3}$

82. $z^{3/5} \cdot z^{-2/5}$

83. $(3x^{-1/3}y^{3/4})^2$

84. $(-2u^{3/5}v^{-1/5})^3$

85. $\dfrac{18y^{4/3}z^{-1/3}}{24y^{-2/3}z}$

86. $\dfrac{a^{3/4} \cdot a^{1/2}}{a^{5/2}}$

87. $\left(\dfrac{x^{1/4}}{x^{1/6}}\right)^3$

88. $\left(\dfrac{3m^{1/6}n^{1/3}}{4n^{-2/3}}\right)^2$

89. $(c^{3/2})^{1/3}$

90. $(k^{-1/3})^{3/2}$

91. $\dfrac{x^{4/3}y^{2/3}}{(xy)^{1/3}}$

92. $\dfrac{(2x^3)^{4/3}}{2^{1/3}x^5}$

93. $\sqrt{\sqrt[4]{y}}$

94. $\sqrt[3]{\sqrt{2x}}$

In Exercises 95–98, determine the domain of the function.

95. $f(x) = 3\sqrt{x}$

96. $h(x) = \sqrt[4]{x}$

97. $g(x) = \dfrac{2}{\sqrt[4]{x}}$

98. $g(x) = \dfrac{10}{\sqrt[3]{x}}$

In Exercises 99–102, use a graphing utility to graph the function. Find the domain of the function algebraically. Did the graphing utility omit part of the domain? If so, complete the graph by hand.

99. $y = \dfrac{5}{\sqrt[4]{x^3}}$

100. $y = 4\sqrt[3]{x}$

101. $g(x) = 2x^{3/5}$

102. $h(x) = 5x^{2/3}$

In Exercises 103–106, perform the multiplication. Use a graphing utility to confirm your result.

103. $x^{1/2}(2x - 3)$

104. $x^{4/3}(3x^2 - 4x + 5)$

105. $y^{-1/3}(y^{1/3} + 5y^{4/3})$

106. $(x^{1/2} - 3)(x^{1/2} + 3)$

Mathematical Modeling   In Exercises 107 and 108, use the formula for the *declining balances method*

$$r = 1 - \left(\frac{S}{C}\right)^{1/n}$$

to find the depreciation rate r. In the formula, n is the useful life of the item (in years), S is the salvage value (in dollars), and C is the original cost (in dollars).

107. A $75,000 truck is depreciated over an 8-year period, as shown in the graph.

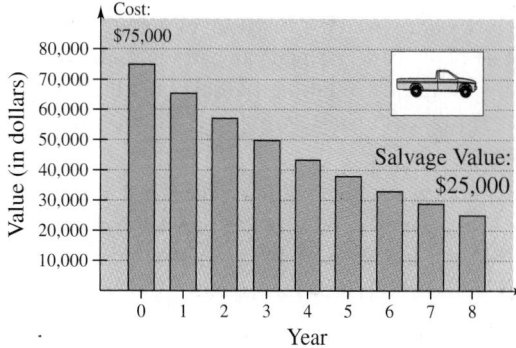

108. A $125,000 printing press is depreciated over a 10-year period, as shown in the graph.

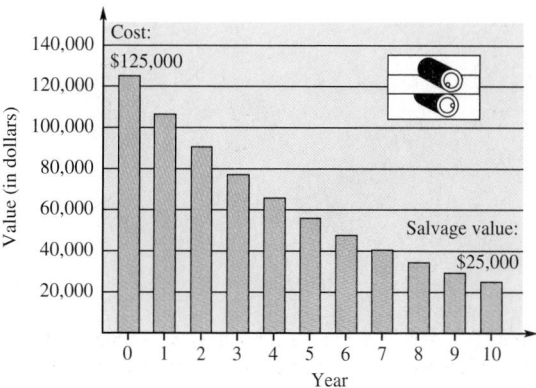

109. *Microwave Oven*   The usable space in a particular microwave oven is in the form of a cube (see figure). The sales brochure indicates that the interior space of the oven is 2197 cubic inches. Find the inside dimensions of the oven.

110. *Dimensions of a Room*   Find the dimensions of a piece of carpet for a classroom with 529 square feet of floor space, assuming the floor is square.

111. *Velocity of a Stream*   A stream of water moving at a rate of v feet per second can carry particles of size $0.03\sqrt{v}$ inches. Find the particle size that can be carried by a stream flowing at a rate of $\frac{3}{4}$ foot per second.

112. *Investigation*   Find all possible "last digits" of perfect squares. (For instance, the last digit of 81 is 1 and the last digit of 64 is 4.) Is it possible that 4,322,788,986 is a perfect square?

113. *Think About It*   Determine the values of x for which $\sqrt{x^2} \neq x$. Explain your answer.

Section Project

Investigation   (a) Choose a positive real number $x > 1$. Enter the number into a calculator and find its square root. Then repeatedly take the square root of the result.

$$\sqrt{x}, \ \sqrt{\sqrt{x}}, \ \sqrt{\sqrt{\sqrt{x}}}, \ldots$$

What real number does the display appear to be approaching?

(b) Repeat part (a) using a positive real number less than 1.

(c) How do the results compare?

5.3	**Simplifying and Combining Radicals**

Simplifying Radicals ▪ Rationalization Techniques ▪
Adding and Subtracting Radicals ▪ Applications of Radicals

Simplifying Radicals

In this section, you will study ways to simplify and combine radicals. For instance, the expression $\sqrt{12}$ can be simplified as

$$\sqrt{12} = \sqrt{4 \cdot 3} = \sqrt{4}\sqrt{3} = 2\sqrt{3}.$$

This rewriting is based on the following rules for multiplying and dividing radicals.

Multiplication and Division Properties of Radicals

Let u and v be real numbers, variables, or algebraic expressions. If the nth roots of u and v are real, the following properties are true.

1. $\sqrt[n]{u}\sqrt[n]{v} = \sqrt[n]{uv}$ Multiplication Property

2. $\dfrac{\sqrt[n]{u}}{\sqrt[n]{v}} = \sqrt[n]{\dfrac{u}{v}}, \qquad v \neq 0$ Division Property

---

You can use these properties of radicals to *simplify* radical expressions as follows.

$$\sqrt{48} = \sqrt{16 \cdot 3} = \sqrt{16}\sqrt{3} = 4\sqrt{3}$$

This simplification process is called **removing perfect square factors from the radical.**

EXAMPLE 1 *Removing Constant Factors from Radicals*

Simplify each radical by removing as many factors as possible.

(a) $\sqrt{75}$ (b) $\sqrt{72}$ (c) $\sqrt{162}$

Solution

(a) $\sqrt{75} = \sqrt{25 \cdot 3} = \sqrt{25}\sqrt{3} = 5\sqrt{3}$

(b) $\sqrt{72} = \sqrt{36 \cdot 2} = \sqrt{36}\sqrt{2} = 6\sqrt{2}$

(c) $\sqrt{162} = \sqrt{81 \cdot 2} = \sqrt{81}\sqrt{2} = 9\sqrt{2}$

When removing *variable* factors from a square root radical, remember that it is not valid to write $\sqrt{x^2} = x$ *unless* you happen to know that x is nonnegative. Without knowing anything about x, the only way you can simplify $\sqrt{x^2}$ is to include absolute value signs when you remove x from the radical.

$$\sqrt{x^2} = |x| \qquad\qquad \text{Restricted by absolute value signs}$$

When simplifying the expression $\sqrt{x^3}$, it is not necessary to include absolute value signs because the domain of this expression does not include negative numbers.

$$\sqrt{x^3} = \sqrt{x^2(x)} = x\sqrt{x} \qquad\qquad \text{Restricted by domain of radical}$$

EXAMPLE 2   Removing Variable Factors from Radicals

Simplify each radical expression.

(a) $\sqrt{25x^2}$ (b) $\sqrt{12x^3}, \quad x \geq 0$ (c) $\sqrt{144x^4}$

Solution

(a) $\sqrt{25x^2} = \sqrt{5^2x^2} = \sqrt{5^2}\sqrt{x^2} = 5|x|$ $\sqrt{x^2} = |x|$

(b) $\sqrt{12x^3} = \sqrt{2^2x^2(3x)} = 2x\sqrt{3x}$ $\sqrt{2^2}\sqrt{x^2} = 2x, \quad x \geq 0$

(c) $\sqrt{144x^4} = \sqrt{12^2(x^2)^2} = 12x^2$ $\sqrt{12^2}\sqrt{(x^2)^2} = 12|x^2| = 12x^2$

In the same way that perfect squares can be removed from square root radicals, perfect nth powers can be removed from nth root radicals.

Students may make mistakes simplifying radicals by confusing whether they are dealing with square roots, cube roots, or fourth roots. Have students compare and simplify the following radicals: $\sqrt{64}$, $\sqrt[3]{64}$, and $\sqrt[4]{64}$.

EXAMPLE 3   Removing Factors from Radicals

Simplify each radical expression.

(a) $\sqrt[3]{40}$ (b) $\sqrt[4]{x^5}, \quad x \geq 0$ (c) $\sqrt[3]{54x^3y^5}$

Solution

(a) $\sqrt[3]{40} = \sqrt[3]{8(5)} = \sqrt[3]{2^3(5)} = 2\sqrt[3]{5}$ $\sqrt[3]{2^3} = 2$

(b) $\sqrt[4]{x^5} = \sqrt[4]{x^4(x)} = x\sqrt[4]{x}$ $\sqrt[4]{x^4} = x, \quad x \geq 0$

(c) $\sqrt[3]{54x^3y^5} = \sqrt[3]{27x^3y^3(2y^2)}$

$\qquad\qquad = \sqrt[3]{3^3x^3y^3(2y^2)}$

$\qquad\qquad = 3xy\sqrt[3]{2y^2}$ $\sqrt[3]{3^3}\sqrt[3]{x^3}\sqrt[3]{y^3} = 3xy$

Remind students to include the index of the radical when simplifying.

Rationalization Techniques

Removing factors from radicals is only one of three techniques that we use to simplify radicals. We summarize all three techniques as follows.

Simplifying Radical Expressions

A radical expression is said to be in simplest form if all three of the following are true.

1. All possible factors have been removed from each radical.

2. No radical contains a fraction.

3. No denominator of a fraction contains a radical.

To meet the last two conditions, you can use a technique called **rationalizing the denominator.** This involves multiplying both the numerator and denominator by a factor that creates a perfect nth power in the denominator.

EXAMPLE 4   *Rationalizing the Denominator*

(a) $\sqrt{\dfrac{3}{5}} = \dfrac{\sqrt{3}}{\sqrt{5}} = \dfrac{\sqrt{3}}{\sqrt{5}} \cdot \dfrac{\sqrt{5}}{\sqrt{5}} = \dfrac{\sqrt{15}}{\sqrt{5^2}} = \dfrac{\sqrt{15}}{5}$

Multiply by $\sqrt{5}/\sqrt{5}$ to create a perfect square in the denominator.

(b) $\dfrac{4}{\sqrt[3]{9}} = \dfrac{4}{\sqrt[3]{9}} \cdot \dfrac{\sqrt[3]{3}}{\sqrt[3]{3}} = \dfrac{4\sqrt[3]{3}}{\sqrt[3]{3^3}} = \dfrac{4\sqrt[3]{3}}{3}$

Multiply by $\sqrt[3]{3}/\sqrt[3]{3}$ to create a perfect cube in the denominator.

(c) $\dfrac{8}{3\sqrt{18}} = \dfrac{8}{3\sqrt{18}} \cdot \dfrac{\sqrt{2}}{\sqrt{2}} = \dfrac{8\sqrt{2}}{3\sqrt{36}} = \dfrac{8\sqrt{2}}{3(6)} = \dfrac{4\sqrt{2}}{9}$

EXAMPLE 5   *Rationalizing the Denominator*

Simplify each expression.

(a) $\sqrt{\dfrac{8x}{12y^5}}$ (b) $\sqrt[3]{\dfrac{54x^6y^3}{5z^2}}$

Solution

(a) $\sqrt{\dfrac{8x}{12y^5}} = \sqrt{\dfrac{2x}{3y^5}} = \dfrac{\sqrt{2x}}{\sqrt{3y^5}} \cdot \dfrac{\sqrt{3y}}{\sqrt{3y}} = \dfrac{\sqrt{6xy}}{\sqrt{9y^6}} = \dfrac{\sqrt{6xy}}{3y^3}$

(b) $\sqrt[3]{\dfrac{54x^6y^3}{5z^2}} = \dfrac{\sqrt[3]{(3^3)(2)(x^6)(y^3)}}{\sqrt[3]{5z^2}} \cdot \dfrac{\sqrt[3]{25z}}{\sqrt[3]{25z}} = \dfrac{3x^2y\sqrt[3]{50z}}{\sqrt[3]{5^3z^3}} = \dfrac{3x^2y\sqrt[3]{50z}}{5z}$

Adding and Subtracting Radicals

Two or more radical expressions are *alike* if they have the same radicand and the same index. For instance, $\sqrt{2}$ and $3\sqrt{2}$ are alike, but $\sqrt{3}$ and $\sqrt[3]{3}$ are not alike. Two radical expressions that are alike can be added or subtracted by adding or subtracting their coefficients.

EXAMPLE 6   Combining Radicals

NOTE   Notice in Example 6(c) that *before* concluding that two radicals cannot be combined, you should check to see that they are written in simplest form.

(a) $\sqrt{7} + 5\sqrt{7} - 2\sqrt{7} = (1 + 5 - 2)\sqrt{7} = 4\sqrt{7}$

(b) $6\sqrt{x} - \sqrt[3]{4} - 5\sqrt{x} + 2\sqrt[3]{4} = 6\sqrt{x} - 5\sqrt{x} - \sqrt[3]{4} + 2\sqrt[3]{4}$

$$= (6 - 5)\sqrt{x} + (-1 + 2)\sqrt[3]{4}$$

$$= \sqrt{x} + \sqrt[3]{4}$$

(c) $3\sqrt[3]{x} + \sqrt[3]{8x} = 3\sqrt[3]{x} + 2\sqrt[3]{x} = (3 + 2)\sqrt[3]{x} = 5\sqrt[3]{x}$

 *Technology*

You can use a graphing utility to check simplifications such as those shown in Example 7. For instance, to check the result of Example 7(a), sketch the graphs of

$$y = \sqrt{45x} + 3\sqrt{20x}$$

and

$$y = 9\sqrt{5x}.$$

Both equations have the same graph, as shown below.

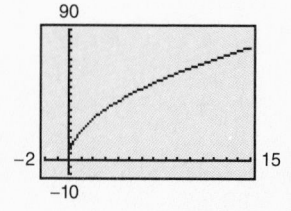

EXAMPLE 7   Simplifying Radical Expressions

(a) $\sqrt{45x} + 3\sqrt{20x} = \sqrt{9 \cdot 5x} + 3\sqrt{4 \cdot 5x} = 3\sqrt{5x} + 6\sqrt{5x} = 9\sqrt{5x}$

(b) $5\sqrt{x^3} - x\sqrt{4x} = 5x\sqrt{x} - 2x\sqrt{x} = 3x\sqrt{x}$

(c) $\sqrt[3]{54y^5} + 4\sqrt[3]{2y^2} = \sqrt[3]{27y^3(2y^2)} + 4\sqrt[3]{2y^2}$

$$= 3y\sqrt[3]{2y^2} + 4\sqrt[3]{2y^2} = (3y + 4)\sqrt[3]{2y^2}$$

In some instances, it may be necessary to rationalize denominators before combining radicals.

EXAMPLE 8   Rationalizing Denominators Before Simplifying

Simplify the expression $\sqrt{7} - \dfrac{5}{\sqrt{7}}$.

Solution

$$\sqrt{7} - \frac{5}{\sqrt{7}} = \sqrt{7} - \left(\frac{5}{\sqrt{7}} \cdot \frac{\sqrt{7}}{\sqrt{7}} \right)$$

$$= \sqrt{7} - \frac{5\sqrt{7}}{7}$$

$$= \left(1 - \frac{5}{7} \right)\sqrt{7}$$

$$= \frac{2}{7}\sqrt{7}$$

Applications of Radicals

EXAMPLE 9   *Finding the Frequencies of Musical Notes*

The musical note A-440 (A above middle C) has a frequency of 440 vibrations per second. The frequency F of any note can be found by the model

$$F = 440 \sqrt[12]{2^n},$$

where n represents the number of notes above or below A-440, as indicated in Figure 5.3. Two notes are said to *harmonize* (sound pleasing to the ear) if the ratio of their frequencies is an integer or a simple rational number. Show that any two notes that are an octave apart harmonize.

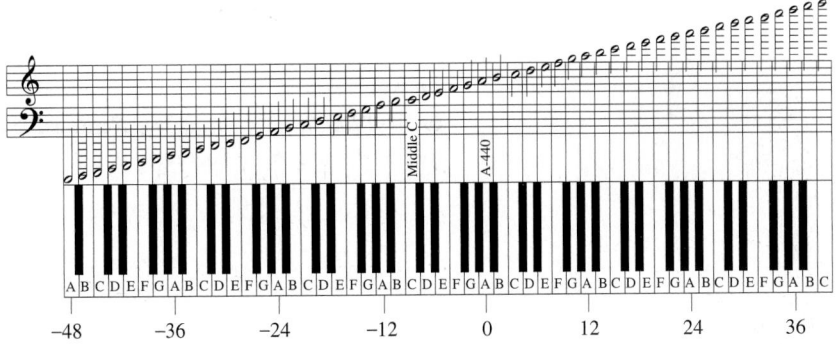

FIGURE 5.3

Solution

Notes that are an octave apart differ by 12 notes. (The "oct" in octave refers to the number of white keys on a piano.) Thus, you can represent the two notes by n and $n + 12$. The ratio of the frequencies of the two notes is as follows.

$$\frac{\text{Frequency of higher note}}{\text{Frequency of lower note}} = \frac{440 \sqrt[12]{2^{n+12}}}{440 \sqrt[12]{2^n}}$$

$$= \frac{\sqrt[12]{2^{n+12}}}{\sqrt[12]{2^n}}$$

$$= \sqrt[12]{\frac{2^{n+12}}{2^n}}$$

$$= \sqrt[12]{2^{12}}$$

$$= 2$$

Thus, the ratio of the two frequencies is 2—which means that the two notes harmonize.

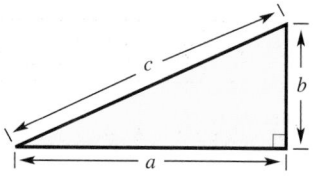

FIGURE 5.4

A common use of radicals occurs in applications involving right triangles. Recall that a right triangle is one that contains a right (or 90°) angle, as shown in Figure 5.4. The relationship among the three sides of a right triangle is described by the *Pythagorean Theorem,* which says that if a and b are the lengths of the legs and c is the length of the hypotenuse, then

$$c = \sqrt{a^2 + b^2}. \qquad \text{Pythagorean Theorem: } a^2 + b^2 = c^2$$

For instance, if $a = 6$ and $b = 9$, then

$$c = \sqrt{6^2 + 9^2} = \sqrt{117} = \sqrt{9}\sqrt{13} = 3\sqrt{13}.$$

EXAMPLE 10 An Application of the Pythagorean Theorem

A softball diamond has the shape of a square with 60-foot sides (see Figure 5.5). The catcher is 5 feet behind home plate. How far does the catcher have to throw to reach second base?

Solution

In Figure 5.5, let x be the hypotenuse of a right triangle with 60-foot legs. Thus, by the Pythagorean Theorem, you have the following.

$$x = \sqrt{60^2 + 60^2} \qquad \text{Pythagorean Theorem}$$
$$x = \sqrt{7200}$$
$$x \approx 84.9 \text{ feet}$$

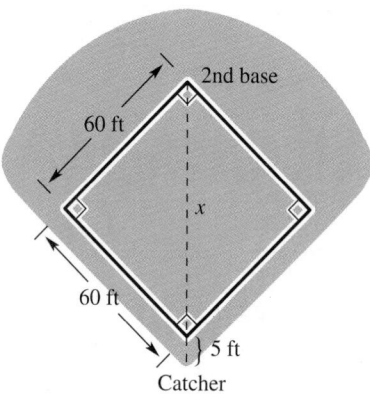

FIGURE 5.5

Historical Note: Pythagoras was a 6th century B.C. philosopher and mathematician. He taught the Pythagorean Theorem, but he didn't discover it. The Theorem was known by the Sumerians as early as 2000 B.C. Note the use of the Pythagorean Theorem in Example 10.

Thus, the distance from home plate to second base is approximately 84.9 feet. Because the catcher is 5 feet behind home plate, the catcher must make a throw of

$$x + 5 \approx 84.9 + 5 = 89.9 \text{ feet}.$$

Group Activities

Extending the Concept

Properties of Radicals In general, $\sqrt{a + b}$ is *not* equal to $\sqrt{a} + \sqrt{b}$. One convenient way to demonstrate this is to let $a = 9$ and $b = 16$. Then

$$\sqrt{9 + 16} = \sqrt{25} = 5, \qquad \text{whereas} \qquad \sqrt{9} + \sqrt{16} = 3 + 4 = 7.$$

In your group, try to find another example in which a, b, and $a + b$ are all perfect squares.

| 5.3 | **Exercises** |

See Warm-Up
Exercises, p. A44

In Exercises 1–8, write the expression as a single radical.

1. $\sqrt{3} \cdot \sqrt{10}$

2. $\sqrt[5]{9} \cdot \sqrt[5]{19}$

3. $\sqrt[3]{11} \cdot \sqrt[3]{10}$

4. $\sqrt[4]{35} \cdot \sqrt[4]{3}$

5. $\dfrac{\sqrt{15}}{\sqrt{31}}$

6. $\dfrac{\sqrt[3]{84}}{\sqrt[3]{9}}$

7. $\dfrac{\sqrt[5]{152}}{\sqrt[5]{3}}$

8. $\dfrac{\sqrt[4]{633}}{\sqrt[4]{5}}$

In Exercises 9–16, write the expression as a product or quotient of radicals and simplify.

9. $\sqrt{9 \cdot 35}$

10. $\sqrt[3]{27 \cdot 4}$

11. $\sqrt[4]{81 \cdot 11}$

12. $\sqrt[5]{100,000 \cdot 3}$

13. $\sqrt{\dfrac{35}{9}}$

14. $\sqrt[4]{\dfrac{165}{16}}$

15. $\sqrt[3]{\dfrac{11}{1000}}$

16. $\sqrt[5]{\dfrac{2}{243}}$

In Exercises 17–32, simplify the radical.

17. $\sqrt{20}$

18. $\sqrt{50}$

19. $\sqrt{27}$

20. $\sqrt{125}$

21. $\sqrt{0.04}$

22. $\sqrt{0.25}$

23. $\sqrt[3]{24}$

24. $\sqrt[3]{54}$

25. $\sqrt[4]{30,000}$

26. $\sqrt[5]{96}$

27. $\sqrt{\dfrac{15}{4}}$

28. $\sqrt{\dfrac{5}{36}}$

29. $\sqrt[3]{\dfrac{35}{64}}$

30. $\sqrt[4]{\dfrac{5}{16}}$

31. $\sqrt[5]{\dfrac{15}{243}}$

32. $\sqrt[3]{\dfrac{1}{1000}}$

In Exercises 33–56, simplify the expression.

33. $\sqrt{4 \times 10^{-4}}$

34. $\sqrt{8.5 \times 10^3}$

35. $\sqrt[3]{2.4 \times 10^6}$

36. $\sqrt[4]{4.4 \times 10^{-4}}$

37. $\sqrt[5]{3 \times 10^5}$

38. $\sqrt[4]{8.1 \times 10^{-3}}$

39. $\sqrt{9x^5}$

40. $\sqrt{64x^3}$

41. $\sqrt{48y^4}$

42. $\sqrt{32x}$

43. $\sqrt[3]{x^4y^3}$

44. $\sqrt[3]{a^5b^6}$

45. $\sqrt[5]{32x^5y^6}$

46. $\sqrt[4]{128u^4v^7}$

47. $\sqrt{\dfrac{13}{25}}$

48. $\sqrt{\dfrac{15}{36}}$

49. $\sqrt[5]{\dfrac{32x^2}{y^5}}$

50. $\sqrt[3]{\dfrac{16z^3}{y^6}}$

51. $\sqrt[3]{\dfrac{54a^4}{b^9}}$

52. $\sqrt[4]{\dfrac{3u^2}{16v^8}}$

53. $\sqrt{\dfrac{32a^4}{b^2}}$

54. $\sqrt{\dfrac{18x^2}{z^6}}$

55. $\sqrt[4]{(3x^2)^4}$

56. $\sqrt[5]{96x^5}$

In Exercises 57–60, use a graphing utility to graph the equations y_1 and y_2 in the same viewing rectangle. Use the graphs to determine if $y_1 = y_2$. If not, explain why.

57. $y_1 = \sqrt{12x^2}$
    $y_2 = 2x\sqrt{3}$

58. $y_1 = \sqrt{12x^3}$
    $y_2 = 2x\sqrt{3x}$

59. $y_1 = \sqrt[3]{16x^3}$
    $y_2 = 2x\sqrt[3]{2}$

60. $y_1 = \sqrt[4]{16x^6}$
    $y_2 = 2x\sqrt[4]{x^2}$

In Exercises 61–80, rationalize the denominator and simplify further, if possible.

61. $\sqrt{\frac{1}{3}}$

62. $\sqrt{\frac{1}{5}}$

63. $\frac{12}{\sqrt{3}}$

64. $\frac{5}{\sqrt{10}}$

65. $\sqrt[4]{\frac{5}{4}}$

66. $\sqrt[3]{\frac{9}{25}}$

67. $\frac{6}{\sqrt[3]{32}}$

68. $\frac{10}{\sqrt[5]{16}}$

69. $\frac{1}{\sqrt{y}}$

70. $\frac{1}{\sqrt{2x}}$

71. $\sqrt{\frac{4}{x}}$

72. $\sqrt{\frac{5}{c}}$

73. $\sqrt{\frac{4}{x^3}}$

74. $\frac{5}{\sqrt{8x^5}}$

75. $\sqrt[3]{\frac{2x}{3y}}$

76. $\sqrt[3]{\frac{20x^2}{9y^2}}$

77. $\frac{a^3}{\sqrt[3]{ab^2}}$

78. $\frac{3u^2}{\sqrt[4]{8u^3}}$

79. $\frac{6}{\sqrt{3b^3}}$

80. $\frac{1}{\sqrt{xy}}$

In Exercises 81 and 82, use a graphing utility to graph the equations in the same viewing rectangle. Use the graphs to verify that the expressions are equivalent. Verify the results algebraically.

81. $y_1 = \sqrt{\frac{3}{x}}$

$y_2 = \frac{\sqrt{3x}}{x}$

82. $y_1 = \frac{4}{\sqrt{2x}}$

$y_2 = \frac{2\sqrt{2x}}{x}$

In Exercises 83–94, combine the radical expressions, if possible.

83. $3\sqrt{2} - \sqrt{2}$

84. $\frac{2}{5}\sqrt{5} - \frac{6}{5}\sqrt{5}$

85. $12\sqrt{8} - 3\sqrt[3]{8}$

86. $4\sqrt{32} + 7\sqrt{32}$

87. $2\sqrt[3]{54} + 12\sqrt[3]{16}$

88. $4\sqrt[4]{48} - \sqrt[4]{243}$

89. $5\sqrt{9x} - 3\sqrt{x}$

90. $3\sqrt{x+1} + 10\sqrt{x+1}$

91. $\sqrt{25y} + \sqrt{64y}$

92. $\sqrt[3]{16t^4} - \sqrt[3]{54t^4}$

93. $10\sqrt[3]{z} - \sqrt[3]{z^4}$

94. $5\sqrt[3]{24u^2} + 2\sqrt[3]{81u^5}$

In Exercises 95 and 96, use a graphing utility to graph the equations in the same viewing rectangle. Use the graphs to verify that the expressions are equivalent. Verify the results algebraically.

95. $y_1 = 7\sqrt{x^3} - 2x\sqrt{4x}$

$y_2 = 3x\sqrt{x}$

96. $y_1 = \sqrt[3]{8x^4} - \sqrt[3]{x^4}$

$y_2 = x\sqrt[3]{x}$

In Exercises 97–100, perform the addition or subtraction and simplify your answer.

97. $\sqrt{5} - \frac{3}{\sqrt{5}}$

98. $\sqrt{10} + \frac{5}{\sqrt{10}}$

99. $\sqrt{20} - \sqrt{\frac{1}{5}}$

100. $\frac{x}{\sqrt{3x}} + \sqrt{27x}$

In Exercises 101–104, place the correct symbol (<, >, or =) between the numbers.

101. $\sqrt{7} + \sqrt{18}$ ▭ $\sqrt{7 + 18}$

102. $\sqrt{10} - \sqrt{6}$ ▭ $\sqrt{10 - 6}$

103. 5 ▭ $\sqrt{3^2 + 2^2}$

104. 5 ▭ $\sqrt{3^2 + 4^2}$

Geometry In Exercises 105 and 106, find the length of the hypotenuse of the right triangle.

105.

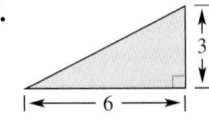

106.

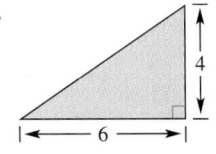

107. *Geometry*   The four corners are cut from a 4-foot by 8-foot sheet of plywood as shown in the figure. Find the perimeter of the remaining piece of plywood.

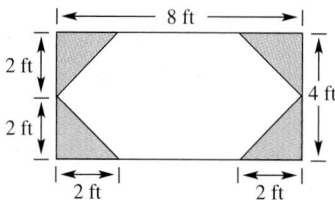

108. *Vibrating String*   The frequency f in cycles per second of a vibrating string is given by

$$f = \frac{1}{100} \sqrt{\frac{400 \times 10^6}{5}}.$$

Use a calculator to approximate this number. (Round the result to two decimal places.)

109. *Period of a Pendulum*   The period T in seconds of a pendulum (see figure) is given by

$$T = 2\pi \sqrt{\frac{L}{32}},$$

where L is the length of the pendulum in feet. Find the period of a pendulum whose length is 4 feet. (Round the result to two decimal places.)

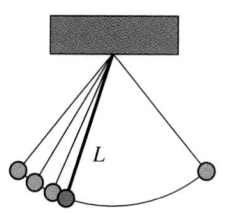

110. *Think About It*   Square the real number $5/\sqrt{3}$ and note that the radical is eliminated from the denominator. Is this equivalent to rationalizing the denominator? Why or why not?

Section Project

The Square Root Spiral   The square root spiral (see figure) is formed by a sequence of right triangles, each with a side whose length is 1. Let r_n be the length of the hypotenuse of the nth triangle.

(a) Each leg of the first triangle has a length of 1 unit. Use the Pythagorean Theorem to show that $r_1 = \sqrt{2}$.

(b) Find r_2, r_3, r_4, r_5, and r_6.

(c) What can you conclude about r_n?

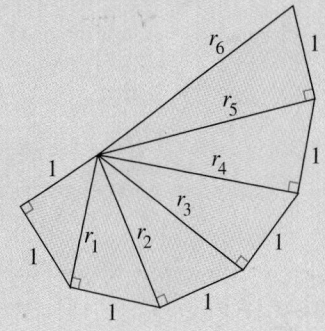

Mid-Chapter Quiz

Take this quiz as you would take a quiz in class. After you are done, check your work against the answers given in the back of the book.

In Exercises 1–4, evaluate the expression.

1. -12^{-2} **2.** $\left(\frac{3}{4}\right)^{-3}$ **3.** $\sqrt{\frac{25}{9}}$ **4.** $(-64)^{2/3}$

In Exercises 5–8, rewrite the expression using only positive exponents and simplify. (Assume any variables in the expression are nonzero.)

5. $(t^3)^{-1/2}(3t^3)$ **6.** $\dfrac{(10x)^0}{(4x^{-2})^{3/2}}$ **7.** $\dfrac{10u^{-2}}{15u}$ **8.** $(3x^2y^{-1})(4x^{-2}y)^{-2}$

9. Write each number in scientific notation.

  (a) 13,400,000   (b) 0.00075

10. Evaluate each quantity without the use of a calculator.

  (a) $(3 \times 10^3)^4$ (b) $\dfrac{3.2 \times 10^4}{16 \times 10^7}$

In Exercises 11–14, simplify each expression.

11. (a) $\sqrt{150}$ (b) $\sqrt[3]{54}$ **12.** (a) $\sqrt{27x^2}$ (b) $\sqrt[4]{81x^6}$

13. (a) $\sqrt[4]{\dfrac{5}{16}}$ (b) $\sqrt{\dfrac{24}{49}}$ **14.** (a) $\sqrt{\dfrac{40u^3}{9}}$ (b) $\sqrt[3]{\dfrac{16}{u^{12}}}$

In Exercises 15 and 16, rationalize the denominator and simplify further, if possible.

15. (a) $\sqrt{\dfrac{2}{3}}$ (b) $\dfrac{24}{\sqrt{12}}$ **16.** (a) $\dfrac{10}{\sqrt{5x}}$ (b) $\sqrt[3]{\dfrac{3}{2a}}$

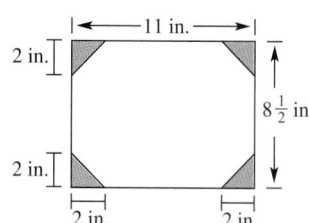

In Exercises 17 and 18, combine the radical expressions, if possible.

17. $\sqrt{200y} - 3\sqrt{8y}$ **18.** $6x\sqrt[3]{5x^2} + 2\sqrt[3]{40x^4}$

19. Explain why $\sqrt{5^2 + 12^2} \neq 17$. Determine the correct value of the radical.

20. The four corners are cut from an $8\frac{1}{2}$-inch by 11-inch sheet of paper as shown in the figure. Find the perimeter of the remaining piece of paper.

Figure for 20

| **5.4** | **Multiplying and Dividing Radicals** |

Multiplying Radical Expressions ▪ Dividing Radical Expressions

Multiplying Radical Expressions

You can multiply radical expressions by using the Distributive Property or the FOIL Method. In both procedures, you also make use of the Multiplication Property of Radicals. Recall from Section 5.3 that the product of two radicals is given by

$$\sqrt[n]{a}\ \sqrt[n]{b} = \sqrt[n]{ab},$$

where a and b are real numbers whose nth roots are also real numbers.

EXAMPLE 1 *Multiplying Radical Expressions*

$$\sqrt{3}\left(2 + \sqrt{5}\right) = 2\sqrt{3} + \sqrt{3}\sqrt{5} \qquad \text{Distributive Property}$$
$$= 2\sqrt{3} + \sqrt{15} \qquad \text{Multiplication Property of Radicals}$$

In Example 1, the product of $\sqrt{3}$ and $\sqrt{5}$ is best left as $\sqrt{15}$. In some cases, however, the product of two radicals can be simplified, as shown in Example 2.

Study Tip

Throughout this chapter, remember that you can use a calculator to check your computations. For instance, in Example 2(b), try using a calculator to evaluate the expressions

$$\sqrt{6}\left(\sqrt{12} - \sqrt{3}\right) \text{ and } 3\sqrt{2}.$$

For each expression, you should obtain a calculator display of about 4.2426.

EXAMPLE 2 *Multiplying Radical Expressions*

Find each product and simplify.

(a) $\sqrt{2}\left(4 - \sqrt{8}\right)$

(b) $\sqrt{6}\left(\sqrt{12} - \sqrt{3}\right)$

Solution

(a) $\sqrt{2}\left(4 - \sqrt{8}\right) = 4\sqrt{2} - \sqrt{2}\sqrt{8}$ Distributive Property
$$= 4\sqrt{2} - \sqrt{16} \qquad \text{Multiplication Property of Radicals}$$
$$= 4\sqrt{2} - 4 \qquad \text{Simplify.}$$

(b) $\sqrt{6}\left(\sqrt{12} - \sqrt{3}\right) = \sqrt{6}\sqrt{12} - \sqrt{6}\sqrt{3}$ Distributive Property
$$= \sqrt{72} - \sqrt{18} \qquad \text{Multiplication Property of Radicals}$$
$$= 6\sqrt{2} - 3\sqrt{2} \qquad \text{Find perfect square factors.}$$
$$= 3\sqrt{2} \qquad \text{Simplify.}$$

Students sometimes have more difficulty using the FOIL method when radicals are involved. You may want to use an additional example such as:

$$\left(4 + 2\sqrt{5}\right)\left(3 - 4\sqrt{5}\right)$$
$$= 12 - 16\sqrt{5} + 6\sqrt{5} - 8(5)$$
$$= -28 - 10\sqrt{5}$$

Have students verify their results using their calculators.

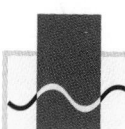

Technology

Remember that you can use a graphing utility to check simplifications. For instance, to check the result of Example 4(b), sketch the graphs of

$$y = \left(\sqrt{3} + \sqrt{x}\right)\left(\sqrt{3} - \sqrt{x}\right)$$

and

$$y = 3 - x.$$

Both equations have the same graph, as shown below.

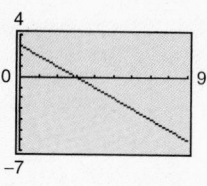

In Examples 1 and 2, the Distributive Property was used to multiply radical expressions. In Example 3, note how the FOIL Method can be used to multiply binomial radical expressions.

EXAMPLE 3   Using the FOIL Method

(a)
$$\left(2\sqrt{7} - 4\right)\left(\sqrt{7} + 1\right) = \overbrace{2\left(\sqrt{7}\right)^2}^{F} + \overbrace{2\sqrt{7}}^{O} - \overbrace{4\sqrt{7}}^{I} - \overset{L}{4} \qquad \text{FOIL Method}$$
$$= 2(7) + (2 - 4)\sqrt{7} - 4 \qquad \text{Combine like radicals.}$$
$$= 10 - 2\sqrt{7} \qquad \text{Simplify.}$$

(b)
$$\left(3 - \sqrt{x}\right)\left(1 + \sqrt{x}\right) = 3 + 3\sqrt{x} - \sqrt{x} - \left(\sqrt{x}\right)^2 \qquad \text{FOIL Method}$$
$$= 3 + 2\sqrt{x} - x \qquad \text{Combine like radicals.}$$

In addition to the FOIL Method, you can also use special product formulas to multiply binomial radical expressions.

EXAMPLE 4   Using a Special Product Formula

(a)
$$\left(2 - \sqrt{5}\right)\left(2 + \sqrt{5}\right) = 2^2 - \left(\sqrt{5}\right)^2 \qquad \text{Special product formula}$$
$$= 4 - 5$$
$$= -1$$

(b)
$$\left(\sqrt{3} + \sqrt{x}\right)\left(\sqrt{3} - \sqrt{x}\right) = \left(\sqrt{3}\right)^2 - \left(\sqrt{x}\right)^2 \qquad \text{Special product formula}$$
$$= 3 - x, \quad x \geq 0$$

In Example 4(a), the expressions $\left(2 - \sqrt{5}\right)$ and $\left(2 + \sqrt{5}\right)$ are called **conjugates** of each other. The product of two conjugates is the difference of two squares, which is given by the special product formula $(a + b)(a - b) = a^2 - b^2$. Here are some other examples.

Expression	Conjugate	Product
$\left(1 - \sqrt{3}\right)$	$\left(1 + \sqrt{3}\right)$	$(1)^2 - \left(\sqrt{3}\right)^2 = 1 - 3 = -2$
$\left(\sqrt{5} + \sqrt{2}\right)$	$\left(\sqrt{5} - \sqrt{2}\right)$	$\left(\sqrt{5}\right)^2 - \left(\sqrt{2}\right)^2 = 5 - 2 = 3$
$\left(\sqrt{10} - 3\right)$	$\left(\sqrt{10} + 3\right)$	$\left(\sqrt{10}\right)^2 - (3)^2 = 10 - 9 = 1$
$\sqrt{x} + 2$	$\sqrt{x} - 2$	$\left(\sqrt{x}\right)^2 - (2)^2 = x - 4, \quad x \geq 0$

Ask students what pattern they can observe (*i.e.*, when you multiply a binomial with radical terms by its conjugate, the result is free of radicals.)

Dividing Radical Expressions

To simplify a *quotient* involving radicals, you rationalize the denominator. For single-term denominators, you can use the rationalizing process described in Section 5.3. To rationalize a denominator involving two terms, multiply both the numerator and denominator by the *conjugate of the denominator,* as demonstrated in Examples 5, 6, and 7.

EXAMPLE 5   *Simplifying Quotients Involving Radicals*

(a)

$$\frac{\sqrt{3}}{1 - \sqrt{5}} = \frac{\sqrt{3}}{1 - \sqrt{5}} \cdot \frac{1 + \sqrt{5}}{1 + \sqrt{5}}$$ Multiply by conjugate of denominator.

$$= \frac{\sqrt{3}\left(1 + \sqrt{5}\right)}{1^2 - \left(\sqrt{5}\right)^2}$$ Special product formula

$$= \frac{\sqrt{3} + \sqrt{15}}{1 - 5}$$ Simplify.

$$= \frac{\sqrt{3} + \sqrt{15}}{-4} \quad \text{or} \quad -\frac{\sqrt{3} + \sqrt{15}}{4}$$ Simplify.

(b)

$$\frac{4}{2 - \sqrt{3}} = \frac{4}{2 - \sqrt{3}} \cdot \frac{2 + \sqrt{3}}{2 + \sqrt{3}}$$ Multiply by conjugate of denominator.

$$= \frac{4\left(2 + \sqrt{3}\right)}{2^2 - \left(\sqrt{3}\right)^2}$$ Special product formula

$$= \frac{8 + 4\sqrt{3}}{4 - 3}$$ Simplify.

$$= 8 + 4\sqrt{3}$$ Simplify.

At this point students may still be tempted to rationalize $4/(2 - \sqrt{3})$ by multiplying $\sqrt{3}/\sqrt{3}$. Ask them why this doesn't rationalize the denominator.

EXAMPLE 6   *Simplifying Quotients Involving Radicals*

$$\frac{5\sqrt{2}}{\sqrt{7} + \sqrt{2}} = \frac{5\sqrt{2}}{\sqrt{7} + \sqrt{2}} \cdot \frac{\sqrt{7} - \sqrt{2}}{\sqrt{7} - \sqrt{2}}$$ Multiply by conjugate of denominator.

$$= \frac{5\left(\sqrt{14} - \sqrt{4}\right)}{\left(\sqrt{7}\right)^2 - \left(\sqrt{2}\right)^2}$$ Special product formula

$$= \frac{5\left(\sqrt{14} - 2\right)}{7 - 2}$$ Simplify.

$$= \frac{5\left(\sqrt{14} - 2\right)}{5}$$ Cancel common factor.

$$= \sqrt{14} - 2$$ Simplest form

EXAMPLE 7   Dividing Radical Expressions

Perform each division and simplify.

(a) $6 \div \left(\sqrt{x} - 2 \right)$ (b) $\left(5 - \sqrt{3} \right) \div \left(\sqrt{6} + \sqrt{2} \right)$

Solution

(a) $\dfrac{6}{\sqrt{x} - 2} = \dfrac{6}{\sqrt{x} - 2} \cdot \dfrac{\sqrt{x} + 2}{\sqrt{x} + 2}$ Multiply by conjugate of denominator.

$= \dfrac{6\left(\sqrt{x} + 2 \right)}{\left(\sqrt{x} \right)^2 - 2^2}$ Special product formula

$= \dfrac{6\sqrt{x} + 12}{x - 4}$ Simplify.

(b) $\dfrac{5 - \sqrt{3}}{\sqrt{6} + \sqrt{2}} = \dfrac{5 - \sqrt{3}}{\sqrt{6} + \sqrt{2}} \cdot \dfrac{\sqrt{6} - \sqrt{2}}{\sqrt{6} - \sqrt{2}}$ Multiply by conjugate of denominator.

$= \dfrac{5\sqrt{6} - 5\sqrt{2} - \sqrt{18} + \sqrt{6}}{\left(\sqrt{6} \right)^2 - \left(\sqrt{2} \right)^2}$ FOIL Method and special product formula

$= \dfrac{6\sqrt{6} - 5\sqrt{2} - 3\sqrt{2}}{6 - 2}$ Simplify.

$= \dfrac{6\sqrt{6} - 8\sqrt{2}}{4}$ Simplify.

$= \dfrac{3\sqrt{6} - 4\sqrt{2}}{2}$ Simplify.

Have students form the ratio of length to width of 3 × 5 and 5 × 8 index cards. Then ask them to convert these ratios to decimal form and compare the results to the golden section.

Group Activities

You Be the Instructor

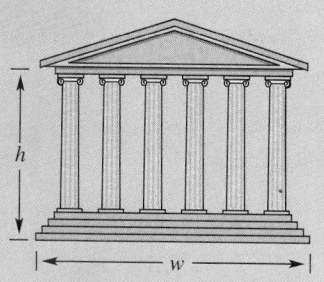

The Golden Section   The ratio of the width of the Temple of Hephaestus to its height is approximately $2/\left(\sqrt{5} - 1 \right)$. This number is called the **golden section.** Early Greeks believed that the most aesthetically pleasing rectangles were those whose sides had this ratio.

(a) Rationalize the denominator for this number. Approximate your answer, rounded to two decimal places.

(b) Use the Pythagorean Theorem, a straightedge, and a compass to construct a rectangle whose sides have the golden section as their ratio.

quiz

$(\sqrt{5} + 2\sqrt{3})(\sqrt{5} - \sqrt{3})$

$\sqrt{5} = \sqrt{15} + 2\sqrt{6} - 6$

5.4 Exercises

See Warm-Up Exercises, p. A44

In Exercises 1–32, multiply and simplify.

1. $\sqrt{2} \cdot \sqrt{8}$
2. $\sqrt{6} \cdot \sqrt{18}$
3. $\sqrt{3} \cdot \sqrt{6}$
4. $\sqrt{5} \cdot \sqrt{10}$
5. $\sqrt{5}(2 - \sqrt{3})$
6. $\sqrt{11}(\sqrt{5} - 3)$
7. $\sqrt{2}(\sqrt{20} + 8)$
8. $\sqrt{7}(\sqrt{14} + 3)$
9. $(\sqrt{3} + 2)(\sqrt{3} - 2)$
10. $(\sqrt{15} + 3)(\sqrt{15} - 3)$
11. $(3 - \sqrt{5})(3 + \sqrt{5})$
12. $(\sqrt{11} + 3)(\sqrt{11} - 3)$
13. $(2\sqrt{2} + \sqrt{4})(2\sqrt{2} - \sqrt{4})$
14. $(4\sqrt{3} + \sqrt{2})(4\sqrt{3} - \sqrt{2})$
15. $(\sqrt{20} + 2)^2$
16. $(4 - \sqrt{20})^2$
17. $(10 + \sqrt{2x})^2$
18. $(5 - \sqrt{3v})^2$
19. $\sqrt{y}(\sqrt{y} + 4)$
20. $\sqrt{x}(5 - \sqrt{x})$
21. $(\sqrt{5} + 3)(\sqrt{3} - 5)$
22. $(\sqrt{30} + 6)(\sqrt{2} + 6)$
23. $(9\sqrt{x} + 2)(5\sqrt{x} - 3)$
24. $(16\sqrt{u} - 3)(\sqrt{u} - 1)$
25. $(\sqrt{x} + \sqrt{y})(\sqrt{x} - \sqrt{y})$
26. $(3\sqrt{u} + \sqrt{3v})(3\sqrt{u} - \sqrt{3v})$
27. $\sqrt[3]{4}(\sqrt[3]{2} - 7)$
28. $(\sqrt[3]{9} + 5)(\sqrt[3]{5} - 5)$
29. $(\sqrt[3]{2x} + 5)^2$
30. $(\sqrt[3]{y} + 2)(\sqrt[3]{y^2} - 5)$
31. $(\sqrt[3]{2y} + 10)(\sqrt[3]{4y^2} - 10)$
32. $(\sqrt[3]{t} + 1)(\sqrt[3]{t^2} + 4\sqrt[3]{t} - 3)$

In Exercises 33–38, complete the statement.

33. $5x\sqrt{3} + 15\sqrt{3} = 5\sqrt{3}(\quad)$
34. $x\sqrt{7} - x^2\sqrt{7} = x\sqrt{7}(\quad)$
35. $4\sqrt{12} - 2x\sqrt{27} = 2\sqrt{3}(\quad)$
36. $5\sqrt{50} + 10y\sqrt{8} = 5\sqrt{2}(\quad)$
37. $6u^2 + \sqrt{18u^3} = 3u(\quad)$
38. $12s^3 - \sqrt{32s^4} = 4s^2(\quad)$

In Exercises 39–42, simplify the expression.

39. $\dfrac{4 - 8\sqrt{x}}{12}$
40. $\dfrac{-3 + 27\sqrt{2y}}{18}$
41. $\dfrac{-2y + \sqrt{12y^3}}{8y}$
42. $\dfrac{-t^2 - \sqrt{2t^3}}{3t}$

In Exercises 43–46, evaluate the function.

43. $f(x) = x^2 - 6x + 1$
   (a) $f(2 - \sqrt{3})$ (b) $f(3 - 2\sqrt{2})$
44. $g(x) = x^2 + 8x + 11$
   (a) $g(-4 + \sqrt{5})$ (b) $g(-4\sqrt{2})$
45. $f(x) = x^2 - 2x - 1$
   (a) $f(1 + \sqrt{2})$ (b) $f(\sqrt{4})$
46. $g(x) = x^2 - 4x + 1$
   (a) $g(1 + \sqrt{5})$ (b) $g(2 - \sqrt{3})$

In Exercises 47–54, find the conjugate of the number. Then multiply the number by its conjugate and simplify.

47. $2 + \sqrt{5}$
48. $\sqrt{2} - 9$
49. $\sqrt{11} - \sqrt{3}$
50. $\sqrt{10} + \sqrt{7}$
51. $\sqrt{x} - 3$
52. $\sqrt{t} + 7$
53. $\sqrt{2u} - \sqrt{3}$
54. $\sqrt{5a} + \sqrt{2}$

In Exercises 55–70, rationalize the denominator of the expression and simplify.

55. $\dfrac{6}{\sqrt{22} - 2}$
56. $\dfrac{3}{2\sqrt{10} - 5}$
57. $\dfrac{8}{\sqrt{7} + 3}$
58. $\dfrac{10}{\sqrt{9} + \sqrt{5}}$
59. $\dfrac{2}{6 + \sqrt{2}}$
60. $\dfrac{44\sqrt{5}}{3\sqrt{5} - 1}$

61. $\left(\sqrt{7}+2\right)\div\left(\sqrt{7}-2\right)$

sec 1 quiz **62.** $\left(5-3\sqrt{3}\right)\div\left(3+\sqrt{3}\right)$

63. $\dfrac{3x}{\sqrt{15}-\sqrt{3}}$

64. $\dfrac{6(y+1)}{y^2+\sqrt{y}}$

65. $\dfrac{2t^2}{\sqrt{5t}-\sqrt{t}}$

66. $\dfrac{5x}{\sqrt{x}-\sqrt{2}}$

67. $\left(\sqrt{x}-5\right)\div\left(2\sqrt{x}-1\right)$

68. $\left(2\sqrt{t}+1\right)\div\left(2\sqrt{t}-1\right)$

69. $\dfrac{\sqrt{u+v}}{\sqrt{u-v}-\sqrt{u}}$

70. $\dfrac{z}{\sqrt{u+z}-\sqrt{u}}$

In Exercises 71–74, use a graphing utility to graph the equations in the same viewing rectangle. Use the graphs to verify that the expressions are equivalent. Verify your results algebraically.

71. $y_1=\dfrac{10}{\sqrt{x}+1}$

$y_2=\dfrac{10\left(\sqrt{x}-1\right)}{x-1}$

72. $y_1=\dfrac{4x}{\sqrt{x}+4}$

$y_2=\dfrac{4x\left(\sqrt{x}-4\right)}{x-16}$

73. $y_1=\dfrac{2\sqrt{x}}{2-\sqrt{x}}$

$y_2=\dfrac{2\left(2\sqrt{x}+x\right)}{4-x}$

74. $y_1=\dfrac{\sqrt{2x}+6}{\sqrt{2x}-2}$

$y_2=\dfrac{x+6+4\sqrt{2x}}{x-2}$

75. *Strength of a Wooden Beam*   The rectangular cross section of a wooden beam cut from a log of diameter 24 inches (see figure) will have maximum strength if its width w and height h are given by

$$w=8\sqrt{3}$$

and

$$h=\sqrt{24^2-\left(8\sqrt{3}\right)^2}.$$

Find the area of the rectangular cross section and express the area in simplest form.

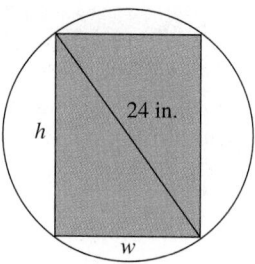
Figure for 75

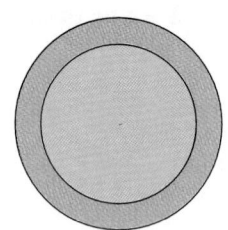
Figure for 76

76. *Geometry*   The areas of the circles in the figure are 15 square centimeters and 20 square centimeters. Find the ratio of the radius of the small circle to the radius of the large circle.

77. *Force to Move a Block*   The force required to slide a 500-pound steel block across a milling machine is

$$\frac{500k}{\sqrt{k^2+1}},$$

where k is the friction constant (see figure). Simplify this expression.

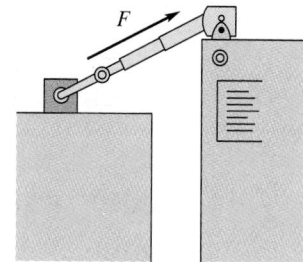

Section Project

Rationalizing Numerators   To simplify a quotient involving radicals, you rationalize the denominator. In the study of calculus, students sometimes rewrite an expression by rationalizing the numerator. For each of the following expressions, rationalize the numerator. (*Note:* Your results will not be in simplest radical form.)

(a) $\dfrac{\sqrt{2}}{7}$ (b) $\dfrac{\sqrt{7}+\sqrt{3}}{5}$ (c) $\dfrac{\sqrt{x}+6}{\sqrt{2}}$

5.5	**Solving Radical Equations**

Solving Radical Equations ▪ Application

Solving Radical Equations

Solving equations involving radicals is somewhat like solving equations that contain fractions—you try to get rid of the radicals and obtain a polynomial equation. Then you solve the polynomial equation using the standard procedures. The following property plays a key role.

Raising Both Sides of an Equation to the *n*th Power

Let u and v be real numbers, variables, or algebraic expressions, and let n be a positive integer. If $u = v$, then it follows that

$$u^n = v^n.$$

This is called **raising both sides of an equation to the *n*th power.**

To use this property to solve an equation, first try to isolate one of the radicals on one side of the equation.

Technology

To use a graphing utility to check the solution in Example 1, graph

$$y = \sqrt{x} - 8,$$

as shown below. Notice that the graph crosses the *x*-axis at $x = 64$, which confirms the solution that was obtained algebraically.

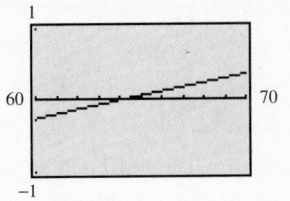

EXAMPLE 1   *Solving an Equation Having One Radical*

Solve $\sqrt{x} - 8 = 0$.

Solution

$\sqrt{x} - 8 = 0$	Original equation
$\sqrt{x} = 8$	Isolate radical.
$\left(\sqrt{x}\right)^2 = 8^2$	Square both sides.
$x = 64$	Simplify.

Check

$\sqrt{x} - 8 = 0$	Original equation
$\sqrt{64} - 8 \stackrel{?}{=} 0$	Substitute 64 for x.
$8 - 8 = 0$	Solution checks. ✓

Therefore, the equation has one solution: $x = 64$.

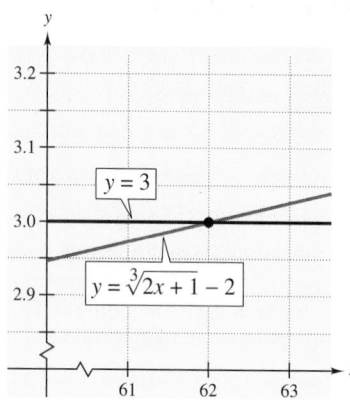

FIGURE 5.6

EXAMPLE 2 *Solving an Equation Having One Radical*

Solve $\sqrt[3]{2x + 1} - 2 = 3$.

Solution

$$\sqrt[3]{2x + 1} - 2 = 3 \qquad \text{Original equation}$$
$$\sqrt[3]{2x + 1} = 5 \qquad \text{Isolate radical.}$$
$$\left(\sqrt[3]{2x + 1}\right)^3 = 5^3 \qquad \text{Cube both sides.}$$
$$2x + 1 = 125 \qquad \text{Simplify.}$$
$$2x = 124 \qquad \text{Subtract 1 from both sides.}$$
$$x = 62 \qquad \text{Divide both sides by 2.}$$

Check this solution in the original equation. You can also check the solution graphically, as shown in Figure 5.6.

The next example demonstrates another reason for checking solutions in the *original* equation. Even with no mistakes in the solution process, it can happen that a "trial solution" does not satisfy the original equation. This type of "solution" is called **extraneous.** An extraneous solution does not satisfy the original equation, and therefore *must not* be listed as an actual solution.

EXAMPLE 3 *An Equation with an Extraneous Solution*

Solve $\sqrt{3x} + 6 = 0$.

Solution

$$\sqrt{3x} + 6 = 0 \qquad \text{Original equation}$$
$$\sqrt{3x} = -6 \qquad \text{Isolate radical.}$$
$$\left(\sqrt{3x}\right)^2 = (-6)^2 \qquad \text{Square both sides.}$$
$$3x = 36 \qquad \text{Simplify.}$$
$$x = 12 \qquad \text{Divide both sides by 3.}$$

Check

$$\sqrt{3x} + 6 = 0 \qquad \text{Original equation}$$
$$\sqrt{3(12)} + 6 \overset{?}{=} 0 \qquad \text{Substitute 12 for } x.$$
$$6 + 6 \neq 0 \qquad \text{Solution does not check.} \quad \text{✗}$$

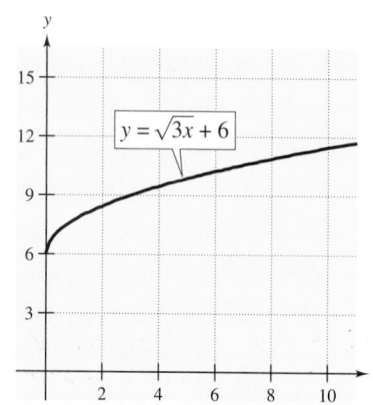

FIGURE 5.7

Therefore, 12 is an *extraneous* solution, which means that the original equation has no solution. You can also check this graphically, as shown in Figure 5.7.

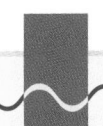

Technology

In Example 4, you can graphically check the solution of the equation by graphing the left side and right side in the same viewing rectangle. That is, by graphing the equations

$$y = \sqrt{5x + 3}$$

and

$$y = \sqrt{x + 11}$$

in the same viewing rectangle, as shown below, you can see that the two graphs intersect at $x = 2$.

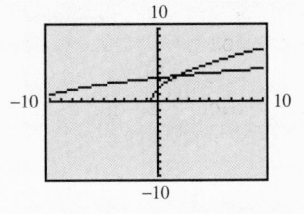

EXAMPLE 4 Solving an Equation Having Two Radicals

Solve $\sqrt{5x + 3} = \sqrt{x + 11}$.

Solution

$\sqrt{5x + 3} = \sqrt{x + 11}$	Original equation
$\left(\sqrt{5x + 3}\right)^2 = \left(\sqrt{x + 11}\right)^2$	Square both sides.
$5x + 3 = x + 11$	Simplify.
$5x = x + 8$	Subtract 3 from both sides.
$4x = 8$	Subtract x from both sides.
$x = 2$	Divide both sides by 4.

Check

$\sqrt{5x + 3} = \sqrt{x + 11}$	Original equation
$\sqrt{5(2) + 3} \overset{?}{=} \sqrt{2 + 11}$	Substitute 2 for x.
$\sqrt{13} = \sqrt{13}$	Solution checks. ✓

Therefore, the equation has one solution: $x = 2$.

EXAMPLE 5 Solving an Equation Having Two Radicals

Solve $\sqrt[4]{3x} - \sqrt[4]{2x - 5} = 0$.

Solution

$\sqrt[4]{3x} - \sqrt[4]{2x - 5} = 0$	Original equation
$\sqrt[4]{3x} = \sqrt[4]{2x - 5}$	Isolate radicals.
$\left(\sqrt[4]{3x}\right)^4 = \left(\sqrt[4]{2x - 5}\right)^4$	Raise both sides to 4th power.
$3x = 2x - 5$	Simplify.
$x = -5$	Subtract $2x$ from both sides.

Check

$\sqrt[4]{3x} - \sqrt[4]{2x - 5} = 0$	Original equation
$\sqrt[4]{3(-5)} - \sqrt[4]{2(-5) - 5} \overset{?}{=} 0$	Substitute -5 for x.
$\sqrt[4]{-15} - \sqrt[4]{-15} \neq 0$	Solution does not check. ✗

Try this simple illustration to show how extraneous roots can be introduced. The equation $x = 2$ has only one solution. Trivially, x is 2. After squaring both sides, $x^2 = 4$ and the resulting equation has two solutions, $x = 2$ and $x = -2$. An "extra" solution was introduced. Stress the checking of solutions in the *original* equation.

The solution does not check because it yields fourth roots of negative radicands. Notice that -5 is not included in the domains of the radicals in this equation. Therefore, the equation has no solution. Try checking this graphically. If you graph both sides of the equation, you will discover that they do not intersect.

EXAMPLE 6   An Equation that Converts to a Quadratic Equation

Solve $\sqrt{x} + 2 = x$.

Solution

$\sqrt{x} + 2 = x$	Original equation
$\sqrt{x} = x - 2$	Isolate radical.
$\left(\sqrt{x}\right)^2 = (x - 2)^2$	Square both sides.
$x = x^2 - 4x + 4$	Simplify.
$0 = x^2 - 5x + 4$	Standard form
$0 = (x - 4)(x - 1)$	Factor.
$x - 4 = 0 \implies x = 4$	Set 1st factor equal to 0.
$x - 1 = 0 \implies x = 1$	Set 2nd factor equal to 0.

Try checking each of these solutions. When you do, you will find that $x = 4$ is a valid solution but that $x = 1$ is extraneous.

When an equation contains two radicals, it may not be possible to isolate both. In such cases, you may have to raise both sides of the equation to a power at *two* different stages in the solution.

EXAMPLE 7   Repeatedly Squaring Both Sides of an Equation

Solve $\sqrt{3t + 1} = 2 - \sqrt{3t}$.

Solution

$\sqrt{3t + 1} = 2 - \sqrt{3t}$	Original equation
$\left(\sqrt{3t + 1}\right)^2 = \left(2 - \sqrt{3t}\right)^2$	Square both sides (1st time).
$3t + 1 = 4 - 4\sqrt{3t} + 3t$	Simplify.
$-3 = -4\sqrt{3t}$	Isolate radical.
$(-3)^2 = \left(-4\sqrt{3t}\right)^2$	Square both sides (2nd time).
$9 = 16(3t)$	Simplify.
$\dfrac{9}{48} = t$	Divide both sides by 48.
$\dfrac{3}{16} = t$	Simplify.

The solution is $t = \frac{3}{16}$. Check this in the original equation.

EXAMPLE 8   *Repeatedly Squaring Both Sides of an Equation*

$$\sqrt{x^2 + 11} - \sqrt{x^2 - 9} = 2 \qquad \text{Original equation}$$

$$\sqrt{x^2 + 11} = 2 + \sqrt{x^2 - 9} \qquad \text{Isolate one of the radicals.}$$

$$\left(\sqrt{x^2 + 11}\right)^2 = \left(2 + \sqrt{x^2 - 9}\right)^2 \qquad \text{Square both sides (1st time).}$$

$$x^2 + 11 = 4 + 4\sqrt{x^2 - 9} + (x^2 - 9)$$

$$16 = 4\sqrt{x^2 - 9}$$

$$4 = \sqrt{x^2 - 9} \qquad \text{Isolate radical.}$$

$$4^2 = \left(\sqrt{x^2 - 9}\right)^2 \qquad \text{Square both sides (2nd time).}$$

$$16 = x^2 - 9$$

$$0 = x^2 - 25$$

$$0 = (x + 5)(x - 5)$$

By setting each factor equal to zero, you can see that this equation has two solutions: 5 and -5. Check these in the original equation.

Additional problem: You may want to have students solve $4 - \sqrt{x - 1} = \sqrt{3x + 3}$ in class. After they have finished, note that there is an extraneous solution. *Answer:* $x = 2$ ($x = 26$ is extraneous.)

Application

EXAMPLE 9   *An Application Involving Electricity*

The power consumed by an electrical appliance is given by

$$I = \sqrt{\frac{P}{R}},$$

where I is the current measured in amps, R is the resistance measured in ohms, and P is the power measured in watts. Find the power used by an electric heater for which $I = 10$ amps and $R = 16$ ohms.

Solution

$$I = \sqrt{\frac{P}{R}} \qquad \text{Original equation}$$

$$10 = \sqrt{\frac{P}{16}} \qquad \text{Substitute for } I \text{ and } R.$$

Try completing this solution. You should discover that the electric heater uses 1600 watts of power.

NOTE   An alternative way to solve the problem in Example 9 would be to first solve the equation for P.

$$I = \sqrt{\frac{P}{R}}$$

$$I^2 = \left(\sqrt{\frac{P}{R}}\right)^2$$

$$I^2 = \frac{P}{R}$$

$$I^2 R = P$$

At this stage, you can substitute the known values of I and R to obtain

$$P = (10)^2 16 = 1600.$$

EXAMPLE 10 *The Velocity of a Falling Object*

The velocity of a free-falling object can be determined from the equation

$$v = \sqrt{2gh},$$

where v is the velocity measured in feet per second, $g = 32$ feet per second per second, and h is the distance (in feet) the object has fallen. Find the height from which a rock has been dropped if it strikes the ground with a velocity of 50 feet per second.

Solution

$v = \sqrt{2gh}$	Original equation
$50 = \sqrt{2(32)h}$	Substitute for v and g.
$(50)^2 = \left(\sqrt{64h}\right)^2$	Square both sides.
$2500 = 64h$	Simplify.
$39 \approx h$	Divide both sides by 64.

Thus, the rock has fallen approximately 39 feet when it hits the ground. Check this solution in the original equation.

Group Activities

Extending the Concept

An Experiment Without using a stopwatch, you can find the length of time an object has been falling by using the following equation from physics:

$$t = \sqrt{\frac{h}{384}},$$

where t is the length of time in seconds and h is the height in inches the object has fallen. How far does an object fall in 0.25 second? in 0.10 second?

Use this equation to test how long it takes members of your group to catch a falling ruler. Hold the ruler vertically while another group member holds his or her hands near the lower end of the ruler ready to catch it. Before releasing the ruler, record the mark on the ruler closest to the top of the catcher's hands. Release the ruler. After it has been caught, again note the mark closest to the top of the catcher's hands. (The difference between these two measurements is h.) Which member of your group reacts most quickly?

5.5   Exercises

See Warm-Up
Exercises, p. A44

In Exercises 1–4, determine whether the values of x are solutions of the radical equation.

Equation	Values of x	
1. $\sqrt{x} - 10 = 0$	(a) $x = -4$	(b) $x = -100$
	(c) $x = \sqrt{10}$	(d) $x = 100$
2. $\sqrt{3x} - 6 = 0$	(a) $x = \frac{2}{3}$	(b) $x = 2$
	(c) $x = 12$	(d) $x = -\frac{1}{3}\sqrt{6}$
3. $\sqrt[3]{y} - 4 = 4$	(a) $x = -60$	(b) $x = 68$
	(c) $x = 20$	(d) $x = 0$
4. $\sqrt[4]{2x} + 2 = 6$	(a) $x = 128$	(b) $x = 2$
	(c) $x = -2$	(d) $x = 0$

In Exercises 5–38, solve the equation. (Some of the equations have no solution.)

5. $\sqrt{x} = 20$ **6.** $\sqrt{x} = 5$

7. $\sqrt{y} - 7 = 0$ **8.** $\sqrt{t} - 13 = 0$

9. $\sqrt{u} + 13 = 0$ **10.** $\sqrt{y} + 15 = 0$

11. $\sqrt{a + 100} = 25$ **12.** $\sqrt{b + 12} = 13$

13. $\sqrt{10x} = 30$ **14.** $\sqrt{8x} = 6$

15. $\sqrt{3y + 5} - 3 = 4$ **16.** $\sqrt{5z - 2} + 7 = 10$

17. $5\sqrt{x + 2} = 8$ **18.** $2\sqrt{x + 4} = 7$

19. $\sqrt{x^2 + 5} = x + 3$ **20.** $\sqrt{x^2 - 4} = x - 2$

21. $\sqrt{2x} = x - 4$ **22.** $\sqrt{x} = x - 6$

23. $\sqrt{3x + 2} + 5 = 0$ **24.** $\sqrt{1 - x} + 10 = 4$

25. $\sqrt{x + 3} = \sqrt{2x - 1}$ **26.** $\sqrt{3t + 1} = \sqrt{t + 15}$

27. $\sqrt{3y - 5} = 3\sqrt{y}$ **28.** $\sqrt{2u + 10} - 2\sqrt{u} = 0$

29. $\sqrt[3]{3x - 4} = \sqrt[3]{x + 10}$

30. $2\sqrt[3]{10 - 3x} = \sqrt[3]{2 - x}$

31. $\sqrt[3]{2x + 15} - \sqrt[3]{x} = 0$

32. $\sqrt[4]{2x} + \sqrt[4]{x + 3} = 0$

33. $\sqrt{8x + 1} = x + 2$ **34.** $\sqrt{3x + 7} = x + 3$

35. $\sqrt{5x - 4} = 2 - \sqrt{5x}$

36. $\sqrt{1 + 4a} = 2\sqrt{a} - 3$

37. $\sqrt{2t + 3} = 3 - \sqrt{2t}$

38. $\sqrt{x + 3} - \sqrt{x - 1} = 1$

In Exercises 39–46, use a graphing utility to graph each side of the equation in the same viewing rectangle. Then use the graphs to approximate the solution.

39. $\sqrt{x} = 2(2 - x)$ **40.** $\sqrt{2x + 3} = 4x - 3$

41. $\sqrt{x^2 + 1} = 5 - 2x$ **42.** $\sqrt{8 - 3x} = x$

43. $\sqrt{x + 3} = 5 - \sqrt{x}$ **44.** $\sqrt[3]{5x - 8} = 4 - \sqrt[3]{x}$

45. $4\sqrt[3]{x} = 7 - x$ **46.** $\sqrt[3]{x + 4} = \sqrt{6 - x}$

Length of a Pendulum   In Exercises 47 and 48, use the equation for the time t in seconds for a pendulum of length L feet to go through one complete cycle (its period). The equation is $t = 2\pi\sqrt{L/32}$.

47. How long is the pendulum of a grandfather clock with a period of 1.5 seconds (see figure)?

48. How long is the pendulum of a mantle clock with a period of 0.75 second?

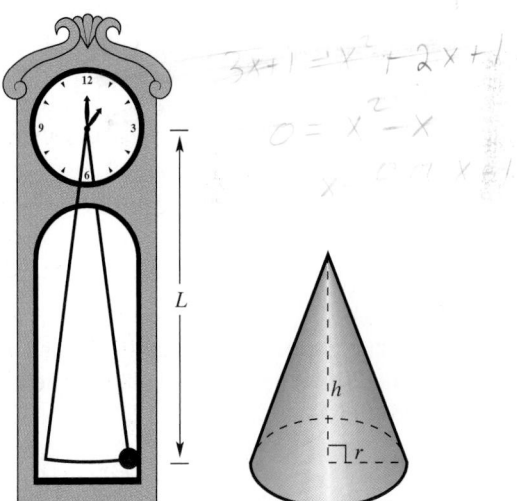

Figure for 47   **Figure for 49**

49. *Surface Area of a Cone*   The surface area of a cone (see figure) is $S = \pi r\sqrt{r^2 + h^2}$. Solve this equation for h^2.

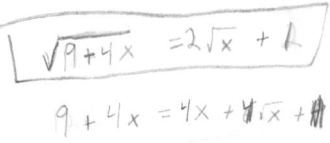

50. *Airline Passengers* An airline offers daily flights between Chicago and Denver. The total monthly cost of the flights is

$$C = \sqrt{0.2x + 1}, \quad x \geq 0,$$

where C is measured in millions of dollars and x is measured in thousands of passengers. The total cost of the flights for a certain month is 2.5 million dollars. Approximately how many passengers flew that month?

Height of an Object In Exercises 51 and 52, use the formula $t = \sqrt{d/16}$, which gives the time t in seconds for a free-falling object to fall d feet.

51. A construction worker drops a nail and observes it strike a water puddle after approximately 2 seconds. Estimate the height of the worker.

52. A construction worker drops a nail and observes it strike a water puddle after approximately 3 seconds. Estimate the height of the worker.

Free-Falling Object In Exercises 53–56, use the equation for the velocity of a free-falling object $(v = \sqrt{2gh})$, as described in Example 10.

53. An object is dropped from a height of 50 feet. Find the velocity of the object when it strikes the ground.

54. An object is dropped from a height of 200 feet. Find the velocity of the object when it strikes the ground.

55. An object that was dropped strikes the ground with a velocity of 60 feet per second. Find the height from which the object was dropped.

56. An object that was dropped strikes the ground with a velocity of 120 feet per second. Find the height from which the object was dropped.

In Exercises 57–60, match the function with its graph. [The graphs are labeled (a), (b), (c), and (d).]

(a)

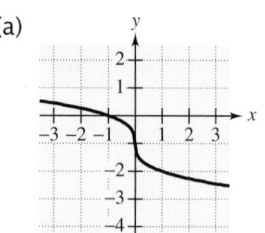

(b)

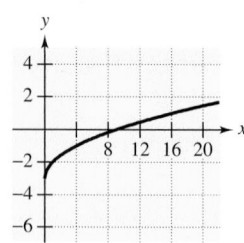

(c)

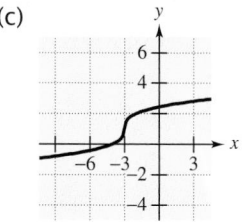

(d)

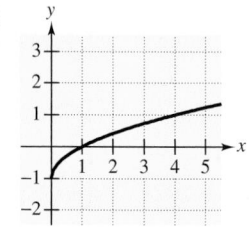

57. $f(x) = \sqrt[3]{x + 3} + 1$ **58.** $f(x) = -\sqrt[3]{x} - 1$

59. $f(x) = \sqrt{x} - 1$ **60.** $f(x) = \sqrt{x} - 3$

61. *Demand for a Product* The demand equation for a certain product is

$$p = 50 - \sqrt{0.8(x - 1)},$$

where x is the number of units demanded per day and p is the price per unit. Find the demand if the price is $30.02.

Section Project

Presidential Elections For 1960 through 1992, the number N (in millions) of votes cast for president can be modeled by the function

$$N = 69.16 + 1.48t - 2.72\sqrt{t},$$

where $t = 0$ represents 1960. *(Source: U.S. Bureau of the Census)*

(a) In what year were approximately 86.5 million votes cast for president?

(b) Use a graphing utility to graph the function. Use the graph to verify the result in part (a).

5.6 | Complex Numbers

The Imaginary Unit i ▪ Complex Numbers ▪ Operations with Complex Numbers ▪ Complex Conjugates

The Imaginary Unit i

In Section 5.2, you learned that a negative number has no *real* square root. For instance, $\sqrt{-1}$ is not real because there is no real number x such that $x^2 = -1$. Thus, as long as you are dealing only with real numbers, the equation

$$x^2 = -1$$

has no solution. To overcome this deficiency, mathematicians have expanded the set of numbers, using the **imaginary unit i,** defined as

$$i = \sqrt{-1}. \qquad \text{Imaginary unit}$$

This number has the property that $i^2 = -1$. Thus, the imaginary unit i is a solution of the equation $x^2 = -1$.

The Square Root of a Negative Number

Let c be a positive real number. Then the square root of $-c$ is given by

$$\sqrt{-c} = \sqrt{c(-1)} = \sqrt{c}\sqrt{-1} = \sqrt{c}\,i.$$

When writing $\sqrt{-c}$ in the *i*-**form**, $\sqrt{c}\,i$, note that i is outside the radical.

EXAMPLE 1 Writing Numbers in i-Form

(a) $\sqrt{-36} = \sqrt{36(-1)} = \sqrt{36}\sqrt{-1} = 6i$

(b) $\sqrt{-\dfrac{16}{25}} = \sqrt{\dfrac{16}{25}(-1)} = \sqrt{\dfrac{16}{25}}\sqrt{-1} = \dfrac{4}{5}i$

(c) $\sqrt{-5} = \sqrt{5(-1)} = \sqrt{5}\sqrt{-1} = \sqrt{5}\,i$

(d) $\sqrt{-54} = \sqrt{54(-1)} = \sqrt{54}\sqrt{-1} = \sqrt{9}\sqrt{6}\sqrt{-1} = 3\sqrt{6}\,i$

(e) $\dfrac{\sqrt{-48}}{\sqrt{-3}} = \dfrac{\sqrt{48}\sqrt{-1}}{\sqrt{3}\sqrt{-1}} = \dfrac{\sqrt{48}\,i}{\sqrt{3}\,i} = \sqrt{\dfrac{48}{3}} = \sqrt{16} = 4$

(f) $\dfrac{\sqrt{-18}}{\sqrt{2}} = \dfrac{\sqrt{18}\sqrt{-1}}{\sqrt{2}} = \dfrac{\sqrt{18}\,i}{\sqrt{2}} = \sqrt{\dfrac{18}{2}}\,i = \sqrt{9}\,i = 3i$

To perform operations with square roots of negative numbers, you must *first* write the numbers in *i*-form. Once the numbers are written in *i*-form, you can add, subtract, and multiply as follows.

$$ai + bi = (a + b)i \qquad \text{Addition}$$
$$ai - bi = (a - b)i \qquad \text{Subtraction}$$
$$(ai)(bi) = ab(i^2) = ab(-1) = -ab \qquad \text{Multiplication}$$

EXAMPLE 2   Adding Square Roots of Negative Numbers

Perform the addition: $\sqrt{-9} + \sqrt{-49}$.

Solution
$$\sqrt{-9} + \sqrt{-49} = \sqrt{9}\sqrt{-1} + \sqrt{49}\sqrt{-1} \qquad \text{Property of radicals}$$
$$= 3i + 7i \qquad \text{Write in } i\text{-form.}$$
$$= 10i \qquad \text{Simplify.}$$

In the next example, notice how you can multiply radicals that involve square roots of negative numbers.

Exploration

When performing operations with numbers in *i*-form, you sometimes need to be able to evaluate powers of the imaginary unit *i*. Evaluate the following powers of *i* using $i^2 = -1$.

$$i^1, i^2, i^3, i^4, i^5, i^6, i^7, i^8$$

Describe the pattern. Use the pattern to evaluate i^{16} and i^{19}.

EXAMPLE 3   Multiplying Square Roots of Negative Numbers

(a) $\sqrt{-15}\sqrt{-15} = (\sqrt{15}i)(\sqrt{15}i) \qquad \text{Write in } i\text{-form.}$
$$= (\sqrt{15})^2 i^2 \qquad \text{Multiply.}$$
$$= 15(-1) \qquad \text{Definition of } i$$
$$= -15 \qquad \text{Simplify.}$$

(b) $\sqrt{-5}(\sqrt{-45} - \sqrt{-4}) = \sqrt{5}i(3\sqrt{5}i - 2i) \qquad \text{Write in } i\text{-form.}$
$$= (\sqrt{5}i)(3\sqrt{5}i) - (\sqrt{5}i)(2i) \qquad \text{Distributive Property}$$
$$= 3(5)(-1) - 2\sqrt{5}(-1) \qquad \text{Multiply.}$$
$$= -15 + 2\sqrt{5} \qquad \text{Simplify.}$$

When multiplying square roots of negative numbers, be sure to write them in *i*-form *before multiplying*. If you do not do this, you can obtain incorrect answers. For instance, in Example 3(a), be sure you see that

$$\sqrt{-15}\sqrt{-15} \neq \sqrt{(-15)(-15)} = \sqrt{225} = 15.$$

Complex Numbers

A number of the form $a + bi$, where a and b are real numbers, is called a **complex number.** The real number a is called the **real part** of the complex number $a + bi$, and the real number b is called the **imaginary part** of the complex number.

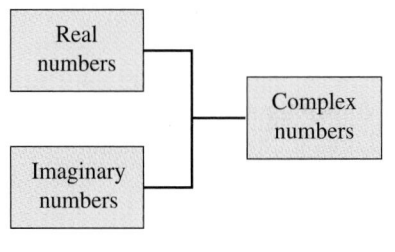

FIGURE 5.8

Definition of a Complex Number

If a and b are real numbers, the number $a + bi$ is a **complex number,** and it is said to be written in **standard form.** If $b = 0$, the number $a + bi = a$ is a real number. If $b \neq 0$, the number $a + bi$ is called an **imaginary number.** A number of the form bi, where $b \neq 0$, is called a **pure imaginary number.**

A number cannot be both real and imaginary. For instance, the numbers -2, 0, 1, $\frac{1}{2}$, and $\sqrt{2}$ are real numbers (but they are *not* imaginary numbers), and the numbers $-3i$, $2 + 4i$, and $-1 + i$ are imaginary numbers (but they are *not* real numbers). The diagram shown in Figure 5.8 further illustrates the relationships among real, complex, and imaginary numbers.

NOTE   Two complex numbers $a + bi$ and $c + di$, in standard form, are equal if and only if $a = c$ and $b = d$.

Point out to students that in $3 - 4\sqrt{3}\,i$, i is *not* under the radical. To avoid carelessly miswriting this expression, students can write $3 - 4\sqrt{3}\,i$ as $3 - 4i\sqrt{3}$.

EXAMPLE 4   Equality of Two Complex Numbers

(a) Are the complex numbers $\sqrt{9} + \sqrt{-48}$ and $3 - 4\sqrt{3}\,i$ equal?

(b) Find x and y such that the equation is valid.
$$3x - \sqrt{-25} = -6 + 3yi$$

Solution

(a) Begin by writing the first number in standard form.
$$\sqrt{9} + \sqrt{-48} = \sqrt{3^2} + \sqrt{4^2(3)(-1)} = 3 + 4\sqrt{3}\,i.$$

From this form, you can see that the two numbers are not equal because they have imaginary parts that differ in sign.

(b) Begin by writing the left side of the equation in standard form.

$$3x - \sqrt{-25} = -6 + 3yi \qquad \text{Original equation}$$
$$3x - 5i = -6 + 3yi \qquad \text{Both sides in standard form}$$

For these two numbers to be equal, their real parts must be equal to each other and their imaginary parts must be equal to each other. Thus, $3x = -6$, which implies that $x = -2$, and $-5 = 3y$, which implies that $y = -\frac{5}{3}$.

Operations with Complex Numbers

To add or subtract two complex numbers, you add (or subtract) the real and imaginary parts separately. This is similar to combining like terms of a polynomial.

$$(a + bi) + (c + di) = (a + c) + (b + d)i \qquad \text{Addition of complex numbers}$$

$$(a + bi) - (c + di) = (a - c) + (b - d)i \qquad \text{Subtraction of complex numbers}$$

EXAMPLE 5 Adding and Subtracting Complex Numbers

(a) $(3 - i) + (-2 + 4i) = (3 - 2) + (-1 + 4)i = 1 + 3i$

(b) $3i + (5 - 3i) = 5 + (3 - 3)i = 5$

(c) $4 - (-1 + 5i) + (7 + 2i) = [4 - (-1) + 7] + (-5 + 2)i$

$$= 12 - 3i$$

Note in part (b) that the sum of two imaginary numbers can be a real number.

The Commutative, Associative, and Distributive Properties of real numbers are also valid for complex numbers.

EXAMPLE 6 Multiplying Complex Numbers

(a) $(1 - i)\left(\sqrt{-9}\right) = (1 - i)(3i)$ Write in i-form.

$$= (1)(3i) - (i)(3i) \qquad \text{Distributive Property}$$

$$= 3i - 3(i^2) \qquad \text{Simplify.}$$

$$= 3i - 3(-1) \qquad \text{Definition of } i$$

$$= 3 + 3i \qquad \text{Simplify.}$$

Stress to students that their answers are not in standard form until they have simplified terms involving i raised to powers greater than 1.

(b) $(2 - i)(4 + 3i) = 8 + 6i - 4i - 3i^2$ FOIL Method

$$= 8 + 6i - 4i - 3(-1) \qquad \text{Definition of } i$$

$$= 11 + 2i \qquad \text{Combine like terms.}$$

(c) $(3 + 2i)(3 - 2i) = 3^2 - (2i)^2$ Special product formula

$$= 9 - 2^2 i^2 \qquad \text{Simplify.}$$

$$= 9 - 4(-1) \qquad \text{Definition of } i$$

$$= 9 + 4 \qquad \text{Simplify.}$$

$$= 13 \qquad \text{Simplify.}$$

Complex Conjugates

In Example 6(c), note that the product of two complex numbers can be a real number. This occurs with pairs of complex numbers of the form $a + bi$ and $a - bi$, called **complex conjugates.** In general, the product of complex conjugates has the following form.

$$(a + bi)(a - bi) = a^2 - (bi)^2$$
$$= a^2 - b^2 i^2$$
$$= a^2 - b^2(-1)$$
$$= a^2 + b^2$$

Here are some examples.

Complex Number	Complex Conjugate	Product
$4 - 5i$	$4 + 5i$	$4^2 + 5^2 = 41$
$3 + 2i$	$3 - 2i$	$3^2 + 2^2 = 13$
$-2 = -2 + 0i$	$-2 = -2 - 0i$	$(-2)^2 + 0^2 = 4$
$i = 0 + i$	$-i = 0 - i$	$0^2 + 1^2 = 1$

Complex conjugates are used to divide one complex number by another. To do this, multiply the numerator and denominator by the *complex conjugate of the denominator,* as shown in Example 7.

EXAMPLE 7   Division of Complex Numbers

Perform the following division and write your answer in standard form.

$$5 \div (3 - 2i)$$

Solution

$$\frac{5}{3 - 2i} = \frac{5}{3 - 2i} \cdot \frac{3 + 2i}{3 + 2i} \qquad \text{Multiply by complex conjugate of denominator.}$$

$$= \frac{5(3 + 2i)}{(3 - 2i)(3 + 2i)} \qquad \text{Multiply fractions.}$$

$$= \frac{5(3 + 2i)}{3^2 + 2^2} \qquad \text{Product of complex conjugates}$$

$$= \frac{15 + 10i}{13} \qquad \text{Simplify.}$$

$$= \frac{15}{13} + \frac{10}{13}i \qquad \text{Standard form}$$

EXAMPLE 8   Division of Complex Numbers

Divide $2 + 3i$ by $4 - 2i$.

Solution

$$\frac{2 + 3i}{4 - 2i} = \frac{2 + 3i}{4 - 2i} \cdot \frac{4 + 2i}{4 + 2i}$$ Multiply by complex conjugate of denominator.

$$= \frac{8 + 16i + 6i^2}{4^2 + 2^2}$$ Multiply fractions.

$$= \frac{8 + 16i + 6(-1)}{4^2 + 2^2}$$ Definition of i

$$= \frac{2 + 16i}{20}$$ Simplify.

$$= \frac{1}{10} + \frac{4}{5}i$$ Standard form

Some students incorrectly rewrite $(2 + 16i)/(20)$ as $(1 + 16i)/(10)$ or as $(2 + 4i)/5$.

EXAMPLE 9   Verifying a Complex Solution of an Equation

Show that $x = 2 + i$ is a solution of the equation $x^2 - 4x + 5 = 0$.

Solution

$$x^2 - 4x + 5 = 0$$ Original equation

$$(2 + i)^2 - 4(2 + i) + 5 \overset{?}{=} 0$$ Substitute $2 + i$ for x.

$$4 + 4i + i^2 - 8 - 4i + 5 \overset{?}{=} 0$$ Expand.

$$i^2 + 1 \overset{?}{=} 0$$ Combine like terms.

$$(-1) + 1 \overset{?}{=} 0$$ Definition of i

$$0 = 0$$ Solution checks. ✔

Therefore, $2 + i$ is a solution of the original equation.

Group Activities   Extending the Concept

Prime Polynomials   The polynomial $x^2 + 1$ is prime *with respect to the integers*. It is not, however, prime *with respect to the complex numbers*. Show how $x^2 + 1$ can be factored using complex numbers. Compare your results with those of the rest of the group.

5.6 **Exercises**

See Warm-Up Exercises, p. A45

In Exercises 1–14, write the number in *i*-form.

1. $\sqrt{-4}$ **2.** $\sqrt{-9}$

3. $-\sqrt{-144}$ **4.** $\sqrt{-49}$

5. $\sqrt{-\frac{4}{25}}$ **6.** $-\sqrt{-\frac{36}{121}}$

7. $\sqrt{-0.09}$ **8.** $\sqrt{-0.0004}$

9. $\sqrt{-8}$ **10.** $\sqrt{-75}$

11. $\sqrt{-27}$ **12.** $\sqrt{-\frac{8}{25}}$

13. $\sqrt{-7}$ **14.** $\sqrt{-15}$

In Exercises 15–28, perform the operations and write the result in standard form.

15. $\sqrt{-16} + 6i$ **16.** $\sqrt{-\frac{1}{4}} - \frac{3}{2}i$

17. $\sqrt{-50} - \sqrt{-8}$ **18.** $\sqrt{-500} + \sqrt{-45}$

19. $\sqrt{-8}\sqrt{-2}$ **20.** $\sqrt{-25}\sqrt{-6}$

21. $\sqrt{-18}\sqrt{-3}$ **22.** $\sqrt{-0.16}\sqrt{-1.21}$

23. $\sqrt{-3}\left(\sqrt{-3} + \sqrt{-4}\right)$

24. $\sqrt{-12}\left(\sqrt{-3} - \sqrt{-12}\right)$

25. $\sqrt{-5}\left(\sqrt{-16} - \sqrt{-10}\right)$

26. $\sqrt{-24}\left(\sqrt{-\frac{4}{9}} + \sqrt{-\frac{1}{4}}\right)$

27. $\left(\sqrt{-16}\right)^2$

28. $\left(\sqrt{-2}\right)^2$

In Exercises 29–36, determine *a* and *b*, where *a* and *b* are real numbers.

29. $3 - 4i = a + bi$

30. $-8 + 6i = a + bi$

31. $5 - 4i = (a + 3) + (b - 1)i$

32. $-10 + 12i = 2a + (5b - 3)i$

33. $-4 - \sqrt{-8} = a + bi$

34. $\sqrt{-36} - 3 = a + bi$

35. $(a + 5) + (b - 1)i = 7 - 3i$

36. $(2a + 1) + (2b + 3)i = 5 + 12i$

In Exercises 37–48, perform the operations and write the result in standard form.

37. $(4 - 3i) + (6 + 7i)$

38. $(-10 + 2i) + (4 - 7i)$

39. $(-4 - 7i) + (-10 - 33i)$

40. $(15 + 10i) - (2 + 10i)$

41. $13i - (14 - 7i)$

42. $(-21 - 50i) + (21 - 20i)$

43. $(30 - i) - (18 + 6i) + 3i^2$

44. $(4 + 6i) + (15 + 24i) - (1 - i)$

45. $\left(\frac{4}{3} + \frac{1}{3}i\right) + \left(\frac{5}{6} + \frac{7}{6}i\right)$

46. $(0.05 + 2.50i) - (6.2 + 11.8i)$

47. $15i - (3 - 25i) + \sqrt{-81}$

48. $(-1 + i) - \sqrt{2} - \sqrt{-2}$

In Exercises 49–70, perform the operations and write the result in standard form.

49. $(3i)(12i)$ **50.** $(-5i)(4i)$

51. $(-6i)(-i)(6i)$ **52.** $\frac{1}{2}(10i)(12i)(-3i)$

53. $(-3i)^3$ **54.** $(8i)^2$

55. $-5(13 + 2i)$ **56.** $10(8 - 6i)$

57. $4i(-3 - 5i)$ **58.** $-3i(10 - 15i)$

59. $(4 + 3i)(-7 + 4i)$ **60.** $(3 + 5i)(2 + 15i)$

61. $(-7 + 7i)(4 - 2i)$ **62.** $(3 + 5i)(2 - 15i)$

63. $(3 - 4i)^2$ **64.** $(7 + i)^2$

65. $(2 + 5i)^2$ **66.** $(8 - 3i)^2$

67. $\left(-2 + \sqrt{-5}\right)\left(-2 - \sqrt{-5}\right)$

68. $\sqrt{-9}\left(1 + \sqrt{-16}\right)$

69. $(2 + i)^3$ **70.** $(3 - 2i)^3$

In Exercises 71–82, multiply the number by its conjugate.

71. $2 + i$ **72.** $3 + 2i$

73. $5 - \sqrt{6}i$ **74.** $10 - 3i$

75. $-2 - 8i$

76. $-4 + \sqrt{2}i$

77. $10i$

78. 20

79. $1 + \sqrt{-3}$

80. $-3 - \sqrt{-5}$

81. $1.5 + \sqrt{-0.25}$

82. $3.2 - \sqrt{-0.04}$

In Exercises 83–92, perform the operations and write the result in standard form.

83. $\dfrac{4}{1 - i}$

84. $\dfrac{20}{3 + i}$

85. $\dfrac{-12}{2 + 7i}$

86. $\dfrac{17i}{5 + 3i}$

87. $\dfrac{20}{2i}$

88. $\dfrac{1 + i}{3i}$

89. $\dfrac{4i}{1 - 3i}$

90. $\dfrac{15}{2(1 - i)}$

91. $\dfrac{2 + 3i}{1 + 2i}$

92. $\dfrac{4 - 5i}{4 + 5i}$

In Exercises 93–96, perform the operations and write the result in standard form.

93. $\dfrac{1}{1 - 2i} + \dfrac{4}{1 + 2i}$

94. $\dfrac{3i}{1 + i} + \dfrac{2}{2 + 3i}$

95. $\dfrac{i}{4 - 3i} - \dfrac{5}{2 + i}$

96. $\dfrac{1 + i}{i} - \dfrac{3}{5 - 2i}$

In Exercises 97–100, decide whether each number is a solution of the equation.

97. $x^2 + 2x + 5 = 0$

   (a) $x = -1 + 2i$ (b) $x = -1 - 2i$

98. $x^2 - 4x + 13 = 0$

   (a) $x = 2 - 3i$ (b) $x = 2 + 3i$

99. $x^3 + 4x^2 + 9x + 36 = 0$

   (a) $x = -4$ (b) $x = -3i$

100. $x^3 - 8x^2 + 25x - 26 = 0$

   (a) $x = 2$ (b) $x = 3 - 2i$

In Exercises 101–104, perform the operations.

101. $(a + bi) + (a - bi)$

102. $(a + bi)(a - bi)$

103. $(a + bi) - (a - bi)$

104. $(a + bi)^2 + (a - bi)^2$

105. *Think About It* Explain why the equation $x^2 = -1$ does not have real number solutions.

106. *Error Analysis* Describe the error.

$$\sqrt{-3}\sqrt{-3} = \sqrt{(-3)(-3)} = \sqrt{9} = 3$$

Section Project

Cube Roots

(a) The principal cube root of 125, $\sqrt[3]{125}$, is 5. Evaluate the expression x^3 for each of the following values of x.

   (i) $x = \dfrac{-5 + 5\sqrt{3}i}{2}$ (ii) $x = \dfrac{-5 - 5\sqrt{3}i}{2}$

(b) The principal cube root of 27, $\sqrt[3]{27}$, is 3. Evaluate the expression x^3 for each of the following values of x.

   (i) $x = \dfrac{-3 + 3\sqrt{3}i}{2}$ (ii) $x = \dfrac{-3 - 3\sqrt{3}i}{2}$

(c) Compare the results in parts (a) and (b). Use the results to find the three cube roots of each number.

   (i) 1   (ii) 8   (iii) 64

   Verify your results algebraically.

Chapter Summary

After studying this chapter, you should have acquired the following skills. These skills are keyed to the Review Exercises that begin on page 402. Answers to odd-numbered Review Exercises are given in the back of the book.

- Evaluate numerical expressions involving integer exponents and scientific notation. *(Section 5.1)* — **Review Exercises 1–8, 23, 24**

- Simplify algebraic expressions involving integer exponents. *(Section 5.1)* — **Review Exercises 9–16**

- Evaluate numerical expressions involving radicals. *(Section 5.2)* — **Review Exercises 17–22, 25, 26, 41, 42**

- Translate numerical expressions from radical form to rational exponent form and vice versa. *(Section 5.2)* — **Review Exercises 27–30**

- Evaluate numerical expressions involving rational exponents. *(Section 5.2)* — **Review Exercises 31–36**

- Simplify algebraic expressions involving rational exponents or radicals. *(Sections 5.2, 5.3)* — **Review Exercises 37–40, 43–46**

- Simplify expressions involving radicals by rationalizing the denominator. *(Section 5.3)* — **Review Exercises 47–54**

- Add, subtract, multiply, and divide expressions involving radicals. *(Sections 5.3, 5.4)* — **Review Exercises 55–64**

- Graph functions using a graphing utility. *(Sections 5.2, 5.3, 5.4)* — **Review Exercises 65–68**

- Solve radical equations. *(Section 5.5)* — **Review Exercises 69–78**

- Approximate solutions of radical equations using a graphing utility. *(Section 5.5)* — **Review Exercises 79, 80**

- Write complex numbers in standard form. *(Section 5.6)* — **Review Exercises 81–86**

- Add, subtract, multiply, and divide complex numbers. *(Section 5.6)* — **Review Exercises 87–96**

- Solve problems using geometry. *(Sections 5.3, 5.5)* — **Review Exercises 97, 98**

- Solve real-life problems modeled by radical equations. *(Section 5.5)* — **Review Exercises 99, 100**

Review Exercises

In Exercises 1–8, evaluate the expression.

1. $(2^3 \cdot 3^2)^{-1}$

2. $(2^{-2} \cdot 5^2)^{-2}$

3. $\left(\dfrac{2}{5}\right)^{-3}$

4. $\left(\dfrac{1}{3^{-2}}\right)^2$

5. $(6 \times 10^3)^2$

6. $(3 \times 10^{-3})(8 \times 10^7)$

7. $\dfrac{3.5 \times 10^7}{7 \times 10^4}$

8. $\dfrac{1}{(6 \times 10^{-3})^2}$

In Exercises 9–16, simplify the expression.

9. $\dfrac{4x^2}{2x}$

10. $4(-3x)^3$

11. $(x^3y^{-4})^2$

12. $5yx^0$

13. $\dfrac{t^{-5}}{t^{-2}}$

14. $\dfrac{a^5 \cdot a^{-3}}{a^{-2}}$

15. $\left(\dfrac{y}{3}\right)^{-3}$

16. $(2x^2y^4)^4(2x^2y^4)^{-4}$

In Exercises 17–22, evaluate the expression.

17. $\sqrt{1.44}$

18. $\sqrt{0.16}$

19. $\sqrt{\tfrac{25}{36}}$

20. $-\sqrt{\tfrac{64}{225}}$

21. $\sqrt{169 - 25}$

22. $\sqrt{16 + 9}$

In Exercises 23–26, evaluate the expression. (Round the result to two decimal places.)

23. $1800(1 + 0.08)^{24}$

24. $0.0024(7,658,400)$

25. $\sqrt{13^2 - 4(2)(7)}$

26. $\dfrac{-3.7 + \sqrt{15.8}}{2(2.3)}$

In Exercises 27–30, fill in the missing description.

Radical Form	Rational Exponent Form
27. $\sqrt{49} = 7$	
28. $\sqrt[3]{0.125} = 0.5$	

Radical Form	Rational Exponent Form
29.	$216^{1/3} = 6$
30.	$16^{1/4} = 2$

In Exercises 31–34, evaluate the expression.

31. $27^{4/3}$

32. $16^{3/4}$

33. $25^{3/2}$

34. $243^{-2/5}$

In Exercises 35 and 36, evaluate the expression. (Round the result to two decimal places.)

35. $75^{-3/4}$

36. $510^{5/3}$

In Exercises 37–46, simplify the expression.

37. $x^{3/4} \cdot x^{-1/6}$

38. $(2y^2)^{3/2}(2y^{-4})^{1/2}$

39. $\dfrac{15x^{1/4}y^{3/5}}{5x^{1/2}y}$

40. $\dfrac{48a^2b^{5/2}}{14a^{-3}b^{-1/2}}$

41. $\sqrt{360}$

42. $\sqrt{\tfrac{50}{9}}$

43. $\sqrt{0.25x^4y}$

44. $\sqrt{0.16s^6t^3}$

45. $\sqrt[3]{48a^3b^4}$

46. $\sqrt[4]{32u^4v^5}$

In Exercises 47–54, rationalize the denominator and simplify further, if possible.

47. $\sqrt{\tfrac{5}{6}}$

48. $\sqrt{\tfrac{3}{20}}$

49. $\dfrac{3}{\sqrt{12x}}$

50. $\dfrac{4y}{\sqrt{10z}}$

51. $\dfrac{2}{\sqrt[3]{2x}}$

52. $\sqrt[3]{\dfrac{16t}{s^2}}$

53. $\dfrac{6}{7 - \sqrt{7}}$

54. $\dfrac{x}{\sqrt{x} + 1}$

In Exercises 55–64, perform the operations and simplify.

55. $3\sqrt{40} - 10\sqrt{90}$ **56.** $9\sqrt{50} - 5\sqrt{8} + \sqrt{48}$

57. $10\sqrt[4]{y+3} - 3\sqrt[4]{y+3}$

58. $\sqrt{25x} + \sqrt{49x} - \sqrt[3]{8x}$

59. $\left(\sqrt{5}+6\right)^2$ **60.** $\left(\sqrt{3}-\sqrt{x}\right)\left(\sqrt{3}+\sqrt{x}\right)$

61. $\left(2\sqrt{3}+7\right)\left(\sqrt{6}-2\right)$

62. $\dfrac{15}{\sqrt{x}+3}$

63. $\left(\sqrt{x}+10\right) \div \left(\sqrt{x}-10\right)$

64. $\left(3\sqrt{s}+4\right) \div \left(\sqrt{s}+2\right)$

In Exercises 65–68, use a graphing utility to graph the function.

65. $y = 3\sqrt[3]{2x}$ **66.** $y = \dfrac{10}{\sqrt[4]{x^2+1}}$

67. $g(x) = 4x^{3/4}$ **68.** $h(x) = \tfrac{1}{2}x^{4/3}$

In Exercises 69–78, solve the equation.

69. $\sqrt{y} = 15$ **70.** $\sqrt{3x}+9=0$

71. $\sqrt{2(a-7)} = 14$ **72.** $\sqrt{3(2x+3)} = \sqrt{x+15}$

73. $\sqrt{2(x+5)} = x+5$

74. $\sqrt{5t} = 1 + \sqrt{5(t-1)}$

75. $\sqrt[3]{5x+2} - \sqrt[3]{7x-8} = 0$

76. $\sqrt[4]{9x-2} - \sqrt[4]{8x} = 0$

77. $\sqrt{1+6x} = 2 - \sqrt{6x}$

78. $\sqrt{2+9b} + 1 = 3\sqrt{b}$

In Exercises 79 and 80, use a graphing utility to approximate the solution of the equation.

79. $4\sqrt[3]{x} = 7 - x$ **80.** $2\sqrt{x} - 1 = \sqrt{10-x}$

In Exercises 81–86, write the complex number in standard form.

81. $\sqrt{-48}$ **82.** $\sqrt{-0.16}$

83. $10 - 3\sqrt{-27}$ **84.** $3 + 2\sqrt{-500}$

85. $\tfrac{3}{4} - 5\sqrt{-\tfrac{3}{25}}$ **86.** $-0.5 + 3\sqrt{-1.21}$

In Exercises 87–96, perform the operations and write the result in standard form.

87. $(-4+5i) - (-12+8i)$

88. $5(3-8i) + (5+12i)$

89. $(-2)(15i)(-3i)$ **90.** $-10i(4-7i)$

91. $(4-3i)(4+3i)$ **92.** $(6-5i)^2$

93. $(12-5i)(2+7i)$ **94.** $\dfrac{4}{5i}$

95. $\dfrac{5i}{2+9i}$ **96.** $\dfrac{2+i}{1-9i}$

97. *Perimeter* The four corners are cut from an $8\tfrac{1}{2}$-inch by 14-inch sheet of paper (see figure). Find the perimeter of the remaining piece of paper.

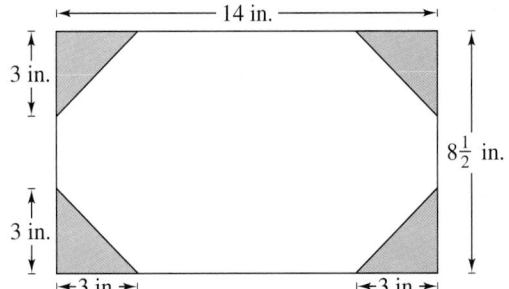

98. *Dimensions of a Rectangle* Determine the length and width of a rectangle with a perimeter of 84 inches and a diagonal of 30 inches.

99. *Length of a Pendulum* The time t in seconds for a pendulum of length L in feet to go through one complete cycle (its period) is

$$t = 2\pi\sqrt{\dfrac{L}{32}}.$$

How long is the pendulum of a grandfather clock with a period of 1.3 seconds?

100. *Height of a Bridge* The time t in seconds for a free-falling object to fall d feet is given by

$$t = \sqrt{\dfrac{d}{16}}.$$

A child drops a pebble from a bridge and observes it strike the water after approximately 4 seconds. Estimate the height of the bridge.

Chapter Test

Take this test as you would take a test in class. After you are done, check your work against the answers given in the back of the book.

In Exercises 1 and 2, evaluate each expression without using a calculator.

1. (a) $2^{-2} + 2^{-3}$

(b) $\dfrac{6.3 \times 10^{-3}}{2.1 \times 10^2}$

2. (a) $27^{-2/3}$

(b) $\sqrt{2}\sqrt{18}$

3. Write 0.000032 in scientific notation.

4. Write 3.04×10^7 in decimal notation.

In Exercises 5–7, simplify each expression.

5. (a) $\dfrac{12t^{-2}}{20t^{-1}}$

(b) $(y^{-2})^{-1}$

6. (a) $\left(\dfrac{x^{1/2}}{x^{1/3}}\right)^2$

(b) $5^{1/4} \cdot 5^{7/4}$

7. (a) $\sqrt{\dfrac{32}{9}}$

(b) $\sqrt[3]{24}$

8. In your own words, explain the meaning of *rationalize* and demonstrate by rationalizing the denominator of (a) $\dfrac{10}{\sqrt{6} - \sqrt{2}}$ and (b) $\dfrac{2}{\sqrt[3]{9y}}$. Simplify the result.

9. Subtract: $5\sqrt{3x} - 3\sqrt{75x}$.

10. Multiply: $\sqrt{5}\left(\sqrt{15x} + 3\right)$.

11. Expand: $\left(4 - \sqrt{2x}\right)^2$.

In Exercises 12–14, solve the equation.

12. $\sqrt{x^2 - 1} = x - 2$

13. $\sqrt{x} - x + 6 = 0$

14. $\sqrt{x - 4} = \sqrt{x + 7} - 1$

In Exercises 15–18, perform the operations and simplify.

15. $(2 + 3i) - \sqrt{-25}$

16. $(2 - 3i)^2$

17. $\sqrt{-16}\left(1 + \sqrt{-4}\right)$

18. $(3 - 2i)(1 + 5i)$

19. Divide $5 - 2i$ by $3 + i$. Write the result in standard form.

20. The velocity v (in feet per second) of an object is given by $v = \sqrt{2gh}$, where $g = 32$ feet per second per second and h is the distance (in feet) the object has fallen. Find the height from which a rock has been dropped if it strikes the ground with a velocity of 80 feet per second.

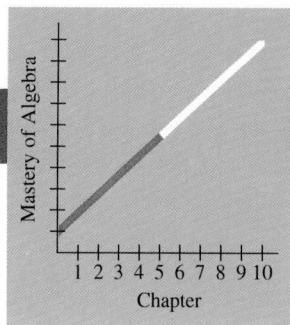

Mastery of Algebra

1 2 3 4 5 6 7 8 9 10

Chapter

Take this test as you would take a test in class. After you are done, check your work against the answers given in the back of the book.

In Exercises 1 and 2, write the statement using inequality notation.

1. y is no more than 45.

2. x is at least 15.

In Exercises 3 and 4, name the property of real numbers that justifies the statement.

3. $3x + 0 = 3x$

4. $-4(x + 10) = -4 \cdot x + (-4)(10)$

In Exercises 5 and 6, write an algebraic expression that represents the quantity in the verbal statement and simplify, if possible.

5. The total hourly wage for an employee when the base pay is $9.35 per hour plus 75 cents for each of q units produced per hour

6. The sum of two consecutive odd integers, the first of which is $2n - 1$

In Exercises 7–10, factor the expression completely.

7. $16x^2 - 121$

8. $9t^2 - 24t + 16$

9. $x(x - 10) - 4(x - 10)$

10. $4x^3 - 12x^2 + 16x$

In Exercises 11–16, solve the equation.

11. $\dfrac{x}{4} - \dfrac{2}{3} = 0$

12. $2x - 3[1 + (4 - x)] = 0$

13. $3x^2 - 13x - 10 = 0$

14. $x(x - 3) = 40$

15. $|2x + 5| = 11$

16. $\sqrt{x - 5} - 6 = 0$

In Exercises 17 and 18, plot the points and find the slope of the line passing through the points.

17. $(-3, 2), (5, -4)$

18. $(2, 8), (7, -3)$

19. Given the function $f(x) = 3x - x^2$, find and simplify $f(2 + t) - f(2)$.

20. An automobile is traveling at an average speed of 48 miles per hour. Express the distance d traveled by the automobile as a function of time t in hours.

21. Use the graph of $f(x) = \sqrt{x}$ to write a function g for each of the graphs.

(a)

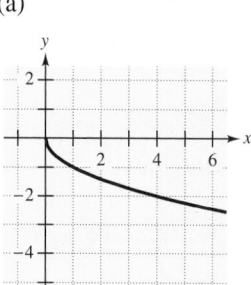

(b)

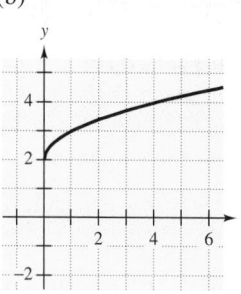

(c)

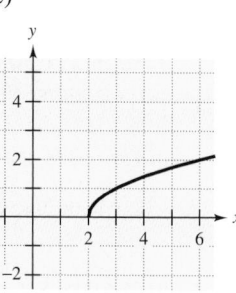

22. Your employer withholds $3\frac{1}{2}\%$ of your gross income for medical insurance coverage. Determine the amount withheld each month if the gross monthly income is \$3100.

In Exercises 23 and 24, solve the system of linear equations.

23. $\begin{cases} 4x - y = 0 \\ -3x + 2y = 2 \end{cases}$

24. $\begin{cases} x - y = -1 \\ x + 2y - 2z = 3 \\ 3x - y + 2z = 3 \end{cases}$

25. The slope and y-intercept of the line $y = mx + b$ that best fits the points in the figure is given by the solution of the following system of linear equations.

$$\begin{cases} 4b + 5m = 11 \\ 5b + 21m = 1 \end{cases}$$

(a) Solve the system and find the equation of the best-fitting line.

(b) Plot the points and graph the best-fitting line.

26. Use a determinant to find the area of the triangle with vertices $(-5, 8)$, $(10, 0)$, and $(3, -4)$.

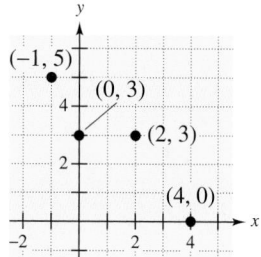

Figure for 25

In Exercises 27–30, simplify the expression.

27. $\sqrt{24x^2y^3}$

28. $\sqrt[3]{9} \cdot \sqrt[3]{15}$

29. $(12a^{-4}b^6)^{1/2}$

30. $(16^{1/3})^{3/4}$

Quadratic Functions, Equations, and Inequalities

Strategies for Success

SKILLS *When you have completed this chapter, make sure you are able to:*

○ Solve quadratic equations by factoring, extracting square roots, completing the square, and using the Quadratic Formula

○ Solve equations of quadratic form and quadratic inequalities

○ Sketch the graph of a quadratic function

○ Model real data using a quadratic function

○ Write quadratic equations that model real-life problems and solve

TOOLS *Use these study tools to master the skills above:*

In Exercise 88 of Section 6.5, you will find the maximum height of a diver when the diver's path is modeled by a quadratic equation.

6.1 The Factoring and Square Root Methods

Solving Equations by Factoring ▪ Extracting Square Roots ▪ Equations with Imaginary Solutions ▪ Equations of Quadratic Form

Solving Equations by Factoring

In this chapter, you will study methods for solving quadratic equations and equations of quadratic form. To begin, let's review the method of factoring that you studied in Section 1.8.

Remember that the first step in solving a quadratic equation by factoring is to write the equation in standard form. Next, factor the left side. Finally, set each factor equal to zero and solve for x.

─────────

EXAMPLE 1 Solving Quadratic Equations by Factoring

Stress to students the need to have the equation set equal to 0 before factoring and setting the factors equal to 0.

(a)

$x^2 + 5x = 24$	Original equation
$x^2 + 5x - 24 = 0$	Standard form
$(x + 8)(x - 3) = 0$	Factor.
$x + 8 = 0 \implies x = -8$	Set 1st factor equal to 0.
$x - 3 = 0 \implies x = 3$	Set 2nd factor equal to 0.

The solutions are -8 and 3. Check these in the original equation.

(b)

$3x^2 = 4 - 11x$	Original equation
$3x^2 + 11x - 4 = 0$	Standard form
$(3x - 1)(x + 4) = 0$	Factor.
$3x - 1 = 0 \implies x = \frac{1}{3}$	Set 1st factor equal to 0.
$x + 4 = 0 \implies x = -4$	Set 2nd factor equal to 0.

The solutions are $\frac{1}{3}$ and -4. Check these in the original equation.

NOTE When the two solutions of a quadratic equation are identical, they are called a **double** or **repeated solution.** This occurs in Example 1(c).

(c)

$9x^2 + 12 = 3 + 12x + 5x^2$	Original equation
$4x^2 - 12x + 9 = 0$	Standard form
$(2x - 3)(2x - 3) = 0$	Factor.
$2x - 3 = 0$	Set factor equal to 0.
$x = \dfrac{3}{2}$	Repeated solution

The only solution is $\frac{3}{2}$. Check this in the original equation.

─────────

EXAMPLE 2 *Solving Quadratic Equations by Factoring*

Solve each equation by factoring.

(a) $(x + 1)(x - 2) = 18$ (b) $5x^2 = 45$

Solution

(a)

$(x + 1)(x - 2) = 18$	Original equation
$x^2 - x - 2 = 18$	Multiply.
$x^2 - x - 20 = 0$	Standard form
$(x - 5)(x + 4) = 0$	Factor.

$x - 5 = 0 \implies x = 5$ Set 1st factor equal to 0.

$x + 4 = 0 \implies x = -4$ Set 2nd factor equal to 0.

The solutions are 5 and -4. Check these in the original equation.

(b)

$5x^2 = 45$	Original equation
$5x^2 - 45 = 0$	Standard form
$5(x^2 - 9) = 0$	Common monomial factor
$5(x + 3)(x - 3) = 0$	Factor as difference of squares.

$5 \neq 0$ Set 1st factor equal to 0.

$x + 3 = 0 \implies x = -3$ Set 2nd factor equal to 0.

$x - 3 = 0 \implies x = 3$ Set 3rd factor equal to 0.

The solutions are -3 and 3. Check these in the original equation.

NOTE Don't be tricked by the quadratic equation in Example 2(a). Even though the left side is factored, the right side is not zero. Thus, you must first multiply the left side, then rewrite the equation in standard form, and finally refactor.

EXAMPLE 3 *Solving Quadratic Equations by Factoring*

Solve $3x^2 - 12x = 0$ by factoring.

Solution

The left side of this equation has a common factor of $3x$.

$3x^2 - 12x = 0$	Original equation
$3x(x - 4) = 0$	Factor.

$3x = 0 \implies x = 0$ Set 1st factor equal to 0.

$x - 4 = 0 \implies x = 4$ Set 2nd factor equal to 0.

The solutions are 0 and 4. Check these in the original equation. You can also check these solutions graphically by observing that the graph of $y = 3x^2 - 12x$ has x-intercepts at $x = 0$ and $x = 4$, as shown in Figure 6.1.

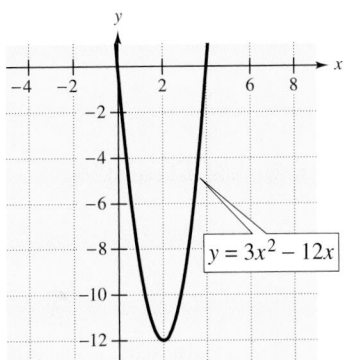

$y = 3x^2 - 12x$

FIGURE 6.1

Extracting Square Roots

Consider the following equation, where $d > 0$ and u is an algebraic expression.

$$u^2 = d$$ Original equation
$$u^2 - d = 0$$ Standard form
$$\left(u + \sqrt{d}\right)\left(u - \sqrt{d}\right) = 0$$ Factor.
$$u + \sqrt{d} = 0 \quad \Longrightarrow \quad u = -\sqrt{d}$$ Set 1st factor equal to 0.
$$u - \sqrt{d} = 0 \quad \Longrightarrow \quad u = \sqrt{d}$$ Set 2nd factor equal to 0.

Because the solutions differ only in sign, they can be written together using a "plus or minus sign"

$$u = \pm\sqrt{d}.$$

This form of the solution is read as "u is equal to plus or minus the square root of d." Solving an equation of the form $u^2 = d$ *without* going through the steps of factoring is called **extracting square roots.**

Technology

To estimate solutions of an equation graphically, write the equation with all of the nonzero terms on the left side and zero on the right side of the equation. Graph the left side of the equation and look at its x-intercepts. For instance, in Example 4(b), you can write the equation as

$$(x - 2)^2 - 10 = 0$$

and then graph

$$y = (x - 2)^2 - 10,$$

as shown below. From the graph, you can zoom in repeatedly to see that the x-intercepts are approximately 5.16 and -1.16. Solving algebraically produces similar results.

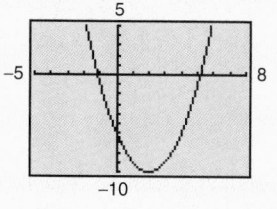

Extracting Square Roots

The equation $u^2 = d$, where $d > 0$, has exactly two solutions:

$$u = \sqrt{d} \quad \text{and} \quad u = -\sqrt{d}.$$

These solutions can also be written as $u = \pm\sqrt{d}$.

EXAMPLE 4   *Extracting Square Roots*

(a) $3x^2 = 15$ Original equation
      $x^2 = 5$ Divide both sides by 3.
      $x = \pm\sqrt{5}$ Extract square roots.

The solutions are $\sqrt{5}$ and $-\sqrt{5}$. Check these in the original equation.

(b) $(x - 2)^2 = 10$ Original equation
      $x - 2 = \pm\sqrt{10}$ Extract square roots.
      $x = 2 \pm \sqrt{10}$ Add 2 to both sides.

The solutions are $2 + \sqrt{10} \approx 5.16$ and $2 - \sqrt{10} \approx -1.16$. Check these in the original equation.

EXAMPLE 5   Extracting Square Roots

Solve the quadratic equation

$$3(5x + 4)^2 - 81 = 0$$

by extracting square roots.

Solution

In this case, some preliminary steps are needed before taking the square roots.

$3(5x + 4)^2 - 81 = 0$	Original equation
$3(5x + 4)^2 = 81$	Add 81 to both sides.
$(5x + 4)^2 = 27$	Divide both sides by 3.
$5x + 4 = \pm\sqrt{27}$	Extract square roots.
$5x = -4 \pm 3\sqrt{3}$	Subtract 4 from both sides.
$x = \dfrac{-4 \pm 3\sqrt{3}}{5}$	Divide both sides by 5.

Thus, the solutions are

$$\frac{-4 + 3\sqrt{3}}{5} \approx 0.24 \qquad \text{and} \qquad \frac{-4 - 3\sqrt{3}}{5} \approx -1.84.$$

An algebraic check of these solutions is possible, but rather messy. With solutions such as these, a graphic check is usually easier. If you have access to a graphing utility, you can easily use a graphic check. To do this, sketch the graph of

$$y = 3(5x + 4)^2 - 81$$

and locate the *x*-intercepts of the graph. From the graph shown in Figure 6.2, you can see that the *x*-intercepts are $x \approx -1.84$ and $x \approx 0.24$. This graphically confirms the two solutions.

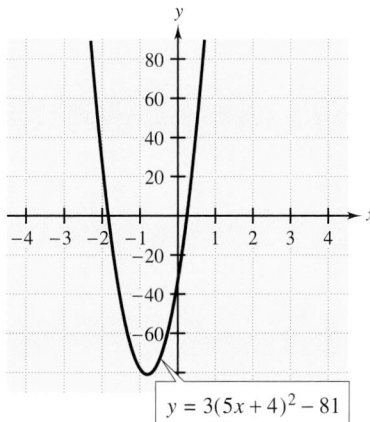

$y = 3(5x + 4)^2 - 81$

FIGURE 6.2

Exploration

Try using a graphing utility to approximate the solutions of

$$2(3x - 2)^2 - 48 = 0.$$

Use the zoom and trace features of the graphing utility to write the approximations with an error of less than 0.01. Then solve the equation algebraically and compare your algebraic solutions with your approximations.

Equations with Imaginary Solutions

Prior to Section 5.6, the only solutions you had been finding had been real numbers. But now that you have studied imaginary numbers, it makes sense to look for other types of solutions. For instance, although the quadratic equation $x^2 + 1 = 0$ has no solutions that are real numbers, it does have two solutions that are imaginary numbers: i and $-i$. To check this, substitute i and $-i$ for x.

$$(i)^2 + 1 = -1 + 1 = 0 \qquad \text{Solution checks.} \quad \checkmark$$
$$(-i)^2 + 1 = -1 + 1 = 0 \qquad \text{Solution checks.} \quad \checkmark$$

One way to find complex solutions of a quadratic equation is to extend the *extraction of square roots* technique to cover the case where d is a negative number.

Extracting Imaginary Square Roots

The equation $u^2 = d$, where $d < 0$, has exactly two solutions:
$$u = \sqrt{|d|}\, i \qquad \text{and} \qquad u = -\sqrt{|d|}\, i.$$
These solutions can also be written as $u = \pm \sqrt{|d|}\, i$.

 *Technology*

When checking solutions graphically, it is important to realize that only the real solutions appear as x-intercepts—the imaginary solutions cannot be estimated from the graph. For instance, in Example 6(a), the graph of

$$y = x^2 + 8$$

has no x-intercepts. This agrees with the fact that both of its solutions are imaginary numbers.

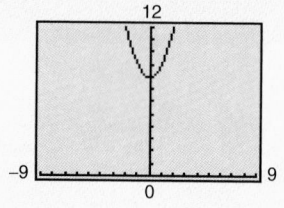

EXAMPLE 6   *Extracting Imaginary Square Roots*

(a) $x^2 + 8 = 0$ Original equation

$\qquad x^2 = -8$ Subtract 8 from both sides.

$\qquad x = \pm \sqrt{8}\, i$ Extract imaginary square roots.

$\qquad x = \pm 2\sqrt{2}\, i$ Simplify.

The solutions are $2\sqrt{2}\, i$ and $-2\sqrt{2}\, i$. Check these in the original equation.

(b) $2(3x - 5)^2 + 32 = 0$ Original equation

$\qquad 2(3x - 5)^2 = -32$ Subtract 32 from both sides.

$\qquad (3x - 5)^2 = -16$ Divide both sides by 2.

$\qquad 3x - 5 = \pm 4i$ Extract imaginary square roots.

$\qquad 3x = 5 \pm 4i$ Add 5 to both sides.

$$x = \frac{5 \pm 4i}{3} \qquad \text{Divide both sides by 3.}$$

The solutions are $(5 + 4i)/3$ and $(5 - 4i)/3$. Check these in the original equation.

Equations of Quadratic Form

Both the factoring and extraction of square roots methods can be applied to nonquadratic equations that are of **quadratic form.** An equation is said to be of quadratic form if it has the form

$$au^2 + bu + c = 0,$$

where u is an algebraic expression. Here are some examples.

Equation	*Written in Quadratic Form*
$x^4 + 5x^2 + 4 = 0$	$(x^2)^2 + 5(x^2) + 4 = 0$
$2x^{2/3} + 3x^{1/3} - 9 = 0$	$2(x^{1/3})^2 + 3(x^{1/3}) - 9 = 0$
$x - 5\sqrt{x} + 6 = 0$	$\left(\sqrt{x}\right)^2 - 5\left(\sqrt{x}\right) + 6 = 0$

Discuss the identifying characteristics of a quadratic-type equation: When written in standard form, an equation has a trinomial of quadratic type if the degree of the leading term is twice the degree of the middle term.

To solve an equation of quadratic form, it helps to make a substitution and rewrite the equation in terms of u, as demonstrated in Examples 7, 8, and 9.

NOTE In Example 7, you can use a graphing utility to graph $y = x^4 - 13x^2 + 36$, as shown in Figure 6.3. The graph indicates that there are four real solutions near $x = \pm 2$ and $x = \pm 3$.

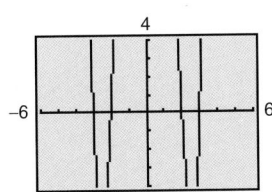

FIGURE 6.3

EXAMPLE 7 Solving an Equation of Quadratic Form

Solve $x^4 - 13x^2 + 36 = 0$.

Solution

This equation is of quadratic form with $u = x^2$.

$x^4 - 13x^2 + 36 = 0$	Original equation
$(x^2)^2 - 13(x^2) + 36 = 0$	Write in quadratic form.
$u^2 - 13u + 36 = 0$	Substitute u for x^2.
$(u - 4)(u - 9) = 0$	Factor.
$u - 4 = 0$ ⟹ $u = 4$	Set 1st factor equal to 0.
$u - 9 = 0$ ⟹ $u = 9$	Set 2nd factor equal to 0.

At this point you have found the "u-solutions." To find the "x-solutions," replace u by x^2, as follows.

$$u = 4 \implies x^2 = 4 \implies x = \pm 2$$
$$u = 9 \implies x^2 = 9 \implies x = \pm 3$$

The solutions are 2, -2, 3, and -3. Check these in the original equation.

NOTE Be sure you see in Example 7 that the u-solutions of 4 and 9 represent only a temporary step. They are not solutions of the original equation.

EXAMPLE 8   Solving an Equation of Quadratic Form

Solve $x - 5\sqrt{x} + 6 = 0$.

Solution

This equation is of quadratic form with $u = \sqrt{x}$.

$$x - 5\sqrt{x} + 6 = 0 \qquad \text{Original equation}$$
$$\left(\sqrt{x}\right)^2 - 5\left(\sqrt{x}\right) + 6 = 0 \qquad \text{Write in quadratic form.}$$
$$u^2 - 5u + 6 = 0 \qquad \text{Substitute } u \text{ for } \sqrt{x}.$$
$$(u - 2)(u - 3) = 0 \qquad \text{Factor.}$$
$$u - 2 = 0 \quad \Longrightarrow \quad u = 2 \qquad \text{Set 1st factor equal to 0.}$$
$$u - 3 = 0 \quad \Longrightarrow \quad u = 3 \qquad \text{Set 2nd factor equal to 0.}$$

Now, using the u-solutions of 2 and 3, you obtain the following x-solutions.

$$u = 2 \quad \Longrightarrow \quad \sqrt{x} = 2 \quad \Longrightarrow \quad x = 4$$
$$u = 3 \quad \Longrightarrow \quad \sqrt{x} = 3 \quad \Longrightarrow \quad x = 9$$

The solutions are 4 and 9. Check these in the original equation.

EXAMPLE 9   Solving an Equation of Quadratic Form

Solve $x^{2/3} - x^{1/3} - 6 = 0$.

Solution

This equation is of quadratic form with $u = x^{1/3}$.

$$x^{2/3} - x^{1/3} - 6 = 0 \qquad \text{Original equation}$$
$$(x^{1/3})^2 - (x^{1/3}) - 6 = 0 \qquad \text{Write in quadratic form.}$$
$$u^2 - u - 6 = 0 \qquad \text{Substitute } u \text{ for } x^{1/3}.$$
$$(u + 2)(u - 3) = 0 \qquad \text{Factor.}$$
$$u + 2 = 0 \quad \Longrightarrow \quad u = -2 \qquad \text{Set 1st factor equal to 0.}$$
$$u - 3 = 0 \quad \Longrightarrow \quad u = 3 \qquad \text{Set 2nd factor equal to 0.}$$

Now, using the u-solutions of -2 and 3, you obtain the following x-solutions.

$$u = -2 \quad \Longrightarrow \quad x^{1/3} = -2 \quad \Longrightarrow \quad x = -8$$
$$u = 3 \quad \Longrightarrow \quad x^{1/3} = 3 \quad \Longrightarrow \quad x = 27$$

The solutions are -8 and 27. Check these in the original equation.

EXAMPLE 10   *The Dimensions of a Room*

You are working on some house plans. You want the living room in the house to have 200 square feet of floor space.

(a) What dimensions should the living room have if you want it to be square?

(b) What dimensions should the living room have if you want it to be a rectangle whose width is two-thirds of its length?

Solution

(a) If the room is square [Figure 6.4(a)], let x represent the length of each side.

$$x^2 = 200 \qquad \text{Formula for area}$$
$$x = \pm\sqrt{200} \qquad \text{Extract square roots.}$$
$$x \approx \pm 14.14 \qquad \text{Use a calculator.}$$

The negative solution makes no sense in this problem, so you can conclude that the living room should have sides of about 14.14 feet.

(b) If the room is rectangular [Figure 6.4(b)], let x represent the length. Because the width is two-thirds of the length, you can represent the width by $\frac{2}{3}x$.

$$x\left(\frac{2}{3}x\right) = 200 \qquad \text{Formula for area}$$
$$\frac{2}{3}x^2 = 200 \qquad \text{Simplify.}$$
$$x^2 = 300 \qquad \text{Divide both sides by } \tfrac{2}{3}.$$
$$x = \pm\sqrt{300} \qquad \text{Extract square roots.}$$
$$x \approx \pm 17.32 \qquad \text{Use a calculator.}$$

From this you can conclude that the room's length is $x \approx 17.32$ feet and its width is $\frac{2}{3}x \approx 11.55$ feet. Check this solution by multiplying 17.32 by 11.55 to see that you obtain approximately 200 square feet of floor space.

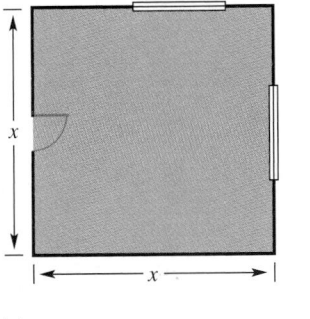

(a)

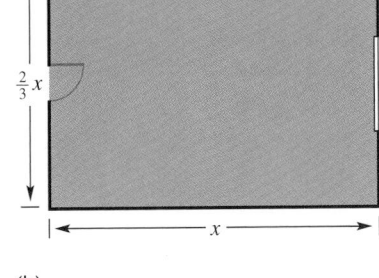

(b)

FIGURE 6.4

Group Activities

Exploring with Technology

Analyzing Solutions of Quadratic Equations Use a graphing utility to graph each of the following equations. How many times does the graph of each equation cross the x-axis?

(a) $y = 2x^2 + x - 15$ (b) $y = 9x^2 + 24x + 16$

(c) $y = -4(x - 2)^2 - 3$ (d) $y = (3x - 1)^2 + 3$

Now set each equation equal to zero and use the techniques of this section to solve the resulting equations. How many of each type of solution (real or imaginary) does each equation have? Summarize the relationship between the number of x-intercepts of the graph of a quadratic equation and the number and type of roots found algebraically.

6.1   Exercises

See Warm-Up Exercises, p. A45

In Exercises 1–26, solve the quadratic equation by factoring.

1. $x^2 - 12x + 35 = 0$ **2.** $x^2 + 15x + 44 = 0$

3. $x^2 + x - 72 = 0$ **4.** $x^2 - 2x - 48 = 0$

5. $x^2 + 4x = 45$ **6.** $x^2 - 7x = 18$

7. $4x^2 - 12x = 0$ **8.** $25y^2 - 75y = 0$

9. $u(u - 9) - 12(u - 9) = 0$

10. $16x(x - 8) - 12(x - 8) = 0$

11. $4x^2 - 25 = 0$ **12.** $16y^2 - 121 = 0$

13. $x^2 - 12x + 36 = 0$ **14.** $9x^2 + 24x + 16 = 0$

15. $x^2 + 60x + 900 = 0$ **16.** $8x^2 - 10x + 3 = 0$

17. $(y - 4)(y - 3) = 6$

18. $(6 + u)(1 - u) = 10$

19. $2x(3x + 2) = 5 - 6x^2$

20. $(2z + 1)(2z - 1) = -4z^2 - 5z + 2$

21. $3x(x - 6) - 5(x - 6) = 0$

22. $3(4 - x) - 2x(4 - x) = 0$

23. $6x^2 = 54$ **24.** $5t^2 = 125$

25. $\dfrac{y^2}{2} = 32$ **26.** $\dfrac{x^2}{6} = 24$

27. *Think About It* Explain the Zero-Factor Property and how it can be used to solve a quadratic equation.

28. *True or False?* The only solution of the equation $x^2 = 25$ is $x = 5$. Explain.

In Exercises 29–44, solve the quadratic equation by extracting square roots.

29. $x^2 = 64$ **30.** $z^2 = 169$

31. $25x^2 = 16$ **32.** $9z^2 = 121$

33. $4u^2 - 225 = 0$ **34.** $16x^2 - 1 = 0$

35. $(x + 4)^2 = 169$ **36.** $(y - 20)^2 = 625$

37. $(x - 3)^2 = 0.25$ **38.** $(x + 2)^2 = 0.81$

39. $(x - 2)^2 = 7$ **40.** $(x + 8)^2 = 28$

41. $(2x + 1)^2 = 50$ **42.** $(3x - 5)^2 = 48$

43. $(4x - 3)^2 - 98 = 0$

44. $(5x + 11)^2 - 300 = 0$

In Exercises 45–62, solve the quadratic equation by extracting imaginary square roots.

45. $z^2 = -36$

46. $x^2 = -9$

47. $x^2 + 4 = 0$

48. $y^2 + 16 = 0$

49. $(t - 3)^2 = -25$

50. $(x + 5)^2 + 81 = 0$

51. $(3z + 4)^2 + 144 = 0$

52. $(2y - 3)^2 + 25 = 0$

53. $9(x + 6)^2 = -121$

54. $4(x - 4)^2 = -169$

55. $(x - 1)^2 = -27$

56. $(2x + 3)^2 = -54$

57. $(x + 1)^2 + 0.04 = 0$

58. $(x - 3)^2 + 2.25 = 0$

59. $\left(c - \frac{2}{3}\right)^2 + \frac{1}{9} = 0$

60. $\left(u + \frac{5}{8}\right)^2 + \frac{49}{16} = 0$

61. $\left(x + \frac{7}{3}\right)^2 = -\frac{38}{9}$

62. $\left(y - \frac{5}{6}\right)^2 = -\frac{4}{5}$

In Exercises 63–72, use a graphing utility to graph the function. Use the graph to approximate any x-intercepts. Set $y = 0$ and solve the resulting equation. Compare the result with the x-intercepts of the graph.

63. $y = x^2 - 9$

64. $y = 5x - x^2$

65. $y = x^2 - 2x - 15$

66. $y = 9 - 4(x - 3)^2$

67. $y = 4 - (x - 3)^2$

68. $y = 4(x + 1)^2 - 9$

69. $y = 2x^2 - x - 6$

70. $y = 4x^2 - x - 14$

71. $y = 3x^2 - 8x - 16$

72. $y = 5x^2 + 9x - 18$

In Exercises 73–78, use a graphing utility to graph the function and observe that the graph has no x-intercepts. Set $y = 0$ and solve the resulting equation. What type of roots does the equation have?

73. $y = (x - 1)^2 + 1$

74. $y = (x + 2)^2 + 3$

75. $y = (x + 3)^2 + 5$

76. $y = (x - 2)^2 + 3$

77. $y = x^2 + 7$

78. $y = x^2 + 5$

In Exercises 79–84, find all real and imaginary solutions of the quadratic equation.

79. $2x^2 - 5x = 0$

80. $3x^2 + 8x - 16 = 0$

81. $x^2 - 100 = 0$

82. $(y + 12)^2 + 400 = 0$

83. $(x - 5)^2 + 100 = 0$

84. $(y + 12)^2 - 400 = 0$

In Exercises 85–88, solve for y in terms of x. Let f and g be functions where f represents the positive square root and g represents the negative square root. Use a graphing utility to graph f and g in the same viewing rectangle.

85. $x^2 + y^2 = 4$

86. $x^2 - y^2 = 4$

87. $x^2 + 4y^2 = 4$

88. $x - y^2 = 0$

In Exercises 89–104, solve the equation of quadratic form. (Find all real *and* imaginary solutions.)

89. $x^4 - 5x^2 + 4 = 0$

90. $4x^4 - 101x^2 + 25 = 0$

91. $x^4 - 5x^2 + 6 = 0$

92. $x^4 - 11x^2 + 30 = 0$

93. $x^4 - 3x^2 - 4 = 0$

94. $x^4 - x^2 - 6 = 0$

95. $(x^2 - 4)^2 + 2(x^2 - 4) - 3 = 0$

96. $(x^2 - 1)^2 + (x^2 - 1) - 6 = 0$

97. $x^{2/3} - x^{1/3} - 6 = 0$

98. $x^{2/3} + 3x^{1/3} - 10 = 0$

99. $2x^{2/3} - 7x^{1/3} + 5 = 0$

100. $3x^{2/3} + 8x^{1/3} + 5 = 0$

101. $x^{2/5} - 3x^{1/5} + 2 = 0$

102. $x^{2/5} + 5x^{1/5} + 6 = 0$

103. $2x^{2/5} - 7x^{1/5} + 3 = 0$

104. $2x^{2/5} + 3x^{1/5} + 1 = 0$

The symbol ▦ indicates an exercise in which you are instructed to use a calculator or graphing utility. The solutions of other exercises may also be facilitated by the use of appropriate technology.

In Exercises 105–108, find the length of the side labeled x. (Round the result to two decimal places.)

105.

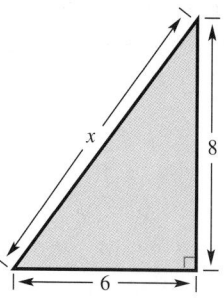

106.

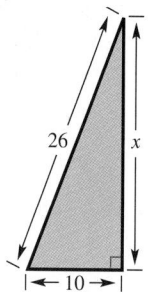

107.

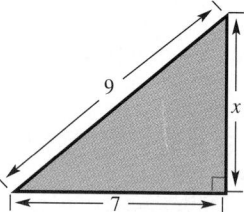

108.

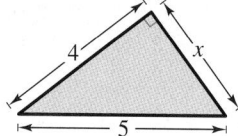

109. *Basketball*   A basketball court is 50 feet wide and 94 feet long. Find the length of the diagonal of the court.

110. *Plumbing*   A house has dimensions 26 feet by 32 feet. The gas hot water heater and furnace are diagonally across the basement from where the natural gas line enters the house. Find the length of the gas line across the basement.

111. *Television Screens*   For a square 25-inch television screen, it is the diagonal measurement of the screen that is 25 inches. What are the dimensions of the screen?

112. *Geometry*   Determine the length and width of a rectangle with a perimeter of 68 inches and a diagonal of length 26 inches.

Free-Falling Object   In Exercises 113 and 114, find the required time for an object to reach the ground if it is dropped from a height of s_0 feet. The height h (in feet) is given by

$$h = -16t^2 + s_0,$$

where t measures the time in seconds from the time the object is released.

113. $s_0 = 128$ **114.** $s_0 = 500$

115. *Free-Falling Object*   The height h (in feet) of an object propelled upward from a building 144 feet high (see figure) is given by

$$h = 144 + 128t - 16t^2,$$

where t measures the time in seconds from when the object is released. Find the time it takes for the object to reach the ground.

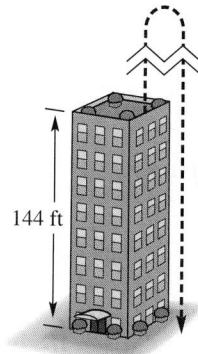

144 ft

116. *Revenue*   The revenue R (in dollars) from sales of x units of a product is given by

$$R = x\left(120 - \frac{1}{2}x\right).$$

Determine the number of units that must be sold to produce a revenue of $7000.

National Health Expenditures   In Exercises 117 and 118, use the following model, which gives the national expenditures for health care in the United States from 1988 through 1993.

$$y = (26.2 + 1.2t)^2, \quad -2 \leq t \leq 3$$

In this model, y represents the expenditures (in billions of dollars) and t represents the year, with $t = 0$ corresponding to 1990 (see figure). *(Source: U.S. Health Care Financing Administration)*

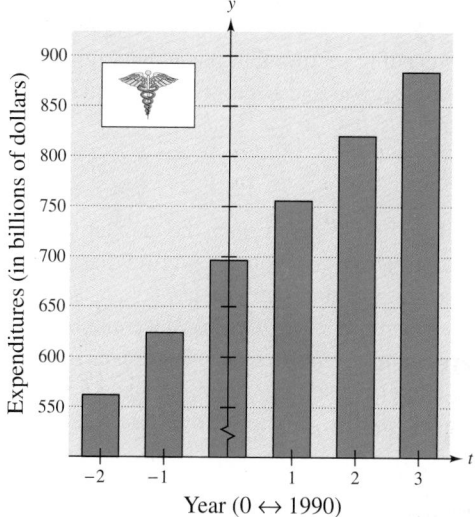

Year (0 ↔ 1990)

117. Analytically determine the year when expenditures were approximately $625 billion. Graphically confirm the result.

118. Analytically determine the year when expenditures were approximately $820 billion. Graphically confirm the result.

119. *Compound Interest*   A principal of $1500 is deposited in an account at an annual interest rate r compounded annually. If the amount after 2 years is $1685.40, the annual interest rate is the solution of the equation

$$1685.40 = 1500(1 + r)^2.$$

Find r.

120. *Think About It*   Is it possible for a quadratic equation to have only one solution? If so, give an example.

Section Project

Constructing a Quadratic Equation from Its Solutions
Suppose that you are tutoring a student in algebra and want to make up several practice equations for the student to solve by factoring. You might want to start by writing several simple equations with easy solutions such as 1 and 2. Then you might want to put in some tougher problems with solutions that include irrational or imaginary numbers.

(a) Find a quadratic equation that has $x = 1$ and $x = 2$ as solutions.

(b) Find a quadratic equation that has $x = 5$ and $x = -2$ as solutions.

(c) Find a quadratic equation that has $x = -3$ and $x = \frac{1}{2}$ as solutions.

(d) Find a quadratic equation that has $x = 1 + \sqrt{2}$ and $x = 1 - \sqrt{2}$ as solutions.

(e) Find a quadratic equation that has $x = 2 + 5i$ and $x = 2 - 5i$ as solutions.

(f) Write a short paragraph describing a general procedure for creating a quadratic equation that has two given numbers as solutions.

6.2	**Completing the Square**

Constructing Perfect Square Trinomials ▪
Solving Equations by Completing the Square

Constructing Perfect Square Trinomials

Study Tip

In Section 1.6, a *perfect square trinomial* was defined as the square of a binomial. For instance, the perfect square trinomial

$$x^2 + 10x + 25 = (x + 5)^2$$

is the square of the binomial $(x + 5)$.

Consider the quadratic equation

$$(x - 2)^2 = 10. \qquad \text{Completed square form}$$

You know from Example 4(b) in the preceding section that this equation has two solutions: $2 + \sqrt{10}$ and $2 - \sqrt{10}$. Suppose you had been given the equation in its standard form

$$x^2 - 4x - 6 = 0. \qquad \text{Standard form}$$

How would you solve this equation if you were given only the standard form? You could try factoring, but after attempting to do so you would find that the left side of the equation is not factorable (using integer coefficients).

In this section, you will study a technique for rewriting an equation in a completed square form. This technique is called **completing the square.** To complete the square, you use the fact that all perfect square trinomials have a similar form. For instance, consider the following perfect square trinomials.

$$x^2 + 6x + 9 = (x + 3)^2$$
$$x^2 - 12x + 36 = (x - 6)^2$$
$$x^2 + 5x + \frac{25}{4} = \left(x + \frac{5}{2}\right)^2$$

In each case, note that the constant term of the perfect square trinomial is the square of half the coefficient of the x-term. That is, a perfect square trinomial has the following form.

Consider adding to a pattern such as this

$$x^2 + 10x + \quad = (x + \quad)^2$$
$$x^2 + 11x + \quad = (x + \quad)^2$$
$$x^2 - 12x + \quad = (x - \quad)^2$$

by asking what should be added to create a perfect square trinomial. (See Examples 1 and 2.)

Perfect Square Trinomial   *Square of Binomial*

$$x^2 + bx + \left(\frac{b}{2}\right)^2 = \left(x + \frac{b}{2}\right)^2$$
$$\underset{(\text{half})^2}{\underline{\qquad\qquad}}$$

Thus, to complete the square for an expression of the form $x^2 + bx$, you must add $(b/2)^2$ to the expression.

Completing the Square

To **complete the square** for the expression $x^2 + bx$, add $(b/2)^2$, which is the square of half the coefficient of x. Consequently,

$$x^2 + bx + \left(\frac{b}{2}\right)^2 = \left(x + \frac{b}{2}\right)^2.$$

EXAMPLE 1   Creating a Perfect Square Trinomial

What term should be added to $x^2 - 8x$ so that it becomes a perfect square trinomial?

Solution

For this expression, the coefficient of the x-term is -8. Divide this term by 2 and square the result to obtain $(-4)^2 = 16$. This is the term that should be added to the expression to make it a perfect square trinomial.

$$\begin{aligned} x^2 - 8x + 16 &= x^2 - 8x + (-4)^2 && \text{Add 16 to the expression.} \\ &= (x - 4)^2 && \text{Completed square form} \end{aligned}$$

EXAMPLE 2   Creating Perfect Square Trinomials

Emphasize that the coefficient of the second-degree term must be 1 before completing the square.

What term should be added to each expression so that it becomes a perfect square trinomial?

(a) $x^2 + 14x$ (b) $x^2 - 9x$

Solution

(a) For this expression, the coefficient of the x-term is 14. By taking half of this and squaring the result, you can see that $7^2 = 49$ should be added to the expression to make it a perfect square trinomial. That is,

$$x^2 + 14x + 49 = x^2 + 14x + 7^2 = (x + 7)^2.$$

(b) For this expression, the coefficient of the x-term is -9. By taking half of this and squaring the result, you can see that $\left(-\frac{9}{2}\right)^2 = \frac{81}{4}$ should be added to the expression to make it a perfect square trinomial. That is,

$$x^2 - 9x + \frac{81}{4} = x^2 - 9x + \left(-\frac{9}{2}\right)^2 = \left(x - \frac{9}{2}\right)^2.$$

Solving Equations by Completing the Square

When completing the square to solve an equation, remember that it is essential to *preserve the equality.* Thus, when you add a constant term to one side of the equation, you must be sure to add the same constant to the other side of the equation.

NOTE In Example 3, completing the square is used for the sake of illustration. This particular equation would be easier to solve by factoring. Try reworking the problem by factoring to see that you obtain the same two solutions. In Example 4, the equation cannot be solved by factoring (using integer coefficients).

EXAMPLE 3 Completing the Square: Leading Coefficient Is 1

Solve $x^2 + 12x = 0$.

Solution

$$x^2 + 12x = 0 \qquad \text{Original equation}$$
$$x^2 + 12x + (6)^2 = 36 \qquad \text{Add } \left(\tfrac{12}{2}\right)^2 = 36 \text{ to both sides.}$$
$$\underbrace{\qquad}_{\text{(half)}^2}$$
$$(x + 6)^2 = 36 \qquad \text{Completed square form}$$
$$x + 6 = \pm\sqrt{36} \qquad \text{Extract square roots.}$$
$$x = -6 \pm 6 \qquad \text{Subtract 6 from both sides.}$$
$$x = 0 \text{ or } x = -12 \qquad \text{Solutions}$$

The solutions are 0 and -12. Check these in the original equation.

EXAMPLE 4 Completing the Square: Leading Coefficient Is 1

Solve $x^2 - 6x + 7 = 0$.

Solution

$$x^2 - 6x + 7 = 0 \qquad \text{Original equation}$$
$$x^2 - 6x = -7 \qquad \text{Subtract 7 from both sides.}$$
$$x^2 - 6x + (-3)^2 = -7 + 9 \qquad \text{Add } (-3)^2 = 9 \text{ to both sides.}$$
$$\underbrace{\qquad}_{\text{(half)}^2}$$
$$(x - 3)^2 = 2 \qquad \text{Completed square form}$$
$$x - 3 = \pm\sqrt{2} \qquad \text{Extract square roots.}$$
$$x = 3 \pm \sqrt{2} \qquad \text{Solutions}$$

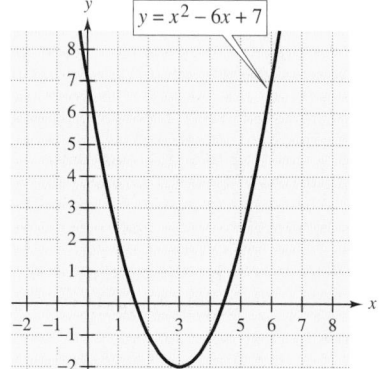

$y = x^2 - 6x + 7$

FIGURE 6.5

The solutions are $3 + \sqrt{2}$ and $3 - \sqrt{2}$. Check these in the original equation. Also try checking the solutions graphically, as shown in Figure 6.5.

EXAMPLE 5   Completing the Square: Leading Coefficient Is Not 1

Solve $3x^2 + 5x = 2$.

Solution

$3x^2 + 5x = 2$	Original equation
$x^2 + \dfrac{5}{3}x = \dfrac{2}{3}$	Divide both sides by 3.
$x^2 + \dfrac{5}{3}x + \left(\dfrac{5}{6}\right)^2 = \dfrac{2}{3} + \dfrac{25}{36}$	Add $\left(\dfrac{5}{6}\right)^2 = \dfrac{25}{36}$ to both sides.
$\left(x + \dfrac{5}{6}\right)^2 = \dfrac{49}{36}$	Completed square form
$x + \dfrac{5}{6} = \pm\dfrac{7}{6}$	Extract square roots.
$x = -\dfrac{5}{6} \pm \dfrac{7}{6}$	Subtract $\frac{5}{6}$ from both sides.
$x = \dfrac{1}{3}$ or $x = -2$	Solutions

The solutions are $\frac{1}{3}$ and -2. Check these in the original equation.

EXAMPLE 6   Completing the Square: Leading Coefficient Is Not 1

Solve $2x^2 - x - 2 = 0$.

Solution

$2x^2 - x - 2 = 0$	Original equation
$2x^2 - x = 2$	Add 2 to both sides.
$x^2 - \dfrac{1}{2}x = 1$	Divide both sides by 2.
$x^2 - \dfrac{1}{2}x + \left(-\dfrac{1}{4}\right)^2 = 1 + \dfrac{1}{16}$	Add $\left(-\dfrac{1}{4}\right)^2 = \dfrac{1}{16}$ to both sides.
$\left(x - \dfrac{1}{4}\right)^2 = \dfrac{17}{16}$	Complete square form
$x - \dfrac{1}{4} = \pm\dfrac{\sqrt{17}}{4}$	Extract square roots.
$x = \dfrac{1}{4} \pm \dfrac{\sqrt{17}}{4}$	Add $\frac{1}{4}$ to both sides.

The solutions are $\frac{1}{4}\left(1 \pm \sqrt{17}\right)$. Check these in the original equation. You can use a calculator to obtain the decimal approximations $x \approx 1.28$ and $x \approx -0.78$.

EXAMPLE 7 A Quadratic Equation with Imaginary Solutions

Solve $x^2 - 4x + 8 = 0$.

Solution

$x^2 - 4x + 8 = 0$	Original equation
$x^2 - 4x = -8$	Subtract 8 from both sides.
$x^2 - 4x + (-2)^2 = -8 + 4$	Add $(-2)^2 = 4$ to both sides.
$(x - 2)^2 = -4$	Completed square form
$x - 2 = \pm 2i$	Extract imaginary square roots.
$x = 2 \pm 2i$	Add 2 to both sides.

The solutions are $2 + 2i$ and $2 - 2i$. The first of these is checked as follows. Try checking the other.

Check

$x^2 - 4x + 8 = 0$	Original equation
$(2 + 2i)^2 - 4(2 + 2i) + 8 \stackrel{?}{=} 0$	Substitute $2 + 2i$ for x.
$4 + 8i - 4 - 8 - 8i + 8 \stackrel{?}{=} 0$	Simplify.
$0 = 0$	Solution checks. ✓

Group Activities   You Be the Instructor

Error Analysis   Suppose you teach an algebra class and one of your students hands in the following solution. Find and correct the error(s). Discuss how to explain the error(s) to your student.

1. Solve $x^2 + 6x - 13 = 0$ by completing the square.

$$x^2 + 6x = 13$$
$$x^2 + 6x + \left(\frac{6}{2}\right)^2 = 13$$
$$(x + 3)^2 = 13$$
$$x + 3 = \pm\sqrt{13}$$
$$x = -3 \pm \sqrt{13}$$

6.2   **Exercises**

See Warm-Up
Exercises, p. A45

1. *Think About It*   Is it possible for a quadratic equation to have no real number solution? If so, give an example and describe what the graph would look like.

2. *Think About It*   When completing the square to solve a quadratic equation, what is the first step if the leading coefficient is not 1? Is the resulting equation equivalent to the original equation? Explain.

In Exercises 3–14, add a term to the expression so that it becomes a perfect square trinomial.

3. $x^2 + 8x +$ ▫
4. $x^2 + 12x +$ ▫
5. $y^2 - 20y +$ ▫
6. $y^2 - 2y +$ ▫
7. $t^2 + 5t +$ ▫
8. $u^2 + 7u +$ ▫
9. $x^2 - \frac{6}{5}x +$ ▫
10. $y^2 + \frac{4}{3}y +$ ▫
11. $y^2 - \frac{3}{5}y +$ ▫
12. $a^2 - \frac{1}{3}a +$ ▫
13. $r^2 - 0.4r +$ ▫
14. $s^2 + 4.5s +$ ▫

In Exercises 15–28, solve the quadratic equation (a) by completing the square and (b) by factoring.

15. $x^2 - 20x = 0$
16. $x^2 + 32x = 0$
17. $x^2 + 6x = 0$
18. $t^2 - 9t = 0$
19. $t^2 - 8t + 7 = 0$
20. $y^2 - 8y + 12 = 0$
21. $x^2 + 2x - 24 = 0$
22. $x^2 + 12x + 27 = 0$
23. $x^2 + 7x + 12 = 0$
24. $z^2 + 3z - 10 = 0$
25. $x^2 - 3x - 18 = 0$
26. $t^2 - 5t - 36 = 0$
27. $2x^2 - 11x + 12 = 0$
28. $3x^2 - 5x - 2 = 0$

In Exercises 29–60, solve the quadratic equation by completing the square. Give the solutions in exact form and in decimal form rounded to two decimal places. (The solutions may be imaginary numbers.)

29. $x^2 - 4x - 3 = 0$
30. $x^2 - 6x + 7 = 0$
31. $x^2 + 4x - 3 = 0$
32. $x^2 + 6x + 7 = 0$
33. $u^2 - 4u + 1 = 0$
34. $a^2 - 10a - 15 = 0$
35. $x^2 + 2x + 3 = 0$
36. $x^2 - 6x + 12 = 0$
37. $x^2 - 10x - 2 = 0$
38. $x^2 + 8x - 4 = 0$
39. $y^2 + 20y + 10 = 0$
40. $y^2 + 6y - 24 = 0$
41. $x^2 - \frac{2}{3}x - 3 = 0$
42. $x^2 + \frac{4}{5}x - 1 = 0$
43. $t^2 + 5t + 3 = 0$
44. $u^2 - 9u - 1 = 0$
45. $v^2 + 3v - 2 = 0$
46. $z^2 - 7z + 9 = 0$
47. $-x^2 + x - 1 = 0$
48. $1 - x - x^2 = 0$
49. $2x^2 + 8x + 3 = 0$
50. $3x^2 - 24x - 5 = 0$
51. $3x^2 + 9x + 5 = 0$
52. $5x^2 - 15x + 7 = 0$
53. $4y^2 + 4y - 9 = 0$
54. $4z^2 - 3z + 2 = 0$
55. $x(x - 7) = 2$
56. $2x\left(x + \frac{4}{3}\right) = 5$
57. $\frac{1}{2}t^2 + t + 2 = 0$
58. $0.1x^2 + 0.5x = -0.2$
59. $0.1x^2 + 0.2x + 0.5 = 0$
60. $0.02x^2 + 0.10x - 0.05 = 0$

In Exercises 61–64, find the real solutions.

61. $\dfrac{x^2}{4} = \dfrac{x + 1}{2}$
62. $\dfrac{x^2 + 2}{24} = \dfrac{x - 1}{3}$
63. $\sqrt{2x + 1} = x - 3$
64. $\sqrt{3x - 2} = x - 2$

In Exercises 65–72, use a graphing utility to graph the function. Use the graph to approximate any x-intercepts. Set $y = 0$ and solve the resulting equation. Compare the result with the x-intercepts of the graph.

65. $y = x^2 + 4x - 1$
66. $y = x^2 + 6x - 4$
67. $y = x^2 - 2x - 5$
68. $y = 2x^2 - 6x - 5$
69. $y = \frac{1}{3}x^2 + 2x - 6$
70. $y = \frac{1}{2}x^2 - 3x + 1$
71. $y = -x^2 - x + 3$
72. $y = \sqrt{x} - x + 2$

73. *Geometric Modeling*

    (a) Find the area of the two adjoining rectangles and large square in the figure.

    (b) Find the area of the small square in the lower right-hand corner of the figure and add it to the area found in part (a).

    (c) Find the dimensions and the area of the entire figure after adjoining the small square in the lower right-hand corner. Note that you have shown completing the square geometrically.

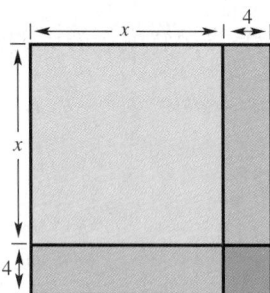

74. *Geometric Modeling*   Repeat Exercise 73 for the model shown below.

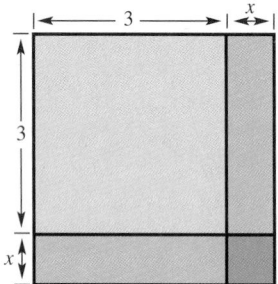

75. *Geometry*   Find the dimensions of the triangle in the figure if its area is 12 square centimeters.

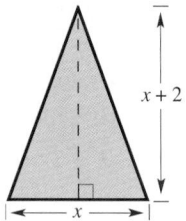

76. *Geometry*   The area of the rectangle in the figure is 160 square feet. Find the rectangle's dimensions.

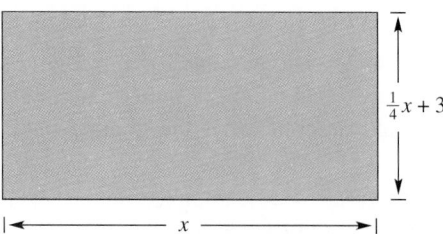

77. *Fencing in a Corral*   You have 200 meters of fencing to enclose two adjacent rectangular corrals (see figure). The total area of the enclosed region is 1400 square meters. What are the dimensions of each corral? (The corrals are the same size.)

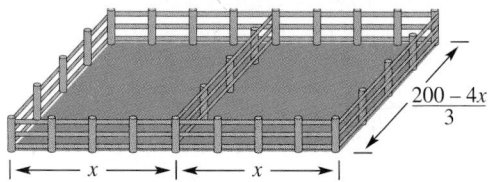

78. *Geometry*   An open box with a rectangular base of x inches by $x + 4$ inches has a height of 6 inches (see figure). Find the dimensions of the box if its volume is 840 cubic inches.

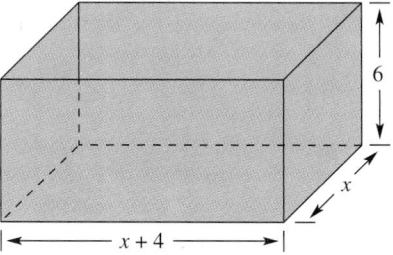

79. *Pulling a Boat to a Dock*   A windlass is used to pull a boat to a dock (see figure). The rope is attached to the boat at a point 15 feet below the level of the windlass. Find the distance from the boat to the dock when the length of the rope is 75 feet.

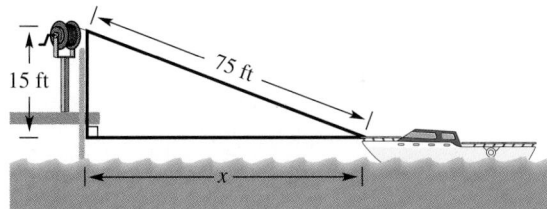

80. *Cutting Across the Lawn*   On the sidewalk, the distance from the dormitory to the cafeteria is 400 meters (see figure). By cutting across the lawn, the walking distance is shortened to 300 meters. How long is each part of the L-shaped sidewalk?

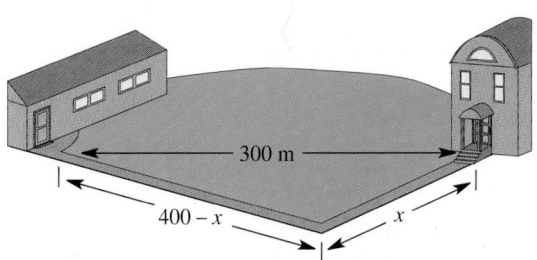

81. *Revenue*   The revenue R from selling x units of a product is

$$R = x\left(50 - \tfrac{1}{2}x\right).$$

Find the number of units that must be sold to produce a revenue of $1218.

82. *Revenue*   The revenue R from selling x units of a product is

$$R = x\left(100 - \tfrac{1}{10}x\right).$$

Find the number of units that must be sold to produce a revenue of $12,000.

Section Project

Completed Square Form   Consider the following quadratic equation.

$$(x - 1)^2 = d$$

(a) What value(s) of d will produce a quadratic equation that has exactly one (repeated) solution?

(b) Describe the values of d that will produce two different solutions, both of which are *rational* numbers.

(c) Describe the values of d that will produce two different solutions, both of which are *irrational* numbers.

(d) Describe the values of d that will produce two different solutions, both of which are *imaginary* numbers.

6.3	**The Quadratic Formula and the Discriminant**

The Quadratic Formula ▪ Solving Equations by the Quadratic Formula ▪ The Discriminant

The Quadratic Formula

Before completing the square for the *general* quadratic equation, consider solving a *specific* quadratic equation by completing the square and leaving it on the board for easy reference. Point out that the derivation of the Quadratic Formula is just a generalization of the technique presented in Section 6.2. Both the completing-the-square method and the Quadratic Formula can be used to solve *any* quadratic equation.

A fourth technique for solving a quadratic equation involves the **Quadratic Formula.** This formula is obtained by completing the square for a general quadratic equation.

$$ax^2 + bx + c = 0 \qquad \text{Standard form, } a \neq 0$$

$$ax^2 + bx = -c \qquad \text{Subtract } c \text{ from both sides.}$$

$$x^2 + \frac{b}{a}x = -\frac{c}{a} \qquad \text{Divide both sides by } a.$$

$$x^2 + \frac{b}{a}x + \left(\frac{b}{2a}\right)^2 = -\frac{c}{a} + \left(\frac{b}{2a}\right)^2 \qquad \text{Add } \left(\frac{b}{2a}\right)^2 \text{ to both sides.}$$

$$\left(x + \frac{b}{2a}\right)^2 = \frac{b^2 - 4ac}{4a^2} \qquad \text{Simplify.}$$

$$x + \frac{b}{2a} = \pm\sqrt{\frac{b^2 - 4ac}{4a^2}} \qquad \text{Extract square roots.}$$

$$x = -\frac{b}{2a} \pm \frac{\sqrt{b^2 - 4ac}}{2|a|} \qquad \text{Subtract } \frac{b}{2a} \text{ from both sides.}$$

$$x = \frac{-b \pm \sqrt{b^2 - 4ac}}{2a} \qquad \text{Simplify.}$$

NOTE Notice in the derivation of the Quadratic Formula that, because $\pm 2|a|$ represents the same numbers as $\pm 2a$, you can omit the absolute value bars.

The Quadratic Formula

The solutions of $ax^2 + bx + c = 0$, $a \neq 0$, are given by the **Quadratic Formula**

$$x = \frac{-b \pm \sqrt{b^2 - 4ac}}{2a}.$$

Suggest that for each homework problem students write the Quadratic Formula before substituting specific values. Repeatedly writing the formula should help students learn it.

The Quadratic Formula is one of the most important formulas in algebra, and you should memorize it. We have found that it helps to try to memorize a verbal statement of the rule. For instance, you might try to remember the following verbal statement of the Quadratic Formula: "The opposite of b, plus or minus the square root of b squared minus $4ac$, all divided by $2a$."

NOTE In Example 1, the solutions are rational numbers, which means that the equation could have been solved by factoring. Try solving the equation by factoring.

Solving Equations by the Quadratic Formula

When using the Quadratic Formula, remember that *before* the formula can be applied, you must first write the quadratic equation in standard form.

EXAMPLE 1   The Quadratic Formula: Two Distinct Solutions

$$x^2 + 6x = 16 \qquad \text{Original equation}$$

$$x^2 + 6x - 16 = 0 \qquad \text{Write in standard form.}$$

$$x = \frac{-b \pm \sqrt{b^2 - 4ac}}{2a} \qquad \text{Quadratic Formula}$$

$$x = \frac{-6 \pm \sqrt{6^2 - 4(1)(-16)}}{2(1)} \qquad \text{Substitute: } a = 1, b = 6, c = -16.$$

$$x = \frac{-6 \pm \sqrt{100}}{2} \qquad \text{Simplify.}$$

$$x = \frac{-6 \pm 10}{2} \qquad \text{Simplify.}$$

$$x = 2 \text{ or } x = -8 \qquad \text{Solutions}$$

The solutions are 2 and -8. Check these in the original equation. Or, try using a graphic check, as shown in Figure 6.6.

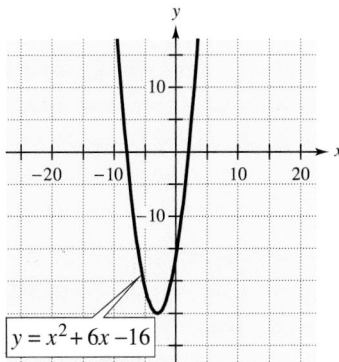

$y = x^2 + 6x - 16$

FIGURE 6.6

EXAMPLE 2   The Quadratic Formula: Two Distinct Solutions

$$-x^2 - 4x + 8 = 0 \qquad \text{Leading coefficient is negative.}$$

$$x^2 + 4x - 8 = 0 \qquad \text{Multiply both sides by } -1.$$

$$x = \frac{-b \pm \sqrt{b^2 - 4ac}}{2a} \qquad \text{Quadratic Formula}$$

$$x = \frac{-4 \pm \sqrt{4^2 - 4(1)(-8)}}{2(1)} \qquad \text{Substitute: } a = 1, b = 4, c = -8.$$

$$x = \frac{-4 \pm \sqrt{48}}{2} \qquad \text{Simplify.}$$

$$x = \frac{-4 \pm 4\sqrt{3}}{2} \qquad \text{Simplify.}$$

$$x = \frac{2(-2 \pm 2\sqrt{3})}{2} \qquad \text{Divide out common factor.}$$

$$x = -2 \pm 2\sqrt{3} \qquad \text{Solutions}$$

The solutions are $-2 + 2\sqrt{3}$ and $-2 - 2\sqrt{3}$. Check these in the original equation.

Study Tip

If the leading coefficient of a quadratic equation is negative, we suggest that you begin by multiplying both sides of the equation by -1, as shown in Example 2. This will produce a positive leading coefficient, which is less cumbersome to work with.

Some students may be tempted to reduce $(-4 \pm 4\sqrt{3})/2$ to $-2 \pm 4\sqrt{3}$. Ask them what is wrong with this reasoning.

EXAMPLE 3 The Quadratic Formula: One Repeated Solution

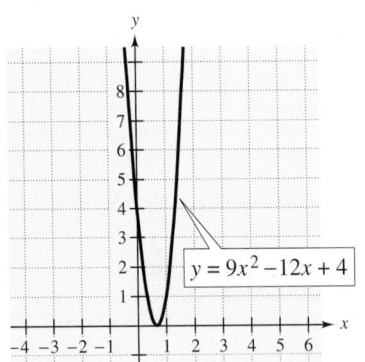

FIGURE 6.7

$$18x^2 - 24x + 8 = 0 \qquad \text{Original equation}$$

$$9x^2 - 12x + 4 = 0 \qquad \text{Divide both sides by 2.}$$

$$x = \frac{-b \pm \sqrt{b^2 - 4ac}}{2a} \qquad \text{Quadratic Formula}$$

$$x = \frac{-(-12) \pm \sqrt{(-12)^2 - 4(9)(4)}}{2(9)} \qquad \begin{array}{l}\text{Substitute:}\\ a = 9, b = -12, c = 4.\end{array}$$

$$x = \frac{12 \pm \sqrt{144 - 144}}{18} \qquad \text{Simplify.}$$

$$x = \frac{12 \pm \sqrt{0}}{18} \qquad \text{Simplify.}$$

$$x = \frac{2}{3} \qquad \text{Solution}$$

The only solution is $\frac{2}{3}$. Check this in the original equation. Or, try using a graphic check, as shown in Figure 6.7. Note that when a quadratic equation has one repeated solution, its graph touches the x-axis at a single point.

EXAMPLE 4 The Quadratic Formula: Imaginary Solutions

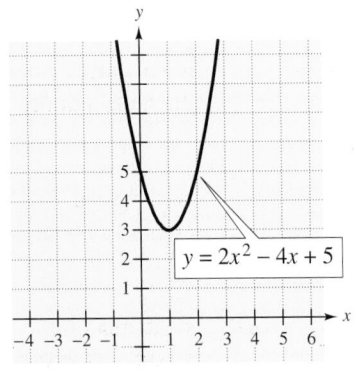

FIGURE 6.8

$$2x^2 - 4x + 5 = 0 \qquad \text{Original equation}$$

$$x = \frac{-b \pm \sqrt{b^2 - 4ac}}{2a} \qquad \text{Quadratic Formula}$$

$$x = \frac{-(-4) \pm \sqrt{(-4)^2 - 4(2)(5)}}{2(2)} \qquad \begin{array}{l}\text{Substitute:}\\ a = 2, b = -4, c = 5.\end{array}$$

$$x = \frac{4 \pm \sqrt{-24}}{4} \qquad \text{Simplify.}$$

$$x = \frac{4 \pm 2\sqrt{6}i}{4} \qquad \text{Write in } i\text{-form.}$$

$$x = \frac{2(2 \pm \sqrt{6}i)}{2 \cdot 2} \qquad \text{Divide out common factor.}$$

$$x = \frac{2 \pm \sqrt{6}i}{2} \qquad \text{Solutions}$$

The solutions are $\frac{1}{2}(2 \pm \sqrt{6}i)$. Check these in the original equation. Or, try using a graphic check, as shown in Figure 6.8. Note that the graph has no x-intercepts. This verifies that the equation has no *real* solutions.

The Discriminant

The radicand in the Quadratic Formula, $b^2 - 4ac$, is called the **discriminant** because it allows you to "discriminate" among different types of solutions.

Study Tip

By reexamining Examples 1 through 4, you can see that the equations with rational or repeated solutions could have been solved by *factoring*. In general, quadratic equations (with integer coefficients) for which the discriminant is either zero or a perfect square are factorable using integer coefficients. Consequently, a quick test of the discriminant will help decide which solution method to use to solve a quadratic equation.

Using the Discriminant

Let a, b, and c be ~~real~~ *rational* numbers such that $a \neq 0$. The **discriminant** of the quadratic equation $ax^2 + bx + c = 0$ is given by $b^2 - 4ac$, and can be used to classify the solutions of the equation as follows.

Discriminant	*Solution Type*
1. Perfect square	Two distinct rational solutions (Example 1)
2. Positive (nonperfect square)	Two distinct irrational solutions (Example 2)
3. Zero	One repeated rational solution (Example 3)
4. Negative number	Two distinct imaginary solutions (Example 4)

Exploration

Use a graphing utility to graph the equations below.

a. $y = x^2 - x + 2$

b. $y = 2x^2 - 3x - 2$

c. $y = x^2 - 2x + 1$

d. $y = x^2 - 2x - 10$

Describe the solution type of each equation and check your results with those shown in Example 5. Why do you think the discriminant is used to determine solution types?

EXAMPLE 5   Using the Discriminant

Equation	*Discriminant*	*Solution Type*
(a) $x^2 - x + 2 = 0$	$b^2 - 4ac = (-1)^2 - 4(1)(2)$ $= 1 - 8$ $= -7$	Two distinct imaginary solutions
(b) $2x^2 - 3x - 2 = 0$	$b^2 - 4ac = (-3)^2 - 4(2)(-2)$ $= 9 + 16$ $= 25$	Two distinct rational solutions
(c) $x^2 - 2x + 1 = 0$	$b^2 - 4ac = (-2)^2 - 4(1)(1)$ $= 4 - 4$ $= 0$	One repeated rational solution
(d) $x^2 - 2x - 1 = 9$	$b^2 - 4ac = (-2)^2 - 4(1)(-10)$ $= 4 + 40$ $= 44$	Two distinct irrational solutions

You have now studied four algebraic ways to solve quadratic equations.

Note this summary.

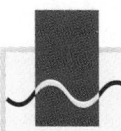

Programming

The *TI-82, TI-83* program below can be used to solve equations with the Quadratic Formula. (See the appendix for other calculator models.) Enter *a*, *b*, and *c* when prompted.

```
PROGRAM:QUADRAT
:Prompt A
:Prompt B
:Prompt C
:B²-4AC→D
:If D≥0
:Then
:(-B+√(D))/(2A)→M
:Disp M
:(-B-√(D))/(2A)→N
:Disp N
:Else
:Disp "NO REAL SOLUTION"
:End
```

When using their calculators, some students err in squaring negative numbers. Be sure students know the difference between $(-17.8)^2$ and -17.8^2.

Summary of Algebraic Methods for Solving Quadratic Equations

1. Factoring   Section 6.1

2. Extracting Square Roots   Section 6.1

3. Completing the Square   Section 6.2

4. The Quadratic Formula   Section 6.3

When choosing one of these methods, first check to see whether the equation is in a form in which you can extract square roots. Next, you can try factoring. If neither of these two methods works, use the Quadratic Formula (or completing the square), which will work for any quadratic equation. For *real* solutions, remember that you can use a graph to approximate the solutions.

EXAMPLE 6   *Using a Calculator with the Quadratic Formula*

Solve $1.2x^2 - 17.8x + 8.05 = 0$.

Solution

Using the Quadratic Formula, you can write

$$x = \frac{-(-17.8) \pm \sqrt{(-17.8)^2 - 4(1.2)(8.05)}}{2(1.2)}.$$

To evaluate these solutions, begin by calculating the square root.

17.8 [+/−] [x²] [−] 4 [×] 1.2 [×] 8.05 [=] [√]   Scientific

[√] [(] [(] [(−)] 17.8 [)] [x²] [−] 4 [×] 1.2   Graphing

[×] 8.05 [)] [ENTER]

The display for either of these keystroke sequences should be 16.67932852. Storing this result and using the recall key, you can find the following two solutions.

$$x \approx \frac{17.8 + 16.67932852}{2.4} \approx 14.366 \qquad \text{Add stored value.}$$

$$x \approx \frac{17.8 - 16.67932852}{2.4} \approx 0.467 \qquad \text{Subtract stored value.}$$

Group Activities   You Be the Instructor

Problem Posing   Suppose you are writing a quiz that covers quadratic equations. Write four quadratic equations, including one with solutions $x = \frac{5}{3}$ and $x = -2$ and one with solutions $x = 4 \pm \sqrt{3}$, and instruct students to use any of the four solution methods: factoring, extracting square roots, completing the square, and using the Quadratic Formula. Trade quizzes with another member of your group and check one another's work.

T. QUIZ #4 solve by compl. sqa
$$x^2 - 6x + 6 = 0$$

6.3   Exercises

See Warm-Up
Exercises, p. A45

In Exercises 1–4, write in standard form.

1. $2x^2 = 7 - 2x$

2. $7x^2 + 15x = 5$

3. $x(10 - x) = 5$

4. $x(3x + 8) = 15$

In Exercises 5–16, solve the quadratic equation by (a) using the Quadratic Formula and (b) factoring.

5. $x^2 - 11x + 28 = 0$

6. $x^2 - 12x + 27 = 0$

7. $x^2 + 6x + 8 = 0$

8. $x^2 + 9x + 14 = 0$

9. $4x^2 + 4x + 1 = 0$

10. $9x^2 + 12x + 4 = 0$

11. $4x^2 + 12x + 9 = 0$

12. $9x^2 - 30x + 25 = 0$

13. $6x^2 - x - 2 = 0$

14. $10x^2 - 11x + 3 = 0$

15. $x^2 - 5x - 300 = 0$

16. $x^2 + 20x - 300 = 0$

In Exercises 17–26, use the discriminant to determine the type of solutions of the quadratic equation.

17. $x^2 + x + 1 = 0$

18. $x^2 + x - 1 = 0$

19. $2x^2 - 5x - 4 = 0$

20. $10x^2 + 5x + 1 = 0$

21. $x^2 + 7x + 15 = 0$

22. $3x^2 - 2x - 5 = 0$

23. $4x^2 - 12x + 9 = 0$

24. $2x^2 + 10x + 6 = 0$

25. $3x^2 - x + 2 = 0$

26. $9x^2 - 24x + 16 = 0$

In Exercises 27–50, solve the quadratic equation by using the Quadratic Formula. (Find all real *and* imaginary solutions.)

27. $x^2 - 2x - 4 = 0$

28. $x^2 - 2x - 6 = 0$

29. $t^2 + 4t + 1 = 0$

30. $y^2 + 6y + 4 = 0$

31. $x^2 + 6x - 3 = 0$

32. $x^2 + 8x - 4 = 0$

33. $x^2 - 10x + 23 = 0$

34. $u^2 - 12u + 29 = 0$

35. $x^2 + 3x + 3 = 0$

36. $2x^2 - x + 1 = 0$

37. $2v^2 - 2v - 1 = 0$

38. $4x^2 + 6x + 1 = 0$

39. $2x^2 + 4x - 3 = 0$

40. $2x^2 + 3x + 3 = 0$

41. $9z^2 + 6z - 4 = 0$

42. $8y^2 - 8y - 1 = 0$

43. $x^2 - 0.4x - 0.16 = 0$

44. $x^2 + 0.6x - 0.41 = 0$

45. $2.5x^2 + x - 0.9 = 0$

46. $0.09x^2 - 0.12x - 0.26 = 0$

47. $4x^2 - 6x + 3 = 0$

48. $-5x^2 - 15x + 10 = 0$

49. $9x^2 = 1 + 9x$

50. $x - x^2 = 1 - 6x^2$

*Quiz
part b) - use discrim $x^2 + 5x - 3 = 0$
a) - find solutions $x^2 + 3x - 1 = 0$*

In Exercises 51–64, solve the quadratic equation by the most convenient method. (Find all real *and* imaginary solutions.)

51. $z^2 - 169 = 0$

52. $t^2 = 150$

53. $y^2 + 15y = 0$

54. $4u^2 + 49 = 0$

55. $25(x - 3)^2 - 36 = 0$

56. $2y(y - 18) + 3(y - 18) = 0$

57. $(x + 4)^2 + 16 = 0$

58. $x^2 - 3x - 4 = 0$

59. $18x^2 + 15x - 50 = 0$

60. $2x^2 - 15x + 225 = 0$

61. $x^2 - 24x + 128 = 0$

62. $1.2x^2 - 0.8x - 5.5 = 0$

63. $x^2 + 8x + 25 = 0$

64. $2x^2 + 8x + 4.5 = 0$

In Exercises 65–72, use a graphing utility to graph the function. Use the graph to approximate any x-intercepts. Set $y = 0$ and solve the resulting equation. Compare the result with the x-intercepts of the graph.

65. $y = 3x^2 - 6x + 1$

66. $y = x^2 + x + 1$

67. $y = -(4x^2 - 20x + 25)$

68. $y = x^2 - 4x + 3$

69. $y = 5x^2 - 18x + 6$

70. $y = 15x^2 + 3x - 105$

71. $y = -0.04x^2 + 4x - 0.8$

72. $y = 3.7x^2 - 10.2x + 3.2$

In Exercises 73–76, use a graphing utility to determine the number of real solutions of the quadratic equation.

73. $2x^2 - 5x + 5 = 0$

74. $2x^2 - x - 1 = 0$

75. $\frac{1}{5}x^2 + \frac{6}{5}x - 8 = 0$

76. $\frac{1}{3}x^2 - 5x + 25 = 0$

Think About It   In Exercises 77–80, describe the value of c such that the equation has (a) two real number solutions, (b) one real number solution, and (c) two imaginary number solutions.

77. $x^2 - 6x + c = 0$

78. $x^2 - 12x + c = 0$

79. $x^2 + 8x + c = 0$

80. $x^2 + 2x + c = 0$

In Exercises 81–84, solve the equation.

81. $\dfrac{2x^2}{5} - \dfrac{x}{2} = 1$

82. $\dfrac{x^2 - 9x}{6} = \dfrac{x - 1}{2}$

83. $\sqrt{x + 3} = x - 1$

84. $\sqrt{2x - 3} = x - 2$

85. *Geometry*   A rectangle has a width of x centimeters, a length of $x + 6.3$ centimeters, and an area of 58.14 square centimeters. Find its dimensions.

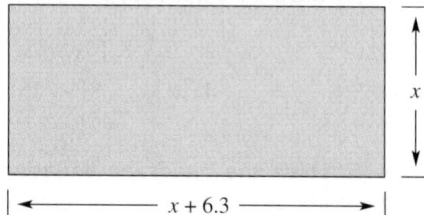

86. *Geometry*   Two circular regions are tangent to each other (see figure). The distance between the centers is 10 feet.

(a) Find the radius of each circle if their combined area is 52π square feet.

(b) Suppose the distance between the centers remained the same but the radii were made equal. Would the combined area of the circles increase or decrease?

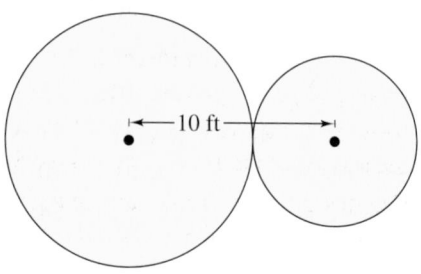

87. *Free-Falling Object*   A ball is thrown upward at a velocity of 40 feet per second from a point that is 50 feet above the level of the water (see figure). The height h (in feet) of the ball at time t (in seconds) after it is thrown is

$$h = -16t^2 + 40t + 50.$$

(a) Find the time when the ball is again 50 feet above the water.

(b) Find the time when the ball strikes the water.

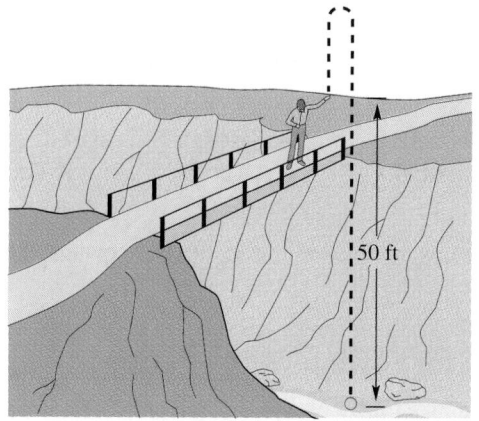

50 ft

88. *Number Problem*   Find two consecutive positive even integers whose product is 288.

89. *Analyzing Data*   Use the following model, which approximates the number of people employed in the aerospace industry in the United States from 1988 through 1994.

$$y = 795.36 - 32.11t - 7.82t^2, \quad -2 \le t \le 4$$

In this model, y represents the number employed in the aerospace industry (in thousands) and t represents the year, with $t = 0$ corresponding to 1990. *(Source: U.S. Department of Commerce)*

(a) Use a graphing utility to graph the model.

(b) Use the graph in part (a) to find the year in which there were approximately 750,000 employed in the aerospace industry in the United States.

(c) Use the model to predict the number employed in the aerospace industry in 1995.

90. *Analyzing Data*   The numbers of cellular phone subscribers (in millions) in the United States for the years 1989 through 1994 are given in the table. *(Source: Cellular Telecommunications Industry Association)*

Year	1989	1990	1991
Subscribers	3.51	5.28	7.56

Year	1992	1993	1994
Subscribers	11.03	16.01	24.13

(a) Create a bar graph for the data.

(b) The data can be approximated by the model

$$s = 0.76t^2 + 1.69t + 4.80, \quad -1 \le t \le 4,$$

where $t = 0$ corresponds to 1990. Use a graphing utility to graph the model.

(c) Use the model to determine the year in which the cellular phone companies had 10 million subscribers.

Section Project

Solutions of a Quadratic Equation   Determine the solutions x_1 and x_2 of each quadratic equation. Use the values of x_1 and x_2 to fill in the boxes.

Equation	x_1, x_2	$x_1 + x_2$	$x_1 x_2$
(a) $x^2 - x - 6 = 0$			
(b) $2x^2 + 5x - 3 = 0$			
(c) $4x^2 - 9 = 0$			
(d) $x^2 - 10x + 34 = 0$			

Consider a general quadratic equation $ax^2 + bx + c = 0$ whose solutions are x_1 and x_2. Can you determine a relationship among the coefficients a, b, and c and the sum $(x_1 + x_2)$ and product $(x_1 x_2)$ of the solutions?

6.4   Applications of Quadratic Equations

Review of Problem-Solving Strategy ■
Applications of Quadratic Equations

Review of Problem-Solving Strategy

You have completed your study of the technical parts of the algebra of quadratic equations. You have also seen some uses of quadratic equations in solving real-life problems. In this section, you will study other real-life problems involving quadratic equations. To construct these equations from verbal problems, you should use the problem-solving strategy developed in Chapter 3.

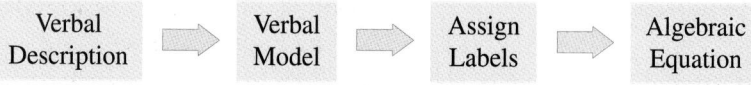

Verbal Description ⟹ Verbal Model ⟹ Assign Labels ⟹ Algebraic Equation

For your convenience, the steps involved in this strategy are reviewed here.

Strategy for Solving Word Problems

1. Ask yourself what you need to know to solve the problem. Then *write a verbal model* that will give you what you need to know.

2. *Assign labels* to each part of the verbal model—numbers to the known quantities and letters (or expressions) to the variable quantities.

3. Use the labels *to write an algebraic model* based on the verbal model.

4. *Solve* the resulting algebraic equation.

5. *Answer* the original question and *check* that your answer satisfies the original problem as stated.

For word problems that involve quadratic equations, the verbal model is often a *product* of two variable quantities. For instance, the area of a rectangular region is the product of the width and length of the rectangle.

Area of rectangle = Width of rectangle · Length of rectangle

On other occasions, the model may be a previously known formula that describes the situation. Watch for these variations in the examples that follow.

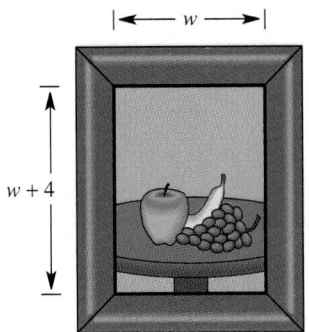

FIGURE 6.9

Point out that the equation has two algebraic solutions, $w = 12$ and $w = -16$. However, $w = -16$ is not a feasible solution to the application. The context of the problem will often dictate the domain.

EXAMPLE 1   Geometry

A picture is 4 inches taller than it is wide and has an area of 192 square inches. What are the dimensions of the picture?

Solution

Begin by drawing a diagram, as shown in Figure 6.9.

Verbal Model: | Area of picture | $=$ | Width | $\cdot$ | Height |

Labels: Picture width $= w$ (inches)
             Picture height $= w + 4$ (inches)
             Area $= 192$ (square inches)

Equation:
$$192 = w(w + 4)$$
$$0 = w^2 + 4w - 192$$
$$0 = (w + 16)(w - 12)$$
$$w + 16 = 0 \implies w = -16$$
$$w - 12 = 0 \implies w = 12$$

Of the two possible solutions, choose the positive value of w and conclude that the picture is $w = 12$ inches wide and $w + 4 = 16$ inches tall. Check these dimensions in the original statement of the problem.

EXAMPLE 2 An Interest Problem

The formula

$$A = P(1 + r)^2$$

represents the amount of money A in an account in which P dollars is deposited for 2 years at an annual interest rate of r (in decimal form). Find the interest rate if a deposit of $6000 increases to $6933.75 over a 2-year period.

Solution

$A = P(1 + r)^2$	Given formula
$6933.75 = 6000(1 + r)^2$	Substitute for A and P.
$1.155625 = (1 + r)^2$	Divide both sides by 6000.
$\pm 1.075 = 1 + r$	Extract square roots.
$0.075 = r$	Choose positive solution.

The annual interest rate is $r = 0.075 = 7.5\%$. Check this result in the original statement of the problem.

Applications of Quadratic Equations

EXAMPLE 3   *A Mathematical Puzzle*

Use algebra to solve the following puzzle.

> *The product of two consecutive positive integers is 10 more than 4 times the smaller integer.*

Solution

You can use the following verbal model.

*Verbal
Model:* $\boxed{\text{Smaller integer}} \cdot \boxed{\text{Larger integer}} = 10 + 4 \cdot \boxed{\text{Smaller integer}}$

Labels: Smaller integer $= n$
Larger integer $= n + 1$

Equation:
$$n(n + 1) = 10 + 4n$$
$$n^2 + n = 10 + 4n$$
$$n^2 - 3n - 10 = 0$$
$$(n - 5)(n + 2) = 0$$
$$n - 5 = 0 \implies n = 5$$
$$n + 2 = 0 \implies n = -2$$

Of the two possible solutions, choose the positive value of n and conclude that the two consecutive integers are $n = 5$ and $n + 1 = 6$. Check these values in the original statement of the problem as follows.

Check

The product of 5 and 6 is 30. Four times 5 is 20, and 10 more than this is 30. Thus, the product of the two integers is 10 more than 4 times the smaller integer.

Exploration

Partner Activity   You and your partner should each write a number puzzle similar to the one described in Example 3. Then, without showing the solutions, trade puzzles and use algebra to solve each other's puzzles. Once you have solved the puzzles, discuss with your partner other ways (that do not involve algebra) of solving them.

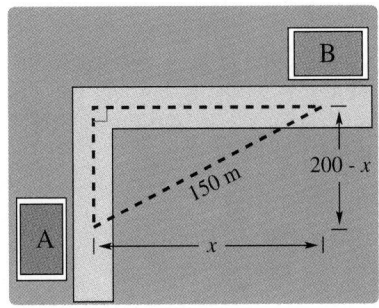

FIGURE 6.10

Remind students that a diagram or sketch is often helpful when solving applications. Also, they should ask themselves if the answers they get are reasonable.

EXAMPLE 4   An Application Involving the Pythagorean Theorem

An L-shaped sidewalk from building A to building B on a college campus is 200 meters long, as shown in Figure 6.10. By cutting diagonally across the grass, students shorten the walking distance to 150 meters. What are the lengths of the two legs of the existing sidewalk?

Solution

Verbal Model: $a^2 + b^2 = c^2$ Pythagorean Theorem

Labels:

Length of one leg $= x$ (meters)
Length of other leg $= 200 - x$ (meters)
Length of diagonal $= 150$ (meters)

Equation:
$$x^2 + (200 - x)^2 = (150)^2$$
$$x^2 + 40{,}000 - 400x + x^2 = 22{,}500$$
$$2x^2 - 400x + 40{,}000 = 22{,}500$$
$$2x^2 - 400x + 17{,}500 = 0$$
$$x^2 - 200x + 8750 = 0 \qquad a = 1, b = -200, c = 8750$$

By the Quadratic Formula, you can find the solutions as follows.

$$x = \frac{-(-200) \pm \sqrt{(-200)^2 - 4(1)(8750)}}{2(1)}$$

$$= \frac{200 \pm \sqrt{5000}}{2}$$

$$= \frac{200 \pm 50\sqrt{2}}{2}$$

$$= 100 \pm 25\sqrt{2}$$

Both solutions are positive, and it does not matter which one you choose. If you let

$$x = 100 + 25\sqrt{2} \approx 135.4 \text{ meters,}$$

the length of the other leg is

$$200 - x \approx 200 - 135.4 \approx 64.6 \text{ meters.}$$

NOTE   In Example 4, notice that you obtain the same dimensions if you choose the other value of x. That is, if the length of one leg is

$$x = 100 - 25\sqrt{2} \approx 64.6 \text{ meters,}$$

the length of the other leg is $200 - x \approx 200 - 64.6 \approx 135.4$ meters.

To help develop intuition about the results produced when numbers are substituted for t in the height equation, ask students to determine the height of the model rocket at $t = 1$ and $t = 4$.

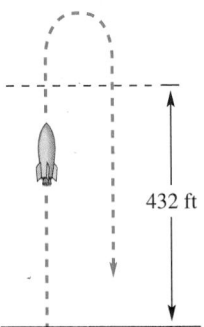

FIGURE 6.11

Technology

Use a graphing utility to estimate graphically the times when the rocket in Example 5 (a) is at a height of 432 feet, (b) reaches its maximum height, and (c) hits the ground.

EXAMPLE 5   The Height of a Model Rocket

A model rocket is projected straight upward from ground level according to the height equation $h = -16t^2 + 192t$, $t \geq 0$, where h is the height in feet and t is the time in seconds. (a) After how many seconds will the height be 432 feet? (b) After how many seconds will the rocket hit the ground?

Solution

(a)

$$h = -16t^2 + 192t \qquad \text{Original equation}$$
$$432 = -16t^2 + 192t \qquad \text{Substitute 432 for } h.$$
$$16t^2 - 192t + 432 = 0 \qquad \text{Standard form}$$
$$t^2 - 12t + 27 = 0 \qquad \text{Divide both sides by 16.}$$
$$(t - 3)(t - 9) = 0 \qquad \text{Factor.}$$
$$t - 3 = 0 \quad \Longrightarrow \quad t = 3 \qquad \text{Set 1st factor equal to 0.}$$
$$t - 9 = 0 \quad \Longrightarrow \quad t = 9 \qquad \text{Set 2nd factor equal to 0.}$$

The rocket obtains a height of 432 feet at two different times—once (going up) after 3 seconds, and again (coming down) after 9 seconds. (See Figure 6.11.)

(b) To find the time it takes for the rocket to hit the ground, let the height be 0.

$$0 = -16t^2 + 192t$$
$$0 = t^2 - 12t$$
$$0 = t(t - 12)$$
$$t = 0 \quad \text{or} \quad t = 12$$

The rocket will hit the ground after 12 seconds. (Note that the time of $t = 0$ seconds corresponds to the time of lift-off.)

Group Activities

Exploring with Technology

Analyzing Quadratic Functions   Use a graphing utility to graph $y_1 = 3x^2 + 2x - 1$ and $y_2 = -x^2 + 5x + 4$. For each function, use the zoom and trace features to find either the maximum or minimum function value, and the time at which it occurs. Discuss other methods that you could use to find these values.

6.4   **Exercises**

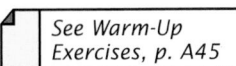
See Warm-Up Exercises, p. A45

1. *Unit Analysis*   Describe the units of the product.

$$\frac{9 \text{ dollars}}{\text{hour}} \cdot (20 \text{ hours})$$

2. *Unit Analysis*   Describe the units of the product.

$$\frac{20 \text{ feet}}{\text{minute}} \cdot \frac{1 \text{ minute}}{60 \text{ seconds}} \cdot (45 \text{ seconds})$$

Number Problems   In Exercises 3–6, find two positive integers that satisfy the requirement.

3. The product of two consecutive integers is 8 less than 10 times the smaller integer.

4. The product of two consecutive integers is 80 more than 15 times the larger integer.

5. The product of two consecutive even integers is 50 more than 3 times the larger integer.

6. The product of two consecutive odd integers is 22 less than 15 times the smaller integer.

Dimensions of a Rectangle   In Exercises 7–16, complete the table of widths, lengths, perimeters, and areas of rectangles.

	Width	Length	Perimeter	Area
7.	$0.75l$	l	42 in.	
8.	w	$1.5w$	40 m	
9.	w	$2.5w$		250 ft^2
10.	w	$1.5w$		216 cm^2
11.	$\frac{1}{3}l$	l		192 in.^{2}
12.	$\frac{3}{4}l$	l		2700 in.^{2}
13.	w	$w + 3$	54 km	
14.	$l - 6$	l	108 ft	
15.	$l - 20$	l		12,000 m^2
16.	w	$w + 5$		500 ft^2

17. *Lumber Storage Area*   A retail lumberyard plans to store lumber in a rectangular region adjoining the sales office (see figure). The region will be fenced on three sides and the fourth side will be bounded by the wall of the office building. Find the dimensions of the region if 350 feet of fencing is available and the area of the region is to be 12,500 square feet.

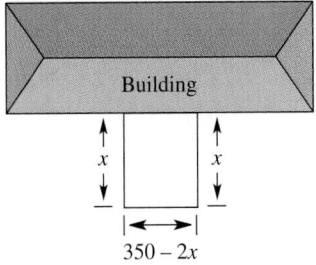

18. *Fencing the Yard*   You have 100 feet of fencing. Do you have enough to enclose a rectangular region whose area is 630 square feet? Is there enough to enclose a circular region of area 630 square feet? Explain.

19. *Fencing the Yard*   A family has built a fence around three sides of their property. In total, they used 550 feet of fencing. By their calculations, the lot is one acre (43,560 square feet). Is this correct? Explain your reasoning.

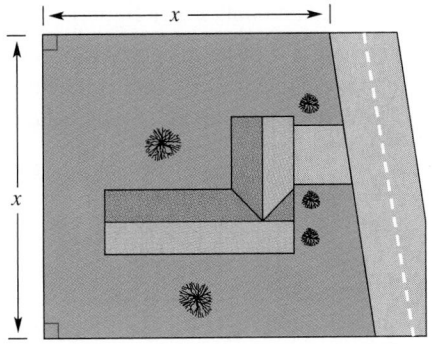

20. *Open Conduit*   An open-topped rectangular conduit for carrying water in a manufacturing process is made by folding up the edges of a sheet of aluminum 48 inches wide (see figure). A cross section of the conduit must have an area of 288 square inches. Find the width and height of the conduit.

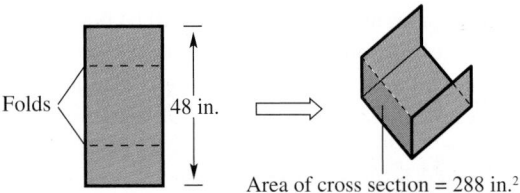

Folds   48 in.   Area of cross section = 288 in.²

21. *More Grass to Mow*   Your home is on a square lot. To add more space to your yard, you purchase an additional 20 feet along the side of your property (see figure). The area of the lot is now 25,500 square feet. What are the dimensions of the new lot?

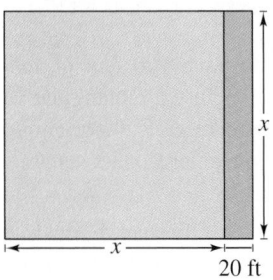

x

x 20 ft

22. *Solving Graphically and Algebraically*   An adjustable rectangular form has minimum dimensions of 3 meters by 4 meters. The length and width can be expanded by equal amounts x (see figure).

(a) Write the length d of the diagonal as a function of x. Use a graphing utility to graph the function. Use the graph to approximate the value of x when $d = 10$ meters.

(b) Find x algebraically when $d = 10$.

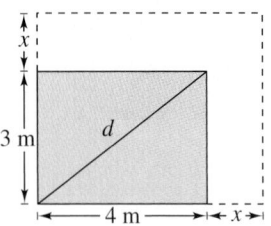

x

3 m d

4 m x

23. *Dimensions of a Rectangle*   The perimeter of a rectangle is 102 inches and the length of the diagonal is 39 inches. Find the dimensions of the rectangle.

24. *Solving Graphically and Numerically*   A meteorologist is positioned 100 feet from the point where a weather balloon is launched (see figure). The instrument package lifted vertically by the balloon transmits data to the meteorologist.

(a) Write the distance d between the balloon and the meteorologist as a function of the height h of the balloon. Use a graphing utility to graph the function. Use the graph to approximate the value of h when $d = 200$ feet.

(b) Complete the table.

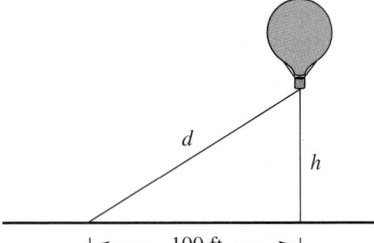

d h

| ← 100 ft → |

h	d
0	
100	
200	
300	

25. *Delivery Route*   You are delivering pizza to offices B and C in your city (see figure), and you are required to keep a log of the mileage between stops. You forget to look at the odometer at stop B, but after getting to stop C you record the total distance traveled from the pizza shop as 18 miles. The return distance from C to A is 16 miles. If the route approximates a right triangle, estimate the distance from A to B.

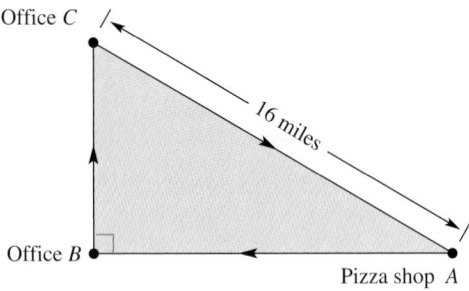

Office C

16 miles

Office B Pizza shop A

26. *Dimensions of a Triangle*   The height of a triangle is twice its base. The area of the triangle is 625 square inches. Find the dimensions of the triangle.

Compound Interest   In Exercises 27–32, find the interest rate r. Use the formula $A = P(1 + r)^2$, where A is the amount after 2 years in an account earning r percent (in decimal form) compounded annually, and P is the original investment.

27. $P = \$3000$
    $A = \$3499.20$

28. $P = \$10,000$
    $A = \$11,990.25$

29. $P = \$250$
    $A = \$280.90$

30. $P = \$500$
    $A = \$572.45$

31. $P = \$8000$
    $A = \$8420.20$

32. $P = \$6500$
    $A = \$7370.46$

Free-Falling Object   In Exercises 33–36, find the time necessary for an object to fall to ground level from an initial height of h_0 feet, if its height h at any time t (in seconds) is given by $h = h_0 - 16t^2$.

33. $h_0 = 144$

34. $h_0 = 625$

35. $h_0 = 1454$ (height of the Sears Tower)

36. $h_0 = 984$ (height of the Eiffel Tower)

37. *Height of a Baseball*   The height h in feet of a baseball hit 3 feet above the ground is given by

$$h = 3 + 75t - 16t^2,$$

where t is the time in seconds. Find the time when the ball hits the ground in the outfield.

38. *Hitting Baseballs*   You are hitting baseballs. When tossing the ball into the air, your hand is 5 feet above the ground (see figure). You hit the ball when it falls back to a height of 4 feet. If you toss the ball with an initial velocity of 25 feet per second, the height h of the ball t seconds after leaving your hand is given by

$$h = 5 + 25t - 16t^2.$$

How much time will pass before you hit the ball?

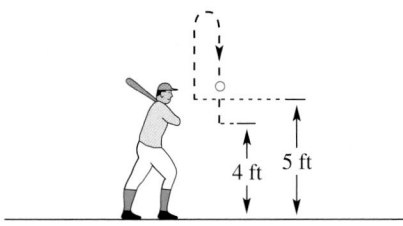

4 ft   5 ft

In Exercises 39 and 40, you are given the cost C for producing x units, the revenue R for selling x units, and the profit P. Find the value of x that will produce the profit P.

39. $C = 100 + 30x$
    $R = x(90 - x)$
    $P = \$800$

40. $C = 4000 - 40x + 0.02x^2$
    $R = x(50 - 0.01x)$
    $P = \$63,500$

41. *Broadcast Range*   A television station claims that it covers a circular region of approximately 25,000 square miles. Assume that the station is located at the center of the circular region. How far is the station from its farthest listener?

Section Project

Geometry   The area A of an ellipse is given by the equation $A = \pi ab$ (see figure). For a certain ellipse it is required that $a + b = 20$.

(a) Show that $A = \pi a(20 - a)$.

(b) Complete the table.

a	4	7	10	13	16
A					

(c) Find two values of a such that $A = 300$.

(d) Use a graphing utility to graph the area function. Then use the graph to verify the results in part (c).

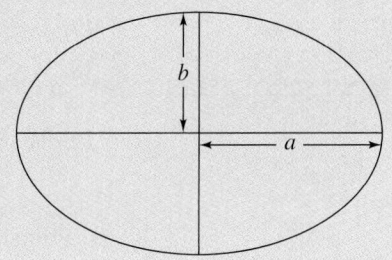

Mid-Chapter Quiz

Take this quiz as you would take a quiz in class. After you are done, check your work against the answers given in the back of the book.

In Exercises 1–8, solve the quadratic equation by the specified method.

1. Factor: $2x^2 - 72 = 0$.

2. Factor: $2x^2 + 3x - 20 = 0$.

3. Extract square roots: $t^2 = 12$.

4. Extract square roots: $(u - 3)^2 - 16 = 0$.

5. Complete the square: $s^2 + 10s + 1 = 0$.

6. Complete the square: $2y^2 + 6y - 5 = 0$.

7. Quadratic Formula: $x^2 + 4x - 6 = 0$

8. Quadratic Formula: $6v^2 - 3v - 4 = 0$

In Exercises 9–16, solve the equation by using the most convenient method. (Find all real *and* imaginary solutions.)

9. $x^2 + 5x + 7 = 0$

10. $36 - (t - 4)^2 = 0$

11. $x(x - 10) + 3(x - 10) = 0$

12. $x(x - 3) = 10$

13. $4b^2 - 12b + 9 = 0$

14. $3m^2 + 10m + 5 = 0$

15. $x^4 + 5x^2 - 14 = 0$

16. $x^{2/3} - 8x^{1/3} + 15 = 0$

In Exercises 17 and 18, use a graphing utility to graph the function. Use the graph to approximate any *x*-intercepts. Set $y = 0$ and solve the resulting equation. Compare the result with the *x*-intercepts of the graph.

17. $y = \frac{1}{2}x^2 - 3x - 1$

18. $y = x^2 + 0.45x - 4$

19. The revenue R from selling x units of a certain product is given by

$$R = x(20 - 0.2x).$$

Find the number of units that must be sold to produce a revenue of $500.

20. The perimeter of a rectangle is 200 meters. Its area A is given by $A = x(100 - x)$ (see figure). Determine the dimensions of the rectangle if its area is 2275 square meters.

x

$100 - x$

Figure for 20

6.5	**Graphing Quadratic Functions**

Graphs of Quadratic Functions ▪ Sketching a Parabola ▪ Finding the Equation of a Parabola ▪ Application

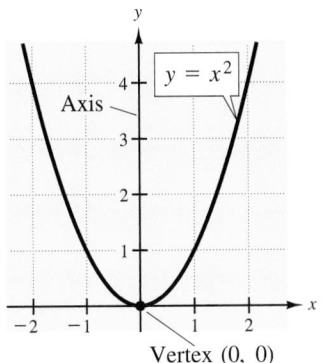

FIGURE 6.12

Historical Note: The Greek mathematician Apollonius of Perga (third century B.C.) was also called the "Great Geometer." He is credited with coining the term *parabola*.

Graphs of Quadratic Functions

In this section, you will study the graphs of quadratic functions of the form

$$f(x) = ax^2 + bx + c. \qquad \text{Quadratic function}$$

Figure 6.12 shows the graph of a simple quadratic function, $y = x^2$. This graph is an example of a **parabola.** The lowest point on this graph, $(0, 0)$, is the **vertex** of the parabola, and the vertical line that passes through the vertex (the y-axis, in this case) is the **axis** of the parabola. Every parabola is *symmetric* about its axis, which means that if it were folded along its axis, the two parts would match.

Equations of Parabolas

The graph of $y = ax^2 + bx + c$, $a \neq 0$, is a **parabola.** The completed square form of the equation

$$y = a(x - h)^2 + k \qquad \text{Standard form}$$

is called the **standard form** of the equation. The **vertex** of the parabola occurs at the point (h, k), and the vertical line passing through the vertex is the **axis** of the parabola.

Parabolas that correspond to equations of the form $y = ax^2 + bx + c$ open *upward* if the leading coefficient a is positive, and open *downward* if the leading coefficient is negative, as shown in Figure 6.13.

You might ask students if they know of objects with this shape. Supplement their answers with radar and satellite dishes, car headlights, and bridge cables.

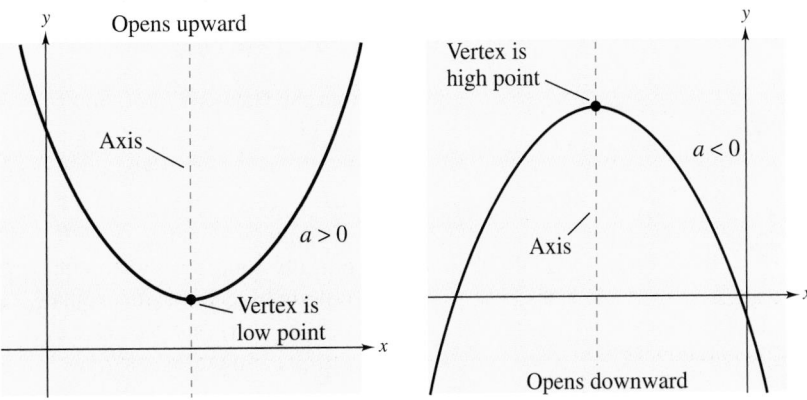

FIGURE 6.13

EXAMPLE 1   Finding the Vertex of a Parabola

Find the vertex of the parabola given by the equation $y = x^2 - 6x + 5$.

Solution

Begin by writing the equation in standard form.

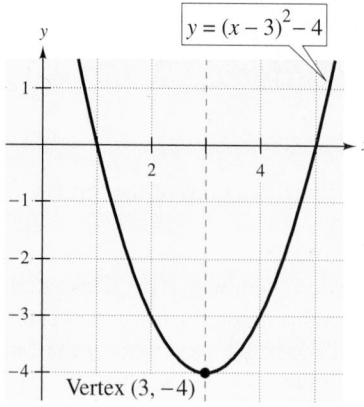

$y = (x - 3)^2 - 4$

$y = x^2 - 6x + 5$	Original equation
$y = x^2 - 6x + (-3)^2 - (-3)^2 + 5$	Complete the square.

half of -6

$y = (x^2 - 6x + 9) - 9 + 5$	Regroup terms.
$y = (x - 3)^2 - 4$	Standard form

Now, from the standard form, you can see that the vertex of the parabola occurs at the point $(3, -4)$, as shown in Figure 6.14.

Vertex $(3, -4)$

FIGURE 6.14

EXAMPLE 2   Finding the Vertex of a Parabola

Find the vertex of the parabola given by the equation $f(x) = x^2 + x$.

Solution

Begin by writing the equation in standard form.

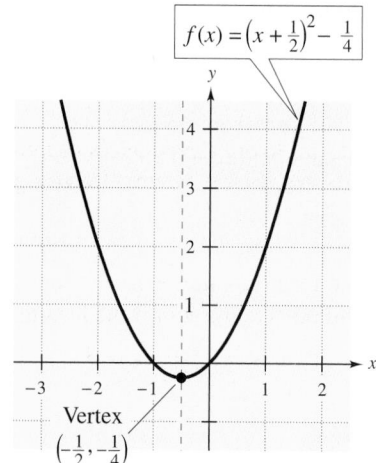

$f(x) = \left(x + \frac{1}{2}\right)^2 - \frac{1}{4}$

$f(x) = x^2 + x$	Original equation
$f(x) = x^2 + x + \left(\frac{1}{2}\right)^2 - \left(\frac{1}{2}\right)^2$	Complete the square.

half of 1

$f(x) = \left(x^2 + x + \frac{1}{4}\right) - \frac{1}{4}$	Regroup terms.
$f(x) = \left(x + \frac{1}{2}\right)^2 - \frac{1}{4}$	Standard form

Now, from the standard form, you can see that the vertex of the parabola occurs at the point $\left(-\frac{1}{2}, -\frac{1}{4}\right)$, as shown in Figure 6.15.

Vertex $\left(-\frac{1}{2}, -\frac{1}{4}\right)$

FIGURE 6.15

NOTE In Figure 6.15, the graph of f can be obtained by shifting the graph of $y = x^2$ to the left $\frac{1}{2}$ unit and down $\frac{1}{4}$ unit, as discussed in Section 2.6.

In Examples 1 and 2, you found the vertex of the given parabola by completing the square *for each* equation. Another technique you could use for this purpose is to complete the square once for the general equation, as follows.

$$y = a\left(x + \frac{b}{2a}\right)^2 + c - \frac{b^2}{4a}$$

From this form, you can see that the vertex occurs when $x = -b/2a$. You can use this method to verify the vertices in Examples 1 and 2.

$$y = x^2 - 6x + 5 \qquad\qquad f(x) = x^2 + x$$
$$x = -b/2a = -(-6)/2(1) = 3 \qquad x = -b/2a = -(1)/2(1) = -1/2$$
$$y = (3)^2 - 6(3) + 5 = -4 \qquad y = (-1/2)^2 + (-1/2) = -1/4$$
$$\text{Vertex: } (3, -4) \qquad\qquad \text{Vertex: } (-1/2, -1/4)$$

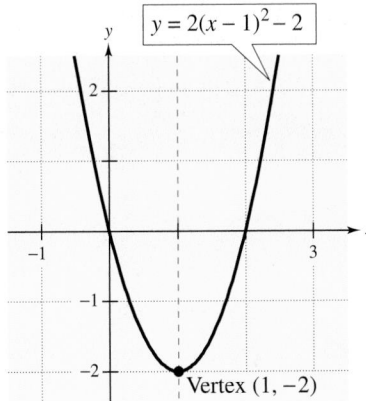

$y = 2(x-1)^2 - 2$

Vertex $(1, -2)$

FIGURE 6.16

Discuss standard form and the signs of the coordinates of the vertex.

EXAMPLE 3   *Finding the Vertex of a Parabola*

Find the vertex of the parabola given by the equation $y = 2x^2 - 4x$.

Solution

Begin by writing the equation in standard form.

$$y = 2x^2 - 4x \qquad\qquad \text{Original equation}$$
$$y = 2(x^2 - 2x) \qquad\qquad \text{Factor out leading coefficient.}$$
$$y = 2[x^2 - 2x + (-1)^2 - (-1)^2] \qquad \text{Complete the square.}$$

half of -2

$$y = 2(x^2 - 2x + 1) - 2(1) \qquad \text{Regroup terms.}$$
$$y = 2(x - 1)^2 - 2 \qquad\qquad \text{Standard form}$$

Now, from the standard form, you can see that the vertex of the parabola occurs at the point $(1, -2)$, as shown in Figure 6.16. Verify the vertex by using $x = -b/2a$.

EXAMPLE 4   *Finding the Vertex of a Parabola*

Find the vertex of the parabola given by the equation $y = -x^2 - 4x - 3$.

Solution

$$y = -x^2 - 4x - 3 \qquad\qquad \text{Original equation}$$
$$y = -1(x^2 + 4x) - 3 \qquad\qquad \text{Factor out leading coefficient.}$$
$$y = -(x^2 + 4x + 2^2 - 2^2) - 3 \qquad \text{Complete the square.}$$

half of 4

$$y = -(x^2 + 4x + 4) + 2^2 - 3 \qquad \text{Regroup terms.}$$
$$y = -(x + 2)^2 + 1 \qquad\qquad \text{Standard form}$$

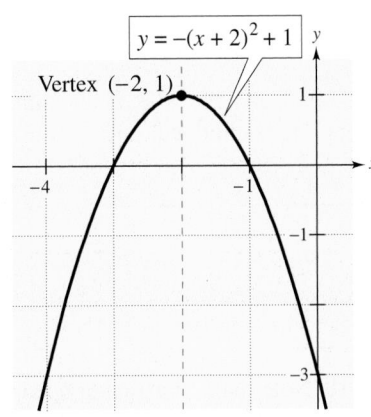

$y = -(x+2)^2 + 1$

Vertex $(-2, 1)$

FIGURE 6.17

Now, from the standard form, you can see that the vertex of the parabola occurs at the point $(-2, 1)$, as shown in Figure 6.17. Notice that the parabola opens downward because the leading coefficient is negative.

Sketching a Parabola

> ### Sketching a Parabola
>
> 1. Write the given equation in the standard form $y = a(x - h)^2 + k$, and determine the vertex and axis of the parabola.
>
> 2. Plot the vertex, axis, and a few additional points of the parabola. (Using the symmetry about the axis can reduce the number of points you need to plot.)
>
> 3. Use the fact that the parabola opens *upward* if $a > 0$ and opens *downward* if $a < 0$ to complete the sketch.

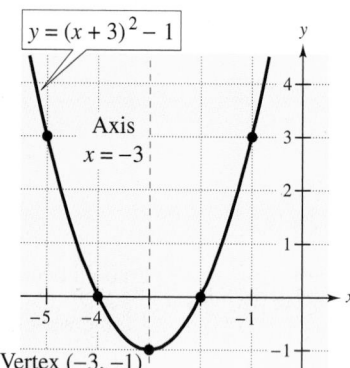

FIGURE 6.18

EXAMPLE 5   Sketching a Parabola

Sketch the graph of the parabola given by the second-degree equation

$$x^2 - y + 6x + 8 = 0.$$

Solution

Begin by writing the equation in standard form.

$$x^2 - y + 6x + 8 = 0 \qquad \text{Original equation}$$
$$-y = -x^2 - 6x - 8$$
$$y = x^2 + 6x + 8$$
$$y = (x^2 + 6x + 3^2 - 3^2) + 8$$

half of 6

$$y = (x^2 + 6x + 9) - 9 + 8$$
$$y = (x + 3)^2 - 1 \qquad \text{Standard form}$$

Therefore, the vertex occurs at the point $(-3, -1)$ and the axis is the line $x = -3$. After plotting this information, calculate a few additional points of the parabola, as shown in the table.

NOTE In Example 5, two of the additional points chosen for the table, $(-4, 0)$ and $(-2, 0)$, are the x-intercepts for this graph. (Remember that you can find any x-intercepts by setting y equal to zero and solving for x. See Example 5 in Section 2.2.)

x	-5	-4	-3	-2	-1
$y = (x + 3)^2 - 1$	3	0	-1	0	3
Solution point	$(-5, 3)$	$(-4, 0)$	$(-3, -1)$	$(-2, 0)$	$(-1, 3)$

The graph of the parabola is shown in Figure 6.18. Note that the parabola opens upward because the leading coefficient (in the standard form) is positive.

EXAMPLE 6   Sketching a Parabola

Sketch the graph of the parabola given by the second-degree equation

$$x^2 + 2y - 4x + 8 = 0.$$

Solution

Begin by writing the equation in standard form.

$$x^2 + 2y - 4x + 8 = 0 \qquad \text{Original equation}$$

$$2y = -x^2 + 4x - 8$$

$$y = -\frac{1}{2}x^2 + 2x - 4$$

$$y = -\frac{1}{2}(x^2 - 4x) - 4$$

$$y = -\frac{1}{2}[x^2 - 4x + (-2)^2 - (-2)^2] - 4$$

half of -4

$$y = -\frac{1}{2}(x^2 - 4x + 4) + 2 - 4$$

$$y = -\frac{1}{2}(x - 2)^2 - 2 \qquad \text{Standard form}$$

Remind students of writing equations of vertical lines in Section 3.1, and relate this to the equation of the axis.

Therefore, the vertex occurs at the point $(2, -2)$ and the axis is the line $x = 2$. After plotting this information, calculate a few additional points of the parabola, as shown in the table.

x	0	1	2	3	4
$y = -\frac{1}{2}(x - 2)^2 - 2$	-4	$-\frac{5}{2}$	-2	$-\frac{5}{2}$	-4
Solution point	$(0, -4)$	$\left(1, -\frac{5}{2}\right)$	$(2, -2)$	$\left(3, -\frac{5}{2}\right)$	$(4, -4)$

The graph of the parabola is shown in Figure 6.19 at the left. Note that the parabola opens downward because the leading coefficient (in the standard form) is negative.

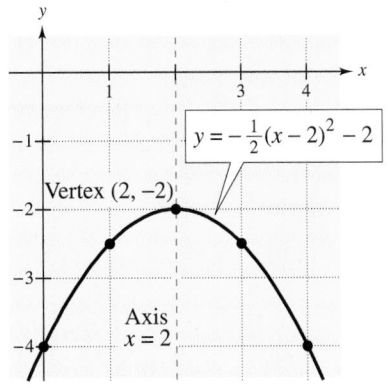

$y = -\frac{1}{2}(x - 2)^2 - 2$

Vertex $(2, -2)$

Axis $x = 2$

FIGURE 6.19

NOTE   In Example 6, from the fifth line of the solution to the sixth line, notice the following.

$$-\frac{1}{2}[-(-2)^2] = -\frac{1}{2}(-4) = 2$$

Finding the Equation of a Parabola

To write the equation of a parabola with a vertical axis, use the fact that its standard equation has the form $y = a(x - h)^2 + k$, where (h, k) is the vertex.

EXAMPLE 7   Finding the Equation of a Parabola

Find the equation of the parabola that has a vertex of $(-2, 1)$ and a y-intercept of $(0, -3)$, as shown in Figure 6.20.

Solution

Because the vertex occurs at $(h, k) = (-2, 1)$, you can write the following.

$$y = a(x - h)^2 + k \qquad \text{Standard form}$$
$$y = a[x - (-2)]^2 + 1$$
$$y = a(x + 2)^2 + 1$$

Now, to find the value of a, use the fact that the y-intercept is $(0, -3)$.

$$y = a(x + 2)^2 + 1 \qquad \text{Standard form}$$
$$-3 = a(0 + 2)^2 + 1 \qquad \text{Substitute } x = 0 \text{ and } y = -3.$$
$$-3 = 4a + 1 \qquad \text{Simplify.}$$
$$-4 = 4a \qquad \text{Subtract 1 from both sides.}$$
$$-1 = a \qquad \text{Divide both sides by 4.}$$

So, the standard form of the equation of the parabola is

$$y = -(x + 2)^2 + 1.$$

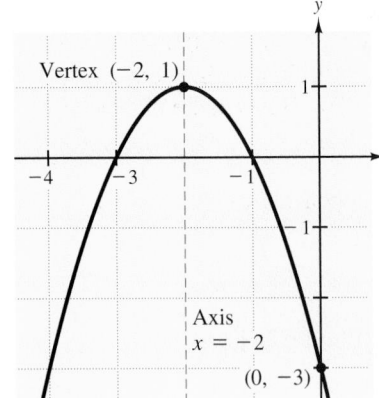

FIGURE 6.20

Earlier it was noted that ". . . if a point is on the line, then it satisfies the equation of that line . . . " Here a similar idea may help students understand the conceptual steps necessary to build an equation of a parabola from some given information.

EXAMPLE 8   Finding the Equation of a Parabola

Find the equation of the parabola that has a vertex of $(0, -4)$ and passes through the point $(3, 1)$, as shown in Figure 6.21.

Solution

Because the vertex occurs at $(h, k) = (0, -4)$, you can write the following.

$$y = a(x - h)^2 + k \qquad \text{Standard form}$$
$$y = a(x - 0)^2 - 4$$
$$y = ax^2 - 4$$

To find the value of a, use the fact that the parabola passes through the point $(3, 1)$. You should get a value of $a = \frac{5}{9}$. Finally, you can conclude that the standard form of the equation of the parabola is $y = \frac{5}{9}x^2 - 4$.

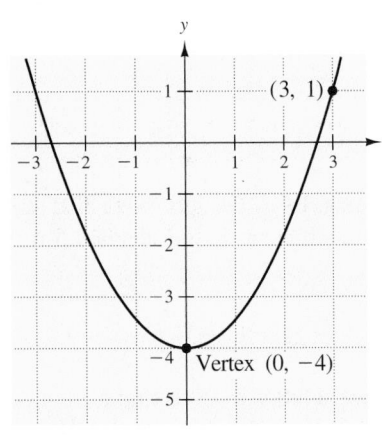

FIGURE 6.21

Application

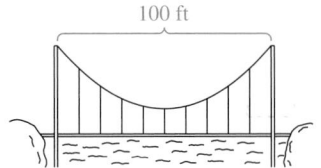

FIGURE 6.22

EXAMPLE 9   An Application Involving a Parabola

A suspension bridge is 100 feet long, as shown in Figure 6.22. The bridge is supported by cables attached to the tops of the towers at each end of the bridge. Each cable hangs in the shape of a parabola given by $y = 0.01x^2 - x + 35$, where x and y are both measured in feet. (a) Find the distance between the lowest point of the cable and the roadbed of the bridge. (b) How tall are the towers?

Solution

(a) By writing the equation of the parabola in standard form,

$$y = 0.01x^2 - x + 35$$
$$y = 0.01(x^2 - 100x) + 35$$
$$y = 0.01[x^2 - 100x + (-50)^2 - (-50)^2] + 35$$
$$y = 0.01(x^2 - 100x + 2500) - 25 + 35$$
$$y = 0.01(x - 50)^2 + 10,$$

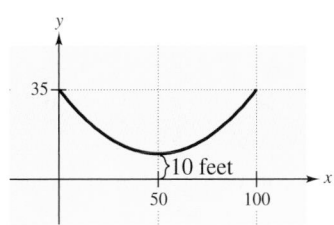

FIGURE 6.23

you can see that the vertex occurs at the point (50, 10). The minimum distance between the cable and the roadbed is 10 feet (see Figure 6.23).

(b) Because the vertex of the parabola occurs at the midpoint of the bridge, the two towers are located at the points where $x = 0$ and $x = 100$. Substituting $x = 0$, you can determine that the corresponding y-value is

$$y = 0.01(0^2) - 0 + 35 = 35 \text{ feet.}$$

Therefore, the towers are each 35 feet tall. (Try substituting $x = 100$ into the equation to see that you obtain the same y-value.)

Group Activities   **Problem Solving**

Quadratic Modeling   The data in the table represents the average monthly temperature y in degrees Fahrenheit in Savannah, Georgia, for the month x, with $x = 1$ corresponding to November. *(Source: National Climate Data Center)*   Plot the data. Find a quadratic model for the data and use it to find the average temperatures for December and February. The actual average temperature for both December and February is 52°F. How well do you think the model fits the data? Use the model to predict the average temperature for June. How useful do you think the model would be for the whole year?

x	1	3	5
y	59	49	59

6.5   **Exercises**

See Warm-Up
Exercises, p. A45

In Exercises 1–6, match the equation with its graph.
[The graphs are labeled (a), (b), (c), (d), (e), and (f).]

(a)
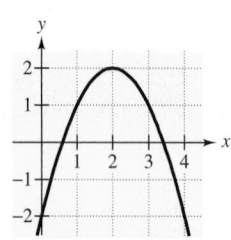

(b)
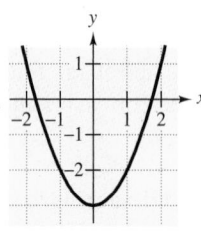

(c)
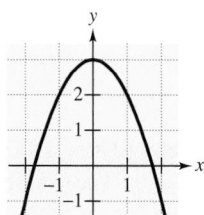

(d)
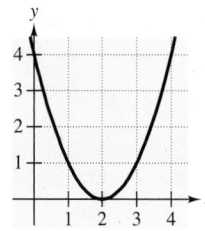

(e)
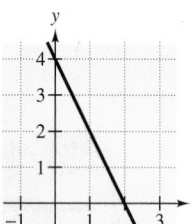

(f)
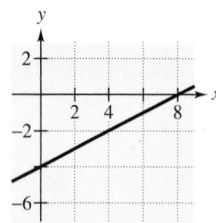

1. $y = 4 - 2x$

2. $y = \frac{1}{2}x - 4$

3. $y = x^2 - 3$

4. $y = -x^2 + 3$

5. $y = (x - 2)^2$

6. $y = 2 - (x - 2)^2$

7. *Think About It*   Explain how to determine whether the graph of a quadratic function opens upward or downward.

8. *Think About It*   Is it possible for the graph of a quadratic function to have two y-intercepts? Explain.

In Exercises 9–16, state whether the graph opens upward or downward and find the vertex.

9. $y = 2(x - 0)^2 + 2$

10. $y = -3(x + 5)^2 - 3$

11. $y = 4 - (x - 10)^2$

12. $y = 2(x - 12)^2 + 3$

13. $y = x^2 - 6$

14. $y = -(x + 1)^2$

15. $y = -(x - 3)^2$

16. $y = x^2 - 6x$

In Exercises 17–24, find the x- and y-intercepts of the graph.

17. $y = 25 - x^2$

18. $y = x^2 - 49$

19. $y = x^2 - 9x$

20. $y = x^2 + 4x$

21. $y = 4x^2 - 12x + 9$

22. $y = 10 - x - 2x^2$

23. $y = x^2 - 3x + 3$

24. $y = x^2 - 3x - 10$

In Exercises 25–36, write the equation in standard form and find the vertex of its graph.

25. $f(x) = x^2 + 2$

26. $g(x) = x^2 + 2x$

27. $y = x^2 - 4x + 7$

28. $y = x^2 + 6x - 5$

29. $y = x^2 + 6x + 5$

30. $y = x^2 - 4x + 5$

31. $y = -x^2 + 6x - 10$

32. $y = 4 - 8x - x^2$

33. $y = -x^2 + 2x - 7$

34. $y = -x^2 - 10x + 10$

35. $y = 2x^2 + 6x + 2$

36. $y = 3x^2 - 3x - 9$

In Exercises 37–60, sketch the graph of the equation. Identify the vertex and any x-intercepts. Use a graphing utility to verify your results.

37. $y = x^2 - 4$

38. $y = x^2 - 9$

39. $y = -x^2 + 4$

40. $y = -x^2 + 9$

41. $y = x^2 - 3x$

42. $y = x^2 - 4x$

43. $y = -x^2 + 3x$

44. $y = -x^2 + 4x$

45. $y = (x - 4)^2$

46. $y = -(x + 4)^2$

47. $y = x^2 - 8x + 15$

48. $y = x^2 + 4x + 2$

49. $y = -(x^2 + 6x + 5)$

50. $y = -x^2 + 2x + 8$

51. $y = -x^2 + 6x - 7$

52. $y = x^2 + 4x + 7$

53. $y = 2(x^2 + 6x + 8)$

54. $y = 3x^2 - 6x + 4$

55. $y = \frac{1}{2}(x^2 - 2x - 3)$ **56.** $y = -\frac{1}{2}(x^2 - 6x + 7)$

57. $y = 5 - \frac{x^2}{3}$ **58.** $y = \frac{x^2}{3} - 2$

59. $y = \frac{1}{5}(3x^2 - 24x + 38)$

60. $y = \frac{1}{5}(2x^2 - 4x + 7)$

In Exercises 61–64, use a graphing utility to approximate the vertex of the graph. Check the result algebraically.

61. $y = \frac{1}{6}(2x^2 - 8x + 11)$

62. $y = -\frac{1}{4}(4x^2 - 20x + 13)$

63. $y = -0.7x^2 - 2.7x + 2.3$

64. $y = 0.75x^2 - 7.50x + 23.00$

In Exercises 65–68, use a graphing utility to graph the two functions in the same viewing rectangle. Do the graphs intersect? If so, approximate the point(s) of intersection.

65. $y_1 = -x^2 + 6$ **66.** $y_1 = x^2 - 6x + 8$

    $y_2 = 2$ $y_2 = 3$

67. $y_1 = \frac{1}{2}x^2 - 3x + \frac{13}{2}$ **68.** $y_1 = -2x^2 - 4x$

    $y_2 = 3$ $y_2 = 1$

In Exercises 69–74, write an equation of the parabola.

69.
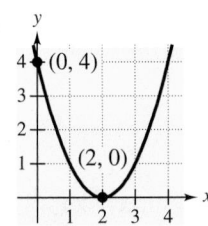

70.
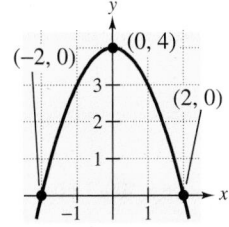

71.
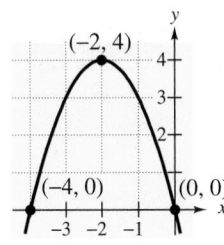

72.
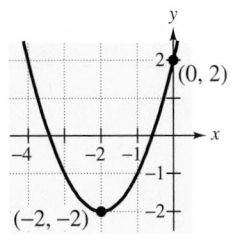

73.
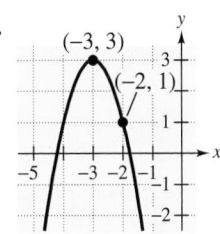

74.
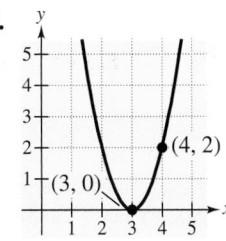

In Exercises 75–86, write an equation of the parabola

$$y = ax^2 + bx + c$$

that satisfies the conditions.

75. Vertex: $(2, 1)$; $a = 1$

76. Vertex: $(-3, -3)$; $a = 1$

77. Vertex: $(-3, 4)$; $a = -1$

78. Vertex: $(3, -2)$; $a = -1$

79. Vertex: $(2, -4)$;   Point on graph: $(0, 0)$

80. Vertex: $(-2, -4)$;   Point on graph: $(0, 0)$

81. Vertex: $(3, 2)$;   Point on graph: $(1, 4)$

82. Vertex: $(-1, -1)$;   Point on graph: $(0, 4)$

83. Vertex: $(-1, 5)$;   Point on graph: $(0, 1)$

84. Vertex: $(5, 10)$;   Point on graph: $(6, 6)$

85. Vertex: $(5, 2)$;   Point on graph: $(10, 3)$

86. Vertex: $(0, 20)$;   Point on graph: $(10, 15)$

87. *Path of a Ball* The height y (in feet) of a ball thrown by a child is given by

$$y = -\frac{1}{12}x^2 + 2x + 4,$$

where x is the horizontal distance (in feet) from where the ball was thrown.

(a) Use a graphing utility to graph the path of the ball.

(b) How high was the ball when it left the child's hand?

(c) How high was the ball when it reached its maximum height?

(d) How far from the child did the ball strike the ground?

88. *Maximum Height of a Diver*   The path of a diver is given by

$$y = -\frac{4}{9}x^2 + \frac{24}{9}x + 10,$$

where y is the height in feet and x is the horizontal distance from the end of the diving board in feet (see figure). What is the maximum height of the diver?

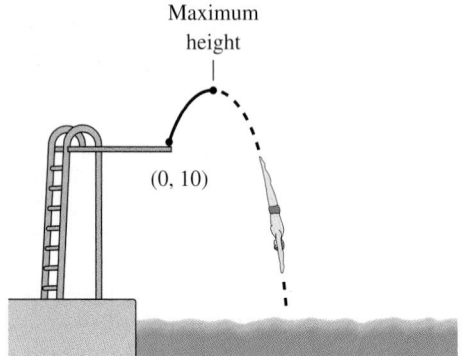

Maximum height

(0, 10)

89. *Graphical Estimation*   The profit (in thousands of dollars) for a company is given by

$$P = 230 + 20s - \frac{1}{2}s^2,$$

where s is the amount (in hundreds of dollars) spent on advertising. Use a graphing utility to graph the profit function and approximate the amount of advertising that yields a maximum profit. Verify the maximum algebraically.

90. *Cost*   The cost of producing x units of a product is given by

$$C = 800 - 10x + \frac{1}{4}x^2, \quad 0 < x < 40.$$

Use a graphing utility to graph the cost function and approximate the value of x when C is minimum.

91. *Area*   The area of a rectangle is given by the equation

$$A = \frac{2}{\pi}(100x - x^2), \quad 0 < x < 100,$$

where x is the length of the rectangle's base in feet. Use a graphing utility to graph the area function and approximate the value of x when A is maximum.

92. *Graphical Interpretation*   The income from state and local governments for public broadcasting in the United States from 1989 through 1993 is approximated by the model

$$I = 480.54 + 20.59t - 7.64t^2.$$

In this model, I is the income in millions of dollars and t is the time in years, with $t = 0$ corresponding to 1990.   *(Source: Corporation for Public Broadcasting)*

(a) Use a graphing utility to graph the income function.

(b) Determine the year when income was greatest. Approximate the income for that year.

93. *Highway Design*   A highway department engineer must design a parabolic arc to create a turn in a freeway around a city. The vertex of the parabola is placed at the origin and must connect with roads represented by the equations

$$y = -0.4x - 10, \quad x < -500$$
$$y = 0.4x - 100, \quad x > 500$$

(see figure). Find an equation for the parabolic arc.

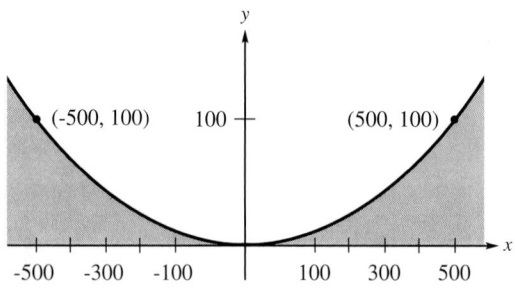

(-500, 100)   100   (500, 100)

-500   -300   -100   100   300   500

94. *Bridge Design*   A bridge is to be constructed over a gorge with the main supporting arch being a parabola (see figure). The equation of the parabola is

$$y = 4\left(100 - \frac{x^2}{2500}\right),$$

where x and y are measured in feet.

(a) Find the length of the road across the gorge.

(b) Find the height of the parabolic arch at the center of the span.

(c) Find the lengths of the vertical girders at intervals of 100 feet from the center of the bridge.

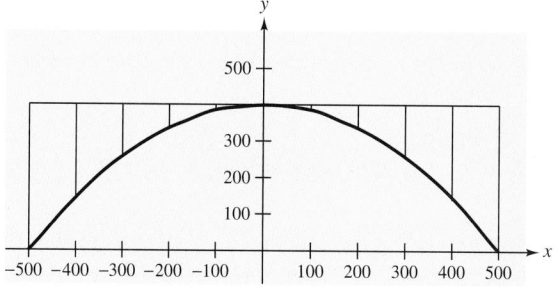

Section Project

Maximum Profit   A company manufactures radios that cost (the company) $60 each. For buyers who purchase 100 or fewer radios, the purchase price is $90 per radio. To encourage large orders, the company will reduce the price *per radio* for orders over 100, as follows. If 101 radios are purchased, the price is $89.85 per radio. If 102 radios are purchased, the price is $89.70 per radio. If $(100 + x)$ radios are purchased, the price per unit is $p = 90 - x(0.15)$, where x is the amount over 100 in the order.

(a) Show that the profit P is given by

$$P = (100 + x)[90 - x(0.15)] - (100 + x)60$$

$$= 3000 + 15x - \frac{3}{20}x^2.$$

(b) Use a graphing utility to graph the profit function.

(c) Find the vertex of the profit curve and determine the order size for maximum profit.

(d) Would you recommend this pricing scheme? Explain your reasoning.

6.6	**Modeling Data with Quadratic Functions**
	Classifying Scatter Plots ▪ Fitting Quadratic Models to Data ▪ Application

Classifying Scatter Plots

In Section 3.2, you saw how to fit linear models to data. In real life, many relationships between two variables are nonlinear. A scatter plot can be used to give you an idea of which type of model can be used to best fit a set of data.

EXAMPLE 1   *Classifying Scatter Plots*

Decide whether the data could be better modeled by a linear model $y = ax + b$ or a quadratic model $y = ax^2 + bx + c$.

(a) (0.2, 2.6), (0.4, 2.3), (0.6, 2.1), (0.8, 2.0), (1.1, 2.1), (1.3, 2.1), (1.4, 2.2), (1.7, 2.5), (1.9, 2.8), (2.2, 3.3), (2.3, 3.8), (2.6, 4.4), (2.7, 5.1), (3.0, 5.9), (3.2, 6.8), (3.4, 7.8)

(b) (0.2, 4.4), (0.4, 4.9), (0.5, 5.2), (0.7, 5.7), (1.1, 6.1), (1.2, 6.5), (1.5, 7.0), (1.7, 7.4), (1.9, 7.8), (2.0, 8.2), (2.3, 8.7), (2.5, 9.0), (2.8, 9.5), (2.9, 9.9), (3.2, 10.4), (3.4, 10.8)

Solution

Begin by entering the data into a graphing utility. You should obtain the scatter plots shown in Figure 6.24.

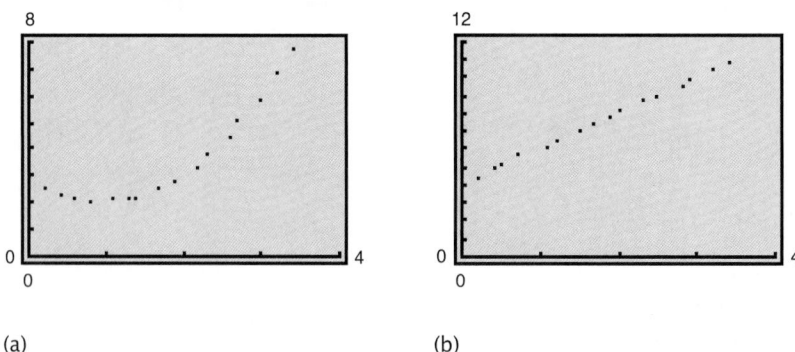

(a)   (b)

FIGURE 6.24

From the scatter plots, it appears that the data in part (a) can be modeled by a quadratic function and the data in part (b) can be modeled by a linear function.

Fitting Quadratic Models to Data

Once you have used a scatter plot to determine the type of model to be fit to a set of data, there are several ways that you can actually find the model. Each method is best used with a computer or calculator, rather than with hand calculations.

EXAMPLE 2   *Fitting a Model to Data*

Fit the following data, from Example 1(a), with a linear model and a quadratic model. Which model do you think fits better?

> (0.2, 2.6), (0.4, 2.3), (0.6, 2.1), (0.8, 2.0), (1.1, 2.1), (1.3, 2.1), (1.4, 2.2), (1.7, 2.5), (1.9, 2.8), (2.2, 3.3), (2.3, 3.8), (2.6, 4.4), (2.7, 5.1), (3.0, 5.9), (3.2, 6.8), (3.4, 7.8)

Solution

Begin by entering the data into a calculator or computer that has least squares regression programs. Then run the regression programs for linear and quadratic models.

Linear: $y = ax + b$ $a = 1.601$ $b = 0.730$

Quadratic: $y = ax^2 + bx + c$ $a = 0.983$ $b = -1.932$ $c = 2.939$

From the graphs in Figure 6.25, you can see that the quadratic model fits better.

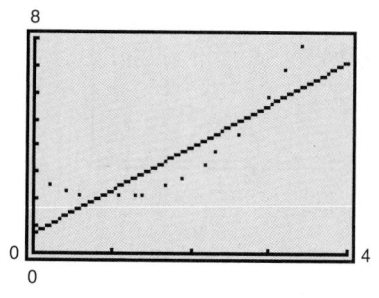
Linear

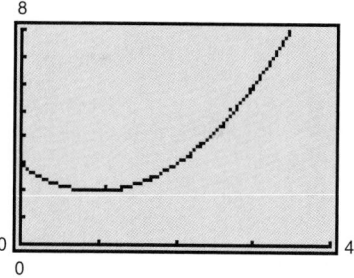
Quadratic

FIGURE 6.25

Students will enter the data into a calculator or computer as they did in Section 3.2.

Technology

Most graphing utilities have built-in programs that will calculate the equation of the best-fitting quadratic for a collection of points. For instance, to find the best-fitting quadratic on a *TI-82* or *TI-83*, enter the *x*- and *y*-values with the (STAT) (Edit) menu so that the *x*-values are in L1 and the *y*-values are in L2. Then, from the (STAT) (Calc) menu, choose "QuadReg." (See the Technology feature on page 200 for an example of fitting a linear model to data.)

NOTE   Deciding which model best fits a set of data is a problem that is studied in detail in statistics. In statistics, how well a model fits a set of data is measured by a number called the *correlation coefficient,* which is denoted by r. The closer $|r|$ is to 1, the better the model fits the data. In Example 2, the correlation coefficient for the linear model is $r = 0.88$, but the correlation coefficient for the quadratic model is $r = 0.99$.

EXAMPLE 3   Fitting a Model to Data

Fit the following data, from Example 1(b), with a linear model and a quadratic model. Which model do you think fits better?

(0.2, 4.4), (0.4, 4.9), (0.5, 5.2), (0.7, 5.7), (1.1, 6.1), (1.2, 6.5), (1.5, 7.0), (1.7, 7.4), (1.9, 7.8), (2.0, 8.2), (2.3, 8.7), (2.5, 9.0), (2.8, 9.5), (2.9, 9.9), (3.2, 10.4), (3.4, 10.8)

Solution

Begin by entering the data into a calculator or computer that has least squares regression programs. Then run the regression programs for linear and quadratic models.

Linear: $y = ax + b$ $a = 1.963$ $b = 4.122$
Quadratic: $y = ax^2 + bx + c$ $a = 0.001$ $b = 1.959$ $c = 4.125$

From the scatter plots, notice that the data appears to follow a linear pattern. Although both models fit the data well, the linear model is better because it is simpler. The graphs of both models are shown in Figure 6.26.

Using Figure 6.26, students will notice that both of these two different models fit the data reasonably well.

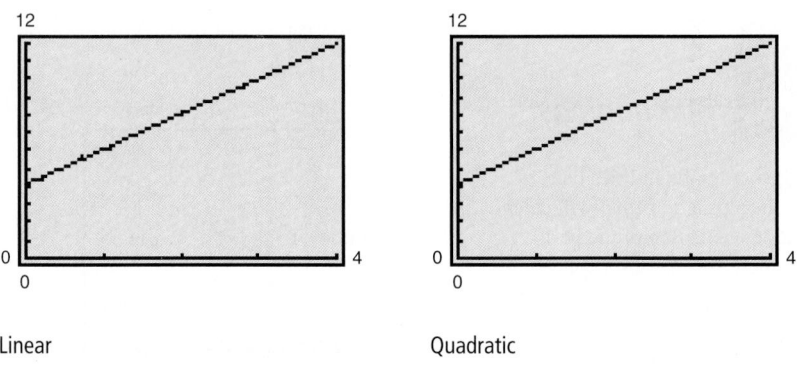

Linear   Quadratic

FIGURE 6.26

Exploration

Write a quadratic function and use it to create a table of values. Then enter the data from the table into a graphing utility and perform the quadratic least squares regression program. You should obtain the original function. Do you?

Application

EXAMPLE 4   *Finding a Quadratic Model*

The total amounts A (in millions of tons) of solid waste materials recycled in the United States in selected years from 1980 through 1993 are shown below. Decide whether a linear model or a quadratic model better fits the data. Then use the model to predict the amount that will be recycled in the year 2000. In the list of data points (t, A), t represents the year, with $t = 0$ corresponding to 1980. *(Source: Franklin Associates, Ltd.)*

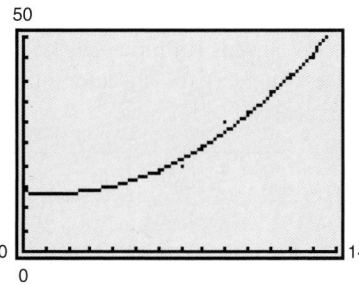

Linear

Quadratic

FIGURE 6.27

$(0, 14.5)$, $(5, 16.4)$, $(6, 18.3)$, $(7, 20.1)$, $(8, 23.5)$, $(9, 29.9)$, $(10, 32.9)$, $(11, 37.3)$, $(12, 41.5)$, $(13, 45.0)$

Solution

Begin by entering the data into a calculator or computer that has least squares regression programs. Then run the regression programs for linear and quadratic models.

$$\text{Linear:} \quad y = ax + b \qquad a = 2.624 \quad b = 6.684$$
$$\text{Quadratic:} \quad y = ax^2 + bx + c \quad a = 0.243 \quad b = -0.655 \quad c = 14.037$$

From the graphs in Figure 6.27, you can see that the quadratic model fits better.

(The correlation coefficient is $r = 0.994$, which implies that the model is a good fit to the data.) From this model, you can predict the amount that will be recycled in the year 2000 to be

$$A = 0.243(20)^2 - 0.655(20) + 14.037 = 98.137 \text{ million tons,}$$

which is more than two times the amount recycled in 1993.

Group Activities   **Problem Solving**

Modeling Data   The amounts y (in gallons) of bottled water consumed in the United States in the years 1980 through 1993 are listed below. The data is given as ordered pairs of the form (t, y), where t is the year, with $t = 0$ representing 1980. Create a scatter plot of the data. With others in your group, decide which type of model best fits the data. Then find the model.

$(0, 2.4)$, $(1, 2.7)$, $(2, 3.0)$, $(3, 3.4)$, $(4, 4.0)$, $(5, 4.5)$, $(6, 5.0)$, $(7, 5.7)$, $(8, 6.5)$, $(9, 7.4)$, $(10, 8.0)$, $(11, 8.0)$, $(12, 8.2)$, $(13, 9.2)$

6.6 Exercises

See Warm-Up Exercises, p. A46

In Exercises 1–4, decide whether you would use the linear model $y = ax + b$ or the quadratic model $y = ax^2 + bx + c$ to model the data.

1.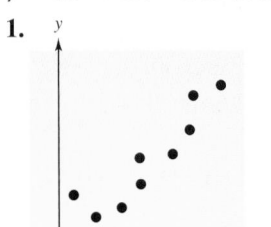

2.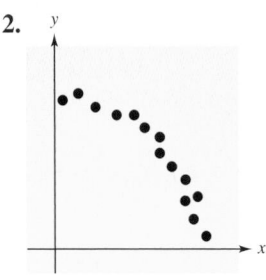

3.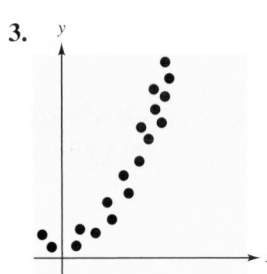

4.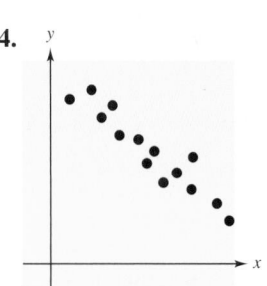

In Exercises 5–14, fit a linear model and a quadratic model to the points. Then plot the data and graph the models. Which model would you use?

5.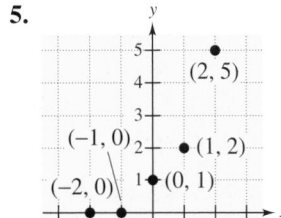

6.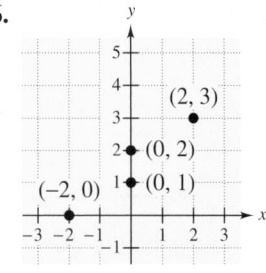

7.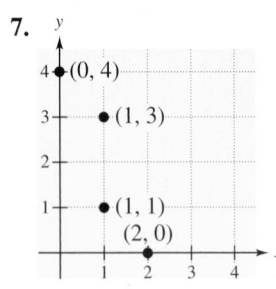

8.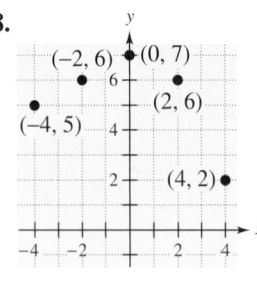

9.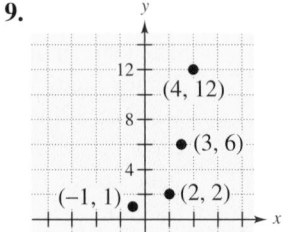

10.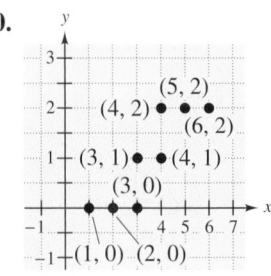

11. $(0, 0)$, $(1, 1)$, $(3, 4)$, $(4, 2)$, $(5, 5)$

12. $(-2, 5)$, $(-1, 5)$, $(0, 4)$, $(1, 3)$, $(2, 0)$

13. $(-4, 5)$, $(-2, 6)$, $(2, 6)$, $(4, 5)$

14. $(0, 6)$, $(4, 3)$, $(5, 0)$, $(8, -4)$, $(10, -5)$

15. *Stopping Distance*   The speeds (in miles per hour) and stopping distances (in feet) for an automobile braking system were recorded as follows.

Speed (x)	30	40	50	60	70
Stopping distance (y)	25	55	105	188	300

(a) Fit a least squares regression quadratic to the data, then plot the data and graph the model in the same viewing rectangle.

(b) Use the model in part (a) to estimate the stopping distance when $x = 70$. Compare this estimate with the actual value given in the table.

16. *Wildlife Management*   A wildlife management team studied the reproduction rates of deer in five tracts of a wildlife preserve. In each tract, the number of females and the percent of females that had offspring were recorded as follows.

Number (x)	80	100	120	140	160
Percent (y)	80	75	68	55	30

(a) Use a graphing utility to fit a least squares regression quadratic to the data.

(b) Use the model in part (a) to predict the percent of females that would have offspring if $x = 170$.

17. *Engine Performance*   After a new turbocharger for an automobile engine was developed, the following experimental data was obtained for speed in miles per hour at 2-second intervals.

Time (x)	0	2	4	6	8	10
Speed (y)	0	15	30	50	65	70

Use a graphing utility to fit a least squares regression quadratic to the data. Then plot the data and graph the model in the same viewing rectangle.

18. *World Population*   The table gives the world population (in billions) for five different years.   *(Source: U.S. Bureau of the Census)*

Year	1960	1970	1980	1990	1996
Population	3.0	3.7	4.5	5.3	5.8

Let $x = 0$ represent 1960.

(a) Use a graphing utility to fit a least squares regression line to the data.

(b) Use a graphing utility to fit a least squares regression quadratic to the data.

(c) Use both models in parts (a) and (b) to forecast the world population for the year 2010. How do the two models differ as you extrapolate into the future? Which do you think is a better model?

19. *Health Maintenance Organizations*   The table gives the number N (in millions) of people enrolled in HMOs in four selected years.   *(Source: National Directory of HMOs)*

Year	1980	1985	1990	1993
Enrollment	9.1	18.9	33.6	45.2

Let $x = 0$ represent 1980.

(a) Use a graphing utility to fit a least squares regression line to the data.

(b) Use a graphing utility to fit a least squares regression quadratic to the data.

(c) Use a graphing utility to plot the data and graph the models in parts (a) and (b).

(d) Use both models in part (c) to forecast the enrollment in HMOs in the year 2010. How do the two models differ as you extrapolate into the future?

20. *Vinyl Singles*   The table gives the number N (in millions) of vinyl single records sold in four selected years.   *(Source: Recording Industry Association of America)*

Year	1980	1985	1990	1994
Singles	164.3	120.7	27.6	11.7

Let $x = 0$ represent 1980.

(a) Use a graphing utility to fit a least squares regression line to the data.

(b) Use a graphing utility to fit a least squares regression quadratic to the data.

(c) Use a graphing utility to plot the data and graph the models in parts (a) and (b). Which do you think is a better model?

Section Project

Mathematical Modeling   After the path of a ball thrown by a baseball player is videotaped, it is analyzed on a television set with a grid covering the screen. The tape is paused three times and the positions of the ball are measured. The coordinates are approximately (0, 5.0), (15, 9.6), and (30, 12.4). (The x-coordinate measures the horizontal distance from the player in feet and the y-coordinate measures the height in feet.)

(a) Use a graphing utility to fit a least squares regression quadratic to the data.

(b) Use a graphing utility to plot the data and graph the model in part (a). How well does the model fit the data? Explain.

(c) Graphically approximate the maximum height of the ball and the point at which it struck the ground.

(d) Algebraically find the maximum height of the ball and the point at which it struck the ground.

6.7	**Quadratic Inequalities in One Variable**

Finding Test Intervals ▪ Quadratic Inequalities ▪
Application

Finding Test Intervals

When you are working with polynomial inequalities, it is important to realize that the value of a polynomial can change signs only at its **zeros.** That is, a polynomial can change signs only at the x-values that make the polynomial zero. For instance, the first-degree polynomial $x + 2$ has a zero at -2, and it changes sign at that zero. You can picture this result on the real number line, as shown in Figure 6.28.

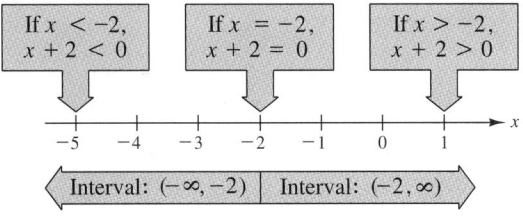

FIGURE 6.28

Note in Figure 6.28 that the zero of the polynomial partitions the real number line into two **test intervals.** The polynomial is negative for every x-value in the first test interval $(-\infty, -2)$, and is positive for every x-value in the second test interval $(-2, \infty)$. You can use the same basic approach to determine the test intervals for any polynomial.

Finding Test Intervals for a Polynomial

1. Find all real zeros of the polynomial, and arrange the zeros in increasing order. The zeros of a polynomial are called its **critical numbers.**

2. Use the critical numbers of the polynomial to determine its **test intervals.**

3. Choose a representative x-value in each test interval and evaluate the polynomial at that value. If the value of the polynomial is negative, the polynomial will have negative values for *every* x-value in the interval. If the value of the polynomial is positive, the polynomial will have positive values for *every* x-value in the interval.

EXAMPLE 1   Finding Test Intervals for a Quadratic

Determine the intervals on which the polynomial $x^2 - x - 6$ is entirely negative or entirely positive.

Solution

By factoring the given quadratic as

$$x^2 - x - 6 = 0$$
$$(x + 2)(x - 3) = 0,$$

you can see that the critical numbers occur at $x = -2$ and $x = 3$. Therefore, the test intervals for the quadratic are

$$(-\infty, -2), \quad (-2, 3), \quad \text{and} \quad (3, \infty). \qquad \text{Test intervals}$$

In each test interval, choose a representative x-value and evaluate the polynomial, as shown in the table.

Test interval	Representative x-value	Value of polynomial	Conclusion
$(-\infty, -2)$	$x = -3$	$(-3)^2 - (-3) - 6 = 6$	Quadratic is positive.
$(-2, 3)$	$x = 0$	$(0)^2 - (0) - 6 = -6$	Quadratic is negative.
$(3, \infty)$	$x = 4$	$(4)^2 - (4) - 6 = 6$	Quadratic is positive.

Therefore, the polynomial has positive values for every x in the intervals $(-\infty, -2)$ and $(3, \infty)$, and negative values for every x in the interval $(-2, 3)$. This result is shown graphically in Figure 6.29.

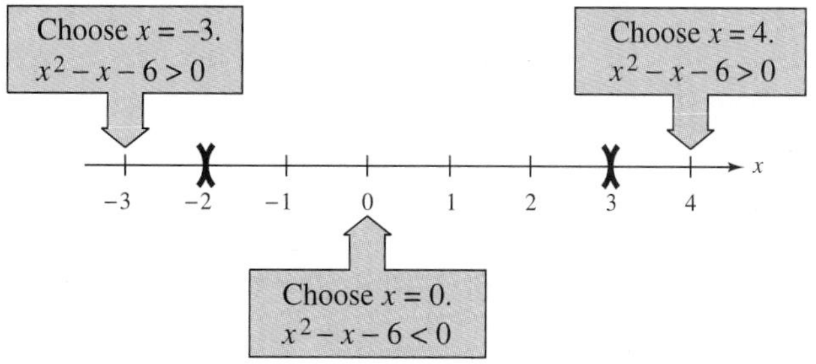

Choose $x = -3$.
$x^2 - x - 6 > 0$

Choose $x = 4$.
$x^2 - x - 6 > 0$

Choose $x = 0$.
$x^2 - x - 6 < 0$

FIGURE 6.29

NOTE   Another way to determine the intervals on which $x^2 - x - 6$ is entirely negative or entirely positive is to sketch the graph of $y = x^2 - x - 6$, as shown in Figure 6.30. The portions of the graph that lie above the x-axis correspond to $x^2 - x - 6 > 0$, and the portion that lies below the x-axis corresponds to $x^2 - x - 6 < 0$.

Discuss how to choose test points. For instance, in Example 1, see if students can come up with an alternate set of test points. Ask if this new selection changes the solution to the problem.

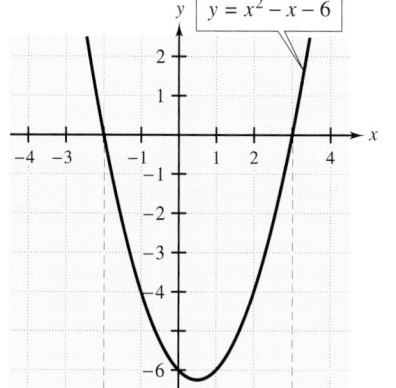

$y = x^2 - x - 6$

$y < 0$ when
$-2 < x < 3$.

FIGURE 6.30

Quadratic Inequalities

The concepts of critical numbers and test intervals can be used to solve nonlinear inequalities, as demonstrated in Examples 2, 3, and 4.

EXAMPLE 2   Solving a Quadratic Inequality

Solve the inequality $x^2 - 5x < 0$.

Solution

Begin by finding the critical numbers of $x^2 - 5x$.

$$x^2 - 5x < 0 \qquad \text{Original inequality}$$
$$x(x - 5) < 0 \qquad \text{Factor.}$$

From this factorization, you can see that the critical numbers of $x^2 - 5x$ are 0 and 5. This implies that the test intervals are

$$(-\infty, 0), \quad (0, 5), \quad \text{and} \quad (5, \infty).$$

To test an interval, choose a convenient number in the interval and compute the sign of $x^2 - 5x$. The results are shown in Figure 6.31.

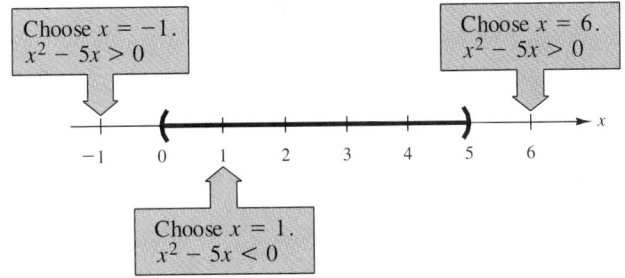

FIGURE 6.31

Because the inequality $x^2 - 5x < 0$ is satisfied only by the middle test interval, you can conclude that the solution set of the inequality is the interval $(0, 5)$.

In Example 2, note that you would have used the same basic procedure if the inequality symbol had been $\leq$, $>$, or $\geq$. For instance, from Figure 6.31, you can see that the solution set of the inequality

$$x^2 - 5x \geq 0$$

consists of the union of the intervals $(-\infty, 0]$ and $[5, \infty)$, which is written as $(-\infty, 0] \cup [5, \infty)$.

Technology

Most graphing utilities can graph a quadratic inequality. For instance, the following steps are used to graph $x^2 - 5x < 0$ on a *TI-82* or *TI-83*.

$\boxed{\text{Y=}}$ Y1=X²–5X<0

$\boxed{\text{GRAPH}}$ (Standard setting)

The graph produced by these steps is shown below. Notice that the graph occurs as an interval *above* the x-axis.

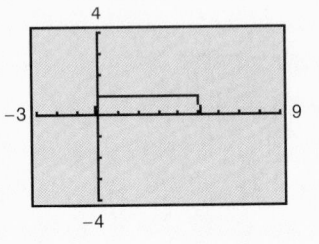

Just as in solving quadratic *equations,* the first step in solving a quadratic *inequality* is to write the inequality in **standard form,** with the polynomial on the left and zero on the right, as demonstrated in Example 3.

EXAMPLE 3   *Solving a Quadratic Inequality*

Solve the inequality $2x^2 + 5x > 12$.

Solution

Begin by writing the inequality in standard form.

$$2x^2 + 5x > 12 \qquad \text{Original inequality}$$
$$2x^2 + 5x - 12 > 0 \qquad \text{Write in standard form.}$$
$$(x + 4)(2x - 3) > 0 \qquad \text{Factor.}$$

From this factorization, you can see that the critical numbers of $2x^2 + 5x - 12$ are -4 and $\frac{3}{2}$. This implies that the test intervals are

$$(-\infty, -4), \quad \left(-4, \tfrac{3}{2}\right), \quad \text{and} \quad \left(\tfrac{3}{2}, \infty\right).$$

To test an interval, choose a convenient number in the interval and compute the sign of $(x + 4)(2x - 3)$. The results are shown in Figure 6.32. From the figure, you can see that the polynomial $2x^2 + 5x - 12$ is positive in the open intervals $(-\infty, -4)$ and $\left(\tfrac{3}{2}, \infty\right)$. Thus, the solution set is $(-\infty, -4) \cup \left(\tfrac{3}{2}, \infty\right)$.

Some students may try to solve the quadratic inequality $(x + 4)(2x - 3) > 0$ by setting each factor greater than 0, as $x + 4 > 0$ and $2x - 3 > 0$, similar to the way they solved quadratic equations by factoring. Point out the error in this reasoning.

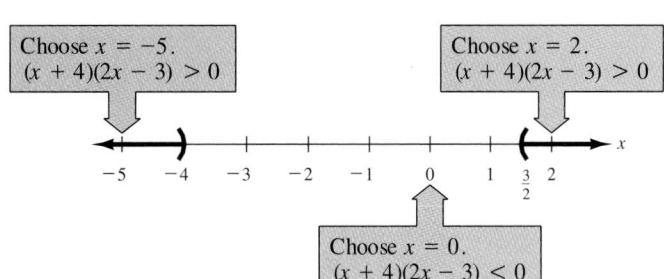

FIGURE 6.32

In Examples 2 and 3, the critical numbers were found by factoring. With quadratic polynomials that do not factor, you can use the Quadratic Formula to find the critical numbers. For instance, to solve the inequality

$$x^2 + 4x + 1 \le 0,$$

you can use the Quadratic Formula to determine that the critical numbers are $-2 - \sqrt{3} \approx -3.732$ and $-2 + \sqrt{3} \approx -0.268$.

Example 4 shows how the Quadratic Formula can be used to solve a quadratic inequality.

EXAMPLE 4   Using the Quadratic Formula

Solve the following quadratic inequality.

$$x^2 - 2x - 1 \leq 0$$

Solution

Point out that "does not factor" does not mean "no solution."

Because the quadratic $x^2 - 2x - 1$ does not factor (using integer coefficients), you can use the Quadratic Formula to find its zeros. That is, the solutions of $x^2 - 2x - 1 = 0$ are given by

$$x = \frac{-(-2) \pm \sqrt{(-2)^2 - 4(1)(-1)}}{2(1)} = \frac{2 \pm \sqrt{8}}{2} = \frac{2 \pm 2\sqrt{2}}{2} = 1 \pm \sqrt{2}.$$

Therefore, the critical numbers of $x^2 - 2x - 1$ are $1 - \sqrt{2}$ and $1 + \sqrt{2}$. This implies that the test intervals are

$$\left(-\infty, 1 - \sqrt{2}\right), \quad \left(1 - \sqrt{2}, 1 + \sqrt{2}\right), \quad \text{and} \quad \left(1 + \sqrt{2}, \infty\right).$$

After testing these intervals, as shown in Figure 6.33, you can see that the polynomial $x^2 - 2x - 1$ is less than or equal to zero in the *closed* interval $\left[1 - \sqrt{2}, 1 + \sqrt{2}\right]$. Therefore, the solution set of the inequality is

$$\left[1 - \sqrt{2}, 1 + \sqrt{2}\right]. \qquad \text{Solution set}$$

Note that $1 - \sqrt{2} \approx -0.414$ and $1 + \sqrt{2} \approx 2.414$.

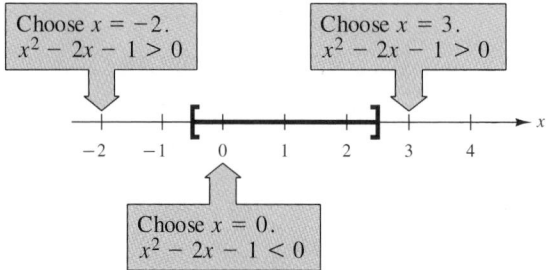

You may suggest that students convert $1 - \sqrt{2}$ and $1 + \sqrt{2}$ to decimal form to help in choosing test points.

FIGURE 6.33

NOTE   When solving a quadratic inequality, be sure you have accounted for the particular type of inequality symbol given in the inequality. For instance, in Example 4, note that the solution is a closed interval because the original inequality contained a "*less than or equal to*" symbol. If the original inequality had been $x^2 - 2x - 1 < 0$, the solution would have been the *open* interval $\left(1 - \sqrt{2}, 1 + \sqrt{2}\right)$.

The solutions of the quadratic inequalities in Examples 2, 3, and 4 consist of a single interval or the union of two intervals. When solving the exercises for this section, you should be on the watch for some unusual solution sets, as illustrated in Example 5.

EXAMPLE 5   *Unusual Solution Sets*

Solve each inequality to verify that the given solution set is correct.

(a) The solution set of the quadratic inequality

$$x^2 + 2x + 4 > 0$$

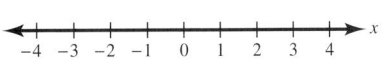

(a)

consists of the entire set of real numbers, $(-\infty, \infty)$. [See Figure 6.34(a).] In other words, the quadratic $x^2 + 2x + 4$ is positive for every real value of x. (Note that this quadratic inequality has no critical numbers. In such a case, there is only one test interval—the entire real line.)

(b) The solution set of the quadratic inequality

$$x^2 + 2x + 1 \le 0$$

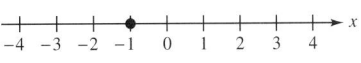

(b)

consists of the single real number $\{-1\}$. [See Figure 6.34(b).]

(c) The solution set of the quadratic inequality

$$x^2 + 3x + 5 < 0$$

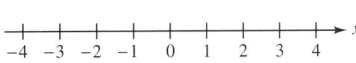

(c)

is empty. [See Figure 6.34(c).] In other words, the quadratic $x^2 + 3x + 5$ is not less than zero for any value of x.

(d) The solution set of the quadratic inequality

$$x^2 - 4x + 4 > 0$$

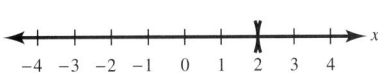

(d)

FIGURE 6.34

consists of all real numbers *except* the number 2. [See Figure 6.34(d).] In interval notation, this solution set can be written as $(-\infty, 2) \cup (2, \infty)$.

Remember that checking the solution set of an inequality is not as straightforward as checking the solutions of an equation, because inequalities tend to have infinitely many solutions. Even so, we suggest that you check several x-values in your solution set to confirm that they satisfy the inequality. Also try checking x-values that are not in the solution set to confirm that they do not satisfy the inequality.

For instance, the solution of the quadratic inequality $x^2 - 5x < 0$ is the interval $(0, 5)$. Try checking some numbers in this interval to verify that they satisfy the inequality. Then check some numbers outside the interval to verify that they do not satisfy the inequality.

Application

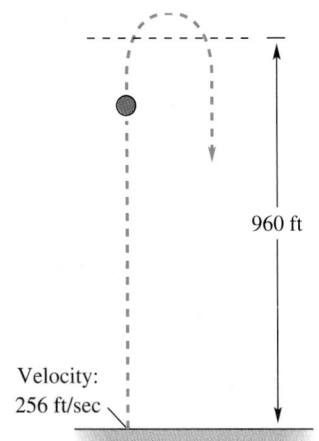

FIGURE 6.35

EXAMPLE 6   The Height of a Projectile

A projectile is fired straight up from ground level with an initial velocity of 256 feet per second, as shown in Figure 6.35, so that its height at any time t is given by

$$h = -16t^2 + 256t,$$

where the height h is measured in feet and the time t is measured in seconds. During what interval of time will the height of the projectile exceed 960 feet?

Solution

To solve this problem, find the values of t for which h is greater than 960.

$$-16t^2 + 256t > 960 \qquad \text{Original inequality}$$

$$-16t^2 + 256t - 960 > 0 \qquad \text{Subtract 960 from both sides.}$$

$$t^2 - 16t + 60 < 0 \qquad \text{Divide both sides by } -16 \text{ and reverse the inequality symbol.}$$

$$(t - 6)(t - 10) < 0 \qquad \text{Factor.}$$

Thus, the critical numbers are $t = 6$ and $t = 10$. A test of the intervals $(-\infty, 6)$, $(6, 10)$, and $(10, \infty)$ shows that the solution interval is $(6, 10)$. Therefore, the height of the projectile will exceed 960 feet for values of t that are greater than 6 seconds and less than 10 seconds.

Group Activities

Exploring with Technology

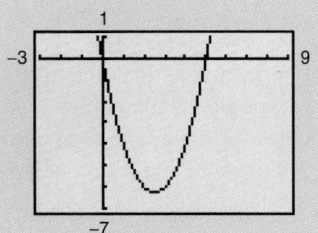

Graphing an Inequality   You can use a graph on a rectangular coordinate system as an alternative method of solving an inequality. For instance, to solve the inequality in Example 2,

$$x^2 - 5x < 0,$$

you can sketch the graph of $y = x^2 - 5x$. Using a graphing utility, you can obtain the graph shown at the left. From the graph, you can see that the only part of the curve that lies below the x-axis is the portion for which $0 < x < 5$. Thus, the solution of $x^2 - 5x < 0$ is $0 < x < 5$. Try using this graphing approach to solve some of the other examples in this section.

| 6.7 | **Exercises** |

See Warm-Up
Exercises, p. A46

1. *Think About It*   Explain the change in an inequality that occurs when both sides are multiplied by a negative real number.

2. *Think About It*   Define the term *critical number* and explain its use in solving quadratic inequalities.

In Exercises 3–10, find the critical numbers.

3. $4x^2 - 81$

4. $y(y - 4) - 3(y - 4)$

5. $x(2x - 5)$

6. $5x(x - 3)$

7. $x^2 - 4x + 3$

8. $3x^2 - 2x - 8$

9. $4x^2 - 20x + 25$

10. $4x^2 - 4x - 3$

In Exercises 11 and 12, determine whether the x-values are solutions of the inequality.

11. $2x^2 - 7x - 4 > 0$

   (a) $x = 0$ (b) $x = 6$

   (c) $x = -\frac{1}{2}$ (d) $x = -\frac{3}{2}$

12. $4x^2 + 3x - 10 \le 0$

   (a) $x = 0$ (b) $x = 3$

   (c) $x = -2$ (d) $x = \frac{1}{2}$

In Exercises 13–20, determine the intervals for which the polynomial is entirely negative and entirely positive.

13. $x - 4$

14. $3 - x$

15. $2x(x - 4)$

16. $7x(3 - x)$

17. $x^2 - 9$

18. $16 - x^2$

19. $x^2 - 4x - 5$

20. $2x^2 - 4x - 3$

In Exercises 21–46, solve the inequality and graph the solution on the real number line. (Some of the inequalities have no solution.)

21. $2x + 6 \ge 0$

22. $5x - 20 < 0$

23. $-\frac{3}{4}x + 6 < 0$

24. $3x - 2 \ge 0$

25. $3x(x - 2) < 0$

26. $2x(x - 6) > 0$

27. $3x(2 - x) < 0$

28. $2x(6 - x) > 0$

29. $x^2 - 4 \ge 0$

30. $z^2 \le 9$

31. $x^2 + 3x \le 10$

32. $t^2 - 15t + 50 < 0$

33. $-2u^2 + 7u + 4 < 0$

34. $-3x^2 - 4x + 4 \le 0$

35. $x^2 + 4x + 5 < 0$

36. $x^2 + 6x + 10 > 0$

37. $x^2 + 2x + 1 \ge 0$

38. $25 \ge (x - 3)^2$

39. $x^2 - 4x + 2 > 0$

40. $-x^2 + 8x - 11 \le 0$

41. $(x - 5)^2 < 0$

42. $(y + 3)^2 \ge 0$

43. $6 - (x - 5)^2 < 0$

44. $(y + 3)^2 - 6 \ge 0$

45. $x^2 - 6x + 9 \ge 0$

46. $x^2 + 8x + 16 < 0$

In Exercises 47–54, use a graphing utility to graph the function and solve the inequality.

47. $y = x^2 - 6x, \quad y < 0$

48. $y = 2x^2 + 5x, \quad y > 0$

49. $y = 0.5x^2 + 1.25x - 3, \quad y > 0$

50. $y = \frac{1}{3}x^2 - 3x, \quad y < 0$

51. $y = x^2 + 4x + 4, \quad y \ge 9$

52. $y = x^2 - 6x + 9, \quad y < 16$

53. $y = 9 - 0.2(x - 2)^2, \quad y < 4$

54. $y = 8x - x^2, \quad y > 12$

55. *Height of a Projectile*   A projectile is fired straight up from ground level with an initial velocity of 128 feet per second, so that its height at any time t is given by

$$h = -16t^2 + 128t,$$

where the height h is measured in feet and the time t is measured in seconds. During what interval of time will the height of the projectile exceed 240 feet?

56. *Height of a Projectile*   A projectile is fired straight up from ground level with an initial velocity of 88 feet per second, so that its height at any time t is given by

$$h = -16t^2 + 88t,$$

where the height h is measured in feet and the time t is measured in seconds. During what interval of time will the height of the projectile exceed 50 feet?

57. *Annual Interest Rate*   You are investing $1000 in a certificate of deposit for 2 years and you want the interest to exceed $150. The interest is compounded annually. What interest rate should you have? [*Hint:* Solve the inequality $1000(1 + r)^2 > 1150$.]

58. *Company Profits*   The revenue and cost equations for a product are given by

$$R = x(50 - 0.0002x)$$
$$C = 12x + 150,000,$$

where R and C are measured in dollars and x represents the number of units sold (see figure). How many units must be sold to obtain a profit of at least $1,650,000?

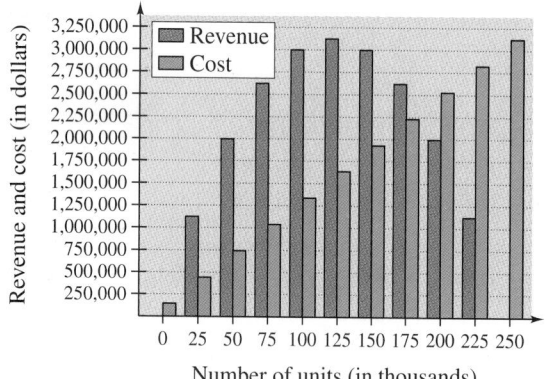

59. *Geometry*   A rectangular playing field with a perimeter of 100 meters must have an area of at least 500 square meters. Within what bounds must the length of the field lie?

60. *Geometry*   You have 64 feet of fencing to enclose a rectangular region. Determine the interval for the length such that the area will exceed 240 square feet.

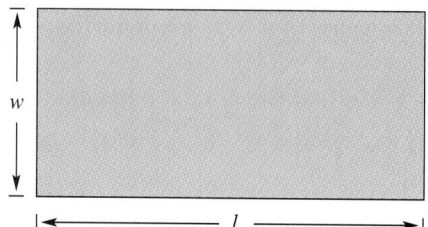

Section Project

Geometry   Two circles are tangent to each other (see figure). The distance between their centers is 12 centimeters.

(a) If x is the radius of one of the circles, express the combined area of the circles as a function of x. What is the domain of the function?

(b) Use a graphing utility to graph the area function of part (a).

(c) Determine the radii of the circles if the combined area of the circles must be at least 300 square centimeters but no more than 400 square centimeters.

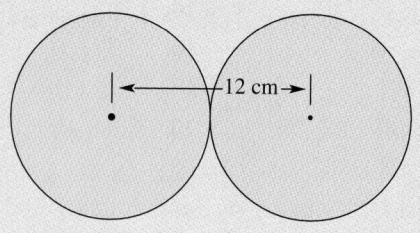

Chapter Summary

After studying this chapter, you should have acquired the following skills. These skills are keyed to the Review Exercises that begin on page 472. Answers to odd-numbered Review Exercises are given in the back of the book.

- Solve quadratic equations by factoring. *(Section 6.1)*   **Review Exercises 1–12**
- Solve quadratic equations by extracting square roots. *(Section 6.1)*   **Review Exercises 13–18**
- Solve quadratic equations by completing the square. *(Section 6.2)*   **Review Exercises 19–24**
- Solve quadratic equations using the Quadratic Formula. *(Section 6.3)*   **Review Exercises 25–30**
- Solve quadratic equations by the most convenient method.   **Review Exercises 31–40**
  (Sections 6.1–6.3)
- Solve equations of quadratic form. *(Section 6.1)*   **Review Exercises 41–44**
- Use a graphing utility to graph functions. Solve the equations when   **Review Exercises 45–50**
  $y = 0$ and interpret the results. *(Sections 6.1–6.3)*
- Determine whether graphs open upward or downward given their equations.   **Review Exercises 51–54**
  (Section 6.5)
- Sketch the graph of an equation and identify the vertex and any x-intercepts. **Review Exercises 55–58**
  Confirm the result using a graphing utility. *(Section 6.5)*
- Find equations of parabolas. *(Section 6.5)*   **Review Exercises 59–62**
- Solve real-life problems modeled by quadratic equations or inequalities.   **Review Exercises 63–65,**
  (Section 6.1–6.5, 6.7)   **70**
- Solve problems involving geometry. *(Sections 6.1–6.4)*   **Review Exercises 66, 67**
- Translate real-life problems into quadratic equations or inequalities and solve. **Review Exercises 68, 69**
  (Sections 6.1–6.7)
- Solve inequalities and graph the solutions on the real number line.   **Review Exercises 71–78**
  (Section 6.7)

Review Exercises

In Exercises 1–12, solve the equation by factoring.

1. $x^2 + 12x = 0$ **2.** $u^2 - 18u = 0$

3. $3z(z + 10) - 8(z + 10) = 0$

4. $7x(2x - 9) + 4(2x - 9) = 0$

5. $4y^2 - 1 = 0$ **6.** $2z^2 - 72 = 0$

7. $4y^2 + 20y + 25 = 0$ **8.** $x^2 + \frac{8}{3}x + \frac{16}{9} = 0$

9. $t^2 - t - 20 = 0$ **10.** $z^2 + \frac{2}{3}z - \frac{8}{9} = 0$

11. $2x^2 - 2x - 180 = 0$ **12.** $15x^2 - 30x - 45 = 0$

In Exercises 13–18, solve the equation by extracting square roots.

13. $x^2 = 10,000$ **14.** $x^2 = 98$

15. $y^2 - 2.25 = 0$ **16.** $y^2 - 8 = 0$

17. $(x - 16)^2 = 400$ **18.** $(x + 3)^2 = 0.04$

In Exercises 19–24, solve the equation by completing the square. (Find all real *and* imaginary solutions.)

19. $x^2 - 6x - 3 = 0$ **20.** $x^2 + 12x + 6 = 0$

21. $x^2 - 3x + 3 = 0$ **22.** $t^2 + \frac{1}{2}t - 1 = 0$

23. $2y^2 + 10y + 3 = 0$ **24.** $3x^2 - 2x + 2 = 0$

In Exercises 25–30, use the Quadratic Formula to solve the equation. (Find all real *and* imaginary solutions.)

25. $y^2 + y - 30 = 0$ **26.** $x^2 - x - 72 = 0$

27. $2y^2 + y - 21 = 0$ **28.** $2x^2 - 3x - 20 = 0$

29. $0.3t^2 - 2t + 5 = 0$ **30.** $-u^2 + 2.5u + 3 = 0$

In Exercises 31–40, solve the equation by the method of your choice. (Find all real *and* imaginary solutions.)

31. $(v - 3)^2 = 250$ **32.** $x^2 - 36x = 0$

33. $-x^2 + 5x + 84 = 0$

34. $9x^2 + 6x + 1 = 0$

35. $(x - 9)^2 - 121 = 0$

36. $60 - (x - 6)^2 = 0$

37. $z^2 - 6z + 10 = 0$

38. $z^2 - 14z + 5 = 0$

39. $2y^2 + 3y + 1 = 0$

40. $x - 5 = \sqrt{x - 2}$

In Exercises 41–44, solve the equation of quadratic form.

41. $x^4 - 4x^2 - 5 = 0$

42. $x + 2\sqrt{x} - 3 = 0$

43. $x^{2/5} + 4x^{1/5} + 3 = 0$

44. $6\left(\dfrac{1}{x}\right)^2 + 7\left(\dfrac{1}{x}\right) - 3 = 0$

In Exercises 45–48, use a graphing utility to graph the function. Use the graph to approximate any x-intercepts. Set $y = 0$ and solve the resulting equation. Compare the result with the x-intercepts of the graph.

45. $y = x^2 - 7x$

46. $y = 12x^2 + 11x - 15$

47. $y = \dfrac{1}{16}x^4 - x^2 + 3$

48. $y = \left(\dfrac{1}{x}\right)^2 + 2\left(\dfrac{1}{x}\right) - 3$

Graphical Reasoning   In Exercises 49 and 50, use a graphing utility to graph the function and observe that the graph has no x-intercepts. Set $y = 0$ and solve the resulting equation. What type of roots does the equation have?

49. $y = x^2 - 8x + 17$

50. $y = -(x + 3)^2 - 2$

In Exercises 51–54, state whether the graph opens upward or downward and find the vertex.

51. $y = 4(x - 3)^2 + 6$ **52.** $y = -(x - 7)^2 - 1$

53. $y = -3(x + 4)^2$ **54.** $y = 5(x + 2)^2 + 9$

In Exercises 55–58, sketch the graph of the equation. Identify the vertex and any x-intercepts. Use a graphing utility to verify your results.

55. $y = x^2 + 8x$ **56.** $y = -x^2 + 1$

57. $y = x^2 - 6x + 5$ **58.** $y = -(x - 2)^2$

In Exercises 59–62, write an equation of the parabola

$$y = ax^2 + bx + c$$

that satisfies the conditions.

59. Vertex: $(3, 5)$; $a = -2$

60. Vertex: $(-2, 3)$; $a = 3$

61. Vertex: $(5, 0)$; Passes through the point $(1, 1)$

62. Vertex: $(-2, 5)$; Passes through the point $(0, 1)$

63. *Vertical Motion*   The height h in feet of an object above the ground is

$$h = 200 - 16t^2, \qquad t \geq 0,$$

where t is the time in seconds.

(a) How high is the object when $t = 0$?

(b) Was the object thrown upward, dropped, or thrown downward? Explain.

(c) Find the time when the object strikes the ground.

64. *Vertical Motion*   The height h in feet of an object above the ground is

$$h = -16t^2 + 64t + 192, \qquad t \geq 0,$$

where t is the time in seconds.

(a) How high is the object when $t = 0$?

(b) Was the object thrown upward, dropped, or thrown downward? Explain.

(c) Find the time when the object strikes the ground.

65. *Path of a Projectile*   The path y of a projectile is

$$y = -\frac{1}{16}x^2 + 5x,$$

where y is the height (in feet) and x is the horizontal distance (in feet) from where the projectile was launched.

(a) Sketch the path of the projectile.

(b) How high is the projectile when it is at its maximum height?

(c) How far from the launch point does the projectile strike the ground?

66. *Geometry*   Find the dimensions of a triangle if its height is $1\frac{2}{3}$ times its base and its area is 3000 square centimeters.

67. *Geometry*   The perimeter of a rectangle of length l and width w is 48 meters (see figure).

$24 - l$

(a) Show that $w = 24 - l$.

(b) Show that the area A is $A = lw = l(24 - l)$.

(c) Use the function in part (b) to complete the table.

l	2	4	6	8	10	12	14	16	18
A									

68. *Solving Graphically, Numerically, and Algebraically*   A launch control team is located 3 miles from the point where a rocket is launched.

(a) Write the distance d between the rocket and the team as a function of the height h of the rocket. Use a graphing utility to graph the function. Use the graph to approximate the value of h when $d = 4$ miles.

(b) Complete the table. Use the table to approximate the value of h when $d = 4$ miles.

h	1	2	3	4	5	6	7
d							

(c) Find h algebraically when $d = 4$ miles.

69. *Median Home Price*   The median sales price S (in thousands of dollars) of privately owned one-family homes in the United States in selected years from 1970 to 1994 is given in the table. The time in years is given by t, with $t = 0$ corresponding to 1970. *(Source: U.S. Department of Housing and Urban Development)*

t	0	5	10	15	20	24
S	23.4	39.3	64.6	84.3	122.9	130.0

(a) Use the regression capabilities of a graphing utility to fit a quadratic model to the data.

(b) Use a graphing utility to plot the data and graph the model in part (a).

(c) Use the model to estimate the median price of a home in the year 2000.

70. *College Completion*   The percent of the American population that graduated from college between 1960 and 1990 is approximated by the model

$$\text{Percent of graduates} = (2.739 + 0.064t)^2, \qquad 0 \le t,$$

where $t = 0$ represents 1960 (see figure). According to this model, when did the percent of college graduates exceed 25% of the population?   *(Source: U.S. Bureau of Census)*

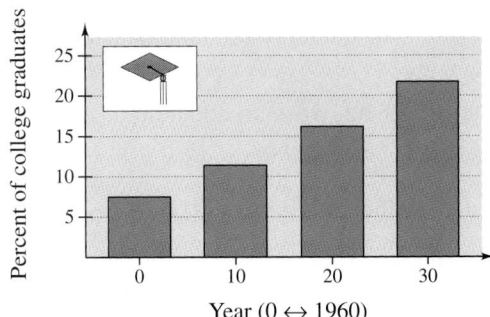

Year (0 ↔ 1960)

In Exercises 71–78, solve the inequality and sketch the graph of the solution on the real number line. (Some of the inequalities have no solution.)

71. $4x - 12 < 0$

72. $3(x + 2) > 0$

73. $5x(7 - x) > 0$

74. $-2x(x - 10) \le 0$

75. $(x - 5)^2 - 36 > 0$

76. $16 - (x - 2)^2 \le 0$

77. $2x^2 + 3x - 20 < 0$

78. $3x^2 - 2x - 8 > 0$

Chapter Test

Take this test as you would take a test in class. After you are done, check your work against the answers given in the back of the book.

In Exercises 1 and 2, solve the quadratic equation by factoring.

1. $x(x + 5) - 10(x + 5) = 0$ **2.** $8x^2 - 21x - 9 = 0$

In Exercises 3 and 4, solve the quadratic equation by extracting square roots.

3. $(x - 2)^2 = 0.09$ **4.** $(x + 3)^2 + 81 = 0$

5. What term should be added to $x^2 - 3x$ so that it becomes a perfect square trinomial?

6. Solve by completing the square: $2x^2 - 6x + 3 = 0$.

7. Find the discriminant and explain how it is used to determine the type of solutions of the quadratic equation $5x^2 - 12x + 10 = 0$.

In Exercises 8 and 9, solve by the Quadratic Formula.

8. $3x^2 - 8x + 3 = 0$ **9.** $2y(y - 2) = 7$

10. Solve: $x^{2/3} - 6x^{1/3} + 8 = 0$.

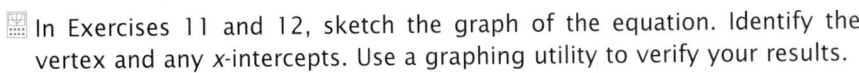 In Exercises 11 and 12, sketch the graph of the equation. Identify the vertex and any *x*-intercepts. Use a graphing utility to verify your results.

11. $y = -x^2 + 16$ **12.** $y = x^2 - 2x - 15$

In Exercises 13 and 14, solve the inequality and sketch the solution on the real number line.

13. $2x(x - 3) < 0$ **14.** $x^2 - 4x \geq 12$

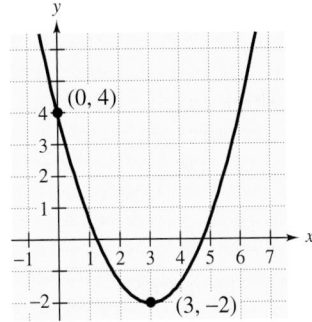

Figure for 15

15. Find the quadratic function, $y = ax^2 + bx + c$, whose graph is shown in the figure.

16. The width of a rectangle is 8 feet less than its length. The area of the rectangle is 240 square feet. Find the dimensions of the rectangle.

17. An object is dropped from a height of 75 feet. Its height (in feet) at any time is given by $h = -16t^2 + 75$, where the time t is measured in seconds. Find the time at which the object has fallen to a height of 35 feet.

18. Use a graphing utility to fit a quadratic model to the points $(0, 0)$, $(2, 2)$, $(3, 6)$, and $(4, 12)$.

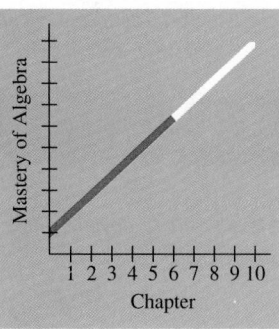

Mastery of Algebra

Chapter

Take this test as you would take a test in class. After you are done, check your work against the answers given in the back of the book.

In Exercises 1–4, factor the expression.

1. $2x^2 + 5x - 7$

2. $11x^2 + 6x - 5$

3. $12x^3 - 27x$

4. $8x^3 + 125$

In Exercises 5–10, solve the equation.

5. $125 - 50x = 0$

6. $t^2 - 8t = 0$

7. $x^2(x + 2) - (x + 2) = 0$

8. $x(10 - x) = 25$

9. $\dfrac{x + 3}{7} = \dfrac{x - 1}{4}$

10. $\dfrac{x}{4} - \dfrac{x + 2}{6} = \dfrac{3}{2}$

In Exercises 11 and 12, write a slope-intercept form of the equation of the line.

11. Passes through $(4, -2)$; $m = \frac{5}{2}$

12. Passes through $(-5, 8)$ and $(1, 2)$

13. Does the graph in the figure represent y as a function of x? Explain.

14. The base of a triangle is $5x$ and its height is $2x + 9$. Write the area A of the triangle as a function of x.

15. A sales representative indicates that if a customer waits another month for a new car that currently costs \$23,500, the price will increase by 4%. However, the customer will pay an interest penalty of \$725 for the early withdrawal of a certificate of deposit if the car is purchased now. Determine whether the customer should buy now or wait another month.

16. Solve the system of equations: $\begin{cases} 2x + 0.5y = 8 \\ 3x + 2y = 22. \end{cases}$

In Exercises 17 and 18, evaluate the determinant of the matrix.

17. $\begin{bmatrix} 3 & 7 \\ -2 & 6 \end{bmatrix}$

18. $\begin{bmatrix} 3 & -2 & 1 \\ 0 & 5 & 3 \\ 6 & 1 & 1 \end{bmatrix}$

19. Simplify the expressions (a) $\sqrt{75x^3y}$ and (b) $40/(3 - \sqrt{5})$.

20. Find an equation of the parabola with a vertex at $(2, -1)$ that passes through the origin.

$x^2 + y^2 = 25$

Figure for 13

476

Rational Expressions and Rational Functions

Strategies for Success

SKILLS *When you have completed this chapter, make sure you are able to:*

○ Determine the domain of a rational expression and simplify the expression
○ Add, subtract, multiply, and divide rational expressions
○ Divide a polynomial by a monomial or a polynomial
○ Sketch the graph of rational functions
○ Model real-life situations using rational equations and inequalities and solve

TOOLS *Use these study tools to master the skills above:*

In Exercise 77 of Section 7.5, you will solve a rational equation to find the year in which total revenues for computer and data services exceeded $100 billion.

7.1	**Simplifying Rational Expressions**

The Domain of a Rational Expression ▪ Simplifying Rational Expressions ▪ Application

The Domain of a Rational Expression

A fraction whose numerator and denominator are polynomials is called a **rational expression.** Some examples are

$$\frac{3}{x+4}, \quad \frac{2x}{x^2-4x+4}, \quad \text{and} \quad \frac{x^2-5x}{x^2+2x-3}.$$

Exploration

Evaluate each of the three rational functions above for all integer values of x from -4 to 4. Organize your results in a table. Then, discuss your results. What do you observe?

Because division by zero is undefined, the denominator of a rational expression cannot be zero. Therefore, in your work with rational expressions, you must assume that all real number values of the variable that make the denominator zero are excluded. For the three fractions above, $x = -4$ is excluded from the first fraction, $x = 2$ from the second, and both $x = 1$ and $x = -3$ from the third. The set of *usable* values of the variable is called the **domain** of the rational expression.

Emphasize the importance of determining the domain of a rational expression.

Definition of a Rational Expression

Let u and v be polynomials. The algebraic expression

$$\frac{u}{v}$$

is a **rational expression.** The **domain** of this rational expression is the set of all real numbers for which $v \neq 0$.

Like polynomials, rational expressions can be used to describe functions. Such functions are called **rational functions.**

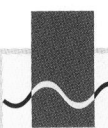

Technology

Sometimes a graphing utility can help you examine the domain of a function. For instance, the graph of $f(x) = 4/(x - 2)$ is shown below. Notice that the graph moves down immediately to the left of $x = 2$ and it moves up immediately to the right of $x = 2$. At $x = 2$, the graph does not exist.

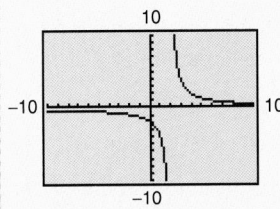

Try graphing the functions in Example 2(a) and (b). What can you conclude?

A rational expression is undefined if the *denominator* is equal to 0. Students often infer that this means the *variable in the denominator is 0*. Distinguish these two rules by the following examples:

(a) $\dfrac{4 - n}{n}$

Domain: $(-\infty, 0) \cup (0, \infty)$

(b) $\dfrac{4 - n}{n + 6}$

Domain: $(-\infty, -6) \cup (-6, \infty)$

EXAMPLE 1   Finding the Domain of a Rational Function

Find the domain of each rational function.

(a) $f(x) = \dfrac{4}{x - 2}$ (b) $g(x) = \dfrac{2x + 5}{8}$ (c) $h(x) = 3x^2 + 2x - 5$

Solution

(a) The denominator is zero when $x = 2$. Therefore, the domain is all real values of x such that $x \neq 2$. In interval notation, you can write the domain as

Domain $= (-\infty, 2) \cup (2, \infty)$.

(b) The denominator, 8, is never zero, and hence the domain is the set of *all* real numbers. In interval notation, you can write the domain as

Domain $= (-\infty, \infty)$.

(c) Note that any polynomial is also a rational expression, because you can consider its denominator to be 1. The domain of this function is the set of all real numbers. In interval notation, you can write the domain as

Domain $= (-\infty, \infty)$.

EXAMPLE 2   Finding the Domain of a Rational Function

Find the domain of each rational function.

(a) $f(x) = \dfrac{3x}{x^2 - 25}$ (b) $g(x) = \dfrac{x^2 + 3x}{x^2 + 5x - 6}$

Solution

(a) The denominator of this rational function is

$x^2 - 25 = (x + 5)(x - 5)$.

Because this denominator is zero when $x = -5$ or when $x = 5$, the domain is all real values of x such that $x \neq -5$ and $x \neq 5$. In interval notation, you can write the domain as

Domain $= (-\infty, -5) \cup (-5, 5) \cup (5, \infty)$.

(b) The denominator of this rational function is

$x^2 + 5x - 6 = (x + 6)(x - 1)$.

Because this denominator is zero when $x = -6$ or when $x = 1$, the domain is all real values of x such that $x \neq -6$ and $x \neq 1$. In interval notation, you can write the domain as

Domain $= (-\infty, -6) \cup (-6, 1) \cup (1, \infty)$.

In applications involving rational functions, it is often necessary to further restrict the domain. To indicate such a restriction, write the domain to the right of the fraction. For instance, the domain of the rational function

$$f(x) = \frac{x^2 + 20}{x + 4}, \quad x > 0$$

is the set of positive real numbers $(0, \infty)$, as indicated by the inequality $x > 0$. (Note that the normal domain of this function would be all real values of x such that $x \neq -4$. However, because "$x > 0$" is listed to the right of the function, the domain is restricted by this inequality.)

EXAMPLE 3   An Application Involving a Restricted Domain

You have started a small manufacturing business. The initial investment for the business is $120,000. The cost of each unit that you manufacture is $15. Thus, your total cost of producing x units is

$$C = 15x + 120,000. \qquad \text{Cost function}$$

Your average cost per unit depends on the number of units produced. For instance, the average cost per unit $\overline{C}$ for producing 100 units is

$$\overline{C} = \frac{15(100) + 120,000}{100} = \$1215. \qquad \text{Average cost per unit for 100 units}$$

The average cost per unit decreases as the number of units increases. For instance, the average cost per unit $\overline{C}$ for producing 1000 units is

$$\overline{C} = \frac{15(1000) + 120,000}{1000} = \$135. \qquad \text{Average cost per unit for 1000 units}$$

In general, the average cost of producing x units is

$$\overline{C} = \frac{15x + 120,000}{x}. \qquad \text{Average cost per unit for } x \text{ units}$$

What is the domain of this rational function?

Solution

Point out that in solving applications we often have additional restrictions on the domain that are based on real-life conditions of the problem.

If you were considering this function from only a mathematical point of view, you would say that the domain is all real values of x such that $x \neq 0$. However, because this fraction is a mathematical model representing a real-life situation, you must consider which values of x make sense in real life. For this model, the variable x represents the number of units that you produce. Assuming that you cannot produce a fractional number of units, you conclude that the domain is the set of positive integers. That is,

$$\text{Domain} = \{1, 2, 3, 4, \ldots\}.$$

Simplifying Rational Expressions

As with numerical fractions, a rational expression is said to be **simplified** or **in reduced form** if its numerator and denominator have no factors in common (other than ± 1). To reduce fractions, you can apply the following rule.

NOTE You can verify the Reduction Rule shown at the right as follows.

$$\frac{uw}{vw} = \frac{u}{v} \cdot \frac{w}{w}$$

$$= \frac{u}{v} \cdot 1$$

$$= \frac{u}{v}$$

Reduction Rule for Fractions

Let u, v, and w represent numbers, variables, or algebraic expressions such that $v \neq 0$ and $w \neq 0$. Then the following Reduction Rule is valid.

$$\frac{u\cancel{w}}{v\cancel{w}} = \frac{u}{v}$$

Be sure you see that this Reduction Rule allows us to divide out only factors, not terms. For instance, consider the following.

$$\frac{2 \cdot 2}{2(x + 5)} \qquad \text{You can divide out common factor 2.}$$

$$\frac{3 + x}{3 + 2x} \qquad \text{You cannot divide out common term 3.}$$

Using the Reduction Rule to simplify a rational expression requires two steps: (1) completely factor the numerator and denominator and (2) apply the Reduction Rule to divide out any *factors* that are common to both the numerator and denominator. Thus, your success in simplifying rational expressions actually lies in your ability to *completely factor* the polynomials in both the numerator and denominator. You may want to review the factoring techniques discussed in Sections 1.5 and 1.6.

EXAMPLE 4 *Simplifying a Rational Expression*

Simplify $\dfrac{2x^3 - 6x}{6x^2}$.

Solution

To begin, completely factor both the numerator and denominator.

$$\frac{2x^3 - 6x}{6x^2} = \frac{2x(x^2 - 3)}{2x(3x)} \qquad \text{Factor numerator and denominator.}$$

$$= \frac{2\cancel{x}\,(x^2 - 3)}{2\cancel{x}(3x)} \qquad \text{Divide out common factor } 2x.$$

$$= \frac{x^2 - 3}{3x} \qquad \text{Simplified form}$$

EXAMPLE 5 Adjusting the Domain After Simplifying

Simplify $\dfrac{x^2 + 2x - 15}{4x - 12}$.

Solution

$$\frac{x^2 + 2x - 15}{4x - 12} = \frac{(x + 5)(x - 3)}{4(x - 3)} \qquad \text{Factor numerator and denominator.}$$

$$= \frac{(x + 5)\cancel{(x - 3)}}{4\cancel{(x - 3)}} \qquad \text{Divide out common factor } (x - 3).$$

$$= \frac{x + 5}{4}, \quad x \neq 3 \qquad \text{Simplified form}$$

Dividing out common factors from the numerator and denominator of a rational expression can change its domain. In Example 5, for instance, the domain of the original expression is all real values of x such that $x \neq 3$. Thus, the original expression is equal to the simplified expression for all real numbers *except* 3.

EXAMPLE 6 Simplifying a Rational Expression

Simplify $\dfrac{x^3 - 16x}{x^2 - 2x - 8}$.

Solution

$$\frac{x^3 - 16x}{x^2 - 2x - 8} = \frac{x(x^2 - 16)}{(x + 2)(x - 4)} \qquad \text{Partially factor.}$$

$$= \frac{x(x + 4)(x - 4)}{(x + 2)(x - 4)} \qquad \text{Factor completely.}$$

$$= \frac{x(x + 4)\cancel{(x - 4)}}{(x + 2)\cancel{(x - 4)}} \qquad \text{Divide out common factor } (x - 4).$$

$$= \frac{x(x + 4)}{x + 2}, \quad x \neq 4 \qquad \text{Simplified form}$$

In this text, when simplifying a rational expression, you follow the convention of listing *by the simplified expression* all values of x that must be specifically excluded from the domain in order to make the domains of the simplified and original expressions agree. In Example 6, for instance, the restriction $x \neq 4$ must be listed with the simplified expression in order to make the two domains agree. (Note that the value of -2 is excluded from both domains, so it is not necessary to list this value.)

Study Tip

Be sure to *completely* factor the numerator and denominator of a rational expression before concluding that there is no common factor. This may involve a change in sign to see if further reduction is possible. Note that the Distributive Property allows you to write $(b - a)$ as $-(a - b)$. Watch for this in Example 7.

EXAMPLE 7 Simplification Involving a Change of Sign

Simplify $\dfrac{2x^2 - 9x + 4}{12 + x - x^2}$.

Solution

$$\frac{2x^2 - 9x + 4}{12 + x - x^2} = \frac{(2x - 1)(x - 4)}{(4 - x)(3 + x)} \qquad \text{Factor numerator and denominator.}$$

$$= \frac{(2x - 1)(x - 4)}{-(x - 4)(3 + x)} \qquad (4 - x) = -(x - 4)$$

$$= \frac{(2x - 1)(x - 4)}{-(x - 4)(3 + x)} \qquad \text{Divide out common factor } (x - 4).$$

$$= -\frac{2x - 1}{3 + x}, \quad x \neq 4 \qquad \text{Simplified form}$$

The simplified form is equivalent to the original expression for all values of x except 4. (Note that -3 is excluded from the domains of both the original and simplified expressions.)

NOTE Here is another way to factor and simplify the expression in Example 7.

$$\frac{2x^2 - 9x + 4}{12 + x - x^2}$$

$$= \frac{2x^2 - 9x + 4}{-x^2 + x + 12}$$

$$= \frac{2x^2 - 9x + 4}{-1(x^2 - x - 12)}$$

$$= \frac{(2x - 1)(x - 4)}{-1(x - 4)(x + 3)}$$

$$= -\frac{2x - 1}{x + 3}, \quad x \neq 4$$

In Example 7, be sure you see that when dividing the numerator and denominator by the common factor of $(x - 4)$, you keep the minus sign. In the simplified form of the fraction, we usually like to move the minus sign out in front of the fraction. However, this is a personal preference. All of the following forms are legitimate.

$$-\frac{2x - 1}{3 + x} = \frac{-(2x - 1)}{3 + x} = \frac{2x - 1}{-3 - x} = \frac{2x - 1}{-(3 + x)}$$

In the next two examples, the Reduction Rule is used to simplify rational expressions that involve more than one variable.

EXAMPLE 8 A Rational Expression Involving Two Variables

Simplify $\dfrac{3xy + y^2}{2y}$.

Solution

$$\frac{3xy + y^2}{2y} = \frac{y(3x + y)}{2y} \qquad \text{Factor numerator and denominator.}$$

$$= \frac{y(3x + y)}{2y} \qquad \text{Divide out common factor } y.$$

$$= \frac{3x + y}{2}, \quad y \neq 0 \qquad \text{Simplified form}$$

On this page you can see various equivalent forms for the simplified rational expression of Example 7. You may want to discuss your preference with students.

EXAMPLE 9 A Rational Expression Involving Two Variables

Simplify $\dfrac{2x^2 + 2xy - 4y^2}{5x^3 - 5xy^2}$.

Solution

$$\frac{2x^2 + 2xy - 4y^2}{5x^3 - 5xy^2} = \frac{2(x^2 + xy - 2y^2)}{5x(x^2 - y^2)}$$ Partially factor.

$$= \frac{2(x - y)(x + 2y)}{5x(x - y)(x + y)}$$ Factor numerator and denominator.

$$= \frac{2(\cancel{x - y})(x + 2y)}{5x(\cancel{x - y})(x + y)}$$ Divide out common factor $(x - y)$.

$$= \frac{2(x + 2y)}{5x(x + y)}, \quad x \neq y$$ Simplified form

Remind students to factor completely both denominator and numerator.

As you study the examples and work the exercises in this and the following three sections, keep in mind that you are *rewriting expressions in simpler forms.* You are not solving equations. Equal signs are used in the steps of the simplification process only to indicate that the new form of the expression is *equivalent* to the previous one.

Application

The geometric model shown in Figure 7.1 can be used to find an algebraic model for the ratio of surface area to volume for a person or animal. The model is created from six rectangular boxes. The surface area S and volume V of each box are as follows.

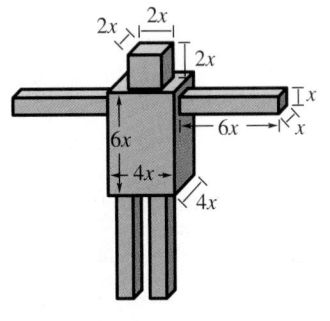

FIGURE 7.1

	Surface Area	Volume
Each Arm and Leg:	$S = x^2 + 4(6x^2)$ $= 25x^2$	$V = 6x^3$
Head:	$S = 5(4x^2)$ $= 20x^2$	$V = 8x^3$
Trunk:	$S = 2(16x^2) + 4(24x^2) - 4x^2 - 4x^2$ $= 120x^2$	$V = 96x^3$

NOTE The air resistance of an animal that is falling through the air depends on the ratio of the animal's surface area to its volume. The greater the ratio, the greater the air resistance. For instance, if a mouse and a human each falls 30 feet, the mouse will encounter a greater air resistance, which means that it will hit the ground at a lower speed.

EXAMPLE 10 Comparing Surface Area to Volume

Find a rational expression that represents the ratio of surface area to volume for the geometric model shown in Figure 7.1 on the previous page. Simplify the expression. Then evaluate the simplified expression for several values of x, where x is measured in feet.

Solution

$$\frac{\text{Surface area}}{\text{Volume}} = \frac{4(25x^2) + 20x^2 + 120x^2}{4(6x^3) + 8x^3 + 96x^3}$$

$$= \frac{240x^2}{128x^3}$$

$$= \frac{15}{8x}$$

In the following table, notice that small animals have large ratios and large animals have small ratios.

Animal	Mouse	Squirrel	Cat	Human	Elephant
x	$\frac{1}{24}$	$\frac{1}{12}$	$\frac{1}{6}$	$\frac{2}{5}$	2
Ratio	45	22.5	11.25	4.7	0.9

Group Activities   You Be the Instructor

Error Analysis Suppose you are the instructor of an algebra course. One of your students turns in the following incorrect simplifications. Find the errors, discuss the student's misconceptions, and construct correct simplifications.

a. $\dfrac{3x^2 + 5x - 4}{x} = 3x + 5 - 4 = 3x + 1$

b. $\dfrac{x^2 + 7x}{x + 7} = \dfrac{x^2}{x} + \dfrac{7x}{7} = x + x = 2x$

7.1   Exercises

(handwritten: $x^2 - 7x + 10$ over $x^2 - 4x - 5$; Quiz $\frac{x^2 + x - 2}{x^2 + 3x + 2}$)

1. *Think About It*   How do you determine whether a rational expression is in reduced form?

2. *Think About It*   Can you divide out common terms of the numerator and denominator of a rational expression?

3. *Error Analysis*   Describe the error.

$$\frac{2x^2}{x^2 + 4} = \frac{2x^2}{x^2 + 4} = \frac{2}{1 + 4} = \frac{2}{5}$$

4. *Think About It*   Is it true that $(6x - 5)/(5 - 6x) = -1$? Explain.

In Exercises 5–18, find the domain of the rational expression.

5. $\dfrac{5}{x - 8}$

6. $\dfrac{9}{x - 13}$

7. $\dfrac{7x}{x + 4}$

8. $x^4 + 2x^2 - 5$

9. $\dfrac{4}{x^2 + 9}$

10. $\dfrac{y^2 - 3}{7}$

11. $\dfrac{5t}{t^2 - 16}$

12. $\dfrac{z + 2}{z(z - 4)}$

13. $\dfrac{u^2}{u^2 - 4u - 5}$

14. $\dfrac{y + 5}{4y^2 - 5y - 6}$

15. $\dfrac{3x + 6}{4}$

16. $\dfrac{x}{x^2 - 4}$

17. $\dfrac{y + 5}{y^2 - 3y}$

18. $\dfrac{3t}{t^2 - 2t - 3}$

In Exercises 19–54, simplify the rational expression.

19. $\dfrac{5x}{25}$

20. $\dfrac{32y}{24}$

21. $\dfrac{12y^2}{2y}$

22. $\dfrac{15z^3}{15z^3}$

23. $\dfrac{18x^2y}{15xy^4}$

24. $\dfrac{16y^2z^2}{60y^5z}$

25. $\dfrac{x^2(x - 8)}{x(x - 8)}$

26. $\dfrac{a^2b(b - 3)}{b^3(b - 3)^2}$

27. $\dfrac{2x - 3}{4x - 6}$

28. $\dfrac{y^2 - 81}{2y - 18}$

29. $\dfrac{5 - x}{3x - 15}$

30. $\dfrac{x^2 - 36}{6 - x}$

31. $\dfrac{3xy^2}{xy^2 + x}$

32. $\dfrac{x + 3x^2y}{3xy + 1}$

33. $\dfrac{y^2 - 64}{5(3y + 24)}$

34. $\dfrac{x^2 - 25z^2}{x + 5z}$ *(handwritten: $\frac{x^2 - 25}{x^2 - 5x}$)*

35. $\dfrac{a + 3}{a^2 + 6a + 9}$

36. $\dfrac{u^2 - 12u + 36}{u - 6}$

37. $\dfrac{x^2 - 7x}{x^2 - 14x + 49}$

38. $\dfrac{z^2 + 22z + 121}{3z + 33}$

39. $\dfrac{y^3 - 4y}{y^2 + 4y - 12}$

40. $\dfrac{x^2 - 7x}{x^2 - 4x - 21}$

41. $\dfrac{3 - x}{2x^2 - 3x - 9}$

42. $\dfrac{2y^2 + 13y + 20}{2y^2 + 17y + 30}$

43. $\dfrac{15x^2 + 7x - 4}{15x^2 + x - 2}$

44. $\dfrac{56z^2 - 3z - 20}{49z^2 - 16}$

45. $\dfrac{5xy + 3x^2y^2}{xy^3}$

46. $\dfrac{4u^2v - 12uv^2}{18uv}$

47. $\dfrac{3m^2 - 12n^2}{m^2 + 4mn + 4n^2}$

48. $\dfrac{x^2 + xy - 2y^2}{x^2 + 3xy + 2y^2}$

49. $\dfrac{u^2 - 4v^2}{u^2 + uv - 2v^2}$

50. $\dfrac{x^2 + 4xy}{x^2 - 16y^2}$

51. $\dfrac{x^3 - 8y^3}{x^2 - 4y^2}$

52. $\dfrac{x^3 + 27z^3}{x^2 + xz - 6z^2}$

53. $\dfrac{uv - 3u - 4v + 12}{v^2 - 5v + 6}$

54. $\dfrac{mn + 3m - n^2 - 3n}{m^2 - n^2}$

In Exercises 55–58, describe the domain.

55. *Inventory Cost*   The inventory cost I when x units of a product are ordered from a supplier is

$$I = \frac{0.25x + 2000}{x}.$$

56. *Pollution Removal* The cost C in dollars of removing $p\%$ of the air pollutants from the stack emissions of a utility company is

$$C = \frac{80{,}000p}{100 - p}.$$

57. *Perimeter* A rectangle of length x inches has an area of 500 square inches. The perimeter P of the rectangle is

$$P = 2\left(x + \frac{500}{x}\right).$$

58. *Cost* The cost C in millions of dollars for the government to seize $p\%$ of a certain illegal drug as it enters the country is

$$C = \frac{528p}{100 - p}.$$

In Exercises 59 and 60, evaluate the rational expression at each value of x. If not possible, state the reason.

Expression   *Values*

59. $\dfrac{x - 10}{4x}$ (a) $x = 10$ (b) $x = 0$

(c) $x = -2$ (d) $x = 12$

60. $\dfrac{x^2 - 4x}{x^2 - 9}$ (a) $x = 0$ (b) $x = 4$

(c) $x = 3$ (d) $x = -3$

In Exercises 61–64, evaluate the rational function as indicated. If not possible, state the reason.

61. $f(x) = \dfrac{4x}{x + 3}$

(a) $f(1)$ (b) $f(-2)$ (c) $f(-3)$ (d) $f(0)$

62. $g(t) = \dfrac{t - 2}{2t - 5}$

(a) $g(2)$ (b) $g\left(\frac{5}{2}\right)$ (c) $g(-2)$ (d) $g(0)$

63. $h(s) = \dfrac{s^2}{s^2 - s - 2}$

(a) $h(10)$ (b) $h(0)$ (c) $h(-1)$ (d) $h(2)$

64. $f(x) = \dfrac{x^3 + 1}{x^2 - 6x + 9}$

(a) $f(-1)$ (b) $f(3)$ (c) $f(-2)$ (d) $f(2)$

In Exercises 65–68, explain how you can show that the two expressions are not equivalent.

65. $\dfrac{x - 4}{4} \neq x - 1$ **66.** $\dfrac{x - 4}{x} \neq -4$

67. $\dfrac{3x + 2}{4x + 2} \neq \dfrac{3}{4}$ **68.** $\dfrac{1 - x}{2 - x} \neq \dfrac{1}{2}$

In Exercises 69–72, write the rational expression in reduced form. (Assume n is a positive integer.)

69. $\dfrac{x^{n+1} - 3x}{x}$ **70.** $\dfrac{x^{2n}}{x^{2n+1} + x^{2n}}$

71. $\dfrac{x^{2n} - 4}{x^n + 2}$ **72.** $\dfrac{x^{2n} + x^n - 12}{x^{n+1} + 4x}$

Geometry In Exercises 73 and 74, find the ratio of the shaded portion of the figure to the total area of the figure.

73.

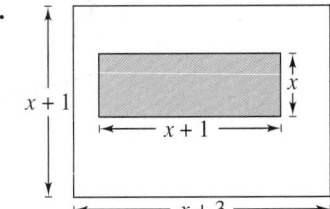

74.

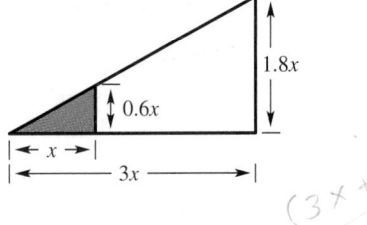

75. *Geometry* One swimming pool is circular and another is rectangular. The rectangular pool's width is three times its depth, and its length is 6 feet more than its width. The circular pool has a diameter that is twice the width of the rectangular pool, and it is 2 feet deeper. Find the ratio of the volume of the circular pool to the volume of the rectangular pool.

76. *Average Cost* A machine shop has a setup cost of $2500 for the production of a new product. The cost for labor and material to produce each unit is $9.25.

(a) Write a rational expression that gives the average cost per unit when x units are produced.

(b) Determine the domain of the expression in part (a).

(c) Find the average cost per unit when $x = 100$ units.

Cost of Medicare In Exercises 77 and 78, use the following models, which give the cost of Medicare and the U.S. population aged 65 and older from 1988 through 1993 (see figures).

$$C = 113.66 + 12.64t \qquad \text{Cost of Medicare}$$

$$P = 31.20 + 0.54t \qquad \text{Population}$$

In these models, C represents the total annual cost of Medicare (in billions of dollars), P represents the U.S. population aged 65 and older (in millions), and t represents the year, with $t = 0$ corresponding to 1990. *(Source: Congressional Budget Office and U.S. Bureau of Census)*

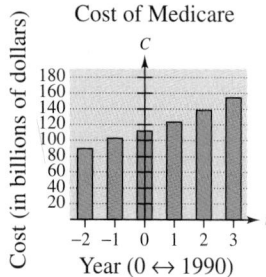

77. Find a rational model that represents the average cost of Medicare per person aged 65 and older during the years 1988 through 1993.

78. Use the model found in Exercise 77 to find the average cost of Medicare per person aged 65 and older for the years 1988 through 1993. Organize your results in a table.

79. *Distance Traveled* A van starts on a trip and travels at an average speed of 45 miles per hour. Three hours later, a car starts on the same trip and travels at an average speed of 60 miles per hour (see figure).

(a) Find the distances d_1 and d_2 each vehicle has traveled when the car has been on the road for t hours.

(b) Use the result in part (a) to determine the ratio of the distance the car has traveled to the distance the van has traveled.

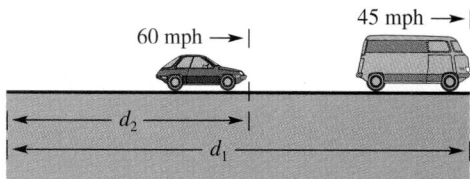

Section Project

(a) Complete the table.

x	-2	-1	0	1	2	3	4
$\dfrac{x^2 - x - 2}{x - 2}$							
$x + 1$							

(b) Write a paragraph describing the equivalence (or nonequivalence) of the two expressions in the table. Support your argument with appropriate algebra from this section *and* a discussion of the domains of the two expressions.

(c) Repeat parts (a) and (b) for the table below.

x	-2	-1	0	1	2	3	4
$\dfrac{x^2 + 5x}{x}$							
$x + 5$							

7.2	**Multiplying and Dividing Rational Expressions**
	Multiplying Rational Expressions ▪ Dividing Rational Expressions ▪ Complex Fractions

Multiplying Rational Expressions

The rule for multiplying rational expressions is the same as the rule for multiplying numerical fractions.

$$\frac{3}{4} \cdot \frac{7}{6} = \frac{21}{24} = \frac{3 \cdot 7}{3 \cdot 8} = \frac{7}{8}$$

That is, you *multiply numerators, multiply denominators, and write the new fraction in reduced form.*

Multiplying Rational Expressions

Let u, v, w, and z be real numbers, variables, or algebraic expressions such that $v \neq 0$ and $z \neq 0$. Then the product of u/v and w/z is

$$\frac{u}{v} \cdot \frac{w}{z} = \frac{uw}{vz}.$$

In order to recognize common factors, write the numerators and denominators in factored form, as demonstrated in Example 1.

NOTE   The result in Example 1 can be read as

$$\frac{4x^3y}{3xy^4} \cdot \frac{-6x^2y^2}{10x^4}$$

is equal to

$$-\frac{4}{5y}$$

for all values of x except 0.

Remind students at the beginning that the restrictions on the original expression are $x \neq 0$ and $y \neq 0$.

EXAMPLE 1   Multiplying Rational Expressions

Multiply the rational expressions.

$$\frac{4x^3y}{3xy^4} \cdot \frac{-6x^2y^2}{10x^4}$$

Solution

$$\frac{4x^3y}{3xy^4} \cdot \frac{-6x^2y^2}{10x^4} = \frac{(4x^3y) \cdot (-6x^2y^2)}{(3xy^4) \cdot (10x^4)}$$ Multiply numerators and denominators.

$$= \frac{-24x^5y^3}{30x^5y^4}$$ Simplify.

$$= \frac{-4(6)(x^5)(y^3)}{5(6)(x^5)(y^3)(y)}$$ Factor and divide out common factors.

$$= -\frac{4}{5y}, \quad x \neq 0$$ Simplified form

EXAMPLE 2   Multiplying Rational Expressions

Multiply the rational expressions.

$$\frac{x}{5x^2 - 20x} \cdot \frac{x - 4}{2x^2 + x - 3}$$

Solution

$$\frac{x}{5x^2 - 20x} \cdot \frac{x - 4}{2x^2 + x - 3}$$

$$= \frac{x \cdot (x - 4)}{(5x^2 - 20x) \cdot (2x^2 + x - 3)} \qquad \text{Multiply numerators and denominators.}$$

$$= \frac{x(x - 4)}{5x(x - 4)(x - 1)(2x + 3)} \qquad \text{Factor.}$$

$$= \frac{\cancel{x}(\cancel{x - 4})}{5\cancel{x}(\cancel{x - 4})(x - 1)(2x + 3)} \qquad \text{Divide out common factors.}$$

$$= \frac{1}{5(x - 1)(2x + 3)}, \quad x \neq 0, x \neq 4 \qquad \text{Simplified form}$$

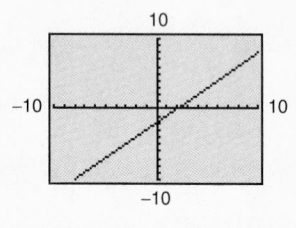

~ **Technology**

You can use a graphing utility to check your results when multiplying rational expressions. In Example 3, for instance, try graphing the equations

$$y_1 = \frac{4x^2 - 4x}{x^2 + 2x - 3} \cdot \frac{x^2 + x - 6}{4x}$$

and

$$y_2 = x - 2$$

in the same viewing rectangle. If the two graphs coincide, as shown below, you can conclude that the solution checks.

EXAMPLE 3   Multiplying Rational Expressions

Multiply the rational expressions.

$$\frac{4x^2 - 4x}{x^2 + 2x - 3} \cdot \frac{x^2 + x - 6}{4x}$$

Solution

$$\frac{4x^2 - 4x}{x^2 + 2x - 3} \cdot \frac{x^2 + x - 6}{4x}$$

$$= \frac{4x(x - 1)(x + 3)(x - 2)}{(x - 1)(x + 3)(4x)} \qquad \text{Multiply and factor.}$$

$$= \frac{4\cancel{x}(\cancel{x - 1})(\cancel{x + 3})(x - 2)}{(\cancel{x - 1})(\cancel{x + 3})(4\cancel{x})} \qquad \text{Divide out common factors.}$$

$$= x - 2, \quad x \neq 0, x \neq 1, x \neq -3 \qquad \text{Simplified form}$$

The rule for multiplying rational expressions can be extended to cover products involving expressions that are not in fractional form. To do this, rewrite the nonfractional expression as a fraction whose denominator is 1. Here is a simple example.

Point out that any polynomial expression can be rewritten as a rational expression with denominator equal to 1.

$$\frac{x + 3}{x - 2} \cdot (5x) = \frac{x + 3}{x - 2} \cdot \frac{5x}{1} = \frac{(x + 3)(5x)}{x - 2} = \frac{5x(x + 3)}{x - 2}$$

In the next example, note how to divide out a factor that differs only in sign. Note the step in which $(y - x)$ is rewritten as $(-1)(x - y)$.

EXAMPLE 4   *Multiplying Rational Expressions*

Multiply the rational expressions.

$$\frac{x - y}{y^2 - x^2} \cdot \frac{x^2 - xy - 2y^2}{3x - 6y}$$

Solution

In Example 4, remind students that $a - b = -(b - a)$. This step is critical in the simplification process.

$$\frac{x - y}{y^2 - x^2} \cdot \frac{x^2 - xy - 2y^2}{3x - 6y}$$

$$= \frac{(x - y)(x - 2y)(x + y)}{(y + x)(y - x)(3)(x - 2y)} \qquad \text{Multiply and factor.}$$

$$= \frac{(x - y)(x - 2y)(x + y)}{(y + x)(-1)(x - y)(3)(x - 2y)} \qquad y - x = -1(x - y)$$

$$= \frac{\cancel{(x - y)}\cancel{(x - 2y)}\cancel{(x + y)}}{\cancel{(x + y)}(-1)\cancel{(x - y)}(3)\cancel{(x - 2y)}} \qquad \begin{array}{l}\text{Divide out common} \\ \text{factors.}\end{array}$$

$$= -\frac{1}{3}, \quad x \neq y, \ x \neq -y, \ x \neq 2y \qquad \text{Simplified form}$$

The rule for multiplying rational expressions can be extended to cover the product of three or more fractions, as shown in Example 5.

EXAMPLE 5   *Multiplying Three Rational Expressions*

Multiply the rational expressions.

$$\frac{x^2 - 3x + 2}{x + 2} \cdot \frac{3x}{x - 2} \cdot \frac{2x + 4}{x^2 - 5x}$$

Solution

$$\frac{x^2 - 3x + 2}{x + 2} \cdot \frac{3x}{x - 2} \cdot \frac{2x + 4}{x^2 - 5x}$$

$$= \frac{(x - 1)(x - 2)(3)(x)(2)(x + 2)}{(x + 2)(x - 2)(x)(x - 5)} \qquad \text{Multiply and factor.}$$

$$= \frac{(x - 1)\cancel{(x - 2)}(3)\cancel{(x)}(2)\cancel{(x + 2)}}{\cancel{(x + 2)}\cancel{(x - 2)}\cancel{(x)}(x - 5)} \qquad \begin{array}{l}\text{Divide out common} \\ \text{factors.}\end{array}$$

$$= \frac{6(x - 1)}{x - 5}, \quad x \neq 0, \ x \neq 2, \ x \neq -2 \qquad \text{Simplified form}$$

Dividing Rational Expressions

To divide two rational expressions, multiply the first fraction by the *reciprocal* of the second. That is, simply *invert the divisor and multiply*. For instance, to perform the following division.

$$\frac{x}{x + 3} \div \frac{4}{x - 1}$$

invert the fraction $4/(x - 1)$ and multiply, as follows.

$$\frac{x}{x + 3} \div \frac{4}{x - 1} = \frac{x}{x + 3} \cdot \frac{x - 1}{4}$$

$$= \frac{x(x - 1)}{(x + 3)(4)}$$

$$= \frac{x(x - 1)}{4(x + 3)}, \quad x \neq 1$$

Dividing Rational Expressions

Let u, v, w, and z be real numbers, variables, or algebraic expressions such that $v \neq 0$, $w \neq 0$, and $z \neq 0$. Then the quotient of u/v and w/z is

$$\frac{u}{v} \div \frac{w}{z} = \frac{u}{v} \cdot \frac{z}{w} = \frac{uz}{vw}.$$

EXAMPLE 6   *Dividing Rational Expressions*

Divide the rational expressions.

$$\frac{2x}{3x - 12} \div \frac{x^2 - 2x}{x^2 - 6x + 8}$$

Solution

$$\frac{2x}{3x - 12} \div \frac{x^2 - 2x}{x^2 - 6x + 8}$$

$$= \frac{2x}{3x - 12} \cdot \frac{x^2 - 6x + 8}{x^2 - 2x} \qquad \text{Invert divisor and multiply.}$$

$$= \frac{(2x)(x - 2)(x - 4)}{(3)(x - 4)(x)(x - 2)} \qquad \text{Factor.}$$

$$= \frac{(2x)(x - 2)(x - 4)}{(3)(x - 4)(x)(x - 2)} \qquad \text{Divide out common factors.}$$

$$= \frac{2}{3}, \quad x \neq 0, \ x \neq 2, \ x \neq 4 \qquad \text{Simplified form}$$

Complex Fractions

Problems involving the division of two rational expressions are sometimes written as **complex fractions.** The rules for dividing fractions still apply in such cases. For instance, consider the following complex fraction.

Complex fractions are given more extensive treatment in the next section where these ideas are extended to the sum or difference of rational expressions.

$$\frac{\left(\dfrac{x-1}{5}\right)\}}{\left(\dfrac{x+3}{x}\right)\}}$$

Numerator fraction

→ Main fraction line

Denominator fraction

(Note that for complex fractions you make the main fraction line slightly longer than the fraction lines in the numerator and denominator.) To perform the division implied by this complex fraction, invert the denominator and multiply:

$$\frac{\left(\dfrac{x-1}{5}\right)}{\left(\dfrac{x+3}{x}\right)} = \frac{x-1}{5} \cdot \frac{x}{x+3} = \frac{x(x-1)}{5(x+3)}, \quad x \neq 0.$$

EXAMPLE 7   *Simplifying a Complex Fraction*

Simplify the complex fraction $\dfrac{\left(\dfrac{x^2+2x-3}{x-3}\right)}{4x+12}$.

Solution

Begin by converting the denominator to fraction form.

$$\frac{\left(\dfrac{x^2+2x-3}{x-3}\right)}{4x+12} = \frac{\left(\dfrac{x^2+2x-3}{x-3}\right)}{\left(\dfrac{4x+12}{1}\right)} \qquad \text{Rewrite denominator.}$$

$$= \frac{x^2+2x-3}{x-3} \cdot \frac{1}{4x+12} \qquad \text{Invert divisor and multiply.}$$

$$= \frac{(x-1)(x+3)}{(x-3)(4)(x+3)} \qquad \text{Factor.}$$

$$= \frac{(x-1)(x+3)}{(x-3)(4)(x+3)} \qquad \text{Divide out common factor.}$$

$$= \frac{x-1}{4(x-3)}, \quad x \neq -3 \qquad \text{Simplified form}$$

Group Activities   **Extending the Concept**

Using a Table Complete the table with the given values of *x*.

x	60	100	1000	10,000	100,000	1,000,000
$\dfrac{x-10}{x+10}$						
$\dfrac{x+50}{x-50}$						
$\dfrac{x-10}{x+10} \cdot \dfrac{x+50}{x-50}$						

What kind of pattern do you see? Try to explain it. Can you see why?

7.2 Exercises

See Warm-Up
Exercises, p. A46

In Exercises 1–26, multiply and simplify.

1. $\dfrac{7x^2}{3} \cdot \dfrac{9}{14x}$

2. $\dfrac{6}{5a} \cdot (25a)$

3. $\dfrac{8s^3}{9s} \cdot \dfrac{6s^2}{32s}$

4. $\dfrac{3x^4}{7x} \cdot \dfrac{8x^2}{9}$

5. $16u^4 \cdot \dfrac{12}{8u^2}$

6. $\dfrac{25}{8x} \cdot \dfrac{8x}{35}$

7. $\dfrac{8}{3+4x} \cdot (9+12x)$

8. $\dfrac{1-3x}{4} \cdot \dfrac{46}{15-45x}$

9. $\dfrac{8u^2v}{3u+v} \cdot \dfrac{u+v}{12u}$

10. $\dfrac{x+25}{8} \cdot \dfrac{8}{x+25}$

11. $\dfrac{12-r}{3} \cdot \dfrac{3}{r-12}$

12. $\dfrac{8-z}{8+z} \cdot \dfrac{z+8}{z-8}$

13. $\dfrac{6r}{r-2} \cdot \dfrac{r^2-4}{33r^2}$

14. $\dfrac{5y-20}{5y+15} \cdot \dfrac{2y+6}{y-4}$

15. $\dfrac{(2x-3)(x+8)}{x^3} \cdot \dfrac{x}{3-2x}$

16. $\dfrac{x+14}{x^3(10-x)} \cdot \dfrac{x(x-10)}{5}$

17. $\dfrac{2t^2-t-15}{t+2} \cdot \dfrac{t^2-t-6}{t^2-6t+9}$

18. $\dfrac{y^2-16}{2y^3} \cdot \dfrac{4y}{y^2-6y+8}$

19. $(x^2-4y^2) \cdot \dfrac{xy}{(x-2y)^2}$

20. $\dfrac{x^2+2xy-3y^2}{(x+y)^2} \cdot \dfrac{x^2-y^2}{x+3y}$

21. $(u-2v)^2 \cdot \dfrac{u+2v}{u-2v}$

22. $\dfrac{x-2y}{x+2y} \cdot \dfrac{x^2+4y^2}{x^2-4y^2}$

23. $\dfrac{x+5}{x-5} \cdot \dfrac{2x^2-9x-5}{3x^2+x-2} \cdot \dfrac{x^2-1}{x^2+7x+10}$

24. $\dfrac{x^3+3x^2-4x-12}{x^3-3x^2-4x+12} \cdot \dfrac{x^2-9}{x}$

25. $\dfrac{xu-yu+xv-yv}{xu+yu-xv-yv} \cdot \dfrac{xu+yu+xv+yv}{xu-yu-xv+yv}$

26. $\dfrac{t^2+4t+3}{2t^2-t-10} \cdot \dfrac{t}{t^2+3t+2} \cdot \dfrac{2t^2+4t^3}{t^2+3t}$

In Exercises 27–40, divide and simplify.

27. $\dfrac{7xy^2}{10u^2v} \div \dfrac{21x^3}{45uv}$

28. $\dfrac{25x^2y}{60x^3y^2} \div \dfrac{5x^4y^3}{16x^2y}$

29. $\dfrac{3(a+b)}{4} \div \dfrac{(a+b)^2}{2}$

30. $\dfrac{x^2+9}{5x+10} \div \dfrac{x+3}{5x^2-20}$

31. $\dfrac{(x^3y)^2}{(x+2y)^2} \div \dfrac{x^2y}{(x+2y)^3}$

32. $\dfrac{x^2-y^2}{2x^2-8x} \div \dfrac{(x-y)^2}{2xy}$

33. $\dfrac{\left(\dfrac{x^2}{12}\right)}{\left(\dfrac{5x}{18}\right)}$

34. $\dfrac{\left[\dfrac{3(u^2v)^2}{6v^3}\right]}{\left[\dfrac{(uv^3)^2}{3uv}\right]}$

35. $\dfrac{\left(\dfrac{25x^2}{x-5}\right)}{\left(\dfrac{10x}{5-x}\right)}$

36. $\dfrac{\left(\dfrac{5x}{x+7}\right)}{\left(\dfrac{10}{x^2+8x+7}\right)}$

37. $\dfrac{16x^2+8x+1}{3x^2+8x-3} \div \dfrac{4x^2-3x-1}{x^2+6x+9}$

38. $\dfrac{x^2-25}{x} \div \dfrac{x^3-5x^2}{x^2+x}$

39. $\dfrac{x(x+3)-2(x+3)}{x^2-4} \div \dfrac{x}{x^2+4x+4}$

40. $\dfrac{t^3+t^2-9t-9}{t^2-5t+6} \div \dfrac{t^2+6t+9}{t-2}$

In Exercises 41–48, perform the operations and simplify. (In Exercises 47 and 48, n is a positive integer.)

41. $\left[\dfrac{x^2}{9} \cdot \dfrac{3(x+4)}{x^2+2x}\right] \div \dfrac{x}{x+2}$

42. $\left(\dfrac{x^2+6x+9}{x^2} \cdot \dfrac{2x+1}{x^2-9}\right) \div \dfrac{4x^2+4x+1}{x^2-3x}$

43. $\left[\dfrac{xy+y}{4x} \div (3x+3)\right] \div \dfrac{y}{3x}$

44. $\left(\dfrac{3u^2-u-4}{u^2}\right)^2 \div \dfrac{3u^2+12u+4}{u^4-3u^3}$

45. $\dfrac{2x^2+5x-25}{3x^2+5x+2} \cdot \dfrac{3x^2+2x}{x+5} \div \left(\dfrac{x}{x+1}\right)^2$

46. $\dfrac{t^2-100}{4t^2} \cdot \dfrac{t^3-5t^2-50t}{t^4+10t^3} \div \dfrac{(t-10)^2}{5t}$

47. $x^3 \cdot \dfrac{x^{2n}-9}{x^{2n}+4x^n+3} \div \dfrac{x^{2n}-2x^n-3}{x}$

48. $\dfrac{x^{n+1}-8x}{x^{2n}+2x^n+1} \cdot \dfrac{x^{2n}-4x^n-5}{x} \div x^n$

In Exercises 49–52, use a graphing utility to graph the two equations in the same viewing rectangle. Use the graphs to verify that the expressions are equivalent. Verify the results algebraically.

49. $y_1 = \dfrac{3x+2}{x} \cdot \dfrac{x^2}{9x^2-4}$

$y_2 = \dfrac{x}{3x-2}, x \neq 0, x \neq -\dfrac{2}{3}$

50. $y_1 = \dfrac{x^2-10x+25}{x^2-25} \cdot \dfrac{x+5}{2}$

$y_2 = \dfrac{x-5}{2}, x \neq \pm 5$

51. $y_1 = \dfrac{3x+15}{x^4} \div \dfrac{x+5}{x^2}$

$y_2 = \dfrac{3}{x^2}, x \neq -5$

52. $y_1 = (x^2+6x+9) \cdot \dfrac{3}{2x(x+3)}$

$y_2 = \dfrac{3(x+3)}{2x}, x \neq -3$

The symbol ▦ indicates an exercise in which you are instructed to use a calculator or graphing utility. The solutions of other exercises may also be facilitated by the use of appropriate technology.

In Exercises 53 and 54, write an expression for the area of the shaded region of the figure. Then simplify the expression.

53.

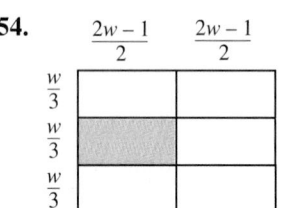

54.

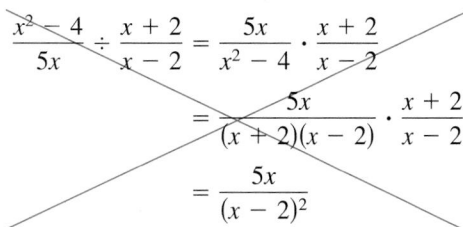

55. *Error Analysis* Describe the error.

$$\frac{x^2 - 4}{5x} \div \frac{x + 2}{x - 2} = \frac{5x}{x^2 - 4} \cdot \frac{x + 2}{x - 2}$$

$$= \frac{5x}{(x + 2)(x - 2)} \cdot \frac{x + 2}{x - 2}$$

$$= \frac{5x}{(x - 2)^2}$$

Section Project

Probability   Consider an experiment in which a marble is tossed into a rectangular box with dimensions x centimeters by $2x + 1$ centimeters. The probability that the marble will come to rest in the *unshaded* portion of the box is equal to the ratio of the unshaded area to the total area of the figure. Find the probability.

(a)

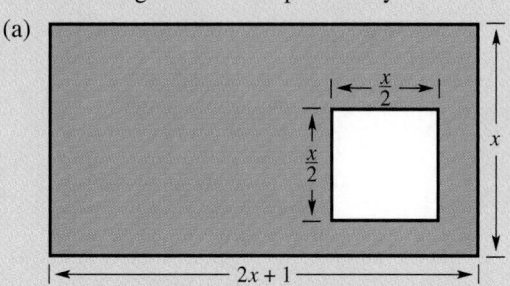

(b)

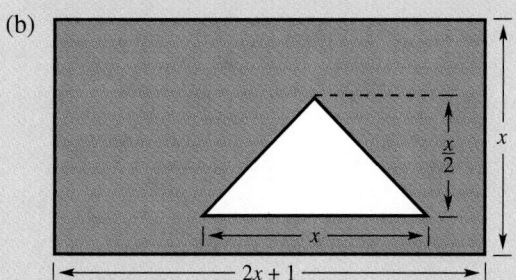

(c)

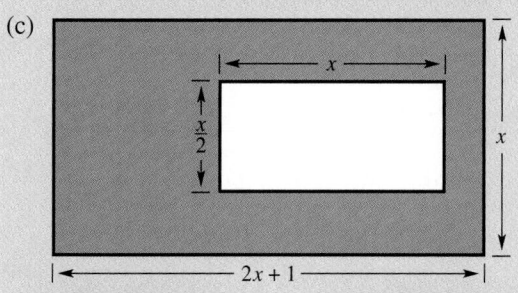

(d)

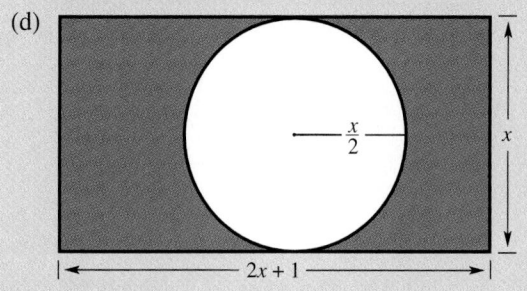

| **7.3** | **Adding and Subtracting Rational Expressions** |

Adding or Subtracting with Like Denominators ▪ Adding or Subtracting with Unlike Denominators ▪ Complex Fractions ▪ Application

Adding or Subtracting with Like Denominators

As with numerical fractions, the procedure used to add or subtract two rational expressions depends on whether the expressions have *like* or *unlike* denominators. To add or subtract two rational expressions with *like* denominators, simply combine their numerators and place the result over the common denominator.

Adding or Subtracting with Like Denominators

If u, v, and w are real numbers, variables, or algebraic expressions, and $w \neq 0$, the following rules are valid.

1. $\dfrac{u}{w} + \dfrac{v}{w} = \dfrac{u + v}{w}$ Add fractions with like denominators.

2. $\dfrac{u}{w} - \dfrac{v}{w} = \dfrac{u - v}{w}$ Subtract fractions with like denominators.

EXAMPLE 1   *Adding and Subtracting with Like Denominators*

(a) $\dfrac{x}{4} + \dfrac{5 - x}{4} = \dfrac{x + (5 - x)}{4} = \dfrac{5}{4}$

(b) $\dfrac{7}{2x - 3} - \dfrac{3x}{2x - 3} = \dfrac{7 - 3x}{2x - 3}$

Study Tip

After adding or subtracting two (or more) rational expressions, check the resulting fraction to see if it can be simplified, as illustrated in Example 2.

EXAMPLE 2   *Subtracting Rational Expressions and Simplifying*

$$\frac{x}{x^2 - 2xy - 3y^2} - \frac{3y}{x^2 - 2xy - 3y^2} = \frac{x - 3y}{x^2 - 2xy - 3y^2} \qquad \text{Subtract.}$$

$$= \frac{x - 3y}{(x - 3y)(x + y)} \qquad \text{Factor.}$$

$$= \frac{(x - 3y)(1)}{(x - 3y)(x + y)} \qquad \text{Divide out common factor.}$$

$$= \frac{1}{x + y}, \quad x \neq 3y \qquad \text{Simplified form}$$

The rule for adding or subtracting rational expressions with like denominators can be extended to cover sums and differences involving three or more rational expressions, as illustrated in Example 3.

EXAMPLE 3   Combining Three Rational Expressions

$$\frac{x^2 - 26}{x - 5} - \frac{2x + 4}{x - 5} + \frac{10 + x}{x - 5} = \frac{(x^2 - 26) - (2x + 4) + (10 + x)}{x - 5}$$

$$= \frac{x^2 - 26 - 2x - 4 + 10 + x}{x - 5}$$

$$= \frac{x^2 - x - 20}{x - 5}$$

$$= \frac{(x - 5)(x + 4)}{x - 5}$$

$$= x + 4, \quad x \neq 5$$

Encourage students to use parentheses in the first step of combining. This will help ensure that parentheses will be used for subtraction when necessary.

Study Tip

To find the least common multiple of polynomials, use the following steps.

1. Factor each polynomial completely.

2. Multiply the highest powers of the factors.

This is demonstrated in Example 4.

Adding or Subtracting with Unlike Denominators

To add or subtract rational expressions with *unlike* denominators, you must first rewrite each expression using the **least common multiple** of the denominators of the individual expressions. The least common multiple of two (or more) polynomials is the simplest polynomial that is a multiple of each of the original polynomials.

EXAMPLE 4   Finding Least Common Multiples

(a) The least common multiple of

$$6x = 2 \cdot 3 \cdot x \quad \text{and} \quad 2x^2 = 2 \cdot x^2$$

is $2 \cdot 3 \cdot x^2 = 6x^2$.

(b) The least common multiple of

$$x^2 - x = x(x - 1) \quad \text{and} \quad 2x - 2 = 2(x - 1)$$

is $2x(x - 1)$.

(c) The least common multiple of

$$3x^2 + 6x = 3x(x + 2) \quad \text{and} \quad x^2 + 4x + 4 = (x + 2)^2$$

is $3x(x + 2)^2$.

Additional problems: Find the least common multiple of each of the following.

(a) $30y^3$ and $45y^2$

(b) $x^2 - 5x + 6$ and $x^2 - 4$

(c) $12x - 24$ and $8x^2 - 8x - 16$

Answers:

(a) $90y^3$

(b) $(x - 2)(x - 3)(x + 2)$

(c) $24(x - 2)(x + 1)$

To add or subtract rational expressions with *unlike* denominators, you must first rewrite the rational expressions so that they have *like* denominators. The like denominator that you use is the least common multiple of the original denominators and is called the **least common denominator** (or **LCD**) of the original rational expressions. Once the rational expressions have been written with like denominators, you can simply add or subtract the rational expressions using the rule given at the beginning of this section.

Technology

You can use a graphing utility to check your results when adding or subtracting rational expressions. In Example 5, for instance, try graphing the equations

$$y_1 = \frac{7}{6x} + \frac{5}{8x}$$

and

$$y_2 = \frac{43}{24x}$$

in the same viewing rectangle. If the two graphs coincide, as shown below, you can conclude that the solution checks.

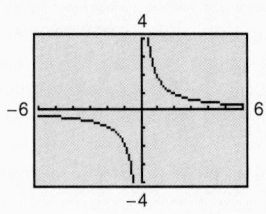

EXAMPLE 5   Adding with Unlike Denominators

Add the rational expressions: $\dfrac{7}{6x} + \dfrac{5}{8x}$.

Solution

By factoring the denominators, $6x = 2 \cdot 3 \cdot x$ and $8x = 2^3 \cdot x$, you can conclude that the least common denominator is $2^3 \cdot 3 \cdot x = 24x$.

$$\frac{7}{6x} + \frac{5}{8x} = \frac{7(4)}{6x(4)} + \frac{5(3)}{8x(3)} \qquad \text{Rewrite fractions using least common denominator.}$$

$$= \frac{28}{24x} + \frac{15}{24x} \qquad \text{Like denominators}$$

$$= \frac{28 + 15}{24x} \qquad \text{Add fractions.}$$

$$= \frac{43}{24x} \qquad \text{Simplified form}$$

EXAMPLE 6   Subtracting with Unlike Denominators

Subtract the rational expressions: $\dfrac{3}{x - 3} - \dfrac{5}{x + 2}$.

Solution

The only factors of the denominators are $(x - 3)$ and $(x + 2)$. Therefore, the least common denominator is $(x - 3)(x + 2)$.

$$\frac{3}{x - 3} - \frac{5}{x + 2} = \frac{3(x + 2)}{(x - 3)(x + 2)} - \frac{5(x - 3)}{(x - 3)(x + 2)}$$

$$= \frac{3x + 6}{(x - 3)(x + 2)} - \frac{5x - 15}{(x - 3)(x + 2)}$$

$$= \frac{(3x + 6) - (5x - 15)}{(x - 3)(x + 2)}$$

$$= \frac{3x + 6 - 5x + 15}{(x - 3)(x + 2)}$$

$$= \frac{-2x + 21}{(x - 3)(x + 2)}$$

EXAMPLE 7   Adding with Unlike Denominators

Add the rational expressions: $\dfrac{6x}{x^2 - 4} + \dfrac{3}{2 - x}$.

Solution

NOTE   In Example 7, note that the factors in the denominator are $x^2 - 4 = (x + 2)(x - 2)$ and $2 - x$. Using the fact that

$$(2 - x) = (-1)(x - 2),$$

you can rewrite the second fraction by multiplying its numerator and denominator by -1.

$$\frac{6x}{x^2 - 4} + \frac{3}{2 - x} = \frac{6x}{x^2 - 4} + \frac{(-1)(3)}{(-1)(2 - x)}$$

Multiply numerator and denominator of second fraction by -1.

$$= \frac{6x}{(x + 2)(x - 2)} + \frac{-3}{x - 2}$$

Simplify.

$$= \frac{6x}{(x + 2)(x - 2)} + \frac{-3(x + 2)}{(x + 2)(x - 2)}$$

Rewrite fractions using least common denominator.

$$= \frac{6x}{(x + 2)(x - 2)} + \frac{-3x - 6}{(x + 2)(x - 2)}$$

Like denominators

$$= \frac{6x - 3x - 6}{(x + 2)(x - 2)}$$

Add fractions.

$$= \frac{3x - 6}{(x + 2)(x - 2)}$$

Combine like terms.

$$= \frac{3(x - 2)}{(x + 2)(x - 2)}$$

Divide out common factor $(x - 2)$.

$$= \frac{3}{x + 2}, \quad x \neq 2$$

Simplified form

EXAMPLE 8   Combining Three Rational Expressions

Point out that you could use any common denominator, but that the use of the *least* common denominator simplifies the process.

$$\frac{2x - 5}{6x + 9} - \frac{4}{2x^2 + 3x} + \frac{1}{x} = \frac{(2x - 5)(x)}{3(2x + 3)(x)} - \frac{(4)(3)}{x(2x + 3)(3)} + \frac{3(2x + 3)}{(x)(3)(2x + 3)}$$

$$= \frac{2x^2 - 5x}{3x(2x + 3)} - \frac{12}{3x(2x + 3)} + \frac{6x + 9}{3x(2x + 3)}$$

$$= \frac{2x^2 - 5x - 12 + 6x + 9}{3x(2x + 3)}$$

$$= \frac{2x^2 + x - 3}{3x(2x + 3)}$$

$$= \frac{(x - 1)(2x + 3)}{3x(2x + 3)}$$

$$= \frac{(x - 1)(2x + 3)}{3x(2x + 3)}$$

$$= \frac{x - 1}{3x}, \quad x \neq -\frac{3}{2}$$

Complex Fractions

NOTE Another way to simplify the complex fraction given in Example 9 is to multiply the numerator and denominator by the least common denominator of *every* fraction in the numerator and denominator. For this fraction, when you multiply the numerator and denominator by $4x$, you get the same result.

Complex fractions can have numerators or denominators that are the sums or differences of fractions. To simplify a complex fraction, first combine its numerator and its denominator into single fractions. Then divide by inverting the denominator and multiplying.

$$\frac{\left(\dfrac{x}{4} + \dfrac{3}{2}\right)}{\left(2 - \dfrac{3}{x}\right)} = \frac{\left(\dfrac{x}{4} + \dfrac{3}{2}\right) \cdot 4x}{\left(2 - \dfrac{3}{x}\right) \cdot 4x}$$

$$= \frac{\dfrac{x}{4}(4x) + \dfrac{3}{2}(4x)}{2(4x) - \dfrac{3}{x}(4x)}$$

$$= \frac{x^2 + 6x}{8x - 12}$$

$$= \frac{x(x + 6)}{4(2x - 3)}, \quad x \neq 0$$

EXAMPLE 9   Simplifying a Complex Fraction

$$\frac{\left(\dfrac{x}{4} + \dfrac{3}{2}\right)}{\left(2 - \dfrac{3}{x}\right)} = \frac{\left(\dfrac{x}{4} + \dfrac{6}{4}\right)}{\left(\dfrac{2x}{x} - \dfrac{3}{x}\right)} \qquad \text{Find least common denominators.}$$

$$= \frac{\left(\dfrac{x + 6}{4}\right)}{\left(\dfrac{2x - 3}{x}\right)} \qquad \begin{array}{l}\text{Add fractions in} \\ \text{numerator and} \\ \text{denominator.}\end{array}$$

$$= \frac{x + 6}{4} \cdot \frac{x}{2x - 3} \qquad \begin{array}{l}\text{Invert divisor} \\ \text{and multiply.}\end{array}$$

$$= \frac{x(x + 6)}{4(2x - 3)}, \quad x \neq 0 \qquad \text{Simplified form}$$

Students who had trouble adding and subtracting rational expressions may find the alternative method, shown above in the Note, useful.

EXAMPLE 10   Simplifying a Complex Fraction

Simplify $\dfrac{\left(\dfrac{2}{x + 2}\right)}{\left(\dfrac{1}{x + 2} + \dfrac{2}{x}\right)}$.

Solution

$$\frac{\left(\dfrac{2}{x + 2}\right)}{\left(\dfrac{1}{x + 2} + \dfrac{2}{x}\right)} = \frac{\left(\dfrac{2}{x + 2}\right)(x)(x + 2)}{\dfrac{1}{x + 2}(x)(x + 2) + \dfrac{2}{x}(x)(x + 2)} \qquad x(x + 2) \text{ is LCD.}$$

$$= \frac{2x}{x + 2(x + 2)}$$

$$= \frac{2x}{3x + 4}, \quad x \neq -2, \ x \neq 0$$

Application

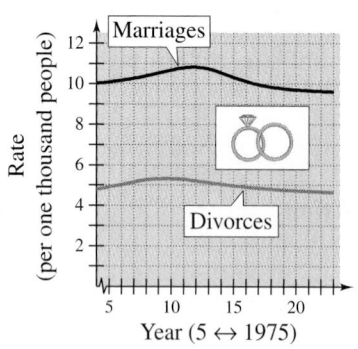

FIGURE 7.2 Marriage (and divorce) rates are often given in terms of the number of marriages (or divorces) per 1000 people. Since 1900, the marriage rate has remained relatively constant: the year with the lowest rate was 1958 (a rate of 8.4), and the year with the highest rate was 1945 (a rate of 12.2). The divorce rate, however, has increased greatly since 1900—from a low of 0.7 in 1900 to a high of 5.3 around 1980. *(Source: U.S. Bureau of Census)*

EXAMPLE 11   *Subtracting Rational Expressions*

For the years 1975 through 1993, the marriage rate M (per 1000 people) and the divorce rate D (per 1000 people) can be modeled by

$$M = 9.6 + \frac{-2t + 48}{t^2 - 25.2t + 178} \quad \text{and} \quad D = 4.4 + \frac{2t + 13}{t^2 - 17t + 106},$$

where $t = 5$ represents 1975. Find a model for the difference R between marriages and divorces (per 1000 people per year). (You do not need to simplify the model.)   *(Source: U.S. National Center for Health Statistics)*

Solution

You can find the model for R by subtracting D from M.

$$R = M - D$$

$$= \left(9.6 + \frac{-2t + 48}{t^2 - 25.2t + 178}\right) - \left(4.4 + \frac{2t + 13}{t^2 - 17t + 106}\right)$$

$$= 5.2 + \frac{-2t + 48}{t^2 - 25.2t + 178} - \frac{2t + 13}{t^2 - 17t + 106}$$

The graphs of the two models are shown in Figure 7.2.

Group Activities   **Extending the Concept**

Comparing Two Methods   Each person in your group should evaluate each of the following expressions for the given variable in two different ways: (1) simplify the rational expressions first and then evaluate the expression at the given value, and (2) substitute the given value for the variable first and then simplify the resulting expression. Do you get the same result with each method? Discuss which method you prefer and why.

a. $\dfrac{1}{m - 4} - \dfrac{1}{m + 4} + \dfrac{3m}{m^2 - 16}$, $m = 2$

b. $\dfrac{x - 2}{x^2 - 9} + \dfrac{3x + 2}{x^2 - 5x + 6}$, $x = 4$

c. $\dfrac{3y^2 + 16y - 8}{y^2 + 2y - 8} - \dfrac{y - 1}{y - 2} + \dfrac{y}{y + 4}$, $y = 3$

7.3   **Exercises**

See Warm-Up
Exercises, p. A46

In Exercises 1–16, combine and simplify.

1. $\dfrac{5x}{8} + \dfrac{7x}{8}$

2. $\dfrac{7y}{12} - \dfrac{5y}{12}$

3. $\dfrac{x}{9} - \dfrac{x+2}{9}$

4. $\dfrac{z^2}{3} + \dfrac{z^2-2}{3}$

5. $\dfrac{4-y}{4} + \dfrac{3y}{4}$

6. $\dfrac{10x^2+1}{3} - \dfrac{10x^2}{3}$

7. $\dfrac{2}{3a} - \dfrac{11}{3a}$

8. $\dfrac{16+z}{5z} - \dfrac{11-z}{5z}$

9. $\dfrac{2x+5}{3} + \dfrac{1-x}{3}$

10. $\dfrac{6x}{13} - \dfrac{7x}{13}$

11. $\dfrac{3y}{3} - \dfrac{3y-3}{3} - \dfrac{7}{3}$

12. $\dfrac{-16u}{9} - \dfrac{27-16u}{9} + \dfrac{2}{9}$

13. $\dfrac{2x-1}{x(x-3)} + \dfrac{1-x}{x(x-3)}$

14. $\dfrac{5x-1}{x+4} + \dfrac{5-4x}{x+4}$

15. $\dfrac{3y-22}{y-6} - \dfrac{2y-16}{y-6}$

16. $\dfrac{7s-5}{2s+5} + \dfrac{3(s+10)}{2s+5}$

In Exercises 17–26, find the least common multiple of the expressions.

17. $5x^2,\ 20x^3$

18. $14t^2,\ 42t^5$

19. $15x^2,\ 3(x+5)$

20. $18y^3,\ 27y(y-3)^2$

21. $9y^3,\ 12y$

22. $6x^2,\ 15x(x-1)$

23. $6(x^2-4),\ 2x(x+2)$

24. $t^3 + 3t^2 + 9t,\ 2t^2(t^2-9)$

25. $8t(t+2),\ 14(t^2-4)$

26. $2y^2 + y - 1,\ 4y^2 - 2y$

In Exercises 27–32, find the missing algebraic expression that makes the two fractions equivalent.

27. $\dfrac{7}{3y} = \dfrac{7x^2}{3y(\quad)},\quad x \neq 0$

28. $\dfrac{2x}{x-3} = \dfrac{14x(x-3)^2}{(x-3)(\quad)}$

29. $\dfrac{3u}{7v} = \dfrac{3u(\quad)}{7v(u+1)},\quad u \neq -1$

30. $\dfrac{3t+5}{t} = \dfrac{(3t+5)(\quad)}{5t^2(3t-5)},\quad t \neq \dfrac{5}{3}$

31. $\dfrac{13x}{x-2} = \dfrac{13x(\quad)}{4-x^2},\quad x \neq -2$

32. $\dfrac{x^2}{10-x} = \dfrac{x^2(\quad)}{x^2-10x},\quad x \neq 0$

In Exercises 33–40, find the least common denominator of the two fractions and rewrite each fraction using the least common denominator.

33. $\dfrac{n+8}{3n-12},\ \dfrac{10}{6n^2}$

34. $\dfrac{8s}{(s+2)^2},\ \dfrac{3}{s^3+s^2-2s}$

35. $\dfrac{v}{2v^2+2v},\ \dfrac{4}{3v^2}$

36. $\dfrac{4x}{(x+5)^2},\ \dfrac{x-2}{x^2-25}$

37. $\dfrac{2}{x^2(x-3)},\ \dfrac{5}{x(x+3)}$

38. $\dfrac{5t}{2t(t-3)^2},\ \dfrac{4}{t(t-3)}$

39. $\dfrac{x-8}{x^2-25},\ \dfrac{9x}{x^2-10x+25}$

40. $\dfrac{3y}{y^2-y-12},\ \dfrac{y-4}{y^2+3y}$

In Exercises 41–74, combine and simplify.

41. $\dfrac{5}{4x} - \dfrac{3}{5}$

42. $\dfrac{10}{b} + \dfrac{1}{10b^2}$

43. $\dfrac{7}{a} + \dfrac{14}{a^2}$

44. $\dfrac{1}{6u^2} - \dfrac{2}{9u}$

(handwritten at top: quiz 3 $\dfrac{3}{x^2+5x+4} - \dfrac{2}{x(x+4)}$ ... $(x+1)(x+3)x$)

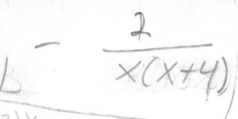

45. $\dfrac{20}{x-4} + \dfrac{20}{4-x}$

46. $\dfrac{15}{2-t} - \dfrac{7}{t-2}$

47. $\dfrac{3x}{x-8} - \dfrac{6}{8-x}$

48. $\dfrac{1}{y-6} + \dfrac{y}{6-y}$

49. $\dfrac{3x}{3x-2} + \dfrac{2}{2-3x}$

50. $\dfrac{y}{5y-3} - \dfrac{3}{3-5y}$

51. $25 + \dfrac{10}{x+4}$

52. $\dfrac{100}{x-10} - 8$

53. $-\dfrac{1}{6x} + \dfrac{1}{6(x-3)}$

54. $\dfrac{1}{x} - \dfrac{1}{x+2}$

55. $\dfrac{x}{x+3} - \dfrac{5}{x-2}$

56. $\dfrac{3t}{t(t+1)} + \dfrac{4}{t}$

57. $\dfrac{3}{x+1} - \dfrac{2}{x}$

58. $\dfrac{5}{x-4} - \dfrac{3}{x}$

59. $\dfrac{3}{x-5y} + \dfrac{2}{x+5y}$

60. $\dfrac{7}{2x-3y} + \dfrac{3}{2x+3y}$

61. $\dfrac{4}{x^2} - \dfrac{4}{x^2+1}$

62. $\dfrac{2}{y} + \dfrac{1}{2y^2}$

63. $\dfrac{x}{x^2-9} + \dfrac{3}{x(x-3)}$

64. $\dfrac{x}{x^2-x-30} - \dfrac{1}{x+5}$

65. $\dfrac{4}{x-4} + \dfrac{16}{(x-4)^2}$

66. $\dfrac{3}{x-2} - \dfrac{1}{(x-2)^2}$

67. $\dfrac{y}{x^2+xy} - \dfrac{x}{xy+y^2}$

68. $\dfrac{5}{x+y} + \dfrac{5}{x-y}$

69. $\dfrac{4}{x} - \dfrac{2}{x^2} + \dfrac{4}{x+3}$

70. $\dfrac{5}{2(x+1)} - \dfrac{1}{2x} - \dfrac{3}{2(x+1)^2}$

71. $\dfrac{3u}{u^2-2uv+v^2} + \dfrac{2}{u-v}$

72. $\dfrac{1}{x} - \dfrac{3}{y} + \dfrac{3x-y}{xy}$

73. $\dfrac{x+2}{x-1} - \dfrac{2}{x+6} - \dfrac{14}{x^2+5x-6}$

74. $\dfrac{x}{x^2+15x+50} + \dfrac{7}{2(x+10)} - \dfrac{3}{2(x+5)}$

In Exercises 75 and 76, use a graphing utility to graph the two equations in the same viewing rectangle. Use the graphs to verify that the expressions are equivalent. Verify the results algebraically.

75. $y_1 = \dfrac{2}{x} + \dfrac{4}{x-2}$

$y_2 = \dfrac{6x-4}{x(x-2)}$

76. $y_1 = 3 - \dfrac{1}{x-1}$

$y_2 = \dfrac{3x-4}{x-1}$

77. *Error Analysis* Describe the error.

$$\dfrac{x-1}{x+4} - \dfrac{4x-11}{x+4} = \dfrac{x-1-4x-11}{x+4}$$

$$= \dfrac{-3x-12}{x+4}$$

$$= \dfrac{-3(x+4)}{x+4}$$

$$= -3$$

78. *Think About It* Is it possible for the least common denominator of two fractions to be the same as one of the fraction's denominators? If so, give an example.

In Exercises 79–94, simplify the complex fraction.

79. $\dfrac{\frac{1}{2}}{\left(3+\frac{1}{x}\right)}$

80. $\dfrac{\frac{2}{3}}{\left(4-\frac{1}{x}\right)}$

81. $\dfrac{\left(\frac{4}{x}+3\right)}{\left(\frac{4}{x}-3\right)}$

82. $\dfrac{\left(\frac{1}{t}-1\right)}{\left(\frac{1}{t}+1\right)}$

83. $\dfrac{\left(16x-\frac{1}{x}\right)}{\left(\frac{1}{x}-4\right)}$

84. $\dfrac{\left(\frac{36}{y}-y\right)}{6+y}$

85. $\dfrac{\left(3+\frac{9}{x-3}\right)}{\left(4+\frac{12}{x-3}\right)}$

86. $\dfrac{\left(x+\frac{2}{x-3}\right)}{\left(x+\frac{6}{x-3}\right)}$

87. $\dfrac{\left(1 - \dfrac{1}{y^2}\right)}{\left(1 - \dfrac{4}{y} + \dfrac{3}{y^2}\right)}$

88. $\dfrac{\left(\dfrac{x+1}{x+2} - \dfrac{1}{x}\right)}{\left(\dfrac{2}{x+2}\right)}$

89. $\dfrac{\left(\dfrac{y}{x} - \dfrac{x}{y}\right)}{\left(\dfrac{x+y}{xy}\right)}$

90. $\dfrac{\left(x - \dfrac{2y^2}{x-y}\right)}{x - 2y}$

91. $\dfrac{\left(\dfrac{3}{x^2} + \dfrac{1}{x}\right)}{\left(2 - \dfrac{4}{5x}\right)}$

92. $\dfrac{\left(16 - \dfrac{1}{x^2}\right)}{\left(\dfrac{1}{4x^2} - 4\right)}$

93. $\dfrac{\left(\dfrac{x}{x-3} - \dfrac{2}{3}\right)}{\left(\dfrac{10}{3x} + \dfrac{x^2}{x-3}\right)}$

94. $\dfrac{\left(\dfrac{1}{2x} - \dfrac{6}{x+5}\right)}{\left(\dfrac{x}{x-5} + \dfrac{1}{x}\right)}$

In Exercises 95–98, simplify the expression.

95. $(u + v^{-2})^{-1}$

96. $x^{-2}(x^2 + y^2)$

97. $\dfrac{a + b}{ba^{-1} - ab^{-1}}$

98. $\dfrac{u^{-1} - v^{-1}}{u^{-1} + v^{-1}}$

99. *Average of Two Numbers*   Determine the average of the two real numbers $x/4$ and $x/6$.

100. *Average of Three Numbers*   Determine the average of the three real numbers x, $x/2$ and $x/3$.

101. Find three real numbers that divide the real number line between $x/6$ and $x/2$ into four equal parts.

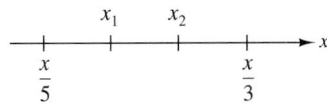

102. Find two real numbers that divide the real number line between $x/5$ and $x/3$ into three equal parts.

103. *Work Rate*   After two workers have worked together for t hours on a common task, the fractional parts of the task done by the two workers are $t/4$ and $t/6$. What fractional part of the task has been completed?

104. *Work Rate*   After two workers have worked together for t hours on a common task, the fractional parts of the task done by the two workers are $t/3$ and $t/5$. What fractional part of the task has been completed?

105. *Parallel Resistance*   When two resistors are connected in parallel (see figure), the total resistance is

$$\dfrac{1}{\left(\dfrac{1}{R_1} + \dfrac{1}{R_2}\right)}.$$

Simplify this complex fraction.

106. *Monthly Payment*   The approximate annual percent rate r of a monthly installment loan is

$$r = \dfrac{\left[\dfrac{24(MN - P)}{N}\right]}{\left(P + \dfrac{MN}{12}\right)},$$

where N is the total number of payments, M is the monthly payment, and P is the amount financed.

(a) Approximate the annual percent rate for a 4-year car loan of $10,000 that has monthly payments of $300.

(b) Simplify the expression for the annual percent rate r, and then rework part (a).

In Exercises 107 and 108, use the function to find and simplify the expression for

$$\frac{f(2 + h) - f(2)}{h}.$$

107. $f(x) = \dfrac{1}{x}$ **108.** $f(x) = \dfrac{x}{x - 1}$

109. *Rewriting a Fraction* The fraction $4/(x^3 - x)$ can be rewritten as a sum of three fractions, as follows.

$$\frac{4}{x^3 - x} = \frac{A}{x} + \frac{B}{x + 1} + \frac{C}{x - 1}$$

The numbers A, B, and C are the solutions of the system

$$A + B + C = 0$$
$$-B + C = 0$$
$$-A = 4.$$

Solve the system and verify that the sum of the three resulting fractions is the original fraction.

110. *Error Analysis* Determine whether the following is correct. If not, find and correct any errors.

$$\frac{2}{x} - \frac{3}{x + 1} + \frac{x + 1}{x^2} = \frac{2x(x + 1) - 3x^2 + (x + 1)^2}{x^2(x + 1)}$$

$$= \frac{2x^2 + x - 3x^2 + x^2 + 1}{x^2(x + 1)}$$

$$= \frac{x + 1}{x^2(x + 1)}$$

$$= \frac{1}{x^2}$$

Section Project

(a) Simplify the complex fraction $\dfrac{\left(1 - \dfrac{1}{x}\right)}{\left(1 - \dfrac{1}{x^2}\right)}$.

(b) Use a graphing utility to complete the table.

x	$\dfrac{\left(1 - \dfrac{1}{x}\right)}{\left(1 - \dfrac{1}{x^2}\right)}$	$\dfrac{x}{x + 1}$
-3		
-2		
-1		
0		
1		
2		
3		

(c) Discuss the domains and the equivalence of the two expressions from the table in part (b).

(d) Simplify the complex fraction $\dfrac{\left(1 + \dfrac{4}{x} + \dfrac{4}{x^2}\right)}{\left(1 - \dfrac{4}{x^2}\right)}$.

(e) Make a table showing the values of the complex fraction in part (d) for integer values of x from -3 through 3. Also show the values of the simplified form of the fraction for these same values of x.

(f) Discuss the domains and the equivalence of the two expressions from the table in part (e).

7.4	**Dividing Polynomials**

Dividing a Polynomial by a Monomial ▪ Long Division ▪
Synthetic Division ▪ Factoring and Division

Dividing a Polynomial by a Monomial

To divide a polynomial by a monomial, reverse the procedure used to add or subtract two rational expressions. Here is an example.

$$2 + \frac{1}{x} = \frac{2x}{x} + \frac{1}{x} = \frac{2x + 1}{x}$$ Add fractions.

$$\frac{2x + 1}{x} = \frac{2x}{x} + \frac{1}{x} = 2 + \frac{1}{x}$$ Divide by monomial.

You might compare these rules to those at the beginning of Section 7.3.

Dividing a Polynomial by a Monomial

Let u, v, and w be real numbers, variables, or algebraic expressions such that $w \neq 0$.

1. $\dfrac{u + v}{w} = \dfrac{u}{w} + \dfrac{v}{w}$ 2. $\dfrac{u - v}{w} = \dfrac{u}{w} - \dfrac{v}{w}$

When dividing a polynomial by a monomial, remember to reduce the resulting expressions to simplest form, as illustrated in Example 1.

EXAMPLE 1   Dividing a Polynomial by a Monomial

Perform the following division and simplify.

$$\frac{12x^2 - 20x + 8}{4x}$$

Solution

$$\frac{12x^2 - 20x + 8}{4x} = \frac{12x^2}{4x} - \frac{20x}{4x} + \frac{8}{4x}$$ Divide each term by $4x$.

$$= \frac{3(4x)(x)}{4x} - \frac{5(4x)}{4x} + \frac{2(4)}{4x}$$ Divide out common factors.

Note that the result is not a poly-nomial.

$$= 3x - 5 + \frac{2}{x}$$ Simplified form

EXAMPLE 2   Dividing a Polynomial by a Monomial

Perform the following division and simplify.

$$\frac{5y^3 + 8y^2 - 10y}{5y^2}$$

Solution

$$\frac{5y^3 + 8y^2 - 10y}{5y^2} = \frac{5y^3}{5y^2} + \frac{8y^2}{5y^2} - \frac{10y}{5y^2} \qquad \text{Divide each term by } 5y^2.$$

$$= \frac{5y^2(y)}{5y^2} + \frac{8y^2}{5y^2} - \frac{2(5y)}{5y(y)} \qquad \text{Divide out common factors.}$$

$$= y + \frac{8}{5} - \frac{2}{y} \qquad \text{Simplified form}$$

EXAMPLE 3   Dividing a Polynomial by a Monomial

Perform the following division and simplify.

$$\frac{-12x^3y^2 + 6x^2y + 9y^3}{3xy^2}$$

Solution

$$\frac{-12x^3y^2 + 6x^2y + 9y^3}{3xy^2} = -\frac{12x^3y^2}{3xy^2} + \frac{6x^2y}{3xy^2} + \frac{9y^3}{3xy^2}$$

$$= -\frac{4(3x)(x^2)(y^2)}{(3x)(y^2)} + \frac{2(3x)(x)(y)}{3x(y)(y)} + \frac{3y(3y^2)}{x(3y^2)}$$

$$= -4x^2 + \frac{2x}{y} + \frac{3y}{x}$$

There are different ways to check division problems such as those shown in Examples 1, 2, and 3. One way is to combine terms. For instance, you can check the result of Example 2 as follows.

$$y + \frac{8}{5} - \frac{2}{y} = \frac{y \cdot 5y^2}{5y^2} + \frac{8 \cdot y^2}{5 \cdot y^2} - \frac{2 \cdot 5y}{y \cdot 5y}$$

$$= \frac{5y^3}{5y^2} + \frac{8y^2}{5y^2} - \frac{10y}{5y^2}$$

$$= \frac{5y^3 + 8y^2 - 10y}{5y^2} \qquad \checkmark$$

Long Division

In Section 7.1, you learned how to divide one polynomial by another by factoring and dividing out common factors. For instance, you can divide $(x^2 - 2x - 3)$ by $(x - 3)$ as follows.

$$(x^2 - 2x - 3) \div (x - 3) = \frac{x^2 - 2x - 3}{x - 3} \qquad \text{Write as fraction.}$$

$$= \frac{(x + 1)(x - 3)}{x - 3} \qquad \text{Factor.}$$

$$= \frac{(x + 1)(x - 3)}{x - 3} \qquad \text{Divide out common factor.}$$

$$= x + 1, \quad x \neq 3 \qquad \text{Simplified form}$$

This procedure works well for polynomials that factor easily. For those that do not, you can use a more general procedure that follows a "long division algorithm" similar to the algorithm used for dividing positive integers. We review that procedure in Example 4.

EXAMPLE 4 *Long Division Algorithm for Positive Integers*

Use the long division algorithm to divide 6584 by 28.

Solution

$$
\begin{array}{r}
\text{Think } \frac{65}{28} \approx 2. \\
\text{Think } \frac{98}{28} \approx 3. \\
\text{Think } \frac{144}{28} \approx 5.
\end{array}
$$

$$
\begin{array}{r}
235 \\
28 \overline{)6584}
\end{array}
$$

$\underline{56}$	Multiply 2 by 28.
98	Subtract and bring down 8.
$\underline{84}$	Multiply 3 by 28.
144	Subtract and bring down 4.
$\underline{140}$	Multiply 5 by 28.
4	Remainder

Thus, you have

$$6584 \div 28 = 235 + \frac{4}{28} = 235 + \frac{1}{7}.$$

In Example 4, the number 6584 is the **dividend,** 28 is the **divisor,** 235 is the **quotient,** and 4 is the **remainder.**

In the next several examples, you will see how the long division algorithm can be extended to cover division of one polynomial by another.

EXAMPLE 5 Long Division Algorithm for Polynomials

Use the long division algorithm to divide $(x^2 + 2x + 4)$ by $(x - 1)$.

Solution

Think $\dfrac{x^2}{x} = x$.

Think $\dfrac{3x}{x} = 3$.

$$
\begin{array}{r}
x + 3 \\
x - 1 \overline{)\, x^2 + 2x + 4} \\
\underline{x^2 - x} \\
3x + 4 \\
\underline{3x - 3} \\
7
\end{array}
$$

Multiply x by $(x - 1)$.

Subtract and bring down 4.

Multiply 3 by $(x - 1)$.

Remainder

Considering the remainder as a fractional part of the divisor, the result is

$$\underbrace{\frac{x^2 + 2x + 4}{x - 1}}_{\text{Divisor}} = \overbrace{x + 3}^{\text{Quotient}} + \frac{\overset{\text{Remainder}}{7}}{\underbrace{x - 1}_{\text{Divisor}}}.$$

Dividend

NOTE Long division of polynomials assists in factoring and in finding zeros of polynomials.

You can check a long division problem by multiplying. For instance, you can check the result of Example 5 as follows.

$$\frac{x^2 + 2x + 4}{x - 1} \overset{?}{=} x + 3 + \frac{7}{x - 1}$$

$$(x - 1)\left(\frac{x^2 + 2x + 4}{x - 1}\right) \overset{?}{=} (x - 1)\left(x + 3 + \frac{7}{x - 1}\right)$$

$$x^2 + 2x + 4 \overset{?}{=} (x + 3)(x - 1) + 7$$

$$x^2 + 2x + 4 \overset{?}{=} (x^2 + 2x - 3) + 7$$

$$x^2 + 2x + 4 = x^2 + 2x + 4 \quad \checkmark$$

If the remainder is 0, the divisor is said to **divide evenly** into the dividend. For instance, $(x + 2)$ divides evenly into $(3x^3 + 10x^2 + 6x - 4)$:

$$\frac{3x^3 + 10x^2 + 6x - 4}{x + 2} = 3x^2 + 4x - 2, \quad x \neq -2.$$

Try checking this result.

When using the long division algorithm for polynomials, be sure that both the divisor and the dividend are written in standard form before beginning the division process.

EXAMPLE 6   *Writing in Standard Form Before Dividing*

Divide $(-13x^3 + 10x^4 + 8x - 7x^2 + 4)$ by $(3 - 2x)$.

Solution

First write the divisor and dividend in standard polynomial form.

Encourage students to perform subtraction carefully.

$$
\begin{array}{r}
-5x^3 - x^2 + 2x - 1 \\
-2x + 3 \overline{)\,10x^4 - 13x^3 - 7x^2 + 8x + 4}
\end{array}
$$

$\underline{10x^4 - 15x^3}$	Multiply $-5x^3$ by $(-2x + 3)$.
$2x^3 - 7x^2$	Subtract and bring down $-7x^2$.
$\underline{2x^3 - 3x^2}$	Multiply $-x^2$ by $(-2x + 3)$.
$-4x^2 + 8x$	Subtract and bring down $8x$.
$\underline{-4x^2 + 6x}$	Multiply $2x$ by $(-2x + 3)$.
$2x + 4$	Subtract and bring down 4.
$\underline{2x - 3}$	Multiply -1 by $(-2x + 3)$.
7	Remainder

This shows that

$$\frac{10x^4 - 13x^3 - 7x^2 + 8x + 4}{-2x + 3} = -5x^3 - x^2 + 2x - 1 + \frac{7}{-2x + 3}.$$

Check this result by multiplying.

Technology

You can check the result of a division problem *algebraically*, as shown at the bottom of page 510. You can check the result *graphically* with a graphing utility by comparing the graphs of the original quotient and the simplified form. For example, the figure at the left shows the graphs of

$$y_1 = \frac{10x^4 - 13x^3 - 7x^2 + 8x + 4}{-2x + 3}$$

and

$$y_2 = -5x^3 - x^2 + 2x - 1 + \frac{7}{-2x + 3}.$$

Because the graphs coincide, you can conclude that the solution checks.

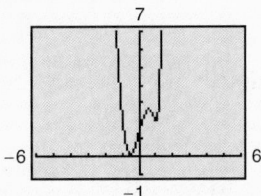

When the dividend is missing one or more powers of x, the long division algorithm requires that you account for the missing powers, as shown in Example 7.

EXAMPLE 7   *Accounting for Missing Powers of x*

Divide $(x^3 - 2)$ by $(x - 1)$.

Solution

Note how the missing x^2- and x-terms are accounted for.

$$
\begin{array}{r}
x^2 + x + 1 \\
x - 1 \overline{\smash{\big)}\ x^3 + 0x^2 + 0x - 2} \\
\end{array}
$$

$\underline{x^3 - x^2}$	Multiply x^2 by $(x - 1)$.
$x^2 + 0x$	Subtract and bring down $0x$.
$\underline{x^2 - x}$	Multiply x by $(x - 1)$.
$x - 2$	Subtract and bring down -2.
$\underline{x - 1}$	Multiply 1 by $(x - 1)$.
-1	Remainder

Stress the use of zero coefficients in terms needed for placeholders. You may want to illustrate this with an arithmetic example such as $1003 \div 20$.

Thus, you have

$$\frac{x^3 - 2}{x - 1} = x^2 + x + 1 - \frac{1}{x - 1}.$$

NOTE In each of the long division examples presented so far, the divisor has been a first-degree polynomial. The long division algorithm works just as well with polynomial divisors of degree 2 or more, as shown in Example 8.

EXAMPLE 8   *A Second-Degree Divisor*

Divide $(x^4 + 6x^3 + 6x^2 - 10x - 3)$ by $(x^2 + 2x - 3)$.

Solution

$$
\begin{array}{r}
x^2 + 4x + 1 \\
x^2 + 2x - 3 \overline{\smash{\big)}\ x^4 + 6x^3 + 6x^2 - 10x - 3} \\
\end{array}
$$

$\underline{x^4 + 2x^3 - 3x^2}$	Multiply x^2 by $(x^2 + 2x - 3)$.
$4x^3 + 9x^2 - 10x$	Subtract and bring down $-10x$.
$\underline{4x^3 + 8x^2 - 12x}$	Multiply $4x$ by $(x^2 + 2x - 3)$.
$x^2 + 2x - 3$	Subtract and bring down -3.
$\underline{x^2 + 2x - 3}$	Multiply 1 by $(x^2 + 2x - 3)$.
0	

Thus, $x^2 + 2x - 3$ divides evenly into $x^4 + 6x^3 + 6x^2 - 10x - 3$:

$$\frac{x^4 + 6x^3 + 6x^2 - 10x - 3}{x^2 + 2x - 3} = x^2 + 4x + 1, \quad x \neq -3, \ x \neq 1.$$

Synthetic Division

There is a nice shortcut for division by polynomials of the form $(x - k)$. It is called **synthetic division** and is outlined for a third-degree polynomial as follows.

Synthetic Division for a Third-Degree Polynomial

Use synthetic division to divide $(ax^3 + bx^2 + cx + d)$ by $(x - k)$ as follows.

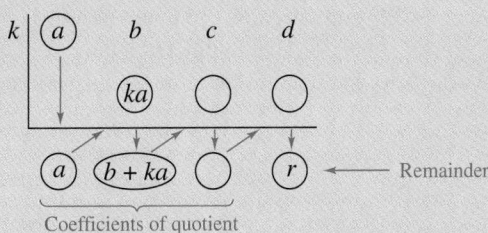

Coefficients of quotient

Vertical Pattern: Add terms.

Diagonal Pattern: Multiply by k.

NOTE Be sure you see that synthetic division works *only* for divisors of the form $(x - k)$. Remember that $x + k = x - (-k)$. Moreover, the degree of the quotient is always 1 less than the degree of the dividend.

EXAMPLE 9   *Using Synthetic Division*

Use synthetic division to divide $(x^3 + 3x^2 - 4x - 10)$ by $(x - 2)$.

Solution

The coefficients of the dividend form the top row of the synthetic division table. Because you are dividing by $(x - 2)$, write 2 at the top left of the table. To begin the algorithm, bring down the first coefficient. Then multiply this coefficient by 2, write the result in the second row of the table, and add the two numbers in the second column. By continuing this pattern, you obtain the following table.

$$
\begin{array}{c|cccc}
2 & 1 & 3 & -4 & -10 \\
  &   & 2 & 10 & 12 \\
\hline
  & 1 & 5 & 6 & 2 \quad \longleftarrow \text{ Remainder}
\end{array}
$$

The bottom row of the table shows that the quotient is

$$(1)x^2 + (5)x + (6)$$

and the remainder is 2. Thus, the result of the division problem is

$$\frac{x^3 + 3x^2 - 4x - 10}{x - 2} = x^2 + 5x + 6 + \frac{2}{x - 2}.$$

If the remainder in a synthetic division problem turns out to be zero, you can conclude that the divisor divides *evenly* into the dividend, as shown in Example 10.

EXAMPLE 10 *Using Synthetic Division*

Use synthetic division to divide $(x^4 - 10x^2 - 2x + 3)$ by $(x + 3)$.

Solution

The coefficients of the dividend form the top row of the synthetic division table, and, because you are dividing by $(x + 3)$, write -3 at the top left of the table. The completed synthetic division table is as follows. (Note that a zero is included in place of the missing term in the dividend.)

$$
\begin{array}{r|rrrrr}
-3 & 1 & 0 & -10 & -2 & 3 \\
   &   & -3 & 9 & 3 & -3 \\
\hline
   & 1 & -3 & -1 & 1 & \boxed{0} \longleftarrow \text{Remainder}
\end{array}
$$

Because the remainder is zero, you can conclude that

$$
\frac{x^4 - 10x^2 - 2x + 3}{x + 3} = x^3 - 3x^2 - x + 1, \quad x \neq -3.
$$

You can check the result of Example 10 by multiplying.

$$
\begin{array}{r}
x^3 - 3x^2 - x + 1 \\
\times \quad\quad\quad x + 3 \\
\hline
3x^3 - 9x^2 - 3x + 3 \\
x^4 - 3x^3 - x^2 + x \quad\quad\quad \\
\hline
x^4 \quad\quad - 10x^2 - 2x + 3
\end{array}
$$

From this, you can conclude that the result of Example 10 is correct because

$$
(x + 3)(x^3 - 3x^2 - x + 1) = x^4 - 10x^2 - 2x + 3. \quad \checkmark
$$

Exploration

Factor the polynomial $x^3 - 64$. Use synthetic division to divide $(x^3 - 64)$ by $(x - 4)$. What can you conclude?

Factoring and Division

One of the uses of synthetic division (or long division) is in factoring of polynomials. This is demonstrated in Example 11.

EXAMPLE 11 *Factoring a Polynomial*

Factor completely the polynomial

$$x^3 - 7x + 6.$$

Use the fact that $(x - 1)$ is one of the factors.

Solution

Because you are given one of the factors, you should divide this factor into the given polynomial. (In this case we use synthetic division, but long division could have been used also.)

Remind students about the place-holder to account for missing powers of x.

$$
\begin{array}{c|cccc}
1 & 1 & 0 & -7 & 6 \\
  &   & 1 & 1 & -6 \\
\hline
  & 1 & 1 & -6 & \widehat{0} \leftarrow \text{Remainder}
\end{array}
$$

Thus, you can conclude that

$$\frac{x^3 - 7x + 6}{x - 1} = x^2 + x - 6, \quad x \neq 1,$$

which implies that

$$x^3 - 7x + 6 = (x - 1)(x^2 + x - 6).$$

Thus, you have used division to factor the polynomial $(x^3 - 7x + 6)$ into the product of a first-degree polynomial and a second-degree polynomial. To complete the factorization, factor the second-degree polynomial as follows.

Remind students to factor completely.

$$
\begin{aligned}
x^3 - 7x + 6 &= (x - 1)(x^2 + x - 6) \\
&= (x - 1)(x + 3)(x - 2)
\end{aligned}
$$

You can check the result of Example 11 as follows.

$$
\begin{aligned}
(x - 1)(x + 3)(x - 2) &= (x - 1)(x^2 + x - 6) \qquad &&\text{Multiply } (x + 3) \text{ by } (x - 2). \\
&= x^3 - 7x + 6 \qquad &&\text{Multiply } (x - 1) \text{ by} \\
& &&(x^2 + x - 6). \; \checkmark
\end{aligned}
$$

Group Activities   **Exploring with Technology**

Investigating Polynomials and Their Factors   Use a graphing utility to graph the following polynomials in the same viewing rectangle using the standard setting. Use the trace feature to find the *x*-intercepts. What can you conclude about the polynomials? Verify your conclusion algebraically.

a. $y = (x - 4)(x - 2)(x + 1)$

b. $y = (x^2 - 6x + 8)(x + 1)$

c. $y = x^3 - 5x^2 + 2x + 8$

Now use your graphing utility to graph the function

$$f(x) = \frac{x^3 - 5x^2 + 2x + 8}{x - 2}.$$

Use the trace feature to find the *x*-intercepts. Why does this function have only two *x*-intercepts? To what other function does the graph of $f(x)$ appear to be equivalent? What is the difference between the two graphs? (*Hint:* Zoom in at $x = 2$.)

7.4   **Exercises**

See Warm-Up
Exercises, p. A46

1. *Error Analysis*   Describe the error.

$$\frac{6x + 5y}{x} = \frac{6x + 5y}{x} = 6 + 5y$$

2. *True or False?*   If the divisor divides evenly into the dividend, the divisor and quotient are factors of the dividend. Explain.

In Exercises 3–52, perform the division.

3. $\dfrac{6z + 10}{2}$

4. $\dfrac{9x + 12}{3}$

5. $\dfrac{10z^2 + 4z - 12}{4}$

6. $\dfrac{4u^2 + 8u - 24}{16}$

7. $(7x^3 - 2x^2) \div x$

8. $(6a^2 + 7a) \div a$

9. $\dfrac{50z^3 + 30z}{-5z}$

10. $\dfrac{18c^4 - 24c^2}{-6c}$

11. $\dfrac{8z^3 + 3z^2 - 2z}{2z}$

12. $\dfrac{6x^4 + 8x^3 - 18x^2}{3x^2}$

13. $\dfrac{m^4 + 2m^2 - 7}{m}$

14. $\dfrac{l^2 - 8}{-l}$

15. $(5x^2y - 8xy + 7xy^2) \div 2xy$

16. $(-14s^4t^2 + 7s^2t^2 - 18t) \div 2s^2t$

17. $\dfrac{4(x + 5)^2 + 8(x + 5)}{x + 5}$

18. $\dfrac{12(r - 9)^3 - 18(r - 9)^2}{6(r - 9)^2}$

19. $\dfrac{x^2 - 8x + 15}{x - 3}$ **20.** $\dfrac{t^2 - 18t + 72}{t - 6}$

21. $(x^2 + 15x + 50) \div (x + 5)$

22. $(y^2 - 6y - 16) \div (y + 2)$

23. $(21 - 4x - x^2) \div (3 - x)$

24. $(5 + 4x - x^2) \div (1 + x)$

25. $(2y^2 + 7y + 3) \div (2y + 1)$

26. $(10t^2 - 7t - 12) \div (2t - 3)$

27. $(12t^2 - 40t + 25) \div (2t - 5)$

28. $(15 - 14u - 8u^2) \div (5 + 2u)$

29. $\dfrac{16x^2 - 1}{4x + 1}$ **30.** $\dfrac{81y^2 - 25}{9y - 5}$

31. $\dfrac{x^3 + 125}{x + 5}$ **32.** $\dfrac{x^3 - 27}{x - 3}$

33. $\dfrac{x^3 - 2x^2 + 4x - 8}{x - 2}$

34. $\dfrac{x^3 - 28x - 48}{x + 4}$

35. $(2x + 9) \div (x + 2)$

36. $(12x - 5) \div (2x + 3)$

37. $\dfrac{x^2 + 16}{x + 4}$ **38.** $\dfrac{y^2 + 8}{y + 2}$

39. $\dfrac{5x^2 + 2x + 3}{x + 2}$ **40.** $\dfrac{2x^2 + 5x + 2}{x + 4}$

41. $\dfrac{12x^2 + 17x - 5}{3x + 2}$ **42.** $\dfrac{8x^2 + 2x + 3}{4x - 1}$

43. $\dfrac{6z^2 + 7z}{5z - 1}$ **44.** $\dfrac{8y^2 - 2y}{3y + 5}$

45. $\dfrac{2x^3 - 5x^2 + x - 6}{x - 3}$

46. $\dfrac{5x^3 + 3x^2 + 12x + 20}{x + 1}$

47. $(x^6 - 1) \div (x - 1)$

48. $x^3 \div (x - 1)$

49. $x^5 \div (x^2 + 1)$ **50.** $x^4 \div (x - 2)$

51. $(x^3 + 4x^2 + 7x + 6) \div (x^2 + 2x + 3)$

52. $(2x^3 + 2x^2 - 2x - 15) \div (2x^2 + 4x + 5)$

In Exercises 53 and 54, perform the division. (Assume n is a positive integer.)

53. $\dfrac{x^{3n} + 3x^{2n} + 6x^n + 8}{x^n + 2}$

54. $\dfrac{x^{3n} - x^{2n} + 5x^n - 5}{x^n - 1}$

In Exercises 55–66, use synthetic division to divide.

55. $\dfrac{x^3 + 3x^2 - 1}{x + 4}$ **56.** $\dfrac{x^4}{x + 2}$

57. $\dfrac{x^4 - 4x^3 + x + 10}{x - 2}$ **58.** $\dfrac{2x^5 - 3x^3 + x}{x - 3}$

59. $\dfrac{5x^3 + 12}{x + 5}$ **60.** $\dfrac{8x + 35}{x - 10}$

61. $\dfrac{5x^3 - 6x^2 + 8}{x - 4}$ **62.** $\dfrac{5x^3 + 6x + 8}{x + 2}$

63. $\dfrac{10x^4 - 50x^3 - 800}{x - 6}$ **64.** $\dfrac{x^5 - 13x^4 - 120x + 80}{x + 3}$

65. $\dfrac{0.1x^2 + 0.8x + 1}{x - 0.2}$ **66.** $\dfrac{x^3 - 0.8x + 2.4}{x + 1}$

In Exercises 67–74, use synthetic division to divide. Use the result to write the dividend in factored form.

67. $\dfrac{15x^2 - 2x - 8}{x - \frac{4}{5}}$

68. $\dfrac{18x^2 - 9x - 20}{x + \frac{5}{6}}$

69. $\dfrac{-3z^3 + 20z^2 - 36z + 16}{z - 4}$

70. $\dfrac{2t^3 + 15t^2 + 19t - 30}{t + 5}$

71. $\dfrac{5t^3 - 27t^2 - 14t - 24}{t - 6}$

72. $\dfrac{y^3 + y^2 - 4y - 4}{y - 2}$ **73.** $\dfrac{x^4 - 16}{x - 2}$

74. $\dfrac{x^4 - x^3 - 3x^2 + 4x - 1}{x - 1}$

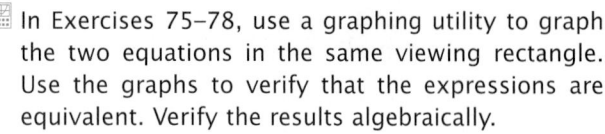 In Exercises 75–78, use a graphing utility to graph the two equations in the same viewing rectangle. Use the graphs to verify that the expressions are equivalent. Verify the results algebraically.

75. $y_1 = \dfrac{x + 4}{2x}$

$y_2 = \dfrac{1}{2} + \dfrac{2}{x}$

76. $y_1 = \dfrac{x^2 + 2}{x + 1}$

$y_2 = x - 1 + \dfrac{3}{x + 1}$

77. $y_1 = \dfrac{x^3 + 1}{x + 1}$

$y_2 = x^2 - x + 1, \; x \neq -1$

78. $y_1 = \dfrac{x^3}{x^2 + 1}$

$y_2 = x - \dfrac{x}{x^2 + 1}$

79. When a polynomial is divided by $x - 6$, the quotient is $x^2 + x + 1$ and the remainder is -4. Find the dividend.

80. When a polynomial is divided by $x + 3$, the quotient is $x^3 + x^2 - 4$ and the remainder is 8. Find the dividend.

In Exercises 81 and 82, find the constant c such that the denominator divides evenly into the numerator.

81. $\dfrac{x^3 + 2x^2 - 4x + c}{x - 2}$

82. $\dfrac{x^4 - 3x^2 + c}{x + 6}$

83. *Geometry*   A rectangle's area is

$$2x^3 + 3x^2 - 6x - 9.$$

Find its width if its length is $2x + 3$ (see figure).

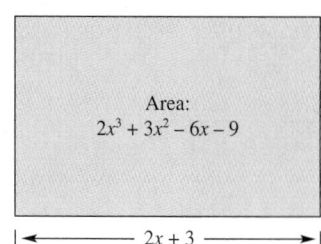

Area:
$2x^3 + 3x^2 - 6x - 9$

|← $2x + 3$ →|

84. *Geometry*   A rectangular house has a volume of

$$x^3 + 55x^2 + 650x + 2000$$

cubic feet (the space in the attic is not included). The height of the house is $x + 5$ (see figure). Find the number of square feet of floor space *on the first floor* of the house.

$x + 5$

In Exercises 85–88, simplify the algebraic expression.

85. $\dfrac{4x^4}{x^3} - 2x$

86. $\dfrac{x^2 + 2x - 3}{x - 1} - (3x - 4)$

87. $\dfrac{15x^3y}{10x^2} + \dfrac{3xy^2}{2y}$

88. $\dfrac{8u^2v}{2u} + \dfrac{3(uv)^2}{uv}$

In Exercises 89–92, determine whether the reduction shown is valid or invalid. If it is invalid, state what is wrong.

89. $\dfrac{5 + 12}{5} = \dfrac{\cancel{5} + 12}{\cancel{5}} = 12$

90. $\dfrac{6 - 3}{6 + 11} = \dfrac{\cancel{6} - 3}{\cancel{6} + 11} = \dfrac{-3}{11}$

91. $\dfrac{9 \cdot 12}{19 \cdot 9} = \dfrac{\cancel{9} \cdot 12}{19 \cdot \cancel{9}} = \dfrac{12}{19}$

92. $\dfrac{28}{83} = \dfrac{2\cancel{8}}{8\cancel{3}} = \dfrac{2}{3}$

Geometry   In Exercises 93 and 94, you are given the expression for the volume of the solid shown. Find the expression for the missing dimension.

93. $V = x^3 + 18x^2 + 80x + 96$

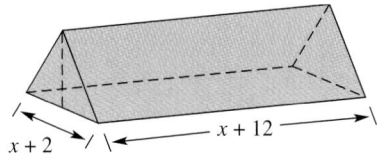

$x + 2$ $x + 12$

94. $V = h^4 + 3h^3 + 2h^2$

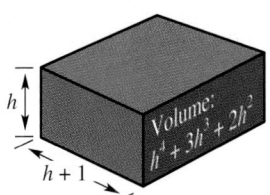

h

Volume:
$h^4 + 3h^3 + 2h^2$

$h + 1$

Section Project

Finding a Pattern   Consider the following function.

$$f(x) = 2x^4 - x^3 - 9x^2 + 4x + 4$$

(a) Use synthetic division to divide $f(x)$ by each of the divisors in the table. Then complete the table.

k	-3	-2	-1	$-\frac{1}{2}$
Divisor $(x - k)$	$x + 3$	$x + 2$	$x + 1$	$x + \frac{1}{2}$
Remainder				

k	0	$\frac{1}{2}$	1	2	3
Divisor $(x - k)$	x	$x - \frac{1}{2}$	$x - 1$	$x - 2$	$x - 3$
Remainder					

(b) Evaluate $f(x)$ at each of the x-values in the table.

x	-3	-2	-1	$-\frac{1}{2}$	0	$\frac{1}{2}$	1	2	3
$f(x)$									

(c) Compare the function values from the table in part (b) with the remainders from the table in part (a). What can you conclude?

(d) Use the complete factorization

$$f(x) = (x - 2)(x + 2)(2x + 1)(x - 1)$$

to find all the solutions of the polynomial equation $2x^4 - x^3 - 9x^2 + 4x + 4 = 0$.

(e) Compare the solutions of the equation $f(x) = 0$ in part (d) with your observations in parts (a) through (c).

(f) Use a graphing utility to graph $f(x)$. What observations can you make about the x-intercepts of the graph?

Mid-Chapter Quiz

Take this quiz as you would take a quiz in class. After you are done, check your work against the answers given in the back of the book.

1. Determine the domain of the rational expression $\dfrac{y + 2}{y(y - 4)}$.

2. Evaluate the rational function $h(x) = (x^2 - 9)/(x^2 - x - 2)$ as indicated. If not possible, state the reason.

    (a) $h(-3)$ (b) $h(0)$ (c) $h(-1)$ (d) $h(5)$

In Exercises 3–8, write the expression in reduced form.

3. $\dfrac{9y^2}{6y}$ **4.** $\dfrac{8u^3v^2}{36uv^3}$ **5.** $\dfrac{4x^2 - 1}{x - 2x^2}$

6. $\dfrac{(z + 3)^2}{2z^2 + 5z - 3}$ **7.** $\dfrac{7ab + 3a^2b^2}{a^2b}$ **8.** $\dfrac{2mn^2 - n^3}{2m^2 + mn - n^2}$

In Exercises 9–18, perform the operations and simplify.

9. $\dfrac{11t^2}{6} \cdot \dfrac{9}{33t}$ **10.** $(x^2 + 2x) \cdot \dfrac{5}{x^2 - 4}$

11. $\dfrac{4}{3(x - 1)} \cdot \dfrac{12x}{6(x^2 + 2x - 3)}$ **12.** $\dfrac{5u}{3(u + v)} \cdot \dfrac{2(u^2 - v^2)}{3v} \div \dfrac{25u^2}{18(u - v)}$

13. $\dfrac{\left(\dfrac{9t^2}{3 - t}\right)}{\left(\dfrac{6t}{t - 3}\right)}$ **14.** $\dfrac{\left(\dfrac{10}{x^2 + 2x}\right)}{\left(\dfrac{15}{x^2 + 3x + 2}\right)}$ **15.** $\dfrac{4x}{x + 5} - \dfrac{3x}{4}$

16. $4 + \dfrac{x}{x^2 - 4} - \dfrac{2}{x^2}$ **17.** $\dfrac{\left(1 - \dfrac{2}{x}\right)}{\left(\dfrac{3}{x} - \dfrac{4}{5}\right)}$ **18.** $\dfrac{\left(\dfrac{3}{x} + \dfrac{x}{3}\right)}{\left(\dfrac{x + 3}{6x}\right)}$

19. Divide $(6x^3 - 16x^2 + 17x - 6)$ by $(3x - 2)$.

20. You start a business with a setup cost of \$6000. The cost of material for producing each unit of your product is \$10.50.

    (a) Write a rational function that gives the average cost per unit when x units are produced. Explain your reasoning.

    (b) Find the average cost per unit when $x = 500$ units are produced.

21. Find the ratio of the area of the shaded portion of the figure to the total area of the figure.

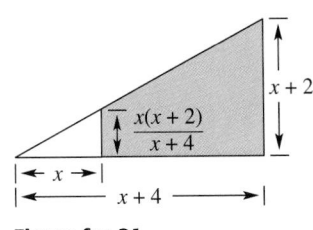

Figure for 21

| **7.5** | **Solving Rational Equations** |

Equations Containing Constant Denominators ▪
Equations Containing Variable Denominators ▪ Application

Equations Containing Constant Denominators

In Section 1.7, you studied a strategy for solving equations that contain fractions with *constant* denominators. We review that procedure here because it is the basis for solving more general equations involving fractions. Recall from Section 1.7 that you can "clear an equation of fractions" by multiplying both sides of the equation by the least common denominator of the fractions in the equation. Note how this is done in Example 1.

EXAMPLE 1   An Equation Containing Constant Denominators

Solve $\dfrac{3}{5} = \dfrac{x}{2}$.

Solution

The least common denominator of the two fractions is 10, so begin by multiplying both sides of the equation by 10.

$$\frac{3}{5} = \frac{x}{2} \qquad \text{Original equation}$$

$$10\left(\frac{3}{5}\right) = 10\left(\frac{x}{2}\right) \qquad \text{Multiply both sides by 10.}$$

$$6 = 5x \qquad \text{Simplify.}$$

$$\frac{6}{5} = x \qquad \text{Divide both sides by 5.}$$

The solution is $\frac{6}{5}$. You can check this as follows.

Check

$$\frac{3}{5} = \frac{x}{2} \qquad \text{Original equation}$$

$$\frac{3}{5} \overset{?}{=} \frac{6/5}{2} \qquad \text{Substitute } \tfrac{6}{5} \text{ for } x.$$

$$\frac{3}{5} \overset{?}{=} \frac{6}{5} \cdot \frac{1}{2} \qquad \text{Invert and multiply.}$$

$$\frac{3}{5} = \frac{3}{5} \qquad \text{Solution checks.} \quad \text{✔}$$

EXAMPLE 2 An Equation Containing Constant Denominators

Solve $\dfrac{x}{6} = 7 - \dfrac{x}{12}$.

Solution

$$\dfrac{x}{6} = 7 - \dfrac{x}{12} \qquad \text{Original equation}$$

$$12\left(\dfrac{x}{6}\right) = 12\left(7 - \dfrac{x}{12}\right) \qquad \text{Multiply both sides by 12.}$$

$$2x = 84 - x \qquad \text{Simplify.}$$

$$3x = 84 \qquad \text{Add } x \text{ to both sides.}$$

$$x = 28 \qquad \text{Divide both sides by 3.}$$

The solution is 28. Check this in the original equation.

Historical Note: The Rhind Papyrus is a major source of knowledge about ancient Egyptian mathematicians. In one of its many documents, we find the *rule of false position*. Problem 24 states "A quantity and its $\frac{1}{7}$ added together become 19. What is the quantity?" It is solved as follows:

$$x + \dfrac{x}{7} = 19$$

Guess that x is, for example, 14.

$$14 + \tfrac{14}{7} = 14 + 2 = 16$$

As $19 = 16\left(\frac{19}{16}\right)$, the solution is $14\left(\frac{16}{19}\right)$ or $\frac{133}{8}$. Try the following from the Rhind Papyrus (using either the rule of false position or an equation involving rational expressions).

Problem 25: A quantity and its $\frac{1}{2}$ added together become 16. What is the quantity? $\left(\text{Solution: } \frac{32}{3}\right)$

EXAMPLE 3 An Equation Containing Constant Denominators

Solve $\dfrac{x+2}{6} - \dfrac{x-4}{8} = \dfrac{2}{3}$.

Solution

$$\dfrac{x+2}{6} - \dfrac{x-4}{8} = \dfrac{2}{3} \qquad \text{Original equation}$$

$$24\left(\dfrac{x+2}{6} - \dfrac{x-4}{8}\right) = 24\left(\dfrac{2}{3}\right) \qquad \text{Multiply both sides by 24.}$$

$$4(x+2) - 3(x-4) = 8(2) \qquad \text{Simplify.}$$

$$4x + 8 - 3x + 12 = 16 \qquad \text{Distributive Property}$$

$$x + 20 = 16 \qquad \text{Combine like terms.}$$

$$x = -4 \qquad \text{Subtract 20 from both sides.}$$

The solution is -4. You can check this as follows.

Check

$$\dfrac{x+2}{6} - \dfrac{x-4}{8} = \dfrac{2}{3} \qquad \text{Original equation}$$

$$\dfrac{-4+2}{6} - \dfrac{-4-4}{8} \stackrel{?}{=} \dfrac{2}{3} \qquad \text{Substitute } -4 \text{ for } x.$$

$$-\dfrac{1}{3} + 1 \stackrel{?}{=} \dfrac{2}{3} \qquad \text{Simplify.}$$

$$\dfrac{2}{3} = \dfrac{2}{3} \qquad \text{Solution checks.} \quad \checkmark$$

EXAMPLE 4   An Equation that Has Two Solutions

Solve $\dfrac{x^2}{3} + \dfrac{x}{2} = \dfrac{5}{6}$.

Solution

First multiply both sides of the equation by the least common denominator, 6.

$$\frac{x^2}{3} + \frac{x}{2} = \frac{5}{6} \qquad\qquad \text{Original equation}$$

$$6\left(\frac{x^2}{3} + \frac{x}{2}\right) = 6\left(\frac{5}{6}\right) \qquad\qquad \text{Multiply both sides by 6.}$$

$$2x^2 + 3x = 5 \qquad\qquad \text{Distribute and simplify.}$$

$$2x^2 + 3x - 5 = 0 \qquad\qquad \text{Write in standard form.}$$

$$(2x + 5)(x - 1) = 0 \qquad\qquad \text{Factor.}$$

$$2x + 5 = 0 \implies x = -\frac{5}{2} \qquad \text{Set 1st factor equal to 0.}$$

$$x - 1 = 0 \implies x = 1 \qquad \text{Set 2nd factor equal to 0.}$$

Therefore, the equation has two solutions: $-\frac{5}{2}$ and 1. You can check these solutions in the original equation as follows.

Check First Solution:

$$\frac{x^2}{3} + \frac{x}{2} = \frac{5}{6} \qquad\qquad \text{Original equation}$$

$$\frac{(-5/2)^2}{3} + \frac{(-5/2)}{2} \stackrel{?}{=} \frac{5}{6} \qquad\qquad \text{Substitute } -\tfrac{5}{2} \text{ for } x.$$

$$\left(\frac{25}{4}\right)\left(\frac{1}{3}\right) + \left(-\frac{5}{2}\right)\left(\frac{1}{2}\right) \stackrel{?}{=} \frac{5}{6} \qquad\qquad \text{Simplify.}$$

$$\frac{25}{12} - \frac{5}{4} \stackrel{?}{=} \frac{5}{6} \qquad\qquad \text{Multiply.}$$

$$\frac{5}{6} = \frac{5}{6} \qquad\qquad \text{Solution checks.} \checkmark$$

Check Second Solution:

$$\frac{x^2}{3} + \frac{x}{2} = \frac{5}{6} \qquad\qquad \text{Original equation}$$

$$\frac{(1)^2}{3} + \frac{(1)}{2} \stackrel{?}{=} \frac{5}{6} \qquad\qquad \text{Substitute 1 for } x.$$

$$\frac{1}{3} + \frac{1}{2} \stackrel{?}{=} \frac{5}{6} \qquad\qquad \text{Simplify.}$$

$$\frac{5}{6} = \frac{5}{6} \qquad\qquad \text{Solution checks.} \checkmark$$

Equations Containing Variable Denominators

Remember that you always *exclude* those values of a variable that make the denominator of a rational expression zero. This is especially critical for solving equations that contain variable denominators. You will see why in the examples that follow.

EXAMPLE 5 An Equation Containing Variable Denominators

Solve $\dfrac{7}{x} - \dfrac{1}{3x} = \dfrac{8}{3}$.

Solution

The least common denominator for this equation is $3x$. Therefore, begin by multiplying both sides of the equation by $3x$.

$$\frac{7}{x} - \frac{1}{3x} = \frac{8}{3} \qquad \text{Original equation}$$

$$3x\left(\frac{7}{x} - \frac{1}{3x}\right) = 3x\left(\frac{8}{3}\right) \qquad \text{Multiply both sides by } 3x.$$

$$\frac{21x}{x} - \frac{3x}{3x} = \frac{24x}{3} \qquad \text{Distributive Property}$$

$$21 - 1 = 8x \qquad \text{Simplify.}$$

$$\frac{20}{8} = x \qquad \text{Divide both sides by 8.}$$

$$x = \frac{5}{2} \qquad \text{Simplify.}$$

The solution is $\frac{5}{2}$. You can check this as follows.

Check

$$\frac{7}{x} - \frac{1}{3x} = \frac{8}{3} \qquad \text{Original equation}$$

$$\frac{7}{5/2} - \frac{1}{3(5/2)} \overset{?}{=} \frac{8}{3} \qquad \text{Substitute } \frac{5}{2} \text{ for } x.$$

$$7\left(\frac{2}{5}\right) - \frac{2}{15} \overset{?}{=} \frac{8}{3} \qquad \text{Invert and multiply.}$$

$$\frac{14}{5} - \frac{2}{15} \overset{?}{=} \frac{8}{3} \qquad \text{Simplify.}$$

$$\frac{40}{15} \overset{?}{=} \frac{8}{3} \qquad \text{Simplify.}$$

$$\frac{8}{3} = \frac{8}{3} \qquad \text{Solution checks. } ✓$$

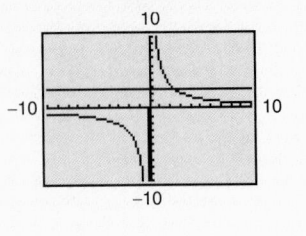

Throughout the text, we have emphasized the importance of checking solutions. Even with no mistakes in the solution process, it can happen that a "trial solution" does not satisfy an original equation. As discussed in Section 5.5, this type of "solution" is called *extraneous*. Recall that an extraneous solution of an equation does not, by definition, satisfy the original equation, and therefore *must not* be listed as an actual solution.

As an aid to checking solutions, you might have students determine the domain *before* beginning the problem.

EXAMPLE 6   An Equation that Has No Solution

Solve $\dfrac{5x}{x-2} = 7 + \dfrac{10}{x-2}$.

Solution

The least common denominator for this equation is $x - 2$. Therefore, begin by multiplying both sides of the equation by $x - 2$.

$$\frac{5x}{x-2} = 7 + \frac{10}{x-2}$$ Original equation

$$(x-2)\left(\frac{5x}{x-2}\right) = (x-2)\left(7 + \frac{10}{x-2}\right)$$ Multiply both sides by $x - 2$.

$$5x = 7(x-2) + 10, \quad x \neq 2$$ Simplify.

$$5x = 7x - 14 + 10$$ Distributive Property

$$5x = 7x - 4$$ Simplify.

$$-2x = -4$$ Subtract $7x$ from both sides.

$$x = 2$$ Divide both sides by -2.

At this point, the solution appears to be 2. However, by performing the following check, you will see that this "trial solution" is extraneous.

Check

$$\frac{5x}{x-2} = 7 + \frac{10}{x-2}$$ Original equation

$$\frac{5(2)}{2-2} \overset{?}{=} 7 + \frac{10}{2-2}$$ Substitute 2 for x.

$$\frac{10}{0} \overset{?}{=} 7 + \frac{10}{0}$$ Solution does not check.

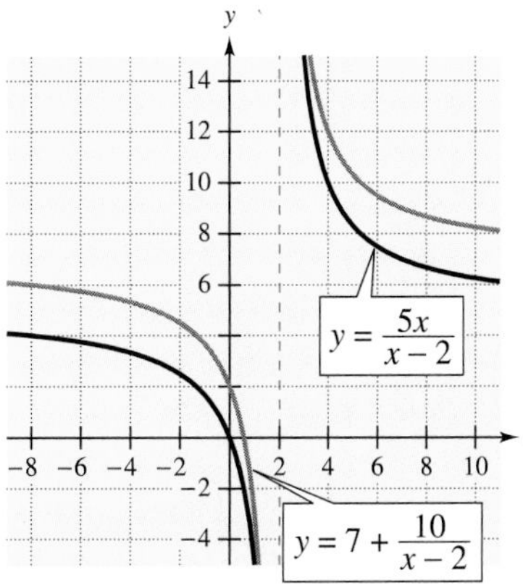

Because the check results in *division by zero*, 2 is extraneous. Therefore, the original equation has no solution. Figure 7.3 shows that the graphs of "each side" of the equation have no point of intersection.

$y = \dfrac{5x}{x-2}$

$y = 7 + \dfrac{10}{x-2}$

FIGURE 7.3

NOTE   In Example 6, can you see why $x = 2$ is extraneous? By looking back at the original equation, you can see that 2 is excluded from the domain of two of the fractions that occur in the equation.

EXAMPLE 7 *An Equation Containing Variable Denominators*

Solve $\dfrac{4}{x-2} + \dfrac{3x}{x+1} = 3$.

Solution

The least common denominator is $(x-2)(x+1)$.

$$\frac{4}{x-2} + \frac{3x}{x+1} = 3$$

$$(x-2)(x+1)\left(\frac{4}{x-2} + \frac{3x}{x+1}\right) = (x-2)(x+1)(3)$$

$$4(x+1) + 3x(x-2) = 3(x^2 - x - 2), \quad x \neq 2, \; x \neq -1$$

$$4x + 4 + 3x^2 - 6x = 3x^2 - 3x - 6$$

$$3x^2 - 2x + 4 = 3x^2 - 3x - 6$$

$$-2x + 4 = -3x - 6$$

$$x + 4 = -6$$

$$x = -10$$

The solution is -10. Check this in the original equation.

Exploration

Use a graphing utility to graph the equation

$$y = \frac{3x}{x+1} - \frac{12}{x^2 - 1} - 2.$$

Then use the zoom and trace features of the graphing utility to determine the x-intercepts. How do the x-intercepts compare with the solutions to Example 8? What can you conclude?

EXAMPLE 8 *An Equation that Has Two Solutions*

Solve $\dfrac{3x}{x+1} = \dfrac{12}{x^2-1} + 2$.

Solution

The least common denominator is $(x+1)(x-1) = x^2 - 1$.

$$\frac{3x}{x+1} = \frac{12}{x^2-1} + 2$$

$$(x^2-1)\left(\frac{3x}{x+1}\right) = (x^2-1)\left(\frac{12}{x^2-1} + 2\right)$$

$$(x-1)(3x) = 12 + 2(x^2-1), \quad x \neq \pm 1$$

$$3x^2 - 3x = 12 + 2x^2 - 2$$

$$x^2 - 3x - 10 = 0$$

$$(x+2)(x-5) = 0$$

$$x + 2 = 0 \quad \Longrightarrow \quad x = -2$$

$$x - 5 = 0 \quad \Longrightarrow \quad x = 5$$

The solutions are -2 and 5. Check these in the original equation.

Application

EXAMPLE 9   A Work-Rate Problem

With only the cold water valve open, it takes 8 minutes to fill the tub of a washer. With both the hot and cold water valves open, it takes only 5 minutes. How long will it take the tub to fill with only the hot water valve open?

Solution

Verbal Model:	Rate for cold water	$+$	Rate for hot water	$=$	Rate for warm water

In answering the question(s) to an application problem in words, use mixed fractions where appropriate. $\frac{40}{3}$ minutes is much more difficult to conceptualize than $13\frac{1}{3}$ minutes.

Labels:	Rate for cold water $= \frac{1}{8}$	(tub per minute)
	Rate for hot water $= 1/t$	(tub per minute)
	Rate for warm water $= \frac{1}{5}$	(tub per minute)

$$\textit{Equation:} \qquad \frac{1}{8} + \frac{1}{t} = \frac{1}{5}$$

$$40t\left(\frac{1}{8} + \frac{1}{t}\right) = 40t\left(\frac{1}{5}\right)$$

$$5t + 40 = 8t$$

$$40 = 3t$$

$$\tfrac{40}{3} = t$$

Thus, it will take about $13\frac{1}{3}$ minutes to fill the tub with hot water alone. Check this in the original statement of the problem.

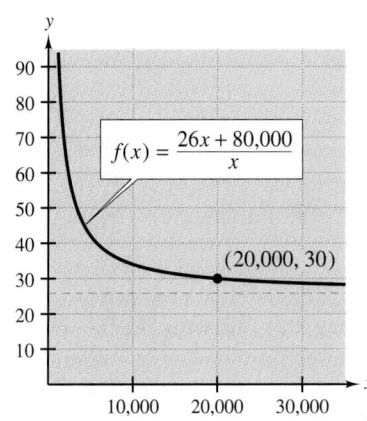

$$f(x) = \frac{26x + 80,000}{x}$$

$(20,000, 30)$

FIGURE 7.4

NOTE   In Figure 7.4, notice that the average cost per unit drops as the number of units increases.

EXAMPLE 10   Average Cost

A manufacturing plant can produce x units of a certain item for $26 per unit *plus* an initial investment of $80,000. How many units must be produced to have an average cost of $30 per unit?

Solution

Verbal Model:	Average cost per unit	$=$	Total cost	$\div$	Number of units

Labels:	Number of units $= x$	(units)
	Average cost per unit $= 30$	(dollars per unit)
	Total cost $= 26x + 80,000$	(dollars)

$$\textit{Equation:} \qquad 30 = \frac{26x + 80,000}{x}$$

Try solving this equation. You should discover that the plant should produce 20,000 units. Check this in the original statement of the problem.

EXAMPLE 11 Batting Average

In this year's playing season, a baseball player has been at bat 140 times and has hit the ball safely 35 times. Thus, the "batting average" for the player is $35/140 = .250$. How many consecutive times must the player hit safely to obtain a batting average of .300?

Solution

Verbal Model:	Batting average	=	Total hits	÷	Total times at bat

Labels: Current times at bat $= 140$
Current hits $= 35$
Additional consecutive hits $= x$

Equation:
$$.300 = \frac{x + 35}{x + 140}$$

$$.300(x + 140) = x + 35$$

$$.3x + 42 = x + 35$$

$$7 = 0.7x$$

$$10 = x$$

The player must hit safely the next 10 times at bat. After that, the batting average will be $45/150 = .300$.

Group Activities Problem Solving

Interpreting Average Cost You buy sand in bulk for a construction project. You find that the total cost of your order depends on the weight of the order. Suppose the total cost C in dollars is given by

$$C = 100 + 50x - 0.2x^2, \quad 1 \le x \le 50,$$

where x is the weight in thousands of pounds. Construct a rational function representing the average cost per thousand pounds. Because of cost constraints, you may proceed with the project only if the average cost of the sand is less than $50 per thousand pounds. What is the smallest order you can place and still proceed with the project?

7.5   **Exercises**

See Warm-Up
Exercises, p. A47

1. *Think About It*   Define the term *extraneous solution*. How do you identify an extraneous solution?

2. *Think About It*   Explain when you can use cross-multiplication to solve a rational equation.

In Exercises 3–6, determine whether the values of x are solutions of the equation.

Equation	Values

3. $\dfrac{x}{3} - \dfrac{x}{5} = \dfrac{4}{3}$ (a) $x = 0$ (b) $x = -1$
   (c) $x = \frac{1}{8}$ (d) $x = 10$

4. $x = 4 + \dfrac{21}{x}$ (a) $x = 0$ (b) $x = -3$
   (c) $x = 7$ (d) $x = -1$

5. $\dfrac{x}{4} + \dfrac{3}{4x} = 1$ (a) $x = -1$ (b) $x = 1$
   (c) $x = 3$ (d) $x = \frac{1}{2}$

6. $5 - \dfrac{1}{x - 3} = 2$ (a) $x = \frac{10}{3}$ (b) $x = -\frac{1}{3}$
   (c) $x = 0$ (d) $x = 1$

In Exercises 7–16, solve the rational equation.

7. $\dfrac{x}{4} = \dfrac{3}{8}$

8. $\dfrac{x}{10} = \dfrac{12}{5}$

9. $\dfrac{t}{2} = \dfrac{1}{8}$

10. $\dfrac{y}{5} = \dfrac{3}{2}$

11. $\dfrac{z + 2}{3} = \dfrac{z}{12}$

12. $5 + \dfrac{y}{3} = y + 2$

13. $\dfrac{4t}{3} = 15 - \dfrac{t}{6}$

14. $\dfrac{x}{3} + \dfrac{x}{6} = 10$

15. $\dfrac{h + 2}{5} - \dfrac{h - 1}{9} = \dfrac{2}{3}$

16. $\dfrac{u - 2}{6} + \dfrac{2u + 5}{15} = 3$

In Exercises 17–58, solve the rational equation. (Check for extraneous solutions.)

17. $\dfrac{7}{x} = 21$

18. $\dfrac{9}{t} = -\dfrac{4}{3}$

19. $\dfrac{9}{25 - y} = -\dfrac{1}{4}$

20. $\dfrac{2}{u + 4} = \dfrac{5}{8}$

21. $5 - \dfrac{12}{a} = \dfrac{5}{3}$

22. $\dfrac{6}{b} + 22 = 24$

23. $\dfrac{12}{y + 5} + \dfrac{1}{2} = 2$

24. $\dfrac{7}{8} - \dfrac{16}{t - 2} = \dfrac{3}{4}$

25. $\dfrac{5}{x} = \dfrac{25}{3(x + 2)}$

26. $\dfrac{10}{x + 4} = \dfrac{15}{4(x + 1)}$

27. $\dfrac{8}{3x + 5} = \dfrac{1}{x + 2}$

28. $\dfrac{500}{3x + 5} = \dfrac{50}{x - 3}$

29. $\dfrac{3}{x + 2} - \dfrac{1}{x} = \dfrac{1}{5x}$

30. $\dfrac{12}{x + 5} + \dfrac{5}{x} = \dfrac{20}{x}$

31. $\dfrac{4}{2x + 3} + \dfrac{17}{5(2x + 3)} = 3$

32. $\dfrac{2}{6q + 5} - \dfrac{3}{4(6q + 5)} = \dfrac{1}{28}$

33. $\dfrac{10}{x(x - 2)} + \dfrac{4}{x} = \dfrac{5}{x - 2}$

34. $\dfrac{x}{x + 4} + \dfrac{4}{x + 4} + 2 = 0$

35. $3\left(\dfrac{1}{x} + 4\right) = 2 + \dfrac{4}{3x}$

36. $\dfrac{x}{x - 2} + \dfrac{1}{x - 4} = \dfrac{2}{x^2 - 6x + 8}$

37. $\dfrac{2}{(x - 4)(x - 2)} = \dfrac{1}{x - 4} + \dfrac{2}{x - 2}$

38. $\dfrac{10}{x + 3} + \dfrac{10}{3} = 6$

39. $\dfrac{1}{x - 5} + \dfrac{1}{x + 5} = \dfrac{x + 3}{x^2 - 25}$

40. $\dfrac{2}{x - 10} - \dfrac{3}{x - 2} = \dfrac{6}{x^2 - 12x + 20}$

41. $\dfrac{1}{2} = \dfrac{18}{x^2}$

42. $\dfrac{1}{6} = \dfrac{150}{z^2}$

43. $\dfrac{32}{t} = 2t$

44. $\dfrac{45}{u} = \dfrac{u}{5}$

45. $x + 1 = \dfrac{72}{x}$

46. $t + 4\left(\dfrac{2t + 5}{t + 4}\right) = 0$

47. $1 = \dfrac{16}{y} - \dfrac{39}{y^2}$

48. $\dfrac{2x}{3x + 10} - \dfrac{5}{x} = 0$

49. $\dfrac{1}{x - 1} + \dfrac{3}{x + 1} = 2$

50. $\dfrac{x + 42}{x} = x$

51. $x - \dfrac{24}{x} = 5$

52. $\dfrac{3x}{2} + \dfrac{4}{x} = 5$

53. $\dfrac{2x}{5} = \dfrac{x^2 - 5x}{5x}$

54. $\dfrac{8(x - 1)}{x^2 - 4} = \dfrac{4}{x - 2}$

55. $\dfrac{x}{2} = \dfrac{2 - \dfrac{3}{x}}{1 - \dfrac{1}{x}}$

56. $\dfrac{2x}{3} = \dfrac{1 + \dfrac{2}{x}}{1 + \dfrac{1}{x}}$

57. $\dfrac{2(x + 7)}{x + 4} - 2 = \dfrac{2x + 20}{2x + 8}$

58. $\dfrac{2x^2 - 5}{x^2 - 4} + \dfrac{6}{x + 2} = \dfrac{4x - 7}{x - 2}$

59. *Number Problem*   Find a number such that the sum of the number and its reciprocal is $\frac{65}{8}$.

60. *Number Problem*   Find a number such that the sum of two times the number and three times its reciprocal is $\frac{97}{4}$.

61. *Wind Speed*   A plane has a speed of 300 miles per hour in still air. Find the speed of the wind if the plane traveled a distance of 680 miles with a tail wind in the same time it took to travel 520 miles into a head wind.

62. *Average Speed*   During the first part of a 6-hour trip, you travel 240 miles at an average speed of r miles per hour. For the next 72 miles of the trip, you increase your speed by 10 miles per hour. What were your two average speeds?

63. *Comparing Two Speeds*   One person runs 2 miles per hour faster than another person. The first person runs 5 miles in the same time the second person runs 4 miles. Find the speed of each person.

64. *Comparing Two Speeds*   The speed of a commuter plane is 150 miles per hour slower than that of a passenger jet. The commuter plane travels 450 miles in the same time the passenger jet travels 1150 miles. Find the speed of each aircraft.

65. *Speed*   A boat travels at a speed of 20 miles per hour in still water. It travels 48 miles upstream and then returns to the starting point in a total of 5 hours. Find the speed of the current.

66. *Speed*   You traveled 72 miles in a certain time period. If you had traveled 6 miles per hour faster, the trip would have taken 10 minutes less time. What was your speed?

67. *Partnership Costs*   A group plans to start a new business that will require $240,000 for start-up capital. The individuals in the group will share the cost equally. If two additional people were to join the group, the cost per person would decrease by $4000. How many people are presently in the group?

68. *Population Growth*   A biologist introduces 100 insects into a culture. The population P of the culture is approximated by the model

$$P = \dfrac{500(1 + 3t)}{5 + t},$$

where t is the time in hours. Find the time required for the population to increase to 1000 insects.

Work Rate   In Exercises 69 and 70, complete the table by finding the time for two individuals to complete a task together. The first two columns in the table give the times required for the two individuals to complete the task working alone. (Assume that when they work together their individual rates do not change.)

69.

Person #1	Person #2	Together
6 hours	6 hours	
3 minutes	5 minutes	

70.

Person #1	Person #2	Together
$5\frac{1}{2}$ hours	3 hours	
a days	b days	

71. *Work Rate*   One landscaper works $1\frac{1}{2}$ times as fast as another landscaper. Find their individual rates if it takes them 9 hours working together to complete a certain job.

72. *Work Rate*   Assume that the slower of the two landscapers in Exercise 71 is given another job after 4 hours. The faster of the two must work an additional 10 hours to complete the job. Find their individual rates.

73. *Swimming Pool*   The flow rate for one pipe is $1\frac{1}{4}$ times that of another pipe. A swimming pool can be filled in 5 hours using both pipes. Find the time required to fill the pool using only the pipe with the lower flow rate.

74. *Swimming Pool*   Assume that the pipe with the higher flow rate in Exercise 73 is shut off after 1 hour and it takes an additional 10 hours to fill the pool. Find the flow rate of each pipe.

75. *Pollution Removal*   The cost C in dollars of removing $p\%$ of the air pollutants in the stack emissions of a utility company is modeled by

$$C = \frac{120,000p}{100 - p}.$$

Determine the percent of the air pollutants in the stack emissions that can be removed for $680,000.

76. *Average Cost*   The average cost of producing x units of a product is given by

$$\overline{C} = 1.50 + \frac{4200}{x}.$$

Determine the number of units that must be produced to have an average cost of $2.90.

77. *Computer and Data Processing Services*   The total revenue y (in billions of dollars) from computer and data services in the United States for the years 1988 through 1993 (see figure) can be modeled by

$$y = \frac{84.981 + 6.457t}{1 - 0.026t}, \quad -2 \le t \le 3.$$

In this model, $t = 0$ represents 1990. *(Source: Current Business Reports)*

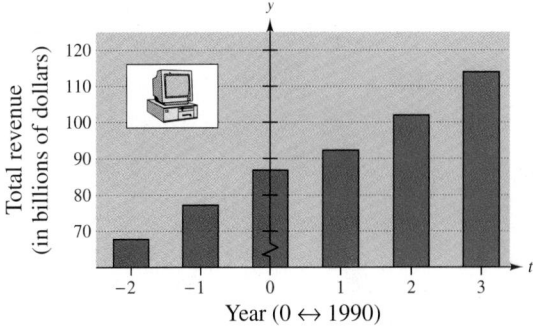

(a) Use the graph to determine *graphically* the year the total revenue first exceeded $100 billion.

(b) Use the model to confirm the result in part (a) *algebraically.*

Section Project

Equations and Expressions   It is important to distinguish between equations and expressions. In parts (a) through (d), if the exercise is an equation, solve it; if the exercise is an expression, simplify it.

(a) $\dfrac{16}{x^2 - 16} + \dfrac{x}{2x - 8} = \dfrac{1}{2}$

(b) $\dfrac{16}{x^2 - 16} + \dfrac{x}{2x - 8} + \dfrac{1}{2}$

(c) $\dfrac{5}{3} - \dfrac{5}{x + 3} + 3$ **(d)** $\dfrac{5}{3} - \dfrac{5}{x + 3} = 3$

(e) Explain the difference between an equation and an expression.

(f) Compare the use of a common denominator in solving rational equations with its use in adding or subtracting rational expressions.

7.6	**Graphing Rational Functions**

Introduction ▪ Horizontal and Vertical Asymptotes ▪ Graphing Rational Functions ▪ Application

Introduction

Recall that the domain of a rational function consists of all values of x for which the denominator is not zero. For instance, the domain of

$$f(x) = \frac{x + 2}{x - 1}$$

is all real numbers except $x = 1$. When graphing a rational function, pay special attention to the shape of the graph near x-values that are not in the domain.

EXAMPLE 1 Point Plotting the Graph of a Rational Function

Use point plotting to sketch the graph of $f(x) = \dfrac{x + 2}{x - 1}$.

Solution

Notice that the domain is all real numbers except $x = 1$. Next, construct a table of values, including x-values that are close to 1 on the left *and* the right.

x-Values to the Left of 1

x	-3	-2	-1	0	0.5	0.9	0.99
$f(x)$	0.25	0	-0.5	-2	-5	-29	-299

x-Values to the Right of 1

x	1.01	1.1	1.5	2	3	4	5
$f(x)$	301	31	7	4	2.5	2	1.75

Plot the points to the left of 1 and connect them with a smooth curve, as shown in Figure 7.5. Do the same for the points to the right of 1. *Do not* connect the two portions of the graph, which are called its **branches.**

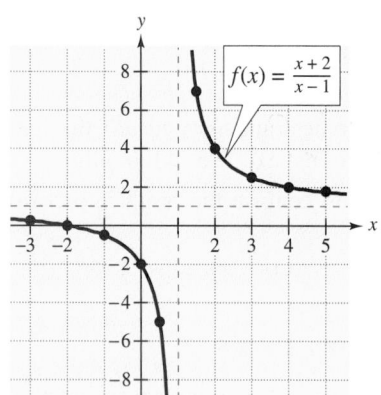

$f(x) = \dfrac{x+2}{x-1}$

FIGURE 7.5

NOTE In Figure 7.5, as x approaches 1 from the left, the values of $f(x)$ approach negative infinity, and as x approaches 1 from the right, the values of $f(x)$ approach positive infinity.

Horizontal and Vertical Asymptotes

An **asymptote** of a graph is a line to which the graph becomes arbitrarily close as $|x|$ or $|y|$ increases without bound. In other words, if a graph has an asymptote, it is possible to move far enough out on the graph so that there is almost no difference between the graph and the asymptote.

The graph in Example 1 has two asymptotes: the line $x = 1$ is a **vertical asymptote,** and the line $y = 1$ is a **horizontal asymptote.** Other examples of asymptotes are shown in Figure 7.6.

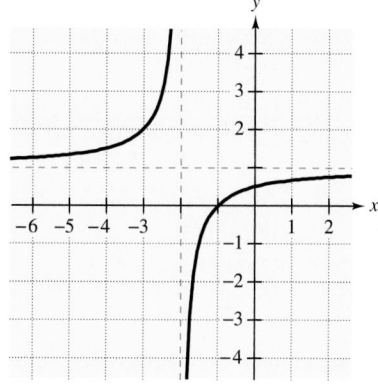

Graph of $y = \dfrac{x + 1}{x + 2}$

Horizontal asymptote: $y = 1$

Vertical asymptote: $x = -2$

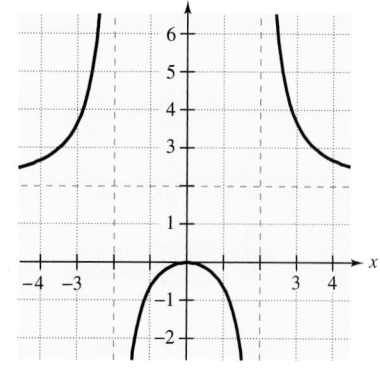

Graph of $y = \dfrac{2x^2}{x^2 - 4}$

Horizontal asymptote: $y = 2$

Vertical asymptotes: $x = \pm 2$

FIGURE 7.6

The graph of a rational function may have no horizontal or vertical asymptotes, or it may have several.

Guidelines for Finding Asymptotes

Let $f(x) = p(x)/q(x)$, where $p(x)$ and $q(x)$ have no common factors.

1. The graph of f has a vertical asymptote at each x-value for which the denominator is zero.

2. The graph of f has at most one horizontal asymptote.

   • If the degree of $p(x)$ is less than the degree of $q(x)$, the line $y = 0$ is a horizontal asymptote.

   • If the degree of $p(x)$ is equal to the degree of $q(x)$, the line $y = a/b$ is a horizontal asymptote, where a is the leading coefficient of $p(x)$ and b is the leading coefficient of $q(x)$.

   • If the degree of $p(x)$ is greater than the degree of $q(x)$, the graph has no horizontal asymptote.

EXAMPLE 2   Finding Horizontal and Vertical Asymptotes

Find all horizontal and vertical asymptotes of the graph of

$$f(x) = \frac{2x}{3x^2 + 1}.$$

Solution

For this rational function, the degree of the numerator is less than the degree of the denominator. This implies that the graph has the line

$$y = 0 \qquad\qquad\qquad \text{Horizontal asymptote}$$

as a horizontal asymptote, as shown in Figure 7.7. To find any vertical asymptotes, set the denominator equal to zero and solve the resulting equation for x.

$$3x^2 + 1 = 0 \qquad\qquad \text{Set denominator equal to zero.}$$

Because this equation has no real solution, you can conclude that the graph has no vertical asymptote.

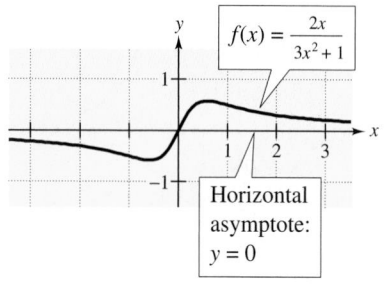

FIGURE 7.7

Remember that the graph of a rational function can have at most one horizontal asymptote, but it can have several vertical asymptotes. For instance, the graph in Example 3 has two vertical asymptotes.

EXAMPLE 3   Finding Horizontal and Vertical Asymptotes

Find all horizontal and vertical asymptotes of the graph of

$$f(x) = \frac{2x^2}{x^2 - 1}.$$

Solution

For this rational function, the degree of the numerator is equal to the degree of the denominator. The leading coefficient of the numerator is 2, and the leading coefficient of the denominator is 1. Thus, the graph has the line

$$y = \frac{2}{1} = 2 \qquad\qquad\qquad \text{Horizontal asymptote}$$

as a horizontal asymptote, as shown in Figure 7.8. Then to find any vertical asymptotes, set the denominator equal to zero and solve the resulting equation for x.

$$x^2 - 1 = 0 \qquad\qquad \text{Set denominator equal to zero.}$$

This equation has two real solutions: -1 and 1. Thus, the graph has two vertical asymptotes: the lines $x = -1$ and $x = 1$.

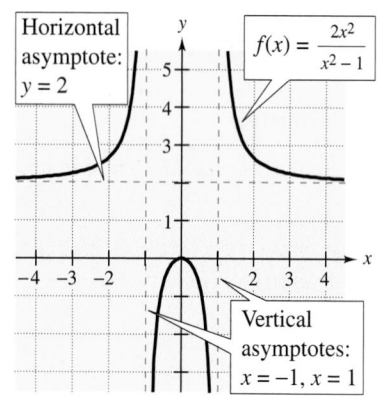

FIGURE 7.8

Graphing Rational Functions

To sketch the graph of a rational function, we suggest the following guidelines.

This organized process will be bene-
ficial to students as the process
becomes more complicated.

Guidelines for Graphing Rational Functions

Let $f(x) = p(x)/q(x)$, where $p(x)$ and $q(x)$ have no common factors.

1. Find and plot the y-intercept (if any) by evaluating $f(0)$.

2. Set the numerator equal to zero and solve the equation for x. The real solutions represent the x-intercepts of the graph. Plot these intercepts.

3. Set the denominator equal to zero and solve the equation for x. The real solutions represent the vertical asymptotes. Sketch these asymptotes.

4. Find and sketch the horizontal asymptote of the graph.

5. Plot at least one point between and one point beyond each x-intercept and vertical asymptote.

6. Use smooth curves to complete the graph between and beyond the vertical asymptotes.

EXAMPLE 4 Sketching the Graph of a Rational Function

Sketch the graph of $f(x) = \dfrac{2}{x-3}$.

Solution

Begin by noting that the numerator and denominator have no common factors. Following the guidelines above produces the following.

- Because $f(0) = -\frac{2}{3}$, the y-intercept is $\left(0, -\frac{2}{3}\right)$.

- Because the numerator is never zero, there are no x-intercepts.

- Because the denominator is zero when $x = 3$, the line $x = 3$ is a vertical asymptote.

- Because the degree of the numerator is less than the degree of the denominator, the line $y = 0$ is a horizontal asymptote.

Plot the intercept, asymptotes, and the additional points from the following table. Then complete the graph by drawing two branches, as shown in Figure 7.9. Note that the two branches are not connected.

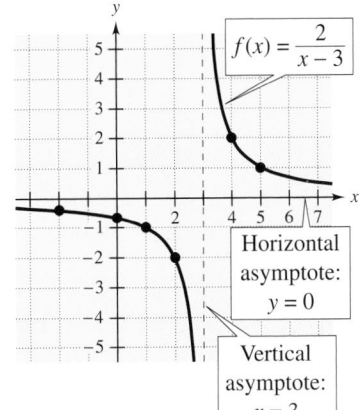

FIGURE 7.9

x	-2	1	2	4	5
$f(x)$	$-\frac{2}{5}$	-1	-2	2	1

EXAMPLE 5   *Sketching the Graph of a Rational Function*

Sketch the graph of $f(x) = \dfrac{2x - 1}{x}$.

Solution

Begin by noting that the numerator and denominator have no common factors.

- Because $x = 0$ is not in the domain of the function, $f(0)$ is undefined and there is no y-intercept.

- Because the numerator is zero when $x = \frac{1}{2}$, the x-intercept is $\left(\frac{1}{2}, 0\right)$.

- Because the denominator is zero when $x = 0$, the line $x = 0$ is a vertical asymptote.

- Because the degree of the numerator is equal to the degree of the denominator and the ratio of the leading coefficients is 2, the line $y = 2$ is a horizontal asymptote.

By plotting the intercept, asymptotes, and the additional points from the following table, you can obtain the graph shown in Figure 7.10.

x	-4	-2	-1	$\frac{1}{4}$	4
$f(x)$	$\frac{9}{4}$	$\frac{5}{2}$	3	-2	$\frac{7}{4}$

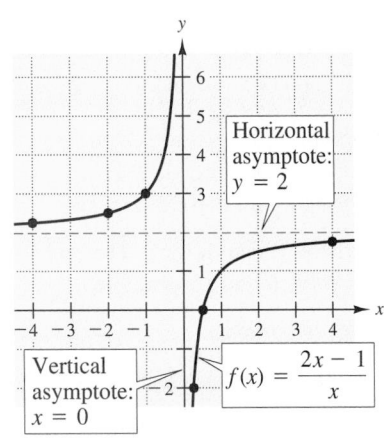

FIGURE 7.10

EXAMPLE 6   *Sketching the Graph of a Rational Function*

Sketch the graph of $f(x) = \dfrac{2}{x^2 + 1}$.

Solution

Begin by noting that the numerator and denominator have no common factors.

- Because $f(0) = 2$, the y-intercept is $(0, 2)$.

- Because the numerator is never zero, there are no x-intercepts.

- Because the equation $x^2 + 1 = 0$ has no real solution, there are no vertical asymptotes.

- Because the degree of the numerator is less than the degree of the denominator, the line $y = 0$ is a horizontal asymptote.

By plotting the intercept, asymptote, and the additional points from the following table, you can obtain the graph shown in Figure 7.11.

x	-2	-1	1	2
$f(x)$	$\frac{2}{5}$	1	1	$\frac{2}{5}$

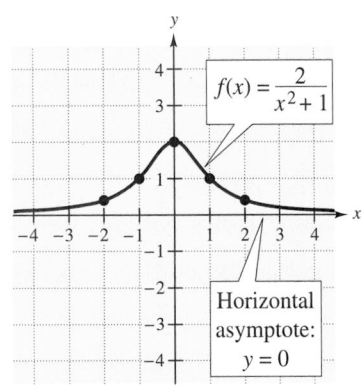

FIGURE 7.11

Application

EXAMPLE 7 *Finding the Average Cost*

As a fund-raising project, a club is publishing a calendar. The costs of photography and typesetting total $850. In addition to these "one-time" costs, the unit cost of printing each calendar is $3.25. Let x represent the number of calendars printed. Write a model that represents the average cost per calendar.

Solution

The total cost C of printing x calendars is

$$C = 3.25x + 850.$$ 	Total cost function

The average cost per calendar $\overline{C}$ for printing x calendars is

$$\overline{C} = \frac{3.25x + 850}{x}.$$ 	Average cost function

From the graph shown in Figure 7.12, notice that the average cost decreases as the number of calendars increases.

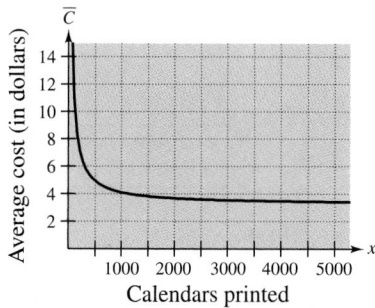

FIGURE 7.12

~~~~ *Technology*

Graphing a Rational Function

A graphing utility can help you sketch the graph of a rational function. With most graphing utilities, however, there are problems with graphs of rational functions. If you use the *connected mode,* the graphing utility will try to connect any branches of the graph. If you use the *dot mode,* the graphing utility will draw a dotted (rather than a solid) graph. Both of these options are shown below for the graph of $y = (x - 1)/(x - 3)$.

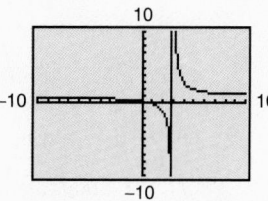

Connected mode

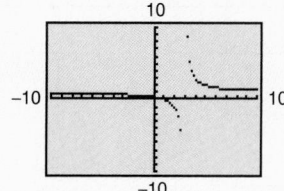

Dot mode

Group Activities **Extending the Concept**

More About the Average Cost In Example 7, what is the horizontal asymptote of the graph of the average cost function? What is the significance of this asymptote in the problem? Is it possible to sell enough calendars to obtain an average cost of $3.00 per calendar? Explain your reasoning.

7.6 Exercises

See Warm-Up Exercises, p. A47

1. *Think About It* In your own words, describe what is meant by an asymptote of a graph.

2. *Think About It* Does every rational function have a vertical asymptote? Explain.

Numerical Analysis In Exercises 3 and 4, (a) complete each table, (b) determine the vertical and horizontal asymptotes of the graph, and (c) find the domain of the function.

x	0	0.5	0.9	0.99	0.999
y					

x	2	1.5	1.1	1.01	1.001
y					

x	2	5	10	100	1000
y					

3. $f(x) = \dfrac{4}{x-1}$

4. $f(x) = \dfrac{2x}{x-1}$

In Exercises 5–8, identify the horizontal and vertical asymptotes of the function. Use the asymptotes to match the rational function with its graph. [The graphs are labeled (a), (b), (c), and (d).]

(a)
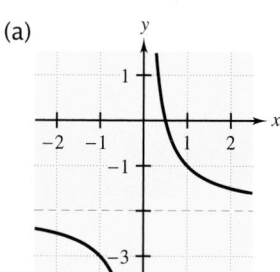

(b)
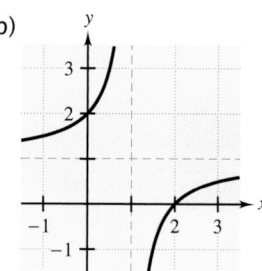

(c)
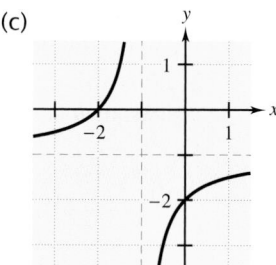

(d)
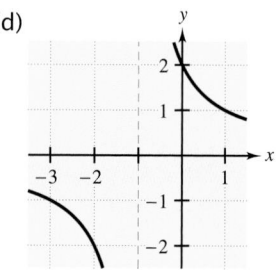

5. $f(x) = \dfrac{2}{x+1}$

6. $f(x) = \dfrac{1-2x}{x}$

7. $f(x) = \dfrac{x-2}{x-1}$

8. $f(x) = -\dfrac{x+2}{x+1}$

In Exercises 9–20, find the domain of the function and identify any horizontal and vertical asymptotes.

9. $f(x) = \dfrac{5}{x^2}$

10. $g(x) = \dfrac{3}{x - 5}$

11. $f(x) = \dfrac{x}{x + 8}$

12. $f(u) = \dfrac{u^2}{u - 10}$

13. $g(t) = \dfrac{3}{t(t - 1)}$

14. $h(x) = \dfrac{4x - 3}{x}$

15. $h(s) = \dfrac{2s^2}{s + 3}$

16. $g(t) = \dfrac{3}{t^2 + 1}$

17. $y = \dfrac{3 - 5x}{1 - 3x}$

18. $y = \dfrac{3x + 2}{2x - 1}$

19. $y = \dfrac{5x^2}{x^2 - 1}$

20. $y = \dfrac{2x^2}{x^2 + 1}$

In Exercises 21–24, match the function with its graph. [The graphs are labeled (a), (b), (c), and (d).]

(a)
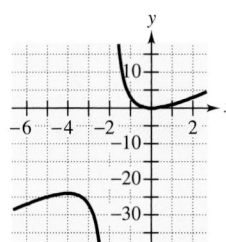

(b)
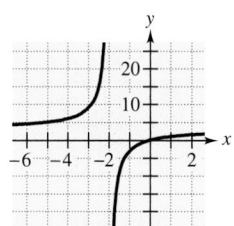

(c)
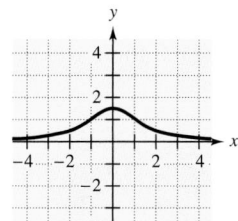

(d)
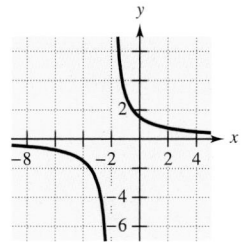

21. $f(x) = \dfrac{3}{x + 2}$

22. $f(x) = \dfrac{3x}{x + 2}$

23. $f(x) = \dfrac{3x^2}{x + 2}$

24. $f(x) = \dfrac{3}{x^2 + 2}$

In Exercises 25–44, sketch the graph of the function. As sketching aids, check for intercepts, vertical asymptotes, and horizontal asymptotes.

25. $g(x) = \dfrac{5}{x}$

26. $g(x) = \dfrac{5}{x - 4}$

27. $f(x) = \dfrac{5}{x^2}$

28. $f(x) = \dfrac{5}{(x - 4)^2}$

29. $f(x) = \dfrac{1}{x - 2}$

30. $f(x) = \dfrac{3}{x + 1}$

31. $g(x) = \dfrac{1}{2 - x}$

32. $g(x) = \dfrac{-3}{x + 1}$

33. $y = \dfrac{2x + 4}{x}$

34. $y = \dfrac{2x}{x + 4}$

35. $y = \dfrac{3x}{x + 4}$

36. $y = \dfrac{x - 2}{x}$

37. $y = \dfrac{2x^2}{x^2 + 1}$

38. $y = \dfrac{10}{(x - 2)^2}$

39. $g(t) = 3 - \dfrac{2}{t}$

40. $g(v) = \dfrac{2v}{v + 1}$

41. $y = \dfrac{4}{x^2 + 1}$

42. $y = \dfrac{4x^2}{x^2 + 1}$

43. $y = -\dfrac{x}{x^2 - 4}$

44. $y = \dfrac{3x^2}{x^2 - x - 2}$

In Exercises 45–52, use a graphing utility to graph the function. Give the domain of the function and identify any horizontal or vertical asymptotes.

45. $h(x) = \dfrac{x - 3}{x - 1}$

46. $h(x) = \dfrac{x^2}{x - 2}$

47. $f(t) = \dfrac{6}{t^2 + 1}$

48. $g(t) = 2 + \dfrac{3}{t + 1}$

49. $y = \dfrac{2(x^2 + 1)}{x^2}$

50. $y = \dfrac{2(x^2 - 1)}{x^2}$

51. $y = \dfrac{3}{x} + \dfrac{1}{x - 2}$

52. $y = \dfrac{x}{2} - \dfrac{2}{x}$

In Exercises 53–56, identify the transformation of the graph of

$$f(x) = \frac{1}{x}$$

and sketch the graph of g. (The graph of f is shown below.)

53. $g(x) = -\dfrac{1}{x}$ **54.** $g(x) = \dfrac{1}{x} + 2$

55. $g(x) = \dfrac{1}{x - 2}$ **56.** $g(x) = \dfrac{1}{x + 3}$

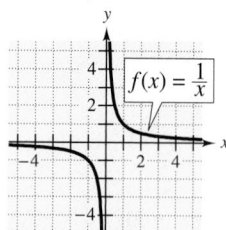
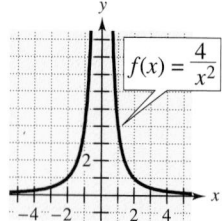

Figure for 53–56   **Figure for 57–60**

In Exercises 57–60, identify the transformation of the graph of

$$f(x) = \frac{4}{x^2}$$

and sketch the graph of g. (The graph of f is shown above.)

57. $g(x) = 2 + \dfrac{4}{x^2}$ **58.** $g(x) = -\dfrac{4}{x^2}$

59. $g(x) = -\dfrac{4}{(x - 2)^2}$ **60.** $g(x) = 5 - \dfrac{4}{x^2}$

Think About It   In Exercises 61 and 62, use a graphing utility to graph the function. Explain why there is no vertical asymptote when a superficial examination of the function may indicate that there should be one.

61. $g(x) = \dfrac{4 - 2x}{x - 2}$ **62.** $h(x) = \dfrac{x^2 - 9}{x + 3}$

In Exercises 63–66, (a) use the graph to determine any x-intercepts of the function, and (b) set $y = 0$ and solve the resulting equation to confirm the result.

63. $y = \dfrac{x + 2}{x - 2}$ **64.** $y = \dfrac{2x}{x + 4}$

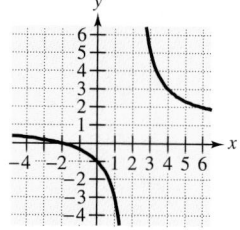
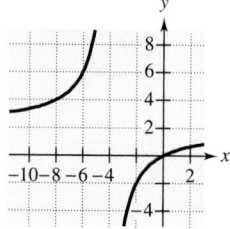

65. $y = x - \dfrac{1}{x}$ **66.** $y = x - \dfrac{2}{x} - 1$

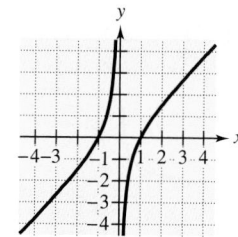
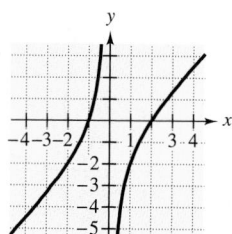

In Exercises 67–74, (a) use a graphing utility to graph the function and determine any x-intercepts of the graph, and (b) set $y = 0$ and solve the resulting equation to confirm the result.

67. $y = \dfrac{1}{x} + \dfrac{4}{x - 5}$ **68.** $y = \dfrac{x}{2} - \dfrac{4}{x} - 1$

69. $y = \dfrac{x - 4}{x + 5}$ **70.** $y = \dfrac{1}{x} - \dfrac{3}{x + 4}$

71. $y = (x + 1) - \dfrac{6}{x}$ **72.** $y = \dfrac{x^2 - 4}{x}$

73. $y = (x - 1) - \dfrac{12}{x}$ **74.** $y = 20\left(\dfrac{2}{x} - \dfrac{3}{x - 1}\right)$

75. *Average Cost* The cost of producing x units is $C = 2500 + 0.50x$, $x > 0$.

(a) Write the average cost $\overline{C}$ as a function of x.

(b) Find the average cost of producing $x = 1000$ and $x = 10,000$ units.

(c) Use a graphing utility to graph the average cost function. Determine the horizontal asymptote of the graph.

76. *Average Cost* The cost of producing x units is $C = 30,000 + 1.25x$, $x > 0$.

(a) Write the average cost $\overline{C}$ as a function of x.

(b) Find the average cost of producing $x = 10,000$ and $x = 100,000$ units.

(c) Use a graphing utility to graph the average cost function. Determine the horizontal asymptote of the graph.

77. *Medicine* The concentration of a certain chemical in the bloodstream t hours after injection into muscle tissue is given by

$$C = \frac{2t}{4t^2 + 25}, \qquad t \geq 0.$$

(a) Determine the horizontal asymptote of the function and interpret its meaning in the context of the problem.

(b) Use a graphing utility to graph the function. Approximate the time when the concentration is the greatest.

78. *Concentration of a Mixture* A 25-liter container contains 5 liters of a 25% brine solution. You add x liters of a 75% brine solution to the container. The concentration C of the resulting mixture is

$$C = \frac{3x + 5}{4(x + 5)}.$$

(a) Determine the domain of the function based on the physical constraints of the problem.

(b) Use a graphing utility to graph the function. As the container is filled, what percent does the concentration of the brine appear to approach?

79. *Geometry* A rectangular region of length x and width y has an area of 400 square meters.

(a) Verify that the perimeter P is given by

$$P = 2\left(x + \frac{400}{x}\right).$$

(b) Determine the domain of the function based on the physical constraints of the problem.

(c) Use a graphing utility to graph the function. Approximate the dimensions of the rectangle that has a minimum perimeter.

80. *Sales* The cumulative number N (in thousands) of units of a product sold over a period of t years on the market is modeled by

$$N = \frac{150t(1 + 4t)}{1 + 0.15t^2}, \qquad t \geq 0.$$

(a) Estimate N when $t = 1$, $t = 2$, and $t = 4$.

(b) Use a graphing utility to graph the function. Determine the horizontal asymptote.

(c) Explain the meaning of the horizontal asymptote in the context of the problem.

Section Project

Think About It Write a rational function satisfying each of the following criteria. Use a graphing utility to verify the results.

(a) Vertical asymptote: $x = 3$

Horizontal asymptote: $y = 2$

Zero of the function: $x = -1$

(b) Vertical asymptote: $x = -2$

Horizontal asymptote: $y = 0$

Zero of the function: $x = 3$

(c) Vertical asymptotes: $x = 1$ and $x = -1$

Horizontal asymptote: $y = 1$

Zero of the function: $x = 0$

(d) Vertical asymptotes: $x = 4$ and $x = -2$

Horizontal asymptote: $y = 0$

Zero of the function: $x = 6$

7.7	**Rational Inequalities in One Variable**

Test Intervals for Rational Inequalities ▪
Solving Rational Inequalities ▪ Application

Test Intervals for Rational Inequalities

In Section 6.7, you studied strategies for solving polynomial inequalities. For instance, to solve the quadratic inequality

$$2x^2 + 5x > 12, \qquad \text{Original inequality}$$

you wrote the inequality in standard form

$$2x^2 + 5x - 12 > 0 \qquad \text{Standard form}$$

and found the solutions of the corresponding equation

$$2x^2 + 5x - 12 = 0. \qquad \text{Corresponding equation}$$

These solutions, $x = -4$ and $x = \frac{3}{2}$, are called **critical numbers** and are used to form test intervals for the inequality.

$$(-\infty, -4), \quad \left(-4, \tfrac{3}{2}\right), \quad \left(\tfrac{3}{2}, \infty\right) \qquad \text{Test intervals}$$

For a review of how to use these test intervals to solve the inequality, see Example 3 on page 465.

The strategy for solving a rational inequality is similar to that for solving a polynomial inequality, and is summarized as follows.

NOTE The value of a rational expression can change sign only at its *real zeros* (the real x-values for which its numerator is zero) and its *undefined values* (the real x-values for which its denominator is zero). These two types of numbers make up the critical numbers of a rational inequality.

You may want students to refer to Section 7.1 and review the definition of rational expression.

Test Intervals for a Rational Inequality

To find the test intervals for a rational inequality, proceed as follows.

1. Write the inequality in standard form, with a single rational expression on the left and zero on the right.

2. Find all real zeros of the numerator of the rational expression *and* all real zeros of the denominator of the rational expression. These zeros are the **critical numbers** of the rational expression.

3. Arrange the critical numbers in increasing order. Then use the critical numbers to determine the **test intervals** of the rational expression.

4. Choose a representative x-value in each test interval and evaluate the rational expression at that value. If the value of the rational expression is negative, it will have negative values for *every* x-value in the interval. If the value of the rational expression is positive, it will have positive values for *every* x-value in the interval.

EXAMPLE 1   Finding Test Intervals for a Rational Expression

Determine the intervals on which the rational expression $\dfrac{x+2}{x-3}$ is entirely negative or entirely positive.

Solution

Begin by finding the critical numbers of the rational expression. To do this, set the numerator equal to zero and the denominator equal to zero and solve the resulting equations.

$$x + 2 = 0 \quad\Longrightarrow\quad x = -2 \qquad \text{Set numerator equal to 0.}$$
$$x - 3 = 0 \quad\Longrightarrow\quad x = 3 \qquad \text{Set denominator equal to 0.}$$

The rational expression has two critical numbers: -2 and 3. These critical numbers determine the following test intervals.

$$(-\infty, -2), \quad (-2, 3), \quad (3, \infty) \qquad \text{Test intervals}$$

NOTE   Another way to determine the intervals on which $(x + 2)/(x - 3)$ is entirely negative or entirely positive is to sketch the graph of $y = (x + 2)/(x - 3)$, as shown in Figure 7.14. The portions of the graph that lie above the x-axis correspond to $(x + 2)/(x - 3) > 0$, and the portion that lies below the x-axis corresponds to $(x + 2)/(x - 3) < 0$.

In each test interval, choose a representative x-value and evaluate the rational expression, as shown in the table.

Test interval	Representative x-value	Value of rational expression	Conclusion
$(-\infty, -2)$	$x = -3$	$\dfrac{-3+2}{-3-3} = \dfrac{1}{6}$	Expression is positive.
$(-2, 3)$	$x = 0$	$\dfrac{0+2}{0-3} = -\dfrac{2}{3}$	Expression is negative.
$(3, \infty)$	$x = 4$	$\dfrac{4+2}{4-3} = 6$	Expression is positive.

Therefore, the rational expression has positive values for every x in the intervals $(-\infty, -2)$ and $(3, \infty)$, and negative values for every x in the interval $(-2, 3)$. This result is shown graphically in Figure 7.13.

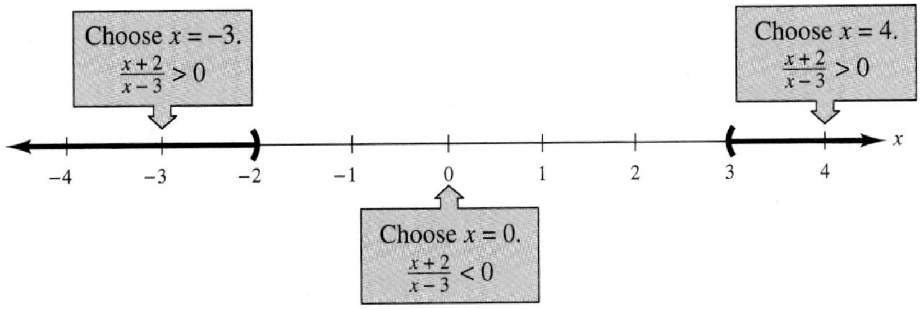

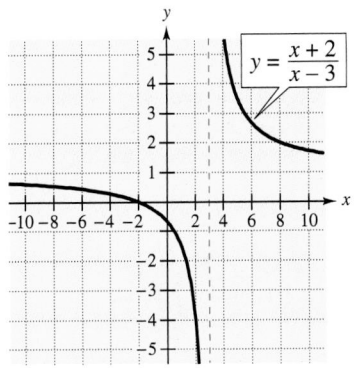

FIGURE 7.14

Choose $x = -3$.
$\dfrac{x+2}{x-3} > 0$

Choose $x = 4$.
$\dfrac{x+2}{x-3} > 0$

Choose $x = 0$.
$\dfrac{x+2}{x-3} < 0$

FIGURE 7.13

Solving Rational Inequalities

EXAMPLE 2   Solving a Rational Inequality

Some students may be tempted to clear the inequality of fractions as a first step. Discuss why this approach is not effective.

Solve the rational inequality $\dfrac{x}{x-2} > 0$.

Solution

The numerator is zero when $x = 0$ and the denominator is zero when $x = 2$. Thus, the two critical numbers are 0 and 2, which implies that the test intervals are

$$(-\infty, 0), \quad (0, 2), \quad \text{and} \quad (2, \infty).$$

A test of these intervals, as shown in Figure 7.15, reveals that the rational expression $x/(x-2)$ is positive in the open intervals $(-\infty, 0)$ and $(2, \infty)$. Therefore, the solution set of the inequality is $(-\infty, 0) \cup (2, \infty)$.

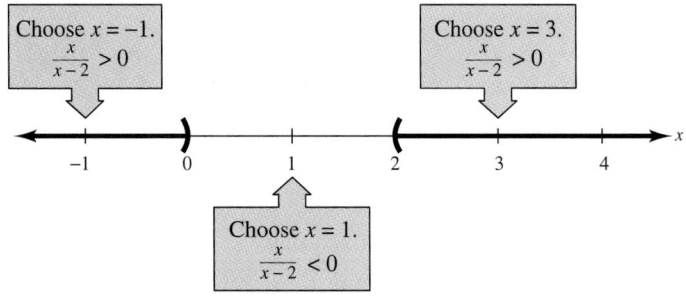

FIGURE 7.15

~ *Technology*

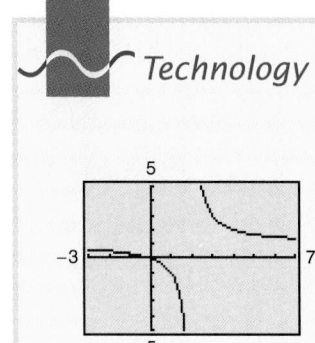

You can use a graphing utility to estimate the solution of a rational inequality (written in standard form). One way to do this is to graph the rational expression and observe the portions of the graph that lie above the x-axis and the portions of the graph that lie below the x-axis. For instance, the graph of

$$y = \frac{x}{x-2}$$

is shown at the left. Notice that the graph lies above the x-axis on the intervals $(-\infty, 0)$ and $(2, \infty)$, and below the x-axis on the interval $(0, 2)$. This observation agrees with the solution of $x/(x-2) > 0$, as found in Example 2.

When solving a rational inequality, you should begin by writing the inequality in standard form, with the rational expression (as a single fraction) on the left and zero on the right. This is demonstrated in Example 3.

EXAMPLE 3   *Writing in Standard Form First*

Solve the rational inequality $\dfrac{2x}{x + 3} \leq 4$.

Solution

Begin by writing the inequality in standard form, as follows.

$$\dfrac{2x}{x + 3} \leq 4 \qquad \text{Original inequality}$$

$$\dfrac{2x}{x + 3} - 4 \leq 0 \qquad \text{Subtract 4 from both sides.}$$

$$\dfrac{2x - 4(x + 3)}{x + 3} \leq 0 \qquad \text{Subtract fractions.}$$

$$\dfrac{-2x - 12}{x + 3} \leq 0 \qquad \text{Simplify.}$$

$$\dfrac{-2(x + 6)}{x + 3} \leq 0 \qquad \text{Factor.}$$

From the factored standard form, you can see that the critical numbers are $x = -6$ and $x = -3$. This implies that the test intervals are

$$(-\infty, -6), \quad (-6, -3), \quad \text{and} \quad (-3, \infty).$$

After testing the intervals, as shown in Figure 7.16, you can see that the rational expression $-2(x + 6)/(x + 3)$ is less than or equal to zero in the intervals $(-\infty, -6]$ and $(-3, \infty)$. Because the rational expression is undefined when $x = -3$, that value is *not* included in the solution set. Thus, the solution set of the inequality is

$$(-\infty, -6] \cup (-3, \infty). \qquad \text{Solution set}$$

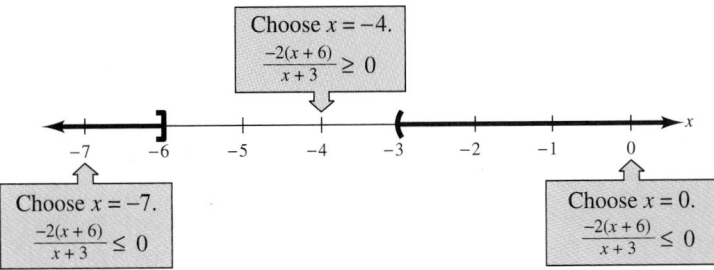

FIGURE 7.16

Application

Applications involving rational inequalities are similar to those involving rational equations. For instance, compare the following example on *average cost* with Example 10 on page 527. You will notice that the two examples have similar solutions.

EXAMPLE 4   Average Cost

A manufacturing plant can produce x units of a certain item for $26 per unit *plus* an initial investment of $80,000. How many units must be produced to have an average cost that is *less than* $30 per unit?

Solution

Verbal Model:

$$\boxed{\text{Average cost per unit}} < \boxed{\$30}$$

$$\boxed{\text{Total cost}} \div \boxed{\text{Number of units}} < \boxed{\$30}$$

Labels: Number of units $= x$ (units)
Total cost $= 26x + 80,000$ (dollars)

Inequality: $\dfrac{26x + 80,000}{x} < 30$

You can solve this inequality as follows.

$$\frac{26x + 80,000}{x} < 30 \qquad \text{Original inequality}$$

$$\frac{26x + 80,000}{x} - 30 < 0 \qquad \text{Subtract 30 from both sides.}$$

$$\frac{26x + 80,000 - 30x}{x} < 0 \qquad \text{Subtract fractions.}$$

$$\frac{-4x + 80,000}{x} < 0 \qquad \text{Simplify.}$$

$$\frac{-4(x - 20,000)}{x} < 0 \qquad \text{Factor.}$$

From the factored standard form, you can see that the critical numbers are $x = 20,000$ and $x = 0$. This implies that the test intervals are $(-\infty, 0)$, $(0, 20,000)$, and $(20,000, \infty)$. Because x cannot be negative in this application, the first test interval cannot be part of the solution. After testing the other two intervals, you can conclude that the plant should produce at least 20,000 units. Check this in the original statement of the problem.

Group Activities   **Exploring with Technology**

Solving a Rational Inequality Graphically   The Technology box on page 544 describes one way to use a graphing utility to estimate (or to solve) a rational inequality. Another way is to use the *inequality grapher* that is built into your graphing utility. For instance, on a TI-82 or TI-83, you can solve the inequality in Example 2 as follows.

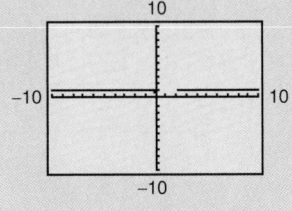

1. Set the calculator in dot mode.

2. In the [Y=] menu, enter X/(X–2) > 0. (Inequality symbols can be found in the TEST menu.)

3. Choose the standard viewing rectangle to obtain the graph shown at the left. You can interpret the graph as saying that the solution is $(-\infty, 0) \cup (2, \infty)$.

With others in your group, try solving the following rational inequalities with a graphing utility. Then check your solutions using the algebraic strategy illustrated in Examples 2 and 3.

a. $\dfrac{x}{x-3} > 0$ b. $\dfrac{x}{x-3} > 2$ c. $\dfrac{x}{x-3} > 4$

7.7   **Exercises**

See Warm-Up
Exercises, p. A47

In Exercises 1 and 2, determine whether the x-values are solutions of the inequality.

1. $\dfrac{2}{3-x} \le 0$ (a) $x = 3$ (b) $x = 4$
   (c) $x = -4$ (d) $x = -\frac{1}{3}$

2. $\dfrac{x-2}{x+1} > 0$ (a) $x = 0$ (b) $x = -1$
   (c) $x = -4$ (d) $x = -\frac{3}{2}$

In Exercises 3–6, find the critical numbers of the rational expression and plot them on the real number line.

3. $\dfrac{5}{x-3}$

4. $\dfrac{-6}{x+2}$

5. $\dfrac{2x}{x+5}$

6. $\dfrac{x-2}{x-10}$

In Exercises 7–22, solve the inequality and sketch the graph of the solution on the real number line.

7. $\dfrac{5}{x-3} > 0$

8. $\dfrac{3}{4-x} > 0$

9. $\dfrac{-5}{x-3} > 0$

10. $\dfrac{-3}{4-x} > 0$

11. $\dfrac{4}{x-3} < 0$

12. $\dfrac{5}{2-x} < 0$

13. $\dfrac{x}{x-4} \le 0$

14. $\dfrac{z-1}{z+3} < 0$

15. $\dfrac{y-3}{2y-11} \ge 0$

16. $\dfrac{x+5}{3x+2} \ge 0$

17. $\dfrac{3(u-3)}{u+1} < 0$

18. $\dfrac{2(4-t)}{4+t} > 0$

19. $\dfrac{6}{x - 4} > 2$

20. $\dfrac{1}{x + 2} > -3$

21. $\dfrac{x}{x - 3} \leq 2$

22. $\dfrac{x + 4}{x - 5} \geq 10$

In Exercises 23–30, use a graphing utility to solve the rational inequality.

23. $\dfrac{1}{x} - x > 0$

24. $\dfrac{1}{x} - 4 < 0$

25. $\dfrac{x + 6}{x + 1} - 2 < 0$

26. $\dfrac{x + 12}{x + 2} - 3 \geq 0$

27. $\dfrac{6x}{x + 5} < 2$

28. $x + \dfrac{1}{x} > 3$

29. $\dfrac{3x - 4}{x - 4} < -5$

30. $4 - \dfrac{1}{x^2} > 1$

Graphical Analysis In Exercises 31–34, use a graphing utility to graph the equation. Use the graph to approximate the values of x that satisfy the specified inequalities.

Equation	Inequalities	
31. $y = \dfrac{3x}{x - 2}$	(a) $y \leq 0$	(b) $y \geq 6$
32. $y = \dfrac{2(x - 2)}{x + 1}$	(a) $y \leq 0$	(b) $y \geq 8$
33. $y = \dfrac{2x^2}{x^2 + 4}$	(a) $y \geq 1$	(b) $y \leq 2$
34. $y = \dfrac{5x}{x^2 + 4}$	(a) $y \geq 1$	(b) $y \geq 0$

35. *Average Speed* During the first 144 miles of a 260-mile trip, you travel at an average speed of r miles per hour. During the last part of the trip, you increase your average speed by 10 miles per hour. Find the interval for the two average speeds if you want to make the trip in less than 5 hours. (Assume that the speed limit is 65 miles per hour.)

36. *Average Cost* The cost of producing x units is $C = 3000 + 0.75x$, $x > 0$.

(a) Write the average cost $\overline{C}$ as a function of x.

(b) Use a graphing utility to graph the average cost function in part (a). Determine the horizontal asymptote of the graph.

(c) How many units must be produced if the average cost per unit is to be less than $2?

37. *Data Analysis* The temperature T (in degrees Fahrenheit) of a metal in a laboratory experiment was recorded every 2 minutes for a period of 16 minutes. The table gives the experimental data, where t is the time in minutes.

t	0	2	4	6	8	10	12	14	16
T	250	290	338	410	498	560	530	370	160

A model for this data is

$$T = \dfrac{244.20 - 13.23t}{1 - 0.13t + 0.005t^2}.$$

(a) Plot the data and graph the model.

(b) Use the graph to approximate the times when the temperature was at least 400°F.

Section Project

The cost of producing x units is $C = 8000 + 0.6x$, $x > 0$.

(a) Write the average cost $\overline{C}$ as a function of x.

(b) Find the number of units that must be produced if $\overline{C} < 3$.

(c) Find the number of units that must be produced if $\overline{C} < 2$.

(d) Find the number of units that must be produced if $\overline{C} < 1$.

(e) Find the number of units that must be produced if $\overline{C} < 0.50$.

(f) Use a graphing utility to graph the average cost function.

(g) Determine the horizontal asymptote of the graph in part (f).

(h) Interpret what the horizontal asymptote in part (g) represents in the context of the problem. Describe how this result is related to the result in part (e).

Chapter Summary

After studying this chapter, you should have acquired the following skills.
These skills are keyed to the Review Exercises that begin on page 550.
Answers to odd-numbered Review Exercises are given in the back of the book.

• Find the domains of rational expressions. *(Section 7.1)*	**Review Exercises 1–4**
• Simplify rational expressions. *(Section 7.1)*	**Review Exercises 5–12**
• Multiply and divide rational expressions. *(Section 7.2)*	**Review Exercises 13–24**
• Add and subtract rational expressions. *(Section 7.3)*	**Review Exercises 25–36**
• Simplify complex fractions. *(Sections 7.2, 7.3)*	**Review Exercises 37–42**
• Divide polynomials. *(Section 7.4)*	**Review Exercises 43–48**
• Divide polynomials using synthetic division. *(Section 7.4)*	**Review Exercises 49–52**
• Use a graphing utility to graph pairs of equations to verify their equivalence. *(Sections 7.2–7.4)*	**Review Exercises 53–56**
• Solve rational equations. *(Section 7.5)*	**Review Exercises 57–70**
• Use a graphing utility to graph rational functions to determine any x-intercepts, and confirm them algebraically. *(Section 7.6)*	**Review Exercises 71, 72**
• Translate real-life situations into rational equations and solve. *(Section 7.5)*	**Review Exercises 73–76**
• Match rational functions with their graphs. *(Section 7.6)*	**Review Exercises 77–80**
• Sketch the graphs of rational functions, and use a graphing utility to confirm the graphs. *(Section 7.6)*	**Review Exercises 81–94**
• Solve real-life problems modeled by rational equations or inequalities. *(Sections 7.6, 7.7)*	**Review Exercises 95, 96, 101, 102**
• Solve rational inequalities and graph their solutions on the real number line. *(Section 7.7)*	**Review Exercises 97–100**

Review Exercises

In Exercises 1–4, find the domain of the rational expression.

1. $\dfrac{3y}{y - 8}$

2. $\dfrac{t + 4}{t + 12}$

3. $\dfrac{u}{u^2 - 7u + 6}$

4. $\dfrac{x - 12}{x(x^2 - 16)}$

In Exercises 5–12, simplify the rational expression.

5. $\dfrac{6x^4y^2}{15xy^2}$

6. $\dfrac{2(y^3z)^2}{28(yz^2)^2}$

7. $\dfrac{5b - 15}{30b - 120}$

8. $\dfrac{4a}{10a^2 + 26a}$

9. $\dfrac{9x - 9y}{y - x}$

10. $\dfrac{x + 3}{x^2 - x - 12}$

11. $\dfrac{x^2 - 5x}{2x^2 - 50}$

12. $\dfrac{x^2 + 3x + 9}{x^3 - 27}$

In Exercises 13–42, perform the operation(s) and simplify.

13. $\dfrac{7}{8} \cdot \dfrac{2x}{y} \cdot \dfrac{y^2}{14x^2}$

14. $\dfrac{15(x^2y)^3}{3y^3} \cdot \dfrac{12y}{x}$

15. $\dfrac{60z}{z + 6} \cdot \dfrac{z^2 - 36}{5}$

16. $\dfrac{1}{6}(x^2 - 16) \cdot \dfrac{3}{x^2 - 8x + 16}$

17. $\dfrac{u}{u - 3} \cdot \dfrac{3u - u^2}{4u^2}$

18. $x^2 \cdot \dfrac{x + 1}{x^2 - x} \cdot \dfrac{(5x - 5)^2}{x^2 + 6x + 5}$

19. $\dfrac{\left(\dfrac{6}{x}\right)}{\left(\dfrac{2}{x^3}\right)}$

20. $\dfrac{0}{\left(\dfrac{5x^2}{2y}\right)}$

21. $25y^2 \div \dfrac{xy}{5}$

22. $\dfrac{6}{z^2} \div 4z^2$

23. $\dfrac{x^2 - 7x}{x + 1} \div \dfrac{x^2 - 14x + 49}{x^2 - 1}$

24. $\left(\dfrac{6x}{y^2}\right)^2 \div \left(\dfrac{3x}{y}\right)^3$

25. $\dfrac{4a}{9} - \dfrac{11a}{9}$

26. $\dfrac{2(3y + 4)}{2y + 1} + \dfrac{3 - y}{2y + 1}$

27. $\dfrac{15}{16x} - \dfrac{5}{24x} - 1$

28. $-\dfrac{3y}{8x} + \dfrac{7y}{6x} - \dfrac{1y}{12x}$

29. $\dfrac{1}{x + 5} + \dfrac{3}{x - 12}$

30. $\dfrac{2}{x - 10} + \dfrac{3}{4 - x}$

31. $5x + \dfrac{2}{x - 3} - \dfrac{3}{x + 2}$

32. $4 - \dfrac{4x}{x + 6} + \dfrac{7}{x - 5}$

33. $\dfrac{6}{x} - \dfrac{6x - 1}{x^2 + 4}$

34. $\dfrac{5}{x + 2} + \dfrac{25 - x}{x^2 - 3x - 10}$

35. $\dfrac{5}{x + 3} - \dfrac{4x}{(x + 3)^2} - \dfrac{1}{x - 3}$

36. $\dfrac{8}{y} - \dfrac{3}{y + 5} + \dfrac{4}{y - 2}$

37. $\dfrac{\left(\dfrac{6x^2}{x^2 + 2x - 35}\right)}{\left(\dfrac{x^3}{x^2 - 25}\right)}$

38. $\dfrac{\left[\dfrac{24 - 18x}{(2 - x)^2}\right]}{\left(\dfrac{60 - 45x}{x^2 - 4x + 4}\right)}$

39. $\dfrac{3t}{\left(5 - \dfrac{2}{t}\right)}$

40. $\dfrac{\left(x - 3 + \dfrac{2}{x}\right)}{\left(1 - \dfrac{2}{x}\right)}$

41. $\dfrac{\left(\dfrac{1}{a^2 - 16} - \dfrac{1}{a}\right)}{\left(\dfrac{1}{a^2 + 4a} + 4\right)}$

42. $\dfrac{\left(\dfrac{1}{x^2} - \dfrac{1}{y^2}\right)}{\left(\dfrac{1}{x} + \dfrac{1}{y}\right)}$

In Exercises 43–48, perform the division.

43. $\dfrac{4x^3 - x}{2x}$

44. $\dfrac{10x^7y^8 - 40x^6y^5 - 25x^3y^3}{5xy^2}$

45. $\dfrac{6x^3 + x^2 - 4x + 2}{3x - 1}$

46. $\dfrac{4x^4 - x^3 - 7x^2 + 18x}{x - 2}$

47. $\dfrac{x^4 - 3x^2 + 2}{x^2 - 1}$

48. $\dfrac{3x^6}{x^2 - 1}$

In Exercises 49–52, use synthetic division to divide.

49. $\dfrac{x^3 + 7x^2 + 3x - 14}{x + 2}$

50. $\dfrac{x^4 - 2x^3 - 15x^2 - 2x + 10}{x - 5}$

51. $(x^4 - 3x^2 - 25) \div (x - 3)$

52. $(2x^3 + 5x - 2) \div \left(x + \frac{1}{2}\right)$

In Exercises 53–56, use a graphing utility to graph the two equations in the same viewing rectangle. Use the graphs to verify that the expressions are equivalent. Verify the results algebraically.

53. $y_1 = \dfrac{x^2 + 6x + 9}{x^2} \cdot \dfrac{x^2 - 3x}{x + 3}, \quad y_2 = \dfrac{x^2 - 9}{x}, \, x \neq -3$

54. $y_1 = \dfrac{1}{x} - \dfrac{3}{x + 3}, \quad y_2 = \dfrac{3 - 2x}{x(x + 3)}$

55. $y_1 = \dfrac{\left(\dfrac{1}{x} - \dfrac{1}{2}\right)}{2x}, \quad y_2 = \dfrac{2 - x}{4x^2}$

56. $y_1 = \dfrac{x^3 - 2x^2 - 7}{x - 2}, \quad y_2 = x^2 - \dfrac{7}{x - 2}$

In Exercises 57–70, solve the rational equation.

57. $\dfrac{3x}{8} = -15$

58. $\dfrac{t + 1}{8} = \dfrac{1}{2}$

59. $3\left(8 - \dfrac{12}{t}\right) = 0$

60. $\dfrac{1}{3y - 4} = \dfrac{6}{4(y + 1)}$

61. $\dfrac{2}{y} - \dfrac{1}{3y} = \dfrac{1}{3}$

62. $8\left(\dfrac{6}{x} - \dfrac{1}{x + 5}\right) = 15$

63. $r = 2 + \dfrac{24}{r}$

64. $\dfrac{3}{y + 1} - \dfrac{8}{y} = 1$

65. $\dfrac{2}{x} - \dfrac{x}{6} = \dfrac{2}{3}$

66. $\dfrac{2x}{x + 3} - \dfrac{3}{x} = 0$

67. $\dfrac{12}{x^2 + x - 12} - \dfrac{1}{x - 3} = -1$

68. $\dfrac{3}{x - 1} + \dfrac{6}{x^2 - 3x + 2} = 2$

69. $\dfrac{5}{x^2 - 4} - \dfrac{6}{x - 2} = -5$

70. $\dfrac{3}{x^2 - 9} + \dfrac{4}{x + 3} = 1$

In Exercises 71 and 72, (a) use a graphing utility to graph the function and determine any x-intercepts of the graph, and (b) set $y = 0$ and solve the resulting equation to confirm the result algebraically.

71. $y = \dfrac{1}{x} - \dfrac{1}{2x + 3}$

72. $y = \dfrac{x}{4} - \dfrac{2}{x} - \dfrac{1}{2}$

73. *Batting Average*   In this year's playing season, a baseball player has been at bat 150 times and has hit the ball safely 45 times. Thus, the "batting average" for the player is $45/150 = 0.300$. How many consecutive times must the player hit safely to obtain a batting average of 0.400?

74. *Average Speed*   You drive 56 miles on a service call for your company. On the return trip, which takes 10 minutes less than the original trip, your average speed is 8 miles per hour faster. What is your average speed on the return trip?

75. *Work Rate*   Suppose that in 12 minutes your supervisor can complete a task that you require 15 minutes to complete. Determine the time required to complete the task if you work together.

76. *Forming a Partnership*   A group of people agree to share equally in the cost of a $60,000 piece of machinery. If they find two more people to join the group, each person's share of the cost will decrease by $5000. How many people are presently in the group?

In Exercises 77–80, match the function with its graph. [The graphs are labeled (a), (b), (c), and (d).]

(a)

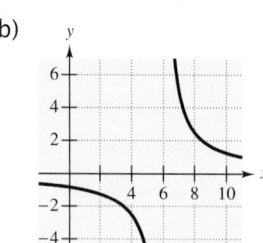

(b)

(c)

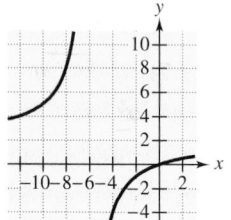

(d)

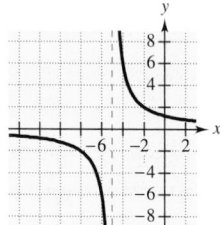

77. $f(x) = \dfrac{5}{x - 6}$

78. $f(x) = \dfrac{6}{x + 5}$

79. $f(x) = \dfrac{6x}{x - 5}$

80. $f(x) = \dfrac{2x}{x + 6}$

In Exercises 81–94, sketch the graph of the equation. Use a graphing utility to confirm the graph.

81. $g(x) = \dfrac{2 + x}{1 - x}$

82. $h(x) = \dfrac{x - 3}{x - 2}$

83. $f(x) = \dfrac{x}{x^2 + 1}$

84. $f(x) = \dfrac{2x}{x^2 + 4}$

85. $P(x) = \dfrac{3x + 6}{x - 2}$

86. $s(x) = \dfrac{2x - 6}{x + 4}$

87. $h(x) = \dfrac{4}{(x - 1)^2}$

88. $g(x) = \dfrac{-2}{(x + 3)^2}$

89. $f(x) = -\dfrac{5}{x^2}$

90. $f(x) = \dfrac{4}{x}$

91. $y = \dfrac{x}{x^2 - 1}$

92. $y = \dfrac{2x}{x^2 - 4}$

93. $y = \dfrac{2x^2}{x^2 - 4}$

94. $y = \dfrac{2}{x + 3}$

95. *Average Cost* A business produces x units at a cost of $C = 0.5x + 500$. The average cost per unit is

$$\overline{C} = \frac{C}{x} = \frac{0.5x + 500}{x}, \qquad x > 0.$$

Find the horizontal asymptote and state what it represents in the model.

96. *Population of Fish* The Parks and Wildlife Commission introduces 80,000 fish into a large lake. The population of the fish (in thousands) is

$$N = \frac{20(4 + 3t)}{1 + 0.05}, \qquad t \geq 0,$$

where t is the time in years.

(a) Find the population when t is 5, 10, and 25.

(b) Find the horizontal asymptote and state what it represents in the model.

In Exercises 97–100, solve the inequality and graph the solution on the real number line.

97. $\dfrac{x}{2x - 7} \geq 0$

98. $\dfrac{2x - 9}{x - 1} \leq 0$

99. $\dfrac{x}{x + 6} + 2 < 0$

100. $\dfrac{3x + 1}{x - 2} > 4$

101. *Seizure of Illegal Drugs* The cost (in millions of dollars) for a government agency to seize $p\%$ of a certain illegal drug as it enters the country is

$$C = \frac{528p}{100 - p}, \qquad 0 \leq p < 100.$$

(a) Find the cost of seizing 25%.

(b) Find the cost of seizing 75%.

(c) According to this model, would it be possible to seize 100% of the drug?

(d) Find the percent of drugs seized if $C < 1500$.

102. *Average Cost* The cost of producing x units is C, and therefore the average cost per unit is

$$\overline{C} = \frac{C}{x} = \frac{100{,}000 + 0.9x}{x}, \qquad x > 0.$$

Find the number of units that must be produced if $\overline{C} < 2$.

Chapter Test

Take this test as you would take a test in class. After you are done, check your work against the answers given in the back of the book.

1. Find the domain of the rational expression. $\dfrac{3y}{y^2 - 25}$

2. Simplify the rational expression. $\dfrac{2 - x}{3x - 6}$

In Exercises 3–11, perform the operation(s) and simplify.

3. $\dfrac{4z^3}{5} \cdot \dfrac{25}{12z^2}$

4. $\dfrac{y^2 + 8y + 16}{2(y - 2)} \cdot \dfrac{8y - 16}{(y + 4)^3}$

5. $(4x^2 - 9) \cdot \dfrac{2x + 3}{2x^2 - x - 3}$

6. $\dfrac{(2xy^2)^3}{15} \div \dfrac{12x^3}{21}$

7. $\dfrac{\left(\dfrac{3x}{x + 2}\right)}{\left(\dfrac{12}{x^3 + 2x^2}\right)}$

8. $\dfrac{\left(9x - \dfrac{1}{x}\right)}{\left(\dfrac{1}{x} - 3\right)}$

9. $2x + \dfrac{1 - 4x^2}{x + 1}$

10. $\dfrac{5x}{x + 2} - \dfrac{2}{x^2 - x - 6}$

11. $\dfrac{3}{x} - \dfrac{5}{x^2} + \dfrac{2x}{x^2 + 2x + 1}$

In Exercises 12–14, perform the division. If possible, use synthetic division.

12. $\dfrac{24a^7 + 42a^4 - 6a^3}{6a^2}$

13. $\dfrac{t^4 + t^2 - 6t}{t^2 - 2}$

14. $\dfrac{2x^4 - 15x^2 - 7}{x - 3}$

15. Sketch the graph of each function. Describe how to find the asymptotes. Use a graphing utility to confirm the results.

   (a) $f(x) = \dfrac{3}{x - 3}$ (b) $g(x) = \dfrac{3x}{x - 3}$

In Exercises 16–18, solve the rational equation.

16. $\dfrac{3}{h + 2} = \dfrac{1}{8}$ **17.** $\dfrac{2}{x + 5} - \dfrac{3}{x + 3} = \dfrac{1}{x}$ **18.** $\dfrac{1}{x + 1} + \dfrac{1}{x - 1} = \dfrac{2}{x^2 - 1}$

In Exercises 19 and 20, solve the inequality and sketch the solution.

19. $\dfrac{3}{x - 2} > 4$

20. $\dfrac{3u + 2}{u - 3} \le 2$

21. One painter works $1\frac{1}{2}$ times as fast as another. Find their individual rates for painting a room if together it takes them 4 hours.

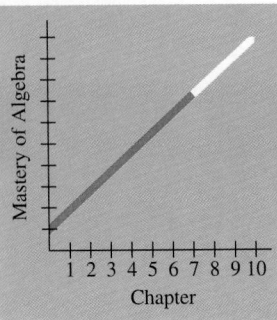

Take this test as you would take a test in class. After you are done, check your work against the answers given in the back of the book.

In Exercises 1 and 2, simplify the expression.

1. $5(x + 2) - 4(2x - 3)$

2. $0.12x + 0.05(2000 - 2x)$

In Exercises 3 and 4, use the function to find and simplify the expression for $f(a + 2)$.

3. $f(x) = x^2 - 3$

4. $f(x) = \dfrac{3}{x + 5}$

In Exercises 5 and 6, simplify the rational expression.

5. $\dfrac{-16x^2}{12x}$

6. $\dfrac{6u^5 v^{-3}}{27uv^3}$

In Exercises 7–9, perform the operation and simplify. (Assume that all variables are positive.)

7. $\left(\sqrt{x} + 3\right)\left(\sqrt{x} - 3\right)$ **8.** $\sqrt{u}\left(\sqrt{20} - \sqrt{5}\right)$ **9.** $\left(2\sqrt{t} + 3\right)^2$

In Exercises 10 and 11, solve the system of equations.

10. $\begin{cases} 5x - 2y = -25 \\ -3x + 7y = 44 \end{cases}$

11. $\begin{cases} 3x - 2y + z = 1 \\ x + 5y - 6z = 4 \\ 4x - 3y + 2z = 2 \end{cases}$

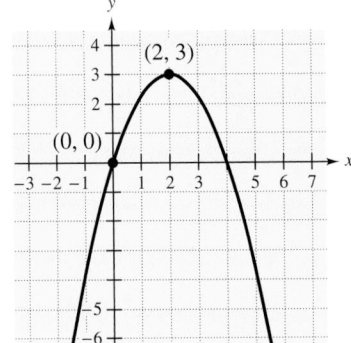

In Exercises 12–14, solve the equation.

12. $x + \dfrac{4}{x} = 4$

13. $\sqrt{x + 10} = x - 2$

14. $(x - 5)^2 + 50 = 0$

15. Use a graphing utility to graph the equation $y = x^2 - 6x - 8$. Use the graph to approximate any x-intercepts. Set $y = 0$ and solve the resulting equation. Compare the results with the x-intercepts of the graph.

16. Find an equation of the parabola shown in the figure.

Figure for 16

17. Sketch the graph of the rational function $y = 4/(x - 2)$.

18. You can mow a lawn in 4 hours, and your friend can mow it in 5 hours. What fractional part of the lawn can each of you mow in 1 hour? How long will it take both of you to mow the lawn?

More About Functions and Relations

Strategies for Success

SKILLS *When you have completed this chapter, make sure you are able to:*

◯ Find combinations, compositions, and inverses of functions
◯ Solve real-life problems involving variation
◯ Sketch and summarize the characteristics of graphs of polynomial functions
◯ Graph circles, ellipses, hyperbolas, and parabolas
◯ Solve nonlinear systems of equations

TOOLS *Use these study tools to master the skills above:*

◯ **Mid-Chapter Quiz** (page 596)
◯ **Chapter Summary** (page 636)
◯ **Review Exercises** (page 637)
◯ **Chapter Test** (page 641)
◯ **Cumulative Test** (page 642)

In Exercise 51 of Section 8.3, you will model an equation using variation to determine the weight of an astronaut on the moon.

8.1	**Combinations of Functions**

Arithmetic Combinations of Functions ▪ Composition of Functions ▪ Application

Exploration

Enter the functions $f(x) = 3x - 1$ and $g(x) = x^2 - 4$ as Y_1 and Y_2 on your graphing utility, and then let $Y_3 = Y_1 + Y_2$. Now, algebraically add $f(x) + g(x)$, and graph the resulting function as Y_4. Are Y_3 and Y_4 equivalent? Do you think they should be? Repeat this using the difference, product, and quotient of the two functions.

Arithmetic Combinations of Functions

In Section 2.4, you were introduced to the concept of a function. In this and the next section, you will learn more about functions.

Just as two real numbers can be combined by the operations of addition, subtraction, multiplication, and division to form other real numbers, two functions can be combined to form new functions. For example, if

$$f(x) = 3x - 1$$

and

$$g(x) = x^2 - 4$$

you can form the sum, difference, product, and quotient of f and g as follows.

$$f(x) + g(x) = (3x - 1) + (x^2 - 4) = x^2 + 3x - 5 \qquad \text{Sum}$$
$$f(x) - g(x) = (3x - 1) - (x^2 - 4) = -x^2 + 3x + 3 \qquad \text{Difference}$$
$$f(x)g(x) = (3x - 1)(x^2 - 4) = 3x^3 - x^2 - 12x + 4 \qquad \text{Product}$$
$$\frac{f(x)}{g(x)} = \frac{3x - 1}{x^2 - 4}, \quad x \neq \pm 2 \qquad \text{Quotient}$$

The domain of an arithmetic combination of functions f and g consists of all real numbers that are common to the domains of f and g. In the case of the quotient $f(x)/g(x)$, there is the further restriction that $g(x) \neq 0$.

Definitions of Sum, Difference, Product, and Quotient of Functions

Let f and g be two functions with overlapping domains. Then, for all x common to both domains, the **sum, difference, product,** and **quotient** of f and g are defined as follows.

1. *Sum:* $\qquad\qquad (f + g)(x) = f(x) + g(x)$

2. *Difference:* $\qquad (f - g)(x) = f(x) - g(x)$

3. *Product:* $\qquad\quad (fg)(x) = f(x) \cdot g(x)$

4. *Quotient:* $\qquad\quad \left(\dfrac{f}{g}\right)(x) = \dfrac{f(x)}{g(x)}, \quad g(x) \neq 0$

Students often think $(fg)(x)$ means f times g times x. Make sure they understand this notation.

EXAMPLE 1   *Finding the Sum and Difference of Two Functions*

Given the functions $f(x) = 4x - 3$ and $g(x) = 2x^2 + x$, find each of the following.

(a) $(f + g)(x)$

(b) $(f - g)(x)$

Then evaluate each of these combinations when $x = 3$.

Solution

(a) The sum of the functions f and g is given by

$$(f + g)(x) = f(x) + g(x)$$
$$= (4x - 3) + (2x^2 + x)$$
$$= 2x^2 + 5x - 3.$$

When $x = 3$, the value of this sum is

$$(f + g)(3) = 2(3)^2 + 5(3) - 3$$
$$= 30.$$

(b) The difference of the functions f and g is given by

$$(f - g)(x) = f(x) - g(x)$$
$$= (4x - 3) - (2x^2 + x)$$
$$= -2x^2 + 3x - 3.$$

When $x = 3$, the value of this difference is

$$(f - g)(3) = -2(3)^2 + 3(3) - 3$$
$$= -12.$$

NOTE   In Example 1, $(f + g)(3)$ also could have been evaluated as

$$(f + g)(3) = f(3) + g(3)$$
$$= [4(3) - 3] + [2(3)^2 + 3]$$
$$= 9 + 21$$
$$= 30.$$

Similarly, $(f - g)(3)$ also could have been evaluated as

$$(f - g)(3) = f(3) - g(3)$$
$$= [4(3) - 3] - [2(3)^2 + 3]$$
$$= 9 - 21$$
$$= -12.$$

In Example 1, both f and g have domains that consist of all real numbers. Thus, the domains of both $(f + g)$ and $(f - g)$ are also the set of all real numbers. In the next example, the domain of the sum of f and g is smaller than the domains of f and g.

EXAMPLE 2   Finding the Domain of the Sum of Two Functions

Given the functions $f(x) = \sqrt{x + 1}$ and $g(x) = \sqrt{1 - x}$, find the sum of f and g. Then find the domain of $(f + g)$.

Solution

The sum of the functions f and g is given by

$$(f + g)(x) = f(x) + g(x)$$
$$= \sqrt{x + 1} + \sqrt{1 - x}.$$

Now, because f is defined for $x + 1 \geq 0$ and g is defined for $1 - x \geq 0$, it follows that the domains of f and g are as follows.

$$(\text{Domain of } f) = [-1, \infty)$$
$$(\text{Domain of } g) = (-\infty, 1]$$

The set that is common to both of these domains is the closed interval $[-1, 1]$, as shown in Figure 8.1. Therefore, the domain of $(f + g)$ is

$$(\text{Domain of } f + g) = [-1, 1].$$

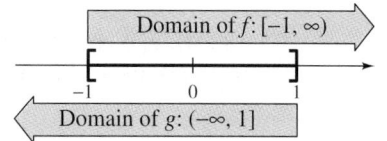

FIGURE 8.1

EXAMPLE 3   Finding the Product of Two Functions

Given the functions $f(x) = x^2$ and $g(x) = x - 3$, find the product of f and g. Then evaluate the product when $x = 4$.

Solution

The product of the functions f and g is given by

$$(fg)(x) = f(x)g(x)$$
$$= (x^2)(x - 3)$$
$$= x^3 - 3x^2.$$

When $x = 4$, the value of this product is

$$(fg)(4) = 4^3 - 3(4)^2$$
$$= 16.$$

Using the functions f and g from Example 4, have students compare the domains of these functions to the domain of $(f/g)(x)$. Ask students if they can think of functions $f(x)$ and $g(x)$ in which the domain of $(f/g)(x)$ doesn't have to be restricted.

EXAMPLE 4 *Finding the Quotient of Two Functions*

Given the functions $f(x) = x^2$ and $g(x) = x - 3$, find the quotient of f and g. Then evaluate the quotient when $x = 5$.

Solution

The quotient of the functions f and g is given by

$$\left(\frac{f}{g}\right)(x) = \frac{f(x)}{g(x)} = \frac{x^2}{x - 3}, \quad x \neq 3.$$

(Note that the domain of the quotient has to be restricted to exclude the number 3.) When $x = 5$, the value of this quotient is

$$\left(\frac{f}{g}\right)(5) = \frac{5^2}{5 - 3} = \frac{25}{2}.$$

Composition of Functions

Another way of combining two functions is to form the **composition** of one with the other. For instance, if $f(x) = 2x^2$ and $g(x) = x - 1$, the composition of f with g is given by

$$f(g(x)) = f(x - 1) = 2(x - 1)^2.$$

This composition is denoted as $f \circ g$.

Definition of Composition of Two Functions

The **composition** of the functions f and g is given by

$$(f \circ g)(x) = f(g(x)).$$

The domain of $(f \circ g)$ is the set of all x in the domain of g such that $g(x)$ is in the domain of f (see Figure 8.2).

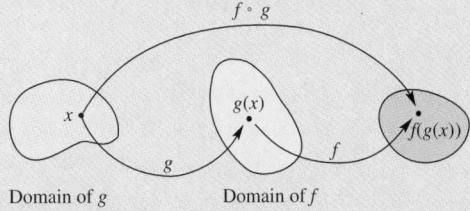

FIGURE 8.2

EXAMPLE 5   *Forming the Composition of Two Functions*

Given the functions $f(x) = 2x + 4$ and $g(x) = 3x - 1$, find the composition of f with g. Then find the value of the composition when $x = 1$ in two ways.

Solution

$$
\begin{aligned}
(f \circ g)(x) &= f(g(x)) \\
&= f(3x - 1) \\
&= 2(3x - 1) + 4 \\
&= 6x - 2 + 4 \\
&= 6x + 2
\end{aligned}
$$

Remind students that composition can be described as the substitution of one function for the variable position of another function. Have students complete a pattern such as the following. Add additional substitutions if necessary.

$$
\begin{aligned}
f(x) &= 4x - 1 \\
f(\ ) &= 4(\ ) - 1 \\
f(2) &= 4(2) - 1 = 7 \\
f(a) &= 4(a) - 1 = 4a - 1 \\
f(x + 2) &= 4(x + 2) - 1 = 4x + 7
\end{aligned}
$$

Note that $f(x + 2)$ is the same as $(f \circ g)(x)$ where $g(x) = x + 2$.

To evaluate this composition when $x = 1$, you can use either of the following two methods. Note that you obtain the same value with either method.

(a) You can use the above result $(f \circ g)(x) = 6x + 2$, and write
$$(f \circ g)(1) = 6(1) + 2 = 8.$$

(b) You can use the fact that $(f \circ g)(x) = f(g(x))$, and write
$$(f \circ g)(1) = f(3(1) - 1) = f(2) = 2(2) + 4 = 8.$$

EXAMPLE 6   *Comparing the Compositions of Functions*

Given the functions $f(x) = 2x - 3$ and $g(x) = x^2 + 1$, find each of the following.

(a) $(f \circ g)(x)$ (b) $(g \circ f)(x)$

NOTE   The composition of f with g is generally *not* the same as the composition of g with f. This is illustrated in Example 6.

Solution

(a)
$$
\begin{aligned}
(f \circ g)(x) &= f(g(x)) && \text{Definition of } f \circ g \\
&= f(x^2 + 1) && \text{Definition of } g(x) \\
&= 2(x^2 + 1) - 3 && \text{Definition of } f(x) \\
&= 2x^2 + 2 - 3 \\
&= 2x^2 - 1
\end{aligned}
$$

(b)
$$
\begin{aligned}
(g \circ f)(x) &= g(f(x)) && \text{Definition of } g \circ f \\
&= g(2x - 3) && \text{Definition of } f(x) \\
&= (2x - 3)^2 + 1 && \text{Definition of } g(x) \\
&= 4x^2 - 12x + 9 + 1 \\
&= 4x^2 - 12x + 10
\end{aligned}
$$

EXAMPLE 7 The Domain of the Composition of Two Functions

Given the functions

$$f(x) = \frac{1}{x - 2} \quad \text{and} \quad g(x) = \sqrt{x},$$

find the composition of f with g. Then find the domain of the composition.

Solution

The composition of f with g is given by

$$(f \circ g)(x) = f(g(x))$$
$$= f\left(\sqrt{x}\right)$$
$$= \frac{1}{\sqrt{x} - 2}.$$

The domain of $(f \circ g)$ is the set of all nonnegative real numbers, except $x = 4$. In interval notation, you can write the domain as

$$(\text{Domain of } f \circ g) = [0, 4) \cup (4, \infty).$$

Application

In real-life applications, the order in which two procedures are performed will often affect the outcome. Examples 8 and 9 illustrate this possibility.

EXAMPLE 8 Forming Composite Functions in Different Orders

In a chemistry lab, you are asked to mix a 2-liter solution that is half water and half sulfuric acid. Describe two ways of doing this. Do they produce the same results?

Solution

(a) One way to obtain the solution is to pour 1 liter of sulfuric acid into a container and then slowly pour 1 liter of water into the acid. *Don't try this!* It produces a violent explosion—much the same way that pouring water into boiling oil produces an explosion.

(b) Another way to obtain the solution is to pour 1 liter of water into a container and then slowly pour 1 liter of sulfuric acid into the water. As long as the solution is stirred constantly, this order of operations will produce the desired 50-50 mixture.

EXAMPLE 9 *Forming Composite Functions in Different Orders*

The regular price of a certain new car is $15,800. The dealership advertises a factory rebate of $1500 *and* a 12% discount. Compare the sale price obtained by subtracting the rebate first and then taking the discount with the sale price obtained by taking the discount first and then subtracting the rebate.

Solution

Using function notation, the price after rebate, $f(x)$, and the price after discount, $g(x)$, can be represented by

$$f(x) = x - 1500 \qquad \text{Rebate of \$1500}$$
$$g(x) = 0.88x \qquad \text{Discount of 12\%}$$

where x is the price of the car.

Rebate First: If you subtract the rebate first, the sale price is given by the composition of g with f.

$$
\begin{aligned}
g(f(15{,}800)) &= g(15{,}800 - 1500) && \text{Subtract rebate first.}\\
&= g(14{,}300) && \text{Simplify.}\\
&= 0.88 \cdot 14{,}300 && \text{Take discount.}\\
&= \$12{,}584 && \text{Simplify.}
\end{aligned}
$$

Discount First: If you take the discount first, the sale price is given by the composition of f with g.

$$
\begin{aligned}
f(g(15{,}800)) &= f(0.88 \cdot 15{,}800) && \text{Take discount first.}\\
&= f(13{,}904) && \text{Simplify.}\\
&= 13{,}904 - 1500 && \text{Subtract rebate.}\\
&= \$12{,}404 && \text{Simplify.}
\end{aligned}
$$

Thus, given an option, you should take the discount first.

Group Activities **Comparing the Compositions of Functions**

You have seen that the composition of f with g is generally not the same as the composition of g with f. There are, however, special cases in which these two compositions are the same. For instance, let $f(x) = 2x$ and $g(x) = \frac{1}{2}x$. Show that for these two functions, the compositions $(f \circ g)$ and $(g \circ f)$ are the same. Can you find other examples of functions f and g such that the compositions $(f \circ g)$ and $(g \circ f)$ are the same?

8.1   Exercises

See Warm-Up
Exercises, p. A47

In Exercises 1–8, use the given functions f and g to find the following combinations.

(a) $(f + g)(x)$ (b) $(f - g)(x)$

(c) $(fg)(x)$ (d) $(f/g)(x)$

Find the domain of each combination.

1. $f(x) = 2x$

$g(x) = x^2$

2. $f(x) = x + 1$

$g(x) = \sqrt{x}$

3. $f(x) = 4x - 3$

$g(x) = x^2 - 9$

4. $f(x) = x^3$

$g(x) = x^2 + 1$

5. $f(x) = \dfrac{1}{x}$

$g(x) = 5x$

6. $f(x) = \dfrac{1}{x + 2}$

$g(x) = \dfrac{1}{x - 2}$

7. $f(x) = \sqrt{x + 4}$

$g(x) = \sqrt{4 - x}$

8. $f(x) = \sqrt{x}$

$g(x) = \sqrt{x^2 - 16}$

In Exercises 9–16, evaluate the combination of the functions f and g at the given x-values. (If not possible, state the reason.)

9. $f(x) = x^2$, $g(x) = 2x + 3$

(a) $(f + g)(2)$ (b) $(f - g)(3)$

10. $f(x) = x^3$, $g(x) = x - 5$

(a) $(f + g)(-2)$ (b) $(f - g)(5)$

11. $f(x) = \sqrt{x}$, $g(x) = x^2 + 1$

(a) $(f + g)(4)$ (b) $(f - g)(9)$

12. $f(x) = \sqrt{x - 3}$, $g(x) = x^2 - 4$

(a) $(f + g)(7)$ (b) $(f - g)(1)$

13. $f(x) = |x|$, $g(x) = 5$

(a) $(fg)(-2)$ (b) $\left(\dfrac{f}{g}\right)(-2)$

14. $f(x) = |x - 3|$, $g(x) = (x - 2)^3$

(a) $(fg)(3)$ (b) $\left(\dfrac{f}{g}\right)\left(\dfrac{5}{2}\right)$

15. $f(x) = \dfrac{1}{x - 4}$, $g(x) = x$

(a) $(fg)(-2)$ (b) $\left(\dfrac{f}{g}\right)(4)$

16. $f(x) = \dfrac{1}{x^2}$, $g(x) = \dfrac{1}{x + 1}$

(a) $(fg)\left(\dfrac{1}{2}\right)$ (b) $\left(\dfrac{f}{g}\right)(0)$

In Exercises 17–24, find the compositions.

17. $f(x) = x - 3$, $g(x) = x^2$

(a) $(f \circ g)(x)$ (b) $(g \circ f)(x)$

(c) $(f \circ g)(4)$ (d) $(g \circ f)(7)$

18. $f(x) = x + 5$, $g(x) = x^3$

(a) $(f \circ g)(x)$ (b) $(g \circ f)(x)$

(c) $(f \circ g)(2)$ (d) $(g \circ f)(-3)$

19. $f(x) = |x - 3|$, $g(x) = 3x$

(a) $(f \circ g)(x)$ (b) $(g \circ f)(x)$

(c) $(f \circ g)(1)$ (d) $(g \circ f)(2)$

20. $f(x) = |x|$, $g(x) = 2x + 5$

(a) $(f \circ g)(x)$ (b) $(g \circ f)(x)$

(c) $(f \circ g)(-2)$ (d) $(g \circ f)(-4)$

21. $f(x) = \sqrt{x}$, $g(x) = x + 5$

(a) $(f \circ g)(x)$ (b) $(g \circ f)(x)$

(c) $(f \circ g)(4)$ (d) $(g \circ f)(9)$

22. $f(x) = \sqrt{x + 6}$, $g(x) = 2x - 3$

(a) $(f \circ g)(x)$ (b) $(g \circ f)(x)$

(c) $(f \circ g)(3)$ (d) $(g \circ f)(-2)$

23. $f(x) = 1/(x - 3)$, $g(x) = \sqrt{x}$

(a) $(f \circ g)(x)$ (b) $(g \circ f)(x)$

(c) $(f \circ g)(49)$ (d) $(g \circ f)(12)$

24. $f(x) = 4/(x^2 - 4)$, $g(x) = 1/x$

(a) $(f \circ g)(x)$ (b) $(g \circ f)(x)$

(c) $(f \circ g)(-2)$ (d) $(g \circ f)(1)$

In Exercises 25–30, find the compositions (a) $f \circ g$ and (b) $g \circ f$. Then find the domain of each composition.

25. $f(x) = x^2 + 1$
$g(x) = 2x$

26. $f(x) = 2 - 3x$
$g(x) = 5x + 3$

27. $f(x) = \sqrt{x}$
$g(x) = x - 2$

28. $f(x) = \sqrt{x - 5}$
$g(x) = x^2$

29. $f(x) = \dfrac{9}{x + 9}$
$g(x) = x^2$

30. $f(x) = \dfrac{x}{x - 4}$
$g(x) = \sqrt{x}$

In Exercises 31–40, find the indicated combination of the functions $f(x) = x^2 - 3x$ and $g(x) = 5x + 3$.

31. $(f - g)(t)$

32. $(f/g)(2)$

33. $(f + g)(t - 2)$

34. $(fg)(z)$

35. $\dfrac{g(x + h) - g(x)}{h}$

36. $\dfrac{f(x + h) - f(x)}{h}$

37. $(f \circ g)(-1)$

38. $(g \circ f)(3)$

39. $(g \circ f)(y)$

40. $(f \circ g)(z)$

In Exercises 41–44, use the graphs to sketch the graph of $h(x) = (f + g)(x)$.

41.

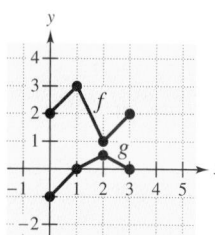

42.

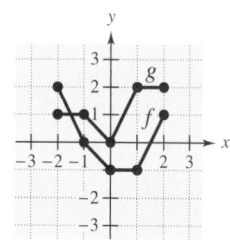

43.

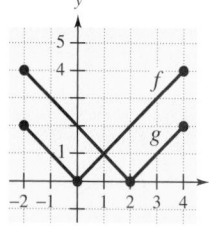

44.

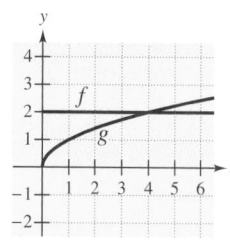

In Exercises 45–48, use a graphing utility to graph the functions f, g, and $f + g$ in the same viewing rectangle.

45. $f(x) = \frac{1}{2}x$, $g(x) = x - 1$

46. $f(x) = \frac{1}{3}x$, $g(x) = -x + 4$

47. $f(x) = x^2$, $g(x) = -2x$

48. $f(x) = 4 - x^2$, $g(x) = x$

In Exercises 49 and 50, use a graphing utility to graph f, g, and $f + g$ in the same viewing rectangle. Observe from the graphs which of the two functions contributes more to the magnitude of the sum for different values of x.

49. $f(x) = x^3$
$g(x) = 3(1 - x^2)$

50. $f(x) = x^3$
$g(x) = -3x$

In Exercises 51–54, use the functions f and g to find the indicated values.

$$f = \{(-2, 3), (-1, 1), (0, 0), (1, -1), (2, -3)\}$$
$$g = \{(-3, 1), (-1, -2), (0, 2), (2, 2), (3, 1)\}$$

51. (a) $f(1)$ (b) $g(-1)$ (c) $(g \circ f)(1)$

52. (a) $g(0)$ (b) $f(2)$ (c) $(f \circ g)(0)$

53. (a) $(f \circ g)(-3)$ (b) $(g \circ f)(-2)$

54. (a) $(f \circ g)(2)$ (b) $(g \circ f)(2)$

In Exercises 55–58, use the functions f and g to find the indicated values.

$$f = \{(0, 1), (1, 2), (2, 5), (3, 10), (4, 17)\}$$
$$g = \{(5, 4), (10, 1), (2, 3), (17, 0), (1, 2)\}$$

55. (a) $f(3)$
(b) $g(10)$
(c) $(g \circ f)(3)$

56. (a) $g(2)$
(b) $f(0)$
(c) $(f \circ g)(10)$

57. (a) $(g \circ f)(4)$
(b) $(f \circ g)(2)$

58. (a) $(f \circ g)(1)$
(b) $(g \circ f)(0)$

The symbol ▦ indicates an exercise in which you are instructed to use a calculator or graphing utility. The solutions of other exercises may also be facilitated by the use of appropriate technology.

59. *Stopping Distance*   A car traveling at x miles per hour stops quickly. The distance the car travels during the driver's reaction time is $R(x) = \frac{3}{4}x$. The distance the car travels while braking is $B(x) = \frac{1}{15}x^2$.

(a) Find a model for the total stopping distance T.

(b) Use a graphing utility to graph the functions R, B, and T in the interval $0 \leq x \leq 60$.

(c) Which function contributes more to the magnitude of the sum at higher speeds?

60. *Data Analysis*   The table gives the variable costs for operating an automobile in the United States in the years 1985 through 1992. The functions y_1, y_2, and y_3 represent the costs in cents per mile for gas and oil, maintenance, and tires, and t represents the year, with $t = 0$ corresponding to 1990. *(Source: American Automobile Manufacturers Association)*

t	-5	-4	-3	-2	-1	0	1	2
y_1	6.16	4.48	4.80	5.20	5.20	5.40	6.70	6.00
y_2	1.23	1.37	1.60	1.60	1.90	2.10	2.20	2.20
y_3	0.65	0.67	0.80	0.80	0.80	0.90	0.90	0.90

Mathematical models for the data are given by

$$y_1 = 0.08t^2 + 0.38t + 5.46$$
$$y_2 = 0.15t + 2.00$$
$$y_3 = 0.04t + 0.86.$$

Use a graphing utility to graph y_1, y_2, y_3, and $y_1 + y_2 + y_3$ in the same viewing rectangle. Use the model to estimate the total variable cost per mile in 1995.

61. *Sales Bonus*   You are a sales representative for a clothing manufacturer. You are paid an annual salary plus a bonus of 2% of your sales over $200,000. Consider the two functions

$$f(x) = x - 200,000 \quad \text{and} \quad g(x) = 0.02x.$$

If x is greater than $200,000, which of the following represents your bonus? Explain.

(a) $f(g(x))$ (b) $g(f(x))$

62. *Daily Production Cost*   The daily cost of producing x units in a manufacturing process is $C(x) = 8.5x + 300$. The number of units produced in t hours during a day is given by $x(t) = 12t$, $0 \leq t \leq 8$. Find, simplify, and interpret $(C \circ x)(t)$.

63. *A Bridge over Troubled Waters*   You are standing on a bridge over a calm pond and drop a pebble, causing ripples of concentric circles in the water (see figure). The radius (in feet) of the outer ripple is given by $r(t) = 0.6t$, where t is the time in seconds after the pebble hits the water. The area of the circle is given by $A(r) = \pi r^2$. Find an equation for the composition $A(r(t))$.

Section Project

Rebate and Discount   The suggested retail price of a new car is p dollars. The dealership advertises a factory rebate of $2000 and a 5% discount.

(a) Write a function R in terms of p, giving the cost of the car after receiving the factory rebate.

(b) Write a function S in terms of p, giving the cost of the car after receiving the dealership discount.

(c) Form the compositions $(R \circ S)(p)$ and $(S \circ R)(p)$ and interpret each.

(d) Find $(R \circ S)(26,000)$ and $(S \circ R)(26,000)$. Which yields the smaller cost for the car? Explain.

8.2 | **Inverse Functions**

The Inverse of a Function ▪ Finding the Inverse of a Function ▪ The Graph of the Inverse of a Function

The Inverse of a Function

When functions were introduced in Section 2.4, you saw that one way to represent a function is by a set of ordered pairs. For instance, the function $f(x) = x + 2$ from the set $A = \{1, 2, 3, 4\}$ to the set $B = \{3, 4, 5, 6\}$ can be written as follows.

$$f(x) = x + 2: \quad \{(1, 3), (2, 4), (3, 5), (4, 6)\}$$

By interchanging the first and second coordinates in each of these ordered pairs, you can form another function that is called the **inverse function** of f. This function is denoted by f^{-1}. It is a function from the set B to the set A, and can be written as follows.

$$f^{-1}(x) = x - 2: \quad \{(3, 1), (4, 2), (5, 3), (6, 4)\}$$

Note that the domain of f is equal to the range of f^{-1}, and vice versa, as shown in Figure 8.3. Also note that the functions f and f^{-1} have the effect of "undoing" each other. In other words, when you form the composition of f with f^{-1}, or the composition of f^{-1} with f, you obtain the identity function, as follows.

$$f(f^{-1}(x)) = f(x - 2) = (x - 2) + 2 = x$$
$$f^{-1}(f(x)) = f^{-1}(x + 2) = (x + 2) - 2 = x$$

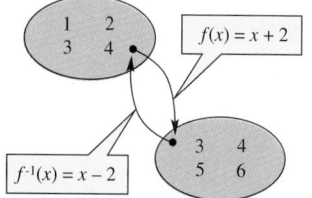

FIGURE 8.3

EXAMPLE 1 *Determining Inverse Functions Informally*

Find the inverse of the function $f(x) = 3x$ and verify that both $f(f^{-1}(x))$ and $f^{-1}(f(x))$ are equal to the identity function.

Solution

The given function is a function that *multiplies* each input by 3. To "undo" this function, you need to *divide* each input by 3. Thus, the inverse of $f(x) = 3x$ is

$$f^{-1}(x) = \frac{x}{3}.$$

You can verify that both $f(f^{-1}(x))$ and $f^{-1}(f(x))$ are equal to the identity function, as follows.

$$f(f^{-1}(x)) = f\left(\frac{x}{3}\right) = 3\left(\frac{x}{3}\right) = x$$

$$f^{-1}(f(x)) = f^{-1}(3x) = \frac{3x}{3} = x$$

Exploration

Consider the functions $f(x) = x + 2$ and $f^{-1}(x) = x - 2$ shown in Figure 8.3. Evaluate $f(f^{-1}(x))$ and $f^{-1}(f(x))$ for the indicated values of x. What can you conclude about the functions?

x	-10	0	7	45
$f(f^{-1}(x))$				
$f^{-1}(f(x))$				

EXAMPLE 2   *Determining Inverse Functions Informally*

Find the inverse of the function $f(x) = x - 4$ and verify that both $f(f^{-1}(x))$ and $f^{-1}(f(x))$ are equal to the identity function.

Solution

The given function is a function that *subtracts* 4 from each input. To "undo" this function, you need to *add* 4 to each input. Thus, the inverse of $f(x) = x - 4$ is

$$f^{-1}(x) = x + 4.$$

Point out that students are already seeing an application of composition. It can be used to verify that two functions are inverses of each other.

You can verify that both $f(f^{-1}(x))$ and $f^{-1}(f(x))$ are equal to the identity function, as follows.

$$f(f^{-1}(x)) = f(x + 4) = (x + 4) - 4 = x$$
$$f^{-1}(f(x)) = f^{-1}(x - 4) = (x - 4) + 4 = x$$

Don't be confused by the use of -1 to denote the inverse function f^{-1}. Whenever we write f^{-1}, we will *always* be referring to the inverse of the function f and *not* to the reciprocal of $f(x)$.

The formal definition of the inverse of a function is given as follows.

Definition of the Inverse of a Function

Let f and g be two functions such that

$$f(g(x)) = x \quad \text{for every } x \text{ in the domain of } g$$

and

$$g(f(x)) = x \quad \text{for every } x \text{ in the domain of } f.$$

Then the function g is called the **inverse** of the function f, and is denoted by f^{-1} (read "f-inverse"). Thus, $f(f^{-1}(x)) = x$ and $f^{-1}(f(x)) = x$. The domain of f must be equal to the range of f^{-1}, and vice versa.

Note from this definition that if the function g is the inverse of the function f, it must also be true that the function f is the inverse of the function g. For this reason, the functions f and g are sometimes called *inverses of each other*.

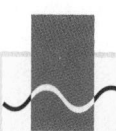

Technology

You can use a graphing utility to get a visual image of two functions being inverses of each other. For example, enter the functions in Example 3 as follows.

Y1=x³+1

Y2=(x−1)$^{1/3}$

Y3=Y1(Y2)

Y4=Y2(Y1)

Y3 and Y4 both yield the line $y = x$, which means that Y1 and Y2 are inverses of each other.

EXAMPLE 3   Verifying Inverse Functions

Show that $f(x) = x^3 + 1$ and $g(x) = \sqrt[3]{x - 1}$ are inverses of each other.

Solution

Begin by noting that the domain and range of both functions are the entire set of real numbers. To show that f and g are inverses of each other, you need to show that $f(g(x)) = x$ and $g(f(x)) = x$, as follows.

$$f(g(x)) = f\left(\sqrt[3]{x - 1}\right)$$
$$= \left(\sqrt[3]{x - 1}\right)^3 + 1$$
$$= (x - 1) + 1$$
$$= x$$

$$g(f(x)) = g(x^3 + 1)$$
$$= \sqrt[3]{(x^3 + 1) - 1}$$
$$= \sqrt[3]{x^3}$$
$$= x$$

Emphasize that *both* conditions, $f(g(x)) = x$ and $g(f(x)) = x$, must be satisfied.

Note that the two functions f and g "undo" each other in the following verbal sense. The function f first cubes the input x and then adds 1, whereas the function g first subtracts 1, and then takes the cube root of the result.

EXAMPLE 4   Verifying Inverse Functions

Which of the following functions is the inverse of the function $f(x) = 1/(5x)$?

$$g(x) = \frac{5}{x} \quad \text{or} \quad h(x) = \frac{1}{5x}$$

Solution

By forming the composition of f with g, you find that

$$f(g(x)) = f\left(\frac{5}{x}\right) = \frac{1}{5\left(\frac{5}{x}\right)} = \frac{1}{\left(\frac{25}{x}\right)} = \frac{x}{25}.$$

NOTE   Some functions are their own inverse. For instance, the function $f(x) = 1/(5x)$ in Example 4 is its own inverse.

Because this composition does not yield x, you can conclude that g *is not* the inverse of f. By forming the composition of f with h, you find that

$$f(h(x)) = f\left(\frac{1}{5x}\right) = \frac{1}{5\left(\frac{1}{5x}\right)} = \frac{1}{\left(\frac{1}{x}\right)} = x.$$

Thus, it appears that h *is* the inverse of f. You can confirm this by showing that the composition of h with f is also equal to the identity function. (Try doing this.)

Finding the Inverse of a Function

For simple functions (such as the ones in Examples 1 and 2), you can find inverse functions by inspection. For instance, the inverse of $f(x) = 10x$ is $f^{-1}(x) = x/10$. For more complicated functions, however, it is best to use the following steps for finding the inverses. The key step in these guidelines is switching the roles of x and y. This step corresponds to the fact that inverse functions have ordered pairs with the coordinates reversed.

Finding the Inverse of a Function

To find the inverse of a function f, use the following steps.

1. In the equation for $f(x)$, replace $f(x)$ by y.

2. Interchange the roles of x and y.

3. If the new equation does not represent y as a function of x, the function f does not have an inverse function. If the new equation does represent y as a function of x, solve the new equation for y.

4. Replace y by $f^{-1}(x)$.

5. Verify that f and f^{-1} are inverses of each other by showing that $f(f^{-1}(x)) = x = f^{-1}(f(x))$.

NOTE Note in Step 3 of the guidelines for finding the inverse of a function that it is possible that a function has no inverse. This possibility is illustrated in Example 6.

EXAMPLE 5 *Finding the Inverse of a Function*

Determine whether $f(x) = 2x + 3$ has an inverse. If it does, find its inverse.

Solution

$$f(x) = 2x + 3 \qquad \text{Original function}$$
$$y = 2x + 3 \qquad \text{Replace } f(x) \text{ by } y.$$
$$x = 2y + 3 \qquad \text{Interchange } x \text{ and } y.$$
$$y = \frac{x - 3}{2} \qquad \text{Solve for } y.$$
$$f^{-1}(x) = \frac{x - 3}{2} \qquad \text{Replace } y \text{ by } f^{-1}(x).$$

Thus, the inverse of $f(x) = 2x + 3$ is

$$f^{-1}(x) = \frac{x - 3}{2}.$$

Verify that $f(f^{-1}(x)) = x = f^{-1}(f(x))$.

EXAMPLE 6   A Function that Has No Inverse

Determine whether $f(x) = x^2$ has an inverse. If it does, find its inverse.

Solution

$$f(x) = x^2 \qquad \text{Original function}$$
$$y = x^2 \qquad \text{Replace } f(x) \text{ by } y.$$
$$x = y^2 \qquad \text{Interchange } x \text{ and } y.$$

Because the equation $x = y^2$ does not represent y as a function of x, you can conclude that the original function f does not have an inverse.

NOTE   The equation $x = y^2$ does not represent y as a function of x because you can find two different y-values that correspond to the same x-value. For example, when $x = 9$, y can be 3 or -3.

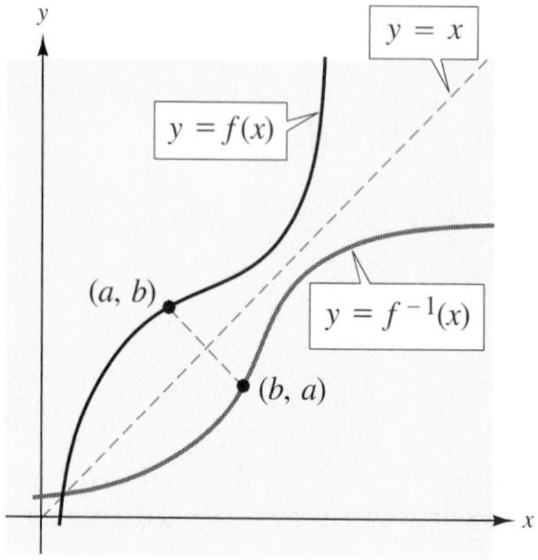

The Graph of the Inverse of a Function

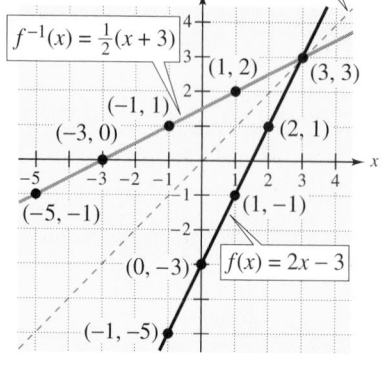

FIGURE 8.4

The graph of f^{-1} is a reflection of the graph of f in the line $y = x$.

The graphs of f and f^{-1} are related to each other in the following way. If the point (a, b) lies on the graph of f, then the point (b, a) must lie on the graph of f^{-1}, and vice versa. This means that the graph of f^{-1} is a reflection of the graph of f in the line $y = x$, as shown in Figure 8.4. This "reflective property" of the graphs of f and f^{-1} is illustrated in Examples 7 and 8.

EXAMPLE 7   The Graphs of f and f^{-1}

Sketch the graphs of the inverse functions $f(x) = 2x - 3$ and $f^{-1}(x) = \frac{1}{2}(x + 3)$ on the same rectangular coordinate system, and show that the graphs are reflections of each other in the line $y = x$.

Solution

The graphs of f and f^{-1} are shown in Figure 8.5. Visually, it appears that the graphs are reflections of each other in the line $y = x$. You can further verify this reflective property by testing a few points on each graph. Note in the following list that if the point (a, b) is on the graph of f, then the point (b, a) is on the graph of f^{-1}.

$f(x) = 2x - 3$	$f^{-1}(x) = \frac{1}{2}(x + 3)$
$(-1, -5)$	$(-5, -1)$
$(0, -3)$	$(-3, 0)$
$(1, -1)$	$(-1, 1)$
$(2, 1)$	$(1, 2)$
$(3, 3)$	$(3, 3)$

FIGURE 8.5

In Example 6, you saw that the function $f(x) = x^2$ has no inverse. A more complete way of saying this is "*assuming that the domain of f is the entire real line*, the function $f(x) = x^2$ has no inverse." If, however, you restrict the domain of f to the nonnegative real numbers, then f does have an inverse, as demonstrated in Example 8.

EXAMPLE 8 The Graphs of f and f⁻¹

Sketch the graphs of the inverse functions

$$f(x) = x^2, \quad x \ge 0 \quad \text{and} \quad f^{-1}(x) = \sqrt{x}$$

on the same rectangular coordinate system, and show that the graphs are reflections of each other in the line $y = x$.

Solution

The graphs of f and f^{-1} are shown in Figure 8.6. Visually, it appears that the graphs are reflections of each other in the line $y = x$. You can further verify this reflective property by testing a few points on each graph. Note in the following list that if the point (a, b) is on the graph of f, then the point (b, a) is on the graph of f^{-1}.

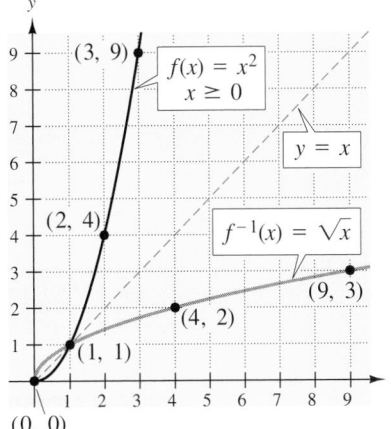

FIGURE 8.6

$f(x) = x^2, \quad x \ge 0$	$f^{-1}(x) = \sqrt{x}$
$(0, 0)$	$(0, 0)$
$(1, 1)$	$(1, 1)$
$(2, 4)$	$(4, 2)$
$(3, 9)$	$(9, 3)$

In the guidelines for finding the inverse of a function, we included an *algebraic* test for determining whether a function has an inverse. The reflective property of the graphs of inverse functions gives you a nice *geometric* test for determining whether a function has an inverse. This is called the **Horizontal Line Test** for inverse functions.

Horizontal Line Test for Inverse Functions

A function f has an inverse function if and only if no *horizontal* line intersects the graph of f at more than one point. Such a function is called **one-to-one**.

The Horizontal Line Test for inverse functions is demonstrated in Example 9.

EXAMPLE 9   *Applying the Horizontal Line Test*

a. The graph of the function $f(x) = 2x - 1$ is shown in Figure 8.7. Because no horizontal line intersects the graph of f at more than one point, you can conclude that f *is* a one-to-one function, and it *does* possess an inverse function.

b. The graph of the function $f(x) = x^2 - 1$ is shown in Figure 8.8. Because it is possible to find a horizontal line that intersects the graph of f at more than one point, you can conclude that f *is not* a one-to-one function, and it *does not* possess an inverse function.

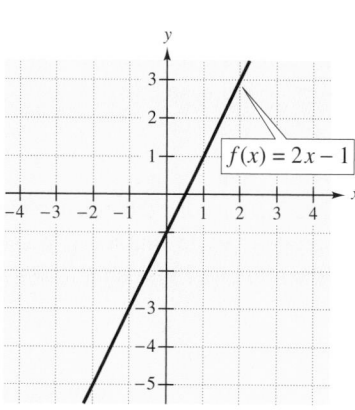

FIGURE 8.7

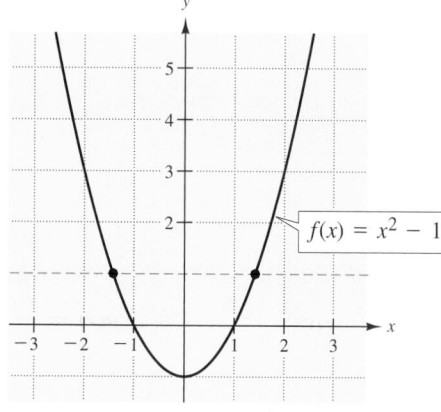

FIGURE 8.8

Group Activities   **The Existence of an Inverse Function**

In your group, describe why the following functions do or do not possess inverse functions.

a. Let x represent the retail price of an item (in dollars), and let $f(x)$ represent the sales tax on the item. Assume that the sales tax is 6% of the retail price *and* that the sales tax is rounded to the nearest cent. Does this function possess an inverse? (*Hint:* Can you undo this function? For instance, if you know that the sales tax is $0.12, can you determine *exactly* what the retail price is?)

b. Let x represent the temperature in degrees Celsius, and let $f(x)$ represent the temperature in degrees Fahrenheit. Does this function possess an inverse? (*Hint:* The formula for converting from degrees Celsius to degrees Fahrenheit is $F = \frac{9}{5}C + 32$.)

8.2 Exercises

See Warm-Up
Exercises, p. A47

In Exercises 1–10, find the inverse of the function f informally. Verify that $f(f^{-1}(x))$ and $f^{-1}(f(x))$ are equal to the identity function.

1. $f(x) = 5x$

2. $f(x) = \frac{1}{3}x$

3. $f(x) = x + 10$

4. $f(x) = x - 5$

5. $f(x) = \frac{1}{2}x$

6. $f(x) = 2x$

7. $f(x) = x^7$

8. $f(x) = x^5$

9. $f(x) = \sqrt[3]{x}$

10. $f(x) = x^{1/5}$

In Exercises 11–22, verify algebraically that the functions f and g are inverses of each other.

11. $f(x) = 10x$

$g(x) = \frac{1}{10}x$

12. $f(x) = \frac{2x}{3}$

$g(x) = \frac{3x}{2}$

13. $f(x) = x + 15$

$g(x) = x - 15$

14. $f(x) = 3 - x$

$g(x) = 3 - x$

15. $f(x) = 1 - 2x$

$g(x) = \frac{1 - x}{2}$

16. $f(x) = 2x - 1$

$g(x) = \frac{1}{2}(x + 1)$

17. $f(x) = 2 - 3x$

$g(x) = \frac{1}{3}(2 - x)$

18. $f(x) = -\frac{1}{4}x + 3$

$g(x) = -4(x - 3)$

19. $f(x) = \sqrt[3]{x + 1}$

$g(x) = x^3 - 1$

20. $f(x) = x^5$

$g(x) = \sqrt[5]{x}$

21. $f(x) = \frac{1}{x}$

$g(x) = \frac{1}{x}$

22. $f(x) = \frac{1}{x - 3}$

$g(x) = 3 + \frac{1}{x}$

(a)
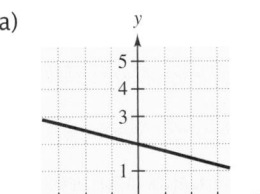

(b)
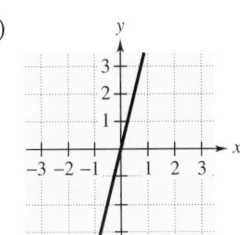

(c)
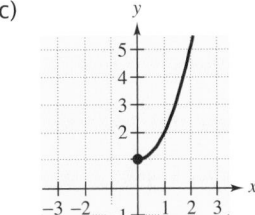

(d)
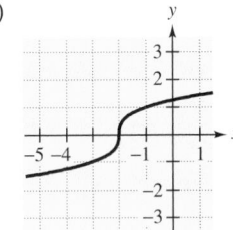

23.
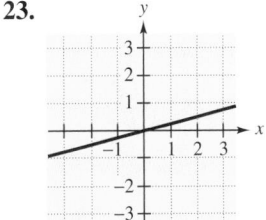

24.
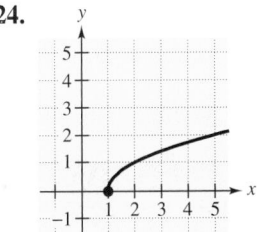

25.
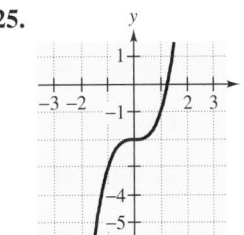

26.
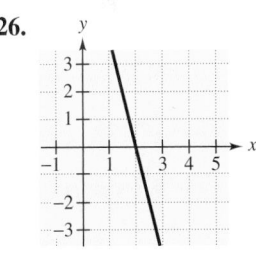

In Exercises 23–26, match the graph with the graph of its inverse. [The graphs of the inverse functions are labeled (a), (b), (c), and (d).]

In Exercises 27–34, use a graphing utility to verify that the functions are inverses of each other.

27. $f(x) = \frac{1}{3}x$

$g(x) = 3x$

28. $f(x) = \frac{1}{5}x - 1$

$g(x) = 5x + 5$

29. $f(x) = \sqrt{x + 1}$

$g(x) = x^2 - 1,\ x \geq 0$

30. $f(x) = \sqrt{4 - x}$

$g(x) = 4 - x^2,\ x \geq 0$

31. $f(x) = \frac{1}{8}x^3$

$g(x) = 2\sqrt[3]{x}$

32. $f(x) = \sqrt[3]{x + 2}$

$g(x) = x^3 - 2$

33. $f(x) = 3x + 4$

$g(x) = \frac{1}{3}(x - 4)$

34. $f(x) = |x - 2|,\ x \geq 2$

$g(x) = x + 2,\ x \geq 0$

53. $g(x) = \sqrt{25 - x^2}$

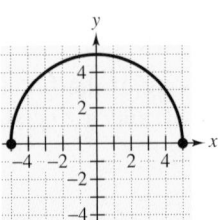

54. $g(x) = |x - 4|$

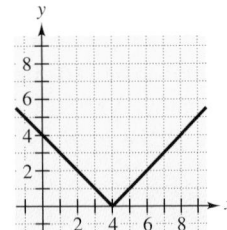

In Exercises 35–48, find the inverse of the function.

35. $f(x) = 8x$

36. $f(x) = \dfrac{x}{10}$

37. $g(x) = x + 25$

38. $f(x) = 7 - x$

39. $g(x) = 3 - 4x$

40. $g(t) = 6t + 1$

41. $g(t) = \frac{1}{4}t + 2$

42. $h(s) = 5 - \frac{3}{2}s$

43. $h(x) = \sqrt{x}$

44. $h(x) = \sqrt{x + 5}$

45. $f(t) = t^3 - 1$

46. $h(t) = t^5$

47. $g(s) = \dfrac{5}{s}$

48. $f(s) = \dfrac{2}{3 - s}$

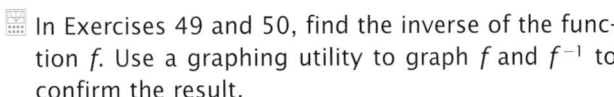 In Exercises 49 and 50, find the inverse of the function f. Use a graphing utility to graph f and f^{-1} to confirm the result.

49. $f(x) = x^3 + 1$

50. $f(x) = \sqrt{x^2 - 4},\ x \geq 2$

In Exercises 51–54, determine whether the function has an inverse.

51. $f(x) = x^2 - 2$

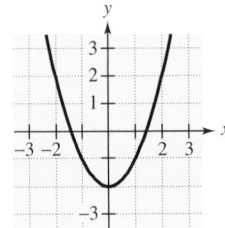

52. $f(x) = \frac{1}{5}x$

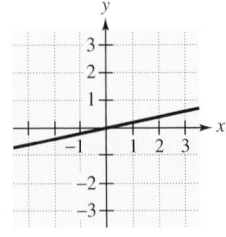

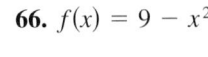 In Exercises 55–64, use a graphing utility to graph the function and determine whether the function is one-to-one.

55. $f(x) = \frac{1}{4}x^3$

56. $f(x) = 9 - x^2$

57. $f(t) = \sqrt[3]{5 - t}$

58. $h(t) = 4 - \sqrt[3]{t}$

59. $g(x) = x^4$

60. $f(x) = (x + 2)^5$

61. $h(t) = \dfrac{5}{t}$

62. $g(t) = \dfrac{5}{t^2}$

63. $f(s) = \dfrac{4}{s^2 + 1}$

64. $f(x) = \dfrac{1}{x - 2}$

In Exercises 65–68, delete part of the graph of the function so that the remaining part is one-to-one. Find the inverse of the remaining part and give the domain of the inverse. (*Note:* There is more than one correct answer.)

65. $f(x) = x^4$

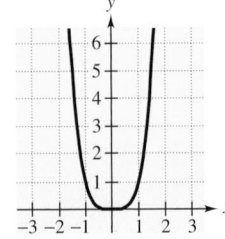

66. $f(x) = 9 - x^2$

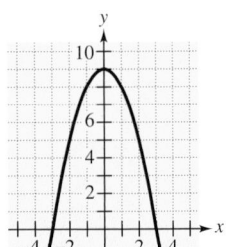

67. $f(x) = (x - 2)^2$ **68.** $f(x) = |x - 2|$

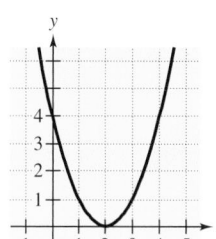   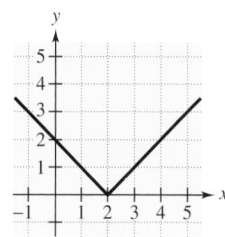

In Exercises 69–72, use the graph of f to sketch the graph of f^{-1}.

69.   **70.**

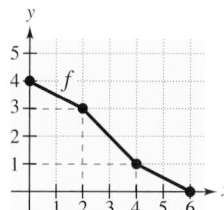   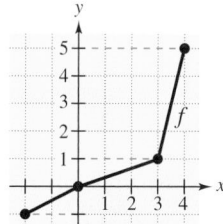

71.   **72.**

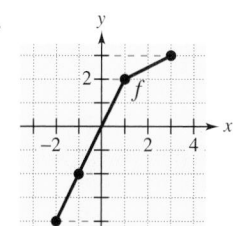   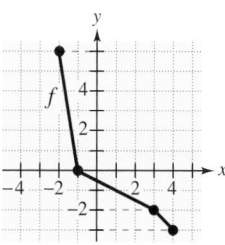

73. *Hourly Wage*   Your wage is $9.00 per hour plus $0.65 for each unit produced per hour. Thus, your hourly wage y in terms of the number of units produced is $y = 9 + 0.65x$.

(a) Find the inverse of the function.

(b) What does each variable represent in the inverse function?

(c) Determine the number of units produced when your hourly wage averages $14.20.

74. *Cost*   Suppose you need 100 pounds of two commodities that cost $0.50 and $0.75 per pound.

(a) Verify that your total cost is $y = 0.50x + 0.75(100 - x)$, where x is the number of pounds of the less expensive commodity.

(b) Find the inverse of the function. What does each variable represent in the inverse function?

(c) Use the context of the problem to determine the domain of the inverse function.

(d) Determine the number of pounds of the less expensive commodity purchased if the total cost is $60.

True or False?   In Exercises 75–78, decide whether the statement is true or false. If true, explain your reasoning. If false, give an example to show why.

75. If the inverse of f exists, the y-intercept of f is an x-intercept of f^{-1}.

76. There exists no function f such that $f = f^{-1}$.

77. If the inverse of f exists, the domains of f and f^{-1} are the same.

78. If the inverse of f exists and its graph passes through the point $(2, 2)$, the graph of f^{-1} also passes through the point $(2, 2)$.

79. Consider the function $f(x) = 3 - 2x$.

(a) Find $f^{-1}(x)$.   (b) Find $(f^{-1})^{-1}(x)$.

Section Project

Consider the functions $f(x) = 4x$ and $g(x) = x + 6$.

(a) Find $(f \circ g)(x)$.

(b) Find $(f \circ g)^{-1}(x)$.

(c) Find $f^{-1}(x)$ and $g^{-1}(x)$.

(d) Find $(g^{-1} \circ f^{-1})(x)$. Compare the result with part (b).

(e) Repeat parts (a) through (d) for $f(x) = x^3 + 1$ and $g(x) = 2x$.

(f) Write two one-to-one functions f and g, and repeat parts (a) through (d) for these functions.

(g) Make a conjecture about $(f \circ g)^{-1}(x)$ and $(g^{-1} \circ f^{-1})(x)$.

8.3	**Variation and Mathematical Models**

Direct Variation ▪ Inverse Variation ▪ Joint Variation

Direct Variation

In this section, you will continue your study of methods for creating mathematical models by looking at models that are related to the concept of **variation.**

Direct Variation

The following statements are equivalent.

1. y **varies directly** as x.

2. y is **directly proportional** to x.

3. $y = kx$ for some constant k.

The number k is called the **constant of proportionality.**

EXAMPLE 1   *Direct Variation*

The total revenue R (in dollars) obtained from selling x units of a given product is directly proportional to the number of units sold x. When 10,000 units are sold, the total revenue is $142,500. (a) Find a mathematical model that relates R to x. (b) Find the total revenue obtained from selling 12,000 units.

Solution

(a) Because the total revenue is directly proportional to the number of units sold, you have the model

$$R = kx.$$

To find the value of k, use the fact that $R = 142,500$ when $x = 10,000$. Substituting these values into the model produces $142,500 = k(10,000)$, which implies that $k = 142,500/10,000 = 14.25$. Thus, the equation relating the total revenue to the total number of units sold is

$$R = 14.25x.$$

(b) When $x = 12,000$, the total revenue is

$$R = 14.25(12,000) = \$171,000.$$

EXAMPLE 2   *Direct Variation*

Hooke's Law for springs states that the distance a spring is stretched (or compressed) is proportional to the force on the spring. A force of 20 pounds stretches a certain spring 5 inches.

(a) Find a mathematical model that relates the distance the spring is stretched to the force applied to the spring.

(b) How far will a force of 30 pounds stretch the spring?

Solution

(a) For this problem, let d represent the distance (in inches) that the spring is stretched and let F represent the force (in pounds) that is applied to the spring. Because the distance d is directly proportional to the force F, you have the model

$$d = kF.$$

To find the value of k, use the fact that $d = 5$ when $F = 20$. Substituting these values into the model produces

$$5 = k(20),$$

which implies that $k = \frac{5}{20} = \frac{1}{4}$. Thus, the equation relating distance and force is

$$d = \frac{1}{4}F.$$

(b) When $F = 30$, the distance is

$$d = \frac{1}{4}(30) = 7.5 \text{ inches.}$$

(See Figure 8.9.)

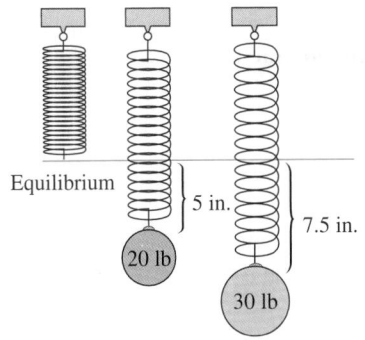

Equilibrium

5 in.

7.5 in.

20 lb

30 lb

FIGURE 8.9

In Examples 1 and 2, the direct variations were such that an *increase* in one variable corresponded to an *increase* in the other variable. For instance, in the model $d = \frac{1}{4}F$, if the force F increases, the distance d also increases. There are, however, other applications of direct variation in which an *increase* in one variable corresponds to a *decrease* in the other variable. For instance, in the model

$$y = -2x,$$

an increase in x will yield a decrease in y.

Another type of direct variation relates one variable to a *power* of another variable.

Direct Variation as *n*th Power

The following statements are equivalent.

1. *y* **varies directly as the *n*th power** of *x*.

2. *y* is **directly proportional to the *n*th power** of *x*.

3. $y = kx^n$ for some constant *k*.

EXAMPLE 3   *Direct Variation as a Power*

The distance a ball rolls down an inclined plane is directly proportional to the square of the time it rolls. During the first second, the ball rolls 6 feet.

(a) Find a mathematical model that relates the distance traveled to the time.

(b) How far will the ball roll during the first 2 seconds?

Solution

(a) Letting *d* be the distance (in feet) that the ball rolls and letting *t* be the time (in seconds), you obtain the model

$$d = kt^2.$$

Because $d = 6$ when $t = 1$, it follows that $6 = k(1)^2$, so $k = 6$. Therefore, the equation relating distance to time is

$$d = 6t^2.$$

(b) When $t = 2$, the distance traveled is

$$d = 6(2)^2 = 6(4) = 24 \text{ feet.}$$

(See Figure 8.10.)

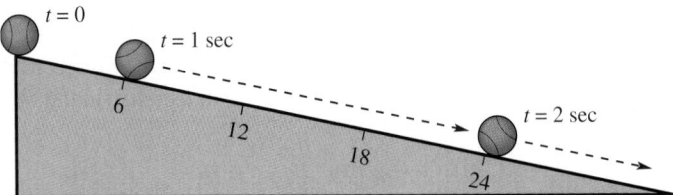

FIGURE 8.10

Inverse Variation

A second type of variation is called **inverse variation.** With this type of variation, one of the variables is said to be inversely proportional to the other variable.

Inverse Variation

The following statements are equivalent.

1. y **varies inversely** as x.

2. y is **inversely proportional** to x.

3. $y = \dfrac{k}{x}$ for some constant k.

EXAMPLE 4   Inverse Variation

The marketing department of a large company has found that the demand for one of its products varies inversely with the price of the product. (When the price is low, more people are willing to buy the product than when the price is high.) When the price of the product is $7.50, the monthly demand is 50,000 units. Approximate the monthly demand if the price is reduced to $6.00.

Solution

Let x represent the number of units sold each month (the demand), and let p represent the price per unit (in dollars). Because the demand is inversely proportional to the price, you obtain the model

$$x = \frac{k}{p}.$$

By substituting $x = 50{,}000$ when $p = 7.50$, you can determine that $k = (7.5)(50{,}000) = 375{,}000$. Thus, the model is

$$x = \frac{375{,}000}{p}.$$

To find the demand that corresponds to a price of $6.00, substitute 6 for p in the equation and obtain a demand of

$$x = \frac{375{,}000}{6} = 62{,}500 \text{ units.}$$

Thus, if the price is lowered from $7.50 per unit to $6.00 per unit, you can expect the monthly demand to increase from 50,000 units to 62,500 units.

Some applications of variation involve problems with *both* direct and inverse variation in the same model. These types of models are said to have **combined variation.** For instance, in the model

$$z = k\left(\frac{x}{y}\right),$$

z is directly proportional to x *and* inversely proportional to y.

EXAMPLE 5 *Direct and Inverse Variation*

A company determines that the demand for one of its products is directly proportional to the amount spent on advertising and inversely proportional to the price of the product. When $40,000 is spent on advertising and the price per unit is $20, the monthly demand is 10,000 units.

Additional problem: Have students write a mathematical model for the statement "*N* varies directly as the square of *r* and inversely as the cube of *s*."

Answer: $N = (kr^2)/(s^3)$

(a) If the amount spent on advertising were increased to $50,000, how much could the price be increased to maintain a monthly demand of 10,000 units?

(b) If you were in charge of the advertising department, would you recommend this increased expenditure for advertising?

Solution

(a) Let x represent the number of units sold each month (the demand), let a represent the amount spent on advertising (in dollars), and let p represent the price per unit (in dollars). Because the demand is directly proportional to the advertising expenditure and inversely proportional to the price, you obtain the model

$$x = \frac{ka}{p}.$$

By substituting 10,000 for x when $a = 40,000$ and $p = 20$, you can determine that $k = (10,000)(20)/(40,000) = 5$. Thus, the model is

$$x = \frac{5a}{p}.$$

To find the price that corresponds to a demand of 10,000 and an advertising expenditure of $50,000, substitute 10,000 for x and 50,000 for a in the model and solve for p.

$$10,000 = \frac{5(50,000)}{p} \quad \Longrightarrow \quad p = \frac{5(50,000)}{10,000} = \$25$$

(b) The total revenue for selling 10,000 units at $20 is $200,000, and the total revenue for selling 10,000 units at $25 is $250,000. Thus, by increasing the advertising expenditure from $40,000 to $50,000, the company can increase its revenue by $50,000. This implies that you should recommend the increased expenditure for advertising.

The graph in Figure 8.11 shows the "banking angles" for a bicycle at a speed of 15 miles per hour. The banking angle and the radius of the turn vary inversely. As the radius of the turn gets smaller, the bicyclist must lean at greater angles to avoid falling over.

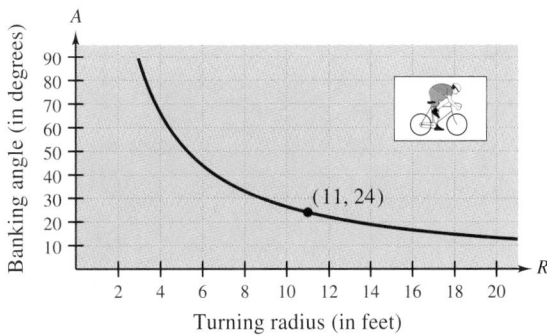

FIGURE 8.11

EXAMPLE 6   *Finding a Model for "Banking Angles"*

Find a model that relates the banking angle A with the turning radius R for a bicyclist at a speed of 15 miles per hour. Use the model to find the banking angle for a turning radius of 8 feet.

Solution

From Figure 8.11, you can see that A is $24°$ when the turning radius R is 11 feet.

$$A = \frac{k}{R} \qquad \text{Model for inverse variation}$$

$$24 = \frac{k}{11} \qquad \text{Substitute 24 for } A \text{ and 11 for } R.$$

$$264 = k \qquad \text{Solve for } k.$$

The model is given by

$$A = \frac{264}{R},$$

where A is measured in degrees and R is measured in feet. When the turning radius is 8 feet, the banking angle is

$$A = \frac{264}{8} = 33°.$$

You can use Figure 8.11 to confirm this result.

The most popular bicycle road race is the Tour de France, lasting 24 days and covering 2500 miles in Europe. Greg LeMond was the first American winner.

Joint Variation

The model used in Example 5 involved both direct and inverse variation, and the word "and" was used to couple the two types of variation together. To describe two different *direct* variations in the same statement, the word "jointly" is used. For instance, the model

$$z = kxy$$

can be described by saying that z is *jointly* proportional to x and y.

Joint Variation

The following statements are equivalent.

1. z **varies jointly** as x and y.

2. z is **jointly proportional** to x and y.

3. $z = kxy$ for some constant k.

If x, y, and z are related by an equation of the form $z = kx^n y^m$, it is said that z varies jointly as the nth power of x and the mth power of y.

EXAMPLE 7   Joint Variation

The *simple interest* for a certain savings account is jointly proportional to the time and the principal. After 1 quarter (3 months), the interest for a principal of $6000 is $120. How much interest would a principal of $7500 earn in 5 months?

Solution

To begin, let I represent the interest earned (in dollars), let P represent the principal (in dollars), and let t represent the time (in years). Then, because the interest is jointly proportional to the time and the principal, you have the following model.

$$I = ktP$$

Because $I = 120$ when $P = 6000$ and $t = \frac{1}{4}$, you can then determine that $k = 120/\left(6000 \cdot \frac{1}{4}\right) = 0.08$. Therefore, the model is

$$I = 0.08tP.$$

To find the interest earned for a principal of $7500 over a 5-month period of time, substitute 7500 for P and $\frac{5}{12}$ for t in the model to obtain an interest of

$$I = 0.08\left(\frac{5}{12}\right)(7500) = \$250.$$

Group Activities   **You Be the Instructor**

Suppose you are teaching an algebra class and are writing a test that covers the material in this section. In your group, write two test questions that give a fair representation of this material. (Assume that your students can spend 10 minutes on each question.)

8.3   Exercises

See Warm-Up Exercises, p. A48

In Exercises 1–12, write a model for the statement.

1. *I* varies directly as *V*.

2. *C* varies directly as *r*.

3. *u* is directly proportional to the square of *v*.

4. *s* varies directly as the cube of *t*.

5. *p* varies inversely as *d*.

6. *S* is inversely proportional to the square of *v*.

7. *P* is inversely proportional to the square root of $1 + r$.

8. *A* varies inversely as the fourth power of *t*.

9. *A* varies jointly as *l* and *w*.

10. *V* varies jointly as *h* and the square of *r*.

11. *Boyle's Law*   If the temperature of a gas is not allowed to change, its absolute pressure *P* is inversely proportional to its volume *V*.

12. *Newton's Law of Universal Gravitation*   The gravitational attraction *F* between two particles of masses m_1 and m_2 is directly proportional to the product of the masses and inversely proportional to the square of the distance *r* between the particles.

In Exercises 13–20, write a verbal sentence using variation terminology to describe the formula.

13. *Area of a Triangle* $A = \frac{1}{2}bh$

14. *Area of a Circle* $A = \pi r^2$

15. *Area of a Rectangle* $A = lw$

16. *Surface Area of a Sphere* $A = 4\pi r^2$

17. *Volume of a Right Circular Cylinder* $V = \pi r^2 h$

18. *Volume of a Sphere* $V = \frac{4}{3}\pi r^3$

19. *Average Speed* $r = \dfrac{d}{t}$

20. *Height of a Cylinder* $h = \dfrac{V}{\pi r^2}$

21. *Think About It*   Suppose the constant of proportionality is positive and *y* varies directly as *x*. If one of the variables increases, how will the other change? Explain.

22. *Think About It*   Suppose the constant of proportionality is positive and *y* varies inversely as *x*. If one of the variables increases, how will the other change? Explain.

In Exercises 23–36, find the constant of proportionality and write an equation that relates the variables.

23. *s* varies directly as *t*, and $s = 20$ when $t = 4$.

24. *h* is directly proportional to *r*, and $h = 28$ when $r = 12$.

25. *F* is directly proportional to the square of *x*, and $F = 500$ when $x = 40$.

26. *v* varies directly as the square root of *s*, and $v = 24$ when $s = 16$.

27. *H* is directly proportional to *u*, and $H = 100$ when $u = 40$.

28. M varies directly as the cube of n, and $M = 0.012$ when $n = 0.2$.

29. n varies inversely as m, and $n = 32$ when $m = 1.5$.

30. q is inversely proportional to p, and $q = \frac{3}{2}$ when $p = 50$.

31. g varies inversely as the square root of z, and $g = \frac{4}{5}$ when $z = 25$.

32. u varies inversely as the square of v, and $u = 40$ when $v = \frac{1}{2}$.

33. F varies jointly as x and y, and $F = 500$ when $x = 15$ and $y = 8$.

34. V varies jointly as h and the square of b, and $V = 288$ when $h = 6$ and $b = 12$.

35. d varies directly as the square of x and inversely as r, and $d = 3000$ when $x = 10$ and $r = 4$.

36. z is directly proportional to x and inversely proportional to the square root of y, and $z = 720$ when $x = 48$ and $y = 81$.

37. *Revenue* The total revenue R is directly proportional to the number of units sold x. When 500 units are sold, the revenue is $3875.

(a) Find the revenue when 635 units are sold.

(b) Interpret the constant of proportionality.

38. *Revenue* The total revenue R is directly proportional to the number of units sold x. When 25 units are sold, the revenue is $300.

(a) Find the revenue when 42 units are sold.

(b) Interpret the constant of proportionality.

39. *Hooke's Law* A force of 50 pounds stretches a spring 3 inches.

(a) How far will a force of 20 pounds stretch the spring?

(b) What force will stretch the spring 1.5 inches?

40. *Hooke's Law* A baby weighing $10\frac{1}{2}$ pounds compresses the spring of a baby scale 7 millimeters. Determine the weight of a baby that compresses the spring 12 millimeters.

41. *Hooke's Law* A force of 50 pounds stretches a spring 5 inches (see figure).

(a) How far will a force of 20 pounds stretch the spring?

(b) What force will stretch the spring 1.5 inches?

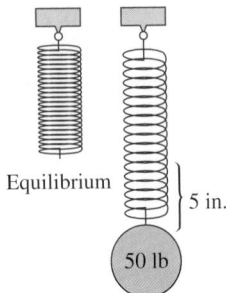

Equilibrium {5 in.

50 lb

42. *Hooke's Law* A force of 50 pounds stretches a spring 1.5 inches.

(a) Write the force F as a function of the distance x the spring is stretched.

(b) Graph the function in part (a) where $0 \le x \le 5$. Identify the graph.

43. *Stopping Distance* The stopping distance d of an automobile is directly proportional to the square of its speed s. On a certain road surface, a car requires 75 feet to stop when its speed is 30 miles per hour. Estimate the stopping distance if the brakes are applied when the car is traveling at 50 miles per hour under similar road conditions.

44. *Free-Falling Object* Neglecting air resistance, the distance d that an object falls varies directly as the square of the time t it has been falling. If an object falls 64 feet in 2 seconds, determine the distance it will fall in 6 seconds.

45. *Free-Falling Object* The velocity v of a free-falling object is directly proportional to the time that it has fallen. The constant of proportionality is the acceleration due to gravity. Find the acceleration due to gravity if the velocity of a falling object is 96 feet per second after the object has fallen for 3 seconds.

46. *Velocity of a Stream* The diameter d of a particle moved by a stream is directly proportional to the square of the velocity v of the stream. A stream with a velocity of $\frac{1}{4}$ mile per hour can move coarse sand particles about 0.02 inch in diameter. What must the velocity be to carry particles with a diameter of 0.12 inch?

47. *Power Generation* The power P generated by a wind turbine varies directly as the cube of the wind speed w. The turbine generates 750 watts of power in a 25-mile-per-hour wind. Find the power it generates in a 40-mile-per-hour wind.

48. *Travel Time* The travel time between two cities is inversely proportional to the average speed. If a train travels between two cities in 3 hours at an average speed of 65 miles per hour, how long would it take at an average speed of 80 miles per hour? What does the constant of proportionality measure in this problem?

49. *Demand Function* A company has found that the daily demand x for its product is inversely proportional to the price p. When the price is $5, the demand is 800 units. Approximate the demand if the price is increased to $6.

50. *Predator-Prey* The number N of prey t months after a natural predator is introduced into a test area is inversely proportional to $t + 1$. If $N = 500$ when $t = 0$, find N when $t = 4$.

51. *Weight of an Astronaut* A person's weight on the moon varies directly with his or her weight on earth. Neil Armstrong, the first man on the moon, weighed 360 pounds on earth, including his heavy equipment. On the moon, he weighed only 60 pounds with equipment. If the first woman in space, Valentina V. Tereshkova, had landed on the moon and weighed 54 pounds with equipment, how much would she have weighed on earth with her equipment?

52. *Weight of an Astronaut* The gravitational force F with which an object is attracted to the earth is inversely proportional to the square of its distance r from the center of the earth. If an astronaut weighs 190 pounds on the surface of the earth ($r \approx 4000$ miles), what will the astronaut weigh 1000 miles above the earth's surface?

53. *Amount of Illumination* The illumination I from a light source varies inversely as the square of the distance d from the light source (see figure). If you raise a study lamp from 18 inches to 36 inches over your desk, the illumination will change by what factor?

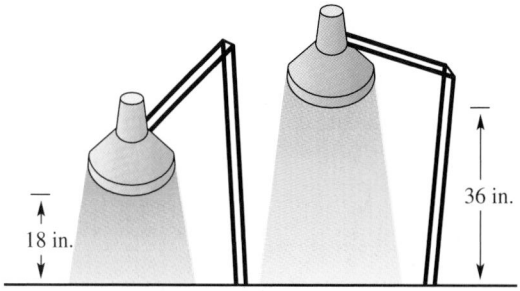

54. *Think About It* If y varies inversely as the square of x and x is doubled, how will y change? Use the properties of exponents to explain your answer.

55. *Frictional Force* The frictional force F between the tires and the road required to keep the car on a curved section of a highway is directly proportional to the square of the speed s of the car. If the speed of the car is doubled, the force will change by what factor?

56. *Best Buy* The prices of 9-inch, 12-inch, and 15-inch diameter pizzas at a certain pizza shop are $6.78, $9.78, and $12.18. One would expect that the price of a certain size pizza would be directly proportional to its surface area. Is this the case for this pizza shop? If not, which size pizza is the best buy?

57. *Snowshoes* When a person walks, the pressure P on each sole varies inversely as the area A of the sole. Denise is trudging through deep snow, wearing boots that have a sole area of 29 square inches each. The sole pressure is 4 pounds per square inch. If Denise were wearing snowshoes, each with an area 11 times that of her boot soles, what would be the pressure on each snowshoe? The constant of proportionality in this problem is Denise's weight. How much does she weigh?

58. *Oil Spill* The graph shows the percent p of oil that remained in Chedabucto Bay, Nova Scotia, after an oil spill. Clean-up was left primarily to natural actions such as wave motion, evaporation, photochemical decomposition, and bacterial de-composition. After about a year, the percent that remained varied inversely as time. Find a model that relates p and t, where t is the number of years since the spill. Then use it to find the amount of oil that remained $6\frac{1}{2}$ years after the spill, and compare the result with the graph.

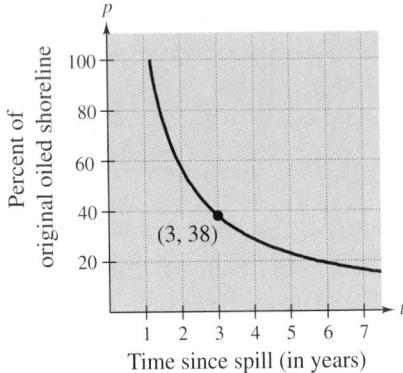

Time since spill (in years)

59. *Ocean Temperatures* The graph shows the temperature of the water in the north central Pacific Ocean. At depths greater than 900 meters, the water temperature varies inversely as the depth. Find a model that relates the temperature T with the depth d. Then use it to find the water temperature at a depth of 4385 meters, and compare the result with the graph.

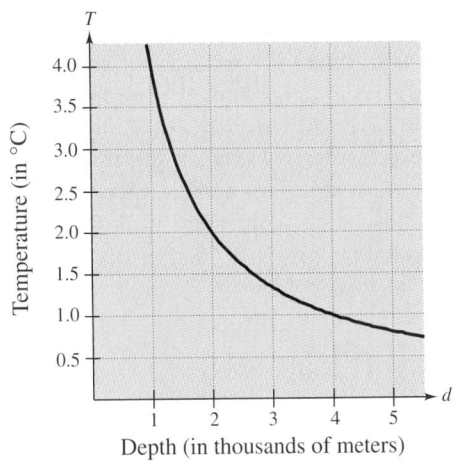

Depth (in thousands of meters)

Think About It In Exercises 60 and 61, determine whether the variation model is of the form $y = kx$ or $y = k/x$, and find k.

60.

x	10	20	30	40	50
y	$\frac{2}{5}$	$\frac{1}{5}$	$\frac{2}{15}$	$\frac{1}{10}$	$\frac{2}{25}$

61.

x	10	20	30	40	50
y	-3	-6	-9	-12	-15

Section Project

Engineering The load P that can be safely supported by a horizontal beam varies jointly as the product of the width W of the beam and the square of its depth D and inversely as its length L (see figure).

(a) Write a model for the statement.

(b) How does P change when the width and length of the beam are doubled?

(c) How does P change when the width and depth of the beam are doubled?

(d) How does P change when all three of the dimensions are doubled?

(e) How does P change when the depth of the beam is cut in half?

(f) A beam of width 3 inches, depth 8 inches, and length 10 feet can safely support 2000 pounds. Determine the safe load of a beam made from the same material if its depth is increased to 10 inches.

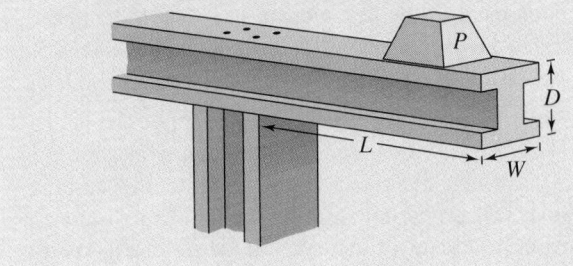

8.4	**Polynomial Functions and Their Graphs**

Graphs of Polynomial Functions ▪ Sketching Graphs of Polynomial Functions ▪ Application

Graphs of Polynomial Functions

Discuss the general description of a polynomial function. Explain that a polynomial function has real number coefficients and that the exponents of the variables are whole numbers.

A function of the form $f(x) = a_n x^n + a_{n-1} x^{n-1} + \cdots + a_1 x + a_0$, $a_n \neq 0$ is a **polynomial function** of degree n. You already know how to sketch the graph of a polynomial function of degree 0, 1, or 2.

Function	Degree	Graph
$f(x) = a$	0	Horizontal line
$f(x) = ax + b$	1	Line of slope a
$f(x) = ax^2 + bx + c$	2	Parabola

EXAMPLE 1   Polynomial Functions of Degrees 0, 1, and 2

(a) The graph of $f(x) = 3$ is a horizontal line, as shown in Figure 8.12(a). This polynomial function has degree 0 and is called a *constant* function.

(b) The graph of $f(x) = -\frac{1}{2}x + 3$ is a line with a slope of $-\frac{1}{2}$ and a y-intercept of 3, as shown in Figure 8.12(b). This polynomial function has degree 1 and is called a *linear* function.

(c) The graph of $f(x) = x^2 + 2x - 3$ is a parabola with a vertex of $(-1, -4)$, as shown in Figure 8.12(c). This polynomial function has degree 2 and is called a *quadratic* function.

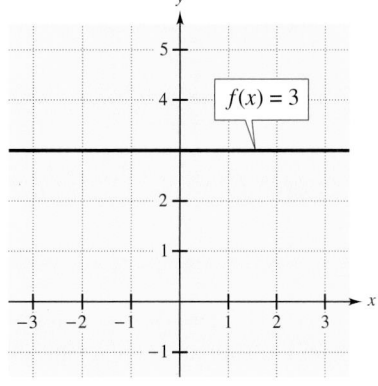

(a)

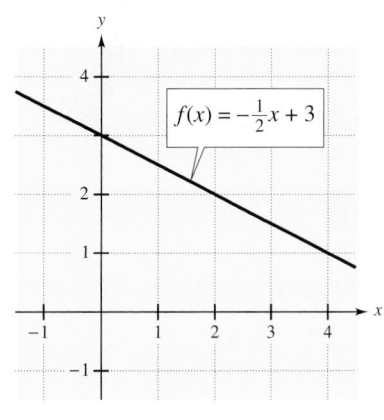

(b)

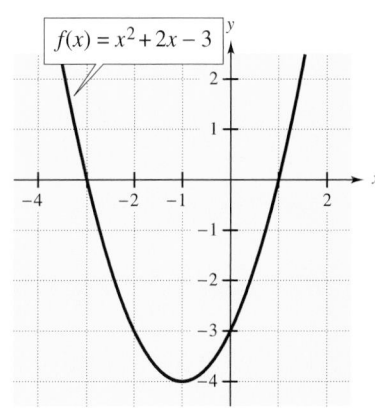

(c)

FIGURE 8.12

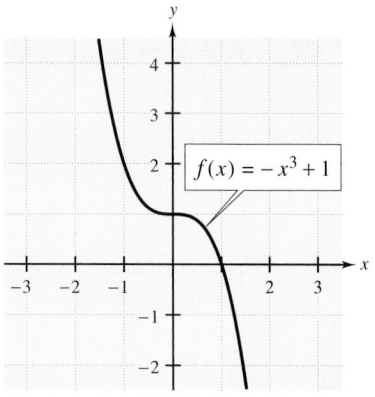
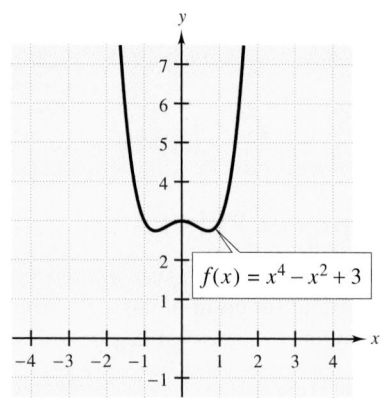
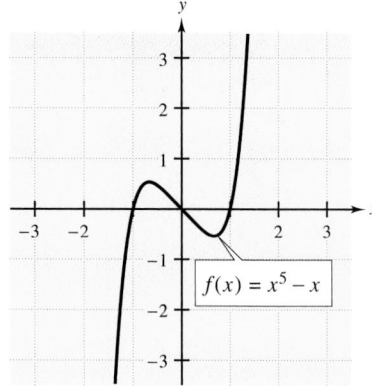

FIGURE 8.13

The graphs of polynomial functions of degree greater than 2 are more difficult to sketch, as illustrated in Figure 8.13. In this section, however, you will learn to recognize some of their basic features. With these features and point plotting, you will be able to make rough sketches *by hand*. (If you go on to take a class in calculus, you will learn to sketch more accurate graphs of polynomial functions of degree greater than 2.)

NOTE   Examine the graphs in Figure 8.13 and notice that the features described at the right hold true.

Features of Graphs of Polynomial Functions

1. The graph of a polynomial function is **continuous.** This means that the graph has no breaks—you could sketch the graph without lifting your pencil from the paper.

2. The graph of a polynomial function has only smooth turns. A function of degree n has *at most $n - 1$* turns.

3. If the leading coefficient of the polynomial function is positive, the graph rises to the right. If the leading coefficient is negative, the graph falls to the right.

The polynomial functions (of degree 2 or greater) that have the simplest graphs are the monomial functions

$$f(x) = a_n x^n.$$

When n is *even*, the graph is similar to the graph of $f(x) = x^2$, as shown in Figure 8.14(a) on the next page. When n is *odd*, the graph is similar to the graph of $f(x) = x^3$, as shown in Figure 8.14(b). Moreover, the greater the value of n, the flatter the graph of a monomial is on the interval $-1 \leq x \leq 1$.

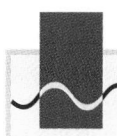 Technology

Use a graphing utility to examine the graphs of x^2, x^4, and x^6. Observe the similarities. Then examine the graphs of x^1, x^3, and x^5. Observe the similarities.

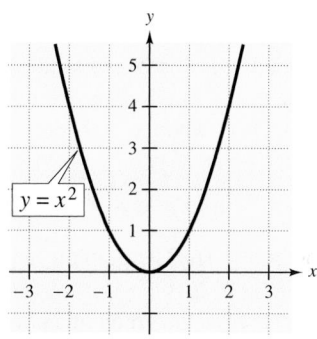

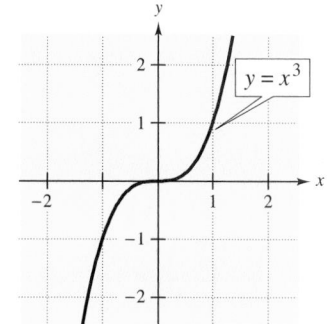

(a) For n even, the graph of $y = x^n$ is similar to the graph of $y = x^2$.

(b) For n odd, the graph of $y = x^n$ is similar to the graph of $y = x^3$.

FIGURE 8.14

EXAMPLE 2 *Sketching Transformations of Monomial Functions*

(a) *Reflection:* To sketch the graph of
$$g(x) = -x^5,$$
reflect the graph of $f(x) = x^5$ in the x-axis, as shown in Figure 8.15(a).

(b) *Vertical Shift:* To sketch the graph of
$$g(x) = x^4 + 1,$$
shift the graph of $f(x) = x^4$ up one unit, as shown in Figure 8.15(b).

(c) *Horizontal Shift:* To sketch the graph of
$$g(x) = (x + 1)^3,$$
shift the graph of $f(x) = x^3$ left one unit, as shown in Figure 8.15(c).

As a classroom demonstration or class exercise, graph $f(x) = x^4$, $g(x) = -x^4$, $h(x) = x^4 - 3$, and $r(x) = (x + 5)^4$ on a single grid using a graphing utility. Compare and contrast the graphs. Help students see the value of technology as a tool for investigation.

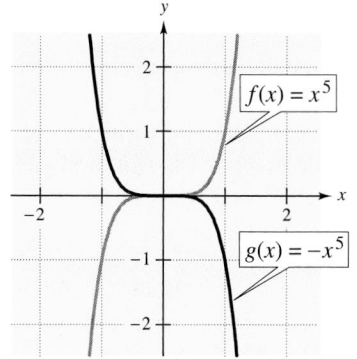

(a)

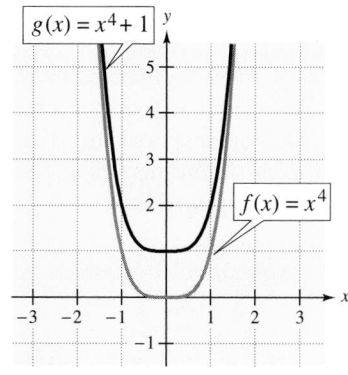

(b)

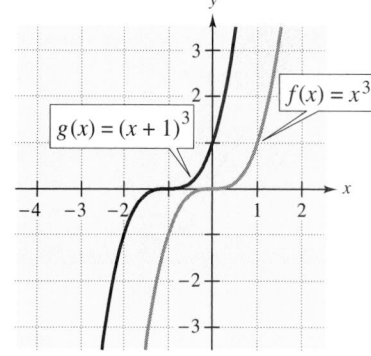

(c)

FIGURE 8.15

Sketching Graphs of Polynomial Functions

In sketching the graph of a polynomial function, it helps to label the intercepts. The *y*-intercept of the graph of a polynomial function is easy to find—it is the value of the function when $x = 0$. For instance, the *y*-intercept of the graph of $f(x) = x^3 - 7x + 6$ is

$$f(0) = 0^3 - 7(0) + 6 = 6, \qquad \text{\small\textit{y}-intercept}$$

as shown in Figure 8.16. The graph of a polynomial function always has exactly one *y*-intercept.

The *x*-intercepts of the graph of a polynomial function are not as easy to find. Moreover, the graph can have no *x*-intercepts, one *x*-intercept, or several *x*-intercepts. The *x*-intercepts of the graph of a polynomial function are the *x*-values for which $f(x) = 0$. You can find these values by solving the equation

$$f(x) = 0. \qquad \text{\small\textit{x}-intercepts}$$

These values are also called the **zeros** of the function. If you go on to take a course in college algebra, you will study techniques for solving polynomial equations of degree 3 or greater. You will also learn that the graph of a polynomial function of degree *n* has *at most n x*-intercepts. For instance, the graph of

$$f(x) = x^3 - 7x + 6$$

has three *x*-intercepts, as shown in Figure 8.16. Try showing that $f(x) = 0$ when $x = -3$, $x = 1$, and $x = 2$.

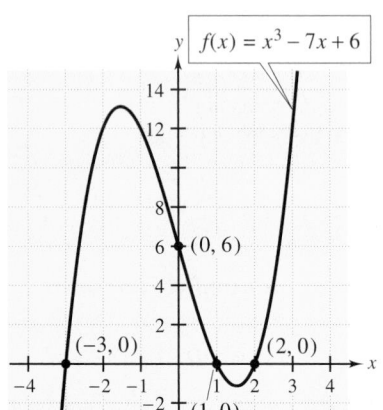

FIGURE 8.16

EXAMPLE 3   Graphing a Cubic Polynomial Function

Sketch the graph of

$$f(x) = x^3 - 9x.$$

Solution

Because the leading coefficient is positive, you know that the graph rises to the right. Also, because the degree of the polynomial function is 3, you know that the graph can have at most two turns. Using these general observations with the values given in the following table, you can sketch the graph, as shown in Figure 8.17. Notice that the graph has three *x*-intercepts: -3, 0, and 3.

x	-4	-3	-2	-1	0	1	2	3	4
$f(x)$	-28	0	10	8	0	-8	-10	0	28

If you have access to a graphing utility, try using it to confirm the graph shown in Figure 8.17.

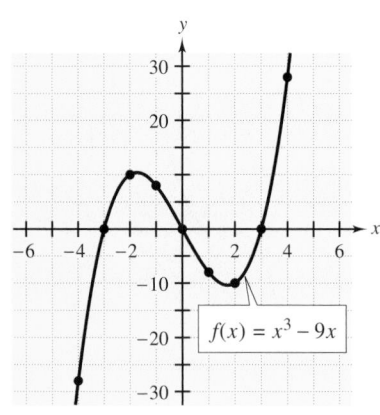

FIGURE 8.17

NOTE The x-intercepts of the graphs of some polynomial functions can be found by setting $f(x) = 0$ and factoring. For the equation in Example 3, this is done as follows.

$$x^3 - 9x = 0$$
$$x(x^2 - 9) = 0$$
$$x(x + 3)(x - 3) = 0$$
$$x = 0$$
$$x + 3 = 0 \implies x = -3$$
$$x - 3 = 0 \implies x = 3$$

Thus, the three x-intercepts are 0, -3, and 3.

Have students discuss why a polynomial function has exactly *one* y-intercept.

In Figure 8.17, note that the graph of the function rises to the right and falls to the left. In general, the graph of a polynomial function of *odd* degree has opposite behaviors to the right and left, whereas the graph of a polynomial function of *even* degree has the same behavior to the right and left.

EXAMPLE 4   Identifying the Left and Right Behavior of a Graph

(a) Because the leading coefficient of $f(x) = -x^3 + 4x$ is negative, the graph falls to the right. Because the degree is odd, the graph rises to the left, as shown in Figure 8.18(a).

(b) Because the leading coefficient of $f(x) = x^4 - 5x^2 + 4$ is positive, the graph rises to the right. Because the degree is even, the graph also rises to the left, as shown in Figure 8.18(b).

(c) Because the leading coefficient of $f(x) = x^5 - x$ is positive, the graph rises to the right. Because the degree is odd, the graph falls to the left, as shown in Figure 8.18(c).

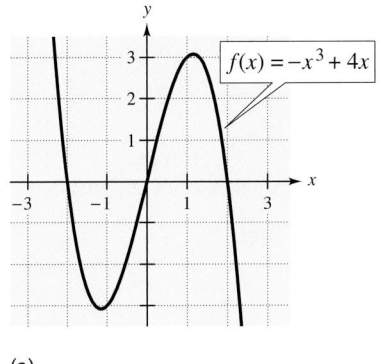

(a)

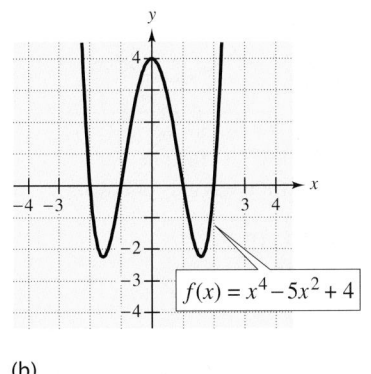

(b)

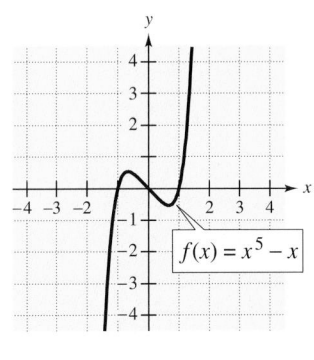

(c)

FIGURE 8.18

Application

EXAMPLE 5 Charitable Contributions

According to a survey conducted by *Independent Sector*, the average percent P (in decimal form) of income that Americans give to charitable organizations is related to their income x (in thousands of dollars) by the model

$$P = 0.000014x^2 - 0.001529x + 0.05855, \qquad 5 \le x \le 100.$$

Use this model to write a model for the average *amount* given.

Solution

The graph of the given model is shown in Figure 8.19. Notice that the vertex of the graph occurs when $x \approx 55$. This means that the income level that corresponds to the smallest percent contributions to charity is the \$55,000 level. To find the average amount A that Americans of each income level give to charity, you can multiply the average percent-of-income model by the income ($1000x$) to obtain the following polynomial.

$$A = 1000x(0.000014x^2 - 0.001529x + 0.05855), \qquad 5 \le x \le 100$$

$$= 0.014x^3 - 1.529x^2 + 58.55x$$

The graph of this third-degree polynomial is shown in Figure 8.20. Note in the graph that the average *amount* given to charity increases as the income level increases.

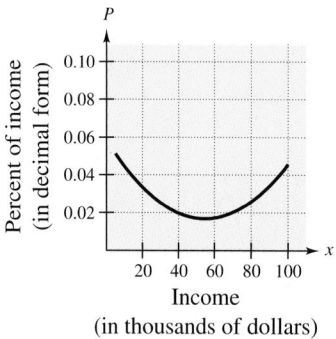

FIGURE 8.19

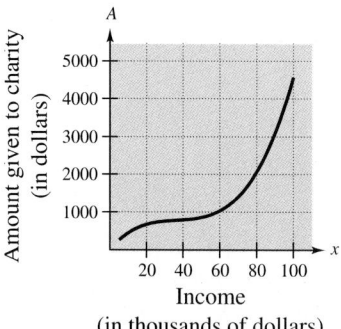

FIGURE 8.20

Group Activities **Finding x-Intercepts of Polynomial Functions**

Choose four integers between -10 and 10. Then, in your group, discuss how to find a fourth–degree polynomial function that has these integers as its zeros. Find the polynomial function and use a graphing utility to confirm your result.

Is there only one fourth-degree polynomial that has these four real numbers as its zeros? If so, explain why. If not, find another and sketch its graph in the same viewing rectangle.

8.4 **Exercises**

See Warm-Up
Exercises, p. A48

In Exercises 1–6, identify and sketch the graph of the polynomial function of degree 0, 1, or 2.

1. $g(x) = -2$
2. $f(x) = 8$
3. $f(x) = 7 - 3x$
4. $h(x) = \frac{1}{2}x + 1$
5. $f(x) = -(x - 3)^2 + 4$
6. $h(t) = t^2 + 2t + 1$

In Exercises 7–12, state the maximum possible number of turns in the graph of the polynomial function.

7. $f(x) = 2x^5 - x + 6$
8. $f(x) = \frac{1}{2}x^3 + 1$
9. $g(s) = -3s^2 + 2s + 4$
10. $h(u) = (u - 3)^2 - 1$
11. $h(x) = x^4 + 9x - 1$
12. $f(x) = -2x^6 + 3x + 7$

In Exercises 13–18, describe the transformation of g that would produce f.

13. $f(x) = (x - 5)^3$
      $g(x) = x^3$
14. $f(x) = (x + 3)^2$
      $g(x) = x^2$
15. $f(x) = x^4 + 3$
      $g(x) = x^4$
16. $f(x) = x^5 - 2$
      $g(x) = x^5$
17. $f(x) = -x^6$
      $g(x) = x^6$
18. $f(x) = -x$
      $g(x) = x$

In Exercises 19 and 20, use the graph to sketch the graphs of the specified transformations.

19. $y = x^3$
(a) $f(x) = (x - 2)^3$
(b) $f(x) = x^3 - 2$
(c) $f(x) = (x - 2)^3 - 2$
(d) $f(x) = -x^3$

20. $y = x^4$
(a) $f(x) = (x + 3)^4$
(b) $f(x) = x^4 - 3$
(c) $f(x) = 4 - x^4$
(d) $f(x) = -(x - 1)^4$

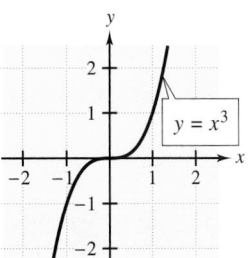

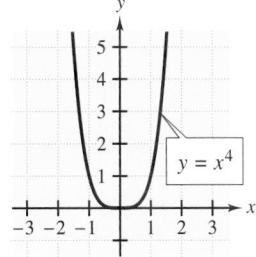

Graphical Analysis   In Exercises 21–24, use a graphing utility to graph the functions f and g in the same viewing rectangle. Zoom out sufficiently far to show that the right and left behavior of f and g appear identical. What is this viewing window?

21. $f(x) = 3x^3 - 9x + 1$ $g(x) = 3x^3$
22. $f(x) = -\frac{1}{3}(x^3 - 3x + 2)$ $g(x) = -\frac{1}{3}x^3$

23. $f(x) = -(x^4 - 4x^3 + 16x)$ $g(x) = -x^4$

24. $f(x) = 3x^4 - 6x^2$ $g(x) = 3x^4$

In Exercises 25–30, determine the right and left behavior of the graph of the polynomial function.

25. $f(x) = 2x^2 - 3x + 1$

26. $g(x) = 5 - \frac{7}{2}x - 3x^2$

27. $f(x) = \frac{1}{3}x^3 + 5x$

28. $f(x) = -2.1x^5 + 4x^3 - 2$

29. $f(x) = 6 - 2x + 4x^2 - 5x^3$

30. $h(x) = 1 - x^6$

In Exercises 31–42, find the x- and y-intercepts of the graph of the polynomial function.

31. $f(x) = x^2 - 25$

32. $f(x) = 49 - x^2$

33. $h(x) = x^2 - 6x + 9$

34. $f(x) = x^2 + 10x + 25$

35. $f(x) = x^2 + x - 2$

36. $f(x) = 3x^2 - 12x + 3$

37. $f(x) = x^3 - 4x^2 + 4x$

38. $f(x) = x^4 - x^3 - 20x^2$

39. $g(x) = \frac{1}{2}x^4 - \frac{1}{2}$

40. $f(x) = 2x^4 - 2x^2 - 40$

41. $f(x) = 5x^4 + 15x^2 + 10$

42. $f(x) = x^3 - 4x^2 - 25x + 100$

▦ *Graphical Analysis* In Exercises 43–50, use a graphing utility to graph the function and approximate any x-intercepts of the graph.

43. $g(t) = \frac{1}{2}t^4 - \frac{1}{2}$ **44.** $f(x) = x^5 + x^3 - 6x$

45. $f(x) = 2x^4 - 2x^2 - 40$

46. $g(t) = t^5 - 6t^3 + 9t$

47. $f(x) = 5x^4 + 15x^2 + 10$

48. $f(x) = x^3 - 4x^2 - 25x + 100$

49. $y = 4x^3 + 4x^2 - 7x + 2$

50. $y = x^5 - 5x^3 + 4x$

In Exercises 51–58, match the polynomial function with the correct graph. [The graphs are labeled (a), (b), (c), (d), (e), (f), (g), and (h).]

(a)
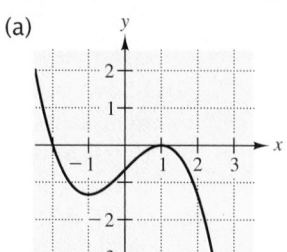

(b)
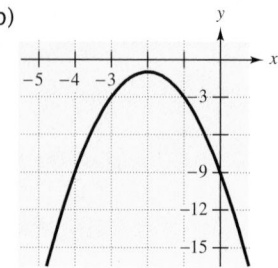

(c)
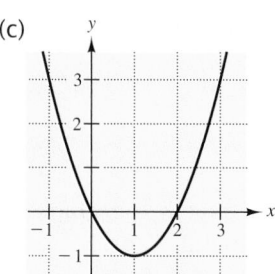

(d)
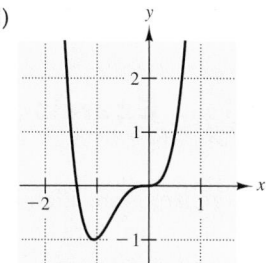

(e)
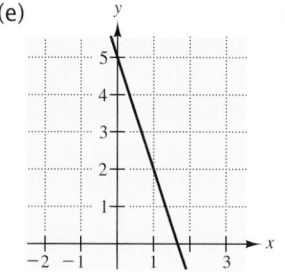

(f)
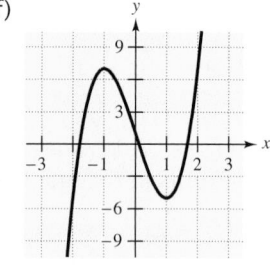

(g)
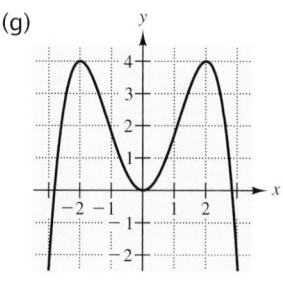

(h)
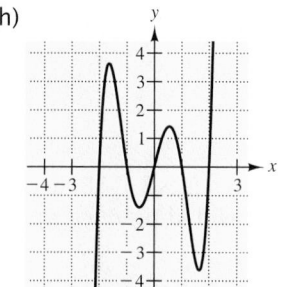

51. $f(x) = -3x + 5$ **52.** $f(x) = x^2 - 2x$

53. $f(x) = -2x^2 - 8x - 9$ **54.** $f(x) = 3x^3 - 9x + 1$

55. $f(x) = -\frac{1}{3}x^3 + x - \frac{2}{3}$ **56.** $f(x) = -\frac{1}{4}x^4 + 2x^2$

57. $f(x) = 3x^4 + 4x^3$ **58.** $f(x) = x^5 - 5x^3 + 4x$

In Exercises 59–70, sketch the graph of the polynomial function.

59. $f(x) = -\frac{3}{2}$

60. $h(x) = \frac{1}{3}x - 3$

61. $f(x) = -x^3$

62. $g(x) = -x^4$

63. $f(x) = x^5 + 1$

64. $f(x) = x^4 - 2$

65. $h(x) = (x + 2)^4$

66. $g(x) = (x - 4)^3$

67. $f(x) = 1 - x^6$

68. $g(x) = 1 - (x + 1)^6$

69. $f(x) = x^3 - 3x^2$

70. $f(x) = x^3 - 4x$

In Exercises 71–76, use a graphing utility to graph the function. Does the graph have the maximum number of turns possible for the degree of the polynomial function?

71. $f(x) = -x^3 + 4x^2$

72. $f(x) = \frac{1}{4}x^4 - 2x^2$

73. $g(t) = -\frac{1}{4}(t - 2)^2(t + 2)^2$

74. $f(x) = x^2(x - 4)$

75. $h(s) = s^5 + s$

76. $g(t) = t^5 + 3t^3 - t$

77. *Volume of a Box* An open box is made from a 12-inch-square piece of material by cutting equal squares from the corners and turning up the sides (see figure).

(a) Verify that the volume of the box is
$V(x) = 4x(6 - x)^2$.

(b) Determine the domain of the function V.

(c) Use a graphing utility to graph the function and use the graph to estimate the value of x for which $V(x)$ is a maximum.

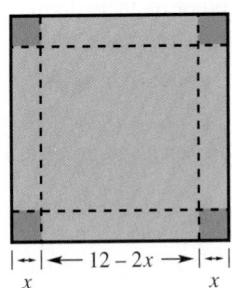
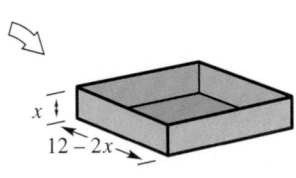

78. *Volume of a Box* An open box with locking tabs is made from a 12-inch-square piece of material. This is done by cutting equal squares from the corners and folding along the dashed lines (see figure).

(a) Verify that the volume of the box is
$V(x) = 8x(3 - x)(6 - x)$.

(b) Determine the domain of the function V.

(c) Use a graphing utility to graph the function and use the graph to estimate the value of x for which $V(x)$ is a maximum.

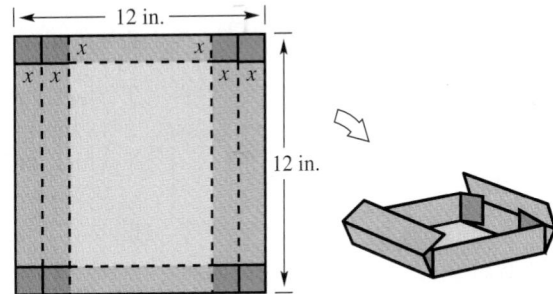

Section Project

Making Connections

(a) Find the zeros of the quadratic function $g(x)$.

(i) $g(x) = x^2 - 4x - 12$ (iv) $g(x) = x^2 - 4x + 4$

(ii) $g(x) = x^2 + 5x$ (v) $g(x) = x^2 - 2x - 6$

(iii) $g(x) = x^2 + 3x - 10$ (vi) $g(x) = x^2 + 3x + 4$

(b) For each function in part (a), use a graphing utility to graph $f(x) = (x - 2) \cdot g(x)$. Verify that $(2, 0)$ is an x-intercept of the graph of $f(x)$. Describe any similarities or differences in the behavior of the six functions at this x-intercept.

(c) For each function in part (b), use the graph of $f(x)$ to approximate the other x-intercepts of the graph.

(d) Describe the connections that you find among the results of parts (a), (b), and (c).

Mid-Chapter Quiz

Take this quiz as you would take a quiz in class. After you are done, check your work against the answers given in the back of the book.

In Exercises 1–8, let $f(x) = 2x - 8$ and $g(x) = x^2$. Find the specified combination of the functions and find its domain.

1. $(f - g)(x)$ **2.** $(f + g)(x)$ **3.** $(fg)(x)$

4. $(g - f)(x)$ **5.** $(f/g)(x)$ **6.** $(g/f)(x)$

7. $(f \circ g)(x)$ **8.** $(g \circ f)(x)$

9. *Weekly Production Cost*   The weekly cost of producing x units in a manufacturing process is given by the function $C(x) = 60x + 750$. The number of units produced in t hours during a week is given by $x(t) = 50t$, $0 \le t \le 40$. Find, simplify, and interpret $(C \circ x)(t)$.

10. Verify algebraically that the functions $f(x) = 100 - 15x$ and $g(x) = \frac{1}{15}(100 - x)$ are inverses of each other.

11. Find the inverse of the function $f(x) = x^3 - 8$. Verify your result algebraically and graphically.

In Exercises 12–14, find the constant of proportionality and give the equation relating the variables.

12. z varies directly as t, and $z = 12$ when $t = 4$.

13. S varies jointly as h and the square of r, and $S = 120$ when $h = 6$ and $r = 2$.

14. N varies directly as the square of t and inversely as s, and $N = 300$ when $t = 10$ and $s = 5$.

15. *Mining Danger*   One of the dangers of coal mining is the methane gas that can leak out of seams in the rock. Methane forms an explosive mixture with air at a concentration of 5% or greater. A steady leak of methane begins in a coal mine so that the concentration of methane gas varies directly as time. Twelve minutes after a leak begins, the concentration of methane in the air is 2%. If the leak continues at the same rate, when could an explosion occur?

16. Does the graph of $h(x) = 3x^5 - 2x^2 + 5x - 7$ rise to the right or fall to the right? Explain your reasoning.

17. Use the graph $y = x^3 - 2x^2$ shown in the figure to sketch the graphs of the specified transformations.

    (a) $f(x) = -x^3 + 2x^2$ (b) $f(x) = (x - 2)^3 - 2(x - 2)^2$

18. Use a graphing utility to graph $g(x) = x^4 - x^3 - 1$ and approximate any x-intercepts of the graph.

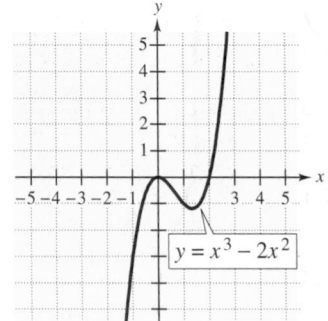

Figure for 17

8.5	**Circles**

The Conics ▪ Circles Centered at the Origin ▪
Circles Centered at *(h, k)* ▪ Application

The Conics

Historical Note: The first notable work of the French mathematician Blaise Pascal (1623–1662) was a paper about conic sections.

In Section 6.5, you saw that the graph of a second-degree equation

$$y = ax^2 + bx + c$$

is a parabola. A parabola is one of four types of curves that are called **conics** or **conic sections.** The other three types are circles, ellipses, and hyperbolas. As indicated in Figure 8.21, the name "conic" relates to the fact that each of these figures can be obtained by intersecting a plane with a double-napped cone.

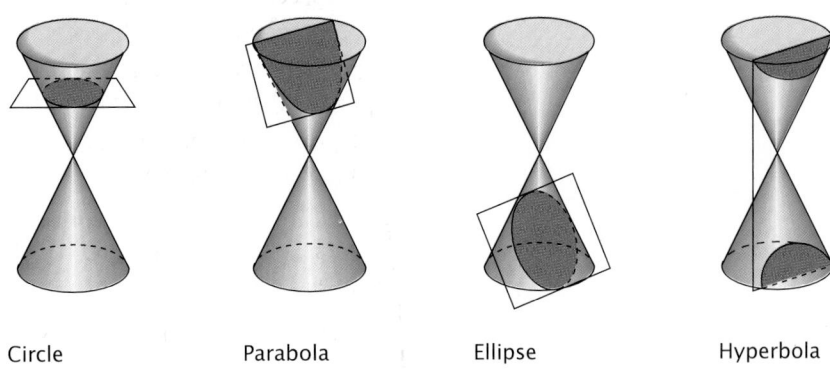

Circle   Parabola   Ellipse   Hyperbola

FIGURE 8.21

Exploration

Practical applications of the conic sections are numerous and varied. Encourage students to suggest other applications of parabolas, ellipses, and hyperbolas.

Imagine that you cut a carrot along its width at a 90° angle. What conic is formed by the cross section of the carrot? What conic is formed when the carrot is cut along its width at a 45° angle? Illustrate your answers with diagrams. Explain how other real-life objects can be used to form cross sections that are conics.

Conic sections occur in many practical applications. Reflective surfaces in satellite dishes, flashlights, and telescopes often have a parabolic shape. The orbits of planets are elliptical, and the orbits of comets are usually either elliptical or hyperbolic. Ellipses and parabolas are also used in the construction of archways and bridges.

Circles Centered at the Origin

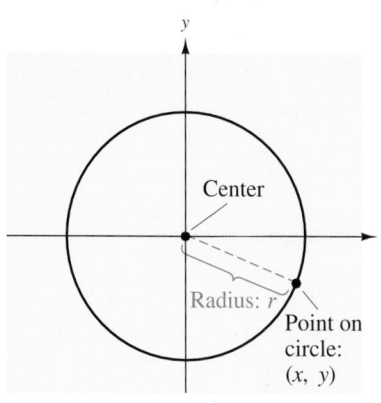

FIGURE 8.22

A **circle** in the rectangular coordinate plane consists of all points (x, y) that are a given positive distance r from a fixed point, called the **center** of the circle. The distance r is called the **radius** of the circle. If the center of the circle is the origin, as shown in Figure 8.22, the relationship between the coordinates of any point (x, y) on the circle and the radius r is given by

$$\text{Radius} = r = \sqrt{(x - 0)^2 + (y - 0)^2} \qquad \text{Distance Formula}$$
$$= \sqrt{x^2 + y^2}.$$

By squaring both sides of this equation, you obtain the one given below, which is called the **standard form of the equation of a circle centered at the origin.**

Standard Equation of a Circle (Center at Origin)

The **standard form of the equation of a circle centered at the origin** is

$$x^2 + y^2 = r^2.$$

The positive number r is called the **radius** of the circle.

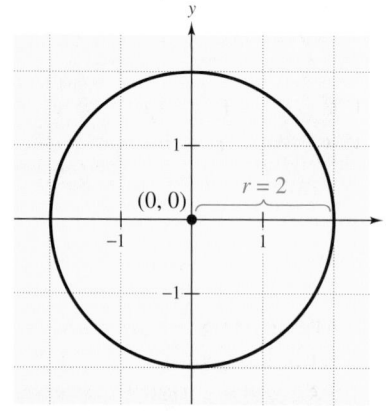

FIGURE 8.23

EXAMPLE 1 Finding an Equation of a Circle

Find an equation of the circle that is centered at the origin and has a radius of 2 (see Figure 8.23).

Solution

Using the standard form of the equation of a circle (with center at the origin) and $r = 2$, you obtain

$$x^2 + y^2 = r^2$$
$$x^2 + y^2 = 2^2$$
$$x^2 + y^2 = 4.$$

Technology

You cannot represent a circle as a single function of x. You can, however, represent a circle by two functions of x—one for the top half of the circle and one for the bottom half. For instance, try using a graphing utility to graph the following. Do you obtain a circle?

$$y_1 = \sqrt{4 - x^2} \qquad \text{Top half of circle}$$
$$y_2 = -\sqrt{4 - x^2} \qquad \text{Bottom half of circle}$$

To sketch the circle for a given equation, first write the equation in standard form. Then, from the standard form, you can identify the center and radius, and sketch the circle.

EXAMPLE 2 Sketching a Circle

Identify the center and radius of the circle given by the following equation, and sketch the circle.

$$4x^2 + 4y^2 - 25 = 0$$

Solution

Begin by writing the equation in standard form.

$$4x^2 + 4y^2 - 25 = 0 \qquad \text{Original equation}$$

$$4x^2 + 4y^2 = 25 \qquad \text{Add 25 to both sides.}$$

$$x^2 + y^2 = \frac{25}{4} \qquad \text{Divide both sides by 4.}$$

$$x^2 + y^2 = \left(\frac{5}{2}\right)^2 \qquad \text{Standard form}$$

Now, from this standard form, you can see that the graph of the equation is a circle that is centered at the origin and has a radius of $\frac{5}{2}$, as shown in Figure 8.24.

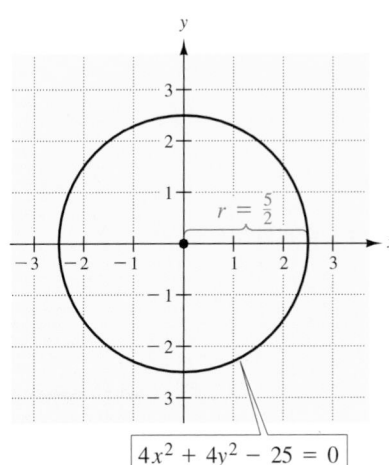

$$4x^2 + 4y^2 - 25 = 0$$

FIGURE 8.24

Circles Centered at (*h*, *k*)

Consider a circle whose radius is r and whose center is the point (h, k), as shown in Figure 8.25. Let (x, y) be any point on the circle. To find an equation for this circle, you can use the Distance Formula and write

$$\text{Radius} = r = \sqrt{(x - h)^2 + (y - k)^2}. \qquad \text{Distance Formula, Section 2.1}$$

By squaring both sides of this equation, you obtain the equation given below, which is called the **standard form of the equation of a circle centered at (*h*, *k*).**

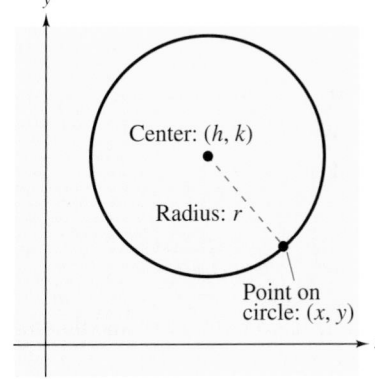

Center: (h, k)

Radius: r

Point on circle: (x, y)

FIGURE 8.25

Standard Equation of a Circle [Center at (*h*, *k*)]

The **standard form of the equation of a circle centered at** (h, k) is

$$(x - h)^2 + (y - k)^2 = r^2.$$

EXAMPLE 3   *Finding an Equation of a Circle*

The point $(2, 5)$ lies on a circle whose center is at $(5, 1)$, as shown in Figure 8.26. Find the standard form of the equation of this circle.

Solution

The radius r of the circle is the distance between $(2, 5)$ and $(5, 1)$.

$$
\begin{aligned}
r &= \sqrt{(2 - 5)^2 + (5 - 1)^2} & &\text{Distance Formula} \\
&= \sqrt{(-3)^2 + 4^2} & &\text{Simplify.} \\
&= \sqrt{9 + 16} & &\text{Simplify.} \\
&= \sqrt{25} & &\text{Simplify.} \\
&= 5 & &\text{Radius}
\end{aligned}
$$

Using $(h, k) = (5, 1)$ and $r = 5$, the equation of the circle is

$$
\begin{aligned}
(x - h)^2 + (y - k)^2 &= r^2 & &\text{Standard form} \\
(x - 5)^2 + (y - 1)^2 &= 5^2 & &\text{Substitute for } h, k, \text{ and } r. \\
(x - 5)^2 + (y - 1)^2 &= 25. & &\text{Equation of circle}
\end{aligned}
$$

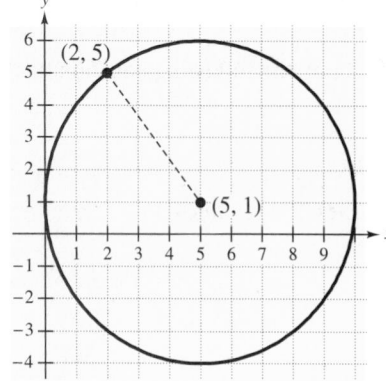

FIGURE 8.26

To write the equation of a circle in standard form, you may need to complete the square, as demonstrated in Example 4.

EXAMPLE 4   *Writing an Equation in Standard Form*

Write the equation $x^2 + y^2 - 2x + 4y - 4 = 0$ in standard form, and sketch the circle represented by the equation.

Solution

$$
\begin{aligned}
x^2 + y^2 - 2x + 4y - 4 &= 0 \\
\left(x^2 - 2x + \right) + \left(y^2 + 4y + \right) &= 4 \\
[x^2 - 2x + (-1)^2] + (y^2 + 4y + 2^2) &= 4 + 1 + 4 \qquad \text{Complete the square.} \\
(\text{half})^2 \qquad\qquad (\text{half})^2 & \\
(x - 1)^2 + (y + 2)^2 &= 3^2
\end{aligned}
$$

From this standard form, you can see that the circle has a radius of 3 and that the center of the circle is $(1, -2)$, as shown in Figure 8.27.

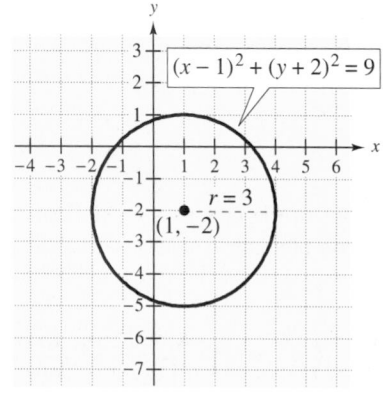

FIGURE 8.27

Application

EXAMPLE 5 *Mechanical Drawing*

You are in a mechanical drawing class and are asked to help program a computer to model the metal piece shown in Figure 8.28. Part of your assignment is to find an equation for the semicircular upper portion of the hole in the metal piece. What is the equation?

Solution

From the drawing, you can see that the center of the circle is $(h, k) = (5, 2)$ and that the radius of the circle is $r = 1.5$. This implies that the equation of the entire circle is

$$(x - h)^2 + (y - k)^2 = r^2 \qquad \text{Standard form}$$
$$(x - 5)^2 + (y - 2)^2 = 1.5^2 \qquad \text{Substitute for } h, k, \text{ and } r.$$
$$(x - 5)^2 + (y - 2)^2 = 2.25 \qquad \text{Equation of circle}$$

To find the equation of the upper portion of the circle, solve this standard equation for y.

$$(x - 5)^2 + (y - 2)^2 = 2.25$$
$$(y - 2)^2 = 2.25 - (x - 5)^2$$
$$y - 2 = \pm\sqrt{2.25 - (x - 5)^2}$$
$$y = 2 \pm \sqrt{2.25 - (x - 5)^2}$$

Finally, take the positive square root to obtain the upper portion of the circle, $y = 2 + \sqrt{2.25 - (x - 5)^2}$. (If you had wanted the lower portion of the circle, you would have taken the negative square root.)

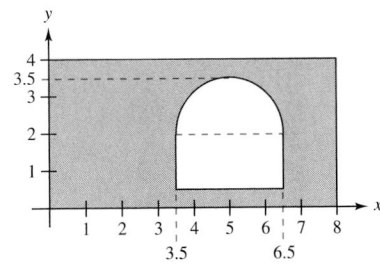

FIGURE 8.28

Group Activities **Finding x-Intercepts of Circles**

The circle given by
$$(x - 2)^2 + (y - 3)^2 = 25$$
intersects the *x*-axis twice. Use a graphing utility to approximate these two *x*-intercepts. Then find the intercepts algebraically and compare your results.

8.5   **Exercises**

See Warm-Up
Exercises, p. A48

1. *Think About It*   Name the four types of conics.
2. *Think About It*   Define a circle and give the standard form of the equation of a circle centered at the origin.

In Exercises 3–8, match the equation with its graph. [The graphs are labeled (a), (b), (c), (d), (e), and (f).]

(a)
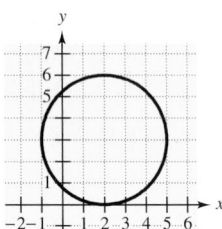

(b)
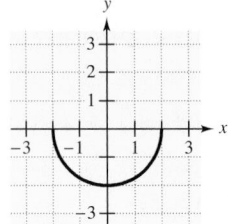

(c)
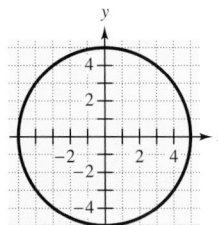

(d)
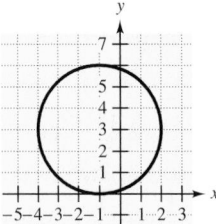

(e)
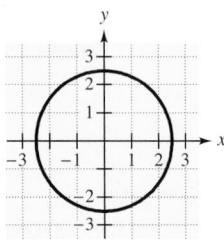

(f)
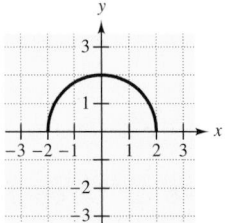

3. $x^2 + y^2 = 25$
4. $4x^2 + 4y^2 = 25$
5. $(x - 2)^2 + (y - 3)^2 = 9$
6. $(x + 1)^2 + (y - 3)^2 = 9$
7. $y = -\sqrt{4 - x^2}$
8. $y = \sqrt{4 - x^2}$

In Exercises 9–16, find an equation of the circle with center at $(0, 0)$ that satisfies the given criterion.

9. Radius: 5
10. Radius: 7
11. Radius: $\frac{2}{3}$
12. Radius: $\frac{5}{2}$
13. Passes through the point $(0, 8)$
14. Passes through the point $(-2, 0)$
15. Passes through the point $(5, 2)$
16. Passes through the point $(-1, -4)$

In Exercises 17–24, find an equation of the circle with center at (h, k) that satisfies the given criteria.

17. Center: $(4, 3)$
    Radius: 10
18. Center: $(-2, 5)$
    Radius: 6
19. Center: $(5, -3)$
    Radius: 9
20. Center: $(-5, -2)$
    Radius: $\frac{5}{2}$
21. Center: $(-2, 1)$
    Passes through the point $(0, 1)$
22. Center: $(8, 2)$
    Passes through the point $(8, 0)$
23. Center: $(3, 2)$
    Passes through the point $(4, 6)$
24. Center: $(-3, -5)$
    Passes through the point $(0, 0)$

In Exercises 25–36, identify the center and radius of the circle and sketch its graph.

25. $x^2 + y^2 = 16$

26. $x^2 + y^2 = 25$

27. $x^2 + y^2 = 36$

28. $x^2 + y^2 = 10$

29. $4x^2 + 4y^2 = 1$

30. $9x^2 + 9y^2 = 64$

31. $25x^2 + 25y^2 - 144 = 0$

32. $\dfrac{x^2}{4} + \dfrac{y^2}{4} - 1 = 0$

33. $(x - 2)^2 + (y - 3)^2 = 4$

34. $(x + 4)^2 + (y - 3)^2 = 25$

35. $\left(x + \frac{5}{2}\right)^2 + (y + 3)^2 = 9$

36. $(x - 5)^2 + \left(y + \frac{3}{4}\right)^2 = 1$

In Exercises 37–40, write the equation in completed square form. Then sketch the graph.

37. $x^2 + y^2 - 4x - 2y + 1 = 0$

38. $x^2 + y^2 + 6x - 4y - 3 = 0$

39. $x^2 + y^2 + 2x + 6y + 6 = 0$

40. $x^2 + y^2 - 2x + 6y - 15 = 0$

In Exercises 41–44, use a graphing utility to graph the equation. Identify the center and radius. (Note: Solve for y. Use the square setting so that the circles appear visually correct.)

41. $x^2 + y^2 = 30$

42. $4x^2 + 4y^2 = 45$

43. $(x - 2)^2 + y^2 = 10$

44. $(x + 3)^2 + y^2 = 15$

45. *Satellite Orbit* Find an equation of the circular orbit of a satellite 500 miles above the surface of the earth. Place the origin of the rectangular coordinate system at the center of the earth and assume that the radius of the earth is 4000 miles.

46. *Height of an Arch* A semicircular arch for a tunnel under a river has a diameter of 100 feet (see figure). Determine the height of the arch 5 feet from the edge of the tunnel.

47. *Architecture* The top portion of a stained-glass window is in the form of a pointed Gothic arch (see figure). Each side of the arch is an arc of a circle of radius 12 feet with its center at the base of the opposite arch. Find an equation of one of the circles and use it to determine the height of the point of the arch above the rectangular portion of the window.

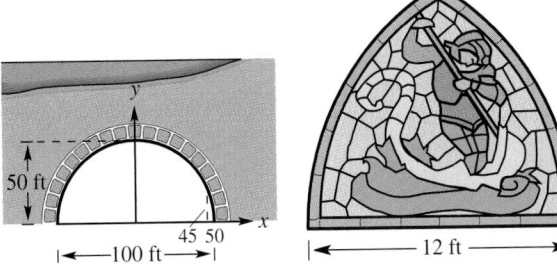

Figure for 46   Figure for 47

Graphical Analysis A rectangle centered at the origin with sides parallel to the coordinate axes is placed in a circle of radius 25 inches centered at the origin (see figure). The length of the rectangle is $2x$ inches. Verify that the width of the rectangle is $2\sqrt{625 - x^2}$. Find a model for the area of the rectangle. Use a graphing utility to graph the model. Approximate the value of x for which the area is a maximum. Repeat this process for a circle of radius 4, and compare the results.

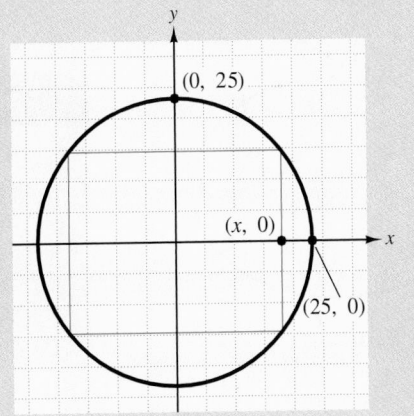

8.6	**_Ellipses and Hyperbolas_**

Ellipses Centered at the Origin ▪ Ellipses Centered at (h, k) ▪
Hyperbolas Centered at the Origin ▪ Hyperbolas Centered at (h, k) ▪
Application

Ellipses Centered at the Origin

An **ellipse** in the rectangular coordinate system consists of all points (x, y) such that the sum of the distances between (x, y) and two distinct fixed points is a constant, as shown in Figure 8.29. Each of the two fixed points is called a **focus** of the ellipse. (The plural of focus is **foci.**)

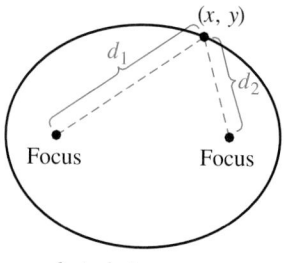
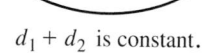
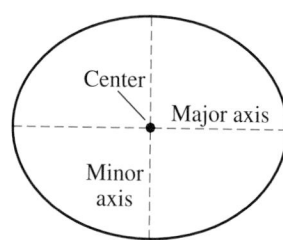

FIGURE 8.29

The line through the foci intersects the ellipse at two points, called the **vertices.** The line segment joining the vertices is called the **major axis,** and its midpoint is called the **center** of the ellipse. The line segment perpendicular to the major axis at the center is called the **minor axis** of the ellipse, and the points at which the minor axis intersects the ellipse are called **co-vertices.**

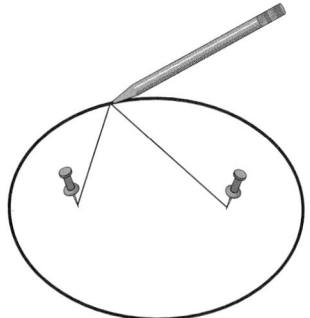

FIGURE 8.30

Exploration

Drawing an Ellipse   You can visualize the definition of an ellipse by imagining two thumbtacks placed at the foci, as shown in Figure 8.30. If the ends of a fixed length of string are fastened to the thumbtacks and the string is drawn taut with a pencil, the path traced by the pencil will be an ellipse. Try doing this. Vary the length of the string and the distance between the thumbtacks. Explain how to obtain ellipses that are almost circular. Explain how to obtain ellipses that are long and narrow.

The standard form of the equation of an ellipse takes one of two forms, depending on whether the major axis is horizontal or vertical.

Standard Equation of an Ellipse (Center at Origin)

The **standard form of the equation of an ellipse centered at the origin** with major and minor axes of lengths $2a$ and $2b$ is

$$\frac{x^2}{a^2} + \frac{y^2}{b^2} = 1 \qquad \text{or} \qquad \frac{x^2}{b^2} + \frac{y^2}{a^2} = 1, \qquad 0 < b < a.$$

The vertices lie on the major axis, a units from the center, and the co-vertices lie on the minor axis, b units from the center, as shown in Figure 8.31.

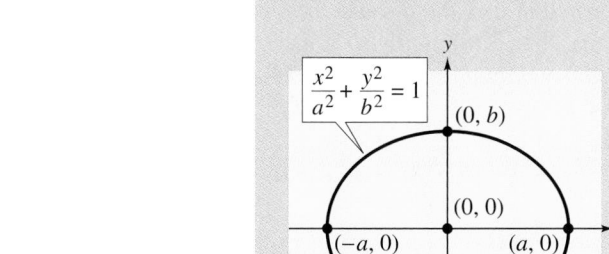 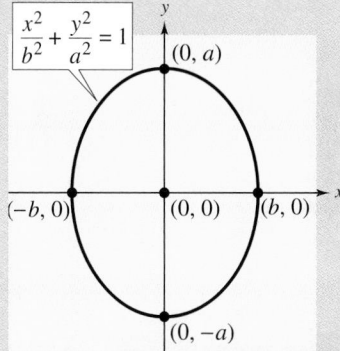

Major axis is horizontal.
Minor axis is vertical.

Major axis is vertical.
Minor axis is horizontal.

FIGURE 8.31

Mention that the major axis is the longer axis.

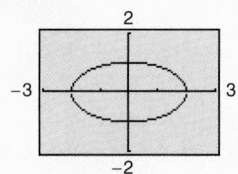

Technology

You can use a graphing utility to graph an ellipse by graphing the upper and lower portions in the same viewing rectangle. For instance, to graph the ellipse $x^2 + 4y^2 = 4$, first solve for y to get

$$y_1 = \tfrac{1}{2}\sqrt{4 - x^2} \qquad \text{and} \qquad y_2 = -\tfrac{1}{2}\sqrt{4 - x^2}.$$

Use a viewing rectangle in which $-3 \le x \le 3$ and $-2 \le y \le 2$. You should obtain the graph shown at the left.

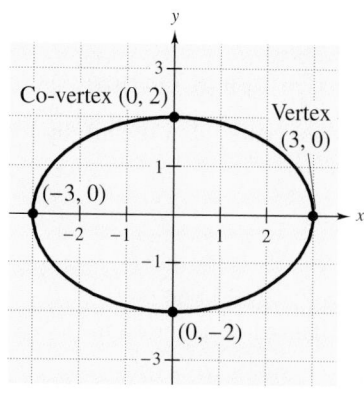

FIGURE 8.32

EXAMPLE 1   Finding the Standard Equation of an Ellipse

Find the standard form of the equation of the ellipse with vertices $(-3, 0)$ and $(3, 0)$ and co-vertices $(0, -2)$ and $(0, 2)$.

Solution

Begin by plotting the two vertices and the two co-vertices, as shown in Figure 8.32. The center of the ellipse is $(0, 0)$, because it is the point that lies halfway between the vertices (or halfway between the co-vertices). From the figure, you can see that the major axis is horizontal. Thus, the equation of the ellipse has the form

$$\frac{x^2}{a^2} + \frac{y^2}{b^2} = 1.$$

In this equation, a is the distance between the center and either vertex, which implies that $a = 3$. Similarly, b is the distance between the center and either co-vertex, which implies that $b = 2$. Thus, the standard form of the equation of the ellipse is

$$\frac{x^2}{3^2} + \frac{y^2}{2^2} = 1.$$

EXAMPLE 2   Sketching an Ellipse

Sketch the graph of the ellipse given by

$$4x^2 + y^2 = 36$$

and identify the vertices and co-vertices.

Solution

Begin by writing the equation in standard form.

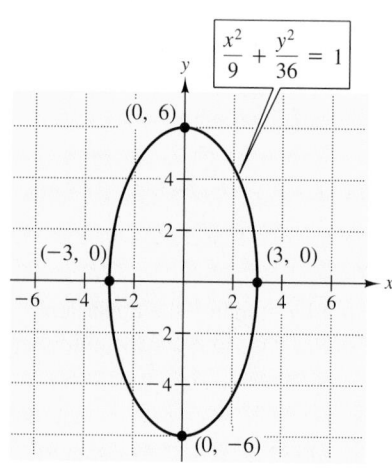

FIGURE 8.33

$4x^2 + y^2 = 36$	Original equation
$\dfrac{4x^2}{36} + \dfrac{y^2}{36} = \dfrac{36}{36}$	Divide both sides by 36.
$\dfrac{x^2}{9} + \dfrac{y^2}{36} = 1$	Simplify.
$\dfrac{x^2}{3^2} + \dfrac{y^2}{6^2} = 1$	Standard form

Because the denominator of the y^2 term is larger than the denominator of the x^2 term, you can conclude that the major axis is vertical. Moreover, because $a = 6$, the vertices are $(0, -6)$ and $(0, 6)$. Finally, because $b = 3$, the co-vertices are $(-3, 0)$ and $(3, 0)$, as shown in Figure 8.33.

Ellipses Centered at (h, k)

Standard Equation of an Ellipse [Center at (h, k)]

The **standard form of the equation of an ellipse centered at (h, k)** with major and minor axes of lengths $2a$ and $2b$, where $0 < b < a$, is

$$\frac{(x - h)^2}{a^2} + \frac{(y - k)^2}{b^2} = 1 \qquad \text{Major axis is horizontal.}$$

or

$$\frac{(x - h)^2}{b^2} + \frac{(y - k)^2}{a^2} = 1. \qquad \text{Major axis is vertical.}$$

The foci lie on the major axis, c units from the center, with $c^2 = a^2 - b^2$.

Figure 8.34 shows both the horizontal and vertical orientations for an ellipse.

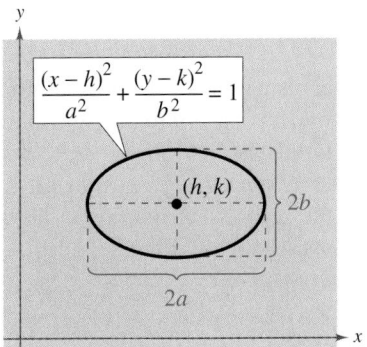 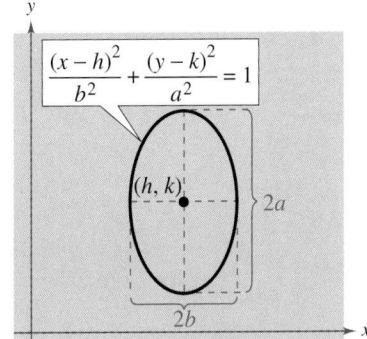

FIGURE 8.34

EXAMPLE 3 *Finding the Standard Equation of an Ellipse*

Find the standard form of the equation of the ellipse with vertices $(-2, 2)$ and $(4, 2)$ and co-vertices $(1, 3)$ and $(1, 1)$, as shown in Figure 8.35.

Solution

Because the vertices are $(-2, 2)$ and $(4, 2)$, the center of the ellipse is $(h, k) = (1, 2)$. The distance from the center to either vertex is $a = 3$, and the distance to either co-vertex is $b = 1$. Because the major axis is horizontal, the standard form of the equation is

$$\frac{(x - 1)^2}{9} + \frac{(y - 2)^2}{1} = 1.$$

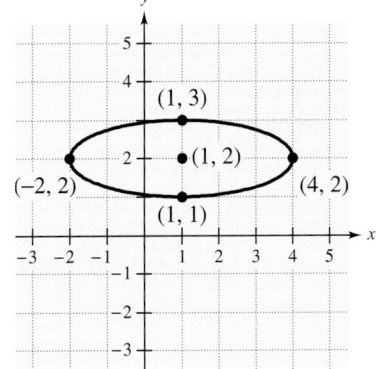

FIGURE 8.35

EXAMPLE 4   Sketching an Ellipse

Sketch the graph of the ellipse given by

$$4x^2 + y^2 - 8x + 6y + 9 = 0.$$

Solution

Begin by writing the equation in standard form. In the fourth step, note that 9 and 4 are added to *both* sides of the equation.

$$4x^2 + y^2 - 8x + 6y + 9 = 0 \qquad \text{Original equation}$$

$$(4x^2 - 8x +) + (y^2 + 6y +) = -9 \qquad \text{Group terms.}$$

$$4(x^2 - 2x +) + (y^2 + 6y +) = -9 \qquad \text{Factor 4 out of } x\text{-terms.}$$

$$4(x^2 - 2x + 1) + (y^2 + 6y + 9) = -9 + 4(1) + 9$$

$$4(x - 1)^2 + (y + 3)^2 = 4 \qquad \text{Completed square form}$$

$$\frac{(x - 1)^2}{1} + \frac{(y + 3)^2}{4} = 1 \qquad \text{Standard form}$$

Now you can see that the center occurs at

$$(h, k) = (1, -3). \qquad \text{Center of ellipse}$$

Because the denominator of the *x*-term is $b^2 = 1^2$, you can locate the endpoints of the minor axis one unit to the right and left of the center. Similarly, because the denominator of the *y*-term is $a^2 = 2^2$, you can locate the endpoints of the major axis two units up and down from the center, as shown in Figure 8.36(a). To complete the graph, sketch an oval shape that is determined by the vertices and co-vertices, as shown in Figure 8.36(b).

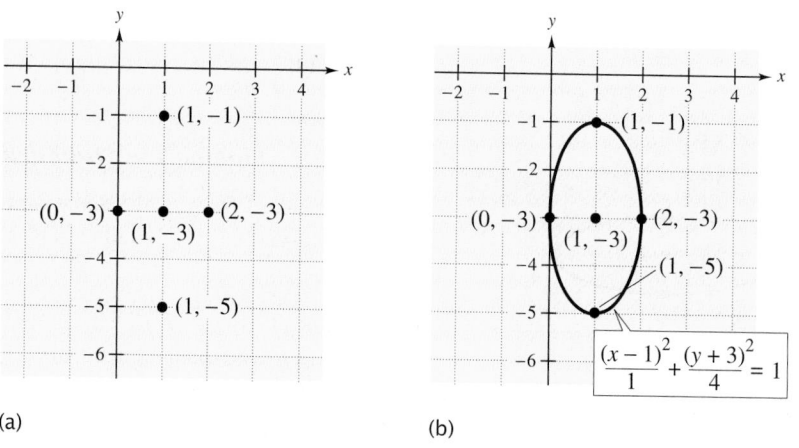

(a)   (b)

FIGURE 8.36

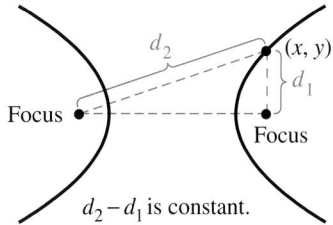

FIGURE 8.37

Hyperbolas Centered at the Origin

The third basic type of conic is called a **hyperbola.** A hyperbola on the rectangular coordinate system consists of all points (x, y) such that the *difference* of the distances between (x, y) and two fixed points is a constant, as shown in Figure 8.37. The two fixed points are called the **foci** of the hyperbola. The line on which the foci lie is called the **transverse axis** of the hyperbola.

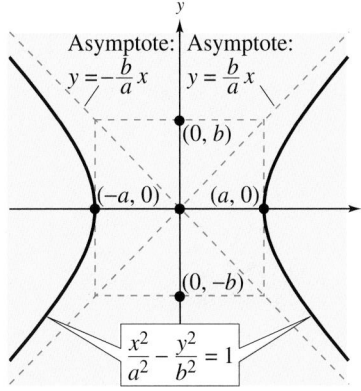

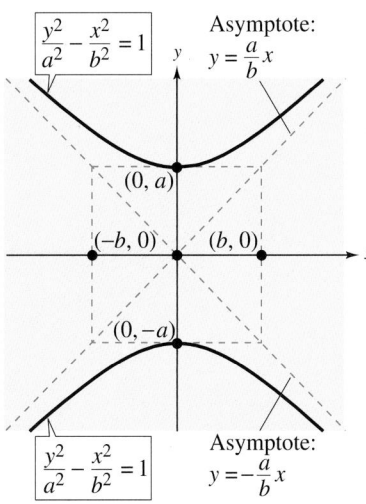

FIGURE 8.39

Standard Equation of a Hyperbola (Center at Origin)

The **standard form of the equation of a hyperbola centered at the origin** is

$$\frac{x^2}{a^2} - \frac{y^2}{b^2} = 1 \qquad \text{Transverse axis is horizontal.}$$

or

$$\frac{y^2}{a^2} - \frac{x^2}{b^2} = 1, \qquad \text{Transverse axis is vertical.}$$

where a and b are positive real numbers. The **vertices** of the hyperbola lie on the transverse axis, a units from the center, as shown in Figure 8.38.

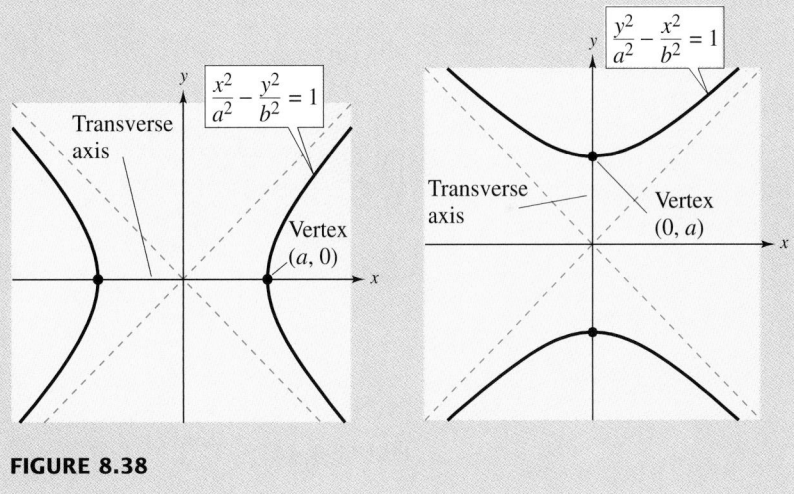

FIGURE 8.38

A hyperbola has two disconnected parts, each of which is called a **branch** of the hyperbola. The two branches approach a pair of intersecting straight lines called the **asymptotes** of the hyperbola. The two asymptotes intersect at the center of the hyperbola. To sketch a hyperbola, form a **central rectangle** that is centered at the origin and has side lengths of $2a$ and $2b$. Note in Figure 8.39 that the asymptotes pass through the corners of the central rectangle and that the vertices of the hyperbola lie at the centers of opposite sides of the central rectangle.

EXAMPLE 5   Sketching a Hyperbola

Identify the vertices of the hyperbola given by the following equation, and sketch the hyperbola.

$$\frac{x^2}{36} - \frac{y^2}{16} = 1$$

Ask students how they can determine from an equation written in standard form whether a graph of the equation is a circle, ellipse, or hyperbola. Have them identify the graph of each of the following equations as one of these conics.

(a) $7x^2 + 7y^2 = 63$

(b) $5y^2 - x^2 = 20$

(c) $-x^2 - 9y^2 = -9$

Answers: (a) circle (b) hyperbola (c) ellipse

Solution

From the standard form of the equation

$$\frac{x^2}{6^2} - \frac{y^2}{4^2} = 1,$$

you can see that the center of the hyperbola is the origin and the transverse axis is horizontal. Therefore, the vertices lie six units to the left and right of the center at the points $(-6, 0)$ and $(6, 0)$. Because $a = 6$ and $b = 4$, you can sketch the hyperbola by first drawing a central rectangle with a width of $2a = 12$ and a height of $2b = 8$. Next, draw the asymptotes of the hyperbola through the corners of the central rectangle and plot the vertices, as shown in Figure 8.40(a). Finally, draw the hyperbola, as shown in Figure 8.40(b).

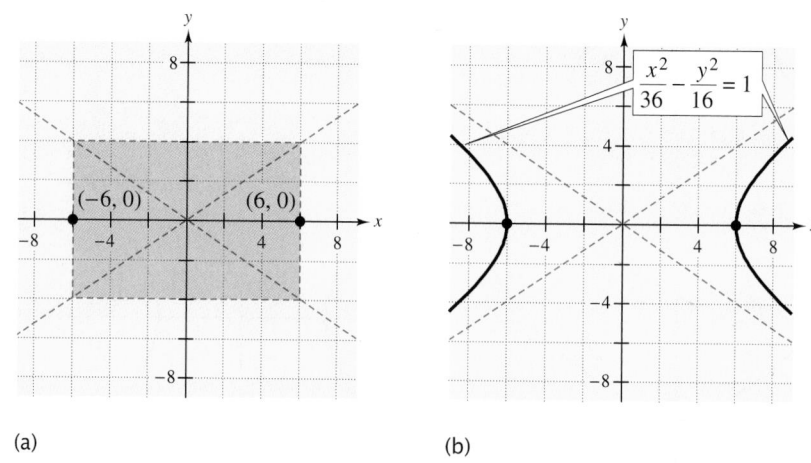

(a)   (b)

FIGURE 8.40

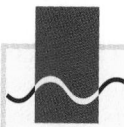 *Technology*

Try using a graphing utility to graph the hyperbola shown in Example 5. To do this, solve the equation for y to obtain two functions of y. Then graph both functions in the same viewing rectangle.

EXAMPLE 6   *Sketching a Hyperbola*

Identify the vertices of the hyperbola given by the following equation, and sketch the hyperbola.

$$4y^2 - 9x^2 = 36$$

Solution

To begin, write the equation in standard form.

$4y^2 - 9x^2 = 36$	Original equation
$\dfrac{4y^2}{36} - \dfrac{9x^2}{36} = \dfrac{36}{36}$	Divide both sides by 36.
$\dfrac{y^2}{9} - \dfrac{x^2}{4} = 1$	Simplify.
$\dfrac{y^2}{3^2} - \dfrac{x^2}{2^2} = 1$	Standard form

Now, from the standard form, you can see that the center of the hyperbola is the origin and the transverse axis is vertical. Thus, the vertices lie three units above and below the center at $(0, 3)$ and $(0, -3)$. Because $a = 3$ and $b = 2$, you can sketch the hyperbola by first drawing a central rectangle with a width of $2b = 4$ and a height of $2a = 6$. Next, draw the asymptotes of the hyperbola through the corners of the central rectangle and plot the vertices, as shown in Figure 8.41(a). Finally, draw the hyperbola, as shown in Figure 8.41(b).

To help students determine whether a transverse axis is horizontal or vertical, have them write the following equations in standard form and then identify the transverse axis:

(a) $3x^2 - 2y^2 = 12$

(b) $8y^2 - 5x^2 = 40$

(c) $6x^2 - 24y^2 = -24$

Answers:

(a) $\dfrac{x^2}{4} - \dfrac{y^2}{6} = 1$ Horizontal

(b) $\dfrac{y^2}{5} - \dfrac{x^2}{8} = 1$ Vertical

(c) $\dfrac{y^2}{1} - \dfrac{x^2}{4} = 1$ Vertical

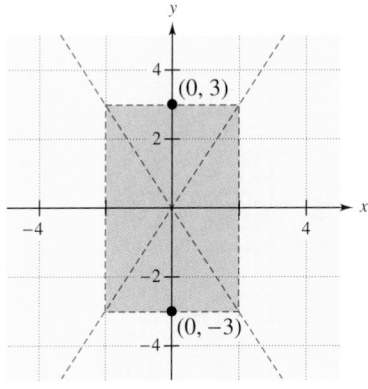

(a)

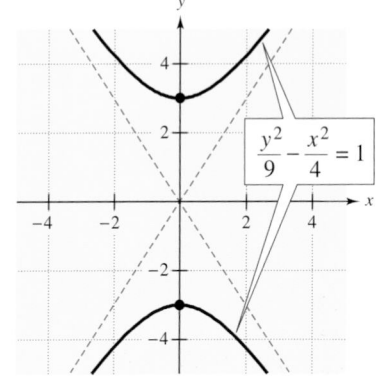

(b)

FIGURE 8.41

Finding the equation of a hyperbola is a little more difficult than finding the equation of one of the other types of conics. However, if you know the vertices and the asymptotes, you can find the values of a and b, which enable you to write the equation. Notice in Example 7 that the key to this procedure is knowing that the central rectangle has a width of $2b$ and a height of $2a$.

EXAMPLE 7   *Finding the Standard Equation of a Hyperbola*

Find the standard form of the equation of the hyperbola with a vertical transverse axis and vertices $(0, 3)$ and $(0, -3)$. The asymptotes of the hyperbola are $y = \frac{3}{5}x$ and $y = -\frac{3}{5}x$.

Solution

To begin, sketch the lines that represent the asymptotes, as shown in Figure 8.42(a). Note that these two lines intersect at the origin, which implies that the center of the hyperbola is $(0, 0)$. Next, plot the two vertices at the points $(0, 3)$ and $(0, -3)$. Because you know where the vertices are located, you can sketch the central rectangle of the hyperbola, as shown in Figure 8.42(a). Note that the corners of the central rectangle occur at the points $(-5, 3)$, $(5, 3)$, $(-5, -3)$, and $(5, -3)$. Because the width of the central rectangle is $2b = 10$, it follows that $b = 5$. Similarly, because the height of the central rectangle is $2a = 6$, it follows that $a = 3$. Thus, you can conclude that the standard form of the equation of the hyperbola is

$$\frac{y^2}{3^2} - \frac{x^2}{5^2} = 1, \qquad \text{or} \qquad \frac{y^2}{9} - \frac{x^2}{25} = 1.$$

The graph is shown in Figure 8.42(b).

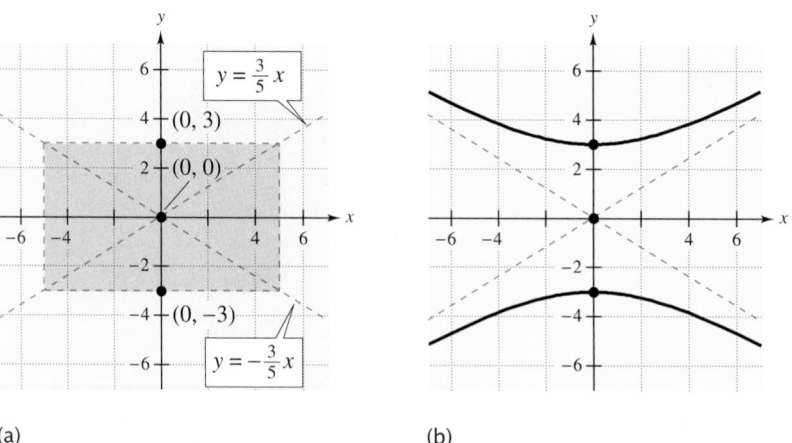

(a)   (b)

FIGURE 8.42

Hyperbolas Centered at (h, k)

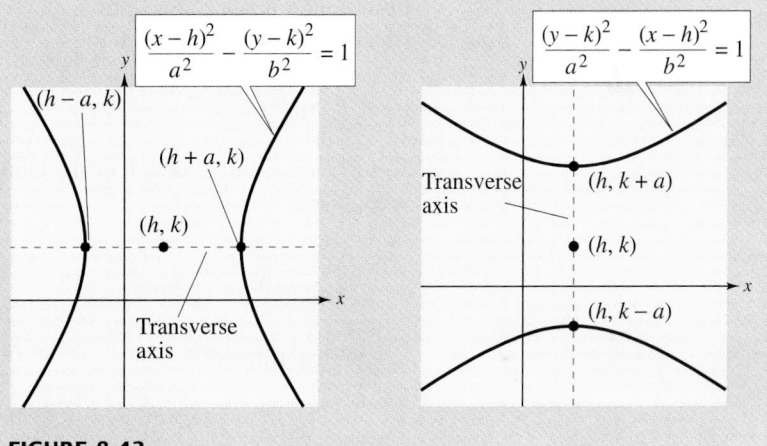

Standard Equation of a Hyperbola [Center at (h, k)]

The **standard form of the equation of a hyperbola centered at (h, k)** is

$$\frac{(x - h)^2}{a^2} - \frac{(y - k)^2}{b^2} = 1 \qquad \text{Transverse axis is horizontal.}$$

or

$$\frac{(y - k)^2}{a^2} - \frac{(x - h)^2}{b^2} = 1, \qquad \text{Transverse axis is vertical.}$$

where a and b are positive real numbers. The vertices lie on the transverse axis, a units from the center, as shown in Figure 8.43.

FIGURE 8.43

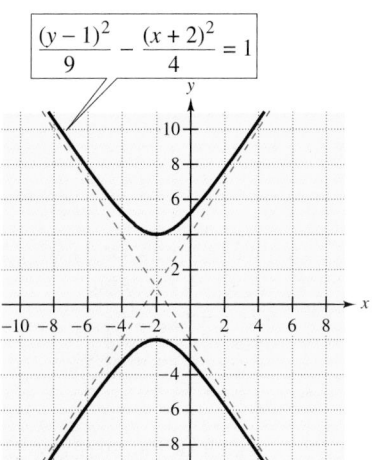

FIGURE 8.44

EXAMPLE 8 Sketching a Hyperbola

Sketch the graph of the hyperbola given by $\dfrac{(y - 1)^2}{9} - \dfrac{(x + 2)^2}{4} = 1$.

Solution

From the form of the equation, you can see that the transverse axis is vertical. The center of the hyperbola is $(h, k) = (-2, 1)$. Because $a = 3$ and $b = 2$, you can begin by sketching a central rectangle that is six units high and four units wide, centered at $(-2, 1)$. Then, sketch the asymptotes by drawing lines through the corners of the central rectangle. Sketch the hyperbola, as shown in Figure 8.44.

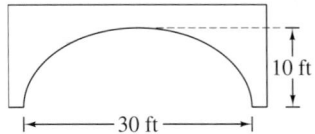

FIGURE 8.45

Application

EXAMPLE 9 An Application Involving an Ellipse

You are responsible for designing a semielliptical archway, as shown in Figure 8.45. The height of the archway is 10 feet, and the width is 30 feet. Find an equation of the ellipse and use the equation to sketch an accurate drawing of the archway.

Solution

To make the equation simple, place the origin at the center of the ellipse. This means that the standard form of the equation is

$$\frac{x^2}{a^2} + \frac{y^2}{b^2} = 1.$$

Because the major axis is horizontal, you can see that the $a = 15$ and $b = 10$, which implies that the equation is

$$\frac{x^2}{15^2} + \frac{y^2}{10^2} = 1.$$

To make an accurate sketch of the ellipse, it is helpful to solve this equation for y, as follows.

$$\frac{x^2}{15^2} + \frac{y^2}{10^2} = 1$$

$$\frac{x^2}{225} + \frac{y^2}{100} = 1$$

$$\frac{y^2}{100} = 1 - \frac{x^2}{225}$$

$$y^2 = 100\left(1 - \frac{x^2}{225}\right)$$

$$y = 10\sqrt{1 - \frac{x^2}{225}}$$

(Note that you use the positive square root because the y-values must be positive.) Finally, calculate several y-values for the archway, as shown in the table.

x-value	± 15	± 12.5	± 10	± 7.5	± 5	± 2.5	0
y-value	0	5.53	7.45	8.66	9.43	9.86	10

Using the values shown in the table, you can make a sketch of the archway, as shown in Figure 8.46.

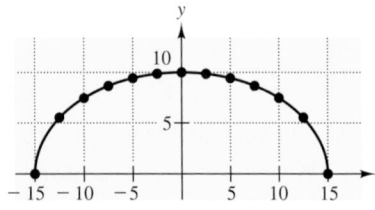

FIGURE 8.46

Group Activities | **Sketching a Hyperbola**

In this section, the only types of hyperbolas that were discussed were those with a horizontal or a vertical transverse axis. There are other types of hyperbolas. For instance, the graph of the rational function

$$f(x) = \frac{1}{x}$$

is a hyperbola. Sketch the graph of this hyperbola. Then label both its vertices and its asymptotes.

8.6 Exercises

See Warm-Up Exercises, p. A48

In Exercises 1–6, match the equation with its graph. [The graphs are labeled (a), (b), (c), (d), (e), and (f).]

(a)
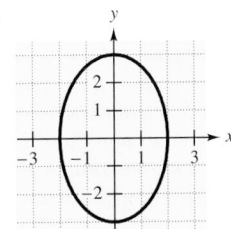

(b)
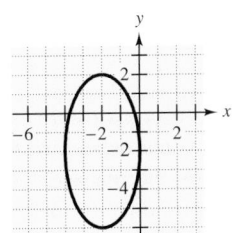

(c)
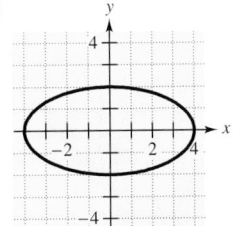

(d)
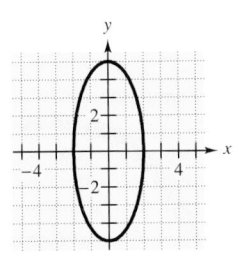

(e)
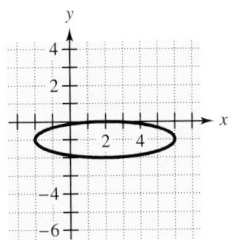

(f)
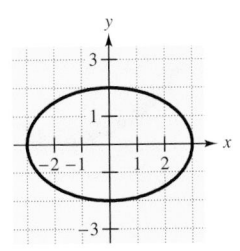

1. $\dfrac{x^2}{4} + \dfrac{y^2}{9} = 1$

2. $\dfrac{x^2}{9} + \dfrac{y^2}{4} = 1$

3. $\dfrac{x^2}{4} + \dfrac{y^2}{25} = 1$

4. $\dfrac{y^2}{4} + \dfrac{x^2}{16} = 1$

5. $\dfrac{(x-2)^2}{16} + \dfrac{(y+1)^2}{1} = 1$

6. $\dfrac{(x+2)^2}{4} + \dfrac{(y+2)^2}{16} = 1$

In Exercises 7–14, sketch the ellipse. Identify its vertices and co-vertices.

7. $\dfrac{x^2}{16} + \dfrac{y^2}{4} = 1$

8. $\dfrac{x^2}{25} + \dfrac{y^2}{9} = 1$

9. $\dfrac{x^2}{4} + \dfrac{y^2}{16} = 1$

10. $\dfrac{x^2}{9} + \dfrac{y^2}{25} = 1$

11. $\dfrac{x^2}{25/9} + \dfrac{y^2}{16/9} = 1$

12. $\dfrac{x^2}{1} + \dfrac{y^2}{1/4} = 1$

13. $4x^2 + y^2 - 4 = 0$

14. $4x^2 + 9y^2 - 36 = 0$

In Exercises 15–18, use a graphing utility to graph the equation. Identify the vertices. (*Note:* Solve for *y*.)

15. $x^2 + 2y^2 = 4$

16. $9x^2 + y^2 = 64$

17. $3x^2 + y^2 - 12 = 0$

18. $5x^2 + 2y^2 - 10 = 0$

In Exercises 19–30, write the standard form of the equation of the ellipse centered at the origin.

Vertices	*Co-vertices*
19. $(-4, 0), (4, 0)$	$(0, -3), (0, 3)$
20. $(-4, 0), (4, 0)$	$(0, -1), (0, 1)$
21. $(-2, 0), (2, 0)$	$(0, -1), (0, 1)$
22. $(-10, 0), (10, 0)$	$(0, -4), (0, 4)$
23. $(0, -4), (0, 4)$	$(-3, 0), (3, 0)$
24. $(0, -5), (0, 5)$	$(-1, 0), (1, 0)$
25. $(0, -2), (0, 2)$	$(-1, 0), (1, 0)$
26. $(0, -8), (0, 8)$	$(-4, 0), (4, 0)$

27. Major axis (vertical) 10 units, minor axis 6 units

28. Major axis (horizontal) 24 units, minor axis 10 units

29. Major axis (horizontal) 20 units, minor axis 12 units

30. Major axis (horizontal) 50 units, minor axis 30 units

In Exercises 31–36, find the center and vertices of the ellipse and sketch its graph. Use a graphing utility to verify the graph.

31. $\dfrac{(x + 5)^2}{16} + y^2 = 1$ **32.** $\dfrac{(x - 2)^2}{4} + \dfrac{(y - 3)^2}{9} = 1$

33. $\dfrac{(x - 1)^2}{9} + \dfrac{(y - 5)^2}{25} = 1$

34. $\dfrac{(x + 2)^2}{1/4} + \dfrac{(y + 4)^2}{1} = 1$

35. $9x^2 + 4y^2 + 36x - 24y + 36 = 0$

36. $9x^2 + 4y^2 - 36x + 8y + 31 = 0$

In Exercises 37–40, find an equation of the ellipse.

37.

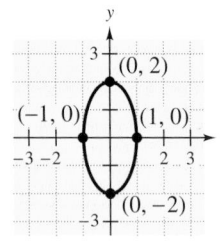

38.

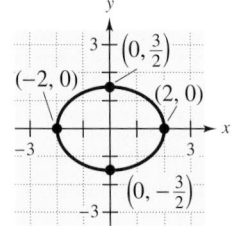

39.

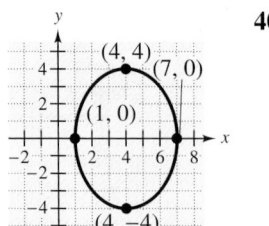

40.

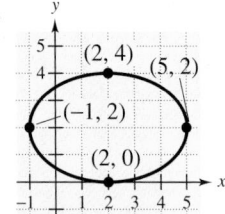

41. *Height of an Arch* A semielliptical arch for a tunnel under a river has a width of 100 feet and a height of 40 feet (see figure). Determine the height of the arch 5 feet from the edge of the tunnel.

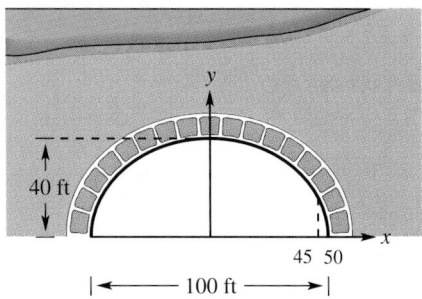

42. *Bicycle Chainwheel* The pedals of a bicycle drive a chainwheel, which drives a smaller sprocket wheel on the rear axle (see figure). Many chainwheels are circular. Some, however, are slightly elliptical, which tends to make pedaling easier. Find an equation of an elliptical chainwheel that is 8 inches at its widest point and $7\frac{1}{2}$ inches at its narrowest point.

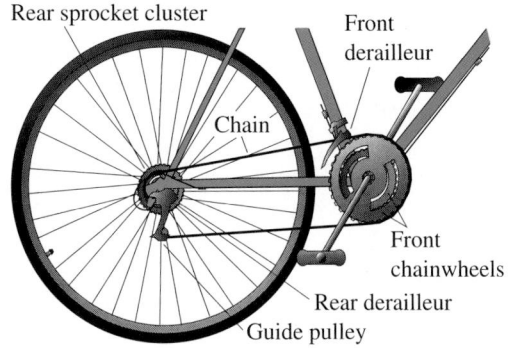

In Exercises 43–48, match the equation with its graph. [The graphs are labeled (a), (b), (c), (d), (e), and (f).]

(a)
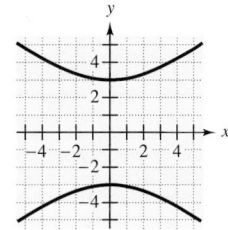

(b)
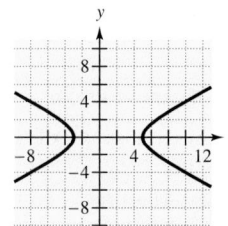

(c)
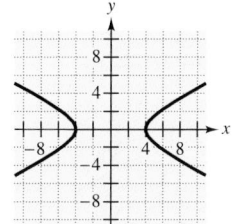

(d)
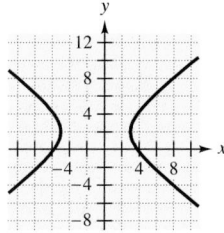

(e)
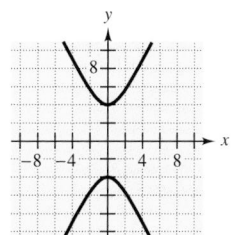

(f)
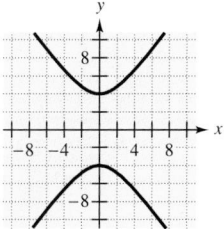

43. $\dfrac{x^2}{16} - \dfrac{y^2}{4} = 1$

44. $\dfrac{y^2}{16} - \dfrac{x^2}{4} = 1$

45. $\dfrac{y^2}{9} - \dfrac{x^2}{16} = 1$

46. $\dfrac{y^2}{16} - \dfrac{x^2}{9} = 1$

47. $\dfrac{(x-1)^2}{16} - \dfrac{y^2}{4} = 1$

48. $\dfrac{(x+1)^2}{16} - \dfrac{(y-2)^2}{9} = 1$

In Exercises 49–60, sketch the hyperbola. Identify its vertices and asymptotes.

49. $x^2 - y^2 = 9$

50. $x^2 - y^2 = 1$

51. $y^2 - x^2 = 9$

52. $y^2 - x^2 = 1$

53. $\dfrac{x^2}{9} - \dfrac{y^2}{25} = 1$

54. $\dfrac{x^2}{4} - \dfrac{y^2}{9} = 1$

55. $\dfrac{y^2}{9} - \dfrac{x^2}{25} = 1$

56. $\dfrac{y^2}{4} - \dfrac{x^2}{9} = 1$

57. $\dfrac{x^2}{1} - \dfrac{y^2}{9/4} = 1$

58. $\dfrac{y^2}{1/4} - \dfrac{x^2}{25/4} = 1$

59. $4y^2 - x^2 + 16 = 0$

60. $4y^2 - 9x^2 - 36 = 0$

In Exercises 61–68, write the standard form of the equation of the hyperbola centered at the origin.

Vertices	Asymptotes	
61. $(-4, 0), (4, 0)$	$y = 2x$	$y = -2x$
62. $(-2, 0), (2, 0)$	$y = \frac{1}{3}x$	$y = -\frac{1}{3}x$
63. $(0, -4), (0, 4)$	$y = \frac{1}{2}x$	$y = -\frac{1}{2}x$
64. $(0, -2), (0, 2)$	$y = 3x$	$y = -3x$
65. $(-9, 0), (9, 0)$	$y = \frac{2}{3}x$	$y = -\frac{2}{3}x$
66. $(-1, 0), (1, 0)$	$y = \frac{1}{2}x$	$y = -\frac{1}{2}x$
67. $(0, -1), (0, 1)$	$y = 2x$	$y = -2x$
68. $(0, -5), (0, 5)$	$y = x$	$y = -x$

In Exercises 69–74, identify the center and vertices of the hyperbola and sketch its graph. Use a graphing utility to verify the graph.

69. $(y + 4)^2 - (x - 3)^2 = 25$

70. $(y + 6)^2 - (x - 2)^2 = 1$

71. $\dfrac{(x-1)^2}{4} - \dfrac{(y+2)^2}{1} = 1$

72. $\dfrac{(x-2)^2}{4} - \dfrac{(y-3)^2}{9} = 1$

73. $9x^2 - y^2 - 36x - 6y + 18 = 0$

74. $x^2 - 9y^2 + 36y - 72 = 0$

In Exercises 75–78, find an equation of the hyperbola.

75.

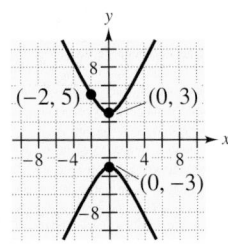

76.

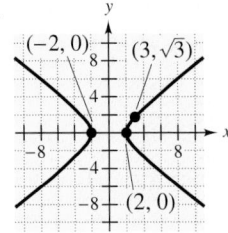

77.

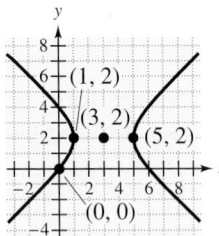

78.

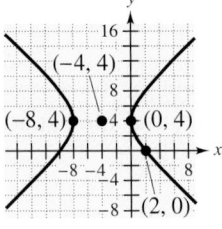

Think About It   In Exercises 79 and 80, describe the part of the hyperbola

$$\frac{(x-3)^2}{4} - \frac{(y-1)^2}{9} = 1$$

given by the equation.

79. $x = 3 - \frac{2}{3}\sqrt{9 + (y-1)^2}$

80. $y = 1 + \frac{3}{2}\sqrt{(x-3)^2 - 4}$

Section Project

(a) Sketch a graph of the ellipse that consists of all points (x, y) such that the sum of the distances between (x, y) and two fixed points is 15 units and the foci are located at the centers of the two sets of concentric circles in the figure.

(b) Sketch a graph of the hyperbola that consists of all points (x, y) such that the difference of the distances between (x, y) and two fixed points is eight units and the foci are located at the centers of the two sets of concentric circles in the figure.

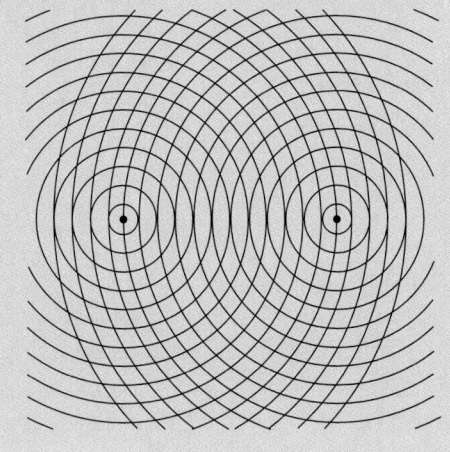

8.7 | Parabolas

Definition of a Parabola ▪ Equations of Parabolas ▪ Application

Definition of a Parabola

The fourth basic type of conic is a **parabola.** In Section 6.5, you studied some of the properties of parabolas. There, you saw that the graph of a quadratic function of the form $y = ax^2 + bx + c$ is a parabola, which opens upward if a is positive and downward if a is negative. You also learned that each parabola has a vertex and that the vertex of the graph of $y = ax^2 + bx + c$ occurs when $x = -b/2a$.

In this section, you will study the technical definition of a parabola, and you will study the equations of parabolas that open to the right and to the left.

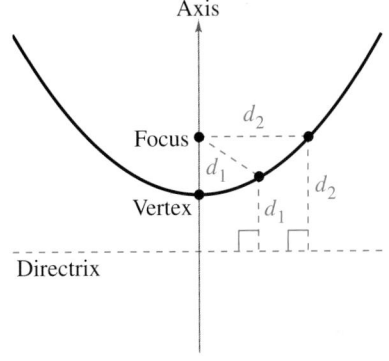

FIGURE 8.47

Definition of a Parabola

A **parabola** is the set of all points (x, y) that are equidistant from a fixed line (**directrix**) and a fixed point (**focus**) not on the line.

NOTE If the focus is above or to the right of the vertex, p is positive. If the focus is below or to the left of the vertex, p is negative.

The midpoint between the focus and the directrix is called the **vertex,** and the line passing through the focus and the vertex is called the **axis** of the parabola. Note in Figure 8.47 that a parabola is symmetric with respect to its axis. Using the definition of a parabola, you can derive the following **standard form of the equation of a parabola** whose directrix is parallel to the x-axis or to the y-axis.

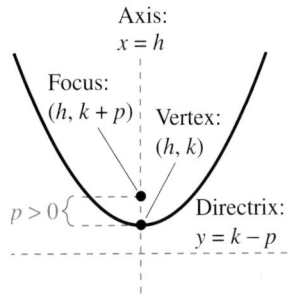

FIGURE 8.48

Standard Equation of a Parabola

The **standard form of the equation of a parabola** with vertex at the origin $(0, 0)$ is

$$x^2 = 4py, \quad p \neq 0 \qquad \text{Vertical axis}$$
$$y^2 = 4px, \quad p \neq 0. \qquad \text{Horizontal axis}$$

The focus lies on the axis p units (*directed distance*) from the vertex. If the vertex is at (h, k), then the standard form of the equation is

$$(x - h)^2 = 4p(y - k), \quad p \neq 0 \qquad \text{Vertical axis; directrix: } y = k - p$$
$$(y - k)^2 = 4p(x - h), \quad p \neq 0. \qquad \text{Horizontal axis; directrix: } x = h - p$$

(See Figure 8.48.)

Equations of Parabolas

EXAMPLE 1   Finding the Standard Equation of a Parabola

Find the standard form of the equation of the parabola with vertex $(0, 0)$ and focus $(0, -2)$, as shown in Figure 8.49.

Solution

Because the vertex is at the origin and the axis of the parabola is vertical, consider the equation

$$x^2 = 4py,$$

where p is the directed distance from the vertex to the focus. Because the focus is two units *below* the vertex, you have $p = -2$. Thus, the equation of the parabola is

$$x^2 = 4py$$
$$x^2 = 4(-2)y$$
$$x^2 = -8y.$$

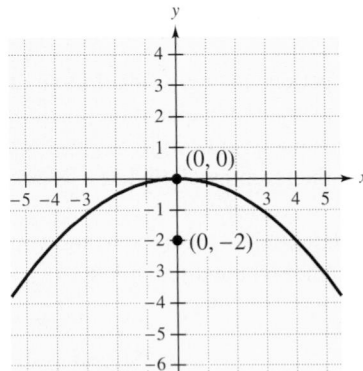

FIGURE 8.49

EXAMPLE 2   Finding the Standard Equation of a Parabola

Find the standard form of the equation of the parabola with vertex $(3, -2)$ and focus $(4, -2)$, as shown in Figure 8.50.

Solution

Because the vertex is at $(h, k) = (3, -2)$ and the axis of the parabola is horizontal, consider the equation

$$(y - k)^2 = 4p(x - h),$$

where $h = 3$, $k = -2$, and $p = 1$. Thus, the equation of the parabola is

$$(y - k)^2 = 4p(x - h)$$
$$[y - (-2)]^2 = 4(1)(x - 3)$$
$$(y + 2)^2 = 4(x - 3).$$

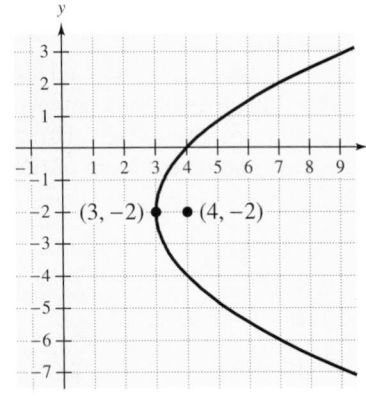

FIGURE 8.50

 *Technology*

Try using a graphing utility to graph the parabola shown in Example 2. To do this, solve the equation for y to obtain two functions of y. Then graph both functions in the same viewing rectangle.

EXAMPLE 3 *Analyzing a Parabola*

Sketch the graph of the parabola $y = 2x^2$ and identify its vertex and focus.

Solution

Because the equation can be written in the standard form $x^2 = 4py$, it is a parabola whose vertex is at the origin. You can identify the focus of the parabola by writing its equation in standard form.

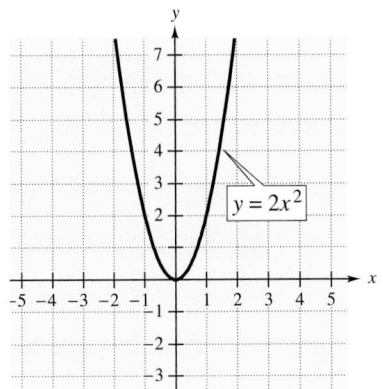

FIGURE 8.51

$$y = 2x^2 \qquad \text{Original equation}$$
$$2x^2 = y \qquad \text{Interchange sides of the equation.}$$
$$x^2 = \tfrac{1}{2}y \qquad \text{Divide both sides by 2.}$$
$$x^2 = 4\left(\tfrac{1}{8}\right)y \qquad \text{Rewrite } \tfrac{1}{2} \text{ in the form } 4p.$$

From this standard form, you can see that $p = \tfrac{1}{8}$. Because the parabola opens upward, as shown in Figure 8.51, you can conclude that the focus lies $p = \tfrac{1}{8}$ unit above the vertex. Thus, the focus is $\left(0, \tfrac{1}{8}\right)$.

EXAMPLE 4 *Analyzing a Parabola*

Sketch the graph of the parabola $x = 2y^2 - 4y + 10$ and identify its vertex and focus.

Solution

This equation can be written in the standard form $(y - k)^2 = 4p(x - h)$. To do this, you can complete the square, as follows.

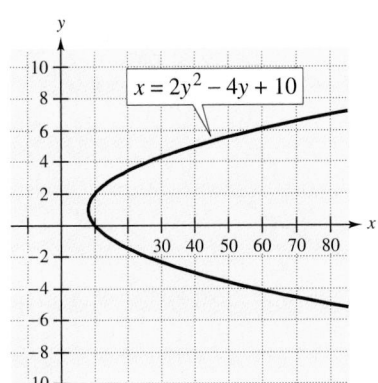

FIGURE 8.52

$$x = 2y^2 - 4y + 10 \qquad \text{Original equation}$$
$$2y^2 - 4y + 10 = x \qquad \text{Interchange sides of the equation.}$$
$$y^2 - 2y + 5 = \tfrac{1}{2}x \qquad \text{Divide both sides by 2.}$$
$$y^2 - 2y = \tfrac{1}{2}x - 5 \qquad \text{Subtract 5 from both sides.}$$
$$y^2 - 2y + 1 = \tfrac{1}{2}x - 5 + 1 \qquad \text{Complete the square on left side.}$$
$$(y - 1)^2 = \tfrac{1}{2}x - 4 \qquad \text{Simplify.}$$
$$(y - 1)^2 = \tfrac{1}{2}(x - 8) \qquad \text{Factor.}$$
$$(y - 1)^2 = 4\left(\tfrac{1}{8}\right)(x - 8) \qquad \text{Rewrite } \tfrac{1}{2} \text{ in the form } 4p.$$

From this standard form, you can see that the vertex is $(h, k) = (8, 1)$ and $p = \tfrac{1}{8}$. Because the parabola opens to the right, as shown in Figure 8.52, you can conclude that the focus lies $p = \tfrac{1}{8}$ unit to the right of the vertex. Thus, the focus is $\left(8\tfrac{1}{8}, 1\right)$.

Application

Parabolas occur in a wide variety of applications. For instance, a parabolic reflector can be formed by revolving a parabola around its axis. The resulting surface has the property that all incoming rays parallel to the axis are reflected through the focus of the parabola; this is the principle behind the construction of the parabolic mirrors used in reflecting telescopes. Conversely, the light rays emanating from the focus of a parabolic reflector used in a flashlight are all parallel to one another, as shown in Figure 8.53.

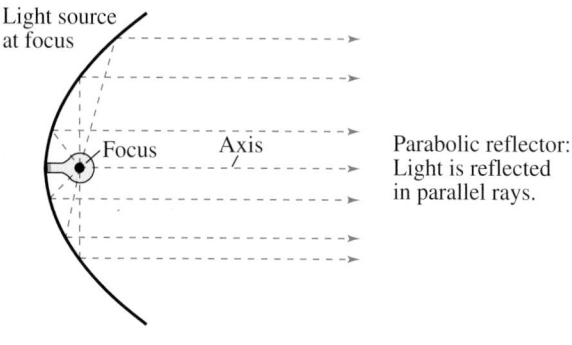

FIGURE 8.53

Group Activities

Satellite Antennas

Satellite television dishes have cross sections that are parabolic, as shown in the diagram below. Use the reflective property of parabolas described above to explain why this parabolic shape helps amplify the television signals received from the satellite.

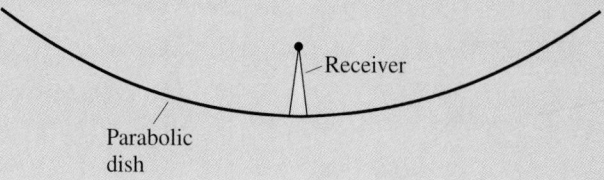

8.7 Exercises

See Warm-Up
Exercises, p. A48

In Exercises 1–6, match the equation with its graph. [The graphs are labeled (a), (b), (c), (d), (e), and (f).]

(a)
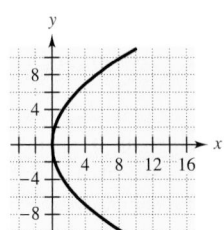

(b)
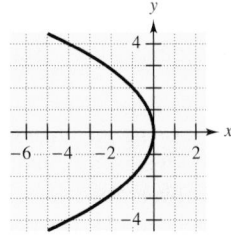

(c)
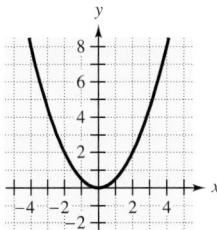

(d)
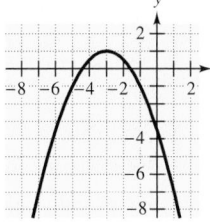

(e)
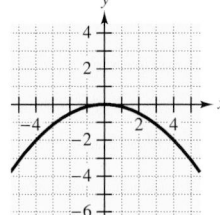

(f)
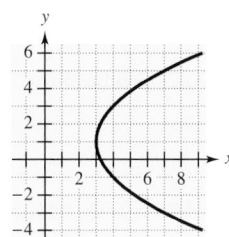

1. $y^2 = -4x$ **2.** $x^2 = 2y$

3. $x^2 = -8y$ **4.** $y^2 = 12x$

5. $(y - 1)^2 = 4(x - 3)$ **6.** $(x + 3)^2 = -2(y - 1)$

In Exercises 7–20, find the vertex and focus of the parabola and sketch its graph.

7. $y = \frac{1}{2}x^2$ **8.** $y = 2x^2$

9. $y^2 = -6x$ **10.** $y^2 = 3x$

11. $x^2 + 8y = 0$ **12.** $x + y^2 = 0$

13. $(x - 1)^2 + 8(y + 2) = 0$

14. $(x + 3) + (y - 2)^2 = 0$

15. $\left(y + \frac{1}{2}\right)^2 = 2(x - 5)$

16. $\left(x + \frac{1}{2}\right)^2 = 4(y - 3)$

17. $y = \frac{1}{4}(x^2 - 2x + 5)$

18. $4x - y^2 - 2y - 33 = 0$

19. $y^2 + 6y + 8x + 25 = 0$

20. $y^2 - 4y - 4x = 0$

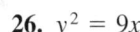 In Exercises 21–24, use a graphing utility to graph the parabola. Identify the vertex and focus.

21. $y = -\frac{1}{6}(x^2 + 4x - 2)$

22. $x^2 - 2x + 8y + 9 = 0$

23. $y^2 + x + y = 0$

24. $y^2 - 4x - 4 = 0$

In Exercises 25 and 26, change the equation so that its graph matches the given graph.

25. $y^2 = -4x$ **26.** $y^2 = 9x$

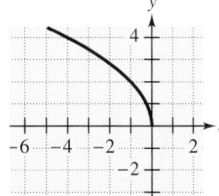

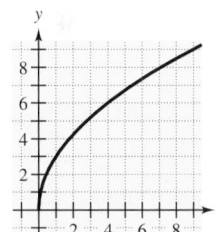

In Exercises 27 and 28, change the equation so that its graph matches the description.

27. Upper half: $(y - 3)^2 = 6(x + 1)$

28. Lower half: $(y + 1)^2 = 2(x - 2)$

In Exercises 29–40, find an equation of the parabola with its vertex at the origin.

29.

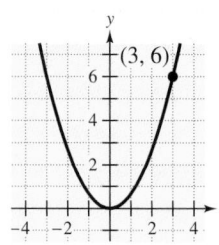

30.

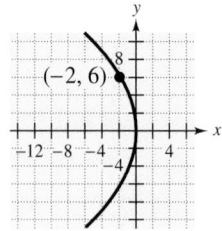

31. Focus: $\left(0, -\frac{3}{2}\right)$ **32.** Focus: $(2, 0)$

33. Focus: $(-2, 0)$ **34.** Focus: $(0, -2)$

35. Focus: $(0, 1)$ **36.** Focus: $(-3, 0)$

37. Focus: $(4, 0)$ **38.** Focus: $(2, 0)$

39. Horizontal axis and passes through the point $(4, 6)$

40. Vertical axis and passes through the point $(-2, -2)$

In Exercises 41–50, find an equation of the parabola with its vertex at (h, k).

41.

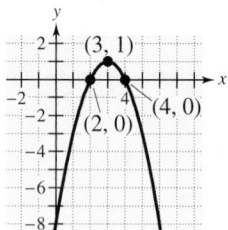

42.

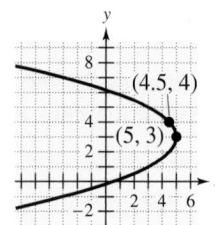

43.

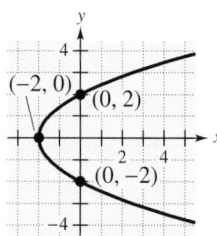

44.

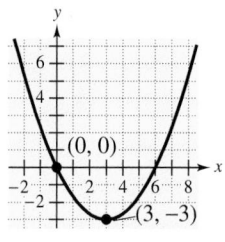

45. Vertex: $(3, 2)$; Focus: $(1, 2)$

46. Vertex: $(-1, 2)$; Focus: $(-1, 0)$

47. Vertex: $(0, 4)$; Focus: $(0, 6)$

48. Vertex: $(-2, 1)$; Focus: $(-5, 1)$

49. Vertex: $(0, 2)$;
Horizontal axis and passes through $(1, 3)$

50. Vertex: $(0, 2)$; Vertical axis and passes through $(6, 0)$

51. *Suspension Bridge*   Each cable of a suspension bridge is suspended (in the shape of a parabola) between two towers that are 120 meters apart, and the top of each tower is 20 meters above the roadway. The cables touch the roadway midway between the towers (see figure).

(a) Find an equation for the parabolic shape of each cable.

(b) Complete the table by finding the height of the suspension cables y over the roadway at a distance of x meters from the center of the bridge.

x	0	20	40	60
y				

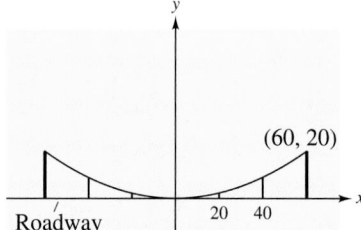

52. *Beam Deflection*   A simply supported beam is 16 meters long and has a load at the center (see figure). The deflection of the beam at its center is 3 centimeters. Assume that the shape of the deflected beam is parabolic.

(a) Find an equation of the parabola. (Assume that the origin is at the center of the deflected beam.)

(b) How far from the center of the beam is the deflection 1 centimeter?

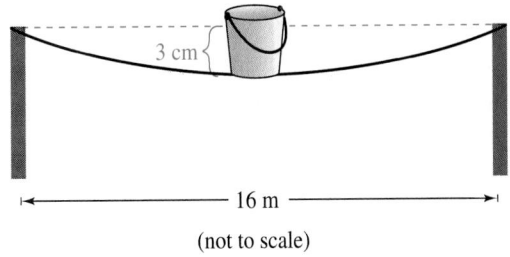

3 cm

← 16 m →

(not to scale)

53. *Revenue*   The revenue R generated by a sale of x units is given by

$$R = 375x - \frac{3}{2}x^2.$$

Use a graphing utility to graph the function and approximate the number of sales that will maximize revenue.

54. *Path of a Projectile*   The path of a softball is given by

$$y = -0.08x^2 + x + 4.$$

The coordinates x and y are measured in feet, with $x = 0$ corresponding to the position from which the ball was thrown.

(a) Use a graphing utility to graph the trajectory of the softball.

(b) Move the cursor along the path to approximate the highest point and the range of the trajectory.

55. *True or False?* y is a function of x in the equation $y^2 = 6x$. Explain.

Section Project

Investigation   Consider the parabola $x^2 = 4py$.

(a) Use a graphing utility to graph the parabola for $p = 1, p = 2, p = 3$, and $p = 4$. Describe the effect on the graph when p increases.

(b) Locate the focus for each parabola in part (a).

(c) For each parabola in part (a), find the length of the chord passing through the focus perpendicular to the axis of the parabola. How can the length of this chord be determined directly from the standard form of the equation of the parabola?

(d) Explain how the result in part (c) can be used as a sketching aid when graphing parabolas.

8.8 Nonlinear Systems of Equations

Solving Nonlinear Systems of Equations by Graphing ▪
Solving Nonlinear Systems of Equations by Substitution ▪
Solving Nonlinear Systems of Equations by Elimination ▪ Application

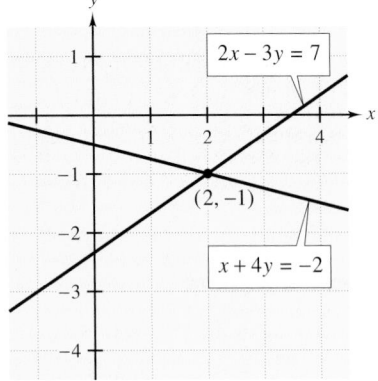

FIGURE 8.54

Review what it means graphically and algebraically for an ordered pair to be a solution to a system of linear equations.

Solving Nonlinear Systems of Equations by Graphing

In Chapter 4, you studied several methods for solving a system of linear equations. For instance, the linear system

$$\begin{cases} 2x - 3y = 7 \\ x + 4y = -2 \end{cases}$$

has one solution: $(2, -1)$. Graphically, this means that $(2, -1)$ is a point of intersection of the two lines represented by the system, as shown in Figure 8.54.

In Chapter 4, you also learned that a linear system can have no solution, exactly one solution, or infinitely many solutions. A **nonlinear system of equations** is a system that contains at least one nonlinear equation. Nonlinear systems of equations can have no solution, one solution, or two or more solutions. For instance, the hyperbola and line in Figure 8.55(a) have no point of intersection, the circle and line in Figure 8.55(b) have one point of intersection, and the parabola and line in Figure 8.55(c) have two points of intersection.

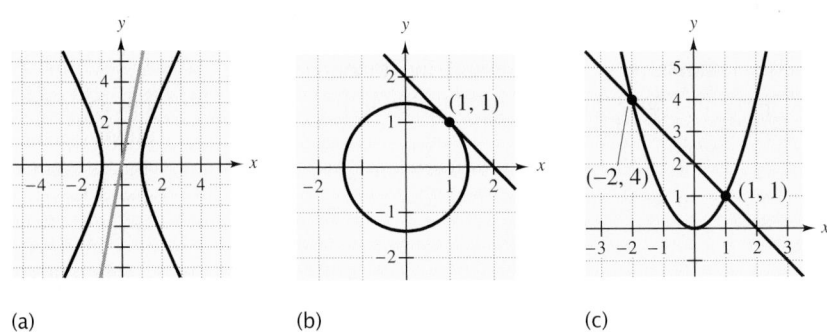

(a)   (b)   (c)

FIGURE 8.55

Solving a Nonlinear System Graphically

1. Sketch the graph of each equation in the system.

2. Locate the point(s) of intersection of the graphs (if any) and graphically approximate the coordinates of the points.

3. Check the coordinate values by substituting them into each equation in the original system. If the coordinate values do not check, you may have to use an algebraic approach, as discussed later in this section.

EXAMPLE 1   Solving a Nonlinear System Graphically

Find all solutions of the following nonlinear system of equations.

$$\begin{cases} x^2 + y^2 = 25 & \text{Equation 1} \\ x \;-\; y \;=\; 1 & \text{Equation 2} \end{cases}$$

Solution

Begin by sketching the graph of each equation. The first equation graphs as a circle centered at the origin and having a radius of 5. The second equation, which can be written as $y = x - 1$, graphs as a line with a slope of 1 and a y-intercept of $(0, -1)$. From the graphs shown in Figure 8.56, you can see that the system appears to have two solutions: $(-3, -4)$ and $(4, 3)$. You can check these solutions as follows.

Check

To check $(-3, -4)$, substitute -3 for x and -4 for y in each equation.

$$(-3)^2 + (-4)^2 = 9 + 16 = 25 \qquad \text{Solution checks in Equation 1.}$$
$$(-3) - (-4) = -3 + 4 = 1 \qquad \text{Solution checks in Equation 2.}$$

To check $(4, 3)$, substitute 4 for x and 3 for y in each equation.

$$4^2 + 3^2 = 16 + 9 = 25 \qquad \text{Solution checks in Equation 1.}$$
$$4 - 3 = 1 \qquad \text{Solution checks in Equation 2.}$$

From the check, you can conclude that both points are actual solutions of the system of equations.

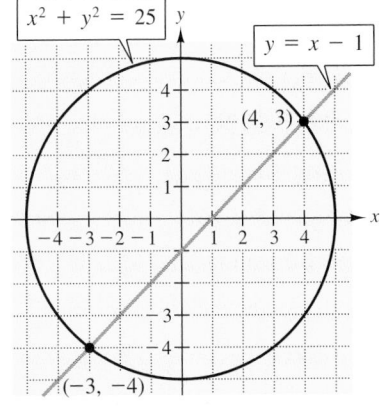

FIGURE 8.56

Technology

Try using a graphing utility to solve the system described in Example 1. When you do this, remember that the circle needs to be entered as two separate equations.

$$y_1 = \sqrt{25 - x^2} \qquad \text{Top half of circle}$$
$$y_2 = -\sqrt{25 - x^2} \qquad \text{Bottom half of circle}$$
$$y_3 = x - 1 \qquad \text{Line}$$

EXAMPLE 2   *Solving a Nonlinear System Graphically*

Find all solutions of the following nonlinear system of equations.

$$\begin{cases} x = (y - 3)^2 & \text{Equation 1} \\ x + y = 5 & \text{Equation 2} \end{cases}$$

Solution

Begin by sketching the graph of each equation. The first equation, which can be written as $y = 3 \pm \sqrt{x}$, graphs as a parabola with its vertex at point $(0, 3)$. The second equation, which can be written as $y = -x + 5$, graphs as a line with a slope of -1 and a y-intercept of $(0, 5)$. From the graphs shown in Figure 8.57, you can see that the system appears to have two solutions: $(4, 1)$ and $(1, 4)$. You can check these solutions as follows.

Check

To check $(4, 1)$, substitute 4 for x and 1 for y in each equation.

$$4 = (1 - 3)^2 = (-2)^2 \qquad \text{Solution checks in Equation 1.}$$
$$4 + 1 = 5 \qquad \text{Solution checks in Equation 2.}$$

To check $(1, 4)$, substitute 1 for x and 4 for y in each equation.

$$1 = (4 - 3)^2 = 1^2 \qquad \text{Solution checks in Equation 1.}$$
$$1 + 4 = 5 \qquad \text{Solution checks in Equation 2.}$$

From the check, you can conclude that both points are actual solutions of the system of equations.

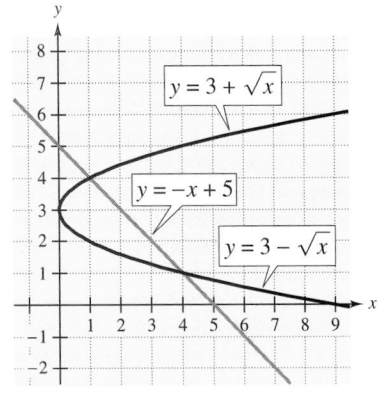

FIGURE 8.57

Solving Nonlinear Systems of Equations by Substitution

The graphical approach to solving any type of system (linear or nonlinear) in two variables is very useful for helping you see the number of solutions and their approximate coordinates. For systems with solutions having messy coordinates, however, a graphical approach is usually not accurate enough to produce exact solutions. In such cases, you should use an algebraic approach. (With an algebraic approach, you should still sketch the graph of each equation in the system.)

As with systems of *linear* equations, there are two basic algebraic approaches: substitution and elimination. Substitution usually works well for systems in which one of the equations is linear, as shown in Example 3.

EXAMPLE 3   Using Substitution to Solve a Nonlinear System

Solve the following nonlinear system of equations.

$$\begin{cases} 4x^2 + y^2 = 4 & \text{Equation 1} \\ -2x + y = 2 & \text{Equation 2} \end{cases}$$

Solution

Begin by solving for y in Equation 2.

$$y = 2x + 2 \qquad\qquad \text{Revised Equation 2}$$

Next, substitute this expression for y into Equation 1.

$$4x^2 + y^2 = 4 \qquad\qquad \text{Equation 1}$$
$$4x^2 + (2x + 2)^2 = 4 \qquad\qquad \text{Replace } y \text{ by } 2x + 2.$$
$$4x^2 + 4x^2 + 8x + 4 = 4 \qquad\qquad \text{Multiply.}$$
$$8x^2 + 8x = 0 \qquad\qquad \text{Simplify.}$$
$$8x(x + 1) = 0 \qquad\qquad \text{Factor.}$$

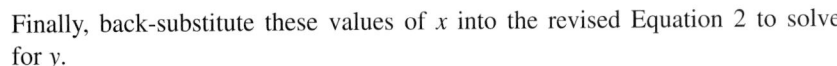

$$8x = 0 \implies x = 0 \qquad \text{Set 1st factor equal to 0.}$$
$$x + 1 = 0 \implies x = -1 \qquad \text{Set 2nd factor equal to 0.}$$

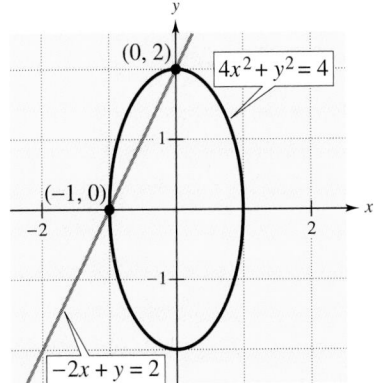

FIGURE 8.58

Finally, back-substitute these values of x into the revised Equation 2 to solve for y.

For $x = 0$: $\qquad y = 2(0) + 2 = 2$

For $x = -1$: $\qquad y = 2(-1) + 2 = 0$

Thus, the system of equations has two solutions: $(0, 2)$ and $(-1, 0)$. Check these solutions in each of the original equations. Figure 8.58 shows the graph of the system.

The steps for using the method of substitution to solve a system of two equations involving two variables are summarized as follows.

Method of Substitution

To solve a system of two equations in two variables, use the following steps.

1. Solve one of the equations for one variable in terms of the other.
2. Substitute the expression found in Step 1 into the other equation to obtain an equation of one variable.
3. Solve the equation obtained in Step 2.
4. Back-substitute the solution from Step 3 into the expression obtained in Step 1 to find the value of the other variable.
5. Check the solution to see that it satisfies *each* of the original equations.

EXAMPLE 4   *Method of Substitution: No-Solution Case*

Solve the following nonlinear system of equations.

$$\begin{cases} x^2 - y = 0 & \text{Equation 1} \\ x - y = 1 & \text{Equation 2} \end{cases}$$

Solution

Begin by solving for y in Equation 2.

$$y = x - 1 \qquad \text{Revised Equation 2}$$

Next, substitute this expression for y into Equation 1.

$$x^2 - y = 0 \qquad \text{Equation 1}$$
$$x^2 - (x - 1) = 0 \qquad \text{Replace } y \text{ by } x - 1.$$
$$x^2 - x + 1 = 0 \qquad \text{Simplify.}$$
$$x = \frac{1 \pm \sqrt{1 - 4}}{2} \qquad \text{Use Quadratic Formula.}$$

Now, because the Quadratic Formula yields a negative number inside the square root sign, you can conclude that the equation $x^2 - x + 1 = 0$ has no (real) solution. Hence, the system has no (real) solution. Figure 8.59 shows the graph of the original system. Notice that the parabola and line have no point of intersection.

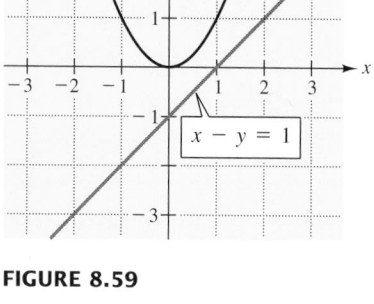

FIGURE 8.59

Stress to students that algebraically "no solution" means graphically "no points of intersection."

Solving Nonlinear Systems of Equations by Elimination

In Section 4.1, you learned how to use the method of elimination to solve a linear system. This method can also be used with special types of nonlinear systems, as demonstrated in Example 5.

EXAMPLE 5   *Using Elimination to Solve a Nonlinear System*

Solve the following nonlinear system of equations.

$$\begin{cases} 4x^2 + y^2 = 64 & \text{Equation 1} \\ x^2 + y^2 = 52 & \text{Equation 2} \end{cases}$$

Solution

Because both equations have y^2 as a term (and no other terms containing y), you can eliminate y by subtracting Equation 2 from Equation 1.

$$\begin{aligned} 4x^2 + y^2 &= 64 \\ -x^2 - y^2 &= -52 \qquad \text{Subtract Equation 2 from Equation 1.} \\ \hline 3x^2 &= 12 \end{aligned}$$

After eliminating y, solve the remaining equation for x.

$$3x^2 = 12$$
$$x^2 = 4$$
$$x = \pm 2$$

To find the corresponding values of y, substitute these values of x into either of the original equations.

$$x^2 + y^2 = 52 \qquad \text{Equation 2}$$
$$(2)^2 + y^2 = 52 \qquad \text{Substitute 2 for } x.$$
$$y^2 = 48 \qquad \text{Subtract 4 from both sides.}$$
$$y = \pm \sqrt{48} \qquad \text{Take square root of both sides.}$$
$$y = \pm 4\sqrt{3} \qquad \text{Simplify.}$$

By substituting $x = -2$, you obtain the same values of y. This implies that the system has four solutions:

$$\left(2, 4\sqrt{3}\right), \quad \left(2, -4\sqrt{3}\right), \quad \left(-2, 4\sqrt{3}\right), \quad \left(-2, -4\sqrt{3}\right).$$

Figure 8.60 shows the graph of the system. Notice that the graph of Equation 1 is an ellipse and the graph of Equation 2 is a circle.

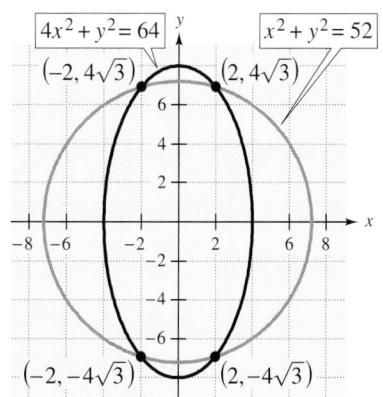

FIGURE 8.60

EXAMPLE 6   *Using Elimination to Solve a Nonlinear System*

Solve the following nonlinear system of equations.

$$\begin{cases} x^2 - 2y = 4 & \text{Equation 1} \\ x^2 - y^2 = 1 & \text{Equation 2} \end{cases}$$

Solution

Because both equations have x^2 as a term (and no other terms containing x), you can eliminate x by subtracting Equation 2 from Equation 1.

$$\begin{aligned} x^2 - 2y &= 4 \\ \underline{-x^2 + y^2} &= -1 \qquad \text{Subtract Equation 2 from Equation 1.} \\ y^2 - 2y &= 3 \end{aligned}$$

Ask students what would occur if they tried to eliminate the y-terms listed instead of the x-terms in Example 6.

After eliminating x, solve the remaining equation for y.

$$y^2 - 2y = 3$$
$$y^2 - 2y - 3 = 0$$
$$(y - 3)(y + 1) = 0$$
$$y - 3 = 0 \implies y = 3$$
$$y + 1 = 0 \implies y = -1$$

When $y = -1$, you have the following.

$$\begin{aligned} x^2 - y^2 &= 1 && \text{Equation 2} \\ x^2 - (-1)^2 &= 1 && \text{Substitute } -1 \text{ for } y. \\ x^2 &= 2 && \text{Add 1 to both sides.} \\ x &= \pm\sqrt{2} && \text{Take square root of both sides.} \end{aligned}$$

When $y = 3$, you have the following.

$$\begin{aligned} x^2 - y^2 &= 1 && \text{Equation 2} \\ x^2 - (3)^2 &= 1 && \text{Substitute 3 for } y. \\ x^2 &= 10 && \text{Add 9 to both sides.} \\ x &= \pm\sqrt{10} && \text{Take square root of both sides.} \end{aligned}$$

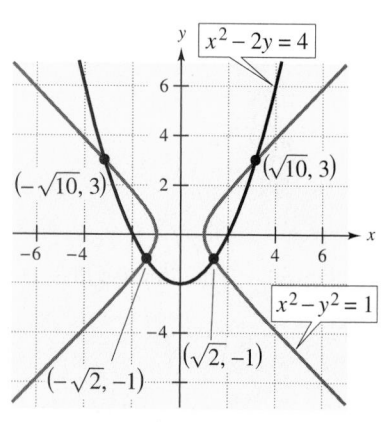

FIGURE 8.61

This implies that the system has four solutions:

$$\left(\sqrt{2}, -1\right), \quad \left(-\sqrt{2}, -1\right), \quad \left(\sqrt{10}, 3\right), \quad \left(-\sqrt{10}, 3\right).$$

Figure 8.61 shows the graph of the system. Notice that the graph of Equation 1 is a parabola and the graph of Equation 2 is a hyperbola.

Application

There are many examples of the use of nonlinear systems of equations in business and science. For instance, the following example uses a nonlinear system of equations to compare the revenues of two companies.

EXAMPLE 7   *Comparing the Revenues of Two Companies*

From 1980 through 1995, the revenues R (in millions of dollars) of Company A and Company B can be modeled by

$$\begin{cases} R = 0.1t + 2.5 & \text{Company A} \\ R = 0.02t^2 - 0.2t + 3.2, & \text{Company B} \end{cases}$$

where t represents the year, with $t = 0$ corresponding to 1980. Sketch the graphs of these two models. During which two years did the companies have approximately equal revenues?

Solution

The graphs of the two models are shown in Figure 8.62. From the graph, you can see that Company A's revenue followed a linear pattern. It had a revenue of $2.5 million in 1980 and had an increase of $0.1 million each year. Company B's revenue followed a quadratic pattern. Between 1980 and 1985, the company's revenue was decreasing. Then, from 1985 through 1995, the revenue was increasing. From the graph, you can see that the two companies had approximately equal revenues in 1983 (Company A had $2.8 million and Company B had $2.78 million) and again in 1992 (Company A had $3.7 million and Company B had $3.68 million).

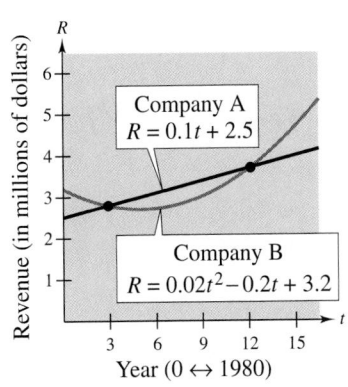

Company A
$R = 0.1t + 2.5$

Company B
$R = 0.02t^2 - 0.2t + 3.2$

Year (0 ↔ 1980)

Revenue (in millions of dollars)

FIGURE 8.62

Group Activities

Creating Examples

Sketch the graph of the circle given by

$$x^2 + y^2 = 4. \qquad \text{Circle}$$

Then find values of C such that the parabola

$$y = x^2 + C \qquad \text{Parabola}$$

intersects the circle at no points, one point, two points, three points, and four points. Use a graphing utility to confirm your results.

8.8 Exercises

See Warm-Up Exercises, p. A48

In Exercises 1–6, graph the equations to determine whether the system has any solutions. Find any solutions that exist.

1. $\begin{cases} x + y = 2 \\ x^2 - y = 0 \end{cases}$

2. $\begin{cases} 2x + y = 10 \\ x^2 + y^2 = 25 \end{cases}$

3. $\begin{cases} y = \sqrt{x - 2} \\ x - 2y = 1 \end{cases}$

4. $\begin{cases} x - 2y = 4 \\ x^2 - y = 0 \end{cases}$

5. $\begin{cases} 9x^2 - 4y^2 = 36 \\ 5x - 2y = 0 \end{cases}$

6. $\begin{cases} 9x^2 + 4y^2 = 36 \\ 3x - 2y + 6 = 0 \end{cases}$

In Exercises 7–12, use a graphing utility to graph the equations and find any solutions of the system.

7. $\begin{cases} y = x \\ y = x^3 \end{cases}$

8. $\begin{cases} y = x^2 \\ y = x + 2 \end{cases}$

9. $\begin{cases} y = x^2 \\ y = -x^2 + 4x \end{cases}$

10. $\begin{cases} y = 8 - x^2 \\ y = 6 - x \end{cases}$

11. $\begin{cases} \sqrt{x} - y = 0 \\ x - 5y = -6 \end{cases}$

12. $\begin{cases} x^2 - y^2 = 12 \\ x - 2y = 0 \end{cases}$

In Exercises 13–28, solve the system by the method of substitution.

13. $\begin{cases} y = 2x^2 \\ y = -2x + 12 \end{cases}$

14. $\begin{cases} y = 5x^2 \\ y = -15x - 10 \end{cases}$

15. $\begin{cases} x^2 + y = 9 \\ x - y = -3 \end{cases}$

16. $\begin{cases} x - y^2 = 0 \\ x - y = 2 \end{cases}$

17. $\begin{cases} x^2 + 2y = 6 \\ x - y = -4 \end{cases}$

18. $\begin{cases} x^2 + y^2 = 100 \\ x = 12 \end{cases}$

19. $\begin{cases} x^2 + y^2 = 25 \\ 2x - y = -5 \end{cases}$

20. $\begin{cases} x^2 + y^2 = 169 \\ x + y = 7 \end{cases}$

21. $\begin{cases} y = \sqrt{4 - x} \\ x + 3y = 6 \end{cases}$

22. $\begin{cases} y = \sqrt[3]{x} \\ y = x \end{cases}$

23. $\begin{cases} 16x^2 + 9y^2 = 144 \\ 4x + 3y = 12 \end{cases}$

24. $\begin{cases} y = 2x^2 \\ y = x^4 - 2x^2 \end{cases}$

25. $\begin{cases} x^2 - y^2 = 9 \\ x^2 + y^2 = 1 \end{cases}$

26. $\begin{cases} x^2 - y^2 = 16 \\ 3x - y = 12 \end{cases}$

27. $\begin{cases} 3x + 2y = 90 \\ xy = 300 \end{cases}$

28. $\begin{cases} x + 2y = 40 \\ xy = 150 \end{cases}$

In Exercises 29–36, solve the system by the method of elimination.

29. $\begin{cases} x^2 + 2y = 1 \\ x^2 + y^2 = 4 \end{cases}$

30. $\begin{cases} x + y^2 = 5 \\ 2x^2 + y^2 = 6 \end{cases}$

31. $\begin{cases} -x + y^2 = 10 \\ x^2 - y^2 = -8 \end{cases}$

32. $\begin{cases} x^2 + y = 9 \\ x^2 - y^2 = 7 \end{cases}$

33. $\begin{cases} x^2 + y^2 = 7 \\ x^2 - y^2 = 1 \end{cases}$

34. $\begin{cases} x^2 + y^2 = 25 \\ y^2 - x^2 = 7 \end{cases}$

35. $\begin{cases} \dfrac{x^2}{4} + y^2 = 1 \\ x^2 + \dfrac{y^2}{4} = 1 \end{cases}$

36. $\begin{cases} x^2 - y^2 = 1 \\ \dfrac{x^2}{2} + y^2 = 1 \end{cases}$

In Exercises 37–44, use a graphing utility to solve the system graphically.

37. $\begin{cases} y = 8 - x^2 \\ y = 6 - x \end{cases}$

38. $\begin{cases} y = \frac{1}{5}(24 - x) \\ y = \sqrt{64 - x^2} \end{cases}$

39. $\begin{cases} x^2 - y^2 = 12 \\ x - 2y = 0 \end{cases}$

40. $\begin{cases} x^2 + y = 4 \\ x + y = 6 \end{cases}$

41. $\begin{cases} y = x^3 \\ y = x^3 - 3x^2 + 3x \end{cases}$

42. $\begin{cases} y = -2(x^2 - 1) \\ y = 2(x^4 - 2x^2 + 1) \end{cases}$

43. $\begin{cases} x - 6y = -8 \\ x^2 - 4y^3 = 0 \end{cases}$

44. $\begin{cases} y = x^3 - 3x^2 + 4 \\ y = -2x + 4 \end{cases}$

45. *Hyperbolic Mirror* In a hyperbolic mirror, light rays directed to one focus are reflected to the other focus. The mirror in the figure has the equation

$$\frac{x^2}{9} - \frac{y^2}{16} = 1.$$

At which point on the mirror will light from the point $(0, 10)$ reflect to the focus?

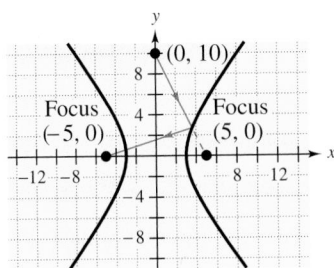

Figure for 45

46. *Miniature Golf* You are playing miniature golf and your golf ball is at $(-15, 25)$ (see figure). A wall at the end of the enclosed area is part of a hyperbola whose equation is

$$\frac{x^2}{19} - \frac{y^2}{81} = 1.$$

Using the reflective property of hyperbolas given in Exercise 45, at which point on the wall must your ball hit for it to go into the hole? (The ball bounces off the wall only once.)

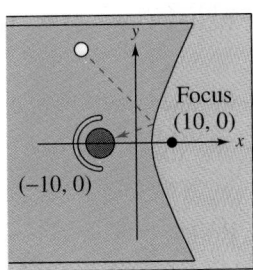

47. *Busing Boundary* To be eligible to ride the school bus to East High School, a student must live at least 1 mile from the school (see figure). Describe the portion of Clarke Street for which the residents are *not* eligible to ride the school bus. Use a coordinate system in which the school is at $(0, 0)$ and each unit represents 1 mile.

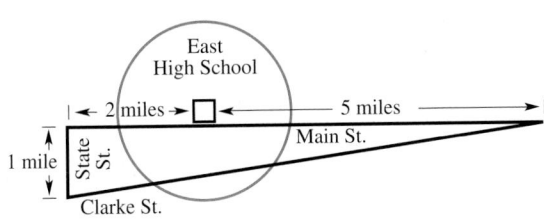

48. *Dimensions of a Corral* Suppose that you have 250 feet of fencing to enclose two corrals of equal size (see figure). The combined area of the corrals is 2400 square feet. Find the dimensions of each corral.

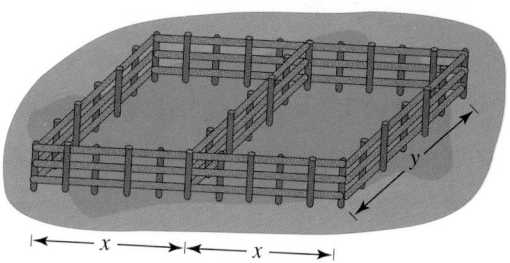

49. *Data Analysis* From 1980 through 1994, the northeastern part of the United States grew at a slower rate than the western part. Two models that represent the populations of the two regions are

$$P = 49{,}152.2 + 164.9t \qquad \text{Northeast}$$
$$P = 43{,}131.7 + 921.0t + 5.3t^2, \qquad \text{West}$$

where P is the population in thousands and t is the calendar year, with $t = 0$ corresponding to 1980. Use a graphing utility to determine when the population of the West overtook the population of the Northeast. *(Source: U.S. Bureau of Census)*

Section Project

Geometry A theorem from geometry states that if a triangle is inscribed in a circle such that one side of the triangle is a diameter of the circle, then the triangle is a right triangle. Show that this theorem is true for the circle

$$x^2 + y^2 = 100$$

and the triangle formed by the lines

$$y = 0, \quad y = \tfrac{1}{2}x + 5, \quad \text{and} \quad y = -2x + 20.$$

(Find the vertices of the triangle and verify that it is a right triangle.)

Chapter Summary

After studying this chapter, you should have acquired the following skills. These skills are keyed to the Review Exercises that begin on page 637. Answers to odd-numbered Review Exercises are given in the back of the book.

- Find combinations of functions. *(Section 8.1)*   **Review Exercises 1–6**

- Find compositions of functions. *(Section 8.1)*   **Review Exercises 7–10**

- Find compositions of functions and determine their domains. *(Section 8.1)*   **Review Exercises 11, 12**

- Use a graphing utility to decide if functions have inverses. *(Section 8.2)*   **Review Exercises 13–16**

- Find the inverses of functions and use a graphing utility to verify the results. *(Section 8.2)*   **Review Exercises 17–22**

- Restrict the domains of functions and determine their inverses. Use a graphing utility to verify the results. *(Section 8.2)*   **Review Exercises 23, 24**

- Find the constants of proportionality and write the equations that relate the variables in variation statements. *(Section 8.3)*   **Review Exercises 25–28**

- Solve real-life problems involving variation. *(Section 8.3)*   **Review Exercises 29–34**

- Determine the right and left behavior of the graphs of polynomial functions. *(Section 8.4)*   **Review Exercises 35–38**

- Graph polynomial functions and identify any intercepts. Use a graphing utility to verify the results. *(Section 8.4)*   **Review Exercises 39–44**

- Match equations of conics with their graphs. *(Sections 8.5–8.7)*   **Review Exercises 45–52**

- Identify and sketch the graphs of conics represented by equations. *(Sections 8.5–8.7)*   **Review Exercises 53–64**

- Determine the equations of circles. *(Section 8.5)*   **Review Exercises 65–68**

- Determine the equations of ellipses. *(Section 8.6)*   **Review Exercises 69–72**

- Determine the equations of hyperbolas. *(Section 8.6)*   **Review Exercises 73–76**

- Determine the equations of parabolas. *(Section 8.7)*   **Review Exercises 77–80**

- Solve real-life problems involving conics. *(Sections 8.6, 8.7)*   **Review Exercises 81, 82, 95**

- Solve nonlinear systems of equations. *(Section 8.8)*   **Review Exercises 83–90**

- Use a graphing utility to solve nonlinear systems of equations. *(Section 8.8)*   **Review Exercises 91–94**

Review Exercises

In Exercises 1–6, evaluate the combination of the functions f and g at the given x-values. (If not possible, state the reason.)

1. $f(x) = x^2$, $g(x) = 4x - 5$
  (a) $(f + g)(-5)$
  (b) $(f - g)(0)$
  (c) $(fg)(2)$
  (d) $\left(\dfrac{f}{g}\right)(1)$

2. $f(x) = \frac{3}{4}x^3$, $g(x) = x + 1$
  (a) $(f + g)(-1)$
  (b) $(f - g)(2)$
  (c) $(fg)\left(\dfrac{1}{3}\right)$
  (d) $\left(\dfrac{f}{g}\right)(2)$

3. $f(x) = \frac{2}{3}\sqrt{x}$, $g(x) = -x^2$
  (a) $(f + g)(1)$
  (b) $(f - g)(9)$
  (c) $(fg)\left(\dfrac{1}{4}\right)$
  (d) $\left(\dfrac{f}{g}\right)(2)$

4. $f(x) = |x|$, $g(x) = 3$
  (a) $(f + g)(-2)$
  (b) $(f - g)(3)$
  (c) $(fg)(-10)$
  (d) $\left(\dfrac{f}{g}\right)(-3)$

5. $f(x) = \dfrac{2}{x - 1}$, $g(x) = x$
  (a) $(f + g)\left(\dfrac{1}{3}\right)$
  (b) $(f - g)(3)$
  (c) $(fg)(-1)$
  (d) $\left(\dfrac{f}{g}\right)(5)$

6. $f(x) = \dfrac{1}{x}$, $g(x) = \dfrac{1}{x - 4}$
  (a) $(f + g)(2)$
  (b) $(f - g)(-2)$
  (c) $(fg)\left(\dfrac{7}{2}\right)$
  (d) $\left(\dfrac{f}{g}\right)(1)$

In Exercises 7–10, find the compositions.

7. $f(x) = x + 2$, $g(x) = x^2$
  (a) $(f \circ g)(x)$
  (b) $(g \circ f)(x)$
  (c) $(f \circ g)(2)$
  (d) $(g \circ f)(-1)$

8. $f(x) = \sqrt[3]{x}$, $g(x) = x + 2$
  (a) $(f \circ g)(x)$
  (b) $(g \circ f)(x)$
  (c) $(f \circ g)(6)$
  (d) $(g \circ f)(64)$

9. $f(x) = \sqrt{x + 1}$, $g(x) = x^2 - 1$
  (a) $(f \circ g)(x)$
  (b) $(g \circ f)(x)$
  (c) $(f \circ g)(5)$
  (d) $(g \circ f)(-1)$

10. $f(x) = \dfrac{1}{x - 5}$, $g(x) = \dfrac{5x + 1}{x}$
  (a) $(f \circ g)(x)$
  (b) $(g \circ f)(x)$
  (c) $(f \circ g)(1)$
  (d) $(g \circ f)\left(\dfrac{1}{5}\right)$

In Exercises 11 and 12, find the domains of the compositions (a) $f \circ g$ and (b) $g \circ f$.

11. $f(x) = \sqrt{x - 4}$, $g(x) = 2x$

12. $f(x) = \dfrac{2}{x - 4}$, $g(x) = x^2$

In Exercises 13–16, use a graphing utility to decide if the function has an inverse. Explain your reasoning.

13. $f(x) = x^2 - 25$
14. $f(x) = \frac{1}{4}x^3$
15. $h(x) = 4\sqrt[3]{x}$
16. $g(x) = \sqrt{9 - x^2}$

In Exercises 17–22, find the inverse of the function. Verify your result by using a graphing utility to graph the function and its inverse. (If not possible, state the reason.)

17. $f(x) = \frac{1}{4}x$
18. $f(x) = 2x - 3$
19. $h(x) = \sqrt{x}$
20. $g(x) = x^2 + 2$, $x \geq 0$
21. $f(t) = |t + 3|$
22. $h(t) = t$

In Exercises 23 and 24, restrict the domain of the function f to an interval over which the function is increasing, and determine f^{-1} over that interval. Use a graphing utility to graph f and f^{-1} in the same viewing rectangle.

23. $f(x) = 2(x - 4)^2$

24. $f(x) = |x - 2|$

In Exercises 25–28, find the constant of proportionality and write an equation that relates the variables.

25. y varies directly as the cube root of x, and $y = 12$ when $x = 8$.

26. r varies inversely as s, and $r = 45$ when $s = \frac{3}{5}$.

27. T varies jointly as r and the square of s, and $T = 5000$ when $r = 0.09$ and $s = 1000$.

28. D is directly proportional to the cube of x and inversely proportional to y, and $D = 810$ when $x = 3$ and $y = 25$.

29. *Power Generation*   The power generated by a wind turbine is given by the function

$$P = kw^3,$$

where P is the number of kilowatts produced at a wind speed of w miles per hour and k is the constant of proportionality.

(a) Find k if $P = 1000$ when $w = 20$.

(b) Find the output for a wind speed of 25 miles per hour.

30. *Hooke's Law*   A force of 100 pounds stretches a spring 4 inches.

(a) How far will a force of 200 pounds stretch the spring?

(b) What force will stretch the spring 2.5 inches?

31. *Hooke's Law*   A force of 100 pounds stretches a spring 4 inches. Find the force required to stretch the spring 6 inches.

32. *Stopping Distance*   The stopping distance d of an automobile is directly proportional to the square of its speed s. How will the stopping distance change if the speed of the car is doubled?

33. *Demand Function*   A company has found that the daily demand x for its product varies inversely as the square root of the price p. When the price is $25, the demand is approximately 1000 units. Approximate the demand if the price is increased to $28.

34. *Weight of an Astronaut*   The gravitational force F with which an object is attracted to the earth is inversely proportional to the square of its distance r from the center of the earth. If an astronaut weighs 200 pounds on the earth's surface ($r \approx 4000$ miles), what will the astronaut weigh 500 miles above the earth's surface?

In Exercises 35–38, determine the right and left behavior of the graph of the polynomial function.

35. $f(x) = -x^2 + 6x + 9$

36. $f(x) = \frac{1}{2}x^3 + 2x$

37. $g(x) = \frac{3}{4}(x^4 + 3x^2 + 2)$

38. $h(x) = -x^5 - 7x^2 + 10x$

In Exercises 39–44, sketch the graph of the polynomial function and identify any intercepts. Use a graphing utility to verify your result.

39. $f(x) = -(x - 2)^3$

40. $f(x) = (x + 1)^3$

41. $g(x) = x^4 - x^3 - 2x^2$

42. $f(x) = x^3 - 4x$

43. $f(x) = x(x + 3)^2$

44. $f(x) = x^4 - 4x^2$

In Exercises 45–52, match the equation with its graph. [The graphs are labeled (a), (b), (c), (d), (e), (f), (g), and (h).]

(a)
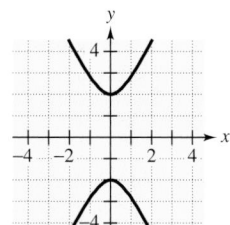

(b)
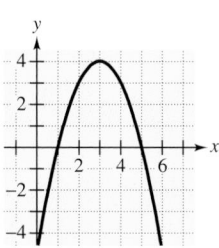

(c)
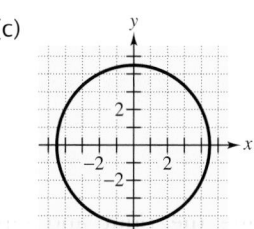

(d)
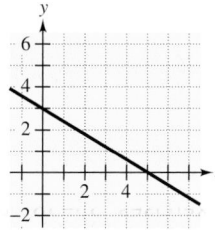

(e)
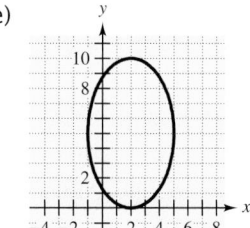

(f)
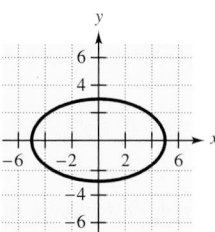

(g)
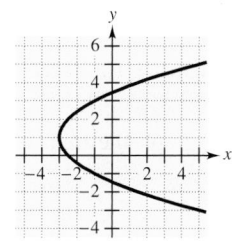

(h)
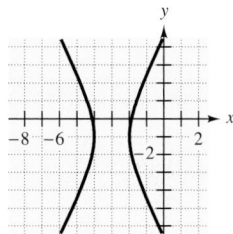

45. $4x^2 + 4y^2 = 81$

46. $3x + 5y = 15$

47. $\dfrac{y^2}{4} - x^2 = 1$

48. $\dfrac{x^2}{25} + \dfrac{y^2}{9} = 1$

49. $y = -x^2 + 6x - 5$

50. $(y - 1)^2 = 2(x + 3)$

51. $(x + 3)^2 - \dfrac{(y + 1)^2}{4} = 1$

52. $\dfrac{(x - 2)^2}{9} + \dfrac{(y - 5)^2}{25} = 1$

In Exercises 53–64, identify and sketch the graph of the conic.

53. $x^2 - 2y = 0$

54. $x^2 + y^2 = 64$

55. $x^2 - y^2 = 64$

56. $x^2 + 4y^2 = 64$

57. $y = x(x - 6)$

58. $y = 9 - (x - 3)^2$

59. $\dfrac{x^2}{25} + \dfrac{y^2}{4} = 1$

60. $\dfrac{x^2}{25} - \dfrac{y^2}{4} = -1$

61. $4x^2 + 4y^2 - 9 = 0$

62. $x^2 + 9y^2 - 9 = 0$

63. $\dfrac{(x - 2)^2}{25} + y^2 = 1$

64. $\dfrac{(y + 1)^2}{4} - \dfrac{(x - 3)^2}{9} = 1$

In Exercises 65–68, find an equation of the circle.

65. Center: $(0, 0)$;   Radius: 12

66. Center: $(0, 0)$;   Radius: 6

67. Center: $(3, 5)$;   Radius: 5

68. Center: $(-2, 3)$;   Passes through the point $(1, 1)$

In Exercises 69–72, find an equation of the ellipse.

69. Vertices: $(0, -5), (0, 5)$:
      Co-vertices: $(-2, 0), (2, 0)$

70. Vertices: $(-10, 0), (10, 0)$:
      Co-vertices: $(0, -6), (0, 6)$

71.
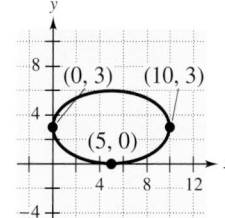

72.
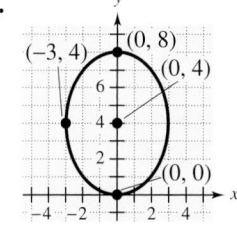

In Exercises 73–76, find an equation of the hyperbola.

73. Vertices: $(\pm 6, 0)$; Asymptotes: $y = \pm\frac{1}{3}x$

74. Vertices: $(0, \pm 4)$; Asymptotes: $y = \pm 2x$

75. 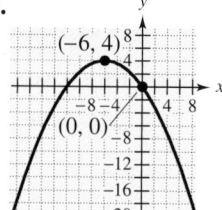   **76.**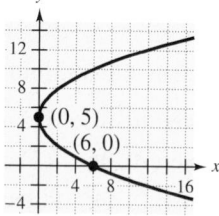

In Exercises 77–80, find an equation of the parabola.

77. Vertex: $(0, 0)$; Focus: $(-2, 0)$

78. Vertex: $(0, 0)$; Focus: $(0, 4)$

79.   **80.**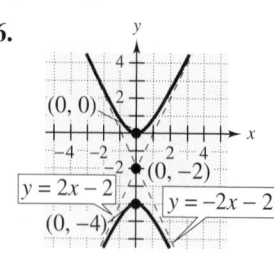

81. *Satellite Orbit*   Find an equation of the circular orbit of a satellite 1000 miles above the surface of the earth. Place the origin of the rectangular coordinate system at the center of the earth and assume the radius of the earth to be 4000 miles.

82. *Satellite Orbit*   Find an equation of the elliptical orbit of a satellite that varies in altitude from 500 miles to 1000 miles above the surface of the earth. Place the origin of the rectangular coordinate system at the center of the earth and assume the radius of the earth to be 4000 miles.

In Exercises 83–90, find any solutions of the nonlinear system of equations.

83. $\begin{cases} y = 5x^2 \\ y = -15x - 10 \end{cases}$ **84.** $\begin{cases} y^2 = 16x \\ 4x - y = -24 \end{cases}$

85. $\begin{cases} x^2 + y^2 = 1 \\ x + y = -1 \end{cases}$ **86.** $\begin{cases} x^2 + y^2 = 100 \\ x + y = 0 \end{cases}$

87. $\begin{cases} \dfrac{x^2}{16} + \dfrac{y^2}{4} = 1 \\ y = x + 2 \end{cases}$ **88.** $\begin{cases} \dfrac{x^2}{100} + \dfrac{y^2}{25} = 1 \\ y = -x - 5 \end{cases}$

89. $\begin{cases} \dfrac{x^2}{25} + \dfrac{y^2}{9} = 1 \\ \dfrac{x^2}{25} - \dfrac{y^2}{9} = 1 \end{cases}$ **90.** $\begin{cases} x^2 + y^2 = 16 \\ -x^2 + \dfrac{y^2}{16} = 1 \end{cases}$

In Exercises 91–94, use a graphing utility to graph the equations and find any solutions of the system.

91. $\begin{cases} y = x^2 \\ y = 3x \end{cases}$ **92.** $\begin{cases} y = 2 + x^2 \\ y = 8 - x \end{cases}$

93. $\begin{cases} x^2 + y^2 = 16 \\ -x + y = 4 \end{cases}$ **94.** $\begin{cases} 2x^2 - y^2 = -8 \\ y = x + 6 \end{cases}$

95. *Path of a Ball*   The height y (in feet) of a ball thrown by a child is

$$y = -\frac{1}{10}x^2 + 3x + 6,$$

where x is the horizontal distance (in feet) from where the ball was thrown.

(a) Use a graphing utility to graph the path of the ball.

(b) How high is the ball when it leaves the child's hand?

(c) How high is the ball when it is at its maximum height?

(d) How far from the child does the ball strike the ground?

Chapter Test

Take this test as you would take a test in class. After you are done, check your work against the answers given in the back of the book.

In Exercises 1–4, use the functions $f(x) = \frac{1}{2}x$ and $g(x) = x^2 - 1$ to find the following.

1. $(f - g)(4)$ **2.** $(fg)(-4)$ **3.** $\left(\dfrac{f}{g}\right)(2)$ **4.** $(f \circ g)(-5)$

5. Find the domain of the composite function $(f \circ g)(x)$ if $f(x) = \sqrt{25 - x}$ and $g(x) = x^2$.

6. Find the inverse of the function $f(x) = \frac{1}{2}x - 1$. Verify that $f(f^{-1}(x)) = x = f^{-1}(f(x))$.

7. Write a mathematical model for the statement "S varies directly as the square of x and inversely as y."

8. Find a mathematical model that relates u and v if v varies directly as the square root of u, and $v = \frac{3}{2}$ when $u = 36$.

9. *Boyle's Law*   If the temperature of a gas is not allowed to change, its absolute pressure P is inversely proportional to its volume V, according to Boyle's Law. A large balloon is filled with 180 cubic meters of helium at atmospheric pressure (1 atm) at sea level. What is the volume of the helium if the balloon rises to an altitude at which the atmospheric pressure is 0.75 atm? (Assume that the temperature does not change.)

10. Determine the right and left behavior of the polynomial function $f(x) = -2x^3 + 3x^2 - 4$.

11. Sketch the graph of the function $f(x) = 1 - (x - 2)^3$. Identify any intercepts of the graph.

12. Write an equation of the circle shown in the figure.

13. Write an equation of the parabola shown in the figure.

14. Find the standard form of the equation of the ellipse with vertices $(0, -10)$ and $(0, 10)$ and co-vertices $(-3, 0)$ and $(3, 0)$.

15. Find the standard form of the equation of the hyperbola with vertices $(-3, 0)$ and $(3, 0)$ and asymptotes $y = \frac{1}{2}x$ and $y = -\frac{1}{2}x$.

16. Sketch the graph of each equation. Use a graphing utility to confirm your results.

(a) $x^2 + y^2 = 9$ (b) $\dfrac{x^2}{9} + \dfrac{y^2}{16} = 1$ (c) $\dfrac{x^2}{9} - \dfrac{y^2}{16} = 1$ (d) $\dfrac{x}{3} - \dfrac{y}{4} = 1$

17. Find any solutions of each nonlinear system of equations.

(a) $\begin{cases} y = -\frac{1}{2}x^2 \\ y = -4x + 6 \end{cases}$ (b) $\begin{cases} x^2 + y^2 = 100 \\ x + y = 14 \end{cases}$

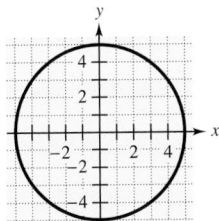

Figure for 12

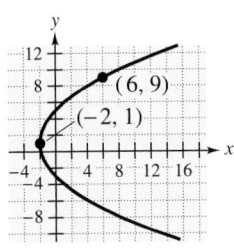

(6, 9)

(−2, 1)

Figure for 13

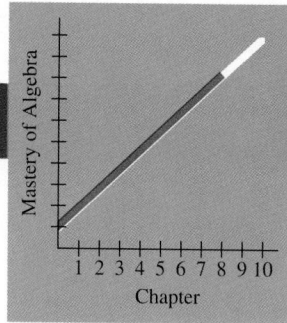

Mastery of Algebra

1 2 3 4 5 6 7 8 9 10

Chapter

Take this test as you would take a test in class. After you are done, check your work against the answers given in the back of the book.

In Exercises 1–3, simplify the expression.

1. $-(-3x^2)^3(2x^4)$ **2.** $-\dfrac{(2u^2v)^2}{-3uv^2}$ **3.** $\dfrac{5}{\sqrt{12}-2}$

In Exercises 4–6, find the distance between the points.

4. $(-2, 4), (-5, 4)$ **5.** $(8, 1), (3, 6)$ **6.** $(-6, 7), (6, -2)$

In Exercises 7 and 8, write an equation of the line through the two points.

7. $(-1, -2), (3, 6)$ **8.** $(1, 5), (6, 0)$

In Exercises 9 and 10, factor the expression completely.

9. $3x^2 - 21x$ **10.** $4t^2 - 169$

In Exercises 11 and 12, simplify the complex fraction.

11. $\dfrac{\left(\dfrac{9}{x}\right)}{\left(\dfrac{6}{x}+2\right)}$ **12.** $\dfrac{\left(1+\dfrac{2}{x}\right)}{\left(x-\dfrac{4}{x}\right)}$

In Exercises 13–15, sketch a graph of the equation.

13. $y = 3 - \dfrac{1}{2}x$ **14.** $(x-5)^2 + (y+2)^2 = 25$ **15.** $\dfrac{x^2}{9} - \dfrac{y^2}{25} = 1$

16. Find an equation of the ellipse shown in the figure.

17. Solve the equation $x^2 + 3x - 2 = 0$.

18. The cost of a long distance telephone call is $1.10 for the first minute and $0.45 for each additional minute. If the total cost of the call cannot exceed $11, find the interval of time that is available for the call.

19. A company produces a product for which the variable cost is $5.35 per unit and the fixed costs are $30,000. The product is sold for $11.60. How many units must be sold before the company breaks even?

20. Find two positive consecutive odd integers such that their sum is 47 less than their product.

Figure for 16

(graph showing points (0, 0), (3, 2), (6, 0), (3, −2))

642

Exponential and Logarithmic Functions and Equations

9

Strategies for Success

SKILLS *When you have completed this chapter, make sure you are able to:*

○ Evaluate and sketch the graphs of exponential and logarithmic functions

○ Write exponential equations in logarithmic form and vice versa

○ Evaluate and rewrite logarithmic expressions using properties of logarithms

○ Solve exponential and logarithmic equations

○ Model data and solve real-life problems using exponential and logarithmic functions

TOOLS *Use these study tools to master the skills above:*

In Exercise 75 of Section 9.1, you will use an exponential function to approximate the time it takes a parachutist to reach the ground.

9.1	**Exponential Functions and Their Graphs**

Exponential Functions ▪ Graphs of Exponential Functions ▪ The Natural Exponential Function ▪ Compound Interest

Exponential Functions

In this section, you will study a new type of function called an **exponential function.** Whereas polynomial and rational functions have terms with variable bases and constant exponents, exponential functions have terms with *constant bases* and *variable exponents*. Here are some examples.

Polynomial or Rational Function   *Exponential Function*

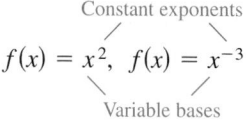

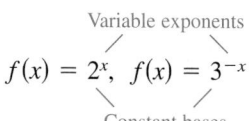

NOTE The base $a = 1$ is excluded because $f(x) = 1^x = 1$ is a constant function, *not* an exponential function.

Definition of Exponential Function

The **exponential function** f **with base** a is denoted by

$$f(x) = a^x,$$

where $a > 0$, $a \neq 1$, and x is any real number.

In Chapter 5, you learned to evaluate a^x for integer and rational values of x. For example, you know that

$$a^3 = a \cdot a \cdot a \quad \text{and} \quad a^{2/3} = \left(\sqrt[3]{a}\right)^2.$$

However, to evaluate a^x for any real number x, you need to interpret forms with *irrational* exponents. For the purpose of this text, it is sufficient to think of a number such as

$$a^{\sqrt{2}},$$

where $\sqrt{2} \approx 1.414214$, as the number that has the successively closer approximations

$$a^{1.4}, a^{1.41}, a^{1.414}, a^{1.4142}, a^{1.41421}, a^{1.414214}, \ldots .$$

The properties of exponents that were discussed in Section 5.1 can be extended to cover exponential functions, as shown on page 645.

Properties of Exponential Functions

Let a be a positive real number, and let x and y be real numbers, variables, or algebraic expressions.

1. $a^x \cdot a^y = a^{x+y}$

2. $(a^x)^y = a^{xy}$

3. $\dfrac{a^x}{a^y} = a^{x-y}$

4. $a^{-x} = \dfrac{1}{a^x} = \left(\dfrac{1}{a}\right)^x$

To evaluate exponential functions with a calculator, you can use the exponential key $\boxed{y^x}$ (where y is the base and x is the exponent) or $\boxed{\wedge}$. For example, to evaluate $3^{-1.3}$, you can use the following keystrokes.

Keystrokes	Display	
3 $\boxed{y^x}$ 1.3 $\boxed{+/-}$ $\boxed{=}$	0.239741	Scientific
3 $\boxed{\wedge}$ $\boxed{(}$ $\boxed{(-)}$ 1.3 $\boxed{)}$ $\boxed{\text{ENTER}}$	0.239741	Graphing

---

EXAMPLE 1   Evaluating Exponential Functions

Evaluate each function at the indicated values of x. Use a calculator only if it is necessary or more efficient.

Function	Values
(a) $f(x) = 2^x$	$x = 3, x = -4, x = \pi$
(b) $g(x) = 12^x$	$x = 3, x = -0.1, x = \frac{6}{7}$
(c) $h(x) = (1.085)^x$	$x = 0, x = -3$

Solution

Evaluation	Comment
(a) $f(3) = 2^3 = 8$	Calculator is not necessary.
$f(-4) = 2^{-4} = \dfrac{1}{2^4} = \dfrac{1}{16}$	Calculator is not necessary.
$f(\pi) = 2^\pi \approx 8.8250$	Calculator is necessary.
(b) $g(3) = 12^3 = 1728$	Calculator is more efficient.
$g(-0.1) = 12^{-0.1} \approx 0.7800$	Calculator is necessary.
$g\left(\dfrac{6}{7}\right) = 12^{6/7} \approx 8.4142$	Calculator is necessary.
(c) $h(0) = (1.085)^0 = 1$	Calculator is not necessary.
$h(-3) = (1.085)^{-3} \approx 0.7829$	Calculator is more efficient.

When using their calculators to evaluate $12^{6/7}$, some students may incorrectly enter 12 $\boxed{\wedge}$ 6 $\boxed{\div}$ 7. Stress the importance of using grouping symbols—even with calculators.

Graphs of Exponential Functions

The basic nature of the graph of an exponential function can be determined by the point-plotting method or by using a graphing utility.

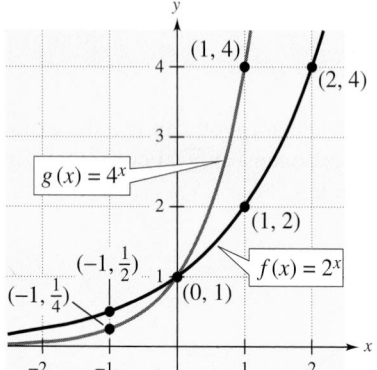

FIGURE 9.1

Have students determine the domain and range of the exponential function $f(x) = a^x$.

EXAMPLE 2   The Graphs of Exponential Functions

In the same coordinate plane, sketch the graphs of the following functions. Determine the domains and ranges.

(a) $f(x) = 2^x$ (b) $g(x) = 4^x$

Solution

The table lists some values of each function, and Figure 9.1 shows their graphs. From the graphs, you can see that the domain of each function is the set of all real numbers and that the range of each function is the set of all positive real numbers.

x	-2	-1	0	1	2	3
2^x	$\frac{1}{4}$	$\frac{1}{2}$	1	2	4	8
4^x	$\frac{1}{16}$	$\frac{1}{4}$	1	4	16	64

You know from your study of functions in Chapter 2 that the graph of $h(x) = f(-x) = 2^{-x}$ is a reflection of the graph of $f(x) = 2^x$ in the y-axis. This is reinforced in the following example.

EXAMPLE 3   The Graphs of Exponential Functions

In the same coordinate plane, sketch the graphs of the following functions.

(a) $f(x) = 2^{-x}$ (b) $g(x) = 4^{-x}$

Solution

The table lists some values of each function, and Figure 9.2 shows their graphs.

x	-3	-2	-1	0	1	2
2^{-x}	8	4	2	1	$\frac{1}{2}$	$\frac{1}{4}$
4^{-x}	64	16	4	1	$\frac{1}{4}$	$\frac{1}{16}$

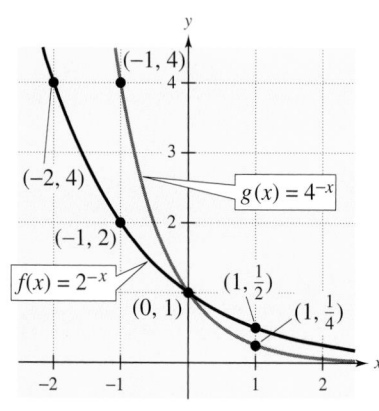

FIGURE 9.2

Point out that the x-axis is an asymptote.

Examples 2 and 3 suggest that for $a > 1$, the values of the function $y = a^x$ increase as x increases and the values of the function $y = a^{-x}$ decrease as x increases. The graphs shown in Figure 9.3 are typical of the graphs of exponential functions. Note that each has a y-intercept at $(0, 1)$ and a horizontal asymptote (the x-axis).

Use a graphing utility to graph $y = a^x$ with $a = 2$, 3, and 5 on the same viewing rectangle. (Use a viewing rectangle in which $-2 \le x \le 1$ and $0 \le y \le 2$.) How do the graphs compare with each other? Repeat this experiment with $y = a^{-x}$, $a = 2$, 3, and 5.

Remind students that 2^{-x}, 5^{-x}, and a^{-x} can also be written as $\dfrac{1}{2^x}$, $\dfrac{1}{5^x}$, and $\dfrac{1}{a^x}$.

Graph of $y = a^x$

- Domain: $(-\infty, \infty)$
- Range: $(0, \infty)$
- Intercept: $(0, 1)$
- Increasing (moves up to the right)
- Asymptote: x-axis

Graph of $y = a^{-x}$

- Domain: $(-\infty, \infty)$
- Range: $(0, \infty)$
- Intercept: $(0, 1)$
- Decreasing (moves down to the right)
- Asymptote: x-axis

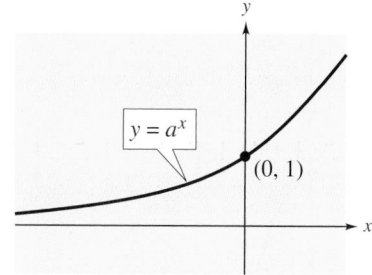

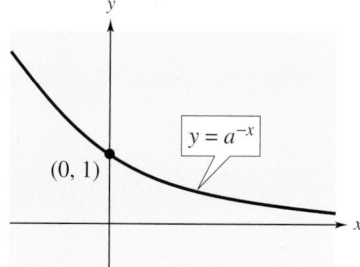

FIGURE 9.3 Characteristics of the exponential functions a^x and a^{-x} $(a > 1)$

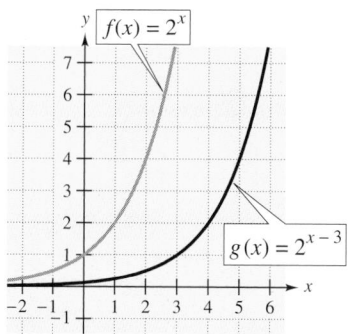

(a)

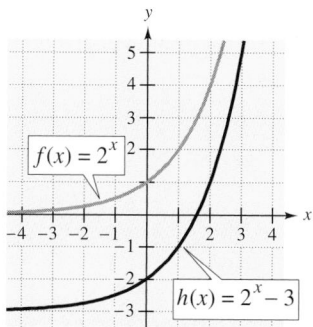

(b)

FIGURE 9.4

EXAMPLE 4   Shifts of Graphs of Exponential Functions

Use transformations to analyze and sketch the graph of each function.

(a) $g(x) = 2^{x-3}$ (b) $h(x) = 2^x - 3$

Solution

Consider the function $f(x) = 2^x$.

(a) The function g is related to f by

   $g(x) = f(x - 3)$.   Horizontal shift (to right)

   To sketch the graph of g, shift the graph of f three units to the right, as shown in Figure 9.4(a).

(b) The function h is related to f by

   $h(x) = f(x) - 3$.   Vertical shift (down)

   To sketch the graph of h, shift the graph of f three units down, as shown in Figure 9.4(b).

EXAMPLE 5   Shifts of Graphs of Exponential Functions

Use transformations to analyze and sketch the graph of each function.

(a) $g(x) = -2^x$ (b) $h(x) = 2 + 2^{-x}$

Solution

(a) The function g is related to $f(x) = 2^x$ by $g(x) = -f(x)$. To sketch the graph of g, reflect the graph of f about the x-axis, as shown in Figure 9.5(a).

(b) The function h is related to $f(x) = 2^x$ by $h(x) = 2 + f(-x)$. To sketch the graph of h, shift the graph of f two units up and reflect it about the y-axis, as shown in Figure 9.5(b).

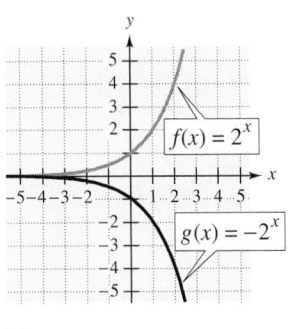

(a)

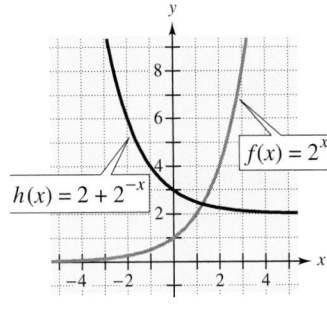

(b)

FIGURE 9.5

EXAMPLE 6   An Application: Radioactive Decay

Let y represent the mass of a particular radioactive element whose half-life is 25 years. The initial mass is 10 grams. After t years, the mass (in grams) is given by

$$y = 10\left(\tfrac{1}{2}\right)^{t/25}, \quad t \geq 0.$$

How much of the initial mass remains after 120 years?

Solution

When $t = 120$, the mass is given by

$$y = 10\left(\tfrac{1}{2}\right)^{120/25} \qquad \text{Substitute 120 for } t.$$
$$= 10\left(\tfrac{1}{2}\right)^{4.8} \qquad \text{Simplify.}$$
$$\approx 0.359 \text{ gram.} \qquad \text{Use a calculator.}$$

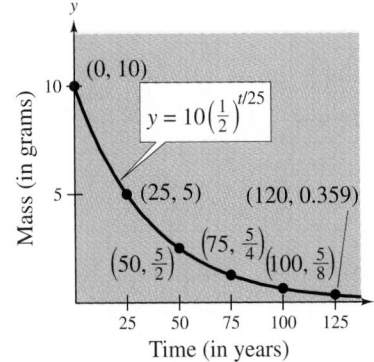

FIGURE 9.6

Thus, after 120 years, the mass has decayed from an initial amount of 10 grams to only 0.359 gram. Note in Figure 9.6 that the graph of the function shows the 25-year half-life. That is, after 25 years the mass is 5 grams (half of the original), after another 25 years the mass is 2.5 grams, and so on.

The Natural Exponential Function

So far, we have used integers or rational numbers as bases of exponential functions. In many applications of exponential functions, the convenient choice for a base is the following irrational number, denoted by the letter "*e.*"

$$e \approx 2.71828. \ldots \qquad \text{Natural base}$$

This number is called the **natural base.** The function

$$f(x) = e^x \qquad \text{Natural exponential function}$$

is called the **natural exponential function.** Be sure you understand that for this function, *e* is the constant number 2.71828. . . , and *x* is a variable. To evaluate the natural exponential function, you need a calculator, preferably one having a natural exponential key $\boxed{e^x}$. Here are some examples of how to use such a calculator to evaluate the natural exponential function.

Value	Keystrokes	Display	
e^2	2 $\boxed{e^x}$	7.3890561	Scientific
e^2	$\boxed{e^x}$ 2 $\boxed{\text{ENTER}}$	7.3890561	Graphing
e^{-3}	3 $\boxed{+/-}$ $\boxed{e^x}$	0.0497871	Scientific
e^{-3}	$\boxed{e^x}$ $\boxed{(}$ $\boxed{(-)}$ 3 $\boxed{)}$ $\boxed{\text{ENTER}}$	0.0497871	Graphing
$e^{0.32}$	.32 $\boxed{e^x}$	1.3771278	Scientific
$e^{0.32}$	$\boxed{e^x}$ .32 $\boxed{\text{ENTER}}$	1.3771278	Graphing

NOTE Some calculators do not have a key labeled $\boxed{e^x}$. If your calculator does not have this key, but does have a key labeled $\boxed{\text{LN}}$, you will have to use the keystroke sequence $\boxed{\text{INV}}$ $\boxed{\text{LN}}$ in place of $\boxed{e^x}$.

After evaluating the natural exponential function at several values, as shown in the table, you can sketch its graph, as shown in Figure 9.7.

x	-1.5	-1.0	-0.5	0.0	0.5	1.0	1.5
$f(x) = e^x$	0.223	0.368	0.607	1.000	1.649	2.718	4.482

From the graph, notice the following properties of the natural exponential function.

- The domain is the set of all real numbers.
- The range is the set of positive real numbers.
- The *y*-intercept is $(0, 1)$.

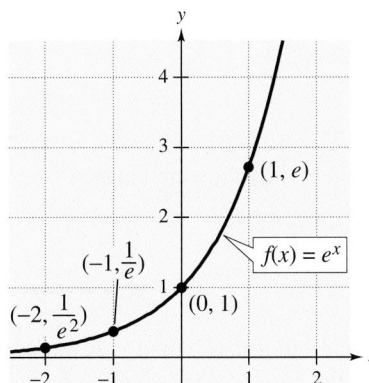

FIGURE 9.7 Natural exponential function

Compound Interest

One of the most familiar uses of exponential functions involves **compound interest.** Suppose a principal P is invested at an annual interest rate r (in decimal form), compounded once a year. If the interest is added to the principal at the end of the year, the balance is

$$A = P + Pr = P(1 + r).$$

This pattern of multiplying the previous principal by $(1 + r)$ is then repeated each successive year, as shown below.

Time in Years	Balance at Given Time
0	$A = P$
1	$A = P(1 + r)$
2	$A = P(1 + r)(1 + r) = P(1 + r)^2$
3	$A = P(1 + r)^2(1 + r) = P(1 + r)^3$
$\vdots$	$\vdots$
t	$A = P(1 + r)^t$

To account for more frequent compounding of interest (such as quarterly or monthly compounding), let n be the number of compoundings per year and let t be the number of years. Then the rate per compounding is r/n and the account balance after t years is

$$A = P\left(1 + \frac{r}{n}\right)^{nt}.$$

EXAMPLE 7   Finding the Balance for Compound Interest

A sum of $10,000 is invested at an annual interest rate of 7.5%, compounded monthly. Find the balance in the account after 10 years.

Solution

Using the formula for compound interest, with $P = 10{,}000$, $r = 0.075$, $n = 12$ (for monthly compounding), and $t = 10$, you obtain the following balance.

$$A = 10{,}000\left(1 + \frac{0.075}{12}\right)^{12(10)} \approx \$21{,}120.65$$

A second method that banks use to compute interest is called **continuous compounding.** The formula for the balance for this type of compounding is

$$A = Pe^{rt}.$$

The formulas for both types of compounding are summarized on page 651.

Additional problem: $1000 is deposited for 9 months at 12%. What is the total amount in the account (a) at simple interest, (b) if interest is compounded quarterly, (c) if interest is compounded continuously?

Answers:
(a) $1088.71
(b) $1092.73
(c) $1094.17

Formulas for Compound Interest

After t years, the balance A in an account with principal P and annual interest rate r (in decimal form) is given by the following formulas.

1. For n compoundings per year: $A = P\left(1 + \dfrac{r}{n}\right)^{nt}$

2. For continuous compounding: $A = Pe^{rt}$

EXAMPLE 8 Comparing Two Types of Compounding

A total of $15,000 is invested at an annual interest rate of 8%. Find the balance after 6 years if the interest is compounded (a) quarterly and (b) continuously.

Solution

(a) Letting $P = 15,000$, $r = 0.08$, $n = 4$, and $t = 6$, the balance after 6 years at quarterly compounding is

$$A = 15,000\left(1 + \frac{0.08}{4}\right)^{4(6)}$$

$$= \$24,126.56.$$

NOTE Example 8 illustrates the following general rule. For a given principal, interest rate, and time, the more often the interest is compounded per year, the greater the balance will be. Moreover, the balance obtained by continuous compounding is larger than the balance obtained by compounding n times per year.

(b) Letting $P = 15,000$, $r = 0.08$, and $t = 6$, the balance after 6 years at continuous compounding is

$$A = 15,000e^{0.08(6)}$$

$$= \$24,241.12.$$

Note that the balance is greater with continuous compounding than with quarterly compounding.

Group Activities Exploring with Technology

Finding a Pattern Use a graphing utility to investigate the function $f(x) = k^x$ for different values of k. Discuss the effect that k has on the shape of the graph.

9.1   Exponential Functions

See Warm-Up Exercises, p. A49

9.1   Exercises

In Exercises 1–4, simplify the expression.

1. $2^x \cdot 2^{x-1}$

2. $\dfrac{3^{2x+3}}{3^{x+1}}$

3. $(2e^x)^3$

4. $\sqrt{4e^{6x}}$

In Exercises 5–10, evaluate the expression. (Round to three decimal places.)

5. $4^{\sqrt{3}}$

6. $6^{-\pi}$

7. $e^{1/3}$

8. $e^{-1/3}$

9. $\dfrac{4e^3}{12e^2}$

10. $(9e^2)^{3/2}$

In Exercises 11–22, evaluate the function as indicated. Use a calculator only if it is necessary. (Round to three decimal places.)

11. $f(x) = 3^x$

  (a) $x = -2$

  (b) $x = 0$

  (c) $x = 1$

12. $F(x) = 3^{-x}$

  (a) $x = -2$

  (b) $x = 0$

  (c) $x = 1$

13. $g(x) = 5^x$

  (a) $x = -1$

  (b) $x = 1$

  (c) $x = 3$

14. $G(x) = 5^{-x}$

  (a) $x = -1$

  (b) $x = 1$

  (c) $x = \sqrt{3}$

15. $f(t) = 500\left(\tfrac{1}{2}\right)^t$

  (a) $t = 0$

  (b) $t = 1$

  (c) $t = \pi$

16. $g(s) = 1200\left(\tfrac{2}{3}\right)^s$

  (a) $s = 0$

  (b) $s = 2$

  (c) $s = 4$

17. $f(x) = 1000(1.05)^{2x}$

  (a) $x = 0$

  (b) $x = 5$

  (c) $x = 10$

18. $P(t) = \dfrac{10{,}000}{(1.01)^{12t}}$

  (a) $t = 2$

  (b) $t = 10$

  (c) $t = 20$

19. $f(x) = e^x$

  (a) $x = -1$

  (b) $x = 0$

20. $A(t) = 200e^{0.1t}$

  (a) $t = 10$

  (b) $t = 20$

21. $g(x) = 10e^{-0.5x}$

  (a) $x = -4$

  (b) $x = 4$

22. $f(z) = \dfrac{100}{1 + e^{-0.05z}}$

  (a) $z = 0$

  (b) $z = 10$

In Exercises 23–30, match the function with its graph. [The graphs are labeled (a), (b), (c), (d), (e), (f), (g), and (h).]

(a)
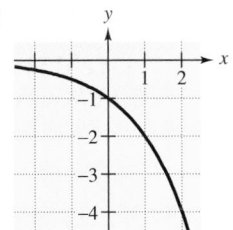

(b)
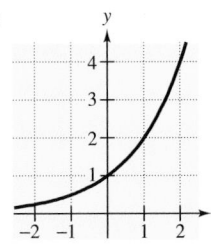

(c)
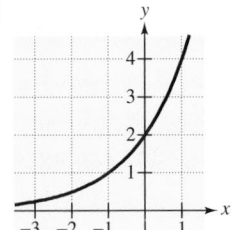

(d)
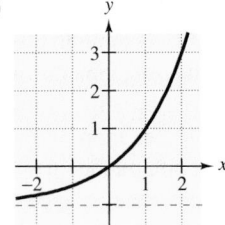

(e)
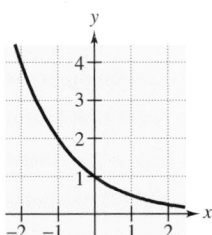

(f)
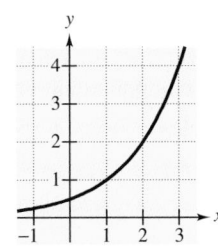

(g)
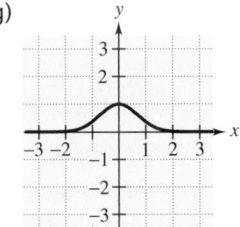

(h)
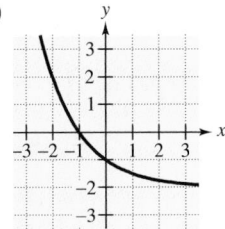

23. $f(x) = 2^x$ **24.** $f(x) = -2^x$

25. $f(x) = 2^{-x}$ **26.** $f(x) = 2^x - 1$

27. $f(x) = 2^{x-1}$ **28.** $f(x) = 2^{x+1}$

29. $f(x) = \left(\frac{1}{2}\right)^x - 2$ **30.** $f(x) = e^{-x^2}$

In Exercises 31–44, sketch the graph of the function.

31. $f(x) = 3^x$ **32.** $f(x) = 3^{-x} = \left(\frac{1}{3}\right)^x$

33. $h(x) = \frac{1}{2}(3^x)$ **34.** $h(x) = \frac{1}{2}(3^{-x})$

35. $g(x) = 3^x - 2$ **36.** $g(x) = 3^x + 1$

37. $f(t) = 2^{-t^2}$ **38.** $f(t) = 2^{t^2}$

39. $f(x) = -2^{0.5x}$ **40.** $h(t) = -2^{-0.5t}$

41. $f(x) = 4^{x-5}$ **42.** $f(x) = 4^{x+1}$

43. $g(t) = 200\left(\frac{1}{2}\right)^t$ **44.** $h(y) = 27\left(\frac{2}{3}\right)^y$

In Exercises 45–56, use a graphing utility to graph the function.

45. $y = 5^{x/3}$ **46.** $y = 5^{(x-2)/3}$

47. $y = 5^{-x/3}$ **48.** $y = 5^{-x/3} + 2$

49. $y = 500(1.06)^t$ **50.** $y = 100(1.06)^{-t}$

51. $y = 3e^{0.2x}$ **52.** $y = 50e^{-0.05x}$

53. $P(t) = 100e^{-0.1t}$ **54.** $A(t) = 1000e^{0.08t}$

55. $y = 6e^{-x^2/3}$ **56.** $g(t) = \dfrac{10}{1 + e^{-0.5t}}$

57. *Think About It*   On the same set of coordinate axes, sketch the graphs of the following functions. Which functions are exponential?

(a) $f(x) = 2x$ (b) $f(x) = 2x^2$

(c) $f(x) = 2^x$ (d) $f(x) = 2^{-x}$

58. *Identifying Graphs*   Identify the graphs of

$$y_1 = e^{0.2x}, \quad y_2 = e^{0.5x}, \quad \text{and} \quad y_3 = e^x$$

in the figure. Describe the effect on the graph of $y = e^{kx}$ when $k > 0$ is changed.

The symbol ▦ indicates an exercise in which you are instructed to use a calculator or graphing utility. The solutions of other exercises may also be facilitated by the use of appropriate technology.

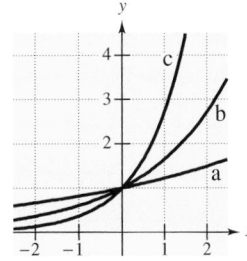

Figure for 58

59. *True or False?* $e = \dfrac{271{,}801}{99{,}990}$. Explain.

60. *Think About It*   Explain why $2^{\sqrt{2}}$ is greater than 2, but less than 4.

In Exercises 61–64, complete the table to determine the balance A for P dollars invested at rate r for t years, compounded n times per year.

n	1	4	12	365	Continuous compounding
A					

Principal	Rate	Time
61. $P = \$100$	$r = 8\%$	$t = 20$ years
62. $P = \$400$	$r = 8\%$	$t = 50$ years
63. $P = \$2000$	$r = 9\%$	$t = 10$ years
64. $P = \$5000$	$r = 10\%$	$t = 40$ years

In Exercises 65–68, complete the table to determine the principal P that yields a balance of A dollars invested at rate r for t years, compounded n times per year.

n	1	4	12	365	Continuous compounding
P					

Balance	Rate	Time
65. $A = \$5000$	$r = 7\%$	$t = 10$ years
66. $A = \$100{,}000$	$r = 9\%$	$t = 20$ years
67. $A = \$1{,}000{,}000$	$r = 10.5\%$	$t = 40$ years
68. $A = \$2500$	$r = 7.5\%$	$t = 2$ years

69. *Graphical Interpretation*   An investment of $500 in two different accounts with interest rates of 6% and 8% is compounded continuously. The balances in the accounts after t years are modeled by

$$A_1 = 500e^{0.06t} \quad \text{and} \quad A_2 = 500e^{0.08t}.$$

(a) Use a graphing utility to graph each of the models in the same viewing rectangle.

(b) Use a graphing utility to graph the function $A_2 - A_1$ in the same viewing rectangle as the graphs in part (a).

(c) Use the graphs to discuss the rates of increase of the balances in the two accounts.

70. *Inflation Rate*   Suppose the annual rate of inflation averages 5% over the next 10 years. With this rate of inflation, the approximate cost C of goods or services during any year in that decade will be given by

$$C(t) = P(1.05)^t, \quad 0 \leq t \leq 10,$$

where t is the time in years and P is the present cost. If the price of an oil change for your car is presently $24.95, estimate the price 10 years from now.

71. *Property Value*   Suppose that the value of a piece of property doubles every 15 years. If you buy the property for $64,000, its value t years after the date of purchase should be

$$V(t) = 64{,}000(2)^{t/15}.$$

Use the model to approximate the value of the property (a) 5 years and (b) 20 years after it is purchased.

72. *Price and Demand*   The daily demand x and the price p for a certain product are related by

$$p = 25 - 0.4e^{0.02x}.$$

Find the prices for demands of (a) $x = 100$ units and (b) $x = 125$ units.

73. *Depreciation*   After t years, the value of a car that originally cost $16,000 depreciates so that each year it is worth $\frac{3}{4}$ of its value for the previous year. Find a model for $V(t)$, the value of the car for year t. Sketch a graph of the model and determine the value of the car 2 years after it was purchased.

74. *Depreciation*   Suppose straight-line depreciation is used to determine the value of the car in Exercise 73. Assume that the car depreciates $3000 per year.

(a) What is $V(t)$, the value of the car for year t?

(b) Sketch a graph of the model in part (a) on the same coordinate axes as the graph in Exercise 73.

(c) If you were selling the car after owning it for 2 years, which depreciation model would you prefer?

(d) If you sell the car after 4 years, which model would be to your advantage?

75. *Parachute Drop*   A parachutist jumps from a plane and opens the parachute at a height of 2000 feet (see figure). The height of the parachutist is

$$h = 1950 + 50e^{-1.6t} - 20t,$$

where h is the height in feet and t is the time in seconds. (The time $t = 0$ corresponds to the time when the parachute is opened.)

(a) Use a graphing utility to graph the function.

(b) Find the height of the parachutist when $t = 0$, 25, 50, and 75.

(c) Approximate the time when the parachutist reaches the ground.

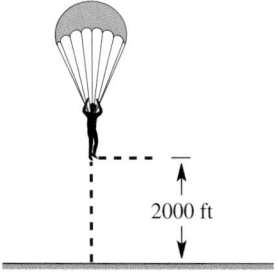

2000 ft

76. *Data Analysis* A meteorologist measures the atmospheric pressure P (in kilograms per square meter) at altitude h (in kilometers). The data is shown in the table.

h	0	5	10	15	20
P	10,332	5583	2376	1240	517

(a) Use a graphing utility to plot the data points.

(b) A model for the data is given by
$$P = 10{,}958e^{-0.15h}.$$
Use a graphing utility to graph the model in the same viewing rectangle as in part (a). How well does the model fit the data?

(c) Use a graphing utility to create a table comparing the model with the data points.

(d) Estimate the atmospheric pressure at a height of 8 kilometers.

(e) Use the graph to estimate the altitude at which the atmospheric pressure is 2000 kilograms per square meter.

77. *Savings Plan* Suppose that you decide to start saving pennies according to the following pattern. You save one penny the first day, two pennies the second day, four pennies the third day, etc. Each day you save twice the number of pennies as the previous day. Which function in Exercise 57 models this problem? How many pennies do you save on the thirtieth day? (In the next chapter you will learn how to find the total number saved.)

78. *Median Home Prices* For the years 1987 through 1992, the median price of a home in the United States is given in the following table. (*Source: Chicago Title Insurance Company*)

Year	1987	1988	1989
Price	$99,260	$121,920	$129,800

Year	1990	1991	1992
Price	$131,200	$134,300	$141,000

A model for this data is given by
$$y = 131{,}368e^{0.0102t^3},$$
where t is the time in years, with $t = 0$ representing 1990.

(a) Use the model to complete the table and compare the results with the actual data.

Year	1987	1988	1989	1990	1991	1992
Price						

(b) Use a graphing utility to graph the model.

(c) If the model were used to predict home prices in the years ahead, would the predictions be increasing at a faster rate or a slower rate with increasing t? Do you think the model would be reliable for predicting the future prices of homes? Explain.

Section Project

Calculator Experiment

(a) Use a calculator to complete the table.

x	1	10	100	1000	10,000
$\left(1 + \dfrac{1}{x}\right)^x$					

(b) Use the table to sketch the graph of the function
$$f(x) = \left(1 + \frac{1}{x}\right)^x.$$
Does this graph appear to be approaching a horizontal asymptote?

(c) From parts (a) and (b), what conclusions can you make about the value of
$$\left(1 + \frac{1}{x}\right)^x$$
as x gets larger and larger?

9.2 Logarithmic Functions and Their Graphs

Logarithmic Functions ▪ Graphs of Logarithmic Functions ▪
The Natural Logarithmic Function ▪ Change of Base

Logarithmic Functions

In Section 8.2, you were introduced to the concept of the inverse of a function. Moreover, you saw that if a function has the property that no horizontal line intersects the graph of the function more than once, the function must have an inverse. By looking back at the graphs of the exponential functions introduced in Section 9.1, you will see that every function of the form

$$f(x) = a^x$$

passes the horizontal line test and therefore must have an inverse. This inverse function is called the **logarithmic function with base a.**

Definition of Logarithmic Function

Let a and x be positive real numbers such that $a \neq 1$. The **logarithm of x with base a** is denoted by $\log_a x$ and is defined as follows.

$$y = \log_a x \qquad \text{if and only if} \qquad x = a^y$$

The function $f(x) = \log_a x$ is the **logarithmic function with base a.**

From the definition of a logarithmic function, you can see that the equations $y = \log_a x$ and $x = a^y$ are equivalent.

Logarithmic Equation   *Exponential Equation*

$$y = \log_a x \qquad\qquad\qquad x = a^y$$

The first equation is in *logarithmic* form and the second equation is in *exponential* form. From these equivalent equations it should also be clear that *a logarithm is an exponent.* For instance, because the exponent in the expression

$$2^3 = 8 \qquad \text{The exponent of } 2^3 \text{ is 3.}$$

is 3, the value of the logarithm $\log_2 8$ is 3. That is,

$$\log_2 8 = 3. \qquad \text{A logarithm is an exponent.}$$

In many applications involving exponential equations, it is helpful to rewrite the equation in logarithmic form.

EXAMPLE 1   *Rewriting Exponential Equations in Logarithmic Form*

Rewrite each exponential equation in logarithmic form.

(a) $4^3 = 64$ (b) $2^0 = 1$ (c) $3^{-1} = \frac{1}{3}$ (d) $9^{1/2} = 3$

Solution

(a) For this equation, the base is 4 and the exponent is 3. Using the fact that a logarithm is an exponent, you can write $3 = \log_4 64$.

(b) For this equation, the base is 2 and the exponent is 0. Using the fact that a logarithm is an exponent, you can write $0 = \log_2 1$.

(c) For this equation, the base is 3 and the exponent is -1. Using the fact that a logarithm is an exponent, you can write $-1 = \log_3 \frac{1}{3}$.

(d) For this equation, the base is 9 and the exponent is $\frac{1}{2}$. Using the fact that a logarithm is an exponent, you can write $\frac{1}{2} = \log_9 3$.

EXAMPLE 2   *Rewriting Logarithmic Equations in Exponential Form*

Rewrite each logarithmic equation in exponential form.

(a) $\log_2 16 = 4$ (b) $\log_{49} 7 = \frac{1}{2}$ (c) $\log_{10} \frac{1}{100} = -2$

Solution

(a) For this equation, the base is 2 and the logarithm is 4. Using the fact that a logarithm is an exponent, you can write $2^4 = 16$.

(b) For this equation, the base is 49 and the logarithm is $\frac{1}{2}$. Using the fact that a logarithm is an exponent, you can write $49^{1/2} = 7$.

(c) For this equation, the base is 10 and the logarithm is -2. Using the fact that a logarithm is an exponent, you can write $10^{-2} = \frac{1}{100}$.

Remember that a logarithm is an exponent. Therefore, to evaluate the logarithmic expression $\log_a x$, you need to ask the question, "To what power must a be raised to obtain x?"

EXAMPLE 3 *Evaluating Logarithms*

Evaluate each logarithm.

(a) $\log_2 16$ (b) $\log_3 9$ (c) $\log_{25} 5$

Solution

In each case you should answer the question, "To what power must the base be raised to obtain the given number?"

Stress that a logarithm is an exponent.

(a) The power to which 2 must be raised to obtain 16 is 4. That is,
$$2^4 = 16 \implies \log_2 16 = 4.$$

(b) The power to which 3 must be raised to obtain 9 is 2. That is,
$$3^2 = 9 \implies \log_3 9 = 2.$$

(c) The power to which 25 must be raised to obtain 5 is $\frac{1}{2}$. That is,
$$25^{1/2} = 5 \implies \log_{25} 5 = \frac{1}{2}.$$

Each of the logarithms in Example 4 involves a special important case.

EXAMPLE 4 *Evaluating Logarithms*

Evaluate each logarithm.

(a) $\log_5 1$ (b) $\log_{10} \dfrac{1}{10}$ (c) $\log_3(-1)$ (d) $\log_4 0$

Solution

(a) The power to which 5 must be raised to obtain 1 is 0. That is,
$$5^0 = 1 \implies \log_5 1 = 0.$$

(b) The power to which 10 must be raised to obtain $\frac{1}{10}$ is -1. That is,
$$10^{-1} = \frac{1}{10} \implies \log_{10} \frac{1}{10} = -1.$$

NOTE Be sure you see that a logarithm can be zero or negative, but that you cannot take the logarithm of zero or a negative number.

(c) There is no power to which 3 can be raised to obtain -1. The reason for this is that for any value of x, 3^x is a positive number. Thus, $\log_3(-1)$ is undefined.

(d) There is no power to which 4 can be raised to obtain 0. Thus, $\log_4 0$ is undefined.

The following properties of logarithms follow directly from the definition of the logarithmic function with base a.

Evaluating Special Logarithms

Let a and x be positive real numbers such that $a \neq 1$. Then the following properties are true.

1. $\log_a 1 = 0$ because $a^0 = 1$.

2. $\log_a a = 1$ because $a^1 = a$.

3. $\log_a a^x = x$ because $a^x = a^x$.

The logarithmic function with base 10 is called the **common logarithmic function.** On most calculators, this function can be evaluated with the common logarithmic key $\boxed{\text{LOG}}$, as illustrated in the next example.

EXAMPLE 5   *Evaluating Common Logarithms*

Evaluate each logarithm. Use a calculator only if necessary.

(a) $\log_{10} 100$ (b) $\log_{10} 0.01$ (c) $\log_{10} 10^{16}$ (d) $\log_{10} 5$

Solution

Inform students that $\log_{10} 100$, $\log_{10} 0.01$, and $\log_{10} \frac{1}{1000}$ can be written as log 100, log 0.01, and log $\frac{1}{1000}$.

(a) The power to which 10 must be raised to obtain 100 is 2. That is,
$$10^2 = 100 \quad \Longrightarrow \quad \log_{10} 100 = 2.$$

(b) The power to which 10 must be raised to obtain 0.01 or $\frac{1}{100}$ is -2. That is,
$$10^{-2} = \frac{1}{100} \quad \Longrightarrow \quad \log_{10} 0.01 = -2.$$

(c) The power to which 10 must be raised to obtain 10^{16} is 16. That is,
$$10^{16} = 10^{16} \quad \Longrightarrow \quad \log_{10} 10^{16} = 16.$$

(d) There is no simple power to which 10 can be raised to obtain 5, so you should use a calculator to evaluate $\log_{10} 5$.

Keystrokes	*Display*	
5 $\boxed{\text{LOG}}$	0.69897	Scientific
$\boxed{\text{LOG}}$ 5 $\boxed{\text{ENTER}}$	0.69897	Graphing

Thus, rounded to three decimal places, $\log_{10} 5 \approx 0.699$.

Graphs of Logarithmic Functions

To sketch the graph of $y = \log_a x$, you can use the fact that the graphs of inverse functions are reflections of each other in the line $y = x$.

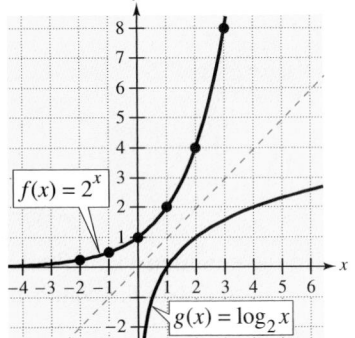

FIGURE 9.8   Inverse functions

EXAMPLE 6   Graphs of Exponential and Logarithmic Functions

On the same rectangular coordinate system, sketch the graphs of the following.

(a) $f(x) = 2^x$ (b) $g(x) = \log_2 x$

Solution

(a) Begin by making a table of values for $f(x) = 2^x$.

x	-2	-1	0	1	2	3
$f(x) = 2^x$	$\frac{1}{4}$	$\frac{1}{2}$	1	2	4	8

By plotting these points and connecting them with a smooth curve, you obtain the graph shown in Figure 9.8.

(b) Because $g(x) = \log_2 x$ is the inverse function of $f(x) = 2^x$, the graph of g is obtained by reflecting the graph of f in the line $y = x$, as shown in Figure 9.8.

Notice from the graph of $g(x) = \log_2 x$, shown in Figure 9.8, that the domain of the function is the set of positive numbers and the range is the set of all real numbers. The basic characteristics of the graph of a logarithmic function are summarized in Figure 9.9. In this figure, note that the graph has one x-intercept, at $(1, 0)$. Also note that the y-axis is a vertical asymptote of the graph.

Study Tip

In Example 6, the inverse property of logarithmic functions was used to sketch the graph of $g(x) = \log_2 x$. You could also use a standard point-plotting approach or a graphing utility. (See the Technology feature on page 663.)

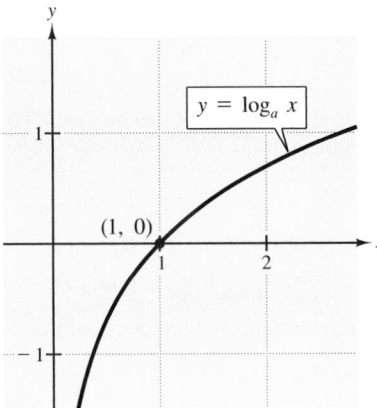

Graph of $y = \log_a x$, $a > 1$

- Domain: $(0, \infty)$
- Range: $(-\infty, \infty)$
- Intercept: $(1, 0)$
- Increasing
- Asymptote: y-axis

FIGURE 9.9   Characteristics of the logarithmic function

Example 7 uses a standard point-plotting approach to sketch a logarithmic function.

EXAMPLE 7   *Sketching the Graph of a Logarithmic Function*

Sketch the graph of the common logarithmic function $f(x) = \log_{10} x$.

Solution

Begin by making a table of values. Note that some of the values can be obtained without a calculator, whereas others require a calculator. Using the points listed in the table, sketch the graph as shown in Figure 9.10. Notice how slowly the graph rises for $x > 1$. In Figure 9.10, you would need to move out to $x = 1000$ before the graph would rise to $y = 3$.

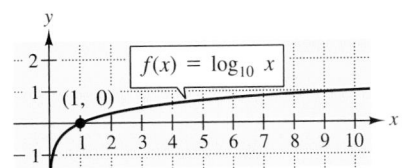

FIGURE 9.10

x	Without calculator				With calculator		
	$\frac{1}{100}$	$\frac{1}{10}$	1	10	2	5	8
$\log_{10} x$	-2	-1	0	1	0.301	0.699	0.903

EXAMPLE 8   *Sketching the Graphs of Logarithmic Functions*

Sketch the graphs of (a) $g(x) = 2 + \log_{10} x$ and (b) $h(x) = \log_{10}(x - 1)$.

Solution

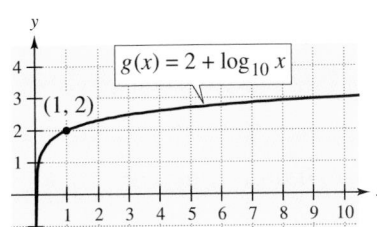

FIGURE 9.11

(a) You can sketch the graph of g by shifting the graph of $f(x) = \log_{10} x$ two units up, as shown in Figure 9.11. Once you have sketched the graph of g, you should check a few points to make sure that you have shifted the graph properly. For instance, when $x = 1$, you have

$$g(1) = 2 + \log_{10} 1 = 2 + 0 = 2.$$

This implies that $(1, 2)$ is a point on the graph of g, as shown in Figure 9.11.

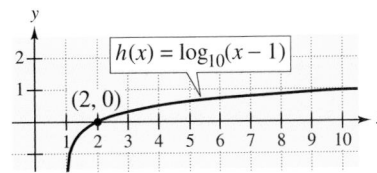

FIGURE 9.12

(b) You can sketch the graph of h by shifting the graph of $f(x) = \log_{10} x$ one unit to the right, as shown in Figure 9.12. After sketching the graph of h, try checking some points to make sure that you have shifted the graph correctly. For instance, when $x = 2$, you have

$$h(2) = \log_{10}(2 - 1) = \log_{10} 1 = 0.$$

This implies that $(2, 0)$ is a point on the graph of h, as shown in Figure 9.12.

The Natural Logarithmic Function

As with exponential functions, the most widely used base for logarithmic functions is the number e. The logarithmic function with base e is the **natural logarithmic function** and is denoted by the special symbol ln x, which is read as "el en of x."

The Natural Logarithmic Function

The function defined by

$$f(x) = \log_e x = \ln x,$$

where $x > 0$, is called the **natural logarithmic function.**

The three properties of logarithms listed earlier in this section are also valid for natural logarithms.

Evaluating Special Natural Logarithms

Let x be a positive real number. Then the following properties are true.

1. $\ln 1 = 0$ because $e^0 = 1$.

2. $\ln e = 1$ because $e^1 = e$.

3. $\ln e^x = x$ because $e^x = e^x$.

Orally or in writing, have students discuss the difference between a common logarithm and a natural logarithm.

EXAMPLE 9   *Evaluating the Natural Logarithmic Function*

Evaluate each of the following. Then incorporate the results into a graph of the natural logarithmic function, $f(x) = \ln x$.

(a) $\ln e^2$ (b) $\ln \dfrac{1}{e}$

Solution

Using the property that $\ln e^x = x$, you obtain the following.

(a) $\ln e^2 = 2$ Note: $e^2 \approx 7.39$

(b) $\ln \dfrac{1}{e} = \ln e^{-1} = -1$ Note: $\dfrac{1}{e} \approx 0.37$

Using the points $(1/e, -1)$, $(1, 0)$, $(e, 1)$, and $(e^2, 2)$, you can sketch the graph of the natural logarithmic function, as shown in Figure 9.13.

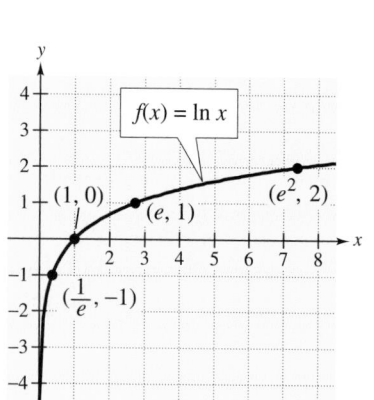

FIGURE 9.13

Change of Base

Although 10 and e are the most frequently used bases, you occasionally need to evaluate logarithms with other bases. In such cases, the following **change-of-base formula** is useful.

You can use a graphing utility to graph logarithmic functions that do not have a base of 10 by using the change-of-base formula. Use the change-of-base formula to rewrite $g(x) = \log_2 x$ in Example 6 on page 660 (with $b = 10$). Use the trace feature to estimate $g(x) = \log_2 x$ when $x = 3$. Verify your estimate arithmetically using a calculator.

Change-of-Base Formula

Let a, b, and x be positive real numbers such that $a \neq 1$ and $b \neq 1$. Then $\log_a x$ is given as follows.

$$\log_a x = \frac{\log_b x}{\log_b a} \quad \text{or} \quad \log_a x = \frac{\ln x}{\ln a}$$

The usefulness of this change-of-base formula is that you can use a calculator that has only the common logarithm key $\boxed{\text{LOG}}$ and the natural logarithm key $\boxed{\text{LN}}$ to evaluate logarithms to any base.

The English mathematician Henry Briggs (1561–1630) was also a founder of calculations by logarithms. He worked with John Napier, and they both published papers on the advantage of having tables of logarithms using the base 10.

EXAMPLE 10   *Changing the Base to Evaluate Logarithms*

(a) Use *common* logarithms to evaluate $\log_3 5$.

(b) Use *natural* logarithms to evaluate $\log_6 2$.

Solution

Using the change-of-base formula, you can convert to common and natural logarithms by writing

$$\log_3 5 = \frac{\log_{10} 5}{\log_{10} 3} \quad \text{and} \quad \log_6 2 = \frac{\ln 2}{\ln 6}.$$

Now, use the following keystrokes.

(a) *Keystrokes*   *Display*

$5\ \boxed{\text{LOG}}\ \boxed{\div}\ 3\ \boxed{\text{LOG}}\ \boxed{=}$ 1.4649735   Scientific

$\boxed{\text{LOG}}\ 5\ \boxed{\div}\ \boxed{\text{LOG}}\ 3\ \boxed{\text{ENTER}}$ 1.4649735   Graphing

Thus, $\log_3 5 \approx 1.465$.

(b) *Keystrokes*   *Display*

$2\ \boxed{\text{LN}}\ \boxed{\div}\ 6\ \boxed{\text{LN}}\ \boxed{=}$ 0.3868528   Scientific

$\boxed{\text{LN}}\ 2\ \boxed{\div}\ \boxed{\text{LN}}\ 6\ \boxed{\text{ENTER}}$ 0.3868528   Graphing

Thus, $\log_6 2 \approx 0.387$.

At this point, you have been introduced to all the basic types of functions that are covered in this course: polynomial functions, radical functions, rational functions, exponential functions, and logarithmic functions. The only other common types of functions are *trigonometric functions,* which you will study if you go on to take a course in trigonometry or precalculus.

Group Activities

Reviewing the Major Concepts

Comparing Models Suppose you work for a research and development firm that deals with a wide variety of disciplines. Your supervisor has asked your group to give a presentation to your department on four basic kinds of mathematical models. Identify each of the models shown below. Develop a presentation describing the types of data sets that each model would best represent. Include distinctions in domain, range, and intercepts, and a discussion of the types of applications to which each model is suited.

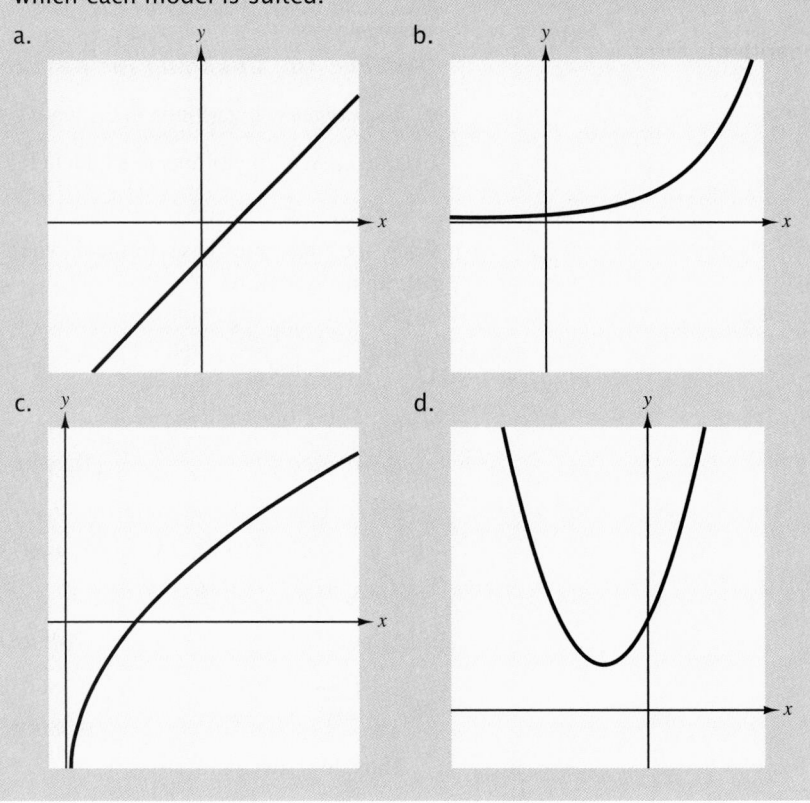

a.

b.

c.

d.

9.2   **Exercises**

See Warm-Up
Exercises, p. A49

1. *Think About It*   Explain the relationship between

$$f(x) = 2^x \quad \text{and} \quad g(x) = \log_2 x.$$

2. *Think About It*   Explain why $\log_a a^x = x$.

In Exercises 3–12, write the logarithmic equation in exponential form.

3. $\log_5 25 = 2$

4. $\log_6 36 = 2$

5. $\log_4 \frac{1}{16} = -2$

6. $\log_8 \frac{1}{8} = -1$

7. $\log_3 \frac{1}{243} = -5$

8. $\log_{10} 10,000 = 4$

9. $\log_{36} 6 = \frac{1}{2}$

10. $\log_{32} 4 = \frac{2}{5}$

11. $\log_8 4 = \frac{2}{3}$

12. $\log_{16} 8 = \frac{3}{4}$

In Exercises 13–22, write the exponential equation in logarithmic form.

13. $7^2 = 49$

14. $6^4 = 1296$

15. $3^{-2} = \frac{1}{9}$

16. $5^4 = 625$

17. $8^{2/3} = 4$

18. $81^{3/4} = 27$

19. $25^{-1/2} = \frac{1}{5}$

20. $6^{-3} = \frac{1}{216}$

21. $4^0 = 1$

22. $10^{0.12} \approx 1.318$

In Exercises 23–40, evaluate the logarithm without a calculator. (If not possible, state the reason.)

23. $\log_2 8$

24. $\log_3 27$

25. $\log_{10} 10$

26. $\log_8 8$

27. $\log_2 \frac{1}{4}$

28. $\log_3 \frac{1}{9}$

29. $\log_2(-3)$

30. $\log_3 1$

31. $\log_{10} 1000$

32. $\log_{10} \frac{1}{100}$

33. $\log_4 1$

34. $\log_5(-6)$

35. $\log_9 3$

36. $\log_{25} 125$

37. $\log_4(-4)$

38. $\log_2 0$

39. $\log_{16} 4$

40. $\log_5 5^3$

In Exercises 41–52, use a calculator to evaluate the logarithm. Round to four decimal places.

41. $\log_{10} 31$

42. $\log_{10} \frac{\sqrt{3}}{2}$

43. $\log_{10} 0.85$

44. $\log_{10} 0.345$

45. $\log_{10}\left(\sqrt{2} + 4\right)$

46. $\log_{10} 5310$

47. $\ln 25$

48. $\ln 6.57$

49. $\ln 0.75$

50. $\ln\left(\sqrt{3} - 1\right)$

51. $\ln\left(\dfrac{1 + \sqrt{5}}{3}\right)$

52. $\ln\left(1 + \dfrac{0.10}{12}\right)$

In Exercises 53–56, state the relationship between the functions f and g.

53. $f(x) = 6^x$

   $g(x) = \log_6 x$

54. $f(x) = 5^x$

   $g(x) = \log_5 x$

55. $f(x) = e^x$

   $g(x) = \ln x$

56. $f(x) = 10^x$

   $g(x) = \log_{10} x$

In Exercises 57–60, sketch the graphs of f and g on the same set of coordinate axes. What can you conclude about the relationship between f and g?

57. $f(x) = \log_3 x$

   $g(x) = 3^x$

58. $f(x) = \log_4 x$

   $g(x) = 4^x$

59. $f(x) = \log_6 x$

   $g(x) = 6^x$

60. $f(x) = \log_{(1/2)} x$

   $g(x) = \left(\frac{1}{2}\right)^x$

In Exercises 61–66, match the function with its graph. [The graphs are labeled (a), (b), (c), (d), (e), and (f).]

(a)
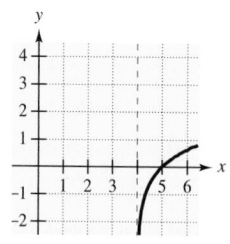

(b)
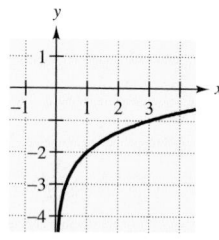

(c)
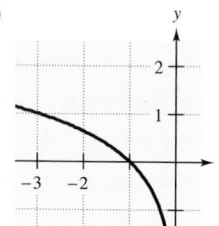

(d)
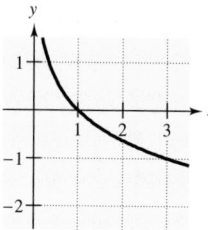

(e)
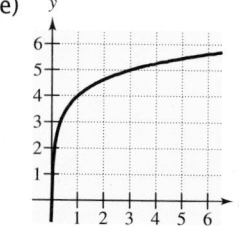

(f)
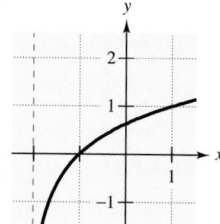

61. $f(x) = 4 + \log_3 x$ **62.** $f(x) = -2 + \log_3 x$

63. $f(x) = -\log_3 x$ **64.** $f(x) = \log_3(-x)$

65. $f(x) = \log_3(x - 4)$ **66.** $f(x) = \log_3(x + 2)$

In Exercises 67–76, sketch the graph of the function.

67. $f(x) = \log_5 x$ **68.** $g(x) = \log_8 x$

69. $g(t) = -\log_2 t$ **70.** $h(s) = -2 \log_3 s$

71. $f(x) = 3 + \log_2 x$ **72.** $f(x) = -2 + \log_3 x$

73. $g(x) = \log_2(x - 3)$ **74.** $h(x) = \log_3(x + 1)$

75. $f(x) = \log_{10}(10x)$ **76.** $g(x) = \log_4(4x)$

In Exercises 77–82, find the domain and vertical asymptote of the logarithmic function. Sketch its graph.

77. $f(x) = \log_4 x$ **78.** $g(x) = \log_6 x$

79. $h(x) = \log_4(x - 3)$ **80.** $f(x) = -\log_6(x + 2)$

81. $y = -\log_3 x + 2$ **82.** $y = \log_5(x - 1) + 4$

In Exercises 83–88, use a graphing utility to graph the function. Determine the domain and identify any vertical asymptote.

83. $y = 5 \log_{10} x$ **84.** $y = 5 \log_{10}(x - 3)$

85. $y = -3 + 5 \log_{10} x$ **86.** $y = 5 \log_{10}(3x)$

87. $y = \log_{10}\left(\dfrac{x}{5}\right)$ **88.** $y = \log_{10}(-x)$

In Exercises 89–94, match the function with its graph. [The graphs are labeled (a), (b), (c), (d), (e), and (f).]

(a)
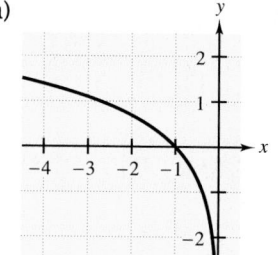

(b)
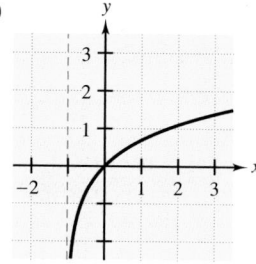

(c)
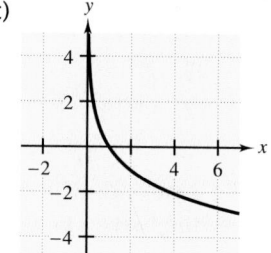

(d)
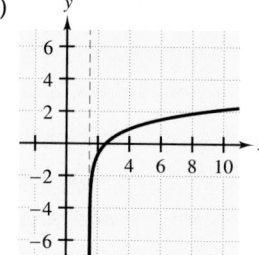

(e)
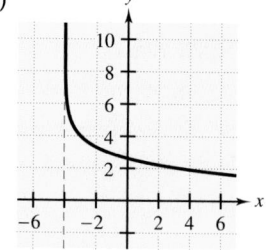

(f)
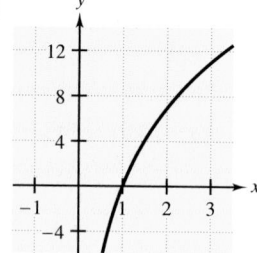

89. $f(x) = \ln(x + 1)$ **90.** $f(x) = 4 - \ln(x + 4)$

91. $f(x) = \ln\left(x - \frac{3}{2}\right)$ **92.** $f(x) = -\frac{3}{2} \ln x$

93. $f(x) = 10 \ln x$ **94.** $f(x) = \ln(-x)$

In Exercises 95–102, sketch the graph of the function.

95. $f(x) = -\ln x$

96. $f(x) = -2 \ln x$

97. $f(x) = 3 \ln x$

98. $h(t) = 4 \ln t$

99. $f(x) = 1 + \ln x$

100. $h(x) = 2 + \ln x$

101. $g(t) = 2 \ln(t - 4)$

102. $g(x) = -3 \ln(x + 3)$

In Exercises 103–106, use a graphing utility to graph the function. Determine the domain and identify any vertical asymptotes.

103. $g(x) = \ln(x + 6)$

104. $h(x) = -\ln(x - 2)$

105. $f(t) = 3 + 2 \ln t$

106. $g(t) = \ln(3 - t)$

In Exercises 107–118, use a calculator to evaluate the logarithm by means of the change-of-base formula. Use (a) the common logarithm key and (b) the natural logarithm key.

107. $\log_8 132$

108. $\log_5 510$

109. $\log_3 7$

110. $\log_7 4$

111. $\log_2 0.72$

112. $\log_{12} 0.6$

113. $\log_{15} 1250$

114. $\log_{20} 125$

115. $\log_{1/2} 4$

116. $\log_{1/3}(0.015)$

117. $\log_4 \sqrt{42}$

118. $\log_3(1 + e^2)$

119. *American Elk*   The antler spread a (in inches) and the shoulder height h (in inches) of an adult male elk are related by the model

$$h = 116 \log_{10}(a + 40) - 176.$$

Approximate the shoulder height of an adult male elk with an antler spread of 55 inches.

120. *Intensity of Sound*   The relationship between the number of decibels B and the intensity of a sound I in watts per meter squared is given by

$$B = 10 \log_{10}\left(\frac{I}{10^{-16}}\right).$$

Determine the number of decibels of a sound with an intensity of 10^{-4} watts per meter squared.

121. *Creating a Table*   The time t in years for an investment to double in value when compounded continuously at annual rate r is given by

$$t = \frac{\ln 2}{r}.$$

Complete the table, which shows the "doubling times" for several annual percent rates.

r	0.07	0.08	0.09	0.10	0.11	0.12
t						

122. *Tractrix*   A person walking along a dock (the y-axis) drags a boat by a 10-foot rope (see figure). The boat travels along a path known as a *tractrix*. The equation of this path is

$$y = 10 \ln\left(\frac{10 + \sqrt{100 - x^2}}{x}\right) - \sqrt{100 - x^2}.$$

(a) Use a graphing utility to graph the function. What is the domain of the function?

(b) Identify any asymptotes of the graph.

(c) Determine the position of the person when the x-coordinate of the position of the boat is $x = 2$.

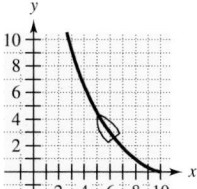

Section Project

Answer the question for the function $f(x) = \log_{10} x$. (Do not use a calculator.)

(a) Describe the values of $f(x)$ for $1000 \le x \le 10{,}000$.

(b) Describe the values of x, given that $f(x)$ is negative.

(c) By what amount will x increase, given that $f(x)$ is increased by one unit?

(d) Find the ratio of a to b, given that $f(a) = 3f(b)$.

9.3	**Properties of Logarithms**

Properties of Logarithms ▪ Rewriting Logarithmic Expressions ▪ Application

Properties of Logarithms

You know from the previous section that the logarithmic function with base a is the *inverse* of the exponential function with base a. Thus, it makes sense that each property of exponents should have a corresponding property of logarithms. For instance, the exponential property

$$a^0 = 1 \qquad \text{Exponential property}$$

has the corresponding logarithmic property

$$\log_a 1 = 0. \qquad \text{Corresponding logarithmic property}$$

In this section, you will study the logarithmic properties that correspond to the following three exponential properties.

Base a *Natural Base*

1. $a^m a^n = a^{m+n}$ $e^m e^n = e^{m+n}$

2. $\dfrac{a^m}{a^n} = a^{m-n}$ $\dfrac{e^m}{e^n} = e^{m-n}$

3. $(a^m)^n = a^{mn}$ $(e^m)^n = e^{mn}$

Properties of Logarithms

Let a be a positive real number such that $a \neq 1$, and let n be a real number. If u and v are real numbers, variables, or algebraic expressions such that $u > 0$ and $v > 0$, the following properties are true.

Logarithm with Base a *Natural Logarithm*

1. $\log_a(uv) = \log_a u + \log_a v$ $\ln(uv) = \ln u + \ln v$

2. $\log_a \dfrac{u}{v} = \log_a u - \log_a v$ $\ln \dfrac{u}{v} = \ln u - \ln v$

3. $\log_a u^n = n \log_a u$ $\ln u^n = n \ln u$

Ask students to use their calculators to evaluate log(2 + 3) and log 2 + log 3 and then to compare the results.

NOTE There is no general property of logarithms that can be used to simplify $\log_a(u + v)$. Specifically,

$$\log_a(u + v) \textit{ does not equal } \log_a u + \log_a v.$$

EXAMPLE 1   Using Properties of Logarithms

Use $\ln 2 \approx 0.693$, $\ln 3 \approx 1.099$, and $\ln 5 \approx 1.609$ to approximate each of the following.

(a) $\ln \dfrac{2}{3}$ (b) $\ln 10$ (c) $\ln 30$

Solution

(a) $\ln \dfrac{2}{3} = \ln 2 - \ln 3$ Property 2

       $\approx 0.693 - 1.099$ Substitute for $\ln 2$ and $\ln 3$.

       $= -0.406$ Simplify.

(b) $\ln 10 = \ln(2 \cdot 5)$ Factor.

       $= \ln 2 + \ln 5$ Property 1

       $\approx 0.693 + 1.609$ Substitute for $\ln 2$ and $\ln 5$.

       $= 2.302$ Simplify.

Point out that Logarithm Property 1 applies even if there are more than two factors.

(c) $\ln 30 = \ln(2 \cdot 3 \cdot 5)$ Factor.

       $= \ln 2 + \ln 3 + \ln 5$ Property 1

       $\approx 0.693 + 1.099 + 1.609$ Substitute for $\ln 2$, $\ln 3$, and $\ln 5$.

       $= 3.401$ Simplify.

NOTE   When using the properties of logarithms, it helps to state the properties *verbally*. For instance, the verbal form of the property $\ln(uv) = \ln u + \ln v$ is: *The log of a product is the sum of the logs of the factors.* Similarly, the verbal form of the property $\ln(u/v) = \ln u - \ln v$ is: *The log of a quotient is the difference of the logs of the numerator and denominator.*

Study Tip

Remember that you can verify results such as those given in Examples 1 and 2 with a calculator.

EXAMPLE 2   Using Properties of Logarithms

Use the properties of logarithms to verify that $-\ln 2 = \ln \frac{1}{2}$.

Solution

Using Property 3, you can write the following.

       $-\ln 2 = (-1)\ln 2$ Rewrite coefficient as -1.

         $= \ln 2^{-1}$ Property 3

         $= \ln \dfrac{1}{2}$ Rewrite 2^{-1} as $\frac{1}{2}$.

Rewriting Logarithmic Expressions

In Examples 1 and 2, the properties of logarithms were used to rewrite logarithmic expressions involving the log of a *constant*. A more common use of the properties of logarithms is to rewrite the log of a *variable expression*.

EXAMPLE 3   Rewriting the Logarithm of a Product

Use the properties of logarithms to rewrite $\log_{10} 7x^3$.

Solution

$$\log_{10} 7x^3 = \log_{10} 7 + \log_{10} x^3 \qquad \text{Property 1}$$
$$= \log_{10} 7 + 3 \log_{10} x \qquad \text{Property 3}$$

NOTE   In Example 3, try confirming the result by substituting values of x.

When you rewrite a logarithmic expression as in Example 3, you are **expanding** the expression. The reverse procedure is demonstrated in Example 4, and is called **condensing** a logarithmic expression.

EXAMPLE 4   Condensing a Logarithmic Expression

Use the properties of logarithms to condense $\ln x - \ln 3$.

Solution

Using Property 2, you can write

$$\ln x - \ln 3 = \ln \frac{x}{3}. \qquad \text{Property 2}$$

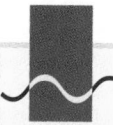

Technology

When you are rewriting a logarithmic expression, remember that you can use a graphing utility to check your result graphically. For instance, in Example 4, try graphing the functions

$$y_1 = \ln x - \ln 3$$

and

$$y_2 = \ln \frac{x}{3}$$

in the same viewing rectangle. You should obtain the same graph for each function.

EXAMPLE 5   Expanding a Logarithmic Expression

Use the properties of logarithms to expand $\log_2 3xy^2$, $x > 0, y > 0$.

Solution

$$\log_2 3xy^2 = \log_2 3 + \log_2 x + \log_2 y^2 \qquad \text{Property 1}$$
$$= \log_2 3 + \log_2 x + 2 \log_2 y \qquad \text{Property 3}$$

Sometimes expanding or condensing logarithmic expressions involves several steps. In the next example, be sure that you can justify each step in the solution. Also, notice how different the expanded expression is from the original.

EXAMPLE 6   *Expanding a Logarithmic Expression*

Use the properties of logarithms to expand $\ln \sqrt{x^2 - 1}, \; x > 1$.

Solution

$$
\begin{aligned}
\ln \sqrt{x^2 - 1} &= \ln(x^2 - 1)^{1/2} && \text{Rewrite using fractional exponent.} \\
&= \tfrac{1}{2} \ln(x^2 - 1) && \text{Property 3} \\
&= \tfrac{1}{2} \ln[(x - 1)(x + 1)] && \text{Factor.} \\
&= \tfrac{1}{2}[\ln(x - 1) + \ln(x + 1)] && \text{Property 1} \\
&= \tfrac{1}{2} \ln(x - 1) + \tfrac{1}{2} \ln(x + 1) && \text{Distributive Property}
\end{aligned}
$$

Stress the importance of grouping symbols, and point out that $\frac{1}{2}$ was distributed to both terms in Example 6.

EXAMPLE 7   *Condensing a Logarithmic Expression*

Use the properties of logarithms to condense $\ln 2 - 2 \ln x$.

Solution

$$
\begin{aligned}
\ln 2 - 2 \ln x &= \ln 2 - \ln x^2, \quad x > 0 && \text{Property 3} \\
&= \ln \frac{2}{x^2}, \quad x > 0 && \text{Property 2}
\end{aligned}
$$

Be aware that some students confuse the property of logarithms that states $\log_b u - \log_b v = \log_b \dfrac{u}{v}$ with $\log_v u = \dfrac{\log_b u}{\log_b v}$. Hence, they may be tempted to rewrite incorrectly $\log_a x = \dfrac{\log_b x}{\log_b a}$ as $\log_b x - \log_b a$.

When you expand or condense a logarithmic expression, it is possible to change the domain of the expression. For instance, the domain of the function

$$f(x) = 2 \ln x \qquad \text{Domain is the set of positive real numbers.}$$

is the set of positive real numbers, whereas the domain of

$$g(x) = \ln x^2 \qquad \text{Domain is the set of nonzero real numbers.}$$

is the set of nonzero real numbers. Thus, when you expand or condense a logarithmic expression, you should check to see whether the rewriting has changed the domain of the expression. In such cases, you should restrict the domain appropriately. For instance, you can write

$$f(x) = 2 \ln x = \ln x^2, \quad x > 0.$$

Exploration

Use the properties of logarithms to rewrite the logarithmic expression $\ln x^4$. Use a graphing utility to graph the two expressions. Should both have the same graph? What can you conclude?

Application

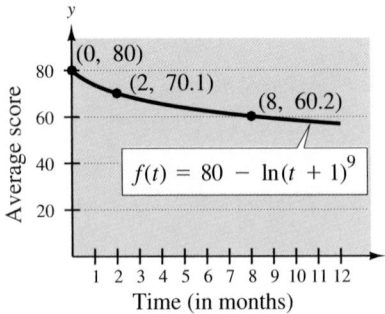

$$f(t) = 80 - \ln(t + 1)^9$$

FIGURE 9.14 Human memory model

Common error: Be aware that some students may incorrectly simplify $80 - 9 \ln 9$ as $71 \ln 9$. Review the order of operations.

EXAMPLE 8   An Application: Human Memory Model

In an experiment, students attended several lectures and then were tested every month for a year to see how much of the material they remembered. The average scores for the group are given by the **human memory model**

$$f(t) = 80 - \ln(t + 1)^9, \quad 0 \le t \le 12,$$

where t is the time in months. Find the average scores for the group after 2 months and after 8 months. The graph of the function is shown in Figure 9.14.

Solution

To make the calculations easier, rewrite the model as

$$f(t) = 80 - 9 \ln(t + 1), \quad 0 \le t \le 12.$$

You can find the average scores as follows.

$$f(2) = 80 - 9 \ln 3 \approx 70.1 \qquad \text{Average score after 2 months}$$
$$f(8) = 80 - 9 \ln 9 \approx 60.2 \qquad \text{Average score after 8 months}$$

EXAMPLE 9   Estimating the Time of Death

At 8:30 A.M., a coroner was called to the home of a person who had died during the night. To estimate the time of death, the coroner took the person's body temperature twice. At 9:00 A.M. the temperature was 85.7°F, and at 9:30 A.M. the temperature was 82.8°F. The room temperature was 70°F. When did the person die?

Solution

From the first temperature reading, the coroner could obtain

$$kt = \ln \frac{T - S}{T_0 - S} = \ln \frac{85.7 - 70}{98.6 - 70} \approx -0.6.$$

From the second temperature reading, the coroner could obtain

$$k\left(t + \frac{1}{2}\right) = \ln \frac{82.8 - 70}{98.6 - 70} \approx -0.8.$$

NOTE In Example 9, the coroner used Newton's Law of Cooling,

$$kt = \ln \frac{T - S}{T_0 - S},$$

where k is a constant, S is the temperature of the surrounding air, and t is the time it takes for the body temperature to cool from $T_0 = 98.6$ to T.

Solving the two equations $kt = -0.6$ and $k(t + 1/2) = -0.8$ for k and t produces $k = -0.4$ and $t = 1.5$. Therefore, the coroner could approximate that the first temperature reading took place 1.5 hours after death, which implies that the death occurred at about 7:30 A.M.

Group Activities

x	5	7	9	11	13
y	522	1682	3370	5375	12,789

Problem Solving

Mathematical Modeling The data in the table represents the annual number y of new AIDS cases in females reported in the United States for the year x from 1985 through 1993, with $x = 5$ corresponding to 1985. *(Source: National Center for Health Statistics)*

a. Plot the data. Would a linear model fit the points well?

b. Add to the table a third row giving the values of $\ln y$.

c. Plot the coordinate pairs $(x, \ln y)$. Would a linear model fit these points well? If so, draw the best-fitting line and find its equation.

d. Describe the shapes of the two scatter plots. Using your knowledge of logarithms, explain why the second scatter plot is so different from the first.

9.3 Exercises

See Warm-Up Exercises, p. A49

1. *Think About It* Is it true that $\log_4(x^2 + 9) = \log_4 x^2 + \log_4 9$? Explain.

2. *Think About It* If $f(x) = \log_a x$, does $f(ax) = 1 + f(x)$? Explain.

In Exercises 3–12, evaluate the expression without a calculator. (If not possible, state the reason.)

3. $\log_5 5^2$

4. $\log_6 2 + \log_6 3$

5. $\ln 1$

6. $\log_{10} 1$

7. $\log_3 9$

8. $\log_6 \sqrt{6}$

9. $\log_2 \frac{1}{8}$

10. $\log_3(-3)$

11. $\ln e^5 - \ln e^2$

12. $\ln e^4$

In Exercises 13–42, use the properties of logarithms to expand the expression.

13. $\log_3 11x$

14. $\ln 5x$

15. $\ln y^3$

16. $\log_7 x^2$

17. $\log_2 \frac{z}{17}$

18. $\log_{10} \frac{7}{y}$

19. $\log_5 x^{-2}$

20. $\log_2 \sqrt{s}$

21. $\log_3 \sqrt[3]{x + 1}$

22. $\log_4 \frac{1}{\sqrt{t}}$

23. $\ln 3x^2 y$

24. $\ln \left[y(y - 1)^2 \right]$

25. $\ln \sqrt{x(x + 2)}$

26. $\ln \frac{5}{x - 2}$

27. $\log_2 \frac{x^2}{x - 3}$

28. $\log_5 \sqrt{\frac{x}{y}}$

29. $\ln \sqrt[3]{x(x + 5)}$

30. $\ln[3x(x - 5)]^2$

31. $\ln \sqrt[3]{\frac{x^2}{x + 1}}$

32. $\ln \left(\frac{x + 1}{x - 1} \right)^2$

33. $\log_4[x^6(x-7)^2]$

34. $\log_3 \dfrac{x^2 y}{z^7}$

35. $\ln \dfrac{a^3(b-4)}{c^2}$

36. $\log_8[(x-y)^4 z^6]$

37. $\log_{10} \dfrac{(4x)^3}{x-7}$

38. $\log_4 \dfrac{\sqrt[3]{a+1}}{(ab)^4}$

39. $\ln \dfrac{x\sqrt[3]{y}}{(wz)^4}$

40. $\log_5[(xy)^2(x+3)^4]$

41. $\ln\left[(x+y)\dfrac{\sqrt[5]{w+2}}{3t}\right]$

42. $\log_6[a\sqrt{b}(c-d)^3]$

In Exercises 43–70, use the properties of logarithms to condense the expression.

43. $\log_2 3 + \log_2 x$

44. $\log_5 2x + \log_5 3y$

45. $\log_{10} 4 - \log_{10} x$

46. $\ln 10x - \ln z$

47. $4 \ln b$

48. $10 \log_4 z$

49. $-2 \log_5 2x$

50. $-5 \ln(x+3)$

51. $\frac{1}{3} \ln(2x+1)$

52. $-\frac{1}{2} \log_3 5y$

53. $\log_3 2 + \frac{1}{2} \log_3 y$

54. $\ln 6 - 3 \ln z$

55. $2 \ln x + 3 \ln y - \ln z$

56. $4 \ln 3 - 2 \ln x - \ln y$

57. $4(\ln x + \ln y)$

58. $2[\ln x - \ln(x+1)]$

59. $\frac{1}{2}(\ln 8 + \ln 2x)$

60. $5[\ln x - \frac{1}{2} \ln(x+4)]$

61. $\log_4(x+8) - 3 \log_4 x$

62. $5 \log_3 x + \log_3(x-6)$

63. $\frac{1}{2} \log_5 x + \log_5(x-3)$

64. $\frac{1}{4} \log_6(x+1) - \log_6 x$

65. $5 \log_6(c+d) - \log_6(m-n)$

66. $2 \log_5(x+y) + 3 \log_5 w$

67. $\frac{1}{5}(3 \log_2 x - \log_2 y)$

68. $\frac{1}{3}[\ln(x-6) - \ln y - 2 \ln z]$

69. $\frac{1}{5} \log_6(x-3) - 2 \log_6 x - 3 \log_6(x+1)$

70. $3[\frac{1}{2} \log_9(a+6) - \log_9(a-1)]$

In Exercises 71–76, use $\log_{10} 3 \approx 0.477$ and $\log_{10} 12 \approx 1.079$ to approximate the logarithm. Use a calculator to verify your result.

71. $\log_{10} 9$

72. $\log_{10} \frac{1}{4}$

73. $\log_{10} 36$

74. $\log_{10} 144$

75. $\log_{10} \sqrt{36}$

76. $\log_{10} 5^0$

In Exercises 77–88, use $\log_4 2 = 0.5000$ and $\log_4 3 \approx 0.7925$ to approximate the logarithm. Do not use a calculator.

77. $\log_4 4$

78. $\log_4 8$

79. $\log_4 6$

80. $\log_4 24$

81. $\log_4 \frac{3}{2}$

82. $\log_4 \frac{9}{2}$

83. $\log_4 \sqrt{2}$

84. $\log_4 \sqrt[3]{9}$

85. $\log_4 (3 \cdot 2^4)$

86. $\log_4 \sqrt{3 \cdot 2^5}$

87. $\log_4 3^0$

88. $\log_4 4^3$

In Exercises 89–94, simplify the expression.

89. $\log_4 \dfrac{4}{x}$

90. $\log_3(3^2 \cdot 4)$

91. $\log_5 \sqrt{50}$

92. $\log_2 \sqrt{22}$

93. $\ln 3e^2$

94. $\ln \dfrac{6}{e^5}$

In Exercises 95–98, use a graphing utility to graph the two equations in the same viewing rectangle. Use the graphs to verify that the expressions are equivalent. (Assume $x > 0$.)

95. $y_1 = \ln\left(\dfrac{10}{x^2+1}\right)^2$, $\quad y_2 = 2[\ln 10 - \ln(x^2+1)]$

96. $y_1 = \ln \sqrt{x(x+1)}$, $\quad y_2 = \frac{1}{2}[\ln x + \ln(x+1)]$

97. $y_1 = \ln[x^2(x+2)]$, $\quad y_2 = 2 \ln x + \ln(x+2)$

98. $y_1 = \ln\left(\dfrac{\sqrt{x}}{x-3}\right)$, $\quad y_2 = \frac{1}{2} \ln x - \ln(x-3)$

True or False? In Exercises 99–104, determine whether the equation is true or false. Explain.

99. $\ln e^{2-x} = 2 - x$

100. $\log_2 8x = 3 + \log_2 x$

101. $\log_8 4 + \log_8 16 = 2$

102. $\log_3(u + v) = \log_3 u + \log_3 v$

103. $\log_3(u + v) = \log_3 u \cdot \log_3 v$

104. $\dfrac{\log_6 10}{\log_6 3} = \log_6 10 - \log_6 3$

105. *Human Memory Model*   Students participating in a psychology experiment attended several lectures on a subject. Every month for a year after that, the students were tested to see how much of the material they remembered. The average scores for the group are given by the human memory model

$$f(t) = 80 - \log_{10}(t + 1)^{12}, \quad 0 \le t \le 12,$$

where t is the time in months.

(a) Find the average scores when $t = 2$ and $t = 8$.

(b) Use a graphing utility to graph the function.

106. *Intensity of Sound*   The relationship between the number of decibels B and the intensity of a sound I in watts per centimeter squared is given by

$$B = 10 \log_{10}\!\left(\dfrac{I}{10^{-16}}\right).$$

Use the properties of logarithms to write the formula in simpler form, and determine the number of decibels of a sound with an intensity of 10^{-10} watts per centimeter squared.

Molecular Transport In Exercises 107 and 108, use the following information. The energy E (in kilocalories per gram molecule) required to transport a substance from the outside to the inside of a living cell is given by

$$E = 1.4(\log_{10} C_2 - \log_{10} C_1),$$

where C_1 and C_2 are the concentrations of the substance outside and inside the cell.

107. Condense the expression.

108. The concentration of a particular substance inside a cell is twice the concentration outside the cell. How much energy is required to transport the substance from outside to inside the cell?

True or False? In Exercises 109–112, determine whether the statement is true or false given that $f(x) = \ln x$. If false, state why or give an example to show that it is false.

109. $f(0) = 0$

110. $\sqrt{f(x)} = \frac{1}{2} \ln x$

111. $f(2x) = \ln 2 + \ln x, \quad x > 0$

112. $f(x - 3) = \ln x - \ln 3, \quad x > 3$

113. *Think About It*   Explain how you can show that

$$\dfrac{\ln x}{\ln y} \neq \ln \dfrac{x}{y}.$$

Section Project

Think About It   Without using a calculator, approximate the natural logarithms of as many integers as possible between 1 and 30 using $\ln 2 \approx 0.6931$, $\ln 3 \approx 1.0986$, $\ln 5 \approx 1.6094$, and $\ln 7 \approx 1.9459$. Explain the method you used. Then verify your results with a calculator and explain differences in the results.

Mid-Chapter Quiz

Take this quiz as you would take a quiz in class. After you are done, check your work against the answers given in the back of the book.

1. Evaluate the function $f(x) = \left(\frac{4}{3}\right)^x$ at the given values of the independent variable. Use a calculator only if it is more efficient.

   (a) $f(2)$ (b) $f(0)$ (c) $f(-1)$ (d) $f(1.5)$

2. Determine the domain and range of the function $g(x) = 2^{-0.5x}$.

In Exercises 3–6, sketch the graph of the function. Use a graphing utility for Exercises 5 and 6.

3. $y = \frac{1}{2}(4^x)$

4. $y = 5(2^{-x})$

5. $f(t) = 12e^{-0.4t}$

6. $g(x) = 100(1.08)^x$

7. You deposit \$750 in an account at an annual interest rate of $7\frac{1}{2}\%$. Complete the table giving the account balance A after 20 years if the interest is compounded n times per year.

n	1	4	12	365	Continuous compounding
A					

8. A gallon of milk costs \$2.23 now. If the price increases by 4% each year, what will the price be after 5 years?

9. Write the logarithmic equation $\log_4\left(\frac{1}{16}\right) = -2$ in exponential form.

10. Write the exponential equation $3^4 = 81$ in logarithmic form.

11. Evaluate $\log_5 125$ without a calculator.

12. Write a paragraph comparing the graphs of $f(x) = \log_5 x$ and $g(x) = 5^x$.

In Exercises 13 and 14, use a graphing utility to graph the function.

13. $f(t) = \frac{1}{2}\ln t$

14. $h(x) = 3 - \ln x$

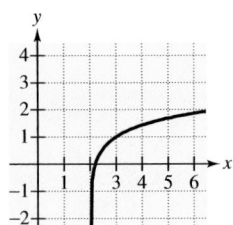

Figure for 15

15. Use the graph of f shown in the figure to determine h and k if $f(x) = \log_5(x - h) + k$.

16. Use a calculator and the change-of-base formula to evaluate $\log_6 450$.

17. Use the properties of logarithms to expand $\ln\left(\dfrac{6x^2}{\sqrt{x^2 + 1}}\right)$.

18. Use the properties of logarithms to condense $2(\ln x - 3\ln y)$.

9.4	**Solving Exponential and Logarithmic Equations**

Exponential and Logarithmic Equations ▪ Solving Exponential
Equations ▪ Solving Logarithmic Equations ▪ Application

Exponential and Logarithmic Equations

So far in this chapter, we have focused on the definitions, graphs, and properties of exponential and logarithmic functions. In this section, you will study procedures for *solving equations* that involve exponential or logarithmic expressions. As a simple example, consider the exponential equation $2^x = 16$. By rewriting this equation in the form $2^x = 2^4$, you can see that the solution is $x = 4$. To solve this equation, you can use one of the following properties.

Properties of Exponential and Logarithmic Equations

Let a be a positive real number such that $a \neq 1$, and let x and y be real numbers. Then the following properties are true.

1. $a^x = a^y$ if and only if $x = y$.

2. $\log_a x = \log_a y$ if and only if $x = y$ $(x > 0,\ y > 0)$.

EXAMPLE 1   Solving Exponential and Logarithmic Equations

Solve each equation.

(a) $4^{x+2} = 4^5$ (b) $\ln(2x - 3) = \ln 11$

Solution

Stress that the bases must be the same before exponents can be equated.

(a) $4^{x+2} = 4^5$ Original equation

     $x + 2 = 5$ Property 1

        $x = 3$ Subtract 2 from both sides.

The solution is 3. Check this in the original equation.

(b) $\ln(2x - 3) = \ln 11$ Original equation

       $2x - 3 = 11$ Property 2

         $2x = 14$ Add 3 to both sides.

           $x = 7$ Divide both sides by 2.

The solution is 7. Check this in the original equation.

Solving Exponential Equations

In Example 1(a), you were able to solve the given equation because both sides of the equation were written in exponential form (with the same base). However, if only one side of the equation is written in exponential form, it is more difficult to solve the equation. For example, how would you solve the following equation?

$$2^x = 7$$

You must find the power to which 2 can be raised to obtain 7. To do this, you can use one of the following inverse properties of exponents and logarithms.

Inverse Properties of Exponents and Logarithms

Base a	Natural Base e
1. $\log_a(a^x) = x$	$\ln(e^x) = x$
2. $a^{(\log_a x)} = x$	$e^{(\ln x)} = x$

Recall that $x = a^y$ if and only if $\log_a x = y$.

EXAMPLE 2   *Solving an Exponential Equation*

Solve $2^x = 7$.

Solution

NOTE In Example 2, you can use the change-of-base formula to evaluate $\log_2 7$, as follows.

$$\log_2 7 = \frac{\log_{10} 7}{\log_{10} 2}$$

$$= \frac{\ln 7}{\ln 2}$$

$$\approx 2.807$$

To isolate the x, take the $\log_2$ of both sides of the equation or write the equation in logarithmic form, as follows.

$2^x = 7$	Original equation
$x = \log_2 7$	Inverse property

The solution is $x = \log_2 7 \approx 2.807$. Check this in the original equation.

EXAMPLE 3   *Solving an Exponential Equation*

Solve $2e^x = 10$.

Solution

$2e^x = 10$	Original equation
$e^x = 5$	Divide both sides by 2.
$x = \ln 5$	Inverse property

The solution is $x = \ln 5 \approx 1.609$. Check this in the original equation.

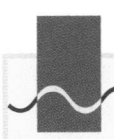

Technology

Graphical Approximation of Solutions

Remember that you can use a graphing utility to solve equations graphically or check solutions that are obtained algebraically. For instance, to check the solutions in Examples 2 and 3, graph both sides of the equations, as shown below.

Graph $y = 2^x$ and $y = 7$. Then approximate the intersection of the two graphs to be $x \approx 2.807$.

Graph $y = 2e^x$ and $y = 10$. Then approximate the intersection of the two graphs to be $x \approx 1.609$.

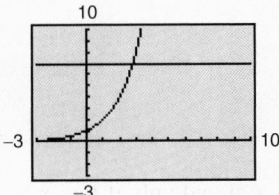

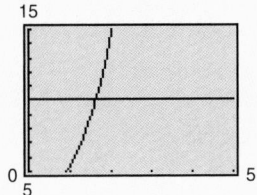

EXAMPLE 4   Solving an Exponential Equation

Solve $5 + e^{x+1} = 20$.

Solution

Point out the need to isolate the exponential term before rewriting the equation in logarithmic form.

$$5 + e^{x+1} = 20 \qquad \text{Original equation}$$

$$e^{x+1} = 15 \qquad \text{Subtract 5 from both sides.}$$

$$x + 1 = \ln 15 \qquad \text{Inverse property}$$

$$x = -1 + \ln 15 \qquad \text{Subtract 1 from both sides.}$$

The solution is $x = -1 + \ln 15 \approx 1.708$. Check this in the original equation, as follows.

Check

$$5 + e^{x+1} = 20 \qquad \text{Original equation}$$

$$5 + e^{1.708+1} \overset{?}{=} 20 \qquad \text{Substitute 1.708 for } x.$$

$$5 + e^{2.708} \overset{?}{=} 20 \qquad \text{Simplify.}$$

$$5 + 14.999 \approx 20 \qquad \text{Solution checks.} \quad$$

Solving Logarithmic Equations

You know how to solve an exponential equation by *taking the logarithms of both sides.* To solve a logarithmic equation, you need to **exponentiate** both sides. For instance, to solve a logarithmic equation such as $\ln x = 2$, you can exponentiate both sides of the equation as follows.

$\ln x = 2$	Original equation
$x = e^2$	Inverse property

Notice that you can obtain the same result by writing the natural logarithmic equation in exponential form. This procedure is demonstrated in the next three examples. We suggest the following guidelines for solving exponential and logarithmic equations.

Solving Exponential and Logarithmic Equations

1. To solve an exponential equation, first isolate the exponential expression, then **take the logarithms of both sides of the equation** (or write the equation in logarithmic form) and solve for the variable.

2. To solve a logarithmic equation, first isolate the logarithmic expression, then **exponentiate both sides of the equation** (or write the equation in exponential form) and solve for the variable.

EXAMPLE 5 Solving a Logarithmic Equation

Solve $2 \ln x = 5$.

Solution

$2 \ln x = 5$	Original equation
$\ln x = \dfrac{5}{2}$	Divide both sides by 2.
$x = e^{5/2}$	Inverse property

The solution is $x = e^{5/2} \approx 12.182$. Check this in the original equation, as follows.

Check

$2 \ln x = 5$	Original equation
$2 \ln(12.182) \overset{?}{=} 5$	Substitute 12.182 for x.
$2(2.49996) \overset{?}{=} 5$	Use a calculator.
$4.99992 \approx 5$	Solution checks. ✓

EXAMPLE 6   *Solving a Logarithmic Equation*

Solve $3 \log_{10} x = 6$.

Solution

$$3 \log_{10} x = 6 \qquad\qquad \text{Original equation}$$
$$\log_{10} x = 2 \qquad\qquad \text{Divide both sides by 3.}$$
$$x = 10^2 \qquad\qquad \text{Inverse property}$$

The solution is $x = 100$. Check this in the original equation.

EXAMPLE 7   *Solving a Logarithmic Equation*

Solve $20 \ln 0.2x = 30$.

Solution

$$20 \ln 0.2x = 30 \qquad\qquad \text{Original equation}$$
$$\ln 0.2x = 1.5 \qquad\qquad \text{Divide both sides by 20.}$$
$$0.2x = e^{1.5} \qquad\qquad \text{Inverse property}$$
$$x = 5e^{1.5} \qquad\qquad \text{Divide both sides by 0.2.}$$

The solution is $x = 5e^{1.5} \approx 22.408$. Check this in the original equation.

The next example uses logarithmic properties as part of the solution.

EXAMPLE 8   *Solving a Logarithmic Equation*

Solve $\log_{10} 2x - \log_{10}(x - 3) = 1$.

Solution

Review the three basic properties of logarithms.

$$\log_{10} 2x - \log_{10}(x - 3) = 1 \qquad\qquad \text{Original equation}$$
$$\log_{10} \frac{2x}{x - 3} = 1 \qquad\qquad \text{Condense the left side.}$$
$$\frac{2x}{x - 3} = 10^1 \qquad\qquad \text{Inverse property}$$
$$2x = 10x - 30 \qquad\qquad \text{Multiply both sides by } x - 3.$$
$$-8x = -30 \qquad\qquad \text{Subtract } 10x \text{ from both sides.}$$
$$x = \frac{15}{4} \qquad\qquad \text{Divide both sides by } -8.$$

The solution is $x = \frac{15}{4}$. Check this in the original equation.

EXAMPLE 9   *Checking for Extraneous Solutions*

Solve $\log_6 x + \log_6(x - 5) = 2$.

Solution

$\log_6 x + \log_6(x - 5) = 2$	Original equation
$\log_6[x(x - 5)] = 2$	Condense the left side.
$x(x - 5) = 6^2$	Inverse property
$x^2 - 5x - 36 = 0$	Standard form
$(x - 9)(x + 4) = 0$	Factor.
$x - 9 = 0$ ⟹ $x = 9$	Set 1st factor equal to 0.
$x + 4 = 0$ ⟹ $x = -4$	Set 2nd factor equal to 0.

From this, it appears that the solutions are 9 and -4. To be sure, you need to check each in the original equation.

Check First Solution:

$\log_6 x + \log_6(x - 5) = 2$	Original equation
$\log_6(9) + \log_6(9 - 5) \overset{?}{=} 2$	Substitute 9 for x.
$\log_6(9 \cdot 4) \overset{?}{=} 2$	Condense the left side.
$\log_6 36 = 2$	Solution checks. ✓

Check Second Solution:

$\log_6 x + \log_6(x - 5) = 2$	Original equation
$\log_6(-4) + \log_6(-4 - 5) \overset{?}{=} 2$	Substitute -4 for x.
$\log_6(-4) + \log_6(-9) \neq 2$	Left side is not defined. ✗

Of the two possible solutions, only $x = 9$ checks. The other possible solution is extraneous.

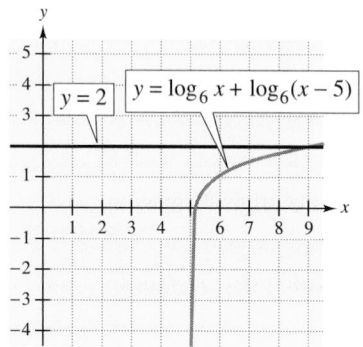

FIGURE 9.15

You can graphically check that the equation in Example 9 has only one real solution by sketching the graphs of the equations in the following nonlinear system.

$$\begin{cases} y = \log_6 x + \log_6(x - 5) & \text{Graph of left side} \\ y = 2 & \text{Graph of right side} \end{cases}$$

From the graph in Figure 9.15, you can see that the only point of intersection occurs when $x = 9$.

Technology

Remember that in order to graph $y = \log_6 x + \log_6(x - 5)$ using a graphing utility, you use the change-of-base formula to enter $y = (\log_{10} x / \log_{10} 6) + [\log_{10}(x - 5) / \log_{10} 6]$.

Application

EXAMPLE 10   An Application: Compound Interest

A deposit of $5000 is placed in a savings account for 2 years. The interest for the account is compounded continuously. At the end of 2 years, the balance in the account is $5867.56. What is the annual interest rate for this account?

Solution

Using the formula for continuously compounded interest, $A = Pe^{rt}$, you have the following solution.

Formula: $A = Pe^{rt}$

Labels: Principal $= P = 5000$ (dollars)
             Amount $= A = 5867.56$ (dollars)
             Time $= t = 2$ (years)
             Annual interest rate $= r$ (percent in decimal form)

Equation: $5867.56 = 5000e^{2r}$

$$\frac{5867.56}{5000} = e^{2r}$$

$$1.1735 \approx e^{2r}$$

$$\ln 1.1735 \approx 2r$$

$$0.16 \approx 2r$$

$$0.08 \approx r$$

The rate is 8%. Check this in the original statement of the problem.

Group Activities   **Reviewing the Major Concepts**

Solving Equations   Solve each equation.

a. $x^2 - 3x - 4 = 0$

b. $e^{2x} - 3e^x - 4 = 0$

c. $(\ln x)^2 - 3 \ln x - 4 = 0$

Explain your strategy. What are the similarities among the three equations? One of the equations has only one solution. Explain why.

9.4 Exinscritial Exercises

See Warm-Up Exercises, p. A49

In Exercises 1–6, determine whether the *x*-values are solutions of the equation.

1. $3^{2x-5} = 27$

  (a) $x = 1$

  (b) $x = 4$

2. $4^{x+3} = 16$

  (a) $x = -1$

  (b) $x = 0$

3. $e^{x+5} = 45$

  (a) $x = -5 + \ln 45$

  (b) $x = -5 + e^{45}$

4. $2^{3x-1} = 324$

  (a) $x \approx 3.1133$

  (b) $x \approx 2.4327$

5. $\log_9(6x) = \frac{3}{2}$

  (a) $x = 27$

  (b) $x = \frac{9}{2}$

6. $\ln(x + 3) = 2.5$

  (a) $x = -3 + e^{2.5}$

  (b) $x \approx 9.1825$

In Exercises 7–30, solve the equation. (Do not use a calculator.)

7. $2^x = 2^5$

8. $5^x = 5^3$

9. $3^{x+4} = 3^{12}$

10. $10^{1-x} = 10^4$

11. $3^{x-1} = 3^7$

12. $4^{2x} = 64$

13. $4^{x-1} = 16$

14. $3^{2x} = 81$

15. $2^{x+2} = \frac{1}{16}$

16. $3^{2-x} = 9$

17. $5^x = \frac{1}{125}$

18. $4^{x+1} = \frac{1}{64}$

19. $\log_2(x + 3) = \log_2 7$

20. $\log_5 2x = \log_5 36$

21. $\log_4(x - 4) = \log_4 12$

22. $\log_3(2 - x) = \log_3 2$

23. $\ln 5x = \ln 22$

24. $\ln(2x - 3) = \ln 17$

25. $\ln(2x - 3) = \ln 15$

26. $\ln 3x = \ln 24$

27. $\log_3 x = 4$

28. $\log_5 x = 3$

29. $\log_{10} 2x = 6$

30. $\log_2(3x - 1) = 5$

In Exercises 31–34, simplify the expression.

31. $\ln e^{2x-1}$

32. $\log_3 3^{x^2}$

33. $10^{\log_{10} 2x}$

34. $e^{\ln(x+1)}$

In Exercises 35–60, solve the exponential equation. (Round the result to two decimal places.)

35. $2^x = 45$

36. $5^x = 212$

37. $10^{2y} = 52$

38. $12^{x-1} = 1500$

39. $4^x = 8$

40. $2^x = 1.5$

41. $\frac{1}{5}(4^{x+2}) = 300$

42. $3(2^{t+4}) = 350$

43. $5(2)^{3x} - 4 = 13$

44. $-16 + 0.2(10)^x = 35$

45. $4 + e^{2x} = 150$

46. $500 - e^{x/2} = 35$

47. $8 - 12e^{-x} = 7$

48. $4 - 2e^x = -23$

49. $23 - 5e^{x+1} = 3$

50. $2e^x + 5 = 115$

51. $300e^{x/2} = 9000$

52. $1000^{0.12x} = 25,000$

53. $6000e^{-2t} = 1200$

54. $10,000e^{-0.1t} = 4000$

55. $250(1.04)^x = 1000$

56. $32(1.5)^x = 640$

57. $\dfrac{1600}{(1.1)^x} = 200$

58. $\dfrac{5000}{(1.05)^x} = 250$

59. $4(1 + e^{x/3}) = 84$

60. $50(3 - e^{2x}) = 125$

In Exercises 61–90, solve the logarithmic equation. (Round the result to two decimal places.)

61. $\log_{10} x = 0$

62. $\ln x = 1$

63. $\log_{10} 4x = \frac{3}{2}$

64. $\log_{10}(x + 3) = \frac{5}{3}$

65. $\log_2 x = 4.5$

66. $\log_4(25x) = 7$

67. $4\log_3 x = 28$

68. $5\log_{10}(x + 2) = 18$

69. $2\log_{10}(x + 5) = 15$

70. $-1 + 3\log_{10}\dfrac{x}{2} = 8$

71. $16\ln x = 30$

72. $\ln x = 2.1$

73. $\ln 2x = 3$

74. $\ln\left(\frac{1}{2}t\right) = \frac{1}{4}$

75. $1 - 2 \ln x = -4$

76. $-5 + 2 \ln 3x = 5$

77. $\frac{2}{3} \ln(x + 1) = -1$

78. $8 \ln(3x - 2) = 1.5$

79. $\ln x^2 = 6$

80. $\ln \sqrt{x} = 6.5$

81. $\log_4 x + \log_4 5 = 2$

82. $\log_5 x - \log_5 4 = 2$

83. $\log_6(x + 8) + \log_6 3 = 2$

84. $\log_7(x - 1) - \log_7 4 = 1$

85. $\log_{10} x + \log_{10}(x - 3) = 1$

86. $\log_{10}(25x) - \log_{10}(x - 1) = 2$

87. $\log_5(x + 3) - \log_5 x = 1$

88. $\log_3(x - 2) + \log_3 5 = 3$

89. $\log_6(x - 5) + \log_6 x = 2$

90. $\log_2 x + \log_2(x + 2) - \log_2 3 = 4$

In Exercises 91–94, use a graphing utility to approximate the *x*-intercept of the graph.

91. $y = 10^{x/2} - 5$

92. $y = 2e^x - 21$

93. $y = 6 \ln(0.4x) - 13$

94. $y = 5 \log_{10}(x + 1) - 3$

In Exercises 95–98, use a graphing utility to approximate the point of intersection of the graphs.

95. $y_1 = 2$
$y_2 = e^x$

96. $y_1 = 2$
$y_2 = \ln x$

97. $y_1 = 3$
$y_2 = 2 \ln(x + 3)$

98. $y_1 = 200$
$y_2 = 1000e^{-x/2}$

99. *Doubling Time*   Solve the exponential equation

$$5000 = 2500e^{0.09t}$$

for *t* to determine the number of years for an investment of \$2500 to double in value when compounded continuously at a rate of 9%.

100. *Doubling Time*   Solve the exponential equation

$$10,000 = 5000e^{10r}$$

for *r* to determine the interest rate required for an investment of \$5000 to double in value when compounded continuously for 10 years.

101. *Intensity of Sound*   The relationship between the number of decibels *B* and the intensity of a sound *I* in watts per centimeter squared is given by

$$B = 10 \log_{10}\left(\frac{I}{10^{-16}}\right).$$

Determine the intensity of a sound *I* if it registers 75 decibels on an intensity meter.

102. *Human Memory Model*   The average score *A* for a group of students who took a test *t* months after the completion of a course is given by the human memory model

$$A = 80 - \log_{10}(t + 1)^{12}.$$

How long after completing the course will the average score fall to $A = 72$?

(a) Answer the question algebraically by letting $A = 72$ and solving the resulting equation.

(b) Answer the question graphically by using a graphing utility to graph the equations $y = 80 - \log_{10}(t + 1)^{12}$ and $y = 72$ and finding their point(s) of intersection.

(c) Which strategy works better for this problem? Explain your reasoning.

103. *Muon Decay*   A muon is an elementary particle that is similar to an electron, but much heavier. Muons are unstable—they quickly decay to form electrons and other particles. In an experiment conducted in 1943, the number of muon decays *m* (of an original 5000 muons) was related to the time *T* (in microseconds) by the model

$$T = 15.7 - 2.48 \ln m.$$

How many decays were recorded when $T = 2.5$?

104. *Friction*   In order to restrain an untrained horse, a person partially wraps a rope around a cylindrical post in a corral (see figure). If the horse is pulling on the rope with a force of 200 pounds, the force F in pounds required by the person is

$$F = 200e^{-0.2\pi\theta/180},$$

where θ is the angle of the wrap in degrees. Find the smallest value of θ if F cannot exceed 80 pounds.

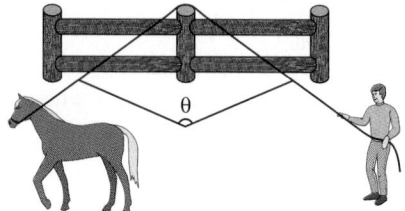

105. *Oceanography*   Oceanographers use the density d (in grams per cubic centimeter) of seawater to obtain information about the circulation of water masses and the rates at which waters of different densities mix. For water with a salinity of 30%, the water temperature T (in degrees Celsius) is related to the density by

$$T = 7.9 \ln(1.0245 - d) + 61.84.$$

Find the densities of the subantarctic water and the antarctic bottom water shown in the figure.

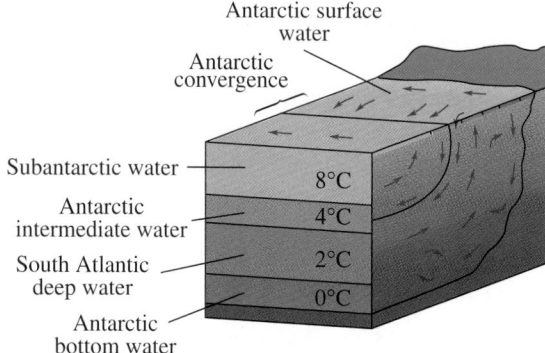

The cross section shows complex currents at various depths in the South Atlantic Ocean off Antarctica.

106. *Military Personnel*   The number N (in thousands) of United States military personnel on active duty in foreign countries for the years 1990 through 1993 is modeled by the equation

$$N = 583e^{-0.2309t}, \quad 0 \le t \le 3,$$

where t is the time in years, with $t = 0$ corresponding to 1990. *(Source: U.S. Department of Defense)*

(a) Use a graphing utility to graph the equation over the specified domain.

(b) Use the graph in part (a) to estimate the value of t when $N = 400$.

107. *Think About It*   Which equation requires logarithms for its solution:

$$2^{x-1} = 32$$

or

$$2^{x-1} = 30?$$

Section Project

Making Ice Cubes   You place a tray of 60°F water into a freezer that is set at 0°F. The water cools according to Newton's Law of Cooling.

$$kt = \ln \frac{T - S}{T_0 - S},$$

where T is the temperature (in °F), t is the number of hours the tray is in the freezer, S is the temperature of the surrounding air, and T_0 is the original temperature of the water.

(a) If the water freezes in 4 hours, what is the constant k?   (*Hint:* Water freezes at 32°F.)

(b) Suppose you lower the temperature in the freezer to -10°F. At this temperature, how long will it take for the ice cubes to form?

(c) Suppose the initial temperature of the water is 50°F. If the freezer temperature is 0°F, how long will it take for the ice cubes to form?

| 9.5 | **Exponential and Logarithmic Applications** |

Compound Interest ▪ Growth and Decay ▪ Intensity Models

Compound Interest

Notice that the formulas for periodic compounding and continuous compounding have five variables and four variables, respectively. Using basic algebraic skills and the properties of exponents and logarithms, we are now able to solve for A, P, r, or t in either formula, given the value of all of the other variables in the formula.

In Section 9.1, you were introduced to the following two formulas for compound interest. Recall that in these formulas, A is the balance, P is the principal, r is the annual interest rate (in decimal form), and t is the time in years.

n Compoundings per Year   *Continuous Compounding*

$$A = P\left(1 + \frac{r}{n}\right)^{nt}$$ $$A = Pe^{rt}$$

EXAMPLE 1   *Finding the Annual Interest Rate*

An investment of $50,000 is made in an account that compounds interest quarterly. After 4 years, the balance in the account is $71,381.07. What is the annual interest rate for this account?

Solution

Formula: $A = P\left(1 + \frac{r}{n}\right)^{nt}$

Labels: Principal $= P = 50{,}000$ (dollars)
              Amount $= A = 71{,}381.07$ (dollars)
              Time $= t = 4$ (years)
              Number of compoundings per year $= n = 4$
              Annual interest rate $= r$ (percent in decimal form)

Equation: $71{,}381.07 = 50{,}000\left(1 + \frac{r}{4}\right)^{(4)(4)}$

NOTE Notice in Example 1 that you can "get rid of" the 16th power on the right side of the equation by raising both sides of the equation to the $\frac{1}{16}$ power.

$1.42762 \approx \left(1 + \frac{r}{4}\right)^{16}$ Divide both sides by 50,000.

$(1.42762)^{1/16} \approx 1 + \frac{r}{4}$ Raise both sides to $\frac{1}{16}$ power.

$1.0225 \approx 1 + \frac{r}{4}$ Simplify.

$0.0225 = \frac{r}{4}$ Subtract 1 from both sides.

$0.09 = r$ Multiply both sides by 4.

The annual interest rate is 9%. Check this in the original problem.

To convince students that they could have chosen the principal arbitrarily, have them repeat Example 2 using $P = \$100$.

EXAMPLE 2 *Doubling Time for Continuous Compounding*

An investment is made in a trust fund at an annual interest rate of 8.75%, compounded continuously. How long will it take for the investment to double?

Solution

$$A = Pe^{rt}$$ — Formula for continuous compounding

$$2P = Pe^{0.0875t}$$ — Substitute known values.

$$2 = e^{0.0875t}$$ — Divide both sides by P.

$$\ln 2 = 0.0875t$$ — Inverse property

$$\frac{\ln 2}{0.0875} = t$$ — Divide both sides by 0.0875.

$$7.92 \approx t$$

NOTE In Example 2, note that you do not need to know the principal to find the doubling time. In other words, the doubling time is the same for *any* principal.

It will take approximately 7.92 years for the investment to double. Check this in the original problem.

EXAMPLE 3 *Finding the Type of Compounding*

You deposit $1000 in an account. At the end of 1 year your balance is $1077.63. If the bank tells you that the annual interest rate for the account is 7.5%, how was the interest compounded?

Solution

If the interest had been compounded continuously at 7.5%, the balance would have been

$$A = 1000e^{(0.075)(1)} = \$1077.88.$$

Because the actual balance is slightly less than this, you should use the formula for interest that is compounded n times per year.

$$A = 1000\left(1 + \frac{0.075}{n}\right)^{n} = 1077.63$$

At this point, it is not clear what you should do to solve the equation for n. However, by completing a table using the most common types of compounding, you can see that $n = 12$. Thus, the interest was compounded monthly.

 *Technology*

You can estimate the solution in Example 3 graphically. Graph

$$Y_1 = 1000(1 + 0.075/x)^x$$

and

$$Y_2 = 1077.63$$

on a graphing utility. By approximating the point of intersection you can determine the value of x.

n	1	4	12	365
$\left(1 + \dfrac{0.075}{n}\right)^{n}$	1.075	1.07714	1.07763	1.07788

In Example 3, notice that an investment of $1000 compounded monthly produced a balance of $1077.63 at the end of 1 year. Because $77.63 of this amount is interest, the **effective yield** for the investment is

$$\text{Effective yield} = \frac{\text{year's interest}}{\text{amount invested}} = \frac{77.63}{1000} = 0.07763 = 7.763\%.$$

In other words, the effective yield for an investment collecting compound interest is the *simple interest rate* that would yield the same balance at the end of 1 year.

EXAMPLE 4 Finding the Effective Yield

NOTE In Example 4, you can find the effective yield *without* substituting a value for P.

$A = Pe^{rt}$

$A = Pe^{(0.0675)(1)}$

$A \approx P(1.06983)$

$A = P + P(0.06983)$

From the last equation, you can see that the principal increased by 6.983%.

An investment is made in an account that pays 6.75% interest, compounded continuously. What is the effective yield for this investment?

Solution

Notice that you do not have to know the principal or the time that the money will be left in the account. Instead, you can choose an arbitrary principal, such as $1000. Then, because effective yield is based on the balance at the end of 1 year, you can use the following formula.

$$A = Pe^{rt} = 1000e^{0.0675(1)} = 1069.83$$

Now, because the account would earn $69.83 in interest after 1 year for a principal of $1000, you can conclude that the effective yield is

$$\text{Effective yield} = \frac{69.83}{1000} = 0.06983 = 6.983\%.$$

Growth and Decay

The balance in an account earning *continuously* compounded interest is one example of a quantity that increases over time according to the **exponential growth model** $y = Ce^{kt}$.

Exponential Growth and Decay

The mathematical model for exponential growth or decay is given by

$$y = Ce^{kt}.$$

For this model, t is the time, C is the original amount of the quantity, and y is the amount after the time t. The number k is a constant that is determined by the rate of growth. If $k > 0$, the model represents **exponential growth,** and if $k < 0$, it represents **exponential decay.**

One common application of exponential growth is in modeling the growth of a population, as shown in Example 5.

EXAMPLE 5   Population Growth

A country's population was 2 million in 1980 and 3 million in 1990. What would you predict the population of the country to be in the year 2000?

Solution

If you assumed a *linear growth model*, you would simply predict the population in the year 2000 to be 4 million. However, social scientists and demographers have discovered that *exponential growth models* are better than linear growth models for representing population growth. Thus, you can use the exponential growth model

$$y = Ce^{kt}.$$

In this model, let $t = 0$ represent 1980. The given information about the population can be described by the following table.

t (year)	0	10	20
Ce^{kt} (million)	$Ce^{k(0)} = 2$	$Ce^{k(10)} = 3$	$Ce^{k(20)} = ?$

To find the population when $t = 20$, you must first find the values of C and k. From the table, you can use the fact that $Ce^{k(0)} = Ce^0 = 2$ to conclude that $C = 2$. Then, using this value of C, you can solve for k as follows.

$$Ce^{k(10)} = 3 \qquad \text{From table}$$

$$2e^{10k} = 3 \qquad \text{Substitute value of } C.$$

$$e^{10k} = \frac{3}{2} \qquad \text{Divide both sides by 2.}$$

$$10k = \ln\frac{3}{2} \qquad \text{Inverse property}$$

$$k = \frac{1}{10}\ln\frac{3}{2} \qquad \text{Divide both sides by 10.}$$

$$k \approx 0.0405 \qquad \text{Simplify.}$$

Finally, you can use this value of k to conclude that the population in the year 2000 is given by

$$2e^{0.0405(20)} \approx 2(2.25) = 4.5 \text{ million.}$$

Figure 9.16 graphically compares the exponential growth model with a linear growth model.

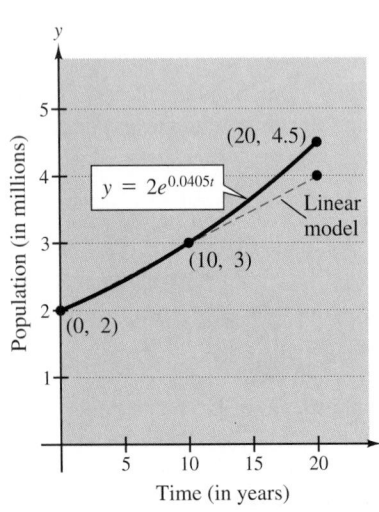

FIGURE 9.16   Population models

EXAMPLE 6 *Radioactive Decay*

Radioactive iodine is a by-product of some types of nuclear reactors. Its **half-life** is 60 days. That is, after 60 days, a given amount of radioactive iodine will have decayed to half the original amount. Suppose a nuclear accident occurs and releases 20 grams of radioactive iodine. How long will it take for the radioactive iodine to decay to a level of 1 gram?

Solution

Discussion project: Have students research and bring in articles relating to the Chernobyl disaster. Note the numerous references to radioactive exposure, the half-life of uranium, and the length of time that the area will be adversely affected.

To solve this problem, use the model for exponential decay.

$$y = Ce^{kt}$$

Next, use the information given in the problem to set up the following table.

t (days)	0	60	?
Ce^{kt} (grams)	$Ce^{k(0)} = 20$	$Ce^{k(60)} = 10$	$Ce^{k(t)} = 1$

Because $Ce^{k(0)} = Ce^0 = 20$, you can conclude that $C = 20$. Then, using this value of C, you can solve for k as follows.

$Ce^{k(60)} = 10$	From table
$20e^{60k} = 10$	Substitute value of C.
$e^{60k} = \dfrac{1}{2}$	Divide both sides by 20.
$60k = \ln\dfrac{1}{2}$	Inverse property
$k = \dfrac{1}{60}\ln\dfrac{1}{2} \approx -0.01155$	Divide both sides by 60 and simplify.

Finally, you can use this value of k to find the time when the amount is 1 gram, as follows.

$Ce^{kt} = 1$	From table
$20e^{-0.01155t} = 1$	Substitute values of C and k.
$e^{-0.01155t} = \dfrac{1}{20}$	Divide both sides by 20.
$-0.01155t = \ln\dfrac{1}{20}$	Inverse property
$t = \dfrac{1}{-0.01155}\ln\dfrac{1}{20} \approx 259.4$ days	Divide both sides by -0.01155 and simplify.

Thus, 20 grams of radioactive iodine will have decayed to 1 gram after about 259.4 days. This solution is shown graphically in Figure 9.17.

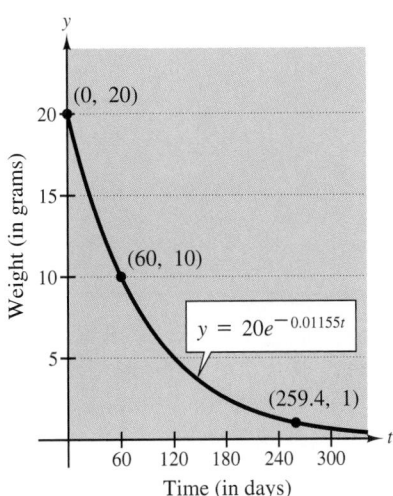

FIGURE 9.17 Radioactive decay

Earthquakes take place along faults in the earth's crust. The 1989 earthquake in California took place along the San Andreas Fault.

Intensity Models

On the **Richter scale,** the magnitude R of an earthquake can be measured by the **intensity model**

$$R = \log_{10} I,$$

where I is the intensity of the shock wave.

EXAMPLE 7   *Earthquake Intensity*

In 1906, San Francisco experienced an earthquake that measured 8.3 on the Richter scale. In 1989, another earthquake, which measured 7.1 on the Richter scale, struck the same area. Compare the intensities of these two earthquakes.

Solution

The intensity of the 1906 earthquake is given as follows.

$$8.3 = \log_{10} I \qquad \text{Given}$$
$$10^{8.3} = I \qquad \text{Inverse property}$$

The intensity of the 1989 earthquake can be found in a similar way.

$$7.1 = \log_{10} I \qquad \text{Given}$$
$$10^{7.1} = I \qquad \text{Inverse property}$$

The ratio of these two intensities is

$$\frac{I \text{ for } 1906}{I \text{ for } 1989} = \frac{10^{8.3}}{10^{7.1}} = 10^{8.3 - 7.1} = 10^{1.2} \approx 15.85.$$

Thus, the 1906 earthquake had an intensity that was about sixteen times greater than that of the 1989 earthquake.

Group Activities   **You Be the Instructor**

Problem Posing   Write a problem that could be answered by investigating the exponential growth model $y = 10e^{0.08t}$ or the exponential decay model $y = 5e^{-0.25t}$. Exchange your problem with that of another group member, and solve one another's problems.

9.5   **Exercises**

See Warm-Up
Exercises, p. A49

1. *Think About It*   If the equation $y = Ce^{kt}$ models exponential growth, what must be true about k?
2. *Think About It*   If the equation $y = Ce^{kt}$ models exponential decay, what must be true about k?

Annual Interest Rate   In Exercises 3–12, find the annual interest rate.

	Principal	Balance	Time	Compounding
3.	$500	$1004.83	10 years	Monthly
4.	$3000	$21,628.70	20 years	Quarterly
5.	$1000	$36,581.00	40 years	Daily
6.	$200	$314.85	5 years	Yearly
7.	$750	$8267.38	30 years	Continuous
8.	$2000	$2718.28	10 years	Continuous
9.	$5000	$22,405.68	25 years	Daily
10.	$10,000	$110,202.78	30 years	Daily
11.	$1500	$24,666.97	40 years	Continuous
12.	$7500	$15,877.50	15 years	Continuous

Doubling Time   In Exercises 13–22, find the time for the investment to double. Use a graphing utility to check the result graphically.

	Principal	Rate	Compounding
13.	$6000	8%	Quarterly
14.	$500	$5\frac{1}{4}\%$	Monthly
15.	$2000	10.5%	Daily
16.	$10,000	9.5%	Yearly
17.	$1500	7.5%	Continuous
18.	$100	6%	Continuous
19.	$300	5%	Yearly
20.	$12,000	4%	Continuous
21.	$6000	7%	Quarterly
22.	$500	9%	Daily

Type of Compounding   In Exercises 23–26, determine the type of compounding. Solve the problem by trying the more common types of compounding.

	Principal	Balance	Time	Rate
23.	$750	$1587.75	10 years	7.5%
24.	$10,000	$73,890.56	20 years	10%
25.	$100	$141.48	5 years	7%
26.	$4000	$4788.76	2 years	9%

Effective Yield   In Exercises 27–36, find the effective yield.

	Rate	Compounding
27.	8%	Continuous
28.	9.5%	Daily
29.	7%	Monthly
30.	8%	Yearly
31.	6%	Quarterly
32.	9%	Quarterly
33.	8%	Monthly
34.	$5\frac{1}{4}\%$	Daily
35.	7.5%	Continuous
36.	4%	Continuous

37. *Doubling Time*   Is it necessary to know the principal P to find the doubling time in Exercises 13–22? Explain.

38. *Effective Yield*
   (a) Is it necessary to know the principal P to find the effective yield in Exercises 27–36? Explain.
   (b) When the interest is compounded more frequently, what inference can you make about the difference between the effective yield and the stated annual percent rate?

Principal   In Exercises 39–48, find the principal that must be deposited in an account to obtain the given balance.

	Balance	Rate	Time	Compounding
39.	$10,000	9%	20 years	Continuous
40.	$5000	8%	5 years	Continuous
41.	$750	6%	3 years	Daily
42.	$3000	7%	10 years	Monthly
43.	$25,000	7%	30 years	Monthly
44.	$8000	6%	2 years	Monthly
45.	$1000	5%	1 year	Daily
46.	$100,000	9%	40 years	Daily
47.	$500,000	8%	25 years	Continuous
48.	$1,000,000	10%	50 years	Continuous

Balance After Monthly Deposits   In Exercises 49–52, you make monthly deposits of *P* dollars in a savings account at an annual interest rate *r*, compounded continuously. Find the balance *A* after *t* years given that

$$A = \frac{P(e^{rt} - 1)}{e^{r/12} - 1}.$$

	Principal	Rate	Time
49.	$P = \$30$	$r = 8\%$	$t = 10$ years
50.	$P = \$100$	$r = 9\%$	$t = 30$ years
51.	$P = \$50$	$r = 10\%$	$t = 40$ years
52.	$P = \$20$	$r = 7\%$	$t = 20$ years

Exponential Growth and Decay   In Exercises 53–56, find the constant *k* such that the graph of $y = Ce^{kt}$ passes through the given points.

53.
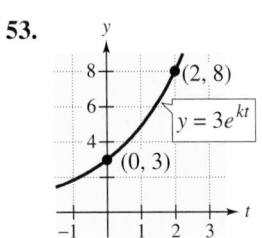

54.
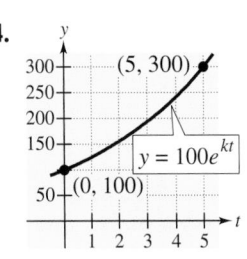

55.
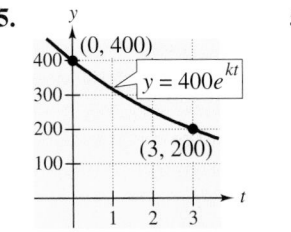

56.
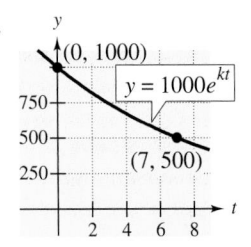

Population of a Region   In Exercises 57–64, the population (in millions) of an urban region for 1994 and the predicted population (in millions) for the year 2015 are given. Find the constants *C* and *k* to obtain the exponential model $y = Ce^{kt}$ for the population. (Let $t = 0$ correspond to 1994.) Use the model to predict the population of the region in the year 2020.   *(Source: United Nations)*

	Region	1994	2015
57.	Los Angeles	12.2	14.3
58.	New York	16.3	17.6
59.	Shanghai, China	14.7	23.4
60.	Jakarta, Indonesia	11.0	21.2
61.	Osaka, Japan	10.5	10.6
62.	Seoul, South Korea	11.5	13.1
63.	Mexico City, Mexico	15.5	18.8
64.	São Paulo, Brazil	16.1	20.8

65. *Rate of Growth*

(a) Compare the values of *k* in Exercises 59 and 61. Which is larger? Explain.

(b) What variable in the continuous compound interest formula is equivalent to *k* in the model for population growth? Use your answer to give an interpretation of *k*.

66. *Radioactive Decay*   Radioactive radium (Ra^{226}) has a half-life of 1620 years. If you start with 5 grams of the isotope, how much remains after 1000 years?

67. *Radioactive Decay*   The isotope Pu^{239} has a half-life of 24,360 years. If you start with 10 grams of the isotope, how much remains after 10,000 years?

68. *Radioactive Decay* Carbon 14 (C^{14}) has a half-life of 5730 years. If you start with 5 grams of the isotope, how much remains after 1000 years?

69. *Carbon 14 Dating* C^{14} dating assumes that the carbon dioxide on earth today has the same radioactive content as it did centuries ago. If this is true, the amount of C^{14} absorbed by a tree that grew several centuries ago should be the same as the amount of C^{14} absorbed by a tree growing today. A piece of ancient charcoal contains only 15% as much of the radioactive carbon as a piece of modern charcoal. How long ago did the tree burn to make the ancient charcoal if the half-life of C^{14} is 5730 years? (Round the result to the nearest 100 years.)

70. *World Population* The figure shows the population P (in billions) of the world as projected by the Population Reference Bureau. The bureau's projection can be modeled by

$$P = \frac{11.14}{1 + 1.101e^{-0.051t}},$$

where $t = 0$ represents 1990. Use the model to estimate the population in 2020.

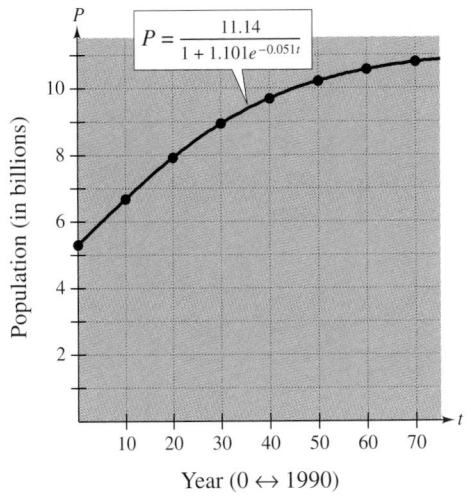

Year ($0 \leftrightarrow 1990$)

71. *Depreciation* A car that cost $22,000 new has a depreciated value of $16,500 after 1 year. Find the value of the car when it is 3 years old by using the exponential model $y = Ce^{kt}$.

72. *Depreciation* After x years, the value y of a truck that cost $32,000 new is given by $y = 32,000(0.8)^x$.

(a) Use a graphing utility to graph the equation.

(b) Graphically approximate the value of the truck after 1 year.

(c) Graphically approximate the time when the truck's value will be $16,000.

Balance After Monthly Deposits In Exercises 73 and 74, you make monthly deposits of $30 in a savings account at an annual interest rate of 8%, compounded continuously (see figure).

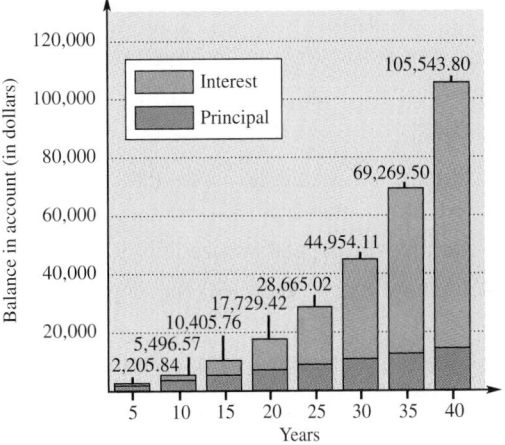

73. Find the total amount that has been deposited in the account in 20 years and the total interest earned.

74. Find the total amount that has been deposited in the account in 40 years and the total interest earned.

Earthquake Intensity In Exercises 75–78, compare the intensities of the two earthquakes.

Location	Date	Magnitude
75. Alaska	3/27/64	8.4
San Fernando Valley	2/9/71	6.6
76. Long Beach, California	3/10/33	6.2
Morocco	2/29/60	5.8
77. Mexico City, Mexico	9/19/85	8.1
Nepal	8/20/88	6.5
78. Chile	8/16/06	8.6
Armenia, USSR	12/7/88	6.8

79. *Population Growth* The population p of a certain species t years after it is introduced into a new habitat is given by

$$p(t) = \frac{5000}{1 + 4e^{-t/6}}.$$

(a) Use a graphing utility to graph the population function.

(b) Determine the population size that was introduced into the habitat.

(c) Determine the population size after 9 years.

(d) After how many years will the population be 2000?

80. *Sales Growth* Annual sales y of a product x years after it is introduced are approximated by

$$y = \frac{2000}{1 + 4e^{-x/2}}.$$

(a) Use a graphing utility to graph the equation.

(b) Graphically approximate annual sales when $x = 4$.

(c) Graphically approximate the time when annual sales are $y = 1100$ units.

(d) Graphically estimate the maximum level that annual sales will approach.

81. *Advertising Effect* The sales S (in thousands of units) of a product after spending x hundred dollars on advertising are given by

$$S = 10(1 - e^{kx}).$$

(a) Find S as a function of x if 2500 units are sold when $500 is spent on advertising.

(b) How many units will be sold if advertising expenditures are raised to $700?

Section Project

Acidity Model Use the acidity model

$$pH = -\log_{10} [H^+],$$

where acidity (pH) is a measure of the hydrogen ion concentration $[H^+]$ (measured in moles of hydrogen per liter) of solution.

(a) Find the pH of a solution that has a hydrogen ion concentration of 9.2×10^{-8}.

(b) Compute the hydrogen ion concentration if the pH of a solution is 4.7.

(c) A certain fruit has a pH of 2.5 and an antacid tablet has a pH of 9.5. The hydrogen ion concentration of the fruit is how many times the concentration of the tablet?

(d) If the pH of a solution is decreased by one unit, the hydrogen ion concentration is increased by what factor?

9.6	**Modeling Data**

Classifying Scatter Plots ▪ Fitting Models to Data ▪ Application

Classifying Scatter Plots

In Sections 3.2 and 6.6, you saw how to fit linear and quadratic models to data. In this section, you will learn how to fit exponential models, logarithmic models, and power models to data.

EXAMPLE 1   *Classifying Scatter Plots*

Decide whether the data could be best modeled by an exponential model, $y = ab^x$, or by a logarithmic model, $y = a + b \ln x$.

(a) (0.4, 2.0), (0.5, 2.2), (1.0, 2.6), (1.4, 3.1), (1.6, 3.4), (2.0, 3.9), (2.3, 4.5), (2.6, 5.1), (3.0, 6.0), (3.3, 6.8), (3.6, 7.6), (3.8, 8.1), (4.0, 9.0), (4.1, 9.3), (4.2, 9.7), (4.3, 10.0)

(b) (0.5, 0.5), (0.7, 1.6), (0.9, 2.7), (1.0, 3.1), (1.2, 3.7), (1.4, 4.4), (1.8, 5.1), (2.0, 5.8), (2.3, 6.4), (2.7, 7.0), (3.2, 7.7), (3.8, 8.3), (4.2, 8.8), (4.4, 8.9), (4.6, 9.1), (4.8, 9.3)

Solution

Begin by entering the data into a graphing utility. You should obtain the scatter plots shown in Figure 9.18.

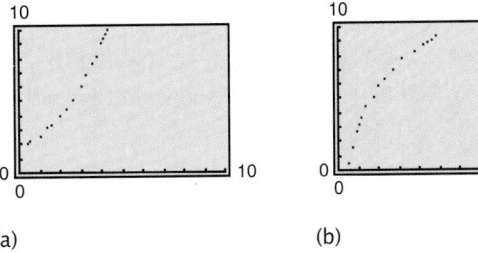

(a)   (b)

FIGURE 9.18

From the scatter plots, it appears that the data in part (a) can be modeled by an exponential function and the data in part (b) can be modeled by a logarithmic function.

Fitting Models to Data

Once you have used a scatter plot to determine the type of model to be fit to a set of data, there are several ways that you can actually find the model. Each method is best used with a computer or calculator, rather than with hand calculations.

Technology

Most graphing utilities have built-in programs that will calculate the equation of the best-fitting model for a collection of points. (For an example of how this is done, see the Technology features on pages 200 and 457.)

EXAMPLE 2 Fitting a Model to Data

Fit the following data, from Example 1a, with a quadratic model, an exponential model, and a power model.

$$\underbrace{y = ax^2 + bx + c,}_{\text{Quadratic}} \quad \underbrace{y = ab^x,}_{\text{Exponential}} \quad \underbrace{y = ax^b}_{\text{Power}}$$

Which model do you think fits the best?

> (0.4, 2.0), (0.5, 2.2), (1.0, 2.6), (1.4, 3.1), (1.6, 3.4), (2.0, 3.9), (2.3, 4.5),
> (2.6, 5.1), (3.0, 6.0), (3.3, 6.8), (3.6, 7.6), (3.8, 8.1), (4.0, 9.0), (4.1, 9.3),
> (4.2, 9.7), (4.3, 10.0)

Solution

Begin by entering the data into a calculator or computer that has least squares regression programs. Then run the regression programs for quadratic, exponential, and power models.

NOTE The question of which model best fits a set of data is one that is studied in detail in statistics. Basically, the model that fits best is the one whose sum of squared differences is the least. In Example 2, the sums of squared differences are 0.102 for the quadratic model, 0.055 for the exponential model, and 4.895 for the power model.

Quadratic	$y = ax^2 + bx + c$	$a = 0.383$	$b = 0.204$	$c = 1.981$
Exponential	$y = ab^x$	$a = 1.765$	$b = 1.499$	
Power	$y = ax^b$	$a = 2.119$	$b = 1.025$	

To decide which model fits the best, you can compare the y-values given by each model with the actual y-values. The model whose y-values are closest to the actual y-values is the one that fits the best. In this case, the best-fitting model is the exponential model (it is slightly better than the quadratic model). The graphs of all three models are shown in Figure 9.19.

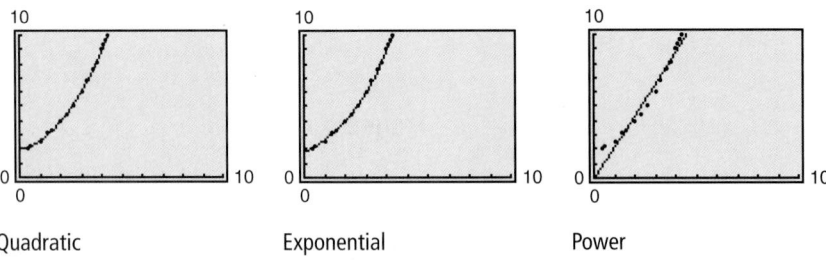

Quadratic   Exponential   Power

FIGURE 9.19

EXAMPLE 3   Fitting a Model to Data

Fit the following data, from Example 1b, with a logarithmic model and a power model.

$$y = a + b \ln x, \qquad y = ax^b$$

Logarithmic   Power

Which model do you think fits the best?

(0.5, 0.5), (0.7, 1.6), (0.9, 2.7), (1.0, 3.1), (1.2, 3.7), (1.4, 4.4), (1.8, 5.1), (2.0, 5.8), (2.3, 6.4), (2.7, 7.0), (3.2, 7.7), (3.8, 8.3), (4.2, 8.8), (4.4, 8.9), (4.6, 9.1), (4.8, 9.3)

Solution

Begin by entering the data into a calculator or computer that has least squares regression programs. Then run the regression programs for logarithmic and power models.

Logarithmic $y = a + b \ln x$ $a = 3.072$ $b = 3.938$
Power $y = ax^b$ $a = 2.364$ $b = 1.008$

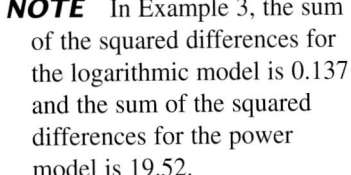

NOTE   In Example 3, the sum of the squared differences for the logarithmic model is 0.137 and the sum of the squared differences for the power model is 19.52.

To decide which model fits the best, compare the y-values given by each model with the actual y-values. The model whose y-values are closest to the actual y-values is the one that fits the best. In this case, the best-fitting model is the logarithmic model. The graphs of both models are shown in Figure 9.20.

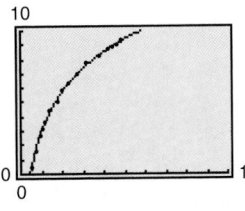

Logarithmic   Power

FIGURE 9.20

Exploration

Try using a calculator or computer to fit the data in Example 3 with a quadratic model of the form $y = ax^2 + bx + c$. Do you think this model fits better than the logarithmic model in the example? Explain your reasoning.

Application

EXAMPLE 4 Finding an Exponential Model

Remind students that they have used these five different models to describe data: linear, quadratic, exponential, logarithmic, and power models.

The amounts by which mail-order sales of computer software and hardware have increased since 1986 for the years 1986 through 1992 are shown below. Find a model for the data. The total amount of computer software and hardware mail-order sales in 1986 was $740 million. Use the model to estimate the total amount in 1997. In the list of data points (t, I), t represents the year, with $t = 6$ corresponding to 1986, and I represents the amount of increase in millions of dollars. *(Source: Fishman, Arnold L., Portable Mail Order Industry Statistics)*

$$(6, 0), (7, 20), (8, 100), (9, 180), (10, 260), (11, 590), (12, 1420)$$

Solution

Begin by entering the data into a computer or calculator that has least squares regression programs. Then, plot the data, as shown in Figure 9.21. From the scatter plot, it appears that an exponential model is a good fit. After running the exponential regression program, you should obtain $I = 0.112(2.195)^t$ or $I = 0.112e^{(0.786)t}$. (The correlation coefficient is $r = 0.997$, which implies that the model is a good fit to the data.) From the model, you can calculate the increase in computer software and hardware mail-order sales since 1986 for the year 1997 to be

$$I = 0.112(2.195)^{17} \approx 71{,}385 \text{ million dollars.}$$

Thus, the total amount of mail-order sales in 1997 would be

$$740 + 71{,}385 = \$72{,}125 \text{ million.}$$

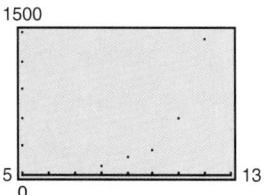

FIGURE 9.21

Group Activities   Fitting a Model to Data

The amounts y (in millions of dollars) of consumer expenditures for paperback books in the United States for the years 1975 through 1993 are listed below. The data is given as ordered pairs of the form (t, y), where t is the year, with $t = 5$ representing 1975. Create a scatter plot of the data. With others in your group, decide which type of model best fits the data. Then find the model. *(Source: Book Industry Study Group, Inc.)*

(5, 1693), (6, 1928), (7, 2269), (8, 2585), (9, 2805), (10, 3123), (11, 3444), (12, 3699), (13, 4438), (14, 4701), (15, 4642), (16, 4853), (17, 5345), (18, 6056), (19, 6793), (20, 7254), (21, 7703), (22, 8178), (23, 8795)

9.6 Exercises

See Warm-Up
Exercises, p. A49

In Exercises 1–6, determine if the scatter plot could best be modeled by a linear model, an exponential model, or a logarithmic model.

1.
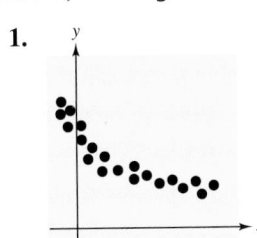

2.
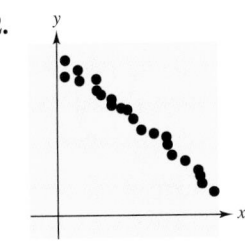

3.
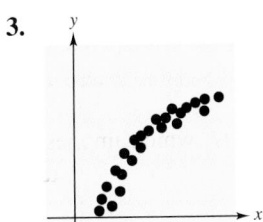

4.
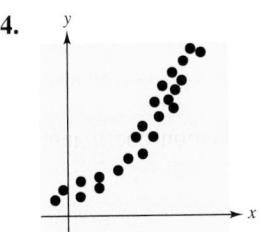

5.
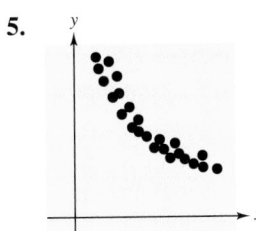

6.
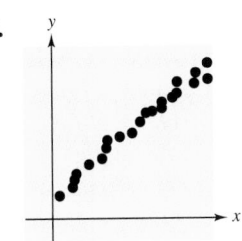

In Exercises 7–12, sketch a scatter plot of the data. Decide whether the data could best be modeled by a linear, an exponential, or a logarithmic model.

7. $(1, 2), (2, 4.1), (4, 6.2), (6, 7.4), (8, 8.2), (10, 8.9)$

8. $(0, 2), (3, 11), (4, 14), (6, 20), (7, 23), (9, 29)$

9. $(-3, 6), (0, 4), \left(4, \frac{4}{3}\right), (6, 0), \left(7, -\frac{2}{3}\right), (9, -2)$

10. $(0, 4), (1, 8.1), (2, 16.2), (3, 32.7), (4, 65.8),$
$(5, 132.5)$

11. $(0, 7), (1, 3.5), (2, 1.7), (3, 0.9), (4, 0.4), (5, 0.2)$

12. $(1, 5), (2, 2.9), (3, 1.7), (5, 0.2), (7, -0.8), (10, -1.9)$

In Exercises 13–16, use a graphing utility to fit an exponential model, $y = ab^x$, to the points. Use the graphing utility to obtain a scatter plot and a graph of the model.

13. $(0, 2), (1, 2.5), (4, 4.5), (6, 6.6), (9, 12.1)$

14. $(1, 45), (2, 201), (3, 910), (4, 4050)$

15. $(1, 12), (2, 7.4), (3, 4.5), (5, 1.5), (8, 0.5), (10, 0.1)$

16. $(0, 500), (4, 390), (6, 345), (8, 300), (10, 270)$

In Exercises 17–20, use a graphing utility to fit a logarithmic model, $y = a + b \ln x$, to the points. Use the graphing utility to obtain a scatter plot and a graph of the model.

17. $(1, 3), (3, 3.5), (5, 3.8), (10, 4)$

18. $\left(\frac{1}{2}, 2.5\right), (2, 9), (3, 11.5), (5, 14), (8, 16)$

19. $(3, 6.7), (4, 5.8), (6, 4.5), (8, 3.6), (10, 3)$

20. $(2, 180), (4, 165), (6, 155), (10, 140), (16, 131),$
$(20, 125)$

In Exercises 21–24, use a graphing utility to fit a power model, $y = ax^b$, to the points. Use the graphing utility to obtain a scatter plot and a graph of the model.

21. $(1, 10), (2, 30), (4, 80), (6, 150), (8, 225)$

22. $(4, 25), (5, 56), (7, 66), (9, 75), (10, 80)$

23. $(3, 58), (4, 50), (5, 45), (8, 35), (10, 32)$

24. $(2, 225), (4, 100), (5, 77), (6, 60), (8, 42)$

25. *Depreciation*   The table gives the predicted value V of a machine t years after it is purchased. Use a graphing utility to fit an exponential model to the data. Use the graphing utility to plot the data and graph the model.

t	0	1	2	3
V	$46,000	$30,500	$20,750	$14,000

26. *Investments*   An advertisement for a stock mutual fund includes the bar graph below, which shows the growth of $10,000 from 1990 through 1996.

(a) Use a graphing utility to fit an exponential model to the data. (Let $t = 0$ correspond to 1990.)

(b) Use part (a) to approximate the continuous annual percent increase in the value of the fund.

(c) Use the model in part (a) to predict the value of the fund in the year 2000.

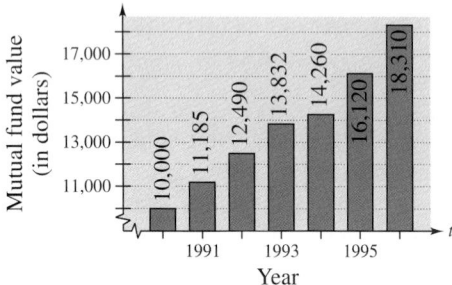

27. *Ventilation*   The table gives the required ventilation rate V (in cubic feet per minute per person) for a room in a public building with an air space of x cubic feet per person.

x	100	200	300	400
V	25	17	12	9

(a) Use a graphing utility to fit a logarithmic model to the data.

(b) Use a graphing utility to plot the data and graph the model in part (a).

(c) Use part (a) to determine V if $x = 250$.

28. *Nails*   The length of a nail is expressed in the "penny" system and is denoted by d. The table gives the length x (in inches) and the number N per pound of selected sizes of common nails. *(Source: Standard Handbook for Mechanical Engineers)*

d	$2d$	$4d$	$6d$	$8d$	$10d$	$16d$	$20d$
x	1	1.5	2	2.5	3	3.5	4
N	740	280	160	88	60	33	23

(a) Use a graphing utility to fit a power model to the data.

(b) Use a graphing utility to plot the data and graph the model in part (a).

(c) Use the model in part (a) to estimate the number of nails per pound of $9d$ nails, which are 2.75 inches long.

29. *Stock Sales*   The table gives the market value V (in billions of dollars) of all sales of stocks and options on registered exchanges for 1991 through 1993. *(Source: U.S. Securities and Exchange Commission)*

Year	1991	1992	1993
V	1903	2149	2734

(a) Use a graphing utility to fit an exponential model and a power model to the data, where $t = 1$ corresponds to 1991.

(b) Use a graphing utility to plot the data and graph the models in part (a).

(c) Select the model that best fits the data and use it to predict V when $t = 4$.

30. *Density of Water*   The figure shows the density D of water at selected temperatures t (in degrees Celsius). *(Source: Standard Handbook for Mechanical Engineers)*

(a) Use the figure to explain why a quadratic model may fit better than an exponential model, a logarithmic model, and a power model.

(b) Use a graphing utility to confirm part (a).

(c) Use the quadratic model to predict the density of water when $t = 80°C$.

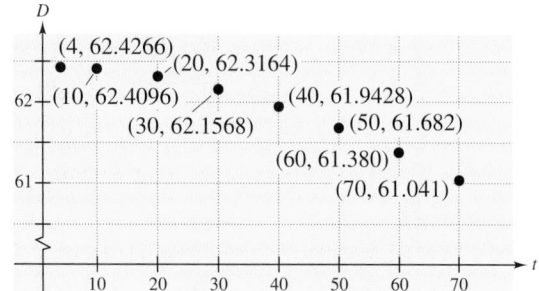

31. *Safe Load* Students in a laboratory experiment determined the force F (in thousands of pounds) required to break a 1-foot span of 2-inch lumber of height x inches. The results are shown in the line graph.

(a) Use a graphing utility to fit a quadratic model, an exponential model, and a power model to the data.

(b) Use a graphing utility to plot the data and graph the models in part (a).

(c) Select the model that best fits the data and use it to predict F when $x = 14$.

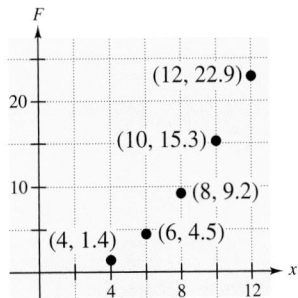

32. *Think About It* A model is needed for the data in the table. After a scatter plot is created, it is determined that a logarithmic model best fits the data. During an attempt to find a logarithmic model, use of the regression capabilities of a graphing utility results in an error. Explain.

x	0	2	4	9
y	2	2.5	2.8	3

Section Project

Comparing Models The rates R of aggravated assault per 100,000 population in the United States from 1984 through 1993 are given in the table. *(Source: U.S. Federal Bureau of Investigation)*

Year	1984	1985	1986	1987	1988
R	290	303	346	351	370

Year	1989	1990	1991	1992	1993
R	383	424	433	442	440

For each of the following, let t be the time in years, with $t = 4$ corresponding to 1984.

(a) Use a graphing utility to fit the following models to the data.

$$y_1 = ax + b$$
$$y_2 = ax^2 + bx + c$$
$$y_3 = a + b \ln x$$
$$y_4 = ab^x$$
$$y_5 = ax^b$$

(b) Use a graphing utility to plot the data and graph the models in part (a). Select the model that best fits the data.

Chapter Summary

After studying this chapter, you should have acquired the following skills. These skills are keyed to the Review Exercises that begin on page 705. Answers to odd-numbered Review Exercises are given in the back of the book.

- Evaluate exponential and logarithmic functions for given values of the variable.   *(Sections 9.1, 9.2)*

  Review Exercises 1–10

- Match exponential and logarithmic functions with their graphs. *(Sections 9.1, 9.2)*

  Review Exercises 11–16

- Sketch the graphs of exponential and logarithmic functions. *(Sections 9.1, 9.2)*

  Review Exercises 17–26

- Graph exponential and logarithmic functions using a graphing utility. *(Sections 9.1, 9.2)*

  Review Exercises 27–34

- Write exponential equations in logarithmic form.   *(Section 9.2)*

  Review Exercises 35, 36

- Write logarithmic equations in exponential form.   *(Section 9.2)*

  Review Exercises 37, 38

- Evaluate logarithmic expressions.   *(Sections 9.2, 9.3)*

  Review Exercises 39–46

- Rewrite logarithmic expressions in expanded form using the properties of logarithms.   *(Section 9.3)*

  Review Exercises 47–54

- Rewrite logarithmic expressions in condensed form using the properties of logarithms.   *(Section 9.3)*

  Review Exercises 55–64

- Graphically verify that two logarithmic expressions are equivalent using a graphing utility.   *(Section 9.3)*

  Review Exercises 65, 66

- Decide whether exponential and logarithmic equations are true or false. *(Section 9.3)*

  Review Exercises 67–72

- Approximate values of logarithmic expressions given the values of specific logarithms.   *(Section 9.3)*

  Review Exercises 73–78

- Evaluate logarithmic expressions using the change-of-base formula. *(Section 9.2)*

  Review Exercises 79–82

- Solve exponential and logarithmic equations.   *(Section 9.4)*

  Review Exercises 83–102

- Solve real-life problems involving exponential and logarithmic functions. *(Sections 9.1–9.6)*

  Review Exercises 103–117

Review Exercises

In Exercises 1–10, evaluate the function as indicated.

1. $f(x) = 2^x$
   (a) $x = -3$ (b) $x = 1$ (c) $x = 2$

2. $g(x) = 2^{-x}$
   (a) $x = -2$ (b) $x = 0$ (c) $x = 2$

3. $g(t) = e^{-t/3}$
   (a) $t = -3$ (b) $t = \pi$ (c) $t = 6$

4. $h(s) = 1 - e^{0.2s}$
   (a) $s = 0$ (b) $s = 2$ (c) $s = \sqrt{10}$

5. $f(x) = \log_3 x$
   (a) $x = 1$ (b) $x = 27$ (c) $x = 0.5$

6. $g(x) = \log_{10} x$
   (a) $x = 0.01$ (b) $x = 0.1$ (c) $x = 30$

7. $f(x) = \ln x$
   (a) $x = e$ (b) $x = \frac{1}{3}$ (c) $x = 10$

8. $h(x) = \ln x$
   (a) $x = e^2$ (b) $x = \frac{5}{4}$ (c) $x = 1200$

9. $g(x) = \ln e^{3x}$
   (a) $x = -2$ (b) $x = 0$ (c) $x = 7.5$

10. $f(x) = \log_2 \sqrt{x}$
   (a) $x = 4$ (b) $x = 64$ (c) $x = 5.2$

In Exercises 11–16, match the function with its graph. [The graphs are labeled (a), (b), (c), (d), (e), and (f).]

(a)
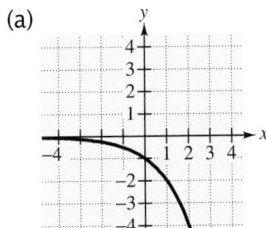

(b)
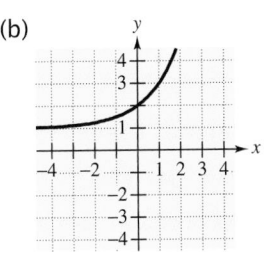

(c)
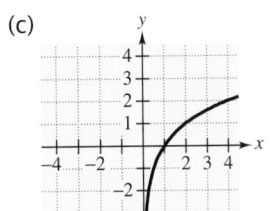

(d)
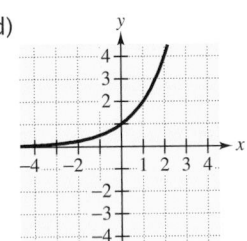

(e)
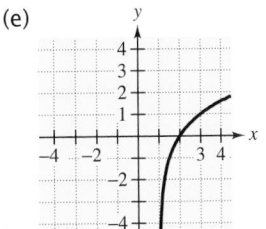

(f)
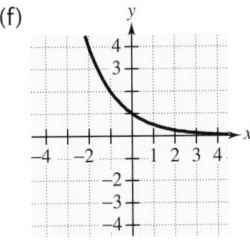

11. $f(x) = 2^x$ **12.** $f(x) = 2^{-x}$

13. $f(x) = -2^x$ **14.** $f(x) = 2^x + 1$

15. $f(x) = \log_2 x$ **16.** $f(x) = \log_2 (x - 1)$

In Exercises 17–26, sketch the graph of the function.

17. $y = 3^{x/2}$ **18.** $y = 3^{x/2} - 2$

19. $f(x) = 3^{-x/2}$ **20.** $f(x) = -3^{-x/2}$

21. $f(x) = 3^{-x^2}$ **22.** $g(t) = 3^{|t|}$

23. $f(x) = -2 + \log_3 x$ **24.** $f(x) = 2 + \log_3 x$

25. $y = \log_2(x - 4)$ **26.** $y = 2 \log_4(x + 1)$

In Exercises 27–34, use a graphing utility to graph the function.

27. $y = 5e^{-x/4}$ **28.** $y = 6 - e^{x/2}$

29. $f(x) = e^{x+2}$ **30.** $h(t) = \dfrac{8}{1 + e^{-t/5}}$

31. $g(x) = \ln 2x$ **32.** $f(x) = 3 + \ln x$

33. $g(t) = 2 - \ln(t - 1)$ **34.** $g(x) = \ln(x - 5)$

In Exercises 35 and 36, write the exponential equation in logarithmic form.

35. $4^3 = 64$

36. $25^{3/2} = 125$

In Exercises 37 and 38, write the logarithmic equation in exponential form.

37. $\ln e = 1$

38. $\log_5 \frac{1}{25} = -2$

In Exercises 39–46, evaluate the expression.

39. $\log_{10} 1000$

40. $\log_9 3$

41. $\log_3 \frac{1}{9}$

42. $\log_4 \frac{1}{16}$

43. $\ln e^7$

44. $\log_a \frac{1}{a}$

45. $\ln 1$

46. $\ln e^{-3}$

In Exercises 47–54, use the properties of logarithms to expand the expression.

47. $\log_4 6x^4$

48. $\log_{10} 2x^{-3}$

49. $\log_5 \sqrt{x + 2}$

50. $\ln \sqrt[3]{\dfrac{x}{5}}$

51. $\ln \dfrac{x + 2}{x - 2}$

52. $\ln [x(x - 3)^2]$

53. $\ln\left[\sqrt{2x}(x + 3)^5\right]$

54. $\log_3 \dfrac{a^2 \sqrt{b}}{cd^5}$

In Exercises 55–64, use the properties of logarithms to condense the expression.

55. $\log_4 x - \log_4 10$

56. $5 \log_2 y$

57. $\log_8 16x + \log_8 2x^2$

58. $4(1 + \ln x + \ln x)$

59. $-2(\ln 2x - \ln 3)$

60. $-\frac{2}{3} \ln 3y$

61. $3 \ln x + 4 \ln y + \ln z$

62. $\frac{1}{3}(\log_8 a + 2 \log_8 b)$

63. $4[\log_2 k - \log_2(k - t)]$

64. $\ln(x + 4) - 3 \ln x - \ln y$

In Exercises 65 and 66, use a graphing utility to verify graphically that the expressions are equivalent.

65. $y_1 = \ln\left(\dfrac{x + 1}{x - 1}\right)^2, \; x > 1$

$y_2 = 2[\ln(x + 1) - \ln(x - 1)]$

66. $y_1 = \ln\sqrt{x^2 - 4}, \; x > 2$

$y_2 = \frac{1}{2}[\ln(x + 2) + \ln(x - 2)]$

True or False? In Exercises 67–72, determine whether the equation is true or false. Explain your reasoning.

67. $\log_2 4x = 2 \log_2 x$

68. $\dfrac{\ln 5x}{\ln 10x} = \ln \dfrac{1}{2}$

69. $\log_{10} 10^{2x} = 2x$

70. $e^{\ln t} = t$

71. $\log_4 \dfrac{16}{x} = 2 - \log_4 x$

72. $e^{2x} - 1 = (e^x + 1)(e^x - 1)$

In Exercises 73–78, approximate the logarithm given that $\log_5 2 \approx 0.43068$ and $\log_5 3 \approx 0.6826$.

73. $\log_5 18$

74. $\log_5 \sqrt{6}$

75. $\log_5 \frac{1}{2}$

76. $\log_5 \frac{2}{3}$

77. $\log_5 (12)^{2/3}$

78. $\log_5 (5^2 \cdot 6)$

In Exercises 79–82, evaluate the logarithm using the change-of-base formula. (Round the result to three decimal places.)

79. $\log_4 9$

80. $\log_{1/2} 5$

81. $\log_{12} 200$

82. $\log_3 0.28$

In Exercises 83–88, solve the equation.

83. $2^x = 64$

84. $3^{x-2} = 81$

85. $4^{x-3} = \frac{1}{16}$

86. $\log_2 2x = \log_2 100$

87. $\log_3 x = 5$

88. $\log_5(x - 10) = 2$

In Exercises 89–102, solve the equation. (Round the result to two decimal places.)

89. $3^x = 500$

90. $8^x = 1000$

91. $2e^{x/2} = 45$

92. $100e^{-0.6x} = 20$

93. $\dfrac{500}{(1.05)^x} = 100$

94. $25(1 - e^t) = 12$

95. $\log_{10} 2x = 1.5$

96. $\frac{1}{3}\log_2 x + 5 = 7$

97. $\ln x = 7.25$

98. $\ln x = -0.5$

99. $\log_2 2x = -0.65$

100. $\log_5(x + 1) = 4.8$

101. $\log_2 x + \log_2 3 = 3$

102. $2 \log_4 x - \log_4(x - 1) = 1$

Creating a Table In Exercises 103–106, complete the table to determine the balance A for P dollars invested at rate r for t years, compounded n times per year.

n	1	4	12	365	Continuous compounding
A					

	Principal	Rate	Time
103.	$P = \$500$	$r = 7\%$	$t = 30$ years
104.	$P = \$100$	$r = 5\frac{1}{4}\%$	$t = 60$ years
105.	$P = \$10,000$	$r = 10\%$	$t = 20$ years
106.	$P = \$2500$	$r = 8\%$	$t = 1$ year

Creating a Table In Exercises 107 and 108, complete the table to determine the principal P that yields a balance of A dollars invested at rate r for t years, compounded n times per year.

n	1	4	12	365	Continuous compounding
P					

	Balance	Rate	Time
107.	$A = \$50,000$	$r = 8\%$	$t = 40$ years
108.	$A = \$1000$	$r = 6\%$	$t = 1$ year

109. *Inflation Rate* If the annual rate of inflation averages 5% over the next 10 years, the approximate cost C of goods or services during any year in that decade will be given by

$$C(t) = P(1.05)^t, \qquad 0 \le t \le 10,$$

where t is the time in years and P is the present cost. If the price of an oil change for your car is presently $19.95, when will it cost $25.00?

110. *Doubling Time* Find the time for an investment of $1000 to double in value when invested at 8% compounded monthly.

111. *Product Demand* The daily demand x and price p for a product are related by $p = 25 - 0.4e^{0.02x}$. Approximate the demand when the price is $16.97.

112. *Sound Intensity* The relationship between the number of decibels B and the intensity of a sound I in watts per centimeter squared is given by

$$B = 10 \log_{10}\left(\dfrac{I}{10^{-16}}\right).$$

Determine the intensity of a sound in watts per centimeter squared if the decibel level is 125.

113. *Population Limit* The population p of a certain species t years after it is introduced into a new habitat is given by

$$p(t) = \dfrac{600}{1 + 2e^{-0.2t}}.$$

Use a graphing utility to graph the function. Use the graph to determine the limiting size of the population in the habitat.

114. *Deer Herd* The state Parks and Wildlife Department releases 100 deer into a wilderness area. The population P of the herd can be modeled by

$$P = \dfrac{500}{1 + 4e^{-0.36t}},$$

where t is measured in years.

(a) Find the population after 5 years.

(b) After how many years will the population be 250?

115. *Data Analysis* The data in the table gives the yield y (in milligrams) of a chemical reaction after t minutes.

t	1	2	3	4
y	1.5	7.4	10.2	13.4

t	5	6	7	8
y	15.8	16.3	18.2	18.3

(a) Use a graphing utility to fit a linear model to the data.

(b) Use a graphing utility to fit a logarithmic model to the data.

(c) Use a graphing utility to plot the data and graph the models in parts (a) and (b).

(d) Which is the better model? Explain your reasoning.

116. *Sporting Goods Sales* The sales y (in billions of dollars) of sporting goods for 1984 through 1993 are given in the table, where $x = 4$ corresponds to 1984. *(Source: National Sporting Goods Association)*

x	4	5	6	7	8
y	26.4	27.4	30.6	33.9	42.1

x	9	10	11	12	13
y	45.2	44.1	42.9	42.4	44.1

(a) Use a graphing utility to fit a power model to the data. Use the graphing utility to plot the data and graph the model.

(b) Use the model in part (a) to predict sales in the year 2000.

117. *Comparing Models* The time t (in seconds) required to attain a speed of s miles per hour from a standing start for a 1995 Dodge Avenger is given in the table. *(Source: Road & Track, March 1995)*

s	30	40	50	60	70	80	90
t	3.4	5.0	7.0	9.3	12.0	15.8	20.0

(a) Use a graphing utility to fit the following models to the data.

$$t_1 = as + b$$
$$t_2 = as^2 + bs + c$$
$$t_3 = a + b \ln s$$
$$t_4 = ab^s$$
$$t_5 = as^b$$

(b) Use a graphing utility to plot the data and graph the models in part (a). Select the model that best fits the data.

Chapter Test

Take this test as you would take a test in class. After you are done, check your work against the answers given in the back of the book.

1. Evaluate $f(t) = 54\left(\frac{2}{3}\right)^t$ when $t = -1, 0, \frac{1}{2}$, and 2.
2. Sketch a graph of the function $f(x) = 2^{x/3}$.
3. Write the logarithmic equation $\log_5 125 = 3$ in exponential form.
4. Write the exponential equation $4^{-2} = \frac{1}{16}$ in logarithmic form.
5. Evaluate $\log_8 2$ without a calculator.
6. Describe the relationship between the graphs of $f(x) = \log_5 x$ and $g(x) = 5^x$.
7. Expand the expression $\log_4\left(5x^2/\sqrt{y}\right)$.
8. Condense the expression $8 \ln a + \ln b - 3 \ln c$.

In Exercises 9–14, solve the equation.

9. $\log_4 x = 3$
10. $10^{3y} = 832$
11. $400e^{0.08t} = 1200$
12. $3 \ln(2x - 3) = 10$
13. $\log_2 x + \log_2(x + 4) = 5$
14. $2 \log_{10} x - \log_{10} 9 = 2$

15. Determine the balance after 20 years if $2000 is invested at 7% compounded (a) quarterly, and (b) continuously.
16. Determine the principal that yields $100,000 invested at 9% compounded quarterly for 25 years.
17. A principal of $500 yields a balance of $1006.88 in 10 years when the interest is compounded continuously. What is the annual interest rate?
18. A car that cost $18,000 new has a depreciated value of $14,000 after 1 year. Use the model $y = Ce^{kt}$ to find the value of the car when it is 3 years old.
19. The population p of a certain species t years after it is introduced into a new habitat is given by

$$p(t) = \frac{2400}{1 + 3e^{-t/4}}.$$

   (a) Determine the population after 4 years.

   (b) After how many years will the population be 1200?

20. The numbers of cellular telephone subscribers y (in millions) for the years 1990 through 1994 are given by $(0, 5.3)$, $(1, 7.6)$, $(2, 11.0)$, $(3, 16.0)$, and $(4, 24.1)$, where x is the time in years, with $x = 0$ corresponding to 1990. A model for the data is $y = 5.23e^{0.3773x}$. Use a graphing utility to plot the data and graph the model in the same viewing rectangle.

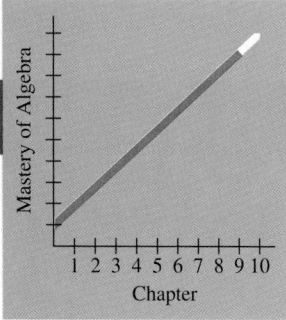

Mastery of Algebra

1 2 3 4 5 6 7 8 9 10
Chapter

Take this test as you would take a test in class. After you are done, check your work against the answers given in the back of the book.

In Exercises 1–3, solve the equation.

1. $4 - \frac{1}{2}x = 6$ **2.** $12(3 - x) = 5 - 7(2x + 1)$ **3.** $x^2 - 10 = 0$

In Exercises 4–6, solve the inequality.

4. $-2 \leq 1 - 2x \leq 2$ **5.** $|x + 7| \geq 2$ **6.** $x^2 + 3x - 10 < 0$

In Exercises 7–9, simplify the expression.

7. $(x^2 \cdot x^3)^4$ **8.** $\left(\dfrac{3x^2}{2y}\right)^{-2}$ **9.** $(9x^8)^{3/2}$

In Exercises 10 and 11, simplify the complex fraction.

10. $\dfrac{\left(\dfrac{4}{x^2 - 9} + \dfrac{2}{x - 2}\right)}{\left(\dfrac{1}{x + 3} + \dfrac{1}{x - 3}\right)}$ **11.** $\dfrac{\left(\dfrac{1}{x + 1} + \dfrac{1}{2}\right)}{\left(\dfrac{3}{2x^2 + 4x + 2}\right)}$

In Exercises 12 and 13, identify the transformation of the graph of $f(x) = \sqrt{x}$. Use a graphing utility to verify the transformation.

12. $h(x) = \sqrt{4 + x}$ **13.** $h(x) = -\sqrt{x}$

In Exercises 14 and 15, evaluate the determinant of the matrix.

14. $\begin{bmatrix} 10 & 25 \\ 6 & -5 \end{bmatrix}$ **15.** $\begin{bmatrix} 4 & 3 & 5 \\ 3 & 2 & -2 \\ 5 & -2 & 0 \end{bmatrix}$

In Exercises 16–18, rewrite the equation in exponential form.

16. $\log_4 64 = 3$ **17.** $\log_3 \frac{1}{81} = -4$ **18.** $\ln 1 = 0$

19. Determine the number of gallons of a 30% solution that must be mixed with a 60% solution to obtain 20 gallons of a 40% solution.

20. After t years, the value of a car that cost $22,000 new is given by $V(t) = 22{,}000(0.8)^t$. Sketch a graph of the function and determine when the value of the car is $15,000.

710

Topics in Discrete Mathematics 10

Strategies for Success

SKILLS *When you have completed this chapter, make sure you are able to:*

◯ Write and evaluate sequences and sums using sequence and sigma notation
◯ Identify, write terms of, and evaluate sums of arithmetic and geometric sequences
◯ Evaluate binomial coefficients and expand binomial expressions
◯ Calculate the number of ways an event can occur using counting principles including combinations and permutations
◯ Find the probabilities of the occurrences of real-life events

TOOLS *Use these study tools to master the skills above:*

In Exercise 111 of Section 10.3, you will use a geometric sequence to determine the total distance traveled by a bungee jumper.

| | 10.1 | **Sequences** |

Sequences ▪ Factorial Notation ▪ Sigma Notation

Sequences

Suppose you were given the following choices of a contract offer for the next 5 years of employment.

Contract A $20,000 the first year and a $2200 raise each year

Contract B $20,000 the first year and a 10% raise each year

Year	Contract A	Contract B
1	$20,000	$20,000
2	$22,200	$22,000
3	$24,400	$24,200
4	$26,600	$26,620
5	$28,800	$29,282
Total	$122,000	$122,102

Which contract offers the largest salary over the 5-year period? The salaries for each contract are shown at the left. The salaries for contract A represent the first five terms of an **arithmetic sequence,** and the salaries for contract B represent the first five terms of a **geometric sequence.** Notice that after 5 years the geometric sequence represents a better contract offer than the arithmetic sequence.

A mathematical **sequence** is simply an ordered list of numbers. Each number in the list is a **term** of the sequence. A sequence can have a finite number of terms or an infinite number of terms. For instance, the sequence of positive odd integers that are less than 15 is a *finite* sequence

$$1, 3, 5, 7, 9, 11, 13 \qquad \text{Finite sequence}$$

whereas the sequence of positive odd integers is an *infinite* sequence.

$$1, 3, 5, 7, 9, 11, 13, \ldots \qquad \text{Infinite sequence}$$

Note that the three dots indicate that the sequence continues and has an infinite number of terms.

Definition of an Infinite Sequence

An **infinite sequence** is an ordered list of real numbers.

$$a_1, a_2, a_3, a_4, a_5, \ldots, a_n, \ldots$$

Each number in the sequence is a **term** of the sequence, and the sequence consists of an infinite number of terms.

NOTE   Sometimes it is convenient to begin subscripting an infinite sequence with 0 instead of 1. In such cases, the terms of the sequence are denoted by

$$a_0, a_1, a_2, a_3, a_4, a_5, \ldots, a_n, \ldots$$

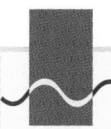

Technology

Most graphing utilities have a "sequence graphing mode" that allows you to plot the terms of a sequence as points on a rectangular coordinate system. For instance, the graph of the first six terms of the sequence given by

$$a_n = n^2 - 1$$

is shown below.

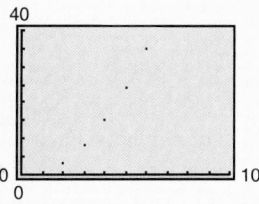

In Example 2, some students may be tempted to write $3(2^n)$ as 6^n. Reinforce the order of operations.

EXAMPLE 1   Finding the Terms of a Sequence

Write the first six terms of the sequence whose nth term is

$$a_n = n^2 - 1. \qquad \text{Begin sequence with } n = 1.$$

Solution

$$a_1 = (1)^2 - 1 = 0 \qquad a_2 = (2)^2 - 1 = 3 \qquad a_3 = (3)^2 - 1 = 8$$
$$a_4 = (4)^2 - 1 = 15 \qquad a_5 = (5)^2 - 1 = 24 \qquad a_6 = (6)^2 - 1 = 35$$

The entire sequence can be written as follows.

$$0, \ 3, \ 8, \ 15, \ 24, \ 35, \ \ldots, \ n^2 - 1, \ \ldots$$

EXAMPLE 2   Finding the Terms of a Sequence

Write the first six terms of the sequence whose nth term is

$$a_n = 3(2^n). \qquad \text{Begin sequence with } n = 0.$$

Solution

$$a_0 = 3(2^0) = 3 \cdot 1 = 3 \qquad a_1 = 3(2^1) = 3 \cdot 2 = 6$$
$$a_2 = 3(2^2) = 3 \cdot 4 = 12 \qquad a_3 = 3(2^3) = 3 \cdot 8 = 24$$
$$a_4 = 3(2^4) = 3 \cdot 16 = 48 \qquad a_5 = 3(2^5) = 3 \cdot 32 = 96$$

The entire sequence can be written as follows.

$$3, \ 6, \ 12, \ 24, \ 48, \ 96, \ \ldots, \ 3(2^n), \ \ldots$$

EXAMPLE 3   A Sequence Whose Terms Alternate in Sign

Write the first six terms of the sequence whose nth term is

$$a_n = \frac{(-1)^n}{2n - 1}. \qquad \text{Begin sequence with } n = 1.$$

Solution

$$a_1 = \frac{(-1)^1}{2(1) - 1} = -\frac{1}{1} \qquad a_2 = \frac{(-1)^2}{2(2) - 1} = \frac{1}{3} \qquad a_3 = \frac{(-1)^3}{2(3) - 1} = -\frac{1}{5}$$
$$a_4 = \frac{(-1)^4}{2(4) - 1} = \frac{1}{7} \qquad a_5 = \frac{(-1)^5}{2(5) - 1} = -\frac{1}{9} \qquad a_6 = \frac{(-1)^6}{2(6) - 1} = \frac{1}{11}$$

The entire sequence can be written as follows.

$$-1, \ \frac{1}{3}, \ -\frac{1}{5}, \ \frac{1}{7}, \ -\frac{1}{9}, \ \frac{1}{11}, \ \ldots, \ \frac{(-1)^n}{2n - 1}, \ \ldots$$

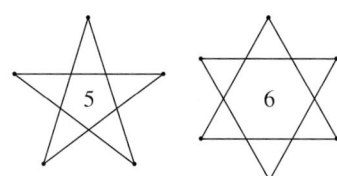

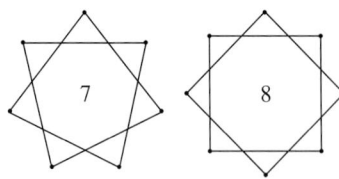

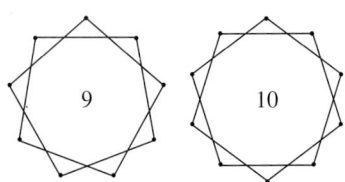

FIGURE 10.1

EXAMPLE 4 *Finding the Terms of a Sequence*

The number of degrees d_n in the angle at each point of each of the six n-pointed stars shown in Figure 10.1 is

$$d_n = \frac{180(n-4)}{n}, \qquad n \geq 5.$$

Write the first six terms of the sequence.

Solution

$$d_5 = \frac{180(5-4)}{5} = 36° \qquad d_6 = \frac{180(6-4)}{6} = 60°$$

$$d_7 = \frac{180(7-4)}{7} \approx 77.14° \qquad d_8 = \frac{180(8-4)}{8} = 90°$$

$$d_9 = \frac{180(9-4)}{9} = 100° \qquad d_{10} = \frac{180(10-4)}{10} = 108°$$

Some sequences are defined **recursively.** To define a sequence recursively, you need to be given one or more of the first few terms. All other terms of the sequence are then defined using previous terms. A well-known example is the Fibonacci Sequence shown in Example 5.

EXAMPLE 5 *A Sequence that Is Defined Recursively*

The Fibonacci Sequence is defined recursively, as follows.

$$a_0 = 1, \ a_1 = 1, \ a_k = a_{k-2} + a_{k-1}, \ \text{where } k \geq 2$$

Write the first six terms of the sequence.

Solution

$$a_0 = 1 \qquad\qquad\qquad \text{0th term is given.}$$
$$a_1 = 1 \qquad\qquad\qquad \text{1st term is given.}$$
$$a_2 = a_0 + a_1 = 1 + 1 = 2 \qquad \text{Use recursive formula.}$$
$$a_3 = a_1 + a_2 = 1 + 2 = 3 \qquad \text{Use recursive formula.}$$
$$a_4 = a_2 + a_3 = 2 + 3 = 5 \qquad \text{Use recursive formula.}$$
$$a_5 = a_3 + a_4 = 3 + 5 = 8 \qquad \text{Use recursive formula.}$$

Factorial Notation

Some very important sequences in mathematics involve terms that are defined with special types of products called **factorials.**

Definition of Factorial

If n is a positive integer, **n factorial** is defined as

$$n! = 1 \cdot 2 \cdot 3 \cdot 4 \cdot \cdot \cdot \cdot \cdot (n - 1) \cdot n.$$

As a special case, zero factorial is defined as $0! = 1$.

The first several factorial values are as follows.

$$0! = 1 \qquad\qquad\qquad 1! = 1$$
$$2! = 1 \cdot 2 = 2 \qquad\qquad 3! = 1 \cdot 2 \cdot 3 = 6$$
$$4! = 1 \cdot 2 \cdot 3 \cdot 4 = 24 \qquad 5! = 1 \cdot 2 \cdot 3 \cdot 4 \cdot 5 = 120$$

Many calculators have a factorial key, denoted by $\boxed{n!}$. If your calculator has such a key, try using it to evaluate $n!$ for several values of n. You will see that the value of n does not have to be very large before the value of $n!$ becomes huge. For instance,

$$10! = 3,628,800.$$

EXAMPLE 6   A Sequence Involving Factorials

Write the first six terms of the sequence whose nth term is

$$a_n = \frac{1}{n!}. \qquad \text{Begin sequence with } n = 0.$$

Solution

$$a_0 = \frac{1}{0!} = \frac{1}{1} = 1 \qquad\qquad a_1 = \frac{1}{1!} = \frac{1}{1} = 1$$

$$a_2 = \frac{1}{2!} = \frac{1}{2} \qquad\qquad a_3 = \frac{1}{3!} = \frac{1}{1 \cdot 2 \cdot 3} = \frac{1}{6}$$

$$a_4 = \frac{1}{4!} = \frac{1}{1 \cdot 2 \cdot 3 \cdot 4} = \frac{1}{24} \qquad a_5 = \frac{1}{5!} = \frac{1}{1 \cdot 2 \cdot 3 \cdot 4 \cdot 5} = \frac{1}{120}$$

The entire sequence can be written as follows.

$$1, \; 1, \; \frac{1}{2}, \frac{1}{6}, \frac{1}{24}, \frac{1}{120}, \cdot \cdot \cdot, \frac{1}{n!}, \cdot \cdot \cdot$$

Technology

Most graphing utilities have a built-in function that will display the terms of a sequence. To display the first six terms of the sequence in Example 6 on a *TI-82* or *TI-83*, use the sequence function in the $\boxed{\text{LIST}}$ (Math) menu and enter the following.

   seq(1/N!,N,0,5,1)

To see the terms displayed in fractional form, add the ▶Frac function (found in the $\boxed{\text{MATH}}$ menu) to the command above.

   seq(1/N!,N,0,5,1) ▶ Frac

After entering these key strokes, press $\boxed{\text{ENTER}}$. Use the right cursor arrow to see the entire display.

Remind students that $0! = 1$, not 0; hence, it is acceptable to have $0!$ in a denominator.

Factorials follow the same conventions for order of operations as do exponents. For instance, $2n!$ means $2(n!)$, not $(2n)!$. Notice how these conventions are used in the next example.

EXAMPLE 7 A Sequence Involving Factorials

Write the first six terms of the sequence whose nth term is

$$a_n = \frac{2n!}{(2n)!}. \qquad \text{Begin sequence with } n = 0.$$

Solution

$$a_0 = \frac{2(0!)}{(2 \cdot 0)!} = \frac{2(0!)}{0!} = \frac{2}{1} = 2$$

$$a_1 = \frac{2(1!)}{(2 \cdot 1)!} = \frac{2(1!)}{2!} = \frac{2}{2} = 1$$

$$a_2 = \frac{2(2!)}{(2 \cdot 2)!} = \frac{2(2!)}{4!} = \frac{4}{24} = \frac{1}{6}$$

$$a_3 = \frac{2(3!)}{(2 \cdot 3)!} = \frac{2(3!)}{6!} = \frac{12}{720} = \frac{1}{60}$$

$$a_4 = \frac{2(4!)}{(2 \cdot 4)!} = \frac{2(4!)}{8!} = \frac{48}{40,320} = \frac{1}{840}$$

$$a_5 = \frac{2(5!)}{(2 \cdot 5)!} = \frac{2(5!)}{10!} = \frac{240}{3,628,800} = \frac{1}{15,120}$$

The entire sequence can be written as follows.

$$2, \; 1, \; \frac{1}{6}, \; \frac{1}{60}, \; \frac{1}{840}, \; \frac{1}{15,120}, \; \cdots \; , \; \frac{2n!}{(2n)!}, \; \cdots$$

In Example 7, the numerators and denominators were multiplied before the fractions were reduced. When you are finding the terms of a sequence, reducing is often easier if you leave the numerator and denominator in factored form. For instance, notice the cancellation in the following fraction.

$$a_5 = \frac{2(5!)}{10!}$$

$$= \frac{2 \cdot 1 \cdot 2 \cdot 3 \cdot 4 \cdot 5}{1 \cdot 2 \cdot 3 \cdot 4 \cdot 5 \cdot 6_3 \cdot 7 \cdot 8 \cdot 9 \cdot 10}$$

$$= \frac{1}{3 \cdot 7 \cdot 8 \cdot 9 \cdot 10}$$

$$= \frac{1}{15,120}$$

Sigma Notation

Many applications involve finding the sum of the first n terms of a sequence. A convenient shorthand notation for such a sum is **sigma notation.** This name comes from the use of the uppercase Greek letter sigma, written as Σ.

Definition of Sigma Notation

The sum of the first n terms of the sequence whose nth term is a_n is

$$\sum_{i=1}^{n} a_i = a_1 + a_2 + a_3 + a_4 + \cdots + a_n,$$

where i is the **index of summation,** n is the **upper limit of summation,** and 1 is the **lower limit of summation.**

NOTE In Example 8, the index of summation is i and the summation begins with $i = 1$. Any letter can be used as the index of summation, and the summation can begin with any integer. For instance, in Example 9, the index of summation is k and the summation begins with $k = 0$.

EXAMPLE 8   Sigma Notation for Sums

Find the sum $\displaystyle\sum_{i=1}^{6} 2i$.

Solution

$$\sum_{i=1}^{6} 2i = 2(1) + 2(2) + 2(3) + 2(4) + 2(5) + 2(6)$$

$$= 2 + 4 + 6 + 8 + 10 + 12$$

$$= 42$$

EXAMPLE 9   Sigma Notation for Sums

Find the sum $\displaystyle\sum_{k=0}^{8} \frac{1}{k!}$.

Solution

Sigma notation expresses a broad idea more concisely. Use several different examples to help students become more comfortable with it. As a first example, consider placing $a_1, a_2, a_3, \ldots, a_n$ above each of the corresponding values of the sequence to help reinforce the meaning of the notation.

$$\sum_{k=0}^{8} \frac{1}{k!} = \frac{1}{0!} + \frac{1}{1!} + \frac{1}{2!} + \frac{1}{3!} + \frac{1}{4!} + \frac{1}{5!} + \frac{1}{6!} + \frac{1}{7!} + \frac{1}{8!}$$

$$= 1 + 1 + \frac{1}{2} + \frac{1}{6} + \frac{1}{24} + \frac{1}{120} + \frac{1}{720} + \frac{1}{5040} + \frac{1}{40,320}$$

$$\approx 2.71828$$

Note that this sum is approximately $e = 2.71828\ldots.$

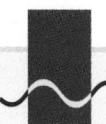

Technology

Most graphing utilities have built-in functions that will calculate the sum of a sequence. For instance, on a *TI-82* or *TI-83*, you can find the sum of the sequence in Example 10(b) by using the [LIST] (Math) and the [LIST] (Ops) menus to produce the following.

sum(seq(1/2 ^ I,I,0,6,1))

After entering these keystrokes and pressing [ENTER], the calculator should display 1.984375, which is equal to 127/64.

Remind students that the process of going from terms of a series to sigma notation requires some trial-and-error, observation, and conjecture. What is the pattern? squares? multiples? cubes less than one?

EXAMPLE 10 Sigma Notation for Sums

(a) $\displaystyle\sum_{i=1}^{4} 5 = 5 + 5 + 5 + 5 = 20$

(b) $\displaystyle\sum_{i=0}^{6} \frac{1}{2^i} = \frac{1}{2^0} + \frac{1}{2^1} + \frac{1}{2^2} + \frac{1}{2^3} + \frac{1}{2^4} + \frac{1}{2^5} + \frac{1}{2^6}$

$\displaystyle = \frac{1}{1} + \frac{1}{2} + \frac{1}{4} + \frac{1}{8} + \frac{1}{16} + \frac{1}{32} + \frac{1}{64}$

$\displaystyle = \frac{127}{64}$

EXAMPLE 11 Writing a Sum in Sigma Notation

Write the sum in sigma notation.

$$\frac{2}{2} + \frac{2}{3} + \frac{2}{4} + \frac{2}{5} + \frac{2}{6}$$

Solution

To write this sum in sigma notation, you must find a pattern for the terms. After examining the terms, you can see that they have numerators of 2 and denominators that range over the integers from 2 to 6. Thus, one possible sigma notation is

$$\sum_{i=1}^{5} \frac{2}{i+1} = \frac{2}{2} + \frac{2}{3} + \frac{2}{4} + \frac{2}{5} + \frac{2}{6}.$$

Group Activities

Communicating Mathematically

Finding a Pattern You learned in this section that a sequence is an ordered list of numbers. Study the following sequence and see if you can guess what its next term should be.

Z, O, T, T, F, F, S, S, E, N, T, E, T, . . .

(*Hint:* You might try to figure out what numbers the letters represent.) Construct another sequence with letters. Can the other members of your group guess the next term?

10.1   Exercises

The symbol ▦ indicates an exercise in which you are instructed to use a calculator or graphing utility. The solutions of other exercises may also be facilitated by the use of appropriate technology.

See Warm-Up Exercises, p. A50

In Exercises 1–22, write the first five terms of the sequence. (Begin with $n = 1$.)

1. $a_n = 2n$
2. $a_n = 3n$
3. $a_n = (-1)^n 2n$
4. $a_n = (-1)^{n+1} 3n$
5. $a_n = \left(\frac{1}{2}\right)^n$
6. $a_n = \left(\frac{1}{3}\right)^n$
7. $a_n = \left(-\frac{1}{2}\right)^{n+1}$
8. $a_n = \left(\frac{2}{3}\right)^{n-1}$
9. $a_n = (-0.2)^{n-1}$
10. $a_n = \left(-\frac{2}{3}\right)^n$
11. $a_n = \frac{1}{n+1}$
12. $a_n = \frac{3}{2n+1}$
13. $a_n = \frac{2n}{3n+2}$
14. $a_n = \frac{5n}{4n+3}$
15. $a_n = \frac{(-1)^n}{n^2}$
16. $a_n = \frac{1}{\sqrt{n}}$
17. $a_n = 5 - \frac{1}{2^n}$
18. $a_n = 7 + \frac{1}{3^n}$
19. $a_n = \frac{2^n}{n!}$
20. $a_n = \frac{n!}{(n-1)!}$
21. $a_n = 2 + (-2)^n$
22. $a_n = \frac{1+(-1)^n}{n^2}$

In Exercises 23 and 24, find the indicated term of the sequence.

23. $a_n = (-1)^n(5n-3)$

$a_{15} = $ ▢

24. $a_n = \frac{n^2}{n!}$

$a_{12} = $ ▢

In Exercises 25–36, simplify the expression.

25. $\frac{5!}{4!}$
26. $\frac{18!}{17!}$
27. $\frac{10!}{12!}$
28. $\frac{5!}{8!}$
29. $\frac{25!}{27!}$
30. $\frac{20!}{15! \cdot 5!}$
31. $\frac{n!}{(n+1)!}$
32. $\frac{(n+2)!}{n!}$
33. $\frac{(n+1)!}{(n-1)!}$
34. $\frac{(3n)!}{(3n+2)!}$
35. $\frac{(2n)!}{(2n-1)!}$
36. $\frac{(2n+2)!}{(2n)!}$

In Exercises 37–52, find the sum.

37. $\sum_{k=1}^{6} 3k$
38. $\sum_{k=1}^{4} 5k$
39. $\sum_{i=0}^{6} (2i+5)$
40. $\sum_{j=3}^{7} (6j-10)$
41. $\sum_{i=0}^{4} (2i+3)$
42. $\sum_{i=2}^{7} (4i-1)$
43. $\sum_{j=1}^{5} \frac{(-1)^{j+1}}{j}$
44. $\sum_{j=0}^{3} \frac{1}{j^2+1}$
45. $\sum_{m=2}^{6} \frac{2m}{2(m-1)}$
46. $\sum_{k=1}^{5} \frac{10k}{k+2}$
47. $\sum_{k=1}^{6} (-8)$
48. $\sum_{n=3}^{12} 10$
49. $\sum_{i=1}^{8} \left(\frac{1}{i} - \frac{1}{i+1}\right)$
50. $\sum_{k=1}^{5} \left(\frac{2}{k} - \frac{2}{k+2}\right)$
51. $\sum_{n=0}^{5} \left(-\frac{1}{3}\right)^n$
52. $\sum_{n=0}^{6} \left(\frac{3}{2}\right)^n$

▦ In Exercises 53–58, use a graphing utility to find the sum.

53. $\sum_{n=1}^{6} n(n+1)$
54. $\sum_{n=0}^{5} 2n^2$
55. $\sum_{j=2}^{6} (j!-j)$
56. $\sum_{j=0}^{4} \frac{6}{j!}$
57. $\sum_{k=1}^{6} \ln k$
58. $\sum_{k=2}^{4} \frac{k}{\ln k}$

In Exercises 59–76, write the sum using sigma notation. (Begin with $k = 0$ or $k = 1$.)

59. $1 + 2 + 3 + 4 + 5$

60. $8 + 9 + 10 + 12 + 13 + 14$

61. $2 + 4 + 6 + 8 + 10$

62. $24 + 36 + 30 + 36 + 42$

63. $\dfrac{1}{2(1)} + \dfrac{1}{2(2)} + \dfrac{1}{2(3)} + \dfrac{1}{2(4)} + \cdots + \dfrac{1}{2(10)}$

64. $\dfrac{3}{1+1} + \dfrac{3}{1+2} + \dfrac{3}{1+3} + \dfrac{3}{1+4} + \cdots + \dfrac{3}{1+50}$

65. $\dfrac{1}{1^2} + \dfrac{1}{2^2} + \dfrac{1}{3^2} + \dfrac{1}{4^2} + \cdots + \dfrac{1}{20^2}$

66. $\dfrac{1}{2^0} + \dfrac{1}{2^1} + \dfrac{1}{2^2} + \dfrac{1}{2^3} + \cdots + \dfrac{1}{2^{12}}$

67. $\dfrac{1}{3^0} - \dfrac{1}{3^1} + \dfrac{1}{3^2} - \dfrac{1}{3^3} + \cdots - \dfrac{1}{3^9}$

68. $\left(-\dfrac{2}{3}\right)^0 + \left(-\dfrac{2}{3}\right)^1 + \left(-\dfrac{2}{3}\right)^2 + \cdots + \left(-\dfrac{2}{3}\right)^{20}$

69. $\dfrac{4}{1+3} + \dfrac{4}{2+3} + \dfrac{4}{3+3} + \cdots + \dfrac{4}{20+3}$

70. $\dfrac{1}{2^3} - \dfrac{1}{4^3} + \dfrac{1}{6^3} - \dfrac{1}{8^3} + \cdots + \dfrac{1}{14^3}$

71. $\frac{1}{2} + \frac{2}{3} + \frac{3}{4} + \frac{4}{5} + \frac{5}{6} + \cdots + \frac{11}{12}$

72. $\frac{2}{4} + \frac{4}{7} + \frac{6}{10} + \frac{8}{13} + \frac{10}{16} + \cdots + \frac{20}{31}$

73. $\frac{2}{4} + \frac{4}{5} + \frac{6}{6} + \frac{8}{7} + \cdots + \frac{40}{23}$

74. $\left(2 + \frac{1}{1}\right) + \left(2 + \frac{1}{2}\right) + \left(2 + \frac{1}{3}\right) + \cdots + \left(2 + \frac{1}{25}\right)$

75. $1 + 1 + 2 + 6 + 24 + 120 + 720$

76. $1 + 1 + \frac{1}{2} + \frac{1}{6} + \frac{1}{24} + \frac{1}{120} + \frac{1}{720}$

Arithmetic Mean In Exercises 77–80, find the arithmetic mean of the set. The *arithmetic mean* of a set of n measurements $x_1, x_2, x_3, \ldots, x_n$ is

$$\bar{x} = \frac{1}{n} \sum_{i=1}^{n} x_i.$$

77. 3, 7, 2, 1, 5

78. 84, 69, 66, 96

79. 0.5, 0.8, 1.1, 0.8, 0.7, 0.7, 1.0

80. −1.0, 4.2, 5.4, −3.2, 3.6

In Exercises 81–84, match the sequence with the graph of its first ten terms. [The graphs are labeled (a), (b), (c), and (d).]

(a)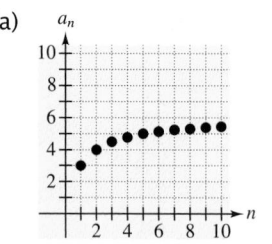

(b)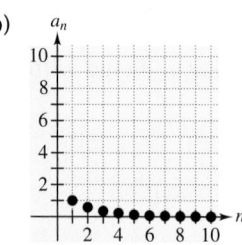

(c)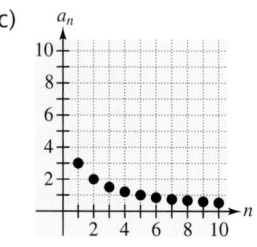

(d)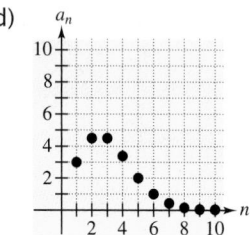

81. $a_n = \dfrac{6}{n+1}$

82. $a_n = \dfrac{6n}{n+1}$

83. $a_n = (0.6)^{n-1}$

84. $a_n = \dfrac{3^n}{n!}$

In Exercises 85–90, use a graphing utility to graph the first ten terms of the sequence.

85. $a_n = (-0.8)^{n-1}$

86. $a_n = \dfrac{2n^2}{n^2+1}$

87. $a_n = \dfrac{1}{2}n$

88. $a_n = \dfrac{n+2}{n}$

89. $a_n = 3 - \dfrac{4}{n}$

90. $a_n = 10\left(\dfrac{3}{4}\right)^{n-1}$

91. *Compound Interest* A deposit of $500 is made in an account that earns 7% interest compounded yearly. The balance in the account after N years is given by

$$A_N = 500(1 + 0.07)^N, \quad N = 1, 2, 3, \ldots$$

(a) Compute the first eight terms of the sequence.

(b) Find the balance in the account after 40 years by computing A_{40}.

(c) Use a graphing utility to graph the first 40 terms of the sequence.

(d) The terms of the sequence are increasing. Is the rate of growth of the terms increasing? Explain.

92. *Depreciation*   At the end of each year, the value of a car with an initial cost of $16,000 is three-fourths what it was at the beginning of the year. Thus, after n years, its value is given by

$$a_n = 16,000 \left(\tfrac{3}{4}\right)^n, \qquad n = 1, 2, 3, \dots$$

(a) Find the value of the car 3 years after it was purchased by computing a_3.

(b) Find the value of the car 6 years after it was purchased by computing a_6. Is this value half of what it was after 3 years?

93. *Soccer Ball*   The number of degrees a_n in each angle of a regular n-sided polygon is

$$a_n = \frac{180(n - 2)}{n}, \qquad n \geq 3.$$

The surface of a soccer ball is made of regular hexagons and pentagons. If the ball is taken apart and flattened, as shown in the figure, the sides don't meet each other. Use the terms a_5 and a_6 to explain why there are gaps between adjacent hexagons.

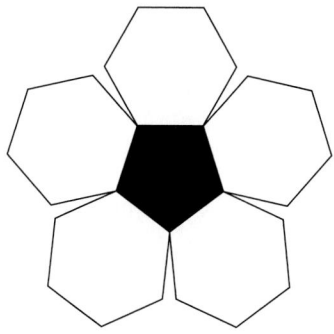

Think About It   In Exercises 94–96, decide whether the statement is true. Explain your reasoning.

94. $\displaystyle\sum_{i=1}^{4} (i^2 + 2i) = \sum_{i=1}^{4} i^2 + \sum_{i=1}^{4} 2i$

95. $\displaystyle\sum_{k=1}^{4} 3k = 3 \sum_{k=1}^{4} k$

96. $\displaystyle\sum_{j=1}^{4} 2^j = \sum_{j=3}^{6} 2^{j-2}$

Section Project

Stars   The stars in Example 4 were formed by placing n equally spaced points on a circle and connecting each point with the second point from it on the circle. The stars in the figure below are formed in a similar way except that each point is connected with the third point from it. For these stars, the number of degrees in each point is given by

$$d_n = \frac{180(n - 6)}{n}, \qquad n \geq 7.$$

(a) Write the first five terms of the sequence.

(b) If you form the stars by connecting each point with the fourth point from it, you obtain stars with the following numbers of points and degrees: 9 points $(20°)$, 10 points $(36°)$, 11 points $\left(49\tfrac{1}{11}°\right)$, 12 points $(60°)$. Find a formula for the number of degrees in each point of an n-pointed star.

<table>
<tr><td>**10.2**</td><td>**Arithmetic Sequences**</td></tr>
</table>

Arithmetic Sequences ▪ The Sum of an Arithmetic Sequence ▪ Application

Arithmetic Sequences

A sequence whose consecutive terms have a common difference is called an **arithmetic sequence.**

Definition of an Arithmetic Sequence

A sequence is called **arithmetic** if the differences between consecutive terms are the same. Thus, the sequence

$$a_1, a_2, a_3, a_4, \ldots, a_n, \ldots$$

is arithmetic if there is a number d such that

$$a_2 - a_1 = d, \quad a_3 - a_2 = d, \quad a_4 - a_3 = d,$$

and so on. The number d is the **common difference** of the sequence.

EXAMPLE 1   *Examples of Arithmetic Sequences*

(a) The sequence whose nth term is $3n + 2$ is arithmetic. For this sequence, the common difference between consecutive terms is 3.

$$5, 8, 11, 14, \ldots, 3n + 2, \ldots \qquad \text{Begin with } n = 1.$$

$$8 - 5 = 3$$

Additional problem: Have students decide whether the sequence 1, 7, 13, 19, 26, 33, 39, . . . is an arithmetic sequence.

Answer: No

(b) The sequence whose nth term is $7 - 5n$ is arithmetic. For this sequence, the common difference between consecutive terms is -5.

$$2, -3, -8, -13, \ldots, 7 - 5n, \ldots \qquad \text{Begin with } n = 1.$$

$$-3 - 2 = -5$$

(c) The sequence whose nth term is $\frac{1}{4}(n + 3)$ is arithmetic. For this sequence, the common difference between consecutive terms is $\frac{1}{4}$.

$$1, \frac{5}{4}, \frac{3}{2}, \frac{7}{4}, \ldots, \frac{n + 3}{4}, \ldots \qquad \text{Begin with } n = 1.$$

$$\tfrac{5}{4} - 1 = \tfrac{1}{4}$$

The nth Term of an Arithmetic Sequence

The nth term of an arithmetic sequence has the form

$$a_n = a_1 + (n - 1)d,$$

where d is the common difference between the terms of the sequence, and a_1 is the first term.

EXAMPLE 2 Finding the nth Term of an Arithmetic Sequence

Find a formula for the nth term of the arithmetic sequence whose common difference is 2 and whose first term is 5.

Solution

You know that the formula for the nth term is of the form $a_n = a_1 + (n - 1)d$. Moreover, because the common difference is $d = 2$, and the first term is $a_1 = 5$, the formula must have the form

$$a_n = 5 + 2(n - 1).$$

Thus, the formula for the nth term is

$$a_n = 2n + 3.$$

The sequence therefore has the following form.

$$5, 7, 9, 11, 13, \ldots, 2n + 3, \ldots$$

EXAMPLE 3 Finding the nth Term of an Arithmetic Sequence

Find a formula for the nth term of the arithmetic sequence whose common difference is 6 and whose *second* term is 11.

Solution

You know that the formula for the nth term is of the form $a_n = a_1 + (n - 1)d$. Moreover, because the common difference is $d = 6$, the formula must have the form

$$a_n = a_1 + 6(n - 1).$$

Because the second term is

Given any two variables, we can easily solve for the third variable by using substitution.

$$a_2 = 11 = a_1 + 6(2 - 1) = a_1 + 6,$$

it follows that $a_1 = 5$. Hence, $a_n = 5 + 6(n - 1) = 6n - 1$.

If you know the nth term and the common difference of an arithmetic sequence, you can find the $(n + 1)$th term by using the following **recursion formula.**

$$a_{n+1} = a_n + d$$

EXAMPLE 4 *Using a Recursion Formula*

The 12th term of an arithmetic sequence is 52 and the common difference is 3. What is the 13th term of the sequence?

Solution

$$\begin{aligned} a_{13} &= a_{12} + 3 \\ &= 52 + 3 \\ &= 55 \end{aligned}$$

Study Tip

Carl Friedrich Gauss's (1777–1855) teacher asked him to add all the integers from 1 to 100. When Gauss returned with the correct answer after only a few moments, the teacher could only look at him in astounded silence. This is what Gauss did:

$$\begin{array}{rcrcrcrcr} 1 &+& 2 &+& 3 &+ \cdots +& 100 \\ 100 &+& 99 &+& 98 &+ \cdots +& 1 \\ \hline 101 &+& 101 &+& 101 &+ \cdots +& 101 \end{array}$$

$$\frac{100 \times 101}{2} = 5050$$

You can use the formula for the nth partial sum of an arithmetic sequence to find the sum of consecutive numbers. For instance,

$$\begin{aligned} \sum_{i=1}^{100} i &= \frac{100}{2}(1 + 100) \\ &= 50(101) \\ &= 5050. \end{aligned}$$

The Sum of an Arithmetic Sequence

The sum of the first n terms of an arithmetic sequence is called the **nth partial sum** of the sequence. For instance, the 5th partial sum of the arithmetic sequence whose nth term is $3n + 4$ is

$$\sum_{i=1}^{5} (3i + 4) = 7 + 10 + 13 + 16 + 19 = 65.$$

A formula for the nth partial sum of an arithmetic sequence is given below.

The *n*th Partial Sum of an Arithmetic Sequence

The nth partial sum of the arithmetic sequence whose nth term is a_n is

$$\begin{aligned} \sum_{i=1}^{n} a_i &= a_1 + a_2 + a_3 + a_4 + \cdots + a_n \\ &= \frac{n}{2}(a_1 + a_n). \end{aligned}$$

In other words, to find the sum of the first n terms of an arithmetic sequence, find the average of the first and nth terms, and multiply by n.

NOTE This summation form of an arithmetic sequence is also called a **series.**

EXAMPLE 5 *Finding the nth Partial Sum*

Find the sum of the first 20 terms of the arithmetic sequence whose nth term is $4n + 1$.

Solution

The first term of this sequence is $a_1 = 4(1) + 1 = 5$ and the 20th term is $a_{20} = 4(20) + 1 = 81$. Therefore, the sum of the first 20 terms is given by

$$\sum_{i=1}^{20}(4i + 1) = \frac{20}{2}(a_1 + a_{20}) \qquad \sum_{i=1}^{n}a_i = \frac{n}{2}(a_1 + a_n)$$

$$= 10(5 + 81)$$

$$= 10(86)$$

$$= 860.$$

EXAMPLE 6 *Finding the nth Partial Sum*

Find the following sum.

$$7 + 10 + 13 + 16 + 19 + 22 + 25 + 28 + 31 + 34 + 37 + 40 + 43$$

Solution

One way to find this sum is simply to add all of the numbers. However, by recognizing that the numbers form an arithmetic sequence that has 13 terms, you can find the sum using the formula for the nth partial sum of an arithmetic sequence.

$$\text{Sum} = \frac{13}{2}(7 + 43) = \frac{13}{2}(50) = 13(25) = 325$$

Check this result on your calculator by actually adding the 13 terms.

EXAMPLE 7 *Finding the nth Partial Sum*

Find the sum of the even integers from 2 to 100.

Solution

Because the integers

$$2, \quad 4, \quad 6, \quad 8, \quad \ldots, \quad 100$$

form an arithmetic sequence, you can find the sum as follows.

$$\sum_{i=1}^{50}2i = \frac{50}{2}(2 + 100) = 25(102) = 2550$$

Application

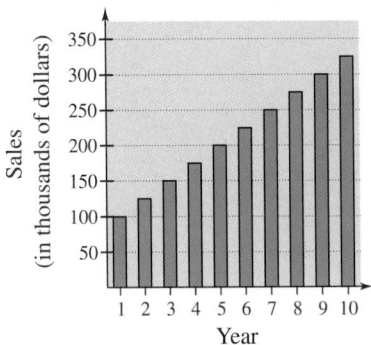

FIGURE 10.2

EXAMPLE 8   An Application: Total Sales

Your business sells $100,000 worth of products during its first year. You have a goal of increasing annual sales by $25,000 each year for 9 years. If you meet this goal, how much will you sell during your first 10 years of business?

Solution

The annual sales during the first 10 years form the following arithmetic sequence.

$100,000, $125,000, $150,000, $175,000, $200,000,

$225,000, $250,000, $275,000, $300,000, $325,000

Using the formula for the nth partial sum of an arithmetic sequence, you can find the total sales during the first 10 years as follows.

$$\text{Total sales} = \frac{10}{2}(100,000 + 325,000) = 5(425,000) = \$2,125,000$$

From the bar graph shown in Figure 10.2, notice that the annual sales for this company follows a *linear growth* pattern. In other words, saying that a quantity increases arithmetically is the same as saying that it increases linearly.

EXAMPLE 9   Seating Capacity

You are organizing a concert at Red Rocks Park. How much should you charge per ticket in order to receive $50,000 in ticket sales for a sold-out performance?

Solution

From the information given in the caption of Figure 10.3, you know that there are an estimated 3320 seats in the last 24 rows. To approximate the number of seats in the first 45 rows, you can use the formula for the nth partial sum of an arithmetic sequence.

$$\sum_{n=1}^{45}\left(87\frac{1}{2} + \frac{3}{2}n\right) = 89 + 90\frac{1}{2} + 92 + 93\frac{1}{2} + \cdots + 155$$

$$= \frac{45}{2}(89 + 155)$$

$$= 5490$$

Thus, the total number of seats is about $3320 + 5490 = 8810$. To bring in $50,000, you should charge about

$$\frac{1}{8810}(50,000) \approx \$5.68 \text{ per ticket.}$$

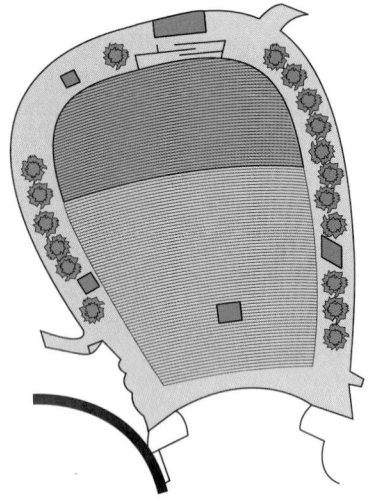

FIGURE 10.3   Red Rocks Park is an open-air amphitheater carved out of rock near Denver, Colorado. The amphitheater has 69 rows of seats. Rows 46 through 69 have seats for an estimated 3320 people. The number of seats in the first 45 rows can be modeled by the arithmetic sequence whose nth term is $87\frac{1}{2} + \frac{3}{2}n$.

Group Activities   **Extending the Concept**

6	1	8
7	5	3
2	9	4

Using Arithmetic Sequences   A *magic square* is a square table of positive integers in which each row, column, and diagonal adds up to the same number. One example is shown at the left. In addition, the values in the middle row, in the middle column, and along both diagonals form arithmetic sequences. See if you can complete the following magic squares.

a.

	11	14
	10	
		15

b.

8		
	9	
	13	

c.

		20
	13	
6		

10.2   **Exercises**

See Warm-Up Exercises, p. A50

1. *Think About It*   In your own words, explain what makes a sequence arithmetic.

2. *Think About It*   Explain what is meant by a recursion formula.

In Exercises 3–10, find the common difference of the arithmetic sequence.

3. $2, 5, 8, 11, \ldots$

4. $-8, 0, 8, 16, \ldots$

5. $100, 94, 88, 82, \ldots$

6. $3200, 2800, 2400, 2000, \ldots$

7. $1, \frac{5}{3}, \frac{7}{3}, 3, \ldots$

8. $\frac{1}{2}, \frac{5}{4}, 2, \frac{11}{4}, \ldots$

9. $\frac{7}{2}, \frac{9}{4}, 1, -\frac{1}{4}, -\frac{3}{2}, \ldots$

10. $\frac{5}{2}, \frac{11}{6}, \frac{7}{6}, \frac{1}{2}, -\frac{1}{6}, \ldots$

In Exercises 11–22, determine whether the sequence is arithmetic. If so, find the common difference.

11. $2, 4, 6, 8, \ldots$

12. $1, 2, 4, 8, 16, \ldots$

13. $32, 16, 0, -16, \ldots$

14. $32, 16, 8, 4, \ldots$

15. $3.2, 4, 4.8, 5.6, \ldots$

16. $8, 4, 2, 1, 0.5, 0.25, \ldots$

17. $2, \frac{7}{2}, 5, \frac{13}{2}, \ldots$

18. $3, \frac{5}{2}, 2, \frac{3}{2}, 1, \ldots$

19. $\frac{1}{3}, \frac{2}{3}, \frac{4}{3}, \frac{8}{3}, \frac{16}{3}, \ldots$

20. $\frac{9}{4}, 2, \frac{7}{4}, \frac{3}{2}, \frac{5}{4}, \ldots$

21. $\ln 4, \ln 8, \ln 12, \ln 16, \ldots$

22. $e, e^2, e^3, e^4, \ldots$

In Exercises 23–30, write the first five terms of the arithmetic sequence. (Begin with $n = 1$.)

23. $a_n = 3n + 4$

24. $a_n = 5n - 4$

25. $a_n = -2n + 8$

26. $a_n = -10n + 100$

27. $a_n = \frac{5}{2}n - 1$

28. $a_n = \frac{2}{3}n + 2$

29. $a_n = -\frac{1}{4}(n - 1) + 4$

30. $a_n = 4(n + 2) + 24$

In Exercises 31–38, write the first five terms of the arithmetic sequence defined recursively.

31. $a_1 = 25$
$a_{k+1} = a_k + 3$

32. $a_1 = 12$
$a_{k+1} = a_k - 6$

33. $a_1 = 9$
$a_{k+1} = a_k - 3$

34. $a_1 = 8$
$a_{k+1} = a_k + 7$

35. $a_1 = -10$
$a_{k+1} = a_k + 6$

36. $a_1 = -20$
$a_{k+1} = a_k - 4$

37. $a_1 = 100$
$a_{k+1} = a_k - 20$

38. $a_1 = 4.2$
$a_{k+1} = a_k + 0.4$

In Exercises 39–44, match the sequence with its graph. [The graphs are labeled (a), (b), (c), (d), (e), and (f).]

(a)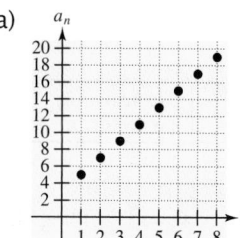

(b)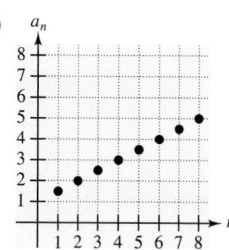

(c)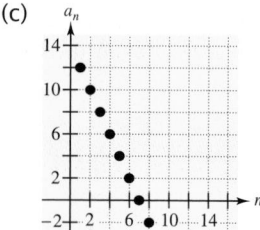

(d)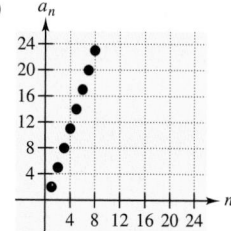

(e)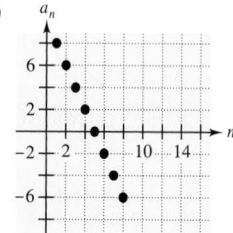

(f)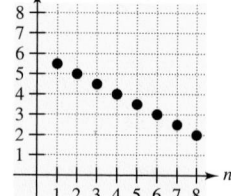

39. $a_n = \frac{1}{2}n + 1$

40. $a_n = -\frac{1}{2}n + 6$

41. $a_n = -2n + 10$

42. $a_n = 2n + 3$

43. $a_1 = 12$
$a_{n+1} = a_n - 2$

44. $a_1 = 2$
$a_{n+1} = a_n + 3$

 In Exercises 45–50, use a graphing utility to graph the first ten terms of the sequence.

45. $a_n = -2n + 21$

46. $a_n = \frac{3}{2}n + 1$

47. $a_n = \frac{3}{5}n + \frac{3}{2}$

48. $a_n = -25n + 500$

49. $a_n = 2.5n - 8$

50. $a_n = 6.2n + 3$

In Exercises 51–68, find a formula for the nth term of the arithmetic sequence.

51. $a_1 = 3,\ d = \frac{1}{2}$

52. $a_1 = -1,\ d = 1.2$

53. $a_1 = 1000,\ d = -25$

54. $a_1 = 64,\ d = -8$

55. $a_1 = 3,\ d = \frac{3}{2}$

56. $a_1 = 12,\ d = -3$

57. $a_3 = 20,\ d = -4$

58. $a_6 = 5,\ d = \frac{3}{2}$

59. $a_1 = 5,\ a_5 = 15$

60. $a_2 = 93,\ a_6 = 65$

61. $a_3 = 16,\ a_4 = 20$

62. $a_5 = 30,\ a_4 = 25$

63. $a_1 = 50,\ a_3 = 30$

64. $a_{10} = 32,\ a_{12} = 48$

65. $a_2 = 10,\ a_6 = 8$

66. $a_7 = 8,\ a_{13} = 6$

67. $a_1 = 0.35,\ a_2 = 0.30$

68. $a_1 = 0.08,\ a_2 = 0.082$

In Exercises 69–78, find the sum.

69. $\sum_{k=1}^{20} k$

70. $\sum_{k=1}^{30} 4k$

71. $\sum_{k=1}^{10} 5k$

72. $\sum_{n=1}^{30} \left(\frac{1}{2}n + 2\right)$

73. $\sum_{k=1}^{50} (k + 3)$

74. $\sum_{k=1}^{100} (4k - 1)$

75. $\sum_{n=1}^{500} \frac{n}{2}$

76. $\sum_{n=1}^{600} \frac{2n}{3}$

77. $\sum_{n=1}^{30} \left(\frac{1}{3}n - 4\right)$

78. $\sum_{n=1}^{75} (0.3n + 5)$

In Exercises 79–84, use a graphing utility to find the sum.

79. $\displaystyle\sum_{j=1}^{25}(750-30j)$

80. $\displaystyle\sum_{i=1}^{60}\left(300-\tfrac{8}{3}i\right)$

81. $\displaystyle\sum_{n=1}^{40}(1000-25n)$

82. $\displaystyle\sum_{n=1}^{20}(500-10n)$

83. $\displaystyle\sum_{n=1}^{50}(2.15n+5.4)$

84. $\displaystyle\sum_{n=1}^{60}(200-3.4n)$

In Exercises 85–96, find the nth partial sum of the arithmetic sequence.

85. $5, 12, 19, 26, 33, \ldots , n = 12$

86. $2, 12, 22, 32, 42, \ldots , n = 20$

87. $2, 8, 14, 20, \ldots , n = 25$

88. $500, 480, 460, 440, \ldots , n = 20$

89. $200, 175, 150, 125, 100, \ldots , n = 8$

90. $800, 785, 770, 755, 740, \ldots , n = 25$

91. $-50, -38, -26, -14, -2, \ldots , n = 50$

92. $-16, -8, 0, 8, 16, \ldots , n = 30$

93. $1, 4.5, 8, 11.5, 15, \ldots , n = 12$

94. $2.2, 2.8, 3.4, 4.0, 4.6, \ldots , n = 12$

95. $0.5, 0.9, 1.3, 1.7, \ldots , n = 10$

96. $a_1 = 15, a_{100} = 307, n = 100$

97. Find the sum of the first 75 positive integers.

98. Find the sum of the integers from 35 to 100.

99. Find the sum of the first 50 positive even integers.

100. Find the sum of the first 100 positive odd integers.

101. *Salary Increases*   In your new job you are told that your starting salary will be $36,000 with an increase of $2000 at the end of each of the first 5 years. How much will you be paid through the end of your first six years of employment with the company?

102. *Would You Accept This Job?*   Suppose that you receive 25 cents the first day of the month, 50 cents the second day, 75 cents the third day, and so on. Determine the total amount that you will receive during a 30-day month.

103. *Ticket Prices*   There are 20 rows of seats on the main floor of a concert hall: 20 seats in the first row, 21 seats in the second row, 22 seats in the third row, and so on (see figure). How much should you charge per ticket in order to obtain $15,000 for the sale of all of the seats on the main floor?

22 seats
21 seats
20 seats

104. *Pile of Logs*   Logs are stacked in a pile as shown in the figure. The top row has 15 logs and the bottom row has 21 logs. How many logs are in the pile?

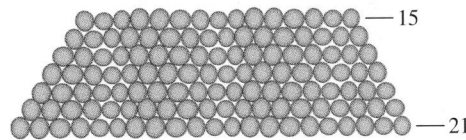

— 15

— 21

105. *Baling Hay*   In the first two trips baling hay around a large field (see figure), a farmer obtains 93 bales and 89 bales. The farmer estimates that the same pattern will continue. Estimate the total number of bales made if there are another six trips around the field.

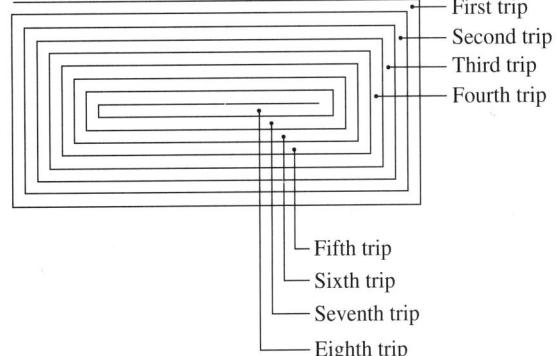

First trip
Second trip
Third trip
Fourth trip

Fifth trip
Sixth trip
Seventh trip
Eighth trip

106. *Clock Chimes*   A clock chimes once at 1:00, twice at 2:00, three times at 3:00, and so on. The clock also chimes once at 15-minute intervals that are not on the hour. How many times does the clock chime in a 12-hour period?

107. *Free-Falling Object*   A free-falling object will fall 16 feet during the first second, 48 more feet during the second, 80 more feet during the third, and so on. What is the total distance the object will fall in 8 seconds if this pattern continues?

108. *Pattern Recognition*

(a) Compute the sums of the positive odd integers.

$$1 + 3 = \boxed{}$$
$$1 + 3 + 5 = \boxed{}$$
$$1 + 3 + 5 + 7 = \boxed{}$$
$$1 + 3 + 5 + 7 + 9 = \boxed{}$$
$$1 + 3 + 5 + 7 + 9 + 11 = \boxed{}$$

(b) Use the sums in part (a) to make a conjecture about the sums of positive odd integers. Check your conjecture for the sum

$$1 + 3 + 5 + 7 + 9 + 11 + 13 = \boxed{}.$$

(c) Verify your conjecture in part (b) analytically.

Section Project

The following sequence of perfect squares is *not* arithmetic.

$$1, \ 4, \ 9, \ 16, \ 25, \ 36, \ 49, \ 64, \ 81, \ \ldots$$

However, you can form a related sequence that is arithmetic by finding the differences of consecutive terms as follows.

$$
\begin{array}{ccccccccc}
 & 5 & & 9 & & 13 & & 17 & \\
1 & 4 & 9 & 16 & 25 & 36 & 49 & 64 & 81 \\
 & 3 & & 7 & & 11 & & 15 &
\end{array}
$$

Thus, the related arithmetic sequence is

$$3, \ 5, \ 7, \ 9, \ 11, \ 13, \ 15, \ 17, \ \ldots$$

(a) Can you find an arithmetic sequence that is related to the following sequence of perfect cubes?

$$1, \ 8, \ 27, \ 64, \ 125, \ 216, \ 343, \ 512, \ 729, \ \ldots$$

(b) Can you find an arithmetic sequence that is related to the following sequence of perfect fourth powers?

$$1, \ 16, \ 81, \ 256, \ 625, \ 1296, \ 2401, \ 4096, \ \ldots$$

| 10.3 | **Geometric Sequences** |

Geometric Sequences ▪ The Sum of a Geometric Sequence ▪ Application

Geometric Sequences

In Section 10.2, you studied sequences whose consecutive terms have a common *difference*. In this section, you will study sequences whose consecutive terms have a common *ratio*.

> ### Definition of a Geometric Sequence
>
> A sequence is called **geometric** if the ratios of consecutive terms are the same. Thus, the sequence $a_1, a_2, a_3, a_4, \ldots, a_n, \ldots$ is geometric if there is a number r, $r \neq 0$, such that
>
> $$\frac{a_2}{a_1} = r, \quad \frac{a_3}{a_2} = r, \quad \frac{a_4}{a_3} = r,$$
>
> and so on. The number r is the **common ratio** of the sequence.

Students often confuse arithmetic and geometric sequences. Remind them that arithmetic sequences have a common *difference* and geometric sequences have a common *ratio* (sometimes called a factor or constant multiplier).

EXAMPLE 1 *Examples of Geometric Sequences*

(a) The sequence whose nth term is 2^n is geometric. For this sequence, the common ratio between consecutive terms is 2.

$$2, 4, 8, 16, \ldots, 2^n, \ldots \qquad \text{Begin with } n = 1.$$

$$\tfrac{4}{2} = 2$$

Additional problem: Have students determine if the sequence 3, −6, 12, −24, −48, 96, . . . is geometric.

Answer: No

(b) The sequence whose nth term is $4(3^n)$ is geometric. For this sequence, the common ratio between consecutive terms is 3.

$$12, 36, 108, 324, \ldots, 4(3^n), \ldots \qquad \text{Begin with } n = 1.$$

$$\tfrac{36}{12} = 3$$

(c) The sequence whose nth term is $\left(-\tfrac{1}{3}\right)^n$ is geometric. For this sequence, the common ratio between consecutive terms is $-\tfrac{1}{3}$.

$$-\frac{1}{3}, \frac{1}{9}, -\frac{1}{27}, \frac{1}{81}, \ldots, \left(-\frac{1}{3}\right)^n, \ldots \qquad \text{Begin with } n = 1.$$

$$\frac{1/9}{-1/3} = -\frac{1}{3}$$

Point out that the factor a_1 remains constant, while the exponent of r increases by 1 in each successive term.

NOTE If you know the nth term of a geometric sequence, the $(n + 1)$th term can be found by multiplying by r. That is,

$$a_{n+1} = ra_n.$$

The nth Term of a Geometric Sequence

The nth term of a geometric sequence has the form

$$a_n = a_1 r^{n-1},$$

where r is the common ratio of consecutive terms of the sequence. Thus, every geometric sequence can be written in the following form.

$$a_1, a_1 r, a_1 r^2, a_1 r^3, a_1 r^4, \ldots, a_1 r^{n-1}, \ldots$$

EXAMPLE 2 *Finding the nth Term of a Geometric Sequence*

Find a formula for the nth term of the geometric sequence whose common ratio is 3 and whose first term is 1. What is the eighth term of the sequence?

Solution

You know that the formula for the nth term is of the form $a_n = a_1 r^{n-1}$. Moreover, because the common ratio is $r = 3$, and the first term is $a_1 = 1$, the formula must have the form

$$a_n = (1)(3)^{n-1} = 3^{n-1}.$$

The sequence therefore has the following form.

$$1, 3, 9, 27, 81, \ldots, 3^{n-1}, \ldots$$

The eighth term of the sequence is $a_8 = 3^{8-1} = 3^7 = 2187.$

EXAMPLE 3 *Finding the nth Term of a Geometric Sequence*

Find a formula for the nth term of the geometric sequence whose first two terms are 4 and 2.

Have students compare Examples 2 and 3. Note that in Example 2 $r > 1$ and the sequence increases; in Example 3, $r < 1$ and the sequence decreases. Ask students to explain why.

Solution

Because the common ratio is

$$\frac{a_2}{a_1} = \frac{2}{4} = \frac{1}{2},$$

the formula for the nth term must be

$$a_n = a_1 r^{n-1} = 4\left(\frac{1}{2}\right)^{n-1}.$$

The sequence therefore has the following form.

$$4, 2, 1, \frac{1}{2}, \frac{1}{4}, \ldots, 4\left(\frac{1}{2}\right)^{n-1}, \ldots$$

The Sum of a Geometric Sequence

In Section 10.2, you saw that there is a simple formula for finding the sum of the first n terms of an arithmetic sequence. There is also a formula for finding the sum of the first n terms of a geometric sequence.

Be sure to have students distinguish between r^{n-1} and $r^n - 1$.

The nth Partial Sum of a Geometric Sequence

The nth partial sum of the geometric sequence whose nth term is $a_n = a_1 r^{n-1}$ is given by

$$\sum_{i=1}^{n} a_1 r^{i-1} = a_1 + a_1 r + a_1 r^2 + a_1 r^3 + \cdots + a_1 r^{n-1} = a_1 \left(\frac{r^n - 1}{r - 1} \right).$$

EXAMPLE 4   *Finding the nth Partial Sum*

Find the following sum.

$$1 + 2 + 4 + 8 + 16 + 32 + 64 + 128$$

Solution

This is a geometric sequence whose common ratio is $r = 2$. Because the first term of the sequence is $a_1 = 1$, it follows that the sum is

$$\sum_{i=1}^{8} 2^{i-1} = (1)\left(\frac{2^8 - 1}{2 - 1} \right) = \frac{256 - 1}{2 - 1} = 255.$$

EXAMPLE 5   *Finding the nth Partial Sum*

Find the sum of the first five terms of the geometric sequence whose nth term is $a_n = \left(\frac{2}{3} \right)^n$.

Solution

Remind students that they can invert and multiply to simplify a complex fraction.

$$
\begin{aligned}
\sum_{i=1}^{5} \left(\frac{2}{3} \right)^i &= \frac{2}{3}\left[\frac{(2/3)^5 - 1}{(2/3) - 1} \right] \qquad \text{Substitute } \tfrac{2}{3} \text{ for } a_1 \text{ and } \tfrac{2}{3} \text{ for } r. \\
&= \frac{2}{3}\left[\frac{(32/243) - 1}{-1/3} \right] \\
&= \frac{2}{3}\left(-\frac{211}{243} \right)(-3) \\
&= \frac{422}{243} \\
&\approx 1.737
\end{aligned}
$$

EXAMPLE 6   *Finding the nth Partial Sum*

Find the following sum.

$$\sum_{i=1}^{7} \frac{3^{i-1}}{2}$$

Solution

By writing out a few terms of the sum,

$$\sum_{i=1}^{7} \frac{3^{i-1}}{2} = \sum_{i=1}^{7} \frac{1}{2}(3^{i-1}) = \frac{1}{2} + \frac{1}{2}(3) + \frac{1}{2}(3^2) + \cdots + \frac{1}{2}(3^6),$$

you can see that $a_1 = \frac{1}{2}$ and $r = 3$. Thus, the sum is as follows.

$$\sum_{i=1}^{n} a_1 r^{i-1} = a_1\left(\frac{r^n - 1}{r - 1}\right)$$

$$\sum_{i=1}^{7} \frac{1}{2}(3^{i-1}) = \frac{1}{2}\left(\frac{3^7 - 1}{3 - 1}\right)$$

$$= \frac{1}{2}\left(\frac{2187 - 1}{2}\right)$$

$$= \frac{2186}{4}$$

$$= 546.5$$

Exploration

Any repeating decimal can be written in rational form (as the ratio of two integers). Here is an example. Consider the repeating decimal

$x = 0.414141\ldots$.

To write this number in rational form, multiply by 100 to obtain an expression for $100x$. Then subtract x from $100x$ to obtain an expression for $99x$. Finally, divide by 99 to obtain a rational expression for x. Try doing this. A repeating decimal can also be written as the sum of an *infinite* geometric sequence. For example,

$x = 0.414141\ldots = 41\left(\frac{1}{100}\right) + 41\left(\frac{1}{100}\right)^2 + 41\left(\frac{1}{100}\right)^3 + \cdots$.

An infinite geometric sequence with common ratio r has the sum $a_1/(1 - r)$ if $|r| < 1$. Try using this formula with $a_1 = 41\left(\frac{1}{100}\right)$ and $r = \frac{1}{100}$ to rewrite x as a rational number.

Application

EXAMPLE 7   An Application: A Lifetime Salary

You have accepted a job that pays a salary of $28,000 the first year. During the next 39 years, suppose you receive a 6% raise each year. What will your total salary be over the 40-year period?

Solution

Using a geometric sequence, your salary during the first year will be

$$a_1 = 28,000.$$

Then, with a 6% raise, your salary during the next 2 years will be as follows.

$$a_2 = 28,000 + 28,000(0.06) = 28,000(1.06)^1$$
$$a_3 = 28,000(1.06) + 28,000(1.06)(0.06) = 28,000(1.06)^2$$

From this pattern, you can see that the common ratio of the geometric sequence is $r = 1.06$. Using the formula for the nth partial sum of a geometric sequence, you will find that the total salary over the 40-year period is given by

$$
\begin{aligned}
\text{Total salary} &= a_1\left(\frac{r^n - 1}{r - 1}\right) \\
&= 28,000\left[\frac{(1.06)^{40} - 1}{1.06 - 1}\right] \\
&= 28,000\left[\frac{(1.06)^{40} - 1}{0.06}\right] \\
&\approx \$4,333,335.
\end{aligned}
$$

The bar graph in Figure 10.4 illustrates your salary during the 40-year period.

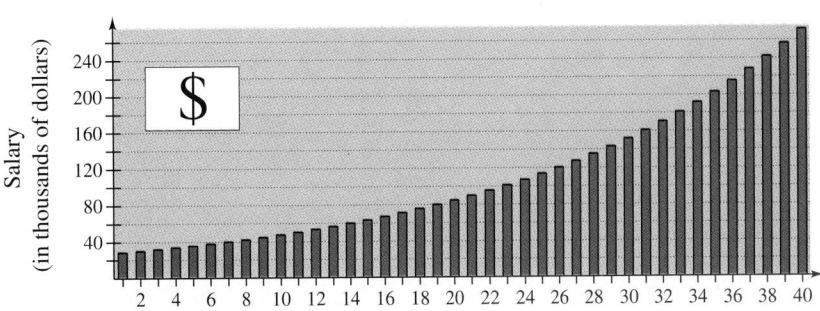

FIGURE 10.4

EXAMPLE 8 An Application: Increasing Annuity

You deposit $100 in an account each month for 2 years. The account pays an annual interest rate of 9%, compounded monthly. What is your balance at the end of 2 years? (This type of savings plan is called an **increasing annuity.**)

Solution

Encourage students to write out the first few terms of the sequence to see "what's going on" and to help determine a_1 and r.

The first deposit would earn interest for the full 24 months, the second deposit would earn interest for 23 months, the third deposit would earn interest for 22 months, and so on. Using the formula for compound interest, you can see that the total of the 24 deposits would be

$$\text{Total} = a_1 + a_2 + \cdots + a_{24}$$

$$= 100\left(1 + \frac{0.09}{12}\right)^1 + 100\left(1 + \frac{0.09}{12}\right)^2 + \cdots + 100\left(1 + \frac{0.09}{12}\right)^{24}$$

$$= 100(1.0075)^1 + 100(1.0075)^2 + \cdots + 100(1.0075)^{24}$$

$$= 100(1.0075)\left(\frac{1.0075^{24} - 1}{1.0075 - 1}\right) \qquad a_1\left(\frac{r^n - 1}{r - 1}\right)$$

$$= \$2638.49.$$

Group Activities

Extending the Concept

Annual Revenue The two bar graphs below show the annual revenues for two companies. One company's revenue grew at an arithmetic rate, whereas the other grew at a geometric rate. Which company had the greatest revenue during the 10-year period? Which company would you rather own? Explain.

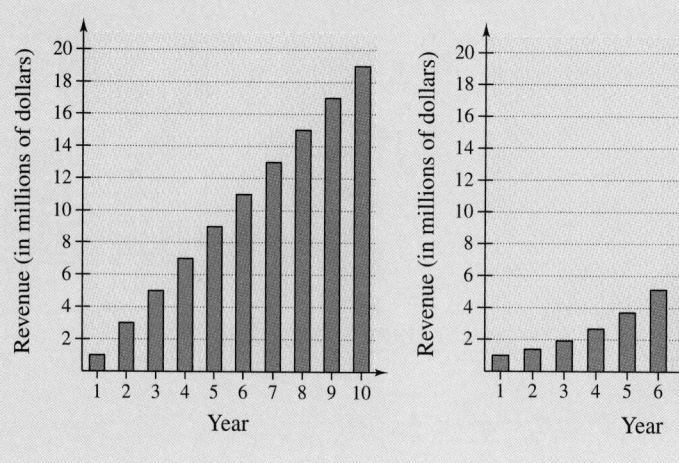

10.3   Exercises

See Warm-Up
Exercises, p. A50

1. *Think About It*   In your own words, explain what makes a sequence geometric.

2. *Think About It*   Give an example of a geometric sequence whose terms alternate in sign.

3. *Think About It*   Explain why the terms of a geometric sequence decrease when $a_1 > 0$ and $0 < r < 1$.

4. *Think About It*   Explain what is meant by the nth partial sum of a sequence.

In Exercises 5–14, find the common ratio of the geometric sequence.

5. $2, 6, 18, 54, \ldots$

6. $5, -10, 20, -40, \ldots$

7. $1, -3, 9, -27, \ldots$

8. $54, 18, 6, 2, \ldots$

9. $1, -\frac{3}{2}, \frac{9}{4}, -\frac{27}{8}, \ldots$

10. $9, 6, 4, \frac{8}{3}, \ldots$

11. $5, -\frac{5}{2}, \frac{5}{4}, -\frac{5}{8}, \ldots$

12. $e, e^2, e^3, e^4, \ldots$

13. $1.1, (1.1)^2, (1.1)^3, (1.1)^4, \ldots$

14. $500(1.06), 500(1.06)^2, 500(1.06)^3, 500(1.06)^4, \ldots$

In Exercises 15–24, determine whether the sequence is geometric. If so, find the common ratio.

15. $64, 32, 16, 8, \ldots$

16. $64, 32, 0, -32, \ldots$

17. $5, 10, 20, 40, \ldots$

18. $54, -18, 6, -2, \ldots$

19. $1, 8, 27, 64, 125, \ldots$

20. $12, 7, 2, -3, -8, \ldots$

21. $1, -\frac{2}{3}, \frac{4}{9}, -\frac{8}{27}, \ldots$

22. $\frac{1}{3}, -\frac{2}{3}, \frac{4}{3}, -\frac{8}{3}, \ldots$

23. $10(1 + 0.02), 10(1 + 0.02)^2, 10(1 + 0.02)^3, \ldots$

24. $1, 0.2, 0.04, 0.008, \ldots$

In Exercises 25–38, write the first five terms of the geometric sequence.

25. $a_1 = 4, r = 2$

26. $a_1 = 3, r = 4$

27. $a_1 = 6, r = \frac{1}{3}$

28. $a_1 = 4, r = \frac{1}{2}$

29. $a_1 = 1, r = -\frac{1}{2}$

30. $a_1 = 32, r = -\frac{3}{4}$

31. $a_1 = 4, r = -\frac{1}{2}$

32. $a_1 = 4, r = \frac{3}{2}$

33. $a_1 = 1000, r = 1.01$

34. $a_1 = 4000, r = \frac{1}{1.01}$

35. $a_1 = 200, r = 1.07$

36. $a_1 = 1000, r = \frac{1}{1.05}$

37. $a_1 = 10, r = \frac{3}{5}$

38. $a_1 = 36, r = \frac{2}{3}$

In Exercises 39–50, find the nth term of the geometric sequence.

39. $a_1 = 6, r = \frac{1}{2}, a_{10} = $

40. $a_1 = 8, r = \frac{3}{4}, a_8 = $

41. $a_1 = 3, r = \sqrt{2}, a_{10} = $

42. $a_1 = 5, r = \sqrt{3}, a_9 = $

43. $a_1 = 200, r = 1.2, a_{12} = $

44. $a_1 = 500, r = 1.06, a_{40} = $

45. $a_1 = 4, a_2 = 3, a_5 = $

46. $a_1 = 1, a_2 = 9, a_7 = $

47. $a_1 = 1, a_3 = \frac{9}{4}, a_6 = $

48. $a_2 = 12, a_3 = 16, a_4 = $

49. $a_3 = 6, a_5 = \frac{8}{3}, a_6 = $

50. $a_4 = 100, a_5 = -25, a_7 = $

In Exercises 51–60, find the formula for the nth term of the geometric sequence. (Begin with $n = 1$.)

51. $a_1 = 2$, $r = 3$

52. $a_1 = 5$, $r = 4$

53. $a_1 = 1$, $r = 2$

54. $a_1 = 1$, $r = -5$

55. $a_1 = 4$, $r = -\frac{1}{2}$

56. $a_1 = 9$, $r = \frac{2}{3}$

57. $a_1 = 8$, $a_2 = 2$

58. $a_1 = 18$, $a_2 = 8$

59. $4, -6, 9, -\frac{27}{2}, \ldots$

60. $1, \frac{3}{2}, \frac{9}{4}, \frac{27}{8}, \ldots$

In Exercises 61–64, match the sequence with its graph. [The graphs are labeled (a), (b), (c), and (d).]

(a)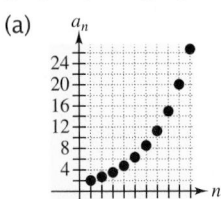

(b)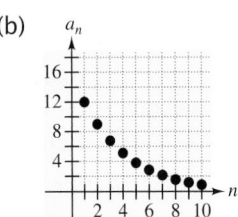

(c)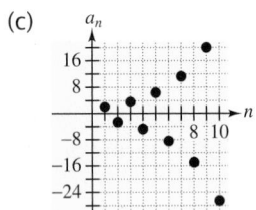

(d)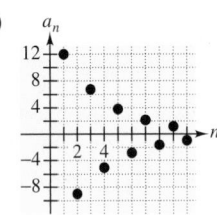

61. $a_n = 12\left(\frac{3}{4}\right)^{n-1}$

62. $a_n = 12\left(-\frac{3}{4}\right)^{n-1}$

63. $a_n = 2\left(\frac{4}{3}\right)^{n-1}$

64. $a_n = 2\left(-\frac{4}{3}\right)^{n-1}$

In Exercises 65–68, use a graphing utility to graph the first ten terms of the sequence.

65. $a_n = 20(-0.6)^{n-1}$

66. $a_n = 4(1.4)^{n-1}$

67. $a_n = 15(0.6)^{n-1}$

68. $a^n = 8(-0.6)^{n-1}$

In Exercises 69–78, find the sum.

69. $\sum_{i=1}^{10} 2^{i-1}$

70. $\sum_{i=1}^{6} 3^{i-1}$

71. $\sum_{i=1}^{12} 3\left(\frac{3}{2}\right)^{i-1}$

72. $\sum_{i=1}^{20} 12\left(\frac{2}{3}\right)^{i-1}$

73. $\sum_{i=1}^{15} 3\left(-\frac{1}{3}\right)^{i-1}$

74. $\sum_{i=1}^{8} 8\left(-\frac{1}{4}\right)^{i-1}$

75. $\sum_{i=1}^{8} 6(0.1)^{i-1}$

76. $\sum_{i=1}^{24} 1000(1.06)^{i-1}$

77. $\sum_{i=1}^{10} 15(0.3)^{i-1}$

78. $\sum_{i=1}^{20} 5(1.25)^{i-1}$

In Exercises 79–82, use a graphing utility to find the sum.

79. $\sum_{i=1}^{30} 100(0.75)^{i-1}$

80. $\sum_{i=1}^{24} 5000(1.08)^{-(i-1)}$

81. $\sum_{i=1}^{20} 100(1.1)^{i}$

82. $\sum_{i=1}^{40} 50(1.07)^{i}$

In Exercises 83–94, find the nth partial sum of the geometric sequence.

83. $1, -3, 9, -27, 81, \ldots$, $n = 10$

84. $3, -6, 12, -24, 48, \ldots$, $n = 12$

85. $8, 4, 2, 1, \frac{1}{2}, \ldots$, $n = 15$

86. $9, 6, 4, \frac{8}{3}, \frac{16}{9}, \ldots$, $n = 10$

87. $4, 12, 36, 108, \ldots$, $n = 8$

88. $\frac{1}{36}, -\frac{1}{12}, \frac{1}{4}, -\frac{3}{4}, \ldots$, $n = 20$

89. $60, -15, \frac{15}{4}, -\frac{15}{16}, \ldots$, $n = 12$

90. $40, -10, \frac{5}{2}, -\frac{5}{8}, \frac{5}{32}, \ldots$, $n = 10$

91. $30, 30(1.06), 30(1.06)^2, \ldots$, $n = 20$

92. $100, 100(1.08), 100(1.08)^2, \ldots$, $n = 40$

93. $500, 500(1.04), 500(1.04)^2, \ldots$, $n = 18$

94. $1, \sqrt{2}, 2, 2\sqrt{2}, 4, \ldots$, $n = 12$

95. *Depreciation* A company buys a machine for $250,000. During the next 5 years, the machine depreciates at a rate of 25% per year. (That is, at the end of each year, the depreciated value is 75% of what it was at the beginning of the year.)

(a) Find a formula for the nth term of the geometric sequence that gives the value of the machine n full years after it was purchased.

(b) Find the depreciated value of the machine at the end of 5 full years.

(c) During which year did the machine depreciate most?

96. *Population Increase*   A city of 500,000 people is growing at a rate of 1% per year. (That is, at the end of each year, the population is 1.01 times what it was at the beginning of the year.)

(a) Find a formula for the nth term of the geometric sequence that gives the population t years from now.

(b) Estimate the population 20 years from now.

97. *Salary Increases*   You accept a job that pays a salary of $30,000 the first year. During the next 39 years, you receive a 5% raise each year. What is your total salary over the 40-year period?

98. *Salary Increases*   You accept a job that pays a salary of $30,000 the first year. During the next 39 years, you receive a 5.5% raise each year.

(a) What is your total salary over the 40-year period?

(b) How much more income did the extra 0.5% provide than the result in Exercise 97?

Increasing Annuity   In Exercises 99–104, find the balance in an increasing annuity in which a principal of P dollars in invested each month for t years, compounded monthly at rate r.

99. $P = \$100$ $t = 10$ years $r = 9\%$

100. $P = \$50$ $t = 5$ years $r = 7\%$

101. $P = \$30$ $t = 40$ years $r = 8\%$

102. $P = \$200$ $t = 30$ years $r = 10\%$

103. $P = \$75$ $t = 30$ years $r = 6\%$

104. $P = \$100$ $t = 25$ years $r = 8\%$

105. *Would You Accept This Job?*   You start work at a company that pays $0.01 the first day, $0.02 the second day, $0.04 the third day, and so on. If the daily wage keeps doubling, what would your total income be for working (a) 29 days and (b) 30 days?

106. *Would You Accept This Job?*   You start work at a company that pays $0.01 the first day, $0.03 the second day, $0.09 the third day, and so on. If the daily wage keeps tripling, what would your total income be for working (a) 25 days and (b) 26 days?

107. *Power Supply*   The electrical power for an implanted medical device decreases by 0.1% each day.

(a) Find a formula for the nth term of the geometric sequence that gives the power n days after the device is implanted.

(b) What percent of the initial power is still available 1 year after the device is implanted?

(c) The power supply needs to be changed when half the power is depleted. Use a graphing utility to graph the first 750 terms of the sequence and estimate when the power source should be changed.

108. *Cooling*   The temperature of water in an ice-cube tray is 70°F when it is placed in a freezer. Its temperature n hours after being placed in the freezer is 20% less than 1 hour earlier.

(a) Find a formula for the nth term of the geometric sequence that gives the temperature of the water n hours after it is placed in the freezer.

(b) Find the temperature of the water 6 hours after it is placed in the freezer.

(c) Use a graphing utility to estimate the time when the water freezes. Explain your reasoning.

109. *Area*   A square has 12-inch sides. A new square is formed by connecting the midpoints of the sides of the square. Then two of the triangles are shaded (see figure). This process is repeated five more times. What is the total area of the shaded region?

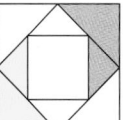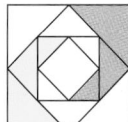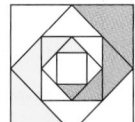

110. *Area* A square has 12-inch sides. The square is divided into nine smaller squares and the center square is shaded (see figure). Each of the eight unshaded squares is then divided into nine smaller squares and each center square is shaded. This process is repeated four more times. What is the total area of the shaded region?

 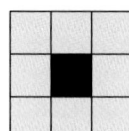 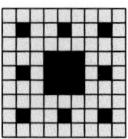

111. *Bungee Jumping* A bungee jumper jumps from a bridge and stretches a cord 100 feet. Successive bounces stretch the cord 75% of each previous length (see figure). Find the total distance traveled by the bungee jumper during 10 bounces.

$$100 + 2(100)(0.75) + \cdots + 2(100)(0.75)^{10}$$

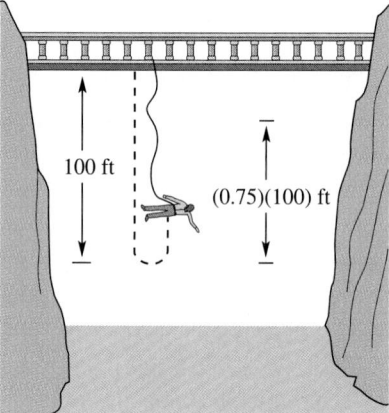

100 ft

(0.75)(100) ft

Section Project

Number of Ancestors The number of direct ancestors a person has had is as follows.

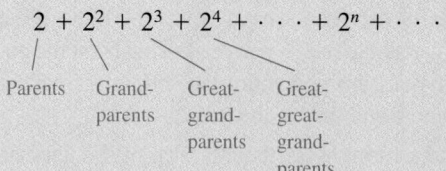

$$2 + 2^2 + 2^3 + 2^4 + \cdots + 2^n + \cdots$$

Parents Grand- Great- Great-
         parents grand- great-
                  parents grand-
                           parents

This formula is valid provided the person has no common ancestors. (A common ancestor is one to whom you are related in more than one way.) During the past 2000 years, suppose your ancestry can be traced through 66 generations. During that time, your total number of ancestors would be

$$2 + 2^2 + 2^3 + 2^4 + \cdots + 2^{66}.$$

Considering the total, do you think that you have had no common ancestors in the past 2000 years? Explain.

Have you had common ancestors?

● = ●

Great-great-grandparents

Great-grandparents

Grandparents

Parents

Mid-Chapter Quiz

Take this quiz as you would take a quiz in class. After you are done, check your work against the answers given in the back of the book.

In Exercises 1–3, write the first five terms of the sequence. (Begin with $n = 1$.)

1. $a_n = 32\left(\frac{1}{4}\right)^{n-1}$

2. $a_n = \dfrac{(-3)^n n}{n + 4}$

3. $a_n = \dfrac{n!}{(n + 1)!}$

In Exercises 4–7, find the sum.

4. $\displaystyle\sum_{k=1}^{4} 10k$

5. $\displaystyle\sum_{i=1}^{10} 4$

6. $\displaystyle\sum_{j=1}^{5} \dfrac{60}{j + 1}$

7. $\displaystyle\sum_{n=1}^{8} 8\left(-\dfrac{1}{2}\right)$

In Exercises 8 and 9, write the sum using sigma notation.

8. $\dfrac{2}{3(1)} + \dfrac{2}{3(2)} + \dfrac{2}{3(3)} + \cdots + \dfrac{2}{3(20)}$

9. $\dfrac{1}{1^3} - \dfrac{1}{2^3} + \dfrac{1}{3^3} - \cdots + \dfrac{1}{25^3}$

In Exercises 10 and 11, find a formula for the nth term of the sequence.

10. *Arithmetic:* $a_1 = 20, a_4 = 11$

11. *Geometric:* $a_1 = 32, r = -\frac{1}{4}$

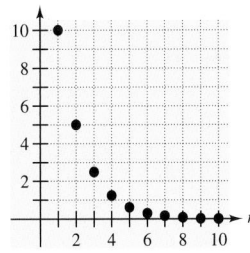

(a)

In Exercises 12–15, find the sum.

12. $\displaystyle\sum_{i=1}^{50} (3i + 5)$

13. $\displaystyle\sum_{j=1}^{300} \dfrac{j}{5}$

14. $\displaystyle\sum_{i=1}^{8} 9\left(\dfrac{2}{3}\right)^{i-1}$

15. $\displaystyle\sum_{j=1}^{20} 500(1.06)^{j-1}$

16. Find the 12th term of the geometric sequence $625, -250, 100, -40, 16, \ldots$.

17. Match $a_n = 10\left(\dfrac{1}{2}\right)^{n-1}$ and $b_n = 10\left(-\dfrac{1}{2}\right)^{n-1}$ with the graphs in the figure.

18. The temperature of a coolant in an experiment decreased by 25.75°F the first hour. For each subsequent hour, the temperature decreased by 2.25°F less than the previous hour. Determine how much the temperature decreased during the tenth hour.

19. This year a company budgets $75,000 for advertising. If this budget is increased by 4.5% per year for 6 years, find the amount that can be spent on advertising 6 years from now.

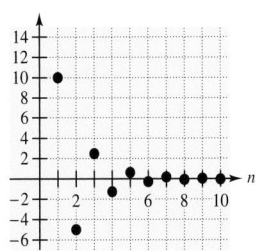

(b)

Figure for 17

10.4	**The Binomial Theorem**

Binomial Coefficients ▪ Pascal's Triangle ▪ Binomial Expansions

Binomial Coefficients

Examine the expansions of $(x + y)^n$ shown at the right. Try to find several patterns and similarities in these expansions.

Remind students to use the distributive property to expand $(x + y)^3$, $(x + y)^4$, $(x + y)^5$, etc.

One strategy is to let students make these observations for the other binomials expanded above. Have students look for the symmetric pattern alluded to in item 4 and see if they can come up with Pascal's Triangle *before* it is introduced.

Recall that a **binomial** is a polynomial that has two terms. In this section, you will study a formula that provides a quick method of raising a binomial to a power. To begin, let's look at the expansion of $(x + y)^n$ for several values of n.

$$(x + y)^0 = 1$$
$$(x + y)^1 = x + y$$
$$(x + y)^2 = x^2 + 2xy + y^2$$
$$(x + y)^3 = x^3 + 3x^2y + 3xy^2 + y^3$$
$$(x + y)^4 = x^4 + 4x^3y + 6x^2y^2 + 4xy^3 + y^4$$
$$(x + y)^5 = x^5 + 5x^4y + 10x^3y^2 + 10x^2y^3 + 5xy^4 + y^5$$

There are several observations you can make about these expansions.

1. In each expansion, there are $n + 1$ terms.

2. In each expansion, x and y have symmetrical roles. The powers of x decrease by 1 in successive terms, whereas the powers of y increase by 1.

3. The sum of the powers of each term is n. For instance, in the expansion of $(x + y)^5$, the sum of the powers of each term is 5.

$$4 + 1 = 5 \qquad 3 + 2 = 5$$
$$(x + y)^5 = x^5 + 5x^4y^1 + 10x^3y^2 + 10x^2y^3 + 5xy^4 + y^5$$

4. The coefficients increase and then decrease in a symmetrical pattern.

The coefficients of a binomial expansion are called **binomial coefficients.** To find them, you can use the following theorem.

NOTE Other notations that are commonly used for $_nC_r$ are $\binom{n}{r}$ and $C(n, r)$.

The Binomial Theorem

In the expansion of $(x + y)^n$,

$$(x + y)^n = x^n + nx^{n-1}y + \cdots + {_nC_r}x^{n-r}y^r + \cdots + nxy^{n-1} + y^n,$$

the coefficient of $x^{n-r}y^r$ is given by

$$_nC_r = \frac{n!}{(n - r)!r!}.$$

EXAMPLE 1   Finding Binomial Coefficients

Find each binomial coefficient.

(a) $_8C_2$ (b) $_{10}C_3$ (c) $_7C_0$ (d) $_8C_8$

Solution

(a) $_8C_2 = \dfrac{8!}{6! \cdot 2!} = \dfrac{(8 \cdot 7) \cdot 6!}{6! \cdot 2!} = \dfrac{8 \cdot 7}{2 \cdot 1} = 28$

(b) $_{10}C_3 = \dfrac{10!}{7! \cdot 3!} = \dfrac{(10 \cdot 9 \cdot 8) \cdot 7!}{7! \cdot 3!} = \dfrac{10 \cdot 9 \cdot 8}{3 \cdot 2 \cdot 1} = 120$

(c) $_7C_0 = \dfrac{7!}{7! \cdot 0!} = 1$

(d) $_8C_8 = \dfrac{8!}{0! \cdot 8!} = 1$

Technology

A graphing utility can be used to find binomial coefficients. For instance, on a *TI-82* or *TI-83*, you can evaluate $_8C_2$ using the following keystrokes.

8 [MATH] ◄ (PRB) 3 (nCr) 2 [ENTER]

Use a graphing utility to find the following binomial coefficients.

$_{10}C_3, \ _7C_0, \ _8C_8$

Check your results against those given in Example 1.

NOTE   When $r \neq 0$ and $r \neq n$, as in parts (a) and (b) of Example 1, there is a simple pattern for evaluating binomial coefficients.

$$\overbrace{\quad}^{2 \text{ factors}} \qquad \overbrace{\quad}^{3 \text{ factors}}$$

$$_8C_2 = \dfrac{8 \cdot 7}{\underbrace{2 \cdot 1}_{2 \text{ factorial}}} \quad \text{and} \quad _{10}C_3 = \dfrac{10 \cdot 9 \cdot 8}{\underbrace{3 \cdot 2 \cdot 1}_{3 \text{ factorial}}}, \text{ so}$$

$$\overbrace{\qquad\qquad\qquad\qquad}^{r \text{ factors}}$$

$$_nC_r = \dfrac{n(n - 1)(n - 2) \ldots (n - r + 1)}{r!}.$$

EXAMPLE 2   Finding Binomial Coefficients

Find each binomial coefficient.

(a) $_7C_3$ (b) $_7C_4$ (c) $_{12}C_1$ (d) $_{12}C_{11}$

Solution

(a) $_7C_3 = \dfrac{7 \cdot 6 \cdot 5}{3 \cdot 2 \cdot 1} = 35$

(b) $_7C_4 = \dfrac{7 \cdot 6 \cdot 5 \cdot 4}{4 \cdot 3 \cdot 2 \cdot 1} = 35$ $\qquad _7C_4 = \ _7C_3$

(c) $_{12}C_1 = \dfrac{12!}{11! \cdot 1!} = \dfrac{(12) \cdot 11!}{11! \cdot 1!} = \dfrac{12}{1} = 12$

(d) $_{12}C_{11} = \dfrac{12!}{1! \cdot 11!} = \dfrac{(12) \cdot 11!}{1! \cdot 11!} = \dfrac{12}{1} = 12$ $\qquad _{12}C_{11} = \ _{12}C_1$

NOTE   In Example 2, it is not a coincidence that the results in parts (a) and (b) are the same, *and* the results in parts (c) and (d) are the same. In general it is true that

$$_nC_r = \ _nC_{n-r}.$$

This shows the symmetric property of binomial coefficients.

Pascal's Triangle

There is a convenient way to remember a pattern for binomial coefficients. By arranging the coefficients in a triangular pattern, you obtain the following array, which is called **Pascal's Triangle.** This triangle is named after the famous French mathematician Blaise Pascal (1623–1662).

NOTE The top row in Pascal's Triangle is called the *zero row* because it corresponds to the binomial expansion

$(x + y)^0 = 1.$

Similarly, the next row is called the *first row* because it corresponds to the binomial expansion

$(x + y)^1 = 1(x) + 1(y).$

In general, the *nth row* in Pascal's Triangle gives the coefficients of $(x + y)^n$.

$$
\begin{array}{ccccccccccccccc}
&&&&&&& 1 &&&&&&& \\
&&&&&& 1 && 1 &&&&&& \\
&&&&& 1 && 2 && 1 &&&&& \\
&&&& 1 && 3 && 3 && 1 &&&& \\
&&& 1 && 4 && 6 && 4 && 1 &&& \\
&& 1 && 5 && 10 && 10 && 5 && 1 && \\
& 1 && 6 && 15 && 20 && 15 && 6 && 1 & \\
1 && 7 && 21 && 35 && 35 && 21 && 7 && 1
\end{array}
$$

The first and last numbers in each row of Pascal's Triangle are 1. Every other number in each row is formed by adding the two numbers immediately above the number. Pascal noticed that numbers in this triangle are precisely the same numbers that are the coefficients of binomial expansions, as follows.

$$(x + y)^0 = 1$$
$$(x + y)^1 = 1x + 1y$$
$$(x + y)^2 = 1x^2 + 2xy + 1y^2$$
$$(x + y)^3 = 1x^3 + 3x^2y + 3xy^2 + 1y^3$$
$$(x + y)^4 = 1x^4 + 4x^3y + 6x^2y^2 + 4xy^3 + 1y^4$$
$$(x + y)^5 = 1x^5 + 5x^4y + 10x^3y^2 + 10x^2y^3 + 5xy^4 + 1y^5$$
$$(x + y)^6 = 1x^6 + 6x^5y + 15x^4y^2 + 20x^3y^3 + 15x^2y^4 + 6xy^5 + 1y^6$$
$$(x + y)^7 = 1x^7 + 7x^6y + 21x^5y^2 + 35x^4y^3 + 35x^3y^4 + 21x^2y^5 + 7xy^6 + 1y^7$$

EXAMPLE 3   *Using Pascal's Triangle*

You can use the seventh row of Pascal's Triangle to find the eighth row.

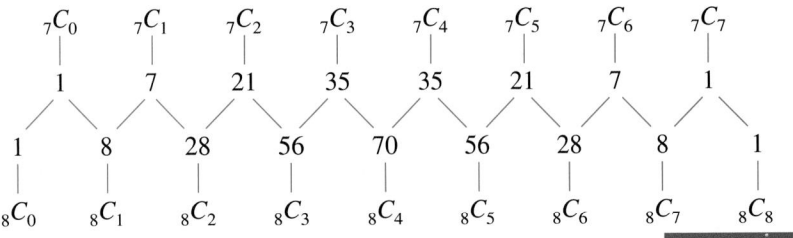

Binomial Expansions

As mentioned at the beginning of this section, when you write out the coefficients for a binomial that is raised to a power, you are **expanding a binomial.** The formulas for binomial coefficients give you an easy way to expand binomials, as demonstrated in the next three examples.

EXAMPLE 4   *Expanding a Binomial*

Write the expansion for the following expression.

$$(x + 1)^3$$

Solution

The binomial coefficients from the third row of Pascal's Triangle are

1, 3, 3, 1.

Therefore, the expansion is as follows.

$$(x + 1)^3 = (1)x^3 + (3)x^2(1) + (3)x(1^2) + (1)(1^3)$$
$$= x^3 + 3x^2 + 3x + 1$$

Stress that when binomials representing differences are expanded, the signs alternate.

To expand binomials representing *differences,* rather than sums, you alternate signs. Here are two examples.

$$(x - 1)^3 = x^3 - 3x^2 + 3x - 1$$
$$(x - 1)^4 = x^4 - 4x^3 + 6x^2 - 4x + 1$$

EXAMPLE 5   *Expanding a Binomial*

Write the expansion for the following expression.

$$(x - 3)^4$$

Solution

The binomial coefficients from the fourth row of Pascal's Triangle are

1, 4, 6, 4, 1.

Therefore, the expansion is as follows.

$$(x - 3)^4 = (1)x^4 - (4)x^3(3) + (6)x^2(3^2) - (4)x(3^3) + (1)(3^4)$$
$$= x^4 - 12x^3 + 54x^2 - 108x + 81$$

EXAMPLE 6   Expanding a Binomial

Write the expansion for $(3x - y)^4$.

Solution

Use the fourth row of Pascal's Triangle, as follows.

$$(3x - y)^4 = (1)(3x)^4 - (4)(3x)^3 y + (6)(3x)^2 y^2 - (4)3xy^3 + (1)y^4$$
$$= 81x^4 - 108x^3 y + 54x^2 y^2 - 12xy^3 + y^4$$

EXAMPLE 7   Finding a Term in the Binomial Expansion

Find the sixth term of $(a + 2b)^8$.

Solution

From the Binomial Theorem, you can see that the $(r + 1)$th term is $_nC_r x^{n-r} y^r$. So in this case, $6 = r + 1$ means that $r = 5$. Because $n = 8$, $x = a$, and $y = 2b$, the sixth term in the binomial expansion is

$$_8C_5 a^{8-5}(2b)^5 = 56 \cdot a^3 \cdot (2b)^5$$
$$= 56(2^5)a^3 b^5$$
$$= 1792a^3 b^5$$

Group Activities   Extending the Concept

Finding a Pattern   By adding the numbers in each of the rows of Pascal's Triangle, you obtain the following.

Row 0: $1 = 1$

Row 1: $1 + 1 = 2$

Row 2: $1 + 2 + 1 = 4$

Row 3: $1 + 3 + 3 + 1 = 8$

Row 4: $1 + 4 + 6 + 4 + 1 = 16$

Find a pattern for the sequence. Then use the pattern to find the sum of the numbers in the 10th row of Pascal's Triangle. Finally, check your answer by actually adding the numbers in the 10th row.

10.4   **Exercises**

See Warm-Up
Exercises, p. A50

1. *Think About It*   How many terms are in the expansion of $(x + y)^n$?

2. *Think About It*   How do the expansions of $(x + y)^n$ and $(x - y)^n$ differ?

3. Which of the following is equal to $_{11}C_5$? Explain.

(a) $\dfrac{11 \cdot 10 \cdot 9 \cdot 8 \cdot 7}{5 \cdot 4 \cdot 3 \cdot 2 \cdot 1}$ (b) $\dfrac{11 \cdot 10 \cdot 9 \cdot 8 \cdot 7}{6 \cdot 5 \cdot 4 \cdot 3 \cdot 2 \cdot 1}$

4. What is the relationship between $_nC_r$ and $_nC_{n-r}$?

In Exercises 5–16, evaluate $_nC_r$.

5. $_6C_4$

6. $_7C_3$

7. $_{10}C_5$

8. $_{12}C_9$

9. $_{20}C_{20}$

10. $_{15}C_0$

11. $_{18}C_{18}$

12. $_{200}C_1$

13. $_{50}C_{48}$

14. $_{75}C_1$

15. $_{25}C_4$

16. $_{18}C_5$

In Exercises 17–26, use a graphing utility to evaluate $_nC_r$.

17. $_{30}C_6$

18. $_{25}C_{10}$

19. $_{12}C_7$

20. $_{40}C_5$

21. $_{52}C_5$

22. $_{100}C_6$

23. $_{200}C_{195}$

24. $_{500}C_4$

25. $_{25}C_{12}$

26. $_{1000}C_2$

In Exercises 27–36, evaluate the binomial coefficient $_nC_r$. Also, evaluate its symmetric coefficient $_nC_{n-r}$.

27. $_{15}C_3$

28. $_9C_4$

29. $_{25}C_5$

30. $_{30}C_3$

31. $_5C_2$

32. $_8C_6$

33. $_{12}C_5$

34. $_{14}C_8$

35. $_{10}C_0$

36. $_{25}C_{25}$

Pascal's Triangle   In Exercises 37–42, use Pascal's Triangle to evaluate $_nC_r$.

37. $_6C_2$

38. $_9C_3$

39. $_7C_3$

40. $_9C_5$

41. $_8C_4$

42. $_{10}C_6$

In Exercises 43–56, use the Binomial Theorem to expand the expression.

43. $(x + 3)^6$

44. $(x - 5)^4$

45. $(x + 1)^5$

46. $(x + 2)^5$

47. $(x - 4)^6$

48. $(x - 8)^4$

49. $(x + y)^4$

50. $(u + v)^6$

51. $(u - 2v)^3$

52. $(2x + y)^5$

53. $(x - y)^5$

54. $(4t - 1)^4$

55. $(3a + 2b)^4$

56. $(4u - 3v)^3$

In Exercises 57–66, use Pascal's Triangle to expand the expression.

57. $(a + 2)^3$

58. $(x + 3)^5$

59. $(2x - 1)^4$

60. $(4 - 3y)^3$

61. $(2y + z)^6$

62. $(2t - s)^5$

63. $(x + y)^8$

64. $(r - s)^7$

65. $(x - 2)^6$

66. $(2x + 3)^5$

In Exercises 67–78, find the coefficient of the given term of the expression.

Expression	*Term*
67. $(x + 1)^{10}$	x^7
68. $(x + 3)^{12}$	x^9
69. $(x - 1)^9$	x^6
70. $(x - 2)^8$	x^4

Expression	Term
71. $(x - y)^{15}$	x^4y^{11}
72. $(x - 3y)^{14}$	x^3y^{11}
73. $(2x + y)^{12}$	x^3y^9
74. $(x + y)^{10}$	x^7y^3
75. $(x^2 - 3)^4$	x^4
76. $(3 - y^3)^5$	y^9
77. $\left(\sqrt{x} + 1\right)^8$	x^2
78. $\left(\dfrac{1}{u} + 2\right)^6$	$\dfrac{1}{u^4}$

Probability In Exercises 79–82, use the Binomial Theorem to expand the expression. In the study of probability, it is sometimes necessary to use the expansion $(p + q)^n$, where $p + q = 1$.

79. $\left(\frac{1}{2} + \frac{1}{2}\right)^5$ **80.** $\left(\frac{2}{3} + \frac{1}{3}\right)^4$

81. $\left(\frac{1}{4} + \frac{3}{4}\right)^4$ **82.** $(0.4 + 0.6)^6$

In Exercises 83–86, use the Binomial Theorem to approximate the quantity accurate to three decimal places. For example,

$$(1.02)^{10} = (1 + 0.02)^{10} \approx 1 + 10(0.02) + 45(0.02)^2.$$

83. $(1.02)^8$ **84.** $(2.005)^{10}$

85. $(2.99)^{12}$ **86.** $(1.98)^9$

Graphical Reasoning In Exercises 87–90, use a graphing utility to graph f and g in the same viewing rectangle. What is the relationship between the two graphs? Use the Binomial Theorem to write the polynomial function g in standard form.

87. $f(x) = -x^2 + 3x + 2$, $\quad g(x) = f(x - 2)$
88. $f(x) = 2x^2 - 4x + 1$, $\quad g(x) = f(x + 3)$
89. $f(x) = x^3 - 4x$, $\quad g(x) = f(x + 4)$
90. $f(x) = -x^4 + 4x^2 - 1$, $\quad g(x) = f(x - 3)$

Section Project

Patterns in Pascal's Triangle Use each encircled group of numbers to form a 2×2 matrix. Find the determinant of each matrix. Describe the pattern.

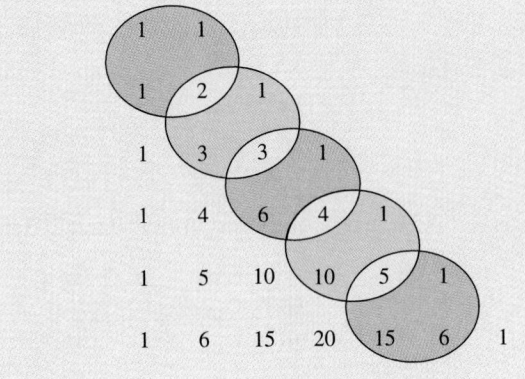

10.5	**Counting Principles**

Simple Counting Problems ▪ Counting Principles ▪
Permutations ▪ Combinations

Simple Counting Problems

The last two sections of this chapter contain a brief introduction to some of the basic counting principles and their application to probability. In the next section, you will see that much of probability has to do with counting the number of ways an event can occur. Examples 1, 2, and 3 describe some simple cases.

EXAMPLE 1   *A Random Number Generator*

A random number generator (on a computer) selects an integer from 1 to 30. Find the number of ways each event can occur.

(a) An even integer is selected.

(b) An integer that is less than 12 is selected.

(c) A prime number is selected.

(d) A perfect square is selected.

Solution

(a) Because half of the integers from 1 to 30 are even, this event can occur in 15 different ways.

(b) The integers from 1 to 30 that are less than 12 are as follows.

$$\{1, 2, 3, 4, 5, 6, 7, 8, 9, 10, 11\}$$

Because this set has 11 members, you can conclude that there are 11 different ways this event can occur.

(c) The prime numbers from 1 to 30 are as follows.

$$\{2, 3, 5, 7, 11, 13, 17, 19, 23, 29\}$$

Because this set has ten members, you can conclude that there are ten different ways this event can occur.

(d) The perfect squares from 1 to 30 are as follows.

$$\{1, 4, 9, 16, 25\}$$

Because this set has five members, you can conclude that there are five different ways this event can occur.

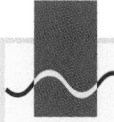 *Programming*

The *TI-82/TI-83* program below can be used to generate ten randomly chosen integers from 0 through 9. (See the appendix for other calculator models.) Repeatedly press ENTER until all ten integers are displayed.

```
PROGRAM:RANDOM
:For(I,1,10)
:int (10*rand)→A
:Disp A
:Pause
:End
```

EXAMPLE 2 *Selecting Pairs of Numbers at Random*

Eight pieces of paper are numbered from 1 to 8 and placed in a box. One piece of paper is drawn from the box, its number is written down, and the piece of paper is replaced in the box. Then, a second piece of paper is drawn from the box, and its number is written down. Finally, the two numbers are added together. How many different ways can a total of 12 be obtained?

Solution

To solve this problem, count the different ways that a total of 12 can be obtained using two numbers from 1 to 8.

$$\boxed{\text{First number}} \ + \ \boxed{\text{Second number}} \ = \ \boxed{12}$$

After considering the various possibilities, you can see that this equation can be solved in the following five ways.

First Number: 4 5 6 7 8
Second Number: 8 7 6 5 4

Thus, a total of 12 can be obtained in five different ways.

Solving counting problems can be tricky. Often, seemingly minor changes in the statement of a problem can affect the answer. For instance, compare the counting problem in the next example with that given in Example 2.

EXAMPLE 3 *Selecting Pairs of Numbers at Random*

NOTE The difference between the counting problems in Examples 2 and 3 can be distinguished by saying that the random selection in Example 2 occurs *with replacement,* whereas the random selection in Example 3 occurs *without replacement,* which eliminates the possibility of choosing two 6's.

Eight pieces of paper are numbered from 1 to 8 and placed in a box. Two pieces of paper are drawn from the box, and the numbers on them are written down and totaled. How many different ways can a total of 12 be obtained?

Solution

To solve this problem, count the different ways that a total of 12 can be obtained *using two different numbers* from 1 to 8.

$$\boxed{\text{First number}} \ + \ \boxed{\text{Second number}} \ = \ \boxed{12}$$

After considering the various possibilities, you can see that this equation can be solved in the following four ways.

First Number: 4 5 7 8
Second Number: 8 7 5 4

Thus, a total of 12 can be obtained in four different ways.

Counting Principles

The first three examples in this section are considered simple counting problems in which you can *list* each possible way that an event can occur. When it is possible, this is always the best way to solve a counting problem. However, some events can occur in so many different ways that it is not feasible to write out the entire list. In such cases, you must rely on formulas and counting principles. The most important of these is called the **Fundamental Counting Principle.**

Fundamental Counting Principle

Let E_1 and E_2 be two events. The first event E_1 can occur in m_1 different ways. After E_1 has occurred, E_2 can occur in m_2 different ways. The number of ways that the two events can occur is

$$m_1 \cdot m_2.$$

EXAMPLE 4   *Applying the Fundamental Counting Principle*

The English alphabet contains 26 letters. Thus, the number of possible "two-letter words" is $26 \cdot 26 = 676$.

EXAMPLE 5   *Applying the Fundamental Counting Principle*

Telephone numbers in the United States have ten digits. The first three are the *area code* and the next seven are the *local telephone number.* How many different telephone numbers are possible within each area code? (A telephone number cannot have 0 or 1 as its first or second digit.)

Solution

There are only eight choices for the first and second digits because neither can be 0 or 1. For each of the other digits, there are 10 choices.

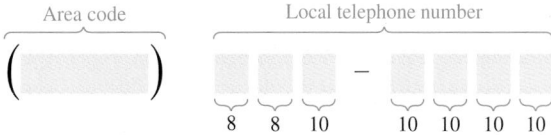

Thus, by the Fundamental Counting Principle, the number of local telephone numbers that are possible within each area code is

$$8 \cdot 8 \cdot 10 \cdot 10 \cdot 10 \cdot 10 \cdot 10 = 6,400,000.$$

Permutations

One important application of the Fundamental Counting Principle is in determining the number of ways that n elements can be arranged (in order). An ordering of n elements is called a **permutation** of the elements.

Definition of Permutation

A **permutation** of n different elements is an ordering of the elements such that one element is first, one is second, one is third, and so on.

EXAMPLE 6   Listing Permutations

The six possible permutations of letters A, B, and C are as follows.

A, B, C	B, A, C	C, A, B
A, C, B	B, C, A	C, B, A

EXAMPLE 7   Finding the Number of Permutations of n Elements

How many permutations are possible for the letters A, B, C, D, E, and F?

Solution

First position:	Any of the *six* letters.
Second position:	Any of the remaining *five* letters.
Third position:	Any of the remaining *four* letters.
Fourth position:	Any of the remaining *three* letters.
Fifth position:	Any of the remaining *two* letters.
Sixth position:	The *one* remaining letter.

Thus, the numbers of choices for the six positions are as follows.

Permutations of six letters

6   5   4   3   2   1

By the Fundamental Counting Principle, the total number of permutations of the six letters is

$$6 \cdot 5 \cdot 4 \cdot 3 \cdot 2 \cdot 1 = 6! = 720.$$

The result obtained in Example 7 is generalized below.

Number of Permutations of *n* Elements

The number of permutations of *n* elements is given by

$$n \cdot (n - 1) \cdot \cdots \cdot 4 \cdot 3 \cdot 2 \cdot 1 = n!.$$

In other words, there are *n*! different ways that *n* elements can be ordered.

EXAMPLE 8 *Finding the Number of Permutations of n Elements*

How many ways can you form a four-digit number using each of the digits 1, 3, 5, and 7 exactly once?

Solution

One way to solve this problem is simply to list the number of ways.

    1357, 1375, 1537, 1573, 1735, 1753

    3157, 3175, 3517, 3571, 3715, 3751

    5137, 5173, 5317, 5371, 5713, 5731

    7135, 7153, 7315, 7351, 7513, 7531

Another way to solve the problem is to use the formula for the number of permutations of four elements. By that formula, there are 4! = 24 permutations.

EXAMPLE 9 *Finding the Number of Permutations of n Elements*

You are a supervisor for 11 different employees. One of your responsibilities is to perform an annual evaluation for each employee, and then rank the performances of the 11 different employees. How many different rankings are possible?

Solution

You have 11 choices for the first ranking. After choosing the first ranking, you can choose any of the remaining ten for the second ranking, and so on.

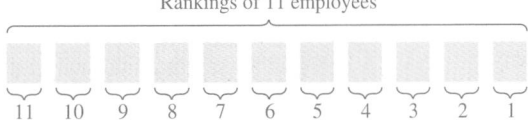

Rankings of 11 employees

11 10 9 8 7 6 5 4 3 2 1

Thus, the number of different rankings is 11! = 39,916,800.

Combinations

When counting the number of possible permutations of a set of elements, order is important. The final topic in this section is a method of selecting subsets of a larger set in which order is *not important.* Such subsets are called **combinations of n elements taken r at a time.** For instance, the combinations

$$\{A, B, C\} \qquad \text{and} \qquad \{B, A, C\}$$

are equivalent because both sets contain the same three elements, and the order in which the elements are listed is *not important.* Hence, you would count only one of the two sets. A common example of how a combination occurs is a card game in which the player is free to reorder the cards after they have been dealt.

Students often confuse a permutation with a combination. Help them see the difference using comparison and contrast. {A, B, C} and {B, A, C} are distinct as a permutation, but identical as a combination.

EXAMPLE 10   Combinations of n Elements Taken r at a Time

In how many different ways can three letters be chosen from the letters A, B, C, D, and E? (The order of the three letters is not important.)

Solution

The following subsets represent the different combinations of three letters that can be chosen from five letters.

$$\{A, B, C\} \qquad \{A, B, D\}$$
$$\{A, B, E\} \qquad \{A, C, D\}$$
$$\{A, C, E\} \qquad \{A, D, E\}$$
$$\{B, C, D\} \qquad \{B, C, E\}$$
$$\{B, D, E\} \qquad \{C, D, E\}$$

From this list, you can conclude that there are 10 different ways that three letters can be chosen from five letters. Because order is not important, sets such as $\{B, C, A\}$ are not chosen, because the set $\{B, C, A\}$ is represented by the set $\{A, B, C\}$.

The formula for the number of combinations of n elements taken r at a time is as follows.

Technology

Use a graphing utility to verify the result in Example 10. (See page 743 for an example of the keystrokes.)

Number of Combinations of n Elements Taken r at a Time

The number of combinations of n elements taken r at a time is

$$_nC_r = \frac{n!}{(n-r)!r!}.$$

Note that the formula for $_nC_r$ is the same one given for binomial coefficients. To see how this formula is used, consider the counting problem given in Example 10. In that problem, you need to find the number of combinations of five elements taken three at a time. Thus, $n = 5$, $r = 3$, and the number of combinations is

$$_5C_3 = \frac{5!}{2!3!}$$
$$= \frac{5 \cdot 4 \cdot 3}{3 \cdot 2 \cdot 1}$$
$$= 10,$$

which is the same as the answer obtained in Example 10.

EXAMPLE 11   Combinations of n Elements Taken r at a Time

A standard poker hand consists of five cards dealt from a deck of 52. How many different poker hands are possible? (After the cards are dealt, the player may reorder them, and therefore order is not important.)

Solution

Use the formula for the number of combinations of 52 elements taken five at a time, as follows.

$$_{52}C_5 = \frac{52!}{47!5!}$$
$$= \frac{52 \cdot 51 \cdot 50 \cdot 49 \cdot 48}{5 \cdot 4 \cdot 3 \cdot 2 \cdot 1}$$
$$= 2{,}598{,}960$$

Thus, there are almost 2.6 million different hands. Use a graphing utility to verify this result.

Remind students that they can use a graphing utility to evaluate $_nC_r$.

Group Activities

Problem Solving

Applying Counting Methods   The Boston Market restaurant chain offers individual rotisserie chicken meals with two side items. Customers can choose either dark or white meat and can select side items from a list of 15. How many different meals are available if two different side items are to be ordered? Of the 15 side items, nine are hot and six are cold. How many different meals are available if a customer wishes to order one hot and one cold side item? *(Source: Boston Market, Inc.)*

10.5   **Exercises**

See Warm-Up
Exercises, p. A50

Random Selection   In Exercises 1–6, find the number of ways the specified event can occur when one or more marbles are selected from a bowl containing ten marbles numbered 0 through 9.

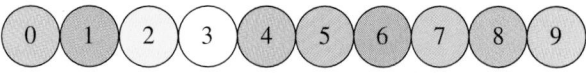

1. One marble is drawn and its number is even.

2. One marble is drawn and its number is prime.

3. Two marbles are drawn one after the other. The first is replaced before the second is drawn. The sum of the numbers is 10.

4. Two marbles are drawn one after the other. The first is replaced before the second is drawn. The sum of the numbers is 7.

5. Two marbles are drawn without replacement. The sum of the numbers is 10.

6. Two marbles are drawn without replacement. The sum of the numbers is 7.

Random Selection   In Exercises 7–16, find the number of ways the specified event can occur when one or more marbles are selected from a bowl containing 20 marbles numbered 1 through 20.

7. One marble is drawn and its number is odd.

8. One marble is drawn and its number is even.

9. One marble is drawn and its number is prime.

10. One marble is drawn and its number is greater than 12.

11. One marble is drawn and its number is divisible by 3.

12. One marble is drawn and its number is divisible by 6.

13. Two marbles are drawn one after the other. The first is replaced before the second is drawn. The sum of the numbers is 8.

14. Two marbles are drawn one after the other. The first is replaced before the second is drawn. The sum of the numbers is 15.

15. Two marbles are drawn without replacement. The sum of the numbers is 8.

16. Three marbles are drawn one after another. Each marble is replaced before the next is drawn. The sum of the numbers is 15.

17. *Staffing Choices*   A small grocery store needs to open another checkout line. Three people who can run the cash register are available and two people are available to bag groceries. How many different ways can the additional checkout line be staffed?

18. *Computer System*   You are in the process of purchasing a new computer system. You must choose one of the three monitors, one of two computers, and one of two keyboards. How many different configurations of the system are available to you?

19. *Identification Numbers*   In a statistical study, each participant was given an identification label consisting of a letter of the alphabet followed by a single digit (0 is a digit). How many distinct identification labels can be made in this way?

20. *Identification Numbers*   How many identification labels (see Exercise 19) can be made consisting of a letter of the alphabet followed by a two-digit number?

21. *License Plates*   How many distinct automobile license plates can be formed using a four-digit number followed by two letters?

22. *Three-Digit Numbers*   How many three-digit numbers can be formed in each of the following situations?

    (a) The hundreds digit cannot be 0.

    (b) No repetition of digits is allowed.

    (c) The number cannot be greater than 400.

23. *Toboggan Ride* Five people line up on a toboggan at the top of a hill. In how many ways can they be seated if only two of the five are willing to sit in the front seat?

24. *Task Assignment* Four people are assigned to four different tasks. In how many ways can the assignments be made if one of the four is not qualified for the first task?

25. *Taking a Trip* Five people are taking a long trip in a car. Two sit in the front seat and three sit in the back seat. Three of the people agree to share the driving. In how many different ways can the five people sit?

26. *Aircraft Boarding* Eight people are boarding an aircraft. Three have tickets for first class and board before those in the economy class. In how many different ways can the eight people board the aircraft?

27. *Permutations* List all the permutations of the letters X, Y, and Z.

28. *Permutations* List all the permutations of the letters A, B, C, and D.

29. *Seating Arrangement* In how many ways can five children be seated in a single row of chairs?

30. *Seating Arrangement* In how many ways can six people be seated in a six-passenger car?

31. *Choosing Officers* From a pool of ten candidates, the offices of president, vice-president, secretary, and treasurer will be filled. In how many ways can the offices be filled if each of the ten candidates can hold any one of the offices?

32. *Time Management Study* There are eight steps in accomplishing a certain task and these steps can be performed in any order. Management wants to test each possible order to determine which is the least time-consuming.

(a) How many different orders will have to be tested?

(b) How many different orders will have to be tested if one step in accomplishing the task must be done first? (The other seven steps can be performed in any order.)

33. *Combination Lock* The combination lock will open when the right choice of three numbers (from 1 to 40, inclusive) is selected in order. How many different lock combinations are possible?

34. *Work Assignments* Out of eight workers five are selected and assigned to different tasks. In how many ways can this be done assuming there are no restrictions in making the assignments?

35. *Number of Subsets* List all the subsets with two elements that can be formed from the set of letters $\{A, B, C, D, E, F\}$.

36. *Number of Subsets* List all the subsets with three elements that can be formed from the set of letters $\{A, B, C, D, E, F\}$.

37. *Committee Selection* Three students are selected from a class of 20 to form a fund-raising committee. Use a graphing utility to determine the number of ways the committee can be formed.

38. *Committee Selection* Use a graphing utility to determine the number of ways a committee of five can be formed from a group of 30 people.

39. *Menu Selection* A group of four people go out to dinner at a restaurant. There are nine entrees on the menu and the four people decide that no two will order the same thing. How many ways can the four order from the nine entrees?

40. *Test Questions* A student is required to answer any nine questions from the 12 questions on an exam. Use a graphing utility to determine the number of ways the student can select the nine questions.

41. *Basketball Lineup* A high school basketball team has 15 players. Use a graphing utility to determine the number of ways the coach can choose the starting lineup. (Assume that each player can play any position.)

42. *Softball League* Six churches form a softball league. If each team must play every other team twice during the season, what is the total number of league games played?

43. *Defective Units*   A shipment of ten microwave ovens contains two defective units. In how many ways can a vending company purchase three of these units and receive all good units?

44. *Job Applicants*   An employer interviews six people for four job openings. Four of the six people are women. If all six are qualified, in how many ways can the employer fill the four positions if (a) the selection is random and (b) exactly two women are chosen?

45. *Group Selection*   Four people are to be selected from four couples.

(a) In how many ways can this be done if there are no restrictions?

(b) In how many ways can this be done if one person from each couple must be selected?

46. *Relationships*   As the size of a group of people increases, the number of relationships increases dramatically (see figure). Determine the numbers of different two-person relationships in groups of the following numbers.

(a) 3   (b) 4   (c) 6   (d) 8   (e) 10   (f) 12

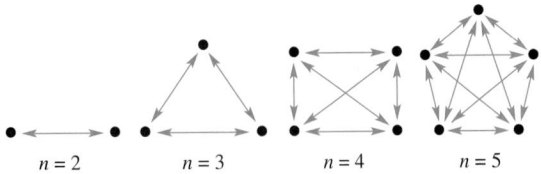

$n = 2$ $n = 3$ $n = 4$ $n = 5$

47. *Number of Triangles*   Eight points are located in the coordinate plane such that no three are collinear. How many different triangles can be formed having three of the eight points as their vertices?

48. *Geometry*   Three points that are not on a line determine three lines (see figure). How many lines are determined by seven points, no three of which are on a line?

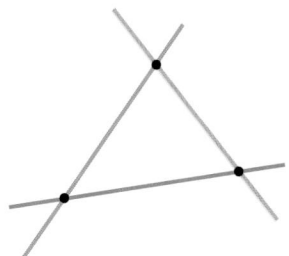

Section Project

Diagonals of a Polygon   Find the number of diagonals of each polygon. (A line segment connecting any two nonadjacent vertices of a polygon is called a *diagonal* of the polygon.)

(a) Pentagon

(b) Hexagon

(c) Octagon

(d) Decagon (ten sides)

Then generalize the results and find the number of diagonals of a polygon with n sides.

10.6	**Probability**

The Probability of an Event ▪
Using Counting Methods to Find Probabilities

The Probability of an Event

The **probability of an event** is a number from 0 to 1 that indicates the likelihood that the event will occur. An event that is certain to occur has a probability of 1. An event that cannot occur has a probability of 0. An event that is equally likely to occur or not occur has a probability of $\frac{1}{2}$ or 0.5.

Stress that the probability of an event must be a number between 0 and 1, inclusive.

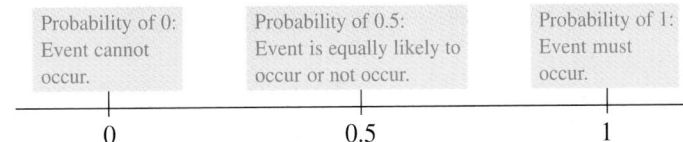

Probability of 0: Event cannot occur.	Probability of 0.5: Event is equally likely to occur or not occur.	Probability of 1: Event must occur.

0   0.5   1

The Probability of an Event

Consider a **sample space** S that is composed of a finite number of outcomes, each of which is equally likely to occur. A subset E of the sample space is an **event**. The probability that an outcome E will occur is

$$P = \frac{\text{number of outcomes in event}}{\text{number of outcomes in sample space}}.$$

EXAMPLE 1   *Finding the Probability of an Event*

(a) You are dialing a friend's phone number but cannot remember the last digit. If you choose a digit at random, the probability that it is correct is

$$P = \frac{\text{number of correct digits}}{\text{number of possible digits}} = \frac{1}{10}.$$

Historical Note: Pierre de Fermat (1601–1675) was a French lawyer whose hobby was mathematics. He made many important contributions to the field of mathematics. He and Blaise Pascal (1623–1662) worked together on probability problems after a professional gambler asked them a number of probability questions.

(b) On a multiple-choice test, you know that the answer to a question is not (a) or (d), but you are not sure about (b), (c), and (e). If you guess, the probability that you are wrong is

$$P = \frac{\text{number of wrong answers}}{\text{number of possible answers}} = \frac{2}{3}.$$

Technology

A well-known experiment in probability is called the "birthday problem." This problem concerns the probability that in a group of *n* people at least two of the people will have the same birthday. The *TI-83* program below is a simulation of the birthday problem. (See the appendix for other calculator models.)

```
PROGRAM:BIRTHDAY
:ClrHome
:ClrList L1
:0→C
:Output(4,3,"COMPUTING...")
:Lbl A
:1+C→C
:randInt(1,365)→D
:D→L1(C)
:If C=1:Goto A
:For(I,1,C-1)
:If D=L1(I):Goto B
:End
:Goto A
:Lbl B
:ClrHome
:Disp "MATCH: NUMBER OF"
:Disp "SELECTIONS IS"
:Disp C
```

Try running the program a few times. How many selections were made each time before two people with the same birthday were found?

EXAMPLE 2   Conducting a Poll

The Centers for Disease Control surveyed 11,631 high school students. The students were asked whether they considered themselves to be a good weight, underweight, or overweight. The results are shown in Figure 10.5.

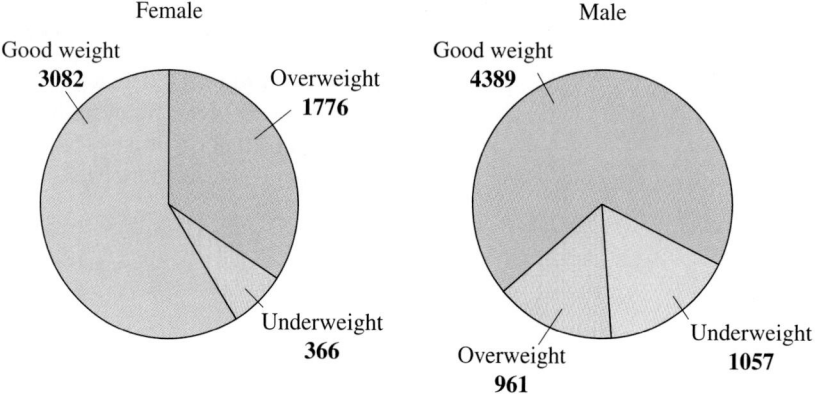

FIGURE 10.5

(a) If you choose a female at random from those surveyed, the probability that she said she was underweight is

$$P = \frac{\text{number of females who answered "underweight"}}{\text{number of females in survey}}$$

$$= \frac{366}{3082 + 366 + 1776}$$

$$= \frac{366}{5224}$$

$$\approx 0.07.$$

(b) If you choose a person who answered "underweight" from those surveyed, the probability that the person is female is

$$P = \frac{\text{number of females who answered "underweight"}}{\text{number in survey who answered "underweight"}}$$

$$= \frac{366}{366 + 1057}$$

$$= \frac{366}{1423}$$

$$\approx 0.26.$$

Polls such as the one described in Example 2 are often used to make inferences about a population that is larger than the sample. For instance, from Example 2, you might infer that 7% of *all* high school girls consider themselves to be underweight. When you make such an inference, it is important that those surveyed are representative of the entire population.

EXAMPLE 3 *Using Area to Find Probability*

You have just stepped into the tub to take a shower when one of your contact lenses falls out. (You have not yet turned on the water.) Assuming the lens is equally likely to land anywhere on the bottom of the tub, what is the probability that it lands in the drain? Use the dimensions in Figure 10.6.

Solution

Because the area of the tub bottom is $(26)(50) = 1300$ square inches and the area of the drain is

$$\pi(1^2) = \pi \qquad \text{Area of drain} = \pi r^2$$

square inches, the probability that the lens lands in the drain is

$$P = \frac{\pi}{1300} \approx 0.0024.$$

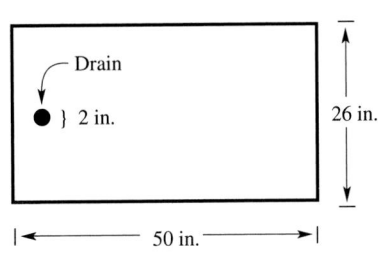

FIGURE 10.6

EXAMPLE 4 *The Probability of Inheriting Certain Genes*

Common parakeets have genes that can produce any one of the following four feather colors.

Green: BBCC, BBCc, BbCC, BbCc   *Blue:* BBcc, Bbcc

Yellow: bbCC, bbCc   *White:* bbcc

Use the *Punnett square* in Figure 10.7 to find the probability that an offspring to two green parents (both with BbCc feather genes) will be yellow. Note that each parent passes along a B or b gene and a C or c gene.

Solution

The probability that an offspring will be yellow is

$$P = \frac{\text{number of yellow possibilities}}{\text{number of possibilities}} = \frac{3}{16}.$$

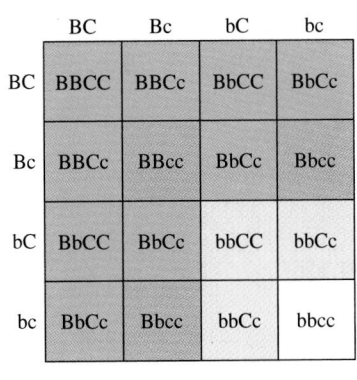

FIGURE 10.7

If P is the probability that an outcome *will* occur, the probability that the outcome *will not* occur is $1 - P$. In Example 3, the probability that the lens does not land in the drain is approximately $1 - 0.0024$, or 0.9976. In Example 4, the probability that an offspring will not be yellow is $1 - \frac{3}{16}$, or $\frac{13}{16}$.

Using Counting Methods to Find Probabilities

Standard 52-Card Deck

A ♠	A ♥	A ♦	A ♣
K ♠	K ♥	K ♦	K ♣
Q ♠	Q ♥	Q ♦	Q ♣
J ♠	J ♥	J ♦	J ♣
10 ♠	10 ♥	10 ♦	10 ♣
9 ♠	9 ♥	9 ♦	9 ♣
8 ♠	8 ♥	8 ♦	8 ♣
7 ♠	7 ♥	7 ♦	7 ♣
6 ♠	6 ♥	6 ♦	6 ♣
5 ♠	5 ♥	5 ♦	5 ♣
4 ♠	4 ♥	4 ♦	4 ♣
3 ♠	3 ♥	3 ♦	3 ♣
2 ♠	2 ♥	2 ♦	2 ♣

FIGURE 10.8

EXAMPLE 5   The Probability of a Royal Flush

Five cards are dealt at random from a standard deck of 52 playing cards (see Figure 10.8). What is the probability that the cards are 10-J-Q-K-A of the same suit?

Solution

On page 755, you saw that the number of possible five-card hands from a deck of 52 cards is $_{52}C_5 = 2,598,960$. Because only four of these five-card hands are 10-J-Q-K-A of the same suit, the probability that the hand contains these cards is

$$P = \frac{4}{2,598,960} = \frac{1}{649,740}.$$

EXAMPLE 6   Conducting a Survey

A survey was conducted of 500 adults who had worn Halloween costumes. Each person was asked how he or she acquired a Halloween costume: created it, rented it, bought it, or borrowed it. The results are shown in Figure 10.9. What is the probability that the first four people who were polled all created their costumes?

Solution

To answer this question, you need to use the formula for the number of combinations *twice*. First, find the number of ways to choose four people from the 360 who created their own costumes.

$$_{360}C_4 = \frac{360 \cdot 359 \cdot 358 \cdot 357}{4 \cdot 3 \cdot 2 \cdot 1} = 688,235,310$$

Next, find the number of ways to choose four people from the 500 who were surveyed.

$$_{500}C_4 = \frac{500 \cdot 499 \cdot 498 \cdot 497}{4 \cdot 3 \cdot 2 \cdot 1} = 2,573,031,125$$

The probability that all of the first four people surveyed created their own costumes is the ratio of these two numbers.

$$P = \frac{\text{number of ways to choose 4 from 360}}{\text{number of ways to choose 4 from 500}}$$

$$= \frac{688,235,310}{2,573,031,125}$$

$$\approx 0.267$$

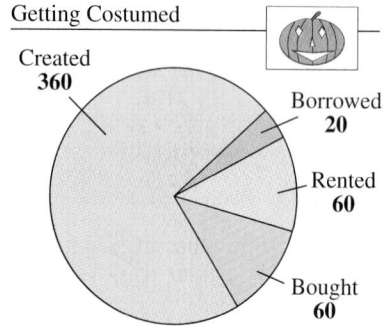

Getting Costumed

Created 360
Borrowed 20
Rented 60
Bought 60

FIGURE 10.9

EXAMPLE 7 Forming a Committee

To obtain input from 200 company employees, the management of a company selected a committee of five. Of the 200 employees, 56 were from minority groups. None of these 56, however, was selected to be on the committee. Does this indicate that the management's selection was biased?

Solution

Part of the solution is similar to that of Example 6. If the five committee members were selected at random, the probability that all five would be nonminority is

$$P = \frac{\text{number of ways to choose 5 from 144 nonminority employees}}{\text{number of ways to choose 5 from 200 employees}}$$

$$= \frac{{}_{144}C_5}{{}_{200}C_5}$$

$$= \frac{481,008,528}{2,535,650,040}$$

$$\approx 0.19.$$

Thus, if the committee were chosen at random (that is, without bias), the likelihood that it would have no minority members is about 0.19. Although this does not *prove* that there was bias, it does suggest it.

Group Activities Extending the Concept

Probability of Guessing Correctly You are taking a chemistry test and are asked to arrange the first ten elements in the order in which they appear on the periodic table of elements. Suppose that you have no idea of the correct order and simply guess. Does the following computation represent the probability that you guess correctly? Explain your reasoning.

Solution

You have ten choices for the first element, nine choices for the second, eight choices for the third, and so on. The number of different orders is 10! = 3,628,800, which means that your probability of guessing correctly is

$$P = \frac{1}{3,628,800}.$$

10.6 Exercises

See Warm-Up Exercises, p. A50

1. *Think About It* The probability of an event is $\frac{3}{4}$. What is the probability that the event *does not* occur? Explain.

2. *Think About It* The weather forecast indicates that the probability of rain is 40%. Explain what this means.

Sample Space In Exercises 3–6, determine the number of outcomes in the sample space for the experiment.

3. One letter from the alphabet is chosen.

4. A six-sided die is tossed twice and the sum is recorded.

5. Two county supervisors are selected from five supervisors, A, B, C, D, and E, to study a recycling plan.

6. A salesperson makes a presentation about a product in three homes. In each home, there may be a sale (denote by Y) or there may be no sale (denote by N).

Sample Space In Exercises 7–10, list the outcomes in the sample space for the experiment.

7. A taste tester must taste and rank three brands of yogurt, A, B, and C, according to preference.

8. A coin and a die are tossed.

9. A basketball tournament between two teams consists of three games. For each game, your team may win (denote by W) or lose (denote by L).

10. Two students are randomly selected from four students, A, B, C, and D.

In Exercises 11 and 12, you are given the probability that an event *will* occur. Find the probability that the event *will not* occur.

11. $p = 0.35$

12. $p = 0.8$

In Exercises 13 and 14, you are given the probability that an event *will not* occur. Find the probability that the event *will* occur.

13. $p = 0.82$

14. $p = 0.13$

Coin Tossing In Exercises 15–18, a coin is tossed three times. Find the probability of the specified event. Use the sample space

$$S = \{HHH,\ HHT,\ HTH,\ HTT,$$
$$THH,\ THT,\ TTH,\ TTT\}.$$

15. The event of getting two heads

16. The event of getting a tail on the second toss

17. The event of getting at least one head

18. The event of getting no more than two heads

Playing Cards In Exercises 19–22, a card is drawn from a standard deck of 52 playing cards. Find the probability of drawing the indicated card.

19. A red card

20. A queen

21. A face card

22. A black face card

Tossing A Die In Exercises 23–26, a six-sided die is tossed. Find the probability of the specified event.

23. The number is a 5.

24. The number is a 7.

25. The number is no more than 5.

26. The number is at least 1.

United States Blood Types In Exercises 27 and 28, use the circle graph, which shows the percent of people in the United States in 1996 with each blood type. *(Source: America's Blood Centers)*

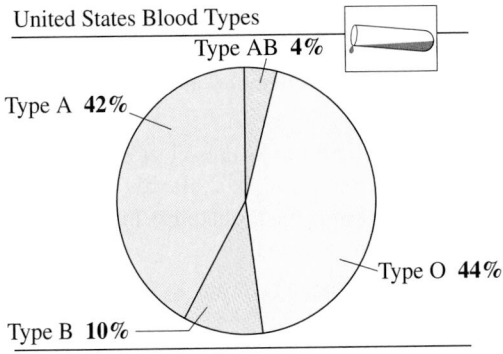

United States Blood Types
Type AB **4%**
Type A **42%**
Type O **44%**
Type B **10%**

27. A person is selected at random from the United States population. What is the probability that the person *does not* have blood type B?

28. What is the probability that a person selected at random from the United States population *does* have blood type B? How is this probability related to the probability found in Exercise 27?

College Freshmen In Exercises 29–32, use the circle graphs, which show the number of incoming college freshmen in each average high school grade category for the years 1970 and 1992.

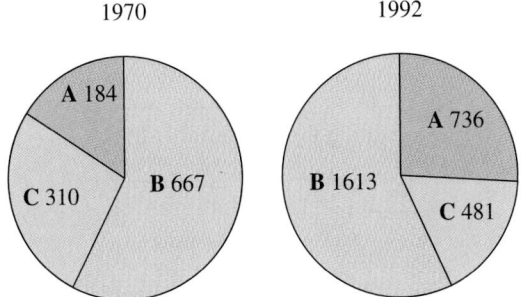

1970
A 184
C 310
B 667

1992
A 736
B 1613
C 481

29. A person is selected at random from the 1970 freshman class of the college. What is the probability that the person's average high school grade was an A?

30. A person is selected at random from the 1992 freshman class of the college. What is the probability that the person's average high school grade was an A?

31. A person is selected at random from the 1970 freshman class of the college. What is the probability that the person's average high school grade was not a C?

32. A person is selected at random from the 1992 freshman class of the college. What is the probability that the person's average high school grade was not a B?

33. *Multiple-Choice Test* A student takes a multiple-choice test in which there are five choices for each question. Find the probability that the first question is answered correctly given the following conditions.

(a) The student has no idea of the answer and guesses at random.

(b) The student can eliminate two of the choices and guesses from the remaining choices.

(c) The student knows the answer.

34. *Multiple-Choice Test* A student takes a multiple-choice test in which there are four choices for each question. Find the probability that the first question is answered correctly given the following conditions.

(a) The student has no idea of the answer and guesses at random.

(b) The student can eliminate two of the choices and guesses from the remaining choices.

(c) The student knows the answer.

35. *Random Selection* Twenty marbles numbered 1 through 20 are placed in a bag, and one is selected. Find the probability of the specified event.

(a) The number is 12.

(b) The number is prime.

(c) The number is odd.

(d) The number is less than 6.

36. *Class Election* Three people are running for class president. It is estimated that the probability that Candidate A will win is 0.5 and the probability that Candidate B will win is 0.3.

(a) What is the probability that *either* Candidate A *or* Candidate B will win?

(b) What is the probability that the third candidate will win?

In Exercises 37–48, some of the sample spaces are large; therefore, you should use the counting principles discussed in Section 10.5.

37. *Game Show*   On a game show, you are given five digits to arrange in the proper order for the price of a car. If you arrange them correctly, you win the car. Find the probability of winning if you know the correct position of only one digit and must guess the positions of the other digits.

38. *Game Show*   On a game show, you are given four digits to arrange in the proper order for the price of a grandfather clock. What is the probability of winning given the following conditions?

(a) You guess the position of each digit.

(b) You know the first digit, but must guess the remaining three.

39. *Lottery*   You buy a lottery ticket inscribed with a five-digit number. On the designated day, five digits (0 though 9 inclusive) are randomly selected. What is the probability that you have a winning ticket?

40. *Shelving Books*   A parent instructs a child to place a five-volume set of books on a bookshelf. Find the probability that the books are placed in the correct order if the child places them at random.

41. *Preparing for a Test*   An instructor gives her class a list of ten study problems, from which she will select eight to be answered on an exam. If you know how to solve eight of the problems, what is the probability that you will be able to answer all eight questions on the exam?

42. *Committee Selection*   A committee of three students is to be selected from a group of three girls and five boys. Find the probability that the committee is composed entirely of girls.

43. *Defective Units*   A shipment of ten food processors to a certain store contains two defective units. If you purchase two of these food processors as birthday gifts for friends, determine the probability that you get both defective units.

44. *Defective Units*   A shipment of 12 compact disc players contains two defective units. A husband and wife buy three of these compact disc players to give to their children as Christmas gifts.

(a) What is the probability that none of the three units is defective?

(b) What is the probability that at least one unit is defective?

45. *Book Selection*   Four books are selected at random from a shelf containing six novels and four autobiographies. Find the probability that the four autobiographies are selected.

46. *Card Selection*   Five cards are selected from a standard deck of 52 cards. Find the probability that four aces are selected.

47. *Card Selection*   Five cards are selected from a standard deck of 52 cards. Find the probability that all hearts are selected.

48. *Card Selection*   Five cards are selected from a standard deck of 52 cards. Find the probability that two aces and three queens are selected.

49. *Girl or Boy?*   The genes that determine the sex of humans are denoted by XX (female) and XY (male). Complete the Punnett square in the figure. Then use the result to explain why it is equally likely that a newborn baby will be a boy or a girl.

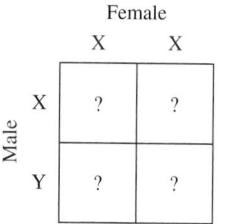
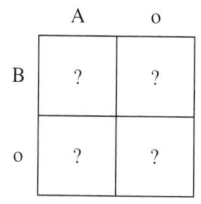

Figure for 49   **Figure for 50**

50. *Blood Types*   There are four basic human blood types: A (AA or Ao), B (BB or Bo), AB (AB), and O (oo). Complete the Punnett square in the figure. What is the blood type of each parent? What is the probability that their offspring will have blood type A? B? AB? O?

51. *Meeting Time*   You and a friend agree to meet at a favorite restaurant between 5:00 P.M. and 6:00 P.M. The one who arrives first will wait 15 minutes for the other, after which the first person will leave (see figure). What is the probability that the two of you actually meet, assuming that your arrival times are random within the hour?

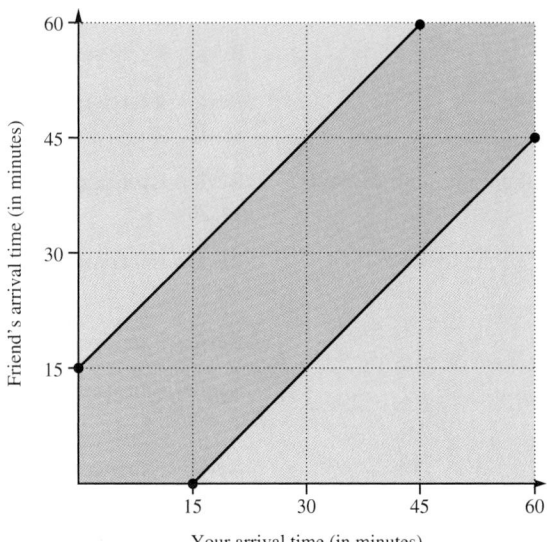

52. *Continuing Education*   In a high school graduating class of 325 students, 255 are going to continue their education. What is the probability that a student selected at random from the class will not be continuing his or her education?

53. *Study Questions*   An instructor gives the class a list of four study questions for the next exam. Two of the four study questions will be on the exam. Find the probability that a student who knows the material relating to three of the four questions will be able to answer both questions selected for the exam.

Section Project

Geometry   A child uses a spring-loaded device to shoot a marble into the square box shown in the figure. The base of the square is horizontal and the marble is equally likely to come to rest at any point on the base. In parts (a)–(d), find the probability that the marble comes to rest in the specified region.

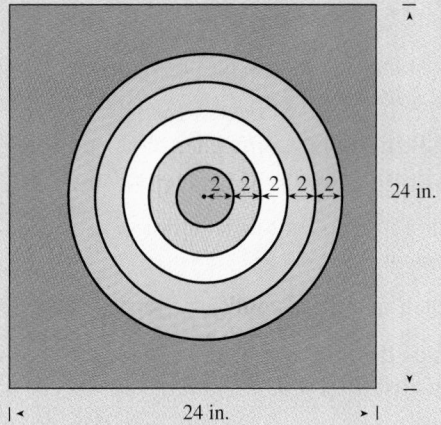

(a) The red center

(b) The blue ring

(c) The purple border

(d) Not in the yellow ring

Chapter Summary

After studying this chapter, you should have acquired the following skills. These skills are keyed to the Review Exercises that begin on page 769. Answers to odd-numbered Review Exercises are given in the back of the book.

• Write sums in sigma notation. *(Section 10.1)*	**Review Exercises 1–4**
• Simplify expressions involving factorials. *(Section 10.1)*	**Review Exercises 5–8**
• Write the first several terms of arithmetic and geometric sequences. *(Sections 10.2, 10.3)*	**Review Exercises 9–12, 15–18**
• Find the common difference of an arithmetic sequence and the common ratio of a geometric sequence. *(Sections 10.2, 10.3)*	**Review Exercises 13, 14, 19, 20**
• Find formulas for the *n*th terms of arithmetic and geometric sequences. *(Sections 10.2, 10.3)*	**Review Exercises 21–30**
• Match sequences with their graphs. *(Sections 10.1–10.3)*	**Review Exercises 31–36**
• Graph the first several terms of sequences using a graphing utility. *(Sections 10.1–10.3)*	**Review Exercises 37–40**
• Evaluate sums expressed in sigma notation. *(Sections 10.1–10.3)*	**Review Exercises 41–56**
• Find and evaluate sums of arithmetic and geometric sequences from verbal statements. *(Sections 10.2, 10.3)*	**Review Exercises 57–62**
• Evaluate binomial coefficients using the Binomial Theorem or Pascal's Triangle. *(Section 10.4)*	**Review Exercises 63–66**
• Evaluate binomial coefficients using a graphing utility. *(Section 10.4)*	**Review Exercises 67–70**
• Expand binomial expressions using the Binomial Theorem. *(Section 10.4)*	**Review Exercises 71–76**
• Find the coefficients of specified terms of binomial expressions. *(Section 10.4)*	**Review Exercises 77, 78**
• Calculate the numbers of ways events can occur using the Fundamental Counting Principle, permutations, or combinations. *(Section 10.5)*	**Review Exercises 79–82**
• Find the probabilities of the occurrences of specified events. *(Section 10.6)*	**Review Exercises 83–88**

Review Exercises

In Exercises 1–4, use sigma notation to write the sum.

1. $[5(1) - 3] + [5(2) - 3] + [5(3) - 3] + [5(4) - 3]$

2. $[9 - 2(1)] + [9 - 2(2)] + [9 - 2(3)] + [9 - 2(4)]$

3. $\dfrac{1}{3(1)} + \dfrac{1}{3(2)} + \dfrac{1}{3(3)} + \dfrac{1}{3(4)} + \dfrac{1}{3(5)} + \dfrac{1}{3(6)}$

4. $\left(-\frac{1}{3}\right)^0 + \left(-\frac{1}{3}\right)^1 + \left(-\frac{1}{3}\right)^2 + \left(-\frac{1}{3}\right)^3 + \left(-\frac{1}{3}\right)^4$

In Exercises 5–8, simplify the expression.

5. $\dfrac{20!}{18!}$

6. $\dfrac{50!}{53!}$

7. $\dfrac{n!}{(n-3)!}$

8. $\dfrac{(n-1)!}{(n+1)!}$

In Exercises 9–12, write the first five terms of the arithmetic sequence. (Begin with $n = 1$.)

9. $a_n = 132 - 5n$

10. $a_n = 2n + 3$

11. $a_n = \frac{3}{4}n + \frac{1}{2}$

12. $a_n = -\frac{3}{5}n + 1$

In Exercises 13 and 14, find the common difference of the arithmetic sequence.

13. $30, 27.5, 25, 22.5, 20, \ldots$

14. $9, 12, 15, 18, 21, \ldots$

In Exercises 15–18, write the first five terms of the geometric sequence.

15. $a_1 = 10, \ r = 3$

16. $a_1 = 2, \ r = -5$

17. $a_1 = 100, \ r = -\frac{1}{2}$

18. $a_1 = 12, \ r = \frac{1}{6}$

In Exercises 19 and 20, find the common ratio of the geometric sequence.

19. $8, 12, 18, 27, \frac{81}{2}, \ldots$

20. $81, -54, 36, -24, 16, \ldots$

In Exercises 21–30, find a formula for the nth term of the specified sequence.

21. Arithmetic sequence: $a_1 = 10, \ d = 4$

22. Arithmetic sequence: $a_1 = 32, \ d = -2$

23. Arithmetic sequence: $a_1 = 1000, \ a_2 = 950$

24. Arithmetic sequence: $a_1 = 12, \ a_2 = 20$

25. Geometric sequence: $a_1 = 1, \ r = -\frac{2}{3}$

26. Geometric sequence: $a_1 = 100, \ r = 1.07$

27. Geometric sequence: $a_1 = 24, \ a_2 = 48$

28. Geometric sequence: $a_1 = 16, \ a_2 = -4$

29. Geometric sequence: $a_1 = 12, \ a_4 = -\frac{3}{2}$

30. Geometric sequence: $a_2 = 1, \ a_3 = \frac{1}{3}$

In Exercises 31–36, match the sequence with its graph. [The graphs are labeled (a), (b), (c), (d), (e), and (f).]

(a)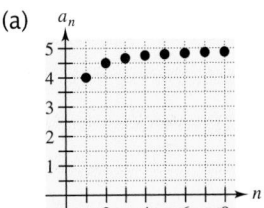

(b)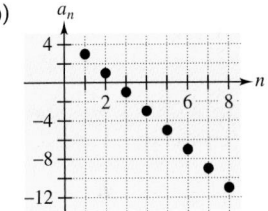

(c)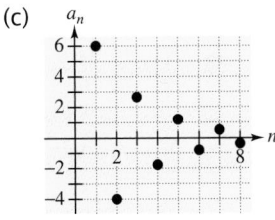

(d)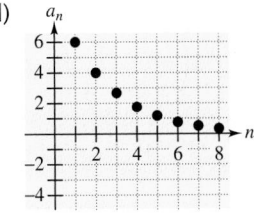

(e)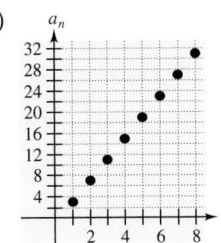

(f)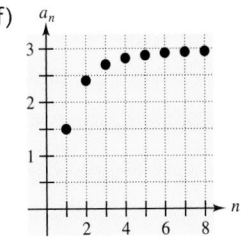

31. $a_n = 5 - \dfrac{1}{n}$

32. $a_n = \dfrac{3n^2}{n^2 + 1}$

33. $a_n = 5 - 2n$

34. $a_n = 4n - 1$

35. $a_n = 6\left(\dfrac{2}{3}\right)^{n-1}$

36. $a_n = 6\left(-\dfrac{2}{3}\right)^{n-1}$

In Exercises 37–40, use a graphing utility to graph the first ten terms of the sequence.

37. $a_n = \dfrac{3n}{n + 1}$

38. $a_n = \dfrac{3}{n + 1}$

39. $a_n = 5\left(\dfrac{3}{4}\right)^{n-1}$

40. $a_n = 5\left(-\dfrac{3}{4}\right)^{n-1}$

In Exercises 41–56, evaluate the sum.

41. $\displaystyle\sum_{k=1}^{4} 7$

42. $\displaystyle\sum_{k=1}^{4} \dfrac{(-1)^k}{k}$

43. $\displaystyle\sum_{n=1}^{4} \left(\dfrac{1}{n} - \dfrac{1}{n + 1}\right)$

44. $\displaystyle\sum_{n=1}^{4} \left(\dfrac{1}{n} - \dfrac{1}{n + 2}\right)$

45. $\displaystyle\sum_{k=1}^{12} (7k - 5)$

46. $\displaystyle\sum_{k=1}^{10} (100 - 10k)$

47. $\displaystyle\sum_{j=1}^{100} \dfrac{j}{4}$

48. $\displaystyle\sum_{j=1}^{50} \dfrac{3j}{2}$

49. $\displaystyle\sum_{n=1}^{12} 2^n$

50. $\displaystyle\sum_{n=1}^{12} (-2)^n$

51. $\displaystyle\sum_{k=1}^{8} 5\left(-\dfrac{3}{4}\right)^k$

52. $\displaystyle\sum_{k=1}^{10} 4\left(\dfrac{3}{2}\right)^k$

53. $\displaystyle\sum_{i=1}^{8} (1.25)^{i-1}$

54. $\displaystyle\sum_{i=1}^{8} (-1.25)^{i-1}$

55. $\displaystyle\sum_{n=1}^{120} 500(1.01)^n$

56. $\displaystyle\sum_{n=1}^{40} 1000(1.1)^n$

57. Find the sum of the first 50 positive integers that are multiples of 4.

58. Find the sum of the integers from 225 to 300.

59. *Auditorium Seating* Each row in a small auditorium has three more seats than the preceding row. Find the seating capacity of the auditorium if the front row seats 22 people and there are 12 rows of seats.

60. *Depreciation* A company pays $120,000 for a machine. During the next 5 years, the machine depreciates at a rate of 30% per year. (That is, at the end of each year, the depreciated value is 70% of what it was at the beginning of the year.)

(a) Find a formula for the *n*th term of the geometric sequence that gives the value of the machine *n* full years after it was purchased.

(b) Find the depreciated value of the machine at the end of 5 full years.

61. *Population Increase* A city of 85,000 people is growing at a rate of 1.2% per year. (That is, at the end of each year, the population is 1.012 times what it was at the beginning of the year.)

(a) Find a formula for the *n*th term of the geometric sequence that gives the population *n* years from now.

(b) Estimate the population 50 years from now.

62. *Salary Increase* You accept a job that pays a salary of $32,000 the first year. During the next 39 years, you receive a 5.5% raise each year. What is your total salary over the 40-year period?

In Exercises 63–66, evaluate $_nC_r$.

63. $_8C_3$

64. $_{12}C_2$

65. $_{12}C_0$

66. $_{100}C_1$

In Exercises 67–70, use a graphing utility to evaluate $_nC_r$.

67. $_{40}C_4$

68. $_{15}C_9$

69. $_{25}C_6$

70. $_{32}C_2$

In Exercises 71–76, use the Binomial Theorem to expand the expression. Simplify the result.

71. $(x + 1)^{10}$ **72.** $(u - v)^9$

73. $(y - 2)^6$ **74.** $(x + 3)^5$

75. $\left(\frac{1}{2} - x\right)^8$ **76.** $(3x - 2y)^4$

In Exercises 77 and 78, find the coefficient of the given term of the expression.

Expression	Term
77. $(x - 3)^{10}$	x^5
78. $(2x - 3y)^5$	$x^2 y^3$

79. *Morse Code*   In Morse code, all characters are transmitted using a sequence of *dots* and *dashes*. How many different characters can be formed using a sequence of three dots and dashes? (These can be repeated. For example, dash-dot-dot represents the letter *d*.)

80. *Forming Line Segments*   How many line segments can be formed by five points of which no three are collinear?

81. *Committee Selection*   Determine the number of ways a committee of five people can be formed from a group of 15 people.

82. *Program Listing*   There are seven participants in a piano recital. In how many orders can their names be listed in the program?

83. *Rolling a Die*   Find the probability of obtaining a number greater than 4 when a six-sided die is rolled.

84. *Coin Tossing*   Find the probability of obtaining at least one head when a coin is tossed four times.

85. *Shelving Books*   A child who does not know how to read carries a four-volume set of books to a bookshelf. Find the probability that the child will put the books on the shelf in the correct order.

86. *Rolling a Die*   Are the chances of rolling a 3 with one six-sided die the same as the chances of rolling a total of 6 with two six-sided dice? If not, which has the greater probability of occurring?

87. *Hospital Inspection*   As part of a monthly inspection at a hospital, the inspection team randomly selects reports from eight of the 84 nurses who are on duty. What is the probability that none of the reports selected will be from the ten most experienced nurses?

88. *Target Shooting*   An archer shoots an arrow at the target shown in the figure. Suppose that the arrow is equally likely to hit any point on the target. What is the probability that the arrow hits the bull's-eye? What is the probability that the arrow hits the blue ring?

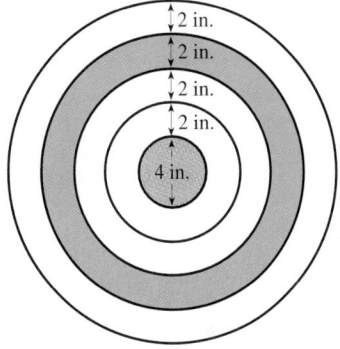

Chapter Test

Take this test as you would take a test in class. After you are done, check your work against the answers given in the back of the book.

1. Write the first five terms of the sequence $a_n = \left(-\frac{2}{3}\right)^{n-1}$. (Begin with $n = 1$.)

In Exercises 2–5, evaluate the sum.

2. $\displaystyle\sum_{j=0}^{4} (3j + 1)$ **3.** $\displaystyle\sum_{n=1}^{5} (3 - 4n)$ **4.** $\displaystyle\sum_{n=1}^{8} 2(2^n)$ **5.** $\displaystyle\sum_{n=1}^{10} 3\left(\frac{1}{2}\right)^n$

6. Use sigma notation to write $\dfrac{2}{3(1) + 1} + \dfrac{2}{3(2) + 1} + \cdots + \dfrac{2}{3(12) + 1}$.

7. Write the first five terms of the arithmetic sequence whose first term is $a_1 = 12$ and whose common difference is $d = 4$.

8. Find a formula for the nth term of the arithmetic sequence whose first term is $a_1 = 5000$ and whose common difference is $d = -100$.

9. Find the sum of the first 50 positive integers that are multiples of 3.

10. Find the common ratio of the geometric sequence $2, -3, \frac{9}{2}, -\frac{27}{4}, \ldots$.

11. Find a formula for the nth term of the geometric sequence whose first term is $a_1 = 4$ and whose common ratio is $r = \frac{1}{2}$.

12. Fifty dollars is deposited each month in an increasing annuity that pays 8%, compounded monthly. What is the balance after 25 years?

13. Evaluate $_{20}C_3$. Use a graphing utility to verify the result.

14. Use Pascal's Triangle to expand $(x - 2)^5$.

15. Find the coefficient of the term $x^3 y^5$ in the expansion of $(x + y)^8$.

16. How many license plates can consist of one letter followed by three digits?

17. Four students are randomly selected from a class of 25 to answer questions from a reading assignment. In how many ways can the four be selected?

18. The weather report indicates that the probability of snow tomorrow is 0.75. What is the probability that it will not snow?

19. A card is drawn from a standard deck of 52 cards. Find the probability that a red face card is drawn.

20. Suppose two spark plugs require replacement in a four-cylinder engine. If the mechanic randomly removes two plugs, find the probability that they are the two defective plugs.

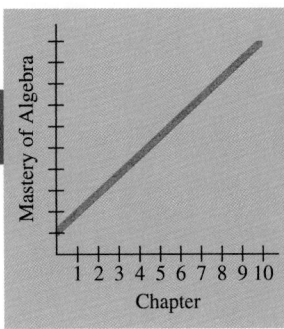

Mastery of Algebra

Chapter

Take this test as you would take a test in class. After you are done, check your work against the answers given in the back of the book.

In Exercises 1–6, simplify the expression.

1. $(-3x^2y^3)^2 \cdot (4xy^2)$

2. $\dfrac{3}{8}x - \dfrac{1}{12}x + 8$

3. $\dfrac{64r^2s^4}{16rs^2}$, $r \neq 0, s \neq 0$

4. $\left(\dfrac{3x}{4y^3}\right)^2$, $y \neq 0$

5. $\dfrac{8}{\sqrt{10}}$

6. $\log_4 64$

In Exercises 7–10, factor the expression completely.

7. $5x - 20x^2$

8. $64 - (x - 6)^2$

9. $15x^2 - 16x - 15$

10. $8x^3 + 1$

In Exercises 11–14, sketch the graph of the equation.

11. $y = 2x - 3$

12. $y = -\dfrac{3}{4}x + 2$

13. $9x^2 + 4y^2 = 36$

14. $(x - 2)y = 5$

15. Write the equation of the line through the points $\left(\frac{3}{2}, 8\right)$ and $\left(\frac{11}{2}, \frac{5}{2}\right)$.

16. Find the domain of the function $h(x) = \sqrt{16 - x^2}$.

17. Find $g(c - 6)$ if $g(x) = \dfrac{x}{x + 10}$.

18. The number N of prey t months after a predator is introduced into an area is inversely proportional to $t + 1$. If $N = 300$ when $t = 0$, find N when $t = 5$.

19. Find an equation of the parabola shown in the figure.

20. Annual sales y of a product x years after it is introduced is approximated by

$$y = \dfrac{10{,}000}{1 + 4e^{-x/3}}.$$

(a) Use a graphing utility to graph the equation.

(b) Use the graph in part (a) to approximate the time when annual sales are $y = 5000$ units.

(c) Use the graph is part (a) to estimate the maximum level that annual sales will approach.

21. Use the Binomial Theorem to expand and simplify $(z - 3)^4$.

22. On a game show, the digits 3, 4, and 5 must be arranged in the proper order to form the price of an appliance and thus win the appliance. What is the probability of winning if the contestant knows that the price is at least $400?

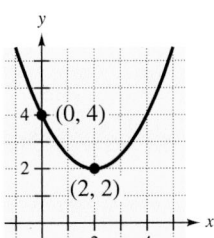

Figure for 19

773

Appendix A   Introduction to Logic

<table>
<tr><td>**A.1**</td><td>**Introduction to Logic**</td></tr>
</table>

Statements • Truth Tables

Statements

In everyday speech and in mathematics you make inferences that adhere to common **laws of logic.** These methods of reasoning allow you to build an algebra of statements by using logical operations to form compound statements from simpler ones. A primary goal of logic is to determine the truth value (true or false) of a compound statement knowing the truth value of its simpler components. For instance, the compound statement "The temperature is below freezing and it is snowing" is true only if both component statements are true.

Definition of a Statement

1. A **statement** is a sentence to which only one truth value (either true or false) can be meaningfully assigned.

2. An **open statement** is a sentence that contains one or more variables and becomes a statement when each variable is replaced by a specific item from a designated set.

NOTE   In this definition, the word *statement* can be replaced by the word *proposition.*

EXAMPLE 1   *Statements, Nonstatements, and Open Statements*

Statement	*Truth Value*
A square is a rectangle.	**T**
-3 is less than -5.	**F**

Nonstatement	*Truth Value*
Do your homework.	No truth value can be meaningfully assigned.
Did you call the police?	No truth value can be meaningfully assigned.

Open Statement	*Truth Value*
x is an irrational number.	We need a value of x.
She is a computer science major.	We need a specific person.

Symbolically, statements are represented by lowercase letters p, q, r, and so on. Statements can be changed or combined to form **compound statements** by means of the three logical operations **and, or,** and **not,** which are represented by $\wedge$ (and), $\vee$ (or), and $\sim$ (not). In logic the word *or* is used in the *inclusive* sense (meaning "and/or" in everyday language). That is, the statement "p or q" is true if p is true, q is true, or both p and q are true. The following list summarizes the terms and symbols used with these three operations of logic.

Operations of Logic

Operation	Verbal Statement	Symbolic Form	Name of Operation
~	not p	$\sim p$	**Negation**
$\wedge$	p and q	$p \wedge q$	**Conjunction**
$\vee$	p or q	$p \vee q$	**Disjunction**

Compound statements can be formed using more than one logical operation, as demonstrated in Example 2.

EXAMPLE 2   Forming Negations and Compound Statements

The statements p and q are as follows.

   $p:$ The temperature is below freezing.

   $q:$ It is snowing.

Write the verbal form for each of the following.

(a) $p \wedge q$ (b) $\sim p$ (c) $\sim(p \vee q)$ (d) $\sim p \wedge \sim q$

Solution

(a) The temperature is below freezing and it is snowing.

(b) The temperature is not below freezing.

(c) It is not true that the temperature is below freezing or it is snowing.

(d) The temperature is not below freezing and it is not snowing.

EXAMPLE 3   Forming Compound Statements

The statements p and q are as follows.

   $p:$ The temperature is below freezing.

   $q:$ It is snowing.

(a) Write the symbolic form for: *The temperature is not below freezing or it is not snowing.*

(b) Write the symbolic form for: *It is not true that the temperature is below freezing and it is snowing.*

Solution

(a) The symbolic form is: $\sim p \vee \sim q$ (b) The symbolic form is: $\sim(p \wedge q)$

Truth Tables

To determine the truth value of a compound statement, you can create charts called **truth tables.** These tables represent the three basic logical operations.

Negation

p	q	$\sim p$	$\sim q$
T	T	F	F
T	F	F	T
F	T	T	F
F	F	T	T

Conjunction

p	q	$p \wedge q$
T	T	T
T	F	F
F	T	F
F	F	F

Disjunction

p	q	$p \vee q$
T	T	T
T	F	T
F	T	T
F	F	F

For the sake of uniformity, all truth tables with two component statements will have **T** and **F** values for p and q assigned in the order shown in the first two columns of each of these three tables. Truth tables for several operations can be combined into one chart by using the same two first columns. For each operation, a new column is added. Such an arrangement is especially useful with compound statements that involve more than one logical operation and for showing that two statements are logically equivalent.

Logical Equivalence

Two compound statements are **logically equivalent** if they have identical truth tables. Symbolically, we denote the equivalence of the statements p and q by writing $p \equiv q$.

EXAMPLE 4 *Logical Equivalence*

Use a truth table to show the logical equivalence of the statements $\sim p \wedge \sim q$ and $\sim(p \vee q)$.

Solution

p	q	$\sim p$	$\sim q$	$\sim p \wedge \sim q$	$p \vee q$	$\sim(p \vee q)$
T	T	F	F	F	T	F
T	F	F	T	F	T	F
F	T	T	F	F	T	F
F	F	T	T	T	F	T

↑——— Identical ———↑

Because the fifth and seventh columns in the table are identical, the two given statements are logically equivalent.

The equivalence established in Example 4 is one of two well-known rules in logic called **DeMorgan's Laws.** Verification of the second of DeMorgan's Laws is left as an exercise.

DeMorgan's Laws	
1. $\sim(p \vee q) \equiv \sim p \wedge \sim q$	2. $\sim(p \wedge q) \equiv \sim p \vee \sim q$

$p \vee \sim p$ **is a tautology**

p	$\sim p$	$p \vee \sim p$
T	F	T
F	T	T

Compound statements that are true, no matter what the truth values of component statements, are called **tautologies.** One simple example is the statement "p or not p," as shown in the table to the left.

A.1   **Exercises**

In Exercises 1–12, classify the sentence as a statement, a nonstatement, or an open statement.

1. All dogs are brown.
2. Can I help you?
3. That figure is a circle.
4. Substitute 4 for x.
5. x is larger than 4.
6. 8 is larger than 4.
7. $x + y = 10$
8. $12 + 3 = 14$
9. Hockey is fun to watch.
10. One mile is greater than 1 kilometer.
11. It is more than 1 mile to the school.
12. Come to the party.

In Exercises 13–20, determine whether the open statement is true for the given values of x.

Open Statement	Values of x	
13. $x^2 - 5x + 6 = 0$	(a) $x = 2$	(b) $x = -2$
14. $x^2 - x - 6 = 0$	(a) $x = 2$	(b) $x = -2$
15. $x^2 \leq 4$	(a) $x = -2$	(b) $x = 0$
16. $|x - 3| = 4$	(a) $x = -1$	(b) $x = 7$
17. $4 - |x| = 2$	(a) $x = 0$	(b) $x = 1$
18. $\sqrt{x^2} = x$	(a) $x = 3$	(b) $x = -3$
19. $\dfrac{x}{x} = 1$	(a) $x = -4$	(b) $x = 0$
20. $\sqrt[3]{x} = -2$	(a) $x = 8$	(b) $x = -8$

In Exercises 21–24, write the verbal form for each of the following.

(a) $\sim p$ (b) $\sim q$ (c) $p \wedge q$ (d) $p \vee q$

21. p: The sun is shining.
    q: It is hot.
22. p: The car has a radio.
    q: The car is red.
23. p: Lions are mammals.
    q: Lions are carnivorous.
24. p: Twelve is less than 15.
    q: Seven is a prime number.

In Exercises 25–28, write the verbal form for each of the following.

(a) $\sim p \wedge q$ (b) $\sim p \vee q$ (c) $p \wedge \sim q$ (d) $p \vee \sim q$

25. p: The sun is shining.
    q: It is hot.
26. p: The car has a radio.
    q: The car is red.
27. p: Lions are mammals.
    q: Lions are carnivorous.
28. p: Twelve is less than 15.
    q: Seven is a prime number.

In Exercises 29–32, write the symbolic form of the given compound statement. In each case, let p represent the statement "It is four o'clock," and let q represent the statement "It is time to go home."

29. It is four o'clock and it is not time to go home.

30. It is not four o'clock or it is not time to go home.

31. It is not four o'clock or it is time to go home.

32. It is four o'clock and it is time to go home.

In Exercises 33–36, write the symbolic form of the given compound statement. In each case, let p represent the statement "The dog has fleas," and let q represent the statement "The dog is scratching."

33. The dog does not have fleas or the dog is not scratching.

34. The dog has fleas and the dog is scratching.

35. The dog does not have fleas and the dog is scratching.

36. The dog has fleas or the dog is not scratching.

In Exercises 37–42, write the negation of the given statement.

37. The bus is not blue.

38. Frank is not 6 feet tall.

39. x is equal to 4.

40. x is not equal to 4.

41. The earth is not flat.

42. The earth is flat.

In Exercises 43–48, construct a truth table for the given compound statement.

43. $\sim p \wedge q$ **44.** $\sim p \vee q$

45. $\sim p \vee \sim q$ **46.** $\sim p \wedge \sim q$

47. $p \vee \sim q$ **48.** $p \wedge \sim q$

In Exercises 49–54, use a truth table to determine whether the given statements are logically equivalent.

49. $\sim p \wedge q,\ \ p \vee \sim q$

50. $\sim (p \wedge \sim q),\ \ \sim p \vee q$

51. $\sim (p \vee \sim q),\ \ \sim p \wedge q$

52. $\sim (p \vee q),\ \ \sim p \vee \sim q$

53. $p \wedge \sim q,\ \ \sim (\sim p \vee q)$

54. $p \wedge \sim q,\ \ \sim (\sim p \wedge q)$

In Exercises 55–58, determine whether the statements are logically equivalent.

55. (a) The house is red and it is not made of wood.

(b) The house is red or it is not made of wood.

56. (a) It is not true that the tree is not green.

(b) The tree is green.

57. (a) The statement that the house is white or blue is not true.

(b) The house is not white and it is not blue.

58. (a) I am not 25 years old and I am not applying for this job.

(b) The statement that I am 25 years old and applying for this job is not true.

In Exercises 59–62, use a truth table to determine whether the given statement is a tautology.

59. $\sim p \wedge p$

60. $\sim p \vee p$

61. $\sim (\sim p) \vee \sim p$

62. $\sim (\sim p) \wedge \sim p$

63. Use a truth table to verify the second of DeMorgan's Laws:

$$\sim (p \wedge q) \equiv \sim p \vee \sim q$$

A.2	**Implications, Quantifiers, and Venn Diagrams**

Implications ▪ Logical Quantifiers ▪
Venn Diagrams

Implications

A statement of the form "If p, then q" is called an **implication** (or a conditional statement) and is denoted by

$$p \rightarrow q.$$

The letter p is the **hypothesis** and the letter q is the **conclusion.** There are many different ways to express the implication $p \rightarrow q$, as shown in the following list.

Different Ways of Stating Implications

The implication $p \rightarrow q$ has the following equivalent verbal forms.

1. If p, then q.   2. p implies q.   3. p, only if q.

4. q follows from p.   5. q is necessary for p.   6. p is sufficient for q.

Normally, you think of the implication $p \rightarrow q$ as having a cause-and-effect relationship between the hypothesis p and the conclusion q. However, you should be careful not to confuse the truth value of the component statements with the truth value of the implication. The following truth table should help you keep this distinction in mind.

Note in the table on the left that the implication $p \rightarrow q$ is false only when p is true and q is false. This is like a promise. Suppose you promise a friend that "If the sun shines, I will take you fishing." The only way your promise can be broken is if the sun shines (p is true) and you do not take your friend fishing (q is false). If the sun doesn't shine (p is false), you have no obligation to go fishing, and hence, the promise cannot be broken.

Implication

p	q	$p \rightarrow q$
T	T	T
T	F	F
F	T	T
F	F	T

EXAMPLE 1   Finding Truth Values of Implications

Give the truth value of each implication.

(a) If 3 is odd, then 9 is odd.   (b) If 3 is odd, then 9 is even.

(c) If 3 is even, then 9 is odd.   (d) If 3 is even, then 9 is even.

Solution

	Hypothesis	Conclusion	Implication
(a)	**T**	**T**	**T**
(b)	**T**	**F**	**F**
(c)	**F**	**T**	**T**
(d)	**F**	**F**	**T**

The next example shows how to write an implication as a disjunction.

EXAMPLE 2   *Identifying Equivalent Statements*

Use a truth table to show the logical equivalence of the following statements.

(a) If I get a raise, I will take my family on a vacation.

$$p \rightarrow q \equiv \sim p \lor q$$

(b) I will not get a raise *or* I will take my family on a vacation.

p	q	$\sim p$	$\sim p \lor q$	$p \rightarrow q$
T	T	F	T	T
T	F	F	F	F
F	T	T	T	T
F	F	T	T	T

$\llcorner$ Identical $\lrcorner$

Solution

Let p represent the statement "I will get a raise," and let q represent the statement "I will take my family on a vacation." Then, you can represent the statement in part (a) as $p \rightarrow q$ and the statement in part (b) as $\sim p \lor q$. The logical equivalence of these two statements is shown at the left in the truth table.

Because the fourth and fifth columns of the truth table are identical, you can conclude that the two statements $p \rightarrow q$ and $\sim p \lor q$ are equivalent.

From the table in Example 2 and the fact that $\sim(\sim p) \equiv p$, you can write the **negation of an implication.** That is, because $p \rightarrow q$ is equivalent to $\sim p \lor q$, it follows that the negation of $p \rightarrow q$ must be $\sim(\sim p \lor q)$, which by DeMorgan's Laws can be written as follows.

$$\sim(p \rightarrow q) \equiv p \land \sim q$$

For the implication $p \rightarrow q$, there are three important associated implications.

1. The **converse** of $p \rightarrow q$: $q \rightarrow p$

2. The **inverse** of $p \rightarrow q$: $\sim p \rightarrow \sim q$

3. The **contrapositive** of $p \rightarrow q$: $\sim q \rightarrow \sim p$

NOTE   The connective "$\rightarrow$" is used to determine the truth values in the last three columns of the table.

From the table below you can see that these four statements yield two pairs of logically equivalent implications.

p	q	$\sim p$	$\sim q$	$p \rightarrow q$	$\sim q \rightarrow \sim p$	$q \rightarrow p$	$\sim p \rightarrow \sim q$
T	T	F	F	T	T	T	T
T	F	F	T	F	F	T	T
F	T	T	F	T	T	F	F
F	F	T	T	T	T	T	T

⌞ Identical ⌟   ⌞ Identical ⌟

NOTE   In Example 3, be sure you see that neither the converse nor the inverse is logically equivalent to the original implication. To see this, consider that the original implication simply states that if you get a B on your test, then you will pass the course. The converse is not true, because knowing that you passed the course does not imply that you got a B on the test. After all, you might have gotten an A on the test.

EXAMPLE 3   Writing the Converse, Inverse, and Contrapositive

Write the converse, inverse, and contrapositive for the implication "If I get a B on my test, then I will pass the course."

Solution

(a) *Converse:* If I pass the course, then I got a B on my test.

(b) *Inverse:* If I do not get a B on my test, then I will not pass the course.

(c) *Contrapositive:* If I do not pass the course, then I did not get a B on my test.

A **biconditional statement,** denoted by $p \leftrightarrow q$, is the conjunction of the implications $p \rightarrow q$ and $q \rightarrow p$. A biconditional statement is written as "p if and only if q," or in shorter form as "p iff q." A biconditional statement is true when both components are true and when both components are false, as shown in the following truth table.

Biconditional Statement: p if and only if q

p	q	$p \rightarrow q$	$q \rightarrow p$	$p \leftrightarrow q$	$(p \rightarrow q) \wedge (q \rightarrow p)$
T	T	T	T	T	T
T	F	F	T	F	F
F	T	T	F	F	F
F	F	T	T	T	T

The following list summarizes some of the laws of logic that have been discussed up to this point.

Laws of Logic

1. For every statement p, either p is true or p is false.	Law of Excluded Middle	2. $\sim(\sim p) \equiv p$	Law of Double Negation
3. $\sim(p \vee q) \equiv \sim p \wedge \sim q$	DeMorgan's Law	4. $\sim(p \wedge q) \equiv \sim p \vee \sim q$	DeMorgan's Law
5. $p \rightarrow q \equiv \sim p \vee q$	Law of Implication	6. $p \rightarrow q \equiv \sim q \rightarrow \sim p$	Law of Contraposition

Logical Quantifiers

Logical quantifiers are words such as *some, all, every, each, one,* and *none.* Here are some examples of statements with quantifiers.

Some isosceles triangles are right triangles.

Every painting on display is for sale.

Not all corporations have male chief executive officers.

All squares are parallelograms.

Being able to recognize the negation of a statement involving a quantifier is one of the most important skills in logic. For instance, consider the statement "All dogs are brown." In order for this statement to be false, we do not have to show that *all* dogs are not brown, we must simply find at least one dog that is not brown. Thus, the negation of the statement is "Some dogs are not brown."

Next we list some of the more common negations involving quantifiers.

Negating Statements with Quantifiers

Statement	*Negation*	*Statement*	*Negation*
1. All p are q.	Some p are not q.	2. Some p are q.	No p is q.
3. Some p are not q.	All p are q.	4. No p is q.	Some p are q.

When logical quantifiers are used, the word *all* can be replaced by the word *each* or the word *every.* For instance, the following are equivalent.

All p are q.   Each p is q.   Every p is q.

Similarly, the word *some* can be replaced by the words *at least one.* For instance, the following are equivalent.

Some p are q.   At least one p is q.

EXAMPLE 4   *Negating Quantifying Statements*

Write the negation of each of the following.

(a) All students study.

(b) Not all prime numbers are odd.

(c) At least one mammal can fly.

(d) Some bananas are not yellow.

Solution

(a) Some students do not study.

(b) All prime numbers are odd.

(c) No mammal can fly.

(d) All bananas are yellow.

Venn Diagrams

Venn diagrams are figures that are used to show relationships between two or more sets of objects. They can help you interpret quantifying statements. Study the following Venn diagrams in which the circle marked *A* represents people over 6 feet tall and the circle marked *B* represents the basketball players.

1. All basketball players are over 6 feet tall.

2. Some basketball players are over 6 feet tall.

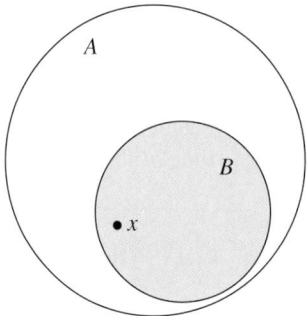

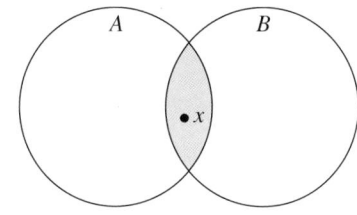

3. Some basketball players are not over 6 feet tall.

4. No basketball player is over 6 feet tall.

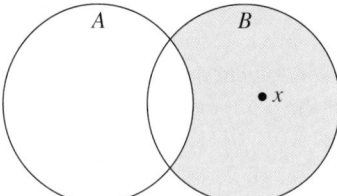

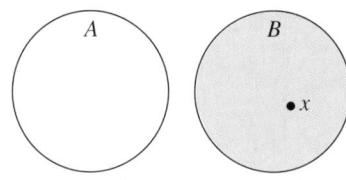

A.2   **Exercises**

In Exercises 1–4, write the verbal form for each of the following.

(a) $p \to q$ (b) $q \to p$ (c) $\sim q \to \sim p$ (d) $p \to \sim q$

1. *p:* The engine is running.

   q: The engine is wasting gasoline.

2. *p:* The student is at school.

   q: It is nine o'clock.

3. *p:* The integer is even.   **4.** *p:* The person is generous.

   q: It is divisible by 2.   *q:* The person is rich.

In Exercises 5–10, write the symbolic form of the compound statement. Let *p* represent the statement "The economy is expanding," and let *q* represent the statement "Interest rates are low."

5. If interest rates are low, then the economy is expanding.

6. If interest rates are not low, then the economy is not expanding.

7. An expanding economy implies low interest rates.

8. Low interest rates are sufficient for an expanding economy.

9. Low interest rates are necessary for an expanding economy.

10. The economy will expand only if interest rates are low.

In Exercises 11–20, give the truth value of the implication.

11. If 4 is even, then 12 is even.

12. If 4 is even, then 2 is odd.

13. If 4 is odd, then 3 is odd.

14. If 4 is odd, then 2 is odd.

15. If $2n$ is even, then $2n + 2$ is odd.

16. If $2n + 1$ is even, then $2n + 2$ is odd.

17. $3 + 11 > 16$ only if $2 + 3 = 5$.

18. $\frac{1}{6} < \frac{2}{3}$ is necessary for $\frac{1}{2} > 0$.

19. $x = -2$ follows from $2x + 3 = x + 1$.

20. If $2x = 224$, then $x = 10$.

In Exercises 21–26, write the converse, inverse, and contrapositive of the statement.

21. If the sky is clear, then you can see the eclipse.

22. If the person is nearsighted, then he is ineligible for the job.

23. If taxes are raised, then the deficit will increase.

24. If wages are raised, then the company's profits will decrease.

25. It is necessary to have a birth certificate to apply for the visa.

26. The number is divisible by 3 only if the sum of its digits is divisible by 3.

In Exercises 27–40, write the negation of the statement.

27. Paul is a junior or a senior.

28. Jack is a senior and he plays varsity basketball.

29. If the temperature increases, then the metal rod will expand.

30. If the test fails, then the project will be halted.

31. We will go to the ocean only if the weather forecast is good.

32. Completing the pass on this play is necessary if we are to win the game.

33. Some students are in extracurricular activities.

34. Some odd integers are not prime numbers.

35. All contact sports are dangerous.

36. All members must pay their dues prior to June 1.

37. No child is allowed at the concert.

38. No contestant is over the age of 12.

39. At least one of the $20 bills is counterfeit.

40. At least one unit is defective.

In Exercises 41–48, construct a truth table for the compound statement.

41. $\sim(p \rightarrow \sim q)$

42. $\sim q \rightarrow (p \rightarrow q)$

43. $\sim(q \rightarrow p) \wedge q$

44. $p \rightarrow (\sim p \vee q)$

45. $[(p \vee q) \wedge (\sim p)] \rightarrow q$

46. $[(p \rightarrow q) \wedge (\sim q)] \rightarrow p$

47. $(p \leftrightarrow \sim q) \rightarrow \sim p$

48. $(p \vee \sim q) \leftrightarrow (q \rightarrow \sim p)$

In Exercises 49–56, use a truth table to show the logical equivalence of the two statements.

49. $q \rightarrow p$ $\sim p \rightarrow \sim q$

50. $\sim p \rightarrow q$ $p \vee q$

51. $\sim(p \rightarrow q)$ $p \wedge \sim q$

52. $(p \vee q) \rightarrow q$ $p \rightarrow q$

53. $(p \rightarrow q) \vee \sim q$ $p \vee \sim p$

54. $q \rightarrow (\sim p \vee q)$ $q \vee \sim q$

55. $p \rightarrow (\sim p \wedge q)$ $\sim p$

56. $\sim(p \wedge q) \rightarrow \sim q$ $p \vee \sim q$

57. Select the statement that is logically equivalent to the statement "If a number is divisible by 6, then it is divisible by 2."

(a) If a number is divisible by 2, then it is divisible by 6.

(b) If a number is not divisible by 6, then it is not divisible by 2.

(c) If a number is not divisible by 2, then it is not divisible by 6.

(d) Some numbers are divisible by 6 and not divisible by 2.

58. Select the statement that is logically equivalent to the statement "It is not true that Pam is a conservative and a Democrat."

(a) Pam is a conservative and a Democrat.

(b) Pam is not a conservative and not a Democrat.

(c) Pam is not a conservative or she is not a Democrat.

(d) If Pam is not a conservative, then she is a Democrat.

59. Select the statement that is *not* logically equivalent to the statement "Every citizen over the age of 18 has the right to vote."

(a) Some citizens over the age of 18 have the right to vote.

(b) Each citizen over the age of 18 has the right to vote.

(c) All citizens over the age of 18 have the right to vote.

(d) No citizen over the age of 18 can be restricted from voting.

60. Select the statement that is *not* logically equivalent to the statement "It is necessary to pay the registration fee to take the course."

(a) If you take the course, then you must pay the registration fee.

(b) If you do not pay the registration fee, then you cannot take the course.

(c) If you pay the registration fee, then you may take the course.

(d) You may take the course only if you pay the registration fee.

In Exercises 61–70, sketch a Venn diagram and shade the region that illustrates the given statement. Let *A* be a circle that represents people who are happy, and let *B* be a circle that represents college students.

61. All college students are happy.

62. All happy people are college students.

63. No college students are happy.

64. No happy people are college students.

65. Some college students are not happy.

66. Some happy people are not college students.

67. At least one college student is happy.

68. At least one happy person is not a college student.

69. Each college student is sad.

70. Each sad person is not a college student.

In Exercises 71–74, state whether the statement follows from the given Venn diagram. Assume that each area shown in the Venn diagram is nonempty. (*Note:* Use only the information given in the diagram. Do not be concerned with whether the statement is actually true or false.)

71. (a) All toads are green.
    (b) Some toads are green.

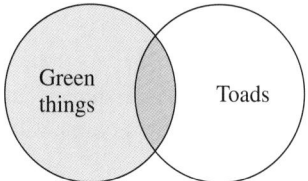

72. (a) All men are company presidents.
    (b) Some company presidents are women.

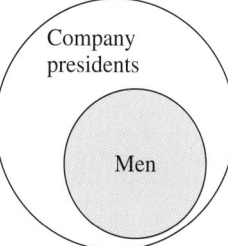

73. (a) All blue cars are old.
    (b) Some blue cars are not old.

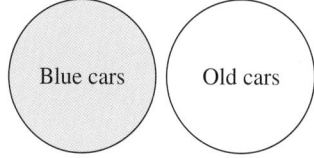

74. (a) No football player is over 6 feet tall.
    (b) Every football player is over 6 feet tall.

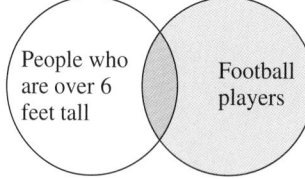

A.3 Logical Arguments

Arguments ▪ Venn Diagrams and Arguments ▪ Proofs

Arguments

An **argument** is a collection of statements, listed in order. The last statement is called the **conclusion** and the other statements are called the **premises.** An argument is **valid** if the conjunction of all the premises implies the conclusion. The most common type of argument takes the following form.

Premise #1: $p \rightarrow q$

Premise #2: p

Conclusion: q

This form of argument is called the **Law of Detachment** or *Modus Ponens.* It is illustrated in the following example.

EXAMPLE 1   *A Valid Argument*

Show that the following argument is valid.

Premise #1:	If Sean is a freshman, then he is taking algebra.
Premise #2:	Sean is a freshman.
Conclusion:	Therefore, Sean is taking algebra.

Solution

Let p represent the statement "Sean is a freshman," and let q represent the statement "Sean is taking algebra." Then the argument fits the Law of Detachment, which can be written as follows.

$$[(p \rightarrow q) \wedge p] \rightarrow q$$

The validity of this argument is shown in the following truth table.

Law of Detachment

p	q	$p \rightarrow q$	$(p \rightarrow q) \wedge p$	$[(p \rightarrow q) \wedge p] \rightarrow q$
T	T	T	T	T
T	F	F	F	T
F	T	T	F	T
F	F	T	F	T

Keep in mind that the validity of an argument has nothing to do with the truthfulness of the premises or conclusion. For instance, the following argument is valid; the fact that it is fanciful does not alter its validity.

Premise #1:	If I snap my fingers, elephants will stay out of my house.
Premise #2:	I am snapping my fingers.
Conclusion:	Therefore, elephants will stay out of my house.

We have discussed the most common form of logical argument. This and three other commonly used forms of valid arguments are summarized in the following list.

Four Types of Valid Arguments

Name	Pattern	Name	Pattern
1. **Law of Detachment or** *Modus Ponens*	Premise #1: $\quad p \to q$ Premise #2: $\quad p$ Conclusion: $\quad q$	2. **Law of Contraposition or** *Modus Tollens*	Premise #1: $\quad p \to q$ Premise #2: $\quad \sim q$ Conclusion: $\quad \sim p$
3. **Law of Transitivity or Syllogism**	Premise #1: $\quad p \to q$ Premise #2: $\quad q \to r$ Conclusion: $\quad p \to r$	4. **Law of Disjunctive Syllogism**	Premise #1: $\quad p \lor q$ Premise #2: $\quad \sim p$ Conclusion: $\quad q$

EXAMPLE 2   An Invalid Argument

Determine whether the following argument is valid.

Premise #1:	If John is elected, the income tax will be increased.
Premise #2:	The income tax was increased.
Conclusion:	Therefore, John was elected.

Solution

This argument has the following form.

	Pattern	Implication
Premise #1:	$p \to q$	$[(p \to q) \land q] \to p$
Premise #2:	q	
Conclusion:	p	

This is not one of the four valid forms of arguments that we listed. You can construct a truth table to verify that the argument is invalid, as follows.

An Invalid Argument

p	q	$p \to q$	$(p \to q) \land q$	$[(p \to q) \land q] \to p$
T	T	T	T	T
T	F	F	F	T
F	T	T	T	F
F	F	T	F	T

An invalid argument, like the one in Example 2, is called a **fallacy.** Other common fallacies are given in the following example.

EXAMPLE 3 Common Fallacies

Each of the following arguments is invalid.

(a) *Arguing from the Converse:* If the football team wins the championship, then students will skip classes. The students skipped classes. Therefore, the football team won the championship.

(b) *Arguing from the Inverse:* If the football team wins the championship, then students will skip classes. The football team did not win the championship. Therefore, the students did not skip classes.

(c) *Arguing from False Authority:* Wheaties are best for you because Joe Montana eats them.

(d) *Arguing from an Example:* Beta Brand products are not reliable because my Beta Brand snowblower does not start in cold weather.

(e) *Arguing from Ambiguity:* If automobile carburetors are modified, the automobile will pollute. Brand X automobiles have modified carburetors. Therefore, Brand X automobiles pollute.

(f) *Arguing by False Association:* Joe was running through the alley when the fire alarm went off. Therefore, Joe started the fire.

EXAMPLE 4 A Valid Argument

Determine whether the following argument is valid.

Premise #1:	You like strawberry pie or you like chocolate pie.
Premise #2:	You do not like strawberry pie.
Conclusion:	Therefore, you like chocolate pie.

Solution

This argument has the following form.

Premise #1:	$p \vee q$
Premise #2:	$\sim p$
Conclusion:	q

This argument is a disjunctive syllogism, which is one of the four common types of valid arguments.

In a valid argument, the conclusion drawn from the premise is called a **valid conclusion.**

EXAMPLE 5   *Making Valid Conclusions*

Given the following two premises, which of the conclusions are valid?

Premise #1:   If you like boating, then you like swimming.
Premise #2:   If you like swimming, then you are a scholar.

(a) Conclusion: If you like boating, then you are a scholar.

(b) Conclusion: If you do not like boating, then you are not a scholar.

(c) Conclusion: If you are not a scholar, then you do not like boating.

Solution

(a) This conclusion is valid. It follows from the Law of Transitivity (or syllogism).

(b) This conclusion is invalid. The fallacy stems from arguing from the inverse.

(c) This conclusion is valid. It follows from the Law of Contraposition.

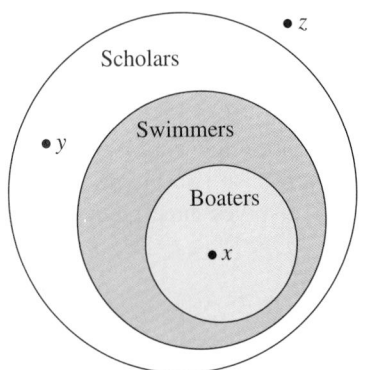

FIGURE A.1

Venn Diagrams and Arguments

Venn diagrams can be used to test informally the validity of an argument. For instance, a Venn diagram for the premises in Example 5 is shown in Figure A.1. In this figure, the validity of conclusion (a) can be seen by choosing a boater x in all three sets. Conclusion (b) can be seen to be invalid by choosing a person y who is a scholar but does not like boating. Finally, person z indicates the validity of conclusion (c).

Venn diagrams work well for testing arguments that involve quantifiers, as shown in the next two examples.

EXAMPLE 6   *Using a Venn Diagram to Show that an Argument Is Not Valid*

Use a Venn diagram to test the validity of the following argument.

Premise #1:   Some plants are green.
Premise #2:   All lettuce is green.
Conclusion: Therefore, lettuce is a plant.

Solution

From the Venn diagram shown in Figure A.2, you can see that this is not a valid argument. Remember that even though the conclusion is true (lettuce is a plant), this does not imply that the argument is true.

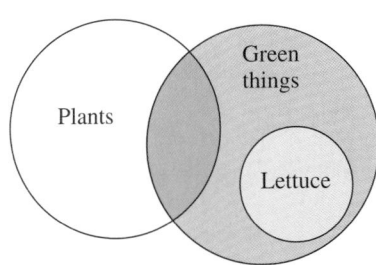

FIGURE A.2

NOTE   In Figure A.2, the circle representing plants is not drawn entirely within the circle representing green things because only *some* plants are green.

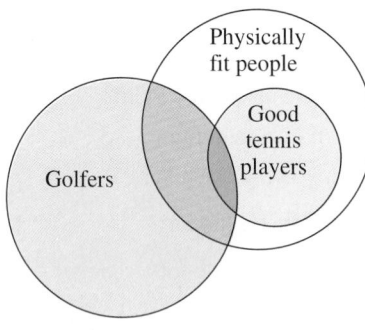

FIGURE A.3

EXAMPLE 7 *Using a Venn Diagram to Show that an Argument Is Valid*

Use a Venn diagram to test the validity of the following argument.

Premise #1:	All good tennis players are physically fit.
Premise #2:	Some golfers are good tennis players.
Conclusion:	Therefore, some golfers are physically fit.

Solution

Because the set of golfers intersects the set of good tennis players, you see from Figure A.3 that the set of golfers must also intersect the set of physically fit people. Therefore, the argument is valid.

Proofs

What does the word *proof* mean to you? In mathematics, the word *proof* means simply a valid argument. Many proofs involve more than two premises and a conclusion. For instance, the proof in Example 8 involves three premises and a conclusion.

EXAMPLE 8 *A Proof by Contraposition*

Use the following three premises to prove that "It is not snowing today."

Premise #1:	If it is snowing today, Greg will go skiing.
Premise #2:	If Greg is skiing today, he is not studying.
Premise #3:	Greg is studying today.

Solution

Let p represent the statement "It is snowing today," let q represent "Greg is skiing," and let r represent "Greg is studying today." Thus, the given premises have the following form.

Premise #1:	$p \rightarrow q$
Premise #2:	$q \rightarrow \sim r$
Premise #3:	r

By noting that $r \equiv \sim(\sim r)$, reordering the premises and writing the contrapositives of the first and second premises, you can obtain the following valid argument.

Premise #3:	r
Contrapositive of Premise #2:	$r \rightarrow \sim q$
Contrapositive of Premise #1:	$\sim q \rightarrow \sim p$
Conclusion:	$\sim p$

Thus, you can conclude $\sim p$. That is, "It is not snowing today."

A.3   Exercises

In Exercises 1–4, use a truth table to show that the given argument is valid.

1. Premise #1: $p \rightarrow \sim q$ **2.** Premise #1: $p \leftrightarrow q$

  Premise #2: q Premise #2: p

  Conclusion: $\sim p$ Conclusion: q

3. Premise #1: $p \vee q$ **4.** Premise #1: $p \wedge q$

  Premise #2: $\sim p$ Premise #2: $\sim p$

  Conclusion: q Conclusion: q

In Exercises 5–8, use a truth table to show that the given argument is invalid.

5. Premise #1: $\sim p \rightarrow q$ **6.** Premise #1: $p \rightarrow q$

  Premise #2: p Premise #2: $\sim p$

  Conclusion: $\sim q$ Conclusion: $\sim q$

7. Premise #1: $p \vee q$ **8.** Premise #1: $\sim (p \wedge q)$

  Premise #2: q Premise #2: q

  Conclusion: p Conclusion: p

In Exercises 9–22, determine whether the argument is valid or invalid.

9. Premise #1: If taxes are increased, then businesses will leave the state.

  Premise #2: Taxes are increased.

  Conclusion: Therefore, businesses will leave the state.

10. Premise #1: If a student does the homework, then a good grade is certain.

  Premise #2: Liza does the homework.

  Conclusion: Therefore, Liza will receive a good grade for the course.

11. Premise #1: If taxes are increased, then businesses will leave the state.

  Premise #2: Businesses are leaving the state.

  Conclusion: Therefore, taxes were increased.

12. Premise #1: If a student does the homework, then a good grade is certain.

  Premise #2: Liza received a good grade for the course.

  Conclusion: Therefore, Liza did her homework.

13. Premise #1: If the doors are kept locked, then the car will not be stolen.

  Premise #2: The car was stolen.

  Conclusion: Therefore, the car doors were unlocked.

14. Premise #1: If Jan passes the exam, she is eligible for the position.

  Premise #2: Jan is not eligible for the position.

  Conclusion: Therefore, Jan did not pass the exam.

15. Premise #1: All cars manufactured by the Ford Motor Company are reliable.

  Premise #2: Lincolns are manufactured by Ford.

  Conclusion: Therefore, Lincolns are reliable cars.

16. Premise #1: Some cars manufactured by the Ford Motor Company are reliable.

  Premise #2: Lincolns are manufactured by Ford.

  Conclusion: Therefore, Lincolns are reliable.

17. Premise #1: All federal income tax forms are subject to the Paperwork Reduction Act of 1980.

  Premise #2: The 1040 Schedule A form is subject to the Paperwork Reduction Act of 1980.

  Conclusion: Therefore, the 1040 Schedule A form is a federal income tax form.

18. Premise #1: All integers divisible by 6 are divisible by 3.

  Premise #2: Eighteen is divisible by 6.

  Conclusion: Therefore, 18 is divisible by 3.

19. Premise #1: Eric is at the store or the handball court.

  Premise #2: He is not at the store.

  Conclusion: Therefore, he must be at the handball court.

20. Premise #1: The book must be returned within 2 weeks or you must pay a fine.

Premise #2: The book was not returned within 2 weeks.

Conclusion: Therefore, you must pay a fine.

21. Premise #1: It is not true that it is a diamond and it sparkles in the sunlight.

Premise #2: It does sparkle in the sunlight.

Conclusion: Therefore, it is a diamond.

22. Premise #1: Either I work tonight or I pass the mathematics test.

Premise #2: I'm going to work tonight.

Conclusion: Therefore, I will fail the mathematics test.

In Exercises 23–30, determine which conclusion is valid from the given premises.

23. Premise #1: If 7 is a prime number, then 7 does not divide evenly into 21.

Premise #2: Seven divides evenly into 21.

(a) Conclusion: Therefore, 7 is a prime number.

(b) Conclusion: Therefore, 7 is not a prime number.

(c) Conclusion: Therefore, 21 divided by 7 is 3.

24. Premise #1: If the fuel is shut off, then the fire will be extinguished.

Premise #2: The fire continues to burn.

(a) Conclusion: Therefore, the fuel was not shut off.

(b) Conclusion: Therefore, the fuel was shut off.

(c) Conclusion: Therefore, the fire becomes hotter.

25. Premise #1: It is necessary that interest rates be lowered for the economy to improve.

Premise #2: Interest rates were not lowered.

(a) Conclusion: Therefore, the economy will improve.

(b) Conclusion: Therefore, interest rates are irrelevant to the performance of the economy.

(c) Conclusion: Therefore, the economy will not improve.

26. Premise #1: It will snow only if the temperature is below 32° at some level of the atmosphere.

Premise #2: It is snowing.

(a) Conclusion: Therefore, the temperature is below 32° at ground level.

(b) Conclusion: Therefore, the temperature is above 32° at some level of the atmosphere.

(c) Conclusion: Therefore, the temperature is below 32° at some level of the atmosphere.

27. Premise #1: Smokestack emissions must be reduced or acid rain will continue as an environmental problem.

Premise #2: Smokestack emissions have not decreased.

(a) Conclusion: Therefore, the ozone layer will continue to be depleted.

(b) Conclusion: Therefore, acid rain will continue as an environmental problem.

(c) Conclusion: Therefore, stricter automobile emission standards must be enacted.

28. Premise #1: The library must upgrade its computer system or service will not improve.

Premise #2: Service at the library has improved.

(a) Conclusion: Therefore, the computer system was upgraded.

(b) Conclusion: Therefore, more personnel were hired for the library.

(c) Conclusion: Therefore, the computer system was not upgraded.

29. Premise #1: If Rodney studies, then he will make good grades.

Premise #2: If he makes good grades, then he will get a good job.

(a) Conclusion: Therefore, Rodney will get a good job.

(b) Conclusion: Therefore, if Rodney doesn't study, then he won't get a good job.

(c) Conclusion: Therefore, if Rodney doesn't get a good job, then he didn't study.

30. Premise #1: It is necessary to have a ticket and an ID card to get into the arena.

Premise #2: Janice entered the arena.

(a) Conclusion: Therefore, Janice does not have a ticket.

(b) Conclusion: Therefore, Janice has a ticket and an ID card.

(c) Conclusion: Therefore, Janice has an ID card.

In Exercises 31–34, use a Venn diagram to test the validity of the argument.

31. Premise #1: All numbers divisible by 10 are divisible by 5.

Premise #2: Fifty is divisible by 10.

Conclusion: Therefore, 50 is divisible by 5.

32. Premise #1: All human beings require adequate rest.

Premise #2: All infants are human beings.

Conclusion: Therefore, all infants require adequate rest.

33. Premise #1: No person under the age of 18 is eligible to vote.

Premise #2: Some college students are eligible to vote.

Conclusion: Therefore, some college students are under the age of 18.

34. Premise #1: Every amateur radio operator has a radio license.

Premise #2: Jackie has a radio license.

Conclusion: Therefore, Jackie is an amateur radio operator.

In Exercises 35–38, use the premises to prove the given conclusion.

35. Premise #1: If Sue drives to work, then she will stop at the grocery store.

Premise #2: If she stops at the grocery store, then she'll buy milk.

Premise #3: Sue drove to work today.

Conclusion: Therefore, Sue will get milk.

36. Premise #1: If Bill is patient, then he will succeed.

Premise #2: Bill will get bonus pay if he succeeds.

Premise #3: Bill did not get bonus pay.

Conclusion: Therefore, Bill is not patient.

37. Premise #1: If this is a good product, then we should buy it.

Premise #2: Either it was made by XYZ Corporation, or we will not buy it.

Premise #3: It is not made by XYZ Corporation.

Conclusion: Therefore, it is not a good product.

38. Premise #1: If the book is returned within 2 weeks, then there is no fine.

Premise #2: You pay a fine or you may not check out another book.

Premise #3: You are allowed to check out another book.

Conclusion: Therefore, the book was not returned within 2 weeks.

Appendix B   Graphing Utilities

B.1   Graphing Utilities

Introduction ▪ Basic Graphing ▪ Special Features

Introduction

In Section 2.2, you studied the point-plotting method for sketching the graph of an equation. One of the disadvantages of the point-plotting method is that, in order to get a good idea about the shape of a graph, you need to plot *many* points. With only a few points, you could badly misrepresent the graph. For instance, consider the equation

$$y = \frac{1}{30}x(39 - 10x^2 + x^4).$$

Suppose you plotted only five points: $(-3, -3)$, $(-1, -1)$, $(0, 0)$, $(1, 1)$, and $(3, 3)$, as shown in Figure B.1. From these five points, you might assume that the graph of the equation is a straight line. This, however, is not correct. By plotting several more points, you can see that the actual graph is not straight at all. (See Figure B.2.)

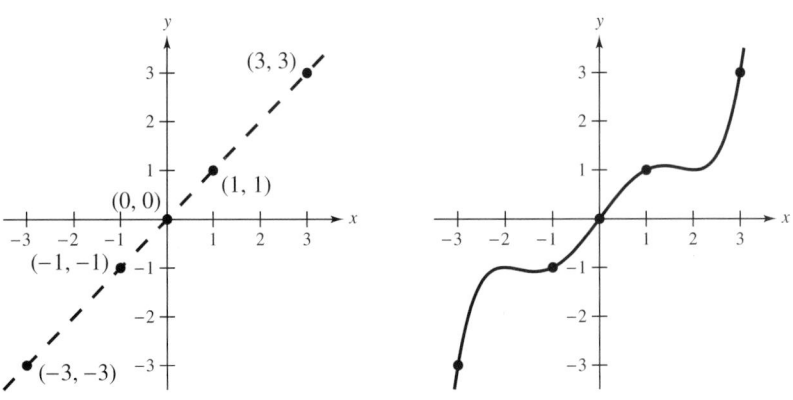

FIGURE B.1   **FIGURE B.2**

Thus, the point-plotting method leaves us with a dilemma. On the one hand, the method can be very inaccurate if only a few points are plotted, but on the other hand, it is very time-consuming to plot a dozen (or more) points.

Technology can help us solve this dilemma. Plotting several (even several hundred) points on a rectangular coordinate system is something that a graphing utility can do easily.

The point-plotting method is the method used by *all* graphing packages for computers and *all* graphing calculators. Each computer or calculator screen is made up of a grid of hundreds or thousands of small areas called **pixels.** Screens that have many pixels per inch are said to have a higher **resolution** than screens that don't have as many. For instance, the screen shown in Figure B.3(a) has a higher resolution than the screen shown in Figure B.3(b). Note that the "graph" of the line on the first screen looks more like a line than the "graph" on the second screen.

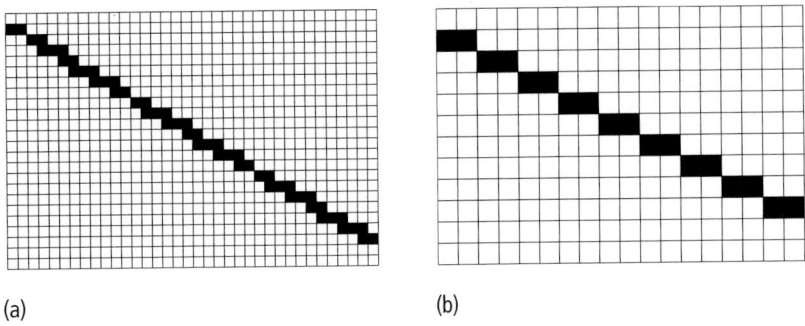

(a)   (b)

FIGURE B.3

Screens on most graphing calculators have 48 pixels per inch. Screens on computer monitors typically have from 32 to 100 pixels per inch.

EXAMPLE 1 *Using Pixels to Sketch a Graph*

Use the grid shown in Figure B.4 to sketch a graph of $y = \frac{1}{2}x^2$. Each pixel on the grid must be either on (shaded black) or off (unshaded).

Solution

To shade the grid, use the following rule. If a pixel contains a plotted point of the graph, it will be "on"; otherwise, the pixel will be "off." Using this rule, the graph of the curve looks like that shown in Figure B.5.

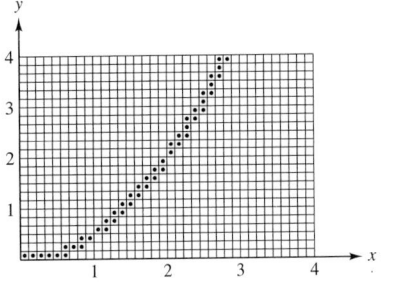

FIGURE B.4

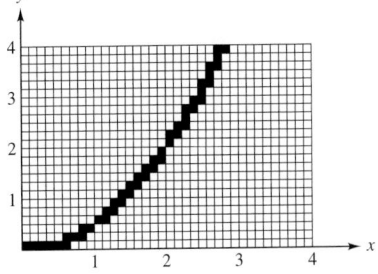

FIGURE B.5

Basic Graphing

There are many different types of graphing utilities/graphing calculators and software packages for computers. The procedures used to draw a graph are similar for most of these utilities.

Basic Graphing Steps for a Graphing Utility

To draw the graph of an equation involving x and y with a graphing utility, use the following steps.

1. Rewrite the equation so that y is isolated on the left side of the equation.

2. Set the boundaries of the viewing rectangle by entering the minimum and maximum x-values and the minimum and maximum y-values.

3. Enter the equation in the form $y =$ (expression involving x). Read the user's guide that accompanies your graphing utility to see how the equation should be entered.

4. Activate the graphing utility.

EXAMPLE 2   Sketching the Graph of an Equation

Sketch the graph of $2y + x^3 = 4x$.

Solution

To begin, solve the original equation for y in terms of x.

$$2y + x^3 = 4x \qquad \text{Original equation}$$
$$2y = -x^3 + 4x \qquad \text{Subtract } x^3 \text{ from both sides.}$$
$$y = -\frac{1}{2}x^3 + 2x \qquad \text{Divide both sides by 2.}$$

Set the viewing rectangle so that $-10 \le x \le 10$ and $-10 \le y \le 10$. (On some graphing utilities, this is the default setting.) Next, enter the equation into the graphing utility.

$$Y_1 = -X \wedge 3/2 + 2 * X$$

Finally, activate the graphing utility. The display screen should look like that shown in Figure B.6.

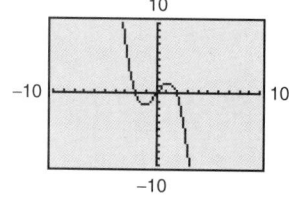

FIGURE B.6

In Figure B.6, notice that the calculator screen does not label the tick marks on the *x*-axis or the *y*-axis. To see what the tick marks represent, check the values in the utility's "range" or "window."

Range or Window

Xmin = −10   The minimum *x*-value is −10.

Xmax = 10   The maximum *x*-value is 10.

Xscl = 1   The *x*-scale is 1 unit per tick mark.

Ymin = −10   The minimum *y*-value is −10.

Ymax = 10   The maximum *y*-value is 10.

Yscl = 1   The *y*-scale is 1 unit per tick mark.

Xres = 1   The *x*-resolution is 1 plotted point per 1 pixel.

These settings are summarized visually in Figure B.7.

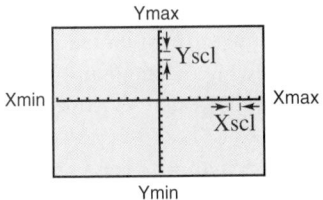

FIGURE B.7

EXAMPLE 3 *Graphing an Equation Involving Absolute Value*

Sketch the graph of $y = |x - 3|$.

Solution

This equation is already written so that *y* is isolated on the left side of the equation, so you can enter the equation as follows.

$$Y_1 = \text{abs}(X - 3)$$

After the graphing utility has been activated, its screen should look like the one shown in Figure B.8.

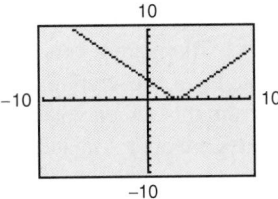

FIGURE B.8

Special Features

In order to be able to use your graphing calculator to its best advantage, you must be able to determine a proper viewing rectangle and use the zoom feature. The next two examples show how this is done.

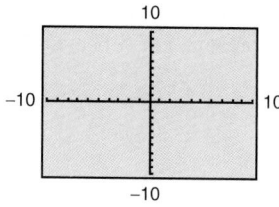

(a)

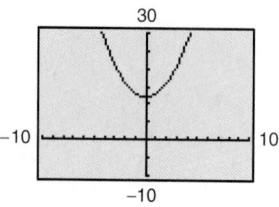

(b)

FIGURE B.9

EXAMPLE 4 Determining a Viewing Rectangle

Sketch the graph of $y = x^2 + 12$ by entering the equation.

$$Y_1 = X \wedge 2 + 12$$

Activate the graphing utility. If you use a viewing rectangle in which $-10 \leq x \leq 10$ and $-10 \leq y \leq 10$, no part of the graph will appear on the screen, as shown in Figure B.9(a). The reason for this is that the lowest point on the graph of $y = x^2 + 12$ occurs at the point $(0, 12)$. With the viewing rectangle in Figure B.9(a), the largest y-value is 10. In other words, none of the graph is visible on a screen whose y-values range between -10 and 10.

To be able to see the graph, change Ymax $= 10$ to Ymax $= 30$ and change Yscl $= 1$ to Yscl $= 5$. Now activate the graphing utility and you will obtain the graph shown in Figure B.9(b). On this graph, note that each tick mark on the y-axis represents five units because you changed the y-scale to 5. Also note that the highest point on the y-axis is now 30 because you changed the maximum value of y to 30.

EXAMPLE 5 Using the Zoom Feature

Sketch the graph of $y = x^3 - x^2 - x$. How many x-intercepts does this graph have?

Solution

Begin by drawing the graph in a "standard" viewing rectangle, as shown in Figure B.10(a). From the display screen, it is clear that the graph has at least one x-intercept (just to the left of $x = 2$), but it is difficult to determine whether the graph has other intercepts. To obtain a better view of the graph near $x = -1$, you can use the zoom feature of the graphing utility. The redrawn screen is shown in Figure B.10(b). From this screen you can tell that the graph has three x-intercepts whose x-coordinates are approximately -0.6, 0, and 1.6.

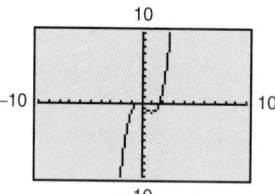

(a)

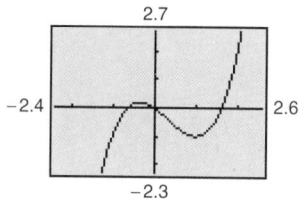

(b)

FIGURE B.10

EXAMPLE 6   *Sketching More than One Graph on the Same Screen*

Sketch the graphs of $y = -\sqrt{36 - x^2}$ and $y = \sqrt{36 - x^2}$ on the same screen.

Solution

To begin, enter both equations in the graphing utility.

$$Y_1 = \sqrt{(36 - X \wedge 2)}$$
$$Y_2 = -\sqrt{(36 - X \wedge 2)}$$

Then, activate the graphing utility to obtain the graph shown in Figure B.11(a). Notice that the graph should be the upper and lower parts of the circle given by $x^2 + y^2 = 6^2$. The reason it doesn't look like a circle is that, with the standard settings, the tick marks on the x-axis are farther apart than the tick marks on the y-axis. To correct this, change the viewing rectangle so that $-15 \le x \le 15$. The redrawn screen is shown in Figure B.11(b). Notice that in this screen the graph appears to be circular.

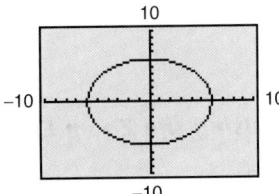

(a)

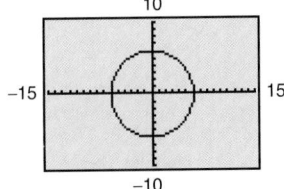

(b)

FIGURE B.11

Group Activities

Extending the Concept

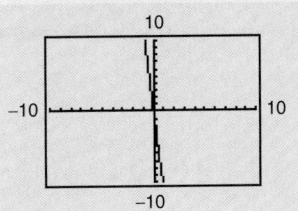

A Misleading Graph   Sketch the graph of $y = x^2 - 12x$, using a standard viewing range of $-10 \le x \le 10$ and $-10 \le y \le 10$. The graph appears to be a straight line, as shown at the left. However, this is misleading because the screen doesn't show an important portion of the graph. Can you find a range setting that reveals a better view of this graph?

B.1   **Exercises**

In Exercises 1–20, use a graphing utility to sketch the graph of the equation. Use a setting on each graph of $-10 \le x \le 10$ and $-10 \le y \le 10$.

1. $y = x - 5$
2. $y = -x + 4$
3. $y = -\frac{1}{2}x + 3$
4. $y = \frac{2}{3}x + 1$
5. $2x - 3y = 4$
6. $x + 2y = 3$
7. $y = \frac{1}{2}x^2 - 1$
8. $y = -x^2 + 6$
9. $y = x^2 - 4x - 5$
10. $y = x^2 - 3x + 2$
11. $y = -x^2 + 2x + 1$
12. $y = -x^2 + 4x - 1$
13. $2y = x^2 + 2x - 3$
14. $3y = -x^2 - 4x + 5$
15. $y = |x + 5|$
16. $y = \frac{1}{2}|x - 6|$
17. $y = \sqrt{x^2 + 1}$
18. $y = 2\sqrt{x^2 + 2} - 4$
19. $y = \frac{1}{5}(-x^3 + 16x)$
20. $y = \frac{1}{8}(x^3 + 8x^2)$

In Exercises 21–30, use a graphing utility to match the equation with its graph. [The graphs are labeled (a), (b), (c), (d), (e), (f), (g), (h), (i), and (j).]

(a)

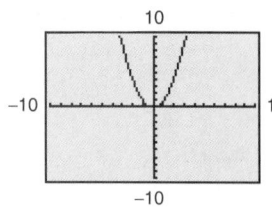

(b)

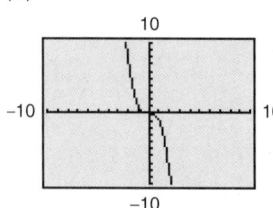

(c)

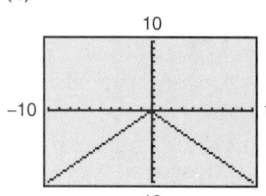

(d)

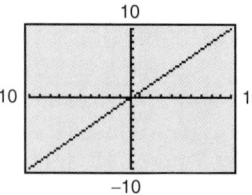

(e)

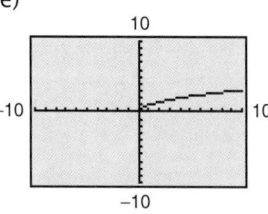

(f)

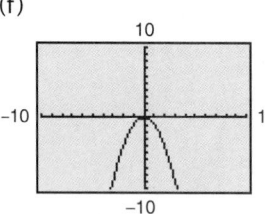

(g)

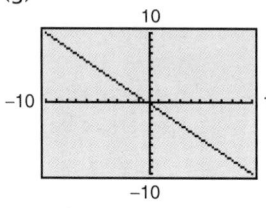

(h)

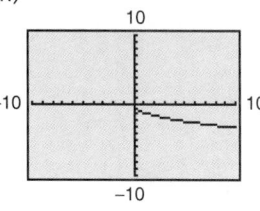

(i)

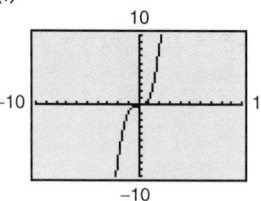

(j)

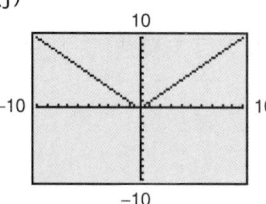

21. $y = x$
22. $y = -x$
23. $y = x^2$
24. $y = -x^2$
25. $y = x^3$
26. $y = -x^3$
27. $y = |x|$
28. $y = -|x|$
29. $y = \sqrt{x}$
30. $y = -\sqrt{x}$

In Exercises 31–34, use a graphing utility to sketch the graph of the equation. Use the indicated setting.

31. $y = -2x^2 + 12x + 14$
32. $y = -x^2 + 5x + 6$

Xmin = -5
Xmax = 10
Xscl = 1
Ymin = -5
Ymax = 35
Yscl = 5
Xres = 1

Xmin = -4
Xmax = 8
Xscl = 1
Ymin = -5
Ymax = 15
Yscl = 5
Xres = 1

33. $y = x^3 + 6x^2$

Xmin = -10
Xmax = 5
Xscl = 1
Ymin = -4
Ymax = 36
Yscl = 3
Xres = 1

34. $y = -x^3 + 16x$

Xmin = -6
Xmax = 6
Xscl = 1
Ymin = -25
Ymax = 25
Yscl = 5
Xres = 1

In Exercises 35–38, find a setting on a graphing utility using the equation to match the graph.

35. $y = -x^2 - 4x + 20$ **36.** $y = x^2 + 12x - 8$

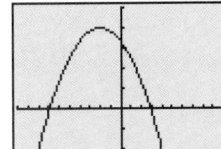 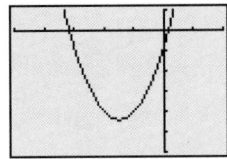

37. $y = -x^3 + x^2 + 2x$ **38.** $y = x^3 + 3x^2 - 2x$

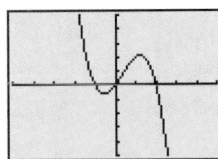 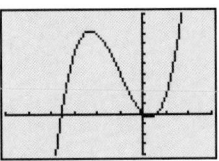

In Exercises 39–42, use a graphing utility to find the number of x-intercepts of the equation.

39. $y = \frac{1}{8}(4x^2 - 32x + 65)$

40. $y = \frac{1}{4}(-4x^2 + 16x - 15)$

41. $y = 4x^3 - 20x^2 - 4x + 61$

42. $y = \frac{1}{4}(2x^3 + 6x^2 - 4x + 1)$

In Exercises 43–46, use a graphing utility to sketch the graphs of the equations on the same screen. Using a "square setting," what geometric shape is bounded by the graphs?

43. $y = |x| - 4$
$y = -|x| + 4$

44. $y = x + |x| - 4$
$y = x - |x| + 4$

45. $y = -\sqrt{25 - x^2}$
$y = \sqrt{25 - x^2}$

46. $y = 6$
$y = -\sqrt{3}x - 4$
$y = \sqrt{3}x - 4$

Ever Been Married? In Exercises 47–50, use the models, which relate ages to the percent of American males and females who have never been married.

Males
$$y = \frac{0.36 - 0.0056x}{1 - 0.0817x + 0.00226x^2}, \quad 20 \le x \le 50$$

Females
$$y = \frac{100}{8.944 - 0.886x + 0.249x^2}, \quad 20 \le x \le 50$$

In these models, y is the percent of the population (in decimal form) who have never been married and x is the age of the person. *(Source: U.S. Bureau of the Census)*

47. Use a graphing utility to sketch the graphs of both equations giving the percent of American males and females who have never been married. Use range settings of $20 \le x \le 50$ and $0 \le y \le 1$.

48. Describe the relationship between the two graphs that were plotted in Exercise 47.

49. If an American male is 25 years old, what is the probability that he has never been married?

50. If an American female is 25 years old, what is the probability that she has never been married?

Earnings and Dividends In Exercises 51–54, use the following model, which approximates the relationship between dividends per share and earnings per share for the Pall Corporation for the years 1980 to1996.

$$y = -0.036 + 0.325x + 0.091x^2, \quad 0.25 \le x \le 1.5$$

In this model, y is the dividends per share (in dollars) and x is the earnings per share (in dollars). *(Source: Value Line)*

51. Use a graphing utility to sketch the graph of the model that gives the dividend per share in terms of the earnings per share. Use range settings of $0 \le x \le 1.5$ and $0 \le y \le 0.7$.

52. Using the model, what size dividend would the Pall Corporation pay if the earnings per share were $1.30?

53. Use the trace feature on your graphing utility to estimate the earnings per share that would produce a dividend per share of $0.25.

54. The **payout ratio** for a stock is the ratio of the dividend per share to earnings per share. Use the model to find the payout ratio for earnings per share of (a) $0.75, (b) $1.00, and (c) $1.25.

B.2	**Programs**

Two-Point Form of a Line (Section 3.1)

This program, shown in the marginal Programming note on page 186, will display the slope and y-intercept of the line that passes through two points, (x_1, y_1) and (x_2, y_2), entered by the user.

TI-80

```
PROGRAM:TWOPTFM
:DISP "ENTER X1, Y1"
:INPUT X
:INPUT Y
:DISP "ENTER X2, Y2"
:INPUT C
:INPUT D
:(D−Y)/(C−X)→M
:M×(-X)+Y→B
:DISP "SLOPE ="
:DISP M
:DISP "Y-INT ="
:DISP B
```

TI-81
TI-82
TI-83

```
PROGRAM:TWOPTFM
:Disp "ENTER X1, Y1"
:Input X
:Input Y
:Disp "ENTER X2, Y2"
:Input C
:Input D
:(D−Y)/(C−X)→M
:M*(-X)+Y→B
:Disp "SLOPE ="
:Disp M
:Disp "Y-INT ="
:Disp B
```

TI-85
TI-86

```
PROGRAM:TWOPTFM
:Disp "Enter X1, Y1"
:Input X
:Input Y
:Disp "Enter X2, Y2"
:Input C
:Input D
:(D−Y)/(C−X)→M
:M*(-X)+Y→B
:Disp "Slope ="
:Disp M
:Disp "Y-int ="
:Disp B
```

TI-92

```
:twoptfm ( )
:Prgm
:Disp "ENTER X1, Y1"
:Input x
:Input y
:Disp "ENTER X2, Y2"
:Input c
:Input d
:(d−y)/(c−x)→m
:m*-x+y→b
:Disp "SLOPE ="
:Disp m
:Disp "Y-INT ="
:Disp b
:EndPrgm
```

Casio fx-7700G

```
TWOPTFORM
"ENTER X1, Y1"?→X:?→Y
"ENTER X2, Y2"?→C:?→D
(D−Y)÷(C−X)→M
M×(-X)+Y→B
"SLOPE =":M ◢
"Y-INT =":B
```

Casio fx-7700GE
Casio fx-9700GE
Casio CFX-9800G

```
TWOPTFORM↵
"ENTER X1, Y1"?→X:?→Y↵
"ENTER X2, Y2"?→C:?→D↵
(D−Y)÷(C−X)→M↵
M×(-X)+Y→B↵
"SLOPE =":M ◢
"Y-INT =":B
```

Casio CFX-9850G

```
======TWOPTFM======
"ENTER X1, Y1"?→X:?→Y↵
"ENTER X2, Y2"?→C:?→D↵
(D−Y)÷(C−X)→M↵
M×(-X)+Y→B↵
"SLOPE =":M ◢
"Y-INT =":B
```

SHARP EL-9200C
SHARP EL-9300C

twoptform
Print "enter x1, y1
Input x
c=x
Input y
d=y
Print "enter x2, y2
Input x
Input y
m=(d−y)/(c−x)
b=m*(-x)+y
Print "slope
Print m
Print "y-int
Print b

HP 38G

TWOPTFM PROGRAM
INPUT X: "ENTER X1, Y1";
"ENTER X1";;1:
INPUT Y: "ENTER X1, Y1";
"ENTER Y1";;1:
INPUT C: "ENTER X2, Y2";
"ENTER X2";;1:
INPUT D: "ENTER X2, Y2";
"ENTER Y2";;1:
(D−Y)/(C−X)▶M
M*-X+Y▶B
DISP 1;"SLOPE ="M:
DISP 3;"Y-INT ="B;
FREEZE:

Simple Interest Program (Section 3.4)

This program, shown in the marginal Programming note on page 227, can be used to find the amount of simple interest earned on a given principal at a given annual interest rate for a certain amount of time.

TI-80

PROGRAM:SIMPINT
:FIX 2
:DISP "PRINCIPAL"
:INPUT P
:DISP "INTEREST RATE"
:DISP "IN DECIMAL FORM"
:INPUT R
:DISP "NUMBER OF YEARS"
:INPUT T
:PRT →I
:DISP "THE INTEREST IS"
:DISP I
:FLOAT

TI-81
TI-82
TI-83

PROGRAM:SIMPINT
:Fix 2
:Disp "PRINCIPAL"
:Input P
:Disp "INTEREST RATE"
:Disp "IN DECIMAL FORM"
:Input R
:Disp "NUMBER OF YEARS"
:Input T
:PRT →I
:Disp "THE INTEREST IS"
:Disp I
:Float

TI-85
TI-86

PROGRAM:SIMPINT
:Fix 2
:Disp "Principal"
:Input P
:Disp "Interest rate"
:Disp "in decimal form"
:Input R
:Disp "Number of years"
:Input T
:P*R*T →I
:Disp "The interest is"
:Disp I
:Float

TI-92

```
:simpint ( )
:Prgm
:setMode("Display Digits", "Fix 2")
:Input "Principal", p
:Input "Interest rate in decimal form", r
:Input "Number of years", t
:p*r*t → i
:Disp "The interest is", i
:setMode("Display Digits", "Float")
:EndPrgm
```

Casio fx-7700G

```
SIMPINT
Fix 2
"PRINCIPAL"? → P
"INTEREST RATE"
"IN DECIMAL FORM"? → R
"NUMBER OF YEARS"? → T
PRT → I
"THE INTEREST IS":I ◢
Norm
```

Casio fx-7700GE
Casio fx-9700GE
Casio CFX-9800G
Casio CFX-9850G

```
SIMPINT ↵
Fix 2 ↵
"PRINCIPAL"? → P ↵
"INTEREST RATE" ↵
"IN DECIMAL FORM"? → R ↵
"NUMBER OF YEARS"? → T ↵
PRT → I ↵
"THE INTEREST IS":I ◢
Norm
```

Sharp EL-9200C
Sharp EL-9300C

```
simpint
──────────REAL
Input principal
Print "Interest rate
Print "in decimal form
Input rate
Print "Number of years
Input time
interest=principal*rate*time
Print interest
```

HP 38G

```
SIMPINT PROGRAM
INPUT P; "SIMPINT";; "ENTER
PRINCIPAL";1:
INPUT R; "SIMPINT";; "INTEREST
RATE IN DECIMAL FORM";1:
INPUT T; "SIMPINT";; "ENTER
NUMBER OF YEARS";1:
P*R*T▶I:
DISP 3; "INTEREST IS" I:
FREEZE:
```

Quadratic Formula Program (Section 6.3)

This program, shown in the Programming note on page 432, will display the solutions to quadratic equations or the words "No Real Solution." To use the program, write the quadratic equation in standard form and enter the values of a, b, and c.

TI-80

PROGRAM:QUADRAT
:Disp "AX^2+BX+C=0"
:Input "ENTER A", A
:Input "ENTER B", B
:Input "ENTER C," C
:B^2−4AC→D
:If D≥0
:Then
:(-B+$\sqrt{\ }$D)/(2A)→M
:Disp M
:(-B−$\sqrt{\ }$D)/(2A)→N
:Disp N
:Else
:Disp "NO REAL SOLUTION"
:End

TI-81

Prgm4: QUADRAT
:Disp "ENTER A"
:Input A
:Disp "ENTER B"
:Input B
:Disp "ENTER C"
:Input C
:B^2−4AC→D
:If D<0
:Goto 1
:((-B+$\sqrt{\ }$D)/(2A))→M
:Disp M
:((-B−$\sqrt{\ }$D)/(2A))→N
:Disp N
:End
:Lbl 1
:Disp "NO REAL"
:Disp "SOLUTION"
:End

TI-82
TI-83

PROGRAM:QUADRAT
:Disp "AX^2+BX+C=0"
:Prompt A
:Prompt B
:Prompt C
:B^2−4AC→D
:If D≥0
:Then
:(-B+$\sqrt{\ }$D)/(2A)→M
:Disp M
:(-B−$\sqrt{\ }$D)/(2A)→N
:Disp N
:Else
:Disp "NO REAL SOLUTION"
:End

TI-85
TI-86

PROGRAM:QUADRAT
:Disp "AX^2+BX+C=0"
:Input "ENTER A", A
:Input "ENTER B", B
:Input "ENTER C", C
:B^2−4*A*C→D
:(-B+$\sqrt{\ }$D)/(2A)→M
:Disp M
:(-B−$\sqrt{\ }$D)/(2A)→N
:Disp N

This program gives both real and complex answers. Solutions to quadratic equations are also available directly by using the POLY function.

TI-92

:quadrat( )
:Prgm
:setMode("Complex Format", "RECTANGULAR")
:Disp "AX$\wedge$2+BX+C=0"
:Input "Enter A.",a
:Input "Enter B.",b
:Input "Enter C.",c
:b$\wedge$2−4*a*c→d
:(-b+$\sqrt{\ }$(d))/(2*a)→m
:(-b−$\sqrt{\ }$(d))/(2*a)→n
:Disp m
:Disp n
:setMode("Complex Format","REAL")
:EndPrgm

This program gives both real and complex answers.

Casio fx-7700G

QUADRATIC
"AX^2+BX+C=0"
"A="?→A
"B="?→B
"C="?→C
B^2−4AC→D
D<0 $\Rightarrow$ Goto 1
"X=":(-B+$\sqrt{\ }$D)÷(2A) ◢
"OR X=":(-B−$\sqrt{\ }$D)÷(2A)
Goto 2
Lbl 1
"NO REAL SOLUTION"
Lbl 2

Casio fx-7700GE

QUADRATIC↵
"AX^2+BX+C=0"↵
"A="?→A↵
"B="?→B↵
"C="?→C↵
B^2−4AC→D↵
D<0 ⟹ Goto 1↵
(-B+$\sqrt{\ }$ D)÷(2A) ◢
(-B−$\sqrt{\ }$ D)÷(2A)↵
Goto 2↵
Lbl 1↵
"NO REAL SOLUTION"↵
Lbl 2

Solutions to quadratic equations are also available directly from the Casio calculator's EQUATION MODE.

Casio fx-9700GE
Casio CFX-9800G

QUADRATIC↵
"AX^2+BX+C=0"↵
"A="?→A↵
"B="?→B↵
"C="?→C↵
B^2−4AC→D↵
(-B+$\sqrt{\ }$ D)÷(2A) ◢
(-B−$\sqrt{\ }$ D)÷(2A)

Both real and complex answers are given. Solutions to quadratic equations are also available directly from the Casio calculator's EQUATION MODE.

Casio CFX-9850G

======QUADRAT======
"AX^2+BX+C=0"↵
"A="?→A↵
"B="?→B↵
"C="?→C↵
B^2−4AC→D↵
(-B+$\sqrt{\ }$ D)÷(2A) ◢
(-B−$\sqrt{\ }$ D)÷(2A)

Both real and complex answers are given. Solutions to quadratic equations are also available directly from the Casio calculator's EQUATION MODE.

Sharp EL-9200C
Sharp EL-9300C

quadratic
——————————COMPLEX
Print "ax^2+bx+c=0"
Input a
Input b
Input c
d=b^2−4a∗c
x1=(-b+$\sqrt{\ }$ d)/(2a)
x2=(-b−$\sqrt{\ }$ d)/(2a)
Print x1
Print x2
End

This program gives both real and complex answers.

HP 38G

QUADRAT PROGRAM
INPUT A;"AX^2+BX+C=0";
"ENTER A";"";1:
INPUT B;"AX^2+BX+C=0";
"ENTER B";"";1:
INPUT C;"AX^2+BX+C=0";
"ENTER C";"";1:
B^2−4AC▶D:
(-B+$\sqrt{\ }$ D)/(2A)▶Z1:
(-B+$\sqrt{\ }$ D)/(2A)▶Z2:
DISP 3;Z1:
DISP 5;Z2:
FREEZE

This program displays the answer in complex form (x, y), where x is the real part and y is the imaginary part.

Random Number Generator (Section 10.5)

This program, shown in the marginal Programming note on page 749, will display ten randomly generated numbers. Repeatedly press ENTER or EXE until all ten integers are displayed.

TI-80

```
PROGRAM:RANDOM
:FOR(I, 1, 10)
:INT(10RAND) → A
:DISP A
:PAUSE
:END
```

TI-81

```
:Prgm2:RANDOM
:1 → I
:Lbl 1
:Int(10Rand) → A
:Disp A
:Pause
:IS>(I,10)
:Goto 1
:Disp "DONE"
:End
```

TI-82
TI-83

```
PROGRAM:RANDOM
:For(I,1,10)
:int(10rand) → A
:Disp A
:Pause
:End
```

TI-85
TI-86

```
PROGRAM:RANDOM
:For(I,1,10)
:iPart(10rand) → A
:Disp A
:Pause
:End
```

TI-92

```
:random ( )
:Prgm
:For i,1,10
:iPart(10rand ( )) → a
:Disp a
:Pause
:EndFor
:Disp "DONE"
:EndPrgm
```

Casio fx-7700G

```
RANDOM
0 → C
Lbl 1
C+1 → C
Int(10Ran#) → A
A ◢
C<10 ⟹ Goto 1
```

Casio fx-7700GE
Casio fx-9700GE
Casio CFX-9800G

```
RANDOM↵
1 → I↵
Lbl 1↵
Int 10Ran# ◢
Isz I↵
I≤10 ⟹ Goto 1↵
"DONE"
```

Casio CFX-9850G

```
======RANDOM======
1 → I↵
Lbl 1↵
Int 10Ran# ◢
Isz I↵
I≤10 ⟹ Goto 1↵
"DONE"
```

Sharp EL-9200C
Sharp EL-9300C

```
random
───────────REAL
i=0
Label 1
i=i+1
X=int 10random
Print X
Wait
If i<10 Goto 1
End
```

HP 38G

```
RANDOM PROGRAM
FOR I=1 TO 10
STEP 1;
INT(10RANDOM)▶A:
DISP 3;A:
FREEZE:
END
```

Birthday Problem Program (Section 10.6)

This program, shown in the marginal Technology note on page 760, is a simulation of the birthday problem. The program displays the number of selections that were made before two people with the same birthday were found. *Note to Casio users:* The command "Defm" allocates additional memory for the storage of the random numbers generated in this program. The number "341" will allow the program to run and store 365 numbers without a memory error. You need to have 2728 bytes of memory available to run this program. If this amount of memory is not available in your graphing calculator, you may lower the number, but be aware you may receive a memory error when running the program. Consult your owner's manual for more information.

TI-80

```
PROGRAM:BIRTHDA
:CLRHOME
:CLRLIST L1
:0 → C
:DISP "COMPUTING . . ."
:LBL A
:1+C → C
:IPART(365RAND) → D
:D → L1(C)
:IF C=1:GOTO A
:For(I,1,C−1)
:If D=L1(I):GOTO B
:END
:GOTO A
:LBL B
:CLRHOME
:DISP "MATCH: NUMBER OF"
:DISP "SELECTIONS IS"
:DISP C
```

TI-81

```
Prgm 3:BIRTHDAY
:ClrHome
:ClrStat
:0 → C
:Disp "COMPUTING . . ."
:Lbl A
:1+C → C
:IPart (365Rand) → D
:D → {x}(C)
:If C=1
:Goto A
:1 → I
:Lbl 1
:If D={x}(I)
:Goto B
:IS>(I,C−1)
:Goto 1
:Goto A
:Lbl B
:ClrHome
:Disp "MATCH, NUMBER OF"
:Disp "SELECTIONS IS"
:Disp C
```

TI-82

```
PROGRAM:BIRTHDAY
:ClrHome
:ClrList L1
:0 → C
:Output(4,3,"COMPUTING . . .")
:Lbl A
:1+C → C
:iPart (365rand) → D
:D → L1(C)
:If C=1:Goto A
:For(I,1,C−1)
:If D=L1(I):Goto B
:End
:Goto A
:Lbl B
:ClrHome
:Disp "MATCH: NUMBER OF"
:Disp "SELECTIONS IS"
:Disp C
```

TI-83

```
PROGRAM:BIRTHDAY
:ClrHome
:ClrList L₁
:0→C
:Output(4,3,"COMPUTING . . .")
:Lbl A
:1+C→C
:randInt(1,365)→D
:D→L₁(C)
:If C=1:Goto A
:For(I,1,C−1)
:If D=L₁(I):Goto B
:End
:Goto A
:Lbl B
:ClrHome
:Disp "MATCH: NUMBER OF"
:Disp "SELECTIONS IS"
:Disp C
```

TI-85
TI-86

```
PROGRAM:BIRTHDAY
:ClLCD
:0→C
:Outpt(4,3,"COMPUTING . . .")
:Lbl A
:1+C→C
:iPart (365rand)→D
:D→TEMP(C)
:If C==1:Goto A
:For(I,1,C−1)
:If D==TEMP(I):Goto B
:End
:Goto A
:End
:Lbl B
:ClLCD
:Disp "MATCH: NUMBER OF"
:Disp "SELECTIONS IS"
:Disp C
```

TI-92

```
:birthday ( )
:Prgm
:ClrIO
:0→c
:Output 4,3,"COMPUTING . . ."
:Lbl a
:1+c→c
:iPart(365rand ( ))→d
:d→list1[c]
:If c=1:Goto a
:For i,1,c−1
:If d=list1[i]:Goto b
:EndFor
:Goto a
:Lbl b
:ClrIO
:Disp "MATCH: NUMBER OF
      SELECTIONS IS"
:Disp c
:EndPrgm
```

Casio fx-7700G

```
BIRTHDAY
Defm 341
0→C
"COMPUTING . . ."
Lbl 1
1+C→C
Int (365Ran#)→D
D→L[C]
C=1⇒Goto 1
C−1→E
Lbl 3
L[E]=D⇒Goto 2
E−1→E
E≠0 ⇒Goto 3
Goto 1
Lbl 2
Defm 0
"MATCH: NUMBER OF"
"SELECTIONS IS":C
```

Casio fx-7700GE
Casio fx-9700GE
Casio CFX-9800G

```
BIRTHDAY↵
Defm 341↵
0→A↵
"COMPUTING . . ."↵
Lbl 1↵
Isz A↵
Int (365ran#)→B↵
B→D[A]↵
A=1⇒Goto 1↵
1→C↵
Lbl 2↵
B=D[C]⇒Goto 3↵
Isz C↵
C<A⇒Goto 2↵
Goto 1↵
Lbl 3↵
Defm 0↵
"MATCH: NUMBER OF"↵
"SELECTIONS IS"↵
A
```

Casio CFX-9850G

```
======BIRTHDAY======
ClrText↵
ClrList↵
Seq(-1,A,1,250,1)→List 1↵
Seq(-1,A,1,115,1)→List 2↵
0→C↵
Locate 4,3,"COMPUTING . . ."↵
Lbl 1↵
Isz C↵
Int (365Ran#)→D↵
If C≤250↵
Then D→List 1[C]↵
Else D→List 2[C−250]↵
IfEnd↵
If C=1↵
Then Goto 1↵
IfEnd↵
For 1→I To C−1 Step 1↵
If I≤250↵
Then List 1[I]→E↵
Else List 2[I−250]→E↵
IfEnd↵
If D=E↵
Then Goto2↵
IfEnd↵
Next↵
Goto1↵
Lbl 2↵
ClrText↵
"MATCH: NUMBER OF"↵
"SELECTIONS IS"↵
C ◢
```

Sharp EL-9200C
Sharp EL-9300C

```
birthday
_____STAT
ClrT
Stat X
c=0
Print "computing . . ."
Label a
c=c+1
X=int 365random
Data X
If c=1 Goto a
i=1
Label 1
If X=St[1,i] Goto b
i=i+1
If i<c Goto 1
Goto a
Label b
ClrT
Print "match, number of"
Print "selections is"
Print c
```

HP 38G

```
BIRTHDAY PROGRAM
DISP 1; "COMPUTING . . .":
MAKELIST(−1,A,1,365,1)▶L1:
0▶C:
RUN "GET NUMBER":
DO
   RUN "DO LOOP"
UNTIL D==E
END:
RUN "PRINT ANSWER":
FREEZE

GET NUMBER PROGRAM
1+C▶C:
INT(365RANDOM)▶D:
D▶L1(C)

DO LOOP PROGRAM
RUN "GET NUMBER":
FOR I=1 TO C−1
STEP 1;
IF D==L1(I)
   THEN L1(I)▶E END
END

PRINT ANSWER PROGRAM
DISP 1;"MATCH: NUMBER OF":
DISP 2;"SELECTIONS IS":
DISP 3;C
```

Enter the four programs (Birthday, Get Number, Do Loop, and Print Answer) into the HP 38G. Then *RUN* the Birthday program.

Warm-Up Exercises

The warm-up exercises for each section listed below involve skills that were covered in previous sections. You will use the skills in the exercise set for each section listed. Before attempting each exercise set, make sure you have mastered the warm-up exercises for that section.

Section 1.2

In Exercises 1–10, perform the indicated operation(s).

1. $625 + (-400)$ **2.** $-360 + 120$

3. $-2(225 - 150)$ **4.** $5(57 - 33)$

5. $7 \times \frac{8}{35}$ **6.** $-\frac{4}{15} \times \frac{15}{16}$

7. $\frac{3}{8} \div \frac{5}{16}$ **8.** $\frac{9}{10} \div 3$

9. $(12 - 15)^3$ **10.** $\left(\frac{5}{8}\right)^2$

Section 1.3

In Exercises 1–6, evaluate the quantity.

1. $6(3 + 5^2)$ **2.** $-5(6 - 10)$

3. $3(-4) + 4\left(\frac{3}{8}\right)$ **4.** $-|35 - 60| + 8(9 - 5)$

5. $\dfrac{360}{3^2 + 4^2}$ **6.** $\dfrac{13}{5} - \dfrac{9}{2}$

In Exercises 7–10, name the property of real numbers that justifies the statement.

7. $8x \cdot \dfrac{1}{8x} = 1$

8. $5 + (-3 + x) = (5 - 3) + x$

9. $-4(x + 10) = -4 \cdot x + (-4)(10)$

10. $3x + 0 = 3x$

Section 1.4

In Exercises 1–4, expand the expression.

1. $23(y - 3)$ **2.** $6(2 - 5z)$

3. $-\frac{2}{3}(15 - 6x)$ **4.** $-5(5x - 4)$

In Exercises 5 and 6, list the terms of the algebraic expression.

5. $32x - 10y - 7$ **6.** $-3s + 4t + 1$

In Exercises 7–10, simplify the algebraic expression.

7. $3(x - 5) + 4x$ **8.** $5(4 - y) + 8(y + 2)$

9. $-3(u - 6) - (u + 18)$ **10.** $(v - 3) - 3(2v + 5)$

Section 1.5

In Exercises 1–10, perform the indicated multiplication.

1. $12x(2x - 3)$ **2.** $7y(4 - 3y)$

3. $-6x(10 - 7x)$ **4.** $-2y(y + 1)$

5. $t(t^2 + 1) - t(t^2 - 1)$ **6.** $2z(z + 5) - 7(z + 5)$

7. $(11 - x)(11 + x)$ **8.** $(6r + 5s)(6r - 5s)$

9. $(x - 2)(x^2 + 2x + 4)$ **10.** $(x + 1)(x^2 - x + 1)$

Section 1.6

In Exercises 1–6, perform the indicated multiplication.

1. $(x + 1)^2$ **2.** $(x - 5)^2$

3. $(2 - y)^2$ **4.** $(2 - y)(3 + 2y)$

5. $(5z + 3)(2z - 7)$ **6.** $t(t - 4)(2t + 3)$

In Exercises 7–10, factor the expression completely.

7. $3x^2 - 21x$ **8.** $-x^3 + 3x^2 - x + 3$

9. $4t^2 - 169$ **10.** $y^3 - 64$

Section 1.7

In Exercises 1–4, evaluate the quantity.

1. $-\dfrac{13}{35} \cdot \dfrac{-25}{104}$ 2. $\dfrac{5}{12} \cdot \dfrac{9}{75}$

3. $\dfrac{14}{3} \div \dfrac{42}{45}$ 4. $\dfrac{-8}{105} \div \dfrac{2}{3}$

In Exercises 5–10, simplify the algebraic expression.

5. $-2(x - 3) - 6$ 6. $6(x + 10) + 4$

7. $4 - 3(2x + 1)$ 8. $5(x + 2) - 4(2x - 3)$

9. $24\left(\dfrac{y}{3} + \dfrac{y}{6}\right)$ 10. $0.12x + 0.05(2000 - 2x)$

Section 1.8

In Exercises 1–4, perform the multiplication.

1. $(x + 1)^2$ 2. $(2 - y)(3 + 2y)$

3. $(4 - 5z)(4 + 5z)$ 4. $t(t - 4)(2t + 3)$

In Exercises 5–10, factor the expression completely.

5. $12x^2 + 7x - 10$ 6. $25x^3 - 60x^2 + 36x$

7. $12z^4 + 17z^3 + 5z^2$ 8. $y^3 + 125$

9. $x^3 + 3x^2 - 4x - 12$

10. $x^3 + 2x^2 + 3x + 6$

Section 2.1

In Exercises 1–6, plot each real number on the real number line and place the correct inequality symbol ($<$ or $>$) between the real numbers.

1. -6 ▨ 2 2. $\dfrac{11}{3}$ ▨ 2

3. -3 ▨ -7 4. -5 ▨ 0

5. $\dfrac{13}{3}$ ▨ $-\dfrac{3}{2}$ 6. $\dfrac{3}{5}$ ▨ $\dfrac{13}{16}$

In Exercises 7–10, evaluate the specified quantity.

7. $5 - |-2|$ 8. $-15 - 3(4 - 18)$

9. $\dfrac{3 - 2(5 - 20)}{4}$ 10. $\dfrac{|3 - 12|}{3}$

Section 2.2

In Exercises 1–4, plot the points on a rectangular coordinate system.

1. $(-3, 2), (5, -4)$ 2. $(2, 8), (7, -3)$

3. $\left(\dfrac{5}{2}, \dfrac{7}{2}\right), \left(\dfrac{7}{3}, -2\right)$ 4. $\left(-\dfrac{9}{4}, -\dfrac{1}{4}\right), \left(-3, \dfrac{9}{2}\right)$

In Exercises 5 and 6, solve the equation.

5. $\dfrac{5}{8}x - 6 = 0$ 6. $14 - \dfrac{7}{3}x = 0$

In Exercises 7–10, find the missing coordinate of the solution point.

7. $y = \dfrac{3}{5}x + 4$, $\left(15, \text{▨}\right)$

8. $y = 3 - \dfrac{5}{9}x$, $\left(12, \text{▨}\right)$

9. $y = 5.5 - 0.95x$, $\left(\text{▨}, -1\right)$

10. $y = 3 + 0.2x$, $\left(\text{▨}, 4.4\right)$

Section 2.3

In Exercises 1–4, evaluate the quantity.

1. $\dfrac{8 - 3}{12 - 2}$ 2. $\dfrac{-3 - (-6)}{11 - 5}$

3. $\dfrac{14 - (-15)}{-13 - (-11)}$ 4. $\dfrac{-3 - 7}{0 - (-20)}$

In Exercises 5 and 6, find $-1/m$ for the given value of m.

5. $m = -\dfrac{5}{16}$ 6. $m = \dfrac{3}{7}$

In Exercises 7–10, solve for y in terms of x.

7. $5x - 8y = 5$ 8. $9x + 2y = 0$

9. $y - (-7) = 2[x - (-3)]$

10. $y - 14 = \dfrac{3}{4}(x - 12)$

Section 2.4

In Exercises 1–6, simplify the expression.

1. $2(x + h) - 3 - (2x - 3)$

2. $3(y + t) + 5 - (3y + 5)$

3. $2(x - 4)^2 - 5$

4. $-3(x + 2)^2 + 10$

5. $4 - 2[3 + 4(x + 1)]$

6. $5x + x[3 - 2(x - 3)]$

In Exercises 7–10, evaluate the expression at the given values of the variable. (If not possible, state the reason.)

7. $3x^2 - 4$ (a) $x = 0$ (b) $x = -2$

                     (c) $x = 2$ (d) $x = 5$

8. $9 - (x - 3)^2$ (a) $x = 0$ (b) $x = 3$

                     (c) $x = 6$ (d) $x = 2.5$

9. $\dfrac{x + 4}{x - 3}$ (a) $x = -4$ (b) $x = 3$

                     (c) $x = 4$ (d) $x = 2$

10. $x + \dfrac{1}{x}$ (a) $x = 1$ (b) $x = -1$

                     (c) $x = \dfrac{1}{2}$ (d) $x = 10$

Section 2.5

In Exercises 1–6, sketch the graph of the equation.

1. $y = 3 - 2x$ **2.** $y = \frac{1}{2}x - 1$

3. $y = (x - 2)^2 - 4$ **4.** $y = 9 - (x + 1)^2$

5. $x - y^2 = 0$ **6.** $y = |x| + 1$

In Exercises 7–10, evaluate the function at the indicated values.

Function	*Function Value*
7. $f(x) = \frac{1}{3}x^2$	(a) $f(6)$ (b) $f\left(\frac{3}{4}\right)$
8. $f(x) = 3 - 2x$	(a) $f(5)$
	(b) $f(x + 3) - f(3)$
9. $f(x) = \dfrac{x}{x + 10}$	(a) $f(5)$ (b) $f(c - 6)$
10. $f(x) = \sqrt{x - 4}$	(a) $f(16)$ (b) $f(t + 3)$

Section 2.6

In Exercises 1–10, simplify the expression.

1. $2(x - 3) + 6$ **2.** $-5(2 - t) - 3$

3. $-3(x + 4) - 8$ **4.** $7(u - 3) + 2(u - 3)$

5. $(x - 1)^2 + 2(x - 1)$ **6.** $5(x + 2) - (x + 2)^2$

7. $4(y + 3) - (y + 3)^2$ **8.** $x(x - 5) + 3(x - 5)$

9. $2(z - 3)^3 + (z - 3)^2$ **10.** $5(t - 4) - (t - 4)^3$

Section 3.1

In Exercises 1–4, determine the slope of the line passing through the points.

1. $(-5, 2), (3, 0)$ **2.** $(-8, -4), (6, 10)$

3. $(-3, -2), (-3, 5)$ **4.** $\left(\frac{5}{6}, \frac{3}{2}\right), \left(\frac{5}{3}, \frac{3}{4}\right)$

In Exercises 5 and 6, sketch the lines through the given point with the indicated slopes. Make the sketches on the same set of coordinate axes.

Point	*Slopes*
5. $(3, 2)$	(a) 0 (b) 14 (c) $\frac{3}{2}$ (d) $-\frac{1}{4}$
6. $(-2, 4)$	(a) 2 (b) -2 (c) $\frac{1}{2}$ (d) Undefined

In Exercises 7–10, use function notation to write y as a linear function of x.

7. $y - 3 = -\frac{3}{5}(x - 2)$ **8.** $y + 5 = \frac{4}{3}(x - 3)$

9. $y - (-2) = \dfrac{5 - (-3)}{4 - 6}(x - 5)$

10. $y - 7 = \dfrac{1 - 3}{0 - 2}[x - (-2)]$

Section 3.2

In Exercises 1–6, solve the equation for k.

1. $8 = k(12)$ **2.** $300 = k(75)$

3. $46 = \dfrac{k}{0.04}$ **4.** $12 = \dfrac{k}{27}$

5. $135 = \dfrac{k(72)}{8}$

6. $70 = \dfrac{k(49)}{2^2}$

In Exercises 7–10, solve for y.

7. $3x + y = 4$

8. $2x + 3y = 2$

9. $5(y + 2) = 4x$

10. $7(y - 4) = 5(x + 2)$

Section 3.3

In Exercises 1–4, use the rules of algebra to simplify the expression.

1. $25 - 6(x - 3)$

2. $-2(u + 8) + u(u + 2)$

3. $3(2n + 1) - 2(2n + 3)$ **4.** $5(2n) + 3(2n + 2)$

In Exercises 5–10, solve the equation. (If not possible, state the reason.)

5. $4 - \dfrac{x}{2} = 6$

6. $\dfrac{x}{3} + 1 = 10$

7. $8(x - 14) = 0$

8. $12(3 - x) = 5 - 7(2x + 1)$

9. $0.75x = 235$

10. $(1 + r)500 = 550$

Section 3.4

In Exercises 1–10, solve the equation.

1. $44 - 16x = 0$

2. $-4(x - 5) = 0$

3. $3[4 + 5(x - 1)] = 6x + 2$

4. $\dfrac{3x}{8} + \dfrac{3}{4} = 2$

5. $\dfrac{x}{3} + \dfrac{x}{2} = \dfrac{1}{3}$

6. $x - \dfrac{x}{4} = 15$

7. $\dfrac{5x}{4} + \dfrac{1}{2} = x - \dfrac{1}{2}$

8. $\dfrac{3}{2}(z + 5) - \dfrac{1}{4}(z + 24) = 0$

9. $0.25x + 0.75(10 - x) = 3$

10. $0.60x + 0.40(100 - x) = 50$

Section 3.5

In Exercises 1–4, place the correct inequality symbol (< or >) between the two real numbers.

1. $-\dfrac{3}{4}$ ▨ -5

2. $-\dfrac{1}{5}$ ▨ $-\dfrac{1}{3}$

3. π ▨ -3

4. 6 ▨ $\dfrac{13}{2}$

In Exercises 5–10, solve the equation.

5. $-2n + 12 = 5$

6. $16 + 7z = 64$

7. $\dfrac{12 + x}{4} = 13$

8. $20 - \dfrac{x}{9} = 3$

9. $55 - 4(3 - y) = -5$

10. $8(t - 24) = 0$

Section 3.6

In Exercises 1–6, evaluate the expression.

1. $|-15|$

2. $|-3.2|$

3. $-|72|$

4. $-|-4|$

5. $|14 - 32|$

6. $-|46.8 - 27|$

In Exercises 7–10, solve the inequality.

7. $-4 < 10x + 1 < 6$

8. $-2 \le 1 - 2x \le 2$

9. $-3 \le \dfrac{x}{2} \le 3$

10. $-5 < x - 25 < 5$

Section 4.1

In Exercises 1–4, sketch the graph of the equation.

1. $y = -\dfrac{1}{4}x + 8$

2. $y = 3(x - 2)$

3. $y - 2 = 2(x + 3)$

4. $y + 3 = \dfrac{1}{3}(x - 4)$

In Exercises 5 and 6, solve the equation.

5. $2y - 3(4y - 2) = 5$

6. $8x + 6(3 - 2x) = 4$

In Exercises 7 and 8, find an equation of the line passing through the two points.

7. $(-1, -2), (3, 6)$

8. $(1, 5), (6, 0)$

In Exercises 9 and 10, determine the slope of the line.

9. $4x + 6y = 4$

10. $-9x + 5y = 10$

Section 4.2

In Exercises 1–6, solve the system of linear equations.

1. $\begin{cases} x + 3y = 17 \\ 2x + y = 14 \end{cases}$

2. $\begin{cases} 4x + 2y = -2 \\ 3x - 2y = -5 \end{cases}$

3. $\begin{cases} -\frac{1}{2}x + \frac{3}{2}y = 2 \\ 2x - 6y = 1 \end{cases}$

4. $\begin{cases} \frac{2}{3}x + \frac{5}{6}y = 3 \\ 4x + 5y = 18 \end{cases}$

5. $\begin{cases} 2x + 5y = -4 \\ 3x + 6y = -3 \end{cases}$

6. $\begin{cases} 4x - 3y = -2 \\ 5x - 4y = -2 \end{cases}$

In Exercises 7–10, simplify the expression.

7. $(2x - 3y + 5z) + 3(x + y - 2z)$

8. $(x + 4y + 3z) - 3(2x - 10y + z)$

9. $-4(6x + 5y - 20z) + 3(8x + y - 15z)$

10. $5(2x - 3y + 4z) - 3(x - 5y - 2z)$

Section 4.3

In Exercises 1–6, solve the system of equations.

1. $\begin{cases} x - 3y = -1 \\ -3x + 2y = -4 \end{cases}$

2. $\begin{cases} 2x + y = 4 \\ -x - y = -1 \end{cases}$

3. $\begin{cases} x - 3y = -7 \\ -2x + y = -6 \end{cases}$

4. $\begin{cases} 2x + 3y = 2 \\ 4x + 9y = 5 \end{cases}$

5. $\begin{cases} x - y = -1 \\ x + 2y - 2z = 3 \\ 3x - y + 2z = 3 \end{cases}$

6. $\begin{cases} 2x + y - 2z = 1 \\ x - z = 1 \\ 3x + 3y + z = 12 \end{cases}$

In Exercises 7–10, evaluate the expression.

7. $3 - \frac{3}{2}(-4)$

8. $-2 + \frac{4}{5}(6)$

9. $5(-2) + 7\left(\frac{3}{4}\right)$

10. $-\frac{2}{3}(5) - \frac{1}{2}(-3)$

Section 4.4

In Exercises 1–10, evaluate the quantity.

1. $2(25) - 5(8)$

2. $6(-8) - 4(2)$

3. $3(-2) - 7(4) + 8(-6)$

4. $10\left(\frac{3}{2}\right) - 6(-8) + (-3)(-30)$

5. $-4\left[-1(8) - 2\left(\frac{3}{4}\right) + (-4)\left(-\frac{3}{8}\right)\right]$

6. $\frac{5}{6}[10(-4) - (-1)(-6) + 4(5)]$

7. $\dfrac{2(-2) - 8(3)}{8\left(\frac{3}{2}\right) - 6\left(-\frac{1}{3}\right)}$

8. $\dfrac{5(3) - 4(-3)}{-6(2) - 5(-9)}$

9. $\dfrac{4(4) - 5(6) + 2(10)}{-5(1) - 10(3) + 8(5)}$

10. $\dfrac{-12\left(\frac{1}{4}\right) - 2(8) + 3(-1)}{3(2) - 4(15) + 10(6)}$

Section 4.5

In Exercises 1–10, sketch the graph of the equation.

1. $x + 2y = 2$

2. $2x - 3y = 4$

3. $y = \frac{1}{2}(x - 2) - 3$

4. $y = -2(x + 1) + 4$

5. $x + 3y = 6$

6. $0.1x - 0.5y = 0.3$

7. $x = -3$

8. $y = 8$

9. $x + y = 0$

10. $9x + 18y = 0$

Section 4.6

In Exercises 1–4, simplify the expression.

1. $(4x + 3y) - 3(5x + y)$

2. $(-15u + 4v) + 5(3u - 9v)$

3. $2x + (2x - 3) + 12x$

4. $y - (y + 2) + 4y$

In Exercises 5 and 6, solve the given equation.

5. $5x - (2x - 5) = 17$

6. $4t - 3(t + 1) = 8$

In Exercises 7–10, determine whether the lines represented by the pair of equations are parallel or perpendicular, or neither.

7. $\begin{cases} 3x - 4y = -10 \\ 4x + 3y = 11 \end{cases}$ **8.** $\begin{cases} 6x - 18y = 8 \\ -2x + 6y = 1 \end{cases}$

9. $\begin{cases} 0.3x - 0.5y = 1 \\ -1.2x + 2y = 2 \end{cases}$ **10.** $\begin{cases} 2x + y = 2 \\ 3x + 4y = 1 \end{cases}$

Section 5.1

In Exercises 1–10, simplify the expression.

1. $a^4 \cdot a^6$

2. $x^2 \cdot y^4 \cdot x^3$

3. $\dfrac{81y^5}{3^2 y^2}$, $y \neq 0$

4. $\dfrac{64r^2 s^4}{16rs^2}$, $r \neq 0, s \neq 0$

5. $(-t^2)^3$

6. $(4z)^2$

7. $(-3x^2 y^3)^2 \cdot (4xy^2)$

8. $(-4uv^2)^3 \cdot (2u^2 v)$

9. $\left(\dfrac{2a^2}{3b}\right)^4$, $b \neq 0$

10. $\left(\dfrac{3x}{4y^3}\right)^2$, $y \neq 0$

Section 5.2

In Exercises 1–6, evaluate the expression.

1. $(-13)^2$

2. -13^2

3. $\left(\frac{2}{3}\right)^3$

4. $\frac{5}{6}\left(\frac{3}{5}\right)^2$

5. $\dfrac{3}{4^{-2}}$

6. $\left(\frac{2}{3}\right)^{-3}$

In Exercises 7–10, simplify the expression.

7. $(x^3 \cdot x^{-2})^{-3}$

8. $(5x^{-4} y^5)(-3x^2 y^{-1})$

9. $\left(\dfrac{2x}{3y}\right)^{-2}$

10. $\left(\dfrac{7u^{-4}}{3v^{-2}}\right)\left(\dfrac{14u}{6v^2}\right)^{-1}$

Section 5.3

In Exercises 1–4, find the indicated *n*th root without using a calculator.

1. $\sqrt{10,000}$

2. $-\sqrt[3]{1000}$

3. $\sqrt[3]{-\frac{27}{125}}$

4. $\sqrt[4]{\frac{16}{625}}$

In Exercises 5–10, simplify the expression.

5. $x^{5/3} \cdot x^{1/6}$

6. $\dfrac{a^{7/4}}{a^{3/2}}$

7. $(2x)^{3/2}(2x)^{5/2}$

8. $(-4x^{1/3} y^{4/3})^6$

9. $\dfrac{32x^{1/4} y^{3/4}}{2x^{1/2} y}$

10. $\dfrac{8y^{3/2} z^{-2}}{y^{-1/2} z^{1/2}}$

Section 5.4

In Exercises 1–6, perform the indicated operations and simplify.

1. $19 - 5(x - 3)$

2. $32 + 6(3x - 10)$

3. $\left(t + \frac{1}{3}\right)^2$

4. $(2x - 5)^2$

5. $(5x - 4)(5x + 4)$

6. $\left(\frac{1}{2} - 3x\right)(4 + x)$

In Exercises 7–10, simplify the expression. (Assume that all variables are positive.)

7. $\sqrt[3]{16x^4 y^3}$

8. $\sqrt{72x^2 y^3}$

9. $\sqrt{3x^2} - 2\sqrt{12}$

10. $\dfrac{14x}{\sqrt{7}}$

Section 5.5

In Exercises 1–6, solve the equation.

1. $13x + 6 = 32$

2. $7x = 18 + 3x$

3. $17 - 5(x - 2) = 0$

4. $32 + 3(2x - 11) = 0$

5. $3(x - 2)(x + 25) = 0$

6. $-2(12 - x)(114 + x) = 0$

In Exercises 7–10, perform the indicated operations and simplify. (Assume that all variables are positive.)

7. $\left(\sqrt{x} + 3\right)\left(\sqrt{x} - 3\right)$

8. $\sqrt{u}\left(\sqrt{20} - \sqrt{5}\right)$

9. $\left(2\sqrt{t} + 3\right)^2$

10. $\dfrac{50x}{\sqrt{2}}$

Section 5.6

In Exercises 1–6, simplify the expression.

1. $\sqrt{48}$ **2.** $\sqrt{108}$

3. $\sqrt{128} + 3\sqrt{50}$ **4.** $3\sqrt{5}\sqrt{500}$

5. $\dfrac{8}{\sqrt{10}}$ **6.** $\dfrac{5}{\sqrt{12} - 2}$

In Exercises 7–10, solve the equation.

7. $x(2x - 35) = 0$ **8.** $(x + 4)\left(x - \frac{5}{4}\right) = 0$

9. $x^2 - 7x - 18 = 0$ **10.** $2x^2 - x - 3 = 0$

Section 6.1

In Exercises 1–6, completely factor the expression.

1. $16x^2 - 121$ **2.** $9t^2 - 24t + 16$

3. $5x^2 - 13x - 6$ **4.** $x(x - 10) - 4(x - 10)$

5. $4y^3 + 4y^2 - 3y$ **6.** $4x^3 - 12x^2 + 16x$

In Exercises 7–10, solve the equation.

7. $5y + 6 = 0$ **8.** $2s - 7 = 0$

9. $(x + 9)(2x - 15) = 0$ **10.** $(5x - 8)(x - 12) = 0$

Section 6.2

In Exercises 1–4, expand and simplify the expression.

1. $(x + 3)^2 - 1$ **2.** $(x + 10)^2 + 50$

3. $\left(u - \frac{1}{2}\right)^2$ **4.** $(v - 12)^2 - 100$

In Exercises 5–10, solve the equation. (Find all real *and* imaginary solutions.)

5. $x^2 = \frac{4}{49}$ **6.** $y^2 = -\frac{9}{16}$

7. $(y - 4)^2 = -36$ **8.** $(z + 2)^2 = 20$

9. $(3x - 2)^2 = 5$

10. $3(12 - x) - 8x(12 - x) = 0$

Section 6.3

In Exercises 1–4, solve the quadratic equation by factoring.

1. $x^2 - 10x + 21 = 0$ **2.** $x^2 + 14x - 32 = 0$

3. $3x^2 + x - 30 = 0$ **4.** $5x^2 - 18x + 9 = 0$

In Exercises 5 and 6, solve the quadratic equation by completing the square. (Find all real *and* imaginary solutions.)

5. $x^2 + 3x - 2 = 0$ **6.** $3x^2 + 6x + 4 = 0$

In Exercises 7–10, simplify the radical.

7. $\sqrt{25 - 4(2)(2)}$ **8.** $\sqrt{9 - 4(1)(5)}$

9. $\sqrt{64 - 4(2)(-4)}$ **10.** $\sqrt{81 - 4(3)(-6)}$

Section 6.4

In Exercises 1–10, solve the equation. (Find all real *and* imaginary solutions.)

1. $3(x + 7) = 0$ **2.** $3x(15x - 2) = 0$

3. $2n(n - 2) + (n - 2) = 3$

4. $2n(n + 2) = 240$

5. $3(x + 8)^2 = 243$ **6.** $t + 3\sqrt{t} - 4 = 0$

7. $x^2 + 3x + 5 = 0$ **8.** $u^2 + 3u - 5 = 0$

9. $t - \dfrac{6}{t} = 5$ **10.** $\sqrt{2s + 3} = s$

Section 6.5

In Exercises 1–4, sketch the graph of the equation.

1. $y = 3x - 2$ **2.** $y = \frac{1}{2}x + 2$

3. $y = -3x + 5$ **4.** $y = -\frac{1}{2}x + 2$

In Exercises 5–8, solve the equation.

5. $10(x - 2) = 25$ **6.** $30 - 3x = 70 - 5x$

7. $10x(x + 8) = 0$ **8.** $3x(2x + 1) = 15$

In Exercises 9 and 10, expand and simplify the expression.

9. $2(x + 5)^2 - 30$ **10.** $-3(x - 2)^2 - 7$

Section 6.6

In Exercises 1–10, sketch the graph of the equation. Identify the vertex.

1. $y = \frac{1}{2}(x - 3)^2$
2. $y = (x + 1)^2$
3. $y = -(x + 4)^2 + 4$
4. $y = -2(x - 3)^2 + 1$
5. $y = x^2 + 4x + 2$
6. $y = x^2 - 10x + 22$
7. $y = -3x^2 - 6x + 1$
8. $y = -2x^2 + 16x - 25$
9. $y = \frac{2}{3}x^2 - 4x + 8$
10. $y = -\frac{3}{4}x^2 - \frac{3}{2}x + \frac{21}{4}$

Section 6.7

In Exercises 1–6, solve the inequality and sketch the graph of the solution on the real number line.

1. $3x - 15 \le 0$
2. $3x + 5 \ge 0$
3. $7 - 3x > 4 - x$
4. $2(x + 6) - 20 < 2$
5. $|x - 3| < 2$
6. $|x - 5| > 3$

In Exercises 7–10, determine whether the product is zero, positive, or negative. (It is not necessary to find the actual product.)

7. $(35 - 10)(13 - 25)$
8. $(254 - 254)(625 - 500)$
9. $(64 - 82)(12 - 15)$
10. $\left(\frac{3}{4} - \frac{3}{5}\right)\left(\frac{1}{6} - \frac{2}{3}\right)$

Section 7.1

In Exercises 1–4, find the missing real number that makes the two fractions equivalent.

1. $-\dfrac{5}{6} = \dfrac{\rule{1cm}{0.15cm}}{24}$
2. $\dfrac{3}{7} = \dfrac{30}{\rule{1cm}{0.15cm}}$
3. $\dfrac{4}{3} = \dfrac{-28}{\rule{1cm}{0.15cm}}$
4. $-\dfrac{9}{16} = \dfrac{\rule{1cm}{0.15cm}}{48}$

In Exercises 5–8, write the fraction in reduced form.

5. $\frac{45}{78}$
6. $\frac{55}{115}$

7. $\frac{160}{256}$
8. $\frac{189}{324}$

In Exercises 9 and 10, factor the algebraic expression.

9. $20ax^2 + 60ax + 80a$
10. $2x^2y^2 - 9xy^2 - 56y^2$

Section 7.2

In Exercises 1–6, evaluate the quantity and write your answer in reduced form.

1. $\dfrac{3}{16} \cdot \dfrac{8}{5}$
2. $\dfrac{5}{12} \cdot \dfrac{9}{75}$
3. $-\dfrac{13}{35} \cdot \dfrac{-25}{104}$
4. $\dfrac{-225}{-448} \div \dfrac{-105}{28}$
5. $\dfrac{14}{3} \div \dfrac{42}{45}$
6. $\dfrac{-8}{105} \div \dfrac{2}{3}$

In Exercises 7–10, factor the algebraic expression.

7. $x^2 - 3x + 2$
8. $2x^2 + 5x - 7$
9. $11x^2 + 6x - 5$
10. $4x^2 - 28x + 49$

Section 7.3

In Exercises 1–10, perform the indicated operations.

1. $\frac{4}{9} + \frac{3}{9}$
2. $\frac{3}{20} + \frac{7}{20}$
3. $\frac{11}{32} - \frac{5}{32}$
4. $\frac{7}{10} - \frac{13}{10}$
5. $\frac{3}{4} + \frac{5}{18}$
6. $\frac{11}{24} + \frac{5}{16}$
7. $-\frac{8}{9} + \frac{18}{75}$
8. $\frac{27}{3} - \frac{8}{21}$
9. $\frac{28}{5} \cdot \frac{65}{14}$
10. $\frac{15}{16} \div 12$

Section 7.4

In Exercises 1–6, rewrite the fraction in reduced form.

1. $\dfrac{36}{144}$
2. $\dfrac{58}{180}$
3. $\dfrac{-16x^2}{12x}$
4. $\dfrac{5t^4}{45t^2}$
5. $\dfrac{6u^2v}{27uv^3}$
6. $\dfrac{14r^4s^2}{-98rs^2}$

In Exercises 7 and 8, find the indicated product.

7. $-3x^3(5x - 4)$ **8.** $10y(7y^2 - 2)$

In Exercises 9 and 10, perform the indicated subtraction.

9. $(3x^3 + 5x^2 - 2x - 1) - (x^3 - 2x^2 + x + 3)$

10. $(7x^2 + 6) - (-x^2 - 10x + 3)$

Section 7.5

In Exercises 1–10, solve the equation.

1. $3x = 16$ **2.** $-5x = 95$

3. $15x + 3 = 48$ **4.** $125 - 50x = 0$

5. $(3x - 2)(x + 8) = 0$ **6.** $(2x + 3)(x - 16) = 0$

7. $x(2x - 21) = 0$ **8.** $x(10 - x) = 25$

9. $x^2 + x - 42 = 0$ **10.** $t^2 - 8t = 0$

Section 7.6

In Exercises 1–4, factor the polynomial.

1. $x^2 - 3x - 10$ **2.** $x^2 - 7x + 10$

3. $x^3 + 4x^2 + 3x$ **4.** $x^3 - 4x^2 - 2x + 8$

In Exercises 5–8, sketch the graph of the equation.

5. $y = 2$ **6.** $x = -1$

7. $y = x - 2$ **8.** $y = -x + 1$

In Exercises 9 and 10, use long division to write the rational expression as the sum of a constant and a rational expression.

9. $\dfrac{5x + 6}{x - 4}$ **10.** $\dfrac{-2x + 3}{x + 2}$

Section 7.7

In Exercises 1–10, solve the inequality.

1. $2x - 12 \geq 0$ **2.** $7 - 3x < 4 - x$

3. $2(1 - 5x) > 10 - 2x$ **4.** $2(x + 6) - 20 < 2$

5. $|x - 3| < 2$ **6.** $|x - 5| > 3$

7. $|x - 3| \geq 2$ **8.** $|x - 5| \leq 3$

9. $x^2 - 5x \geq 0$ **10.** $3x - x^2 > 0$

Section 8.1

In Exercises 1–4, find the domain of the function.

1. $f(x) = x^3 + 3x^2 - 2$ **2.** $A(x) = \sqrt{10 - x}$

3. $H(x) = \dfrac{12}{x - 5}$ **4.** $g(x) = \dfrac{25}{x^2 - 25}$

In Exercises 5–10, sketch the graph of the function.

5. $f(x) = \frac{1}{2}x$ **6.** $f(x) = -2x + 3$

7. $f(x) = x^2 - 4$ **8.** $f(x) = 2x^2 - x$

9. $f(x) = |x| + 1$ **10.** $f(x) = x^3 + 3$

Section 8.2

In Exercises 1–4, find the domain of the function.

1. $f(x) = x^3 - 2x$ **2.** $g(x) = \sqrt[3]{x}$

3. $h(x) = \sqrt{16 - x^2}$ **4.** $A(x) = \dfrac{3}{36 - x^2}$

In Exercises 5–10, use the functions f and g to find the combinations $(f + g)(x)$, $(f - g)(x)$, $(fg)(x)$, $(f/g)(x)$, $(f \circ g)(x)$, and $(g \circ f)(x)$. Find the domain of each combination.

5. $f(x) = x^2 - 9$
$g(x) = 3x$

6. $f(x) = 2x - 3$
$g(x) = \frac{1}{2}x^2$

7. $f(x) = \sqrt[3]{x - 2}$
$g(x) = x^2$

8. $f(x) = \sqrt{x}$
$g(x) = x^2 - 4$

9. $f(x) = \dfrac{1}{x}$
$g(x) = \dfrac{1}{x - 2}$

10. $f(x) = x + 2$
$g(x) = \dfrac{1}{x + 2}$

Section 8.3

In Exercises 1–6, solve the equation for k.

1. $4 = k(12)$

2. $250 = k(75)$

3. $46 = \dfrac{k}{0.02}$

4. $19 = \dfrac{k}{27}$

5. $360 = \dfrac{k(72)}{8}$

6. $700 = \dfrac{k(49)}{2^2}$

In Exercises 7–10, sketch the graph of $y = kx^2$ for the value of k.

7. $k = \frac{1}{4}$

8. $k = 2$

9. $k = -4$

10. $k = -\frac{1}{2}$

Section 8.4

In Exercises 1–6, factor the expression completely.

1. $12x^2 + 7x - 10$

2. $25x^3 - 60x^2 + 36x$

3. $12z^4 + 17z^3 + 5z^2$

4. $y^3 + 125$

5. $x^3 + 3x^2 - 4x - 12$

6. $x^3 + 2x^2 + 3x + 6$

In Exercises 7–10, find all real solutions of the equation.

7. $5x^2 + 8 = 0$

8. $x^2 - 6x + 4 = 0$

9. $4x^2 + 4x - 11 = 0$

10. $x^4 - 18x^2 + 81 = 0$

Section 8.5

In Exercises 1–4, solve the equation for y.

1. $2x + 3y - 9 = 0$

2. $x^2 + y^2 = 36$

3. $2x^2 + 3y^2 - 6 = 0$

4. $2x^2 - y^2 - 3 = 0$

In Exercises 5 and 6, simplify the radical.

5. $\sqrt{450}$

6. $\sqrt{1350}$

In Exercises 7–10, sketch the graph of the equation.

7. $y = 2x - 3$

8. $y = -\frac{3}{4}x + 2$

9. $y = x^2 - 4x + 4$

10. $y = 9 - x^2$

Section 8.6

In Exercises 1 and 2, find the distance between the points.

1. $(4, 1), (10, 6)$

2. $(-1, 5), (3, -2)$

In Exercises 3–6, graph the lines on the same set of coordinate axes.

3. $y = \pm\frac{1}{2}x$

4. $y = 3 \pm \frac{1}{2}x$

5. $y = 3 \pm \frac{1}{2}(x - 4)$

6. $y = \pm\frac{1}{2}(x - 4)$

In Exercises 7–10, find the unknown in the equation $c^2 = a^2 - b^2$. (Assume that a, b, and c are positive.)

7. $a = 13, b = 5$

8. $a = \sqrt{10}, c = 3$

9. $b = 6, c = 8$

10. $a = 7, b = 5$

Section 8.7

In Exercises 1–4, expand and simplify the expression.

1. $(x - 5)^2 - 20$

2. $(x + 3)^2 - 1$

3. $10 - (x + 4)^2$

4. $4 - (x - 2)^2$

In Exercises 5–8, complete the square for the quadratic expression.

5. $x^2 + 6x + 8$

6. $x^2 - 10x + 21$

7. $-x^2 + 2x + 1$

8. $-2x^2 + 4x - 2$

In Exercises 9 and 10, find an equation of the line passing through the point with the specified slope.

9. $(1, 6), m = -\frac{2}{3}$

10. $(3, -2), m = \frac{3}{4}$

Section 8.8

In Exercises 1–6, sketch the graph of the equation.

1. $y = \sqrt{x - 3}$

2. $x^2 + y^2 = 49$

3. $x^2 + 4y^2 = 16$

4. $y^2 - x^2 = 16$

5. $y = (x - 2)^2$

6. $y = 3 - x^3$

In Exercises 7–10, solve the system of linear equations.

7. $\begin{cases} x - 4y = 0 \\ 3x + 4y = 8 \end{cases}$ **8.** $\begin{cases} 4x + 5y = 6 \\ 3x - y = 14 \end{cases}$

9. $\begin{cases} x + y + z = 0 \\ 4x + y - 3z = -9 \\ 2x - 3y + 2z = 20 \end{cases}$ **10.** $\begin{cases} 2x - 3y + 2z = 20 \\ 3x + y - z = 6 \\ 2x + 3y - z = 2 \end{cases}$

Section 9.1

In Exercises 1–6, evaluate the expression.

1. $\left(\frac{3}{4}\right)^3$ **2.** $(3 \cdot 10)^2$ **3.** 2^{-2}

4. 3^{-4} **5.** $25^{1/2}$ **6.** $81^{-1/2}$

In Exercises 7–10, use the properties of exponents to simplify the expression.

7. z^0 **8.** $y^3 \cdot y^2$ **9.** $\dfrac{26x^4}{2x^3}$ **10.** $\left(-\dfrac{2}{y}\right)^3$

Section 9.2

In Exercises 1–10, use the properties of exponents to simplify the expression.

1. $x^5 \cdot x^3$ **2.** $(x + 10)^0, \quad x \neq -10$

3. $\dfrac{z^7}{z^4}, \quad z \neq 0$ **4.** $\dfrac{18(x - 3)^5}{24(x - 3)}, \quad x \neq 3$

5. $(x^2y)^4$ **6.** $\left(\dfrac{x}{2y}\right)^5, \quad y \neq 0$

7. $x^4y^{-3}, \quad y \neq 0$ **8.** $(a^2)^{-4}, \quad a \neq 0$

9. $(b^6)^{1/2}$ **10.** $(8x^3)^{1/3}$

Section 9.3

In Exercises 1–6, use the properties of exponents to simplify the expression.

1. $(x^2 \cdot x^3)^4$ **2.** $4^{-2} \cdot x^2$ **3.** $\dfrac{15y^{-3}}{10y^2}$

4. $\left(\dfrac{3x^2}{2y}\right)^{-2}$ **5.** $\dfrac{3x^2y^3}{18x^{-1}y^2}$ **6.** $(x^2 + 1)^0$

In Exercises 7–10, rewrite the equation in exponential form.

7. $\log_4 64 = 3$ **8.** $\log_3 \frac{1}{81} = -4$

9. $\ln e^{2x} = 2x$ **10.** $\ln 1 = 0$

Section 9.4

In Exercises 1–6, solve for x.

1. $10x - 3 = 22$ **2.** $3 - 2(x - 1) = 12$

3. $(x - 4)^2 - 36 = 0$ **4.** $x^2 - 5x + 6 = 0$

5. $-28 + x \ln 5 = 0$ **6.** $5xe^2 = 0$

In Exercises 7–10, simplify the expression.

7. $\log_{10} 10^3$ **8.** $\log_5 5^6$

9. $\log_2 1$ **10.** $\ln e^4$

Section 9.5

In Exercises 1–10, solve the equation. (Round your answer to two decimal places.)

1. $6 + 3(x - 5) = 27 - 5(x - 6)$

2. $5x^2 + 2x - 3 = 0$

3. $\sqrt{x - 5} = 6$ **4.** $\dfrac{4}{t} + \dfrac{3}{2t} = 1$

5. $\log_6 2x = 2$ **6.** $\log_2(x - 5) = 6$

7. $e^{x/2} = 8$ **8.** $e^{x-5} = 550$

9. $300(1.01)^t = 1200$ **10.** $100(1 - e^t) = 75$

Section 9.6

In Exercises 1–6, evaluate the expression. Use a calculator only if it is necessary or more efficient.

1. $\log_{10} 10,000$ **2.** $\log_4 64$ **3.** $\log_3 \frac{1}{81}$

4. $\ln e^3$ **5.** e^{-1} **6.** $e^{1/2}$

In Exercises 7–10, describe the relationship between the graphs of f and g.

7. $g(x) = f(x) - 4$ **8.** $g(x) = f(x - 4)$

9. $g(x) = -f(x)$ **10.** $g(x) = f(-x)$

Section 10.1

In Exercises 1–10, perform the indicated operations and simplify.

1. $\dfrac{1}{3} \cdot \dfrac{3}{5} \cdot \dfrac{5}{7}$ **2.** $\dfrac{3 \cdot 6 \cdot 9 \cdot 12}{3^4}$

3. $\dfrac{1}{1 \cdot 2} + \dfrac{1}{2 \cdot 3} + \dfrac{1}{3 \cdot 4}$ **4.** $\dfrac{1}{2} + \dfrac{1}{3} + \dfrac{1}{4}$

5. $\dfrac{n(n-1)(n-2)}{(n-1)(n-2)}$ **6.** $\dfrac{n^2 - 1}{n + 1}$

7. $\dfrac{2n^2 + 7n - 15}{n^2 - 25}$ **8.** $\dfrac{n^2 + 3n + 2}{n^2 + 2n}$

9. $\dfrac{1}{n} + \dfrac{2}{n^2}$ **10.** $\dfrac{1}{n-1} + \dfrac{1}{n+1}$

Section 10.2

In Exercises 1–4, find the sum.

1. $\displaystyle\sum_{i=1}^{6} \dfrac{1}{i}$ **2.** $\displaystyle\sum_{i=0}^{6} \dfrac{5}{i+1} - \dfrac{5}{i+3}$

3. $\displaystyle\sum_{n=1}^{4} 3^n$ **4.** $\displaystyle\sum_{n=1}^{5} (3n - 1)$

In Exercises 5–10, evaluate the expression.

5. $15\left(\dfrac{5 + 33}{2}\right)$ **6.** $35\left(\dfrac{5 + 107}{2}\right)$

7. $11\left(\dfrac{\frac{16}{3} + \frac{26}{3}}{2}\right)$ **8.** $20\left(\dfrac{\frac{7}{4} + 16}{2}\right)$

9. $\frac{7}{2}[2(-3) + 6(4)]$ **10.** $\frac{9}{2}[2(4) + 8(3)]$

Section 10.3

In Exercises 1–4, evaluate the quantity.

1. $\left(\dfrac{2}{3}\right)^4$ **2.** $\left(\dfrac{3}{5}\right)^3$ **3.** $\dfrac{2^{-3}}{3^2}$ **4.** $\dfrac{8^2}{4^3}$

In Exercises 5–10, simplify the expression.

5. $\dfrac{n!}{(n-1)!}$ **6.** $\dfrac{(n+1)!}{n!}$

7. $(3n)(4n^2)(5n^3)$ **8.** $n(2n)^4$

9. $\dfrac{(3n)^2}{18n}$ **10.** $\dfrac{18n}{10n^{-1}}$

Section 10.4

In Exercises 1–6, perform the indicated operations and simplify.

1. $-3x(x - 2)$ **2.** $(x + 3)(x - 3)$

3. $(x - 2)^2$ **4.** $(2x + 3)^2$

5. $x^2y(5xy^{-3})$ **6.** $(-2x^2y)^3$

In Exercises 7–10, evaluate the quantity.

7. $4!$ **8.** $\dfrac{8!}{7!}$ **9.** $\dfrac{12!}{9!}$ **10.** $\dfrac{10!}{2!8!}$

Section 10.5

In Exercises 1–4, evaluate $_nC_r$.

1. $_5C_2$ **2.** $_{10}C_3$ **3.** $_9C_6$ **4.** $_{20}C_{20}$

In Exercises 5–10, evaluate the quantity.

5. $8 \cdot 2^3 \cdot 3^2$ **6.** $6^2 \cdot 5 \cdot 4$

7. $\dfrac{30!}{28!}$ **8.** $\dfrac{9!}{2!(4!)(3!)}$

9. $\dfrac{20!}{16!4!}$ **10.** $\dfrac{2 \cdot 4 \cdot 6 \cdot 8}{2^4}$

Section 10.6

In Exercises 1–10, evaluate the quantity.

1. $\frac{1}{6} + \frac{3}{4}$ **2.** $\frac{3}{8} + \frac{1}{4}$

3. $\frac{3}{5} + \frac{7}{10} - \frac{1}{15}$ **4.** $\frac{5}{9} + \frac{5}{12} - \frac{5}{16}$

5. $\dfrac{6 \cdot 5}{6!}$ **6.** $\dfrac{8 \cdot 7 \cdot 6}{8!}$

7. $\dfrac{_4C_2}{_8C_2}$ **8.** $\dfrac{_6C_3}{_{10}C_3}$

9. $\frac{7}{8} \cdot \frac{6}{7} \cdot \frac{5}{6}$ **10.** $1 - \left(\frac{2}{5}\right)^3$

Answers to Warm-Up Exercises

Section 1.2

1. 225   **2.** -240 **3.** -150 **4.** 120   **5.** $\frac{8}{5}$
6. $-\frac{1}{4}$ **7.** $\frac{6}{5}$ **8.** $\frac{3}{10}$ **9.** -27 **10.** $\frac{25}{64}$

Section 1.3

1. 168   **2.** 20   **3.** $-\frac{21}{2}$ **4.** 7   **5.** $\frac{72}{5}$ **6.** $-\frac{19}{10}$
7. Multiplicative Inverse Property
8. Associative Property of Addition
9. Distributive Property   **10.** Additive Identity Property

Section 1.4

1. $23y - 69$ **2.** $12 - 30z$ **3.** $-10 + 4x$
4. $-25x + 20$ **5.** $32x, -10y, -7$ **6.** $-3s, 4t, 1$
7. $7x - 15$ **8.** $3y + 36$ **9.** $-4u$ **10.** $-5v - 18$

Section 1.5

1. $24x^2 - 36x$ **2.** $28y - 21y^2$ **3.** $-60x + 42x^2$
4. $-2y^2 - 2y$ **5.** $2t$ **6.** $2z^2 + 3z - 35$
7. $121 - x^2$ **8.** $36r^2 - 25s^2$ **9.** $x^3 - 8$
10. $x^3 + 1$

Section 1.6

1. $x^2 + 2x + 1$ **2.** $x^2 - 10x + 25$ **3.** $4 - 4y + y^2$
4. $6 + y - 2y^2$ **5.** $10z^2 - 29z - 21$
6. $2t^3 - 5t^2 - 12t$ **7.** $3x(x - 7)$
8. $-(x - 3)(x^2 + 1)$ **9.** $(2t - 13)(2t + 13)$
10. $(y - 4)(y^2 + 4y + 16)$

Section 1.7

1. $\frac{5}{56}$ **2.** $\frac{1}{20}$ **3.** 5   **4.** $-\frac{4}{35}$ **5.** $-2x$
6. $6x + 64$ **7.** $1 - 6x$ **8.** $-3x + 22$ **9.** $12y$
10. $0.02x + 100$

Section 1.8

1. $x^2 + 2x + 1$ **2.** $6 + y - 2y^2$ **3.** $16 - 25z^2$
4. $2t^3 - 5t^2 - 12t$ **5.** $(3x - 2)(4x + 5)$
6. $x(5x - 6)^2$ **7.** $z^2(12z + 5)(z + 1)$
8. $(y + 5)(y^2 - 5y + 25)$ **9.** $(x + 3)(x + 2)(x - 2)$

10. $(x + 2)(x^2 + 3)$

Section 2.1

1. $-6 < 2$ **2.** $\frac{11}{3} > 2$

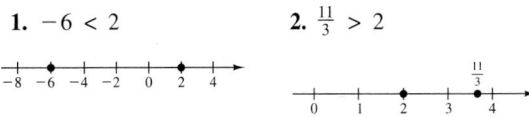

3. $-3 > -7$ **4.** $-5 < 0$

5. $\frac{13}{3} > -\frac{3}{2}$ **6.** $\frac{3}{5} < \frac{13}{16}$

7. 3   **8.** 27   **9.** $\frac{33}{4}$ **10.** 3

Section 2.2

1.   **2.**

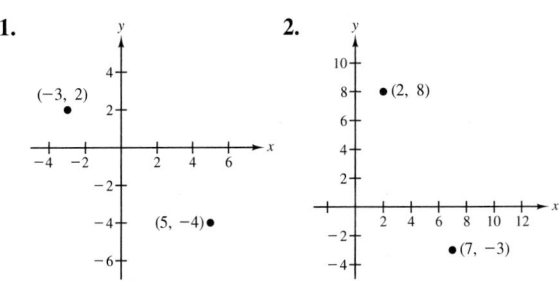

3.   **4.**

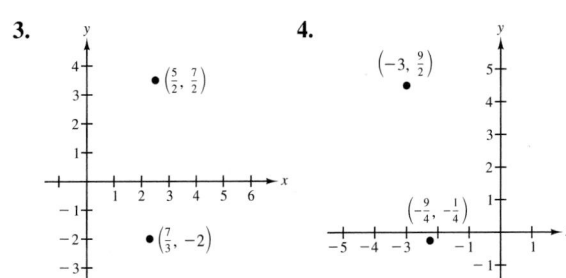

5. $\frac{48}{5}$ **6.** 6   **7.** 13   **8.** $-\frac{11}{3}$ **9.** $\frac{130}{19} \approx 6.84$
10. 7

Section 2.3

1. $\frac{1}{2}$ **2.** $\frac{1}{2}$ **3.** $-\frac{29}{2}$ **4.** $-\frac{1}{2}$ **5.** $\frac{16}{5}$ **6.** $-\frac{7}{3}$
7. $y = \frac{5}{8}(x - 1)$ **8.** $y = -\frac{9}{2}x$ **9.** $y = 2x - 1$
10. $y = \frac{1}{4}(3x + 20)$

Section 2.4

1. $2h$ **2.** $3t$ **3.** $2x^2 - 16x + 27$
4. $-3x^2 - 12x - 2$ **5.** $-8x - 10$ **6.** $-2x^2 + 14x$
7. **(a)** -4 **(b)** 8 **(c)** 8 **(d)** 71
8. **(a)** 0 **(b)** 9 **(c)** 0 **(d)** 8.75
9. **(a)** 0 **(b)** Undefined **(c)** 8 **(d)** -6
10. **(a)** 2 **(b)** -2 **(c)** $\frac{5}{2}$ **(d)** $\frac{101}{10}$

Section 2.5

1.   **2.**
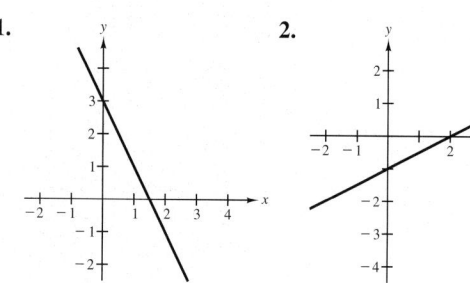

3.   **4.**
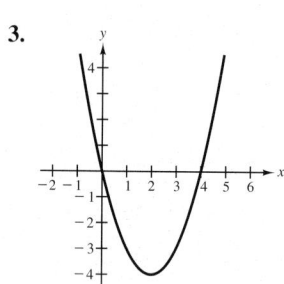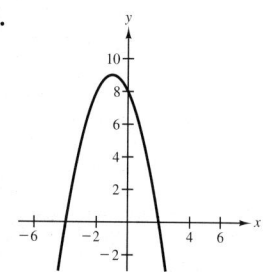

5.   **6.**
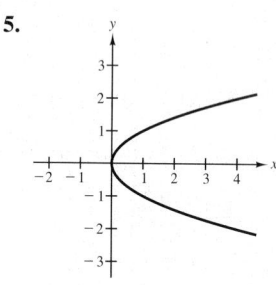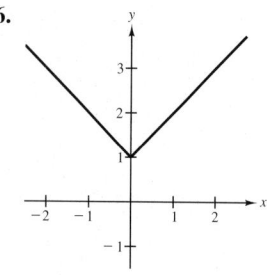

7. **(a)** 12 **(b)** $\frac{3}{16}$ **8.** **(a)** -7 **(b)** $-2x$

9. **(a)** $\frac{1}{3}$ **(b)** $\frac{c - 6}{c + 4}$ **10.** **(a)** $2\sqrt{3}$ **(b)** $\sqrt{t - 1}$

Section 2.6

1. $2x$ **2.** $5t - 13$ **3.** $-3x - 20$ **4.** $9u - 27$
5. $x^2 - 1$ **6.** $-x^2 + x + 6$ **7.** $y^2 - 2y + 3$
8. $x^2 - 2x - 15$ **9.** $2z^3 - 17z^2 + 48z - 45$
10. $-t^3 + 12t^2 - 43t + 44$

Section 3.1

1. $-\frac{1}{4}$ **2.** 1 **3.** Undefined   **4.** $-\frac{9}{10}$
5.   **6.**
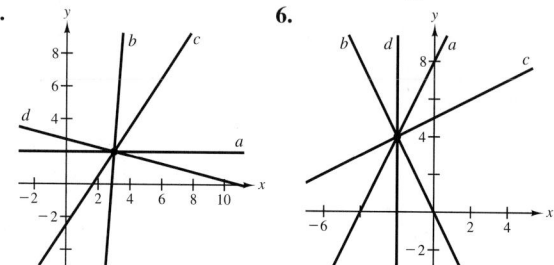

7. $y = -\frac{3}{5}x + \frac{21}{5}$ **8.** $y = \frac{4}{3}x - 9$ **9.** $y = -4x + 18$
10. $y = x + 9$

Section 3.2

1. $\frac{2}{3}$ **2.** 4 **3.** 1.84 **4.** 324 **5.** 15 **6.** $\frac{40}{7}$
7. $y = -3x + 4$ **8.** $y = -\frac{2}{3}x + \frac{2}{3}$ **9.** $y = \frac{4}{5}x - 2$
10. $y = \frac{5}{7}x + \frac{38}{7}$

Section 3.3

1. $43 - 6x$ **2.** $u^2 - 16$ **3.** $2n - 3$ **4.** $16n + 6$
5. -4 **6.** 27 **7.** 14 **8.** -19 **9.** $\frac{940}{3}$
10. $\frac{1}{10}$

Section 3.4

1. $\frac{11}{4}$ **2.** 5 **3.** $\frac{5}{9}$ **4.** $\frac{10}{3}$ **5.** $\frac{2}{5}$ **6.** 20
7. -4 **8.** $-\frac{6}{5}$ **9.** 9 **10.** 50

Section 3.5

1. $-\frac{3}{4} > -5$ **2.** $-\frac{1}{5} > -\frac{1}{3}$ **3.** $\pi > -3$
4. $6 < \frac{13}{2}$ **5.** $\frac{7}{2}$ **6.** $\frac{48}{7}$ **7.** 40 **8.** 153
9. -12 **10.** 24

Section 3.6

1. 15   **2.** 3.2   **3.** -72 **4.** -4 **5.** 18
6. -19.8 **7.** $-\frac{1}{2} < x < \frac{1}{2}$ **8.** $-\frac{1}{2} \le x \le \frac{3}{2}$
9. $-6 \le x \le 6$ **10.** $20 < x < 30$

Section 4.1

1. 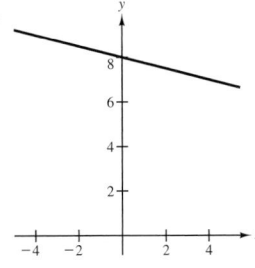   **2.**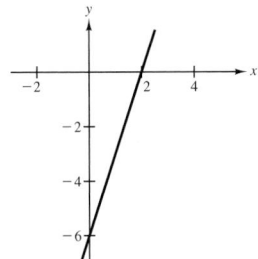

3. 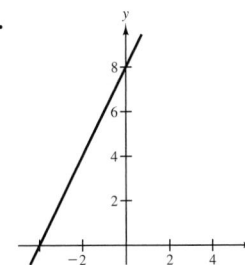   **4.**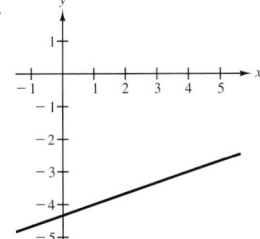

5. $\frac{3}{5}$ **6.** $\frac{7}{2}$ **7.** $2x - y = 0$ **8.** $x + y - 6 = 0$
9. $-\frac{2}{3}$ **10.** $\frac{9}{5}$

Section 4.2

1. $(5, 4)$ **2.** $(-1, 1)$ **3.** Inconsistent
4. Infinite number of solutions   **5.** $(3, -2)$
6. $(-2, -2)$ **7.** $5x - z$ **8.** $34y - 5x$
9. $35z - 17y$ **10.** $7x + 26z$

Section 4.3

1. $(2, 1)$ **2.** $(3, -2)$ **3.** $(5, 4)$ **4.** $\left(\frac{1}{2}, \frac{1}{3}\right)$
5. $(1, 2, 1)$ **6.** $(4, -1, 3)$ **7.** 9   **8.** $\frac{14}{5}$ **9.** $-\frac{19}{4}$
10. $-\frac{11}{6}$

Section 4.4

1. 10   **2.** -56 **3.** -82 **4.** 153   **5.** 32
6. $-\frac{65}{3}$ **7.** -2 **8.** $\frac{9}{11}$ **9.** $\frac{6}{5}$ **10.** $-\frac{11}{3}$

Section 4.5

1. 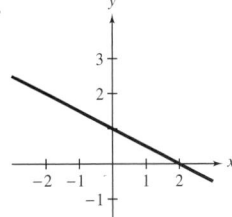   **2.**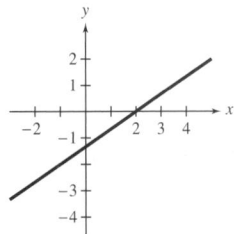

3. 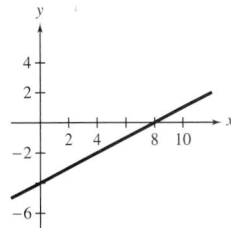   **4.**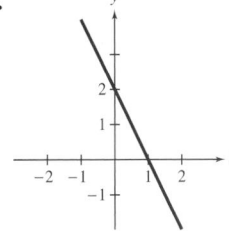

5. 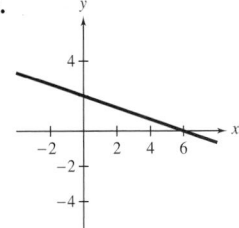   **6.**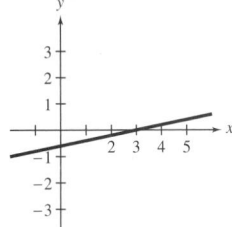

7. 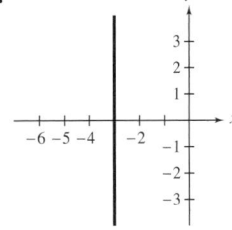   **8.**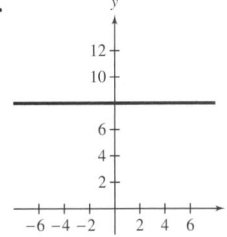

9. 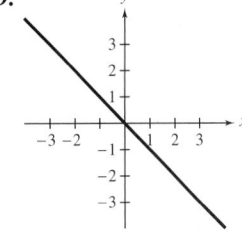   **10.**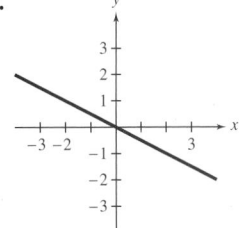

Section 4.6

1. $-11x$ **2.** $-41v$ **3.** $16x - 3$ **4.** $4y - 2$
5. 4 **6.** 11 **7.** Perpendicular **8.** Parallel
9. Parallel **10.** Neither

Section 5.1

1. a^{10} **2.** $x^5 y^4$ **3.** $9y^3$ **4.** $4rs^2$ **5.** $-t^6$
6. $16z^2$ **7.** $36x^5 y^8$ **8.** $-128u^5 v^7$ **9.** $\dfrac{16a^8}{81b^4}$ **10.** $\dfrac{9x^2}{16y^6}$

Section 5.2

1. 169 **2.** -169 **3.** $\frac{8}{27}$ **4.** $\frac{3}{10}$ **5.** 48
6. $\dfrac{27}{8}$ **7.** $\dfrac{1}{x^3}$ **8.** $-\dfrac{15y^4}{x^2}$ **9.** $\dfrac{9y^2}{4x^2}$ **10.** $\dfrac{v^4}{u^5}$

Section 5.3

1. 100 **2.** -10 **3.** $-\frac{3}{5}$ **4.** $\frac{2}{5}$ **5.** $x^{11/6}$
6. $a^{1/4}$ **7.** $16x^4$ **8.** $4096x^2 y^8$ **9.** $\dfrac{16}{x^{1/4} y^{1/4}}$ **10.** $\dfrac{8y^2}{z^{5/2}}$

Section 5.4

1. $34 - 5x$ **2.** $18x - 28$ **3.** $t^2 + \frac{2}{3}t + \frac{1}{9}$
4. $4x^2 - 20x + 25$ **5.** $25x^2 - 16$ **6.** $2 - \frac{23}{2}x - 3x^2$
7. $2xy\sqrt[3]{2x}$ **8.** $6xy\sqrt{2y}$ **9.** $(x - 4)\sqrt{3}$
10. $2\sqrt{7}x$

Section 5.5

1. 2 **2.** $\frac{9}{2}$ **3.** $\frac{27}{5}$ **4.** $\frac{1}{6}$ **5.** $2, -25$
6. $12, -114$ **7.** $x - 9$ **8.** $\sqrt{5u}$
9. $4t + 12\sqrt{t} + 9$ **10.** $25\sqrt{2}x$

Section 5.6

1. $4\sqrt{3}$ **2.** $6\sqrt{3}$ **3.** $23\sqrt{2}$ **4.** 150 **5.** $\dfrac{4\sqrt{10}}{5}$
6. $\dfrac{5(\sqrt{3} + 1)}{4}$ **7.** $0, \dfrac{35}{2}$ **8.** $-4, \dfrac{5}{4}$ **9.** $-2, 9$
10. $-1, \dfrac{3}{2}$

Section 6.1

1. $(4x + 11)(4x - 11)$ **2.** $(3t - 4)^2$
3. $(x - 3)(5x + 2)$ **4.** $(x - 10)(x - 4)$

5. $y(2y - 1)(2y + 3)$ **6.** $4x(x^2 - 3x + 4)$ **7.** $-\frac{6}{5}$
8. $\frac{7}{2}$ **9.** $-9, \frac{15}{2}$ **10.** $\frac{8}{5}, 12$

Section 6.2

1. $x^2 + 6x + 8$ **2.** $x^2 + 20x + 150$ **3.** $u^2 - u + \frac{1}{4}$
4. $v^2 - 24v + 44$ **5.** $\pm\frac{2}{7}$ **6.** $\pm\frac{3}{4}i$ **7.** $4 \pm 6i$
8. $-2 \pm \sqrt[2]{5}$ **9.** $\dfrac{2}{3} \pm \dfrac{\sqrt{5}}{3}$ **10.** $\dfrac{3}{8}, 12$

Section 6.3

1. $3, 7$ **2.** $2, -16$ **3.** $3, -\frac{10}{3}$ **4.** $\frac{3}{5}, 3$
5. $\dfrac{-3 \pm \sqrt{17}}{2}$ **6.** $-1 \pm \dfrac{\sqrt{3}}{3}i$ **7.** 3 **8.** $\sqrt{11}\,i$
9. $4\sqrt{6}$ **10.** $3\sqrt{17}$

Section 6.4

1. -7 **2.** $0, \frac{2}{15}$ **3.** $\frac{5}{2}, -1$ **4.** $10, -12$
5. $1, -17$ **6.** 1 **7.** $-\dfrac{3}{2} \pm \dfrac{\sqrt{11}}{2}i$ **8.** $\dfrac{-3 \pm \sqrt{29}}{2}$
9. $6, -1$ **10.** $3, -1$

Section 6.5

1.
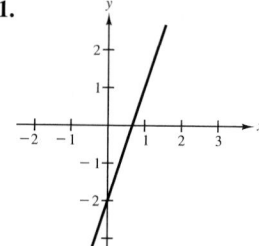

2.
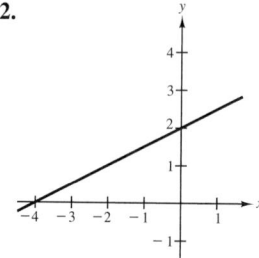

3. **4.**
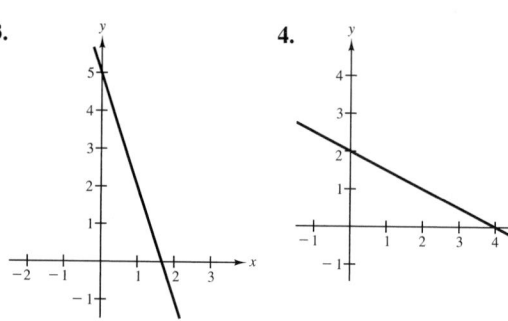

5. $\frac{9}{2}$ **6.** 20   **7.** $0, -8$ **8.** $\dfrac{-1 \pm \sqrt{41}}{4}$

9. $2x^2 + 20x + 20$ **10.** $-3x^2 + 12x - 19$

Section 6.6

1.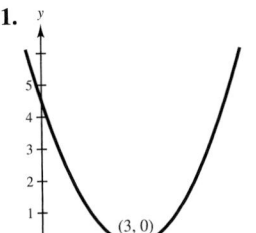

2.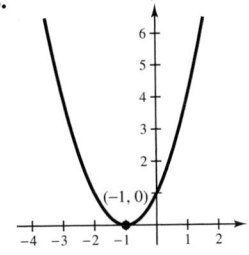

3.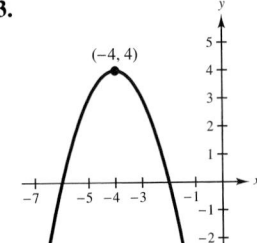

4.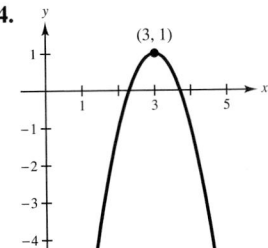

5.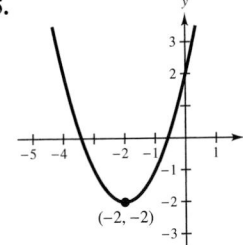

6.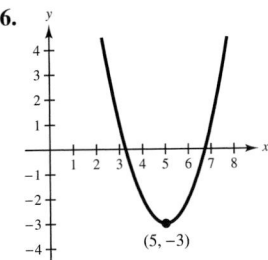

7.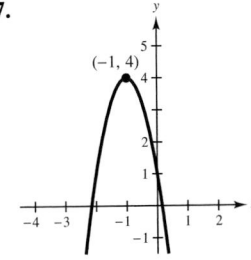

8.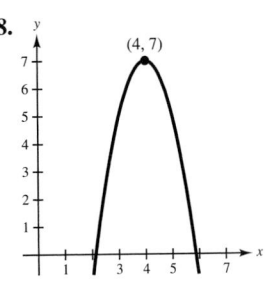

9.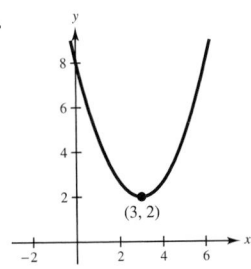

10.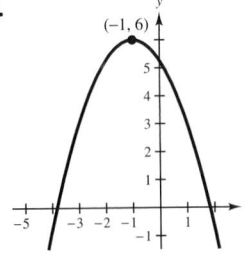

Section 6.7

1. $x \leq 5$

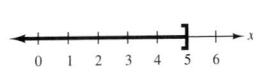

2. $x \geq -\frac{5}{3}$

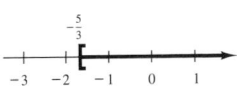

3. $x < \frac{3}{2}$

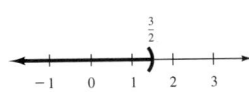

4. $x < 5$

5. $1 < x < 5$

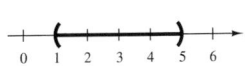

6. $x < 2, \ x > 8$

7. Negative   **8.** Zero   **9.** Positive   **10.** Negative

Section 7.1

1. -20 **2.** 70   **3.** -21 **4.** -27 **5.** $\frac{15}{26}$

6. $\frac{11}{23}$ **7.** $\frac{5}{8}$ **8.** $\frac{7}{12}$ **9.** $20a(x^2 + 3x + 4)$

10. $y^2(x - 8)(2x + 7)$

Section 7.2

1. $\frac{3}{10}$ **2.** $\frac{1}{20}$ **3.** $\frac{5}{56}$ **4.** $-\frac{15}{112}$ **5.** 5   **6.** $-\frac{4}{35}$

7. $(x - 1)(x - 2)$ **8.** $(x - 1)(2x + 7)$

9. $(x + 1)(11x - 5)$ **10.** $(2x - 7)^2$

Section 7.3

1. $\frac{7}{9}$ **2.** $\frac{1}{2}$ **3.** $\frac{3}{16}$ **4.** $-\frac{3}{5}$ **5.** $\frac{37}{36}$ **6.** $\frac{37}{48}$

7. $-\frac{146}{225}$ **8.** $\frac{181}{21}$ **9.** 26   **10.** $\frac{5}{64}$

Section 7.4

1. $\frac{1}{4}$ **2.** $\frac{29}{90}$ **3.** $-\frac{4x}{3}$ **4.** $\frac{t^2}{9}$ **5.** $\frac{2u}{9v^2}$

6. $-\frac{r^3}{7}$ **7.** $-15x^4 + 12x^3$ **8.** $70y^3 - 20y$

9. $2x^3 + 7x^2 - 3x - 4$ **10.** $8x^2 + 10x + 3$

Section 7.5

1. $\frac{16}{3}$ **2.** -19 **3.** 3 **4.** $\frac{5}{2}$ **5.** $\frac{2}{3}, -8$
6. $16, -\frac{3}{2}$ **7.** $0, \frac{21}{2}$ **8.** 5 **9.** $6, -7$ **10.** $0, 8$

Section 7.6

1. $(x - 5)(x + 2)$ **2.** $(x - 5)(x - 2)$
3. $x(x + 1)(x + 3)$ **4.** $(x^2 - 2)(x - 4)$

5. 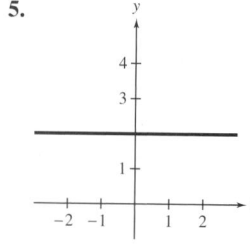   **6.**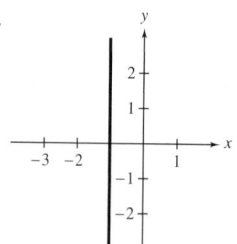

7. 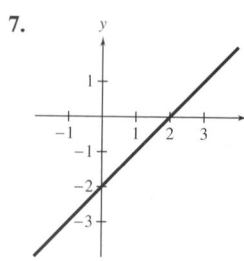   **8.**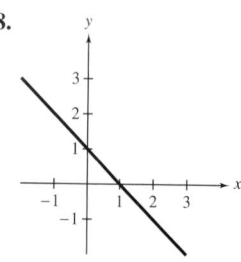

9. $5 + \dfrac{26}{x - 4}$ **10.** $-2 + \dfrac{7}{x + 2}$

Section 7.7

1. $(6, \infty)$ **2.** $\left(\frac{3}{2}, \infty\right)$ **3.** $(-\infty, -1)$ **4.** $(-\infty, 5)$
5. $(1, 5)$ **6.** $(-\infty, 2) \cup (8, \infty)$ **7.** $(-\infty, 1] \cup [5, \infty)$
8. $[2, 8]$ **9.** $(-\infty, 0] \cup [5, \infty)$ **10.** $(0, 3)$

Section 8.1

1. All real values x
2. All real values x such that $x \leq 10$

3. All real values x such that $x \neq 5$
4. All real values x such that $x \neq 5$ and $x \neq -5$
5. 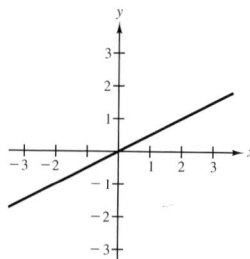   **6.**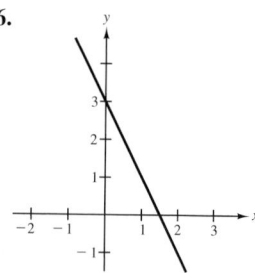

7. 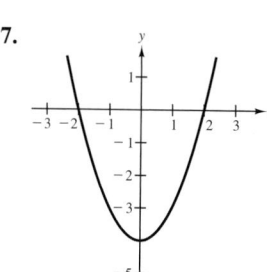   **8.**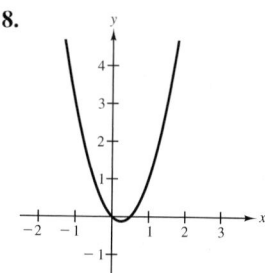

9. 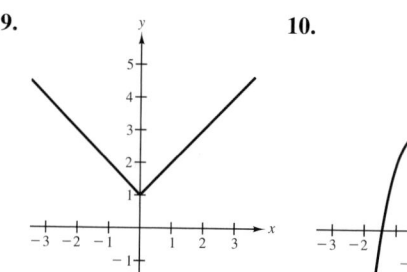   **10.**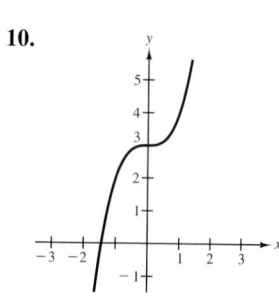

Section 8.2

1. All real values x **2.** All real values x
3. All real values x such that $-4 \leq x \leq 4$
4. All real values x such that $x \neq -6$ and $x \neq 6$
5. $(f + g)(x) = x^2 + 3x - 9, \ (-\infty, \infty)$
$(f - g)(x) = x^2 - 3x - 9, \ (-\infty, \infty)$
$(fg)(x) = 3x^3 - 27x, \ (-\infty, \infty)$
$\left(\dfrac{f}{g}\right)(x) = \dfrac{x^2 - 9}{3x}, \ (-\infty, 0) \cup (0, \infty)$
$(f \circ g)(x) = 9x^2 - 9, \ (-\infty, \infty)$
$(g \circ f)(x) = 3(x^2 - 9), \ (-\infty, \infty)$

6. $(f + g)(x) = \frac{1}{2}x^2 + 2x - 3$, $(-\infty, \infty)$
$(f - g)(x) = -\frac{1}{2}x^2 + 2x - 3$, $(-\infty, \infty)$
$(fg)(x) = x^3 - \frac{3}{2}x^2$, $(-\infty, \infty)$
$\left(\dfrac{f}{g}\right)(x) = \dfrac{4x - 6}{x^2}$, $(-\infty, 0) \cup (0, \infty)$
$(f \circ g)(x) = x^2 - 3$, $(-\infty, \infty)$
$(g \circ f)(x) = \frac{1}{2}(2x - 3)^2$, $(-\infty, \infty)$

7. $(f + g)(x) = \sqrt[3]{x - 2} + x^2$, $(-\infty, \infty)$
$(f - g)(x) = \sqrt[3]{x - 2} - x^2$, $(-\infty, \infty)$
$(fg)(x) = x^2\sqrt[3]{x - 2}$, $(-\infty, \infty)$
$\left(\dfrac{f}{g}\right)(x) = \dfrac{\sqrt[3]{x - 2}}{x^2}$, $(-\infty, 0) \cup (0, \infty)$
$(f \circ g)(x) = \sqrt[3]{x^2 - 2}$, $(-\infty, \infty)$
$(g \circ f)(x) = \sqrt[3]{(x - 2)^2}$, $(-\infty, \infty)$

8. $(f + g)(x) = \sqrt{x} + x^2 - 4$, $[0, \infty)$
$(f - g)(x) = \sqrt{x} - x^2 + 4$, $[0, \infty)$
$(fg)(x) = \sqrt{x}(x^2 - 4)$, $[0, \infty)$
$\left(\dfrac{f}{g}\right)(x) = \dfrac{\sqrt{x}}{x^2 - 4}$, $[0, 2) \cup (2, \infty)$
$(f \circ g)(x) = \sqrt{x^2 - 4}$, $(-\infty, -2] \cup [2, \infty)$
$(g \circ f)(x) = x - 4$, $[0, \infty)$

9. $(f + g)(x) = \dfrac{2(x - 1)}{x(x - 2)}$, $(-\infty, 0) \cup (0, 2) \cup (2, \infty)$

$(f - g)(x) = \dfrac{2}{x(2 - x)}$, $(-\infty, 0) \cup (0, 2) \cup (2, \infty)$

$(fg)(x) = \dfrac{1}{x(x - 2)}$, $(-\infty, 0) \cup (0, 2) \cup (2, \infty)$

$\left(\dfrac{f}{g}\right)(x) = \dfrac{x - 2}{x}$, $(-\infty, 0) \cup (0, 2) \cup (2, \infty)$

$(f \circ g)(x) = x - 2$, $(-\infty, 2) \cup (2, \infty)$

$(g \circ f)(x) = \dfrac{x}{1 - 2x}$, $\left(-\infty, \frac{1}{2}\right) \cup \left(\frac{1}{2}, 2\right) \cup (2, \infty)$

10. $(f + g)(x) = \dfrac{x^2 + 4x + 5}{x + 2}$, $(-\infty, -2) \cup (-2, \infty)$

$(f - g)(x) = \dfrac{x^2 + 4x + 3}{x + 2}$, $(-\infty, -2) \cup (-2, \infty)$

$(fg)(x) = 1$, $(-\infty, -2) \cup (-2, \infty)$

$\left(\dfrac{f}{g}\right)(x) = (x + 2)^2$, $(-\infty, -2) \cup (-2, \infty)$

$(f \circ g)(x) = \dfrac{2x + 5}{x + 2}$, $(-\infty, -2) \cup (-2, \infty)$

$(g \circ f)(x) = \dfrac{1}{x + 4}$, $(-\infty, -4) \cup (-4, \infty)$

Section 8.3

1. $\frac{1}{3}$ **2.** $\frac{10}{3}$ **3.** 0.92   **4.** 513   **5.** 40   **6.** $\frac{400}{7}$

7. 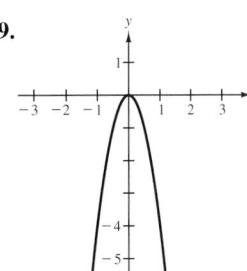   **8.**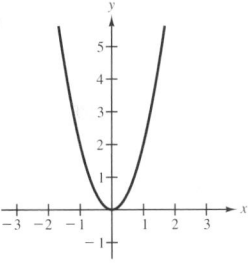

9. 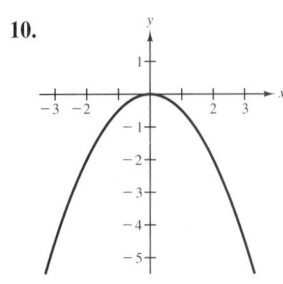   **10.**

Section 8.4

1. $(3x - 2)(4x + 5)$ **2.** $x(5x - 6)^2$
3. $z^2(12z + 5)(z + 1)$ **4.** $(y + 5)(y^2 - 5y + 25)$
5. $(x + 3)(x + 2)(x - 2)$ **6.** $(x + 2)(x^2 + 3)$
7. No real solution   **8.** $3 \pm \sqrt{5}$ **9.** $-\frac{1}{2} \pm \sqrt{3}$
10. ± 3

Section 8.5

1. $y = \frac{1}{3}(9 - 2x)$ **2.** $y = \pm\sqrt{36 - x^2}$

3. $y = \pm \sqrt{\dfrac{6 - 2x^2}{3}}$ **4.** $y = \pm \sqrt{2x^2 - 3}$ **5.** $15\sqrt{2}$

6. $15\sqrt{6}$

7.

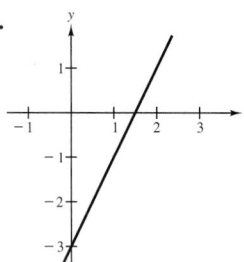

8.

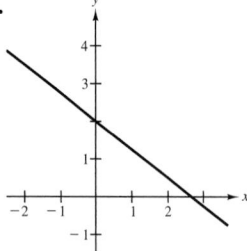

9.

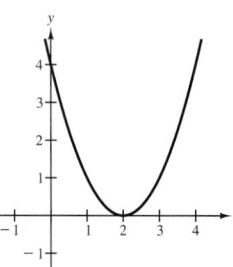

10.

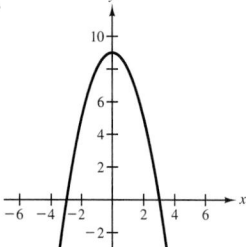

Section 8.6

1. $\sqrt{61}$ **2.** $\sqrt{65}$

3.

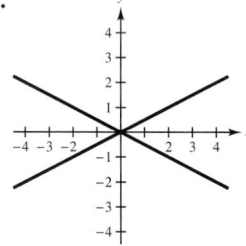

4.

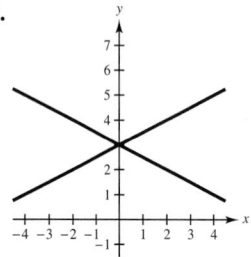

5.

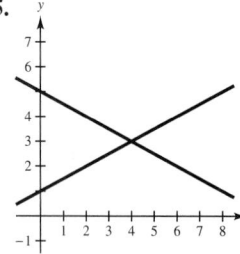

6.

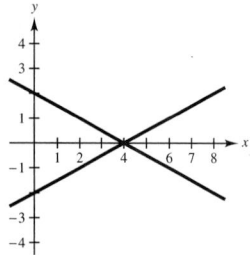

7. $c = 12$ **8.** $b = 1$ **9.** $a = 10$ **10.** $c = 2\sqrt{6}$

Section 8.7

1. $x^2 - 10x + 5$ **2.** $x^2 + 6x + 8$ **3.** $-x^2 - 8x - 4$

4. $-x^2 + 4x$ **5.** $(x + 3)^2 - 1$ **6.** $(x - 5)^2 - 4$

7. $2 - (x - 1)^2$ **8.** $-2(x - 1)^2$

9. $2x + 3y - 20 = 0$ **10.** $3x - 4y - 17 = 0$

Section 8.8

1.

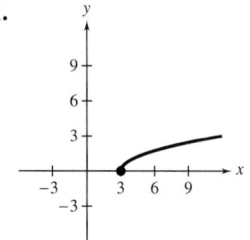

2.

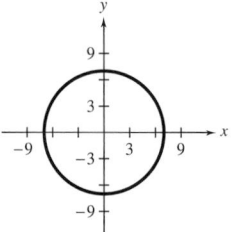

3.

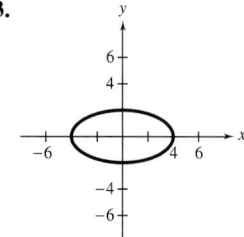

4.

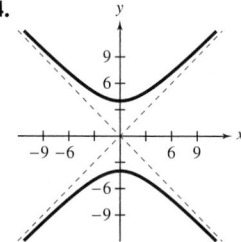

5.

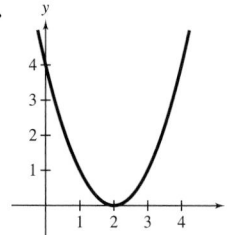

6.

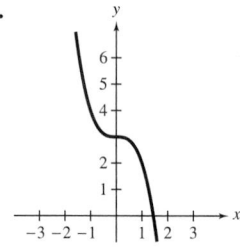

7. $\left(2, \tfrac{1}{2}\right)$ **8.** $(4, -2)$ **9.** $(1, -4, 3)$ **10.** $(4, 0, 6)$

Section 9.1

1. $\dfrac{27}{64}$ **2.** 900 **3.** $\dfrac{1}{4}$ **4.** $\dfrac{1}{81}$ **5.** 5 **6.** $\dfrac{1}{9}$

7. 1 **8.** y^5 **9.** $13x$ **10.** $-\dfrac{8}{y^3}$

Section 9.2

1. x^8 **2.** 1 **3.** z^3 **4.** $\dfrac{3}{4}(x - 3)^4$ **5.** $x^8 y^4$

6. $\dfrac{x^5}{32y^5}$ **7.** $\dfrac{x^4}{y^3}$ **8.** $\dfrac{1}{a^8}$ **9.** b^3 **10.** $2x$

Section 9.3

1. x^{20} **2.** $\dfrac{x^2}{16}$ **3.** $\dfrac{3}{2y^5}$ **4.** $\dfrac{4y^2}{9x^4}$ **5.** $\dfrac{1}{6}x^3y$

6. 1 **7.** $4^3 = 64$ **8.** $3^{-4} = \frac{1}{81}$ **9.** $e^{2x} = e^{2x}$
10. $e^0 = 1$

Section 9.4

1. $\frac{5}{2}$ **2.** $-\frac{7}{2}$ **3.** $10, -2$ **4.** $2, 3$ **5.** $\dfrac{28}{\ln 5}$

6. 0 **7.** 3 **8.** 6 **9.** 0 **10.** 4

Section 9.5

1. $\frac{33}{4}$ **2.** $\frac{3}{5}, -1$ **3.** 41 **4.** $\frac{11}{2}$ **5.** 18 **6.** 69
7. 4.16 **8.** 11.31 **9.** 139.32 **10.** -1.39

Section 9.6

1. 4 **2.** 3 **3.** -3 **4.** 3 **5.** 0.368
6. 1.649
7. g is a vertical translation of f four units downward.
8. g is a horizontal translation of f four units to the right.
9. g is a reflection of f in the x-axis.
10. g is a reflection of f in the y-axis.

Section 10.1

1. $\frac{1}{7}$ **2.** 24 **3.** $\frac{3}{4}$ **4.** $\frac{13}{12}$ **5.** n **6.** $n - 1$
7. $\dfrac{2n - 3}{n - 5}$ **8.** $\dfrac{n + 1}{n}$ **9.** $\dfrac{n + 2}{n^2}$ **10.** $\dfrac{2n}{n^2 - 1}$

Section 10.2

1. $\frac{49}{20}$ **2.** $\frac{455}{72}$ **3.** 120 **4.** 40 **5.** 285
6. 1960 **7.** 77 **8.** $\frac{355}{2}$ **9.** 63 **10.** 144

Section 10.3

1. $\frac{16}{81}$ **2.** $\frac{27}{125}$ **3.** $\frac{1}{72}$ **4.** 1 **5.** n **6.** $n + 1$
7. $60n^6$ **8.** $16n^5$ **9.** $\dfrac{n}{2}$ **10.** $\dfrac{9}{5}n^2$

Section 10.4

1. $-3x^2 + 6x$ **2.** $x^2 - 9$ **3.** $x^2 - 4x + 4$
4. $4x^2 + 12x + 9$ **5.** $\dfrac{5x^3}{y^2}$ **6.** $-8x^6y^3$ **7.** 24
8. 8 **9.** 1320 **10.** 45

Section 10.5

1. 10 **2.** 120 **3.** 84 **4.** 1 **5.** 576 **6.** 720
7. 870 **8.** 1260 **9.** 4845 **10.** 24

Section 10.6

1. $\frac{11}{12}$ **2.** $\frac{5}{8}$ **3.** $\frac{37}{30}$ **4.** $\frac{95}{144}$ **5.** $\frac{1}{24}$ **6.** $\frac{1}{120}$
7. $\frac{3}{14}$ **8.** $\frac{1}{6}$ **9.** $\frac{5}{8}$ **10.** $\frac{117}{125}$

Answers to Exercises

CHAPTER 1

Section 1.1 (page 14)

1. **(a)** $1, 4, 6$ **(b)** $-10, 0, 1, 4, 6$
   (c) $-10, -\frac{2}{3}, -\frac{1}{4}, 0, \frac{5}{8}, 1, 4, 6$ **(d)** $-\sqrt{5}, \sqrt{3}, 2\pi$
2. **(a)** $8, 245$ **(b)** $0, 8, 245$ **(c)** $-\frac{7}{2}, -\frac{3}{8}, 0, \frac{10}{3}, 8, 245$
   (d) $-\sqrt{6}, -\frac{\pi}{2}, \sqrt{15}$
3. $-5, -4, -3, -2, -1, 0, 1, 2, 3$
4. $-2, 0, 2, 4, 6, 8, 10$ **5.** $1, 3, 5, 7, 9$
6. $2, 3, 5, 7, 11, 13, 17, 19, 23$
7. **(a)**  **(b)**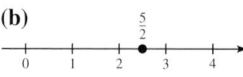
   (c) $-\frac{7}{2}$ **(d)** -5.2
8. **(a)** **(b)** $\frac{4}{3}$
   (c) -6.75 **(d)** $-\frac{9}{2}$
9. $-1 < 3$ **10.** $-\frac{9}{2} < -2$
11. $2 < 5$ **12.** $8 > 3$

13. $-7 < -2$ **14.** $-2 > -5$

15. $-\frac{2}{3} > -\frac{10}{3}$ **16.** $\frac{11}{4} < \pi$

17. Yes. The nonnegative real numbers include 0.
18. 6. The distance between -4 and -8 is 4 units, but the distance between -4 and 6 is 10 units.
19. $x < 0$ **20.** $y > 25$ **21.** $u \geq 16$
22. $2 < z \leq 10$ **23.** $p < 225$ **24.** $30 \leq p \leq 35$
25. 6 **26.** 55 **27.** 50 **28.** 86 **29.** 35

30. 35 **31.** 10 **32.** 62 **33.** -3.5 **34.** 6
35. -25 **36.** $-\frac{3}{4}$ **37.** $|-6| > |2|$
38. $|150| < |-310|$ **39.** $\left|-\frac{3}{4}\right| > -\left|\frac{4}{5}\right|$
40. $-|-16.8| = -|16.8|$ **41.** $-14, 14$
42. $22.5, 22.5$ **43.** $\frac{5}{4}, \frac{5}{4}$ **44.** $-\pi, \pi$
45.  46.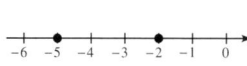

47. $-\frac{5}{3}$ $\frac{5}{3}$ 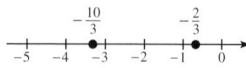 **48.** $-\frac{3}{4}$ $\frac{3}{4}$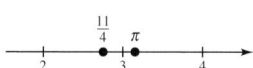

49. 45 **50.** 100 **51.** 19 **52.** -68 **53.** -22
54. -57 **55.** 0.7 **56.** -5.27 **57.** $\frac{1}{2}$ **58.** 2
59. $\frac{1}{24}$ **60.** $-\frac{49}{22}$ **61.** $\frac{105}{8}$ **62.** $-\frac{97}{6}$ **63.** 45.95
64. -38.53 **65.** -28 **66.** -28 **67.** -30
68. -21 **69.** 32.13 **70.** 14.08 **71.** $\frac{1}{2}$ **72.** 2
73. 6 **74.** 2 **75.** $-\frac{5}{2}$ **76.** $-\frac{22}{5}$ **77.** $\frac{46}{17}$
78. $\frac{23}{16}$ **79.** 4.03 **80.** 10.37
81. $(-3)(-3)(-3)(-3)$ **82.** $\left(\frac{2}{3}\right)\left(\frac{2}{3}\right)\left(\frac{2}{3}\right)$ **83.** $(-5)^4$
84. -5^6 **85.** -64 **86.** 81 **87.** -25
88. -243 **89.** $\frac{49}{64}$ **90.** $\frac{16}{81}$ **91.** 9 **92.** 10
93. 5 **94.** 4 **95.** -2 **96.** -1 **97.** 5
98. 2 **99.** 36 **100.** 40 **101.** $-\frac{17}{6}$ **102.** $1\frac{1}{10}$
103. 72.2 **104.** 22 **105.** 161 **106.** $-126,630$
107. 14,425 **108.** -82.88 **109.** 171.36
110. 10.37 **111.** 11.72 **112.** 6.35
113. No. $3 \cdot 4^2 = 3 \cdot 16 = 48$

114.

Year	1993	1994	1995	1996	1997
Yearly gain or loss (in millions)	$+0.5$	-0.2	$+1.3$	-0.2	$+0.9$

115. \$2533.56 **116.** \$10,800 **117.** $\frac{17}{180}$ **118.** $\frac{11}{70}$
119. 15 m^2 **120.** 20 in.^{2} **121.** 36 ft^2
122. 128 cm^2 **123.** 6.125 ft^3 **124.** 40, 2940 ft^3
125. True **126.** False. $\frac{2}{3}$ is not an integer.
127. False. The reciprocal of the integer 2 is $\frac{1}{2}$, which is not an integer.
128. True **129.** False. $\frac{2}{3} + \frac{3}{2} = \frac{13}{6}$
130. False. A negative number raised to an odd power is negative.

131. The calculator steps correspond to $523 - 145 - 136$ instead of $523 - (145 - 136)$. The correct calculator steps are:

523 $\boxed{-}$ $\boxed{(}$ 145 $\boxed{-}$ 136 $\boxed{)}$ $\boxed{\text{ENTER}}$

Display: 514

132. The calculator steps correspond to $126 + \frac{37}{4}$ instead of $\frac{126 + 37}{4}$. The correct calculator steps are:

$\boxed{(}$ 126 $\boxed{+}$ 37 $\boxed{)}$ $\boxed{\div}$ 4 $\boxed{\text{ENTER}}$

Display: 40.75

Section Project 1.1

(a) $(0 + 6 + 3 + 2 + 0 + 2)(3) + (7 + 7 + 7 + 0 + 1) = 61$, $70 - 61 = 9$, which is the check digit.

(b) $(0 + 1 + 0 + 4 + 7 + 0)(3) + (4 + 8 + 0 + 8 + 0) = 56$, $60 - 56 = 4$, which is not the check digit.

Section 1.2   (page 25)

1. Commutative Property of Addition

2. Commutative Property of Multiplication

3. Associative Property of Multiplication

4. Additive Identity Property

5. Associative Property of Addition

6. Distributive Property

7. Commutative Property of Multiplication

8. Associative Property of Addition

9. Commutative Property of Addition

10. Multiplicative Identity Property

11. Associative Property of Multiplication

12. Distributive Property   **13.** Additive Identity Property

14. Multiplicative Identity Property

15. Additive Inverse Property

16. Multiplicative Inverse Property

17. Commutative Property of Multiplication

18. Associative Property of Multiplication

19. Associative Property of Multiplication

20. Additive Identity Property

21. Multiplicative Inverse Property

22. Multiplicative Inverse Property

23. Multiplicative Identity Property

24. Additive Inverse Property   **25.** Distributive Property

26. Associative Property of Addition   **27.** $(3 \cdot 6)y$

28. $-6 + 10$ **29.** $-3(15)$ **30.** $(6 + 5) + y$

31. $5 \cdot 6 + 5 \cdot z$ or $30 + 5z$

32. $-3 \cdot 4 + (-3)x$ or $-12 - 3x$ **33.** $-x + 25$

34. 0 **35.** $x + 8$ **36.** $8x$

37. True, Additive Inverse Property

38. False, $-9 + 5 = -4 \neq -5 + 9 = 4$

39. False, $6(7 + 2) = 6(7) = 6(2)$

40. True, Distributive Property

41. No. Zero does not have a multiplicative inverse.

42. The additive inverse of a real number a is the number $-a$. The sum of a number and its additive inverse is the additive identity 0. For example, $8 + (-8) = 0$.

43. (a) -10 (b) $\frac{1}{10}$ **44.** (a) -18 (b) $\frac{1}{18}$

45. (a) 16 (b) $-\frac{1}{16}$ **46.** (a) 52 (b) $-\frac{1}{52}$

47. (a) $-6z$ (b) $\dfrac{1}{6z}$ **48.** (a) $-2y$ (b) $\dfrac{1}{2y}$

49. (a) $-x - 1$ or $-(x + 1)$ (b) $\dfrac{1}{x + 1}$

50. (a) $-(y - 4)$ or $-y + 4$ (b) $\dfrac{1}{y - 4}$

51. $x + (5 - 3)$ **52.** $z + (-6 + 10)$

53. $(32 - 4) + y$ **54.** $(15 + 3) + x$ **55.** $(3 \cdot 4)5$

56. $10 \cdot (8 \cdot 5)$ **57.** $(6 \cdot 2)y$ **58.** $(8 \cdot 3)x$

59. Given

Addition Property of Equality

Associative Property of Addition

Additive Inverse Property

Additive Identity Property

60. Given

Addition Property of Equality

Associative Property of Addition

Additive Inverse Property

Additive Identity Property

61. Given

Addition Property of Equality

Associative Property of Addition

Additive Inverse Property

Additive Identity Property

Multiplication Property of Equality

Associative Property of Multiplication

Multiplicative Inverse Property

Multiplicative Identity Property

62. Given

Addition Property of Equality

Associative Property of Addition

Additive Inverse Property

Additive Identity Property

Multiplication Property of Equality

Associative Property of Multiplication

Multiplicative Inverse Property

Multiplicative Identity Property

63. $20 \cdot a + 20 \cdot 5$ or $20a + 100$

64. $-3 \cdot y + (-3)8$ or $-3y - 24$

65. $5 \cdot 3x + 5 \cdot 4$ or $15x + 20$

66. $6 \cdot 2x + 6 \cdot 5$ or $12x + 30$

67. $x \cdot (-2) + 6 \cdot (-2)$ or $-2x - 12$

68. $z \cdot 12 + 10 \cdot 12$ or $12z + 120$ **69.** $3x + 15$

70. $4x + 8$ **71.** $-2x - 16$ **72.** $-9x - 36$ **73.** 0

74. 1 **75.** 28 **76.** 25 **77.** 434 **78.** 245

79. 62.82 **80.** 239.4 **81.** Answers may vary.

82. Answers may vary.

83. (a) No. $4 \odot 7 = 15 \neq 18 = 7 \odot 4$

(b) No. $3 \odot (4 \odot 7) = 21$
$$\neq 27 = (3 \odot 4) \odot 7$$

84. $\$0.86$ **85.** $\$0.08$ **86.** $\$1.18$ **87.** Same

Section Project 1.2

(a) $a(b + c) = ab + ac$ (b) $a(b - c) = ab - ac$

Section 1.3   (page 37)

1. Terms are those parts separated by addition. Factors are separated by multiplication.

2. Constants are real numbers and variables are letters used to represent numbers.

3. $10x, 5$ **4.** $-16t^2, 48$ **5.** $-3y^2, 2y, -8$

6. $25z^3, -4.8z^2$ **7.** $4x^2, -3y^2, -5x, 2y$

8. $14u^2, 25uv, -3v^2$ **9.** $x^2, -2.5x, -\dfrac{1}{x}$ **10.** $\dfrac{3}{t^2}, -\dfrac{4}{t}, 6$

11. 5 **12.** 4 **13.** $-\dfrac{3}{4}$ **14.** -8.4

15. Commutative Property of Addition

16. Associative Property of Addition

17. Associative Property of Multiplication

18. Commutative Property of Multiplication

19. Multiplicative Inverse Property

20. Additive Inverse Property   **21.** Distributive Property

22. Additive Identity Property

23. Multiplicative Identity Property

24. Distributive Property

25. (a) $5x + 5 \cdot 6$ or $5x + 30$ (b) $(x + 6)5$

26. (a) $6(x + 1)$ (b) $6 + 6x$ **27.** (a) $(xy)6$ (b) $(6x)y$

28. (a) $3ab$ (b) $0 + 3ab$ **29.** (a) 0 (b) $(-4t^2) + 4t^2$

30. (a) $3 + (6 - 9)$ (b) 0

31. True, Associative Property of Multiplication

32. True, Associative Property of Addition

33. False, $5(y^3 + 2) = 5y^3 + 5(2)$

34. True, Distributive Property

35. $(x \cdot x \cdot x) \cdot (x \cdot x \cdot x \cdot x)$

36. $(z \cdot z) \cdot (z \cdot z \cdot z \cdot z \cdot z)$ **37.** $(-2x)(-2x)(-2x)$

38. $(2y)(2y)(2y)$ **39.** $\left(\dfrac{y}{5}\right)\left(\dfrac{y}{5}\right)\left(\dfrac{y}{5}\right)\left(\dfrac{y}{5}\right)$

40. $\left(\dfrac{3}{t}\right)\left(\dfrac{3}{t}\right)\left(\dfrac{3}{t}\right)\left(\dfrac{3}{t}\right)\left(\dfrac{3}{t}\right)$ **41.** $(5x)^4$ **42.** y^3y^4

43. x^3y^3 **44.** $(-9t)^6$ **45.** $27y^6$ **46.** $36x^8$

47. $16x^2$ **48.** $-64x^3$ **49.** $-125z^6$ **50.** $25z^6$

51. $6x^3y^4$ **52.** $-10a^3b^7$ **53.** $81x^2$ **54.** $4y^2$

55. $\dfrac{16}{3}x^2y^2$ **56.** $16a^4b^4$ **57.** $-25y^6$ **58.** $54y^5$

59. $-3125z^8$ **60.** $18n^3$ **61.** $16a^4$ **62.** $16a^3$

63. $8xy$ **64.** $-z^4$ **65.** $\dfrac{2}{3}a^7b$ **66.** $-\dfrac{9}{8}c^7d^3$

67. $-\dfrac{4x^8}{25y^2}$ **68.** $-\dfrac{27a^9}{8b^{15}}$ **69.** x **70.** a^m

71. x^{4n} **72.** a^{3k} **73.** x^{n+4} **74.** y^m **75.** rs^{m+3}

76. $x^{2n}y^{2m+2}$

77. $a^m \cdot a^n = a^{m+n}$
$(ab)^m = a^m \cdot b^m$
$(a^m)^n = a^{mn}$
$\dfrac{a^m}{a^n} = a^{m-n}, \; m > n, \; a \neq 0$
$\left(\dfrac{a}{b}\right)^m = \dfrac{a^m}{b^m}, \; b \neq 0$

78. The base of the exponent 3 in the expression $(2x)^3$ is $2x$. The base of the exponent 3 in the expression $2x^3$ is x.

79. $7x$ **80.** $2x^2$ **81.** $8y$ **82.** $14y$

83. $8x + 18y$ **84.** $-9a - \dfrac{2}{3}b$ **85.** $\dfrac{11}{2}z^2 + \dfrac{3}{2}z + 10$

86. $3y^3 - 6y^2 + 4y - 4$ **87.** $4u^2v^2 + uv$

88. $5m^2n^2 - 9mn$ **89.** $5a^2b^2 - 2ab$ **90.** $2xy + 8$

91. $12x - 35$ **92.** $4x^2 - 3$ **93.** $-3y^2 - 7y - 7$

94. $-4a^2 + 13a + 34$ **95.** $-12x^2 - 100$

96. $9x^3 - 5$ **97.** $-2b^2 + 4b - 36$ **98.** $x^3 - 12$

99. $2y^3 + y^2 + y$ **100.** $ab^3 - 8ab$

101. $-x^2y^2 - xy$ **102.** $z^6 + 3z^4 + 4z^3$ **103.** $-51a^7$

104. 0 **105.** $6x$ **106.** $\dfrac{1}{4}b^2 + \dfrac{1}{2}b$

107. (a) 3 (b) -10 **108.** (a) 7 (b) $-\dfrac{13}{7}$

109. (a) 7 (b) 7 **110.** (a) 0 (b) 0

111. (a) 0 (b) $\dfrac{3}{10}$ **112.** (a) Undefined   (b) $\dfrac{11}{2}$

113. (a) 13 (b) -36 **114.** (a) 7 (b) 7

115. (a) 0 (b) Undefined   **116.** (a) 3 (b) 0

117. (a) 210 (b) 140 **118.** (a) $\$4250$ (b) $\$157.50$

119. $\dfrac{1}{2}b(b - 3)$, 90 **120.** $\left(\dfrac{5}{4}h + 10\right)h$, 300

121. Graph: $\$1450$ million   **122.** Graph: $\$2500$ million
Model: $\$1441$ million   Model: $\$2450$ million

123. $b_1h + \frac{1}{2}(b_2 - b_1)h = b_1h + \left(\frac{1}{2}b_2 - \frac{1}{2}b_1\right)h$
$$= b_1h + \tfrac{1}{2}b_2h - \tfrac{1}{2}b_1h$$
$$= \left(b_1h - \tfrac{1}{2}b_1h\right) + \tfrac{1}{2}b_2h$$
$$= \tfrac{1}{2}b_1h + \tfrac{1}{2}b_2h$$
$$= \tfrac{1}{2}h(b_1 + b_2)$$
$$= \frac{h}{2}(b_1 + b_2)$$

124. $\frac{57}{2}$ **125.** 1440 ft^2

Section Project 1.3

(a)

x	-1	0	1	2	3	4
$2x - 5$	-7	-5	-3	-1	1	3

(b) 2

(c)

x	-1	0	1	2	3	4
$3x + 2$	-1	2	5	8	11	14

(d) 3

(e) 7

x	-1	0	1	2	3	4
$7x + 4$	-3	4	11	18	25	32

(f) -3

x	-1	0	1	2	3	4
$-3x + 1$	4	1	-2	-5	-8	-11

(g) The change in the value of the expression for a one unit change in x.

Section 1.4   (page 50)

1. $\sqrt{x}$ is not an integer power of x.

2. The degree of the term ax^k is k. The term of highest degree in a polynomial has the same degree as the polynomial.

3. Addition (subtraction) separates terms. Multiplication separates factors.

4. No. $x^3 + 2x + 3$ **5.** $10x - 4$, 1, 10

6. $3x^2 + 8$, 2, 3   **7.** $-3y^4 + 5$, 4, -3

8. $-3x^3 - 2x^2 - 3$, 3, -3 **9.** $4t^5 - t^2 + 6t + 3$, 5, 4

10. $-16z^2 + 8z$, 2, -16 **11.** -4, 0, -4

12. $-16t^2 + v_0t$, 2, -16 **13.** Binomial

14. Monomial   **15.** Trinomial   **16.** Binomial

17. ax^3 where $a \neq 0$ **18.** $-2x^4 + x^3 + 1$

19. $8x^2 + 4$ **20.** Any constant

21. (a) 16   **(b)** 0   **(c)** -16 **(d)** 16

22. (a) -4 **(b)** 0   **(c)** -4 **(d)** $\frac{9}{4}$

23. (a) -27 **(b)** -16 **(c)** 0   **(d)** $\frac{9}{16}$

24. (a) -1 **(b)** $-\frac{16}{27}$ **(c)** 0   **(d)** 7   **25.** $7x^2 + 3$

26. $3x^3 + x + 3$ **27.** $6x^2 - 7x + 8$

28. $4y^2 - y + 3$ **29.** $-2x^2 + 15$

30. $3s^2 + 26s - 32$ **31.** $2x^2 - 3x$

32. $4x^3 + 2x^2 + 9x - 6$ **33.** 4   **34.** $3v^2 + 2v - 2$

35. Yes.
$$(x^3 - 2x^2 + 4) + (-x^3 + x^2 + 3x) = -x^2 + 3x + 4$$

36. To subtract one polynomial from another, add the opposite. You can do this by changing the sign of each term of the polynomial that is being subtracted and then adding the resulting like terms.

37. $x^2 - 3x + 2$ **38.** $2y^4 - 4$ **39.** $7x^3 + 2x$

40. $7y^2 - 9y + 2$ **41.** $4x^3 + x + 4$ **42.** $2y^4 + 2y^2$

43. $x^2 - 2x + 5$ **44.** $1.2t^4 - 0.8t^2 + 1.4$

45. $-4x^3 - 2x + 13$ **46.** 0   **47.** $2x^2 + 9x - 11$

48. $8x^3 - 9x^2 - 3x - 8$ **49.** $7x^3 + 22x^2 + 4$

50. $26y^4 - 10y - 10$ **51.** $29s + 8$

52. $5x^2 - 6x - 18$ **53.** $3t^2 + 29$ **54.** $-5v - 1$

55. $16a^3$ **56.** $-18n^3$ **57.** $10y - 2y^2$

58. $10z^2 - 35z$ **59.** $8x^5 - 12x^4 + 20x^3$

60. $-9y^4 + 21y^3 - 9y^2$ **61.** $-10x^2 - 6x^4 + 14x^5$

62. $-33a^3 + 9a^2$ **63.** $x^2 + 3x - 28$

64. $y^2 + y - 6$ **65.** $6x^2 + 7xy + 2y^2$

66. $6x^2 - 7xy + 2y^2$ **67.** $48y^2 + 32y - 3$

68. $10t^2 - \frac{163}{2}t + 12$ **69.** $75x^3 + 30x^2$

70. $-12t^4 + 12t^2$ **71.** $-a^2 + 19a$

72. $2t^2 + 7t - 16$ **73.** $x^4 - 2x^3 - 3x^2 + 8x - 4$

74. $t^3 - 2t^2 - 14t + 3$ **75.** $2u^3 + 13u^2 + 11u - 20$

76. $x^4 - 2x^3 - 8x - 16$ **77.** $4x^4 - 4x^3 + 6x - 9$

78. $x^3 - 5x^2 + 10x - 6$ **79.** $a^3 + 15a^2 + 75a + 125$

80. $y^3 - 6y^2 + 12y - 8$ **81.** $7x^3 + 7x^2 - 33x + 27$

82. $8x^6 + 27$ **83.** $-2x^3 + 3x^2 - 1$

84. $6s^3 - 23s^2 + 38s - 24$ **85.** $a^2 - 36c^2$

86. $64n^2 - m^2$ **87.** $4x^2 - \frac{1}{16}$ **88.** $\frac{4}{9}x^2 - 49$

89. $0.04t^2 - 0.25$ **90.** $16a^2 - 0.01b^2$

91. $x^2 + 10x + 25$ **92.** $x^2 + 18x + 81$

93. $25x^2 - 20x + 4$ **94.** $9x^2 - 48x + 64$

95. $4a^2 + 12ab + 9b^2$ **96.** $16x^2 - 40xy + 25y^2$

97. $x^2 - 2xy + y^2 + 4x - 4y + 4$

98. $x^2 + 2xy + y^2 - 8x - 8y + 16$

99. $u^2 - 2uv + 6u + v^2 - 6v + 9$

100. $4z^2 + 4yz + 4z + y^2 + 2y + 1$ **101.** $-8x - 7$

102. $k - 67$ **103.** $6y^2 - 32y + 36$

104. $-3b^2 - 18b + 1$ **105.** $12t$ **106.** $2a^2 + 72$

107. $m + n$

108. (a) True (b) False. $(x + 2)(x - 3) = x^2 - x - 6$

109. $8x^2 + 26x$ **110.** $1.2x^2$

111. $(x + a)(x + b)$, $x^2 + ax + bx + ab$;

$(x + a)(x + b) = x^2 + ax + bx + ab$

This statement illustrates the FOIL Method.

112. $(x + a)^2 = x^2 + 2ax + a^2$

$(x + a)^2 = x^2 + 2ax + a^2$

This statement illustrates the special product for the square root of a binomial.

113. (a) $5w$ (b) $\frac{3}{2}w^2$ **114.** $\frac{3}{2}x^2 + \frac{15}{2}x$

115. $1000 + 2000r + 1000r^2$ **116.** $1199.03

117. (a) $x^2 - 1$ (b) $x^3 - 1$ (c) $x^4 - 1, x^5 - 1$

118. $49 \neq 25$ **119.** Dropped, 100 ft

120. Thrown upward, 0 ft   **121.** Thrown downward, 50 ft

122. Thrown upward, 300 ft   **123.** 224 ft, 216 ft, 176 ft

124. $4.29x^2$ **125.** $-14.14y^3$

126. $-7.148a^2 + 15.691a$ **127.** $12.823k^2 - 15.204k$

Section Project 1.4

(a) $y = 25.93 - 0.13t - 0.01t^2$ (b) Decreasing

1985: 25.03 gallons

1990: 23.63 gallons

Mid-Chapter Quiz   (page 54)

1. -68 **2.** $\frac{1}{24}$ **3.** 71.03   **4.** $\frac{2}{5}$ **5.** $\frac{27}{8}$

6. -2 **7.** Distributive Property

8. Multiplicative Identity Property

9. Additive Inverse Property   **10.** $(5x)^4$ **11.** $4x^5$

12. $16x^4$ **13.** $\frac{y^6}{27}$ **14.** $\frac{3}{2}xy^2$ **15.** $-x^2 + 2xy$

16. $-6x^3 - 3x$ **17.** $7x - 15$ **18.** $-5v - 18$

19. $36r^2 - 25s^2$ **20.** $x^3 + 1$ **21.** $2z^2 + 3z - 35$

22. $-12v$ **23.** (a) 225 (b) 5   **24.** $6x + 16$

25. Loss of $498,833.36

Section 1.5   (page 62)

1. It is written as a product of polynomials.

2. *Noun:* Any one of the expressions that, when multiplied together, yields the product

Verb: The process of finding expressions that, when multiplied together, yield the given product

3. 6   **4.** 2   **5.** $3x$ **6.** $9x^3$ **7.** $6z^2$ **8.** $15y$

9. $14b^2$ **10.** $4xy$ **11.** $21(x + 8)^2$ **12.** $22(3 - y)$

13. $x(1 - z)^2$ **14.** $2(x + 5)$ **15.** $8(z - 1)$

16. $5(x + 1)$ **17.** $6(4x^2 - 3)$ **18.** $7(2z^3 + 3)$

19. $x(2x + 1)$ **20.** $-a(a^2 + 4)$ **21.** $7u(3u - 2)$

22. $12y^2(3y^2 + 2)$ **23.** No common factor other than 1

24. $1(16x^2 - 3y^3)$ **25.** $3y(x^2y - 5)$ **26.** $2uv(2 + 3uv)$

27. $4(7x^2 + 4x - 2)$ **28.** $3(3 - 9y - 5y^2)$

29. $x^2(14x^2 + 21x + 9)$ **30.** $y^2(17x^5y - x + 34)$

31. $-5(x - 2)$ **32.** $-4(x^4 - 8)$ **33.** $-7(2x - 1)$

34. $-5(x - 3)$ **35.** $-2(x^2 - 2x - 4)$

36. $-6(x^2 - 2x + 3)$ **37.** $-1(4t^2 - 2t + 15)$

38. $-1(5s^4 - 32s - 16)$ **39.** $6x + 5$ **40.** $2x - 5$

41. $2x + y$ **42.** $2u - 9v$ **43.** $(y - 3)(2y + 5)$

44. $(s + 9)(7t - 6)$ **45.** $(t^2 + 1)(5t - 4)$

46. $(a^2 - 3)(3a + 10)$ **47.** $a(a + 6)(1 - a)$

48. $y(x - y)$ **49.** $(y - 6)(y + 2)$ **50.** $(y + 3)(y + 4)$

51. $(x + 25)(x + 1)$ **52.** $(x - 7)(x + 1)$

53. $(x + 2)(x^2 + 1)$ **54.** $(t - 11)(t^2 + 1)$

55. $(a - 4)(a^2 + 2)$ **56.** $(s + 2)(3s^2 + 5)$

57. $(z + 3)(z^3 - 2)$ **58.** $(2u - 1)(2u^3 - 3)$

59. $(c - 3)(d + 3)$ **60.** $(u + v)(u - 4)$

61. $(x + 8)(x - 8)$ **62.** $(y - 12)(y + 12)$

63. $(4y + 3z)(4y - 3z)$ **64.** $(3z - 5w)(3z + 5w)$

65. $(x + 2y)(x - 2y)$ **66.** $(9a + b^3)(9a - b^3)$

67. $(a^4 + 6)(a^4 - 6)$ **68.** $(y^5 - 8)(y^5 + 8)$

69. $(10 - 3y)(10 + 3y)$ **70.** $(25 - 7x)(25 + 7x)$

71. $(ab - 4)(ab + 4)$ **72.** $(uv - 5)(uv + 5)$

73. $(a + 11)(a - 3)$ **74.** $(x - 5)(x - 1)$

75. $(4 - z)(14 + z)$ **76.** $(13 - y)(7 + y)$

77. $(x - 2)(x^2 + 2x + 4)$ **78.** $(t - 3)(t^2 + 3t + 9)$

79. $(y + 4z)(y^2 - 4yz + 16z^2)$

80. $(z + 5w)(z^2 - 5wz + 25w^2)$

81. $(2t - 3)(4t^2 + 6t + 9)$

82. $(3s + 4)(9s^2 - 12s + 16)$ **83.** $2(2 - 5x)(2 + 5x)$

84. $a(a - 4)(a + 4)$ **85.** $(y + 3x)(y - 3x)(y^2 + 9x^2)$

86. $(u - 4v)(u + 4v)(u^2 + 16v^2)$

87. $2(x - 3)(x^2 + 3x + 9)$

88. $5(y - 5)(y^2 + 5y + 25)$ **89.** 6399   **90.** 396

91. $(2x^n - 5)(2x^n + 5)$ **92.** $(3 - 2y^n)(3 + 2y^n)(9 + 4y^{2n})$

93. $x^2(3x + 4) - (3x + 4) = (x - 1)(x + 1)(3x + 4)$

$3x(x^2 - 1) + 4(x^2 - 1) = (x - 1)(x + 1)(3x + 4)$

94. $2x^2(3x - 4) + 3(3x - 4) = (3x - 4)(2x^2 + 3)$

$3x(2x^2 + 3) - 4(2x^2 + 3) = (2x^2 + 3)(3x - 4)$

95. $P(1 + rt)$ **96.** $p = 800 - 0.25x$

CHAPTER 1

97. $w = 45 - l$ **98.** $l = 32 - w$

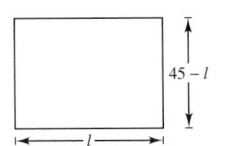

99. $S = 2\pi r(r + h)$ **100.** $\pi(R - r)(R + r)$
101. $kx(Q - x)$

Section Project 1.5

(a)

Solid	Length	Width	Height	Volume
Entire cube	a	a	a	a^3
Solid I	a	a	$a - b$	$a^2(a - b)$
Solid II	a	$a - b$	b	$ab(a - b)$
Solid III	$a - b$	b	b	$b^2(a - b)$
Solid IV	b	b	b	b^3

(b) $a^2(a - b) + ab(a - b) + b^2(a - b) =$
$(a - b)(a^2 + ab + b^2)$

(c) If the smaller cube is removed from the larger, the remaining solid has a volume of $a^3 - b^3$ and is composed of the three rectangular boxes labeled Solid I, Solid II, and Solid III. From part (b) we have
$$a^3 - b^3 = (a - b)(a^2 + ab + b^2).$$

Section 1.6   (page 74)

1. $9x^2 - 9x - 54 = 9(x^2 - x - 6)$
$= 9(x - 3)(x + 2)$
2. No. $x(x + 2) - 2(x + 2) = (x + 2)(x - 2)$
3. $x + 1$ **4.** $a - 2$ **5.** $y - 5$ **6.** $y + 2$
7. $z - 2$ **8.** $z + 6$ **9.** $(x + 3)(x + 1)$
10. $(x - 6)(x - 4)$ **11.** $(y + 10)(y - 3)$
12. $(m - 5)(m + 2)$ **13.** $(t - 7)(t + 3)$
14. $(x + 6)(x - 2)$ **15.** $(x - 12)(x - 8)$
16. $(u + 3v)(u + 2v)$ **17.** $(x - 7y)(x + 5y)$
18. $(a - 10b)(a - 11b)$ **19.** $5x + 3$ **20.** $5x + 4$
21. $5a - 3$ **22.** $5c - 4$ **23.** $2y - 9$ **24.** $3y - 10$
25. $(3x + 1)(x + 1)$ **26.** $(5x + 2)(x + 1)$
27. $(2t + 1)(4t - 5)$ **28.** $(3b - 1)(2b + 7)$
29. $(2a - 5)(a - 4)$ **30.** $(6x + y)(4x - 3y)$
31. $(5x + 4)(4x - 3)$ **32.** $(5x - 3y)(2x + 3y)$
33. $(2u - 5v)(u + 7v)$ **34.** Not factorable
35. $(-1)(2x - 3)(x + 2)$ **36.** $-(3x + 2)(2x - 3)$
37. $(-1)(4x + 1)(15x - 1)$ **38.** $(-1)(4x + 1)(3x - 2)$
39. $(3x + 4)(x + 2)$ **40.** $(2x + 3)(x + 3)$

41. $(2x - 1)(3x + 2)$ **42.** $(2x + 3)(3x - 5)$
43. $(3x - 1)(5x - 2)$ **44.** $(2x - 3)(6x - 5)$
45. $(x + 2)^2$ **46.** $(z + 3)^2$ **47.** $(a - 6)^2$
48. $(y - 7)^2$ **49.** $(5y - 1)^2$ **50.** $(2z + 7)^2$
51. $(3b + 2)^2$ **52.** $(2x - 1)^2$ **53.** $(2x - y)^2$
54. $(m + 3n)^2$ **55.** $(u + 4v)^2$ **56.** $(2y + 5z)^2$
57. $3x^3(x - 2)(x + 2)$ **58.** $5(2y - 3)(2y + 3)$
59. $2t(5t - 9)(t + 2)$ **60.** $(4z - 7)^2$
61. $(3x - 2)(7x - 2)$ **62.** $(1 - x)(1 + x)(6 - x)$
63. $(3 - z)(9 + z)$ **64.** $3(t - 2)(t^2 + 2t + 4)$
65. $2(3x - 1)(9x^2 + 3x + 1)$ **66.** $v(v^2 + 3v + 5)$
67. $3a(3a - b)(3a + b)$ **68.** $4mn(m + 4n)(2m - 3n)$
69. $(x + 2)(x + 4)(x - 4)$ **70.** $(x + 2)(x - 2)(x - 7)$
71. $(x + 3)(x - 3)(x - 6)$ **72.** $(x + 4)(x - 4)(x + 10)$
73. $(x - 5 + y)(x - 5 - y)$
74. $(3y + 2 + z)(3y + 2 - z)$
75. $(a - b + 4)(a - b - 4)$
76. $(x + 7y + 2a)(x + 7y - 2a)$
77. $(x^4 + 1)(x^2 + 1)(x + 1)(x - 1)$
78. $(x^2 + 4y^2)(x + 2y)(x - 2y)$
79. $b(b - 6)(b^2 + 6b + 36)$
80. $3(y - 4)(y^2 + 4y + 16)$ **81.** ±18 **82.** ±8
83. ±6 **84.** ±12 **85.** ±12 **86.** ±40 **87.** 16
88. 36   **89.** 9   **90.** 100   **91.** 25   **92.** 4
93. 14   **94.** (-3) **95.** $\pm9, \pm11, \pm19$
96. $\pm9, \pm15$ **97.** $\pm4, \pm20$ **98.** ±6
99. $\pm13, \pm14, \pm22, \pm41$ **100.** $\pm1, \pm7, \pm13, \pm29$
101. $8, -16$ **102.** $18, -36$ **103.** $2, -40$
104. $35, -28$ **105.** $-5, -32$ **106.** $-12, -6$
107. 2704   **108.** 1521   **109.** $4(6 + x)(6 - x)$
110. $\frac{5}{8}(x - 1)(x + 7)$ **111.** (c)   **112.** (b)   **113.** (a)
114. (d)

Section Project 1.6

(a) $(x + 3)(x + 1)$

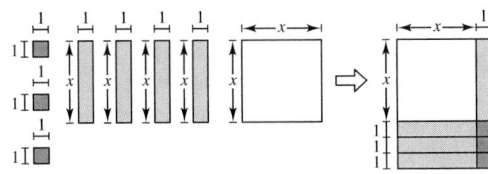

(b) $(x + 4)(x + 1)$

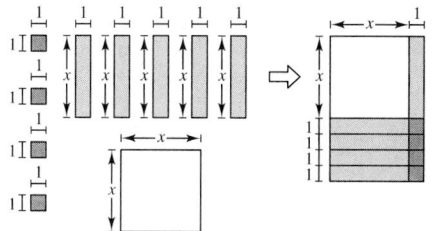

(c) $(x + 2)(x + 4)$

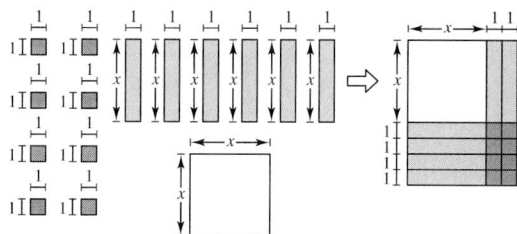

(d) $(x + 2)(x + 3)$

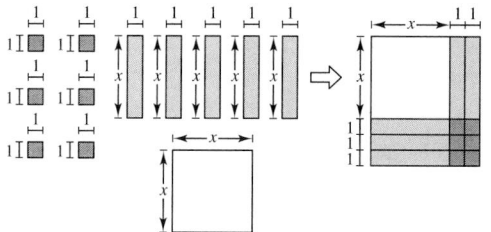

Section 1.7   (page 85)

1. (a) No **(b)** Yes   **2. (a)** Yes **(b)** No
3. (a) No **(b)** Yes   **4. (a)** No **(b)** No
5. (a) Yes **(b)** No   **6. (a)** No **(b)** No
7. (a) An expression is a collection of variables and constants using the operations of addition, subtraction, multiplication, and division. An equation is a statement of equality of two expressions.
  (b) An equation whose solution set is not the entire set of real numbers is called a conditional equation. The solution set of an identity is all real numbers.
8. Equivalent equations have the same solution set. For example, $3x + 4 = 10$ and $3x - 6 = 0$ are equivalent.
9. No solution   **10.** Conditional   **11.** Identity
12. Conditional   **13.** Linear
14. Not linear, because it is a second-degree equation
15. Not linear, because the variable is in the denominator

16. Linear
17. Original equation
  Subtract 15 from both sides.
  Combine like terms.
  Divide both sides by 3.
  Simplify.
18. Original equation
  Add 21 to both sides.
  Combine like terms.
  Divide both sides by 7.
  Simplify.
19. Original equation.
  Subtract 5 from both sides.
  Combine like terms.
  Divide both sides by -2.
  Simplify.
20. Original equation
  Subtract 25 from both sides.
  Combine like terms.
  Divide both sides by -3.
  Simplify.
21. 4   **22.** $\frac{5}{4}$ **23.** 2   **24.** 3   **25.** $\frac{1}{3}$ **26.** $-\frac{1}{2}$
27. 0   **28.** Not possible because $2 \neq 0$
29. Not possible because $7 \neq 0$ **30.** 0   **31.** 6
32. -8 **33.** -2 **34.** $\frac{1}{2}$ **35.** 11   **36.** 3
37. -3 **38.** 10   **39.** $\frac{6}{5}$ **40.** -3 **41.** -2
42. $\frac{3}{4}$ **43.** 100   **44.** 1   **45.** 50   **46.** -14
47. $\frac{19}{10}$ **48.** $-\frac{11}{30}$ **49.** 0   **50.** $-\frac{2}{11}$ **51.** $-\frac{10}{3}$
52. $\frac{24}{7}$ **53.** 72   **54.** $-\frac{48}{11}$ **55.** 23   **56.** 46
57. 3.89   **58.** 14.58   **59.** 30.28   **60.** 0.16
61. (a) $x = \dfrac{6 + 3y}{2}$ **(b)** $y = \dfrac{2x - 6}{3}$
62. (a) $x = \dfrac{3y - 15}{5}$ **(b)** $y = \dfrac{5x + 15}{3}$
63. (a) $x = \dfrac{10y - 11}{7}$ **(b)** $y = \dfrac{7x + 11}{10}$
64. (a) $x = \dfrac{42 - 7y}{12}$ **(b)** $y = \dfrac{42 - 12x}{7}$
65. (a) $x = -6y$ **(b)** $y = -\frac{1}{6}x$
66. (a) $x = -\dfrac{1 + 27y}{3}$ **(b)** $y = -\dfrac{1 + 3x}{27}$
67. (a) $x = \dfrac{10 - 2y}{5}$ **(b)** $y = \dfrac{10 - 5x}{2}$
68. (a) $x = \dfrac{60 - 2y}{9}$ **(b)** $y = \dfrac{60 - 9x}{2}$

69. – 74. Answers may vary.

75. $R = \dfrac{E}{I}$ **76.** $C = \dfrac{S}{1 + r}$ **77.** $L = \dfrac{S}{1 - r}$

78. $h = \dfrac{2A}{b}$ **79.** $b = \dfrac{2A - ah}{h}$ **80.** $h = \dfrac{V}{\pi r^2}$

81. $r = \dfrac{A - P}{Pt}$ **82.** $P = \dfrac{A}{(1 + r/n)^{nt}}$

83. $n = \dfrac{2S}{a_1 + a_n}$ **84.** $n = \dfrac{L + d - a}{d}$

85. $r = \dfrac{S - a_1}{S}$ **86.** $m_2 = \dfrac{Fr^2}{km_1}$

87. (a)

t	1	1.5	2
Width	300	240	200
Length	300	360	400
Area	90,000	86,400	80,000

t	3	4	5
Width	150	120	100
Length	450	480	500
Area	67,500	57,600	50,000

(b) In a rectangle of fixed perimeter with length l equal to t times width w and $t \geq 1$, as t increases, w decreases, l increases, and the area A decreases. The maximum area occurs when the length and width are equal (when $t = 1$).

88. 1.5 sec

89. 2 sec. The velocity is zero when the object is at its maximum height.

90. 6 hr   **91.** 1988

Section Project 1.7

(a)

t	0	1	2	3	4	5
y	4193.0	5292.4	6391.8	7491.2	8590.6	9690.0

t	6	7	8	9	10
y	10,789.4	11,888.8	12,988.2	14,087.6	16,369.0

t	11	12	13	14	15
y	19,579.0	22,789.0	25,999.0	29,209.0	32,419.0

t	16	17	18	19	20
y	35,629.0	38,839.0	42,049.0	45,259.0	48,469.0

t	21	22	23
y	51,679.0	54,889.0	58,099.0

(b)

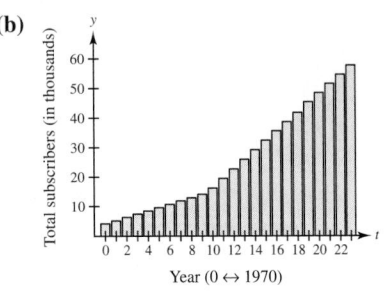

(c) 1976

(d) 1988

(e) The table was the easiest to use.

Section 1.8   (page 95)

1. 0, 8   **2.** $-6, 0$ **3.** $-10, 3$ **4.** $-15, 16$

5. $-4, 2$ **6.** $-8, 3$ **7.** $-4, -\frac{1}{2}, 3$ **8.** $-\frac{4}{3}, 0, 2$

9. $-\frac{25}{2}, 0, \frac{3}{2}$ **10.** $-12, -\frac{7}{2}, 39$ **11.** $-2, 5$

12. $-4, 3$ **13.** 4, 5   **14.** $-9, -3$ **15.** $-3, 0$

16. 0, 5   **17.** ± 5 **18.** ± 9 **19.** ± 10 **20.** 0, ± 3

21. 4   **22.** -2 **23.** $\frac{3}{2}$ **24.** -6 **25.** $-\frac{1}{2}, 7$

26. $-\frac{1}{3}, 11$ **27.** $-2, 5$ **28.** $-12, 8$ **29.** $-12, 6$

30. $-3, 5$ **31.** $-2, 10$ **32.** $-3, 15$ **33.** $-2, 6$

34. $-11, 3$ **35.** $-\frac{1}{3}, 0, \frac{1}{2}$ **36.** $-\frac{1}{3}, 0, 2$

37. 0, 7, 12   **38.** $-15, -3, 0$ **39.** $\pm 4, 25$

40. $-250, \pm 1$ **41.** ± 2 **42.** $\pm 4, 3$ **43.** ± 3

44. $-4, \pm 2$ **45.** $-2, \pm 3$ **46.** ± 2 **47.** 0, 5, ± 3

48. $\pm 5, -3, 0$

49. False. It is not an application of the Zero-Factor Property because there is an unlimited number of factors whose product is 1.

50. Yes. The only solution of the equation $x^2 + 2x + 1 = (x + 1)^2 = 0$ is $x = -1$.

51. Maximum number: n. The third-degree equation $(x + 1)^3 = 0$ has one solution: $x = -1$.

52. (a) and **(b)** $-5, -\frac{10}{3}$

   (c) Answers may vary.

53. $-\frac{20}{3}, -2$ **54.** $-\frac{5}{4}, -\frac{1}{2}$ **55.** 0, $-\frac{b}{a}$ **56.** 0, 1

57. $x^2 - 2x - 15 = 0$ **58.** $x^2 - 7x + 6 = 0$ **59.** 15
60. 8   **61.** 20 sec   **62.** 4 sec   **63.** 15 ft × 22 ft
64. $\frac{5}{2}$ cm   **65.** $b = 8, h = 12$ **66.** $b = 14, h = 10$
67. 20 in. × 20 in.   **68.** 10 units

69. (a)

x	3	4	5	6	7	8
P	-8	0	10	22	36	52

(b) P increases by $2(x + 1)$.   **(c)** 9

Section Project 1.8

(a) Volume = (length)(width)(height) = $(5 - 2x)(4 - 2x)x$
(b) 0, 2, 2.5

(c)

x	0.25	0.50	0.75	1.00	1.25	1.50	1.75
V	3.94	6	6.56	6	4.69	3	1.31

(d) 1.50

Review Exercises   (page 99)

1. $-\frac{1}{8} < 3$ **2.** $-2 > -8$

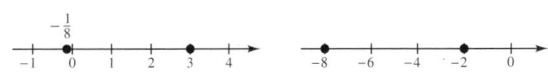

3. $-\frac{8}{5} < -\frac{2}{5}$ **4.** $8.4 > -\pi$

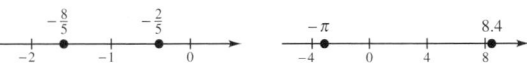

5. 230   **6.** -38 **7.** -4200 **8.** -960 **9.** 14
10. Undefined   **11.** $\frac{11}{21}$ **12.** $\frac{1}{2}$ **13.** $\frac{1}{6}$ **14.** $\frac{107}{96}$
15. $\frac{17}{8}$ **16.** $\frac{9}{4}$ **17.** $-\frac{1}{20}$ **18.** 1   **19.** 2
20. $-\frac{5}{2}$ **21.** -216 **22.** -81 **23.** 20   **24.** 40
25. Additive Inverse Property
26. Associative Property of Addition
27. Commutative Property of Multiplication
28. Additive Identity Property
29. Multiplicative Inverse Property
30. Associative Property of Multiplication   **31.** $3y^2 - 10y$
32. $3x^2 + 4xy$ **33.** $u - 3v$ **34.** $-20 + 12j$
35. $5x - y$ **36.** $20x - 80$ **37.** $18b - 15a$
38. $-2t^2 + t$ **39.** x^6 **40.** $-2y^5$ **41.** $-3x^3y^4$
42. $12u^3v^5$ **43.** $-64a^7$ **44.** $2(a - b)^6$
45. $4u^7v^3$ **46.** $36x^4y^5$ **47.** $8u^2v^2$ **48.** $\frac{4}{3}x^3y^4$
49. $P = 8x + 8$ **50.** $P = 30a - 60$
    $A = 15x$ $A = 30(a - 2)^2$

51. 6910   **52.** \$1203   **53.** 1 hr   **54.** 1986–1990
55. $6x - x^2$ **56.** $\frac{9}{2}x + 1$ **57.** $-9x^3 + 9x - 4$
58. $-8z^2$ **59.** $6y^2 - 2y + 15$ **60.** $6a^3 + 5$
61. $-8x^4 - 32x^3$ **62.** $-12y^3 + 24y^2$
63. $6z^2 - z - 15$ **64.** $6t^2 - 65t - 11$
65. $15x^2 - 11x - 12$ **66.** $12y^4 - 7y^2 - 10$
67. $4x^3 - 5x + 6$ **68.** $20s^3 - 9s^2 - 32s + 15$
69. $16x^2 - 56x + 49$ **70.** $64 - 48x + 9x^2$
71. $25u^2 - 64$ **72.** $49a^2 - 16$
73. $u^2 - v^2 - 6u + 9$
74. $m^2 - 10m + 25 + 2mn - 10n + n^2$
75. $3x^2(2 + 5x)$ **76.** $4y(2 - 3y^3)$ **77.** $(u - 5v)^2$
78. $(u - 1)(u^2 + u + 1)$ **79.** $(3a + 10)(3a - 10)$
80. $(x - 8)(x - 3)$ **81.** $(a - 2)(a + 9)$
82. $(3x - 1)(x + 8)$ **83.** $(3x + 4)(9x^2 - 12x + 16)$
84. $(y + 1)(y - 1)(y + 4)$ **85.** $t(t + 2)(4t - 1)$
86. $(x - 20)^2$ **87.** $(v + 2)(v - 2)^2$
88. $(y + 1)(y - 7)$ **89.** $(2u - 7)^2$
90. $-14(x + 5)(5x + 23)$ **91.** $h(2h + 1)(3h - 13)$
92. $5y(y - 1)(7y + 5)$ **93.** $x(x + 3)(x - 3)(x + 7)$
94. $u(u - 9v)$ **95.** $[(x + 9) + 2y][(x + 9) - 2y]$
96. $2(5a - b)(25a^2 + 5ab + b^2)$ **97.** 14   **98.** $-\frac{6}{5}$
99. -8.2 **100.** 2   **101.** 2   **102.** $\frac{17}{2}$ **103.** $\frac{20}{17}$
104. $\frac{5}{7}$ **105.** $h = \dfrac{V}{\pi r^2}$ **106.** $h = \dfrac{S - 2\pi r^2}{2\pi r}$
107. 0, 3   **108.** $0, -\frac{7}{4}$ **109.** $-10, 10$ **110.** $-8, 2$
111. $-\frac{4}{3}, 2$ **112.** $-4, 0, 3$ **113.** $-4, 9$ **114.** $\pm 1, 6$
115. -2 **116.** 3   **117.** ± 4 **118.** ± 11
119. $-\frac{1}{2}, 7$ **120.** 2   **121.** $\frac{25}{3}$ **122.** $-\frac{1}{3}, 11$
123. -20 **124.** 12   **125.** $-6, 5$ **126.** $-12, 10$
127. 3 sec   **128.** 38,416; (x^4) is the same as $(x^2)^2$.
129. 24 in. × 18 in.
130. (a) $x^2 - y^2$ **(b)** Resulting rectangle = $(x + y)(x - y)$

Chapter 1 Test   (page 102)

1. $-\frac{1}{2}$ **2.** $\frac{4}{27}$ **3.** $-\frac{27}{125}$ **4.** 11
5. Associative Property of Multiplication   **6.** $-5x$
7. $3x^4y^3$ **8.** $-2x^2 + 5x - 1$ **9.** $-2y^2 - 2y$
10. $8x^2 - 4x + 10$ **11.** $11t + 7$ **12.** $\dfrac{y^3}{8}$
13. $2x^2 + 7xy - 15y^2$ **14.** $6s^3 - 17s^2 + 26s - 21$
15. $16x^2 - 24x + 9$ **16.** $16 - a^2 - 2ab - b^2$
17. $6y(3y - 2)$ **18.** $(x - 2)(5x^2 - 6)$ **19.** $(3u - 1)^2$
20. $2(3x + 2)(x - 5)$ **21.** 4   **22.** 4   **23.** 24
24. $-\frac{4}{3}, 3$ **25.** $-5, 1$ **26.** $-4, -1, 0$
27. $x(x + 26)$ **28.** $b = \frac{1}{2}(3a + 17)$

CHAPTER 2

Section 2.1   (page 114)

1. Order is significant because each number in the pair has a particular interpretation. The first measures horizontal distance and the second measures vertical distance.

2. The x-coordinate of any point on the y-axis is 0. The y-coordinate of any point on the x-axis is 0.

3. 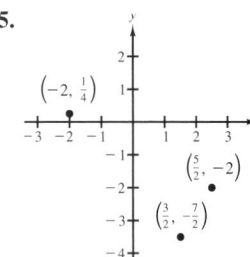   4.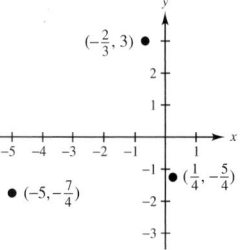

5. 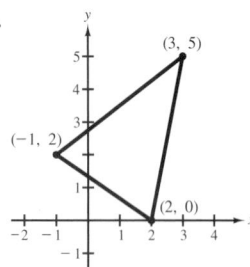   6.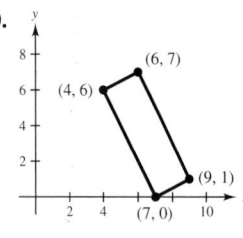

7. $A: (4, -2)$ $B: (-3, -2.5)$ $C: (3, 0.5)$
8. $A: (-3, 2)$ $B: (4, -1)$ $C: \left(-\frac{1}{2}, -2\right)$

9. 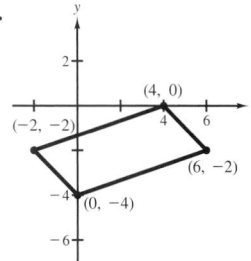   10.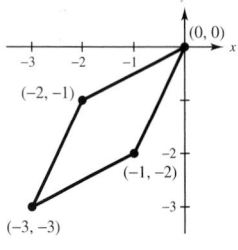

11. 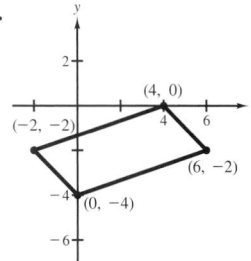   12.

13. Quadrant III   14. Quadrant IV
15. Quadrants I and II   16. Quadrants II and III
17. Quadrants II and IV   18. Quadrant I   19. $(-5, 2)$
20. $(10, -4)$ 21. $(10, 0)$ 22. $(0, -5)$

23.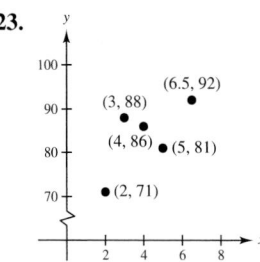

24.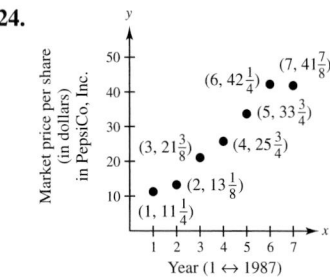

25.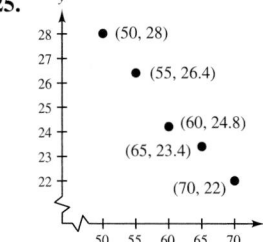

26.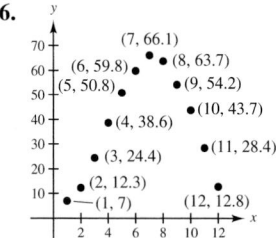

27. (a) Yes   (b) No   (c) No   (d) Yes
28. (a) Yes   (b) No   (c) Yes   (d) No
29. (a) Yes   (b) Yes   (c) Yes   (d) No
30. (a) No   (b) No   (c) No   (d) No
31. (a) No   (b) Yes   (c) Yes   (d) No
32. (a) No   (b) Yes   (c) Yes   (d) No

33.

x	-2	0	2	4	6
$y = 5x - 1$	-11	-1	9	19	29

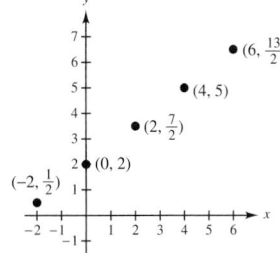

34.

x	-2	0	2	4	6
$y = \frac{3}{4}x + 2$	$\frac{1}{2}$	2	$\frac{7}{2}$	5	$\frac{13}{2}$

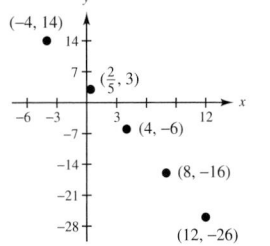

35.

x	-4	$\frac{2}{5}$	4	8	12
$y = -\frac{5}{2}x + 4$	14	3	-6	-16	-26

36.

x	-6	-3	0	$\frac{3}{4}$	10
$y = \frac{4}{3}x - \frac{1}{3}$	$-\frac{25}{3}$	$-\frac{13}{3}$	$-\frac{1}{3}$	$\frac{2}{3}$	13

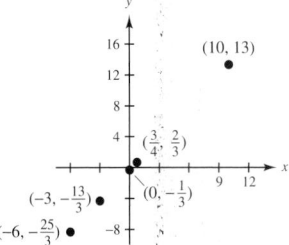

37.

x	-2	0	2	4	6
$y = 4x^2 + x - 2$	12	-2	16	66	148

38.

x	-2	0	2	4	6
$y = \frac{4}{3}x - \frac{1}{3}$	-3	$-\frac{1}{3}$	$\frac{7}{3}$	5	$\frac{23}{3}$

39. Reflection about the y-axis

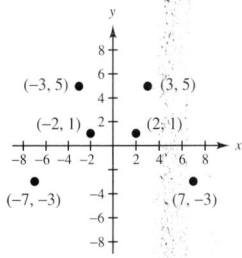

40. Reflection about the x-axis

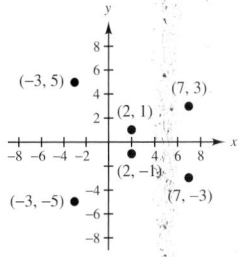

41.

x	100	150	200	250	300
$C = 28x + 3000$	5800	7200	8600	10,000	11,400

For each one-unit increase in x, C increases 28 units.

42.

x	2	4	8	10	20
$y = 0.75x + 8$	9.5	11	14	15.5	23

For each one-unit increase in x, y increases 0.75 unit.

43. $(-1, 1), (3, 2), (0, 4)$

44. $(1, -1), (3, -4), (5, -1), (3, 2)$

45. 7   **46.** 7

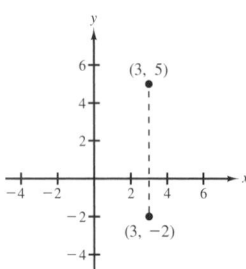   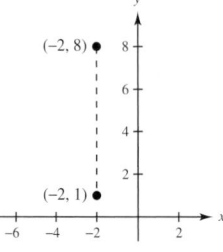

Vertical line   Vertical line

47. 5   **48.** 11

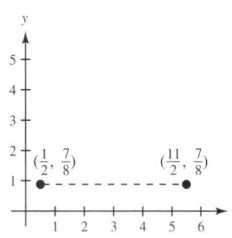   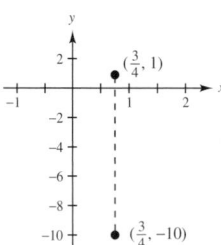

Horizontal line   Vertical line

49. 7   **50.** 250

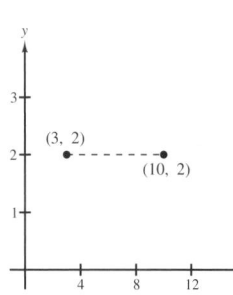

Horizontal line   Horizontal line

51. $\frac{3}{4}$ **52.** $\frac{13}{2}$

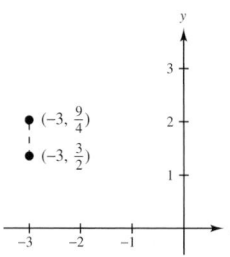   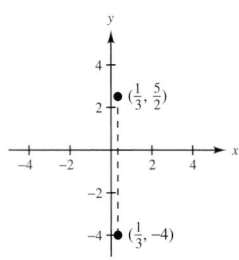

Vertical line   Vertical line

53. 5   **54.** 13   **55.** 15   **56.** 17   **57.** 7.81

58. 16.55   **59.** 5.39   **60.** 1.41   **61.** Right triangle

62. Right triangle   **63.** Not a right triangle

64. Not a right triangle   **65.** $3 + \sqrt{26} + \sqrt{29} \approx 13.48$

66. $\sqrt{41} + \sqrt{52} + \sqrt{65} \approx 21.68$

67. $(1, 4)$ **68.** $(2, 0)$

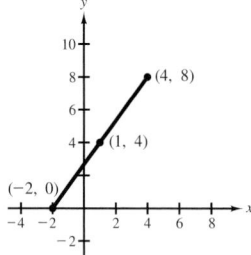   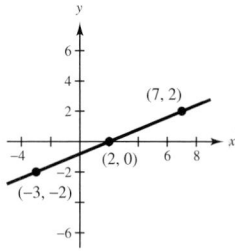

69. $\left(\frac{7}{2}, \frac{9}{2}\right)$ **70.** $\left(\frac{11}{2}, 3\right)$

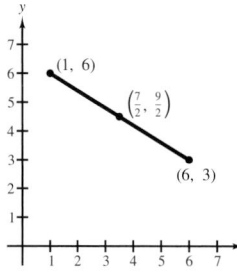   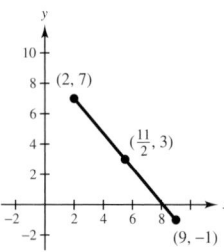

71. 15   **72.** 22 and 23

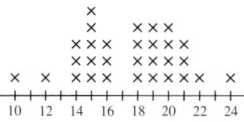   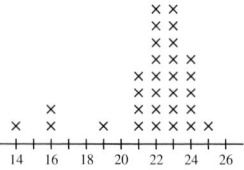

73. 81 and 85

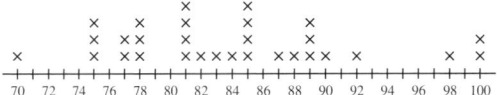

74. 68

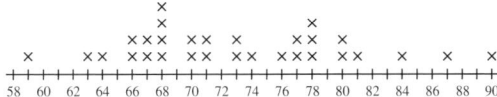

75. Total number

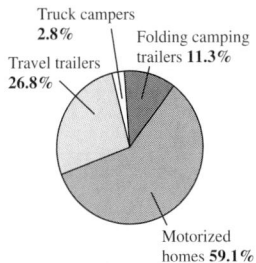

Retail value

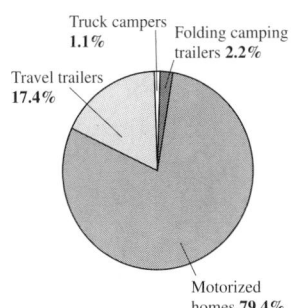

76.

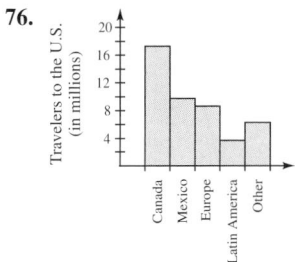

77.

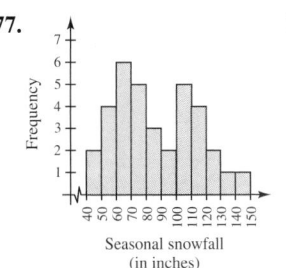

78.

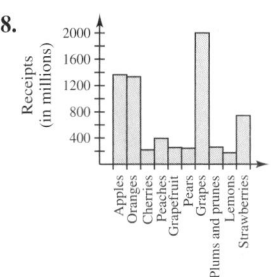

79.

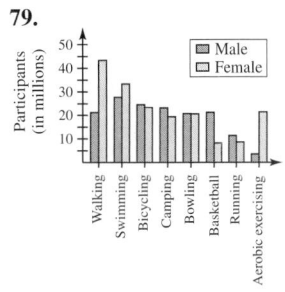

80.

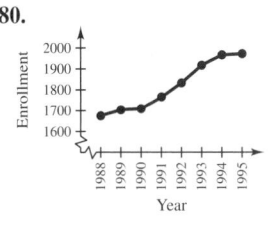

81.

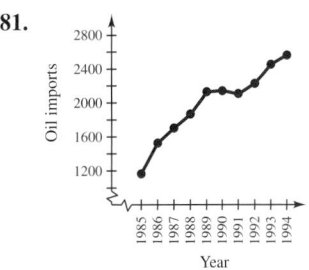

82.

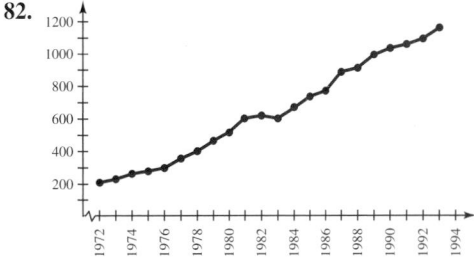

83. 18.55 ft

Section Project 2.1

(a) $(x, y) = (4, -4)$; 8, 7; $\sqrt{113}$

(b) $(x, y) = (6, 6)$; 8, 6; 10

(c) $(x, y) = (10, 2)$; 6, 8; 10

(d) $(x, y) = (10, 1)$; 5, 12; 13

CHAPTER 2

Section 2.2   (page 128)

1. (e)   **2.** (b)   **3.** (f)   **4.** (a)   **5.** (d)   **6.** (c)

7.

x	-4	-2	0	2	4
y	11	7	3	-1	-5

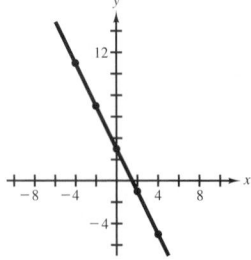

8.

x	-3	0	2	3	$\frac{21}{2}$
y	-4	-2	$-\frac{2}{3}$	0	5

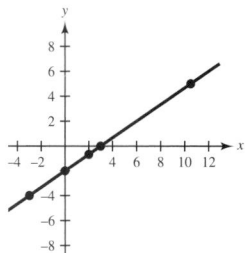

9.

x	± 2	-1	0	2	± 3
y	0	3	4	0	-5

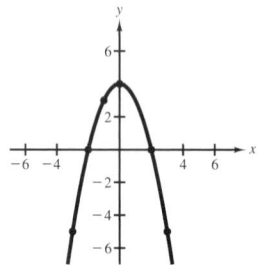

10.

x	-1	0	1	2	3
y	$-\frac{9}{2}$	-4	$-\frac{7}{2}$	0	$\frac{19}{2}$

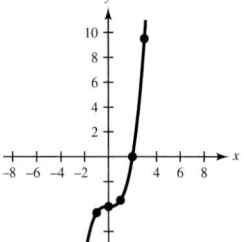

11.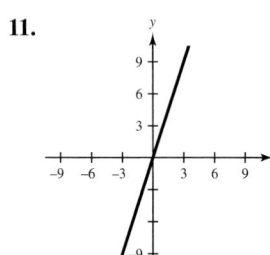

12.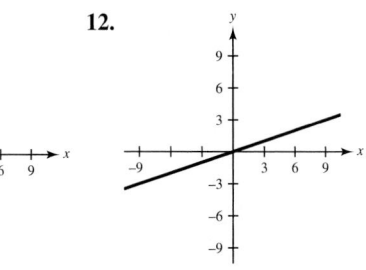

13.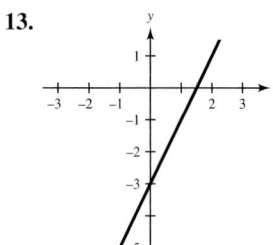

14.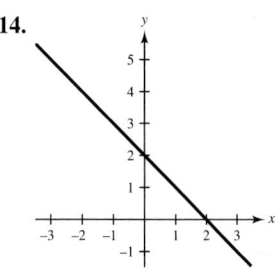

15.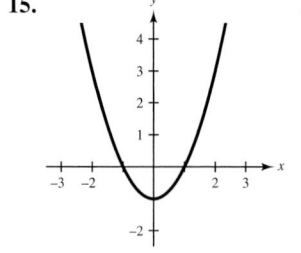

16.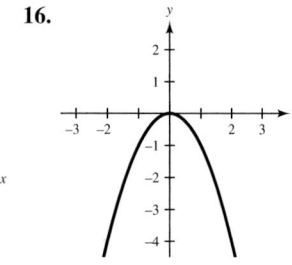

17. 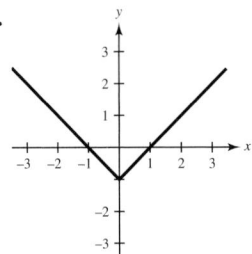   **18.**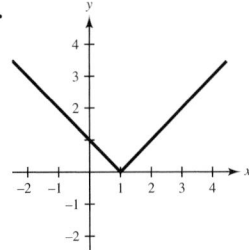

19. $(0, 3)$ **20.** $(3, 0)$ **21.** $(2, 0), (0, 2)$
22. $(3, 0), (4, 0), (0, 12)$ **23.** $(10, 0), (0, 5)$
24. $(-4, 0), (0, 6)$ **25.** $(-1, 0), (0, 1)$
26. $(-5, 0), (5, 0), (0, -25)$ **27.** $(-20, 0), (0, 15)$
28. $(30, 0), (0, 12)$ **29.** $(0, 0)$ **30.** $(0, 0)$

31. 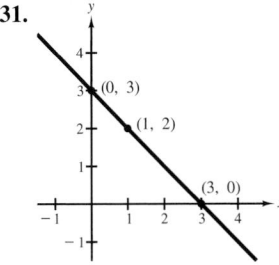   **32.**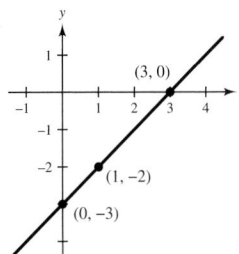

33. 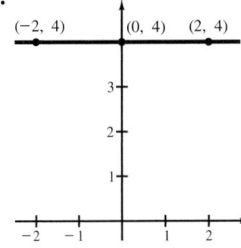   **34.**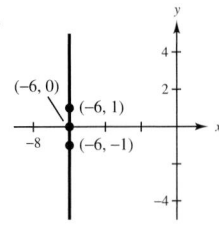

35. 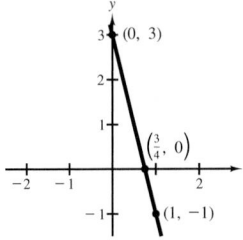   **36.**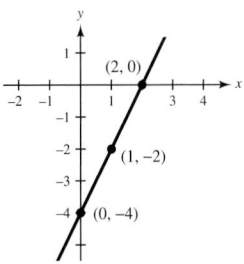

37. 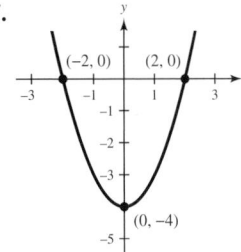   **38.**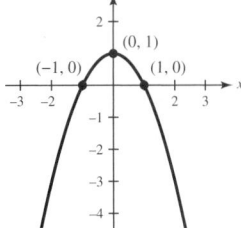

39. 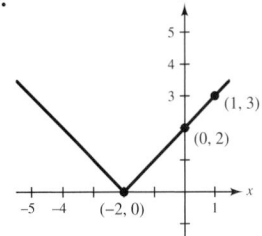   **40.**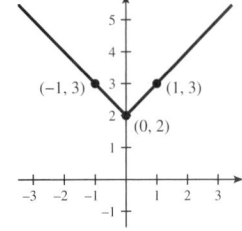

41. 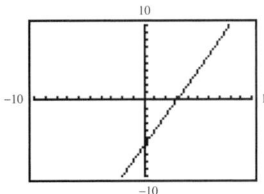   **42.**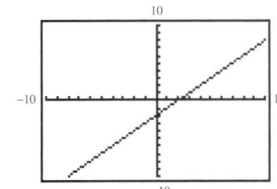

43. 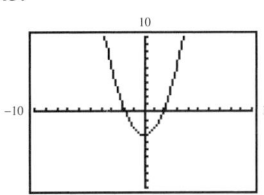   **44.**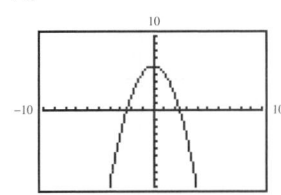

45. 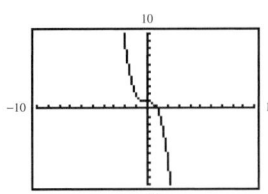   **46.**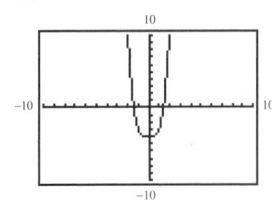

47.
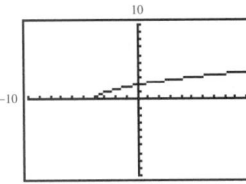

48.
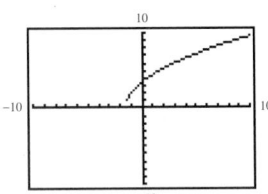

49.
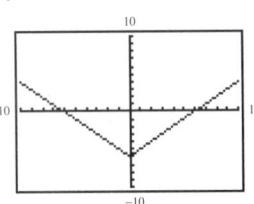

50.
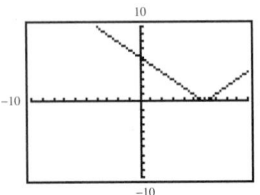

51. A more complete graph with the specified setting

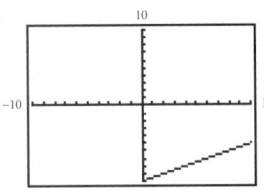

52. A more complete graph with the specified setting

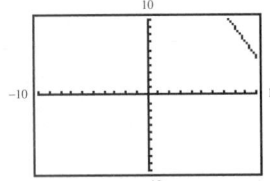

53. A closer look at the important characteristics of the graph

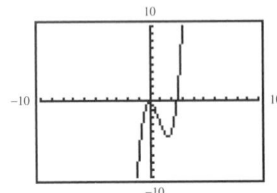

54. A more complete graph with the specified setting

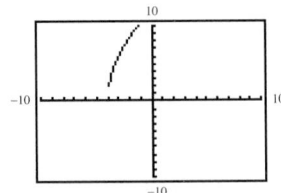 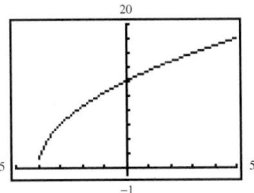

55.
Xmin = -2
Xmax = 8
Xscl = 1
Ymin = -1
Ymax = 17
Yscl = 1

56.
Xmin = -1
Xmax = 8
Xscl = 1
Ymin = -6
Ymax = 10
Yscl = 1

57.
Xmin = -5
Xmax = 5
Xscl = 1
Ymin = -3
Ymax = 6
Yscl = 1

58.
Xmin = -3
Xmax = 3
Xscl = 1
Ymin = -1
Ymax = 18
Yscl = 1

59. Identical graphs
Distributive Property

60. Identical graphs
Commutative Property of Addition

61. Identical graphs
Associative Property of Addition

62. Identical graphs
Additive Identity Property

63. x-intercepts: $(\pm 3, 0)$
Solutions: $x = \pm 3$

64. x-intercept: $(2, 0)$
Solution: $x = 2$

65. x-intercepts: $(-1, 0), (3, 0)$
Solutions: $x = -1, x = 3$

66. x-intercepts: $(-1, 0), (1, 0), (3, 0)$
Solutions: $x = -1, x = 1, x = 3$

67.
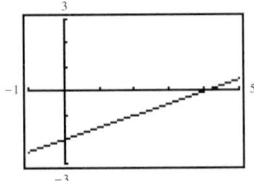

x-intercept: $(4, 0)$

68.

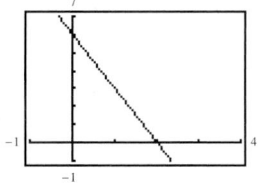

x-intercept: $(2, 0)$

69.

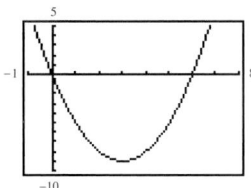

x-intercepts: $(0, 0), (6, 0)$

70.

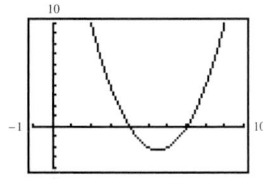

x-intercepts: $(4, 0), (7, 0)$

71. $\frac{9}{2}$ **72.** -4 **73.** ± 2 **74.** $-2, 0$ **75.** 1
76. $1, 3$ **77.** $-4, \frac{3}{2}$ **78.** $-1, 5$ **79.** $\pm 2, 0$
80. $-2, \pm 1$

81. (a)

x	0	3	6	9	12
F	0	4	8	12	16

(b)

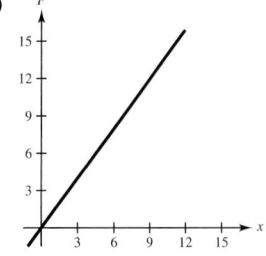

(c) The length is also doubled.

82. (a)

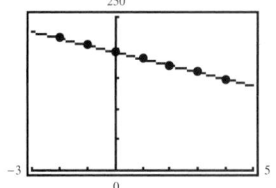

(b) The model fits the data well.
(c) 115,900
(d) The model decreases to negative values of *N*.

83.

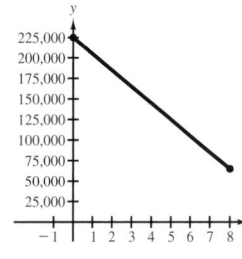

84. (a) Answers may vary.
(b)

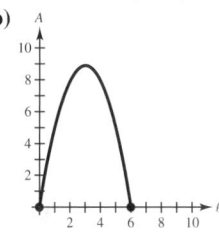

(c) 3 m × 3 m

85.

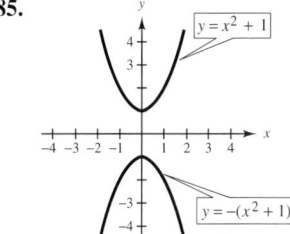

Reflection in the *x*-axis

86.

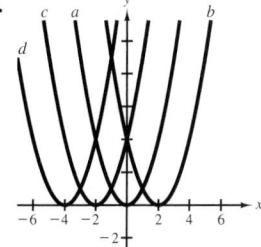

The graph of $y = (x + c)^2$, $c > 0$, is obtained by shifting the graph of $y = x^2$ to the *left c* units.
The graph of $y = (x - c)^2$, $c > 0$, is obtained by shifting the graph of $y = x^2$ to the *right c* units.

87.

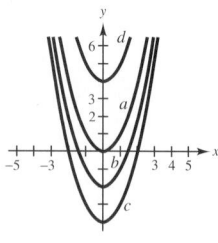

The graph of $y = x^2 + c$, $c > 0$, is obtained by shifting the graph of $y = x^2$ *upward* c units.

The graph of $y = x^2 - c$, $c > 0$, is obtained by shifting the graph of $y = x^2$ *downward* c units.

Section Project 2.2

(a) The graph on the left is misleading because it is difficult to detect any increase in sales.

(b) The graph on the left is misleading because it is difficult to detect any change in profits.

(c) Answers may vary.

Section 2.3   (page 141)

1. $\frac{2}{3}$ **2.** $-\frac{1}{2}$ **3.** -2 **4.** 4   **5.** Undefined   **6.** 0

7. (a) L_3 **(b)** L_2 **(c)** L_1 **8. (a)** L_2 **(b)** L_3 **(c)** L_1

9. $\frac{5}{7}$, rises   **10.** $\frac{4}{3}$, rises

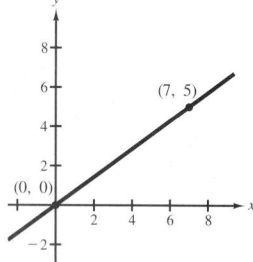 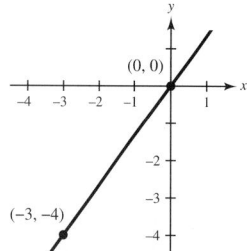

11. $\frac{1}{2}$, rises   **12.** $\frac{2}{3}$, rises

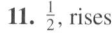

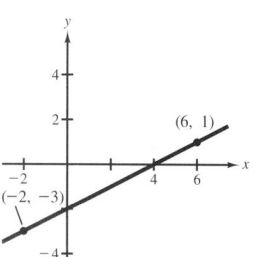 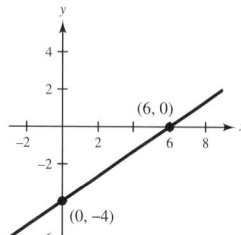

13. $-\frac{3}{2}$, falls

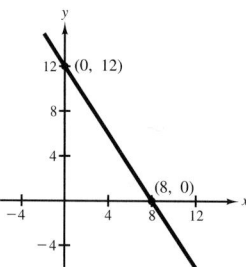

14. -4, falls

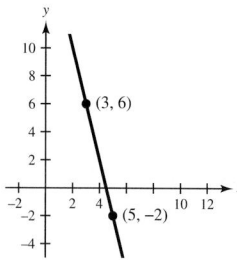

15. Undefined, vertical

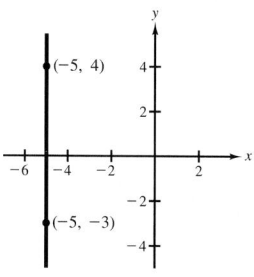

16. $-\frac{8}{5}$, falls

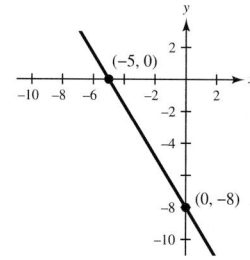

17. 0, horizontal

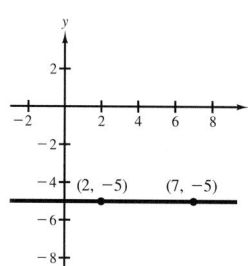

18. 2, rises

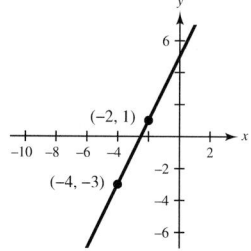

19. $-\frac{18}{17}$, falls

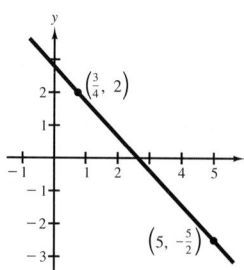

20. $\frac{8}{17}$, rises

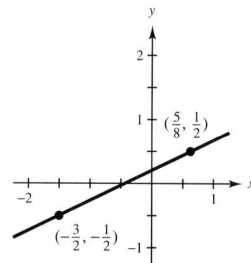

21. $-\frac{5}{6}$, falls

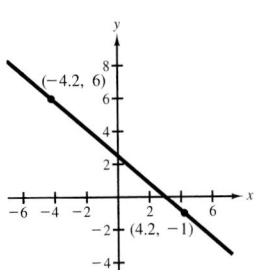

22. Undefined, vertical

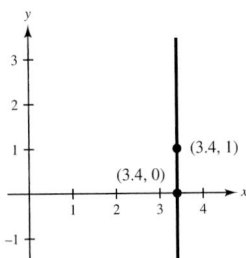

23. 0, horizontal

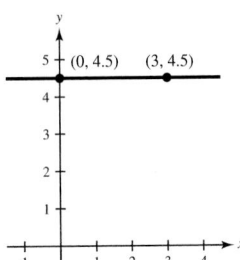

24. $\frac{29}{9}$, rises

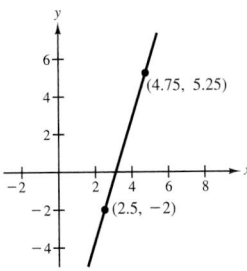

25. $\frac{5}{3}$, rises

26. $\frac{6}{5}$, rises

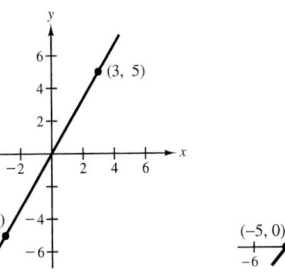

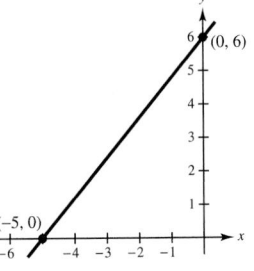

27. Negative slope: line falls to the right.
Zero slope: line is horizontal.
Positive slope: line rises to the right.

28. Yes. When different pairs of points are selected, the change in y and the change in x are lengths of the sides of similar triangles. Corresponding sides of similar triangles are proportional.

29. $x = 1$ **30.** $x = \frac{7}{3}$ **31.** $y = -15$ **32.** $y = -10$

33. $(6, 2), (10, 2)$ **34.** $(-4, 2), (-4, -8)$

35. $(4, -1), (5, 2)$ **36.** $(0, -3), (1, -1)$

37. $(1, 2), (2, 1)$ **38.** $(-1, 3), (0, 0)$ **39.** $(-2, 4), (1, 8)$

40. $(3, -2), (7, -5)$ **41.** $(4, 0), (4, -1)$

42. $(1, -2), (0, -2)$

43.

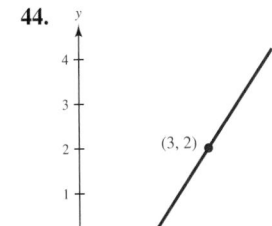

44.

45.

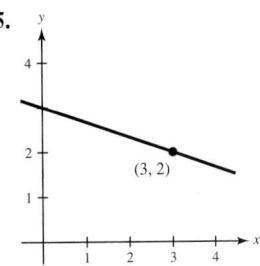

46.

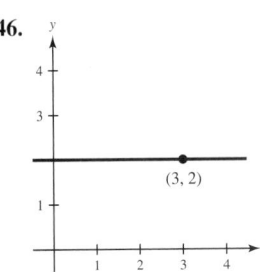

47.

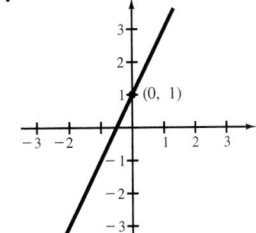

48.

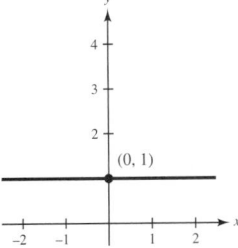

49.

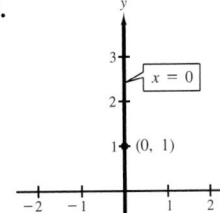

50.

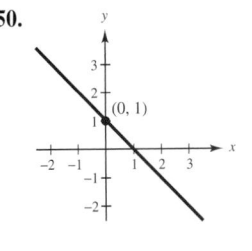

51.

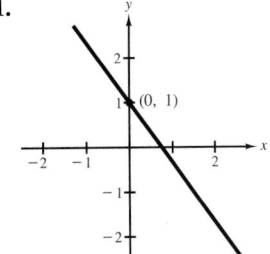

52.

53. Line with slope m_2 **54.** Line with slope m_2

55.

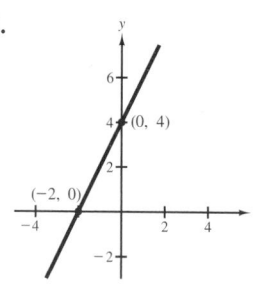

56.

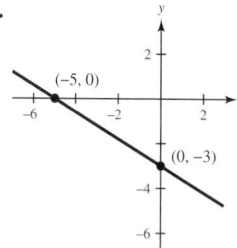

57.

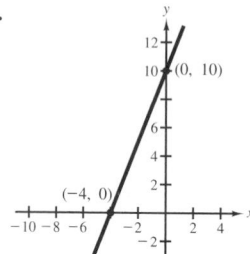

58.

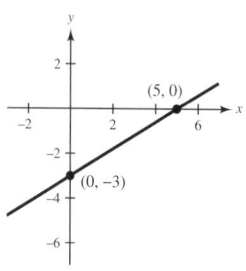

59. $y = 3x - 2$

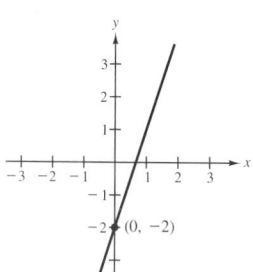

60. $y = x - 5$

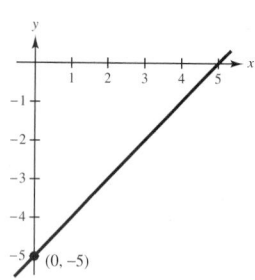

61. $y = -x$

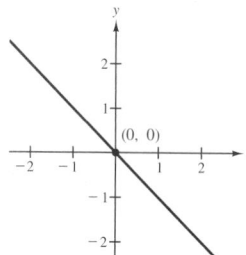

62. $y = x$

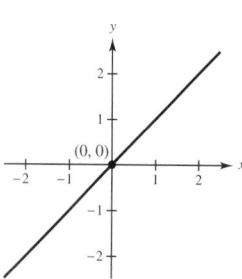

63. $y = -\frac{3}{2}x + 1$

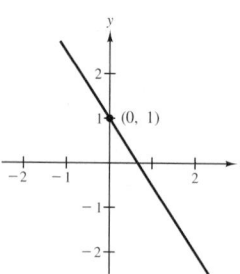

64. $y = \frac{1}{2}x + 1$

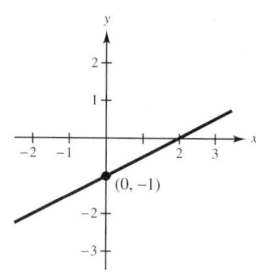

65. $y = \frac{1}{4}x + \frac{1}{2}$

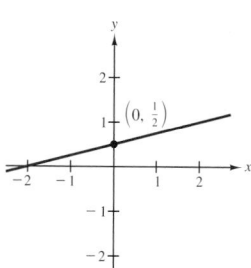

66. $y = -\frac{4}{3}x + \frac{1}{2}$

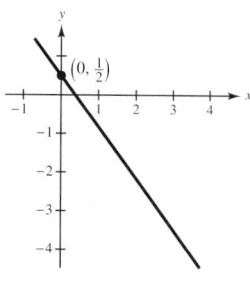

67. $y = 2$

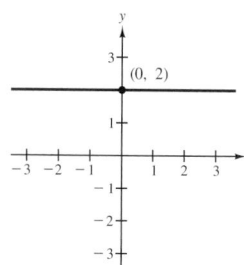

68. $y = -4$

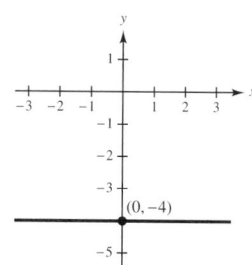

69. $y = 5x - 5$

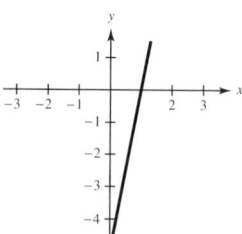

70. $y = -\frac{5}{6}x + 5$

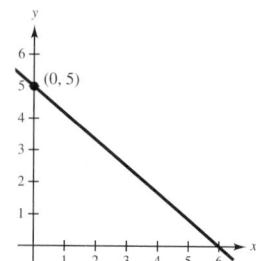

71. Perpendicular **72.** Neither
73. Parallel **74.** Perpendicular

75. Parallel

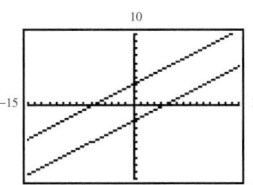

76. Parallel

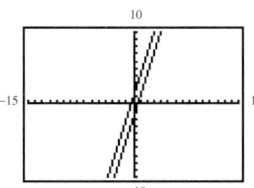

77. Perpendicular

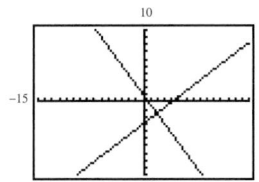

78. Perpendicular

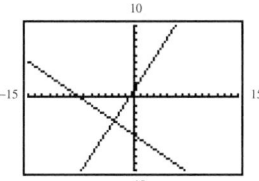

79.

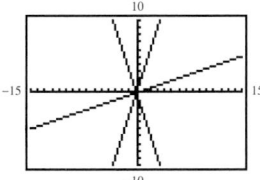

y_2 and y_3 are perpendicular.

80.

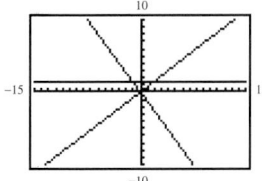

y_1 and y_2 are perpendicular.

81.

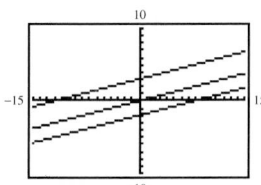

The lines are parallel.

82.

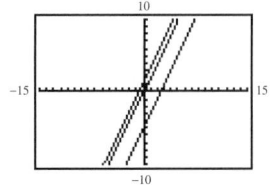

The lines are parallel.

83. No. Their slopes must be negative reciprocals of each other.

84. (a) $y = 8x + 140$

(b)

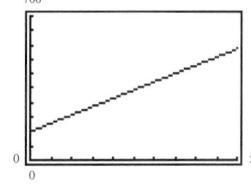

(c) $m = 8$, 8 ft

85. 16,667 ft

86. (a) $8000 - x$

(b) $y = 0.06x + 0.075(8000 - x)$

(c)

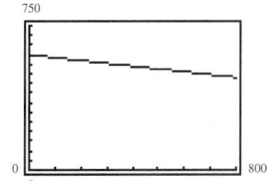

(d) As the amount invested at 6% increases, the interest decreases.

87. 1991 and 1995   **88.** 11.25   **89.** 11.25 ft

Section Project 2.3

(a)

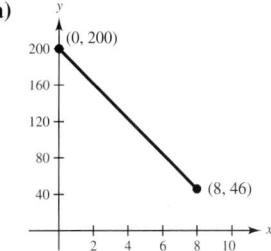

(b) -19.25 (c) $19.25

(d)

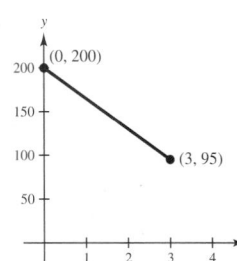

(e) -35.00 **(f)** 35.00
(g) The slope is the average rate at which your pocket money decreases per day.

Mid-Chapter Quiz (page 145)

1. (a)

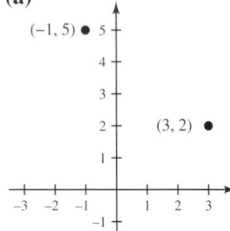

(b) 5
(c) $-\frac{3}{4}$

2. (a)

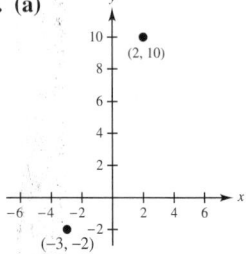

(b) 13
(c) $\frac{12}{5}$

3. $(10, -3)$ **4. (a)** No **(b)** Yes **(c)** Yes **(d)** Yes

5.

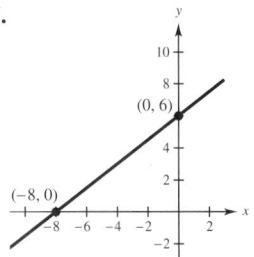

6.

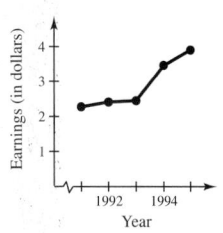

7.

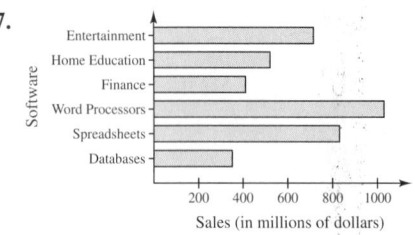

8.

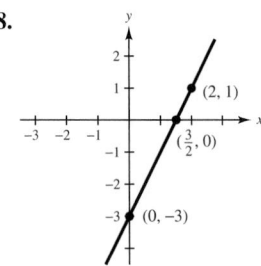

9.

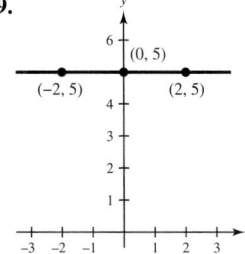

10.

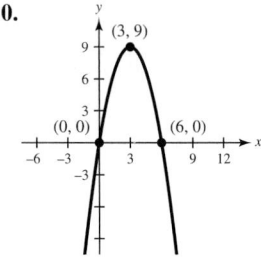

11.

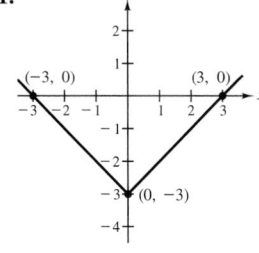

12. $y = -3x + 6$

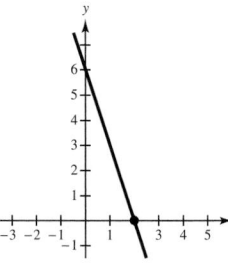

13. $y = \frac{4}{3}x - 5$

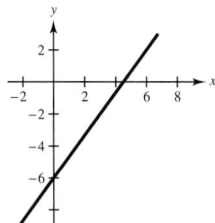

14.

```
Xmin = -4
Xmax = 4
Xscl = 1
Ymin = -3
Ymax = 12
Yscl = 2
```

15.

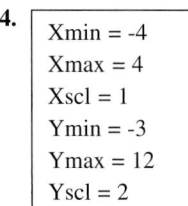

16. $y = -\frac{5}{3}x + 3, \ \frac{3}{5}$

Section 2.4   (page 154)

1. Domain: $\{-2, 0, 1\}$
    Range: $\{-1, 0, 1, 4\}$

2. Domain: $\{3, 4, 6, 8\}$
    Range: $\{-2, 5, 3, 10\}$

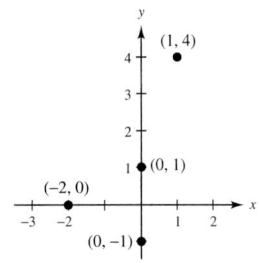

3. Domain: $\{0, 2, 4, 5, 6\}$
    Range: $\{-3, 0, 5, 8\}$

4. Domain: $\{-3\}$
    Range: $\{2, 5, 6\}$

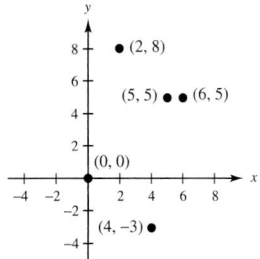
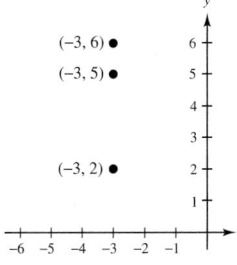

5. $(3, 150), (2, 100), (8, 400), (6, 300), \left(\frac{1}{2}, 25\right)$

6. $(1, 1), (2, 8), (3, 27), (4, 64), (5, 125), (6, 216), (7, 343)$

7. (1992, Toronto), (1993, Toronto), (1995, Atlanta), (1996, New York)

8. (1969, Nixon), (1973, Nixon), (1977, Carter), (1981, Reagan), (1989, Bush), (1993, Clinton), (1997, Clinton)

9. Not a function   **10.** Function   **11.** Not a function

12. Function   **13.** Function   **14.** Not a function

15. Not a function   **16.** Function

17. (a) Function from A to B
    (b) Not a function from A to B
    (c) Function from A to B
    (d) Not a function from A to B

18. (a) Not a function from A to B
    (b) Function from A to B
    (c) Not a function from A to B
    (d) Function from A to B

19. A relation is any set of ordered pairs. A function is a relation in which no two ordered pairs have the same first component and different second components.

20. No. $\{(4, 3), (4, -2)\}$ is a relation, but not a function. Every function is a relation.

21. There are two values of y associated with one value of x.

22. There are two values of y associated with one value of x.

23. There are two values of y associated with one value of x.

24. There are two values of y associated with one value of x.

25. Both high school enrollment and college enrollment are functions of the year. For each there is exactly one enrollment for each year.

26. 16,300,000   **27.** (a) 2   (b) -2 (c) k (d) $k + 1$

28. (a) 0   (b) -3 (c) m (d) $t + 2$

29. (a) 3, 3   (b) $-4, -4$ (c) s, s (d) $s - 2, s - 2$

30. (a) 1   (b) -4 (c) h (d) $h - 8$

31. (a) 29   (b) 11   (c) $12a - 2$ (d) $12a + 5$

32. (a) 10   (b) $-\frac{1}{2}$ (c) $20 - 7t$ (d) $3 - 7w$

33. (a) 2   (b) 3   (c) $\sqrt{\frac{31}{3}}$ (d) $\sqrt{5z + 5}$

34. (a) 4   (b) 4   (c) -7 (d) $8 - |x - 6|$

35. (a) 0   (b) $-\dfrac{3}{2}$ (c) $-\dfrac{5}{2}$ (d) $\dfrac{3x + 12}{x - 1}$

36. (a) 1   (b) 1   (c) -1 (d) 0

37. (a) 2   (b) -2 (c) 10   (d) -8

38. (a) 0   (b) $\frac{3}{2}$ (c) 4   (d) 552

39. (a) 2   (b) $\dfrac{2x - 12}{x}$ **40.** (a) $x^2 + 2x$ (b) $-2x + 5$

41. (a) -3.84 (b) -4.2

42. (a) $x^3 + 6x^2 + 12x + 8$ (b) $3x^2 + 6x + 4$

43. Domain: $\{0, 2, 4, 6\}$ **44.** Domain: $\{-3, -1, 4, 10\}$
    Range: $\{0, 1, 8, 27\}$ Range: $\left\{-\frac{17}{2}, -\frac{5}{2}, 2, 11\right\}$

45. Domain: All real numbers r such that $r > 0$
    Range: All real numbers C such that $C > 0$

46. Domain: All real numbers s such that $s > 0$
    Range: All real numbers A such that $A > 0$

47. All real numbers x such that $x \neq 3$

48. All real numbers x such that $x \neq -4$

49. All real numbers x such that $x \neq 2, 1$

50. All real numbers x

51. All real numbers t such that $t \neq 0, -2$

52. All real numbers s such that $s \neq 6, 10$

53. All real numbers x

54. All real numbers x such that $x \neq 2, -2$

55. All real numbers x such that $x \geq 2$

56. All real numbers x such that $x \geq 5$

57. All real numbers x such that $x \geq 0$

58. All real numbers x **59.** All real numbers x

60. All real numbers x such that $x \geq 0$

61. All real numbers t **62.** All real numbers x

63. $P = 4x$ **64.** $S = 6x^2$

65. $V = x(24 - x)^2$
$= 4x(12 - x)^2$

66. $A = (32 - 2x)^2$
$= 4(16 - x)^2$

67. $A = (32 - x)^2$

68. $C(x) = 1.95x + 8000, \ x > 0$

69. (a) \$700   **(b)** \$750   **70.** $d(t) = 230t, \ t > 0$

71. (a) 10,680 lb   **(b)** 8010 lb

72. (a) Correct   **(b)** Not correct

73. (a) Not correct   **(b)** Correct

Section Project 2.4

(a) \$360   **(b)** \$480   **(c)** \$570   **(d)** \$660

$h < 0$ is not in the domain of W.

$$W(h) = \begin{cases} 12h, & 0 < h \le 45 \\ 18(h - 45) + 540 \end{cases}$$

Section 2.5   (page 164)

1. $f(x) = 3x + 10$ **2.** $f(x) = -\frac{1}{2}x + 4$

3. $f(x) = -\frac{1}{4}x + \frac{45}{8}$ **4.** $f(x) = \frac{8}{9}x$

5. $f(x) = 4$ **6.** $f(x) = \frac{2}{3}x + \frac{13}{3}$

7. $(0, -6), (3, 0)$ **8.** $(0, 6), (2, 0)$

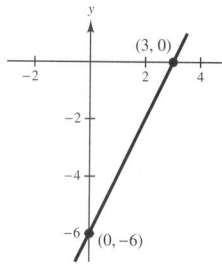   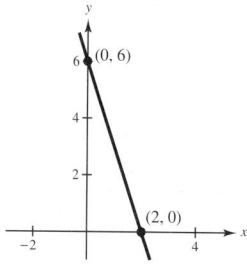

9. $(0, 1), \left(\frac{4}{3}, 0\right)$ **10.** $(0, 2), \left(-\frac{5}{2}, 0\right)$

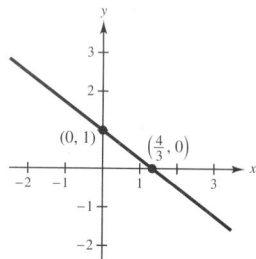   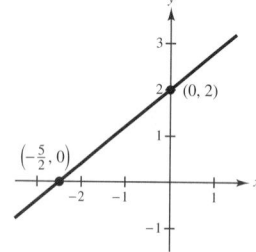

11. Function   **12.** Function   **13.** Not a function

14. Function   **15.** Not a function   **16.** Not a function

17. Function. For each value of y there corresponds one value of x.

18. Not a function. For each value of $y > -1$ there correspond two values of x.

19. y is a function of x.

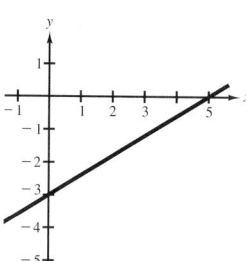

20. y is a function of x.

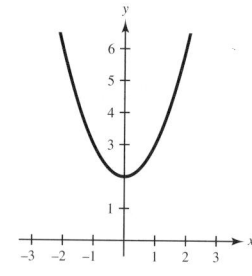

21. y is not a function of x.

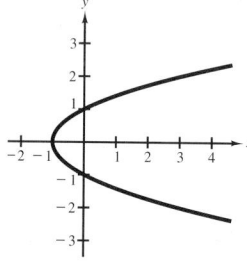

22. y is not a function of x.

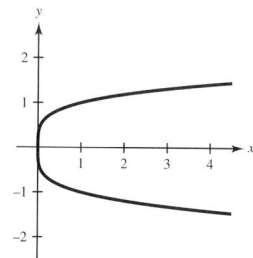

23. Domain: $-\infty < x \le -2, \ 2 \le x < \infty$
Range: $0 \le y < \infty$

24. Domain: $-\infty < x < \infty$
Range: $0 \le y < \infty$

25. Domain: $-3 \le x \le 3$
Range: $0 \le y \le 3$

26. Domain: $-\infty < x < 3, 3 < x < \infty$
Range: $y = -3, \ y = 3$

27.

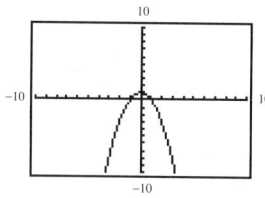

28.

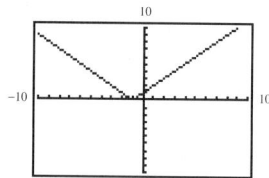

Domain: $-\infty < x < \infty$
Range: $-\infty < y \le 1$

Domain: $-\infty < x < \infty$
Range: $0 < y \le 1$

29.

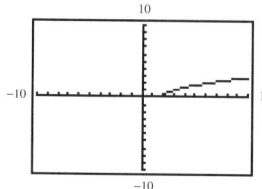

30.

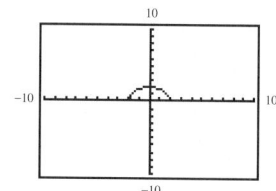

Domain: $2 \le x < \infty$
Range: $0 \le y < \infty$

Domain: $-2 \le t \le 2$
Range: $0 \le y \le 2$

31. (c)   **32.** (b)   **33.** (e)   **34.** (a)   **35.** (f)   **36.** (d)

37.

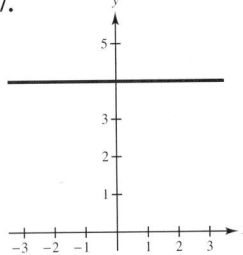

Domain: $-\infty < x < \infty$
Range: $y = 4$

38.

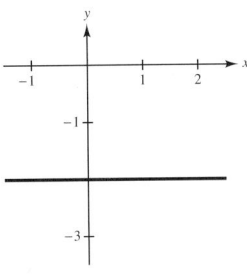

Domain: $-\infty < x < \infty$
Range: $y = -2$

39.

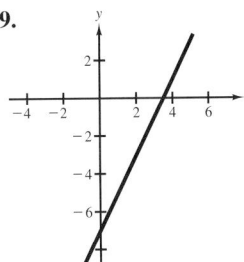

Domain: $-\infty < x < \infty$
Range: $-\infty < y < \infty$

40.

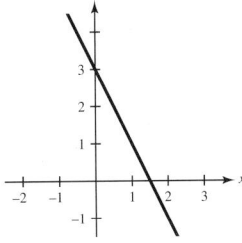

Domain: $-\infty < x < \infty$
Range: $-\infty < y < \infty$

41.

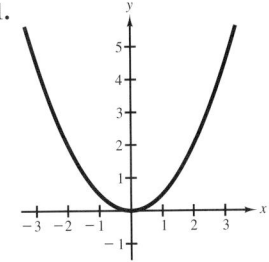

Domain: $-\infty < x < \infty$
Range: $0 \le y < \infty$

42.

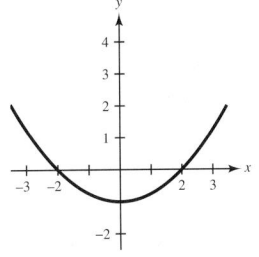

Domain: $-\infty < x < \infty$
Range: $-1 \le y < \infty$

43.

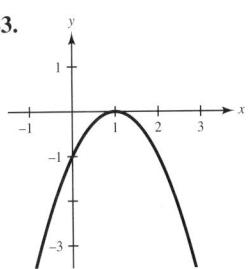

Domain: $-\infty < x < \infty$
Range: $-\infty < y \le 0$

44.

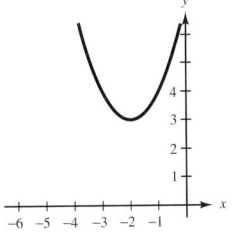

Domain: $-\infty < x < \infty$
Range: $3 \le y < \infty$

45.

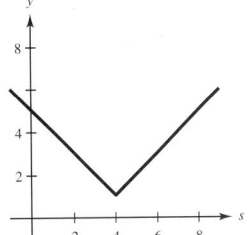

Domain: $-\infty < s < \infty$
Range: $1 \le y < \infty$

46.

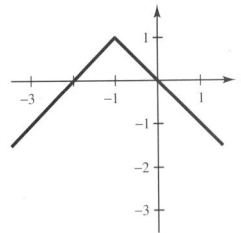

Domain: $-\infty < t < \infty$
Range: $-\infty \le y \le 1$

47.

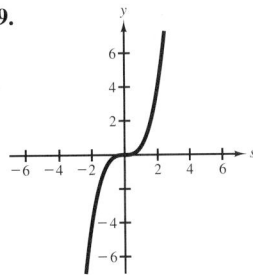

Domain: $2 \le t < \infty$
Range: $0 \le y < \infty$

48.

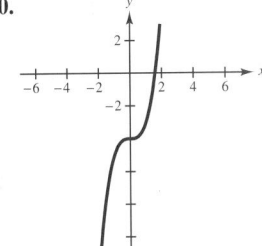

Domain: $-\infty < x \le 4$
Range: $0 \le y < \infty$

49.

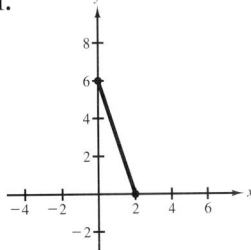

Domain: $-\infty < s < \infty$
Range: $-\infty < y < \infty$

50.

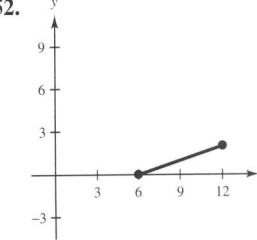

Domain: $-\infty < x < \infty$
Range: $-\infty < y < \infty$

51.

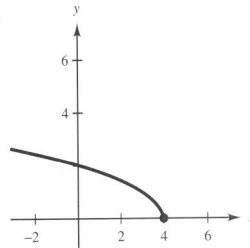

Domain: $0 \le x \le 2$
Range: $0 \le y \le 6$

52.

Domain: $6 \le x \le 12$
Range: $0 \le y \le 2$

53.

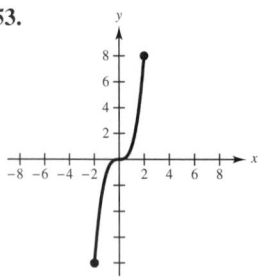

Domain: $-2 \leq x \leq 2$
Range: $-8 \leq y \leq 8$

54.

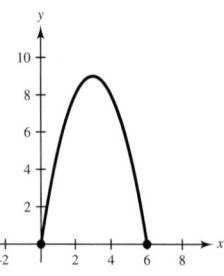

Domain: $0 \leq x \leq 6$
Range: $0 \leq y \leq 9$

Domain: $-4 \leq t \leq 4$
Range: $0 \leq y \leq 2$

55.

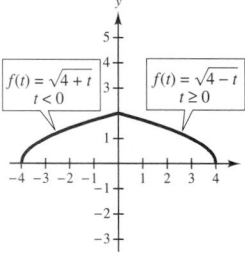

56.

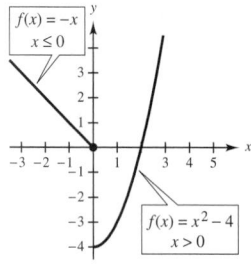

Domain: $-\infty < x < \infty$
Range: $-4 \leq y < \infty$

57.

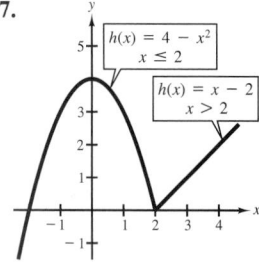

Domain: $-\infty < x < \infty$
Range: $-\infty \leq y < \infty$

58.

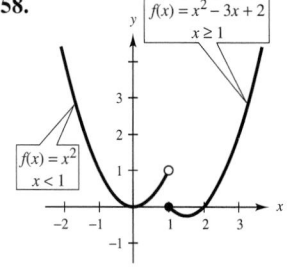

Domain: $-\infty < x < \infty$
Range: $-\frac{1}{4} \leq y < \infty$

59.

Xmin = 0
Xmax = 20
Xscl = 2
Ymin = -10
Ymax = 60
Yscl = 6

60.

Xmin = -3
Xmax = 12
Xscl = 1
Ymin = -1200
Ymax = 400
Yscl = 100

61. (a) Answers may vary.

(b)

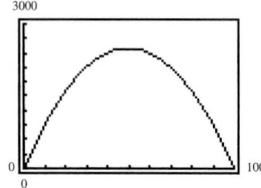

(c) $x = 50$. The figure is a square.

62. (a)

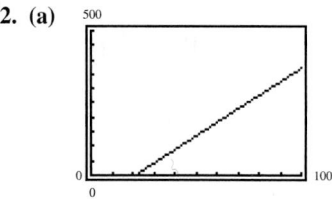

(b) 213   **(c)** 851

Section Project 2.5

(a) 9, 10, 11, 12, 13, 14

(b)

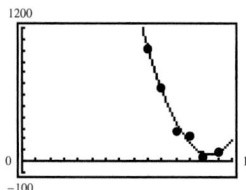

(c) Most accurate: 1994
Least accurate: 1991

CHAPTER 2

Section 2.6 (page 173)

1.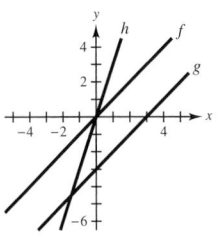

2.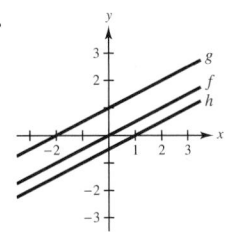

3.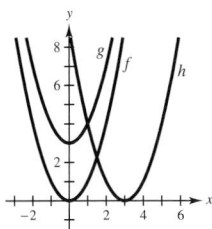

4.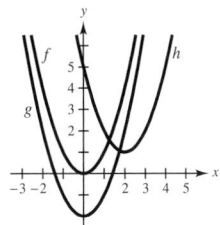

5.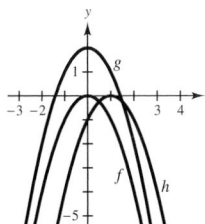

6.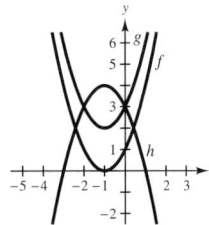

7.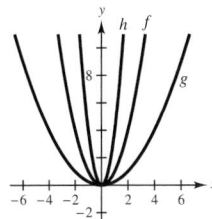

8.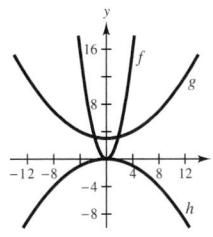

9.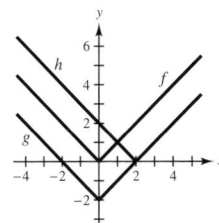

10.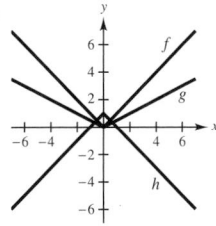

11.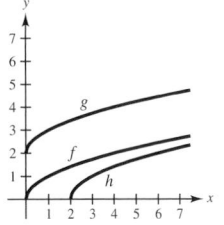

12.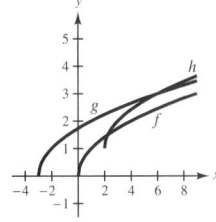

13. Horizontal shift of $y = x^3$
$y = (x - 2)^3$

14. Vertical shift of $y = |x|$
$y = |x| + 3$

15. Reflection in the x-axis of $y = x^2$
$y = -x^2$

16. Constant function $y = 7$

17. Reflection in the x-axis and a vertical shift of $y = \sqrt{x}$
$y = 1 - \sqrt{x}$

18. Horizontal shift of $y = |x|$
$y = |x + 2|$

19. Vertical shift of $y = x^2$
$y = x^2 - 1$

20. Reflection in the x-axis and a horizontal and vertical shift of $y = x^2$
$y = -(x + 1)^2 + 1$

21. Reflection in the x- or y-axis and a vertical shift of $y = x^3$
$y = -x^3 + 1$

22. Horizontal and vertical shifts of $y = x^3$
$y = (x - 1)^3 + 1$

23. Vertical or horizontal shift of $y = x$
$y = x + 3$

24. Reflection in the y-axis and a horizontal shift of $y = \sqrt{x}$
$y = \sqrt{4 - x}$

25. Horizontal and vertical shifts of $y = |x|$
$y = |x - 2| - 2$

26. Horizontal and vertical shifts of $y = |x|$
$y = |x - 1| + 3$

27. $y = -\sqrt{x}$ **28.** $y = \sqrt{x} + 1$ **29.** $y = \sqrt{x + 2}$
30. $y = \sqrt{x - 3}$ **31.** $y = \sqrt{-x}$
32. $y = \sqrt{x + 3} + 2$

33. Vertical shift three units upward

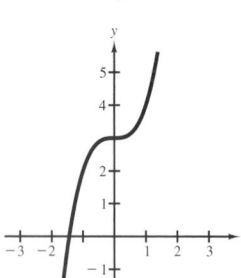

34. Vertical shift five units downward

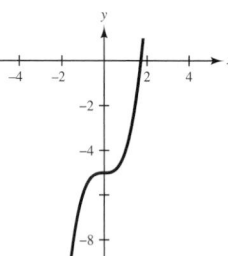

35. Horizontal shift three units to the right

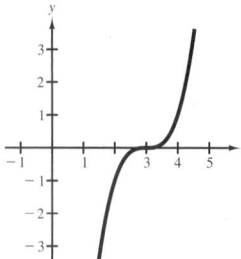

36. Horizontal shift two units to the left

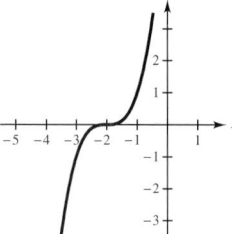

37. Reflection in the y-axis

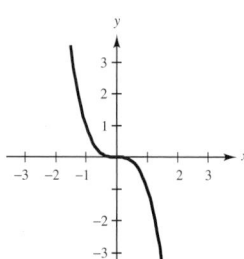

38. Reflection in the x-axis

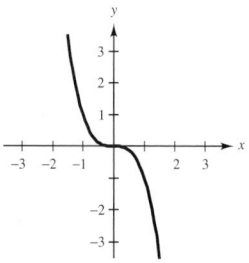

39. Reflection in the x-axis followed by a horizontal shift one unit to the right followed by a vertical shift two units upward

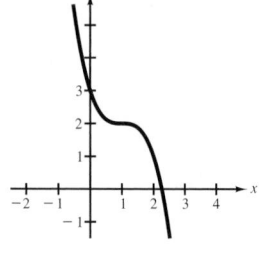

40. Horizontal shift two units to the left followed by a vertical shift three units downward

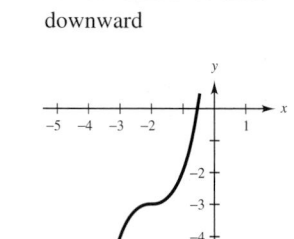

41. Horizontal shift five units to the right

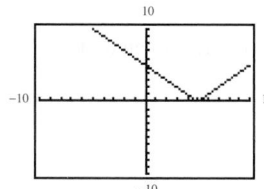

42. Horizontal shift three units to the left

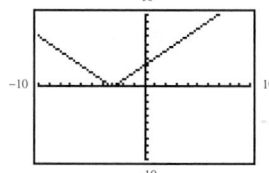

43. Vertical shift five units downward

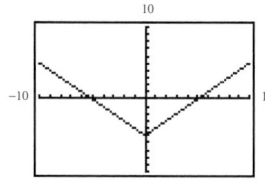

44. Reflection in the y-axis

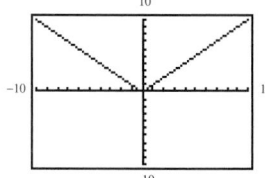

45. Reflection in the x-axis

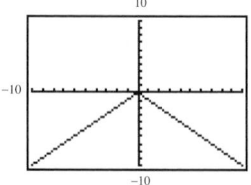

46. Reflection in the x-axis followed by a vertical shift five units upward

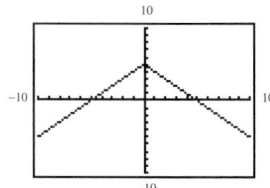

47.

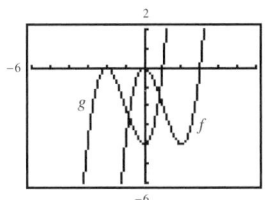

g is a horizontal shift two units to the left.

48.

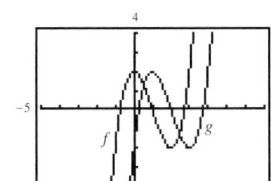

g is a horizontal shift one unit to the right.

(e)

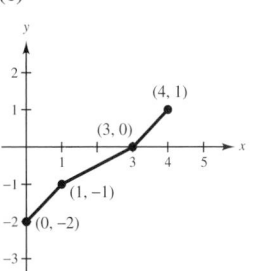

(f)

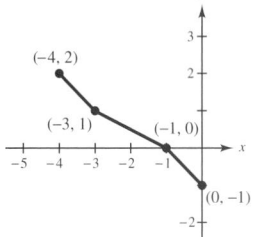

49.

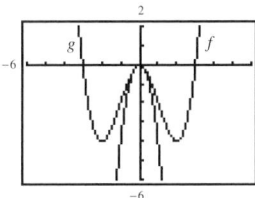

g is a reflection in the y-axis.

50.

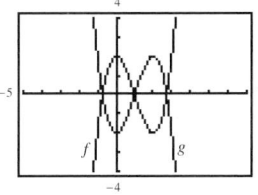

g is a reflection in the x-axis.

54. (a)

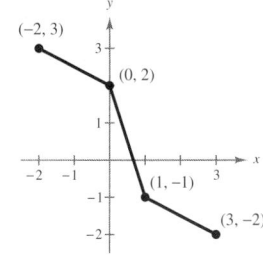

(b)

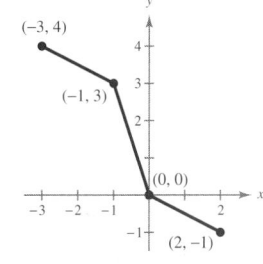

51. $g(x) = -x^3 + 3x^2 + 1$

52. $g(x) = (x - 2)^3 - 3(x - 2)^2 + 1$

53. (a)

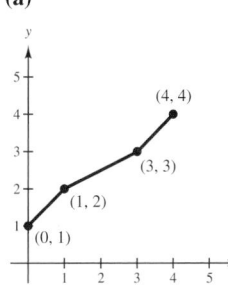

(b)

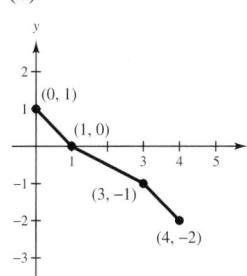

(c)

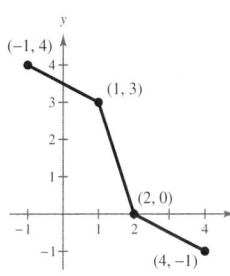

(d)

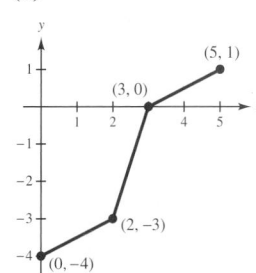

(c)

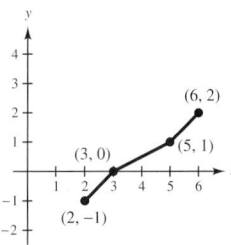

(d)

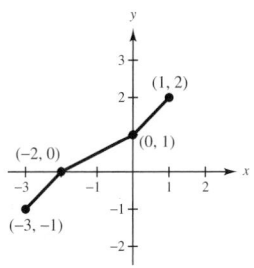

(e)

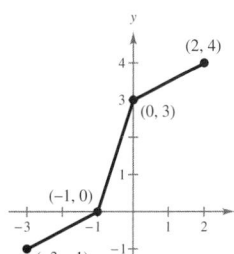

(f)

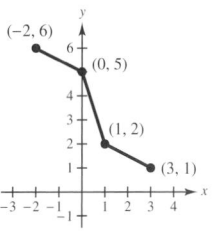

55. (a)

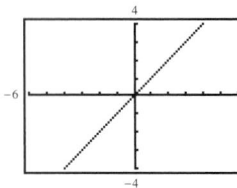

(b)

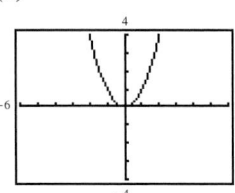

(c)

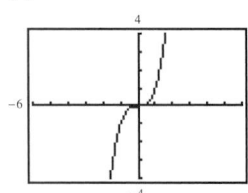

(d)

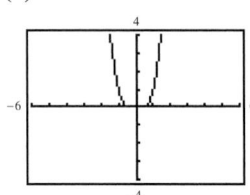

(e)

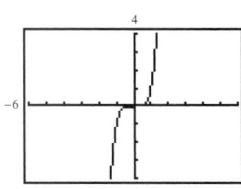

(f)

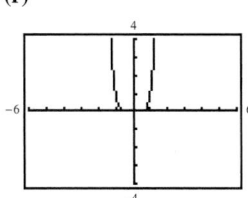

All the graphs pass through the origin. The graphs of the odd powers of x resemble the graphs of the cubing function and the graphs of the even powers resemble the graphs of the squaring function. As the powers increase, the graphs become flatter in the interval $-1 < x < 1$.

56.

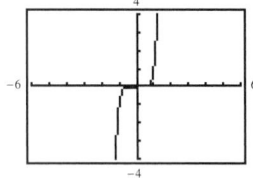

57.

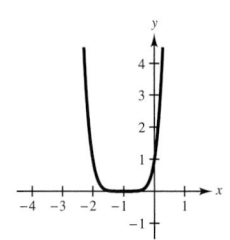

58.

59.

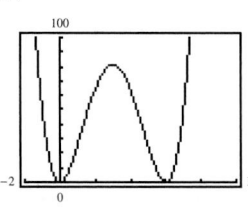

60.

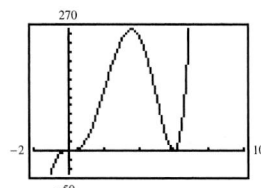

61.

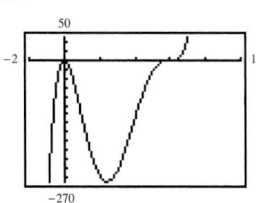

62.

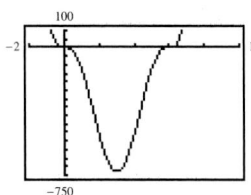

63. (a)

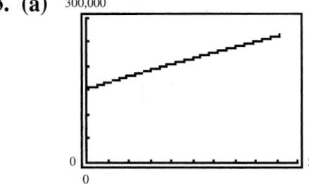

(b) 1970. P_2 is a horizontal shift 20 units to the left of P_1.

(c)

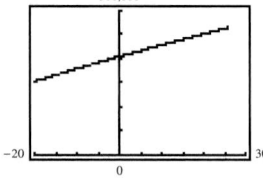

Section Project 2.6

(a) To each time t there corresponds one and only one temperature T.

(b) $60°, 72°$

(c) All the temperature changes would occur 1 hour later.

(d) The temperature would be decreased by 1 degree.

Review Exercises (page 177)

1.
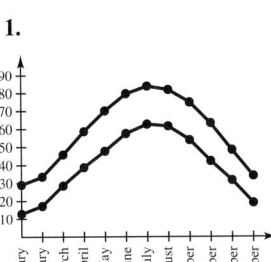

June

2.
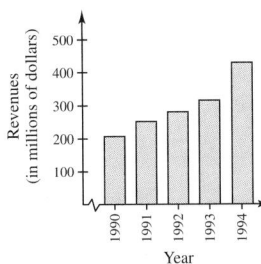

3.
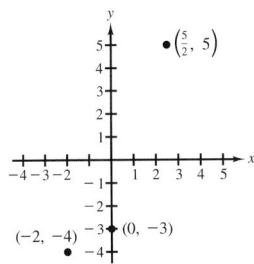

Triangle

4.
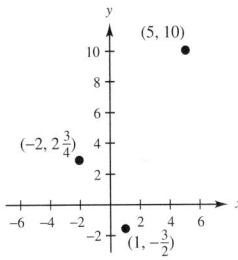

Triangle

5.
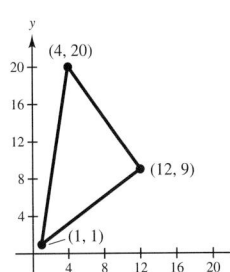

6.
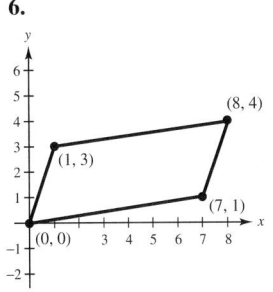

7. Quadrant IV   **8.** Quadrant III

9. Quadrants I and IV   **10.** Quadrants I and III

11. (a) Yes **(b)** No **(c)** No **(d)** Yes

12. (a) No **(b)** Yes **(c)** Yes **(d)** No

13. (c)   **14.** (b)   **15.** (a)   **16.** (d)

17.
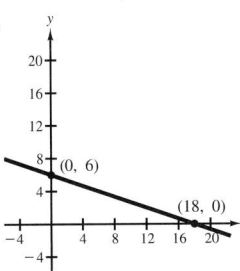

18.
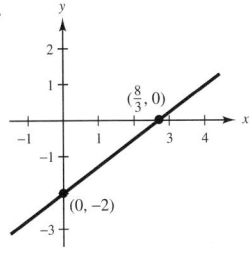

19.
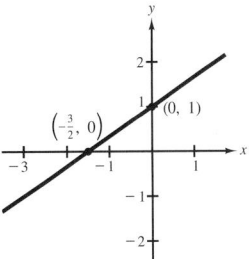

20.
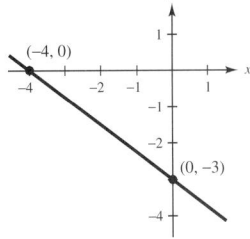

21.
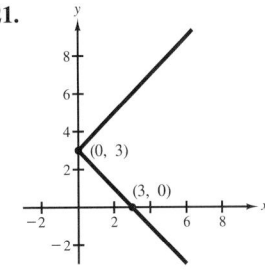

22.
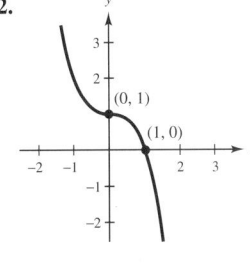

23. $\frac{2}{7}$ **24.** $-\frac{13}{5}$ **25.** 0   **26.** Undefined   **27.** $-\frac{3}{4}$

28. $\frac{12}{7}$ **29.** $\frac{3}{2}$ **30.** $\frac{2}{3}$ **31.** $(1, -1), (0, 2)$

32. $\left(-3, \frac{5}{2}\right), \left(-2, \frac{9}{2}\right)$ **33.** $(7, 6), (11, 11)$

34. $\left(0, -\frac{5}{2}\right), \left(3, -\frac{7}{2}\right)$ **35.** $(3, 0), (3, 5)$

36. $(0, -2), (-3, -2)$

37. $y = \frac{5}{2}x - 2$ **38.** $y = \frac{1}{3}x - 2$

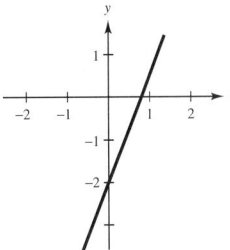

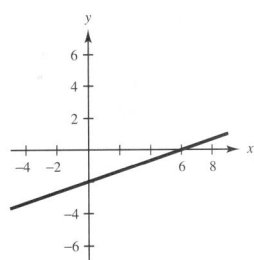

39. $y = -\frac{1}{2}x + 1$ **40.** $y = 6$

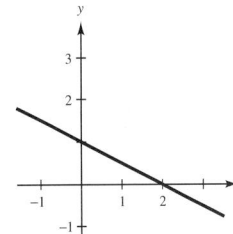

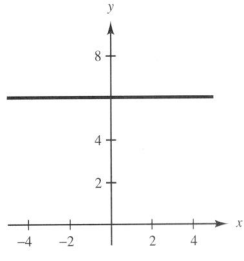

41. Neither

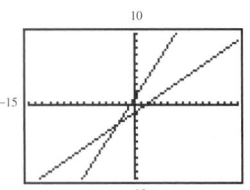

42. Parallel

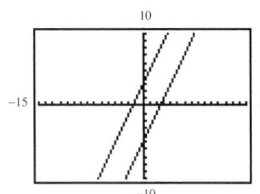

43. Perpendicular

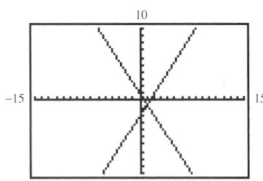

44. Neither

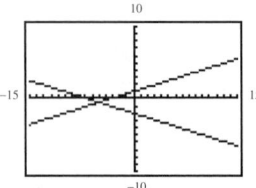

45. Not a function **46.** Function

47. Function **48.** Not a function

49. Intercept: $(0, 0)$ **50.** Intercepts: $(0, 0), (4, 0)$
Not a function Function

51. Intercepts: $(0, 0), (3, 0)$ **52.** Intercept: $(0, 0)$
Function Not a function

53.

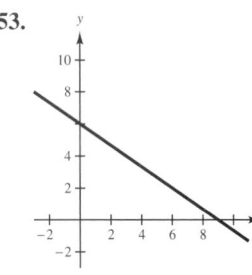

Function

54.

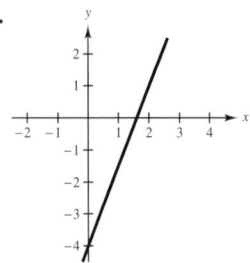

Function

55.

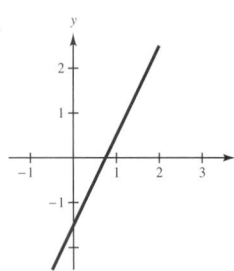

Function

56.

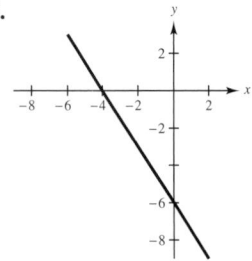

Function

57.

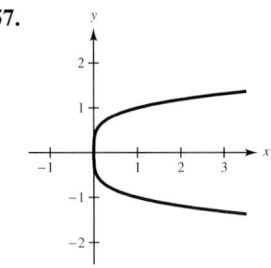

Not a function

58.

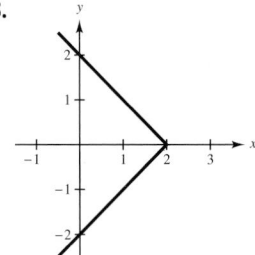

Not a function

59. (a) 29 (b) 3 (c) $\dfrac{36 - 5t}{2}$ (d) $4 - \dfrac{5}{2}(x + h)$

60. (a) 0 (b) 20 (c) 33 (d) $t^2 - 16$

61. (a) 3 (b) 0 (c) $\sqrt{2}$ (d) $\sqrt{5 - 5z}$

62. (a) 1 (b) 1 (c) $\dfrac{5}{4}$ (d) $\dfrac{|x + 2|}{4}$

63. (a) -3 (b) 2 (c) 0 (d) -7

64. (a) 2 (b) $-\dfrac{1}{8}$ (c) 0 (d) 5

65. (a) -2 (b) $\dfrac{12 - 2x}{x}$ **66.** (a) 7 (b) $\dfrac{7x - 70}{x}$

67. $-\infty < x < \infty$

68. $-\infty < x < -5, -5 < x < 1, 1 < x < \infty$

69. $2 \le x < \infty$ **70.** $-\infty < x < \infty$

71.

Xmin = -3
Xmax = 3
Xscl = 1
Ymin = -3
Ymax = 5
Yscl = 1

72.

Xmin = -6
Xmax = 6
Xscl = 1
Ymin = -50
Ymax = 50
Yscl = 10

73.

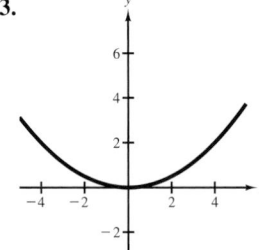

74.

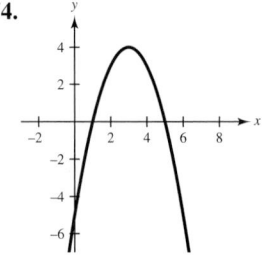

75.
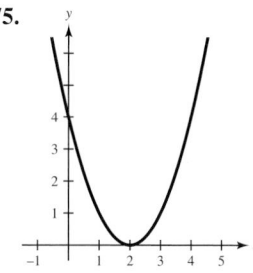

76.
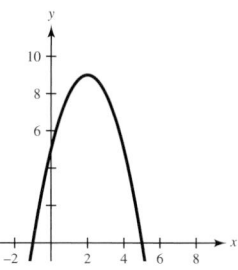

77.
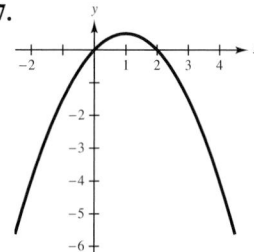

78.
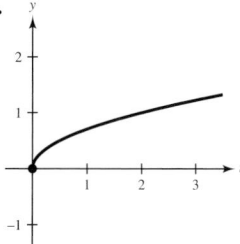

79.
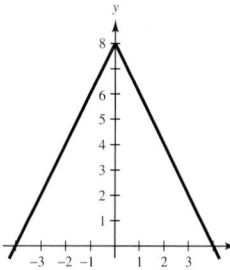

80.
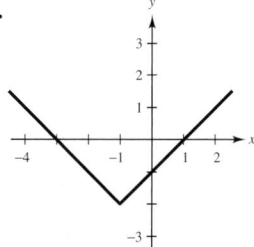

81.
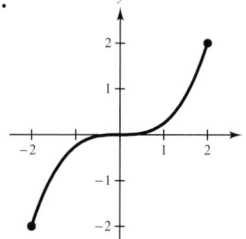

82.
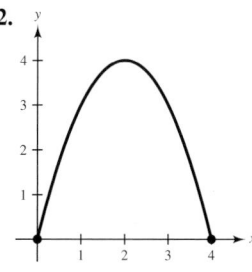

83.
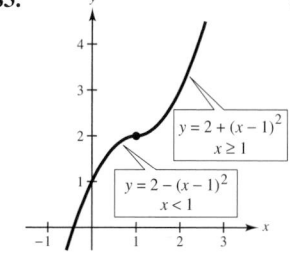

84.
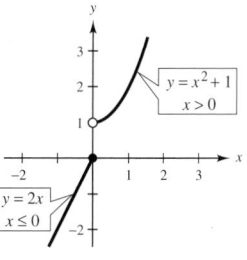

85. Reflection in the *x*-axis

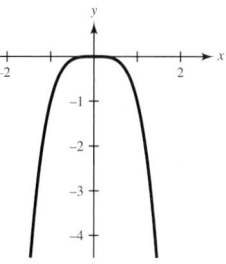

86. Vertical shift two units upward

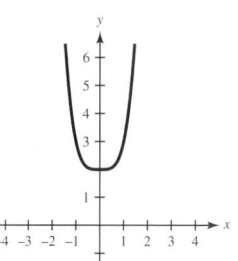

87. Horizontal shift one unit to the right

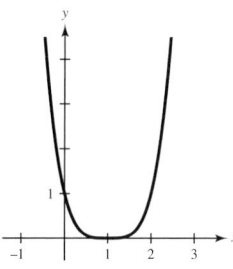

88. Reflection in the *x*-axis followed by a vertical shift one unit upward

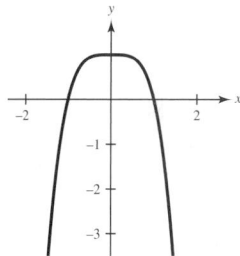

89. (a)

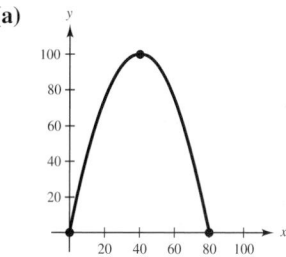

(b) 80 ft

90. (a) 16 ft/sec   **(b)** 2.5 sec   **(c)** -16 ft/sec

91. (a) $k = \frac{1}{8}$ **(b)** 1953.1 kw

92. $A = x(50 - x)$

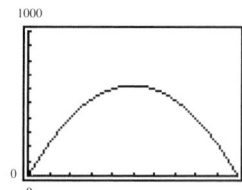

Chapter 2 Test (page 181)

1. 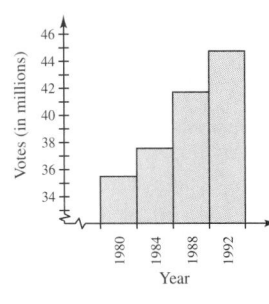   **2.** Quadrant IV

3. $\sqrt{73} \approx 8.54$ **4.** $(-1, 0), (0, -3)$

5. 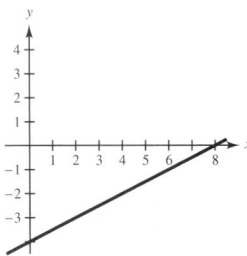   **6. (a)** $-\frac{2}{3}$ **(b)** Undefined

7.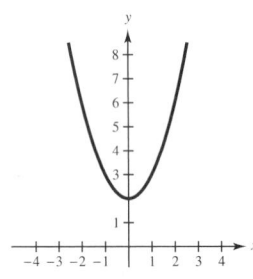

8.
Slope of line that is
perpendicular: -2

9. Not a function

10. (a) -2 **(b)** 7 **(c)** Undefined **(d)** $\dfrac{x + 2}{x - 1}$

11. (a) 104 **(b)** -25 **(c)** 29 **(d)** -1

12. (a) $-9 \leq t < \infty$ **(b)** $-\infty < x < 4,\ \ 4 < x < \infty$

13.

14. (a) $y = |x - 2|$
(b) $y = |x| - 2$
(c) $y = 2 - |x|$

Cumulative Test: Chapters 1 and 2   (page 182)

1. $-\frac{10}{27}$ **2.** 24 **3. (a)** $8a^8b^7$ **(b)** $2x^3 - 11x$

4. (a) $t^2 - 9t$ **(b)** $x^2 - 2xy + y^2 + 4x - 4y + 4$

5. (a) $\frac{3}{2}$ **(b)** $-\frac{3}{2}$ **6. (a)** ± 8 **(b)** $-\frac{1}{2}, 3$

7. $y = \frac{2}{3}x + 3$ **8. (a)** -4 **(b)** -4 **(c)** $\frac{1}{4}$

9. (a) Multiplicative Inverse Property
  (b) Associative Property of Addition

10. $8x + 12$ **11.** $(y + 3)(y - 3)^2$

12. $(x - 5)(3x + 7)$ **13.** Function   **14.** $2 \leq x < \infty$

15. (a) 4 **(b)** $c^2 + 3c$ **16.** $\frac{3}{4}$

17. 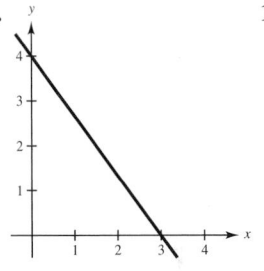   **18.**

19. 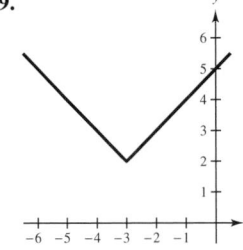   **20.**

CHAPTER 3

Section 3.1   (page 192)

1. Slope: $\frac{2}{3}$; y-intercept: -2 **2.** Slope: -5; y-intercept: 12

3. Slope: $\frac{3}{2}$; y-intercept: 0 **4.** Slope: -12; y-intercept: 2

5. Slope: $\frac{5}{2}$; y-intercept: 12 **6.** Slope: $-\frac{3}{4}$; y-intercept: 4

7. $y = -\frac{1}{2}x$ **8.** $y = -2x$

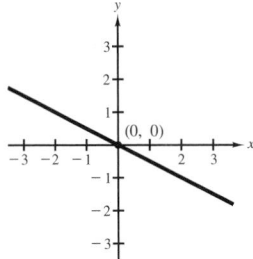   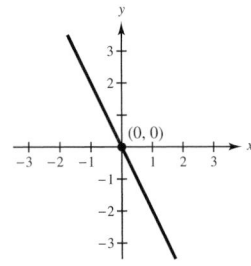

9. $y = 3x - 4$

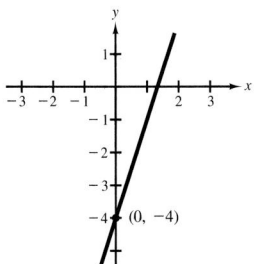

10. $y = -\frac{1}{3}x + 9$

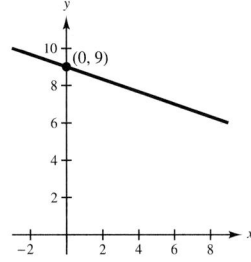

11. $y = 2x - 4$

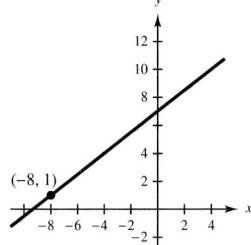

12. $y = 4x - 5$

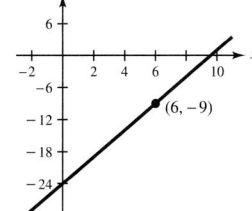

13. $y = \frac{3}{4}x + 7$

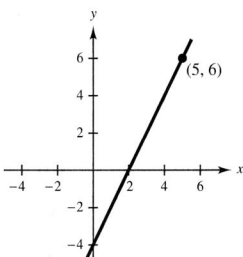

14. $y = \frac{5}{2}x - 24$

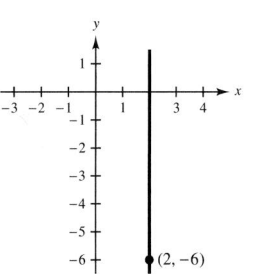

15. $y = \frac{2}{3}x - \frac{19}{3}$

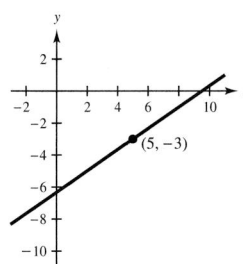

16. $x = 2$

17. $y = 5$

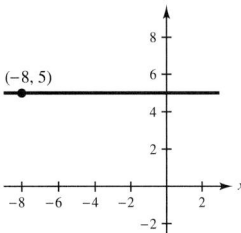

18. $y = \frac{5}{4}x - \frac{13}{4}$

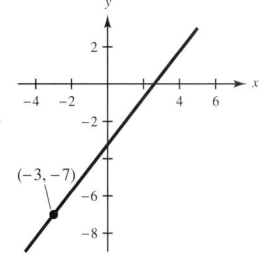

19. $x = 2$

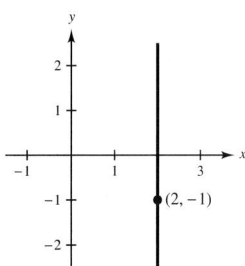

20. $y = -1$

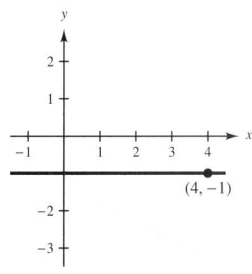

21. $y = \frac{4}{3}x + \frac{3}{2}$

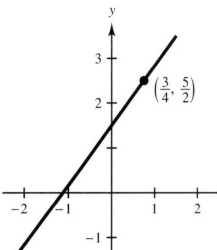

22. $y = -3x - 4$

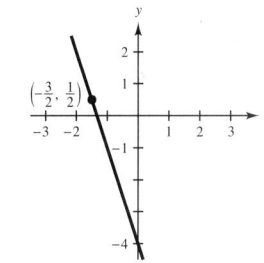

23. (b)   **24.** (d)   **25.** (a)   **26.** (c)   **27.** $y = \frac{3}{2}x$

28. $y = -\frac{5}{3}x$ **29.** $y = 4x - 25$ **30.** $y = 3x + 28$

31. $y = -\frac{2}{5}x$ **32.** $y = -3$ **33.** $y = 12$

34. $y = -\frac{2}{3}x - 4$ **35.** $y = -\frac{3}{7}x + \frac{15}{7}$

36. $y = -\frac{1}{8}x + \frac{37}{8}$ **37.** $x = 1$ **38.** $y = \frac{1}{3}x + 1$

39. $y = -\frac{1}{2}x - \frac{19}{10}$ **40.** $x = 7.5$ **41.** $y = -\frac{7}{3}x + \frac{13}{2}$

42. $y = \frac{8}{3}x - \frac{40}{3}$

43. $f(x) = \frac{1}{2}x + 3$ **44.** $f(x) = -2x + 10$

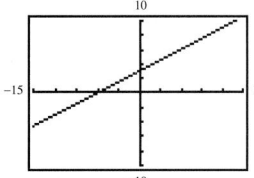

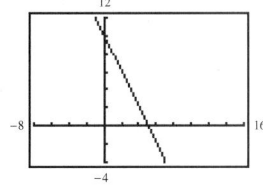

45. $f(x) = 3$

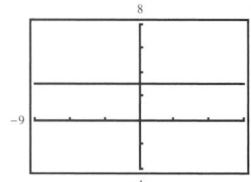

46. $f(x) = \frac{3}{5}x + \frac{3}{5}$

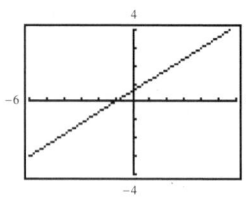

47. Yes. When different pairs of points are selected, the change in y and the change in x are the lengths of the sides of similar triangles. Corresponding sides of similar triangles are proportional.

48. Point-slope form: $y - y_1 = m(x - x_1)$
Slope-intercept form: $y = mx + b$
General form: $ax + by + c = 0$

49. (a) $y = 3x - 5$ **(b)** $y = -\frac{1}{3}x + \frac{5}{3}$
50. (a) $y = 4x + 35$ **(b)** $y = -\frac{1}{4}x + 1$
51. (a) $y = -\frac{3}{2}x - 7$ **(b)** $y = \frac{2}{3}x + 6$
52. (a) $y = -\frac{1}{6}x + \frac{7}{2}$ **(b)** $y = 6x + 22$
53. (a) $y = \frac{4}{3}x - \frac{25}{3}$ **(b)** $y = -\frac{3}{4}x - \frac{25}{4}$
54. (a) $y = -\frac{5}{4}x - \frac{9}{4}$ **(b)** $y = \frac{4}{5}y + 8$
55. (a) $y = 2$ **(b)** $x = -1$
56. (a) $x = 3$ **(b)** $y = -4$

57.

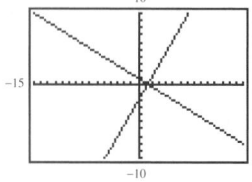

Perpendicular

58.

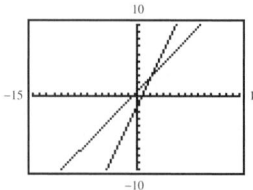

Neither

59.

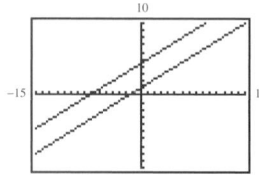

Parallel

60.

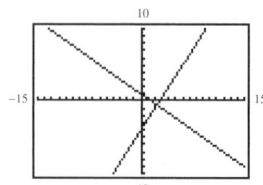

Perpendicular

61.

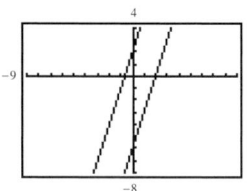

Parallel

62.

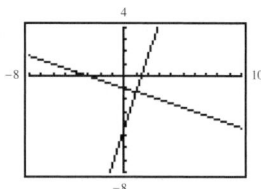

Perpendicular

63.

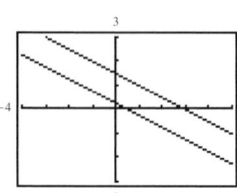

Parallel

64.

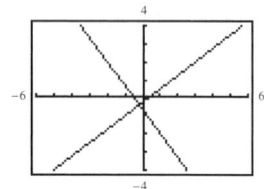

Perpendicular

65.

x	0	50	100	500	1000
C	5000	6000	7000	15,000	25,000

66.

F	$-20°$	$-0.04°$	$20°$	$32°$	$100°$	$212°$
C	$-\frac{260}{9}°$	$-17.8°$	$-\frac{20}{3}°$	$0°$	$\frac{340}{9}°$	$100°$

67. $S = 2500 + 0.03m$ **68.** $C = 175 + 0.32x$
69. (a) $S = 0.70L$ **(b)** \$94.50
70. (a) $V = 7400 - 1475t$ **(b)** \$4450
71. (a) $x = 80 - \frac{1}{15}p$

(b)

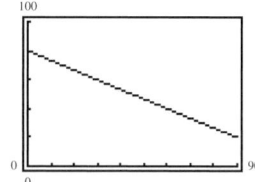

(c) 42   **(d)** 48
72. (a) $x = 14{,}000 - 10{,}000p$

(b)

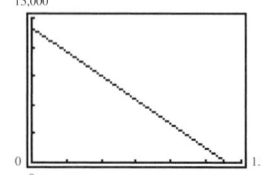

(c) 3000   **(d)** 5000

73. (a) $N = 1500 + 60t$

(b)

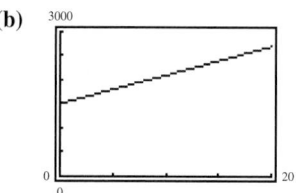

(c) 2700   **(d)** 1800

Section Project 3.1

(a) $-\dfrac{b}{a}$ **(b)** $y = -\dfrac{b}{a}x + b$

(c) Answers may vary.   **(d)** Answers may vary.

(i) $\dfrac{x}{3} + \dfrac{y}{2} = 1$ **(ii)** $-\dfrac{x}{6} + \dfrac{y}{2} = 1$

(iii) $-\dfrac{6x}{5} - \dfrac{3y}{7} = 1$ **(iv)** $-\dfrac{3x}{8} - \dfrac{y}{4} = 1$

Section 3.2   (page 202)

1. (a)

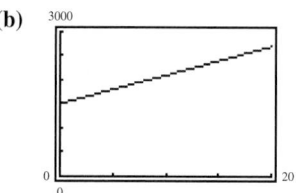

(b)

t	0	1	2	3	4
s	0	11.0	19.4	29.2	39.4
Approx. s	0.4	10.1	19.8	29.5	39.2

(c) 48.9 m/sec

(d) The increase in speed for each one-unit increase in time.

2. (a)

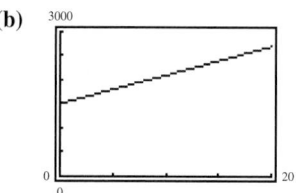

(b)

Year	1989	1990	1991
R	15.21	16.78	18.10
Approx. R	15.44	16.61	17.78

Year	1992	1993	1994
R	19.08	19.39	21.62
Approx. R	18.95	20.12	21.29

(c) $28.31

(d) The average increase in monthly rate for each 1-yr increase in time.

3. (a) $T = 0.0625v$ **(b)** $718.75   **(c)** 6.25%

4. (a) $R = 249x$ **(b)** $7968   **(c)** $249

5. (a) $I = \frac{50}{127}C$ **(b)** 38.1 cm   **(c)** 255.9 in.

(d)

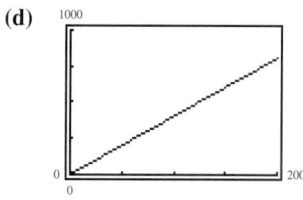

6. (a) $l = 0.02957$ **(b)** 913 oz   **(c)** 0.946 L

(d)

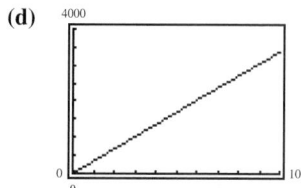

7. $y = 1.77x + 106.28$, 138.1 million

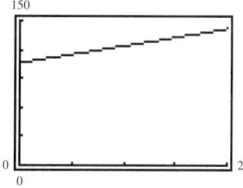

8. 5.56%   **9.** Yes; negative   **10.** No   **11.** No

12. Yes; positive

13. $V = 125(t - 8) + 2540$ **14.** $V = 4.50(t - 8) + 156$

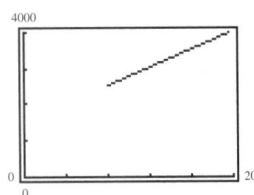

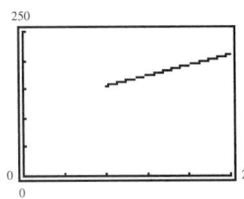

15. $V = 2000(t - 8) + 20,400$

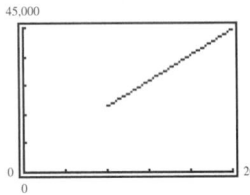

16. $V = 5600(t - 8) + 245,000$

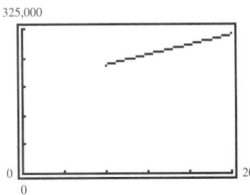

17. (b)   **18.** (c)   **19.** (a)   **20.** (d)
21. (a) and (b)

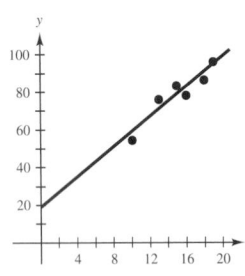

(c) $y = 4x + 19$
(d) 87
(e) Vertical shift four units upward

22. (a) and (b)

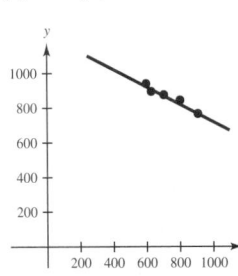

(c) $y = -0.5x + 1221$
(d) The average change in mortgage debt held in commercial banks for each one-unit change in debt held by savings institutions

23. (a) and (b)

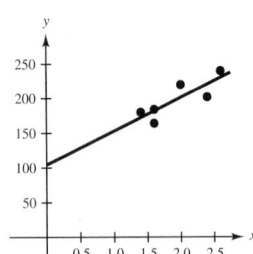

(c) $y = 48.5x + 104.5$
(d) The average change in sales for each one-unit change in advertising expenditures

24. $y = 0.42x + 1.54$ **25.** $y = -1.25x + 3$
26. $y = 0.95x + 0.92$ **27.** $y = -1.15x + 6.85$

Section Project 3.2

(a)

x	0	1	2	3	4	5
$y = 5x - 3$	-3	2	7	12	17	22
Differences		5	5	5	5	5

(b)

x	0	1	2	3	4	5
$y = -2x + 7$	7	5	3	1	-1	-3
Differences		-2	-2	-2	-2	-2

(c) $y = 4x + 3$; m is the slope and b is the y-intercept.
(d) $y = -6x - 1$; m is the slope and b is the y-intercept.

Section 3.3 (page 215)

1. $8 + n$ **2.** $n - 6$ **3.** $15 - 3n$ **4.** $4n - 6$
5. $\frac{1}{3}n$ **6.** $\frac{7n}{5}$ **7.** $0.30L$ **8.** $0.40C$ **9.** $\frac{x}{6}$
10. $\frac{y}{3}$ **11.** $\frac{3 + 4x}{8}$ **12.** $10y - 35$ **13.** $|n - 5|$
14. $\left|\frac{y}{4}\right|$ **15.** The sum of three times a number and 2
16. The difference of four times a number and 5
17. Eight times the difference of a number and 5
18. Negative three times the sum of a number and 2
19. The ratio of a number to 8
20. Four-fifths of a number
21. The sum of a number and 10, all divided by 3
22. Twenty-five more than the quotient of a number and 6
23. $0.25n$ **24.** $10m + 25n$ **25.** $55t$ **26.** $5r$
27. $\frac{100}{r}$ **28.** $\frac{360}{t}$ **29.** $0.45y$ **30.** $0.65q$

31. $0.0125I$　**32.** $0.06L$　**33.** $0.80L$　**34.** $18 + 3n$
35. $8.25 + 0.60q$　**36.** $11.65 + 0.80q$　**37.** 4
38. $\frac{216}{7}$　**39.** $\frac{15}{2}$　**40.** $\frac{7}{4}$　**41.** 16　**42.** $\frac{20}{3}$　**43.** $\frac{3}{4}$
44. $\frac{10}{3}$　**45.** $\frac{1}{25}$　**46.** $\frac{5}{8}$　**47.** Perimeter: $6w$; Area: $2w^2$
48. Perimeter: $12 + 4x$; Area: $9x$　**49.** s^2　**50.** $\frac{3}{8}b^2$
51. $6w^2\,\text{m}^2$　**52.** $l^2 - 6l\,\text{ft}^2$　**53.** $44, 46, 48$　**54.** 10
55. 42　**56.** 52　**57.** 62　**58.** $15, 17$　**59.** 43
60. $8\text{ in.} \times 24\text{ in.}$　**61.** 3　**62.** 88　**63.** $\$420$
64. $\$247$　**65.** 15.625%　**66.** 7%　**67.** 200
68. $\$22,391$
69. Taxes: 7.12%　　　　**70.** Monroe: 30.84%
　　Wages: 59.35%　　　　　Washington: 6.01%
　　Employee benefits: 10.98%　　Howard: 11.30%
　　Miscellaneous: 12.46%　　　Spring: 17.44%
　　Insurance: 2.37%　　　　West: 12.77%
　　Supplies: 2.67%　　　　Clark: 21.66%
　　Utilities: 1.19%
　　Rent: 3.86%
71. **(a)** 1980: 62,400; 1990: 66,144
　　(b) 10.24%; Using different bases
72. No. $\frac{1}{2}\% = 0.5\% = 0.005$　**73.** $\frac{85}{4}$　**74.** $\frac{9}{1}$
75. 2%　**76.** $\frac{3}{2}$　**77.** $\$1800$　**78.** 46,400 votes
79. 20 pt　**80.** 10.5 cups　**81.** 360 mi　**82.** 48 lb

Section Project 3.3

Geometry: $x = 3\frac{1}{3}$
Tree Height: $\frac{516}{11} \approx 46.9$ feet
Shadow Length: $\frac{45}{7} \approx 6.4$ feet

Mid-Chapter Quiz　(page 220)

1. $y = 2x - \frac{3}{2}$　**2.** $y = \frac{1}{2}x + 5$
3. $y = -\frac{3}{4}x + \frac{63}{8}$　**4.** $y = 3x + \frac{87}{10}$
5. $y = 2x - 3$　**6.** $y = -\frac{31}{30}x + \frac{4}{5}$

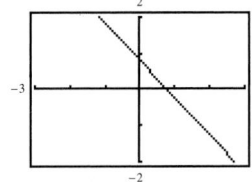

7. $y = -1$　　　　　　**8.** $x = 4$

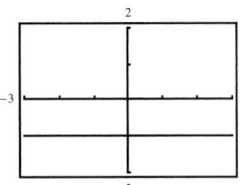　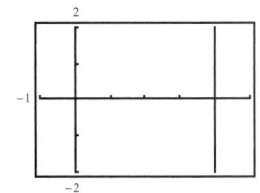

9. **(a)** $y = \frac{2}{3}x + 3$　**(b)** $y = -\frac{3}{2}x + \frac{19}{2}$
10. $(3, 0), \left(0, -\frac{9}{2}\right)$　　　　**11.** $P = 3.6x - 24,000$

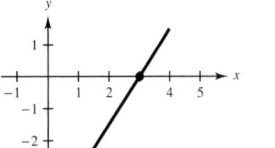

12.　　　　　　　　　　**13.** $5n - 8$

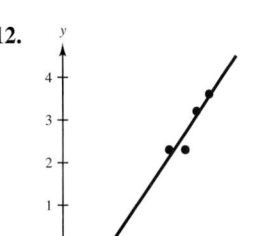

$y = 0.057x - 1.58, 3.3$
14. Perimeter: $3.2l$; Area: $0.6l^2$　**15.** $3n + 6$
16. 2000

Section 3.4　(page 230)

1. $\$18.36$, 40%　**2.** $\$29.47$, 35%　**3.** $\$152.00$, 65%
4. $\$419.25$, 44%　**5.** $\$22,250.00$, 21%
6. $\$12,550.00$, 31%　**7.** $\$416.70$, $\$191.70$
8. $\$976.00$, $\$244.00$　**9.** $\$23.76$, 48%
10. $\$39.27$, 33%　**11.** $\$111.00$, 63%　**12.** $\$210.45$, 39%
13. $\$33.25$, $\$61.75$　**14.** $\$19.90$, $\$3.98$
15. $\$1145.00$, 22%　**16.** $\$459.50$, 44%
17. 19.65 quadrillion Btu　**18.** 9.3%
19. Department store　**20.** Mail order　**21.** 2.5 hr
22. 2 hr　**23.** 9 min, $\$2.06$　**24.** 8 hr　**25.** 18.3%
26. 15.2%　**27.** 9%　**28.** $\$1034.40$
29. Tax $= \$267$; Total bill $= \$4717$;
　　Amount financed $= \$3717$

30. $580   **31.** 12 $0.20 stamps; 58 $0.32 stamps
32. 15 dimes; 35 quarters   **33.** 100
34. 75 lb at $3.88; 25 lb at $4.88
35. (a) 8%: $9600; 10%: $12,000   **(b)** $15,000
36. (a) $6\frac{1}{2}$%: $780; 9%: $1080   **(b)** $1600
37. 50 gal at 20%; 50 gal at 60%
38. 6 L at 50%; 4 L at 75%
39. 8 qt at 15%; 16 qt at 60%
40. 13.75 gal at 60%; 41.25 gal at 80%   **41.** $\frac{5}{6}$ gal
42. 700   **43.** 2275 mi   **44.** 450 ft   **45.** $\frac{100}{11}$ hr
46. $\frac{125}{16}$ sec   **47.** $\frac{2000}{3}$ ft/sec   **48.** 55 mph
49. 1440 mi   **50.** 375 m/min   **51.** 17.65 min
52. 3 hr, $2\frac{3}{4}$ hr
53. (a) $T = \dfrac{x}{63} + \dfrac{280 - x}{54}$

(b)
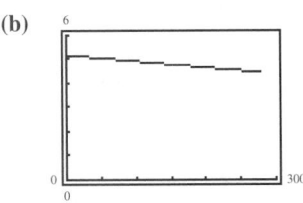

$0 \le x \le 280$

(c) 164.5 mi
54. 8.32 min   **55. (a)** 8 pages/min   **(b)** $\frac{15}{4}$ units/hr
56. $\frac{1}{5}, \frac{1}{8}, \frac{40}{13} = 3\frac{1}{13}$ hr   **57.** 10 min   **58.** 6 hr
59. 0.926 ft   **60.** 81 in.^{2}
61. (a) 1989   **(b)** $0.37, slope
62. (a) 1988   **(b)** $0.282, slope
63. Bus drivers. The slope is greater in the model for bus drivers.
64. (a) $A = 12x$

(b)
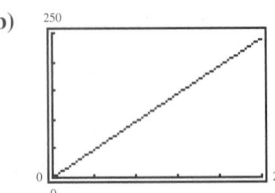
(c) $16\frac{2}{3}$ units

65. (a) $A = \frac{1}{3}x^2 + \frac{1}{2}x$

(b)
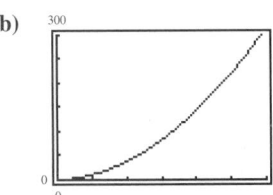
(c) 23.76 units

Section Project 3.4

Oats x	Corn $500 - x$	Price/bushel of the mixture
0	500	$4.00
100	400	$3.74
200	300	$3.48
300	200	$3.22
400	100	$2.96
500	0	$2.70

(a) Decreases **(b)** Decreases
(c) Average of the prices of all components

Section 3.5   (page 243)

1. The multiplication and division properties differ. The inequality symbol is reversed if both sides of the inequality are multiplied or divided by a negative real number.
2. $3x - 2 \le 4$, $-(3x - 2) \ge -4$
3. (a) Yes **(b)** No **(c)** Yes **(d)** No
4. (a) No **(b)** No **(c)** Yes **(d)** No
5. (a) No **(b)** Yes **(c)** Yes **(d)** No
6. (a) Yes **(b)** Yes **(c)** No **(d)** No
7. (d)   **8.** (a)   **9.** (c)   **10.** (b)

11. $(-5, 3]$ **12.** $[1, 4)$

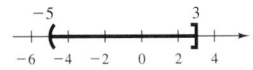 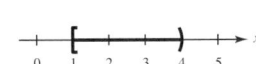

13. $\left(0, \frac{3}{2}\right]$ **14.** $(-7, -3]$

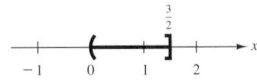 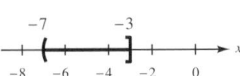

15. $\left(-\frac{15}{4}, -\frac{5}{2}\right)$ **16.** $(-5, -\pi)$

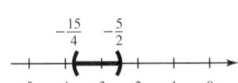 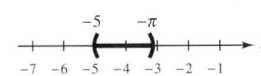

17. $(-\infty, -2]$

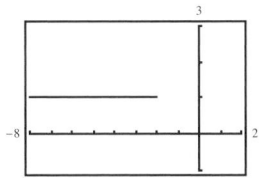

18. $(6, \infty)$

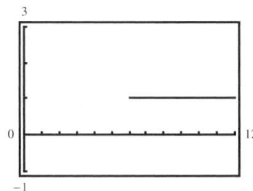

19. $(-2, 4]$

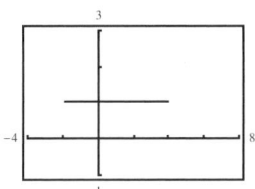

20. $(-10, -5)$

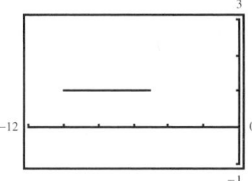

21. $[3, 9]$

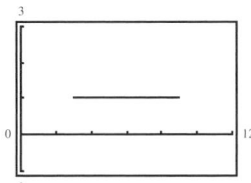

22. $(\infty, -1], (1, \infty)$

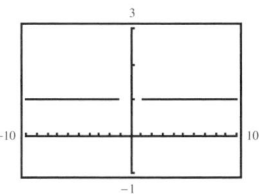

23. $x \le 2$

24. $z > 4$

25. $x < \frac{11}{2}$

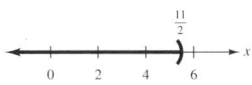

26. $x > \frac{5}{2}$

27. $x \le -4$

28. $x \ge -4$

29. $x > 8$

30. $x < 10$

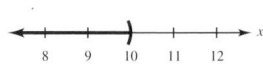

31. $x > 7$

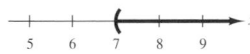

32. $x \le 6$

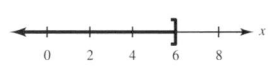

33. $x \ge 7$

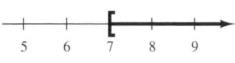

34. $y \le 6$

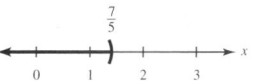

35. $x > -\frac{2}{3}$

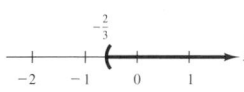

36. $x < \frac{7}{5}$

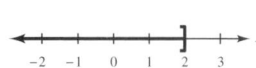

37. $x > \frac{9}{2}$

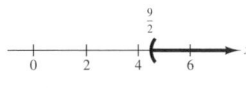

38. $x \le 2$

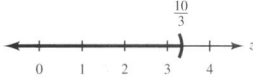

39. $y > 2$

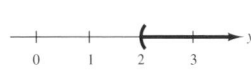

40. $z < \frac{10}{3}$

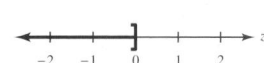

41. $y \le -10$

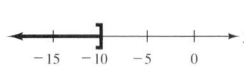

42. $z \le 0$

43. $x \ge -12$

44. $x \ge \frac{12}{7}$

45. $\frac{5}{2} < x < 7$

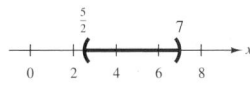

46. $-5 < x \le 0$

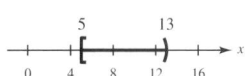

47. $-\frac{3}{2} < x < \frac{9}{2}$

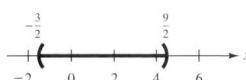

48. $5 \le x < 13$

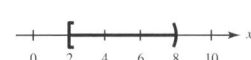

49. $1 < x < 10$

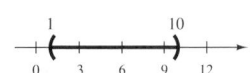

50. $2 \le x < 8$

51. $-\frac{3}{5} < x < -\frac{1}{5}$ **52.** $-6 < x < 6$

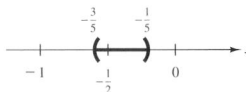

53. $x \geq 0$ **54.** $y > -2$ **55.** $z \geq 2$ **56.** $m \geq 4$
57. $n \leq 16$ **58.** $450 \leq x \leq 500$ **59.** x is at least $\frac{5}{2}$.
60. t is less than 4.
61. z is greater than 0 and no more than π.
62. t is at least -4 but no more than 4.   **63.** $-t > -8$
64. $-10 \leq -x < 3$ **65.** $3 \leq n \leq \frac{15}{2}$
66. $3 \leq n \leq 9$ **67.** $2600   **68.** $C \leq \$850$
69. The average temperature in Miami is greater than the average temperature in New York.
70. The elevation of San Francisco is less than the elevation of Denver.
71. $m \leq 25,357$ **72.** 29,655 mi
73. $x \geq 31$ **74.** $x \geq 942$

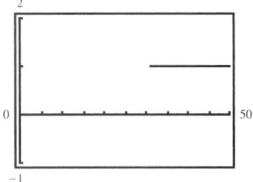 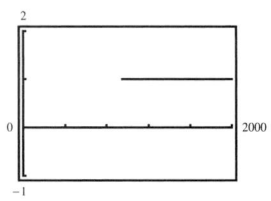

75. The call must be less than 6.38 min. If a portion of a minute is billed as a full minute, the call must be less than or equal to 6 min.

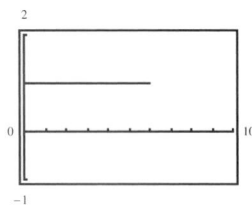

76. The call must be less than 15.26 min. If a portion of a minute is billed as a full minute, the call must be less than or equal to 15 min.

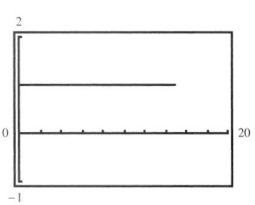

77. $2 \leq d \leq 8$ **78.** $5 \leq d \leq 15$
79. $12.50 < 8 + 0.75n; \quad n > 6$
80. $3000 < 1000 + 0.04S; \quad S > \$50,000$
81. 1987, 1988, 1989   **82.** 1993

Section Project 3.5

(a) $10,000 \leq b \leq 120,000$ **(b)** $1000 \leq B \leq 120,000$
(c) No   **(d)** Yes

Section 3.6 (page 253)

1. Not a solution   **2.** Solution   **3.** Solution
4. Not a solution   **5.** $x - 10 = 17; \; x - 10 = -17$
6. $7 - 2t = 5; \; 7 - 2t = -5$
7. $4x + 1 = \frac{1}{2}; \; 4x + 1 = -\frac{1}{2}$
8. $22k + 6 = 9; \; 22k + 6 = -9$ **9.** $45, -45$
10. $16, -16$ **11.** 0 **12.** No solution   **13.** $21,11$
14. $200, 0$ **15.** $11, -14$ **16.** $\frac{2}{7}, -2$ **17.** $\frac{16}{3}, 16$
18. $\frac{16}{5}, -2$ **19.** No solution   **20.** $-6, 14$
21. $-\frac{11}{5}, \frac{17}{5}$ **22.** $-\frac{11}{4}, -\frac{29}{4}$ **23.** $\frac{4}{3}$ **24.** No solution
25. $-\frac{39}{2}, \frac{15}{2}$ **26.** $\frac{5}{8}, \frac{25}{8}$ **27.** $18.75, -6.25$
28. $1.14, 4.93$ **29.** $-3, 7$ **30.** $\frac{3}{4}, \frac{17}{2}$
31. $11, 13$ **32.** $\frac{21}{2}, -\frac{29}{8}$ **33.** $\frac{3}{2}, -\frac{1}{4}$ **34.** $13, \frac{17}{3}$
35. $|2x + 3| = 5$ **36.** $|4x - 6| = 7$
37. **(a)** Yes **(b)** No **(c)** No **(d)** Yes
38. **(a)** No **(b)** Yes **(c)** Yes **(d)** No
39. **(a)** No **(b)** Yes **(c)** Yes **(d)** No
40. **(a)** Yes **(b)** No **(c)** No **(d)** Yes
41. **(a)** Yes **(b)** No **(c)** No **(d)** Yes
42. **(a)** No **(b)** Yes **(c)** Yes **(d)** No
43. **(a)** No **(b)** Yes **(c)** Yes **(d)** No
44. **(a)** Yes **(b)** No **(c)** No **(d)** Yes
45. $-3 < y + 5 < 3$ **46.** $-5 \leq 6x + 7 \leq 5$
47. $7 - 2h \geq 9; 7 - 2h \leq -9$
48. $8 - x > 25; 8 - x < -25$

49. 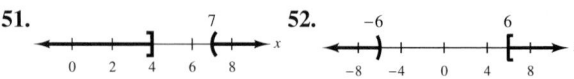 **50.**

51.  **52.**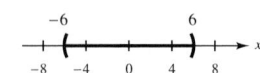

53. $-4 < y < 4$ **54.** $-6 < x < 6$

55. $-2 \le y \le 6$

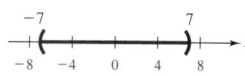

56. $-3 \le x \le 9$

57. $-6 < y < 2$

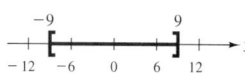

58. $-9 < x < 3$

59. $y \le -4$ or $y \ge 4$

60. $x \le -6$ or $x \ge 6$

61. $y < -2$ or $y > 6$

62. $x < -3$ or $x > 9$

63. $y \le -6$ or $y \ge 2$

64. $x \le -9$ or $x \ge 3$

65. $-7 < x < 7$

66. $-\frac{9}{4} \le z \le \frac{9}{4}$

67. $-9 \le y \le 9$

68. $-8 < t < 8$

69. $-2 < x < \frac{2}{3}$ **70.** $-1 \le x \le 2$ **71.** $3 \le x \le 7$
72. $-5 < a < 3$ **73.** $x > 3$ or $x < -6$
74. $r > 2$ or $r < -\frac{8}{7}$ **75.** $-5 < x < 35$
76. $-5.\overline{3} \le t \le 16$ **77.** $-82 \le x \le 78$
78. $-104 < y < 136$ **79.** $t \le -\frac{15}{2}$ or $t \ge \frac{5}{2}$
80. $t < -2$ or $t > \frac{4}{3}$ **81.** $s > 23$ or $s < -17$
82. $a \le -38$ or $a \ge 26$ **83.** $z < -50$ or $z > 110$
84. No solution **85.** (d) **86.** (c) **87.** (b)
88. (a) **89.** $|x| \le 2$ **90.** $|x| < 4$
91. $|x - 10| < 3$ **92.** $|x - 75| \le 3$
93. $|x - 19| < 3$ **94.** $|x + 11| \le 2$

95.

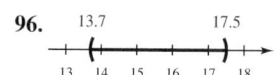

96.

97. $|t - 98.6| < 1$
98. (a) $|s - x| \le \frac{3}{16}$ (b) $\frac{79}{16} \le x \le \frac{85}{16}$
99. 66 oz; 62 oz; Maximum error for each bag is $\frac{1}{2}$ ounce. So for 4 bags the maximum error is $4\left(\frac{1}{2}\right) = 2$ ounces.

Section Project 3.6

$|x - 4| < 1$: All real numbers that are within 1 unit of 4.
$|y + 2| > 4$: All real numbers that are greater than 4 units from -2.

(a) $|x| < 3$

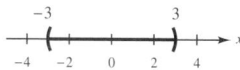

(b) $|x| > 2$

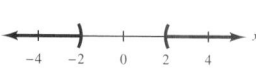

(c) $|x - 5| > 6$

(d) $|x - 16| < 5$

Review Exercises (page 257)

1. $2x - y - 6 = 0$ **2.** $3x - y + 10 = 0$
3. $4x + y = 0$ **4.** $2x + y - 8 = 0$
5. $2x + 3y - 17 = 0$ **6.** $9x - 6y + 10 = 0$
7. $x - 7 = 0$ **8.** $y - 5 = 0$
9. $x + 2y + 6 = 0$ **10.** $3x - 2y = 0$
11. $y - 10 = 0$ **12.** $9x + 14y + 62 = 0$
13. $9x - 24y - 8 = 0$ **14.** $2x - 5 = 0$
15. (a) $x + 2y - 9 = 0$ (b) $2x - y + 7 = 0$
16. (a) $3x + y - 1 = 0$ (b) $x - 3y - 3 = 0$
17. (a) $8x - 6y + 15 = 0$ (b) $24x + 32y - 105 = 0$
18. (a) $x - 12 = 0$ (b) $y - 1 = 0$
19. **20.**

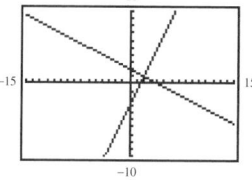

Perpendicular

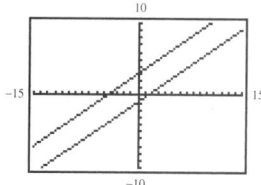

Parallel

21.

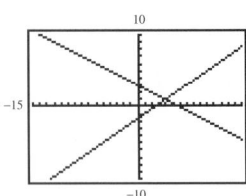

22.

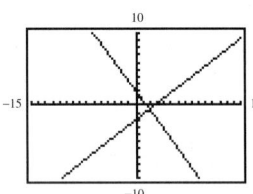

Neither   Perpendicular

23. (a) $C = 8.55x + 25{,}000$ **(b)** $P = 4.05x - 25{,}000$

24. (a) 16 ft/sec   **(b)** 2.5 sec   **(c)** -16 ft/sec

25. (a)

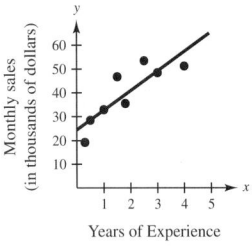

(b)

x	1.5	1.0	0.3	3.0
y	46.7	32.9	19.2	48.4
Approx. y	36.8	32.7	26.9	49.1

x	4.0	0.5	2.5	1.8
y	51.2	28.5	53.4	35.5
Approx. y	57.4	28.6	45.0	39.3

(c) $53,200

(d) Rate of increase (in thousands of dollars) of sales for each 1-yr increase in years of experience

26. (a)

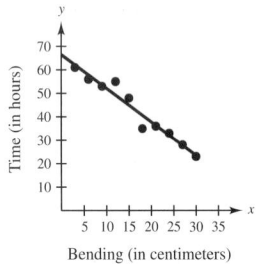

(b)

x	3	6	9	12	15
y	61	56	53	55	48
Approx. y	62.2	57.9	53.6	49.3	45.0

x	18	21	24	27	30
y	35	36	33	28	23
Approx. y	40.7	36.4	32.1	27.8	23.5

(c) 37.8 hr

(d) The average decrease in time until failure for each 1 cm increase in bending distance

27. (a) $F = 15d$ **28. (a)** $o = 0.0353g$
   (b) $5\frac{1}{3}$ cm   **(b)** 907 g
   (c) 48 kg   **(c)** 18 oz

29. (a)

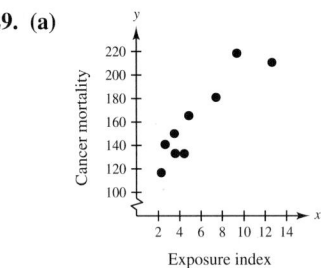

(b) Yes. The cancer mortality appears to increase linearly with increased exposure to the carcinogenic substance.

30. (a)

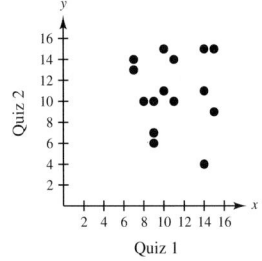

(b) No. Quiz scores are dependent on several variables, such as study time, class attendance, etc. These variables may change from one quiz to the next.

31. $y = -0.5x + 3$ **32.** $y = x + 2$

33. (a) and **(b)**

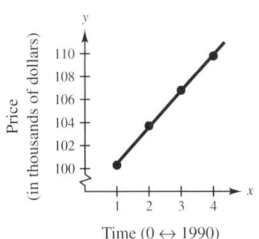

Time (0 ↔ 1990)

(c) $y = 3.16x + 97.25$

(d) The average increase (in thousands of dollars) of the median price of homes for each 1-yr increase in time

(e) $128,850

34. (a) and **(b)**

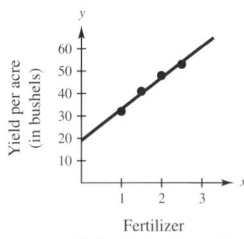

Fertilizer
(in hundreds of pounds)

(c) $y = 14x + 19$

(d) The average increase (in bushels) of wheat for each 100 pound increase per acre of fertilizer

(e) 35.8 bushels

35. $200 - 3n$ **36.** $100 + 15n$ **37.** $n^2 + 49$

38. $|n + 10|$ **39.** $\left|\dfrac{n}{5}\right|$ **40.** $n + n^2$

41. The sum of twice a number and 7

42. Three less than five times a number

43. The difference of a number and 5, all divided by 4

44. Negative three times the difference of a number and 10

45. $0.18l$ **46.** $8r$ **47.** $l(l - 5) = l^2 - 5l$

48. $3n + 6$ **49.** $30p$ **50.** $\dfrac{z}{8}$ **51.** $\dfrac{4}{3}$ **52.** $\dfrac{6}{5}$

53. $\dfrac{3}{20}$ **54.** $\dfrac{2}{1}$ **55.** $y = \dfrac{7}{2}$ **56.** $x = \dfrac{20}{3}$

57. $b = \dfrac{25}{2}$ **58.** $x = 5$ **59.** $1856.25

60. $\dfrac{15}{4}$ or $3\dfrac{3}{4}$ cups   **61.** 487.5 mi

62. 25 pt or 3.125 gal   **63.** 3   **64.** $\dfrac{14}{3}$ **65.** 80 ft

66. 50 ft   **67.** $\approx 40\%$, $\approx \$2,625,000,000$ **68.** 5.5%

69. 10.2%, 9.6%, 9.9%, 13.2%   **70.** $27,166.25

71. $350.93   **72.** 84.2%   **73.** $194.25

74. Department store   **75.** $8333.33   **76.** 480

77. 30% solution: $3\dfrac{1}{3}$ L; 60% solution: $6\dfrac{2}{3}$ L

78. 25% solution: $3\dfrac{1}{5}$ gal; 50% solution: $4\dfrac{4}{5}$ gal

79. 6.35 hr   **80.** $\dfrac{45}{124}$ km/min   **81.** 2800 mi

82. $53\dfrac{1}{3}$ mph   **83.** $\dfrac{20}{9} \approx 2.22$ hr   **84.** $\dfrac{18}{7} \approx 2.57$ hr

85. $340   **86.** 7.5%   **87.** $210,526.32

88. $4,030,000   **89.** $30,000   **90.** $71.23

91. 8 in × 6 in.   **92.** 5   **93.** $-4, 8$ **94.** $-5, 2$

95. $\dfrac{1}{2}, 3$ **96.** $-\dfrac{7}{3}, -\dfrac{5}{7}$

97. $x > 3$ **98.** $x \le -4$

99. $y > -\dfrac{70}{3}$ **100.** $y \le \dfrac{4}{5}$

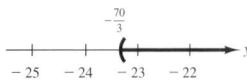

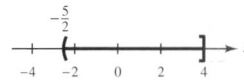

101. $-20 < x \le 20$ **102.** $-\dfrac{5}{2} < x \le 4$

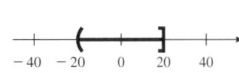

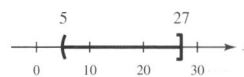

103. $-16 < x < -1$ **104.** $5 < x \le 27$

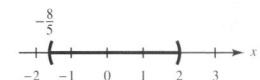

105. $-4 < x < 11$ **106.** $-\dfrac{8}{5} < x < 2$

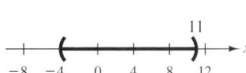

107. $x > 7$ or $x < 1$ **108.** $t < -5$ or $t > -1$

109. $b \ge 5$ or $b \le -9$ **110.** $-3 < t < 3$

111. $z \le 10$ **112.** $x \ge 0$ **113.** $7 \le y < 14$

114. $V < 27$ **115.** $|x - 3| < 2$ **116.** $|x + 15| \le 3$

Chapter 3 Test   (page 263)

1. $y = \dfrac{1}{2}x - \dfrac{55}{2}$ **2.** $y = \dfrac{3}{5}x - 4$

CHAPTER 4

3. $V = -4000t + 26,000$; 2.5 years

4.

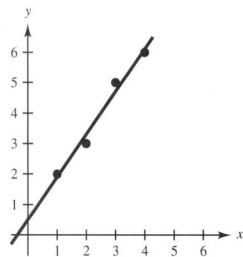

$y = 1.4x + 0.5$

5. $T = 3t + 72$ **6.** $4x + 14$ **7.** 7

8. 26, 28   **9.** \$1466.67   **10.** $2\frac{1}{2}$ hr

11. 10% solution: $33\frac{1}{3}$ L; 40% solution: $66\frac{2}{3}$ L   **12.** 40 min

13. \$2000   **14.** $t \geq 8$ **15.** $-17, 5$ **16.** $-1, 4$

17. $x > 2$ **18.** $-7 < x \leq 1$

   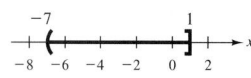

19. $1 \leq x \leq 5$ **20.** $x > -3$ or $x < -5$

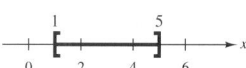   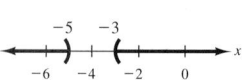

Cumulative Test: Chapters 1–3   (page 264)

1. 120   **2.** $-\frac{1}{4}$ **3.** -27 **4.** $\frac{25}{64}$ **5.** 10

6. $-8x - 10$ **7.** $(x-2)(x-1)$ **8.** $(2x-7)^2$

9. $-4, 1$ **10. (a)** -7 **(b)** $-2x$ **11.** $\frac{7}{3}$

12. $y = -\frac{2}{7}x + \frac{45}{7}$ **13.** $C = 5.75x + 12,000$

14. \$1408.75   **15.** 6.5

16. 7.5%: \$15,000; 9%: \$9000

17. $x < \frac{3}{2}$ **18.** $x \leq -1$ or $x \geq 5$

   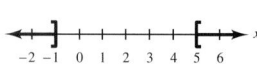

CHAPTER 4

Section 4.1   (page 278)

1. The system has no solution.

2. (a) The lines intersect at a point and only one line lies inside the viewing rectangle.

(b) The lines coincide.

(c) The lines are parallel and only one line lies inside the viewing rectangle.

3. (a) Yes   **(b)** No   **4. (a)** No   **(b)** Yes

5. (a) No   **(b)** Yes   **6. (a)** Yes   **(b)** No

7. No solution   **8.** $(-2, 3)$ **9.** $\left(1, \frac{1}{3}\right)$

10. No solution   **11.** Infinite number of solutions

12. $(5, -3)$

13. $(-1, -1)$ **14.** $(0, 5)$

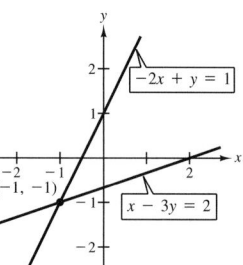   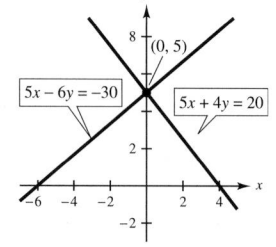

15. $(10, 0)$ **16.** $(6, -2)$

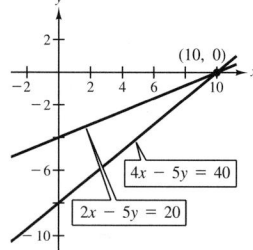   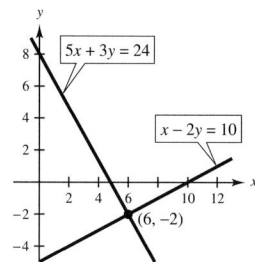

17. $(9, 12)$ **18.** $(4, 3)$

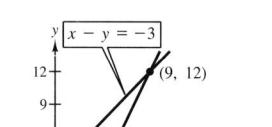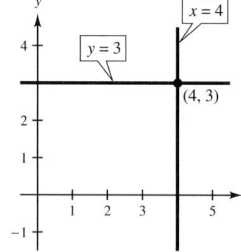

19. $\left(-\frac{3}{2}, \frac{5}{2}\right)$ **20.** $\left(1, -\frac{1}{4}\right)$

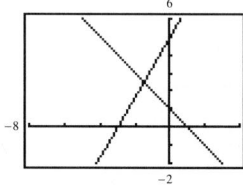   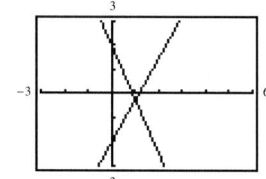

21. $(3, 5)$

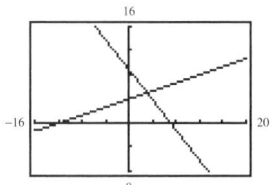

22. $\left(2, \frac{5}{2}\right)$

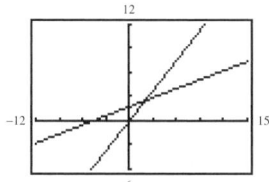

23. $\left(\frac{1}{2}, -1\right)$

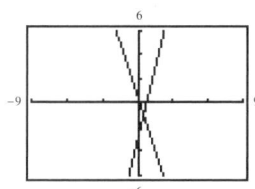

24. $\left(4, \frac{1}{3}\right)$

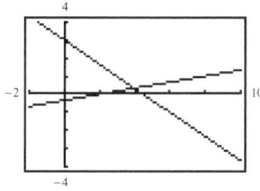

25. Consistent   **26.** Inconsistent
27. Consistent   **28.** Consistent

29.

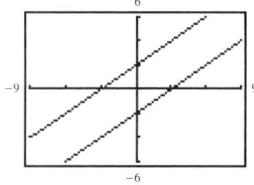

Inconsistent

30.

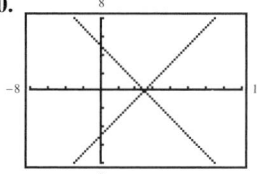

One solution

31.

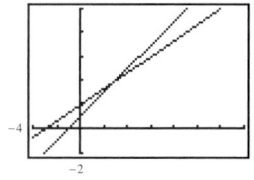

One solution

32.

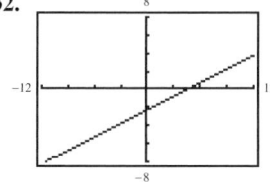

Infinitely many solutions

33.

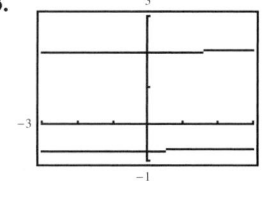

$\left(2992, \frac{798}{25}\right)$

34.

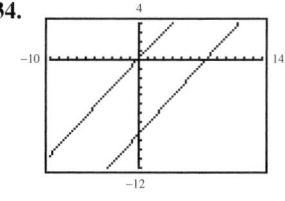

$\left(276, \frac{3220}{11}\right)$

35.

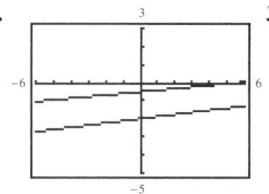

$(50, 4)$

36.

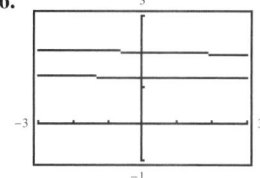

$\left(60, \frac{1}{2}\right)$

37. $(2, 1)$ **38.** $(2, 2)$ **39.** $(4, 3)$ **40.** $(6, 2)$
41. $\left(4, -\frac{1}{2}\right)$ **42.** $\left(\frac{5}{2}, \frac{3}{2}\right)$ **43.** $(1, 2)$ **44.** $\left(-\frac{20}{3}, -\frac{5}{3}\right)$
45. $(4, -2)$ **46.** $(-3, -1)$ **47.** $(7, 2)$ **48.** $(4, 2)$
49. $\left(\frac{1}{3}, \frac{17}{3}\right)$ **50.** $(-2, -1)$ **51.** $\left(\frac{3}{2}, \frac{3}{2}\right)$ **52.** $\left(4, \frac{4}{3}\right)$
53. $(5, 3)$ **54.** $(2, -3)$

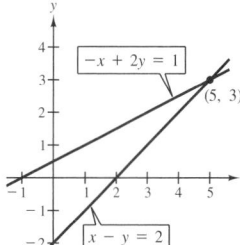

55. $(2, -2)$ **56.** $(4, 3)$

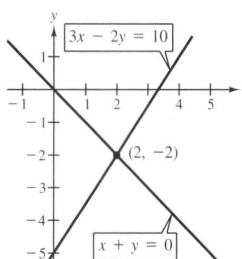
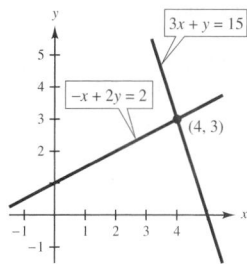

57. No solution   **58.** No solution

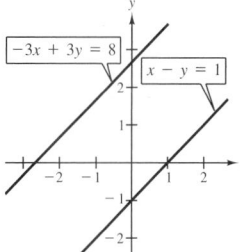
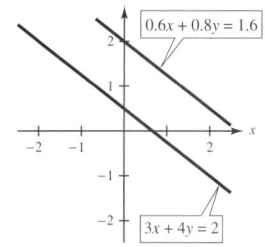

CHAPTER 4

59. Infinite number of solutions   **60.** $\left(2, -\frac{3}{4}\right)$

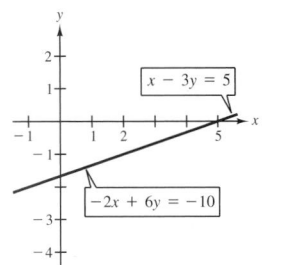

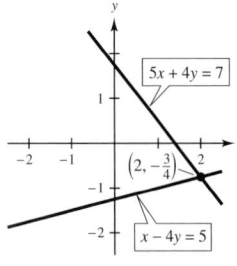

61. $(3, 2)$ **62.** $(1, 5)$ **63.** $(-2, 5)$ **64.** $(1, -4)$
65. $(-1, -1)$ **66.** $(-3, 4)$ **67.** $(7, -2)$
68. $(10, 5)$ **69.** No solution
70. Infinite number of solutions   **71.** $\left(\frac{3}{2}, 1\right)$ **72.** $\left(\frac{1}{4}, \frac{3}{4}\right)$
73. $(-2, -1)$ **74.** $(20, 10)$
75. Infinite number of solutions   **76.** No solution
77. $(12.5, 4.948)$ **78.** $\left(\frac{47}{6}, \frac{11}{3}\right)$ **79.** No solution
80. Infinite number of solutions   **81.** $(4, 3)$
82. $(-2, 6)$ **83.** $(2, 7)$ **84.** $(-3, 7)$ **85.** $(15, 10)$
86. $(-2, 7)$ **87.** $(6, 11)$ **88.** $\left(1, \frac{3}{2}\right)$
89. $(3000, -2000)$ **90.** $(-2500, 1800)$ **91.** $k = 4$
92. $k = 27$
93. False. The lines represented by a consistent system of linear equations intersect at one point (one solution) or are identical lines (infinite number of solutions).
94. Answers may vary, depending on the method used to solve the system.
95. $x + 2y = 0$ **96.** $x + y = 4$
    $4x + 2y = 9$ $2x + 3y = 20$
97. 50 m × 60 m   **98.** 60 ft × 80 ft
99. Regular: $1.11; Premium: $1.22
100. 280 student tickets; 520 adult tickets
101. 70 t of the $75-per-ton hay; 30 t of the $125-per-ton hay
102. 40% solution; 12 L; 65% solution: 8 L
103. $15,000 at 8%; $5000 at 9.5%   **104.** 10,000 units
105. 32 in., 128 in.   **106.** 225 mi, 75 mi
107. Speed of plane: 540 mph; Speed of wind: 60 mph

108. (a) $y = \frac{3}{2}x - \frac{1}{6}$

(b)

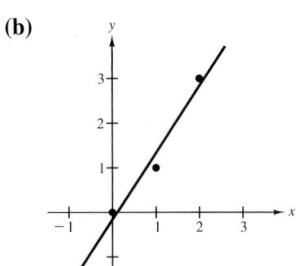

109. (a)

(b) $y = -85x + \frac{2777}{3}$

(c) The average decrease in employment (in thousands) per year

Section Project 4.1

Depth: 10 feet
Length of sections: 122 feet, 125 feet

Section 4.2   (page 293)

1. (a) Not a solution   **2. (a)** Solution
   (b) Solution   **(b)** Not a solution
   (c) Solution   **(c)** Not a solution
   (d) Not a solution   **(d)** Solution
3. $(22, -1, -5)$ **4.** $(-1, 2, 3)$ **5.** $(14, 3, -1)$
6. $(10, -14, 2)$
7. The system has a "stair-step" pattern with leading coefficients of 1.
8. No. When the first equation was multiplied by -2 and added to the second equation, the constant term should have been -11.
9. $x - 2y + 3z = 5$ **10.** $x - 2y + 3z = 5$
      $y - 2z = 9$ $-x + y + 5z = 4$
     $2x \quad\quad - 3z = 0$ $4y - 9z = -10$
   Eliminated the x-term   Eliminated the x-term
   in Equation 2   in Equation 3
11. $(1, 2, 3)$ **12.** $(3, 2, 1)$ **13.** $(1, 2, 3)$

14. $(0, 4, -2)$ **15.** $(2, -3, -2)$ **16.** $(-2, 4, 1)$

17. No solution   **18.** $(5, 2, -4)$ **19.** $(-4, 8, 5)$

20. $(5, -2, 0)$ **21.** No solution   **22.** $\left(\frac{3}{10}, \frac{2}{5}, 0\right)$

23. $(1, 0, -2)$ **24.** $(2, -1, 1)$ **25.** $(-4, 2, 3)$

26. $(0, -4, 5)$ **27.** $(-1, 5, 5)$ **28.** No solution

29. $\left(\frac{7}{5}a + 1, \frac{1}{5}a - 1, a\right)$ **30.** No solution

31. $\left(-\frac{1}{2}a + \frac{1}{4}, \frac{1}{2}a + \frac{5}{4}, a\right)$ **32.** No solution

33. $(1, -1, 2)$ **34.** $\left(\frac{1}{2}, \frac{13}{11}, \frac{35}{22}\right)$

35. $\left(\frac{6}{13}a + \frac{10}{13}, \frac{5}{13}a + \frac{4}{13}, a\right)$ **36.** $\left(-\frac{1}{2}a + \frac{1}{2}, \frac{3}{5}a + \frac{2}{5}, a\right)$

37.
$$
\begin{aligned}
x + 2y - z &= -4 \\
y + 2z &= 1 \\
3x + y + 3z &= 15
\end{aligned}
$$

38.
$$
\begin{aligned}
x - y + z &= -12 \\
x + y + z &= 2 \\
-2x - 4y + 3z &= -68
\end{aligned}
$$
   39. $y = 2x^2 + 3x - 4$

40. $y = -x^2 + 2x + 5$ **41.** $y = x^2 - 4x + 3$

42. $y = -2x^2 + 5x - 1$ **43.** $y = -x^2 + 2x$

44. $y = -2x^2 + x + 2$ **45.** $x^2 + y^2 - 4x = 0$

46. $x^2 + y^2 - 6y = 0$ **47.** $x^2 + y^2 - 6x - 8y = 0$

48. $x^2 + y^2 - 3x - 2y = 0$

49. $x^2 + y^2 - 2x - 4y - 20 = 0$

50. $x^2 + y^2 - 10x - 144 = 0$ **51.** $s = -16t^2 + 144$

52. $s = -16t^2 + 64t$ **53.** $s = -16t^2 + 48t$

54. $s = -16t^2 + 60t + 10$

55. 20 gal of spray X; 18 gal of spray Y; 16 gal of spray Z

56. $\frac{1}{2}x^2 - \frac{1}{2}x$; Yes

57. Strings: 50; Wind: 20; Percussion: 8

Section Project 4.2

(a) 10% solution: 0 liters   **(b)** 10% solution: $6\frac{1}{4}$ liters
20% solution: $8\frac{1}{3}$ liters   20% solution: 0 liters
50% solution: $1\frac{2}{3}$ liters   50% solution: $3\frac{3}{4}$ liters

(c) 10% solution: 1 liter
20% solution: 7 liters
50% solution: 2 liters

Section 4.3   (page 304)

1. 3×2 **2.** 3×3 **3.** 2×3 **4.** 2×2

5. $\begin{bmatrix} 4 & -5 & \vdots & -2 \\ -1 & 8 & \vdots & 10 \end{bmatrix}$ **6.** $\begin{bmatrix} 8 & 3 & \vdots & 25 \\ 3 & -9 & \vdots & 12 \end{bmatrix}$

7. $\begin{bmatrix} 1 & 10 & -3 & \vdots & 2 \\ 5 & -3 & 4 & \vdots & 0 \\ 2 & 4 & 0 & \vdots & 6 \end{bmatrix}$

8. $\begin{bmatrix} 9 & -3 & 1 & \vdots & 13 \\ 12 & 0 & -8 & \vdots & 5 \end{bmatrix}$

9. $\begin{cases} 4x + 3y = 8 \\ x - 2y = 3 \end{cases}$ **10.** $\begin{cases} 9x - 4y = 0 \\ 6x + y = -4 \end{cases}$

11. $\begin{cases} x \quad + 2z = -10 \\ 3y - z = 5 \\ 4x + 2y \quad = 3 \end{cases}$

12. $\begin{cases} 5x + 8y + 2z \quad = -1 \\ -2x + 15y + 5z + w = 9 \\ x + 6y - 7z \quad = -3 \end{cases}$

13. $\begin{bmatrix} 1 & 4 & 3 \\ 0 & 2 & -1 \end{bmatrix}$ **14.** $\begin{bmatrix} 3 & 6 & 8 \\ 1 & -9 & -2 \end{bmatrix}$

15. $\begin{bmatrix} 1 & 1 & 4 & -1 \\ 0 & 5 & -2 & 6 \\ 0 & 3 & 20 & 4 \end{bmatrix}$ $\begin{bmatrix} 1 & 1 & 4 & -1 \\ 0 & 1 & -\frac{2}{5} & \frac{6}{5} \\ 0 & 3 & 20 & 4 \end{bmatrix}$

16. $\begin{bmatrix} 1 & 2 & 4 & \frac{3}{2} \\ 1 & -1 & -3 & 2 \\ 2 & 6 & 4 & 9 \end{bmatrix}$ $\begin{bmatrix} 1 & 2 & 4 & \frac{3}{2} \\ 0 & -3 & -7 & \frac{1}{2} \\ 0 & 2 & -4 & 6 \end{bmatrix}$

17. $\begin{bmatrix} 1 & 2 & 3 \\ 0 & 1 & 2 \end{bmatrix}$ **18.** $\begin{bmatrix} 1 & 3 & 6 \\ 0 & 1 & 9 \end{bmatrix}$

19. $\begin{bmatrix} 1 & -1 & -\frac{5}{2} \\ 0 & 1 & \frac{11}{10} \end{bmatrix}$ **20.** $\begin{bmatrix} 1 & \frac{2}{3} & 2 \\ 0 & 1 & -\frac{21}{5} \end{bmatrix}$

21. $\begin{bmatrix} 1 & 1 & 0 & 5 \\ 0 & 1 & 2 & 0 \\ 0 & 0 & 1 & -1 \end{bmatrix}$ **22.** $\begin{bmatrix} 1 & 2 & -1 & 3 \\ 0 & 1 & -2 & 5 \\ 0 & 0 & 1 & -1 \end{bmatrix}$

23. $\begin{bmatrix} 1 & -1 & -1 & 1 \\ 0 & 1 & 6 & 3 \\ 0 & 0 & 1 & \frac{4}{5} \end{bmatrix}$ **24.** $\begin{bmatrix} 1 & -3 & 0 & -7 \\ 0 & 1 & 1 & 2 \\ 0 & 0 & 0 & 0 \end{bmatrix}$

25. $\begin{bmatrix} 1 & 1 & -1 & 3 \\ 0 & 1 & -4 & 1 \\ 0 & 0 & 0 & 0 \end{bmatrix}$ **26.** $\begin{bmatrix} 1 & -3 & -2 & -8 \\ 0 & 1 & 0 & \frac{25}{6} \\ 0 & 0 & 0 & 1 \end{bmatrix}$

27. $\begin{cases} x - 2y = 4 \\ y = -3 \end{cases}$ **28.** $\begin{cases} x + 5y = 0 \\ y = -1 \end{cases}$

$(-2, -3)$ $(5, -1)$

29. $\begin{cases} x - y + 2z = 4 \\ y - z = 2 \\ z = -2 \end{cases}$ **30.** $\begin{cases} x + 2y - 2z = -1 \\ y + z = 9 \\ z = -3 \end{cases}$

$(8, 0, -2)$ $(-31, 12, -3)$

31. $\left(\frac{9}{5}, \frac{13}{5}\right)$ **32.** $(-1, 3)$ **33.** $(1, 1)$ **34.** $(0.2, 0.5)$

35. No solution   **36.** $(3a + 5, a)$ **37.** $(2, -3, 2)$

38. $(4, -3, 2)$ **39.** $(2a + 1, 3a + 2, a)$

40. No solution **41.** $(1, 2, -1)$ **42.** $(8, 10, 6)$

43. $(1, -1, 2)$ **44.** $\left(-\frac{3}{2}a + \frac{3}{2}, \frac{1}{3}a + \frac{1}{3}, a\right)$

45. $(34, -4, -4)$ **46.** $(-1, -2, 3)$ **47.** No solution

48. No solution **49.** $(-12a - 1, 4a + 1, a)$

50. $\left(-\frac{1}{2}a + 5, -1, a\right)$

51. There will be a row in the matrix with all zero entries except in the last column.

52. The row-echelon form of the matrix will have fewer rows with nonzero entries than there are variables in the system.

53. 8%: $800,000; 9%: $500,000; 12%: $200,000

54. $4000 invested at 5%, $5000 invested at 6%, $7000 invested at 7%

55. $y = x^2 + 2x + 4$ **56.** $y = -x^2 + 2x + 10$

57. The one matrix can be obtained from the other by using the elementary row operations.

58. $\begin{bmatrix} 1 & -2 & 6 \\ 0 & 1 & 5 \\ 0 & 0 & 0 \end{bmatrix}$

59. (a) $y = -\frac{1}{250}x^2 + \frac{3}{5}x + 6$

(b) 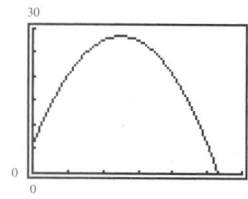 Maximum height:28.5 ft The ball struck the ground at approximately $(159.4, 0)$.

60. (a) $y = \frac{1}{20}t^2 + \frac{3}{20}t + \frac{33}{5}$ (c) 12 million short tons

(b)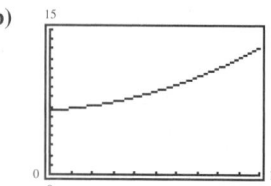

61. (a) $y = 14t^2 - 47t + 1018$

(b)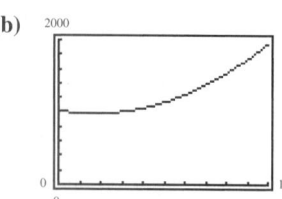

(c) About 1948 billion dollars

62. $\dfrac{2x^2 - 9x}{(x-2)^3} = \dfrac{2}{x-2} - \dfrac{1}{(x-2)^2} - \dfrac{10}{(x-2)^3}$

63. $\dfrac{1}{x^3 - x} = \dfrac{-1}{x} + \dfrac{1/2}{x+1} + \dfrac{1/2}{x-1}$

Section Project 4.3

(a) Certificates of deposit: $250,000 - \frac{1}{2}s$
Municipal bonds: $125,000 + \frac{1}{2}s$
Blue-chip stocks: $125,000 - s$
Growth stocks: s
If $s = \$100,000$, then
    Certificates of deposit: $200,000
    Municipal bonds: $175,000
    Blue-chip stocks: $25,000
    Growth stocks: $100,000

(b) Certificates of deposit: $406,250 - \frac{1}{2}s$
Municipal bonds: $-31,250 + \frac{1}{2}s$
Blue-chip stocks: $125,000 - s$
Growth stocks: s where $s \geq 62,500$
If $s = \$100,000$, then
    Certificates of deposit: $356,250
    Municipal bonds: $18,750
    Blue-chip stocks: $25,000
    Growth stocks: $100,000

Mid-Chapter Quiz (page 308)

1. $(2, 1)$ **2.** $(2, -1)$

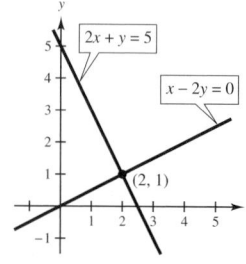

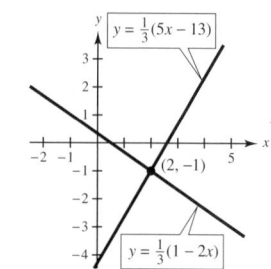

3. $\left(\frac{90}{13}, \frac{34}{13}\right)$ **4.** $(5, 10)$ **5.** $(8, 1)$ **6.** $(-2, 4)$

7. $(5, -1, 3)$ **8.** $(2, 1, -2)$

9. $\begin{aligned} x + y &= -2 \\ 2x + y &= 8 \end{aligned}$ **10.** $\begin{aligned} x + y + z &= 0 \\ x - y + z &= 6 \\ x - y - 3z &= 2 \end{aligned}$

11. $\begin{aligned} x + y &= 300 \\ x - 3y &= 0 \end{aligned}$ **12.** $\begin{aligned} x + y &= 20 \\ 0.2x + 0.5y &= 6 \end{aligned}$

13. $\begin{aligned} a + b + c &= 2 \\ a - b + c &= -4 \\ 4a + 2b + c &= 8 \end{aligned}$

Section 4.4 (page 317)

1. 5 **2.** -11 **3.** 27 **4.** 14 **5.** 0 **6.** 0

7. 6 **8.** 0 **9.** -24 **10.** -13 **11.** -0.16

12. -0.72 **13.** -24 **14.** -48 **15.** -2 **16.** 6

17. -30 **18.** -20 **19.** 3 **20.** -2 **21.** 0

22. 0 **23.** -75 **24.** 151 **25.** -58 **26.** 0

27. -0.22 **28.** 0.002 **29.** $x - 5y + 2$

30. $-7x + 3y - 8$ **31.** 248 **32.** 105.625

33. 19,185 **34.** -32 **35.** 0 **36.** -28

37. $-\frac{167}{8}$ **38.** $\frac{196}{3}$ **39.** 77 **40.** 0 **41.** -6.37

42. 4,561,250 **43.** $(1, 2)$ **44.** $(-3, 4)$ **45.** $(2, -2)$

46. $\left(\frac{1}{2}, \frac{1}{3}\right)$ **47.** $\left(\frac{3}{4}, -\frac{1}{2}\right)$

48. $D = 0$, cannot use Cramer's Rule

49. $D = 0$, cannot use Cramer's Rule **50.** $\left(\frac{32}{7}, \frac{30}{7}\right)$

51. $\left(\frac{2}{3}, \frac{1}{2}\right)$ **52.** $\left(-\frac{1}{13}, \frac{8}{13}\right)$ **53.** $(-1, 3, 2)$

54. $(5, 8, -2)$ **55.** $\left(1, \frac{1}{2}, \frac{3}{2}\right)$ **56.** $\left(\frac{3}{4}, \frac{25}{28}, -\frac{73}{28}\right)$

57. $(1, -2, 1)$ **58.** $D = 0$, cannot use Cramer's Rule

59. $\left(\frac{22}{27}, \frac{22}{9}\right)$ **60.** $\left(-\frac{1}{13}, \frac{6}{13}\right)$ **61.** $\left(\frac{1}{2}, 4\right)$ **62.** $\left(\frac{1}{10}, \frac{7}{10}\right)$

63. $\left(\frac{1}{3}, 1, -\frac{2}{3}\right)$ **64.** $\left(-\frac{2}{3}, -34, -12\right)$ **65.** $\left(2, \frac{1}{2}, -\frac{1}{2}\right)$

66. $(1, -1, 2)$ **67.** $\left(\frac{51}{16}, -\frac{7}{16}, -\frac{13}{16}\right)$ **68.** $\left(-2, \frac{3}{2}, 2\right)$

69. No. The matrix must be square.

70. The determinant of the coefficient matrix must be a nonzero real number.

71. 16 **72.** 27 **73.** $\frac{31}{2}$ **74.** $\frac{33}{8}$ **75.** $\frac{53}{2}$

76. 15 **77.** Collinear **78.** Not collinear

79. Not collinear **80.** Collinear **81.** $3x - 5y = 0$

82. $2x + 6y - 10 = 0$ **83.** $7x - 6y - 28 = 0$

84. $2x + 3y - 8 = 0$

85. $y = 2x^2 - 6x + 1$ **86.** $y = x^2 - \frac{3}{2}x + 2$

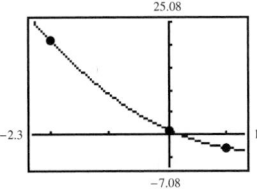 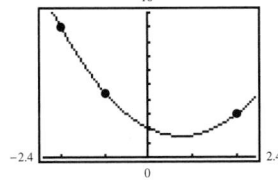

87. $y = -3x^2 + 2x$ **88.** $y = -x^2 + 10$

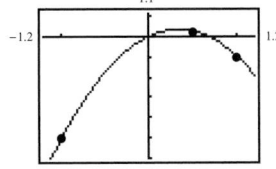 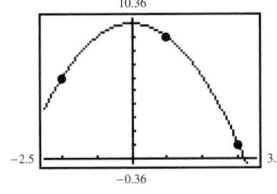

89. (a) $y_1 = 15.35t^2 - 59.85t + 506.5$

(b) $y_2 = 17.55t^2 - 39.75t + 542.0$

(c)
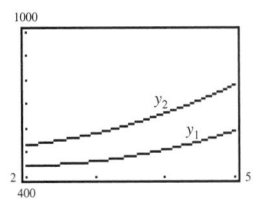

(d) $y_1 - y_2 = -2.20t^2 - 20.10t - 35.5$

90. $I_1 = 1, I_2 = 2, I_3 = 1$

91. (a) $\left(\dfrac{4k - 3}{2k - 1}, \dfrac{4k - 1}{2k - 1}\right)$ (b) $\dfrac{1}{2}$ **92.** 1, 6 **93.** 2, 3

Section Project 4.4

(a) 250 square miles (b) 250 square miles

(c) No, the triangle has been translated on the coordinate plane, but its size has not changed.

Section 4.5 (page 326)

1. (b) **2.** (a) **3.** (d) **4.** (e) **5.** (f) **6.** (c)

7. (a) Yes (b) No (c) Yes (d) Yes

8. (a) No (b) No (c) Yes (d) Yes

9. (a) No (b) No (c) Yes (d) Yes

10. (a) Yes (b) Yes (c) No (d) Yes

11. (a) Yes (b) No (c) No (d) Yes

12. (a) No (b) No (c) Yes (d) Yes

13. (a) No (b) Yes (c) No (d) Yes

14. (a) No (b) Yes (c) Yes (d) No

15. **16.**

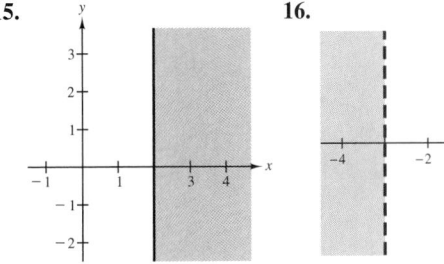

17. **18.**

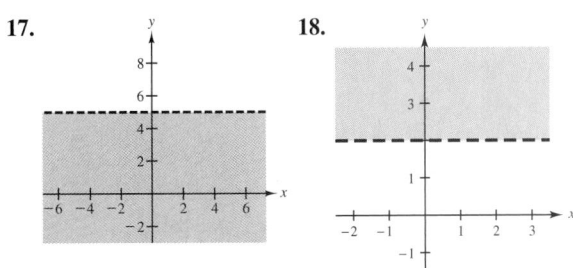

19.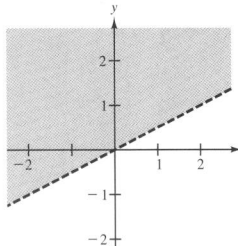

20.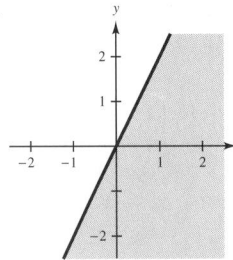

21.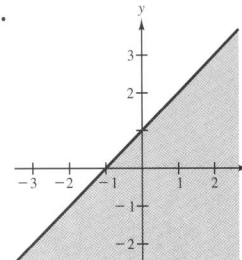

22.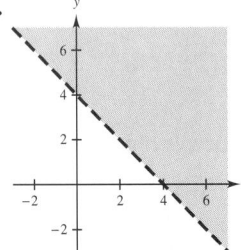

23.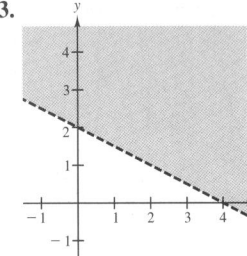

24.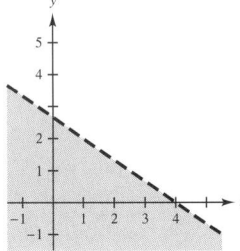

25.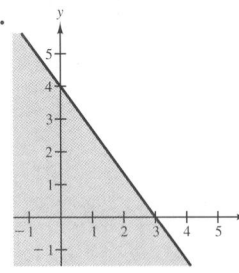

26.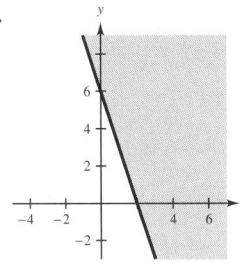

27.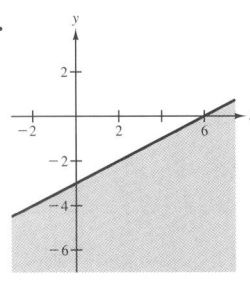

28.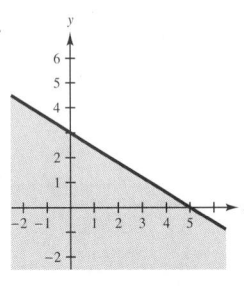

29.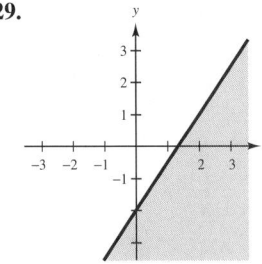

30.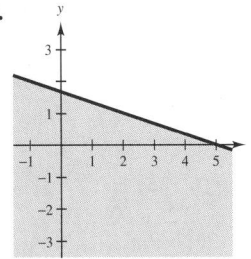

31.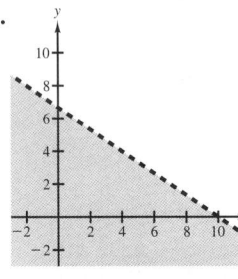

32.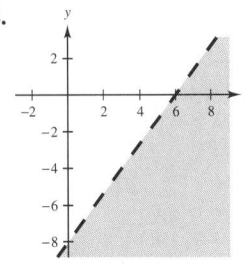

33.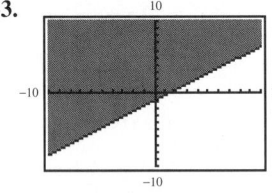

34.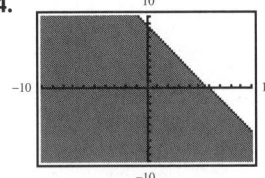

35.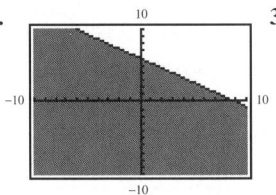

36.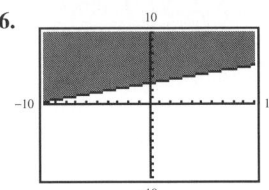

37.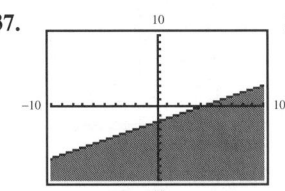

38.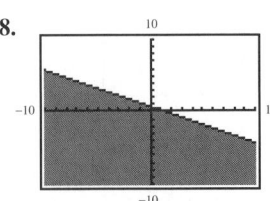

39.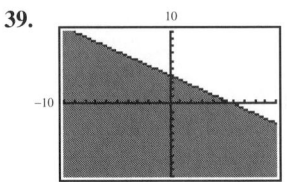

40.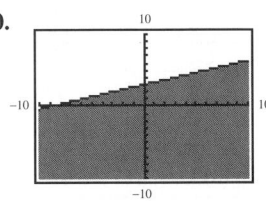

41. $3x + 4y > 17$ **42.** $-x + y < 2$ **43.** $y < 2$

44. $x < 2$ **45.** $x - 2y < 0$ **46.** $x + y < 0$

47. The graph of a line divides the plane into two half-planes. In graphing a linear inequality, graph the corresponding linear equation. The solution to the inequality will be one of the half-planes determined by the line. Example: $x - 2y > 1$.

48. The solution to $x - y > 1$ does not include the points on the line $x - y = 1$. The solution to $x - y \geq 1$ does include the points on the line $x - y = 1$.

49. Test a point in one of the half-planes.

50. On the real number line, the solution to $x \leq 3$ is an unbounded interval. In the rectangular plane, the solution is a half-plane.

51. $2x + 2y \leq 500$ or $y \leq -x + 250$
(*Note:* x and y cannot be negative.)

52. $10x + 15y \leq 1000$ or $y \leq -\frac{2}{3}x + \frac{200}{3}$
(*Note:* x and y cannot be negative.)

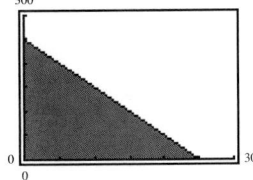

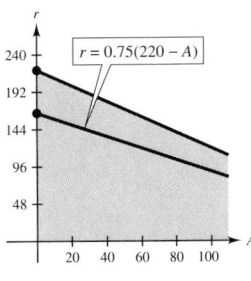

53. $9x + 6y \geq 150$
(*Note:* x and y cannot be negative.)

54. (a) and **(b)**

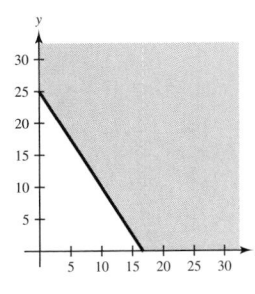

$(10, 10), (8, 14), (20, 0)$

Section Project 4.5

$2(8) + 0.40x + 0.80y \leq 26$
$0.40x + 0.80y \leq 10$
$(6, 6)$ is a solution point.

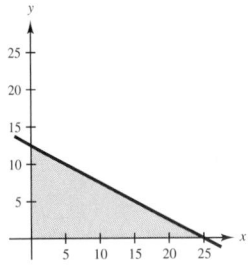

Section 4.6   (page 335)

1. (c)   **2.** (b)   **3.** (f)   **4.** (e)   **5.** (a)   **6.** (d)

7.

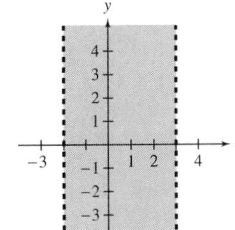

8.

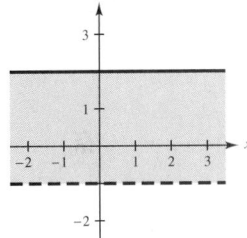

9.

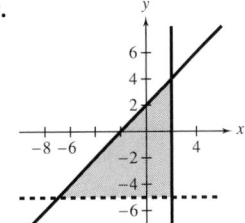

10.

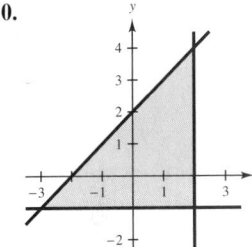

11.

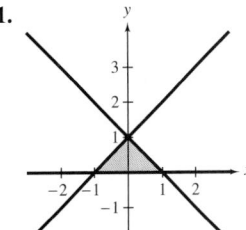

12.

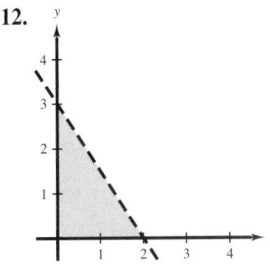

13.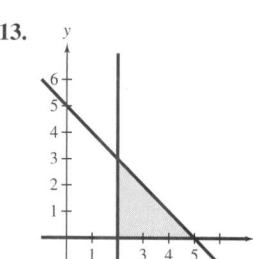

14.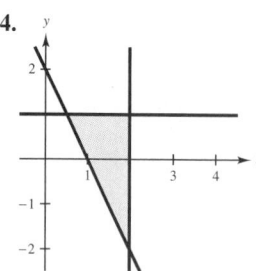

15.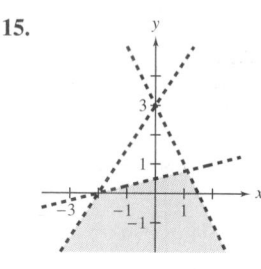

16.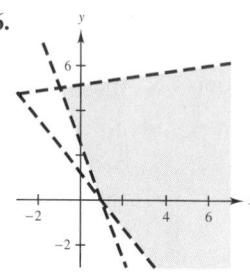

17.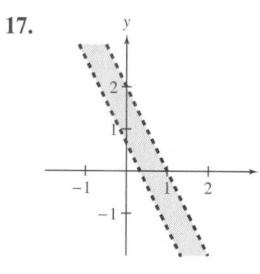

18.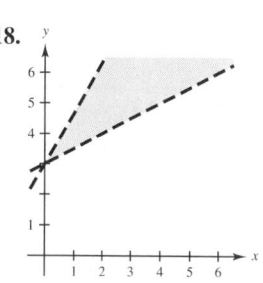

19.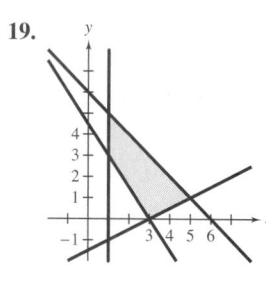

20.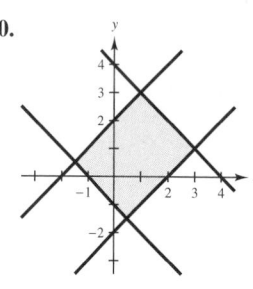

21. **22.**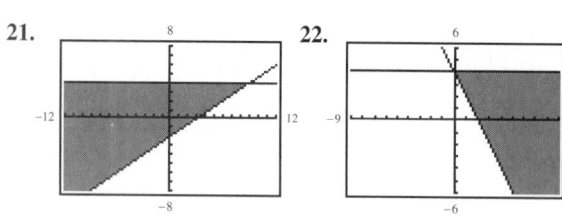

23. **24.**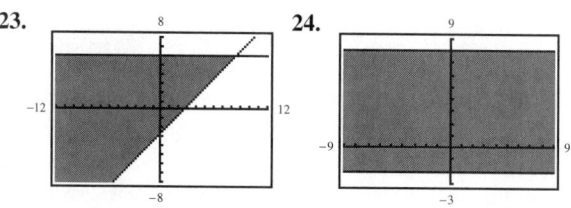

25. **26.**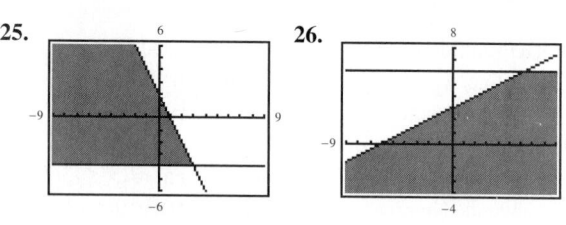

27. **28.**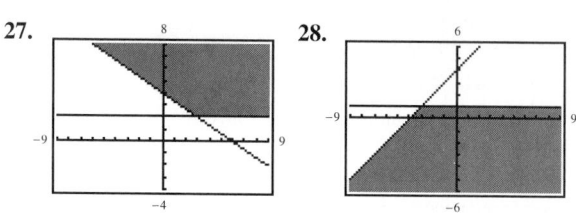

29. $\begin{cases} x \geq 1 \\ x \leq 8 \\ y \geq -5 \\ y \leq 3 \end{cases}$

30. $\begin{cases} y \leq 3 \\ y \geq -1 \\ y \geq x - 6 \\ y \leq x + 3 \end{cases}$

31. $\begin{cases} y \leq \frac{9}{10}x + \frac{42}{5} \\ y \geq 3x \\ y \geq \frac{2}{3}x + 7 \end{cases}$

32. $\begin{cases} y \geq 2x - 4 \\ y \leq 3x - 2 \\ y \leq \frac{5}{4}x - \frac{1}{4} \end{cases}$

33. $\begin{cases} x < 90 \\ y \leq 0 \\ y \geq -10 \\ y \geq -\frac{1}{7}x \end{cases}$

34. $\begin{cases} y \leq 22 \\ y \geq 10 \\ y \geq 2x - 24 \\ y \geq -2x - 24 \end{cases}$

35. $\begin{cases} x + y \leq 20{,}000 \\ x \geq 5000 \\ y \geq 5000 \\ y \geq 2x \end{cases}$

36.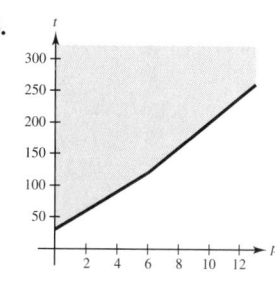

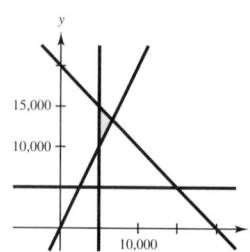

(12, 220), Yes

37. $\begin{cases} x + y \geq 15{,}000 \\ 15x + 25y \geq 275{,}000 \\ x \geq 8000 \\ y \geq 4000 \end{cases}$

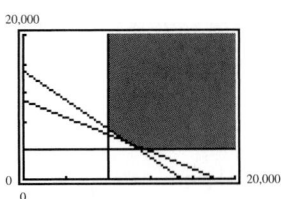

38. $\begin{cases} 20x + 10y \geq 280 \\ 15x + 10y \geq 160 \\ 10x + 20y \geq 180 \\ x \geq 0 \\ y \geq 0 \end{cases}$

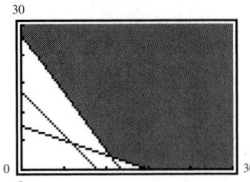

39. Minimum value at $(0, 0)$: 0
Maximum value at $(0, 6)$: 30
40. Minimum value at $(0, 0)$: 0
Maximum value at $(0, 4)$: 32
41. Minimum value at $(0, 0)$: 0
Maximum value at $(6, 0)$: 60
42. Minimum value at $(0, 0)$: 0
Maximum value at $(2, 0)$: 14
43. Minimum value at $(0, 0)$: 0
Maximum value at $(3, 4)$: 17
44. Minimum value at $(0, 2)$: 6
Maximum value at $(5, 3)$: 29
45. Minimum value at $(0, 0)$: 0
Maximum value at $(4, 0)$: 20
46. Minimum value at $(3, 0)$: 3
Maximum value at $(0, 4)$: 24
47. Minimum value at $(0, 0)$: 0
Maximum value at $(60, 20)$: 740
48. Minimum value at $(0, 600)$: 21,000
Maximum value at $(900, 0)$: 45,000
49. Minimum value at $(0, 0)$: 0
Maximum value at $(30, 45)$: 2325
50. Minimum value at $(675, 0)$: 10,800
Maximum value at $(0, 800)$: 16,000

51. Minimum value at
$(4, -1)$: -9
Maximum value at
$(-5, 3)$: 13

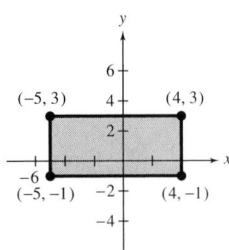

52. Minimum value at
$(-4, 1)$: -16
Maximum value at
$(2, 6)$: 34

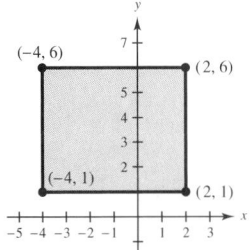

53. Minimum value at
$(12, 0)$: 12
Maximum value at
$(12, 10)$: 52

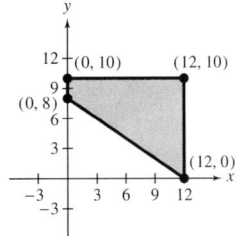

54. Minimum value at
$\left(\frac{1}{4}, 0\right)$: $\frac{3}{2}$
Maximum value at
$(5, 19)$: 68

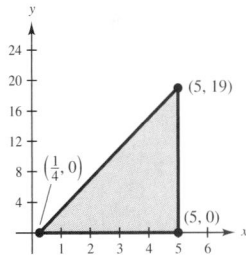

55. Minimum value at
$(0, 0)$: 0
Maximum value at
$(5, 0)$: 30

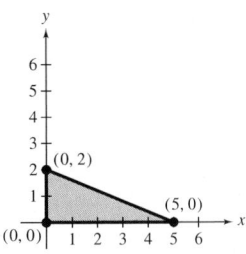

56. Minimum value at
$(0, 0)$: 0
Maximum value at
$(0, 8)$: 64

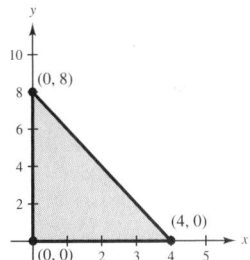

57. Minimum value at
(0, 0): 0
Maximum value at
(5, 0): 45

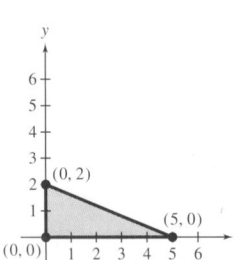

58. Minimum value at
(0, 0): 0
Maximum value at
(4, 0): 28

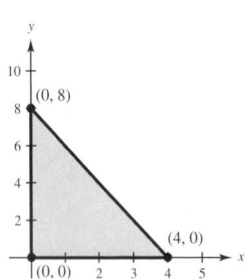

59. Minimum value at
(5, 3): 35
No maximum value

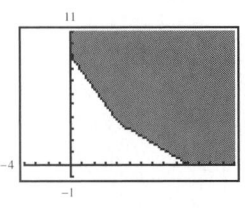

60. Minimum value at
(0, 0): 0
Maximum value at
(4, 1): 21

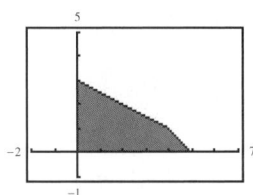

61. Minimum value at
(10, 0): 20
No maximum value

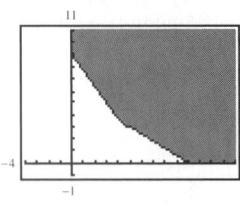

62. Minimum value at
(0, 3): -3
Maximum value at
(5, 0): 10

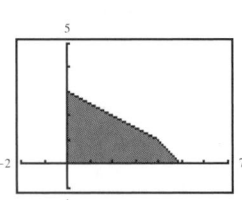

63. 60 acres of crop A; 90 acres of crop B
64. $225,000 in type A; $225,000 in type B
65. 3 bags of brand X; 6 bags of brand Y

Section Project 4.6

1000 units of Model A
500 units of Model B
Maximum profit: $76,000

Review Exercises　(page 341)

1. (1, 1)

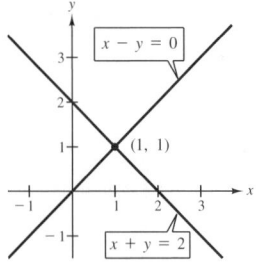

2. (3, 3)

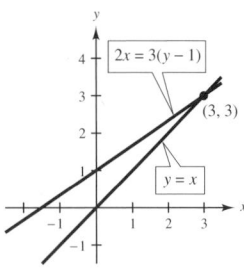

3. No solution

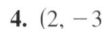
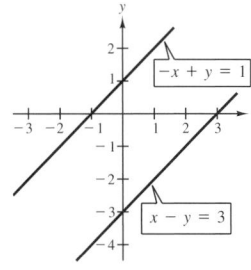

4. (2, -3)

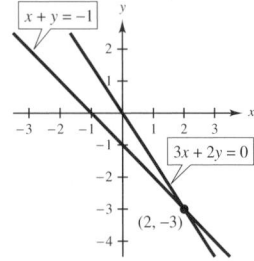

5. (4, 8)

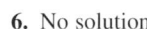
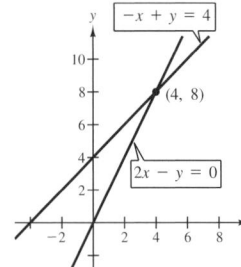

6. No solution

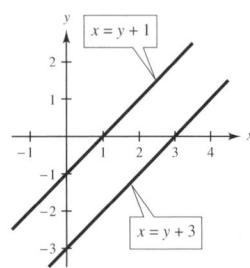

7. (3, 4)

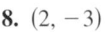
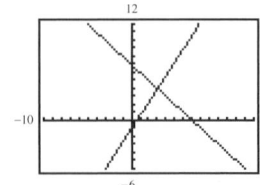

8. (2, -3)

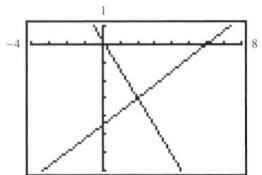

9. $(2, 1)$

10. $(3, -2)$

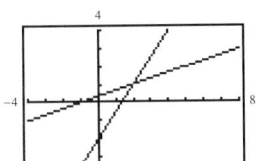

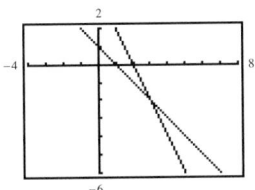

11. $(2, -1)$ **12.** $(8, 2)$ **13.** No solution

14. $(-1, 4)$ **15.** $(-10, -5)$

16. Infinite number of solutions   **17.** $(0, 0)$

18. $(1, -3)$ **19.** $\left(\frac{5}{2}, 3\right)$ **20.** $\left(-\frac{1}{2}, \frac{4}{5}\right)$ **21.** $(2, -3, 3)$

22. $(5, 2, -6)$ **23.** $(10, -12)$ **24.** $(-9, -4)$

25. $\left(\frac{24}{5}, \frac{22}{5}, -\frac{8}{5}\right)$ **26.** $\left(\frac{1}{2}, -\frac{1}{3}, 1\right)$ **27.** $(0.6, 0.5)$

28. $(-0.2, 0.7)$ **29.** $\left(\frac{5}{16}a + \frac{13}{16}, \frac{19}{16}a + \frac{11}{16}, a\right)$

30. $(0, -3, 2)$ **31.** 5 **32.** -2.5 **33.** -51

34. 5 **35.** 1 **36.** 360 **37.** 102 **38.** 88

39. $(-3, 7)$ **40.** $\left(\frac{1}{3}, -\frac{1}{2}\right)$

41. $D = 0$, cannot use Cramer's Rule   **42.** $\left(3, \frac{1}{3}\right)$

43. $(2, -3, 3)$ **44.** $D = 0$, cannot use Cramer's Rule

45. 16 **46.** 24 **47.** 7 **48.** $\frac{25}{8}$

49. $x - 2y + 4 = 0$ **50.** $3x + 2y - 16 = 0$

51. $2x + 6y - 13 = 0$ **52.** $-3x + 1.5y - 2.7 = 0$

53. $\begin{cases} 3x + y = -2 \\ 6x + y = 0 \end{cases}$ **54.** $\begin{cases} x + y = 2 \\ 2x - y = -32 \end{cases}$

55. $16{,}667$ units   **56.** $96 \text{ m} \times 144 \text{ m}$

57. 75% solution: 40 gal; 50% solution: 60 gal

58. Shorter piece: 32 in.;   Longer piece: 96 in.

59. $\$9.95$ tapes: 400; $\$14.95$ tapes: 250

60. $218\frac{3}{4}$ mph, $193\frac{3}{4}$ mph

61. $\$8000$ at 7%; $\$5000$ at 9%; $\$7000$ at 11%

62. $16, 20, 32$ **63.** $y = 2x^2 + x - 6$

64. **(a)** $y = -\frac{1}{45}x^2 + \frac{2}{3}x + 11$

(b)   **(c)** $7\frac{1}{4}$ ft

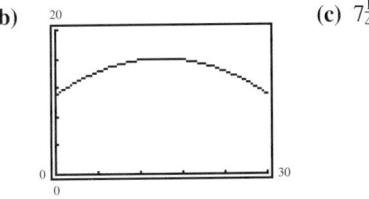

65. $x^2 + y^2 - 4x + 2y - 4 = 0$

66. $x^2 + y^2 - 2x + 4y - 20 = 0$

67.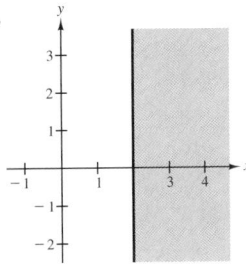

68.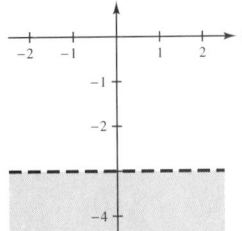

69.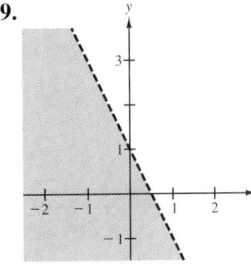

70.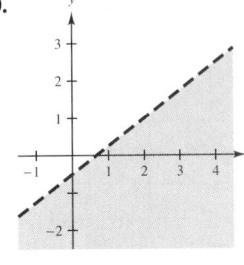

71.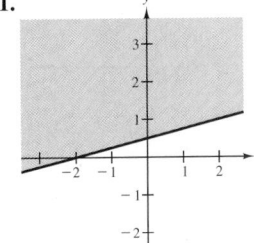

72.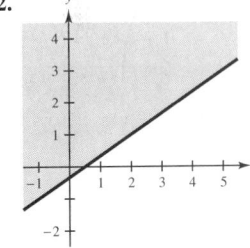

73.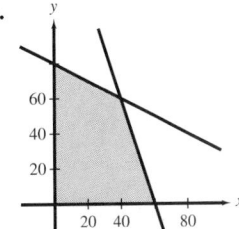

74.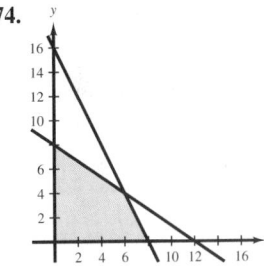

75.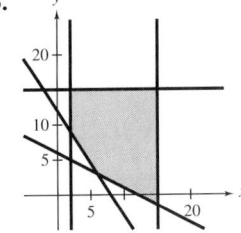

76.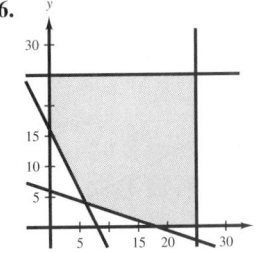

77. $\begin{cases} 2x + y \geq 7 \\ x - y \leq 2 \\ 2x + y \leq 22 \\ x - y \geq -4 \end{cases}$ **78.** $\begin{cases} x - y \geq -1 \\ 3x + y \leq 25 \\ x + 7y \geq 15 \end{cases}$

79. $\begin{cases} x + y \leq 1500 \\ x \geq 400 \\ y \geq 600 \end{cases}$ **80.** $\begin{cases} 20x + 30y \leq 24,000 \\ 12x + 8y \leq 12,400 \\ x \geq 0 \\ y \geq 0 \end{cases}$

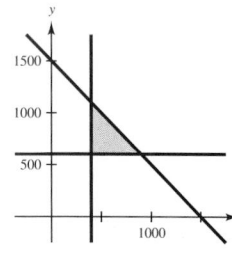   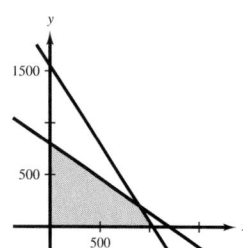

81. Maximum at $(5, 8)$: 47
82. Minimum at $(25, 50)$: 600

83. 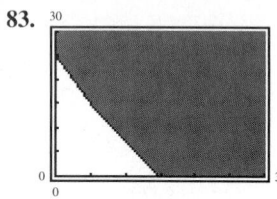   **84.**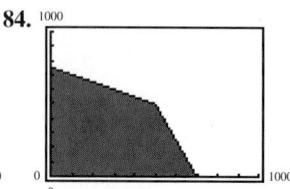

Minimum at   Maximum at
$(15, 0)$: 26.25 $(500, 500)$: 60,000

85. Five units of product A;   two units of product B
86. Three bags of brand X;   two bags of brand Y

Chapter 4 Test   (page 345)

1. $(2, 4)$ **2.** $(5, 4)$ **3.** $(2, 2a - 1, a)$
4. $(2, 1, -2)$ **5.** $\left(4, \frac{1}{7}\right)$ **6.** $\left(-\frac{11}{5}, \frac{56}{25}, \frac{32}{25}\right)$ **7.** -62
8. $y = 2x^2 - 3x + 4$ **9.** 12

10. 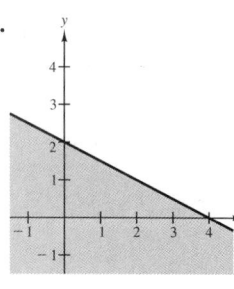   **11.**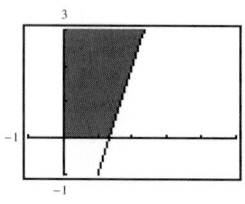

12. 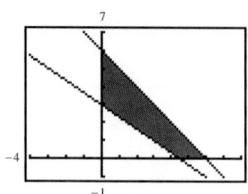   **13.** Minimum at
$(0, 0)$: 0
Maximum at
$(3, 3)$: 48

14. $x + y = 200$
$4x - y = 0$
40 mi, 160 mi

Cumulative Test: Chapters 1–4   (page 346)

1. $-54a^{18}$ **2.** $\frac{8}{3}x^{10}y$ **3.** $(x + 1)(x - 1)(x - 3)$
4. $(y - 4)(y^2 + 4y + 16)$ **5.** $\frac{8}{3}$ **6.** 3   **7.** $-7, 6$
8. (a)   **(b)**

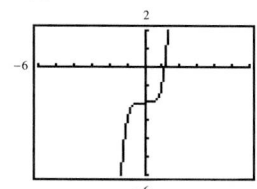   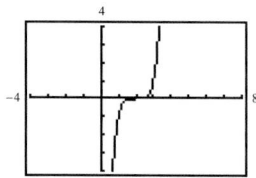

Vertical shift two units   Horizontal shift two units
downward.   to the right.

(c)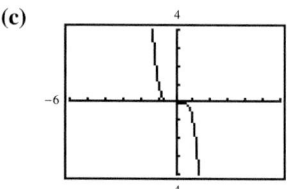

Reflection in the x-axis
9. (a) $y = 250 - x$ **(b)** $A = x(250 - x)$
10. $y = 5x - 37$

11. $-3 < x \leq \frac{1}{2}$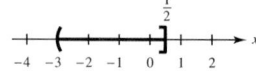

12. $\left(3, \frac{5}{2}\right)$ **13.** $(-5, 7)$

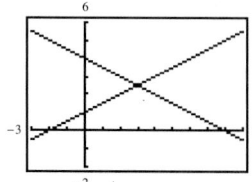   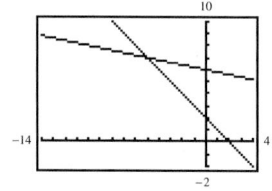

14. $(4, -1, 3)$

CHAPTER 5

Section 5.1 (page 354)

1. $(-2x)^4 = (-2)^4 x^4 = 16x^4 \neq -2x^4$

2. False. The number 32.5 is not in the interval $[0, 10)$.

3. $\frac{1}{25}$ **4.** $\frac{1}{16}$ **5.** $-\frac{1}{1000}$ **6.** $-\frac{1}{400}$ **7.** $-\frac{1}{243}$

8. 1 **9.** 64 **10.** -64 **11.** -32 **12.** $-\frac{1}{36}$

13. $\frac{3}{2}$ **14.** $\frac{125}{64}$ **15.** 1 **16.** $\frac{64}{25}$ **17.** 1 **18.** $\frac{1}{4}$

19. 729 **20.** $\frac{1}{125}$ **21.** 100,000 **22.** 10 **23.** $\frac{1}{16}$

24. 125 **25.** $\frac{1}{64}$ **26.** 16 **27.** $\frac{3}{16}$ **28.** $\frac{35}{9}$

29. $\frac{64}{121}$ **30.** -6 **31.** $\frac{16}{15}$ **32.** 1 **33.** y^2

34. $\frac{1}{x^7}$ **35.** z^2 **36.** $\frac{1}{t^7}$ **37.** x^6 **38.** $\frac{y}{x^3}$

39. a **40.** $\frac{2}{5u}$ **41.** t^2 **42.** 1 **43.** $\frac{1}{4x^4}$

44. $\frac{a^6}{64b^9}$ **45.** $-\frac{12}{xy^3}$ **46.** $\frac{3s^3}{10t^4}$ **47.** $\frac{y^4}{9x^4}$

48. $-\frac{y^9}{64z^3}$ **49.** $\frac{3x^5}{y^4}$ **50.** $\frac{1}{y^2}$ **51.** $\frac{81v^8}{u^6}$

52. $\frac{3x}{25y^3}$ **53.** $\frac{x^{12}y^8}{16}$ **54.** $\frac{64y^6}{x^{12}}$ **55.** $36m^5$

56. $\frac{x^2}{9z^4}$ **57.** $125x^3y^{27}$ **58.** 1 **59.** $\frac{10}{x}$ **60.** $\frac{z^2}{16}$

61. $\frac{b^5}{a^5}$ **62.** 1 **63.** x^8y^{12} **64.** $\frac{1}{a^4b^4}$ **65.** $\frac{2b^{10}}{25a^{12}}$

66. $\frac{x^3y}{10}$ **67.** $\frac{1}{2x^8y^3}$ **68.** $3x^5y^4$ **69.** $4x^8$

70. $36\pi x^8$ **71.** 3.6×10^6 **72.** 9.81×10^7

73. 3.81×10^{-3} **74.** 7.384×10^{-4} **75.** 5.75×10^7

76. 1.394×10^8 **77.** 9.461×10^{15} **78.** 1×10^{-7}

79. 8.99×10^{-5} **80.** 3.937×10^{-5} **81.** 60,000,000

82. 5,050,000,000,000 **83.** 0.0000001359

84. 0.0000000086 **85.** \$93,600,000,000

86. 350,000,000 **87.** 13,000,000 **88.** 0.000000009

89. 0.00000000048 **90.** 0.0009 **91.** 6.8×10^5

92. 1.25×10^{14} **93.** 4×10^3 **94.** 5×10^{-6}

95. 9×10^{15} **96.** 1.6×10^{12} **97.** 1.3×10^{11}

98. 1.25×10^{26} **99.** 6×10^6 **100.** 6×10^{14}

101. 4.70×10^{11} **102.** 3.68×10^{34} **103.** 3.46×10^{10}

104. 2.74×10^{20} **105.** 1.58×10^{-5} yr ≈ 8.3 min

106. 8.505×10^{17} m or \$850,500,000,000,000,000

107. 3.33×10^5 or 333,000 **108.** \$19,699.25

Section Project 5.1

$k \approx 1$ for each planet.

Section 5.2 (page 366)

1. 7 **2.** 24.5 **3.** 4.2 **4.** 6

5. Square root **6.** Cube root

7. If a and b are real numbers, n is an integer greater than or equal to 2, and $a = b^n$, then b is an nth root of a.

8. No. $\sqrt{2}$ is an irrational number. Its decimal representation is a nonterminating, nonrepeating decimal.

9. 8 **10.** -10 **11.** Not a real number **12.** 12

13. 9 **14.** Not a real number **15.** $-\frac{2}{3}$ **16.** $\frac{3}{4}$

17. 0.3 **18.** 0.6 **19.** 0.4 **20.** -0.03 **21.** 13

22. 5 **23.** 5 **24.** -2 **25.** 10 **26.** 4

27. $-\frac{1}{4}$ **28.** -0.2 **29.** 3 **30.** 2

31. Not a real number **32.** $-\frac{1}{5}$ **33.** -0.3 **34.** 2

35. Irrational **36.** Rational **37.** Rational

38. Irrational **39.** $16^{1/2} = 4$ **40.** $81^{1/4} = 3$

41. $27^{2/3} = 9$ **42.** $\sqrt[3]{125} = 5$ **43.** $\sqrt[4]{256^3} = 64$

44. $\sqrt[3]{27^2} = 9$ **45.** 5 **46.** 7 **47.** -6

48. -11 **49.** 8 **50.** $-\frac{1}{25}$ **51.** $\frac{1}{4}$ **52.** $\frac{1}{27}$

53. $\frac{4}{9}$ **54.** $\frac{4}{5}$ **55.** $\frac{3}{11}$ **56.** $\frac{10,000}{81}$ **57.** 8.5440

58. Not a real number **59.** 1.0420 **60.** 5.1701

61. 0.0038 **62.** 97.4503 **63.** 4.3004 **64.** 5.4175

65. 66.7213 **66.** -8.4419 **67.** $|t|$ **68.** z **69.** y^3

70. a^2 **71.** t^2 **72.** $|z|$ **73.** x^4 **74.** y^3 **75.** 3

76. $\sqrt[3]{2}$ **77.** $\frac{1}{2}$ **78.** $\frac{1}{5^{7/4}}$ **79.** $\frac{4}{9}$ **80.** $4^{3/4}$

81. x^3 **82.** $z^{1/5}$ **83.** $\frac{9y^{3/2}}{x^{2/3}}$ **84.** $\frac{-8u^{9/5}}{v^{3/5}}$

85. $\frac{3y^2}{4z^{4/3}}$ **86.** $\frac{1}{a^{5/4}}$ **87.** $x^{1/4}$ **88.** $\frac{9m^{1/3}n^2}{16}$

89. $c^{1/2}$ **90.** $\frac{1}{k^{1/2}}$ **91.** $xy^{1/3}$ **92.** $\frac{2}{x}$ **93.** $\sqrt[8]{y}$

94. $\sqrt[6]{2x}$ **95.** All real numbers $x \geq 0$

96. All real numbers $x \geq 0$ **97.** All real numbers $x > 0$

98. All real numbers $x \neq 0$

99. **100.**

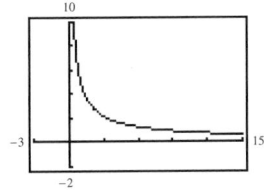

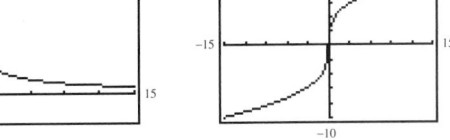

Domain: $(0, \infty)$ Domain: $(-\infty, \infty)$

101.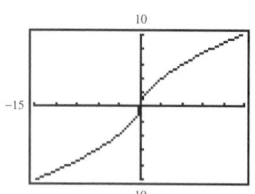

102.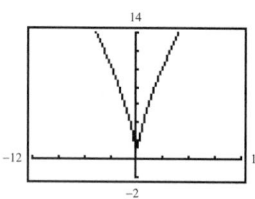

Domain: $(-\infty, \infty)$ Domain: $(-\infty, \infty)$

103. $2x^{3/2} - 3x^{1/2}$ **104.** $3x^{10/3} - 4x^{7/3} + 5x^{4/3}$

105. $1 + 5y$ **106.** $x - 9$ **107.** 12.8%

108. 14.9%   **109.** 13 in. × 13 in. × 13 in.

110. 23 ft × 23 ft   **111.** 0.026 in.   **112.** 1, 4, 5, 6, 9; No

113. $x < 0$. If $x < 0$, then $\sqrt{x^2} = -x$.

Section Project 5.2

(a) 1   **(b)** 1   **(c)** They both approach 1.

Section 5.3   (page 375)

1. $\sqrt{30}$ **2.** $\sqrt[5]{171}$ **3.** $\sqrt[3]{110}$ **4.** $\sqrt[4]{105}$

5. $\sqrt{\frac{15}{31}}$ **6.** $\sqrt[4]{\frac{85}{9}}$ **7.** $\sqrt[5]{\frac{152}{3}}$ **8.** $\sqrt[4]{\frac{633}{5}}$

9. $3\sqrt{35}$ **10.** $3\sqrt[3]{4}$ **11.** $3\sqrt[4]{11}$ **12.** $10\sqrt[5]{3}$

13. $\dfrac{\sqrt{35}}{3}$ **14.** $\dfrac{\sqrt[4]{165}}{2}$ **15.** $\dfrac{\sqrt[3]{11}}{10}$ **16.** $\dfrac{\sqrt[5]{2}}{3}$

17. $2\sqrt{5}$ **18.** $5\sqrt{2}$ **19.** $3\sqrt{3}$ **20.** $5\sqrt{5}$

21. 0.2   **22.** 0.5   **23.** $2\sqrt[3]{3}$ **24.** $3\sqrt[3]{2}$

25. $10\sqrt[4]{3}$ **26.** $2\sqrt[5]{3}$ **27.** $\dfrac{\sqrt{15}}{2}$ **28.** $\dfrac{\sqrt{5}}{6}$

29. $\dfrac{\sqrt[3]{35}}{4}$ **30.** $\dfrac{\sqrt[4]{5}}{2}$ **31.** $\dfrac{\sqrt[5]{15}}{3}$ **32.** $\dfrac{1}{10}$

33. 0.02   **34.** $10\sqrt{85}$ **35.** $20\sqrt[3]{300}$ **36.** $\frac{1}{10}\sqrt[4]{4.4}$

37. $10\sqrt[5]{3}$ **38.** 0.3   **39.** $3x^2\sqrt{x}$ **40.** $8x\sqrt{x}$

41. $4y^2\sqrt{3}$ **42.** $4\sqrt{2x}$ **43.** $xy\sqrt[3]{x}$ **44.** $ab^2\sqrt[3]{a^2}$

45. $2xy\sqrt[5]{y}$ **46.** $2|uv|\sqrt[4]{8v^3}$ **47.** $\dfrac{\sqrt{13}}{5}$ **48.** $\dfrac{\sqrt{15}}{6}$

49. $\dfrac{2\sqrt[5]{x^2}}{y}$ **50.** $\dfrac{2z\sqrt[3]{2}}{y^2}$ **51.** $\dfrac{3a\sqrt[3]{2a}}{b^3}$ **52.** $\dfrac{\sqrt[4]{3u^2}}{2v^2}$

53. $\dfrac{4a^2\sqrt{2}}{|b|}$ **54.** $\dfrac{3|x|\sqrt{2}}{z^3}$ **55.** $3x^2$ **56.** $2x\sqrt[5]{3}$

57.

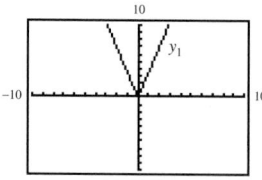 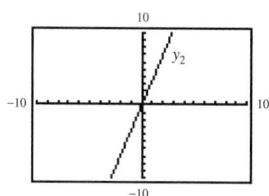

$y_1 \neq y_2$ because the range of y_1 is $[0, \infty)$ and the range of y_2 is $(-\infty, \infty)$.

58.

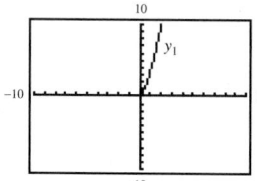 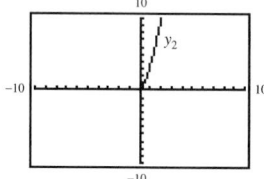

$y_1 = y_2$

59.

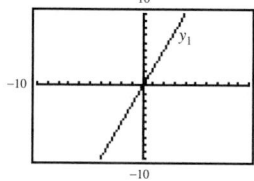 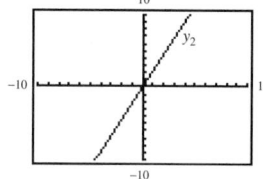

$y_1 = y_2$

60.

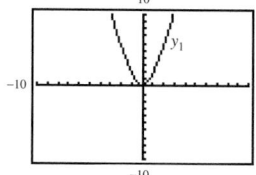 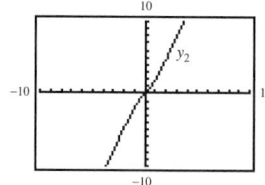

$y_1 \neq y_2$ because the range of y_1 is $[-\infty, 0)$ and the range of y_2 is $(-\infty, \infty)$.

61. $\dfrac{\sqrt{3}}{3}$ **62.** $\dfrac{\sqrt{5}}{5}$ **63.** $4\sqrt{3}$ **64.** $\dfrac{\sqrt{10}}{2}$

65. $\dfrac{\sqrt[4]{20}}{2}$ **66.** $\dfrac{\sqrt[3]{45}}{5}$ **67.** $\dfrac{3\sqrt[3]{2}}{2}$ **68.** $5\sqrt[5]{2}$

69. $\dfrac{\sqrt{y}}{y}$ **70.** $\dfrac{\sqrt{2x}}{2x}$ **71.** $\dfrac{2\sqrt{x}}{x}$ **72.** $\dfrac{\sqrt{5c}}{c}$

73. $\dfrac{2\sqrt{x}}{x^2}$ **74.** $\dfrac{5\sqrt{2x}}{4x^3}$ **75.** $\dfrac{\sqrt[3]{18xy^2}}{3y}$ **76.** $\dfrac{\sqrt[3]{60x^2y}}{3y}$

77. $\dfrac{a^2\sqrt[3]{a^2b}}{b}$ **78.** $\dfrac{3u\sqrt[4]{2u}}{2}$ **79.** $\dfrac{2\sqrt{3b}}{b^2}$ **80.** $\dfrac{\sqrt{xy}}{xy}$

81.   **82.**

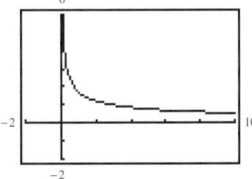   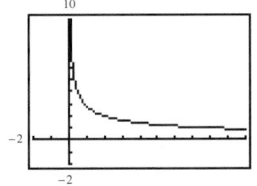

83. $2\sqrt{2}$ **84.** $-\dfrac{4}{5}\sqrt{5}$ **85.** $24\sqrt{2}-6$

86. $44\sqrt{2}$ **87.** $30\sqrt[3]{2}$ **88.** $5\sqrt[4]{3}$ **89.** $12\sqrt{x}$

90. $13\sqrt{x+1}$ **91.** $13\sqrt{y}$ **92.** $-t\sqrt[3]{2t}$

93. $(10-z)\sqrt[3]{z}$ **94.** $(10+6u)\sqrt[3]{3u^2}$

95.   **96.**

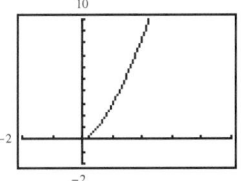   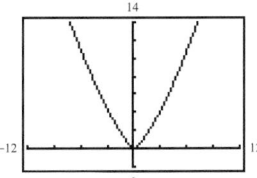

97. $\dfrac{2\sqrt{5}}{5}$ **98.** $\dfrac{3\sqrt{10}}{2}$ **99.** $\dfrac{9\sqrt{5}}{5}$ **100.** $\dfrac{10\sqrt{3x}}{3}$

101. $\sqrt{7}+\sqrt{18}>\sqrt{7+18}$

102. $\sqrt{10}-\sqrt{6}<\sqrt{10-6}$ **103.** $5>\sqrt{3^2+2^2}$

104. $5=\sqrt{3^2+4^2}$ **105.** $3\sqrt{5}$ **106.** $2\sqrt{13}$

107. $8+8\sqrt{2}$ ft   **108.** 89.44 cps

109. 2.22 sec   **110.** No. $\dfrac{5\sqrt{3}}{3}\neq\dfrac{25}{3}$

Section Project 5.3

(a) Answers may vary.   **(b)** $\sqrt{2}$, $\sqrt{3}$, $\sqrt{4}$, $\sqrt{5}$, $\sqrt{6}$, $\sqrt{7}$

(c) $r_n=\sqrt{n+1}$

Mid-Chapter Quiz (page 378)

1. $-\dfrac{1}{144}$ **2.** $\dfrac{64}{27}$ **3.** $\dfrac{5}{3}$ **4.** 16   **5.** $3t^{3/2}$

6. $\dfrac{x^3}{8}$ **7.** $\dfrac{2}{3u^3}$ **8.** $\dfrac{3x^6}{16y^3}$

9. (a) 1.34×10^7 (b) 7.5×10^{-4}

10. (a) 8.1×10^{13} (b) 2×10^{-4}

11. (a) $5\sqrt{6}$ (b) $3\sqrt[3]{2}$ **12.** (a) $3|x|\sqrt{3}$ (b) $3|x|\sqrt{x}$

13. (a) $\dfrac{\sqrt[4]{5}}{2}$ (b) $\dfrac{2\sqrt{6}}{7}$ **14.** (a) $\dfrac{2u\sqrt{10u}}{3}$ (b) $\dfrac{2\sqrt[3]{2}}{u^4}$

15. (a) $\dfrac{\sqrt{6}}{3}$ (b) $4\sqrt{3}$ **16.** (a) $\dfrac{2\sqrt{5x}}{x}$ (b) $\dfrac{\sqrt[3]{12a^2}}{2a}$

17. $4\sqrt{2y}$ **18.** $6x\sqrt[3]{5x^2}+4x\sqrt[3]{5x}$

19. $\sqrt{5^2+12^2}=\sqrt{169}=13$ **20.** $\left(23+8\sqrt{2}\right)$ in.

Section 5.4 (page 383)

1. 4   **2.** $6\sqrt{3}$ **3.** $3\sqrt{2}$ **4.** $5\sqrt{2}$

5. $2\sqrt{5}-\sqrt{15}$ **6.** $\sqrt{55}-3\sqrt{11}$

7. $2\sqrt{10}+8\sqrt{2}$ **8.** $7\sqrt{2}+3\sqrt{7}$ **9.** -1

10. 6   **11.** 4   **12.** 2   **13.** 4   **14.** 46

15. $8\sqrt{5}+24$ **16.** $36-16\sqrt{5}$

17. $2x+20\sqrt{2x}+100$ **18.** $25-10\sqrt{3v}+3v$

19. $y+4\sqrt{y}$ **20.** $5\sqrt{x}-x$

21. $\sqrt{15}+3\sqrt{3}-5\sqrt{5}-15$

22. $2\sqrt{15}+6\sqrt{30}+6\sqrt{2}+36$

23. $45x-17\sqrt{x}-6$ **24.** $16u-19\sqrt{u}+3$

25. $x-y$ **26.** $9u-3v$ **27.** $2-7\sqrt[3]{4}$

28. $\sqrt[3]{45}-5\sqrt[3]{9}+5\sqrt[3]{5}-25$

29. $\sqrt[3]{4x^2}+10\sqrt[3]{2x}+25$

30. $y-5\sqrt[3]{y}+2\sqrt[3]{y^2}-10$

31. $2y-10\sqrt[3]{2y}+10\sqrt[3]{4y^2}-100$

32. $t+5\sqrt[3]{t^2}+\sqrt[3]{t}-3$ **33.** $(x+3)$

34. $(1-x)$ **35.** $(4-3x)$ **36.** $(5+4y)$

37. $\left(2u+\sqrt{2u}\right)$ **38.** $\left(3s-\sqrt{2}\right)$ **39.** $\dfrac{1-2\sqrt{x}}{3}$

40. $\dfrac{-1+9\sqrt{2y}}{6}$ **41.** $\dfrac{-1+\sqrt{3y}}{4}$ **42.** $-\dfrac{t+\sqrt{2t}}{3}$

43. (a) $2\sqrt{3}-4$ (b) 0   **44.** (a) 0   (b) $43-32\sqrt{2}$

45. (a) 0   (b) -1 **46.** (a) $3-2\sqrt{5}$ (b) 0

47. $2-\sqrt{5},-1$ **48.** $\sqrt{2}+9,-79$

49. $\sqrt{11}+\sqrt{3},8$ **50.** $\sqrt{10}-\sqrt{7},3$

51. $\sqrt{x}+3,x-9$ **52.** $\sqrt{t}-7,t-49$

53. $\sqrt{2u}+\sqrt{3},2u-3$ **54.** $\sqrt{5a}-\sqrt{2},5a-2$

55. $\dfrac{\sqrt{22}+2}{3}$ **56.** $\dfrac{2\sqrt{10}+5}{5}$ **57.** $-4\sqrt{7}+12$

58. $\dfrac{15-5\sqrt{5}}{2}$ **59.** $\dfrac{6-\sqrt{2}}{17}$ **60.** $15+\sqrt{5}$

61. $\dfrac{4\sqrt{7}+11}{3}$ **62.** $\dfrac{12-7\sqrt{3}}{3}$ **63.** $\dfrac{x\sqrt{15}+x\sqrt{3}}{4}$

64. $\dfrac{6(y+1)\left(y^2-\sqrt{y}\right)}{y(y^3-1)}$ **65.** $\dfrac{t\sqrt{5t}+t\sqrt{t}}{2}$

66. $\dfrac{5x\sqrt{x}+5x\sqrt{2}}{x-2}$ **67.** $\dfrac{2x-9\sqrt{x}-5}{4x-1}$

68. $\dfrac{4t+4\sqrt{t}+1}{4t-1}$ **69.** $-\dfrac{\sqrt{u+v}\left(\sqrt{u-v}+\sqrt{u}\right)}{v}$

70. $\sqrt{u+z}+\sqrt{u}$

71. 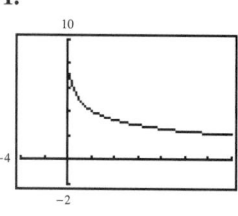   **72.**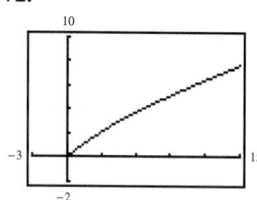

73. 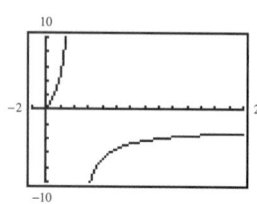   **74.**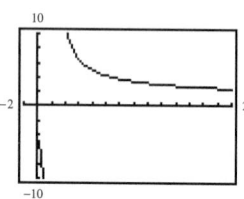

75. $192\sqrt{2}$ in.²   **76.** $\dfrac{\sqrt{3}}{2}$ **77.** $\dfrac{500k\sqrt{k^2+1}}{k^2+1}$

Section Project 5.4

(a) $\dfrac{2}{7\sqrt{2}}$ **(b)** $\dfrac{4}{5\left(\sqrt{7}-\sqrt{3}\right)}$ **(c)** $\dfrac{x-36}{\sqrt{2}\left(\sqrt{x}-6\right)}$

Section 5.5   (page 391)

1. (a) Not a solution
  (b) Not a solution
  (c) Not a solution
  (d) Solution
3. (a) Not a solution
  (b) Solution
  (c) Not a solution
  (d) Not a solution

2. (a) Not a solution
  (b) Not a solution
  (c) Solution
  (d) Not a solution
4. (a) Solution
  (b) Not a solution
  (c) Not a solution
  (d) Not a solution

5. 400   **6.** 25   **7.** 49   **8.** 169   **9.** No solution
10. No solution   **11.** 525   **12.** 157   **13.** 90
14. $\dfrac{9}{2}$ **15.** $\dfrac{44}{3}$ **16.** $\dfrac{11}{5}$ **17.** $\dfrac{14}{25}$ **18.** $\dfrac{33}{4}$
19. $-\dfrac{2}{3}$ **20.** 2   **21.** 8   **22.** 9   **23.** No solution
24. No solution   **25.** 4   **26.** 7   **27.** No solution
28. 5   **29.** 7   **30.** $\dfrac{78}{23}$ **31.** -15 **32.** No solution
33. 1, 3   **34.** $-1, -2$ **35.** $\dfrac{4}{5}$ **36.** No solution
37. $\dfrac{1}{2}$ **38.** $\dfrac{13}{4}$

39. 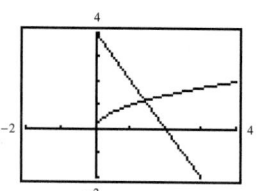   **40.**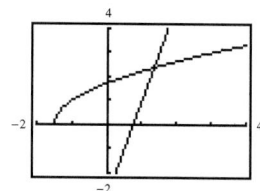

1.407   1.347

41. 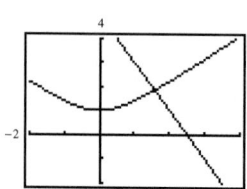   **42.**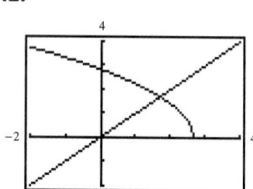

1.569   1.702

43. 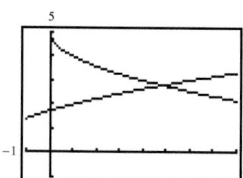   **44.**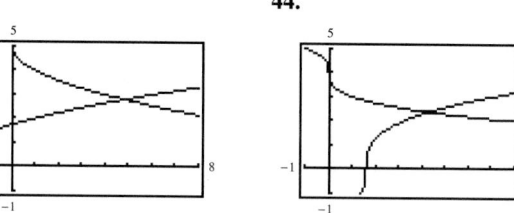

4.840   4.283

45. 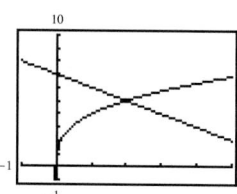   **46.**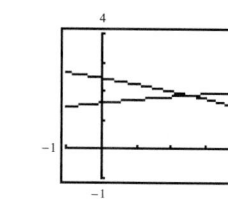

1.978   2.513

47. 1.82 ft   **48.** 0.46 ft   **49.** $h^2 = \dfrac{S^2 - \pi^2 r^4}{\pi^2 r^2}$
50. 26,250 passengers   **51.** 64 ft   **52.** 144 ft
53. 56.57 ft/sec   **54.** 113.14 ft/sec   **55.** 56.25 ft
56. 225 ft   **57.** (c)   **58.** (a)   **59.** (d)   **60.** (b)
61. 500 units

Section Project 5.5

(a) 1980 (b)

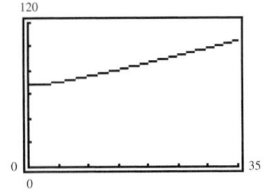

Section 5.6 (page 399)

1. $2i$ **2.** $3i$ **3.** $-12i$ **4.** $7i$ **5.** $\frac{2}{5}i$

6. $-\frac{6}{11}i$ **7.** $0.3i$ **8.** $0.02i$ **9.** $2\sqrt{2}i$

10. $5\sqrt{3}i$ **11.** $3\sqrt{3}i$ **12.** $\frac{2\sqrt{2}i}{5}$ **13.** $\sqrt{7}i$

14. $\sqrt{15}i$ **15.** $10i$ **16.** $-i$ **17.** $3\sqrt{2}i$

18. $13\sqrt{5}i$ **19.** -4 **20.** $-5\sqrt{6}$ **21.** $-3\sqrt{6}$

22. -0.44 **23.** $-2\sqrt{3}-3$ **24.** 6

25. $5\sqrt{2}-4\sqrt{5}$ **26.** $-\frac{7}{3}\sqrt{6}$ **27.** -16 **28.** -2

29. $a=3, b=-4$ **30.** $a=-8, b=6$

31. $a=2, b=-3$ **32.** $a=-5, b=3$

33. $a=-4, b=-2\sqrt{2}$ **34.** $a=-3, b=6$

35. $a=2, b=-2$ **36.** $a=2, b=\frac{9}{2}$ **37.** $10+4i$

38. $-6-5i$ **39.** $-14-40i$ **40.** 13

41. $-14+20i$ **42.** $-70i$ **43.** $9-7i$

44. $18+31i$ **45.** $\frac{13}{6}+\frac{3}{2}i$ **46.** $-6.15-9.3i$

47. $-3+49i$ **48.** $\left(-1-\sqrt{2}\right)+\left(1-\sqrt{2}\right)i$

49. -36 **50.** 20 **51.** $-36i$ **52.** $180i$

53. $27i$ **54.** -64 **55.** $-65-10i$ **56.** $80-60i$

57. $20-12i$ **58.** $-45-30i$ **59.** $-40-5i$

60. $-69+55i$ **61.** $-14+42i$ **62.** $81-35i$

63. $-7-24i$ **64.** $48+14i$ **65.** $-21+20i$

66. $55-48i$ **67.** 9 **68.** $-12+3i$ **69.** $2+11i$

70. $-9-46i$ **71.** 5 **72.** 13 **73.** 31 **74.** 109

75. 68 **76.** 18 **77.** 100 **78.** 400 **79.** 4

80. 14 **81.** 2.5 **82.** 10.28 **83.** $2+2i$

84. $6-2i$ **85.** $-\frac{24}{53}+\frac{84}{53}i$ **86.** $\frac{3}{2}+\frac{5}{2}i$ **87.** $-10i$

88. $\frac{1}{3}-\frac{1}{3}i$ **89.** $-\frac{6}{5}+\frac{2}{5}i$ **90.** $\frac{15}{4}+\frac{15}{4}i$

91. $\frac{8}{5}-\frac{1}{5}i$ **92.** $-\frac{9}{41}-\frac{40}{41}i$ **93.** $1-\frac{6}{5}i$

94. $\frac{47}{26}+\frac{27}{26}i$ **95.** $-\frac{53}{25}+\frac{29}{25}i$ **96.** $\frac{14}{29}-\frac{35}{29}i$

97. (a) Solution (b) Solution

98. (a) Solution (b) Solution

99. (a) Solution (b) Solution

100. (a) Solution (b) Solution

101. $2a$ **102.** a^2+b^2 **103.** $2bi$ **104.** $2a^2-2b^2$

105. The square of any real number is nonnegative.

106. $\sqrt{-3}\,\sqrt{-3}=\left(\sqrt{3}i\right)\left(\sqrt{3}i\right)$
$=3i^2=-3$

Section Project 5.6

(a) (i) $\left(\dfrac{-5+5\sqrt{3}i}{2}\right)^3=125$ (ii) $\left(\dfrac{-5-5\sqrt{3}i}{2}\right)^3=125$

(b) (i) $\left(\dfrac{-3+3\sqrt{3}i}{2}\right)^3=27$ (ii) $\left(\dfrac{-3-3\sqrt{3}i}{2}\right)^3=27$

(c) (i) $1, \dfrac{-1+\sqrt{3}i}{2}, \dfrac{-1-\sqrt{3}i}{2}$

(ii) $2, \dfrac{-2+2\sqrt{3}i}{2}=-1+\sqrt{3}i, \dfrac{-2-2\sqrt{3}i}{2}$
$=-1-\sqrt{3}i$

(iii) $4, \dfrac{-4+4\sqrt{3}i}{2}=-2+2\sqrt{3}i, \dfrac{-4-4\sqrt{3}i}{2}$
$=-2-2\sqrt{3}i$

Review Exercises (page 402)

1. $\frac{1}{72}$ **2.** $\frac{16}{625}$ **3.** $\frac{125}{8}$ **4.** 81 **5.** 3.6×10^7

6. 2.4×10^5 **7.** 500 **8.** $\frac{250{,}000}{9}$ **9.** $2x$

10. $-108x^3$ **11.** $\frac{x^6}{y^8}$ **12.** $5y$ **13.** $\frac{1}{t^3}$ **14.** a^4

15. $\frac{27}{y^3}$ **16.** 1 **17.** 1.2 **18.** 0.4 **19.** $\frac{5}{6}$

20. $-\frac{8}{15}$ **21.** 12 **22.** 5 **23.** $11{,}414.13$

24. $18{,}380.16$ **25.** 10.63 **26.** 0.06 **27.** $49^{1/2}=7$

28. $0.125^{1/3}=0.5$ **29.** $\sqrt[3]{216}=6$ **30.** $\sqrt[4]{16}=2$

31. 81 **32.** 8 **33.** 125 **34.** $\frac{1}{9}$ **35.** 0.04

36. $32{,}554.94$ **37.** $x^{7/12}$ **38.** $4y$ **39.** $\frac{3}{x^{1/4}y^{2/5}}$

40. $\frac{24}{7}a^5b^3$ **41.** $6\sqrt{10}$ **42.** $\frac{5}{3}\sqrt{2}$ **43.** $0.5x^2\sqrt{y}$

44. $0.4s^3t\sqrt{t}$ **45.** $2ab\sqrt[3]{6b}$ **46.** $2uv\sqrt[4]{2v}$

47. $\frac{\sqrt{30}}{6}$ **48.** $\frac{\sqrt{15}}{10}$ **49.** $\frac{\sqrt{3x}}{2x}$ **50.** $\frac{2y\sqrt{10z}}{5z}$

51. $\frac{\sqrt[3]{4x^2}}{x}$ **52.** $\frac{2\sqrt[3]{2st}}{s}$ **53.** $\frac{7+\sqrt{7}}{7}$

54. $\frac{x\left(\sqrt{x}-1\right)}{x-1}$ **55.** $-24\sqrt{10}$ **56.** $35\sqrt{2}+4\sqrt{3}$

57. $7\sqrt[4]{y}+3$ **58.** $12\sqrt{x}-2\sqrt[3]{x}$ **59.** $12\sqrt{5}+41$

60. $3-x$ **61.** $6\sqrt{2}+7\sqrt{6}-4\sqrt{3}-14$

62. $\frac{15\sqrt{x}-45}{x-9}$ **63.** $\frac{x+20\sqrt{x}+100}{x-100}$

64. $\frac{3s-2\sqrt{s}-8}{s-4}$

65. 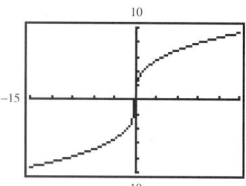   **66.**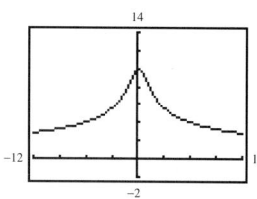

67. 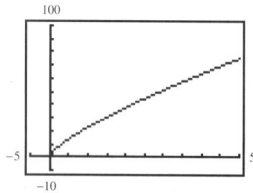   **68.**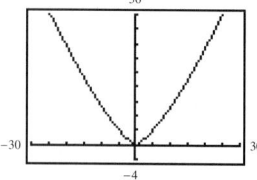

69. 225   **70.** No solution   **71.** 105   **72.** $\frac{6}{5}$

73. $-3, -5$ **74.** $\frac{9}{5}$ **75.** 5   **76.** 2   **77.** $\frac{3}{32}$

78. No solution   **79.** 1.978   **80.** 3.240   **81.** $4\sqrt{3}i$

82. $0.4i$ **83.** $10 - 9\sqrt{3}i$ **84.** $3 + 20\sqrt{5}i$

85. $\frac{3}{4} - \sqrt{3}i$

86. $-0.5 + 3.3i$ **87.** $8 - 3i$

88. $20 - 28i$ **89.** -90 **90.** $-70 - 40i$ **91.** 25

92. $11 - 60i$ **93.** $59 + 74i$ **94.** $-\frac{4}{5}i$

95. $\frac{9}{17} + \frac{2}{17}i$ **96.** $-\frac{7}{82} + \frac{19}{82}i$ **97.** $21 + 12\sqrt{2}$

98. 24 in. × 18 in.   **99.** 1.37 ft   **100.** 256 ft

Chapter 5 Test   (page 404)

1. (a) $\frac{3}{8}$ (b) 3.0×10^{-5} **2.** (a) $\frac{1}{9}$ (b) 6

3. 3.2×10^{-5} **4.** 30,400,000   **5.** (a) $\dfrac{3}{5t}$ (b) y^2

6. (a) $x^{1/3}$ (b) 25   **7.** (a) $\frac{4}{3}\sqrt{2}$ (b) $2\sqrt[3]{3}$

8. (a) $\dfrac{5\left(\sqrt{6} + \sqrt{2}\right)}{2}$ (b) $\dfrac{2\sqrt[3]{3y^2}}{3y}$

To multiply the numerator and denominator of a fraction by a factor so that no radical contains a fraction and no denominator of a fraction contains a radical.

9. $-10\sqrt{3x}$ **10.** $5\sqrt{3x} + 3\sqrt{5}$

11. $16 - 8\sqrt{2x} + 2x$ **12.** No solution   **13.** 9

14. 29   **15.** $2 - 2i$ **16.** $-5 - 12i$ **17.** $-8 + 4i$

18. $13 + 13i$ **19.** $\frac{13}{10} - \frac{11}{10}i$ **20.** 100 ft

Cumulative Test: Chapters 1–5   (page 405)

1. $y \le 45$ **2.** $x \ge 15$ **3.** Additive Identity Property

4. Distributive Property   **5.** $9.35 + 0.75q$ **6.** $4n$

7. $(4x + 11)(4x - 11)$ **8.** $(3t - 4)^2$

9. $(x - 10)(x - 4)$ **10.** $4x(x^2 - 3x + 4)$ **11.** $\frac{8}{3}$

12. 3   **13.** $-\frac{2}{3}, 5$ **14.** $-5, 8$ **15.** $-8, 3$ **16.** 41

17. 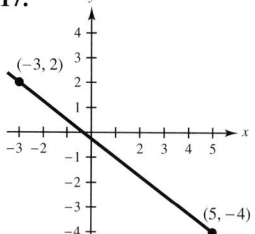   **18.**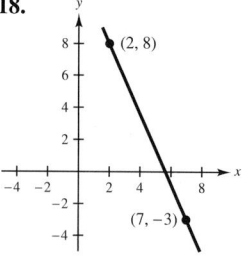

$m = -\frac{3}{4}$ $m = -\frac{11}{5}$

19. $-t^2 - t$ **20.** $d = 48t$

21. (a) $g(x) = -\sqrt{x}$ (b) $g(x) = \sqrt{x} + 2$
(c) $g(x) = \sqrt{x - 2}$

22. \$108.50   **23.** $\left(\frac{2}{5}, \frac{8}{5}\right)$ **24.** $(1, 2, 1)$

25. (a) $y = -\frac{51}{59}x + \frac{226}{59}$ (b)

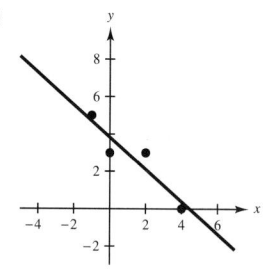

26. 58   **27.** $2|x|y\sqrt{6y}$ **28.** $3\sqrt[3]{5}$

29. $\dfrac{2\sqrt{3}b^3}{a^2}$ **30.** 2

CHAPTER 6

Section 6.1   (page 416)

1. 5, 7   **2.** $-11, -4$ **3.** $-9, 8$ **4.** $-6, 8$

5. $-9, 5$ **6.** $-2, 9$ **7.** 0, 3   **8.** 0, 3   **9.** 9, 12

10. $8, \frac{3}{4}$ **11.** $\pm\frac{5}{2}$ **12.** $\pm\frac{11}{4}$ **13.** 6   **14.** $-\frac{4}{3}$

15. -30 **16.** $\frac{1}{2}, \frac{3}{4}$ **17.** 1, 6   **18.** $-1, -4$

19. $\frac{1}{2}, -\frac{5}{6}$ **20.** $\frac{3}{8}, -1$ **21.** $\frac{5}{3}, 6$ **22.** $\frac{3}{2}, 4$

23. ± 3 **24.** ± 5 **25.** ± 8 **26.** ± 12

27. Factoring and the Zero-Factor Property allow you to solve a quadratic equation by converting it into two linear equations that you already know how to solve.

28. False. The solutions are $x = 5$ and $x = -5$.

29. ± 8 **30.** ± 13 **31.** $\pm\frac{4}{5}$ **32.** $\pm\frac{11}{3}$ **33.** $\pm\frac{15}{2}$

34. $\pm\frac{1}{4}$ **35.** 9, -17 **36.** 45, -5 **37.** 2.5, 3.5

38. $-1.1, -2.9$ **39.** $2 \pm \sqrt{7}$ **40.** $-8 \pm 2\sqrt{7}$

41. $\dfrac{-1 \pm 5\sqrt{2}}{2}$ **42.** $\dfrac{5 \pm 4\sqrt{3}}{3}$ **43.** $\dfrac{3 \pm 7\sqrt{2}}{4}$

44. $\dfrac{-11 \pm 10\sqrt{3}}{5}$ **45.** $\pm 6i$ **46.** $\pm 3i$ **47.** $\pm 2i$

48. $\pm 4i$ **49.** $3 \pm 5i$ **50.** $-5 \pm 9i$ **51.** $-\dfrac{4}{3} \pm 4i$

52. $\dfrac{3}{2} \pm \dfrac{5}{2}i$ **53.** $-6 \pm \dfrac{11}{3}i$ **54.** $4 \pm \dfrac{13}{2}i$

55. $1 \pm 3\sqrt{3}i$ **56.** $-\dfrac{3}{2} \pm \dfrac{3}{2}\sqrt{6}i$ **57.** $-1 \pm 0.2i$

58. $3 \pm 1.5i$ **59.** $\dfrac{2}{3} \pm \dfrac{1}{3}i$ **60.** $-\dfrac{5}{8} \pm \dfrac{7}{4}i$

61. $-\dfrac{7}{3} \pm \dfrac{\sqrt{38}}{3}i$ **62.** $\dfrac{5}{6} \pm \dfrac{2}{5}\sqrt{5}i$

63.
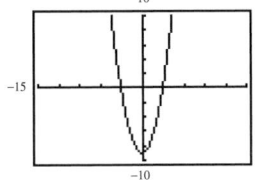
$(-3, 0), (3, 0)$

64.
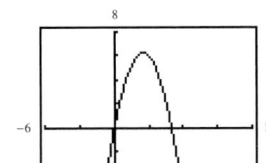
$(0, 0), (5, 0)$

65.
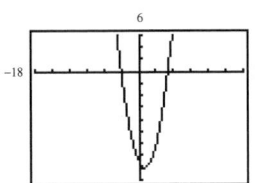
$(-3, 0), (5, 0)$

66.
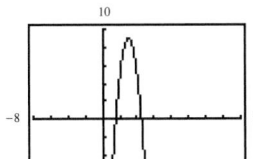
$\left(\dfrac{3}{2}, 0\right), \left(\dfrac{9}{2}, 0\right)$

67.
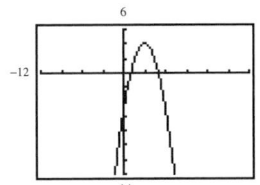
$(1, 0), (5, 0)$

68.
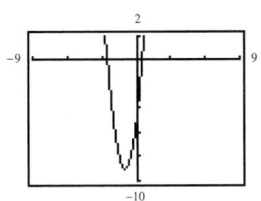
$\left(-\dfrac{5}{2}, 0\right), \left(\dfrac{1}{2}, 0\right)$

69.
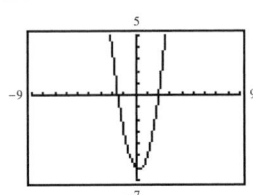
$(2, 0), \left(-\dfrac{3}{2}, 0\right)$

70.
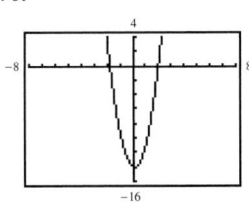
$\left(-\dfrac{7}{4}, 0\right), (2, 0)$

71.
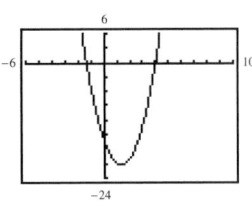
$\left(-\dfrac{4}{3}, 0\right), (4, 0)$

72.
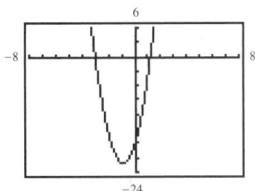
$(-3, 0), \left(\dfrac{6}{5}, 0\right)$

73.
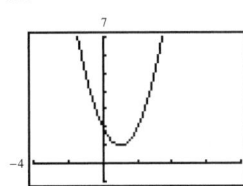
$1 \pm i$

74.
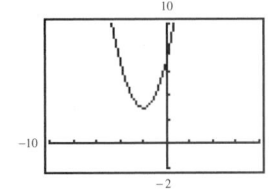
$-2 \pm \sqrt{3}i$

75.
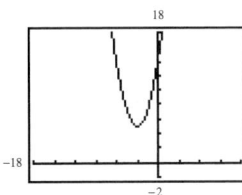
$-3 \pm \sqrt{5}i$

76.
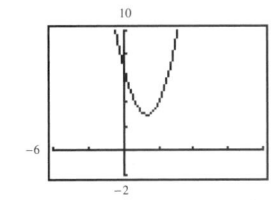
$2 \pm \sqrt{3}i$

77.
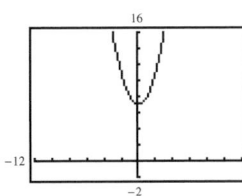
$\pm \sqrt{7}i$

78.
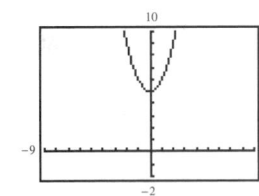
$\pm \sqrt{5}i$

79. $0, \dfrac{5}{2}$ **80.** $-4, \dfrac{4}{3}$ **81.** ± 10 **82.** $-12 \pm 20i$

83. $5 \pm 10i$ **84.** $8, -32$

85. $f(x) = \sqrt{4 - x^2}$
$g(x) = -\sqrt{4 - x^2}$

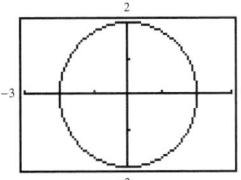

86. $f(x) = \sqrt{x^2 - 4}$
$g(x) = -\sqrt{x^2 - 4}$

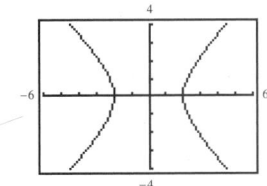

87. $f(x) = \frac{1}{2}\sqrt{4 - x^2}$
$g(x) = -\frac{1}{2}\sqrt{4 - x^2}$

88. $f(x) = \sqrt{x}$
$g(x) = -\sqrt{x}$

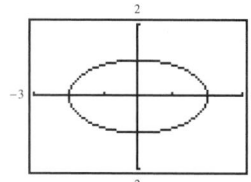
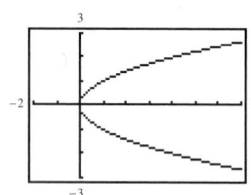

89. $\pm 1, \pm 2$ **90.** $\pm\frac{1}{2}, \pm 5$ **91.** $\pm\sqrt{2}, \pm\sqrt{3}$

92. $\pm\sqrt{5}, \pm\sqrt{6}$ **93.** $\pm 2, \pm i$ **94.** $\pm\sqrt{3}, \pm\sqrt{2}i$

95. $\pm 1, \pm\sqrt{5}$ **96.** $\pm\sqrt{3}, \pm\sqrt{2}i$ **97.** $-8, 27$

98. $-125, 8$ **99.** $1, \frac{125}{8}$ **100.** $-\frac{125}{27}, -1$

101. $1, 32$ **102.** $-243, -32$ **103.** $\frac{1}{32}, 243$

104. $-1, -\frac{1}{32}$ **105.** 10 **106.** 24 **107.** 5.66

108. 3 **109.** $2\sqrt{2834} \approx 106.47$ ft

110. $10\sqrt{17} \approx 41.23$ ft   **111.** $\frac{25\sqrt{2}}{2}$ in. $\times$ $\frac{25\sqrt{2}}{2}$ in.

112. 24 in. $\times$ 10 in.   **113.** $2\sqrt{2} \approx 2.83$ sec

114. $\frac{5\sqrt{5}}{2} \approx 5.59$ sec   **115.** 9 sec

116. 100 units, 140 units   **117.** 1989   **118.** 1992

119. $0.06 = 6\%$

120. Yes. If $(x - 1)^2 = 0$, the only solution is $x = 1$.

Section Project 6.1

(a) $(x - 1)(x - 2) = x^2 - 3x + 2 = 0$

(b) $(x - 5)(x + 2) = x^2 - 3x - 10 = 0$

(c) $(x + 3)(x - \frac{1}{2}) = \frac{1}{2}(2x^2 + 5x - 3) = 0$

(d) $\left[x - \left(1 + \sqrt{2}\right)\right]\left[x - \left(1 - \sqrt{2}\right)\right] = x^2 - 2x - 1 = 0$

(e) $[x - (2 + 5i)][x - (2 - 5i)] = x^2 - 4x + 29 = 0$

(f) A quadratic equation with solutions $x = r_1$ and $x = r_2$ is $(x - r_1)(x - r_2) = 0$.

Section 6.2   (page 425)

1. Yes. $x^2 + 1 = 0$

2. Divide both sides of the equation by the leading coefficient. Dividing both sides of an equation by a nonzero constant yields an equivalent equation.

3. 16   **4.** 36   **5.** 100   **6.** 1   **7.** $\frac{25}{4}$ **8.** $\frac{49}{4}$

9. $\frac{9}{25}$ **10.** $\frac{4}{9}$ **11.** $\frac{9}{100}$ **12.** $\frac{1}{36}$ **13.** 0.04

14. 5.0625   **15.** 0, 20   **16.** 0, -32 **17.** 0, -6

18. 0, 9   **19.** 1, 7   **20.** 2, 6   **21.** 4, -6

22. $-3, -9$ **23.** $-3, -4$ **24.** 2, -5 **25.** $-3, 6$

26. 9, -4 **27.** $\frac{3}{2}, 4$ **28.** $-\frac{1}{3}, 2$

29. $2 + \sqrt{7} \approx 4.65$
$2 - \sqrt{7} \approx -0.65$

30. $3 + \sqrt{2} \approx 4.41$
$3 - \sqrt{2} \approx 1.59$

31. $-2 + \sqrt{7} \approx 0.65$
$-2 - \sqrt{7} \approx -4.65$

32. $-3 + \sqrt{2} \approx -1.59$
$-3 - \sqrt{2} \approx -4.41$

33. $2 + \sqrt{3} \approx 3.73$
$2 - \sqrt{3} \approx 0.27$

34. $5 + 2\sqrt{10} \approx 11.32$
$5 - 2\sqrt{10} \approx -1.32$

35. $-1 + \sqrt{2}i \approx -1 + 1.41i$
$-1 - \sqrt{2}i \approx -1 - 1.41i$

36. $3 + \sqrt{3}i \approx 3 + 1.73i$
$3 - \sqrt{3}i \approx 3 - 1.73i$

37. $5 + 3\sqrt{3} \approx 10.20$
$5 - 3\sqrt{3} \approx -0.20$

38. $-4 + 2\sqrt{5} \approx 0.47$
$-4 - 2\sqrt{5} \approx -8.47$

39. $-10 + 3\sqrt{10} \approx -0.51$
$-10 - 3\sqrt{10} \approx -19.49$

40. $-3 + \sqrt{33} \approx 2.74$
$-3 - \sqrt{33} \approx -8.74$

41. $\frac{1 + 2\sqrt{7}}{3} \approx 2.10$
$\frac{1 - 2\sqrt{7}}{3} \approx -1.43$

42. $\frac{-2 + \sqrt{29}}{5} \approx 0.68$
$\frac{-2 - \sqrt{29}}{5} \approx -1.48$

43. $\frac{-5 + \sqrt{13}}{2} \approx -0.70$
$\frac{-5 - \sqrt{13}}{2} \approx -4.30$

44. $\frac{9 + \sqrt{85}}{2} \approx 9.11$
$\frac{9 - \sqrt{85}}{2} \approx -0.11$

45. $\frac{-3 + \sqrt{17}}{2} \approx 0.56$
$\frac{-3 - \sqrt{17}}{2} \approx -3.56$

46. $\frac{7 + \sqrt{13}}{2} \approx 5.30$
$\frac{7 - \sqrt{13}}{2} \approx 1.70$

47. $\frac{1}{2} + \frac{\sqrt{3}}{2}i \approx 0.5 + 0.87i$
$\frac{1}{2} - \frac{\sqrt{3}}{2}i \approx 0.5 - 0.87i$

48. $\frac{-1 + \sqrt{5}}{2} \approx 0.62$
$\frac{-1 - \sqrt{5}}{2} \approx -1.62$

49. $\frac{-4 + \sqrt{10}}{2} \approx -0.42$
$\frac{-4 - \sqrt{10}}{2} \approx -3.58$

50. $\frac{12 + \sqrt{159}}{3} \approx 8.20$
$\frac{12 - \sqrt{159}}{3} \approx -0.20$

51. $\frac{-9 + \sqrt{21}}{6} \approx -0.74$
$\frac{-9 - \sqrt{21}}{6} \approx -2.26$

52. $\frac{15 + \sqrt{85}}{10} \approx 2.42$
$\frac{15 - \sqrt{85}}{10} \approx 0.58$

53. $\frac{-1 + \sqrt{10}}{2} \approx 1.08$
$\frac{-1 - \sqrt{10}}{2} \approx -2.08$

54. $\frac{3}{8} + \frac{\sqrt{23}}{8}i \approx 0.38 + 0.60i$
$\frac{3}{8} - \frac{\sqrt{23}}{8}i \approx 0.38 - 0.60i$

55. $\dfrac{7 + \sqrt{57}}{2} \approx 7.27$

$\dfrac{7 - \sqrt{57}}{2} \approx -0.27$

56. $\dfrac{-4 + \sqrt{106}}{6} \approx 1.05$

$\dfrac{-4 - \sqrt{106}}{6} \approx -2.38$

57. $-1 + \sqrt{3}\,i \approx -1 + 1.73i$

$-1 - \sqrt{3}\,i \approx -1 - 1.73i$

58. $\dfrac{-5 + \sqrt{17}}{2} \approx -0.44$

$\dfrac{-5 - \sqrt{17}}{2} \approx -4.56$

59. $-1 \pm 2i$

60. $\dfrac{-5 + \sqrt{35}}{2} \approx 0.46$

$\dfrac{-5 - \sqrt{35}}{2} \approx -5.46$

61. $1 \pm \sqrt{3}$ **62.** $4 \pm \sqrt{6}$ **63.** $4 + 2\sqrt{2}$ **64.** 6

65.

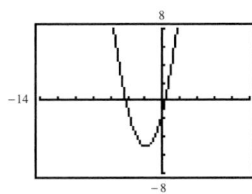

$\left(-2 \pm \sqrt{5},\, 0\right)$

66.

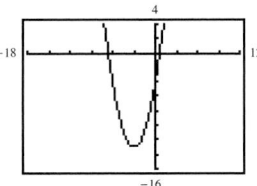

$\left(-3 \pm \sqrt{13},\, 0\right)$

67.

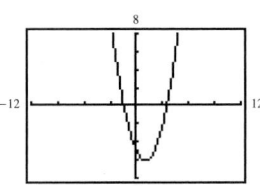

$\left(1 \pm \sqrt{6},\, 0\right)$

68.

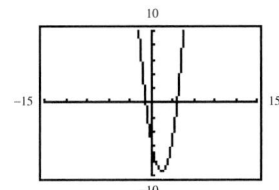

$\left(\dfrac{3 \pm \sqrt{19}}{2},\, 0\right)$

69.

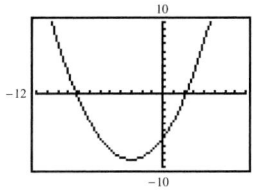

$\left(-3 \pm 3\sqrt{3},\, 0\right)$

70.

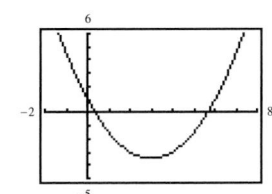

$\left(3 \pm \sqrt{7},\, 0\right)$

71.

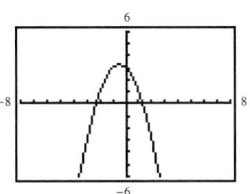

$\left(\dfrac{-1 \pm \sqrt{13}}{2},\, 0\right)$

72.

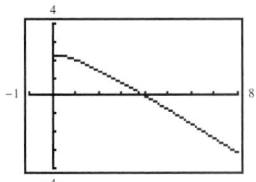

$(4, 0)$

73. **(a)** $x^2 + 8x$ **(b)** $x^2 + 8x + 16$ **(c)** $(x + 4)^2$

74. **(a)** $6x + 9$ **(b)** $x^2 + 6x + 9$ **(c)** $(x + 3)^2$

75. 4 cm, 6 cm **76.** 8 ft $\times$ 20 ft

77. 15 ft $\times$ $46\frac{2}{3}$ ft or 20 ft $\times$ 35 ft

78. 6 in. $\times$ 10 in. $\times$ 14 in. **79.** 73.5 ft

80. 271 m, 129 m **81.** 42 units, 58 units

82. 139 units, 861 units

Section Project 6.2

(a) $d = 0$ **(b)** d is positive and a perfect square.

(c) d is positive and is not a perfect square **(d)** $d < 0$

Section 6.3 (page 433)

1. $2x^2 + 2x - 7 = 0$ **2.** $7x^2 + 15x - 5 = 0$

3. $-x^2 + 10x - 5 = 0$ **4.** $3x^2 + 8x - 15 = 0$

5. 4, 7 **6.** 9, 3 **7.** $-2, -4$ **8.** $-2, -7$

9. $-\frac{1}{2}$ **10.** $-\frac{2}{3}$ **11.** $-\frac{3}{2}$ **12.** $\frac{5}{3}$ **13.** $-\frac{1}{2}, \frac{2}{3}$

14. $\frac{3}{5}, \frac{1}{2}$ **15.** $-15, 20$ **16.** $-30, 10$

17. Two distinct imaginary solutions

18. Two distinct irrational solutions

19. Two distinct irrational solutions

20. Two distinct imaginary solutions

21. Two distinct imaginary solutions

22. Two distinct rational solutions

23. One (repeated) rational solution

24. Two distinct irrational solutions

25. Two distinct imaginary solutions

26. One (repeated) rational solution

27. $1 \pm \sqrt{5}$ **28.** $1 \pm \sqrt{7}$ **29.** $-2 \pm \sqrt{3}$

30. $-3 \pm \sqrt{5}$ **31.** $-3 \pm 2\sqrt{3}$ **32.** $-4 \pm 2\sqrt{5}$

33. $5 \pm \sqrt{2}$ **34.** $6 \pm \sqrt{7}$ **35.** $-\dfrac{3}{2} \pm \dfrac{\sqrt{3}}{2}i$

36. $\dfrac{1}{4} \pm \dfrac{\sqrt{7}}{4}i$ **37.** $\dfrac{1 \pm \sqrt{3}}{2}$ **38.** $\dfrac{-3 \pm \sqrt{5}}{4}$

39. $\dfrac{-2 \pm \sqrt{10}}{2}$ **40.** $-\dfrac{3}{4} \pm \dfrac{\sqrt{15}}{4}i$ **41.** $\dfrac{-1 \pm \sqrt{5}}{3}$

42. $\dfrac{2 \pm \sqrt{6}}{4}$ **43.** $\dfrac{1 \pm \sqrt{5}}{5}$ **44.** $\dfrac{-0.6 \pm \sqrt{2}}{2}$

45. $\dfrac{-1 \pm \sqrt{10}}{5}$ **46.** $\dfrac{-0.02 \pm \sqrt{0.003}}{0.03}$ **47.** $\dfrac{3}{4} \pm \dfrac{\sqrt{3}}{4}i$

48. $\dfrac{-3 \pm \sqrt{17}}{2}$ **49.** $\dfrac{3 \pm \sqrt{13}}{6}$ **50.** $\dfrac{-1 \pm \sqrt{21}}{10}$

51. ± 13 **52.** $\pm 5\sqrt{6}$ **53.** $0, -15$ **54.** $\pm\dfrac{7}{2}i$

55. $\dfrac{9}{5}, \dfrac{21}{5}$ **56.** $-\dfrac{3}{2}, 18$ **57.** $-4 \pm 4i$ **58.** $-1, 4$

59. $\dfrac{-5 \pm 5\sqrt{17}}{12}$ **60.** $\dfrac{15}{4} \pm \dfrac{15\sqrt{7}}{4}i$ **61.** $8, 16$

62. $\dfrac{5}{2}, -\dfrac{11}{6}$ **63.** $-4 \pm 3i$ **64.** $-2 \pm \dfrac{\sqrt{7}}{2}$

65.
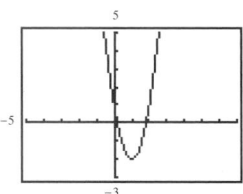
$(0.18, 0), (1.82, 0)$

66.
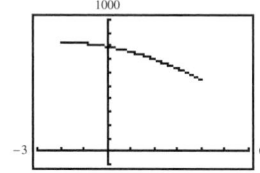

No x-intercepts

67.
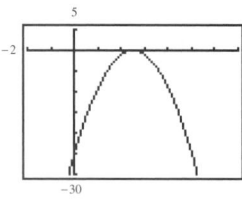
$(2.50, 0)$

68.
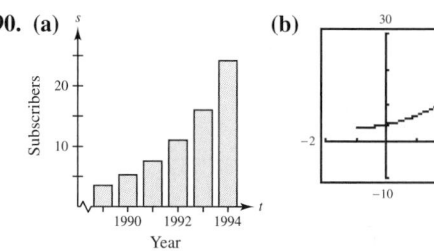
$(1, 0), (3, 0)$

69.
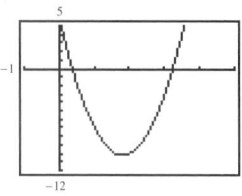
$(3.23, 0), (0.37, 0)$

70.
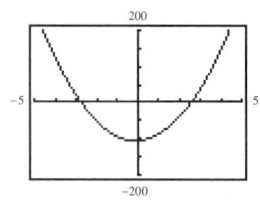
$(2.55, 0), (-2.75, 0)$

71.
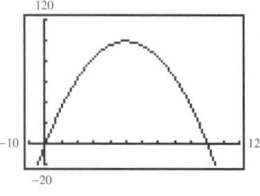
$(99.80, 0), (0.20, 0)$

72.
$(2.40, 0), (0.36, 0)$

73. No real solutions   **74.** Two real solutions
75. Two real solutions   **76.** No real solutions
77. (a) $c < 9$ (b) $c = 9$ (c) $c > 9$
78. (a) $c < 36$ (b) $c = 36$ (c) $c > 36$
79. (a) $c < 16$ (b) $c = 16$ (c) $c > 16$

80. (a) $c < 1$ (b) $c = 1$ (c) $c > 1$ **81.** $\dfrac{5 \pm \sqrt{185}}{8}$

82. $6 \pm \sqrt{33}$ **83.** $\dfrac{3 + \sqrt{17}}{2}$ **84.** $3 + \sqrt{2}$

85. $5.1 \text{ cm} \times 11.4 \text{ cm}$ **86.** (a) 6 ft, 4 ft   (b) Decrease

87. (a) 2.5 sec   (b) $\dfrac{5 + 5\sqrt{3}}{4} \approx 3.4$ sec   **88.** 16, 18

89. (a)   (b) 1991
                (c) 439,000

90. (a)   (b)

(c) 1991

Section Project 6.3

	x_1	x_2	$x_1 + x_2$	$x_1 x_2$
(a)	-2	3	1	-6
(b)	-3	$\frac{1}{2}$	$-\frac{5}{2}$	$-\frac{3}{2}$
(c)	$-\frac{3}{2}$	$\frac{3}{2}$	0	$-\frac{9}{4}$
(d)	$5 + 3i$	$5 - 3i$	10	34

For the general quadratic equation $ax^2 + bx + c = 0$ with solutions x_1 and x_2, $x_1 + x_2 = -b/a$ and $x_1 x_2 = c/a$.

Section 6.4   (page 441)

1. Dollars   **2.** Feet   **3.** 1, 2 or 8, 9   **4.** 19, 20
5. 8, 10   **6.** 11, 13   **7.** 108 in.^{2} **8.** 96 m^2
9. 70 ft   **10.** 60 cm   **11.** 64 in.   **12.** 210 in.
13. 180 km^2 **14.** 720 ft^2 **15.** 440 m   **16.** 90 ft

17. 50 ft × 250 ft or 100 ft × 125 ft

18. Rectangular region: No
    Circular region: Yes

19. No.
    Area $= \frac{1}{2}(b_1 + b_2)h = \frac{1}{2}x[x + (550 - 2x)] = 43{,}560$
    This equation has no real solution.

20. Height: 12 in.; Width: 24 in.   **21.** 150 ft × 170 ft

22. (a) $d = \sqrt{(3 + x)^2 + (4 + x)^2}$

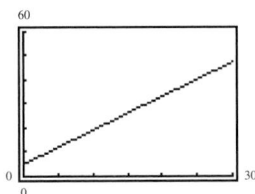

(b) $\dfrac{-7 + \sqrt{199}}{2} \approx 3.55$ m

23. 15 in. × 36 in.

24. (a) $d = \sqrt{100^2 + h^2}$

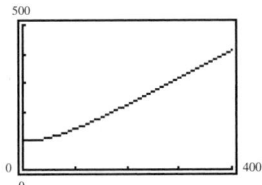

(b)

h	0	100	200	300
d	100	141.4	223.6	316.2

25. 15.86 mi or 2.14 mi   **26.** Base: 25 in.; Height: 50 in.

27. 8%   **28.** 9.5%   **29.** 6%   **30.** 7%

31. 2.59%   **32.** 6.5%   **33.** 3 sec   **34.** $6\frac{1}{4}$ sec

35. 9.5 sec   **36.** 7.8 sec   **37.** 4.7 sec   **38.** 1.6 sec

39. 30 units   **40.** 1500 units   **41.** 90 mi

Section Project 6.4

(a) $b = 20 - a$
    $A = \pi ab$
    $A = \pi a(20 - a)$

(b)

a	4	7	10	13	16
A	201.1	285.9	314.2	285.9	201.1

(c) 7.9, 12.1   (d)

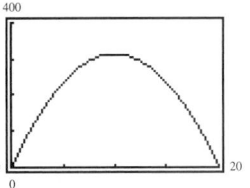

Mid-Chapter Quiz (page 444)

1. ± 6 **2.** $-4, \frac{5}{2}$ **3.** $\pm 2\sqrt{3}$ **4.** $-1, 7$

5. $-5 \pm 2\sqrt{6}$ **6.** $\dfrac{-3 \pm \sqrt{19}}{2}$ **7.** $-2 \pm \sqrt{10}$

8. $\dfrac{3 \pm \sqrt{105}}{12}$ **9.** $-\dfrac{5}{2} \pm \dfrac{\sqrt{3}}{2}i$ **10.** $-2, 10$

11. $-3, 10$ **12.** $-2, 5$ **13.** $\dfrac{3}{2}$ **14.** $\dfrac{-5 \pm \sqrt{10}}{3}$

15. $\pm\sqrt{2}, \pm\sqrt{7}i$ **16.** 27, 125

17.   **18.**

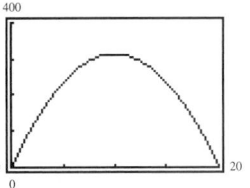

$(-0.32, 0), (6.32, 0)$ $(-2.24, 0), (1.79, 0)$

19. 50 units   **20.** 35 m × 65 m

Section 6.5   (page 452)

1. (e)   **2.** (f)   **3.** (b)   **4.** (c)   **5.** (d)   **6.** (a)

7. If $a > 0$, the graph of $f(x) = ax^2 + bx + c$ opens upward, and if $a < 0$, it opens downward.

8. No. The relationship $f(x) = ax^2 + bx + c$ is a function, and therefore any vertical line will intersect the graph, at most, once.

9. Upward, $(0, 2)$ **10.** Downward, $(-5, -3)$

11. Downward, $(10, 4)$ **12.** Upward, $(12, 3)$

13. Upward, $(0, -6)$ **14.** Downward, $(-1, 0)$

15. Downward, $(3, 0)$ **16.** Upward, $(3, -9)$

17. $(\pm 5, 0), (0, 25)$ **18.** $(\pm 7, 0), (0, -49)$

19. $(0, 0), (9, 0)$ **20.** $(0, 0), (-4, 0)$ **21.** $\left(\frac{3}{2}, 0\right), (0, 9)$

22. $\left(\frac{5}{2}, 0\right), (2, 0), (0, 10)$ **23.** $(0, 3)$

24. $(-2, 0), (5, 0), (0, -10)$

25. $f(x) = (x - 0)^2 + 2, \ (0, 2)$

26. $g(x) = (x + 1)^2 - 1, \ (-1, -1)$

27. $y = (x - 2)^2 + 3, \ (2, 3)$

28. $y = (x + 3)^2 - 14$, $(-3, -14)$

29. $y = (x + 3)^2 - 4$, $(-3, -4)$

30. $y = (x - 2)^2 + 1$, $(2, 1)$

31. $y = -(x - 3)^2 - 1$, $(3, -1)$

32. $y = -(x + 4)^2 + 20$, $(-4, 20)$

33. $y = -(x - 1)^2 - 6$, $(1, -6)$

34. $y = -(x + 5)^2 + 35$, $(-5, 35)$

35. $y = 2\left(x + \frac{3}{2}\right)^2 - \frac{5}{2}$, $\left(-\frac{3}{2}, -\frac{5}{2}\right)$

36. $y = 3\left(x - \frac{1}{2}\right)^2 - \frac{39}{4}$, $\left(\frac{1}{2}, -\frac{39}{4}\right)$

37.
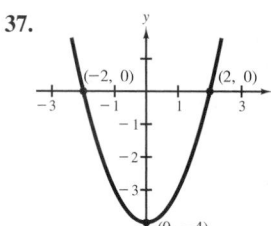

38.
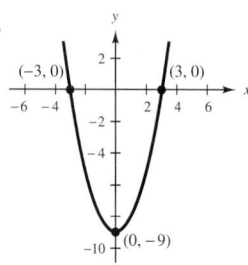

39.
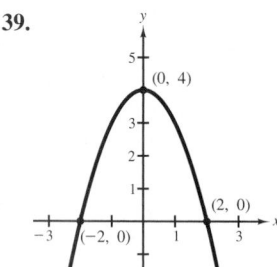

40.
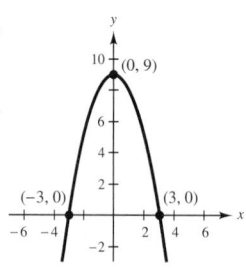

41.
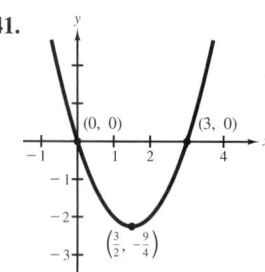

42. **43.** **44.**
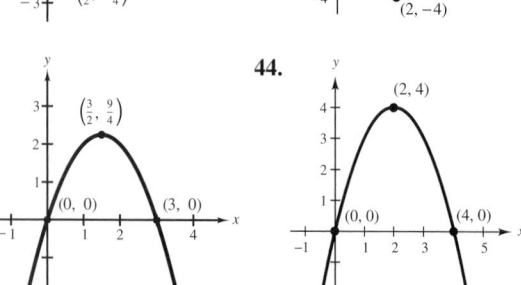

45.
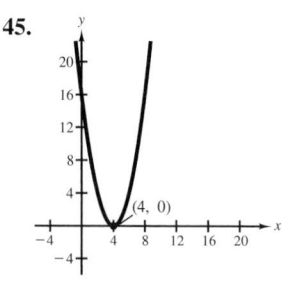

46.
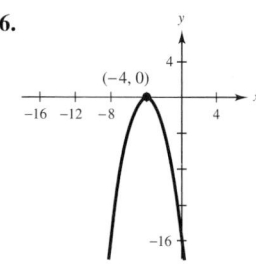

47.
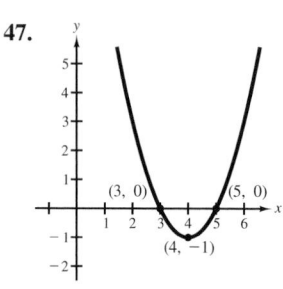

48.
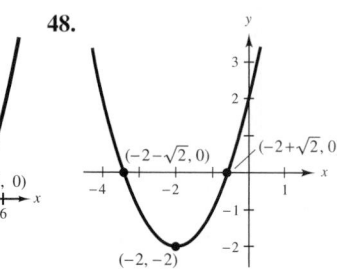

49.
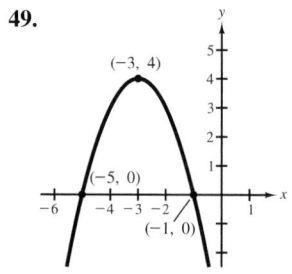

50.
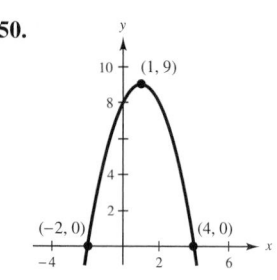

51.
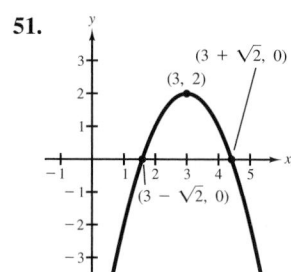

52.
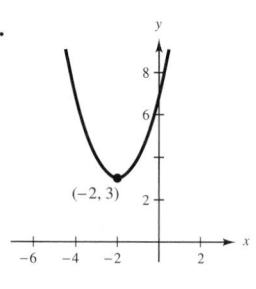

53.
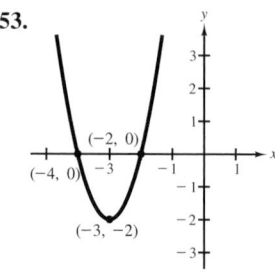

54.
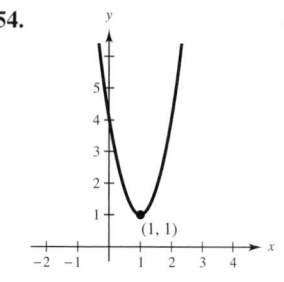

55.

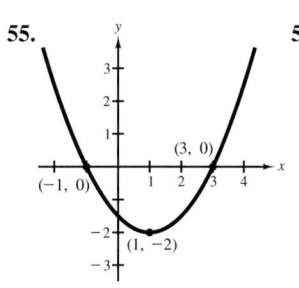

56.

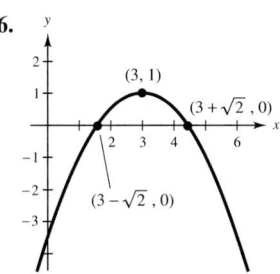

57.

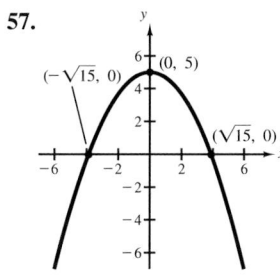

58.

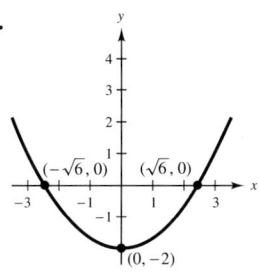

59.

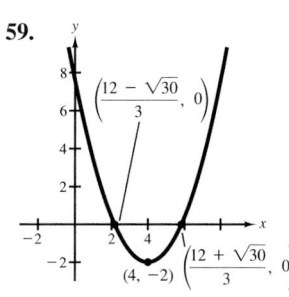

60.

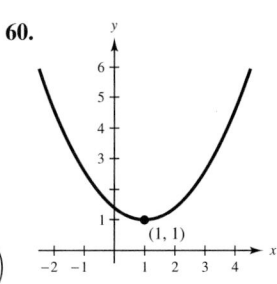

61.

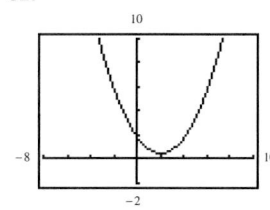

Vertex: $(2, 0.5)$

62.

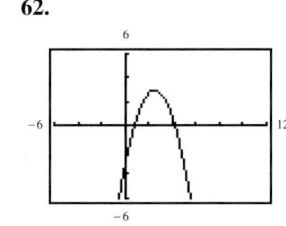

Vertex: $(2.5, 3)$

63.

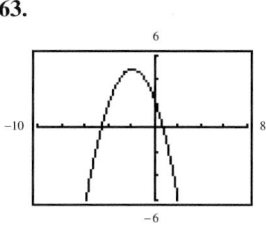

Vertex: $(-1.9, 4.9)$

64.

Vertex: $(5, 4.25)$

65.

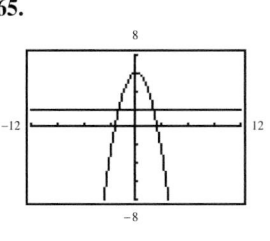

$(\pm 2, 2)$

66.

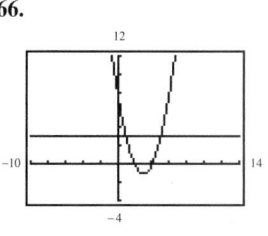

$(1, 3), (5, 3)$

67.

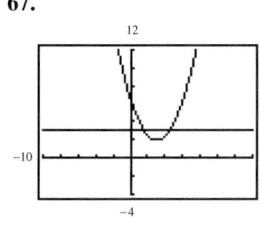

$(4.4, 3), (1.6, 3)$

68.

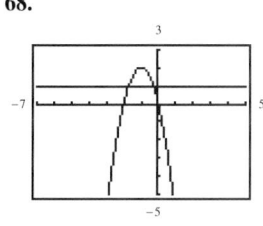

$(-0.3, 1), (-1.7, 1)$

69. $y = x^2 - 4x + 4$ **70.** $y = -x^2 + 4$
71. $y = -x^2 - 4x$ **72.** $y = x^2 + 4x + 2$
73. $y = -2x^2 - 12x - 15$ **74.** $y = 2x^2 - 12x + 18$
75. $y = x^2 - 4x + 5$ **76.** $y = x^2 + 6x + 6$
77. $y = -x^2 - 6x - 5$ **78.** $y = -x^2 + 6x - 11$
79. $y = x^2 - 4x$ **80.** $y = x^2 + 4x$
81. $y = \frac{1}{2}x^2 - 3x + \frac{13}{2}$ **82.** $y = 5x^2 + 10x + 4$
83. $y = -4x^2 - 8x + 1$ **84.** $y = -4x^2 + 40x - 90$
85. $y = \frac{1}{25}x^2 - \frac{2}{5}x + 3$ **86.** $y = -\frac{1}{20}x^2 + 20$

87. (a)

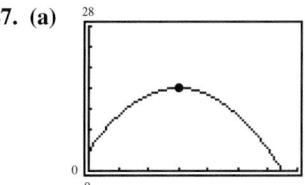

(b) 4 ft
(c) 16 ft
(d) $12 + 8\sqrt{3} \approx 25.9$ ft

88. 14 ft
89. $2000   **90.** 20

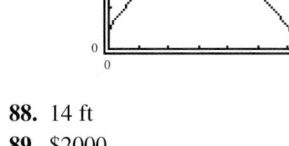

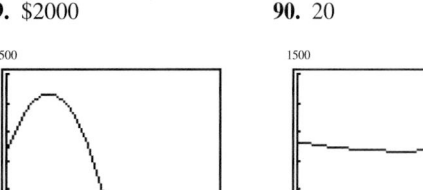

91. 50

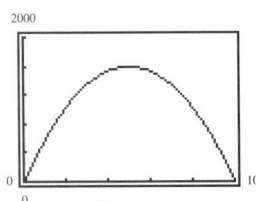

92. (a)

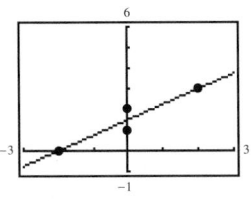

(b) 1991,
494.4 million dollars

93. $y = \frac{1}{2500}x^2$

94. (a) 1000 ft **(b)** 400 ft

(c)

x	± 100	± 200	± 300	± 400	± 500
y	16	64	144	256	400

Section Project 6.5

(a) Answers may vary.

(b)

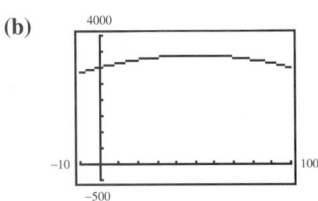

(c) (50, 3375), 150 radios
(d) Recommend for orders of 150 radios or less

Section 6.6 (page 460)

1. Linear **2.** Quadratic **3.** Quadratic **4.** Linear
5.

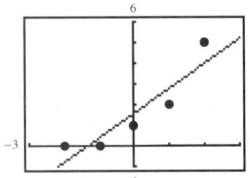
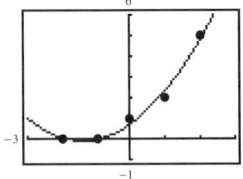

$y = 1.2x + 1.6$ $y = 0.429x^2 + 1.200x + 0.743$

6.

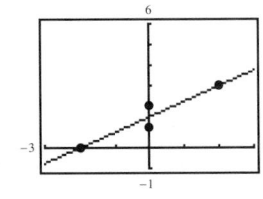

$y = 0.75x + 1.5$ $y = 0x^2 + 0.75x + 1.5$

7.

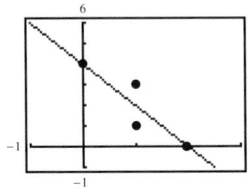
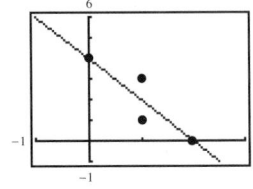

$y = -2x + 4$ $y = 0x^2 - 2x + 4$

8.

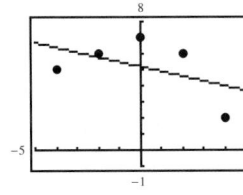
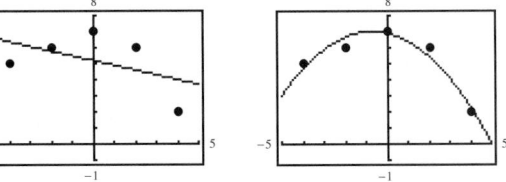

$y = -0.3x + 5.2$ $y = -0.214x^2 - 0.300x + 6.914$

9.

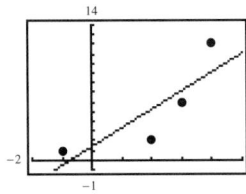
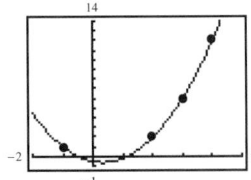

$y = 1.929x + 1.393$ $y = 0.936x^2 - 0.613x - 0.547$

10.

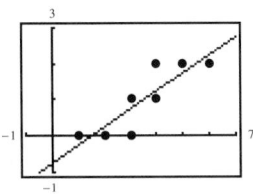
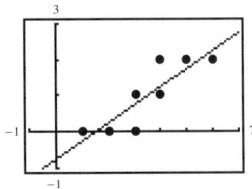

$y = 0.5x - 0.75$ $y = 0x^2 + 0.5x - 0.75$

11.

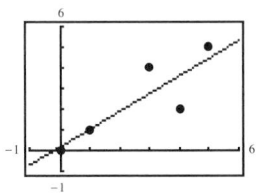 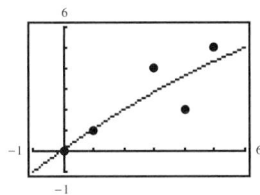

$y = 0.860x + 0.163$ $y = -0.036x^2 + 1.036x + 0.071$

12.

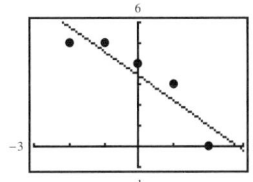 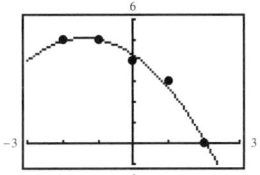

$y = -1.2x + 3.4$ $y = -0.429x^2 - 1.200x + 4.257$

13.

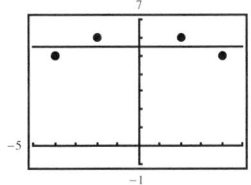 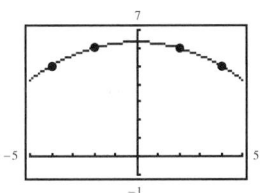

$y = 5.5$ $y = -0.083x^2 + 0x + 6.333$

14.

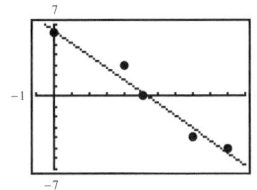 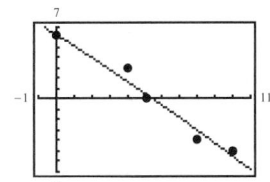

$y = -1.182x + 6.385$ $y = -0.009x^2 - 1.090x + 6.264$

15. (a)   **(b)** 301

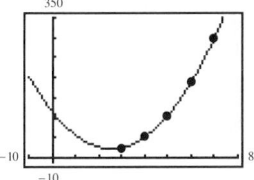

$y = 0.141x^2 - 7.241x + 116.743$

16. (a)   **(b)** 23.8%

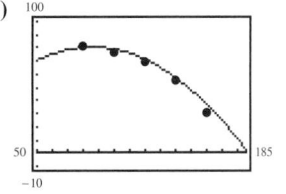

$y = -0.008x^2 + 1.371x + 21.886$

17.

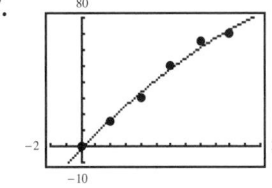

$y = -0.223x^2 + 9.661x - 1.786$

18. (a) $y = 0.078x + 2.960$
(b) $y = 0.0002x^2 + 0.070x + 2.993$
(c) Linear: 6.87
Quadratic: 7.05
The quadratic will increase at a faster rate.

19. (a) $N = 2.760x + 7.379$
(b) $N = 0.104x^2 + 1.418x + 9.126$
(c)

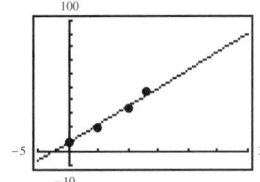 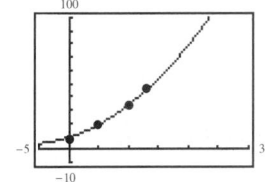

(d) Linear: 90.2; Quadratic: 145.3
The quadratic will increase at a faster rate.

20. (a) $N = -11.809x + 166.692$
(b) $N = 0.139x^2 - 13.743x + 169.564$
(c)

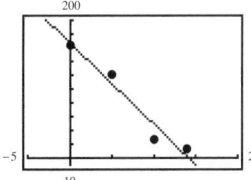 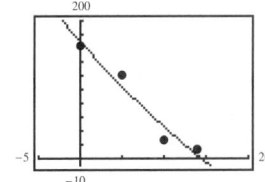

Quadratic

Section Project 6.6

(a) $y = -0.004x^2 + 0.367x + 5$

(b)

(c) Maximum height: $y \approx 13.4$ feet
     Range: $x \approx 103.7$ feet
(d) Maximum height: $y \approx 13.4$ feet
     Range: $x \approx 103.7$ feet

Section 6.7   (page 469)

1. The direction of the inequality is reversed.
2. A polynomial can change signs only at the x-values that make the polynomial zero. The zeros of the polynomial are the critical numbers, and they are used to determine the test intervals in solving polynomial inequalities.
3. $\pm\frac{9}{2}$ 4. $3, 4$ 5. $0, \frac{5}{2}$ 6. $0, 3$ 7. $1, 3$
8. $-\frac{4}{3}, 2$ 9. $\frac{5}{2}$ 10. $-\frac{1}{2}, \frac{3}{2}$
11. (a) Not a solution   12. (a) Solution
     (b) Solution   (b) Not a solution
     (c) Not a solution   (c) Solution
     (d) Solution   (d) Solution
13. Negative: $(-\infty, 4)$ 14. Positive: $(-\infty, 3)$
     Positive: $(4, \infty)$ Negative: $(3, \infty)$
15. Negative: $(0, 4)$
     Positive: $(-\infty, 0) \cup (4, \infty)$
16. Negative: $(-\infty, 0) \cup (3, \infty)$
     Positive: $(0, 3)$
17. Negative: $(-3, 3)$
     Positive: $(-\infty, -3) \cup (3, \infty)$
18. Negative: $(-\infty, -4) \cup (4, \infty)$
     Positive: $(-4, 4)$
19. Negative: $(-1, 5)$
     Positive: $(-\infty, -1) \cup (5, \infty)$
20. Positive: $\left(-\infty, \dfrac{2 - \sqrt{10}}{2}\right) \cup \left(\dfrac{2 + \sqrt{10}}{2}, \infty\right)$

     Negative: $\left(\dfrac{2 - \sqrt{10}}{2}, \dfrac{2 + \sqrt{10}}{2}\right)$
21. $[-3, \infty)$ 22. $(-\infty, 4)$

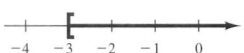 

23. $(8, \infty)$ 24. $\left[\frac{2}{3}, \infty\right)$

25. $(0, 2)$ 26. $(-\infty, 0) \cup (6, \infty)$

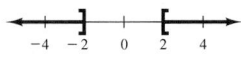

27. $(-\infty, 0) \cup (2, \infty)$ 28. $(0, 6)$

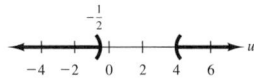 

29. $(-\infty, -2] \cup [2, \infty)$ 30. $[-3, 3]$

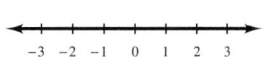

31. $[-5, 2]$ 32. $(5, 10)$

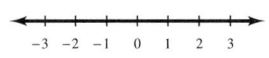 

33. $\left(-\infty, -\frac{1}{2}\right) \cup (4, \infty)$ 34. $(-\infty, -2] \cup \left[\frac{2}{3}, \infty\right)$

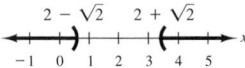

35. No solution   36. $(-\infty, \infty)$

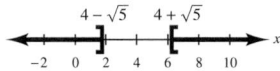

37. $(-\infty, \infty)$ 38. $[-2, 8]$

39. $\left(-\infty, 2 - \sqrt{2}\right) \cup \left(2 + \sqrt{2}, \infty\right)$

40. $\left(-\infty, 4 - \sqrt{5}\right] \cup \left[4 + \sqrt{5}, \infty\right)$

41. No solution   42. $(-\infty, \infty)$

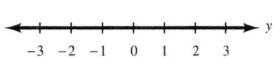

43. $\left(-\infty, 5 - \sqrt{6}\right) \cup \left(5 + \sqrt{6}, \infty\right)$

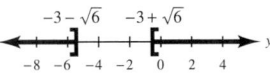

44. $\left(-\infty, -3 - \sqrt{6}\right] \cup \left[-3 + \sqrt{6}, \infty\right)$

45. $(-\infty, \infty)$ **46.** No solution

47.

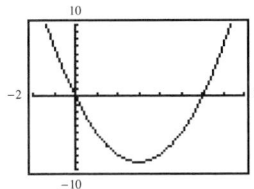

$(0, 6)$

48.

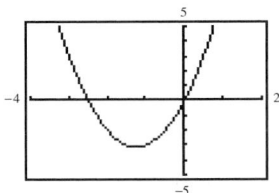

$\left(-\infty, -\frac{5}{2}\right) \cup (0, \infty)$

49.

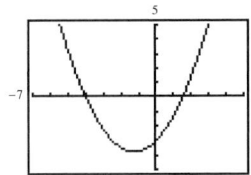

$(-\infty, -4) \cup \left(\frac{3}{2}, \infty\right)$

50.

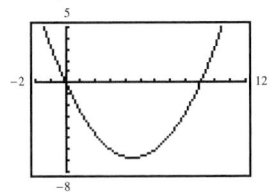

$(0, 9)$

51.

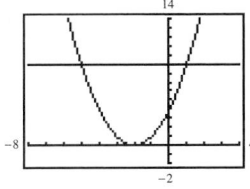

$(-\infty, -5] \cup [1, \infty)$

52.

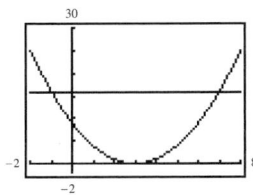

$(-1, 7)$

53.

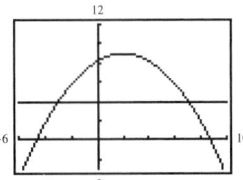

54.

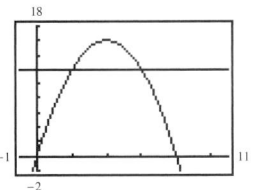

$(-\infty, -3) \cup (7, \infty)$ $(2, 6)$

55. $(3, 5)$ **56.** $\left(\dfrac{11 - \sqrt{71}}{4}, \dfrac{11 + \sqrt{71}}{4}\right)$

57. $r > 7.24\%$ **58.** $90,000 \leq x \leq 100,000$
59. $\left[25 - 5\sqrt{5}, 25 + 5\sqrt{5}\right]$ **60.** $(12, 20)$

Section Project 6.7

(a) $A = \pi x^2 + \pi(12 - x)^2, \ 0 < x < 12$

(b)

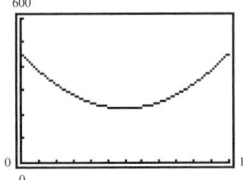

(c) $0.74 < x < 2.57$

Review Exercises   (page 472)

1. $0, -12$ **2.** $0, 18$ **3.** $-10, \frac{8}{3}$ **4.** $\frac{9}{2}, -\frac{4}{7}$
5. $\pm\frac{1}{2}$ **6.** ± 6 **7.** $-\frac{5}{2}$ **8.** $-\frac{4}{3}$ **9.** $-4, 5$
10. $-\frac{4}{3}, \frac{2}{3}$ **11.** $-9, 10$ **12.** $3, -1$ **13.** ± 100
14. $\pm 7\sqrt{2}$ **15.** ± 1.5 **16.** $\pm 2\sqrt{2}$ **17.** $-4, 36$
18. $-2.8, -3.2$ **19.** $3 \pm 2\sqrt{3}$ **20.** $-6 \pm \sqrt{30}$
21. $\frac{3}{2} \pm \frac{\sqrt{3}}{2}i$ **22.** $-\frac{1}{4} \pm \frac{\sqrt{17}}{4}$ **23.** $\frac{-5 \pm \sqrt{19}}{2}$
24. $\frac{1}{3} \pm \frac{\sqrt{5}}{3}i$ **25.** $5, -6$ **26.** $9, -8$ **27.** $3, -\frac{7}{2}$
28. $4, -\frac{5}{2}$ **29.** $\frac{10}{3} \pm \frac{5\sqrt{2}}{3}i$ **30.** $\frac{-2.5 \pm 0.5\sqrt{73}}{-2}$
31. $3 \pm 5\sqrt{10}$ **32.** $0, 36$ **33.** $-7, 12$ **34.** $-\frac{1}{3}$
35. $-2, 20$ **36.** $6 \pm 2\sqrt{15}$ **37.** $3 \pm i$
38. $7 \pm 2\sqrt{11}$ **39.** $-\frac{1}{2}, -1$ **40.** $\frac{11 + \sqrt{13}}{2}$
41. $\pm\sqrt{5}, \pm i$ **42.** 1 **43.** $-243, -1$ **44.** $-\frac{2}{3}, 3$

45.
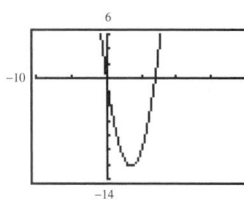
$(0, 0), (7, 0)$

46.
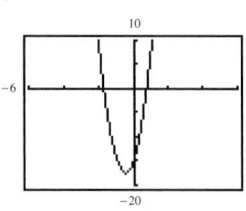
$\left(-\frac{5}{3}, 0\right), \left(\frac{3}{4}, 0\right)$

47.
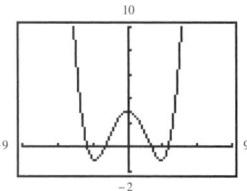
$(\pm 2, 0), \left(\pm 2\sqrt{3}, 0\right)$

48.
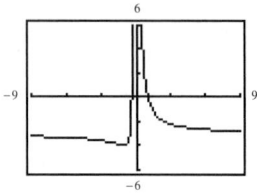
$\left(-\frac{1}{3}, 0\right), (1, 0)$

49.
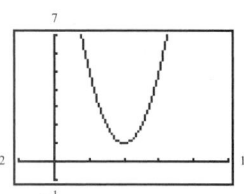
$4 \pm i$

50.
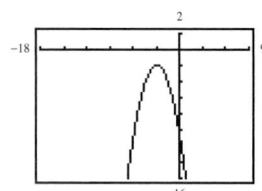
$-3 \pm \sqrt{2}i$

51. Upward; $(3, 6)$ **52.** Downward; $(7, -1)$
53. Downward; $(-4, 0)$ **54.** Upward; $(-2, 9)$

55.
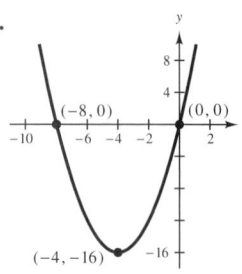

56.
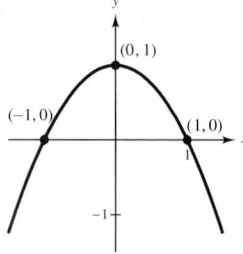

57.
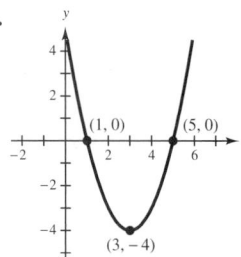

58.
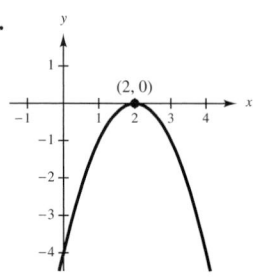

59. $y = -2x^2 + 12x - 13$ **60.** $y = 3x^2 + 12x + 15$
61. $y = \frac{1}{16}x^2 - \frac{5}{8}x + \frac{25}{16}$ **62.** $y = -x^2 - 4x + 1$
63. (a) 200 ft
  (b) Dropped. The coefficient of the first-degree term is 0.
  (c) $\dfrac{5\sqrt{2}}{2} \approx 3.54$ sec

64. (a) 192 ft
  (b) Thrown upward. The coefficient of the first-degree term is positive.
  (c) 6 sec

65. (a)
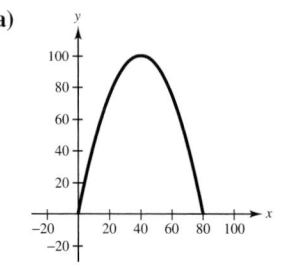
  (b) 100 ft
  (c) 80 ft

66. 60 in. $\times$ 100 in.
67. (a) $P = 2l + 2w = 48$ **(b)** $A = lw$
  $l + w = 24$ $w = 24 - l$
  $w = 24 - l$ $A = l(24 - l)$
  (c) 44, 80, 108, 128, 140, 144, 140, 128, 108
68. (a) $d = \sqrt{9 + h^2}$

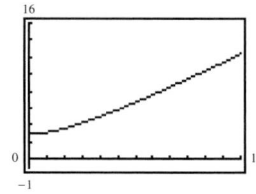

$h \approx 2.6$

(b)

h	1	2	3	4	5	6	7
d	3.16	3.61	4.24	5	5.83	6.71	7.62

$h \approx 2.5$
(c) $h = \sqrt{7} \approx 2.646$
69. (a) $S = 0.031t^2 + 3.977t + 21.444$

(b)
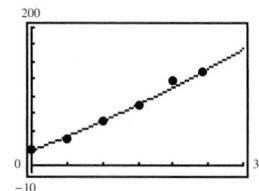
(c) \$169,000

70. 1995

71. $(-\infty, 3)$

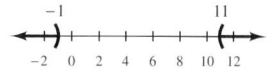

72. $(-2, \infty)$

73. $(0, 7)$

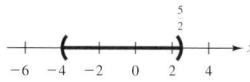

74. $(-\infty, 0] \cup [10, \infty)$

75. $(-\infty, -1) \cup (11, \infty)$

76. $(-\infty, -2] \cup [6, \infty)$

77. $\left(-4, \frac{5}{2}\right)$

78. $\left(-\infty, -\frac{4}{3}\right) \cup (2, \infty)$

Chapter 6 Test (page 475)

1. $-5, 10$ **2.** $-\frac{3}{8}, 3$ **3.** $1.7, 2.3$

4. $-3 \pm 9i$ **5.** $\frac{9}{4}$ **6.** $\frac{3 \pm \sqrt{3}}{2}$

7. -56; If the discriminant is a perfect square, there are two distinct rational solutions. If the discriminant is a positive nonperfect square, there are two distinct irrational solutions. If the discriminant is zero, there is one repeated rational solution. If the discriminant is a negative number, there are two distinct imaginary solutions.

8. $\frac{4 \pm \sqrt{7}}{3}$ **9.** $\frac{2 \pm 3\sqrt{2}}{2}$ **10.** $8, 64$

11.

12.

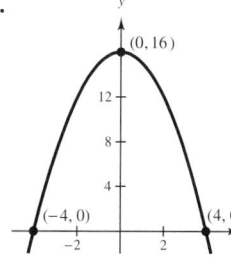

13. $(0, 3)$

14. $(-\infty, -2] \cup [6, \infty)$

15. $y = \frac{2}{3}x^2 - 4x + 4$ **16.** $12 \text{ ft} \times 20 \text{ ft}$

17. $\frac{\sqrt{10}}{2} \approx 1.58$ sec **18.** $y = x^2 - x$

Cumulative Test: Chapters 1–6 (page 476)

1. $(2x + 7)(x - 1)$ **2.** $(11x - 5)(x + 1)$

3. $3x(2x + 3)(2x - 3)$ **4.** $(2x + 5)(4x^2 - 10x + 25)$

5. $\frac{5}{2}$ **6.** $0, 8$ **7.** $-2, \pm 1$ **8.** 5 **9.** $\frac{19}{3}$

10. 22 **11.** $y = \frac{5}{2}x - 12$ **12.** $y = -x + 3$

13. No. For some values of x there correspond two values of y.

14. $A = \frac{5}{2}x(2x + 9)$ **15.** Buy now. **16.** $(2, 8)$

17. 32 **18.** -60 **19.** (a) $5x\sqrt{3xy}$ (b) $10(3 + \sqrt{5})$

20. $y = \frac{1}{4}x^2 - x$

CHAPTER 7

Section 7.1 (page 486)

1. The rational expression is in reduced form if the numerator and denominator have no common factors other than ± 1.

2. No. You can divide out only common factors.

3. You can divide out only common factors.

4. True. $\dfrac{6x - 5}{5 - 6x} = \dfrac{-(5 - 6x)}{5 - 6x} = -1$

5. $(-\infty, 8) \cup (8, \infty)$ **6.** $(-\infty, 13) \cup (13, \infty)$

7. $(-\infty, -4) \cup (-4, \infty)$ **8.** $(-\infty, \infty)$

9. $(-\infty, \infty)$ **10.** $(-\infty, \infty)$

11. $(-\infty, -4) \cup (-4, 4) \cup (4, \infty)$

12. $(-\infty, 0) \cup (0, 4) \cup (4, \infty)$

13. $(-\infty, -1) \cup (-1, 5) \cup (5, \infty)$

14. $\left(-\infty, -\frac{3}{4}\right) \cup \left(-\frac{3}{4}, 2\right) \cup (2, \infty)$

15. $(-\infty, \infty)$ **16.** $(-\infty, -2) \cup (-2, 2) \cup (2, \infty)$

17. $(-\infty, 0) \cup (0, 3) \cup (3, \infty)$

18. $(-\infty, -1) \cup (-1, 3) \cup (3, \infty)$ **19.** $\dfrac{x}{5}$ **20.** $\dfrac{4y}{3}$

21. $6y, \ y \neq 0$ **22.** $1, \ z \neq 0$ **23.** $\dfrac{6x}{5y^3}, \ x \neq 0$

24. $\dfrac{4z}{15y^3}, \ z \neq 0$ **25.** $x, \ x \neq 8, x \neq 0$ **26.** $\dfrac{a^2}{b^2(b - 3)}$

27. $\dfrac{1}{2}, \ x \neq \dfrac{3}{2}$ **28.** $\dfrac{y + 9}{2}, \ y \neq 9$ **29.** $-\dfrac{1}{3}, \ x \neq 5$

30. $-(x + 6)$ or $-x - 6, \ x \neq 6$ **31.** $\dfrac{3y^2}{y^2 + 1}, \ x \neq 0$

32. $x, \ 3xy \neq -1$ **33.** $\dfrac{y - 8}{15}, \ y \neq -8$

34. $x - 5z$, $x \neq -5z$ **35.** $\dfrac{1}{a + 3}$ **36.** $u - 6$, $u \neq 6$

37. $\dfrac{x}{x - 7}$ **38.** $\dfrac{z + 11}{3}$, $z \neq -11$ **39.** $\dfrac{y(y + 2)}{y + 6}$, $y \neq 2$

40. $\dfrac{x}{x + 3}$, $x \neq 7$ **41.** $\dfrac{-1}{2x + 3}$, $x \neq 3$

42. $\dfrac{y + 4}{y + 6}$, $y \neq -\dfrac{5}{2}$ **43.** $\dfrac{5x + 4}{5x + 2}$, $x \neq \dfrac{1}{3}$

44. $\dfrac{8z - 5}{7z - 4}$, $z \neq -\dfrac{4}{7}$ **45.** $\dfrac{5 + 3xy}{y^2}$, $x \neq 0$

46. $\dfrac{2(u - 3v)}{9}$, $u \neq 0$, $v \neq 0$ **47.** $\dfrac{3(m - 2n)}{m + 2n}$

48. $\dfrac{x - y}{x + y}$, $x \neq -2y$ **49.** $\dfrac{u - 2v}{u - v}$, $u \neq -2v$

50. $\dfrac{x}{x - 4y}$, $x \neq -4y$ **51.** $\dfrac{x^2 + 2xy + 4y^2}{x + 2y}$, $x \neq 2y$

52. $\dfrac{x^2 - 3xz + 9z^2}{x - 2z}$, $x \neq -3z$ **53.** $\dfrac{u - 4}{v - 2}$, $v \neq 3$

54. $\dfrac{n + 3}{m + n}$, $m \neq n$ **55.** $\{1, 2, 3, 4, \ldots\}$

56. $[0, 100)$ **57.** $(0, \infty)$ **58.** $[0, 100)$

59. (a) 0   (b) $\dfrac{x - 10}{4x}$ is undefined for $x = 0$.

   (c) $\dfrac{3}{2}$ (d) $\dfrac{1}{24}$

60. (a) 0   (b) 0   (c) $\dfrac{x^2 - 4x}{x^2 - 9}$ is undefined for $x = 3$.

   (d) $\dfrac{x^2 - 4x}{x^2 - 9}$ is undefined for $x = -3$.

61. (a) 1   (b) -8 (c) Undefined   (d) 0
62. (a) 0   (b) Undefined   (c) $\dfrac{4}{9}$ (d) $\dfrac{2}{5}$
63. (a) $\dfrac{25}{22}$ (b) 0   (c) Undefined   (d) Undefined
64. (a) 0   (b) Undefined   (c) $-\dfrac{7}{25}$ (d) 9
65. Evaluating both sides when $x = 10$ yields $\dfrac{3}{2} \neq 9$.
66. Evaluating both sides when $x = 10$ yields $\dfrac{3}{5} \neq -4$.
67. Evaluating both sides when $x = 0$ yields $1 \neq \dfrac{3}{4}$.
68. Evaluating both sides when $x = 1$ yields $0 \neq \dfrac{1}{2}$.

69. $x^n - 3$, $x \neq 0$ **70.** $\dfrac{1}{x + 1}$, $x \neq 0$

71. $x^n - 2$, $x^n \neq -2$ **72.** $\dfrac{x^n - 3}{x}$, $x^n \neq -4$

73. $\dfrac{x}{x + 3}$, $x > 0$ **74.** $\dfrac{1}{9}$, $x > 0$ **75.** π

76. (a) $\dfrac{2500 + 9.25x}{x}$ (b) $\{1, 2, 3, 4, \ldots\}$ (c) \$34.25

77. $\dfrac{C}{P} = \dfrac{1000(113.66 + 12.64t)}{31.20 + 0.54t}$

78.

Year	1988	1989	1990	1991	1992	1993
Average cost	\$2934	\$3295	\$3643	\$3979	\$4304	\$4619

79. (a) van: $45(t + 3)$, car: $60t$ (b) $\dfrac{4t}{3(t + 3)}$

Section Project 7.1

(a)

x	-2	-1	0	1	2		3	4
$\dfrac{x^2 - x - 2}{x - 2}$	-1	0	1	2	Undef.		4	5
$x + 1$	-1	0	1	2	3		4	5

(b) $\dfrac{x^2 - x - 2}{x - 2} = \dfrac{(x - 2)(x + 1)}{x - 2} = x + 1$, $x \neq 2$

(c)

x	-2	-1	0		1	2	3	4
$\dfrac{x^2 + 5x}{x}$	3	4	Undef.		6	7	8	9
$x + 5$	3	4	5		6	7	8	9

$\dfrac{x^2 + 5x}{x} = \dfrac{x(x + 5)}{x} = x + 5$, $x \neq 0$

Section 7.2   (page 494)

1. $\dfrac{3x}{2}$, $x \neq 0$ **2.** 30, $a \neq 0$ **3.** $\dfrac{s^3}{6}$, $s \neq 0$

4. $\dfrac{8x^5}{21}$, $x \neq 0$ **5.** $24u^2$, $u \neq 0$ **6.** $\dfrac{5}{7}$, $x \neq 0$

7. 24, $x \neq -\dfrac{3}{4}$ **8.** $\dfrac{23}{30}$, $x \neq \dfrac{1}{3}$ **9.** $\dfrac{2uv(u + v)}{3(3u + v)}$, $u \neq 0$

10. 1, $x \neq -25$ **11.** -1, $r \neq 12$

12. -1, $z \neq 8$, $z \neq -8$ **13.** $\dfrac{2(r + 2)}{11r}$, $r \neq 2$

14. 2, $y \neq -3$, $y \neq 4$ **15.** $-\dfrac{x + 8}{x^2}$, $x \neq \dfrac{3}{2}$

16. $-\dfrac{x + 14}{5x^2}$, $x \neq 10$ **17.** $2t + 5$, $t \neq 3$, $t \neq -2$

18. $\dfrac{2(y + 4)}{y^2(y - 2)}$, $y \neq 4$ **19.** $\dfrac{xy(x + 2y)}{(x - 2y)}$

20. $\dfrac{(x - y)^2}{x + y}$, $x \neq -3y$ **21.** $(u - 2v)(u + 2v)$, $u \neq 2v$

22. $\dfrac{x^2 + 4y^2}{(x + 2y)^2}$, $x \neq 2y$

23. $\dfrac{(x - 1)(2x + 1)}{(3x - 2)(x + 2)}$, $x \neq 5, x \neq -5, x \neq -1$

24. $\dfrac{(x + 3)^2}{x}$, $x \neq 3, x \neq 4$ **25.** $\dfrac{(u + v)^2}{(u - v)^2}$, $x \neq y, x \neq -y$

26. $\dfrac{2t^2(1 + 2t)}{(2t - 5)(t + 2)^2}$, $t \neq -1, t \neq -3, t \neq 0$

27. $\dfrac{3y^2}{2ux^2}$, $v \neq 0$ **28.** $\dfrac{4}{3x^3y^3}$ **29.** $\dfrac{3}{2(a + b)}$

30. $\dfrac{(x^2 + 9)(x - 2)}{x + 3}$, $x \neq -2, x \neq 2$

31. $x^4y(x + 2y)$, $x \neq 0, y \neq 0, x \neq -2y$

32. $\dfrac{y(x + y)}{(x - 4)(x - y)}$, $x \neq 0, y \neq 0$ **33.** $\dfrac{3x}{10}$, $x \neq 0$

34. $\dfrac{9u^3}{2v^6}$, $u \neq 0$ **35.** $-\dfrac{5x}{2}$, $x \neq 0, x \neq 5$

36. $\dfrac{x(x + 1)}{2}$, $x \neq -7, x \neq -1$

37. $\dfrac{(x + 3)(4x + 1)}{(3x - 1)(x - 1)}$, $x \neq -3, x \neq -\dfrac{1}{4}$

38. $\dfrac{(x + 5)(x + 1)}{x^2}$, $x \neq 5, x \neq -1$

39. $\dfrac{(x + 2)(x + 3)}{x}$, $x \neq -2, x \neq 2$

40. $\dfrac{t + 1}{t + 3}$, $t \neq 3, t \neq 2$ **41.** $\dfrac{x + 4}{3}$, $x \neq -2, x \neq 0$

42. $\dfrac{x + 3}{x(2x + 1)}$, $x \neq -3, x \neq 3$

43. $\frac{1}{4}$, $x \neq -1, x \neq 0, y \neq 0$

44. $\dfrac{(u - 3)(3u - 4)^2(u + 1)^2}{u(3u^2 + 12u + 4)}$, $u \neq 3$

45. $\dfrac{(x + 1)(2x - 5)}{x}$, $x \neq -1, x \neq -5, x \neq -\dfrac{2}{3}$

46. $\dfrac{5(t + 5)}{4t^3}$, $t \neq -10, t \neq 10$

47. $\dfrac{x^4}{(x^n + 1)^2}$, $x^n \neq -3, x^n \neq 3, x \neq 0$

48. $\dfrac{(x^n - 8)(x^n - 5)}{x^n(x^n + 1)}$

49. 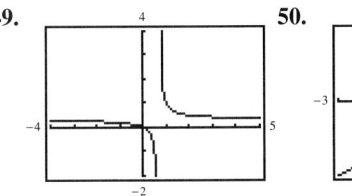 **50.**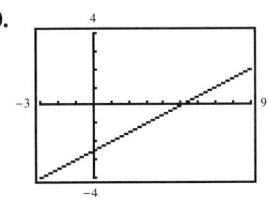

51. 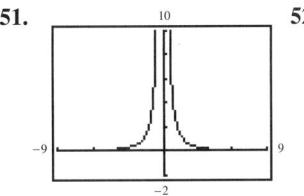 **52.**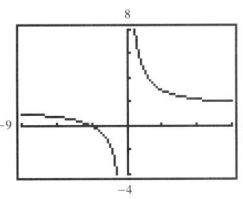

53. $\dfrac{2w^2 + 3w}{6}$ **54.** $\dfrac{2w^2 - w}{6}$

55. Invert the divisor, not the dividend.

Section Project 7.2

(a) $\dfrac{x}{4(2x + 1)}$ (b) $\dfrac{x}{4(2x + 1)}$

(c) $\dfrac{x}{2(2x + 1)}$ (d) $\dfrac{\pi x}{4(2x + 1)}$

Section 7.3 (page 503)

1. $\dfrac{3x}{2}$ **2.** $\dfrac{y}{6}$ **3.** $-\dfrac{2}{9}$ **4.** $\dfrac{2z^2 - 2}{3}$ **5.** $\dfrac{2 + y}{2}$

6. $\dfrac{1}{3}$ **7.** $-\dfrac{3}{a}$ **8.** $\dfrac{5 + 2z}{5z}$ **9.** $\dfrac{x + 6}{3}$ **10.** $-\dfrac{x}{13}$

11. $-\dfrac{4}{3}$ **12.** $-\dfrac{25}{9}$ **13.** $\dfrac{1}{x - 3}$, $x \neq 0$

14. 1, $x \neq -4$ **15.** 1, $y \neq 6$ **16.** 5, $s \neq -\dfrac{5}{2}$

17. $20x^3$ **18.** $42t^5$ **19.** $15x^2(x + 5)$

20. $54y^3(y - 3)^2$ **21.** $36y^3$ **22.** $30x^2(x - 1)$

23. $6x(x + 2)(x - 2)$ **24.** $2t^2(t - 3)(t + 3)(t^2 + 3t + 9)$

25. $56t(t + 2)(t - 2)$ **26.** $2y(y + 1)(2y - 1)$ **27.** x^2

28. $7(x - 3)^2$ **29.** $(u + 1)$ **30.** $5t(3t - 5)$

31. $-(x + 2)$ **32.** $-x$ **33.** $\dfrac{2n^2(n + 8)}{6n^2(n - 4)}, \dfrac{10(n - 4)}{6n^2(n - 4)}$

34. $\dfrac{8s^2(s - 1)}{s(s + 2)^2(s - 1)}, \dfrac{3(s + 2)}{s(s + 2)^2(s - 1)}$

35. $\dfrac{3v^2}{6v^2(v + 1)}, \dfrac{8(v + 1)}{6v^2(v + 1)}$

36. $\dfrac{4x(x - 5)}{(x + 5)^2(x - 5)}, \dfrac{(x - 2)(x + 5)}{(x + 5)^2(x - 5)}$

37. $\dfrac{2(x + 3)}{x^2(x + 3)(x - 3)}, \dfrac{5x(x - 3)}{x^2(x + 3)(x - 3)}$

38. $\dfrac{5t}{2t(t - 3)^2}, \dfrac{8(t - 3)}{2t(t - 3)^2}$

39. $\dfrac{(x - 8)(x - 5)}{(x + 5)(x - 5)^2}, \dfrac{9x(x + 5)}{(x + 5)(x - 5)^2}$

40. $\dfrac{3y^2}{y(y + 3)(y - 4)}, \dfrac{(y - 4)^2}{y(y + 3)(y - 4)}$ **41.** $\dfrac{25 - 12x}{20x}$

42. $\dfrac{100b + 1}{10b^2}$ **43.** $\dfrac{7(a + 2)}{a^2}$ **44.** $\dfrac{3 - 4u}{18u^2}$

45. $0,\ x \ne 4$ **46.** $\dfrac{22}{2 - t}$ **47.** $\dfrac{3(x + 2)}{x - 8}$

48. $\dfrac{1 - y}{y - 6}$ **49.** $1,\ x \ne \dfrac{2}{3}$ **50.** $\dfrac{y + 3}{5y - 3}$

51. $\dfrac{5(5x + 22)}{x + 4}$ **52.** $\dfrac{4(45 - 2x)}{x - 10}$ **53.** $\dfrac{1}{2x(x - 3)}$

54. $\dfrac{2}{x(x + 2)}$ **55.** $\dfrac{x^2 - 7x - 15}{(x + 3)(x - 2)}$ **56.** $\dfrac{7 + 4t}{t(t + 1)}$

57. $\dfrac{x - 2}{x(x + 1)}$ **58.** $\dfrac{2(x + 6)}{x(x - 4)}$ **59.** $\dfrac{5(x + y)}{(x + 5y)(x - 5y)}$

60. $\dfrac{4(5x + 3y)}{(2x + 3y)(2x - 3y)}$ **61.** $\dfrac{4}{x^2(x^2 + 1)}$ **62.** $\dfrac{4y + 1}{2y^2}$

63. $\dfrac{x^2 + 3x + 9}{x(x - 3)(x + 3)}$ **64.** $\dfrac{6}{(x + 5)(x - 6)}$

65. $\dfrac{4x}{(x - 4)^2}$ **66.** $\dfrac{3x - 7}{(x - 2)^2}$ **67.** $\dfrac{y - x}{xy},\ x \ne -y$

68. $\dfrac{10x}{(x + y)(x - y)}$ **69.** $\dfrac{2(4x^2 + 5x - 3)}{x^2(x + 3)}$

70. $\dfrac{4x^2 - 1}{2x(x + 1)^2}$ **71.** $\dfrac{5u - 2v}{(u - v)^2}$ **72.** $0,\ x \ne 0,\ y \ne 0$

73. $\dfrac{x}{x - 1},\ x \ne -6$ **74.** $\dfrac{6x + 5}{2(x + 10)(x + 5)}$

75. 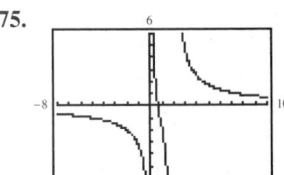 **76.**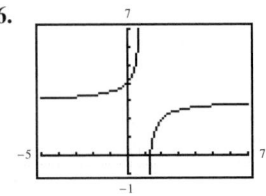

77. When the numerators are subtracted, the result should be $(x - 1) - (4x - 11) = x - 1 - 4x + 11.$

78. Yes. $\dfrac{3}{2(x + 2)} + \dfrac{x}{x + 2}$ **79.** $\dfrac{x}{2(3x + 1)},\ x \ne 0$

80. $\dfrac{2x}{3(4x - 1)},\ x \ne 0$ **81.** $\dfrac{4 + 3x}{4 - 3x},\ x \ne 0$

82. $\dfrac{1 - t}{1 + t},\ t \ne 0$ **83.** $-4x - 1,\ x \ne 0,\ x \ne \dfrac{1}{4}$

84. $\dfrac{6 - y}{y},\ y \ne -6$ **85.** $\dfrac{3}{4},\ x \ne 0,\ x \ne 3$

86. $\dfrac{(x - 2)(x - 1)}{x^2 - 3x + 6},\ x \ne 3$ **87.** $\dfrac{y + 1}{y - 3},\ y \ne 0,\ y \ne 1$

88. $\dfrac{x^2 - 2}{2x},\ x \ne -2$ **89.** $y - x,\ x \ne 0,\ y \ne 0,\ x \ne -y$

90. $\dfrac{x + y}{x - y},\ x \ne 2y$ **91.** $\dfrac{5(x + 3)}{2x(5x - 2)}$

92. $-4,\ x \ne 0,\ x \ne \pm\dfrac{1}{4}$

93. $\dfrac{x(x + 6)}{3x^3 + 10x - 30},\ x \ne 0,\ x \ne 3$

94. $\dfrac{(5 - x)(11x - 5)}{2(x + 5)(x^2 + x - 5)},\ x \ne 0,\ x \ne 5$

95. $\dfrac{v^2}{uv^2 + 1},\ v \ne 0$ **96.** $\dfrac{x^2 + y^2}{x^2}$

97. $\dfrac{ab}{b - a},\ a \ne 0,\ b \ne 0$ **98.** $\dfrac{v - u}{u + v},\ u \ne 0,\ v \ne 0$

99. $\dfrac{5x}{24}$ **100.** $\dfrac{11x}{18}$ **101.** $\dfrac{x}{4},\dfrac{x}{3},\dfrac{5x}{12}$ **102.** $\dfrac{11x}{45},\dfrac{13x}{45}$

103. $\dfrac{5t}{12}$ **104.** $\dfrac{8t}{15}$ **105.** $\dfrac{R_1 R_2}{R_1 + R_2}$

106. (a) 19.6% (b) $\dfrac{288(MN - P)}{N(MN + 12P)}$ **107.** $-\dfrac{1}{2(h + 2)}$

108. $-\dfrac{1}{h + 1}$ **109.** $A = -4,\ B = 2,\ C = 2;$
$$\dfrac{4}{x^3 - x} = -\dfrac{4}{x} + \dfrac{2}{x + 1} + \dfrac{2}{x - 1}$$

110. $\dfrac{4x + 1}{x^2(x + 1)}$

Section Project 7.3

(a) $\dfrac{x}{x + 1}$

(b)

x	-3	-2	-1	0	1	2	3
$\dfrac{\left(1 - \dfrac{1}{x}\right)}{\left(1 - \dfrac{1}{x^2}\right)}$	$\dfrac{3}{2}$	2	Undef.	Undef.	Undef.	$\dfrac{2}{3}$	$\dfrac{3}{4}$
$\dfrac{x}{x + 1}$	$\dfrac{3}{2}$	2	Undef.	0	$\dfrac{1}{2}$	$\dfrac{2}{3}$	$\dfrac{3}{4}$

(c) Domain of the complex fraction: $(-\infty, -1), (-1, 0),$ $(0, 1), (1, \infty).$ Domain of the simplified fraction: $(-\infty, -1), (-1, \infty).$ The two expressions are equivalent except at $x = 0$ and $x = 1.$

(d) $\dfrac{x + 2}{x - 2}$

(e)

x	-3	-2	-1	0	1	2	3
$\dfrac{\left(1 + \dfrac{4}{x} + \dfrac{4}{x^2}\right)}{\left(1 - \dfrac{4}{x^2}\right)}$	$\dfrac{1}{5}$	Undef.	$-\dfrac{1}{3}$	Undef.	-3	Undef.	5
$\dfrac{x + 2}{x - 2}$	$\dfrac{1}{5}$	0	$-\dfrac{1}{3}$	-1	-3	Undef.	5

(f) Domain of the complex fraction: $(-\infty, -2)$, $(-2, 0)$, $(0, 2)$, $(2, \infty)$. Domain of the simplified fraction: $(-\infty, 2)$, $(2, \infty)$. The two expressions are equivalent except at $x = -2$ and $x = 0$.

Section 7.4 (page 516)

1. x is not a factor of the numerator.

2. True. If $\dfrac{n(x)}{d(x)} = q(x)$, then $n(x) = d(x) \cdot q(x)$.

3. $3z + 5$ **4.** $3x + 4$ **5.** $\frac{5}{2}z^2 + z - 3$

6. $\dfrac{u^2}{4} + \dfrac{u}{2} - \dfrac{3}{2}$ **7.** $7x^2 - 2x$, $x \ne 0$

8. $6a + 7$, $a \ne 0$ **9.** $-10z^2 - 6$, $z \ne 0$

10. $-3c^3 + 4c$, $c \ne 0$ **11.** $4z^2 + \frac{3}{2}z - 1$, $z \ne 0$

12. $2x^2 + \dfrac{8}{3}x - 6$, $x \ne 0$ **13.** $m^3 + 2m - \dfrac{7}{m}$

14. $-l + \dfrac{8}{l}$ **15.** $\dfrac{5}{2}x - 4 + \dfrac{7}{2}y$, $x \ne 0, y \ne 0$

16. $-7s^2t + \dfrac{7}{2}t - \dfrac{9}{s^2}$, $t \ne 0$ **17.** $4(x + 7)$, $x \ne -5$

18. $2r - 21$, $r \ne 9$ **19.** $x - 5$, $x \ne 3$

20. $t - 12$, $t \ne 6$ **21.** $x + 10$, $x \ne -5$

22. $y - 8$, $y \ne -2$ **23.** $x + 7$, $x \ne 3$

24. $-x + 5$, $x \ne -1$ **25.** $y + 3$, $y \ne -\frac{1}{2}$

26. $5t + 4$, $t \ne \frac{3}{2}$ **27.** $6t - 5$, $t \ne \frac{5}{2}$

28. $-4u + 3$, $u \ne -\frac{5}{2}$ **29.** $4x - 1$, $x \ne -\frac{1}{4}$

30. $9y + 5$, $y \ne \frac{5}{9}$ **31.** $x^2 - 5x + 25$, $x \ne -5$

32. $x^2 + 3x + 9$, $x \ne 3$ **33.** $x^2 + 4$, $x \ne 2$

34. $x^2 - 4x - 12$, $x \ne -4$ **35.** $2 + \dfrac{5}{x + 2}$

36. $6 - \dfrac{23}{2x + 3}$ **37.** $x - 4 + \dfrac{32}{x + 4}$

38. $y - 2 + \dfrac{12}{y + 2}$ **39.** $5x - 8 + \dfrac{19}{x + 2}$

40. $2x - 3 + \dfrac{14}{x + 4}$ **41.** $4x + 3 - \dfrac{11}{3x + 2}$

42. $2x + 1 + \dfrac{4}{4x - 1}$ **43.** $\dfrac{6}{5}z + \dfrac{41}{25} + \dfrac{41}{25(5z - 1)}$

44. $\dfrac{8}{3}y - \dfrac{46}{9} + \dfrac{230}{9(3y + 5)}$ **45.** $2x^2 + x + 4 + \dfrac{6}{x - 3}$

46. $5x^2 - 2x + 14 + \dfrac{6}{x + 1}$

47. $x^5 + x^4 + x^3 + x^2 + x + 1$, $x \ne 1$

48. $x^2 + x + 1 + \dfrac{1}{x - 1}$ **49.** $x^3 - x + \dfrac{x}{x^2 + 1}$

50. $x^3 + 2x^2 + 4x + 8 + \dfrac{16}{x - 2}$ **51.** $x + 2$

52. $x - 1 + \dfrac{-3x - 10}{2x^2 + 4x + 5}$ **53.** $x^{2n} + x^n + 4$, $x^n \ne -2$

54. $x^{2n} + 5$, $x^n \ne 1$ **55.** $x^2 - x + 4 - \dfrac{17}{x + 4}$

56. $x^3 - 2x^2 + 4x - 8 + \dfrac{16}{x + 2}$

57. $x^3 - 2x^2 - 4x - 7 - \dfrac{4}{x - 2}$

58. $2x^4 + 6x^3 + 15x^2 + 45x + 136 + \dfrac{408}{x - 3}$

59. $5x^2 - 25x + 125 - \dfrac{613}{x + 5}$ **60.** $8 + \dfrac{115}{x - 10}$

61. $5x^2 + 14x + 56 + \dfrac{232}{x - 4}$

62. $5x^2 - 10x + 26 - \dfrac{44}{x + 2}$

63. $10x^3 + 10x^2 + 60x + 360 + \dfrac{1360}{x - 6}$

64. $x^4 - 16x^3 + 48x^2 - 144x + 312 - \dfrac{856}{x + 3}$

65. $0.1x + 0.82 + \dfrac{1.164}{x - 0.2}$ **66.** $x^2 - x + 0.2 + \dfrac{2.2}{x + 1}$

67. $5(3x + 2)\left(x - \frac{4}{5}\right)$ **68.** $6(3x - 4)\left(x + \frac{5}{6}\right)$

69. $(-3z + 2)(z - 2)$ **70.** $(2t^2 + 5t - 6)(t + 5)$

71. $(5t^2 + 3t + 4)(t - 6)$ **72.** $(y + 2)(y + 1)(y - 2)$

73. $(x + 2)(x^2 + 4)(x - 2)$ **74.** $(x^3 - 3x + 1)(x - 1)$

75.
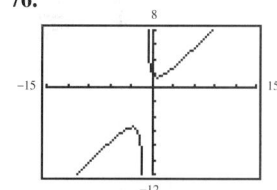

76.

77.
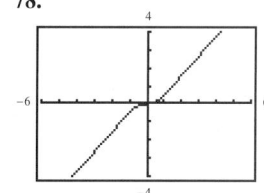

78.

79. $x^3 - 5x^2 - 5x - 10$ **80.** $x^4 + 4x^3 + 3x^2 - 4x - 4$

81. -8 **82.** -1188 **83.** $x^2 - 3$

84. $x^2 + 50x + 400$ **85.** $2x, x \neq 0$

86. $-2x + 7, x \neq 1$ **87.** $3xy, x \neq 0, y \neq 0$

88. $7uv, u \neq 0, v \neq 0$ **89.** Invalid. 5's are terms, not factors.

90. Invalid. 6's are terms, not factors.   **91.** Valid

92. Invalid. 8's are digits of the integers, not factors.

93. $2x + 8$ **94.** $h^2 + 2h$

Section Project 7.4

(a)

k	-3	-2	-1	$-\frac{1}{2}$
Divisor $(x - k)$	$x + 3$	$x + 2$	$x + 1$	$x + \frac{1}{2}$
Remainder	100	0	-6	0

k	0	$\frac{1}{2}$	1	2	3
Divisor $(x - k)$	x	$x - \frac{1}{2}$	$x - 1$	$x - 2$	$x - 3$
Remainder	4	$\frac{15}{4}$	0	0	70

(b)

x	-3	-2	-1	$-\frac{1}{2}$	0	$\frac{1}{2}$	1	2	3
$f(x)$	100	0	-6	0	4	$\frac{15}{4}$	0	0	70

(c) $f(k)$ equals the remainder when $f(x)$ is divided by $x - k$.

(d) $2, -2, -\frac{1}{2}, 1$

(e) If $f(k) = 0$, then the remainder is 0 when $f(x)$ is divided by $x - k$.

(f)

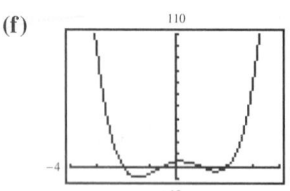

The x-intercepts are the zeros of the function.

Mid-Chapter Quiz   (page 520)

1. $(-\infty, 0) \cup (0, 4) \cup (4, \infty)$

2. (a) 0   **(b)** $\frac{9}{2}$ **(c)** Undefined   **(d)** $\frac{8}{9}$

3. $\frac{3}{2}y, y \neq 0$ **4.** $\frac{2u^2}{9v}, u \neq 0$ **5.** $-\frac{2x + 1}{x}, x \neq \frac{1}{2}$

6. $\frac{z + 3}{2z - 1}, z \neq -3$ **7.** $\frac{7 + 3ab}{a}, b \neq 0$

8. $\frac{n^2}{m + n}, n \neq 2m$ **9.** $\frac{t}{2}, t \neq 0$ **10.** $\frac{5x}{x - 2}, x \neq -2$

11. $\frac{8x}{3(x - 1)^2(x + 3)}$

12. $\frac{4(u - v)^2}{5uv}, u \neq v, u \neq -v$ **13.** $-\frac{3t}{2}, t \neq 0, t \neq 3$

14. $\frac{2(x + 1)}{3x}, x \neq -2, x \neq -1$ **15.** $\frac{-3x^2 + x}{4(x + 5)}$

16. $\frac{4x^4 + x^3 - 18x^2 + 8}{x^2(x^2 - 4)}$ **17.** $\frac{5(2 - x)}{4x - 15}, x \neq 0$

18. $\frac{2(x^2 + 9)}{x + 3}, x \neq 0$ **19.** $2x^2 - 4x + 3, x \neq \frac{2}{3}$

20. (a) $\frac{6000 + 10.50x}{x}$ **(b)** \$22.50   **21.** $\frac{8(x + 2)}{(x + 4)^2}$

Section 7.5   (page 529)

1. It is an extra solution found by multiplying both sides of the original equation by an expression containing the variable. It is identified by checking all solutions in the original equation.

2. When the equation involves only two fractions, one on each side of the equation, the equation can be solved by cross-multiplication.

3. (a) No   **(b)** No   **(c)** No   **(d)** Yes

4. (a) No   **(b)** Yes   **(c)** Yes   **(d)** No

5. (a) No   **(b)** Yes   **(c)** Yes   **(d)** No

6. (a) Yes   **(b)** No   **(c)** No   **(d)** No

7. $\frac{3}{2}$ **8.** 24   **9.** $\frac{1}{4}$ **10.** $\frac{15}{2}$ **11.** $-\frac{8}{3}$ **12.** $\frac{9}{2}$

13. 10   **14.** 20   **15.** $\frac{7}{4}$ **16.** 10   **17.** $\frac{1}{3}$

18. $-\frac{27}{4}$ **19.** 61   **20.** $-\frac{4}{5}$ **21.** $\frac{18}{5}$ **22.** 3

23. 3   **24.** 130   **25.** 3   **26.** $\frac{4}{5}$ **27.** $-\frac{11}{5}$

28. 5   **29.** $\frac{4}{3}$ **30.** -25 **31.** $-\frac{4}{15}$ **32.** 5

33. No solution   **34.** No solution   **35.** $-\frac{1}{6}$ **36.** -1

37. No solution   **38.** $\frac{3}{4}$ **39.** 3   **40.** 20   **41.** ± 6

42. ± 30 **43.** ± 4 **44.** ± 15 **45.** 8, -9

46. $-2, -10$ **47.** 3, 13   **48.** $-\frac{5}{2}, 10$ **49.** 0, 2

50. $-6, 7$ **51.** 8, -3 **52.** $\frac{4}{3}, 2$ **53.** -5 **54.** 4

55. 2, 3   **56.** $-\frac{3}{2}, 2$ **57.** No solution

58. $\frac{3}{2}, 1$ **59.** 8, $\frac{1}{8}$ **60.** 12, $\frac{1}{8}$ **61.** 40 mph

62. 50 mph, 60 mph   **63.** 8 mph, 10 mph

64. 96 mph, 246 mph   **65.** 4 mph   **66.** 48 mph

67. 10   **68.** 9 hr   **69.** 3 hr; $1\frac{7}{8}$ min

70. $1\frac{16}{17}$ hr; $\frac{ab}{a + b}$ days   **71.** 15 hr; $22\frac{1}{2}$ hr

72. $16\frac{2}{3}$ hr; 25 hr   **73.** $11\frac{1}{4}$ hr   **74.** $9\frac{4}{5}$ hr; $12\frac{1}{4}$ hr

75. 85%   **76.** 3000 units

77. (a) 1992 **(b)** Answers may vary.

Section Project 7.5

(a) $x = -12$ **(b)** $\dfrac{x^2 + 2x + 8}{(x + 4)(x - 4)}$

(c) $\dfrac{14x + 27}{3(x + 3)}$ **(d)** $x = -\dfrac{27}{4}$

(e) An equation is a statement of equality of two expressions.

(f) To add or subtract rational expressions with unlike denominators, you must first rewrite the rational expressions so that they have like denominators. The like denominators that you use is the least common denominator. When solving rational equations, multiply both sides of the equation by the least common denominator.

Section 7.6 (page 538)

1. An asymptote of a graph is a line to which the graph becomes arbitrarily close as $|x|$ or $|y|$ increases without bound.

2. No. $f(x) = \dfrac{1}{x^2 + 1}$

3. (a)

x	0	0.5	0.9	0.99	0.999
y	-4	-8	-40	-400	-4000

x	2	1.5	1.1	1.01	1.001
y	4	8	40	400	4000

x	2	5	10	100	1000
y	4	1	0.4444	0.0404	0.0040

(b) Vertical asymptote: $x = 1$;
Horizontal asymptote: $y = 0$
(c) $(-\infty, 1) \cup (1, \infty)$

4. (a)

x	0	0.5	0.9	0.99	0.999
y	0	-2	-18	-198	-1998

x	2	1.5	1.1	1.01	1.001
y	4	6	22	202	2002

x	2	5	10	100	1000
y	4	2.5	2.2222	2.0202	2.0020

(b) Vertical asymptote: $x = 1$;
Horizontal asymptote: $y = 2$
(c) $(-\infty, 1) \cup (1, \infty)$
5. (d) **6. (a)** **7. (b)** **8. (c)**
9. Domain: $(-\infty, 0) \cup (0, \infty)$; Horizontal asymptote: $y = 0$; Vertical asymptote: $x = 0$
10. Domain: $(-\infty, 5) \cup (5, \infty)$; Horizontal asymptote: $y = 0$; Vertical asymptote: $x = 5$
11. Domain: $(-\infty, -8) \cup (-8, \infty)$; Horizontal asymptote: $y = 1$; Vertical asymptote: $x = -8$
12. Domain: $(-\infty, 10) \cup (10, \infty)$; Horizontal asymptotes: None; Vertical asymptote: $u = 10$
13. Domain: $(-\infty, 0) \cup (0, 1) \cup (1, \infty)$; Horizontal asymptote: $y = 0$; Vertical asymptotes: $t = 0, t = 1$
14. Domain: $(-\infty, 0) \cup (0, \infty)$; Horizontal asymptote: $y = 4$; Vertical asymptote: $x = 0$
15. Domain: $(-\infty, -3) \cup (-3, \infty)$; Horizontal asymptotes: None; Vertical asymptote: $s = -3$
16. Domain: $(-\infty, \infty)$; Horizontal asymptote: $y = 0$; Vertical asymptotes: None
17. Domain: $\left(-\infty, \frac{1}{3}\right), \cup \left(\frac{1}{3}, \infty\right)$; Horizontal asymptote: $y = \frac{5}{3}$; Vertical asymptote: $x = \frac{1}{3}$
18. Domain: $\left(-\infty, \frac{1}{2}\right) \cup \left(\frac{1}{2}, \infty\right)$; Horizontal asymptote: $y = \frac{3}{2}$; Vertical asymptote: $x = \frac{1}{2}$
19. Domain: $(-\infty, -1) \cup (-1, 1) \cup (1, \infty)$; Horizontal asymptote: $y = 5$; Vertical asymptotes: $x = -1, x = 1$
20. Domain: $(-\infty, \infty)$; Horizontal asymptote: $y = 2$; Vertical asymptotes: None
21. (d) **22. (b)** **23. (a)** **24. (c)**

25.

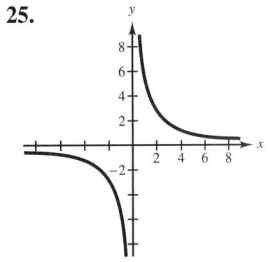

26.

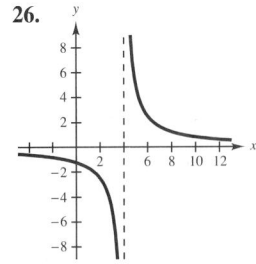

27.

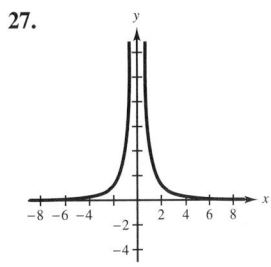

28.

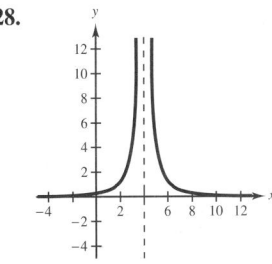

29.

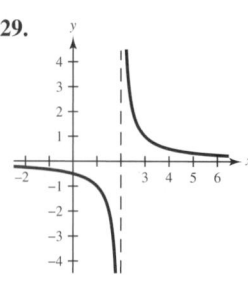

30.

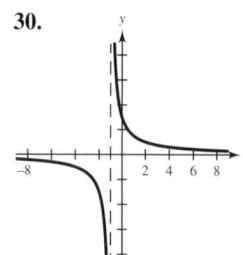

31.

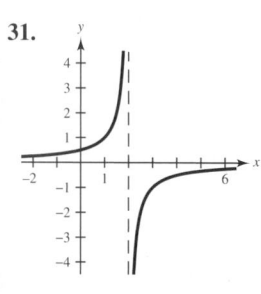

32.

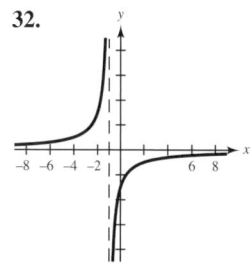

33.

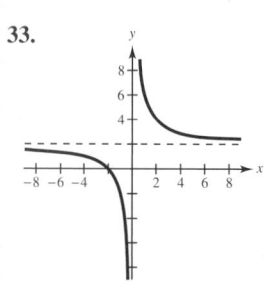

34.

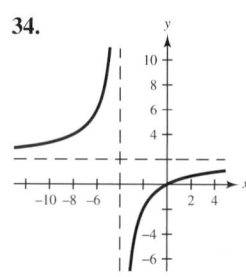

35.

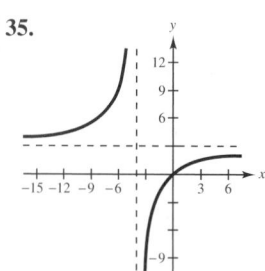

36.

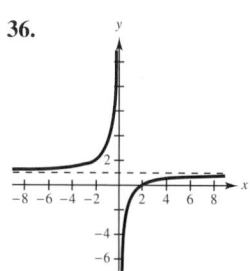

37.

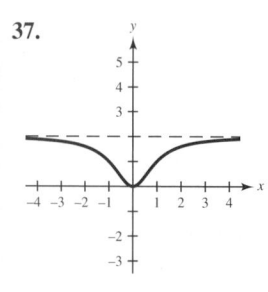

38.

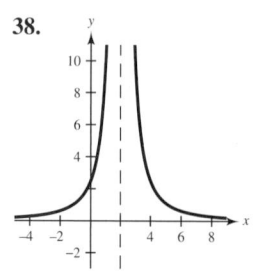

39.

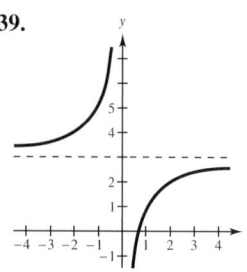

40.

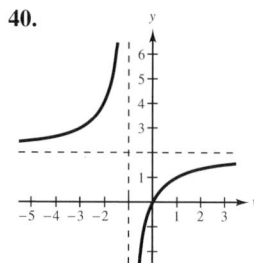

41.

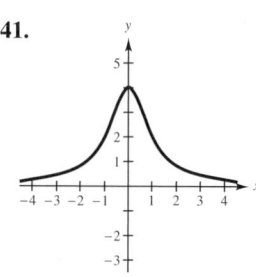

42.

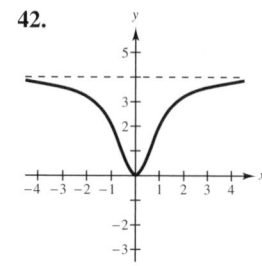

43.

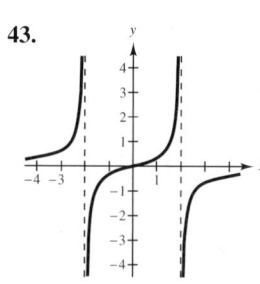

44.

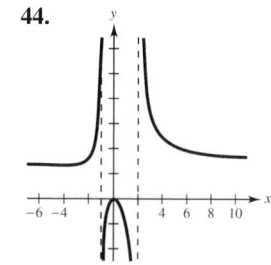

45.

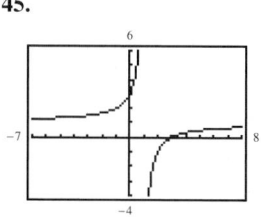

46.

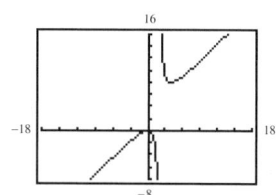

Domain: $(-\infty, 1) \cup (1, \infty)$
Horizontal asymptote: $y = 1$
Vertical asymptote: $x = 1$

Domain: $(-\infty, 2) \cup (2, \infty)$
Horizontal asymptotes: None
Vertical asymptote: $x = 2$

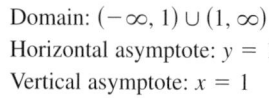

47.

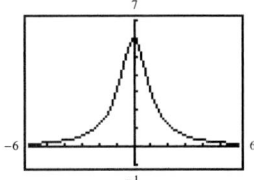

Domain: $(-\infty, \infty)$
Horizontal asymptote: $y = 0$
Vertical asymptotes: None

48.

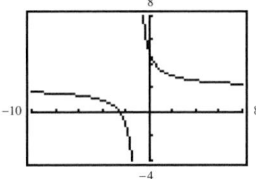

Domain:
$(-\infty, -1) \cup (-1, \infty)$
Horizontal asymptote: $y = 2$
Vertical asymptote: $x = -1$

49.

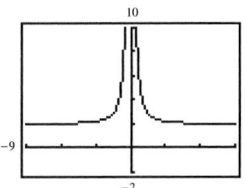

Domain: $(-\infty, 0) \cup (0, \infty)$
Horizontal asymptote: $y = 2$
Vertical asymptote: $x = 0$

50.

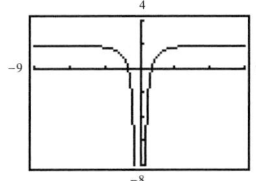

Domain: $(-\infty, 0) \cup (0, \infty)$
Horizontal asymptote: $y = 2$
Vertical asymptote: $x = 0$

51.

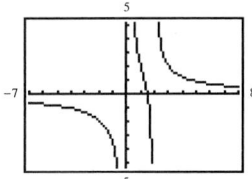

Domain:
$(-\infty, 0) \cup (0, 2) \cup (2, \infty)$
Horizontal asymptote:
$y = 0$
Vertical asymptotes:
$x = 0, x = 2$

52.

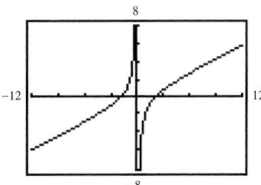

Domain:
$(-\infty, 0) \cup (0, \infty)$
Horizontal asymptotes:
None
Vertical asymptote: $x = 0$

53.

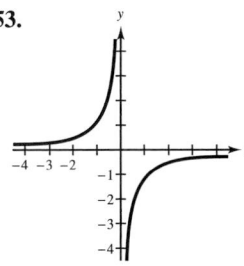

Reflection in the x-axis

54.

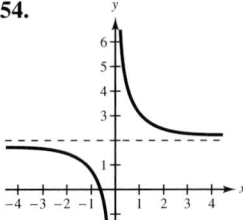

Vertical shift 2 units upward

55.

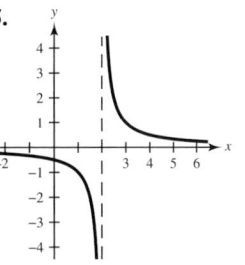

Horizontal shift 2 units
right

56.

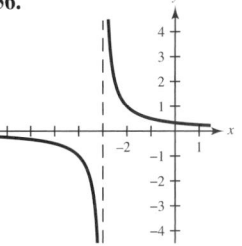

Horizontal shift 3 units left

57.

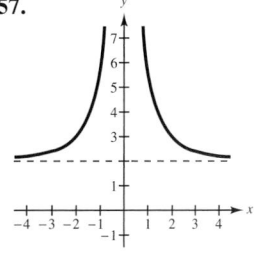

Vertical shift 2 units
upward

58.

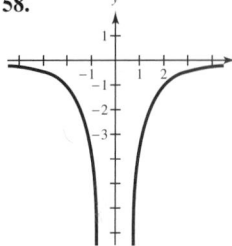

Reflection in the x-axis

59.

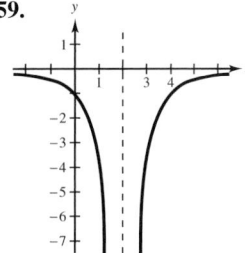

Reflection in the x-axis
and a horizontal shift 2
units right

60.

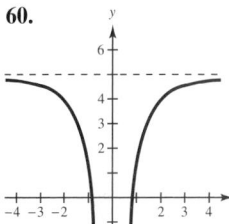

Reflection in the x-axis and
a vertical shift 5 units upward

61.

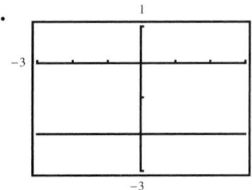

The fraction is not reduced
to lowest terms.

62.

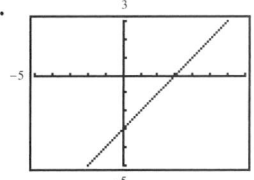

The fraction is not reduced
to lowest terms.

63. $(-2, 0)$ **64.** $(0, 0)$ **65.** $(-1, 0), (1, 0)$

66. $(-1, 0), (2, 0)$

67. (a)

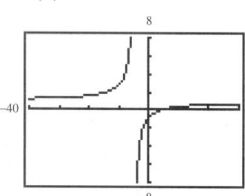

68. (a)

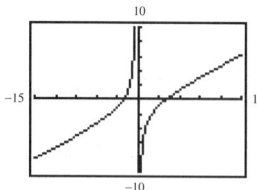

(b) $(1, 0)$ **(b)** $(-2, 0), (4, 0)$

69. (a)

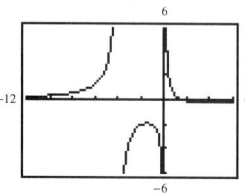

70. (a)

(b) $(4, 0)$ **(b)** $(2, 0)$

71. (a)

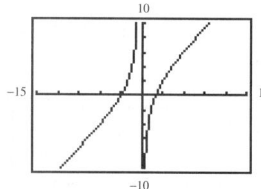

72. (a)

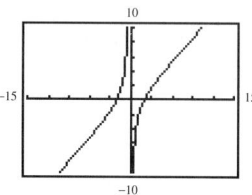

(b) $(-3, 0), (2, 0)$ **(b)** $(-2, 0), (2, 0)$

73. (a)

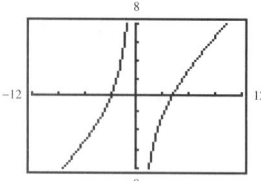

74. (a)

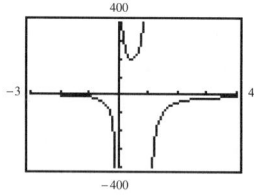

(b) $(-3, 0), (4, 0)$ **(b)** $(-2, 0)$

75. (a) $\overline{C} = \dfrac{2500 + 0.50x}{x}$ **(b)** $3, \$0.75$

(c)

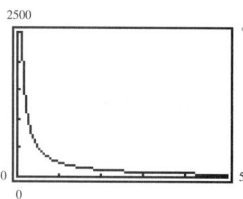

$\overline{C} = \$0.50$

76. (a) $\overline{C} = \dfrac{30,000 + 1.25x}{x}$ **(b)** $\$4.25, \$1.55$

(c)

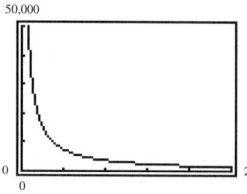

$\overline{C} = \$1.25$

77. (a) $C = 0$. The chemical is eliminated from the body.

(b)

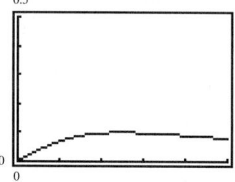

$t \approx 2.5$ hours

78. (a) $[0, 20]$

(b)

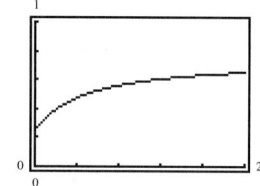

0.75

79. (a) Answers may vary.

(b) $(0, \infty)$

(c)

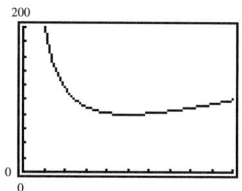

20 units $\times$ 20 units

80. (a) $652,170; 1,687,500; 3,000,000$

(b)

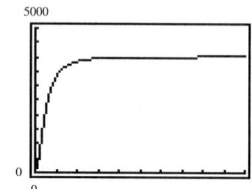

4000 thousand or 4,000,000

(c) Upper limit on sales

Section Project 7.6

(a) $\dfrac{2(x + 1)}{x - 3}$ (b) $\dfrac{x - 3}{(x + 2)(x^2 + 1)}$

(c) $\dfrac{x^2}{x^2 - 1}$ (d) $\dfrac{x - 6}{(x - 4)(x + 2)}$

Section 7.7   (page 547)

1. (a) Not a solution
   (b) Solution
   (c) Not a solution
   (d) Not a solution

2. (a) Not a solution
   (b) Not a solution
   (c) Solution
   (d) Solution

3. 3

4. -2

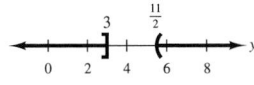

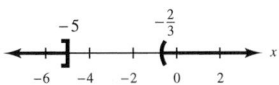

5. $0, -5$

6. $2, 10$

7. $(3, \infty)$

8. $(-\infty, 4)$

9. $(-\infty, 3)$

10. $(4, \infty)$

11. $(-\infty, 3)$

12. $(2, \infty)$

13. $[0, 4)$

14. $(-3, 1)$

15. $(-\infty, 3] \cup \left(\frac{11}{2}, \infty\right)$

16. $(-\infty, -5] \cup \left(-\frac{2}{3}, \infty\right)$

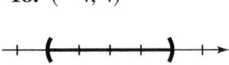

17. $(-1, 3)$

18. $(-4, 4)$

19. $(4, 7)$

20. $\left(-\infty, -\frac{7}{3}\right) \cup (-2, \infty)$

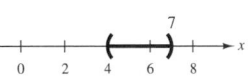

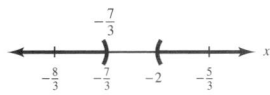

21. $(-\infty, 3) \cup [6, \infty)$

22. $(5, 6]$

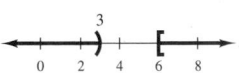

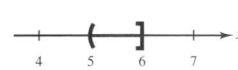

23. $(-\infty, -1) \cup (0, 1)$

24. $(-\infty, 0) \cup \left(\frac{1}{4}, \infty\right)$

25. $(-\infty, -1) \cup (4, \infty)$

26. $[-2, 3]$ 27. $\left(-5, \frac{5}{2}\right)$

28. $(0, 0.382) \cup (2.618, \infty)$ 29. $(3, 4)$

30. $(-\infty, -0.58) \cup (0.58, \infty)$

31.

32.

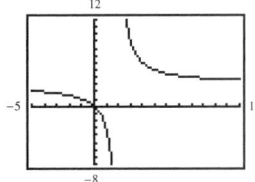

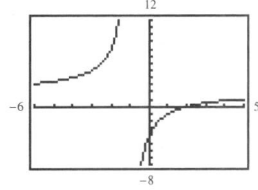

(a) $[0, 2)$ (b) $(2, 4]$

(a) $(-1, 2]$ (b) $[-2, -1)$

33.

34.

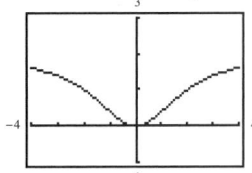

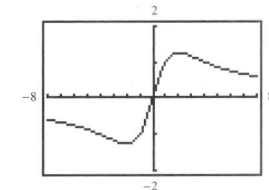

(a) $(-\infty, -2] \cup [2, \infty)$
(b) $(-\infty, \infty)$

(a) $[1, 4]$ (b) $[0, \infty)$

35. $48 < r < 55$; $58 < r + 10 < 65$

36. (a) $\overline{C} = \dfrac{3000}{x} + 0.75$, $x > 0$

(b)

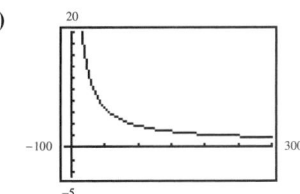

Horizontal asymptote: $\overline{C} = 0.75$

(c) $x > 2400$ units

37. (a)

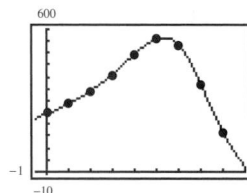

(b) $5.7 \le t \le 13.6$

Section Project 7.7

(a) $\overline{C} = \dfrac{8000}{x} + 0.6$ **(b)** $x > 3333$ **(c)** $x > 5714$

(d) $x > 20{,}000$ **(e)** No solution

(f)

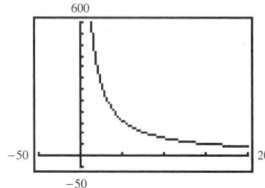

(g) $\overline{C} = 0.6$

(h) With increasing production, the average cost per unit decreases toward the average cost of $\overline{C} = 0.6$.

Review Exercises (page 550)

1. $(-\infty, 8) \cup (8, \infty)$ **2.** $(-\infty, -12) \cup (-12, \infty)$

3. $(-\infty, 1) \cup (1, 6) \cup (6, \infty)$

4. $(-\infty, -4) \cup (-4, 0) \cup (0, 4) \cup (4, \infty)$

5. $\dfrac{2x^3}{5}, \; x \ne 0, y \ne 0$ **6.** $\dfrac{y^4}{14z^2}, \; y \ne 0$ **7.** $\dfrac{b - 3}{6(b - 4)}$

8. $\dfrac{2}{5a + 13}, \; a \ne 0$ **9.** $-9, \; x \ne y$

10. $\dfrac{1}{x - 4}, \; x \ne -3$ **11.** $\dfrac{x}{2(x + 5)}, \; x \ne 5$ **12.** $\dfrac{1}{x - 3}$

13. $\dfrac{y}{8x}, \; y \ne 0$ **14.** $60x^5y, \; x \ne 0, y \ne 0$

15. $12z(z - 6), \; z \ne -6$ **16.** $\dfrac{x + 4}{2(x - 4)}$

17. $-\frac{1}{4}, \; u \ne 0, u \ne 3$

18. $\dfrac{25x(x - 1)}{x + 5}, \; x \ne 0, x \ne 1, x \ne -1$

19. $3x^2, \; x \ne 0$ **20.** $0, \; x \ne 0, y \ne 0$ **21.** $\dfrac{125y}{x}, \; y \ne 0$

22. $\dfrac{3}{2z^4}$ **23.** $\dfrac{x(x - 1)}{x - 7}, \; x \ne -1, x \ne 1$ **24.** $\dfrac{4}{3xy}$

25. $-\dfrac{7a}{9}$ **26.** $\dfrac{5y + 11}{2y + 1}$ **27.** $\dfrac{-48x + 35}{48x}$ **28.** $\dfrac{17y}{24x}$

29. $\dfrac{4x + 3}{(x + 5)(x - 12)}$ **30.** $\dfrac{x - 22}{(x - 10)(4 - x)}$

31. $\dfrac{5x^3 - 5x^2 - 31x + 13}{(x + 2)(x - 3)}$ **32.** $\dfrac{31x - 78}{(x + 6)(x - 5)}$

33. $\dfrac{x + 24}{x(x^2 + 4)}$ **34.** $\dfrac{4x}{(x + 2)(x - 5)}$

35. $\dfrac{6(x - 9)}{(x + 3)^2(x - 3)}$ **36.** $\dfrac{9y^2 + 50y - 80}{y(y + 5)(y - 2)}$

37. $\dfrac{6(x + 5)}{x(x + 7)}, \; x \ne 5, x \ne -5$ **38.** $\dfrac{2}{5}, \; x \ne 2, x \ne \dfrac{4}{3}$

39. $\dfrac{3t^2}{5t - 2}, \; t \ne 0$ **40.** $x - 1, \; x \ne 0, x \ne 2$

41. $\dfrac{-a^2 + a + 16}{(4a^2 + 16a + 1)(a - 4)}, \; a \ne 0, a \ne -4$

42. $\dfrac{y - x}{xy}, \; x \ne -y$ **43.** $2x^2 - \dfrac{1}{2}, \; x \ne 0$

44. $2x^6y^6 - 8x^5y^3 - 5x^2y, \; x \ne 0, y \ne 0$

45. $2x^2 + x - 1 + \dfrac{1}{3x - 1}$

46. $4x^3 + 7x^2 + 7x + 32 + \dfrac{64}{x - 2}$

47. $x^2 - 2, \; x \ne 1, x \ne -1$

48. $3x^4 + 3x^2 + 3 + \dfrac{3}{x^2 - 1}$ **49.** $x^2 + 5x - 7, \; x = -2$

50. $x^3 + 3x^2 - 2$ **51.** $x^3 + 3x^2 + 6x + 18 + \dfrac{29}{x - 3}$

52. $2x^2 - x + \dfrac{11}{2} - \dfrac{\frac{19}{4}}{x + \frac{1}{2}}$

53.

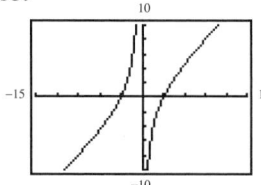

54.

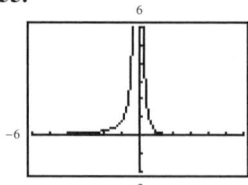

55.

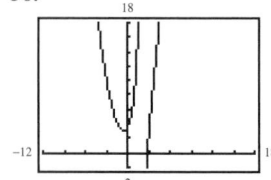

56.

57. -40 **58.** 3 **59.** $\frac{3}{2}$ **60.** 2 **61.** 5

62. $3, -\frac{16}{3}$ **63.** $6, -4$ **64.** $-2, -4$

65. $2, -6$ **66.** $3, -\frac{3}{2}$ **67.** $-2, 2$ **68.** $\frac{1}{2}, 4$

69. $-\frac{9}{5}, 3$ **70.** $0, 4$

71. $(-3, 0)$

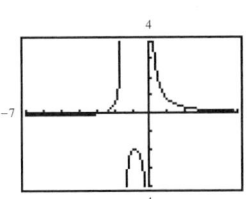

72. $(-2, 0), (4, 0)$

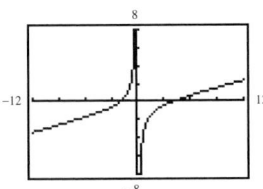

87.

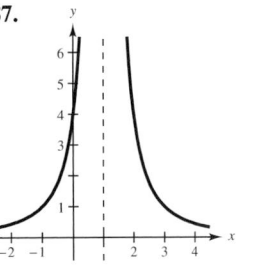

88.

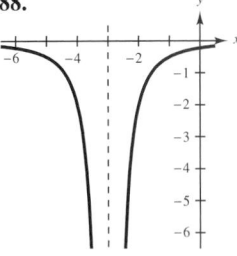

73. 25 **74.** 56 mph **75.** $6\frac{2}{3}$ min **76.** 4
77. (b) **78.** (d) **79.** (a) **80.** (c)

81.

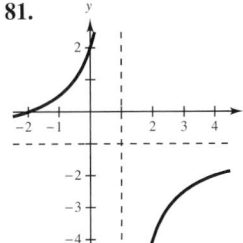

82.

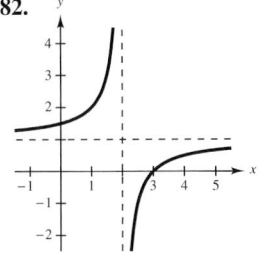

89.

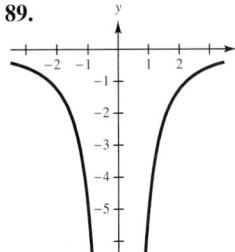

90.

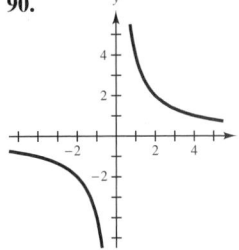

83.

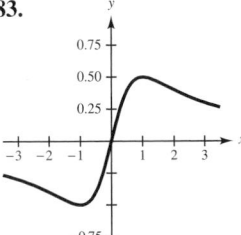

84.

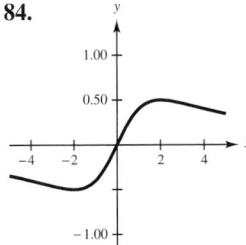

91.

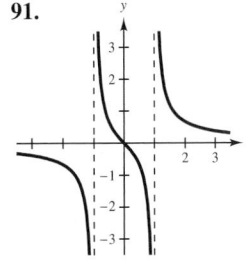

92.

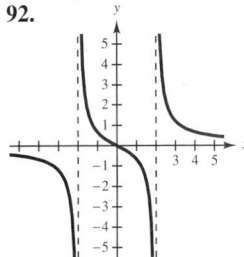

85.

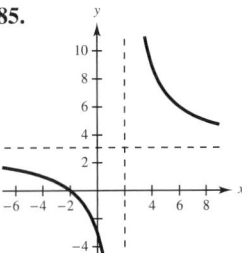

86.

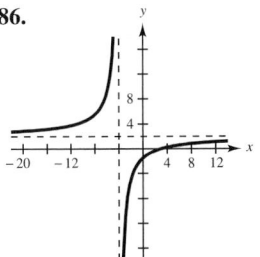

93.

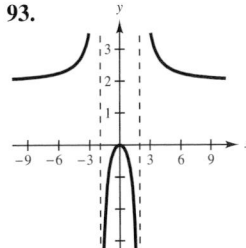

94.

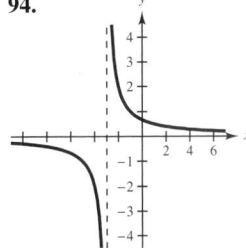

95. Horizontal asymptote: $\overline{C} = \frac{1}{2}$; As x gets larger, the average cost per unit approaches \$0.50.

96. (a) $N(5) = 304$ thousand fish; $N(10) \approx 453.3$ thousand fish; $N(25) \approx 702.2$ thousand fish

(b) The population is limited by the horizontal asymptote $N = 1200$ thousand fish.

97. $(-\infty, 0] \cup \left(\frac{7}{2}, \infty\right)$ **98.** $\left(1, \frac{9}{2}\right]$

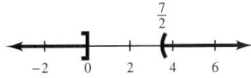

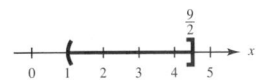

99. $(-6, -4)$ **100.** $(2, 9)$

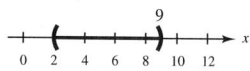

101. (a) \$176 million   (b) \$1584 million
     (c) No. $p = 100$ is a vertical asymptote.
     (d) $[0, 74)$

102. $x > 90,909$

Chapter 7 Test   (page 553)

1. $(-\infty, -5) \cup (-5, 5) \cup (5, \infty)$ **2.** $-\frac{1}{3}, \ x \neq 2$

3. $\frac{5z}{3}, \ z \neq 0$ **4.** $\frac{4}{y + 4}, \ y \neq 2$ **5.** $\frac{(2x + 3)^2}{x + 1}, \ x \neq \frac{3}{2}$

6. $\frac{14y^6}{15}, \ x \neq 0$ **7.** $\frac{x^3}{4}, \ x \neq 0, x \neq -2$

8. $-(3x + 1), \ x \neq 0, \ x \neq \frac{1}{3}$ **9.** $\frac{-2x^2 + 2x + 1}{x + 1}$

10. $\frac{5x^2 - 15x - 2}{(x - 3)(x + 2)}$ **11.** $\frac{5x^3 + x^2 - 7x - 5}{x^2(x + 1)^2}$

12. $4a^5 + 7a^2 - a, a \neq 0$ **13.** $t^2 + 3 - \frac{6t - 6}{t^2 - 2}$

14. $2x^3 + 6x^2 + 3x + 9 + \frac{20}{x - 3}$

15. (a)   (b)

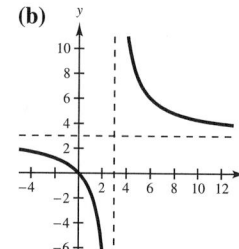

16. 22   **17.** $-1, -\frac{15}{2}$ **18.** No solution
19. $\left(2, \frac{11}{4}\right)$ **20.** $[-8, 3)$

 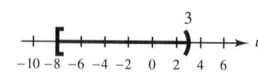

21. $6\frac{2}{3}$ hr, 10 hr

Cumulative Test: Chapters 1–7   (page 554)

1. $-3x + 22$ **2.** $0.02x + 100$ **3.** $a^2 + 4a + 1$

4. $\frac{3}{a + 7}$ **5.** $-\frac{4x}{3}, \ x \neq 0$ **6.** $\frac{2u^4}{9v^6}, \ u \neq 0$

7. $x - 9$ **8.** $\sqrt{5u}$ **9.** $4t + 12\sqrt{t} + 9$

10. $(-3, 5)$ **11.** $(2, 4, 3)$ **12.** 2   **13.** 6
14. $5 \pm 5\sqrt{2}\,i$

15.

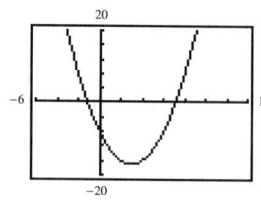

$x = 3 \pm \sqrt{17}$

16. $y = -\frac{3}{4}x^2 + 3x$

17.

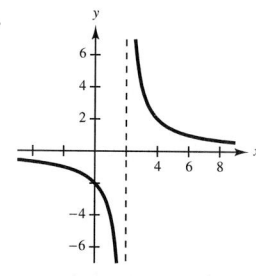

18. $\frac{1}{4}, \frac{1}{5}, \frac{20}{9}$

CHAPTER 8

Section 8.1   (page 563)

1. (a) $(f + g)(x) = x^2 + 2x, \ (-\infty, \infty)$
     (b) $(f - g)(x) = 2x - x^2, \ (-\infty, \infty)$
     (c) $(fg)(x) = 2x^3, \ (-\infty, \infty)$
     (d) $\left(\dfrac{f}{g}\right)(x) = \dfrac{2}{x}, \ (-\infty, 0) \cup (0, \infty)$

2. (a) $(f + g)(x) = x + 1 + \sqrt{x}, \ [0, \infty)$
     (b) $(f - g)(x) = x + 1 - \sqrt{x}, \ [0, \infty)$
     (c) $(fg)(x) = x^{3/2} + x^{1/2}, \ [0, \infty)$
     (d) $\left(\dfrac{f}{g}\right)(x) = \dfrac{x + 1}{\sqrt{x}}, \ (0, \infty)$

3. (a) $(f + g)(x) = x^2 + 4x - 12, \ (-\infty, \infty)$
     (b) $(f - g)(x) = -x^2 + 4x + 6, \ (-\infty, \infty)$
     (c) $(fg)(x) = 4x^3 - 3x^2 - 36x + 27, \ (-\infty, \infty)$
     (d) $\left(\dfrac{f}{g}\right)(x) = \dfrac{4x - 3}{x^2 - 9}, \ (-\infty, -3) \cup (-3, 3) \cup (3, \infty)$

4. (a) $(f + g)(x) = x^3 + x^2 + 1, \ (-\infty, \infty)$
     (b) $(f - g)(x) = x^3 - x^2 - 1, \ (-\infty, \infty)$
     (c) $(fg)(x) = x^5 + x^3, \ (-\infty, \infty)$
     (d) $\left(\dfrac{f}{g}\right)(x) = \dfrac{x^3}{x^2 + 1}, \ (-\infty, \infty)$

5. (a) $(f + g)(x) = \dfrac{1}{x} + 5x,\ (-\infty, 0) \cup (0, \infty)$

  (b) $(f - g)(x) = \dfrac{1}{x} - 5x,\ (-\infty, 0) \cup (0, \infty)$

  (c) $(fg)(x) = 5,\ (-\infty, 0) \cup (0, \infty)$

  (d) $\left(\dfrac{f}{g}\right)(x) = \dfrac{1}{5x^2},\ (-\infty, 0) \cup (0, \infty)$

6. (a) $(f + g)(x) = \dfrac{2x}{(x + 2)(x - 2)},$
  $(-\infty, -2) \cup (-2, 2) \cup (2, \infty)$

  (b) $(f - g)(x) = \dfrac{-4}{(x + 2)(x - 2)},$
  $(-\infty, -2) \cup (-2, 2) \cup (2, \infty)$

  (c) $(fg)(x) = \dfrac{1}{(x + 2)(x - 2)},$
  $(-\infty, -2) \cup (-2, 2) \cup (2, \infty)$

  (d) $\left(\dfrac{f}{g}\right)(x) = \dfrac{x - 2}{x + 2},\ (-\infty, -2) \cup (-2, 2) \cup (2, \infty)$

7. (a) $(f + g)(x) = \sqrt{x + 4} + \sqrt{4 - x},\ [-4, 4]$
  (b) $(f - g)(x) = \sqrt{x + 4} - \sqrt{4 - x},\ [-4, 4]$
  (c) $(fg)(x) = \sqrt{16 - x^2},\ [-4, 4]$

  (d) $\left(\dfrac{f}{g}\right)(x) = \dfrac{\sqrt{x + 4}}{\sqrt{4 - x}},\ [-4, 4)$

8. (a) $(f + g)(x) = \sqrt{x} + \sqrt{x^2 - 16},\ [4, \infty)$
  (b) $(f - g)(x) = \sqrt{x} - \sqrt{x^2 - 16},\ [4, \infty)$
  (c) $(fg)(x) = \sqrt{x^3 - 16x},\ [4, \infty)$

  (d) $\left(\dfrac{f}{g}\right)(x) = \dfrac{\sqrt{x}}{\sqrt{x^2 - 16}},\ (4, \infty)$

9. (a) 11 **(b)** 0   **10. (a)** -15 **(b)** 125
11. (a) 19 **(b)** -79
12. (a) 47
  (b) $(f - g)(1)$ is not real [1 is not in the domain of $f(x)$].
13. (a) 10 **(b)** $\frac{2}{5}$ **14. (a)** 0 **(b)** 4
15. (a) $\frac{1}{3}$ **(b)** Undefined (division by zero)
16. (a) $\frac{8}{3}$ **(b)** Undefined (division by zero)
17. (a) $x^2 - 3$ **(b)** $(x - 3)^2$ **(c)** 13 **(d)** 16
18. (a) $x^3 + 5$ **(b)** $(x + 5)^3$ **(c)** 13 **(d)** 8
19. (a) $3|x - 1|$ **(b)** $3|x - 3|$ **(c)** 0 **(d)** 3
20. (a) $|2x + 5|$ **(b)** $2|x| + 5$ **(c)** 1 **(d)** 13
21. (a) $\sqrt{x + 5}$ **(b)** $\sqrt{x} + 5$ **(c)** 3 **(d)** 8
22. (a) $\sqrt{2x + 3}$ **(b)** $2\sqrt{x + 6} - 3$ **(c)** 3 **(d)** 1

23. (a) $\dfrac{1}{\sqrt{x} - 3}$ **(b)** $\dfrac{1}{\sqrt{x - 3}}$ **(c)** $\dfrac{1}{4}$ **(d)** $\dfrac{1}{3}$

24. (a) $\dfrac{4x^2}{1 - 4x^2}$ **(b)** $\dfrac{x^2 - 4}{4}$ **(c)** $-\dfrac{16}{15}$ **(d)** $-\dfrac{3}{4}$

25. $(f \circ g)(x) = 4x^2 + 1,\ (-\infty, \infty)$
  $(g \circ f)(x) = 2(x^2 + 1),\ (-\infty, \infty)$
26. $(f \circ g)(x) = -15x - 7,\ (-\infty, \infty)$
  $(g \circ f)(x) = -15x + 13,\ (-\infty, \infty)$
27. $(f \circ g)(x) = \sqrt{x} - 2,\ [2, \infty)$
  $(g \circ f)(x) = \sqrt{x} - 2,\ [0, \infty)$
28. $(f \circ g)(x) = \sqrt{x^2 - 5},\ (-\infty, -\sqrt{5}\,] \cup [\sqrt{5}, \infty)$
  $(g \circ f)(x) = x - 5,\ [5, \infty)$

29. $(f \circ g)(x) = \dfrac{9}{x^2 + 9},\ (-\infty, \infty)$

  $(g \circ f)(x) = \left(\dfrac{9}{x + 9}\right)^2,\ (-\infty, -9) \cup (-9, \infty)$

30. $(f \circ g)(x) = \dfrac{\sqrt{x}}{\sqrt{x} - 4},\ [0, 16) \cup (16, \infty)$

  $(g \circ f)(x) = \sqrt{\dfrac{x}{x - 4}},\ (-\infty, 0] \cup (4, \infty)$

31. $t^2 - 8t - 3$ **32.** $-\frac{2}{13}$ **33.** $t^2 - 2t + 3$
34. $5z^3 - 12z^2 - 9z$ **35.** 5 **36.** $2x + h - 3$
37. 10 **38.** 3 **39.** $5y^2 - 15y + 3$
40. $25z^2 + 15z$

41.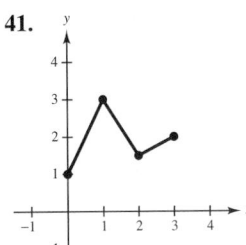

42.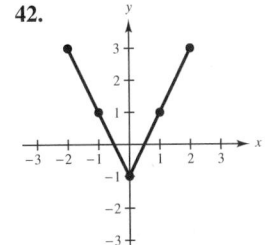

43.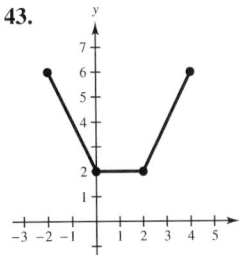

44.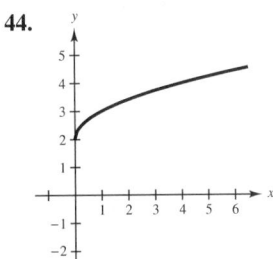

45.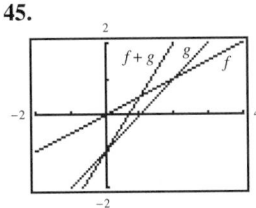

46.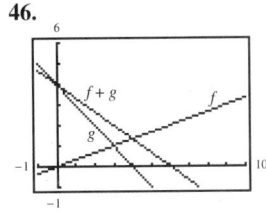

47.

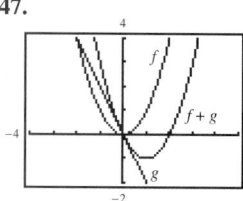

48.

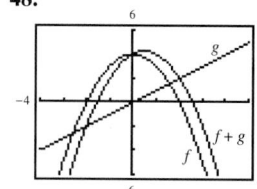

49.

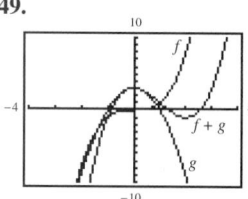

50.

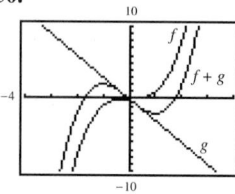

f is more significant for large x.

f is more significant for large x.

51. (a) -1 (b) -2 (c) -2

52. (a) 2 (b) -3 (c) -3

53. (a) -1 (b) 1

54. (a) -3 (b) 1

55. (a) 10 (b) 1 (c) 1

56. (a) 3 (b) 1 (c) 2

57. (a) 0 (b) 10

58. (a) 5 (b) 2

59. (a) $T(x) = \frac{3}{4}x + \frac{1}{15}x^2$

(b)

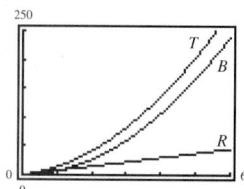

(c) B

60.

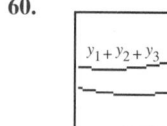

13.17 cents per mile

61. (a) $f(g(x)) = 0.02x - 200,000$

(b) $g(f(x)) = 0.02(x - 200,000)$

This part represents the bonus, because it gives 2% of sales over $200,000.

62. $(C \circ x)t = 102t + 300$

Production cost after t hours of operation

63. $A(r(t)) = 0.36\pi t^2$

Section Project 8.1

(a) $R = p - 2000$ (b) $S = 0.95p$

(c) $(R \circ S)(p) = 0.95p - 2000$; 5% discount followed by the $2000 rebate

$(S \circ R)(p) = 0.95(p - 2000)$; 5% discount after the price is reduced by the rebate

(d) $(R \circ S)(26,000) = 22,700$, $(S \circ R)(26,000) = 22,800$.

$R \circ S$ yields the smaller cost because the dealer discount is calculated on a larger base.

Section 8.2 (page 573)

1. $f^{-1}(x) = \dfrac{x}{5}$ **2.** $f^{-1}(x) = 3x$ **3.** $f^{-1}(x) = x - 10$

4. $f^{-1}(x) = x + 5$ **5.** $f^{-1}(x) = 2x$

6. $f^{-1}(x) = \frac{1}{2}x$ **7.** $f^{-1}(x) = \sqrt[3]{x}$

8. $f^{-1}(x) = \sqrt[5]{x}$ **9.** $f^{-1}(x) = x^3$ **10.** $f^{-1}(x) = x^5$

11. $f(g(x)) = f\left(\frac{1}{10}x\right) = 10\left(\frac{1}{10}x\right) = x$

$g(f(x)) = g(10x) = \frac{1}{10}(10x) = x$

12. $f(g(x)) = f\left(\dfrac{3x}{2}\right) = \dfrac{2}{3}\left(\dfrac{3x}{2}\right) = x$

$g(f(x)) = g\left(\dfrac{2x}{3}\right) = \dfrac{3}{2}\left(\dfrac{2x}{3}\right) = x$

13. $f(g(x)) = f(x - 15) = (x - 15) + 15 = x$

$g(f(x)) = g(x + 15) = (x + 15) - 15 = x$

14. $f(g(x)) = f(3 - x) = 3 - (3 - x) = x$

$g(f(x)) = g(3 - x) = 3 - (3 - x) = x$

15. $f(g(x)) = f\left(\dfrac{1 - x}{2}\right) = 1 - 2\left(\dfrac{1 - x}{2}\right)$

$= 1 - (1 - x) = x$

$g(f(x)) = g(1 - 2x) = \dfrac{1 - (1 - 2x)}{2} = \dfrac{2x}{2} = x$

16. $f(g(x)) = f\left[\frac{1}{2}(x + 1)\right] = 2\left[\frac{1}{2}(x + 1)\right] - 1 = x$

$g(f(x)) = g(2x - 1) = \frac{1}{2}[(2x - 1) + 1] = x$

17. $f(g(x)) = f\left[\frac{1}{3}(2 - x)\right] = 2 - 3\left[\frac{1}{3}(2 - x)\right]$

$= 2 - (2 - x) = x$

$g(f(x)) = g(2 - 3x) = \frac{1}{3}[2 - (2 - 3x)]$

$= \frac{1}{3}(3x) = x$

18. $f(g(x)) = f[-4(x - 3)] = -\frac{1}{4}[-4(x - 3)] + 3 = x$

$g(f(x)) = g\left(-\frac{1}{4}x + 3\right) = -4\left[\left(-\frac{1}{4}x + 3\right) - 3\right] = x$

19. $f(g(x)) = f(x^3 - 1) = \sqrt[3]{(x^3 - 1) + 1} = \sqrt[3]{x^3} = x$

$g(f(x)) = g(\sqrt[3]{x + 1}) = (\sqrt[3]{x + 1})^3 - 1$

$= x + 1 - 1 = x$

20. $f(g(x)) = f(\sqrt[5]{x}) = (\sqrt[5]{x})^5 = x$

$g(f(x)) = g(x^5) = \sqrt[5]{x^5} = x$

21. $f(g(x)) = f\left(\dfrac{1}{x}\right) = \dfrac{1}{(1/x)} = x$

$\quad g(f(x)) = g\left(\dfrac{1}{x}\right) = \dfrac{1}{(1/x)} = x$

22. $f(g(x)) = f\left(3 + \dfrac{1}{x}\right) = \dfrac{1}{(3 + 1/x) - 3} = x$

$\quad g(f(x)) = g\left(\dfrac{1}{x + 3}\right) = 3 + \dfrac{1}{1/(x - 3)} = x$

23. (b)　**24.** (c)　**25.** (d)　**26.** (a)

27.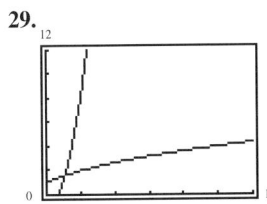

28.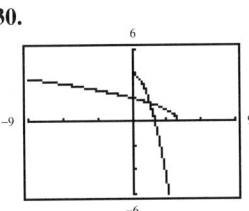

29.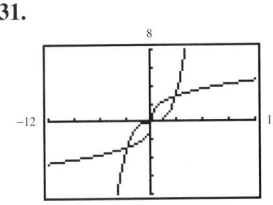

30.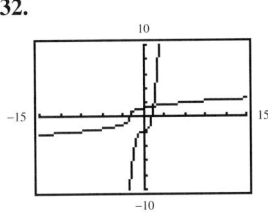

31.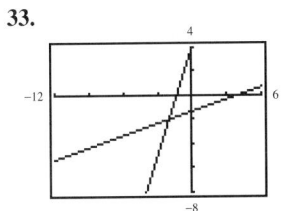

32.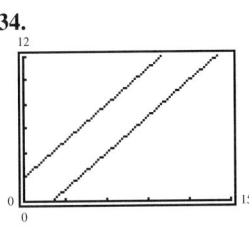

33.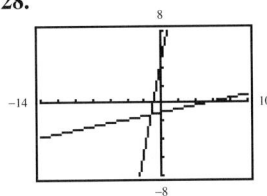

34.

35. $f^{-1}(x) = \dfrac{x}{8}$　**36.** $f^{-1}(x) = 10x$

37. $g^{-1}(x) = x - 25$　**38.** $f^{-1}(x) = 7 - x$

39. $g^{-1}(x) = \dfrac{3 - x}{4}$　**40.** $g^{-1}(t) = \dfrac{t - 1}{6}$

41. $g^{-1}(t) = 4t - 8$　**42.** $h^{-1}(s) = \dfrac{2}{3}(5 - s)$

43. $h^{-1}(x) = x^2,\ x \ge 0$　**44.** $h^{-1}(x) = x^2 - 5,\ x \ge 0$

45. $f^{-1}(t) = \sqrt[3]{t + 1}$　**46.** $h^{-1}(t) = \sqrt[5]{t}$

47. $g^{-1}(s) = \dfrac{5}{s}$　**48.** $f^{-1}(s) = 3 - \dfrac{2}{s}$

49. $f^{-1}(x) = \sqrt[3]{x - 1}$　**50.** $f^{-1}(x) = \sqrt{x^2 + 4},\ x \ge 0$

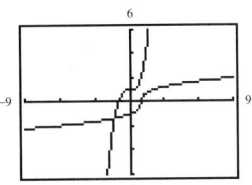　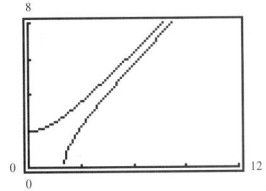

51. No　**52.** Yes　**53.** No　**54.** No

55. 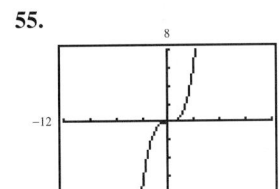　**56.**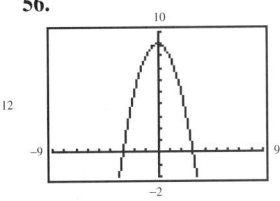

Yes　　　　　No

57.　**58.**

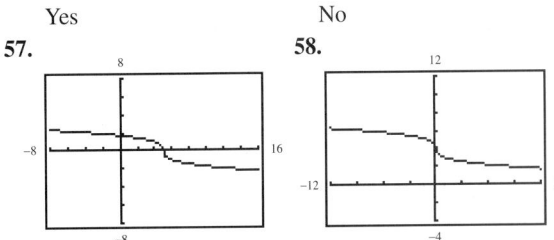

Yes　　　　　Yes

59.　**60.**

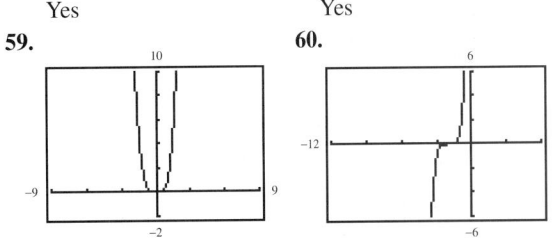

No　　　　　Yes

61.　**62.**

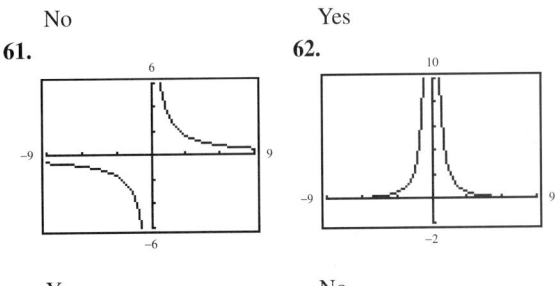

Yes　　　　　No

63.

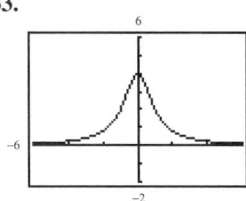

64.

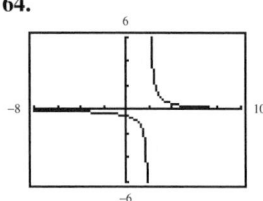

No   Yes

65. $x \geq 0$; $f^{-1}(x) = \sqrt[4]{x}$ **66.** $x \geq 0$; $f^{-1}(x) = \sqrt{9 - x}$

67. $x \geq 2$; $f^{-1}(x) = \sqrt{x} + 2$

68. $x \geq 2$; $f^{-1}(x) = x + 2, x \geq 0$

69.

x	0	1	3	4
f^{-1}	6	4	2	0

70.

x	-1	0	1	5
f^{-1}	-2	0	3	4

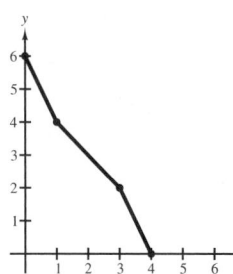

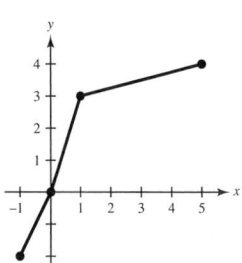

71.

x	-4	-2	2	3
f^{-1}	-2	-1	1	3

72.

x	-3	-2	0	6
f^{-1}	4	3	-1	-2

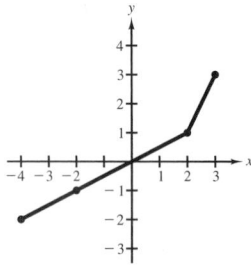

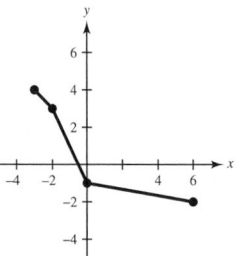

73. (a) $y = \frac{20}{13}(x - 9)$

(b) x: hourly wage; y: number of units produced

(c) 8

74. (a) Total cost = cost of $0.50 compound
+ cost of $0.75 compound
$y = 0.50x + 0.75(100 - x)$

(b) $y = 4(75 - x)$

x: total cost

y: number of pounds at $0.50 per pound

(c) $0 \leq x \leq 75$ **(d)** 60 lb

75. True   **76.** False. $f(x) = \dfrac{1}{x}$

77. False.

$f(x) = \sqrt{x - 1}$; Domain: $[1, \infty)$
$f^{-1}(x) = x^2 + 1, x \geq 0$; Domain: $[0, \infty)$

78. True   **79. (a)** $f^{-1}(x) = \dfrac{3 - x}{2}$

(b) $(f^{-1})^{-1}(x) = 3 - 2x$

Section Project 8.2

(a) $(f \circ g)(x) = 4x + 24$ **(b)** $(f \circ g)^{-1}(x) = \dfrac{x - 24}{4}$

(c) $f^{-1}(x) = \dfrac{x}{4}$, $g^{-1}(x) = x - 6$

(d) $(g^{-1} \circ f^{-1})(x) = \dfrac{x - 24}{4} = (f \circ g)^{-1}(x)$

(e) $(f \circ g)(x) = 8x^3 + 1$

$(f \circ g)^{-1}(x) = \dfrac{\sqrt[3]{x - 1}}{2}$

$f^{-1}(x) = \sqrt[3]{x - 1}$

$g^{-1}(x) = \dfrac{x}{2}$

$(g^{-1} \circ f^{-1})(x) = \dfrac{\sqrt[3]{x - 1}}{2} = (f \circ g)^{-1}(x)$

(f) Answers may vary.   **(g)** $(f \circ g)^{-1}(x) = (g^{-1} \circ f^{-1})(x)$

Section 8.3   (page 583)

1. $I = kV$ **2.** $C = kr$ **3.** $u = kv^2$ **4.** $s = kt^3$

5. $p = \dfrac{k}{d}$ **6.** $S = \dfrac{k}{v^2}$ **7.** $P = \dfrac{k}{\sqrt{1 + r}}$ **8.** $A = \dfrac{k}{t^4}$

9. $A = klw$ **10.** $V = khr^2$ **11.** $P = \dfrac{k}{V}$

12. $F = \dfrac{km_1m_2}{r^2}$ **13.** A varies jointly as the base and height.

14. A varies directly as the square of the radius.

15. A varies jointly as the length and the width.

16. A is proportional to the square of the radius.

17. V varies jointly as the square of the radius and the height.

18. V varies directly as the cube of the radius.

19. r varies directly as the distance and inversely as the time.

20. h varies directly as the volume and inversely as the square of the radius.

21. Increase. Because $y = kx$ and $k > 0$, the variables increase or decrease together.

22. Decrease. Because $y = k/x$ and $k > 0$, one variable decreases when the other increases.

23. $s = 5t$ **24.** $h = \frac{7}{3}r$ **25.** $F = \frac{5}{16}x^2$

26. $v = 6\sqrt{s}$ **27.** $H = \frac{5}{2}u$ **28.** $M = \frac{3}{2}n^3$

29. $n = \dfrac{48}{m}$ **30.** $q = \dfrac{75}{p}$ **31.** $g = \dfrac{4}{\sqrt{z}}$ **32.** $u = \dfrac{10}{v^2}$

33. $F = \dfrac{25}{6}xy$ **34.** $V = \dfrac{1}{3}hb^2$ **35.** $d = \dfrac{120x^2}{r}$

36. $z = \dfrac{125x}{\sqrt{y}}$ **37. (a)** $4921.25   **(b)** Price per unit

38. (a) $504   **(b)** Price per unit

39. (a) 1.2 in.   **(b)** 25 lb   **40.** 18 lb

41. (a) 2 in.   **(b)** 15 lb

42. (a) $F = \frac{100}{3}x$

(b)

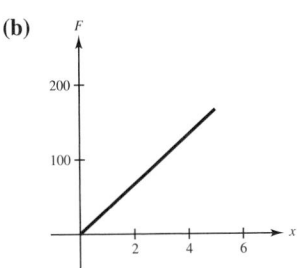

The graph is a line with slope $\frac{100}{3}$ and a y-intercept at $(0, 0)$.

43. $208\frac{1}{3}$ ft   **44.** 576 ft   **45.** 32 ft/sec²

46. 0.61 mph   **47.** 3072 W   **48.** 2.44 hr; distance

49. 667 units   **50.** 100   **51.** 324 lb   **52.** 121.6 lb

53. $\frac{1}{4}$

54. y will be one-fourth as great. If $y = k/x^2$ and x is replaced by $2x$, you have $y = k/(2x)^2 = k/(4x^2)$.

55. 4

56. No, k is different for each pizza. The 15-inch pizza is the best buy.

57. 0.36 psi; 116 lb   **58.** $p = \dfrac{114}{t}$, 17.5%

59. $T = \dfrac{4000}{d}$, 0.91°C   **60.** $y = \dfrac{4}{x}$ **61.** $y = -\dfrac{3}{10}x$

Section Project 8.3

(a) $P = \dfrac{kWD^2}{L}$ **(b)** Unchanged

(c) Increases by a factor of 8   **(d)** Increases by a factor of 4

(e) Changes by a factor of $\frac{1}{4}$ **(f)** 3125 pounds

Section 8.4   (page 593)

1. Horizontal line

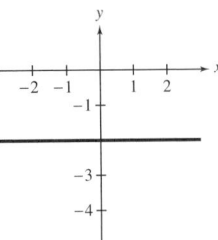

2. Horizontal line

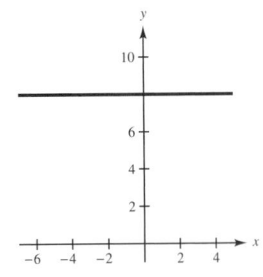

3. Line with slope -3

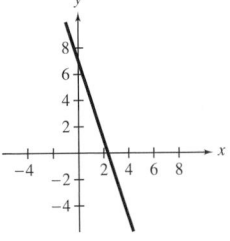

4. Line with slope $\frac{1}{2}$

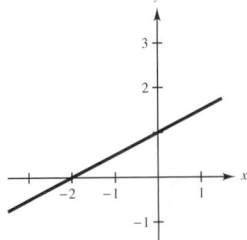

5. Parabola

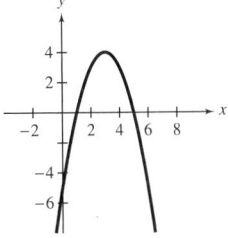

6. Parabola

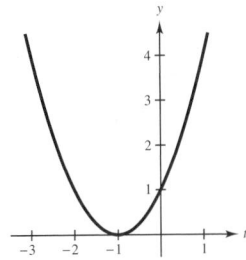

7. 4   **8.** 2   **9.** 1   **10.** 1   **11.** 3   **12.** 5

13. Horizontal translation five units to the right

14. Horizontal translation three units to the left

15. Vertical translation three units upward

16. Vertical translation two units downward

17. Reflection in the x-axis   **18.** Reflection in the x-axis

19. (a)

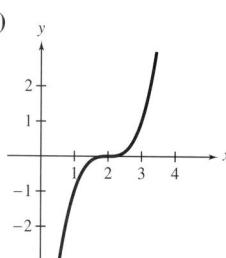

(b)

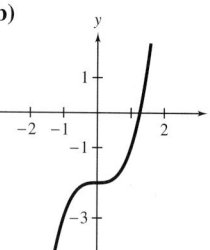

(c)

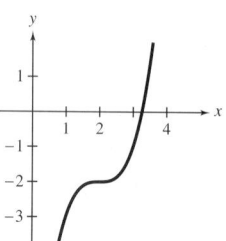

(d)

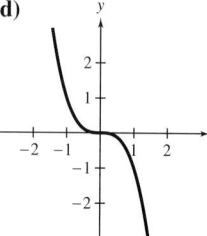

20. (a)

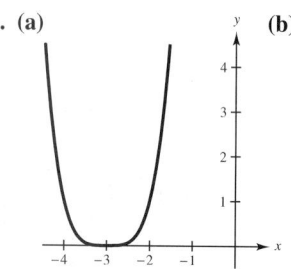

(b)

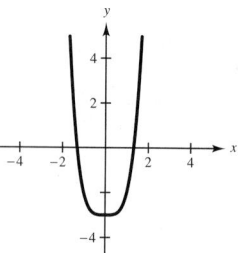

(c)

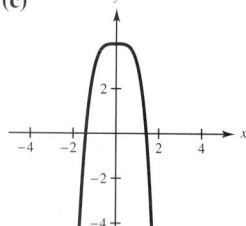

(d)

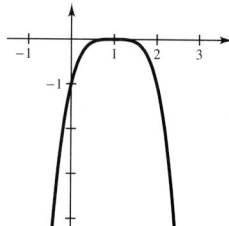

21.

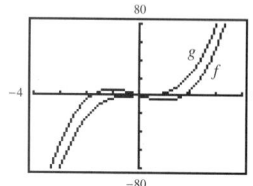

22.

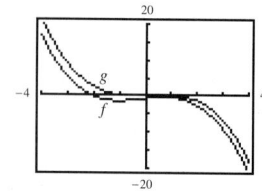

23.

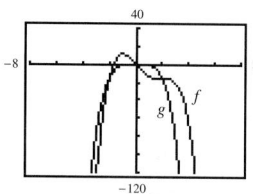

24.

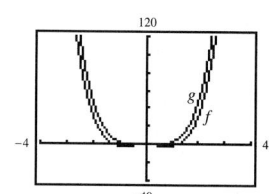

25. Rises to the left; Rises to the right
26. Falls to the left; Falls to the right
27. Falls to the left; Rises to the right
28. Rises to the left; Falls to the right
29. Rises to the left; Falls to the right
30. Falls to the left; Falls to the right
31. $(\pm 5, 0), (0, -25)$ **32.** $(\pm 7, 0), (0, 49)$
33. $(3, 0), (0, 9)$ **34.** $(-5, 0), (0, 25)$
35. $(-2, 0), (1, 0), (0, -2)$ **36.** $\left(2 \pm \sqrt{3}, 0\right), (0, 3)$
37. $(2, 0), (0, 0)$ **38.** $(-4, 0), (0, 0), (5, 0)$
39. $(\pm 1, 0), \left(0, -\frac{1}{2}\right)$ **40.** $\left(\pm \sqrt{5}, 0\right), (0, -40)$
41. $(0, 10)$ **42.** $(4, 0), (\pm 5, 0), (0, 100)$
43.

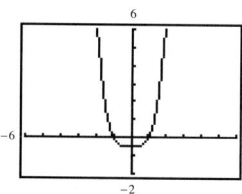

44.

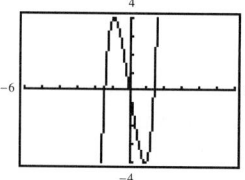

$(-1, 0), (1, 0)$

$(0, 0), (-1.414, 0), (1.414, 0)$

45.

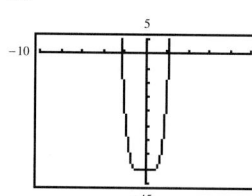

46.

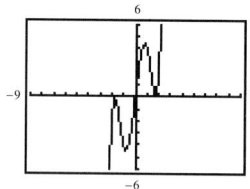

$(-2.236, 0), (2.236, 0)$

$(0, 0), (-1.732, 0), (1.732, 0)$

47.

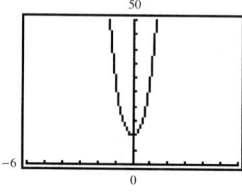

48.

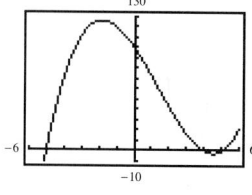

No x-intercepts

$(4, 0), (5, 0), (-5, 0)$

49.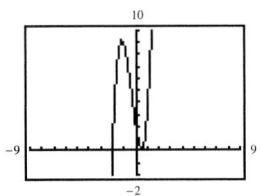

50.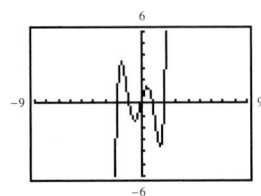

$(-2, 0), \left(\frac{1}{2}, 0\right)$ $(0, 0), (\pm 1, 0), (\pm 2, 0)$

51. (e)   **52.** (c)   **53.** (b)   **54.** (f)   **55.** (a)

56. (g)   **57.** (d)   **58.** (h)

59.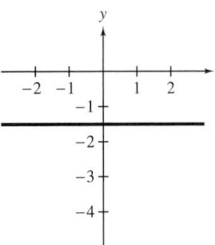

60.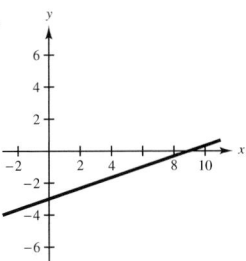

61.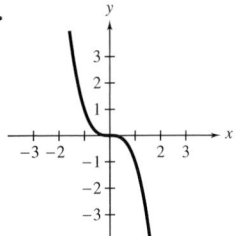

62.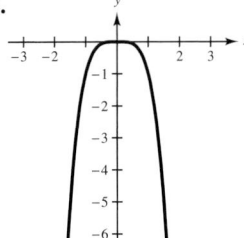

63.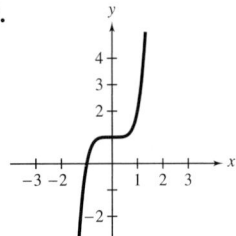

64.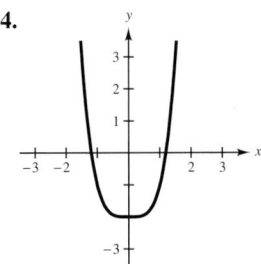

65.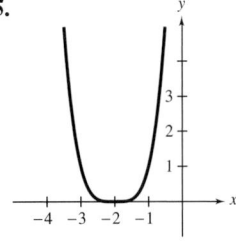

66.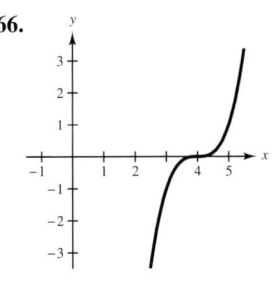

67.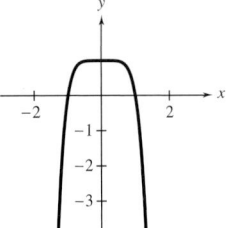

68.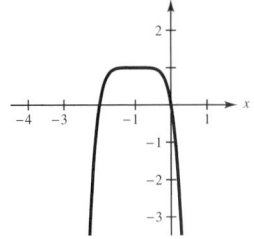

69.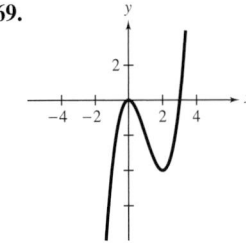

70.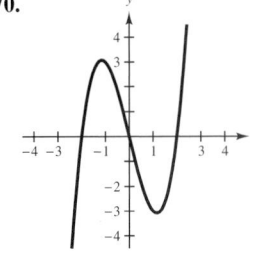

71.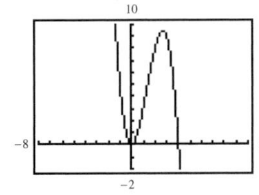

Yes

72.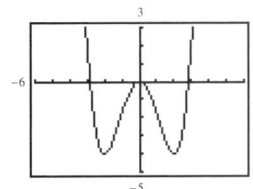

Yes

73.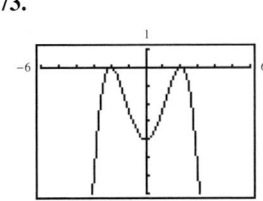

Yes

74.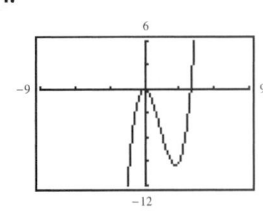

Yes

75.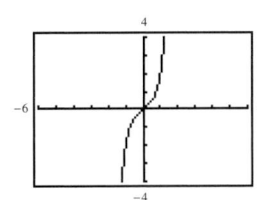

No

76.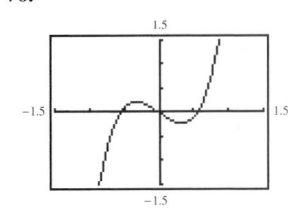

No

77. (a) Answers may vary.
(b) $(0, 6)$
(c)

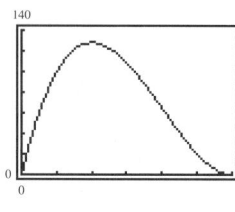

$x = 2$

78. (a) Answers may vary.
(b) $(0, 3)$
(c)

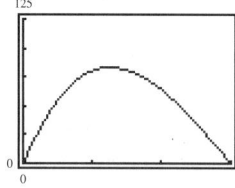

$x = 1.27$

(c) (i) $-2, 2, 6$ **(ii)** $-5, 0, 2$
(iii) $-5, 2$ **(iv)** 2
(v) $2, 1 \pm \sqrt{7}$ **(vi)** $2, \dfrac{3 \pm \sqrt{7}i}{2}$

(d) Two is always a zero of f. If 2 is also a zero of g, then f has more than one factor of $x - 2$.

Section Project 8.4

(a) (i) $-2, 6$ **(ii)** $-5, 0$
(iii) $-5, 2$ **(iv)** 2
(v) $1 \pm \sqrt{7}$ **(vi)** $\dfrac{3 \pm \sqrt{7}i}{2}$

(b) (i)

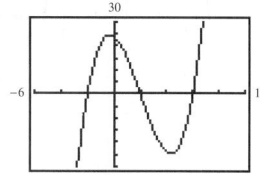

Crosses the x-axis

(ii)

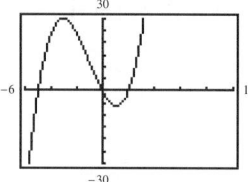

Crosses the x-axis

(iii)

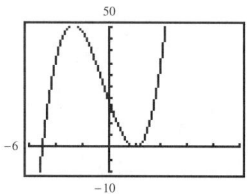

Tangent to the x-axis

(iv)

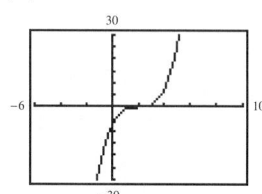

Crosses the x-axis

(v)

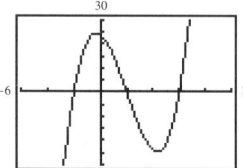

Crosses the x-axis

(vi)

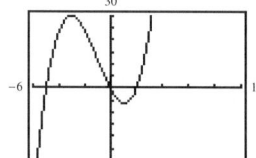

Crosses the x-axis

Mid-Chapter Quiz   (page 596)

1. $-x^2 + 2x - 8,\ (-\infty, \infty)$
2. $x^2 + 2x - 8,\ (-\infty, \infty)$ **3.** $2x^3 - 8x^2,\ (-\infty, \infty)$
4. $x^2 - 2x + 8,\ (-\infty, \infty)$
5. $\dfrac{2x - 8}{x^2},\ (-\infty, 0) \cup (0, \infty)$
6. $\dfrac{x^2}{2x - 8},\ (-\infty, 4) \cup (4, \infty)$ **7.** $2x^2 - 8,\ (-\infty, \infty)$
8. $(2x - 8)^2,\ (-\infty, \infty)$
9. $C(x(t)) = 3000t + 750,\ 0 \le t \le 40$
Cost of units produced in t hours
10. $f(g(x)) = 100 - 15\left[\frac{1}{15}(100 - x)\right]$
$\qquad = 100 - (100 - x) = x$
$\quad g(f(x)) = \frac{1}{15}[100 - (100 - 15x)]$
$\qquad = \frac{1}{15}(15x) = x$
11. $f^{-1}(x) = \sqrt[3]{x + 8}$ **12.** $z = 3t$

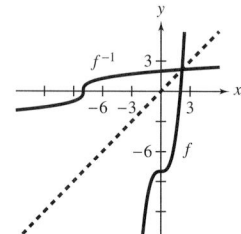

13. $S = 5hr^2$ **14.** $N = \dfrac{15t^2}{s}$ **15.** 30 min
16. Rises to the right because the leading coefficient is positive
17. (a)

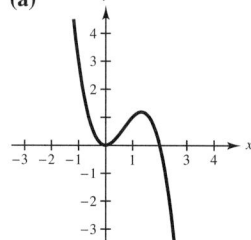

(b)

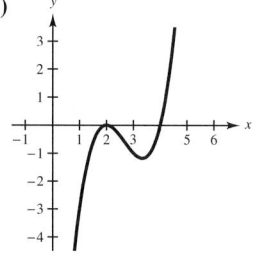

18.

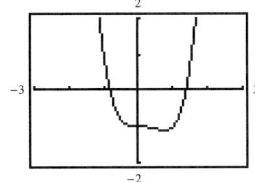

$(-0.819, 0), (1.380, 0)$

Section 8.5 (page 602)

1. Circles, parabolas, ellipses, and hyperbolas
2. A circle consists of all points (x, y) in a plane that are a given positive distance r from a fixed point (h, k) called the center.
$x^2 + y^2 = r^2$
3. (c)   **4.** (e)   **5.** (a)   **6.** (d)   **7.** (b)   **8.** (f)
9. $x^2 + y^2 = 25$ **10.** $x^2 + y^2 = 49$
11. $x^2 + y^2 = \frac{4}{9}$ **12.** $x^2 + y^2 = \frac{25}{4}$
13. $x^2 + y^2 = 64$ **14.** $x^2 + y^2 = 4$
15. $x^2 + y^2 = 29$ **16.** $x^2 + y^2 = 17$
17. $(x - 4)^2 + (y - 3)^2 = 100$
18. $(x + 2)^2 + (y - 5)^2 = 36$
19. $(x - 5)^2 + (y + 3)^2 = 81$
20. $(x + 5)^2 + (y + 2)^2 = \frac{25}{4}$
21. $(x + 2)^2 + (y - 1)^2 = 4$
22. $(x - 8)^2 + (y - 2)^2 = 4$
23. $(x - 3)^2 + (y - 2)^2 = 17$
24. $(x + 3)^2 + (y + 5)^2 = 34$
25. Center: $(0, 0)$ **26.** Center: $(0, 0)$
$r = 4$ $r = 5$

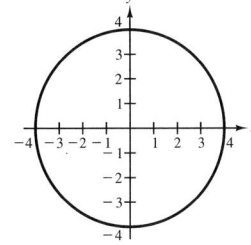   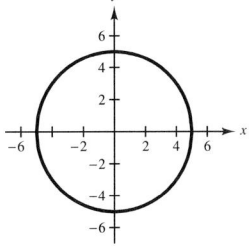

27. Center: $(0, 0)$ **28.** Center: $(0, 0)$
$r = 6$ $r = \sqrt{10}$

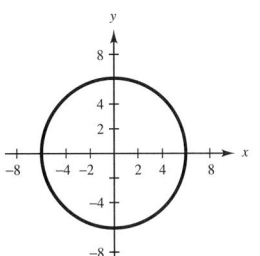   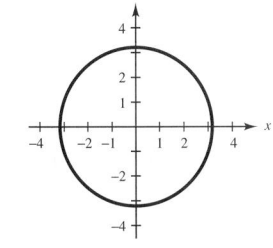

29. Center: $(0, 0)$ **30.** Center: $(0, 0)$
$r = \frac{1}{2}$ $r = \frac{8}{3}$

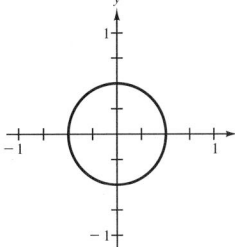   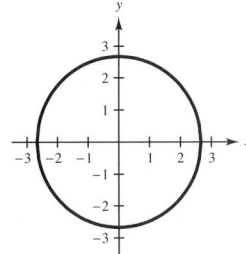

31. Center: $(0, 0)$ **32.** Center: $(0, 0)$
$r = \frac{12}{5}$ $r = 2$

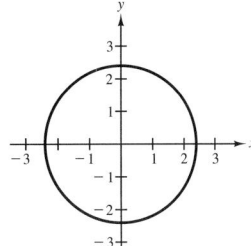   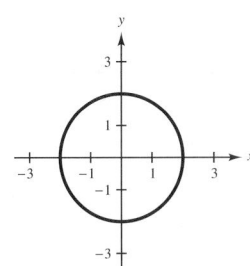

33. Center: $(2, 3)$ **34.** Center: $(-4, 3)$
$r = 2$ $r = 5$

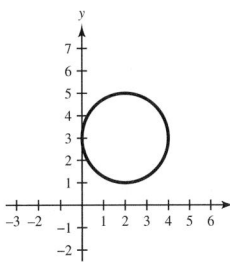   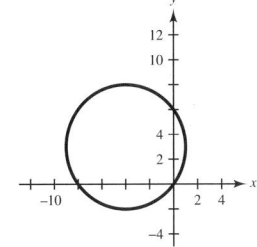

35. Center: $\left(-\frac{5}{2}, -3\right)$
   $r = 3$

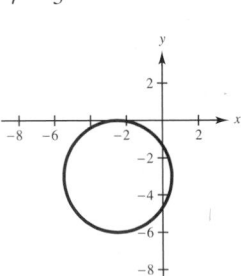

36. Center: $\left(5, -\frac{3}{4}\right)$
   $r = 1$

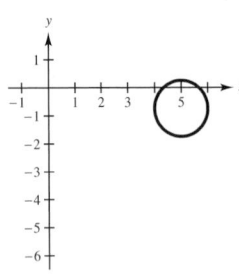

37. $(x - 2)^2 + (y - 1)^2 = 4$

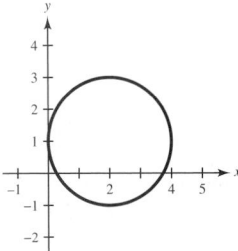

38. $(x + 3)^2 + (y - 2)^2 = 16$

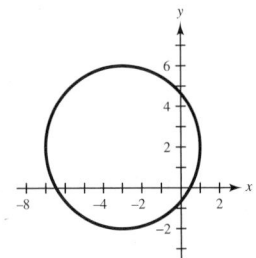

39. $(x + 1)^2 + (y + 3)^2 = 4$

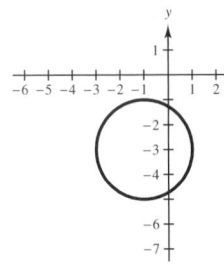

40. $(x - 1)^2 + (y + 3)^2 = 25$

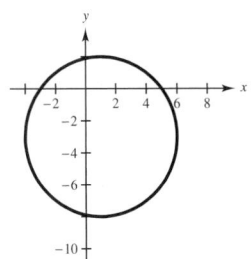

41.

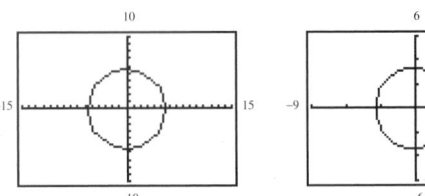

Center: $(0, 0)$

Radius: $\sqrt{30}$

42.

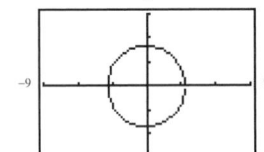

Center: $(0, 0)$

Radius: $\dfrac{3\sqrt{5}}{2}$

43.

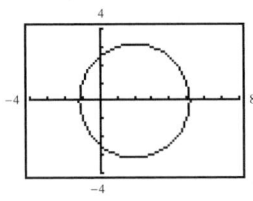

Center: $(2, 0)$
Radius: $\sqrt{10}$

44.

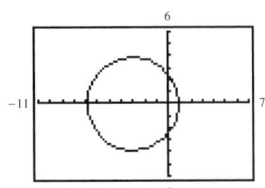

Center: $(-3, 0)$
Radius: $\sqrt{15}$

45. $x^2 + y^2 = 4500^2$ **46.** $\sqrt{475} \approx 21.8$ ft

47. $x^2 + y^2 = 144$,   10.4 ft

Section Project 8.5

Circle of radius 25: $A = 4x\sqrt{625 - x^2}$

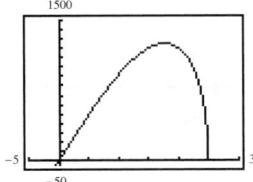

Maximum when $x \approx 17.68$

Circle of radius 4: $A = 4x\sqrt{16 - x^2}$

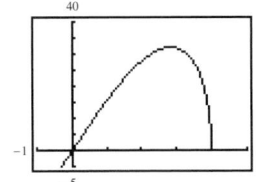

Maximum when $x \approx 2.83$

Section 8.6   (page 615)

1. (a)   **2.** (f)   **3.** (d)   **4.** (c)   **5.** (e)   **6.** (b)

7. Vertices: $(\pm 4, 0)$
   Co-vertices: $(0, \pm 2)$

8. Vertices: $(\pm 5, 0)$
   Co-vertices: $(0, \pm 3)$

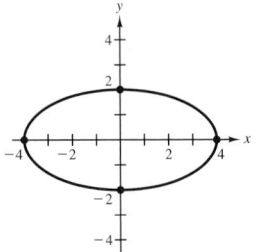

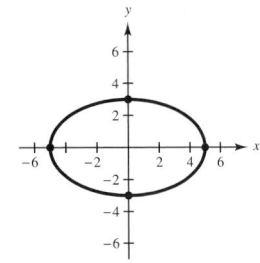

9. Vertices: $(0, \pm 4)$
Co-vertices: $(\pm 2, 0)$

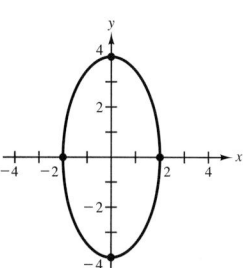

10. Vertices: $(0, \pm 5)$
Co-vertices: $(\pm 3, 0)$

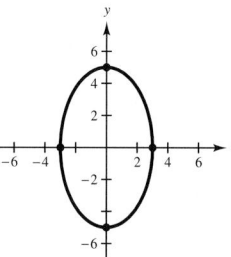

11. Vertices: $\left(\pm \frac{5}{3}, 0\right)$
Co-vertices: $\left(0, \pm \frac{4}{3}\right)$

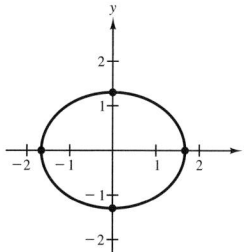

12. Vertices: $(\pm 1, 0)$
Co-vertices: $\left(0, \pm \frac{1}{2}\right)$

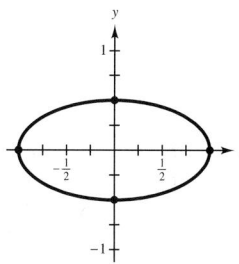

13. Vertices: $(0, \pm 2)$
Co-vertices: $(\pm 1, 0)$

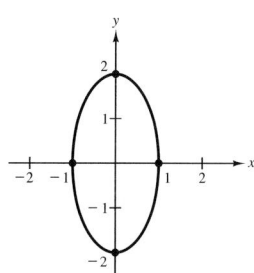

14. Vertices: $(\pm 3, 0)$
Co-vertices: $(0, \pm 2)$

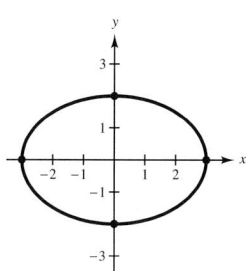

15. Vertices: $(\pm 2, 0)$

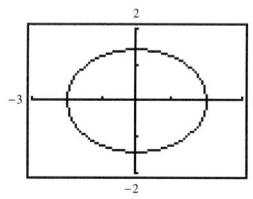

16. Vertices: $(0, \pm 8)$

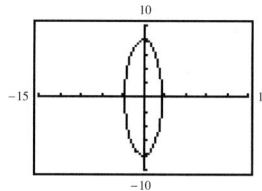

17. Vertices: $\left(0, \pm 2\sqrt{3}\right)$

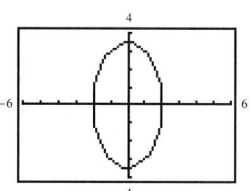

18. Vertices: $\left(0, \pm \sqrt{5}\right)$

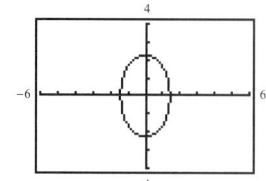

19. $\dfrac{x^2}{16} + \dfrac{y^2}{9} = 1$ **20.** $\dfrac{x^2}{16} + \dfrac{y^2}{1} = 1$ **21.** $\dfrac{x^2}{4} + \dfrac{y^2}{1} = 1$

22. $\dfrac{x^2}{100} + \dfrac{y^2}{16} = 1$ **23.** $\dfrac{x^2}{9} + \dfrac{y^2}{16} = 1$

24. $\dfrac{x^2}{1} + \dfrac{y^2}{25} = 1$ **25.** $\dfrac{x^2}{1} + \dfrac{y^2}{4} = 1$ **26.** $\dfrac{x^2}{16} + \dfrac{y^2}{64} = 1$

27. $\dfrac{x^2}{9} + \dfrac{y^2}{25} = 1$ **28.** $\dfrac{x^2}{144} + \dfrac{y^2}{25} = 1$

29. $\dfrac{x^2}{100} + \dfrac{y^2}{36} = 1$ **30.** $\dfrac{x^2}{625} + \dfrac{y^2}{225} = 1$

31. Center: $(-5, 0)$
Vertices:
$(-9, 0), (-1, 0)$

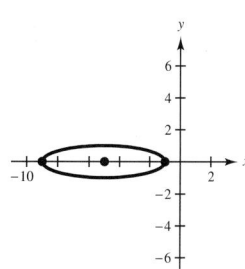

32. Center: $(2, 3)$
Vertices: $(2, 0), (2, 6)$

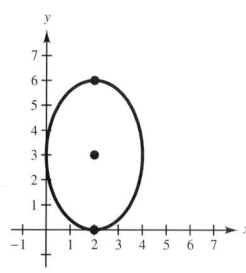

33. Center: $(1, 5)$
Vertices: $(1, 0), (1, 10)$

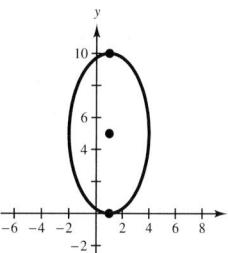

34. Center: $(-2, -4)$
Vertices:
$(-2, -5), (-2, -3)$

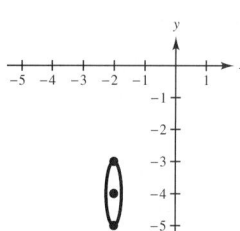

35. Center: $(-2, 3)$
Vertices:
$(-2, 6), (-2, 0)$

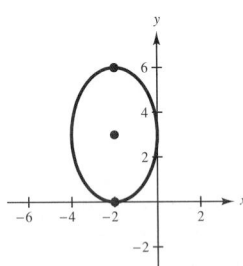

36. Center: $(2, -1)$
Vertices:
$\left(2, -\frac{5}{2}\right), \left(2, \frac{1}{2}\right)$

37. $\dfrac{x^2}{1} + \dfrac{y^2}{4} = 1$ **38.** $\dfrac{x^2}{4} + \dfrac{4y^2}{9} = 1$

39. $\dfrac{(x-4)^2}{9} + \dfrac{y^2}{16} = 1$ **40.** $\dfrac{(x-2)^2}{9} + \dfrac{(y-2)^2}{4} = 1$

41. $\sqrt{304} \approx 17.4$ ft

42. $\dfrac{x^2}{16} + \dfrac{16y^2}{225} = 1$ or $\dfrac{16x^2}{225} + \dfrac{y^2}{16} = 1$ **43.** (c)

44. (e)   **45.** (a)   **46.** (f)   **47.** (b)   **48.** (d)

49. Vertices: $(\pm 3, 0)$
Asymptotes: $y = \pm x$

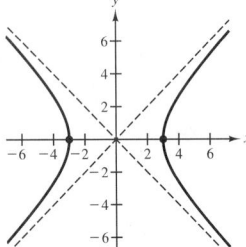

50. Vertices: $(\pm 1, 0)$
Asymptotes: $y = \pm x$

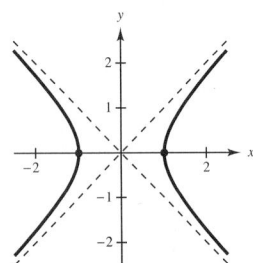

51. Vertices: $(0, \pm 3)$
Asymptotes: $y = \pm x$

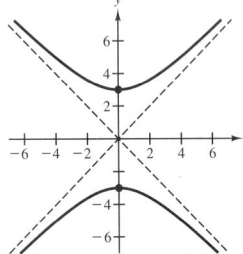

52. Vertices: $(0, \pm 1)$
Asymptotes: $y = \pm x$

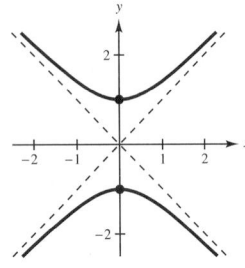

53. Vertices: $(\pm 3, 0)$
Asymptotes: $y = \pm\frac{5}{3}x$

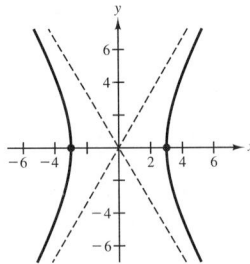

54. Vertices: $(\pm 2, 0)$
Asymptotes: $y = \pm\frac{3}{2}x$

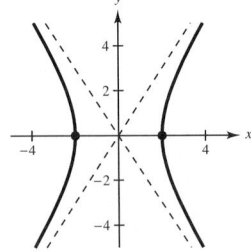

55. Vertices: $(0, \pm 3)$
Asymptotes: $y = \pm\frac{3}{5}x$

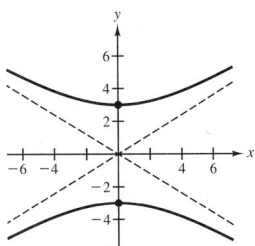

56. Vertices: $(0, \pm 2)$
Asymptotes: $y = \pm\frac{2}{3}x$

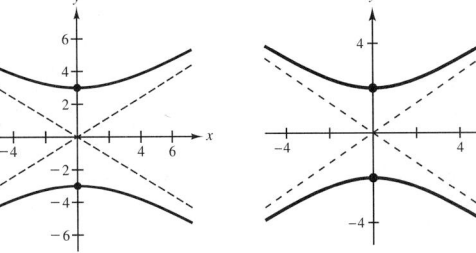

57. Vertices: $(\pm 1, 0)$
Asymptotes: $y = \pm\frac{3}{2}x$

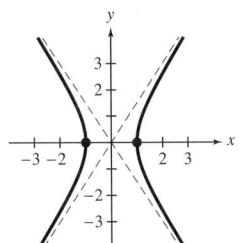

58. Vertices: $\left(0, \pm\frac{1}{2}\right)$
Asymptotes: $y = \pm\frac{1}{5}x$

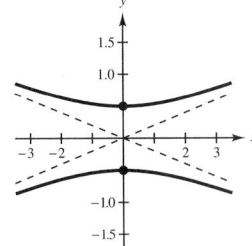

59. Vertices: $(\pm 4, 0)$
Asymptotes: $y = \pm\frac{1}{2}x$

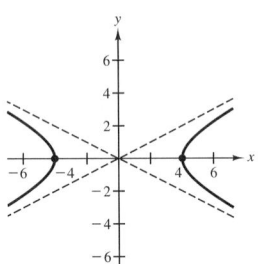

60. Vertices: $(0, \pm 3)$
Asymptotes: $y = \pm\frac{3}{2}x$

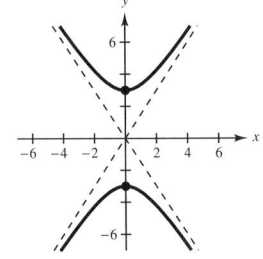

61. $\dfrac{x^2}{16} - \dfrac{y^2}{64} = 1$ **62.** $\dfrac{x^2}{4} - \dfrac{9y^2}{4} = 1$

63. $\dfrac{y^2}{16} - \dfrac{x^2}{64} = 1$ **64.** $\dfrac{y^2}{4} - \dfrac{9x^2}{4} = 1$

65. $\dfrac{x^2}{81} - \dfrac{y^2}{36} = 1$ **66.** $\dfrac{x^2}{1} - \dfrac{y^2}{1/4} = 1$

67. $\dfrac{y^2}{1} - \dfrac{x^2}{1/4} = 1$ **68.** $\dfrac{y^2}{25} - \dfrac{x^2}{25} = 1$

69. Center: $(3, -4)$
Vertices: $(3, 1), (3, -9)$

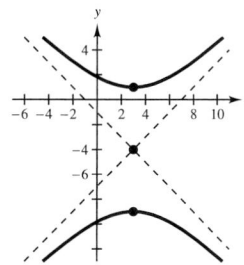

70. Center: $(2, -6)$
Vertices: $(2, -5), (2, -7)$

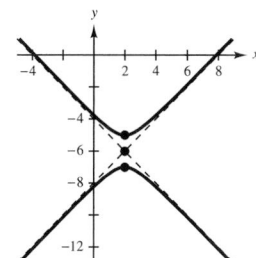

71. Center: $(1, -2)$
Vertices:
$(-1, -2), (3, -2)$

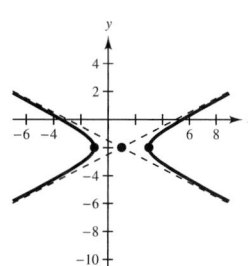

72. Center: $(2, 3)$
Vertices: $(0, 3), (4, 3)$

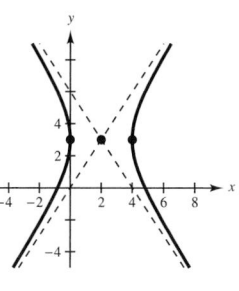

73. Center: $(2, -3)$
Vertices: $(3, -3), (1, -3)$

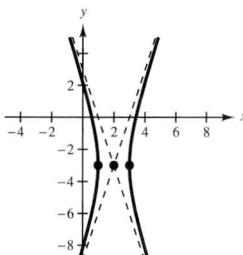

74. Center: $(0, 2)$
Vertices: $(-6, 2), (6, 2)$

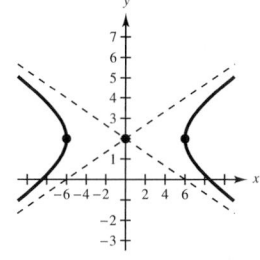

75. $\dfrac{y^2}{9} - \dfrac{x^2}{9/4} = 1$ **76.** $\dfrac{x^2}{4} - \dfrac{y^2}{12/5} = 1$

77. $\dfrac{(x-3)^2}{4} - \dfrac{(y-2)^2}{16/5} = 1$

78. $\dfrac{(x+4)^2}{16} - \dfrac{(y-4)^2}{64/5} = 1$

79. Left half   **80.** Top half

Section Project 8.6

(a)

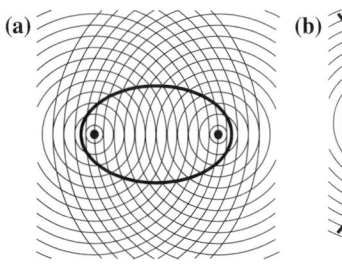

(b)

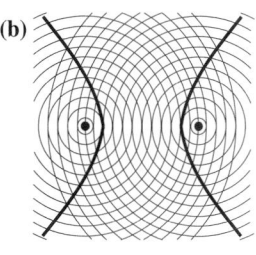

Section 8.7   **(page 623)**

1. (b)   **2.** (c)   **3.** (e)   **4.** (a)   **5.** (f)   **6.** (d)

7. Vertex: $(0, 0)$
Focus: $\left(0, \frac{1}{2}\right)$

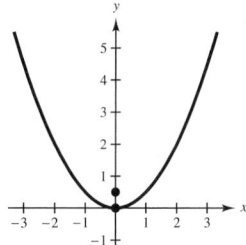

8. Vertex: $(0, 0)$
Focus: $\left(0, \frac{1}{8}\right)$

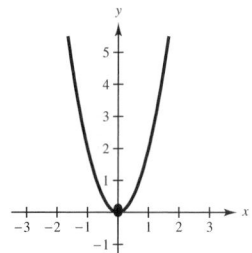

9. Vertex: $(0, 0)$
Focus: $\left(-\frac{3}{2}, 0\right)$

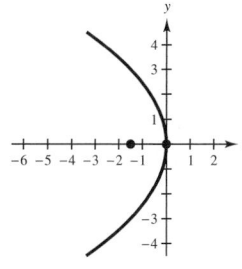

10. Vertex: $(0, 0)$
Focus: $\left(\frac{3}{4}, 0\right)$

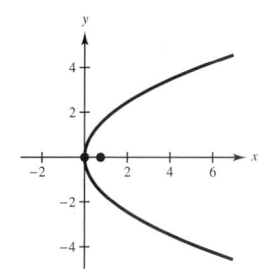

11. Vertex: $(0, 0)$
Focus: $\left(0, -2\right)$

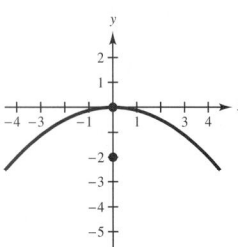

12. Vertex: $(0, 0)$
Focus: $\left(-\frac{1}{4}, 0\right)$

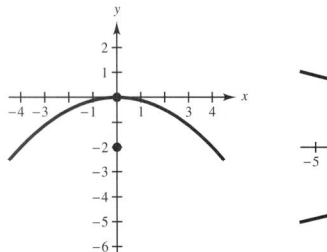

19. Vertex: $(-2, -3)$
Focus: $(-4, -3)$

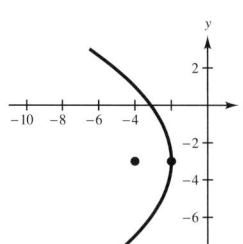

20. Vertex: $(-1, 2)$
Focus: $(0, 2)$

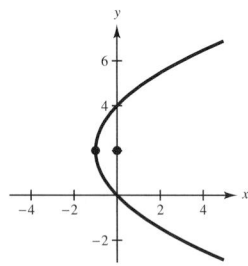

13. Vertex: $(1, -2)$
Focus: $(1, -4)$

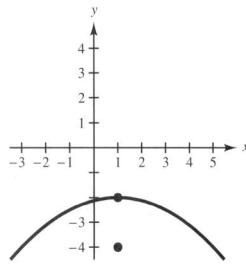

14. Vertex: $(-3, 2)$
Focus: $\left(-\frac{13}{4}, 2\right)$

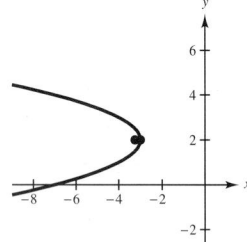

21. Vertex: $(-2, 1)$
Focus: $\left(-2, -\frac{1}{2}\right)$

22. Vertex: $(1, -1)$
Focus: $(1, -3)$

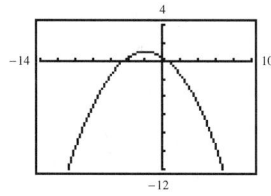 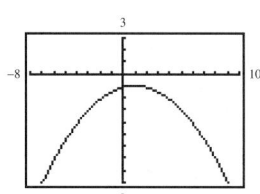

23. Vertex: $\left(\frac{1}{4}, -\frac{1}{2}\right)$
Focus: $\left(0, -\frac{1}{2}\right)$

24. Vertex: $(-1, 0)$
Focus: $(0, 0)$

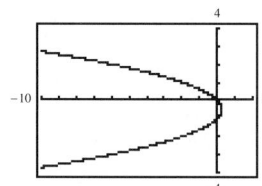 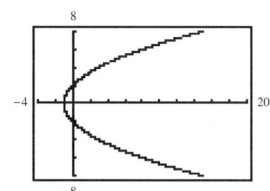

15. Vertex: $\left(5, -\frac{1}{2}\right)$
Focus: $\left(\frac{11}{2}, -\frac{1}{2}\right)$

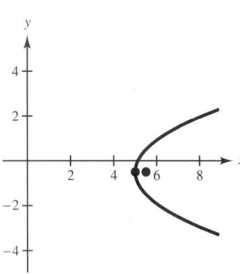

16. Vertex: $\left(-\frac{1}{2}, 3\right)$
Focus: $\left(-\frac{1}{2}, 4\right)$

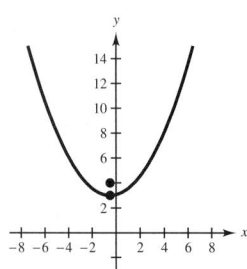

25. $y = 2\sqrt{-x}$ **26.** $y = 3\sqrt{x}$
27. $y = 3 + \sqrt{6(x + 1)}$ **28.** $y = -1 - \sqrt{2(x - 2)}$
29. $x^2 = \frac{3}{2}y$ **30.** $y^2 = -18x$ **31.** $x^2 = -6y$
32. $y^2 = 8x$ **33.** $y^2 = -8x$ **34.** $x^2 = -8y$
35. $x^2 = 4y$ **36.** $y^2 = -12x$ **37.** $y^2 = 16x$
38. $y^2 = 8x$ **39.** $y^2 = 9x$ **40.** $x^2 = -2y$
41. $(x - 3)^2 = -(y - 1)$ **42.** $(y - 3)^2 = -2(x - 5)$
43. $y^2 = 2(x + 2)$ **44.** $(x - 3)^2 = 3(y + 3)$
45. $(y - 2)^2 = -8(x - 3)$ **46.** $(x + 1)^2 = -8(y - 2)$
47. $x^2 = 8(y - 4)$ **48.** $(y - 1)^2 = -12(x + 2)$
49. $(y - 2)^2 = x$ **50.** $x^2 = -18(y - 2)$

17. Vertex: $(1, 1)$
Focus: $(1, 2)$

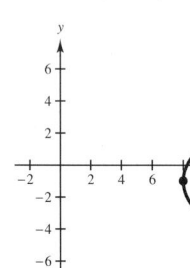

18. Vertex: $(8, -1)$
Focus: $(9, -1)$

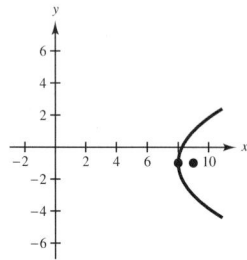

51. (a) $y = \dfrac{x^2}{180}$ **(b)**

x	0	20	40	60
y	0	$2\frac{2}{9}$	$8\frac{8}{9}$	20

52. (a) $y = \dfrac{3x^2}{640,000}$ (b) 653 cm

53.

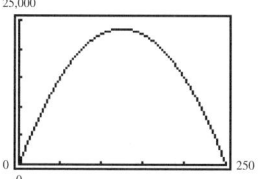

$x = 125$

54. (a)

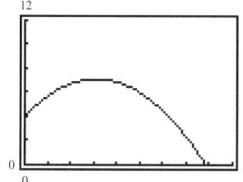

(b) Highest point:
(6.25, 7.125)
Range: 15.69 ft

55. False. There correspond two values of y for each $x > 0$.

Section Project 8.7

(a)

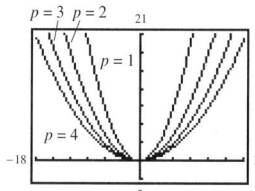

As p increases, the parabola opens wider.

(b) For $p = 1$, the focus is $(0, 1)$.
For $p = 2$, the focus is $(0, 2)$.
For $p = 3$, the focus is $(0, 3)$.
For $p = 4$, the focus is $(0, 4)$.

(c) For $p = 1$, chord length $= 4$.
For $p = 2$, chord length $= 8$.
For $p = 3$, chord length $= 12$.
For $p = 4$, chord length $= 16$.
For any parabola, the length of the chord passing through the focus perpendicular to the axis is $4p$.

(d) By sketching the chord, you can plot two additional points on the parabola.

Section 8.8   (page 634)

1.

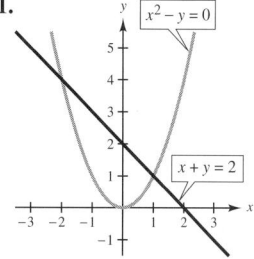

2.

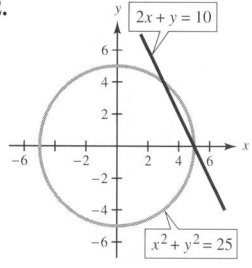

$(-2, 4), (1, 1)$

$(3, 4), (5, 0)$

3.

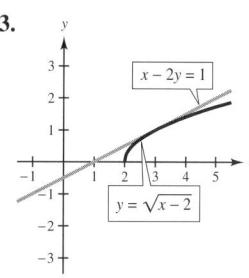

4.

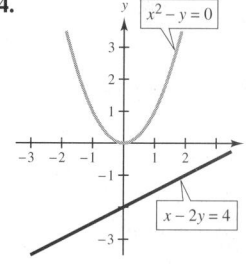

$(3, 1)$

No real solution

5.

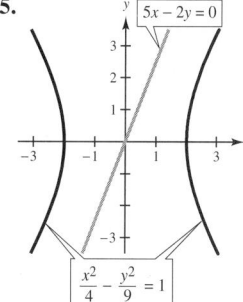

6.

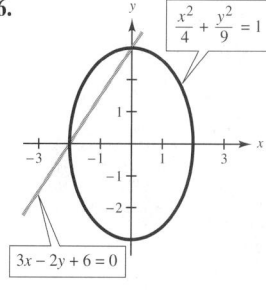

No real solution

$(-2, 0), (0, 3)$

7. $(0, 0), (1, 1), (-1, -1)$

8. $(2, 4), (-1, 1)$

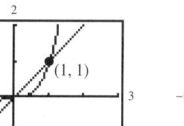

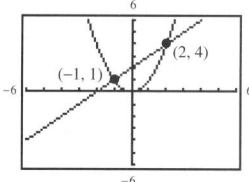

9. $(0, 0), (2, 4)$

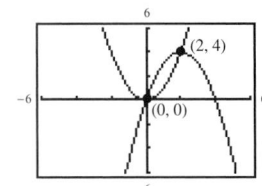

10. $(2, 4), (-1, 7)$

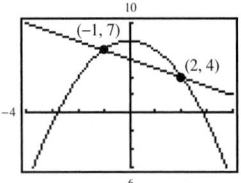

11. $(4, 2), (9, 3)$

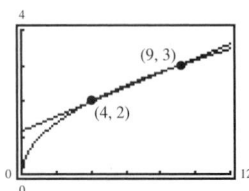

12. $(4, 2), (-4, -2)$

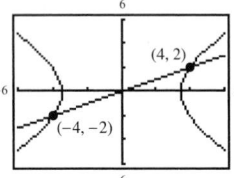

13. $(-3, 18), (2, 8)$ **14.** $(-1, 5), (-2, 20)$
15. $(2, 5), (-3, 0)$ **16.** $(4, 2), (1, -1)$
17. No real solution   **18.** No real solution
19. $(0, 5), (-4, -3)$ **20.** $(-5, 12), (12, -5)$
21. $(0, 2), (3, 1)$ **22.** $(-1, -1), (0, 0), (1, 1)$
23. $(0, 4), (3, 0)$ **24.** $(0, 0), (2, 8), (-2, 8)$
25. No real solution   **26.** $(5, 3), (4, 0)$
27. $(10, 30), (20, 15)$ **28.** $(10, 15), (30, 5)$
29. $\left(\pm\sqrt{3}, -1\right)$ **30.** $\left(1, \pm 2\right), \left(-\frac{1}{2}, \pm\frac{\sqrt{22}}{2}\right)$
31. $\left(2, \pm 2\sqrt{3}\right), (-1, \pm 3)$ **32.** $\left(\pm\sqrt{11}, -2\right), \left(\pm 2\sqrt{2}, 1\right)$
33. $\left(\pm 2, \pm\sqrt{3}\right)$ **34.** $(\pm 3, \pm 4)$
35. $\left(\pm\frac{2\sqrt{5}}{5}, \pm\frac{2\sqrt{5}}{5}\right)$ **36.** $\left(\pm\frac{2\sqrt{3}}{3}, \pm\frac{\sqrt{3}}{3}\right)$

37.

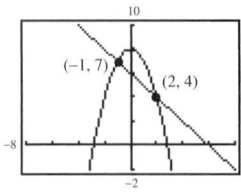

$(-1, 7), (2, 4)$

38.

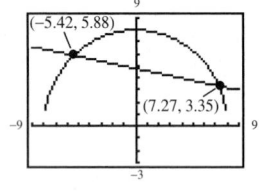

$(7.27, 3.35), (-5.42, 5.88)$

39.

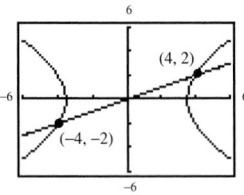

$(4, 2), (-4, -2)$

40.

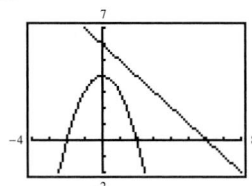

No solution

41.

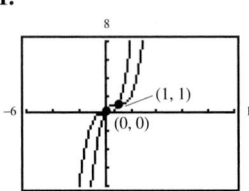

$(0, 0), (1, 1)$

42.

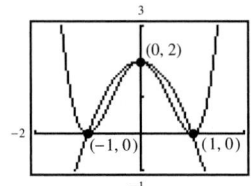

$(\pm 1, 0), (0, 2)$

43.

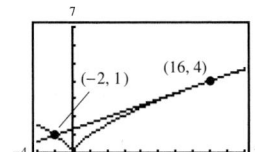

$(-2, 1), (16, 4)$

44.

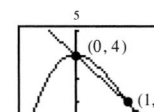

$(0, 4), (1, 2), (2, 0)$

45. $(3.633, 2.733)$
46. $\left(\dfrac{-190 + 81\sqrt{38}}{62}, \dfrac{810 - 81\sqrt{38}}{62}\right) \approx (4.989, 5.011)$
47. Between points $\left(-\frac{3}{5}, -\frac{4}{5}\right)$ and $\left(\frac{4}{5}, -\frac{3}{5}\right)$
48. 40 ft $\times$ 30 ft   **49.** 1987
$22\frac{1}{2}$ ft $\times 53\frac{1}{3}$ ft

Section Project 8.8

Answers may vary.

Review Exercises   (page 637)

1. (a) 0   **(b)** 5   **(c)** 12   **(d)** -1
2. (a) $-\frac{3}{4}$ **(b)** 3   **(c)** $\frac{1}{27}$ **(d)** 2
3. (a) $-\dfrac{1}{3}$ **(b)** 83   **(c)** $-\dfrac{1}{48}$ **(d)** $-\dfrac{\sqrt{2}}{6}$
4. (a) 5   **(b)** 0   **(c)** 30   **(d)** 1
5. (a) $-\frac{8}{3}$ **(b)** -2 **(c)** 1   **(d)** $\frac{1}{10}$
6. (a) 0   **(b)** $-\frac{1}{3}$ **(c)** $-\frac{4}{7}$ **(d)** -3
7. (a) $x^2 + 2$ **(b)** $(x + 2)^2$ **(c)** 6   **(d)** 1

8. (a) $\sqrt[3]{x+2}$ **(b)** $\sqrt[3]{x}+2$ **(c)** 2 **(d)** 6
9. (a) $|x|$ **(b)** $x,\ x \geq -1$ **(c)** 5 **(d)** -1
10. (a) x **(b)** x **(c)** 1 **(d)** $\frac{1}{5}$
11. (a) $[2, \infty)$ **(b)** $[4, \infty)$
12. (a) $(-\infty, -2) \cup (-2, 2) \cup (2, \infty)$
  (b) $(-\infty, 4) \cup (4, \infty)$
13.   **14.**

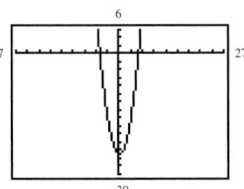 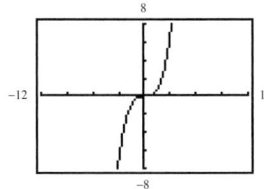

No; function is not one-to-one.   Yes; function is one-to-one.
15.   **16.**

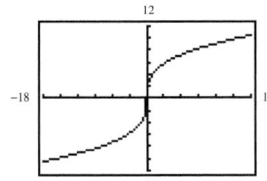 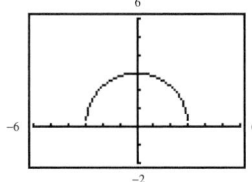

Yes; function is one-to-one.   No; function is not one-to-one.
17. $f^{-1}(x) = 4x$ **18.** $f^{-1}(x) = \frac{1}{2}(x+3)$
19. $h^{-1}(x) = x^2,\ x \geq 0$ **20.** $g^{-1}(x) = \sqrt{x-2}$
21. f is not one-to-one. **22.** $h^{-1}(t) = t$
23. $f(x) = 2(x-4)^2,\ x \geq 4$

$$f^{-1}(x) = \frac{\sqrt{2x}}{2} + 4$$

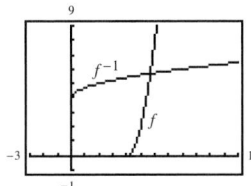

24. $f(x) = |x-2|,\ x \geq 2$
  $f^{-1}(x) = x+2,\ x \geq 0$

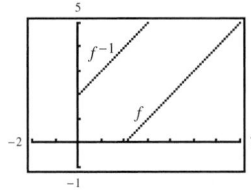

25. $k = 6,\ y = 6\sqrt[3]{x}$ **26.** $k = 27,\ r = \dfrac{27}{s}$

27. $k = \dfrac{1}{18},\ T = \dfrac{1}{18}rs^2$ **28.** $k = 750,\ D = \dfrac{750x^3}{y}$

29. (a) $k = \frac{1}{8}$ **(b)** 1953.125 kW
30. (a) 8 in. **(b)** 62.5 lb   **31.** 150 lb
32. d will increase by a factor of 4.   **33.** 945 units
34. 158 lb   **35.** Falls to the left; Falls to the right
36. Falls to the left; Rises to the right
37. Rises to the left; Rises to the right
38. Rises to the left; Falls to the right
39. $(2, 0), (0, 8)$ **40.** $(-1, 0), (0, 1)$

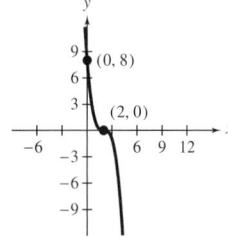 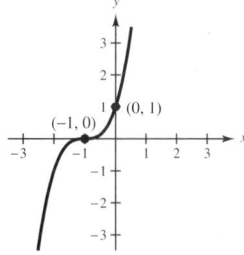

41. $(0, 0), (-1, 0), (2, 0)$ **42.** $(0, 0), (\pm 2, 0)$

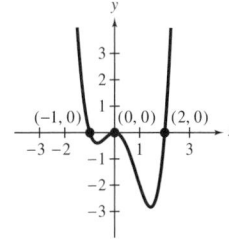 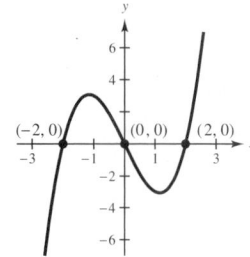

43. $(0, 0), (-3, 0)$ **44.** $(0, 0), (\pm 2, 0)$

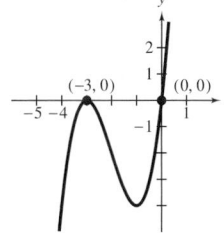 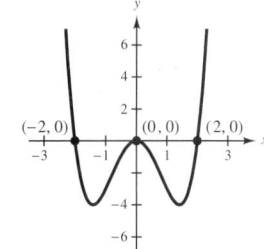

45. (c) **46.** (d) **47.** (a) **48.** (f) **49.** (b)
50. (g) **51.** (h) **52.** (e)

53. Parabola

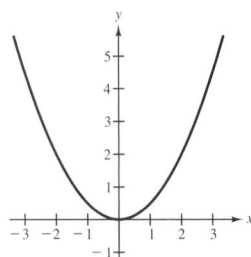

54. Circle

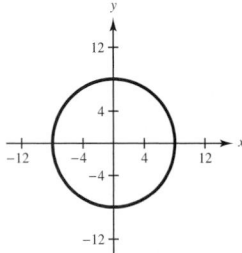

61. Circle

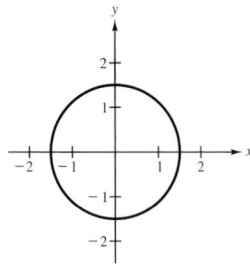

62. Ellipse

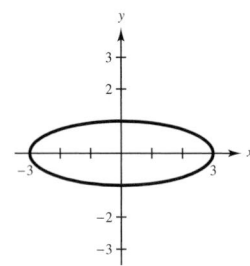

55. Hyperbola

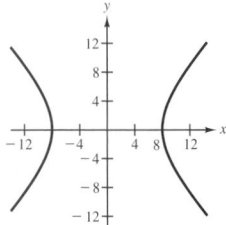

56. Ellipse

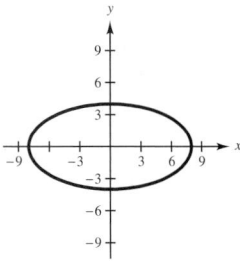

63. Ellipse

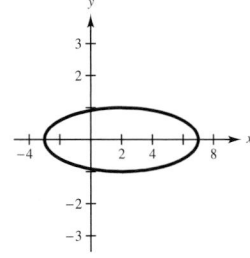

64. Hyperbola

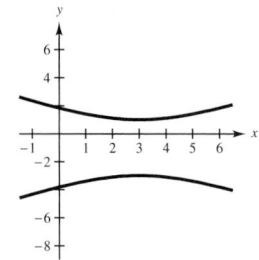

57. Parabola

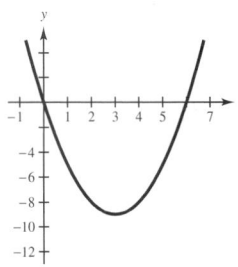

58. Parabola

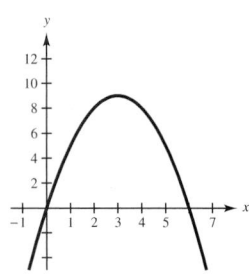

65. $x^2 + y^2 = 144$ **66.** $x^2 + y^2 = 36$

67. $(x - 3)^2 + (y - 5)^2 = 25$

68. $(x + 2)^2 + (y - 3)^2 = 13$ **69.** $\dfrac{x^2}{4} + \dfrac{y^2}{25} = 1$

70. $\dfrac{x^2}{100} + \dfrac{y^2}{36} = 1$ **71.** $\dfrac{(x - 5)^2}{25} + \dfrac{(y - 3)^2}{9} = 1$

72. $\dfrac{x^2}{9} + \dfrac{(y - 4)^2}{16} = 1$ **73.** $\dfrac{x^2}{36} - \dfrac{y^2}{4} = 1$

74. $\dfrac{y^2}{16} - \dfrac{x^2}{4} = 1$ **75.** $\dfrac{x^2}{1} - \dfrac{y^2}{4} = 1$

76. $\dfrac{(y + 2)^2}{4} - \dfrac{x^2}{1} = 1$ **77.** $y^2 = -8x$ **78.** $x^2 = 16y$

79. $(x + 6)^2 = -9(y - 4)$ **80.** $(y - 5)^2 = \frac{25}{6}x$

81. $x^2 + y^2 = 25{,}000{,}000$

82. $\dfrac{x^2}{20{,}250{,}000} + \dfrac{y^2}{25{,}000{,}000} = 1$ **83.** $(-1, 5), (-2, 20)$

84. No real solution   **85.** $(-1, 0), (0, -1)$

86. $\left(-5\sqrt{2}, 5\sqrt{2}\right), \left(5\sqrt{2}, -5\sqrt{2}\right)$

87. $(0, 2), \left(-\frac{16}{5}, -\frac{6}{5}\right)$ **88.** $(0, -5), (-8, 3)$

89. $(\pm 5, 0)$ **90.** $(0, \pm 4)$

59. Ellipse

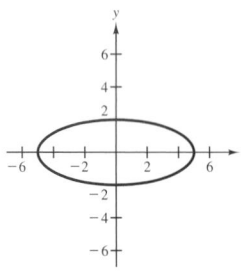

60. Hyperbola

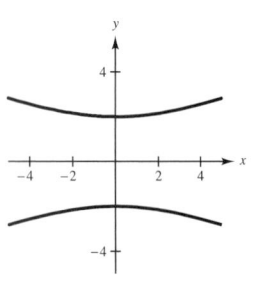

91.

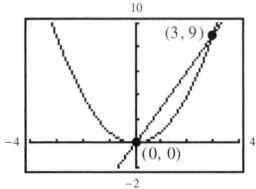

$(0, 0), (3, 9)$

92.

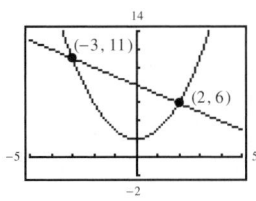

$(-3, 11), (2, 6)$

93.

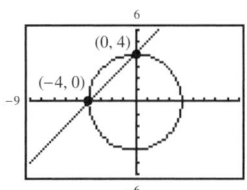

$(-4, 0), (0, 4)$

94.

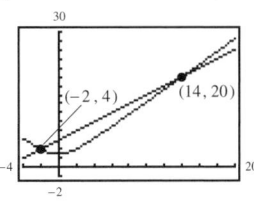

$(-2, 4), (14, 20)$

95. (a)

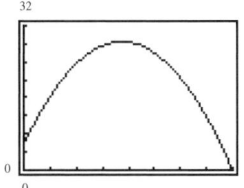

(b) 6 ft
(c) 28.5 ft
(d) 31.9 ft

Chapter 8 Test (page 641)

1. -13 **2.** -30 **3.** $\frac{1}{3}$ **4.** 12   **5.** $[-5, 5]$

6. $f^{-1}(x) = 2x + 2$ **7.** $S = \dfrac{kx^2}{y}$ **8.** $v = \dfrac{1}{4}\sqrt{u}$

9. 240 cm³   **10.** Rises to the left; Falls to the right
11. $(3, 0), (0, 9)$ **12.** $x^2 + y^2 = 25$

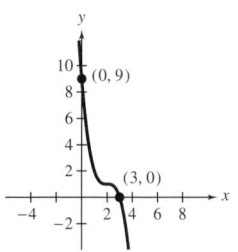

13. $(y - 1)^2 = 8(x + 2)$ **14.** $\dfrac{x^2}{9} + \dfrac{y^2}{100} = 1$

15. $\dfrac{x^2}{9} - \dfrac{y^2}{\frac{9}{4}} = 1$

16. (a)

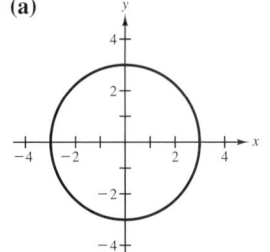

(b)

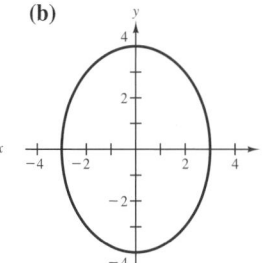

(c)

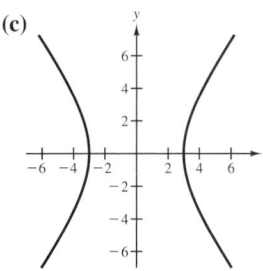

(d)

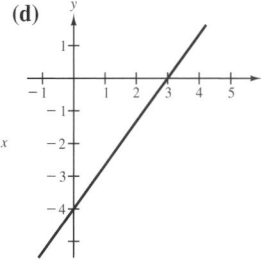

17. (a) $(2, -2),(6, -18)$ **(b)** $(6, 8), (8, 6)$

Cumulative Test: Chapters 1–8   (page 642)

1. $54x^{10}$ **2.** $\dfrac{4u^3}{3}$, $v \ne 0$, $u \ne 0$ **3.** $\dfrac{5(1 + \sqrt{3})}{4}$

4. 3   **5.** $5\sqrt{2}$ **6.** 15   **7.** $y = 2x$

8. $y = -x + 6$ **9.** $3x(x - 7)$ **10.** $(2t + 13)(2t - 13)$

11. $\dfrac{9}{2(x + 3)}$, $x \ne 0$ **12.** $\dfrac{1}{x - 2}$, $x \ne 0, x \ne -2$

13.

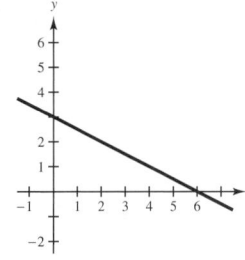

14.

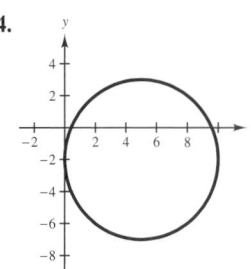

15.

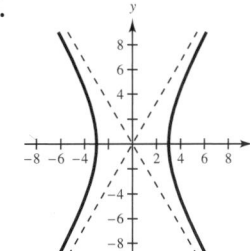

16. $\dfrac{(x - 3)^2}{9} + \dfrac{y^2}{4} = 1$

17. $\dfrac{-3 \pm \sqrt{17}}{2}$

18. $0 < t \le 23$ min
19. 4800 units
20. 7, 9

CHAPTER 9

CHAPTER 9

Section 9.1   (page 652)

1. 2^{2x-1} **2.** 3^{x+2} **3.** $8e^{3x}$ **4.** $2e^{3x}$ **5.** 11.036

6. 0.004   **7.** 1.396   **8.** 0.717   **9.** 0.906

10. 542.309   **11. (a)** $\frac{1}{9}$ **(b)** 1   **(c)** 3

12. (a) 9   **(b)** 1   **(c)** $\frac{1}{3}$ **13. (a)** $\frac{1}{5}$ **(b)** 5   **(c)** 125

14. (a) 5   **(b)** $\frac{1}{5}$ **(c)** 0.062

15. (a) 500   **(b)** 250   **(c)** 56.657

16. (a) 1200   **(b)** 533.333   **(c)** 237.037

17. (a) 1000   **(b)** 1628.895   **(c)** 2653.298

18. (a) 7875.661   **(b)** 3029.948   **(c)** 918.058

19. (a) 0.368   **(b)** 1   **20. (a)** 543.656   **(b)** 1477.811

21. (a) 73.891   **(b)** 1.353   **22. (a)** 50   **(b)** 62.246

23. (b)   **24.** (a)   **25.** (e)   **26.** (d)   **27.** (f)

28. (c)   **29.** (h)   **30.** (g)

31.
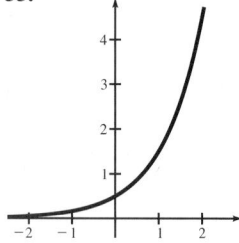

32.
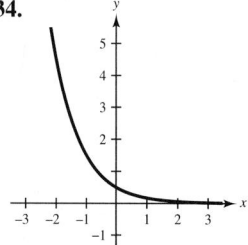

33.
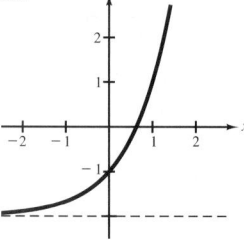

34.
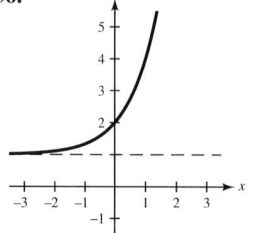

35.
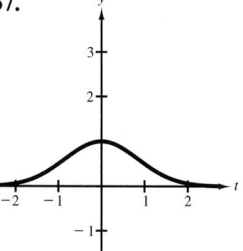

36.
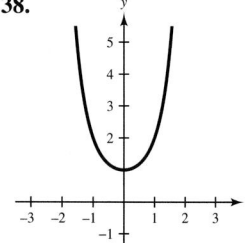

37.
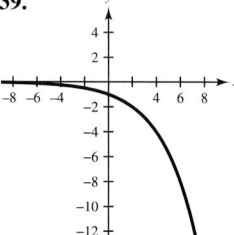

38.
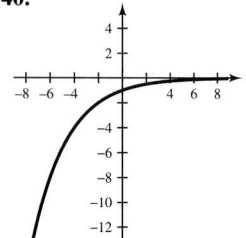

39.
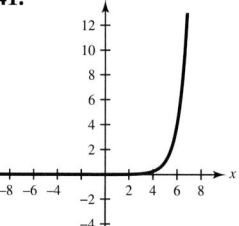

40.
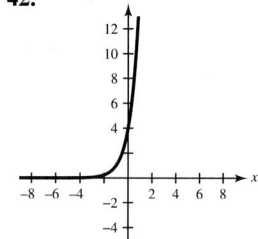

41.
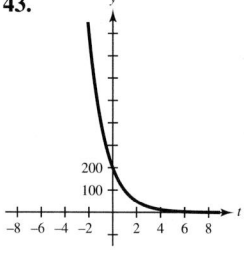

42.
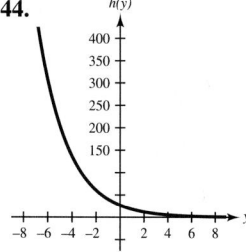

43.
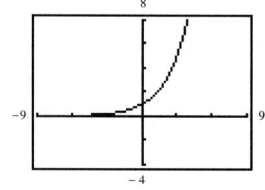

44.
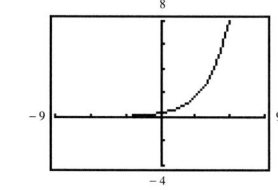

45.

46.

47.

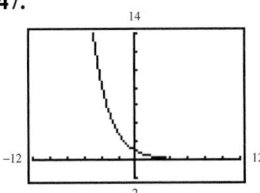

48.

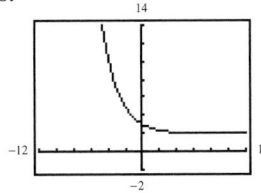

49.

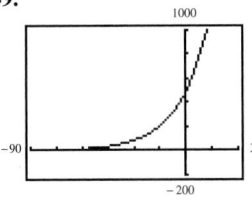

50.

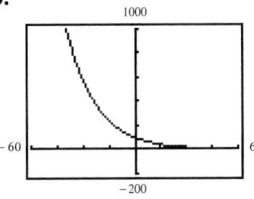

51.

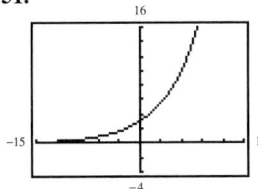

52.

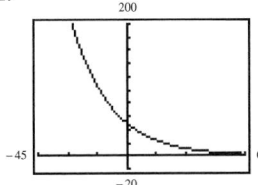

53.

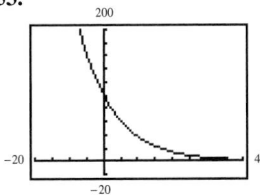

54.

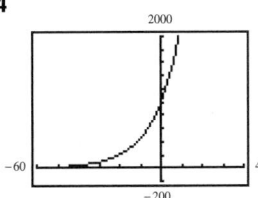

55.

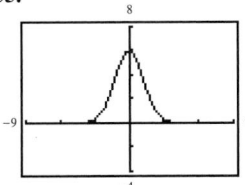

56.

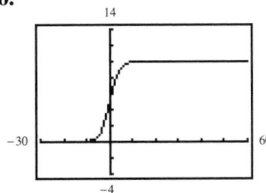

57.

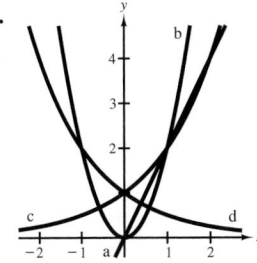

$f(x) = 2^x$ and $f(x) = 2^{-x}$
are exponential.

58. $y_1 \leftrightarrow a$; $y_2 \leftrightarrow b$; $y_3 \leftrightarrow c$
The graph increases at a greater rate as k is increased.

59. No. e is an irrational number.

60. Because $1 < \sqrt{2} < 2$, $2^1 < 2^{\sqrt{2}} < 2^2$.

61.

n	1	4	12
A	\$466.10	\$487.54	\$492.68

n	365	Continuous compounding
A	\$495.22	\$495.30

62.

n	1	4	12
A	\$18,760.65	\$20,993.96	\$21,551.27

n	365	Continuous compounding
A	\$21,829.69	\$21,839.26

63.

n	1	4	12
A	\$4734.73	\$4870.38	\$4902.71

n	365	Continuous compounding
A	\$4918.66	\$4919.21

64.

n	1	4	12
A	\$226,296.28	\$259,889.34	\$268,503.32

n	365	Continuous compounding
A	\$272,841.24	\$272,990.75

65.

n	1	4	12
P	\$2541.75	\$2498.00	\$2487.98

n	365	Continuous compounding
P	\$2483.09	\$2482.93

66.

n	1	4	12
P	$17,843.09	$16,862.99	$16,641.28

n	365	Continuous compounding
P	$16,533.56	$16,529.89

67.

n	1	4	12
P	$18,429.30	$15,830.43	$15,272.04

n	365	Continuous compounding
P	$15,004.64	$14,995.58

68.

n	1	4	12
P	$2163.33	$2154.76	$2152.77

n	365	Continuous compounding
P	$2151.80	$2151.77

69. (a)   **(b)**

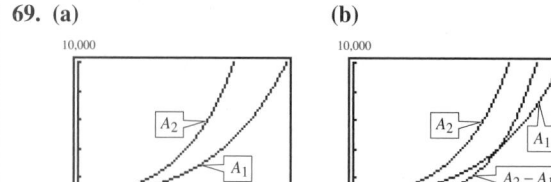

(c) The difference between the functions increases at an increasing rate.

70. $40.64   **71. (a)** $80,634.95   **(b)** $161,269.89
72. (a) $22.04   **(b)** $20.13
73. $V(t) = 16,000\left(\frac{3}{4}\right)^t$ **74. (a)** $V(t) = 16,000 - 3000t$

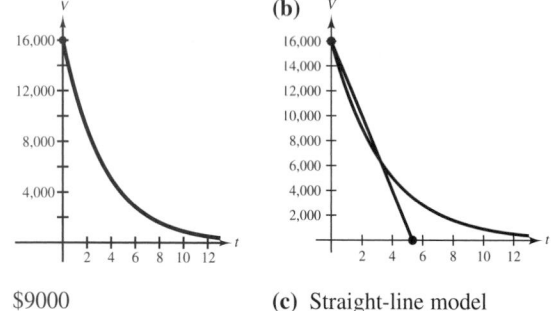

$9000

(c) Straight-line model
(d) Exponential model

75. (a)

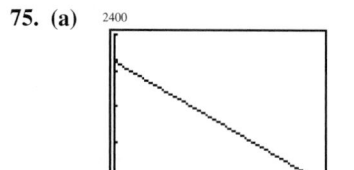

(b)

t	0	25	50	75
h	2000 ft	1450 ft	950 ft	450 ft

(c) Ground level: 97.5 sec
76. (a) and **(b)**

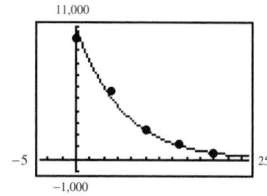

(c)

h	0	5	10	15	20
P	10,332	5583	2376	1240	517
Approx.	10,958	5176	2445	1155	546

(d) 3300   **(e)** 11.3
77. $f(x) = 2^x$; $f(30) = 1{,}073{,}741{,}824$ pennies

78. (a)

Year	1987	1988	1989
Price	$99,744	$121,074	$130,035

Year	1990	1991	1992
Price	$131,368	$132,715	$142,537

(b)

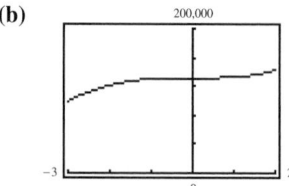

(c) Increasing at a higher rate. Home prices probably will not increase at a higher rate indefinitely.

Section Project 9.1

(a)

x	1	10	100	1000	10,000
$\left(1 + \dfrac{1}{x}\right)^x$	2	2.5937	2.7048	2.7169	2.7181

(b)

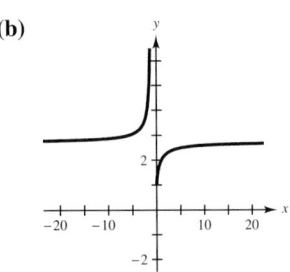

The graph appears to be approaching a horizontal asymptote.

(c) The value approaches e.

Section 9.2   (page 665)

1. $g = f^{-1}$
2. Writing the logarithmic equation in exponential form yields $a^x = a^x$.
3. $5^2 = 25$ **4.** $6^2 = 36$ **5.** $4^{-2} = \frac{1}{16}$
6. $8^{-1} = \frac{1}{8}$ **7.** $3^{-5} = \frac{1}{243}$ **8.** $10^4 = 10,000$
9. $36^{1/2} = 6$ **10.** $32^{2/5} = 4$ **11.** $8^{2/3} = 4$
12. $16^{3/4} = 8$ **13.** $\log_7 49 = 2$ **14.** $\log_6 1296 = 4$
15. $\log_3 \frac{1}{9} = -2$ **16.** $\log_5 625 = 4$ **17.** $\log_8 4 = \frac{2}{3}$
18. $\log_{81} 27 = \frac{3}{4}$ **19.** $\log_{25} \frac{1}{5} = -\frac{1}{2}$
20. $\log_6 \frac{1}{216} = -3$ **21.** $\log_4 1 = 0$
22. $\log_{10} 1.318 \approx 0.12$ **23.** 3   **24.** 3   **25.** 1
26. 1   **27.** -2 **28.** -2
29. There is no power to which 2 can be raised to obtain -3.
30. 0   **31.** 3   **32.** -2 **33.** 0
34. There is no power to which 5 can be raised to obtain -6.
35. $\frac{1}{2}$ **36.** $\frac{3}{2}$
37. There is no power to which 4 can be raised to obtain -4.
38. There is no power to which 2 can be raised to obtain 0.
39. $\frac{1}{2}$ **40.** 3   **41.** 1.4914   **42.** -0.0625
43. -0.0706 **44.** -0.4622 **45.** 0.7335
46. 3.7251   **47.** 3.2189   **48.** 1.8825   **49.** -0.2877
50. -0.3119 **51.** 0.0757   **52.** 0.0083
53. f and g are inverse functions.
54. f and g are inverse functions.
55. f and g are inverse functions.
56. f and g are inverse functions.

57. f and g are inverse functions.

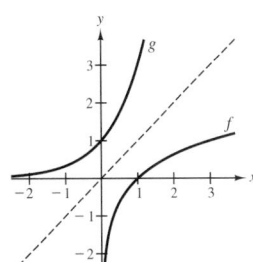

58. f and g are inverse functions.

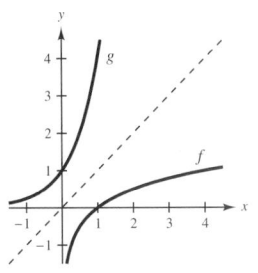

59. f and g are inverse functions.

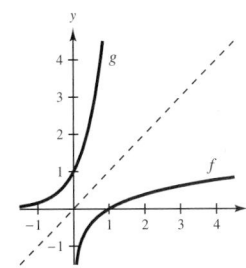

60. f and g are inverse functions.

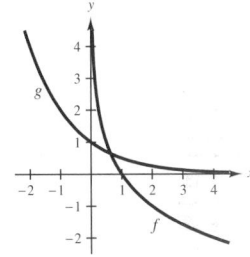

61. (e)   **62.** (b)   **63.** (d)   **64.** (c)   **65.** (a)   **66.** (f)

67.

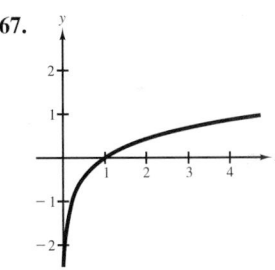

68.

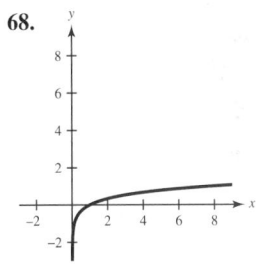

69.

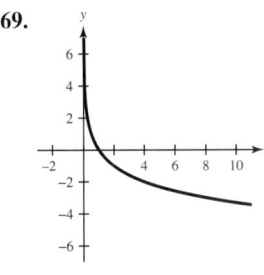

70.

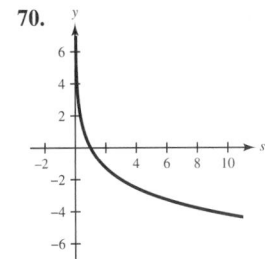

71.

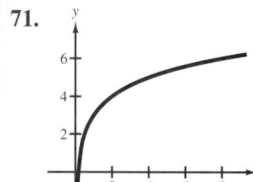

72.

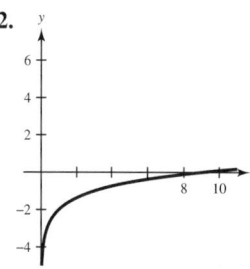

79. Domain: $(3, \infty)$
Vertical asymptote: $x = 3$

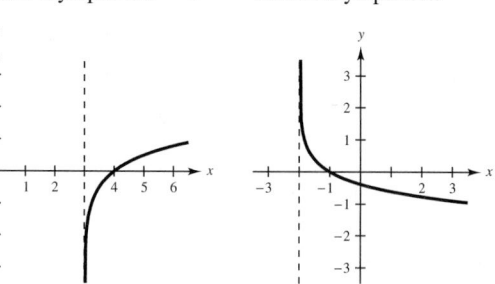

80. Domain: $(-2, \infty)$
Vertical asymptote: $x = -2$

73.

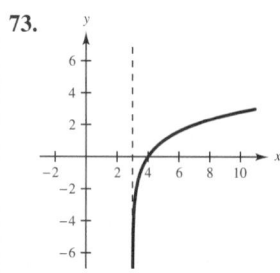

74.

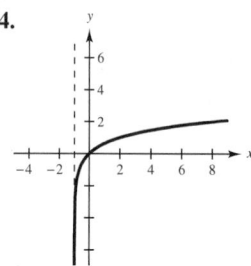

81. Domain: $(0, \infty)$
Vertical asymptote: $x = 0$

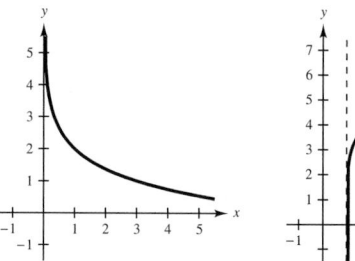

82. Domain: $(1, \infty)$
Vertical asymptote: $x = 1$

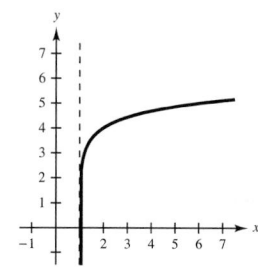

75.

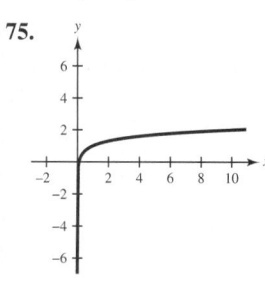

76.

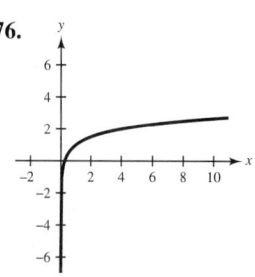

83. Domain: $(0, \infty)$
Vertical asymptote: $x = 0$

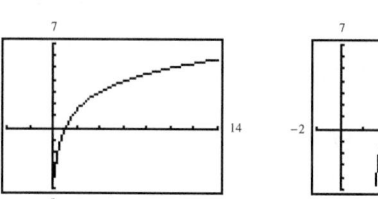

84. Domain: $(3, \infty)$
Vertical asymptote: $x = 3$

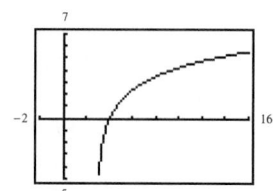

77. Domain: $(0, \infty)$
Vertical asymptote: $x = 0$

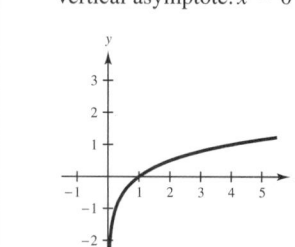

78. Domain: $(0, \infty)$
Vertical asymptote: $x = 0$

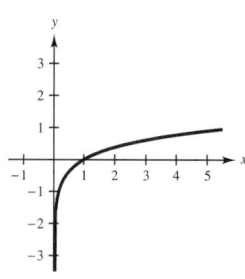

85. Domain: $(0, \infty)$
Vertical asymptote: $x = 0$

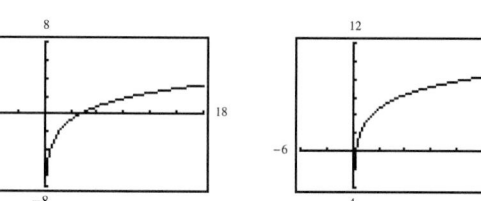

86. Domain: $(0, \infty)$
Vertical asymptote: $x = 0$

87. Domain: $(0, \infty)$
Vertical asymptote: $x = 0$

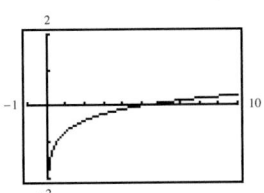

88. Domain: $(-\infty, 0)$
Vertical asymptote: $x = 0$

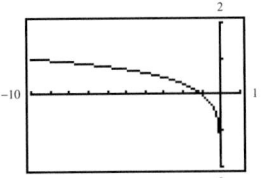

89. (b)   **90.** (e)   **91.** (d)   **92.** (c)   **93.** (f)   **94.** (a)

95.

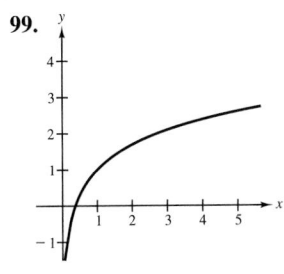

96.

97.

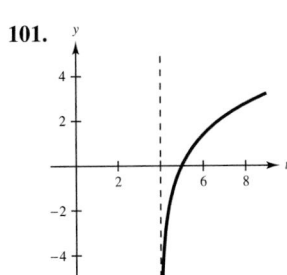

98.

99.

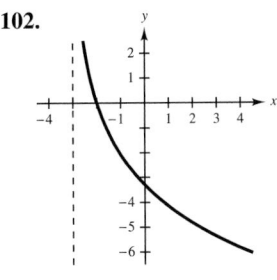

100.

101.

102.

103. Domain: $(-6, \infty)$
Vertical asymptote:
$x = -6$

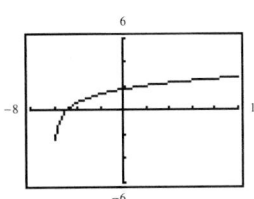

104. Domain: $(2, \infty)$
Vertical asymptote:
$x = 2$

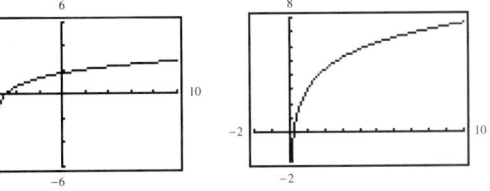

105. Domain: $(0, \infty)$
Vertical asymptote:
$t = 0$

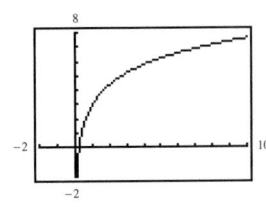

106. Domain: $(-\infty, 3)$
Vertical asymptote:
$t = 3$

107. 2.3481   **108.** 3.8737   **109.** 1.7712
110. 0.7124   **111.** -0.4739 **112.** -0.2056
113. 2.6332   **114.** 1.6117   **115.** -2
116. 3.8227   **117.** 1.3481   **118.** 1.9360
119. 53.4 in.   **120.** 120 db

121.

r	0.07	0.08	0.09	0.10	0.11	0.12
t	9.9	8.7	7.7	6.9	6.3	5.8

122. (a)

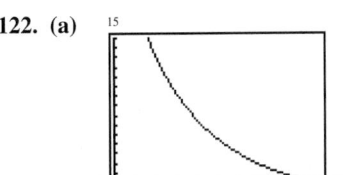

Domain: $(0, 10]$
(b) $x = 0$
(c) $(0, y) \approx (0, 22.9)$

Section Project 9.2

(a) $3 \le f(x) \le 4$ (b) $0 < x < 1$
(c) A factor of 10   (d) 10^3

Section 9.3   (page 673)

1. No. The log of a sum is not the sum of the logs of the terms.

2. Yes. $f(ax) = \log_a(ax) = \log_a a + \log_a x = 1 + f(x)$

3. 2   **4.** 1   **5.** 0   **6.** 0   **7.** 2   **8.** $\frac{1}{2}$ **9.** -3

10. There is no power to which 3 can be raised to obtain -3.

11. 3   **12.** 4   **13.** $\log_3 11 + \log_3 x$ **14.** $\ln 5 + \ln x$

15. $3 \ln y$ **16.** $2 \log_7 x$ **17.** $\log_2 z - \log_2 17$

18. $\log_{10} 7 - \log_{10} y$ **19.** $-2 \log_5 x$ **20.** $\frac{1}{2} \log_2 s$

21. $\frac{1}{3} \log_3(x + 1)$ **22.** $-\frac{1}{2} \log_4 t$

23. $\ln 3 + 2 \ln x + \ln y$ **24.** $\ln y + 2 \ln(y - 1)$

25. $\frac{1}{2}[\ln x + \ln(x + 2)]$ **26.** $\ln 5 - \ln(x - 2)$

27. $2 \log_2 x - \log_2(x - 3)$ **28.** $\frac{1}{2}(\log_5 x - \log_5 y)$

29. $\frac{1}{3}[\ln x + \ln(x + 5)]$ **30.** $2[\ln 3 + \ln x + \ln(x - 5)]$

31. $\frac{1}{3}[2 \ln x - \ln(x + 1)]$ **32.** $2[\ln(x + 1) - \ln(x - 1)]$

33. $6 \log_4 x + 2 \log_4(x - 7)$

34. $2 \log_3 x + \log_3 y - 7 \log_3 z$

35. $3 \ln a + \ln(b - 4) - 2 \ln c$

36. $4 \log_8(x - y) + 6 \log_8 z$

37. $3(\log_{10} 4 + \log_{10} x) - \log_{10}(x - 7)$

38. $\frac{1}{3} \log_4(a + 1) - 4(\log_4 a + \log_4 b)$

39. $\ln x + \frac{1}{3} \ln y - 4(\ln w + \ln z)$

40. $2(\log_5 x + \log_5 y) + 4 \log_5(x + 3)$

41. $\ln(x + y) + \frac{1}{5} \ln(w + 2) - \ln 3 - \ln t$

42. $\log_6 a + \frac{1}{2} \log_6 b + 3 \log_6(c - d)$ **43.** $\log_2 3x$

44. $\log_5 6xy$ **45.** $\log_{10} \dfrac{4}{x}$ **46.** $\ln \dfrac{10x}{z}$

47. $\ln b^4, \; b > 0$ **48.** $\log_4 z^{10}, \; z > 0$

49. $\log_5(2x)^{-2}, \; x > 0$ **50.** $\ln(x + 3)^{-5}$

51. $\ln \sqrt[3]{2x + 1}$ **52.** $\log_3 \dfrac{1}{\sqrt{5y}}$ **53.** $\log_3 2\sqrt{y}$

54. $\ln \dfrac{6}{z^3}$ **55.** $\ln \dfrac{x^2 y^3}{z}, \; x > 0, y > 0, z > 0$

56. $\ln \dfrac{3^4}{x^2 y}, \; x > 0$ **57.** $\ln(xy)^4, \; x > 0, y > 0$

58. $\ln \left(\dfrac{x}{x + 1}\right)^2, \; x > 0$ **59.** $\ln\left(4\sqrt{x}\right)$

60. $\ln \dfrac{x^5}{(x + 4)^{5/2}}$ **61.** $\log_4 \dfrac{x + 8}{x^3}, \; x > 0$

62. $\log_3[x^5(x - 6)], \; x > 6$ **63.** $\log_5\left[\sqrt{x}\,(x - 3)\right]$

64. $\log_6 \dfrac{\sqrt[4]{x + 1}}{x}$

65. $\log_6 \dfrac{(c + d)^5}{m - n}, \; c > -d, m > n$

66. $\log_5[w^3(x + y)^2], \; x + y > 0$

67. $\log_2 \sqrt[5]{\dfrac{x^3}{y}}, \; x > 0, y > 0$

68. $\ln \sqrt[3]{\dfrac{x - 6}{yz^2}}, \; x > 6, y > 0, z > 0$

69. $\log_6 \dfrac{\sqrt[5]{x - 3}}{x^2(x + 1)^3}, \; x > 3$ **70.** $\log_9\left[\dfrac{\sqrt{a + 6}}{a - 1}\right]^3, \; a > 1$

71. 0.954   **72.** -0.602 **73.** 1.556   **74.** 2.158

75. 0.778   **76.** 0   **77.** 1   **78.** 1.5000

79. 1.2925   **80.** 2.2925   **81.** 0.2925   **82.** 1.0850

83. 0.2500   **84.** 0.5283   **85.** 2.7925   **86.** 1.6463

87. 0   **88.** 3   **89.** $1 - \log_4 x$ **90.** $2 + \log_3 4$

91. $1 + \frac{1}{2} \log_5 2$ **92.** $\frac{1}{2} + \frac{1}{2} \log_2 11$ **93.** $2 + \ln 3$

94. $\ln 6 - 5$

95. 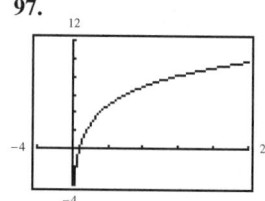   **96.**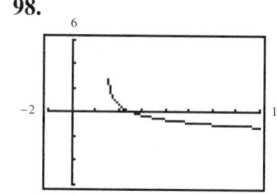

97.   **98.**

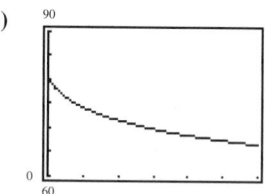

99. True   **100.** True   **101.** True   **102.** False

103. False   **104.** False

105. (a) $f(2) = 74.27, f(8) = 68.55$

(b)

106. $B = 10(\log_{10} I + 16)$, 60 db   **107.** $E = 1.4 \log_{10} \dfrac{C_2}{C_1}$

108. 0.4214   **109.** False. 0 is not in the domain of f.

110. False. $f\left(\sqrt{x}\right) = \frac{1}{2} \ln x$ **111.** True

112. False. $f(x - 3) = \ln(x - 3)$

113. Evaluate when $x = e$ and $y = e$.

Section Project 9.3

$\ln 1 = 0$ $\ln 9 \approx 2.1972$
$\ln 2 \approx 0.6931$ $\ln 10 \approx 2.3025$
$\ln 3 \approx 1.0986$ $\ln 12 \approx 2.4848$
$\ln 4 \approx 1.3862$ $\ln 14 \approx 2.6390$
$\ln 5 \approx 1.6094$ $\ln 15 \approx 2.7080$
$\ln 6 \approx 1.7917$ $\ln 16 \approx 2.7724$
$\ln 7 \approx 1.9459$ $\ln 18 \approx 2.8903$
$\ln 8 \approx 2.0793$ $\ln 20 \approx 2.9956$

Any differences are due to round-off errors.

Mid-Chapter Quiz   (page 676)

1. (a) $\frac{16}{9}$ (b) 1 (c) $\frac{3}{4}$ (d) 1.54

2. Domain: $(-\infty, \infty)$; Range: $(0, \infty)$

3.   **4.**

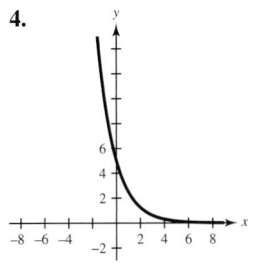

5.   **6.**

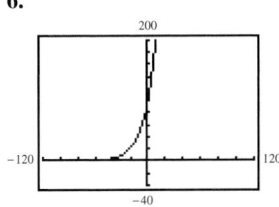

7.

n	1	4	12
A	$3185.89	$3314.90	$3345.61

n	365	Continuous
A	$3360.75	$3361.27

8. $2.71   **9.** $4^{-2} = \frac{1}{16}$ **10.** $\log_3 81 = 4$ **11.** 3

12. $f^{-1}(x) = g(x)$, because their graphs are reflections in the line $y = x$.

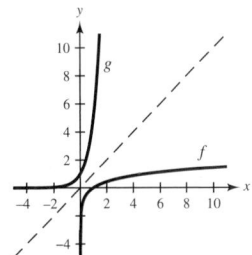

13.   **14.**

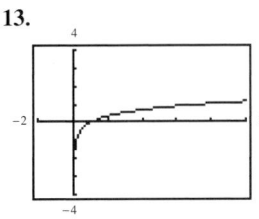   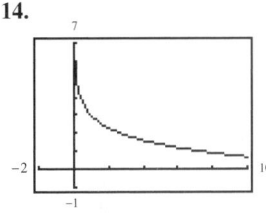

15. $h = 2, k = 1$ **16.** 3.4096

17. $\ln 6 + 2 \ln x - \dfrac{1}{2} \ln(x^2 + 1)$ **18.** $\ln\left(\dfrac{x}{y^3}\right)^2$

Section 9.4   (page 684)

1. (a) Not a solution (b) Solution
2. (a) Solution (b) Not a solution
3. (a) Solution (b) Not a solution
4. (a) Solution (b) Not a solution
5. (a) Not a solution (b) Solution
6. (a) Solution (b) Solution   **7.** 5   **8.** 3   **9.** 8
10. -3 **11.** 8   **12.** $\frac{3}{2}$ **13.** 3   **14.** 2   **15.** -6
16. 0   **17.** -3 **18.** -4 **19.** 4   **20.** 18
21. 16   **22.** 0   **23.** $\frac{22}{5}$ **24.** 10   **25.** 9
26. 8   **27.** 81   **28.** 125   **29.** 500,000   **30.** 11
31. $2x - 1$ **32.** x^2 **33.** $2x, x > 0$
34. $x + 1, x > -1$ **35.** 5.49   **36.** 3.33   **37.** 0.86
38. 3.94   **39.** 1.5   **40.** 0.58   **41.** 3.28   **42.** 2.87
43. 0.59   **44.** 2.41   **45.** 2.49   **46.** 12.28
47. 2.48   **48.** 2.60   **49.** 0.39   **50.** 4.01
51. 6.80   **52.** 12.22   **53.** 0.80   **54.** 9.16
55. 35.35   **56.** 7.39   **57.** 21.82   **58.** 61.40
59. 8.99   **60.** -0.35 **61.** 1   **62.** e **63.** 7.91
64. 43.42   **65.** 22.63   **66.** 655.36   **67.** 2187
68. 3979.07   **69.** 31,622,771.60   **70.** 2000
71. 6.52   **72.** 8.17   **73.** 10.04   **74.** 2.57
75. 12.18   **76.** 49.47   **77.** -0.78 **78.** 1.07

79. ±20.09 **80.** 442,413.39 **81.** 3.20 **82.** 100
83. 4 **84.** 29 **85.** 5 **86.** 1.33 **87.** 0.75
88. 7.40 **89.** 9 **90.** 6
91. **92.**

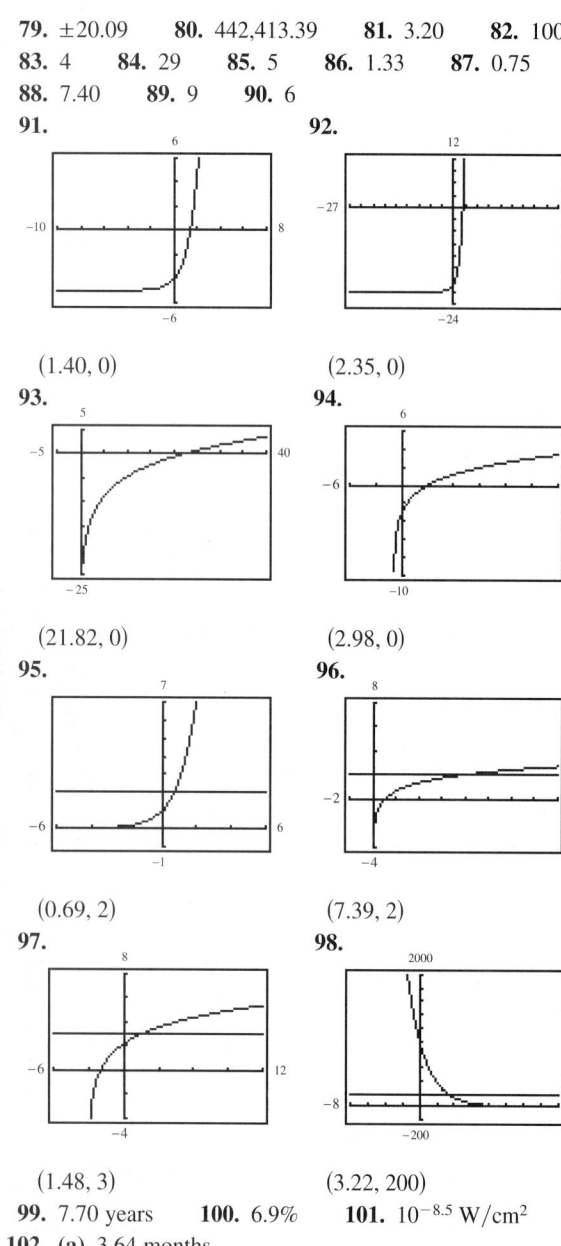

(1.40, 0)   (2.35, 0)
93. **94.**

(21.82, 0)   (2.98, 0)
95. **96.**

(0.69, 2)   (7.39, 2)
97. **98.**

(1.48, 3)   (3.22, 200)
99. 7.70 years **100.** 6.9% **101.** $10^{-8.5}$ W/cm^2
102. (a) 3.64 months

(b)

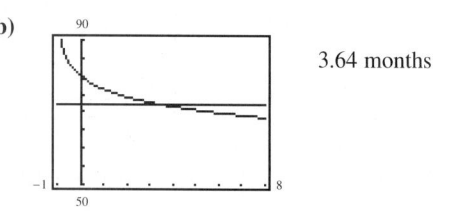

3.64 months

(c) Answers may vary.

103. 205 **104.** 262.5°
105. 1.0234 g/cm^3; 1.0241 g/cm^3

106. (a)

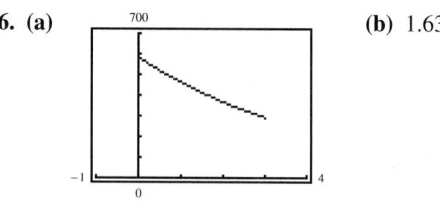

(b) 1.63

107. $2^{x-1} = 30$

Section Project 9.4

(a) $k = \frac{1}{4} \ln \frac{8}{15} \approx -0.1572$ **(b)** ≈ 3.25 hours
(c) ≈ 2.84 hours

Section 9.5 (page 693)

1. $k > 0$ **2.** $k < 0$ **3.** 7% **4.** 10% **5.** 9%
6. 9.5% **7.** 8% **8.** 7.5% **9.** 6% **10.** 8%
11. 7% **12.** 5% **13.** 8.75 years **14.** 13.23 years
15. 6.60 years **16.** 7.64 years **17.** 9.24 years
18. 11.55 years **19.** 14.21 years **20.** 17.33 years
21. 9.99 years **22.** 7.71 years **23.** Continuous
24. Continuous **25.** Quarterly **26.** Daily
27. 8.33% **28.** 9.97% **29.** 7.23% **30.** 8%
31. 6.136% **32.** 9.31% **33.** 8.30% **34.** 5.39%
35. 7.79% **36.** 4.08% **37.** No
38. (a) No **(b)** Becomes greater **39.** $1652.99
40. $3351.60 **41.** $626.46 **42.** $1492.79
43. $3080.15 **44.** $7097.49 **45.** $951.23
46. $2733.59 **47.** $67,667.64 **48.** $6737.95
49. $5496.57 **50.** $184,369.97 **51.** $320,250.81
52. $10,444.45 **53.** $k = \frac{1}{2} \ln \frac{8}{3} \approx 0.4904$
54. $k = \frac{1}{5} \ln 3 \approx 0.2197$ **55.** $k = \frac{1}{3} \ln \frac{1}{2} \approx -0.2310$
56. $k = \frac{1}{7} \ln \frac{1}{2} \approx -0.0990$ **57.** $y = 12.2e^{0.0076t}$; 14.9
58. $y = 16.3e^{0.0037t}$; 17.9 **59.** $y = 14.7e^{0.0221t}$; 26.1
60. $y = 11.0e^{0.0312t}$; 24.8 **61.** $y = 10.5e^{0.0005t}$; 10.6
62. $y = 11.5e^{0.0062t}$; 13.5 **63.** $y = 15.5e^{0.0092t}$; 19.7
64. $y = 16.1 e^{0.0122t}$; 22.1
65. (a) k is larger in Exercise 59, because the population of
Shanghai is increasing faster than the population of
Osaka.
(b) k corresponds to r; k gives the annual percentage rate
of growth.
66. 3.3 g **67.** 7.5 g **68.** 4.43 g **69.** 15,700 years
70. 8.995 billion **71.** $9281

72. (a)

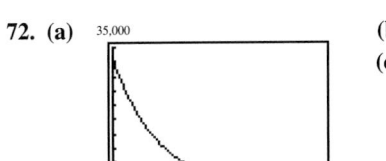

(b) $25,600

(c) 3.1 years

73. Total deposits: $7200.00; Total interest: $10,529.42

74. Total deposits: $14,400.00; Total interest: $91,143.80

75. The one in Alaska is 63 times as great.

76. The one in California is 2.5 times as great.

77. The one in Mexico is 40 times as great.

78. The one in Chile is 63 times as great.

79.

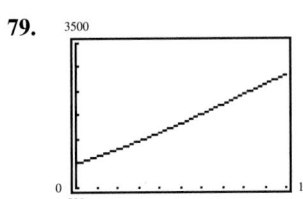

(b) 1000

(c) 2642

(d) 5.88 years

80. (a)

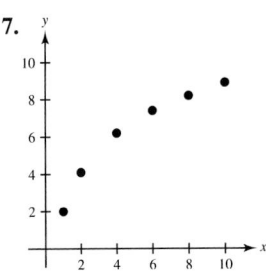

(b) 1298 units

(c) 3.2 years

(d) 2000 units

81. (a) $S = 10(1 - e^{-0.0575x})$ **(b)** 3300 units

Section Project 9.5

(a) 7.04 **(b)** 2.0×10^{-5} **(c)** 10^7 times **(d)** 10

Section 9.6 (page 701)

1. Exponential **2.** Linear **3.** Logarithmic

4. Exponential **5.** Exponential **6.** Linear

7.

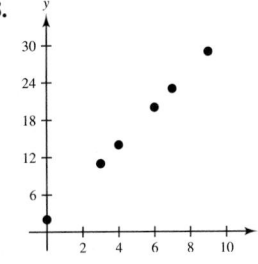

Logarithmmic

8.

Linear

9.

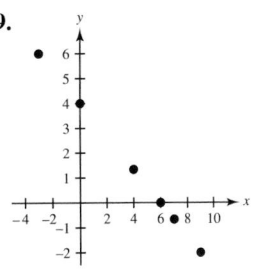

Linear

10.

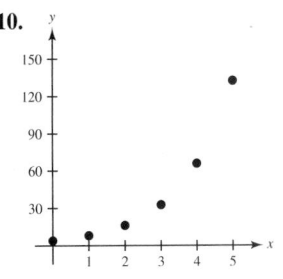

Exponential

11.

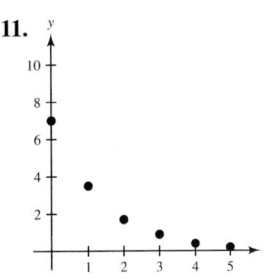

Exponential

12.

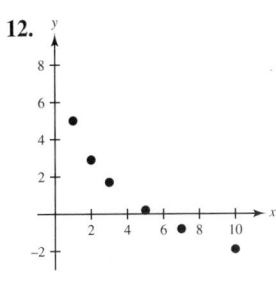

Logarithmic

13.

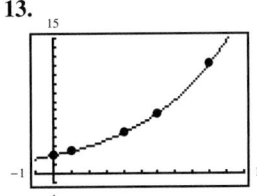

$y = 2.0234(1.2196^x)$

14.

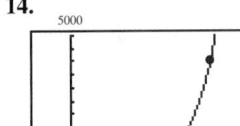

$y = 10.0250(4.4860^x)$

15.

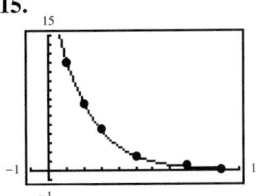

$y = 20.6257(0.6011^x)$

16.

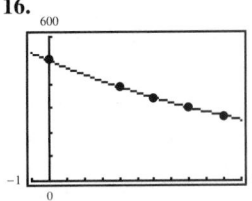

$y = 499.78(0.9396^x)$

17.

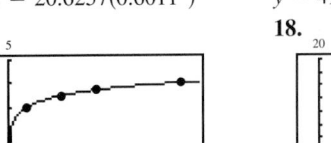

$y = 3.0165 + 0.4458 \ln x$

18.

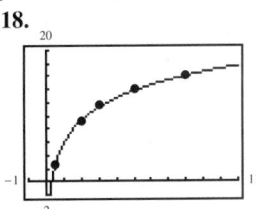

$y = 5.8755 + 4.9343 \ln x$

19.

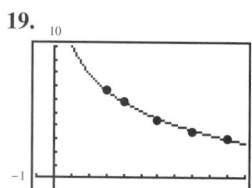

$y = 10.0928 - 3.1025 \ln x$

20.

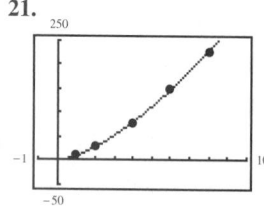

$y = 197.4759 - 24.188 \ln x$

21.

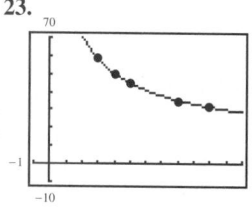

$y = 10.2420x^{1.4923}$

22.

$y = 7.2011x^{1.0869}$

23.

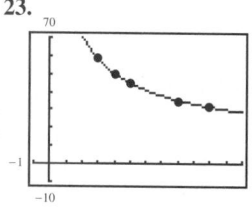

$y = 100.2415x^{-0.4999}$

24.

$y = 526.7657x^{-1.2085}$

25.

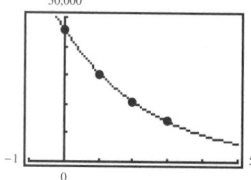

$V = 45,722.15(0.6734^t)$

26. (a) $y = 10,125.60(1.1004^t)$ **(b)** 10% **(c)** \$26,352

27. (a) $V = 78.56 - 11.6314 \ln x$

(b)

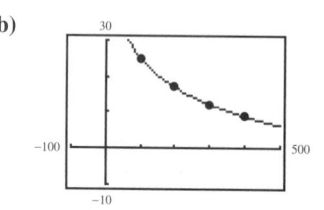

(c) 14.3

28. (a) $N = 793.067x^{-2.4704}$

(b)

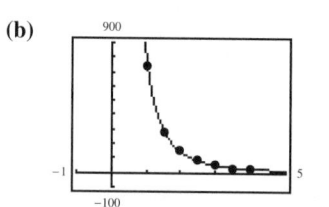

(c) 65

29. (a) $V = 1556.44(1.1986^t)$ $V = 1854.63t^{0.3132}$

(b)

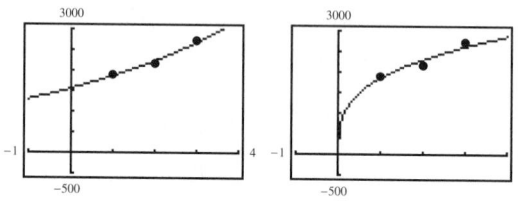

(c) Exponential: 3213 billion dollars

30. (a) and **(b)** Answers may vary. **(c)** 60.623

31. (a) $F = 0.186x^2 - 0.281x - 0.460$
   $F = 0.476(1.4059^x)$
   $F = 0.0443\, x^{2.5393}$

(b)

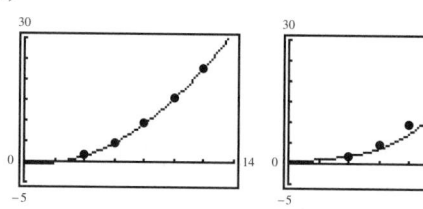

Quadratic   Exponential

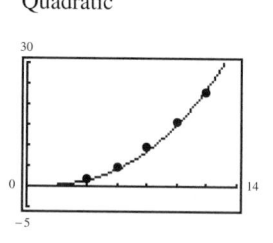

Power

(c) Quadratic: 32,000lb

32. Answers may vary.

Section Project 9.6

(a) $y_1 = 18.12x + 224.17$
$y_2 = -0.93x^2 + 33.96x + 164.53$
$y_3 = 84.36 + 141.54 \ln x$
$y_4 = 246.39(1.0504^x)$
$y_5 = 167.31x^{0.3879}$

(b)

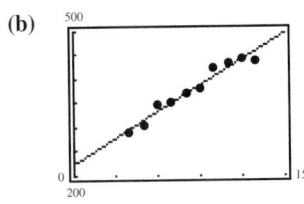

Linear

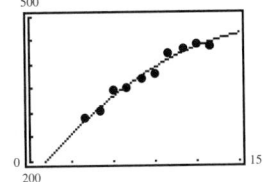

Quadratic

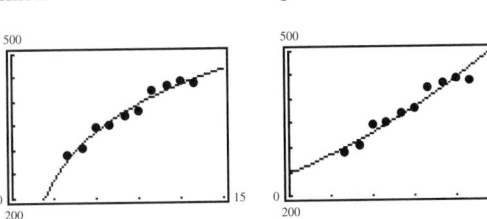

Logarithmic   Exponential

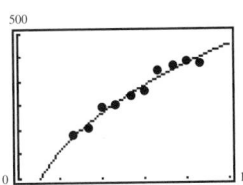

Power

The fit of the quadratic, logarithmic, and power models are similar.

Review Exercises   (page 705)

1. (a) $\frac{1}{8}$ (b) 2 (c) 4   **2.** (a) 4 (b) 1 (c) $\frac{1}{4}$
3. (a) 2.718 (b) 0.351 (c) 0.135
4. (a) 0 (b) -0.492 (c) -0.882
5. (a) 0 (b) 3 (c) -0.631
6. (a) -2 (b) -1 (c) 1.477
7. (a) 1 (b) -1.099 (c) 2.303
8. (a) 2 (b) 0.233 (c) 7.090
9. (a) -6 (b) 0 (c) 22.5
10. (a) 1 (b) 3 (c) 1.189   **11.** (d)   **12.** (f)
13. (a)   **14.** (b)   **15.** (c)   **16.** (e)

17.

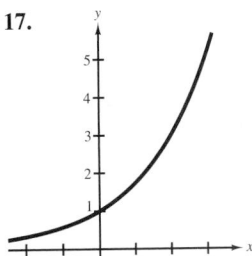

18.

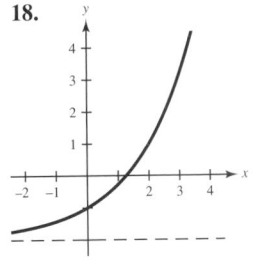

19.

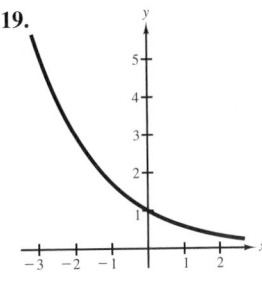

20.

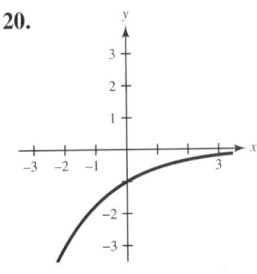

21.

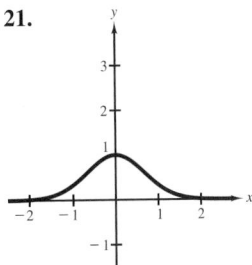

22.

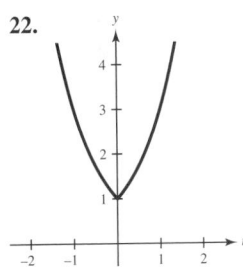

23.

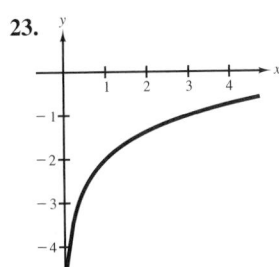

24.

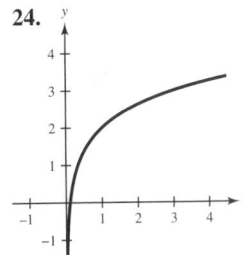

25.

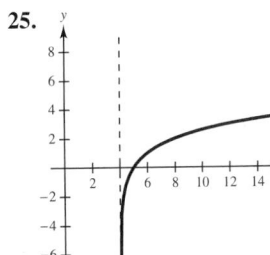

26.

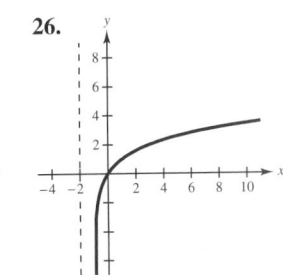

27.
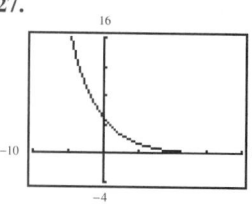

28.
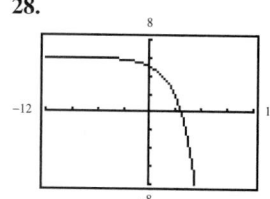

29.
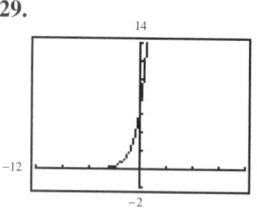

30.
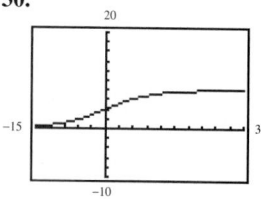

31.
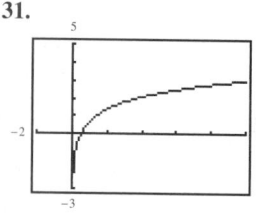

32.
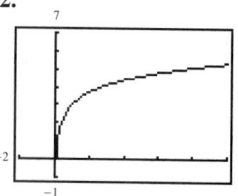

33.
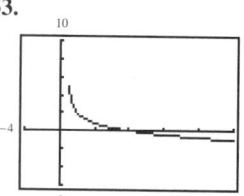

34.
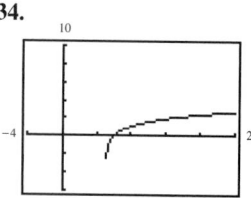

35. $\log_4 64 = 3$ **36.** $\log_{25} 125 = \frac{3}{2}$ **37.** $e^1 = e$
38. $5^{-2} = \frac{1}{25}$ **39.** 3   **40.** $\frac{1}{2}$ **41.** -2 **42.** -2
43. 7   **44.** -1 **45.** 0   **46.** -3
47. $\log_4 6 + 4\log_4 x$ **48.** $\log_{10} 2 - 3\log_{10} x$
49. $\frac{1}{2}\log_5(x+2)$ **50.** $\frac{1}{3}(\ln x - \ln 5)$
51. $\ln(x+2) - \ln(x-2)$ **52.** $\ln x + 2\ln(x-3)$
53. $\frac{1}{2}(\ln 2 + \ln x) + 5\ln(x+3)$
54. $2\log_3 a + \frac{1}{2}\log_3 b - \log_3 c - 5\log_3 d$ **55.** $\log_4 \frac{x}{10}$
56. $\log_2 y^5$ **57.** $\log_8 32x^3$ **58.** $4 + \ln x^8,\ x > 0$
59. $\ln \frac{9}{4x^2},\ x > 0$ **60.** $\ln\left(\frac{1}{3y}\right)^{2/3}$ **61.** $\ln(x^3\,y^4 z)$
62. $\log_8 \sqrt[3]{ab^2}$ **63.** $\log_2\left(\frac{k}{k-t}\right)^4$ **64.** $\ln \frac{x+4}{x^3 y}$

65.
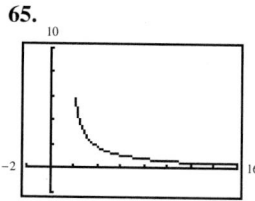

66.
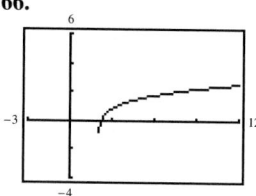

67. False. $\log_2 4x = 2 + \log_2 x$ **68.** False. $\ln \frac{5x}{10x} = \ln \frac{1}{2}$

69. True   **70.** True   **71.** True   **72.** True
73. 1.79588   **74.** 0.55664   **75.** -0.43068
76. -0.25192 **77.** 1.02931   **78.** 3.11328
79. 1.585   **80.** -2.322 **81.** 2.132   **82.** -1.159
83. 6   **84.** 6   **85.** 1   **86.** 50   **87.** 243
88. 35   **89.** 5.66   **90.** 3.32   **91.** 6.23   **92.** 2.68
93. 32.99   **94.** -0.65 **95.** 15.81   **96.** 64
97. 1408.10   **98.** 0.61   **99.** 0.32   **100.** 2263.94
101. 2.67   **102.** 2

103.

n	1	4	12
A	\$3806.13	\$4009.59	\$4058.25

n	365	Continuous compounding	
A	\$4082.26	\$4083.08	

104.

n	1	4	12
A	\$2154.40	\$2286.27	\$2317.63

n	365	Continuous compounding	
A	\$2333.08	\$2333.61	

105.

n	1	4	12
A	\$67,275.00	\$72,095.68	\$73,280.74

n	365	Continuous compounding	
A	\$73,870.32	\$73,890.56	

106.

n	1	4	12
A	\$2700	\$2706.08	\$2707.50

n	365	Continuous compounding
A	\$2708.19	\$2708.22

107.

n	1	4	12
P	\$2301.55	\$2103.50	\$2059.87

n	365	Continuous compounding
P	\$2038.82	\$2038.11

108.

n	1	4	12
P	\$943.40	\$942.18	\$941.91

n	365	Continuous compounding
P	\$941.77	\$941.76

109. 4.6 years   **110.** 8.7 years   **111.** 150 units

112. 0.000316 W/m²   **113.**

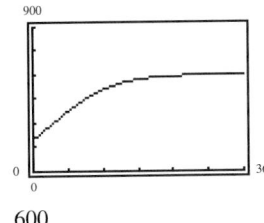

114. (a) 301   **(b)** 3.85 years
115. (a) $y = 2.29t + 2.34$ **(b)** $y = 1.54 + 8.37 \ln t$

(c)

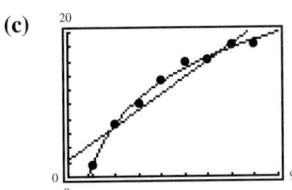

(d) Logarithmic model

116. (a) $y = 13x^{0.5062}$ **(b)** 59.2

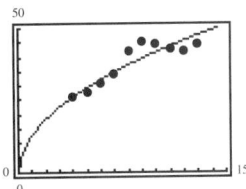

117. (a) $t_1 = 0.2729s - 6.0143$
$t_2 = 0.0027s^2 - 0.0529s + 2.6714$
$t_3 = -47.9558 + 14.4611 \ln s$

$t_4 = 1.5385\,(1.0296^s)$
$t_5 = 0.0136(s^{1.6058})$

(b)

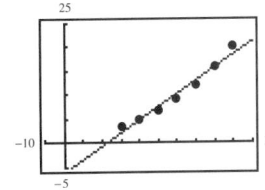

Linear

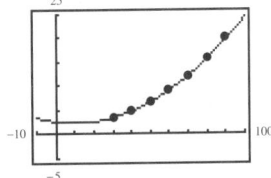

Quadratic

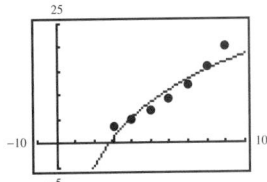

Logarithmic

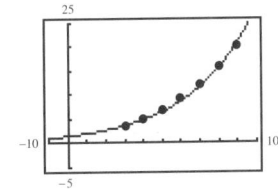

Exponential

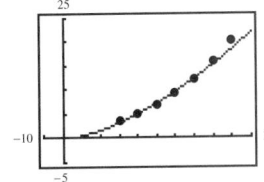

Power
Quadratic

Chapter 9 Test   (page 709)

1. $f(-1) = 81$
$f(0) = 54$
$f(\tfrac{1}{2}) = 18\sqrt{6} \approx 44.09$
$f(2) = 24$

2.

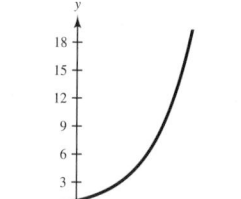

3. $5^3 = 125$ **4.** $\log_4 \tfrac{1}{16} = -2$ **5.** $\tfrac{1}{3}$
6. f is the inverse of g.   **7.** $\log_4 5 + 2\log_4 x - \tfrac{1}{2}\log_4 y$

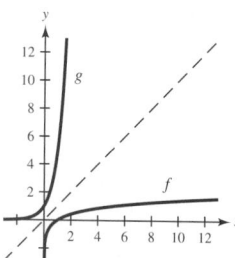

8. $\ln \dfrac{a^8 b}{c^3}$ **9.** 64   **10.** 0.973   **11.** 13.733

12. 15.516   **13.** 4   **14.** 30
15. (a) \$8012.78 (b) \$8110.40   **16.** \$10,806.08
17. 7%   **18.** \$8469.14   **19.** (a) 1141 (b) 4.4 years

20.

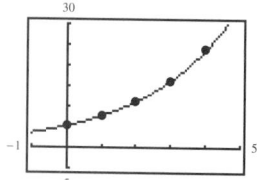

Cumulative Test: Chapters 1–9   (page 710)

1. -4 **2.** -19 **3.** $\pm\sqrt{10}$ **4.** $-\tfrac{1}{2} \le x \le \tfrac{3}{2}$
5. $x \le -9$ or $x \ge -5$ **6.** $-5 < x < 2$ **7.** x^{20}
8. $\dfrac{4y^2}{9x^4}$ **9.** $27x^{12}$ **10.** $\dfrac{x^2 + 2x - 13}{x(x-2)}, x \ne -3, x \ne 3$
11. $\dfrac{(x+1)(x+3)}{3}, x \ne -1$ **12.** Horizontal shift
13. Reflection in the x-axis   **14.** -200 **15.** -126
16. $4^3 = 64$ **17.** $3^{-4} = \tfrac{1}{81}$ **18.** $e^0 = 1$
19. 30% solution: $13\tfrac{1}{3}$ gal; 60% solution: $6\tfrac{2}{3}$ gal

20.

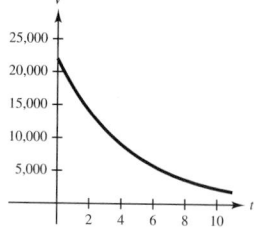

$t = 1.7$

CHAPTER 10

Section 10.1   (page 719)

1. 2, 4, 6, 8, 10   **2.** 3, 6, 9, 12, 15
3. $-2, 4, -6, 8, -10$ **4.** $3, -6, 9, -12, 15$
5. $\tfrac{1}{2}, \tfrac{1}{4}, \tfrac{1}{8}, \tfrac{1}{16}, \tfrac{1}{32}$ **6.** $\tfrac{1}{3}, \tfrac{1}{9}, \tfrac{1}{27}, \tfrac{1}{81}, \tfrac{1}{243}$
7. $\tfrac{1}{4}, -\tfrac{1}{8}, \tfrac{1}{16}, -\tfrac{1}{32}, \tfrac{1}{64}$ **8.** $1, \tfrac{2}{3}, \tfrac{4}{9}, \tfrac{8}{27}, \tfrac{16}{81}$
9. $1, -0.2, 0.04, -0.008, 0.0016$
10. $-\tfrac{2}{3}, \tfrac{4}{9}, -\tfrac{8}{27}, \tfrac{16}{81}, -\tfrac{32}{243}$ **11.** $\tfrac{1}{2}, \tfrac{1}{3}, \tfrac{1}{4}, \tfrac{1}{5}, \tfrac{1}{6}$
12. $1, \tfrac{3}{5}, \tfrac{3}{7}, \tfrac{1}{3}, \tfrac{3}{11}$ **13.** $\tfrac{2}{5}, \tfrac{1}{2}, \tfrac{6}{11}, \tfrac{4}{7}, \tfrac{10}{17}$ **14.** $\tfrac{5}{7}, \tfrac{10}{11}, 1, \tfrac{20}{19}, \tfrac{25}{23}$
15. $-1, \tfrac{1}{4}, -\tfrac{1}{9}, \tfrac{1}{16}, -\tfrac{1}{25}$ **16.** $1, \tfrac{1}{\sqrt{2}}, \tfrac{1}{\sqrt{3}}, \tfrac{1}{2}, \tfrac{1}{\sqrt{5}}$
17. $\tfrac{9}{2}, \tfrac{19}{4}, \tfrac{39}{8}, \tfrac{79}{16}, \tfrac{159}{32}$ **18.** $\tfrac{22}{3}, \tfrac{64}{9}, \tfrac{190}{27}, \tfrac{568}{81}, \tfrac{1702}{243}$
19. $2, 2, \tfrac{4}{3}, \tfrac{2}{3}, \tfrac{4}{15}$ **20.** 1, 2, 3, 4, 5
21. $0, 6, -6, 18, -30$ **22.** $0, \tfrac{1}{2}, 0, \tfrac{1}{8}, 0$ **23.** -72
24. $\tfrac{1}{3,326,400}$ **25.** 5   **26.** 18   **27.** $\tfrac{1}{132}$ **28.** $\tfrac{1}{336}$
29. $\dfrac{1}{702}$ **30.** 15,504   **31.** $\dfrac{1}{n+1}$
32. $(n+2)(n+1)$ **33.** $n(n+1)$
34. $\dfrac{1}{(3n+2)(3n+1)}$ **35.** $2n$ **36.** $(2n+2)(2n+1)$
37. 63   **38.** 50   **39.** 77   **40.** 100   **41.** 35
42. 102   **43.** $\tfrac{47}{60}$ **44.** $\tfrac{9}{5}$ **45.** $\tfrac{437}{60}$ **46.** 28.143
47. -48 **48.** 100   **49.** $\tfrac{8}{9}$ **50.** $\tfrac{50}{21}$ **51.** $\tfrac{182}{243}$
52. $\tfrac{2059}{64}$ **53.** 112   **54.** 110   **55.** 852   **56.** $\tfrac{65}{4}$
57. 6.5793   **58.** 8.5015   **59.** $\displaystyle\sum_{k=1}^{5} k$ **60.** $\displaystyle\sum_{k=1}^{6} (k+7)$
61. $\displaystyle\sum_{k=1}^{5} 2k$ **62.** $\displaystyle\sum_{k=1}^{4} 6(k+3)$ **63.** $\displaystyle\sum_{k=1}^{10} \tfrac{1}{2k}$
64. $\displaystyle\sum_{k=1}^{50} \tfrac{3}{1+k}$ **65.** $\displaystyle\sum_{k=1}^{20} \tfrac{1}{k^2}$ **66.** $\displaystyle\sum_{k=0}^{12} \tfrac{1}{2^k}$
67. $\displaystyle\sum_{k=0}^{9} \tfrac{1}{(-3)^k}$ **68.** $\displaystyle\sum_{k=0}^{20} \left(-\tfrac{2}{3}\right)^k$ **69.** $\displaystyle\sum_{k=1}^{20} \tfrac{4}{k+3}$
70. $\displaystyle\sum_{k=1}^{7} \tfrac{(-1)^{k+1}}{(2k)^3}$ **71.** $\displaystyle\sum_{k=1}^{11} \tfrac{k}{k+1}$ **72.** $\displaystyle\sum_{k=0}^{9} \tfrac{2k+2}{3k+4}$

73. $\sum_{k=1}^{20} \dfrac{2k}{k+3}$ **74.** $\sum_{k=1}^{25}\left(2+\dfrac{1}{k}\right)$ **75.** $\sum_{k=0}^{6} k!$

76. $\sum_{k=0}^{6} \dfrac{1}{k!}$ **77.** 3.6   **78.** 78.75   **79.** 0.8   **80.** 1.8

81. (c)   **82.** (a)   **83.** (b)   **84.** (d)

85.   **86.**

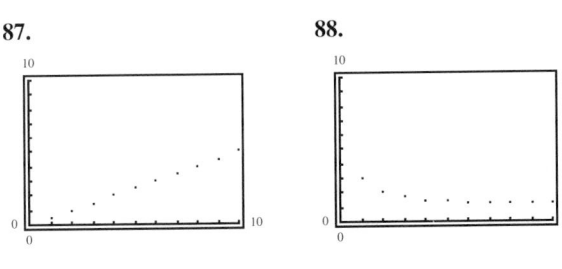

87.   **88.**

89.   **90.**

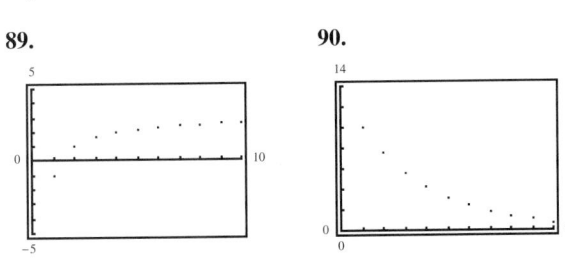

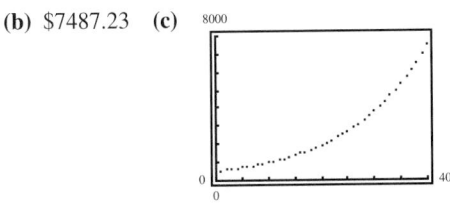

91. (a) \$535, \$572.45, \$612.52, \$655.40, \$701.28, \$750.37, \$802.89, \$859.09

(b) \$7487.23   **(c)**

(d) Yes. Investment earning compound interest increases at an increasing rate.

92. (a) \$6750   **(b)** \$2847.66; No

93. $a_5 = 108°, a_6 = 120°$
At the point where any two hexagons and a pentagon meet, the sum of the three angles is $a_5 + 2a_6 = 348° < 360°$. Therefore, there is a gap of $12°$.

94. True   **95.** True   **96.** True

Section Project 10.1

(a) $25.7°, 45°, 60°, 72°, 81.8°$ **(b)** $\dfrac{180(n-8)}{n}$

Section 10.2   (page 727)

1. A sequence is arithmetic if the differences between consecutive terms are the same.

2. The formula gives the relationship between the terms a_{n+1} and a_n.

3. 3   **4.** 8   **5.** -6 **6.** -400 **7.** $\frac{2}{3}$ **8.** $\frac{3}{4}$

9. $-\frac{5}{4}$ **10.** $-\frac{2}{3}$ **11.** Arithmetic, 2

12. Not arithmetic   **13.** Arithmetic, -16

14. Not arithmetic   **15.** Arithmetic, 0.8

16. Not arithmetic   **17.** Arithmetic, $\frac{3}{2}$

18. Arithmetic, $-\frac{1}{2}$ **19.** Not arithmetic

20. Arithmetic, $-\frac{1}{4}$ **21.** Not arithmetic

22. Not arithmetic   **23.** 7, 10, 13, 16, 19

24. 1, 6, 11, 16, 21   **25.** 6, 4, 2, 0, -2

26. 90, 80, 70, 60, 50   **27.** $\frac{3}{2}, 4, \frac{13}{2}, 9, \frac{23}{2}$

28. $\frac{8}{3}, \frac{10}{3}, 4, \frac{14}{3}, \frac{16}{3}$ **29.** $4, \frac{15}{4}, \frac{7}{2}, \frac{13}{4}, 3$

30. 36, 40, 44, 48, 52   **31.** 25, 28, 31, 34, 37

32. 12, 6, 0, -6, -12 **33.** 9, 6, 3, 0, -3

34. 8, 15, 22, 29, 36   **35.** $-10, -4, 2, 8, 14$

36. $-20, -24, -28, -32, -36$ **37.** 100, 80, 60, 40, 20

38. 4.2, 4.6, 5.0, 5.4, 5.8   **39.** (b)   **40.** (f)   **41.** (e)

42. (a)   **43.** (c)   **44.** (d)

45.   **46.**

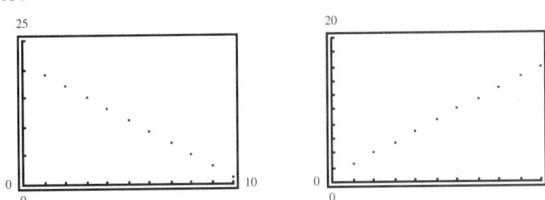

47.   **48.**

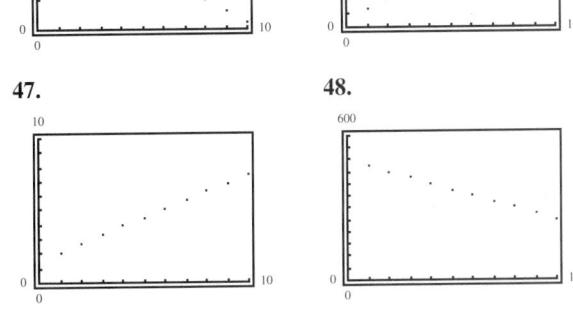

49. **50.**

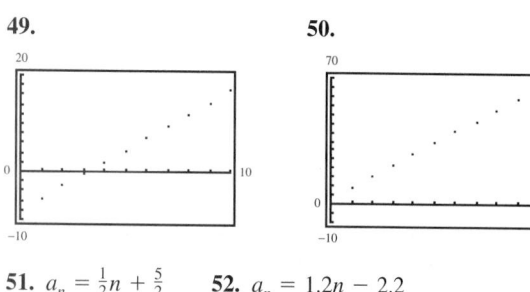

51. $a_n = \frac{1}{2}n + \frac{5}{2}$ **52.** $a_n = 1.2n - 2.2$

53. $a_n = -25n + 1025$ **54.** $a_n = -8n + 72$

55. $a_n = \frac{3}{2}n + \frac{3}{2}$ **56.** $a_n = -3n + 15$

57. $a_n = -4n + 32$ **58.** $a_n = \frac{3}{2}n - 4$

59. $a_n = \frac{5}{2}n + \frac{5}{2}$ **60.** $a_n = -7n + 107$

61. $a_n = 4n + 4$ **62.** $a_n = 5n + 5$

63. $a_n = -10n + 60$ **64.** $a_n = 8n - 48$

65. $a_n = -\frac{1}{2}n + 11$ **66.** $a_n = -\frac{1}{3}n + \frac{31}{3}$

67. $a_n = -0.05n + 0.40$ **68.** $a_n = 0.002n + 0.078$

69. 210 **70.** 1860 **71.** 275 **72.** $\frac{585}{2}$ **73.** 1425

74. 20,100 **75.** 62,625 **76.** 120,200 **77.** 35

78. 1230 **79.** 9000 **80.** 13,120 **81.** 19,500

82. 7900 **83.** 3011.25 **84.** 5778 **85.** 522

86. 1940 **87.** 1850 **88.** 6200 **89.** 900

90. 15,500 **91.** 12,200 **92.** 3000 **93.** 243

94. 66 **95.** 23 **96.** 16,100 **97.** 2850

98. 4455 **99.** 2550 **100.** 10,000 **101.** $246,000

102. $116.25 **103.** $25.43 **104.** 126

105. 632 bales **106.** 114 **107.** 1024 ft

108. **(a)** 4, 9, 16, 25, 36 **(b)** 49

(c) $\sum_{k=1}^{n} (2k - 1) = n^2$

Section Project 10.2

(a) 12, 18, 24, 30, 36, 42, 48

To obtain this arithmetic sequence, find the difference of consecutive terms of the sequence of perfect cubes. Then find the difference of consecutive terms of this sequence.

(b) 60, 84, 108, 132, 156

This is the third sequence obtained by taking differences of consecutive terms in consecutive sequences.

Section 10.3 (page 737)

1. A sequence is geometric if the ratios of consecutive terms are the same.

2. $a_n = \left(-\frac{2}{3}\right)^{n-1}$

3. Raising a real number between 0 and 1 to higher powers yields smaller real numbers.

4. The nth partial sum of a sequence is the sum of the first n terms of the sequence.

5. 3 **6.** -2 **7.** -3 **8.** $\frac{1}{3}$ **9.** $-\frac{3}{2}$ **10.** $\frac{2}{3}$

11. $-\frac{1}{2}$ **12.** e **13.** 1.1 **14.** 1.06

15. Geometric, $\frac{1}{2}$ **16.** Not geometric **17.** Geometric, 2

18. Geometric, $-\frac{1}{3}$ **19.** Not geometric

20. Not geometric **21.** Geometric, $-\frac{2}{3}$

22. Geometric, -2 **23.** Geometric, 1.02

24. Geometric, 0.2 **25.** 4, 8, 16, 32, 64

26. 3, 12, 48, 192, 768 **27.** 6, 2, $\frac{2}{3}, \frac{2}{9}, \frac{2}{27}$

28. 4, 2, 1, $\frac{1}{2}, \frac{1}{4}$ **29.** 1, $-\frac{1}{2}, \frac{1}{4}, -\frac{1}{8}, \frac{1}{16}$

30. 32, -24, 18, $-\frac{27}{2}, \frac{81}{8}$ **31.** 4, -2, 1, $-\frac{1}{2}, \frac{1}{4}$

32. 4, 6, 9, $\frac{27}{2}, \frac{81}{4}$

33. 1000, 1010, 1020.1, 1030.30, 1040.60

34. 4000, 3960.40, 3921.18, 3882.36, 3843.92

35. 200, 214, 228.98, 245.01, 262.16

36. 1000, 952.38, 907.03, 863.84, 822.70

37. 10, 6, $\frac{18}{5}, \frac{54}{25}, \frac{162}{125}$ **38.** 36, 24, 16, $\frac{32}{3}, \frac{64}{9}$ **39.** $\frac{3}{256}$

40. $\frac{2187}{2048}$ **41.** $48\sqrt{2}$ **42.** 405 **43.** 1486.02

44. $500(1.06)^{39}$ **45.** $\frac{81}{64}$ **46.** 531,441 **47.** $\pm\frac{243}{32}$

48. $\frac{64}{3}$ **49.** $\pm\frac{16}{9}$ **50.** $-\frac{25}{16}$ **51.** $a_n = 2(3)^{n-1}$

52. $a_n = 5(4)^{n-1}$ **53.** $a_n = 2^{n-1}$ **54.** $a_n = (-5)^{n-1}$

55. $a_n = 4\left(-\frac{1}{2}\right)^{n-1}$ **56.** $a_n = 9\left(\frac{2}{3}\right)^{n-1}$ **57.** $a_n = 8\left(\frac{1}{4}\right)^{n-1}$

58. $a_n = 18\left(\frac{4}{9}\right)^{n-1}$ **59.** $a_n = 4\left(-\frac{3}{2}\right)^{n-1}$ **60.** $a_n = \left(\frac{3}{2}\right)^{n-1}$

61. (b) **62.** (d) **63.** (a) **64.** (c)

65. **66.**

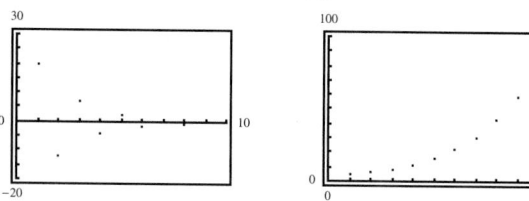

67. **68.**

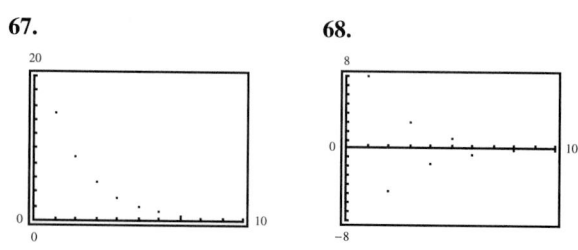

69. 1023 **70.** 364 **71.** 772.48 **72.** 35.99

73. 2.25 **74.** 6.40 **75.** 6.67 **76.** 50,815.58

77. 21.43 **78.** 1714.72 **79.** 399.93 **80.** 56,855.29

81. 6300.25 **82.** 9981.76 **83.** $-14,762$

84. -4095 **85.** 16.00 **86.** 26.53 **87.** 13,120

88. $-24,213,780.56$ **89.** 48.00 **90.** 32.00

91. 1103.57   **92.** 25,905.65

93. 12,822.71   **94.** 152.10

95. (a) $250,000(0.75)^n$ **(b)** \$59,326.17 **(c)** The first year

96. (a) $500.000(1.01)^n$ **(b)** 610,095   **97.** \$3,623,993

98. (a) \$4,098,168 **(b)** \$474,175   **99.** \$19,496.56

100. \$3600.53   **101.** \$105,428.44   **102.** \$455,865.06

103. \$75,715.32   **104.** \$95,736.66

105. (a) \$5,368,709.11 **(b)** \$10,737,418.23

106. (a) \$4,236,400,000 **(b)** \$12,709,000,000

107. (a) $P = (0.999)^n$ **(b)** 69.4%

(c) 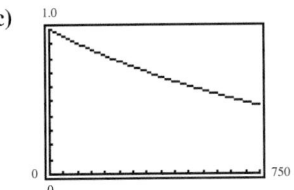   693 days

108. (a) $70(0.8)^n$ **(b)** 18.4° **(c)** 3.5 hr

109. 70.875 in.²   **110.** 72.969 in.²   **111.** 666.21 ft

Section Project 10.3

No. The sum is 1.476×10^{20}.

Mid-Chapter Quiz (page 741)

1. $32, 8, 2, \frac{1}{2}, \frac{1}{8}$ **2.** $-\frac{3}{5}, 3, -\frac{81}{7}, \frac{81}{2}, -135$

3. $\frac{1}{2}, \frac{1}{3}, \frac{1}{4}, \frac{1}{5}, \frac{1}{6}$ **4.** 100   **5.** 40   **6.** 87   **7.** -32

8. $\sum_{k=1}^{20} \frac{2}{3k}$ **9.** $\sum_{k=1}^{25} \frac{(-1)^{k-1}}{k^3}$ **10.** $a_n = -3n + 23$

11. $a_n = 32\left(-\frac{1}{4}\right)^{n-1}$ **12.** 4075   **13.** 9030

14. 25.947   **15.** 18,392.796   **16.** -0.026

17. a_n: (a); b_n: (b)   **18.** 5.5°   **19.** \$97,670

Section 10.4 (page 747)

1. $n + 1$

2. The signs of the terms alternate in the expansion of $(x - y)^n$.

3. $_{11}C_5 = \dfrac{11!}{6!5!} = \dfrac{11 \cdot 10 \cdot 9 \cdot 8 \cdot 7}{5 \cdot 4 \cdot 3 \cdot 2 \cdot 1}$

4. They are the same.   **5.** 15   **6.** 35   **7.** 252

8. 220   **9.** 1   **10.** 1   **11.** 1   **12.** 200

13. 1225   **14.** 75   **15.** 12,650   **16.** 8568

17. 593,775   **18.** 3,268,760   **19.** 792   **20.** 658,008

21. 2,598,960   **22.** 1,192,052,400

23. 2,535,650,040   **24.** 2,573,031,125

25. 5,200,300   **26.** 499,500   **27.** 455   **28.** 126

29. 53,130   **30.** 4060   **31.** 10   **32.** 28   **33.** 792

34. 3003   **35.** 1   **36.** 1   **37.** 15   **38.** 84

39. 35   **40.** 126   **41.** 70   **42.** 210

43. $x^6 + 18x^5 + 135x^4 + 540x^3 + 1215x^2 + 1458x + 729$

44. $x^4 - 20x^3 + 150x^2 - 500x + 625$

45. $x^5 + 5x^4 + 10x^3 + 10x^2 + 5x + 1$

46. $x^5 + 10x^4 + 40x^3 + 80x^2 + 80x + 32$

47. $x^6 - 24x^5 + 240x^4 - 1280x^3 + 3840x^2 - 6144x + 4096$

48. $x^4 - 32x^3 + 384x^2 - 2048x + 4096$

49. $x^4 + 4x^3y + 6x^2y^2 + 4xy^3 + y^4$

50. $u^6 + 6u^5v + 15u^4v^2 + 20u^3v^3 + 15u^2v^4 + 6uv^5 + v^6$

51. $u^3 - 6u^2v + 12uv^2 - 8v^3$

52. $32x^5 + 80x^4y + 80x^3y^2 + 40x^2y^3 + 10xy^4 + y^5$

53. $x^5 - 5x^4y + 10x^3y^2 - 10x^2y^3 + 5xy^4 - y^5$

54. $256t^4 - 256t^3 + 96t^2 - 16t + 1$

55. $81a^4 + 216a^3b + 216a^2b^2 + 96ab^3 + 16b^4$

56. $64u^3 - 144u^2v + 108uv^2 - 27v^3$

57. $a^3 + 6a^2 + 12a + 8$

58. $x^5 + 15x^4 + 90x^3 + 270x^2 + 405x + 243$

59. $16x^4 - 32x^3 + 24x^2 - 8x + 1$

60. $64 - 144y + 108y^2 - 27y^3$

61. $64y^6 + 192y^5z + 240y^4z^2 + 160y^3z^3 + 60y^2z^4$
$\qquad + 12yz^5 + z^6$

62. $32t^5 - 80st^4 + 80s^2t^3 - 40s^3t^2 + 10s^4t - s^5$

63. $x^8 + 8x^7y + 28x^6y^2 + 56x^5y^3 + 70x^4y^4 + 56x^3y^5$
$\qquad + 28x^2y^6 + 8xy^7 + y^8$

64. $r^7 - 7r^6s + 21r^5s^2 - 35r^4s^3 + 35r^3s^4 - 21r^2s^5$
$\qquad + 7rs^6 - s^7$

65. $x^6 - 12x^5 + 60x^4 - 160x^3 + 240x^2 - 192x + 64$

66. $32x^5 + 240x^4 + 720x^3 + 1080x^2 + 810x + 243$

67. 120   **68.** 5940   **69.** -84 **70.** 1120

71. -1365 **72.** $-64,481,508$ **73.** 1760   **74.** 120

75. 54   **76.** -90 **77.** 70   **78.** 60

79. $\frac{1}{32} + \frac{5}{32} + \frac{10}{32} + \frac{10}{32} + \frac{5}{32} + \frac{1}{32}$

80. $\frac{16}{81} + \frac{32}{81} + \frac{24}{81} + \frac{8}{81} + \frac{1}{81}$

81. $\frac{1}{256} + \frac{12}{256} + \frac{54}{256} + \frac{108}{256} + \frac{81}{256}$

82. $\frac{8}{125} + \frac{36}{125} + \frac{54}{125} + \frac{27}{125}$ **83.** 1.172   **84.** 1049.890

85. 510,568.785   **86.** 467.721

87.   **88.**

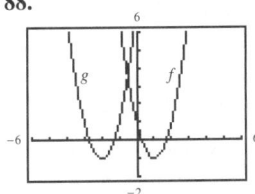

g shifted two units to the right of f

$g(x) = -x^2 + 7x - 8$

g shifted three units to the left of f

$g(x) = 2x^2 + 8x + 7$

89.

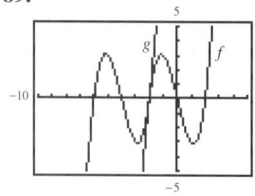

g shifted four units to the left of f
$g(x) = x^3 + 12x^2 + 44x + 48$

90.

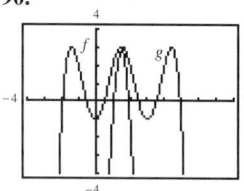

g shifted three units to the right of f
$g(x) = -x^4 + 12x^3 - 50x^2 + 84x - 46$

Section Project 10.4

1, 3, 6, 10, 15

The difference between consecutive determinants increases by 1.

Section 10.5 (page 756)

1. 5 **2.** 4 **3.** 9 **4.** 8 **5.** 8 **6.** 8 **7.** 10
8. 10 **9.** 8 **10.** 8 **11.** 6 **12.** 3 **13.** 7
14. 14 **15.** 6 **16.** 91 **17.** 6 **18.** 12
19. 260 **20.** 2600 **21.** 6,760,000
22. (a) 900 **(b)** 720 **(c)** 400 **23.** 48 **24.** 18
25. 72 **26.** 720
27. XYZ, XZY, YXZ, YZX, ZXY, ZYX
28. ABCD, ABDC, ACBD, ACDB, ADBC, ADCB, BACD, BADC, BCAD, BCDA, BDAC, BDCA, CABD, CADB, CBAD, CBDA, CDAB, CDBA, DABC, DACB, DBAC, DBCA, DCAB, DCBA
29. 120 **30.** 720 **31.** 5040
32. (a) 40,320 **(b)** 5040 **33.** 64,000 **34.** 6720
35. {A, B}, {A, C}, {A, D}, {A, E}, {A, F}, {B, C}, {B, D}, {B, E}, {B, F}, {C, D}, {C, E}, {C, F}, {D, E}, {D, F}, {E, F}
36. {A, B, C}, {A, B, D}, {A, B, E}, {A, B, F}, {A, C, D}, {A, C, E}, {A, C, F}, {A, D, E}, {A, D, F}, {A, E, F}, {B, C, D}, {B, C, E}, {B, C, F}, {B, D, E}, {B, D, F}, {B, E, F}, {C, D, E}, {C, D, F}, {C, E, F}, {D, E, F}
37. 1140 **38.** 142,506 **39.** 126 **40.** 220
41. 3003 **42.** 30 **43.** 56
44. (a) 15 **(b)** 6 **45. (a)** 70 **(b)** 16
46. (a) 3 **(b)** 6 **(c)** 15 **(d)** 28 **(e)** 45 **(f)** 66
47. 56 **48.** 21

Section Project 10.5

(a) 5 **(b)** 9 **(c)** 20 **(d)** 35
Polygon with n sides: $_nC_2 - n$

Section 10.6 (page 764)

1. The probability that the event does not occur is $1 - \frac{3}{4} = \frac{1}{4}$.
2. Over an extended period, it will rain 40% of the time under the given weather conditions.
3. 26 **4.** 11 **5.** 10 **6.** 8
7. {ABC, ACB, BAC, BCA, CAB, CBA}
8. {$(h, 1), (h, 2), (h, 3), (h, 4), (h, 5), (h, 6), (t, 1), (t, 2), (t, 3), (t, 4), (t, 5), (t, 6)$}
9. {WWW, WWL, WLW, WLL, LWW, LWL, LLW, LLL}
10. {AB, AC, AD, BC, BD, CD} **11.** 0.65 **12.** 0.2
13. 0.18 **14.** 0.87 **15.** $\frac{3}{8}$ **16.** $\frac{1}{2}$ **17.** $\frac{7}{8}$
18. $\frac{7}{8}$ **19.** $\frac{1}{2}$ **20.** $\frac{1}{13}$ **21.** $\frac{3}{13}$ **22.** $\frac{3}{26}$ **23.** $\frac{1}{6}$
24. 0 **25.** $\frac{5}{6}$ **26.** 1 **27.** 0.9 **28.** $0.1 = 1 - 0.9$
29. 0.158 **30.** 0.260 **31.** 0.733 **32.** 0.430
33. (a) $\frac{1}{5}$ **(b)** $\frac{1}{3}$ **(c)** 1 **34. (a)** $\frac{1}{4}$ **(b)** $\frac{1}{2}$ **(c)** 1
35. (a) $\frac{1}{20}$ **(b)** $\frac{2}{5}$ **(c)** $\frac{1}{2}$ **(d)** $\frac{1}{4}$ **36. (a)** 0.8 **(b)** 0.2
37. $\frac{1}{24}$ **38. (a)** $\frac{1}{24}$ **(b)** $\frac{1}{6}$ **39.** $\frac{1}{100,000}$ **40.** $\frac{1}{120}$
41. $\frac{1}{45}$ **42.** $\frac{1}{56}$ **43.** $\frac{1}{45}$ **44. (a)** $\frac{6}{11}$ **(b)** $\frac{5}{11}$
45. $\frac{1}{210}$ **46.** $\frac{1}{54,145}$ **47.** $\frac{33}{66,640}$ **48.** 0.000009

49.

	Female	
	X	X
Male X	XX	XX
Y	XY	XY

Probability of a girl $= \frac{2}{4} = \frac{1}{2}$
Probability of a boy $= \frac{2}{4} = \frac{1}{2}$

50.

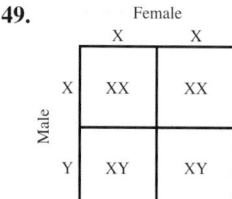

	A	o
B	AB	Bo
o	Ao	oo

Type A: $\frac{1}{4}$ Type B: $\frac{1}{4}$
Type AB: $\frac{1}{4}$ Type O: $\frac{1}{4}$

51. 0.4375 **52.** $\frac{14}{65}$ **53.** $\frac{1}{2}$

Section Project 10.6

(a) 0.022 **(b)** 0.196 **(c)** 0.455 **(d)** 0.891

Review Exercises (page 769)

1. $\sum_{k=1}^{4} (5k - 3)$ **2.** $\sum_{k=1}^{4} (9 - 2k)$ **3.** $\sum_{k=1}^{6} \frac{1}{3k}$

4. $\sum_{k=0}^{4} \left(-\frac{1}{3}\right)^k$ **5.** 380 **6.** $\frac{1}{140,556}$

7. $n(n - 1)(n - 2)$ **8.** $\frac{1}{(n + 1)n}$

9. 127, 122, 117, 112, 107 **10.** 5, 7, 9, 11, 13
11. $\frac{5}{4}, 2, \frac{11}{4}, \frac{7}{2}, \frac{17}{4}$ **12.** $\frac{2}{5}, -\frac{1}{5}, -\frac{4}{5}, -\frac{7}{5}, -2$ **13.** -2.5
14. 3 **15.** 10, 30, 90, 270, 810
16. 2, -10, 50, -250, 1250
17. 100, -50, 25, -12.5, 6.25 **18.** 12, 2, $\frac{1}{3}, \frac{1}{18}, \frac{1}{108}$

19. $\frac{3}{2}$ **20.** $-\frac{2}{3}$ **21.** $4n + 6$ **22.** $-2n + 34$

23. $-50n + 1050$ **24.** $8n + 4$ **25.** $a_n = \left(-\frac{2}{3}\right)^{n-1}$

26. $a_n = 100(1.07)^{n-1}$ **27.** $a_n = 24(2)^{n-1}$

28. $a_n = 16\left(-\frac{1}{4}\right)^{n-1}$ **29.** $a_n = 12\left(-\frac{1}{2}\right)^{n-1}$

30. $a_n = 3\left(\frac{1}{3}\right)^{n-1}$ **31.** (a)   **32.** (f)   **33.** (b)

34. (e)   **35.** (d)   **36.** (c)

37.   **38.**

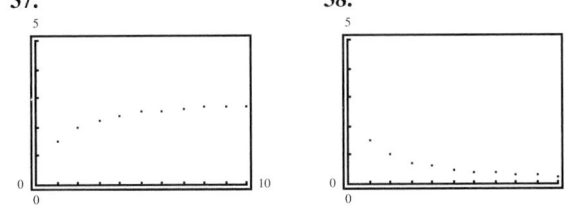

39.   **40.**

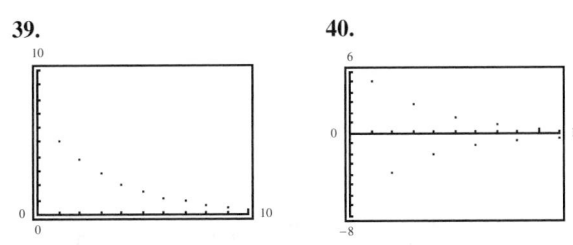

41. 28   **42.** $-\frac{7}{12}$ **43.** $\frac{4}{5}$ **44.** $\frac{17}{15}$ **45.** 486

46. 450   **47.** $\frac{2525}{2}$ **48.** $\frac{3825}{2}$ **49.** 8190   **50.** 2730

51. -1.928 **52.** 453.320   **53.** 19.842   **54.** -2.205

55. 116,169.54   **56.** 486,851.81   **57.** 5100

58. 19,950   **59.** 462

60. (a) $a_n = 120,000(0.70)^n$ (b) \$20,168.40

61. (a) $a_n = 85,000(1.012)^n$ (b) 154,328

62. \$4,371,379.65   **63.** 56   **64.** 66   **65.** 1   **66.** 100

67. 91,390   **68.** 5005   **69.** 177,100   **70.** 496

71. $x^{10} + 10x^9 + 45x^8 + 120x^7 + 210x^6 + 252x^5 + 210x^4$
$+ 120x^3 + 45x^2 + 10x + 1$

72. $u^9 - 9u^8v + 36u^7v^2 - 84u^6v^3 + 126u^5v^4 - 126u^4v^5$
$+ 84u^3v^6 - 36u^2v^7 + 9uv^8 - v^9$

73. $y^6 - 12y^5 + 60y^4 - 160y^3 + 240y^2 - 192y + 64$

74. $x^5 + 15x^4 + 90x^3 + 270x^2 + 405x + 243$

75. $x^8 - 4x^7 + 7x^6 - 7x^5 + \frac{35}{8}x^4 - \frac{7}{4}x^3 + \frac{7}{16}x^2 - \frac{1}{16}x + \frac{1}{256}$

76. $81x^4 - 216x^3y + 216x^2y^2 - 96xy^3 + 16y^4$

77. $-61,236$ **78.** -1080 **79.** 8   **80.** 21

81. 3003   **82.** 5040   **83.** $\frac{1}{3}$ **84.** $\frac{15}{16}$ **85.** $\frac{1}{24}$

86. No, rolling a 3 with one six-sided die has the greater probability of occurring.

87. 0.346   **88.** $\frac{1}{25}, \frac{7}{25}$

Chapter 10 Test   (page 772)

1. $1, -\frac{2}{3}, \frac{4}{9}, -\frac{8}{27}, \frac{16}{81}$ **2.** 35   **3.** -45 **4.** 1020

5. $\frac{3069}{1024}$ **6.** $\sum_{k=1}^{12} \frac{2}{3k+1}$ **7.** 12, 16, 20, 24, 28

8. $a_n = -100n + 5100$ **9.** 3825   **10.** $-\frac{3}{2}$

11. $a_n = 4\left(\frac{1}{2}\right)^{n-1}$ **12.** \$47,868.33   **13.** 1140

14. $x^5 - 10x^4 + 40x^3 - 80x^2 + 80x - 32$ **15.** 56

16. 26,000   **17.** 12,650   **18.** 0.25   **19.** $\frac{3}{26}$ **20.** $\frac{1}{6}$

Cumulative Test: Chapters 1–10   (page 773)

1. $36x^5y^8$ **2.** $\frac{7}{24}x + 8$ **3.** $4rs^2$ **4.** $\frac{9x^2}{16y^6}$

5. $\frac{4\sqrt{10}}{5}$ **6.** 3   **7.** $5x(1 - 4x)$

8. $(2 + x)(14 - x)$ **9.** $(3x - 5)(5x + 3)$

10. $(2x + 1)(4x^2 - 2x + 1)$

11.   **12.**

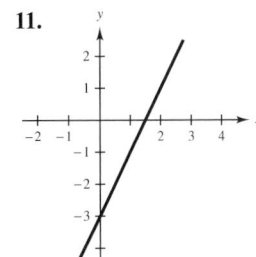 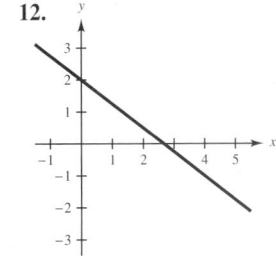

13.   **14.**

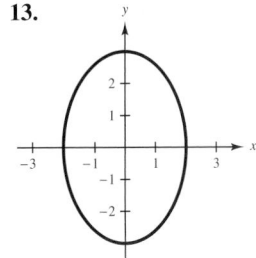 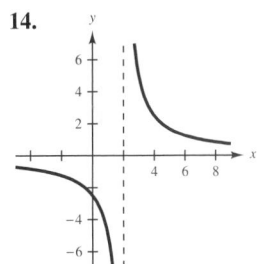

15. $y = -\frac{11}{8}x + \frac{161}{16}$ **16.** $[-4, 4]$ **17.** $\frac{c - 6}{c + 4}$

18. 50   **19.** $y = \frac{1}{2}(x - 2)^2 + 2$

20. (a)   (b) $t \approx 4.16$
   (c) 10,000

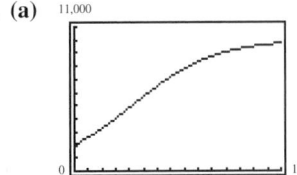

21. $z^4 - 12z^3 + 54z^2 - 108z + 81$ **22.** $\frac{1}{4}$

APPENDIX A

Section A.1   (page A4)

1. Statement   2. Nonstatement   3. Open statement
4. Nonstatement   5. Open statement   6. Statement
7. Open statement   8. Statement   9. Nonstatement
10. Statement   11. Open statement   12. Nonstatement
13. (a) True (b) False   14. (a) False (b) True
15. (a) True (b) True   16. (a) True (b) True
17. (a) False (b) False   18. (a) True (b) False
19. (a) True (b) False   20. (a) False (b) True
21. (a) The sun is not shining.
  (b) It is not hot.
  (c) The sun is shining and it is hot.
  (d) The sun is shining or it is hot.
22. (a) The car does not have a radio.
  (b) The car is not red.
  (c) The car has a radio and the car is red.
  (d) The car has a radio or the car is red.
23. (a) Lions are not mammals.
  (b) Lions are not carnivorous.
  (c) Lions are mammals and lions are carnivorous.
  (d) Lions are mammals or lions are carnivorous.
24. (a) Twelve is not less than fifteen.
  (b) Seven is not a prime number.
  (c) Twelve is less than fifteen and seven is a prime number.
  (d) Twelve is less than fifteen or seven is a prime number.
25. (a) The sun is not shining and it is hot.
  (b) The sun is not shining or it is hot.
  (c) The sun is shining and it is not hot.
  (d) The sun is shining or it is not hot.
26. (a) The car does not have a radio and the car is red.
  (b) The car does not have a radio or the car is red.
  (c) The car has a radio and the car is not red.
  (d) The car has a radio or the car is not red.
27. (a) Lions are not mammals and lions are carnivorous.
  (b) Lions are not mammals or lions are carnivorous.
  (c) Lions are mammals and lions are not carnivorous.
  (d) Lions are mammals or lions are not carnivorous.
28. (a) Twelve is not less than fifteen and seven is a prime number.
  (b) Twelve is not less than fifteen or seven is a prime number.
  (c) Twelve is less than fifteen and seven is not a prime number.
  (d) Twelve is less than fifteen or seven is not a prime number.

29. $p \wedge \sim q$ 30. $\sim p \vee \sim q$ 31. $\sim p \vee q$
32. $p \wedge q$ 33. $\sim p \vee \sim q$ 34. $p \wedge q$ 35. $\sim p \wedge q$
36. $p \vee \sim q$ 37. The bus is blue.
38. Frank is six feet tall.   39. x is not equal to 4.
40. x is equal to 4.   41. The Earth is flat.
42. The Earth is not flat.

43.

p	q	$\sim p$	$\sim p \wedge q$
T	T	F	F
T	F	F	F
F	T	T	T
F	F	T	F

44.

p	q	$\sim p$	$\sim p \vee q$
T	T	F	T
T	F	F	F
F	T	T	T
F	F	T	T

45.

p	q	$\sim p$	$\sim q$	$\sim p \vee \sim q$
T	T	F	F	F
T	F	F	T	T
F	T	T	F	T
F	F	T	T	T

46.

p	q	$\sim p$	$\sim q$	$\sim p \wedge \sim q$
T	T	F	F	F
T	F	F	T	F
F	T	T	F	F
F	F	T	T	T

47.

p	q	$\sim p$	$p \vee \sim q$
T	T	F	T
T	F	T	T
F	T	F	F
F	F	T	T

48.

p	q	$\sim p$	$p \wedge \sim q$
T	T	F	F
T	F	T	T
F	T	F	F
F	F	T	F

49. Not logically equivalent   **50.** Logically equivalent
51. Logically equivalent   **52.** Not logically equivalent
53. Logically equivalent   **54.** Not logically equivalent
55. Not logically equivalent   **56.** Logically equivalent
57. Logically equivalent   **58.** Not logically equivalent
59. Not tautology   **60.** A tautology   **61.** A tautology
62. Not a tautology

63.

┌── Identical ──┐

p	q	$\sim p$	$\sim q$	$p \wedge q$	$\sim(p \wedge q)$	$\sim p \vee \sim q$
T	T	F	F	T	F	F
T	F	F	T	F	T	T
F	T	T	F	F	T	T
F	F	T	T	F	T	T

Section A.2 (page A11)

1. **(a)** If the engine is running, then the engine is wasting gasoline.
   (b) If the engine is wasting gasoline, then the engine is running.
   (c) If the engine is not wasting gasoline, then the engine is not running.
   (d) If the engine is running, then the engine is not wasting gasoline.
2. **(a)** If the student is at school, then it is nine o'clock.
   (b) If it is nine o'clock, then the student is at school.
   (c) If it is not nine o'clock, then the student is not at school.
   (d) If the student is at school, then it is not nine o'clock.
3. **(a)** If the integer is even, then it is divisible by two.
   (b) If it is divisible by two, then the integer is even.
   (c) If it is not divisible by two, then the integer is not even.
   (d) If the integer is even, then it is not divisible by two.
4. **(a)** If the person is generous, then the person is rich.
   (b) If the person is rich, then the person is generous.

   (c) If the person is not rich, then the person is not generous.
   (d) If the person is generous, then the person is not rich.
5. $q \to p$ **6.** $\sim q \to \sim p$ **7.** $p \to q$ **8.** $q \to p$
9. $p \to q$ **10.** $p \to q$ **11.** True   **12.** False
13. True   **14.** True   **15.** False   **16.** True
17. True   **18.** True   **19.** True   **20.** False
21. Converse: If you can see the eclipse, then the sky is clear.
   Inverse: If the sky is not clear, then you cannot see the eclipse.
   Contrapositive: If you cannot see the eclipse, then the sky is not clear.
22. Converse: If he is ineligible for the job, then the person is nearsighted.
   Inverse: If the person is not nearsighted, then he is eligible for the job.
   Contrapositive: If he is eligible for the job, then the person is not nearsighted.
23. Converse: If the deficit increases, then taxes were raised.
   Inverse: If taxes are not raised, then the deficit will not increase.
   Contrapositive: If the deficit will not increase, then the taxes are not raised.
24. Converse: If the company's profits will decrease, then wages are raised.
   Inverse: If wages are not raised, then the company's profit will not decrease.
   Contrapositive: If the company's profits will not decrease, then wages are not raised.
25. Converse: It is necessary to apply for the visa to have a birth certificate.
   Inverse: It is not necessary to have a birth certificate to apply for the visa.
   Contrapositive: It is not necessary to apply for the visa to not have a birth certificate.
26. Converse: The sum of its digits is divisible by three only if the number is divisible by three.
   Inverse: The number is not divisible by three only if the sum of its digits is not divisible by three.
   Contrapositive: The sum of its digits is not divisible by three only if the number is not divisible by three.
27. Paul is not a junior and not a senior.
28. Jack is not a senior or he does not play varsity basketball.
29. The temperature increases and the metal rod will not expand.
30. The test fails and the project will not be halted.

31. We will go to the ocean and the weather forecast is not good.

32. We are going to win the game and not complete the pass on this play.

33. No student is in an extracurricular activity.

34. No odd integer is not a prime number.

35. Some contact sports are not dangerous.

36. Some members must not pay their dues to prior June 1.

37. Some children are allowed at the concert.

38. Some contestants are over the age of twelve.

39. No $20 bills are counterfeit.

40. No units are defective.

41.

p	q	$\sim q$	$p \to \sim q$	$\sim(p \to \sim q)$
T	T	F	F	T
T	F	T	T	F
F	T	F	T	F
F	F	T	T	F

42.

p	q	$\sim q$	$p \to q$	$\sim q \to (p \to q)$
T	T	F	T	T
T	F	T	F	F
F	T	F	T	T
F	F	T	T	T

43.

p	q	$q \to p$	$\sim(q \to p)$	$\sim(q \to p) \wedge q$
T	T	T	F	F
T	F	T	F	F
F	T	F	T	T
F	F	T	F	F

44.

p	q	$\sim p$	$\sim p \vee q$	$p \to (\sim p \vee q)$
T	T	F	T	T
T	F	F	F	F
F	T	T	T	T
F	F	T	T	T

45.

p	q	$\sim p$	$p \vee q$	$(p \vee q) \wedge (\sim p)$	$[(p \vee q) \wedge (\sim p)] \to q$
T	T	F	T	F	T
T	F	F	T	F	T
F	T	T	T	T	T
F	F	T	F	F	T

46.

p	q	$\sim p$	$p \to q$	$(p \to q) \wedge (\sim q)$	$[(p \to q) \wedge (\sim q)] \to p$
T	T	F	T	F	T
T	F	T	F	F	T
F	T	F	T	F	T
F	F	T	T	T	T

47.

p	q	$\sim p$	$\sim q$	$p \leftrightarrow (\sim q)$	$(p \leftrightarrow \sim q) \to \sim p$
T	T	F	F	F	T
T	F	F	T	T	F
F	T	T	F	T	T
F	F	T	T	F	T

48.

p	q	$\sim p$	$\sim q$	$p \vee \sim q$	$q \to \sim p$	$(p \vee \sim q) \leftrightarrow (q \to \sim p)$
T	T	F	F	T	F	F
T	F	F	T	T	T	T
F	T	T	F	F	T	F
F	F	T	T	T	T	T

49.

p	q	$\sim p$	$\sim q$	$q \to p$	$\sim p \to \sim q$
T	T	F	F	T	T
T	F	F	T	T	T
F	T	T	F	F	F
F	F	T	T	T	T

$\llcorner$ Identical $\lrcorner$

50.

p	q	~p	~p→q	p∨q
T	T	F	T	T
T	F	F	T	T
F	T	T	T	T
F	F	T	F	F

Identical (between ~p→q and p∨q)

51.

p	q	~q	p→q	~(p→q)	p∧~q
T	T	F	T	F	F
T	F	T	F	T	T
F	T	F	T	F	F
F	F	T	T	F	F

Identical (between ~(p→q) and p∧~q)

52.

p	q	p∨q	(p∨q)→q	p→q
T	T	T	T	T
T	F	T	F	F
F	T	T	T	T
F	F	F	T	T

Identical (between (p∨q)→q and p→q)

53.

p	q	~p	~q	p→q	(p→q)∨~q	p∨~p
T	T	F	F	T	T	T
T	F	F	T	F	T	T
F	T	T	F	T	T	T
F	F	T	T	T	T	T

Identical (between (p→q)∨~q and p∨~p)

54.

p	q	~p	~q	~p∨q	q→(~p∨q)	q∨~q
T	T	F	F	T	T	T
T	F	F	T	F	T	T
F	T	T	F	T	T	T
F	F	T	T	T	T	T

Identical (between q→(~p∨q) and q∨~q)

55.

p	q	~p	~p∧q	p→(~p∧q)
T	T	F	F	F
T	F	F	F	F
F	T	T	T	T
F	F	T	F	T

Identical (between ~p∧q and p→(~p∧q))

56.

p	q	~q	p∧q	~(p∧q)	~(p∧q)→~q	p∨~q
T	T	F	T	F	T	T
T	F	T	F	T	T	T
F	T	F	F	T	F	F
F	F	T	F	T	T	T

Identical (between ~(p∧q)→~q and p∨~q)

57. (c)   **58.** (c)   **59.** (a)   **60.** (c)

61.

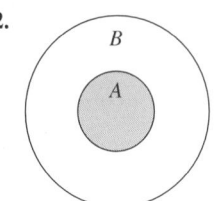

62.

63.

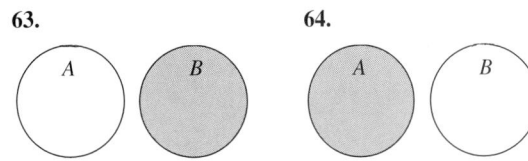

64.

65.

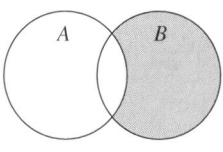

66.

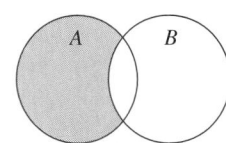

67.

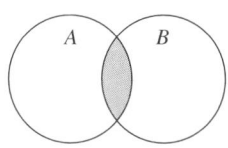

68.

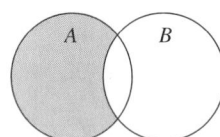

69.

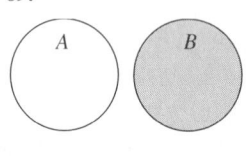

70.

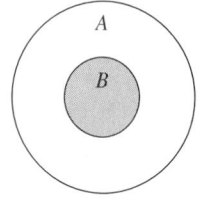

71. (a) Statement does not follow.
    (b) Statement follows.
72. (a) Statement follows.
    (b) Statement follows.
73. (a) Statement does not follow.
    (b) Statement does not follow.
74. (a) Statement does not follow.
    (b) Statement does not follow.

Section A.3   (page A19)

1.

p	q	$\sim p$	$\sim q$	$p \to \sim q$	$(p \to \sim q) \wedge q$	$[(p \to \sim q) \wedge q] \to \sim p$
T	T	F	F	F	F	T
T	F	F	T	T	F	T
F	T	T	F	T	T	T
F	F	T	T	T	F	T

2.

p	q	$p \leftrightarrow q$	$(p \leftrightarrow q) \wedge p$	$[(p \leftrightarrow q) \wedge p] \to q$
T	T	T	T	T
T	F	F	F	T
F	T	F	F	T
F	F	T	F	T

3.

p	q	$\sim p$	$p \vee q$	$(p \vee q) \wedge \sim p$	$[(p \vee q) \wedge \sim p] \to q$
T	T	F	T	F	T
T	F	F	T	F	T
F	T	T	T	T	T
F	F	T	F	F	T

4.

p	q	$\sim p$	$p \wedge q$	$(p \wedge q) \wedge \sim p$	$[(p \wedge q) \wedge \sim p] \to q$
T	T	F	T	F	T
T	F	F	F	F	T
F	T	T	F	F	T
F	F	T	F	F	T

5.

p	q	$\sim p$	$\sim q$	$(\sim p \to q)$	$(\sim p \to q) \wedge p$	$[(\sim p \to q) \wedge p] \to \sim q$
T	T	F	F	T	T	F
T	F	F	T	T	T	T
F	T	T	F	T	F	T
F	F	T	T	F	F	T

6.

p	q	$\sim p$	$\sim q$	$p \to q$	$(p \to q) \wedge \sim p$	$[(p \to q) \wedge \sim p] \to \sim q$
T	T	F	F	T	F	T
T	F	F	T	F	F	T
F	T	T	F	T	T	F
F	F	T	T	T	T	T

7.

p	q	$p \vee q$	$(p \vee q) \wedge q$	$[(p \vee q) \wedge q] \to p$
T	T	T	T	T
T	F	T	F	T
F	T	T	T	F
F	F	F	F	T

8.

p q	$p \wedge q$	$\sim(p \wedge q)$	$\sim(p \wedge q) \wedge q$	$[\sim(p \wedge q) \wedge q] \rightarrow p$
T T	T	F	F	T
T F	F	T	F	T
F T	F	T	T	F
F F	F	T	F	T

9. Valid   **10.** Valid   **11.** Invalid   **12.** Invalid
13. Valid   **14.** Valid   **15.** Valid   **16.** Invalid
17. Invalid   **18.** Valid   **19.** Valid   **20.** Valid
21. Invalid   **22.** Invalid   **23.** (b)   **24.** (c)   **25.** (c)
26. (c)   **27.** (b)   **28.** (a)   **29.** (c)   **30.** (b)
31. Valid

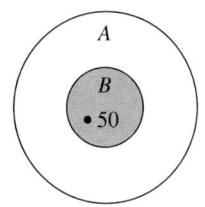

A: All numbers divisible by five
B: All numbers divisible by ten

32. Valid

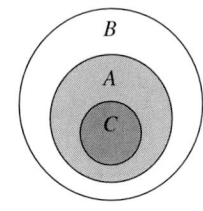

A: All human beings
B: Things that require adequate rest
C: All infants

33. Invalid

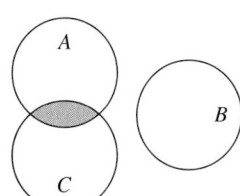

A: People eligible to vote
B: People under the age of 18
C: College students

34. Invalid

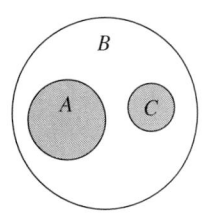

A: Every amateur radio operator
B: People who have a radio license
C: Jackie

35. Let p be the statement "Sue drives to work," let q represent "She will stop at the grocery store," and let r represent "She'll buy milk."

First write: Reorder the premises:
     Premise #1: $p \rightarrow q$ Premise #3: p
     Premise #2: $q \rightarrow r$ Premise #1: $p \rightarrow q$
     Premise #3: p Premise #2: $q \rightarrow r$
                       Conclusion: r

Then we can conclude r. That is, "Sue will get milk."

36. Let p represent "Bill is patient," let q represent "He will succeed," let r represent "Bill will get bonus pay."
First write:
     Premise #1: $p \rightarrow q$
     Premise #2: $q \rightarrow r$
     Premise #3: $\sim r$
Conclusion from Premise #1, Premise #2: $p \rightarrow r$
Conclusion from $p \rightarrow r$ and Premise #3: $\sim p$
That is, "Bill is not patient."

37. Let p represent "This is a good product," let q represent "We will buy it," and let r represent "The product was made by XYZ Corporation."
First write:
     Premise #1: $p \rightarrow q$
     Premise #2: $r \vee \sim q$
     Premise #3: $\sim r$
Note that $p \rightarrow q \equiv \sim q \rightarrow \sim p$, and reorder the premises:
     Premise #2: $r \vee \sim q$
     Premise #3: $\sim r$
     (Conclusion from Premise #2, Premise #3: $\sim q$)
     Premise #1: $\sim q \rightarrow \sim p$
     Conclusion: $\sim p$
Then we can conclude $\sim p$. That is, "It is not a good product."

38. Let p represent "The book is returned in two weeks," let q represent "There is a fine," and let r represent "You may not check out another book."

First write:

Premise #1: $p \rightarrow \sim q$

Premise #2: $q \lor r$

Premise #3: $\sim r$

Conclusion from Premise #2, Premise #3: q

Note $q \equiv \sim(\sim q)$

Conclusion from q, Premise #1: $\sim p$

That is, "The book was not returned within two weeks."

APPENDIX B

Section B.1 (page A28)

1.

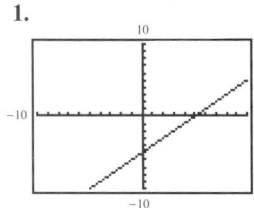

2.

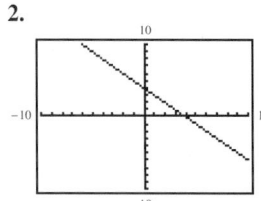

3.

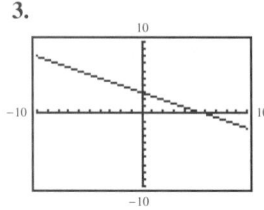

4.

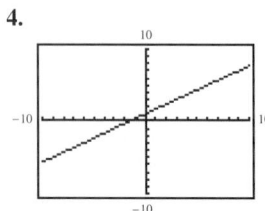

5.

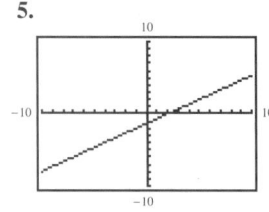

6.

7.

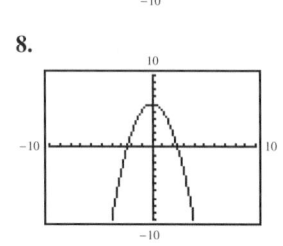

8.

9.

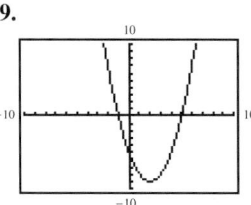

10.

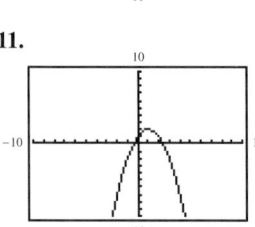

11.

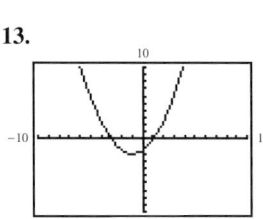

12.

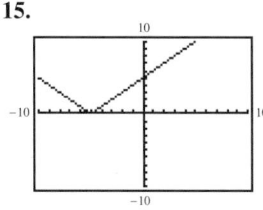

13.

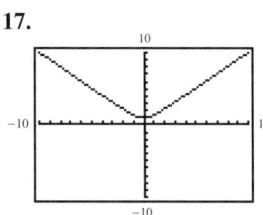

14.

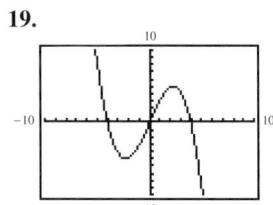

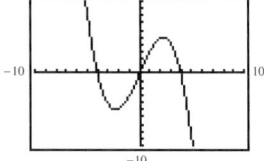

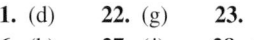

15.

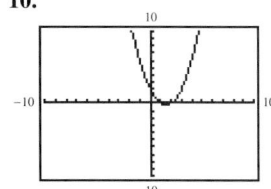

16.

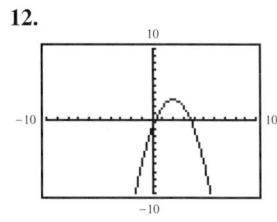

17.

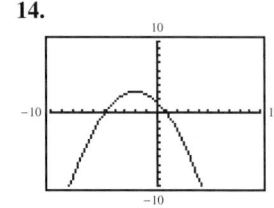

18.

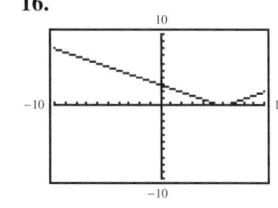

19.

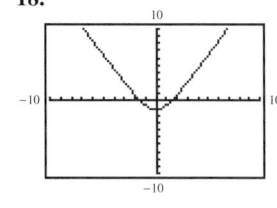

20.

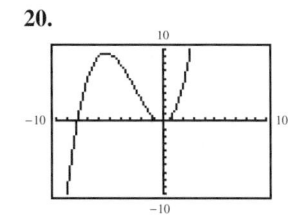

21. (d)   **22.** (g)   **23.** (a)   **24.** (f)   **25.** (i)

26. (b)   **27.** (j)   **28.** (c)   **29.** (e)   **30.** (h)

31.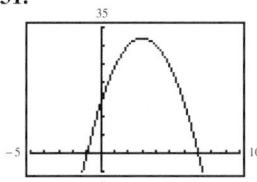

32.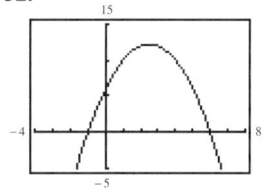

33.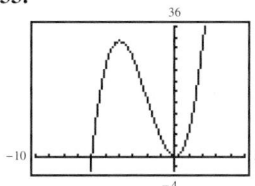

34.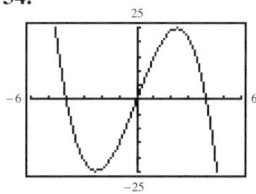

35.

Xmin = -10
Xmax = 10
Xscl = 1
Ymin = -12
Ymax = 30
Yscl = 6
Xres = 1

36.

Xmin = -20
Xmax = 8
Xscl = 4
Ymin = -60
Ymax = 10
Yscl = 10
Xres = 1

37.

Xmin = -5
Xmax = 5
Xscl = 1
Ymin = -5
Ymax = 5
Yscl = 1
Xres = 1

38.

Xmin = -6
Xmax = 3
Xscl = 1
Ymin = -4
Ymax = 10
Yscl = 1
Xres = 1

39. No intercepts   **40.** Two x-intercepts

41. Three x-intercepts   **42.** One x-intercept

43. Square   **44.** Parallelogram

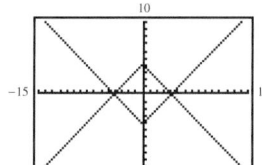

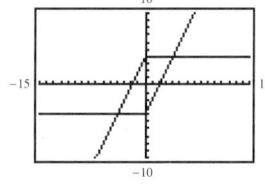

45. Circle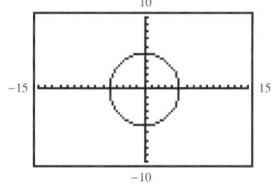

46. Triangle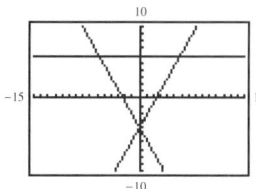

47.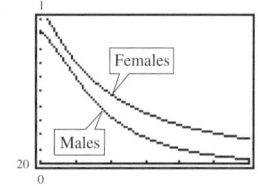

48. Answers may vary.

49. 0.59   **50.** 0.70

51.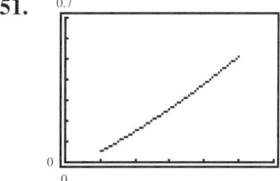

52. $0.54   **53.** $0.73

54. **(a)** 0.345

(b) 0.380

(c) 0.410

Section 1.3   (page 30)

(a) 2^5 **(b)** 2^5 **(c)** 2^7 **(d)** 2^7 **(e)** 2^6

When multiplying exponential expressions that have the same base, you add the exponents.

Section 1.5   (page 60)

$u^6 - v^6 = (u + v)(u - v)(u^2 - uv + v^2)(u^2 + uv + v^2)$

$x^6 - 1 = (x + 1)(x - 1)(x^2 - x + 1)(x^2 + x + 1)$

$x^6 - 64 = (x + 2)(x - 2)(x^2 - 2x + 4)(x^2 + 2x + 4)$

Section 2.1   (page 112)

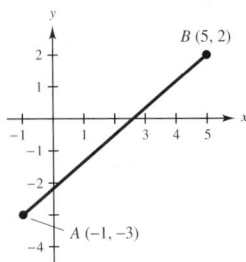

Show that the distance from A to C is equal to the distance from C to B and that the sum of those two distances is the same as the distance from A to B.

C may not lie on the same line as points A and B.

Midpoint of a line segment joining (x_1, y_1) and (x_2, y_2) is $\left(\dfrac{x_1 + x_2}{2}, \dfrac{y_1 + y_2}{2} \right)$.

Section 2.2   (page 126)

Because of the resolution of the pixels on the screen of a graphing utility, you may not be able to read exact solutions to an equation. You would then have to algebraically check that the solution is correct.

Section 2.3   (page 136)

(a)

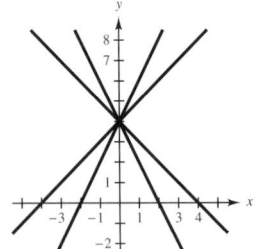

The equations all have a constant term of 4. The graphs all have a y-intercept at 4. The constant term is the y-intercept.

(b)

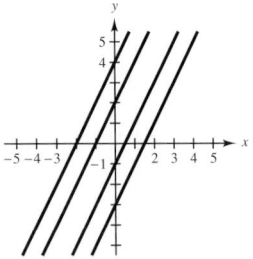

The equations all have an x-coefficient of 2. The graphs all have the same slope. The coefficient of the x-term is the slope.

Section 2.5   (page 158)

(a) Xmin $= -10$, Xmax $= 10$, Xscl $= 1$,
      Ymin $= -5$, Ymax $= 15$, Yscl $= 1$,
(b) Xmin $= -10$, Xmax $= 10$, Xscl $= 1$,
      Ymin $= -10$, Ymax $= 10$, Yscl $= 1$
(c) Xmin $= -15$, Xmax $= 15$, Xscl $= 5$,
      Ymin $= -30$, Ymax $= 10$, Yscl $= 5$

Section 2.6 (page 169)

(a) Use the graph of the squaring function with a right shift of three units and a downward shift of one unit.

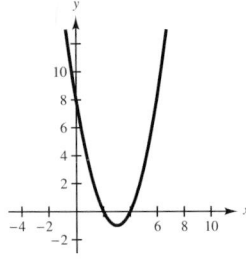

(b) Use the graph of the absolute value function with a left shift of one unit and a downward shift of two units.

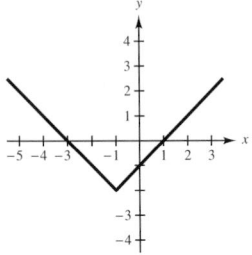

(c) Use the graph of the cubing function with a right shift of one unit and an upward shift of two units.

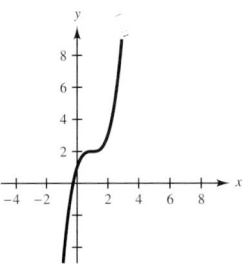

(d) Use the graph of the square root function with a left shift of four units and an upward shift of one unit.

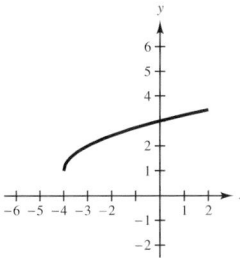

(e) Use the graph of the squaring function with a left shift of two units and an upward shift of two units.

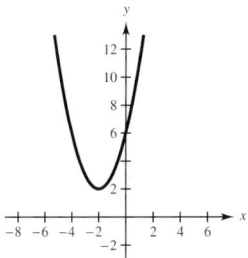

(f) Use the graph of the absolute value function with a right shift of one unit and a downward shift of four units.

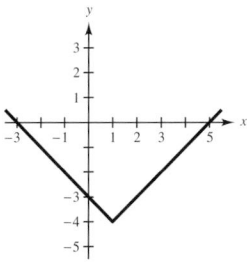

Section 3.6 (page 247)

Any results will not check. The expression $|5x - 6|$ cannot equal a negative number.

Section 3.6 (page 249)

(a) All numbers less than or equal to -3 *or* all numbers greater than or equal to 3

(b) All numbers greater than -3 *and* all numbers less than 3

Section 4.1 (page 267)

Systems of equations in two variables may have infinitely many solutions, one solution, or no solutions.

Section 4.3 (page 302)

Stock X: $6000; stock Y: $6000; stock Z: $18,000

Section 4.5 (page 321)

Points to the left satisfy $4x - 3y < 12$. Points to the right satisfy $4x - 3y > 12$. The graph of the solution of a linear inequality in two variables consists of a half-plane.

Section 5.1 (page 348)

$$a^0 = 1; a^{-4} = \frac{1}{a^4}$$

Section 5.2 (page 360)

≈ 6.2693; ≈ 6.2693; The expressions are equal.

Section 5.5 (page 388)

One solution: $c = -\frac{1}{4}, c > 0$; two solutions: $-\frac{1}{4} < c \le 0$; no solutions: $c < \frac{1}{4}$

Section 5.6 (page 394)

$i^1 = i, i^2 = -1, i^3 = -i, i^4 = 1, i^5 = i, i^6 = -1, i^7 = -i,$ $i^8 = 1$; The pattern is $i, -1, -i, 1$. $i^{16} = 1, i^{19} = -i$

Section 6.1 (page 411)

$\approx -0.97, \approx 2.30$

Section 6.3 (page 431)

(a) No real solution   (b) Two real solutions
(c) One real solution   (d) Two real solutions
The discriminant can determine the difference between two rational solutions, two irrational solutions, or one repeated rational solution.

Section 6.4 (page 438)

Answers may vary.

Section 6.6 (page 458)

Answers may vary.

Section 6.7 (page 462)

Negative: $-2 < x < 3$; positive: $x < -2$ or $x > 3$

Section 7.1 (page 478)

	-4	-3	-2	-1	0
$\dfrac{3}{x+4}$	Undef.	3	$\frac{3}{2}$	1	$\frac{3}{4}$
$\dfrac{2x}{x^2-4x+4}$	$-\frac{2}{9}$	$-\frac{6}{25}$	$-\frac{1}{4}$	$-\frac{2}{9}$	0
$\dfrac{x^2-5x}{x^2+2x-3}$	$\frac{36}{5}$	Undef.	$-\frac{14}{3}$	$-\frac{3}{2}$	0

	1	2	3	4
$\dfrac{3}{x+4}$	$\frac{3}{5}$	$\frac{1}{2}$	$\frac{3}{7}$	$\frac{3}{16}$
$\dfrac{2x}{x^2-4x+4}$	2	Undef.	6	2
$\dfrac{x^2-5x}{x^2+2x-3}$	Undef.	$-\frac{6}{5}$	$-\frac{1}{2}$	$-\frac{4}{21}$

$x = -4$ is not valid for the first expression. $x = 2$ is not valid for the second expression. $x = -3$ and $x = 1$ is not valid for the third expression.

Section 7.4 (page 514)

$x^3 - 64 = (x - 4)(x^2 + 4x + 16)$; Synthetic division can be used to factor polynomials.

Section 7.5 (page 526)

The x-intercepts are $(-2, 0)$ and $(5, 0)$. The x-intercepts are the solutions to the equations. To find solutions graphically, rewrite the equation so that all the terms are on the left side of the equal sign (zero will be on the right side of the equal sign). Set the expression on the left side equal to y and graph the resulting equation. The x-intercepts of the graph are the solution to the original equation.

Section 7.6 (page 533)

$f(x)$ has a vertical asymptote at $x = 0$ and $g(x)$ has a vertical asymptote at $x = 4$. This occurs at the x-value where the denominator is zero.
$f(x)$ and $g(x)$ have horizontal asymptotes at $y = 0$. As x increases or decreases without bound, the values of the functions approach zero.

Section 7.7 (page 545)

$(-\infty, -6] \cup (-3, \infty)$; yes

Section 8.1 (page 556)

Yes, yes

Section 8.2 (page 566)

x	-10	0	7	45
$f(f^{-1}(x))$	-10	0	7	45
$f^{-1}(f(x))$	-10	0	7	45

The functions are inverses

Section 8.5 (page 597)

Circle, ellipse

Section 8.6 (page 604)

When the foci are close together, the ellipse is almost circular. When the foci are far apart, the ellipse is long and narrow.

Section 8.8 (page 628)

(a) Zero, one, two, three, or four points of intersection
(b) Zero, one, two, three, or four points of intersection
(c) Zero or two points of intersection

Section 9.3 (pages 671)

$\ln x^4 = 4 \ln x$; no; The domains of the expressions are different.

Section 9.6 (pages 699)

$y = -0.435x^2 + 4.16x - 0.9$; The logarithmic model is better.

Section 10.3 (page 734)

$x = \frac{41}{99}$

Section 10.4 (page 742)

The coefficients form Pascal's Triangle (see page 744). From left to right the exponents of the x-variable decrease and the exponents of the y-variable increase.

TECHNOLOGY NOTES

Section 1.7 (page 80)

$x = -3$

Section 3.4 (page 227)

$P = \$18,500$

Section 4.1 (page 268)

(**a**) Infinitely many solutions   (**b**) One solution
(**c**) No solutions

Section 4.5 (page 324)

(**a**)

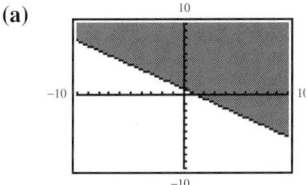

(**b**)

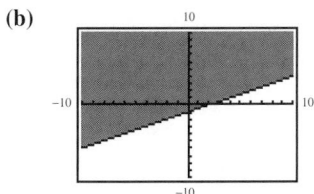

Section 4.6 (page 329)

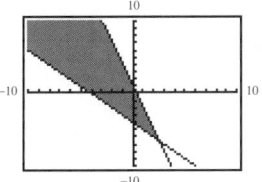

Section 6.4 (page 440)

(**a**) 3 seconds and 9 seconds   (**b**) 6 seconds   (**c**) 12 seconds

Section 7.1 (page 479)

(**a**) The graph does not exist at $x = -5$ and $x = 5$, therefore the domain is $(-\infty, -5) \cup (-5, 5) \cup (5, \infty)$.
(**b**) The graph does not exist at $x = -6$ and $x = 1$, therefore the domain is $(-\infty, -6) \cup (-6, 1) \cup (1, \infty)$.

Section 8.4 (page 589)

The graphs of x^2, x^4, and x^6 rise to the left and rise to the right. The graphs of x^1, x^3, and x^5 fall to the left and rise to the right.

Section 8.5 (page 598)

To obtain a circle use a square setting for the viewing rectangle.

Section 9.2 (page 663)

$g(x) = \log_2 x = \dfrac{\log_{10} x}{\log_{10} 2} = \dfrac{\ln x}{\ln 2}$; $g(3) \approx 1.585$

GROUP ACTIVITIES

Section 1.1 (page 13)

Answers may vary.

Section 1.2 (page 24)

Answers may vary.

Section 1.3 (page 37)

Multiply the units of each factor. Cancel "millions of books" and "book." The remaining unit is "millions of dollars per year".

Section 1.4 (page 49)

(a) $-16t^2 - 80t + 30$; $v_0 = -80$ which indicates that the object was thrown downward.

(b) $-16t^2 + 30$; $v_0 = 0$ which indicates that the object was dropped, and $s_0 = 30$ which indicates that the initial height was 30 feet.

(c) $-16t^2 + 30t$; $v_0 = 30$ which indicates that the object was thrown upward, and $s_0 = 0$ which indicates that the initial height was at ground level.

Section 1.5 (page 62)

Answers may vary.

Section 1.6 (page 74)

$(a + b)^2 - 7(a + b) + 12 = [(a + b) - 4][(a + b) - 3]$
$(m - n)^2 - 4(m - n) - 32 = [(m - n) - 8][(m - n) + 4]$
$3(w + k)^2 + 19(w + k) - 14 = [3(w + k) - 2][(w + k) + 7]$
$x^2 + 2xy + y^2 + 6x + 6y + 8$
$\quad = (x + y)^2 + 6(x + y) + 8$
$\quad = [(x + y) + 4][(x + y) + 2]$
$c^2 - 2cd + d^2 + 8c - 8d - 20$
$\quad = (c - d)^2 + 8(c - d) - 20$
$\quad = [(c - d) + 10][(c - d) - 2]$
$2r^2 + 4rt + 2t^2 - 15r - 15t + 18$
$\quad = 2(r + t)^2 - 15(r + t) + 18$
$\quad = [2(r + t) - 3][(r + t) - 6]$

Section 1.7 (page 84)

(a) No solution   **(b)** Conditional   **(c)** Identity

Section 1.8 (page 95)

(a) $x = -3, x = -1$ **(b)** $x = -\frac{2}{3}, x = 1$
(c) $x = -6, x = \frac{1}{2}$

Section 2.1 (page 113)

(a) For the first set of points, $d_1 = \sqrt{13}$, $d_2 = \sqrt{52} = 2\sqrt{13}$, and $d_3 = \sqrt{117} = 3\sqrt{13}$; relationship: $d_1 + d_2 = d_3$ because $\sqrt{13} + 2\sqrt{13} = 3\sqrt{13}$. For the second set of points, $d_1 = \sqrt{5}$, $d_2 = \sqrt{2}$, and $d_3 = \sqrt{13}$; relationship: $d_1 + d_2 \geq d_3$.

(b) First set of points   Second set of points

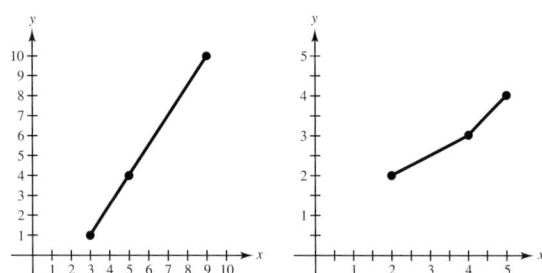

All points in the first set appear to lie on the same line, but the points in the second set do not.

(c) In parts (a) and (b), an exact relationship among the distances exists when the points all fall on the same line when graphed, but when the points do not fall on the same line, there exists an inexact relationship. In general, a set of three points is collinear if the sum of two distances among the points is exactly equal to the third distance.

Section 2.2 (page 127)

Depreciation: $y = 65{,}000 - 7500t$

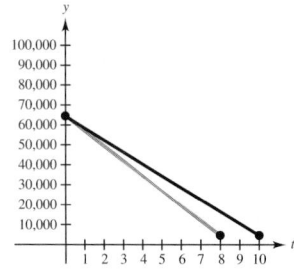

Section 2.3 (page 140)

Examples will vary. Possibilities include:

1. Let y = total pay per hour, x = number pieces, then $y = 0.35x + 6$, where slope is 0.35.
   For every additional piece completed within an hour, the hourly pay increases by $0.35.

2. Let y = total cost, x = list price, then $y = 0.05x + x = 1.05x$, where the slope is 1.05.
   For every one dollar increase in list price, the total price increases by $1.05.

Section 2.4 (page 153)

Answers may vary.

Section 2.5 (page 164)

Answers may vary.

Section 2.6 (page 172)

(a) $y = \sqrt{x} - 3$ **(b)** $y = |x + 2|$ **(c)** $y = \sqrt{x - 4}$
(d) $y = -|x|$ **(e)** $y = x^2 - 1$ **(f)** $y = (x + 4)^3 + 1$
(g) $y = -(x - 5)^2$ **(h)** $y = -x^3 + 2$ **(i)** $y = -\sqrt{x + 1}$

Section 3.1 (page 191)

A linear model should work reasonably well. Models will vary depending on how the "best-fitting" line is drawn by the student. One linear model that fits the data well is $y = 2.6x + 0.8$. The slope of this equation indicates that the percent of households with personal computers increases 2.6% each year. According to this model, the percent in 1999 will be 50.2%.

Section 3.2 (page 201)

Answers may vary.

Section 3.3 (page 214)

(a) A life expectancy of 124 years is too long; 70 years would be more reasonable.
(b) A temperature well above freezing on one scale should not convert to a temperature well below freezing on another scale; the correct answer is 26.7°C.
(c) A diameter of 5.0 inches is too large; 0.5 inch would be more reasonable.
(d) An area of 300 square feet is too small; something like 5500 square feet would be more reasonable (this would depend on the type of gymnasium).

Section 3.4 (page 229)

Let a represent age, and let i represent desired intensity level. The 10-second pulse count formula is $\dfrac{(220 - a)i}{6}$.

Section 3.5 (page 242)

Answers may vary.

Section 3.6 (page 252)

Because $|3x - 4|$ is always nonnegative, the inequality is always true for *all* values of x. The solution given eliminates the values $-\frac{1}{3} \le x \le 3$.

Section 4.1 (page 277)

There are infinitely many systems of equations that have $(3, -2)$ as a solution.

Section 4.2 (page 292)

The system is

$$
\begin{aligned}
c &= 467 \\
a + b + c &= 468 \\
4a + 2b + c &= 476
\end{aligned}
$$

The model is $y = 3.5x^2 - 2.5x + 467$. For 1993, the model gives $491 billion, which is quite different from the actual value of $510 billion—the model gives a conservative estimate. It could be argued that underestimating government income for budgeting purpose is a better situation than over estimating income, because counting on (and subsequently spending) income that is never received will contribute to the federal deficit.

Section 4.3 (page 303)

When Gaussian elimination is used, a system with a unique solution has leading 1 in each row, a system with infinitely many solutions has at least one row of all zeros, and a system with no solution has least one row with a nonzero entry in the last column and all zeros in the other columns.

Section 4.4 (page 316)

Answers may vary.

Section 4.5 (page 325)

Answers may vary.

Section 4.6 (page 335)

$$
\begin{cases}
y < -2x + 8 \\
y \ge 0 \\
y \le 3 \\
x \ge 0
\end{cases}
$$

One possible objective function: $P = x + y$

Section 5.1 (page 354)

Larger: 7×10^{51}

Section 5.2 (page 365)

(a) Domain: $x \geq 0$; range: $y \geq 0$
(b) Domain: all real numbers; range: $y \geq 0$
(c) Domain: all real numbers; range: all real numbers
(d) Domain: $x \geq 0$; range: $y \geq 0$
(e) Domain: all real numbers except 0: range: $y > 0$
(f) Domain: all real numbers; range: all real numbers

Section 5.3 (page 374)

$a = 36, b = 64, a + b = 100$

Section 5.4 (page 382)

(a) $\dfrac{\sqrt{5}+1}{2} \approx 1.62$

(b) Many rectangles are possible.

Section 5.5 (page 390)

When $t = 0.25$ second, $h = 24$ inches. When $t = 0.10$ second, $h = 3.84$ inches.

Section 5.6 (page 398)

$x^2 + 1 = (x + i)(x - i)$

Section 6.1 (page 416)

(a) Crosses x-axis twice; $x = \frac{5}{2}$ and $x = -3$; two real solutions
(b) Does not cross x-axis; $x = 2 \pm i\sqrt{\frac{3}{4}}$; two complex solutions
(c) Crosses x-axis once; $x = -\frac{4}{3}$ one (repeated) real solution
(d) Does not cross x-axis; $x = \dfrac{1 + i\sqrt{3}}{3}$; two complex solutions

If the graph crosses the x-axis twice, the equation has two real solutions; if the graph crosses the x-axis only once, the equation has one repeated real solution; if the graph does not cross the x-axis, the graph has no real solutions but two complex solutions.

Section 6.2 (page 424)

Correct solution:
$$x^2 + 6x = 13$$
$$x^2 + 6x + \left(\tfrac{6}{2}\right)^2 = 13 + 3^2$$
$$(x + 3)^2 = 22$$
$$x + 3 = \pm\sqrt{22}$$
$$x = -3 \pm \sqrt{22}$$

Section 6.3 (page 433)

Answers may vary.

Section 6.4 (page 440)

Minimum function value is $y_1 \approx -1.333$; maximum value is $y_2 = 10.25$

Section 6.5 (page 451)

Quadratic model: $y = 2.5(x - 3)^2 + 49$. Average temperature for December and February from the model is 51.5°F. The model seems to fit the data from November to March quite well. Average temperature for June from the model is 111.5°F. Because an average temperature of 111.5°F does not seem reasonable, the model is probably not the best choice to use for the rest of the year.

Section 6.6 (page 459)

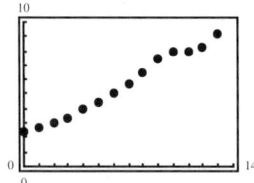

Linear model: $y = 0.549x + 2.00$
Quadratic model: $y = 0.0046x^2 + 0.489x + 2.12$

Section 7.1 (page 485)

(a) Student forgot to divide each term of the numerator by the denominator.
Correct solution:
$$\frac{3x^2 + 5x - 4}{x} = \frac{3x^2}{x} + \frac{5x}{x} - \frac{4}{x} = 3x + 5 - \frac{4}{x}$$
(b) Student incorrectly canceled; the denominator may not be split up.
Correct solution:
$$\frac{x^2 + 7x}{x + 7} = \frac{x(x + 7)}{x + 7} = x$$

Section 7.2 (page 494)

x	60	100	1000	10,000
$\dfrac{x-10}{x+10}$	0.71429	0.81818	0.98020	0.99800
$\dfrac{x+50}{x-50}$	11	3	1.10526	1.01005
$\dfrac{x-10}{x+10}\cdot\dfrac{x+50}{x-50}$	7.85714	2.45455	1.08338	1.00803

x	100,000	1,000,000
$\dfrac{x-10}{x+10}$	0.99980	0.99998
$\dfrac{x+50}{x-50}$	1.00100	1.00010
$\dfrac{x-10}{x+10}\cdot\dfrac{x+50}{x-50}$	1.00080	1.00008

The value of the first row gets larger and closer to 1 as the value of x increases (because as x becomes larger, the value of 10 becomes much smaller in comparison). The value of the second row gets smaller and closer to 1 as the value of x increases (because as x becomes larger, the value of 50 becomes much smaller in comparison). The value of the third row is in between the values of the other two rows and gets smaller and closer to 1 as the value of x increases.

Section 7.3 (page 502)

(a) $-\frac{7}{6}$ (b) $\frac{51}{7}$ (c) 8

Section 7.4 (page 516)

The polynomials of parts (a), (b), and (c) are all equivalent. The x-intercepts are $(-1, 0)$, $(2, 0)$, and $(4, 0)$.
The function has x-intercepts $(-1, 0)$ and $(4, 0)$. It has only two x-intercepts because the numerator factors and it is possible to cancel the same factor in the numerator and denominator, leaving a second-degree polynomial having two factors. The function appears to be equivalent to $f(x) = (x-4)(x+1)$. The difference is at $x = 2$.

Section 7.5 (page 528)

Average cost per thousand pounds $= \dfrac{100 + 50x - 0.2x^2}{x}$;
to proceed with project, place order of $x = \sqrt{500} \approx 22.36$ thousand pounds.

Section 7.6 (page 538)

Horizontal asymptote: $y = 3.25$. The average cost of producing the calendars becomes arbitrarily close to 3.25 through values greater than 3.25 as the number of calendars increases. It is impossible to obtain an average cost of $3 per calendar because the graph will never cross its asymptote of $y = 3.25$.

Section 7.7 (page 547)

(a) $(-\infty, 0) \cup (3, \infty)$ (b) $(3, 6)$ (c) $(3, 4)$

Section 8.1 (page 562)

One example: $f(x) = x^2$, $g(x) = \sqrt{x}$

Section 8.2 (page 572)

(a) Does not possess an inverse (b) Possesses an inverse

Section 8.3 (page 583)

Answers may vary.

Section 8.4 (page 593)

For integers a, b, c and d, one fourth-degree polynomial is $(x-a)(x-b)(x-c)(x-d) = 0$. Another fourth-degree polynomial can be found by multiplying the first polynomial by a constant factor.

Section 8.5 (page 601)

$(-2, 0), (6, 0)$

Section 8.6 (page 615)

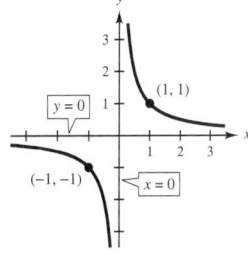

Section 8.7 (page 622)

All television signals striking the satellite's parabolic dish will be reflected to the receiver located at the focus of the dish.

Section 8.8   (page 633)

No points: $c < -\frac{17}{4}$ or $c > 2$; One point: $c = 2$;
Two points: $c = -\frac{17}{4}$ or $-2 < c < 2$; Three points: $c = -2$;
Four points: $-\frac{17}{4} < c < -2$

Section 9.1   (page 651)

The larger the value of k, the steeper the graph.

Section 9.2   (page 664)

(a) Linear model   (b) Exponential model
(c) Logarithmic model   (d) Quadratic model

Section 9.3   (page 673)

(a) No, not very well.

(b)

x	5	7	9	11	13
y	522	1682	3370	5375	12,789
$\ln y$	6.258	7.428	8.123	8.590	9.456

(c) Yes, rather well. The equation of the line that best fits this data will vary depending on how the line is drawn. One possibility is $\ln y = 0.38x + 4.6$.

(d) The first scatter plot has the shape of an exponential model; the second is more linear. Taking the logarithm of exponentially distributed data "straightens" it, explaining the difference between the graphs:

$$y = e^x$$
$$\ln y = \ln(e^x)$$
$$\ln y = x$$

Section 9.4   (page 683)

(a) $x = 4$ and $x = -1$ (b) $x = \ln 4$
(c) $x = e^4$ and $x = e^{-1}$
All three equations can be solved by factoring. Part (b) has only one solution because the other would be $x = \ln(-1)$, which is undefined.

Section 9.5   (page 692)

Answers may vary.

Section 9.6   (page 700)

An exponential model fits the data the best.
$y = 1242.5(1.092)^x$

Section 10.1   (page 718)

Next term in the sequence is T. (The given sequence consists of the first letters of the word forms of the integers 0, 1, 2, 3, 4, 5, 6, 7, 8, 9, 10, 11, 12, . . .)

Section 10.2   (page 727)

(a)

5	11	14
19	10	1
6	9	15

(b)

8	5	14
15	9	3
4	13	10

(c)

9	10	20
24	13	2
6	16	17

Section 10.3   (page 736)

The company whose revenue grew arithmetically had the greatest revenue in the 10-year period. If the trends continue for both companies, the one with the geometric growth will earn increasingly more all the time and will consistently outperform the other company in the future.

Section 10.4   (page 746)

The sum of each row is double the sum of the previous row. Sum of row 10: 1024.

Section 10.5   (page 755)

Number of meals with two different side items:
$$2 \cdot {}_{15}C_2 = 2 \cdot 105 = 210$$
Number of meals with one hot side item and one cold side item:
$$2 \cdot 9 \cdot 6 = 108$$

Section 10.6   (page 763)

Yes

Index of Applications

Biology and Life Science Applications

Business Applications

Miscellaneous Applications

Time and Distance Applications

U.S. Demographics Applications

Index

Basic Rules of Algebra

Commutative Property of Addition

$a + b = b + a$

Commutative Property of Multiplication

$ab = ba$

Associative Property of Addition

$(a + b) + c = a + (b + c)$

Associative Property of Multiplication

$(ab)c = a(bc)$

Left Distributive Property

$a(b + c) = ab + ac$

Right Distributive Property

$(a + b)c = ac + bc$

Additive Identity Property

$a + 0 = a$

Multiplicative Identity Property

$a \cdot 1 = 1 \cdot a = a$

Additive Inverse Property

$a + (-a) = 0$

Multiplicative Inverse Property

$a \cdot \dfrac{1}{a} = 1, \quad a \neq 0$

Properties of Equality

Addition Property of Equality

If $a = b$, then $a + c = b + c$.

Multiplication Property of Equality

If $a = b$, then $ac = bc$.

Cancellation Property of Addition

If $a + c = b + c$, then $a = b$.

Cancellation Property of Multiplication

If $ac = bc$ and $c \neq 0$, then $a = b$.

Zero-Factor Property

If $ab = 0$, then $a = 0$ or $b = 0$.

Properties of Negation

Multiplication by -1

$(-1)a = -a \qquad (-1)(-a) = a$

Placement of Minus Signs

$(-a)(b) = -(ab) = (a)(-b)$

Product of Two Opposites

$(-a)(-b) = ab$

Operations with Fractions

$$\frac{a}{b} \cdot \frac{c}{d} = \frac{a \cdot c}{b \cdot d} \qquad\qquad \frac{\left(\dfrac{a}{b}\right)}{\left(\dfrac{c}{d}\right)} = \frac{a}{b} \cdot \frac{d}{c}$$

$$\frac{a}{b} + \frac{c}{d} = \frac{ad + bc}{bd} \qquad\qquad \frac{a}{b} - \frac{c}{d} = \frac{ad - bc}{bd}$$

Properties of Exponents

$a^0 = 1$ $\qquad\qquad$ $a^m \cdot a^n = a^{m+n}$

$(ab)^m = a^m \cdot b^m$ $\qquad$ $(a^m)^n = a^{mn}$

$\dfrac{a^m}{a^n} = a^{m-n}, \quad a \neq 0$ $\qquad$ $\left(\dfrac{a}{b}\right)^m = \dfrac{a^m}{b^m}, \quad b \neq 0$

$a^{-n} = \dfrac{1}{a^n}, \quad a \neq 0$ $\qquad$ $\left(\dfrac{a}{b}\right)^{-n} = \dfrac{b^n}{a^n}$

Special Products

Square of a Binomial $\qquad (u + v)^2 = u^2 + 2uv + v^2$

$\qquad\qquad\qquad\qquad\qquad (u - v)^2 = u^2 - 2uv + v^2$

Difference of Two Squares

$u^2 - v^2 = (u + v)(u - v)$

Difference of Two Cubes

$u^3 - v^3 = (u - v)(u^2 + uv + v^2)$

Sum of Two Cubes

$u^3 + v^3 = (u + v)(u^2 - uv + v^2)$

The Quadratic Formula

Solutions of $ax^2 + bx + c = 0$

$$x = \frac{-b \pm \sqrt{b^2 - 4ac}}{2a}$$